Numerical Methods for Nonlinear Engineering Models

John R. Hauser

Numerical Methods for Nonlinear Engineering Models

 Springer

John R. Hauser
Department of Electrical & Computer Engineering
North Carolina State University
Raleigh, NC 27695
USA

ISBN-13: 978-94-017-7707-0   ISBN-13: 978-1-4020-9920-5 (eBook)
DOI 10.1007/978-1-4020-9920-5

Printed on acid-free paper

9 8 7 6 5 4 3 2 1

springer.com

Contents

12 Partial Differential Equations: Finite Difference Approaches   705

13 Partial Differential Equations: The Finite Element Method   883

Preface

There are many books on the use of numerical methods for solving engineering problems and for modeling of engineering artifacts. In addition there are many styles of such presentations ranging from books with a major emphasis on theory to books with an emphasis on applications. The purpose of this book is hopefully to present a somewhat different approach to the use of numerical methods for engineering applications.

Engineering models are in general nonlinear models where the response of some appropriate engineering variable depends in a nonlinear manner on the application of some independent parameter. It is certainly true that for many types of engineering models it is sufficient to approximate the real physical world by some linear model. However, when engineering environments are pushed to extreme conditions, nonlinear effects are always encountered. It is also such extreme conditions that are of major importance in determining the reliability or failure limits of engineering systems. Hence it is essential than engineers have a toolbox of modeling techniques that can be used to model nonlinear engineering systems. Such a set of basic numerical methods is the topic of this book. For each subject area treated, nonlinear models are incorporated into the discussion from the very beginning and linear models are simply treated as special cases of more general nonlinear models. This is a basic and fundamental difference in this book from most books on numerical methods.

The exposition of this book is at a fairly basic level. The material should be understandable to a reader with a basic undergraduate education in mathematics and some knowledge of engineering models. As compared with many numerical methods books this exposition may be viewed as somewhat light on the theory of numerical methods. The method of presentation is more that of "learning by example" as opposed to presenting a large body of theory. For each chapter and topic area covered, some theory is discussed and computer code is developed and presented such that the reader can conveniently apply the developed code to practical engineering problems. For each topic area not all valid approaches are discussed but for each topic covered at least one approach is extensively developed leading to working computer code that can be applied to nonlinear problems. This book is not a collection of various approaches to numerical analysis but an introduction to one (or more) of the most appropriate approaches to the use of numerical methods for nonlinear models in various applications. One should consult other books for more general overviews of other numerical methods as applied to engineering models.

Each chapter presents computer code segments that can be used to solve a range of engineering problems. The exposition assumes only a basic knowledge of some computer programming language. The code is presented in a modern scripting language that is easy to program and understand, although it is not the most well known computer language. Many readers may wish that another language such as C (or C++) or MATLAB had been chosen for the programming language. However, the Lua language was selected because of its ease of understanding, rapid execution speed and scripting nature that allows the user to readily modify code and execute programs. Many of the simple programming approaches used in the code are not possible in conventional programming languages. Finally, languages such as C and MATLAB incur an additional cost to the uses whereas all of the code used in this book is readily available at no expense to the user of this book. Only public domain programs are used in developing the numerical code segments in this book.

The book is organized into chapters on basic numerical techniques and examples in the beginning with more advanced topics in later chapters building on the early chapters. Very basic information on numerical approaches and the Lua programming language is contained in Chapter 2. For the reader with knowledge of some programming language, this should be sufficient to understand the coded examples in the book.

Chapters 3 and 4 on Nonlinear Equations and Solution of Sets of Equations provide an introductory approach to the methods used throughout the book for approaching nonlinear engineering models. This is referred to as the "linearize and iterate" approach and forms the basis for all the nonlinear examples in the book. These chapters include useful computer code that can be used to solve systems of nonlinear equations and these programs provide a basis for much of the remainder of the book.

Chapter 5 on Numerical Differentiation and Integration is another fundamental aspect of the books approach to nonlinear models. As opposed to most books on numerical methods, numerical differential is extensively used throughout the book for converting nonlinear models into linearized approximations which are then solved by the "linearize and iterate" approach. Numerical differentiation, when properly coded, is an important key for use in solving nonlinear problems.

Chapters 6, 7, 8 and 9 provide important applications of nonlinear models to example engineering problems. Topics covered include data fitting to nonlinear models using least squares techniques, various statistical methods used in the analysis of data and parameter estimation for nonlinear engineering models. These important topics build upon the basic nonlinear analysis techniques of the preceding chapters. These topics are not extensively covered by most introductory books on numerical methods.

Chapters 10 through 13 are devoted to the solution of nonlinear differential equations using numerical techniques. The chapters build in complexity from differential equations in single variables to coupled systems of nonlinear differential equations to nonlinear partial differential equations. Both initial value and boundary value differential equations are discussed with many examples. The discussion emphasizes the reuse of computer code developed in previous chapters in solving systems of coupled nonlinear equations.

In all the applications of numerical methods, considerable attention is given to the accuracy obtained in solving various engineering models. In all the sections, techniques are developed and discussed that allow the user to estimate the accuracy of the numerical techniques when applied to nonlinear problems. This is a very important topic since exact solutions can not be obtained for the majority of interesting nonlinear engineering models. Most books on numerical methods provide little guidance for the user on the accuracy of various methods especially for nonlinear problems.

This book is appropriate for an advanced engineering course at the senior level in numerical methods or for a first year graduate level course. For a one semester course the material in Chapters 1 through 7 should be covered along with selected material in Chapters 10 through 13. If time allows selected sections of Chapters 8 and 9 could also be covered. The book should also prove valuable to engineers and scientists as a reference source for computer code segments that can be readily applied to a wide range of nonlinear models encountered in their work.

In summary, this book emphasizes the general area of nonlinear engineering models with an in-depth discussion of selected computer algorithms for solving nonlinear problems coupled with appropriate examples of such nonlinear engineering problems. The book provides a balance of numerical theory applied to nonlinear or linear engineering models with many example problems. This book is for the engineer who wants to learn the basics of the theory of numerical methods but who wishes to rapidly attain knowledge sufficient for applying modern computer methods to real nonlinear engineering models.

1 Introduction to Nonlinear Engineering Problems and Models

This book emphasizes the general area of nonlinear engineering problems with an in depth discussion of selected computer algorithms for solving nonlinear problems coupled with appropriate examples of such nonlinear engineering models. This introductory chapter provides a general discussion of the theme of the book as well as laying the foundation for subsequent chapters. Many books covering somewhat similar themes to this book have somewhere in their title the term "Numerical Methods". It is certainly true that this book discusses, develops and uses many numerical methods. However, numerical techniques are viewed here as a means to an end. They are simply the techniques used by modern digital computers to solve engineering problems. The end objective of this work is to provide tools and techniques in order for the practicing engineer to understand and model engineering problems. In general such problems are nonlinear in nature; hence the title Numerical Methods for Nonlinear Engineering Models.

1.1 Science and Engineering

Many people in the general public (and even in science and engineering) do not appreciate the fundamental differences between science activities and engineering activities. The standard definitions of science are usually somewhat along the following lines (http://www.sciencemadesimple.com/science-definition.html):

Science refers to a system of acquiring knowledge. This system uses observation and experimentation to describe and explain natural phenomena. The term science also refers to the organized body of knowledge people have gained using that system. Less formally, the word science often describes any systematic field of study or the knowledge gained from it.

The objective of science is typically understood to be the development of a knowledge base sufficient to understand physical reality in terms of a small set of fundamental physical laws and physical principles. Science seeks to understand nature as one finds it. Methods for accomplishing this are typically referred to as the Scientific Method.

Engineering on the other hand (as practiced by Engineers) is an entirely different and separate activity. Engineering seeks to use a set of basic knowledge and guiding principles to create new artifacts of various types that are useful and desired by civilizations. In this pursuit engineers may and do use scientific knowledge. However, they are not limited to the known base of scientific principles but also frequently rely upon a base of experience and empirical knowledge gleamed

1

J.R. Hauser, *Numerical Methods for Nonlinear Engineering Models*, 1–15.
© Springer Science + Business Media B.V. 2009

over years of practice. Engineers create new artifacts that have never before been seen by man. In many cases it is only after such artifacts have existed for some time that science fully understands the underlying fundamental principles upon which the artifacts are based. Engineers employ methods known as the "Engineering Method" which relies on not only scientific knowledge but also empirical knowledge and so called "rules of thumb". The distinction of science and engineering has been succinctly summarized by Theodore von Karman as:

Scientists study the world as it is; Engineers create the world that has never been.

Man's engineering activities preceded his scientific activities on earth. The first engineers – the builders of stone houses and later wood houses – would probably today have been called Civil engineers. Mechanical engineers certainly appeared by the time of the invention of the wheel (probably the 5th millennium BC). Materials engineers emerged during the Bronze Age (3500 – 1200 BC) and the Iron Age (1300 – 900 BC) with the development of new metals and alloys which Mechanical engineers used for new tools and weapons. Later the wide scale use of cements (new materials) by the Romans revolutionized building (Civil engineering). In modern times Engineers have truly revolutionized the way we live especially with the development of Electrical and Computer Engineers in the last two centuries.

The early engineering activities were accomplished long before the development of the scientific base of understanding of the new artifacts. However these accomplishments certainly fit the definition of engineering accomplishments and they provided new artifacts for use by man which had never been seen before and which do not naturally occur in nature. For the early design and building of such artifacts the builders (or engineers) relied upon empirically generated data and empirically developed rules-of-thumb. In engineering such a collection of information is frequently referred to as the state-of-art in a field.

The ability to engineer new artifacts took giant leaps forward with the development of mathematics and new scientific discoveries in modern times. Perhaps the single most important period was that of the late 1600 and early 1700, around the time of Isaac Newton. The development of calculus and Newton's Laws of Motion provided the fundamental mathematical and science basis for the explosion of engineering activity since that time with the development of a marvelous array on new artifacts for our pleasure and use. It's difficult to envision our understanding today of engineering systems and artifacts without the use of calculus.

Some readers may feel that the distinction between science and engineering is more blurred that that indicated above – that these is a more continuous spectrum between science and engineering. However, this is not the case. The major objectives of science and engineering are clearly very different as the statement by von Karman above indicates. It is certainly true that some persons trained in the science disciplines spend their careers doing primarily engineering work and some persons trained in the engineering disciplines have even been known to contribute to fundamental science, but the two disciplines are far apart in their fundamental objectives.

1.2 The Engineering Method

Since this book is about engineering problems it is perhaps useful to understand a little about the engineering method that gives rise to engineering problems to be studied herein. The engineering method shares some similarities with the scientific method but differs in important areas. Engineering assumes that there is a set of scientific principles and laws which can be used by the discipline. However, engineering would be severely limited in developing new artifacts if it had to solely rely upon fundamental physical laws. For example if a structural engineer had to relay on the calculation from fundamental physical laws of the weight for failure of a concrete column, he/she would be severely limited in designing safe bridges across rivers. However, the engineer can construct empirical tables relating column failure to weight on columns and treat this data as just as important as other fundamental physical laws. Science may later relate such failure to an understanding of more fundamental microscopic features of the concrete. However, this does not invalidate the engineer's tables, but may provide more insight into the failure process. Thus the engineer consults not only fundamental physical laws and principles in artifact development but sets of empirically generated data over past experience – the so called state-of-the-art. The typical major steps in the engineering design of a new artifact are shown in Figure 1.1. The process typically begins with the concept of a new engineering artifact – such as a bridge, automobile, computer chip, cell phone, etc. The engineer then has at his disposal a collection of physical and empirical laws and relationships on which to draw for his/her design. He/she can even draw upon the properties, look, feel, etc. of similar artifacts if they exist.

Past history or state-of-the-art is an important part of the engineer's toolbox. Taking prior art and physical laws and algorithms into account, the engineer can then produce various mathematical models, equations and algorithms for describing the expected performance of the new artifact. This also includes typically models for the performance limits of the artifact. Engineering models may also consider economic issues and environmental issues – issues not usually of concern to the pure scientist seeking to discover new laws. In many engineering works, not all the data or models required to design the new artifact is known so the engineer must then perform experiments to collect new data and develop required relationships. This is indicated by the side box in the figure. This is one of the steps where engineering activity and scientific activity appears similar. In fact the engineer may employ a similar methodology to the scientific method to collect the required information. However, the objective of the engineer is not foremost to discover new physical laws, although this might occur, but to generate needed relationships between physical variables. After various stages of model building, the next step is then typically the solution of or the exercise of the various models to evaluate the expected performance of the artifact – load limits for a bridge, speed of a computer chip, throughput of a transmission line, etc. For modern day engineering, this typically involves extensive calculation work on a digital computer.

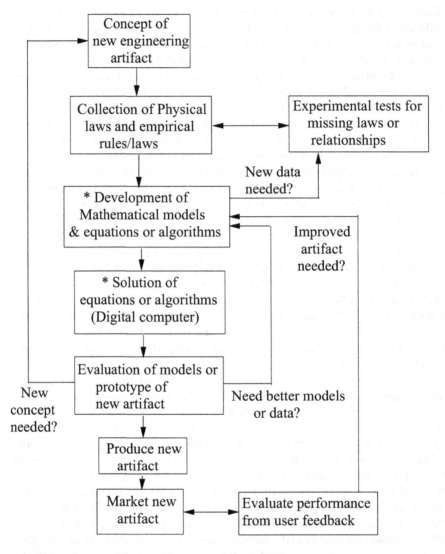

Figure 1.1. Typical major steps in the engineering method for developing a new artifact.

The last two steps mentioned above are the main subject areas of this book – see the boxes in Figure 1.1 with a star (*). In fact the bulk of the discussion will involve the solving of engineering models (or equations). Also some discussion is included about model building and how to determine parameters of models using experimental data. The purpose of computer solutions of the various models is to gain knowledge about the expected performance of the proposed artifact. From the engineer's prospective the numerical methods are simply a means to the end of generating accurate simulations of various aspects of a desired artifact. To this end the engineer needs to be somewhat familiar with various numerical techniques and their limitations and accuracy. It is hoped that this work will provide that knowledge for a range of engineers.

The results of computer solutions for various aspect of an artifact are used by the engineer to then access the feasibility of a new design. At this point several conclusions may be reached. The conclusion may be that improved models or data are needed and the process loops back to a previous design point as indicated by the right branch on the figure. The conclusion may be that the proposed concept is not feasible either from a performance standpoint or from an economic standpoint or even from an environmental standpoint. In such case the process may need to loop back all the way to the design concept stage as indicated by the left branch in the figure. It is not uncommon to iterate through such design loops many times before a satisfactory new artifact reaches the production stage. Thomas Edison, perhaps the US's greatest engineering inventor, cycled through thousands of materials before hitting upon the carbon filament for obtaining a practical incandescent lamp. At this evaluation stage the engineer may also produce a prototype of the artifact which may be a close approximation to the final artifact or a scale model of the final artifact. Part of the design evaluation process may then be an evaluation of the performance of the prototype artifact. In many engineering designs it is simply impractical to produce a full size prototype – for example a large bridge or a large building. In such cases it is essential that the engineer has performed extensive model testing and that means in today's world extensive computer modeling and simulations.

To complete the engineering process the designed artifact is then produced and marketed to a select customer (such as a state for a bridge) or to the mass consumer market. Hopefully the engineer or company marketing the artifact is able to make a monetary profit on the artifact so he/she can continue with the design of newer and better artifacts. A final step in which engineers are also involved is typically the performance evaluation of the artifact with feedback from users. Such feedback can in some cases necessitate modifications in the design for safety or performance issues. Feedback may also lead to the decision to start the design process anew to produce the next generation model of artifact or to produce an entirely new artifact.

The development of modern day high speed computers (on engineer's desktops) has drastically changed the design process for engineers. Before the computer age engineers had to rely much more heavily on design prototypes and scale models to access the quality and expected performance of a new artifact. With computers and the accumulated knowledge (state-of-the-art) in many fields, it is possible to predict with a high degree of accuracy the performance of a new artifact before it is ever produced. In a vast number of cases physical prototypes are no longer necessary before committing to a new design. Hopefully, the material in this book will aid engineers in the use of digital techniques as applied to engineering problems.

1.3 Some General Features of Engineering Models

In the design of artifacts, engineers typically build models for a set of physical performance variables in terms of a set of other physical variables. In the broadest sense an engineering model is an equation (or set of equations) that expresses some features of a physical system in terms of other parameters such as shown in Eq. (1.1)

$$\left. \begin{array}{l} \text{Dependent} \\ \text{variable} \end{array} \right\} = f(\text{independent variables, parameters, forcing functions}), \qquad (1.1)$$

where the system variables are typically characterized as "independent variables" "parameters" and "forcing functions". To have some simple examples of these concepts one might have the elongation of a rod (the dependent variable, strain) dependent on the force applied to the rod (the forcing function, stress) with the modulus of elasticity as the material parameter – i.e. the stress-strain relationship. Or one might have the current (dependent variable) flowing in an electric circuit due to the applied voltage (or forcing function) with the resistance as the material parameter – i.e. the I-V relationship. In mathematical form such relationships might be expressed as:

$$
\begin{aligned}
U_1 &= F_1(x_1, x_2 \cdots x_n), \\
U_2 &= F_2(x_1, x_2 \cdots x_n), \\
&\vdots \\
U_m &= F_m(x_1, x_2 \cdots x_n).
\end{aligned}
\qquad (1.2)
$$

In this the x parameters are assumed to be the independent parameters of the model and may include the forcing function parameters and the U parameters are the dependent parameters. The F functions represent some functional relationship between the independent parameters and the dependent parameters. This relationship may be explicitly given by some closed form mathematical expression or it may be given only by some algorithm which allows one to numerically compute the relationship. In general the relationships may also involve various derivatives of the dependent or independent variables.

Exactly what distinguishes a dependent variable from an independent variable is somewhat fluid in many cases. For example for some features of a model, temperature may be considered an independent variable determined by the environment in which the artifact is embedded. However, a closer examination (or more detailed model) may indicate that at a more accurate level, temperature depends on other variables in the modeling space. Variables may thus change roles as more detailed and complete models are developed. For more concrete examples, general spatial variables and time typically fall into the independent category. However, the spatial variables associated with an artifact or part of an artifact may be important dependent variables depending on forces present on the object. Engineers frequently speak in terms of forcing functions which are external influences acting on an engineering system. For the purpose here these can be considered simply as part of the independent variable set.

In addition to independent variables, an artifact typically has a set of intrinsic parameters (material, chemical, spatial, etc.) that also help determine the model equations. For the example relationships in Eq. (1.2), each dependent variable is shown as a function of only the independent variables. In the most general case these relationships can involve very complicated nonlinear relationships and very complicated coupled relationships between the dependent and independent variables, time derivatives of the dependent variables or integrals of the time dependent variables. Such a set of functional relationships between independent and dependent parameters is what one frequently refers to as a computer model or engineering model of the artifact.

This book is about using modern computer languages to solve engineering equations or models as briefly described above. Only a little discussion will be devoted here to how one develops such engineering models. One means of developing model equations is from physical constraints such as:

1. Newton's laws of motion
2. Conservation laws
3. Chemical reaction laws
4. Electric circuit laws
5. Electromagnetic laws
6. Stress-strain laws
7. Gas and fluid dynamic laws

In addition to these physical laws, other model relationships and equations typically result from (a) empirically known relationships among variables and (b) economic constraints. From this array of laws and relationships, the engineer formulates as complete a set of system equations or models as possible describing the artifact and then uses computer techniques to evaluate and simulate a set of desired response variables.

For a specific example the designer of a new bridge must be able to confidently predict that the bridge will not collapse and must be able to estimate the cost of the new bridge. An engineering design usually involves some level of uncertainty and some level of tradeoffs. In the case of a bridge, the engineer can not predict exactly at what load level the bridge will collapse. In fact he does not want to push his design to the edge of collapse, although this would probably be the lowest cost solution. So the engineer approaches the uncertainty of the load level for collapse by designing in a safety margin factor, setting the predicted load for collapse at a multiplicative factor above the expected maximum load of traffic under the most severe weather conditions. By handling the uncertainty in this manner, the engineer must specify larger size concrete supports, for example, than a marginal design. This in turn increases the cost of the bridge and this is one of the tradeoffs in the design cycle. The engineer can be more confident in the design (reduce the uncertainty) at the expense of a larger cost for construction. If the engineer is with a construction company bidding for a bridge design, he/she must carefully consider the tradeoff as too large a multiplicative factor may result in lose of the contract due to too high a cost and too low a multiplicative factor may result in a questionable safety margin. Such engineering tradeoffs are ideally suited for extensive computer modeling coupled with a base of extensive empirically based knowledge.

It is hoped that this brief discussion will give the reader a little appreciation of how sets of model equations or relationships are developed for engineering design and how important such models are in designing and building new engineering artifacts. The work here will simply assume that such engineering models are known and concentrate on methods of analysis using numerical computer techniques.

1.4 Linear and Nonlinear Models

The equations and models used by engineers can be grouped into two broad categories as either "linear" or "nonlinear" models. While there are almost minimal differences in many cases between the formulation of such models, approaches to simulating or solving such model equations are considerable different. To discuss some of the differences, consider the example in Figure 1.2 of some engineering system excited by some external force. In this it is assumed that some external force, $f(t)$, produces a response of some system variable, $y(t)$. For a simple example, force on the concrete support pillar for a bridge will produce some change in the length of the support. Or voltage applied to an electrical circuit element may produce some current in the element or some power dissipated in the element. If the model of the engineering system is a linear model it satisfies the following properties:

If $f_1(t)$ and $f_2(t)$ produce responses $y_1(t)$ and $y_2(t)$ then the external force
$f(t) = C_1 f_1(t) + C_2 f_2(t)$ will produce the response $y(t) = C_1 y_1(t) + C_2 y_2(t)$.

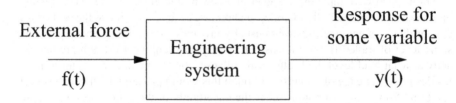

Figure 1.2. Illustration of some engineering system with an external force producing a response of some system variable.

For such a system, knowledge of the response of the system to an external force of some given magnitude implies that one also knows the response of the system to a similar force of any desired magnitude. There are many other important implications of dealing with linear systems that are too extensive to enumerate here and the reader is referred to the many texts devoted to linear system analysis.

For a nonlinear system, no such general properties exist between the responses of a system from one external force to another external force. Knowing the response to a force of one magnitude does not imply knowledge of the response to a force of the same type but of different magnitude. This makes nonlinear systems considerably more difficult to simulate and analyze. Simple examples of nonlinear models are model equations involving powers of the external force or differential equations involving products of derivatives and variables such as the following:

$$y = C_1 f + C_2 f^2 + C_3 f^3$$

$$C_1 \frac{d^2 y}{dt^2} + (C_2 + C_3 y)\frac{dy}{dt} + C_4 y + f(t) = 0 \tag{1.3}$$

In the first case, the nonlinearity is in the powers of the f terms while in the second case the nonlinear term is the $C_3 y$ term multiplying the first derivative. In terms of functional form there is very little difference between the nonlinear differential equation with a nonzero C_3 term and one in which C_3 is zero. However, the mathematical tools available for solving even the simplest nonlinear equations are considerable different from the vast array of tools available for studying linear models.

The emphasis of this book is on developing tools for the analysis of nonlinear models and equations. There are several reasons for this. First and foremost is the fact that essentially all engineering models are fundamentally nonlinear models. Some may argue with this view. However, if one pushes all engineering systems to extremes of excitation, the models and equations in general become nonlinear models. Take the case of stress on the concrete support for a bridge design. If pushed to too large a load, the concrete support will become a nonlinear element giving way as the bridge collapses. In fact the yield point of the bridge support is an important design consideration so that the engineer can provide an appropriate margin of safety. The response of engineering systems to limits of performance is of major interest and this generally means that one is considering the presence of nonlinear terms in one's models. Other important classes of engineering systems directly involve nonlinear models in the basic operation. For example, all classes of digital electronic circuit elements involve the nonlinear operation of electronic devices. It is the basic inherent nonlinearities of such devices that allow one to implement digital computing elements dealing with two stable logic states.

Thus the world of nonlinear models and equations is very important to the field of engineering. An advantage of approaching everything directly from the start as a nonlinear model is that cases of linear models become simply special cases of general nonlinear models. A brief discussion of how the analysis approach for nonlinear models differs from that for linear models is thus appropriate. A vast array of books and analysis tools exist for linear models and linear equations. No such array of material exists for nonlinear models. Even proofs that solutions exist for nonlinear models are in the vast majority of cases not available. So what are the general principles available for solving nonlinear models as opposed to linear models? There is one overriding principle that has survived the test of time for approaching the solution of nonlinear models. This is the principle of "linearize and iterate" (the L&I technique) as illustrated in Figure 1.3. Starting with some general nonlinear models (or sets of equations) one approximates the models by a set of linear models using some initial approximations to the variables in the models. This set of linear equations is then solved for a new and hopefully better approximation to the model variables by use of any of the array of techniques available for solving linear models. After updating the model variables, a test of some type is then made of the new variables to determine if sufficient accuracy is

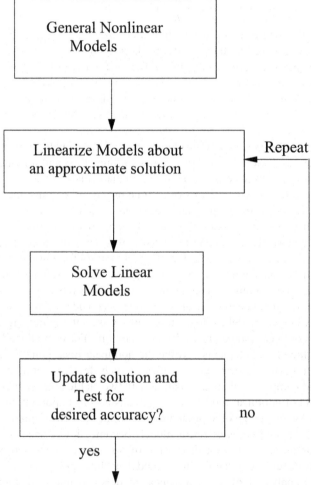

Figure 1.3. Illustration of the Linearize & Iterate (L&I) technique.

achieved in the solution set. If not the process is repeated in an iterative loop. If sufficient accuracy is achieved the process is terminated with hopefully an accurate set of modeled parameters.

Such a linearization and iterative approach typically goes by the name of the Newton method or the Newton-Raphson method. This approach was originally developed by Newton (in 1669 and published in 1685 in *A Treatist of Algebra both Historical and Practical* by John Wallis) for application to finding the roots of polynomial equations. It was based upon earlier work by Heron of Alexandria for finding the square root of numbers. This work was extended by Joseph Raphson in 1690 in *Analysis Aequationum Universalis* but again purely for roots of polynomials. In modern times, extensions of this basic linearization approach have been applied to more general functional approaches involving systems of general equations and nonlinear differential equations. This general L&I approach forms the basis for solving all types of nonlinear models in this book.

From this brief discussion one can see that the solution of nonlinear models is closely tied to a toolbox of methods for solving linear models. A linear set of model equations requires only one loop through the L&I iterative loop and thus can be considered a special case of solving nonlinear models. Thus of necessity considerable discussion will be presented in subsequent chapters of solution methods for linear models. However, this will be considered only as an intermediate step in embedding such linear models within an iterative loop for the solution of more general nonlinear models. Considerable effort will be given to the linearization of various models, on the testing of solution accuracy and on the final accuracy of various solution methods for nonlinear models.

The Linearize and Iterate approach is essentially the only general approach available for solving a broad range of nonlinear engineering models. When properly applied, it is a very powerful method for nonlinear problem solving. In a large number of applications it can be applied in such a manner that the iterative loop converges to a solution in what is known as "quadratic" convergence. This means that the error in the solution decreases in a quadratic manner with each iterative step. For example if the relative error in some modeled variable is of order 1.e-2 in one iterative step it will be of order 1.e-4 in the next loop and 1.e-8 in the next loop and 1.e-16 in the next loop. For such a convergence rate, only a few L&I loops are needed for converged solutions and in the majority of cases less than 20 such L&I loops are required if convergence is to be achieved. Convergence meaning that one obtains the same solution set (at least to some level of accuracy) with each transit through the iterative loop. Of course this does not guarantee that the converged solutions are the only solution or the sought after solution, since for many nonlinear models multiple solution sets may exist. Considerable discussion is given to accessing the accuracy of L&I solution sets.

An iterative loop such as shown in Figure 1.3 may converge to a stable solution set or it may never terminate with a stable solution set. For many simple applications of the Newton-Ralpson technique it is known that a stable solution set can only be obtained if the initial approximation to the solution set is sufficiently close to the final converged solution set. However, no general rules can be given for how close the initial approximation must be for general nonlinear models. The more information one has about solutions to one's nonlinear models, the more successful one is likely to be in obtaining accurate solutions to nonlinear models.

The question of proving that a solution set actually exists for a given model is much discussed in the theory of linear models by mathematicians. However, for many engineering models, this is more of academic interest than practical interest. If one is attempting to model a physical artifact, then one knows, or strongly expects, that a real solution exists because one can construct the artifact and some manifestation of it will exist. Thus engineers really expect that a solution set exists for the models that they construct. In any given model of a system, the defining equation set may lack sufficient detail to give a solution or the equations may not be sufficiently accurate to give an accurate solution, but with sufficient detail in the model, the engineer expects that a solution exists. If the engineer is unable to obtain a converged, solution to a nonlinear model, it may be due to (a) an insufficiently detailed model description or (b) an initial approximation that is too far removed from the actual solution. It's not always easy to determine which of these cases have occurred, although one can help to eliminate the latter by exploring different initial approximations. This text contains a large number of examples that hopefully will give the reader a much deeper appreciation of the L&I method and how it is applied to a variety of nonlinear models.

Another important limitation related to using computers to solve engineering models relates to the fact that computers can only operate on a finite number of data values. Consider the relatively simple problem of numerically computing the value of the integral of a function such as shown below:

$$C = \int_a^b f(x)dx \tag{1.4}$$

Because x is a continuous variable, there are an infinite number of possible x values over any finite range such as the range a to b. However a computer evaluation of this integral can only consider a finite number of spatial points (in a finite time). This limitation of approximating the range of parameters by finite numbers of points introduces errors in any numerical results and this applies to either linear or nonlinear models. One general numerical technique that will be used throughout the book to improve the results of such calculations is that of "value extrapolation" as illustrated in Figure 1.4. In this technique calculations are made with varying numbers of parameter values that differ by some fixed ratio (such as a factor of 2). Within an iterative loop the results of calculations with varying numbers of parameters can be extrapolated to predict the results with ever increasing numbers of parameters. In the case of evaluating an integral, for example, the number of points used to evaluate the function can be increased in each iterative loop. Under rather general conditions on the numerical algorithm, the accuracy of the extrapolated value can be greatly increased for a given computational effort. This will become clearer in subsequent chapters which use this technique, but this is a general numerical approach that can be used to advantage in numerical analysis. This and the L&I technique provide two general analysis techniques for use with nonlinear numerical models.

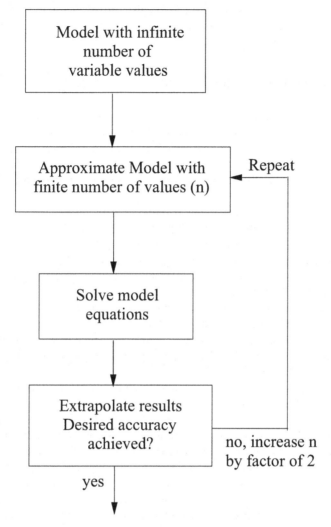

Figure 1.4. Illustration of technique of "value extrapolation" to improve the accuracy of numerical results.

When a solution to a nonlinear model is obtained, an important question always concerns the accuracy of the achieved solution. For numerical solutions to linear models, much is known about the accuracy of such techniques. However, such is not the case with regard to nonlinear problems. The topic of the accuracy of various numerical methods is frequently given little coverage in books on introductory numerical methods. In this work considerable effort is devoted to discussing the accuracy of all the numerical techniques developed. This is essential for nonlinear models as for almost all such problems exact closed form solutions are not known. A variation of the approach shown in Figure 1.4 using results from model evaluation with varying numbers of system parameters also provides an important algorithm that is repeatedly used to approximate the computed accuracy of many nonlinear engineering models.

1.5 A Brief Look Ahead

Before jumping into the details of nonlinear models, it is probably useful to summarize the general type of models and problems to be addressed in this book. The next chapter continues to provide some necessary background material on basic numerical methods and on basic computer programming techniques. The approach to solving nonlinear models begins with Chapter 3 and the major topic areas are:

> Single nonlinear equations in one variable – Chapter 3
> Sets of nonlinear equations in multiple variables – Chapter 4
> Numerical differentiation and integration – Chapter 5
> Interpolation of nonlinear data arrays – Chapter 6
> Fitting of nonlinear data and data plotting – Chapter 7
> Statistical methods for data – Chapter 8
> Parameter estimation for nonlinear functions – Chapter 9
> Nonlinear differential equations of initial value type – Chapter 10
> Nonlinear differential equations of the boundary value type – Chapter 11
> Nonlinear partial differential equations by finite difference methods
>   – Chapter 12
> Nonlinear partial differential equations by finite element methods
>   – Chapter 13

The material builds from the simplest nonlinear cases to the more complex cases of partial differential equations in two spatial dimensions and one time dimension.

The approach in each chapter will be to provide sufficient discussion and background material for the reader to understand the techniques developed with a minimum of previous knowledge required. This is then followed by the development of complete computer code needed to implement the algorithms discussed. Finally each chapter contains examples of the used of the generated methods and code for solving typical nonlinear models in each problem domain. The approach presented here however, differs from that of many similar texts in the following ways:

> The chapters do not cover all possible approaches to solving problems in a given area. Only certain selected approaches are used in developing detailed computer programs. The selected approaches are believed by this author to be those either most easily understood or most practical for a wide range of nonlinear problems. In most cases the approaches most easily implemented are the approaches selected for detailed implementation. One of the philosophies used in guiding algorithm development is the principle that a simple algorithm applied many times is to be preferred to a more complex algorithm applied a fewer number of times.

These guiding principles will not find favor with all readers. While in most chapters some mention is made of other approaches, the reader interested in a broad coverage of numerical methods for nonlinear problems is encouraged to supplement this work other texts that may provide a broader overview of solution methods. The advantage of the approach selected here is that much more usable computer code is presented herein than in most texts addressing numerical methods. The other major difference is the emphasis on nonlinear models from the very beginning and then treating linear problems as simply special cases of the more general algorithms developed.

2 Numerical Fundamentals and Computer Programming

The previous chapter has briefly discussed the concept of nonlinear engineering models and their solution which is the major topic of this work. To review, the term engineering models refers to a collection of relationships between some set of parameters associated with an engineering artifact and some desired set of performance parameters. The relationships may be explicitly given but in most cases are implicitly expressed through sets of nonlinear equations, differential equations and/or algorithms which specify how one generates the associated relationships. An engineer is typically interested in exploring the space of performance parameters as a function of some set (or subset) of artifact parameters. In modern day practice, this exploration of responses is typically done by use of a digital computer executing some type of computer program written in some type of computer language. Some of the fundamental background expected of the reader to implement such numerical computer programs is reviewed in this chapter.

Not many years ago engineers primarily used large centrally located mainframe computers for any extensive modeling work. However, this has now changed to the primary use of desktop computers or workstations for all except the most numerically intensive calculations which may still involve supercomputers. For the discussion here it will be assumed that the code examples will be implemented on typical desktop computers (typically called PCs).

2.1 Computer Programming Languages

The component (or we might say artifact) that interacts between the engineer and the computer hardware is typically a computer program written in some type of computer language. A vast array of computer programming languages has been developed over the years. As of late 2007, a ranking of the usage of over 100 such languages (http://www.tiobe.com/tpci.htm) lists the top 20 most web accessed languages as Java, C, Basic, C^{++}, PHP, Perl, C#, Python, JavaScript, Ruby, PL/SQL, SAS, D, Delphi, Lua, COBOL, ABAP, Lisp, SQL and Ada. For numerical engineering work the programming languages have traditionally been FORTRAN and C (or C^{++}). However, FORTRAN has not been in much favor in recent years and probably should not be considered for new applications. For the program examples in this book a programming language must be selected -- but which one?

Let's briefly consider the desired attributes of the language to be selected for this work. (a) First and foremost should be the ability to clearly and succinctly

J.R. Hauser, *Numerical Methods for Nonlinear Engineering Models*, 17–41.

express numerical algorithms. (b) The language should be readily available to users of this book. (c) The language should have modern features such as object oriented features. (d) The execution speed should be reasonable fast. Several of the languages listed above can be eliminated because of some of these desired features. Some of the languages are designed primarily for web applications such as PHP, ABAP and some such as SQL are designed for data base management. The most obvious choice from the 20 languages listed above is probably C or C^{++}. However, in recent years a class of programming languages collectively known as "scripting" languages has become very popular. Languages in this category include Java, Perl, Python, JavaScript, Ruby and Lua. Another language with somewhat similar features that finds considerable usage among engineers is MATLAB. These languages are typically built on top of a conventional language such as C or C^{++}, they are typically interpreted languages, or in some cases compiled to an intermediate code, and have good high level language features. They can be classified as "code and execute" languages as compared with the classical "code, compile, link and execute" languages such as C. By eliminating the compile and link steps they make program development much simpler and faster.

In deciding on a computer language for this work, three of these scripting languages were examined in detail: Python, Ruby and Lua. In addition MATLAB was also considered. All of these are somewhat similar in design philosophy and usage. Some of the important features of these languages are:

> No compile or link steps – just code and execute
> No type declarations – variable names can refer to any type variable
> Automatic memory management – great simplification in programming
> Simple high level data types – simplifies programming
> Object oriented features – simplifies modern program development
> Embeddable in C programs – useful for mixed language execution
> Dynamic loading of C or source programs – easily modified code
> Widely available code – source code freely available, except for
>   MATLAB

Of these features the first three of no compile/link step, no type declarations and automatic memory management greatly simplify the program development cycle and make the reading and understanding of source code much simpler than equivalent C or C^{++} programs. These languages also have extendibility features that greatly simplify the ability to deal with complex data structures.

Of all these scripting languages, the language with the fastest execution speed is Lua and because of this it is finding increasing applications in computer games. Lua also probably has the simplest and most straightforward syntax and the simplest data structure as only one data type, a table, is supported by the native language. It is also probably the most extendible as to new data types and it can be used as a callable routine from C code or one can call C routines from Lua code. The Lua language is written in ANSI C, is freely available and has been compiled for a wide range of computer hardware and operating systems.

Some readers will obviously find the selection of Lua instead or C (or C^{++}) as a distinct stumbling block at first. However, one should first just give it a try. In this author's opinion, one should use a modern scripting language where ever possible as opposed to conventional languages. One will program considerable faster and write much cleaner code that will be much more readable by ones self or by others. The only drawback to Lua (or another scripting language) is possibly some sacrifice in execution speed. However, in most of the examples in this text the time saved by eliminating compile and link steps more than makes up for the interpreted nature of the language. The code examples in this text can always be converted to C if desired. However, some of the higher level, object oriented features that are used can not be easily expressed in C. It should also be noted that Lua has no "pointer" variable type as in C thus simplifying programming, especially for inexperienced programmers. Finally Lua is freely available.

2.2 Lua as a Programming Language

This section discusses some of the general features of Lua as a programming language. A more detailed discussion of Lua is contained in Appendix A. For this discussion the reader is assumed to be familiar with at lease one conventional programming language such as C, BASIC or Java. The basic constructs provided by all programming languages are somewhat similar and Lua is no exception. Lua is a case sensitive, dynamically typed language with each data value carrying its own type. Thus one can start using variables and values without any type definitions. The following words are reserved in Lua:

and, break, do, else, elseif, end, false, for, function, if, in, local, nil, not, or, repeat, return, then, true, until and *while*.

Most of these should be familiar from other languages. The data types of Lua are:

nil, Boolean, number, string, userdata, function, thread and *table*.

If some of these are unfamiliar, they will become clear when used in examples.

Two of the data types should be discussed as their usage in Lua is somewhat different from most other computer languages. First Lua has only one numerical type, *number*, used to represent both integers and double precision floating point numbers. As such Lua can represent larger integers than most languages which use special integer types. Second, Lua has only one data structuring type, the *table* type which implements general "associative" arrays that can be as large as computer memory allows and that are dynamically managed by the internal garbage collection feature. Any table may have associated with it another special table called a "metatable" through which one can implement powerful language extensions and object oriented features.

Lua has the standard array of assignment statements, arithmetic operators, relational operators and logical operators with the standard precedence. Lua logic control is implemented with:

(a) *if ... then ... else* (or *elseif*) *... end* statements,

(b) *while ... end* statements,

(c) *repeat ... until* statements

(d) numeric and generic *for ... end* statements.

Finally *break* and *return* allow one to jump out of a logic control block.

Functions are implemented in Lua with the syntax:

> *function fx(argument_list) ... end* or with
>
> *fx = function(argument_list) ... end.*

One difference with many languages is that functions may return multiple values as for example *x, y = fx()*. Another feature is that functions may receive a variable number of arguments with the function determining how many arguments are present. Also the number of arguments used in calling a function may differ from the number used in defining the function. These features will become clearer as the language is used in examples.

In addition to the basic programming language, Lua provides several libraries of functions for various standard functions. These are:

1. the mathematical library (math[] with sin, cos, etc.),
2. the table library (table[] with insert, remove, sort, etc.),
3. the string library (string[] with string matching, substituting, etc),
4. the I/O library (io[] with read, write, etc.),
5. the operating system library (os[] with system calls) and
6. the debug library (debug[] for program development).

In addition a Lua program may add additional libraries of user defined functions through the *require("library_name")* function. In the above naming of the libraries, a pair of square brackets (such as math[]) have been added at the end of the name to indicate that the named libraries refer to a table of elements. As an example for the math library, the sin() function would be accessed as math.sin(). In this example, a pair of parentheses (such as sin()) have been added to the sin function to indicate that this name refers to a Lua function. This notation to identify tables and functions will be extensively used in this work.

For a more detailed description of the Lua language, the reader is referred to Appendix A. After learning a few basics of a programming language probably the most effective learning method is through program examples, of which a few will now be given to demonstrate the major features of the Lua language. Listing 2.1 gives a very simple code example illustrating some of the basic operations of the language. The code should be readily understandable to anyone familiar with a previous programming language. However, several important features of the example will be briefly discussed.

First, comments (text beginning with a double hyphen) are liberally dispersed throughout the code. In addition to single line comments with the double hyphen, one can use block comments by enclosing any number of lines of comments between the characters --[[and the characters]]. Lines 4 through 14 illustrate a function definition. This function attempts to evaluate the largest integer value that can be represented in the computer language. It contains a "for" loop from lines 6 to 13 that double an x value each time through the loop. The initial value of x is either input to the function or set to 1 on line 5 as the first statement of the function if the input value is "nil" or not defined. The function exits on line 11 when adding and subtracting 1 from the loop variable can no longer be accurately computed. This provides an estimate (to within a factor of 2) of the maximum integer value that can be used with the computer language. This value is returned by the function on line 11 and set to the variable i on line 16. The reader will note the consistent use of the *end* keyword to terminate control blocks and function definitions. The printed output shows that the value returned by the function is $9007199254740992 = 2^{53}$. This illustrates that the language has slightly better than 15 digits of precision in representing integers. The 53 printed out by line 18 of the code also shows that 53 binary digits are being used to store the mantissa of a floating point number. This version of Lua was compiled with Microsoft's Visual C and this is in fact the maximum value of the mantissa for a double precision number using this compiler.

```
 1 : -- File list2_1.lua -- Simple Lua operations
 2 :
 3 : -- Test for maximum integer value
 4 : function tst(n) -- function definition
 5 :     x = n or 1
 6 :     for i=1,400 do -- Do loop
 7 :       x = x + x -- double x value
 8 :       local y = x+1 -- Local variable
 9 :       z = i -- new variable
10 :       if y-1~=x then -- Simple if statement
11 :          return i -- Return from function
12 :       end -- End if test
13 :     end -- End for loop
14 : end -- end function
15 :
16 : i = tst() -- Call function with return value
17 :
18 :  print("i, x, y, z = ",i,x,y,z) -- Simple print
19 :
20 : -- Formatted print using io and string libraries
21 : io.write(string.format('x = %40.10f \nx-1 =%40.10f \n',x,x-1))
22 :
23 : -- Use of math library
24 : x,y = math.sin(math.pi/4), math.cos(math.pi/4)
25 : print('x, y = ',x,y)
26 :
27 : -- Illustration of table
28 : t = {n = 4, 1,3,5}
29 : table.foreach(t,print) -- print table using library
Selected Output:
i, x, y, z =    53      9.007199254741e+015    nil    53
x = 9007199254740992.0000000000
x-1 = 9007199254740991.0000000000
x, y = 0.70710678118655    0.70710678118655
1       1
2       3
3       5
n       4
```

Listing 2.1. Simple example of Lua computer code.

Associated with this result are various print statements on lines 18 and 21 using the simple print() function and functions from the string and io libraries. The format for calling various functions from a library can be noted as for example: string.format() or io.write() in calling format() from the string library or write() from the io library. A final feature of the function usage is that on line 16 with i = tst(), no argument is supplied for the function call. In this case Lua will automatically provide the keyword *nil* for the argument and when called this will result in the assignment of 1 to the variable x on line 5 of the function.

The code in this example illustrates several language features such as the "do .... end" loop, the "if ... then ... end" loop and a function definition. The code also illustrate some features of the lexical scope of Lua variables with the print() statement on line 18. It is noted that variables x, y and z are first introduced within the body of the tst() function definition. However, y is declared as a "local" variable and is thus known only within the scope of the code block in which it is defined, which in this case is within lines 6 through 13 of the code. The attempt to print the value of y on line 18 thus results in the printed value of 'nil' indicating an unknown value. This illustrates the fact that in Lua, variables are known globally unless restricted to a code block by the local keyword, i.e. the default is that all variables have global scope.

Multiple assignment on a line and two functions from the math library are illustrated on line 24. Multiple assignment is a very useful feature and allows the one line reversing of values in variables as in the statement x, y = y, x which is perfectly valid in Lua. A simple illustration of defining a table of values is given on line 28 with the printing of the values by a table library function on line 29 using the function table.foreach(). Tables in Lua are associative arrays meaning that values can be indexed not only by numbers but by any strings or any other language value, except *nil*. In this example on line 28, one element is indexed by the value "n" (which is treated as a string). The other table values of 1, 3 and 5 are given default integer index values of 1, 2 and 3 as can be verified from the printed output. The value associated with n may be accessed by either t.n or by t["n"] while the integer indexed values may be accessed by t[1], t[2] and t[3]. The use of the form t.n is considered syntactic sugar for t["n"]. It can be noted that the access to library functions such as string.format() on line 20 simply accesses an entry in the string table which is an array of the names of functions and in this case an entry with the name "format". Values stored in a table may be any valid Lua data type, including other tables and functions. Of course only the name associated with the data types is actually stored in the table with the actual table or function definition stored in other locations in Lua memory. A matrix of values is then stored in Lua as a table of tables with one table pointing to the tables storing the rows (or columns) of the matrix. Tables are one of the key features of Lua and require a little getting familiar with as they are somewhat different from arrays in many languages. By combining associative arrays and integer indexed arrays into one table form, Lua provides a very powerful data structuring feature.

2.3 Data Representation and Associated Limitations

Internally, digital computers store and process data in the form of binary digits. A computer language such as Lua must store numbers in terms of some finite number of binary digits and this leads to certain limitations on the representation of numerical data. Since Lua is written and compiled in C, it has inherently the same data representation limitations as the base language (ANSI C). For numerical work, several limiting values are important to know. It is not essential for most numerical work to know the exact format in which a computer language internally stores binary digits. However, it very important to know the maximum and minimum number values that can be stored in the language. From Listing 2.1 it has been shown that 9007199254740992 is the largest integer value that can be stored in a floating point number and that 53 binary digits are used to store the mantissa of a floating point number.

```
 1 : -- /* File list2_2.lua */
 2 : -- test of fundamental arithmetic limits with Lua
 4 : -- Test for relative accuracy
 5 : eps = 1
 6 : while 1 do
 7 :     eps = eps/2
 8 :     b = 1 + eps
 9 :     if b==1 then break end
10 : end
11 : print("Machine eps = ", 2*eps, math.log(2*eps)/math.log(2))
12 :
13 : -- Test for smallest floating point number
14 : nmn,a = 1,1
15 : while 1 do
16 :     nmn = nmn/2
17 :     if nmn==0 then break end
18 :     a = nmn
19 : end
20 : print("Smallest floating point number = ",a,nmn)
21 : print("Values around smallest number = ",1.4*a,1.5*a,1.9*a)
22 :
23 : -- Test for largest floating point number
24 : nmx,a,inf = 1,1,1/0
25 : while 1 do
26 :     nmx = nmx*2
27 :     if nmx==inf then break end
28 :     a = nmx
29 : end
30 : print("Largest floating point number = ",a,nmx)
31 : print("Values around largest number =",2*(1-eps)*a)
Output:
Machine eps =   2.2204460492503e-016   -52
Smallest floating point number =      4.9406564584125e-324   0
Values around smallest number =       4.9406564584125e-324
9.8813129168249e-324   9.8813129168249e-324
Largest floating point number = 8.9884656743116e+307   1.#INF
Values around largest number = 1.7976931348623e+308
```

Listing 2.2. Simple experimental tests of floating point number limits.

The code in Listing 2.2 further explores the limits of floating point numbers through the technique of numerical experimentation. The first loop from lines 5 through 11 tests for what is known as the machine epsilon (ε_M) which is the smallest value that can be added to unity which results in a stored value not equal to unity. This is also a measure of the smallest relative accuracy that can be achieved in any numerical calculation. In this and all of the tests in this example, the test value is stepped by factors of 2, so that the results may only be accurate to within a factor of 2. However, the tests are simple to perform and give results close to the exact answers (to within a factor of 2). For the machine epsilon the printed results give:

$$\varepsilon_M = 2.2204460492503\text{e-}016 = 2^{-52} \qquad (2.1)$$

While the code only tests to within a factor of 2, this value is in fact the exact value of the machine epsilon and is in fact the value specified by the IEEE standard for double precision floating point numbers:

(http://steve.hollasch.net/cgindex/coding/ieeefloat.html).

A quantity frequently used in describing the error in a number is the relative error defined as

$$\text{Relative error} = \varepsilon_R = \frac{x - \overline{x}}{x}, \qquad (2.2)$$

where $\overline{x}$ is some approximation to x. For the number 1 we can then say that it can be represented with a relative error of the machine epsilon or 2.22e-16. Another term frequently used is the number of significant digits in representing a number. For this a number $\overline{x}$ is considered to be an approximation to a true value x to d significant digits if d is the largest positive integer for which

$$\left| \frac{x - \overline{x}}{x} \right| < 10^{-d}. \qquad (2.3)$$

Applying this to the intrinsic machine relative error we would say that the language has 15 significant digits of machine precision. The machine epsilon is very important to know as one can not expect any computed numerical results using double precision floating point calculations to have a smaller relative accuracy or to have more than 15 significant digits of accuracy. This will be important in setting limits on relative accuracy within iterative loops such as Newton loops represented in Figure 1.3.

Before looking at the smallest number let's discuss the probing of the largest number on lines 24 through 29 of the code. The loop starts with 1 and multiplies by a factor of 2 until the internal number representation agrees with 1/0 which signifies that it exceeds the number representation capability of the language. The result obtained to within a factor of 2 is 8.9884656743116e+307. The printed result from line 31 indeed shows that a larger value (by almost a factor of 2) can be represented. This result is in fact the correct answer and the result is

$$\text{Largest number} = 1.7976931348623e + 308 = (1 - 2^{-53})2^{1024} \qquad (2.4)$$

This number with a precision of 15 digits should be sufficient for a broad range of engineering problems, especially considering the estimation that there are only of the order of 10^{81} or fewer atoms in the known universe.

The code in Listing 2.2 from line 14 through 21 explores the smallest number that can be represented as non-zero. The loop begins with unity and divides by 2 until the number representation is zero as tested on line 17. The resulting number obtained and printed as output is

$$\text{Smallest representable number} = 4.9406564584125e\text{-}324 = 2^{-1074} \qquad (2.5)$$

This is in fact the correct smallest number that can be represented in double precision and agrees with the IEEE standards as given by 2 to the -1074 power. However, this is probably not the lower limit that one would wish to consider for most applications. As the values printed out from line 21 indicate, one can only represent integer multiples of this smallest number. Thus a number this small has very poor precision or has only one significant digit. More importantly is the smallest number that can be represented with the same number of significant digits as represented by the largest number. This value can not be determined from a simple probing calculation as in Listing 2.2. However the IEEE standard for double precision numbers gives the answer as:

$$\text{Smallest number} = 2.225073858507e-308 = 2^{-1022} \qquad (2.6)$$

This then results in an almost symmetrical range of largest and smallest practical numbers from about 2e308 to 2e-308 in value.

Values less than the smallest value given by Eq. (2.5) will be set to zero and values larger than the value given by Eq. (2.4) will be set to a special value identified as infinity and printed as shown in Listing 2.2 as 1.#INF. Mathematical operations with a number represented as infinity are well defined in the IEEE standard. Division of any finite number by infinity gives zero while all other operations (+,-,*) give infinity. Not a number (NaN) is also a special value reserved for certain indeterminable mathematical operations such as 0/0, infinity/infinity or infinity +/- infinity. Such results will print as 1.#IND.

In determining the accuracy of numerical techniques, the relative error as defined by Eq. (2.2) will be frequently used. A relative error criterion is frequently used to terminate an iterative calculation process. In order to discuss some of the terms used in such a calculation, consider the evaluation of an exponential value by a series expansion as:

$$f = \exp(x) = 1 + x + \frac{x^2}{2!} + \frac{x^3}{3!} + ...$$
$$f \simeq f_n = 1 + x + \frac{x^2}{2!} + \frac{x^3}{3!} + ... \frac{x^n}{n!} \qquad (2.7)$$

At some nth approximation to the value of the exponential, some value f_n has been calculated as an approximation to the exact value. At such an iterative step the known quantities are only the approximate value of the parameter and the previous corrections to the parameter value. In general the final exact value of the parameter is not known. Thus instead of the exact relative error of Eq. (2.2) only an approximation to the relative error can be evaluated as:

$$\text{Relative error} = \varepsilon_R = \frac{f - f_n}{f}$$

$$\text{Approximate Relative error} = \frac{f_n - f_{n-1}}{f_n} = \frac{\delta f_n}{f_n}$$

(2.8)

The best approximation to the relative error can only be obtained from the ratio of the last correction value ($x^n / n!$ in this example) divided by the best approximation know for the parameter value. In some cases this can give a reasonable approximation to the relative error while in other cases it can only provide an approximation that hopefully approaches zero as the number of iterative terms increases.

Figure 2.1 shows some results for evaluation of terms in the infinite series of Eq. (2.7) for x = 0.5. Plotted in the figure as the two upper curves are the approximate values of the sum and the correction term at each of the iterative steps from n = 1 to n = 20. The bottom two curves then show the estimated relative and the exact relative error using Eq. (2.8). In this particular example it is seen that the estimated relative error is about a factor of 10 larger than the actual relative error at each iterative step.

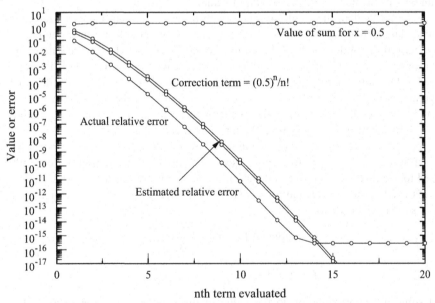

Figure 2.1. Some calculated results for a finite series approximation to the exponential function (at x = 0.5).

In the developments of numerical methods in subsequent chapters, many computational procedures will involve iterative loops to evaluate a final numerical answer. The basic formulation of most procedures for nonlinear problems involves an iterative loop as discussed in Chapter 1 in connection with Figure 1.3. In almost all of these calculations an improved solution will be obtained at each iterative loop that is hopefully converging to an exact solution. Conceptually the calculations will be very similar to that of calculating the value of an exponential value with a finite number of iterative terms. In the case of Eq. (2.7), the exact answer is known so an exact relative error can be evaluated at each iterative step. For most calculations of this type in this work, only the approximate value of the result and the corrective values are known. Also the corrective term is in many cases calculated by an involved computer algorithm so that no exact formula is available for the corrective term. In such a case only the approximate relative error evaluation is available for estimating the accuracy of the calculated term (or terms). For this example it can be seen that the estimated relative error could be used to provide a limit to the number of terms evaluated and in this case, the actually achieved accuracy would be about a factor of 10 better that that from the estimated relative accuracy. However, no such general conclusions can be drawn as to whether the actual accuracy is better or worse than the estimated relative accuracy.

Another feature seen in the data of Figure 2.1 is the saturation of the actual relative error at a value of about 2e-16 as seen for terms beyond around the 15th term. It can be seen that this is very close to the machine epsilon as previously given in Eq. (2.1). This is just one of the manifestations of limits due to the intrinsic numerical accuracy of number representations in any numerical computer calculation. Although the value of each computer corrective term continues to decrease as shown in the figure, there is no improvement in the accuracy of the computed results. The conclusion is that it is fruitless to try to obtain a relative accuracy of computed results better than the intrinsic machine epsilon, or another way of expressing it is that numerical answers can not be obtained to an accuracy of better than about 15 digits.

In any numerical modeling of a real physical system there are several important sources of error such as:

(1) Machine errors – Errors such as discussed above due to the representation of numbers by a finite number of binary digits with a resulting finite number of significant digits. Also truncation and rounding errors in machine calculations.

(2) Mathematical modeling errors – Errors in formulation the models used to represent physical systems. Mathematical or algorithmic models of physical reality are always approximations even with the most complicated of models. At the extreme of accuracy for example one would need to model the individual motion of atoms in a material which is impossible for most artifacts because of the large number of atoms.

(3) Errors in algorithms used to solve models – For computer solutions the generated models must be translated into algorithms which can be used by a computer to obtain a solution. For example the solution

of a problem involving space and time requires that solutions be obtained at a finite grid of spatial and time values – otherwise one has an infinite set of points. Such approximations to models by computer algorithms introduce errors. Much will be discussed about this source of error in subsequent chapters.

(4) Truncation errors in iterative loops – The majority of algorithms used for nonlinear models involve an iterative loop such as shown in Figure 1.3. Even if an algorithm gives an exact result for an infinite number of loops, iterative loops must be terminated after some finite number of iterations, introducing some errors in the results. Again this will be much discussed herein.

(5) Errors in input data – The term garbage in, garbage out is well known and computer calculations are only as good as the input data. Data is always subject to truncation errors again due to the use of finite precision in input data. It is certainly possible to formulate models of selected physical systems that are very sensitive to small variations in input data. Hopefully the models for artifacts one is interested in do not exhibit such features. However, one should always explore the sensitivity of any computer results to the accuracy of one's input data.

(6) Propagation errors – In a series of calculations or in an iterative loop, errors produced at each step of a calculation can propagate to subsequent steps causing even larger errors to accumulate. Some models such as the numerical solution of differential equations are especially susceptible to errors of this type. This type of error is extensively explored in subsequent sections.

(7) Errors in computer programming – This category of errors are human errors introduced in translating correctly formed physical models into algorithms for computer solution. In many cases these are the most insidious of all errors. In some cases they can lie seemingly dormant in a section of computer code only to become evident at the most inappropriate times. These are the ones that can be eliminated with appropriate care and exercising of programmed algorithms. Hopefully these have been eliminated in the coding examples in this book. However, if history is any indicator, there are still some hidden errors lurking somewhere in the example programs.

In addition to numerical data, the implementation of complex data structures in Lua is through the table mechanism. While Lua has only a simple single type of data structure it has proven to be very flexible for implementing all types of complex data structures. One such example is presented in the next section regarding complex numbers. Character data is also of use in documenting numerical work. Lua has standard provisions for character data and for the manipulation of character data through a set of string functions associated with a string library. This provides normal character manipulation capabilities as well as pattern matching with substring capture and replacement capabilities.

2.4 Language Extensibility

As used here the term extensibility refers to the ability to easily extend the semantics of the computer language. Languages known as scripting languages are much more amenable to language extensions than are conventional compile-link-execute languages. This extensibility of Lua is one of the important features of the language. This is due to the interpreted, dynamic typed nature of the language coupled with an ability to easily overload standard mathematical operations for table defined objects. As an example of extensibility, Lua has no built-in capability for complex numbers which are important in many engineering models. However the language is easily extended to support complex numbers as will be shown in this section.

Complex numbers, as are all objects in Lua, are implemented as entries in tables. In this case one needs a table with two real number entries to contain the real and imaginary parts of the complex number, for example C = {3, 4} with 3 as the real part and 4 as the imaginary part. One of the powerful features of Lua is the ability to associate another table known as a "metatable" with any given table. While a metatable is in fact no different in structure from any other table, it takes on a special meaning and relationship when associated with another table through the Lua function setmetatable(otable, mtable) where otable is the name of the original table and mtable is the name of the metatable to be associated with the otable. The function getmetatable(otable) is used to return the metatable if it exists for otable. When a table such as the C table above has a metatable associated with it, all the standard mathematical and logical operations associated with the C table or similar objects can be redefined by functions entered into the metatable.

If all this is a little confusion, it is perhaps best understood with an example. Listing 2.3 shown code for defining simple metatable functions for the four arithmetic operations between two complex numbers and a function for defining new objects of complex type. The function Cmp.new() defined on lines 25 through 28 describes code for setting up a table of real, imaginary values and returning a table with an associated metatable (the mtcmp table). Terms like r = r on line 26 are at first somewhat confusing to those new to Lua. In case this is true, it should be recalled that this is shorthand for the associative table assignment c1['r'] = r and the first r is treated as a character string while the second r is the number passed to the function. The mtcmp metatable is defined on lines 4 through 24 and consists of 6 entries which are function names with corresponding definitions (__add, __sub, __mul, __div, __unm and __tostring). These functions define how addition, subtraction, multiplication and division should be performed for two complex numbers. Each function returns a new complex object after performing the associated mathematical operations. These definitions should be clear to anyone familiar with complex numbers as the operations are defined in terms of real parts (c1.r and c2.r) and imaginary parts (c1.i and c2.i) of the complex numbers. The functions __unm defines how the unitary minus is defined and finally the __tostring defines how a complex object is converted to a returned string using the concatenation operator '..'.

```
 1 :  -- File list2_3.lua -- Simple code for complex numbers
 2 :
 3 :  Cmp = {} -- Complex table
 4 :  mtcmp = { -- metatable for Complex numbers
 5 :     __add = function(c1,c2) -- Add two Complex numbers
 6 :        return Cmp.new(c1.r+c2.r, c1.i+c2.i)
 7 :     end,
 8 :     __sub = function(c1,c2) -- Subtract Complex numbers
 9 :        return Cmp.new(c1.r-c2.r, c1.i-c2.i)
10 :     end,
11 :     __mul = function(c1,c2) -- Multiple Complex numbers
12 :        return Cmp.new(c1.r*c2.r-c1.i*c2.i,
                 c1.i*c2.r+c2.i*c1.r)
13 :     end,
14 :     __div = function(c1,c2) -- Divide Complex numbers
15 :        local d = c2.r*c2.r+c2.i*c2.i
16 :        return Cmp.new((c1.r*c2.r+c1.i*c2.i)/d,
                 (c1.i*c2.r-c2.i*c1.r)/d)
17 :     end,
18 :     __unm = function(c1) -- Negative of complex number
19 :        return Cmp.new(-c1.r, -c1.i)
20 :     end,
21 :     __tostring = function(c1)
22 :        return '('..c1.r..') + j('..c1.i..')'
23 :     end
24 :  } -- End metatable functions
25 :  Cmp.new = function(r,i) -- Define new complex number
26 :     local c1 = {r = r, i = i}
27 :     return setmetatable(c1, mtcmp)
28 :  end
```

Listing 2.3. Simple Lua extension for math operations for complex numbers.

With these operations defined in a file named list2.3.lua, the code in Listing 2.4 illustrates some simple operations on complex numbers. This simple example loads the previously defined metatable operations for complex numbers on line 3 with the require"list2.3" statement. This is the same as if the code in file list2.3.lua were inserted at line 3 in the listing. Lines 6, 7 and 8 define three new complex numbers. Simple math operations on the complex numbers (objects defined by tables) are then performed and the results printed on lines 10 through 15. The function __tostring() in Listing 2.3 formats a complex variable for printing and this is used by the Lua print() function on line 15 as seen by the listed output from executing the code. The reader can easily verify the accuracy of the printed results for the indicated complex math operations.

In addition to the "metamethods" illustrated in Listing 2.3, Lua has a broader range of possible user defined operators for tables. For completeness they are
(a) Arithmetic metamethods: __add (for addition), __sub (for subtraction), __mul (for multiplication), __div (for division), __unm (for negation), __mod (for modulo), __pow (for exponentiation) and __concat (for concatenation),

(b) Relational metamethods: __eq (equal to), __lt (less than) and __le (less than or equal) and

(c) Other metamethods: __tostring (to string), __metamethod (returned by get-metatable()), __index (access an absent table field) and __newindex (set a new table field).

```
 1 : -- File list2_4.lua -- Simple complex number operations
 2 :
 3 : require"list2.3" -- Load defined complex extensions
 4 : Cnew = Cmp.new -- Local define
 5 :
 6 : c1 = Cnew(3,5) -- Define complex numbers
 7 : c2 = Cnew(2,8)
 8 : j = Cnew(0,1) -- pure imaginary of magnitude unity
 9 :
10 : print(c1*c2) -- Test math operations
11 : print(c1/c2)
12 : print(c1+c2)
13 : print(-c1+c2)
14 : print(j*c1)
15 : print(tostring(c1))
Output:
(-34) + j(34)
(0.67647058823529) + j(-0.20588235294118)
(5) + j(13)
(-1) + j(3)
(-5) + j(3)
(3) + j(5)
```

Listing 2.4. Illustration of Complex number operations using extensibility of Lua.

All of these possible operations may not have valid meaning for every user defined object such as the complex object in the above example. For example, the meaning of relational operators between complex numbers is not well defined (such as $c1 < c2$). The user is free to define the meaning if one can determine what is meant by one complex number being less than another complex number. In the course of this work it will be useful to define several new objects and metamethods for the newly defined objects.

The simple metamethods for complex numbers defined in Listing 2.3 are incomplete in many aspects. They work appropriately as long as one is using arithmetic operations between two complex numbers, such as $c1*c2$. However, how about the operation $c1*4$, an operation between a real number and a complex number? In Lua this will trigger a call to the addition metamethod with $c1$ as one argument and 4 as the second argument. From the code for __mul() in Listing 2.3, it can be readily seen that this will cause an error because the code assumes that both function arguments are tables. Thus a complete Lua extension for complex numbers must be able to handle cases of arithmetic operations between real numbers and complex numbers by various tests in the metamethods. Also one might wish to have traditional math functions such as sin(), cos(), tan() etc. that work with complex number arguments. All of this is relatively straightforward and a Lua file with all this has been provided named "Complex.lua", but the details will not be presented here. The user can look at the details for experience in the Lua language and in understanding its easy extendibility.

```
 1 : -- File list2_5.lua -- Complex numbers using Complex.lua
 2 :
 3 : require"Complex" --Load Complex extensions
 4 : Cnew = Complex.new
 5 :
 6 : c1 = Cnew(3,5) -- Define complex numbers
 7 : c2 = Cnew(2,8)
 8 :
 9 : print(c1*c2); print(4*c1) -- Test math operations
10 : print(c1/c2); print(c1+c2); print(-c1+c2); print(j*c1)
11 : print(tostring(c1))
12 : print('sin =',Complex.sin(c1))
13 : print('sin =',c1:sin()); print('tan =',c1:tan())
Output:
(-34) + j(34)
(12) + j(20)
(0.67647058823529) + j(-0.20588235294118)
(5) + j(13)
(-1) + j(3)
(-5) + j(3)
(3) + j(5)
sin =    (10.47250853394) + j(-73.460621695674)
sin =    (10.47250853394) + j(-73.460621695674)
tan =    (-2.5368676207676e-005) + j(0.99991282015135)
```

Listing 2.5. Example of complex number operations using Complex.lua extensions.

Some examples of complex number operations using the Complex.lua extension are illustrated in Listing 2.5. The extensions are loaded by the require"Complex" statement on line 3. The same complex numbers are defined as in Listing 2.4. This only tests a small subset of the complex number operations but with the expanded metafunction definitions, any mixture of real and complex math operations can be easily performed. Note that the purely imaginary, j, is used on line 10 to multiply c1 although it is not defined in the listed code. It is predefined by the Complex extensions for use when the Complex extensions are loaded. Also the final two lines of code (lines 13 and 14) illustrate the complex sin() and tan() functions. Two different methods are illustrated for calling the complex sin() function: Complex.sin(c1) and c1:sin(). The printed output shows that the results are identical. In the first case the function sin() located in the Complex table of functions (or values) is called while in the second case it is not obvious how Lua accesses the proper sin() function. In Lua the notation obj:fn(...) is simply shorthand for obj.fn(obj, ...) so in the above case the second form is really c1.sin(c1). But how does this work since the sin() function is not listed as an entry in the c1 object table? The magic is in the __index setting of the metatable associated with each complex variable. When Lua looks for the sin() function in the object table (the c1 table) the function is not found. Lua then looks for an __index entry in the metatable associated with the object. Lua then looks in the __index table which is set to the Complex table for the desired function. It then finds the appropriate sin() function to execute. This object oriented calling notation will be familiar to those with experience using other object oriented languages. One can also chain __index tables so an object can inherit methods associated with previously defined tables.

This is a simple example of how easily the functionality of the Lua language can be extended for user defined objects – giving it object oriented features. In the course of this book several other language extensions will be developed and used. The most important of these are:

1. Complex – Extensions for complex number operations and complex functions.
2. Matrix – Extensions for standard matrix operations.
3. Polynomial – Extensions for easy manipulation of polynomial functions.
4. ExE – Extensions for element by element table operations (similar to MATLAB's element by element array operations).

Complex numbers and matrices are important in many areas of nonlinear models and these extensions are frequently used in subsequent chapters.

2.5 Some Language Enhancement Functions

This section discusses some Lua language enhancement functions that have been found very useful in developing code for solving nonlinear models in this book. These functions are discussed here because they are not directly related to nonlinear models but are frequently used in subsequent chapters. The most useful functions can be grouped into the following categories:

1. help() and whatis() functions – Debugging functions for printing the type and properties of other Lua objects.
2. printf(), fprintf() and sprintf() functions – Formatted print functions equivalent to similarly named functions in the C language.
3. write_data() and read_data() functions – Functions to easily write Lua tables to files and read Lua tables from saved files.
4. plot(), stem(), scatterplot(), splot(), cplot() – Functions to easily produce pop-up plots of data (these use a freeware program gnuplot).
5. metatable(), globals() and makeglobal() – Functions for looking at metatables and globals of objects and making table entries globally available.

These functions are coded in Lua in a supplied table named init.lua (Lua basic initialization functions). This table of functions is automatically loaded when using the recommended code editor (see Appendix B for more details).

Listing 2.6 illustrates the usage of the debugging functions help() and whatis(). The function help() is executed on line 6 and produces the output seen beginning with the _G line of output. This function is in fact the same as whatis(_G) and results in the printing of all globally known objects at the time it is executed. Lua, by default, maintains a table, known as _G, of all known objects within a section of Lua code. This is the table printed by the help() statement. The entries in any table such as the _G table may be of any of the seven basic Lua object types. The output produced by whatis() and shown on the lines following the "With members" statement in the output consists of the name of the table entry as well as a final two characters (added by whatis()) that identifies the type of object in the table. The identifying character strings are: '<$' for string, '<#' for number, '()' for function, '[]' for table, '<@' for userdata, '<>' for thread and '<&' for boolean. In this work the userdata and thread objects will not be encountered, but the other types will. In the printed output examples of strings, tables, and functions can be seen. For example _VERSION<$ indicates that the variable _VERSION is a character string, math[] indicates that the variable math is a table and print() indicates that the variable print is a function. All of the entry names within the _G[] table can be used by simply typing the table name within Lua code.

```
 1 : -- File list2_6.lua -- Examples of some functions in init.lua
 2 : -- Uncomment next line if needed
 3 : -- require"init" -- Load basic Lua functions if needed
 4 :
 5 : -- List known objects and type (in _G table)
 6 : help() -- Same as whatis(_G)
 7 : -- List entries in math table -- names and types
 8 : whatis(math)
 9 : -- Probe object type of math.sin and math.pi
10 : whatis(math.sin); whatis(math.pi)
11 : -- Make sin a globally known function
12 : sin = math.sin; whatis(sin)
13 : -- Probe a string
14 : whatis(_VERSION)
Output:
_G is a global table
  With members {_G[], _VERSION<$, arg[], assert(), collectgarbage(),
     coroutine[], cplot(), debug[],dofile(),engr(),error(), fprintf(),
     gcinfo(), getfenv(), getmetatable(), globals(), help(), io[],
     ipairs(), load(), loadfile(), loadstring(), makeglobal(), math[],
     metatable(), module(), newproxy(), next(), os[], package[],
     pairs(), pcall(), plot(),print(),printf(),printnow(), rawequal(),
     rawget(), rawset(),rawtype(), read_data(), require(), requires(),
     scatterplot(), select(), setfenv(), setmetatable(), splot(),
     sprintf(),stem(),string[],table[],testme(),tonumber(),tostring(),
     type(), unload(), unpack(), whatis(), write_data(), xpcall()}
math is a global table
  With members {abs(), acos(), asin(), atan(),atan2(), ceil(), cos(),
     cosh(), deg(), exp(), floor(), fmod(), frexp(), huge<#, ldexp(),
     log(), log10(), max(), min(), mod(), modf(), pi<#, pow(), rad(),
     random(), randomseed(), sin(), sinh(), sqrt(), tan(), tanh()}
Variable is a non-global function
Variable is a non-global number = 3.1415926535898
sin is a global function
Variable is a string = Lua 5.1
```

Listing 2.6. Code example for help() and whatis() functions.

Following the _G output is the output printed by the whatis(math) function execution. This begins at the "math is a global table" line. The entries in the math[] table are printed and one readily sees the expected functions such as abs(), atan(), cos(), sin(), sqrt(), etc. In addition an entry pi (listed as pi<#) can be seen with the '<#' indicating that this is a number entry. The object nature of math.sin and math.pi are probed on line 10 with additional calls to whatis(). The output from print() for these calls indicates, as expected that math.sin is a 'non-global function' and that math.pi is a 'non-global number = 3.1415926535898'. The non-global refers to the fact that the sin and pi names are not globally known but must be accessed as math.sin() and math.pi. On line 12 the variable sin is set to the math.sin function and thereafter the variable sin is known globally as the next to last printed output indicates. Finally whatis() is used on line 14 to probe the value of a table entry known to be of type string and the resulting output is identified as 'Lua 5.1'. The whatis() function can be invoked anywhere within a section of Lua code to determine the nature of any desired variable name. This is a very useful function in probing the internal workings of the Lua language and in program debugging.

The second category of functions printf(), sprintf() and fprintf() are simply formatting print statements which are used with exactly the same calling arguments as the functions by the same name in the C programming language. The standard printing functions in Lua are slightly different from the C language functions requiring that one pass the arguments through the string.format() function to obtain formatted printing. These are supplied in the init.lua package to make Lua programming easier for those familiar with C. If the reader is not familiar with these C functions, later examples will make these familiar.

The write_data() and read_data() functions are supplied in order to make the writing and reading of tabular data to files easier. Lua supplies basic low level file operations through the io[] set of library functions. By using these two functions, the user does not have to be involved with these low level file operations, for reading and writing data stored in tables. Listing 2.7 shows a simple example of writing to and reading from a disk file for data defined in Lua tables. The for ... loop in the code from line 4 through line 7 generates x and y arrays of data with integer indexes of 1 to n. The generated data is obviously 3 cycles of a sin wave. The two data tables are written to a file named 'list2.7.dat' on line 9 using the write_data() function. While two data arrays are shown here, any number of tables may be passed as arguments to the function. In addition the arguments may be tables of tables such as {{},{}, .. {}} and the data will still be written correctly. However, only one level of table within a table is supported. The lengths of the tables passed may also be different and the data will still be written correctly to the file. The file will contain a character representation of the numerical data so one can readily read the stored data files. For this example the stored format is two columns of data. For reading the stored data a table of tables is defined on line 11 with no entries in the two internal tables. The read_data() call then on line 12 fills in the two tables within the xy table at xy[1] and xy[2]. The whatis(xy) call on line 15 produces the given output indicating that the resulting table consists of two tables, with indices 1 and 2. Extensive use of these two functions is made in subsequent chapters to read and write tabular data.

```
 1 : -- File list2_7.lua -- Examples of reading and writing tables
 2 :
 3 : x,y,n = {},{},1000
 4 : for i=1,n do -- Define tables of values
 5 :     x[i] = (i-1)/(n-1)
 6 :     y[i] = math.sin(6*math.pi*x[i])
 7 : end
 8 : plot(x,y) -- pop-up plot
 9 : write_data('list2.7.dat', x,y) -- Save data to disk file
10 :
11 : xy = {{},{}} -- New table of tables for reading data
12 : read_data('list2.7.dat',xy) -- Retreive two column data
13 : plot(xy[1],xy[2]) -- pop-up plot of retreived data
14 : -- xy[1] is same as x and xy[2] is same as y data
15 : whatis(xy)
Output:
xy is a global table
  With members {1[], 2[]}
```

Listing 2.7 Code segment for a simple example of writing and reading of table data to disk file using write_data() and read_data() functions.

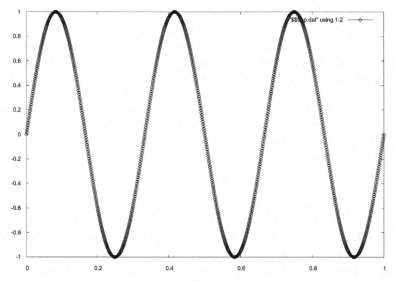

Figure 2.2. Copy of pop-up plot produced by plot() code in Listing 2.7 (uses gnuplot for basic plotting features).

The code in Listing 2.7 also demonstrates the use of the pop-up plots with the plot() functions on line 8 and 13. In the first case the plot is of the data before storing in the data file and the second call is of the retrieved data. The user is encouraged to execute the code and observe the pop-up plots and thereby verify the writing and reading of the data arrays. Figure 2.2 shows a copy of the pop-up plot produced by the code. The plots can be saved to a clipboard for later import into the user's plotting software. This feature of rapidly and easily producing a pop-up plot to observe some calculated result will be frequently used in subsequent chapters. This feature makes use of the freely available plotting program "gnuplot" and the user must have this package installed to use this feature incorporated into the init.lua functions. (see Appendix B for properly installing the software).

The final functions metatable(), makeglobal() and globals() functions listed above are for use with tables as calling arguments. The first function explores and print the entries in any metatable associated with a table as an argument. Makeglobal(tbl) will make the entries in a table given as the argument (tbl) known on a global basis. For example makeglobal(math) with make all the math functions sin(), cos(), tan() etc. usable without adding the math prefix to the names, i.e. using sin(x) instead of math.sin(x). Finally the global() function takes a function as an argument and prints a list of the global objects known to the function. The global environment of a function can be limited and controlled by the setfenv() function to be discussed later. The first and last of these three functions are used primarily as debugging aids to probe the properties of tables and functions. The reader can explore the use of these functions with various tables and functions such as those associated with complex variables in Listing 2.5.

These basic Lua functions in this section are extensively used in subsequent chapters and it is recommended that these be loaded by any coding example developed in connection with this work. (See Appendix B for setting up an appropriate programming environment.)

2.6 Software Coding Practices

The emphasis of this work is the solution of nonlinear models associated with physical problems. Computer software programs are simply viewed as the modern means of achieving that end. While this is not intended to be a manual about computer programming, there are certain relatively simple software coding practices that can greatly aid in the time and effort expended in developing software. Many books have been written about this subject so this simply contains some of the present author's suggestion for use with Lua and the software in this work.

1. Use a modern text editor – Computer program files are simply text files and can be composed by any appropriate text editor. However, the text editor used should support language highlighting in the language being used. Highlighting allows one to readily verify the proper closure of brackets and parentheses as well as checking the correct spelling for reserved words. One such example is the freely available text editor SciTE (http://scintilla.sourceforge.net/SciTE.html). This editor not only can be used for editing program files, but has a built-in code execution feature with output window. Figure 2.3 shows a screen shot of SciTE with the code of Listing 2.5. On the left is the text editor window and on the right is the output produced by execution the Lua code. With SciTE one can readily edit code in the editor window, execute the code and see the re-

sults without leaving the program – it's very good and highly recommended.

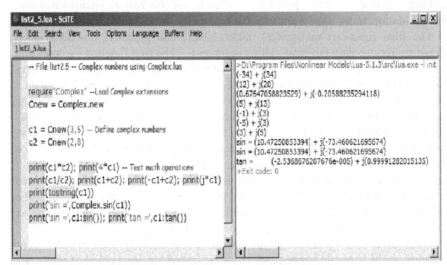

Figure 2.3. Picture of SciTE editor with code of Listing 2.5.

2. Modularize computer programs – Computer code of almost any length
    should be modularized into smaller code blocks usually implemented
    with appropriately defined callable functions. In Lua, callable functions
    can return any number of variables or objects which aids in modularizing
    programs. As a general rule, the modules should be no longer than can
    be printed on a single page. In general this rule is followed in this work
    although some of the most complicated functions are about two pages in
    length. Finally use short descriptive names for coded functions. Coded
    functions should be developed with the goal of reuse in larger programs.
    This requires some thought with respect to objects passed to the functions
    and objects returned by the function.

3. Collect functions into modules – As used here a code module is a collec-
    tion of related functions and variable or object definitions. Related code
    should be collected together into a single file that can be loaded into
    other Lua programs. A simple example so far given is the Complex.lua
    file containing all the code related to defining and using complex nu-
    merical numbers.

4. Use indentation of control blocks – One of the simplest and most effec-
    tive means of producing correct computer code is to use a consistent
    practice of indenting code contained within the scope of a control block (
    begin ...end, for ...end, etc. blocks). This is illustrated by the code in
    Listing 2.7 which shows a for ...end block from lines 4 through 7. By
    indenting all code lines between lines 4 and 7 one can readily see the
    scope of the block and readily see that a proper end statement has been

used to terminate the block. Lua imposes no restrictions on the placement of code on a line, but it is highly recommended that code within all control blocks be indented by some fixed number of spaces. Example code in this work will follow this practice.

5.   Use meaningful variable and object names – Give some thought to the names used for variables and objects in coding. The names should ideally convey some indication of the meaning for the names. For example names beginning with a, b, c and d are frequently used for constants. Names beginning with x, y, z and t are frequently used for spatial variables and time. Also names beginning with i, j, k, l, m and n are frequently used for integer constants used in array indexing and loop counting. In this author's opinion shorter names are usually preferred to longer names although single letter names frequently do no convey much information as to the meaning of the variable (except for simple constants).

6.   Limit scope of variables – In Lua all variable names, by default, have global scope. This can be a source of hard to find errors. For this reason one should carefully consider the scope of all named variables. The use of the 'local' keyword is very useful to appropriately limit the scope of variables. This is very important when developing modules that may define a large number of variables needed in the function. A very useful technique for functions is the use of the setfenv() function that allows one to set the environment of variables known by a function. When the environment of a defined function is set in this manner, any variables defined within the function are not known outside the function. Without this technique, a variable defined in some called function can inadvertently pop up in a main program with an unexpected value. This technique is extensively used in the code developed in this work.

7.   Use appropriate comments – Well placed comments in computer code greatly aid others reading the code but are also very helpful to one reading his/her own code at some later time. Major blocks of code and major control loops probably need some brief comment. However, one can easily overdo the comments. Commenting all lines is just as bad as having no comments. The statement a=a+1 does not need the comment -- increment a by 1.

8.   Code for clarity, then speed – One's first concern in code development should be program simplicity and clarity of algorithm expression. While execution speed is always of some concern in numerical work, this should not be one's first concern especially when using a scripting language such as Lua. After proper algorithm development one can go back and recode sections for speed if necessary. For the Lua language, one can recode critical parts in C for improved speed and gain the best of the world of a scripting language and a compiled language. There are only a few examples in this work where it would be nice to have faster executing code.

9. Test, test and test some more – Obtaining correctly executing computer algorithms is a matter of having well thought out algorithms, carefully crafted code and extensive testing with known results. Computers always execute the code that one has coded correctly. It's just that the code produced is in many cases not the code one thought he/she were producing. There are errors either in one's algorithms or errors in coding the algorithms. The easiest to find errors are language errors (missing brackets, missing end statements, etc.). Lua provides very good language checks and error messages usually indicating exactly which line of code contains the error. Missing brackets and end statements are more difficult to find because the error message may occur many lines from the error. Incorrectly typed variable names are some of the more difficult errors to catch as Lua does not require variable type definitions. While Lua has some special debug capabilities, all the code in this work was debugged by the simple technique of inserting appropriately placed print() statements and whatis() statements within the code. After one's code is apparently executing, the really important testing of algorithms can begin. This requires that one exercise code with a wide range of input data and hopefully with many cases for which the expected results are known. Testing all cases of code for nonlinear problems is somewhat challenging since the exact solution of only a few nonlinear problems is known. The question of algorithm accuracy will be repeatedly addresses in this book.

These are some of the most important software practices that have been found by this author to greatly aid in the development of correctly executing algorithms for solving nonlinear models.

2.7 Summary

This chapter has addressed some fundamental aspects of using computer software and numerical methods for solving nonlinear engineering models. The major reasons for selecting Lua as the language for this work have been discussed. The major features of Lua have then been presented with some very simple examples of the language. The ease with which the language is extensible to general data objects has been stressed and demonstrated with the case of implementing complex numbers as a new data object.

Important code segments developed (or simply discussed) in this chapter are:

1. Complex – A code module for complex number arithmetic along with extending the normal trig functions to handle complex numbers.
2. init – A set of functions for implementing some basic functions useful for further development of Lua code. The most important of these are: help0 and whatis() for debugging and language exploration; printf(), fprintf() and sprintf() for C-like printing; write_data() and read_data() for simple data writing to and reading from files; plot(), stem(), scatterplot(), cplot()

and splot() for pop-up plots of generated data and metatable() globals() and makeglobal() for exploring properties of functions.

For the work in subsequent chapters it is highly recommended that the reader has available a computer with the Lua programming language so the code developed can be exercised and extended. In addition it is highly recommended that the user has the SciTE text editor and the gnuplot software available for use with the example code. All of this software is freely available and Appendix B discusses the software supplied on CD with this book.

With this background we are now ready to hasten on to the meat of this work on solving nonlinear models. Let the show begin!

3 Roots of Nonlinear Equations

A large number of engineering and scientific problems can be formulated in terms of finding the value, or values, of some variable x which results in a zero value of some function of that variable. Mathematically, this is represented by the equation
$$F(x) = 0 , \tag{3.1}$$
where $F(x)$ is some given function of x. Examples are polynomial equations such as
$$F(x) = 5x^4 - 4x^2 + 2x - 3 = 0 , \tag{3.2}$$
or transcendental equations such as
$$F(x) = \tan(x) - 1/x . \tag{3.3}$$
For the case of polynomial equations, the solution values of x which satisfy the equation are frequently called "roots" of the polynomial. In general these may be real numbers or complex numbers. For the case of transcendental equations such as Eq. (3.3) the solution values are typically called "zeros" of the function. Mathematically the terms roots and zeros are used interchangeably.

As a matter of definition a transcendental function is a function for which the value of the function can not be obtained by a finite number of additions, subtractions, multiplications or divisions. Exponential, trigonometric, logarithmic and hyperbolic functions are all examples of transcendental functions. Such functions play extremely important roles in engineering *problems* and are the source of many of the nonlinear equations of interest in this book. For engineering models an important feature of transcendental functions is that their argument must be a dimensionless mathematical variable.

When solving for the zeros of an equation, it is usually important to have some idea of the general behavior of the function through a graphical representation of the function. Such a plot is shown in Figure 3.1 for the polynomial of Eq. (3.2). Graphically it can be seen that the polynomial has zeros at x near $+1$ and near $x = -1.25$. However only two real roots are seen on the graph and since the equation is a fourth degree polynomial and must have four roots, it can be inferred that there must be two complex roots.

Another useful technique in viewing the solution of a nonlinear equation is to separate the equation into two equal parts such as representing Eq. (3.3) as
$$F_1(x) = \tan(x) = F_2(x) = 1/x . \tag{3.4}$$
Each of these two functions can be separately graphed as in Figure 3.2 and the intersections of the two graphs give the solution values as shown by the circled intersection points in the figure. Only the principle quadrant values of the $\tan(x)$ function are shown in the figure. As is well known the $\tan(x)$ function repeats with period π, so there will be an infinite number of intersection points between the $1/x$ function and the $\tan(x)$ function.

J.R. Hauser, *Numerical Methods for Nonlinear Engineering Models*, 43–76.

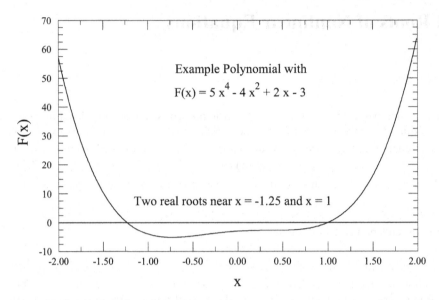

Figure 3.1 Graphical plot of Eq. (3.2).

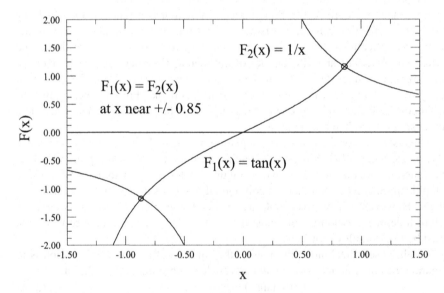

Figure 3.2 Graphical plot of two parts of Eq. (3.4).

These two functions illustrate some important features of the zeros of nonlinear equations. There may be no solutions, a small finite number of solutions or an infinite number of solutions. In working with such problems and in finding solution, it is always wise to have as much general knowledge about the solution values as possible. This can often be obtained by graphically exploring the anticipated range of a function and having a general idea as to where solutions exist. In many real physical problems only a limited number of solutions have physical significance. For example, only the lowest valued (x wise) solutions in Fig 3.2 might be physically significant in some real physical problem, or only the positive roots might be physically significant in some other real physical problem.

One feature of possible solutions to nonlinear equations not illustrated by the two examples is the possibility of multiple zeros at the same value of x. For example the function:

$$F(x) = x \tan(x) , \tag{3.5}$$

has multiple zeros at $x = 0$ since both x and $\tan(x)$ are zero at $x = 0$. Multiple roots are characterized by the fact that not only the function but one or more derivative is also zero at the solution point.

This chapter is devoted to discussing some general solution methods for solving nonlinear equations such as those of Eqs. (3.1) and (3.2). Finding all the roots of a polynomial of given order is a fairly specialized task that will be addressed again in a later chapter, after some pertinent numerical tools have been developed. The major numerical technique to be developed in this chapter, known as Newton's method, requires that one has an initial guess "reasonably" close to the actual solution. One's initial guess can be obtained from the physical context of the problem being addressed or can be obtained from a graphical exploration of the function over an appropriate range of the independent variable as illustrated in the previous two figures.

3.1 Successive Substitutions or Fixed Point Iteration

The method of successive substitutions (also called fixed point iteration) is perhaps the simplest method of obtaining a solution to a nonlinear equation. This technique begins by rearranging the basic $F(x) = 0$ equation so that the variable x is given as some new function of the same variable x. The original equation is thus converted into an equation of the form:

$$x = G(x) . \tag{3.6}$$

Starting with an initial guess, an iterative procedure is then initiated where the value of x is substituted into the right hand side $G(x)$ function and a new value of x is calculated. The procedure is iterated until there is no further change in x or until the change is less than some desired accuracy. Let's use Eq. (3.2) as an example to clarify the procedure. Eq. (3.2) can be rearranged in several ways to obtain an equation in the form of Eq. (3.6). Two of the possibilities are:

$$x = (3 + 4x^2 - 5x^4)/2 , \tag{3.7}$$

$$x = \pm((4x^2 - 2x + 3)/5)^{1/4} . \tag{3.8}$$

There is one other possibility that is not enumerated. This illustrates one of the features, or perhaps problems, with the successive substitutions method which is the fact that for most equations there is more than one way of converting the equation into the form needed for the iterations. Which equation should be used and why? This may become a little clearer after completing this example. Let's select Eq. (3.8) (from experience) and implement the algorithm. Computer code for this is shown in Listing 3.1 along with selected output from executing the code. In fact two iterations are shown, one for the positive ¼ root and one for the negative root as indicated in Eq. (3.8). In each case an initial guess of $x = 0$ is taken. One could of course start with a better guess, since from Figure 3.1 it can be seen that one solution is near 1.0 and another solution is near -1.25. However, it is seen that the algorithm converges, even with the zero initial value. Iteration values are shown for the first 10 iterations or successive substitutions, and for selected values up to 40 iterations. As can be seen in the output table, after about 30 iterations, the solution has converged to the values 1.0 for the positive root and -1.2326393307127 for the negative root. Both of these are accurate to within the computer accuracy of about 14 digits. It is seen that the root near -1.25 is not at exactly that value but at a value close to this value.

```
1 : --/* File list3_1.lua Successive substitution code */
2 :
3 : x1 = 0 -- Initial guess
4 : x2 = 0
5 : for i=1,40 do
6 :     x1 = ((3-2*x1+4*x1^2)/5)^.25
7 :     x2 = -((3-2*x2+4*x2^2)/5)^.25
8 :     print(i,x1,i,x2)
9 : end

Selected Output:

      1      0.88011173679339      1  -0.88011173679339
      2      0.96512590009482      2  -1.1196800369364
      3      0.98962052568194      3  -1.1966904312367
      4      0.99689325658256      4  -1.2212366586237
      5      0.99906860691387      5  -1.2290266417362
      6      0.99972063853033      6  -1.2314951534342
      7      0.99991619663375      7  -1.232277000416
      8      0.99997485944667      8  -1.232524594724
      9      0.99999245787509      9  -1.2326029986961
     10      0.99999773736622     10  -1.2326278259489
---
     20      0.99999999998664     20  -1.2326393305961
---
     25      0.99999999999997     25  -1.2326393307123
---
     30      1                    30  -1.2326393307127
---
     40      1                    40  -1.2326393307127
```

Listing 3.1. Code segment for the successive substitution method applied to Eq. (3.2).

The successive substitution method has successfully calculated the two real roots of the fourth order polynomial in this case. However, the convergence is slow, and follows what is known as "linear" convergence. This means that as the true solution is approached, the error at each iteration is some linear fraction of the error at the previous iteration. Let's now study the convergence rate and the condition for convergence of the successive substitution method. Let's let $x_1 = x_0 + \delta x_1$, where x_0 is the true solution value and $\delta x_1 = x_1 - x_0$ is the error in the solution value at step 1 in the iteration. The next approximation to the solution is obtained from

$$x_2 = G(x_1) = G(x_0 + \delta x_1) = G(x_0) + G'(x_0)\delta x_1,$$

where the first term in a Taylor series has been used to approximate the value of the function $G(x_0 + \delta x_1)$. Since the new calculated value must be of the form

$$x_2 = x_0 + \delta x_2 \text{ and } x_0 = G(x_0),$$

it is seen that the error in the value at the next iteration is:

$$\delta x_2 = G'(x_0)\delta x_1 \text{ or}$$

$$\varepsilon_2 = \frac{\delta x_2}{x_0} = G'(x_0)\varepsilon_1 \tag{3.9}$$

Thus the error or relative error at each iteration is linearly proportional to the error or relative error at the previous iteration and the criteria for a smaller error with each iteration is that

$$|G'(x_0)| < 1. \tag{3.10}$$

The magnitude of the derivative of the successive substitution function evaluated at the solution point must be less than unity for the error to decrease with each iteration. If the derivative is negative, the calculated values will oscillate about the solution point while if it is positive the calculated values will monotonically approach the solution point. As the magnitude of the derivative approaches unity, the convergence of the technique becomes very slow and the successive substitution technique works best when the substitution function is a slowly varying function of x.

Selecting the appropriate function to iterate on is very important to the success of the successive substitution technique. Although both Eqs. (3.7) and (3.8) represent the same relationship, if one attempts to apply the successive substitution technique to Eq. (3.7), it rapidly diverges. From the requirement of Eq. (3.10) this is to be expected as the x^4 term on the right hand side is a rapidly varying function. In many cases, a rearrangement of an equation can be performed such that the successive substitution technique will converge to a correct solution value. In any case the algorithm is extremely simple to apply and in many cases can rapidly give a solution to a nonlinear equation.

Since nonlinear problems in general require some type of iteration to approach a solution, the question of when one stops an iterative process typically occurs especially if one wishes to fully automate a solution such as the successive substitution approach. One wants to perform no more work than is needed to achieve a desired degree of accuracy. In general this is a very difficult subject and the ap-

proach taken in this book will be to use some engineering rules regarding accuracy
of solutions and the approach will not be that of a rigorous and complete discussion. From the printed iteration values in Listing 3.1 it is seen that the difference
in value between any two iterations provides some estimate of the error in the solution. The relative error in a solution value can be expressed as:

$$\varepsilon = \left| \frac{x - x_0}{x_0} \right|, \tag{3.11}$$

where x_0 is the true solution value and x is an approximation to the true value. In
an iterative solution process, x_0 is typically not known. At some nth iterative step,
the best estimate of the relative error that can be formed is:

$$\varepsilon = \left| \frac{x_n - x_{n-1}}{x_n} \right|, \tag{3.12}$$

where x_0 is approximated by x_n. Another estimate can be formed by

$$\varepsilon = \left| \frac{x_n - x_{n-1}}{0.5*(x_n + x_{n-1})} \right|, \tag{3.13}$$

where x_0 is approximated by $0.5*(x_n + x_{n-1})$. In most cases either of these can be
used as an estimate of the relative error in a solution value.

Listing 3.2 shows computer code with Eq. (3.12) implemented in a more general ssroot() function which can be called for any input function and initial guess.
The code to terminate the iterative loop in on line 15 and uses the form of Eq.
(3.12) with the estimated relative error set to less than 1.e-10. Executing the code
gives the output seen at the end of the listing. The last number printed on each
line is the actually achieved relative error in the solution and for both roots it is in
deed less than the value of 1.E-10 used to terminate the iterations. While it is difficult to generalize to all possible functions from this simple example, it is seen
that using an approximation for the relative error provides an excellent criterion
for terminating the iterative loop for this example.

The second printed number in each case indicates the number of iterations
needed to achieve the specified accuracy (20 and 21 iterations for this example).
As a general programming practice it is good to always look at the number of iterations, as a value of 100 in this case would indicate that the algorithm did not
converge to the accuracy specified by the convergence criteria. One might think
that a better approach would be to simply use an infinite loop such as a while(1)
… end code loop on lines 13 to 17, for example, and iterate until the desired accuracy is achieved. The problem with this is that the algorithm may never converge
and one then has a code loop which never terminates. So the best approach is to
set an upper limit on the number of iterations and then check to see that the algorithm did not reach the upper limit. The ssroot() function given in Listing 3.2 can
then be used with any defined function to apply the successive substitution algorithm for finding the roots of an equation. As the discussion progresses in this
book many such code fragments will be developed which can be used in larger
programs to solve increasingly more difficult problems.

```
 1 : --/* File 3_2.lua Successive substitution code */
 2 :
 3 : function g1(x) -- Positive root function
 4 :    return ((3-2*x+4*x^2)/5)^.25
 5 : end
 6 : function g2(x) -- Negative root function
 7 :    return -((3-2*x+4*x^2)/5)^.25
 8 : end
 9 :
10 : function ssroot(g,x) -- Successive Substitution function
11 :    local xold = x
12 :    local itt=NT
13 :    for i=1,NT do
14 :        x = g(x)
15 :        if abs((xold-x)/x)<err then itt=i; break end
16 :        xold = x
17 :    end
18 :    return x,itt
19 : end
20 : setfenv(ssroot,{abs=(Complex or math).abs,NT=200,err=1.e-10})
21 :
22 : root1,n1 = ssroot(g1,0)
23 : print(root1,n1,(1-root1)/root1)
24 : root2,n2 = ssroot(g2,0)
25 : print(root2,n2,(-1.2326393307127-root2)/root2)
Output:
0.99999999998664      20      1.3360645923123e-011
-1.2326393306758      21      2.9945522644232e-011
```

Listing 3.2. Successive substitution algorithm with termination criteria using approximate relative error.

The code in Listing 3.2 introduces some additional programming concepts that will be used extensively in this work. Note the use of the setfenv() (set function environment) function on line 20 of the code. This function takes the name of a function as the first argument (ssroot) and a table of environmental variables as the second argument. This not only provides an interface for variables known to exist between the function and a calling environment, but also limits the scope of any variables defined within the function. In this example a local variable xold is defined on line 11 and thus is known locally to the ssroot() function. However, even if the local keyword is not used, because of the setfenv() function, this variable would not become a global variable and would still be known only internally to the ssroot() function. Note the use of the expression abs = (Complex or math).abs as the first table entry in the setfenv() table. This defines the abs() function used in the ssroot() function as being either the Complex.abs() function or the math.abs() function. If the Complex.lua file has been loaded in a program before this statement is encountered, then the absolute value function (abs()) on line 15 will use the complex variable form of the absolute value. If the Complex.lua file has not been loaded then Complex with return nil and the abs function will be set to equal the math.abs() function. In this manner, the ssroot() function can be set to use either real math or complex math. In addition, the setfenv() table defines two variables, NT=200 and err=1.e-10, that are used in the function. A reason for defining these within the setfenv() table is that these parameters can subsequently be accessed and changed by programming code without changing the code defining the ssroot() function. The environmental table is returned by the getfenv() function so one can subsequently change the value of NT for example by the code statement: getfenv(ssroot).NT = 400, which would change the iteration limit to 400. The setfenv() function is a very valuable tool for isolating the range of function parameter values and for defining function parameters that one does not want to include in the list of arguments passed to the function but which one might want to change for particular problems. This technique will be used extensively in code segments developed in this work.

With respect to terminating an iterative loop, in almost all cases it is best to evaluate a solution's accuracy in terms of the relative error and not in terms of an absolute error. This allows solution values to vary greatly in magnitude, but to obtain the same number of significant digits of accuracy in the results. A relative error of 1.E-10 will result in approximately 10 digits of accuracy in the results regardless of the magnitude of the answer. In this particular case, since the answers are not far from unity, the difference between relative error and absolute error is small, but this will not be the case when solution values are far from unity. In subsequent algorithms, termination criteria will essentially always be specified in terms of relative error and not absolute error. However, this does give a problem if the solution value happens to be exactly zero, in which case dividing by the solution value as on line 15 of the code will obviously give problems. For the special case of zero solution, one would need to fall back on an absolute error criterion. Computer code to be presented in later sections will show how this can be combined with the relative error criteria. In the interest of simplicity, this has not been included in the code of Listing 3.2.

For possible use in other programs, the ssroot() function in Listing 3.2 has been saved in a file named ssroot.lua. Another example of finding the roots of a simple example by using this file is shown in Listing 3.3. In this example the Complex and ssroot files are loaded on line 3 of the code. Three functions are defined on lines 5 through 13 with g2() being the same as the g2() function in Listing 3.3. The first printed line under the output: shows that the calculated root is as expected the value near -1.23. For the g3() function consider again Eq. (3.2). Not only can this be rearranged as in Eq. (3.8), but another equally valid form is:

$$x = \pm j((4x^2 - 2x + 3)/5)^{1/4}, \qquad (3.14)$$

where j is the imaginary value (valid since $j^4 = 1$). Using this function after the Complex file is loaded on line 3 will cause all the algebra to use complex algebra and the obtained root as printed on the second line of output is (0.1163) + j (0.6879) which is one of the two complex roots of the polynomial. The complex root is obtained even though the initial guess on line 16 is a real number.

```
 1 : --/* File 3_3.lua Successive substitution example */
 2 :
 3 : require"Complex"; require"ssroot"
 4 :
 5 : function g2(x) -- Negative root function
 6 :     return -((3-2*x+4*x^2)/5)^.25
 7 : end
 8 : function g3(x) -- Complex root
 9 :     return j*((3-2*x+4*x^2)/5)^.25
10 : end
11 : function g4(x)
12 :     return math.atan(1/x)
13 : end
14 :
15 : print(ssroot(g2,0)) -- Real initial guess
16 : print(ssroot(g3,1)) -- Real initial guess
17 : print(ssroot(g4,1)) -- Real initial guess
Output:
-1.2326393306758    21
(0.11631966532609) + j(0.68791723656191)    122
0.86033358900115    41
```

Listing 3.3 Example of calculation roots using the successive substitution (ssroot()) function.

The reader can re-execute the code using the negative sign of Eq. (3.14) to obtain the other complex root but one should know the answer without executing the code. The reader can also use the printed complex root value back in Eq. (3.2) to verify that the result is zero to within about 10 decimal digits of accuracy which is the relative error criteria used to terminate the iterative loop in ssroot(). It can be seen that 122 iterations of the ssroot() loop are required for this complex answer, so the convergence is slow but a correct answer is obtained. Finally the g4() function defined on line 12 is used to solve for a zero of Eq. (3.3). For this function there are two possible iterative equations:

$$G(x) = 1/\tan(x) \text{ or}$$
$$G(x) = \tan^{-1}(1/x) \tag{3.15}$$

While the first form appears simpler, one finds that the successive substitution method only converges when using the second form. The printed output shows that the method converges to a solution value of $x = 0.86033$, this time achieved in 41 iterations. Now, let's hasten on to a more general solution method known as Newton's Method.

3.2 Newton's Method or Newton-Raphson Method

The next method to be presented for solving nonlinear equations is known as Newton's method. This is probably the most important numerical method available for analyzing nonlinear phenomena. This method and extensions of the method are used throughout this book to approach the nonlinear modeling of physical phenomena. The method was published by Newton in *Method of*

Fluxions in 1736 based upon work done in 1671. The method was also described by Joseph Raphson in *Analysis Aequationum* in 1690. Hence the technique is also frequently known as the Newton-Raphson method.

Newton's method is closely related to Taylor's series (first published in 1715). Given a function $f(x)$ the Taylor series of the function about some point x_1 is:

$$f(x) = f(x_1) + f'(x_1)(x - x_1) + \frac{f''(x_1)}{2!}(x - x_1)^2 + \frac{f'''(x_1)}{3!}(x - x_1)^3 + - - . \quad (3.16)$$

When x is sufficiently close to x_1 only a few terms in the series may be sufficient to give a good approximation to the function. In Newton's method one keeps only the first two terms of the series to give:

$$f(x) = f(x_1) + f'(x_1)(x - x_1) . \quad (3.17)$$

If this is being applied to the problem of finding a root of an equation, one assumes that one has a value x_1 that is close to the value for which $f(x) = 0$ and the first order term can be used to approximate the solution value as:

$$x = x_1 - \frac{f(x_1)}{f'(x_1)} . \quad (3.18)$$

Thus given a value x_1 close to the solution, Eq. (3.18) is applied and an improved solution value is calculated. The technique can be repeated to further improve the solution value and the iteration algorithm is:

$$x_{n+1} = x_n - \frac{f(x_n)}{f'(x_n)} . \quad (3.19)$$

The algorithm is repeated until some desired degree of accuracy is achieved. A graphical representation of the algorithm such as seen in Figure 3.3 can help to understand the technique. Eq. (3.17) is the defining equation for a straight line passing through the function at point x_1 and having the slope of the function at the same point. The intersection of this line with the horizontal axis then gives the new value as represented by Eq. (3.18). Beginning at point x_1, this new value of $x = x_2$ then defines a new point on the curve and a new slope and the process is repeated giving a new x_3, etc. until the solution value is approached as closely as desired. At each step the algorithm is essentially approximating the nonlinear function by a straight line segment and solving for the zero of the straight line segment.

In terms of a general approach to nonlinear problems, it's important to keep some general features of the Newton method in mind. The method of approximating a nonlinear problem by a related linear problem when close to a solution value and iterating on the formulated linear problem is a general approach used in almost all types of nonlinear modeling of physical effects. Newton's method has some very important convergence properties. First, in order for convergence to occur, the initial starting point must be "sufficiently close" to the solution value. It is very difficult to give general criteria for how close the initial guess must be because it is so dependent on the exact nature of the function being solved. No attempt will be made here to provide such general guidance. One can certainly conceive of many cases where one can show that a solution can not be obtained by some range of initial guesses. It has been the author's experience in applying Newton's method to a range of physical problems that one usually knows the solution value to within sufficient accuracy that Newton's method tends to converge for real physical problems. One can also worry about the possible existence of any solution and mathematicians worry a lot about such problems. Again the author's experience has been such that a correctly formulated physical problem will typically have a solution. If one's equations do not have a solution then one hasn't formulated the physical problem correctly. One potential problem can be seen if the derivative term in the denominator of Eq. (3.19) ever becomes zero. This is discussed later in this chapter.

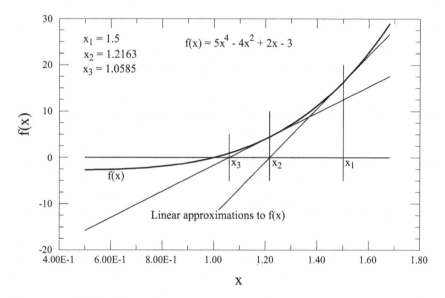

Figure 3.3 Graphical illustration of Newton's method.

If one makes a Taylor series expansion of the function about the x_n point and evaluates the equation at the zero point one has

$$f(x_0) = 0 = f(x_n) + f'(x_n)(x_0 - x_n) + f''(x_n)(x_0 - x_n)^2 / 2 + \ldots. \qquad (3.20)$$

One also has the Newton iteration equation of

$$0 = f(x_n) + f'(x_n)(x_{n+1} - x_n). \qquad (3.21)$$

Now subtracting Eq. (3.21) from Eq. (3.20) and keeping results to second order gives

$$0 = f'(x_n)(x_0 - x_{n+1}) + f''(x_n)(x_0 - x_n)^2 / 2. \qquad (3.22)$$

This can now be converted into an equation for relative error as:

$$\varepsilon_{n+1} = -\left[\frac{f''(x_n)x_0}{2f'(x_n)}\right]\varepsilon_n^2 .$$

(3.23)

What this means is that if the relative error is ε at one iteration step, then it is proportional to ε^2 at the next iteration step. This means that convergence occurs very rapidly as one approaches the solution provided that the quantity in brackets is not too different from unity. For example if one has two digits of accuracy at some iterative step, one has four digits at the next step and eight digits at the next step, etc. The quadratic convergence of Newton's method is one of the main features that make it very attractive as a solution technique. However, some of this rapid convergence is lost if the bracketed term in Eq. (3.23) is very large compared with unity. This does occur with some types of nonlinear equations.

At each iterative step in the Newton method, one must evaluate the function and its derivative at the existing approximation to the solution. A question arises as to how one obtains the derivative term, especially for a very complicated function. There are basically only two ways: either one calculates derivatives analytically or the computer is used to evaluate then numerically. For simple functions, taking an analytical derivative may be a valid approach. However, for complicated functions, considerable effort may be required and in some cases the function is only known through a computer algorithm from which an analytical evaluation of the derivative is not possible. It has been the author's experience that a numerical derivative works fine for a wide range of physical problems and this allows one to much more readily apply Newton's method to a wide range of physical problems. This will be the approach taken in this book, although for simple functions one could consider taking an analytical derivative for use in Newton's formula.

```
 1 :  --/* File newton_1.lua */
 2 :  -- Solves a nonlinear equation f(x) = 0 for the value x
 3 :
 4 :  function newton(f,x)
 5 :     local nend,cx,fx = NTT
 6 :     for i=1,NTT do
 7 :        local dx = x*FACT
 8 :        if dx==0.0 then dx = FACT end -- Protect against zero
 9 :        fx = f(x)
10 :        cx = fx - f(x+dx)
11 :        cx = fx*dx/cx -- Correction to x value
12 :        x = x + cx
13 :        dx = abs(cx) + abs(x)
14 :        if dx==0.0 or abs(cx/dx)<ERR then nend = i; break end
15 :     end
16 :     return x,nend,cx,fx
17 :  end
18 :  setfenv(newton, {abs=(Complex or math).abs, FACT=1.e-6,
19 :  ERR=1.e-8, NTT=200}) -- Make all terms local to newton
```

Listing 3.4. Code segment for implementing Newton's method.

A general code segment for implementing Newton's method is shown in Listing 3.4. The main iterative loop extends from line 6 through line 15 with a maximum iteration number of NTT = 200. Newton's formula is implemented on lines 10 and 11 with the correction value on line 11. Lines 7 through 11 are used to perform a numerical derivative for use in Newton's method and a simple first-order numerical approximation is used in the form:

$$f'(x) = \frac{f(x+dx) - f(x)}{dx}. \tag{3.24}$$

In evaluating the function derivative according to Eq. (3.24) one must select an appropriate value for the increment in x i.e. dx. The question of how to best implement numerical derivatives is extensively covered in Chapter 5, where it is shown that making the dx increment some fraction of the actual x value is a much more robust technique than simply using a fixed displacement. This approach compensates for the possibility that the magnitude of x may vary over many orders of magnitude from one physical problem to another. One might also expect that a double sided derivative where one considers the function value evaluated at $+dx$ and $-dx$ would provide a more accurate estimate of the derivative. However, when Newton's method converges, the converged value and converged accuracy doesn't depend on the accuracy of the derivative term as the numerator term in Eq. (3.19) is driven to zero even if the derivative is slightly inaccurate. Also for some problems where the derivative is actually zero at the root point, it is better to have a slightly inaccurate derivative and avoid a zero in the denominator term. For this chapter, the reader should just take as a given that the numerical derivative is a good approach and consult Chapter 5 for further verification. Also justification for taking the relative displacement factor as 1.e-6 is also found in Chapter 5. With this approach to dx, one must protect against an $x = 0$ root value and line 8 performs this function by using a fixed displacement if x is ever exactly zero. Finally, the function returns the solution value, the number of iterations, the relative error at convergence and the value of the function at convergence on line 16.

The code in Listing 3.4 is saved in a file, newton_1.lua (newton.lua will contain an improved version) and Listing 3.5 shows some examples of using Newton's method to solve some simple nonlinear equations. On line 3 both the Complex number package and the Newton code is loaded by the require statements. The code is arranged in function definition and solution pairs with the answer printed for each solution along with the number of iterations taken for convergence. The $f1$ function defines the cube root of 1.e20 and with an initial guess of 1.e10, the solution takes 24 iterations for an accuracy of better than 10 digits with an answer of 4641588.8336128.

The $f2$ function defines the fourth degree equation defined in Eq. (3.2) for which real approximate solutions exist near 1.0 and -1.25. From the output generated by the code (shown at the bottom of Listing 3.5), it is seen that starting at +2, Newton's method takes 8 iterations to obtain the correct solution value of 1.0 with an accuracy of better than 10 digits. If the successive substitutions algorithm

```
 1 : -- /* list3_5.lua */
 2 :
 3 : require"Complex"; require"newton_1"
 4 : -- Cube root of 1X10^20
 5 : function f1(x) return x^3 - 1.E20 end
 6 : print('f1 ',newton(f1,1.e10))
 7 : -- Real and complex roots of 4th degree polynomial
 8 : function f2(x) return 5*x^4 - 4*x^2 + 2*x -3 end
 9 : print('f2 ',newton(f2,2))
10 : print('f2 ',newton(f2,j))
11 : print('f2 ',newton(f2,-j))
12 : -- Function that varies rapidly with x
13 : function f3(x) return math.exp(10*x)*(1-.5*x) -1 end
14 : print('f3 ',newton(f3,4))
15 : -- Rearranged f3 for more linear function
16 : function f4(x) return (1-.5*x) - math.exp(-10*x) end
17 : print('f4 ',newton(f4,4))
18 : -- Linear equation
19 : function f5(x) return 5*x - math.pi end
20 : print('f5 ',newton(f5,0))
21 : -- Complex roots of 2nd degree polynomial
22 : function f6(x) return x^2 -4*x + 13 end
23 : print('f6 ',newton(f6,j))
24 : print('f6 ',newton(f6,-j))
Selected Output:
f1   4641588.8336128     24
f2   1            8
f2   (0.11631966535636) + j(0.68791723656161)     6
f2   (0.11631966535636) + j(-0.68791723656161)    6
f3   1.9999999958777          28
f4   1.9999999958777          2
f5   0.62831853071796         2
f6   (2) + j(3)          7
f6   (2) + j(-3)         7
```

Listing 3.5. Several examples of Newton's method with nonlinear equations.

is executed starting with the same initial guesses, about 20 iterations are required for the same level of accuracy. The newton() function is called twice again on lines 10 and 11 but starting in these cases with initial guesses of $+j$ and $-j$. This is an attempt to evaluate the two complex roots that one knows must exist for the fourth order equation since only two real roots exist. Since the Complex package is loaded on line 3 before the newton package, the newton() function will in this case be perfectly happy to use complex algebra in all its calculations. The second and third lines of f2 printed output do in fact verify that two complex roots have been evaluated at (0.11631966535636) + j(0.68791723656161) and (0.11631966535636) + j(-0.68791723656161) after 6 Newton iterations. The reader can verify the accuracy of these solutions.

The $f3$ function defines the equation

$$f3(x) = \exp(10x)(1-.5x) - 1 = 0 .$$

(3.25)

This function has an obvious root near the value 2.0. With an initial guess of 4.0, Newton's method requires 28 iterations to reach convergence for 10 digit accuracy.

However if we write the function as

$$f4(x) = (1 - .5x) - \exp(-10x) = 0 , \qquad (3.26)$$

it is seen that Newton's method converges to the same answer in only 2 iterations. Both the $f3$ and $f4$ functions obviously have the same zero values, but much different convergence rates for Newton's method. This illustrates a very important point when using Newton's method. It is frequently possible to rewrite an equation in more than one form with vastly different convergence rates for the various forms of the equation. This is similar to the case encountered with the successive substitution algorithm. Any rearrangement of the equation which results in a more linear form of the equation will result in faster convergence as a linear equation can be solved in only two iterations. The lesson to be learned here is to experiment with different functional forms for nonlinear equations to be solved for roots.

This brings up another feature of Newton's method and this is the fact that it will solve not only a nonlinear equation, but also a linear equation. This is illustrated by the $f5(x) = 5x - \pi = 0$ equation and the resulting two iterations required for convergence as seen in the printed output. Actually, the true solution is obtained in only one Newton loop, but Newton's method requires a second iteration to recognize that the correction error is then near zero. Of course there are simpler and faster ways of solving a linear equation. However, many times it is simpler to use a general solution technique that works for nonlinear and linear equations for a range of problems rather than switching between one technique that works for linear equations and one that works for nonlinear equations and Newton's method is one such technique that can be used for both linear and nonlinear equations. With today's fast computers the additional time taken by Newton's method over a linear equation solver is not significant for many engineering problems. One can also cut the computational time in half in Newton's method by setting the maximum number of iterations to one for linear equations.

A final function in Listing 3.5 is a quadratic equation $f(x) = x^2 - 4x + 13 = 0$ that has no real roots but a pair of complex roots at $x = 2 +/- j3$. The code and output show that Newton's method again works in the complex math domain and from an initial guess of $+/- j$ converges to the correct solution in 7 iterations. In order for the newton() function to use complex algebra, the initial approximation must simply be a complex number.

A typical nonlinear problem from Electrical Engineering is the diode circuit shown in Figure 3.4. Fundamental circuit equations are:

$$v_s = v_d + i_d R \text{ and } i_d = I_s(\exp(v_d / v_t) - 1) \text{ where ,} \qquad (3.27)$$

I_s and v_t are diode parameters. Combining the two circuit parameters results in the single nonlinear equation:

$$v_s = v_d + RI_s(\exp(v_d / v_t) - 1) . \qquad (3.28)$$

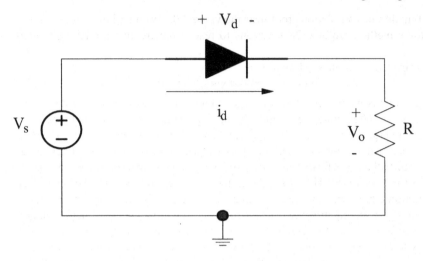

Figure 3.4 Simple diode circuit.

This can be used in Newton's method to solve for the diode voltage v_d. Before plunging ahead, let's review briefly what is known about the diode and the circuit. When the voltage source is positive, current flows through the diode with a small voltage drop (~.6 to .8 volts) across the diode and the difference between this voltage and the source voltage appears across the resistor. When the voltage source is negative, little current flows in the circuit and essentially the entire voltage source is dropped across the diode with a negative diode voltage. A second form of the circuit equation can also be obtained by solving for the exponential term in Eq. (3.28) giving the equivalent equation:

$$v_d = v_t \log((v_s - v_d)/RI_s + 1). \tag{3.29}$$

This could also form the basis for a Newton's method solution for the circuit.

Listing 3.6 shows code for both of these equations and some example applications of Newton's method for both a positive source voltage of 15 volts and a negative source voltage of -15 volts. Functions f1 and f2 implement Eqs. (3.28) and (3.29) respectively. For the positive source voltage, lines 14, 15 and 16 call Newton's method for function f1() with three different initial guesses at the solution, 0, .4 and .8 volts. From the output, it is seen that the first two guesses result in 200 iterations with the final voltage value of around 9.8 volts which is obviously incorrect. However, an initial guess of .8 volts converges in 8 iterations to the correct value of 0.70966 volts. Similar calculations with the f2() function used in lines 17, 18 and 19 with the same initial guesses converge to the correct solutions in only 3 iterations. One might expect the f2() function to converge rapidly since it involves a log() function as opposed to the f1() function which involves an exp() function. However, the lack of convergence for initial guesses of 0 and .4 is perhaps unexpected for the f1 function. Without going into great detail about the solution, with an initial guess of 0 or .4 volts, the initial derivative is very small

```
 1 : --/* File list3_6.lua */ -- Diode equation example
 3 : require "newton_1"
 5 : Is,vt = 1.0e-14,0.026 -- Diode parameters
 6 : R = 2.0e3 -- Resistance value
 7 :
 8 : function f1(vd) return Is*R*(math.exp(vd/vt)-1) + vd - vs end
 9 :
10 : function f2(vd) return vd - vt*math.log((vs - vd)/(Is*R) + 1)
      end
11 :
12 : vs = 15 -- Positive voltage source
13 : print("Positive voltage solutions")
14 : print(newton(f1,0))
15 : print(newton(f1,.4))
16 : print(newton(f1,.8))
17 : print(newton(f2,0))
18 : print(newton(f2,.4))
19 : print(newton(f2,.8))
20 :
21 : vs = -15 -- Negative voltage source
22 : print("Negative voltage solutions")
23 : print(newton(f1,0))
24 : print(newton(f1,-1))
25 : print(newton(f1,.8))
26 : print(newton(f2,0))
27 : print(newton(f2,-1))
28 : print(newton(f2,-14.999))
Output:
Positive voltage solutions
9.827236337169   200   -0.025995073695858    7.6877704188719e+153
9.7733993511304  200   -0.025995100610812    9.6945476347504e+152
0.70966667615268   8   -1.9488814683518e-009  1.0731238031525e-006
0.70966667615265   3   -6.0951370705418e-014  6.1062266354384e-014
0.70966667615265   3   -1.9947721321771e-015  1.9984014443253e-015
0.70966667615265   3   -0          0
Negative voltage solutions
-14.99999999998    2   -1.5399347574227e-008  1.5399347574885e-008
-14.99999999998    2   1.0477606338356e-008   -1.04776063381e-008
-14.99999999998    8   1.1457501614551e-012   -1.1457501614132e-012
-1.#IND   200   -1.#IND   -1.#IND
-1.#IND   200   -1.#IND   -1.#IND
-1.#IND   200   -1.#IND   -1.#IND
```

Listing 3.6. Newton's method for diode circuit example with Eqs. (3.28) and (3.29).

and the next approximation using Newton's method and the f1() function is close to the source voltage of 15 volts which is very far from the solution value near 0.7volts. With such a large second approximation, and the exponential factor is dominant in the f1() equation and the solution slowly works back toward the true solution, but only decreases by about 0.026 volts (v_t) per iterations. It can also be noted that a v_d value of 15.0 in the exponential term has a value of $\exp(15/.026) = 3.6 \times 10^{250}$ which is approaching the largest number that can be stored in the language. The solution will eventually converge, but requires many more iterations than the 200 limit used here. The reader can verify that the solutions will eventually converge by setting the maximum number of allowed iterations to say 1000 (the solution actually requires 556 iterations).

These differences in the convergence rates could have been anticipated if graphs of the two functions had simply been plotted before executing the code. Figure 3.5 shows plots of the f1() and f2() functions over the interval of voltage from 0 to 1.0 Volt. In order to see the results on the same graph, the f2() function has been multiplied by a factor of 100. It is readily seen that the f2() function is very close to being a linear graph over this range and one expects very rapid convergence as observed in Listing 3.6. Also it can be seen that the f1() function has a very small derivative for voltages below about 0.6 Volts and an initial guess within this range of voltages will result in a second voltage value that is very large (near +15) and very far from the zero solution point. While Newton's method will eventually converge, many iterative loops are required. This illustrates the importance of looking at a graph of a function before attempting a solution or taking an initial guess at a solution value.

Next consider the negative source voltage and Newton's method on lines 23 through 28 using both the f1() and f2() function and several initial guesses. For this case the solution value is almost exactly -15 volts, which is the source voltage, and Newton's method with the f1() function is seen to converge to the correct solution in a small number of iterations (2 to 8) with almost any reasonable initial guess. However, with the f2() function on lines 26 through 28, the solution doesn't converge for any of the three initial guesses given in the example. In fact it is almost impossible to get the f2() function to converge unless one used the exact solution value. Even the initial guess of -14.999 used on line 28 does not give a converged solution. The iteration terminates after 200 iterations and gives the answer -1.#IND. This is the printed result when the calculation can not properly be performed. Thus the f2() equation does not provide a reliable function for solving the circuit equation for negative, v_s, although it converges very rapidly for positive v_s. The reason for this lack of convergence for the f2() function is not too difficult to ascertain. Consider the math.log() term in the f2() equation with a negative v_s. The denominator term Is*R is very small making the factor multiplying the (vs - vd) term very large (5e10 in fact). Then for values of vd slightly less negative than vd, the argument of the log() term becomes negative and this can not be computed except in terms of complex numbers and the desired solution is a real number. In fact if one attempts to calculate values of the function for a negative vs, one readily encounters the negative arguments to the log() function. When the argument goes negative, Newton's method will continue, but with values equal to -1.#IND. This is a case where a graph (or attempted graph) can greatly aid in understanding nonlinear equations and save one much effort in attempting an impossible solution. However, a graph of the valid function range is very difficult to obtain as the range over which the log argument has a valid value is only from -15 to -15 +Is*R (or -15 + 2x10^{-11}). Also a graph of the f1() function is not easily obtained as the zero crossing point again occurs very close to the -15 value.

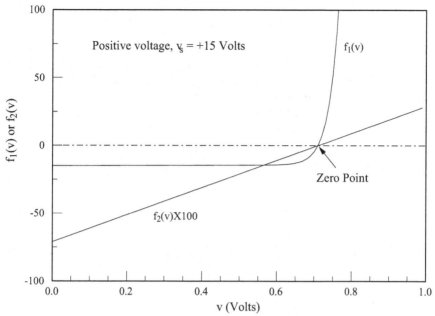

Figure 3.5. Graph of the two functions f1(v) and f2(v) over the range of 0 to 1.0 Volts.

Even when using the f1() equation, the only initial guess in this example that works for both the positive and negative source voltages is the initial 0.8 volts guess, although a range of values from about 0.7 to about 0.9 will also work with varying numbers of iterations. Although for negative source voltages, an initial guess of a positive 0.8 is known to be far from the solution value, Newton's method is readily able to converge to the correct negative voltage values. An exponential term such as is present in this example is one of the more difficult physical models as the value and derivative of the function changes by many orders of magnitude over the range of voltages typically present. The reader is encouraged to execute the code with different initial approximations and observe the results.

With this as a background a sinusoidal voltage source can now be considered and the solution obtained for a rectified sine wave. For this the source voltage is taken as:

$$v_s = V_s \sin(x), \tag{3.30}$$

where x varies from 0 to 4π (for two cycles of a sin wave). Listing 3.7 shows the code for a sinusoidal voltage source of magnitude 15 volts. Initial parameters and data arrays are defined on lines 5 and 6. The function fv() (same as previous f1()) is defined on line 9. A loop is defined from line 12 to 18 that steps the angle value over 400 intervals, evaluates the source voltage on line 14 and calling the newton() function on line 15 to return the solved diode voltage. Based upon the previ-

ous example, an initial guess of 0.8 volts is used for all voltage values in the new-
ton() function. It should be noted that since variables are globally known unless
limited in scope, the vs value on line 9 of the fv() definition is exactly the same
variable as the vs value set on line 14 of the voltage increment iterative loop. Being able to set the value of a variable in a subsequent loop and have it known in a
previously defined function is an unexpected result to some people new to the Lua
language, but is a very convenient feature of the language if properly used. One
just has to know where variables are defined and where they are used. Keeping
code segments short greatly aids this knowledge.

```
 1 : -- /* list3_7.lua */ -- Newton's method for diode circuit
 2 :
 3 : require "newton_1"
 4 :
 5 : Vs,R,Is,vt = 15.00,1.e3,1.e-14,.026 -- device parameters
 6 : xa,vo,vsa = {},{},{} -- Define empty arrays for values
 7 :
 8 : -- Rectifier circuit equation to be solved
 9 : function fv(v) return vs - v - R*Is*(math.exp(v/vt) - 1) end
10 :
11 : nmx = 0
12 : for i=1,401 do -- Select 400 intervals for calculation
13 :    x = (i-1)*4*math.pi/(400) -- Angle value
14 :    vs = Vs*math.sin(x) -- source voltage
15 :    vd,n,err = newton(fv, 0.8) --diode voltage Newton's method
16 :    nmx = math.max(nmx,n) -- Check on convergence
17 :    xa[i], vo[i], vsa[i] = x, vs-vd, vs -- Save arrays
18 : end
19 : print('Maximum number of iterations =',nmx)
20 : write_data('list3_7.dat',xa,vsa,vo)
21 : plot(xa,vsa,vo)
Output:
Maximum number of iterations = 25
```

Listing 3.7 Solution of rectifier circuit by Newton's method.

As is good programming practice not only the value of the returned solution but
the number of iterations for each call to Newton's method is captured on line 15.
The code does not check each returned number of iterations against the newton()
maximum (200 default) but the maximum value of the number of iterations is captured by line 16 of the code using the math.max(nmx,n) function. After the calculation is completed this number is printed and the printed output results shows that
the newton() function took at most 25 iterations which is well below the coded
maximum. One can then be assured that all the returned values were obtained
with a properly converged Newton iterative loop. If this number had been the
NTT value (200) then more extensive testing would be called for to determine the
problem area and to determine how to formulate the equation and solution to give
a properly converged solution for all desired source values.

Finally three array values are saved by the write_data() call on line 20 and then
a graph of the source voltage and the output voltage will popup at the end of the
calculation from the plot() function on line 21. A graph of this calculated output is

shown in Figure 3.6. As can be seen from the graph, the output voltage is about 0.7 volt below the source voltage for positive source voltages and essentially zero for negative source voltages.

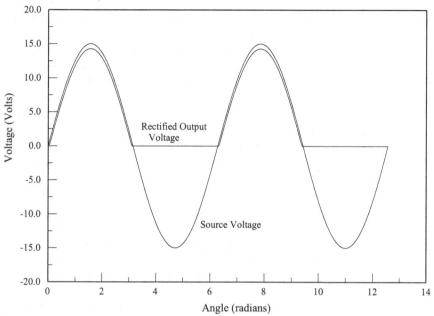

Figure 3.6. Source and output voltage for diode rectifier of Figure 3.4.

This example illustrates several important facts when solving nonlinear equations representing physical phenomena. First the form of the equation to be solved is important – some forms work best over certain solution values and other forms work best over other solution values. Second, the initial guess is an important key in achieving a valid solution. Often different initial guesses are needed for different parameter values or for different function values. The solution shown in Figure 3.6 and Listing 3.7 uses a single equation and single initial value over the range of positive and negative source voltage values. One can envision other strategies for a solution. For example one could use different equations for the positive and negative voltage ranges, such as the f1() and f2() functions of Eqs. (3.28) and (3.29) since it was shown that the solution converges rapidly with f2() for positive voltages and converges rapidly with f1() for negative voltages. However, the solution is simpler if only a single function can be used as in this example even though more Newton iterations may be required for the solution. Different guesses can be used for the different voltage ranges – for example 0.8 volts for positive source voltages and the source voltage for negative source voltages. Such an implementation can be achieved by replacing the 0.8 initial guess in line 15 of the listing by the statement, math.min(vs,0.8) -- Give it a try. This will give a slightly faster execution of the code as the maximum required number of iterations is 11.

Another very useful technique with iterative solutions is to limit the maximum change that can occur between iteration steps in some way. In discussing the code in Listing 3.6 and the initial guesses of 0 and 0.4 for lines 14 and 15, it was pointed out that the application of Newton's method with these values, give subsequent approximation values that are very far from the actual solution points. In fact Newton's method goes from the initial guesses to a value very close to the source voltage value of 15.0, while the solution point is at about 0.7, a value much closer to the initial guess. Cases such as these can be prevented by limiting in some way the step size that can be taken at each iterative loop. One can assume that Newton's method indicates the proper direction to change the variable, but not necessarily the proper magnitude of the change. This is especially useful when one has a function that is very nonlinear such as the exp() function in this example and where the function derivative at some point can be very small. There are at least two useful ways to limit the change in a variable in an iterative step. One can limit the magnitude of the change or one can limit the percentage change in the variable. If a variable can be either positive or negative and one does not know which value it should be then only the magnitude of the change can be limited. On the other hand if one knows that a variable is either positive or negative, then the percentage change per step can be very useful, especially for functions involving terms like log() or sqrt() where a negative value will terminate the calculation. However, if one limits the percentage change, then one must begin with some non-zero initial value.

Code to implement a Newton's method with step size limits is given is Listing 3.8. This is again named newton() but is located in a new file named newton.lua. The code to implement limits is contained in line 5 and lines 13-24, essentially doubling the length of the code. Both the absolute and percentage change limits are implemented. The function argument list on line 4 contains a third variable, step, in addition to the function name and the initial value. If step is negative, it is interpreted as being a +/- limit. For example -.5 will limit changes in the solution value at each iterative step to the previous value + or − 0.5 (see line 15 of listing). If the step input value is positive, it is taken as a limit on the relative change. For example a positive step value will limit the iterative solution values as follows:

$$x_n / step < x_{n+1} < x_n \cdot step .\tag{3.31}$$

This is implemented on lines 16 through 23 of the code. As an example, if x_n is 4.0 and step is 2.0 the next iteration value will be limited to the range 2.0 and 8.0 a factor of 2 (step) below the value and a factor of 2 above the value. If the step parameter is 1.1 the new solution values will be limited to a change of approximately 10% around the previous value. Also a step size of less than unity is converted on line 7 to a value greater than unity so that step sizes of 0.5 and 2.0 have exactly the same effect. In addition to step size limits, the new listing also includes a print capability on line 26 for each iterative loop if the nprint variable is set to any value different from nil.

```
 1 : --/* File newton.lua */
 2 : -- Solves a nonlinear equation f(x)=0 with step limits
 3 :
 4 : function newton(f,x,step) -- step may be omited in calling
 5 :    if step then if step>0.0 and step<1.0 then step=1./step end
          end
 6 :    nend = NTT
 7 :    for i=1,NTT do
 8 :        dx = x*FACT
 9 :        if dx==0.0 then dx = FACT end -- Protect against zero
10 :        fx = f(x)
11 :        cx = fx - f(x+dx)
12 :        cx = fx*dx/cx -- Correction to x value
13 :        if step and step~=0.0 then -- Skip if no step specified
14 :            if step<0.0 then -- Limit absolute step size
15 :                if abs(cx)>-step then cx = -step*(cx/abs(cx)) end
16 :            else -- Limit percentage increase in solution value
17 :                dx = 1. + cx/x
18 :                if dx>step then
19 :                    cx = (step-1)*x
20 :                elseif dx<1/step then
21 :                    cx = (1/step-1)*x
22 :                end
23 :            end
24 :        end -- End of step adjustment
25 :        x = x + cx
26 :        if nprint then print("Iteration ",i,x,dx) end - Print
27 :        dx = abs(cx) + abs(x)
28 :        if dx==0.0 or abs(cx/dx)<ERR then nend = i; break end
29 :    end
30 :    return x,nend,cx/dx
31 : end
32 : -- Make all variables and constants local to newtonl
33 : setfenv(newton, {abs=(Complex or math).abs,FACT=1.e-6,
          ERR=1.e-9,NTT=200,nprint=nil,print=print})
```

Listing 3.8 Newton's method with limits on change at each iterative step.

Examples of using this new code are shown in Listing 3.9. This is a partial repeat of the diode solution of Listing 3.6 but for only the f1() function and now with limits imposed on the maximum change per iteration. The use of limits does not solve the negative log() argument for the f2() function. The switch to using the new improved Newton code is on line 3 which now inputs the newton() function from the new file "newton". It will be recalled that without limits, newton()would not converge with the f1() function for a positive source voltage of 15 and an initial guess of 0 or 0.4. The new code on line 12 now sets limits of +/- 1 on the voltage change per iteration and on line 13 sets a factor of 2.0 for the voltage change beginning with 0.4. The first two lines of output show that the method now converges in 17 and 10 iterations to the correct values. Setting either a +/- limit or a limit on the relative change is seen to be an effective means to achieve a converged solution value.

It is instructive to look at the convergence rate for the examples in Listing 3.9 for different initial guesses. Figure 3.7 shows this for the three cases of a positive value of vs. The relative accuracy of the solution is plotted in each of the three

```
 1 : --/* File list3_9.lua */   -- Diode equation with iter limits
 2 :
 3 : require "newton"
 4 :
 5 : Is,vt = 1.0e-14,0.026 -- Diode parameters
 6 : R = 2.0e3 -- Resistance value
 7 :
 8 : function f1(vd) return Is*R*(math.exp(vd/vt)-1) + vd - vs end
 9 :
10 : vs = 15 -- Positive source voltage
11 : print("Positive voltage solutions")
12 : print(newton(f1,0,-1))
13 : print(newton(f1,.4,2))
14 : print(newton(f1,.8,-1))
15 :
16 : vs = -15 -- Negative voltage source
17 : print("Negative voltage solutions")
18 : print(newton(f1,0,-1))
19 : print(newton(f1,-1,2))
20 : print(newton(f1,.8,-2))
Output:
Positive voltage solutions
0.70966667615265    17    -8.4352506710307e-011
0.70966667615265    10    -3.7525674557288e-014
0.70966667615265     9    -3.7525674557288e-014
Negative voltage solutions
-14.99999999998     16    -8.5963828651149e-013
-14.99999999998      5     3.8391097705182e-011
-14.99999999998     14    -4.8045715555562e-012
```

Listing 3.9. Application of Newton's method with limits at each iterative step.

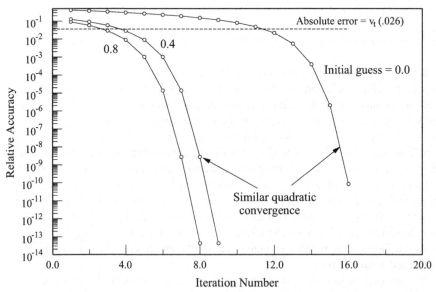

Figure 3.7. Observed relative convergence rate for the three positive voltage cases of Listing 3.9.

cases and the rapid convergence of the solution to the solution value is observed as the solution approaches the converged value. For example the relative error for the 0.8 and 0.4 initial guesses are in the range of 10^{-9} in one of the iterations and in the range of 10^{-14} in the next and final iteration shown for these two cases. For the initial guess of 0.0, the solution takes longer to approach the exact value because the solution is initially (on the second iterative value) forced further from the solution value. However, it is seen that all three solutions exhibit the desired quadratic convergence behavior as the zero point is approached. It simply takes more iterative steps if the initial guess is far from the true solution. Also shown by the dotted line is the point where the absolute error equals v_t and it is seen that this is about the point were the relative accuracy begins to rapidly improve with each iterative step.

Lines 18 through 20 of Listing 3.9 explore the effects of limits on the convergence of the f1() function for a negative source voltage where the solution is known to be close to the negative source voltage. Line 18 begins the Newton loop with a value of zero and sets a limit of +/- 1.0 volt. In order to reach -15, it is readily seen that it will take a minimum of 15 iterative loops and the actual number taken is 16 as indicated by the printed output. Line 19 begins at an initial value of -1 and limits the solution steps by a factor of 2 (-2, -4, -8, etc.). This case takes 5 iterative steps for convergence. Note that one can not begin with a positive initial value and achieve the correct solution with a step parameter of 2 as the value will always be limited to positive values. Finally for the f1() function, line 20 begins at a positive value of 0.8 and sets a +/- limit of 2. The output shows that this takes 14 iterations for the final solution. The limits have little effect for the final solution values with negative source voltages except to increase the number of iterations for the f1() function. However, the limits are seen to be very useful for achieving convergence with a positive voltage source. The convergence rate for the negative voltages is determined primarily by how large a voltage change is allowed in the calling argument to the newton() function.

It should be noted that calling a function with the same number of arguments as used in the defining code is not required. For example if one uses the statement newton(f,x) and omits a value for the step variable, Lua will simply substitute nil for the value. In this case all the code related to limiting the step size will be skipped and the newton() function in Listing 3.8 will be essentially the same as the newton() function in Listing 3.4. Thus this more general function can replace the original newton() function for all our applications. The ability to omit calling arguments for a simpler function application will be used in many function definitions in subsequent chapters.

A fairly thorough look has been taken at Newton's method for solving a single nonlinear (or linear) equation for a root (or roots) of a function. If a function has multiple roots, Newton's method must be started close to each root in order to find more than one solution. It is also seen that for some formulation of a physical problem, it is essentially impossible to get Newton's method to converge to a solution while a rearrangement of the basic equation can rapidly lead to valid solutions with Newton's method. The more one knows about a particular physical problem the easier it is to formulate an equation and obtain a valid solution with Newton's method.

While Newton's method is not the only approach to solving nonlinear problems, it is in this author's opinion the single most important algorithm for approaching nonlinear problems. One of its most important features in that of quadratic convergence which leads to very rapid convergence as a solution is approached. It forms the basis for almost all the techniques to be discussed in subsequent chapters for more advanced and complicated nonlinear problems involving many variables and for obtaining solutions to nonlinear differential equations. Thus a thorough understanding of Newton's method is essential to the remainder of this text.

3.3 Halley's Iteration Method

Newton's iterative method considers a first order Taylor series approximation to a function near a root of the function. Halley's method extends this to a consideration of second order terms as in:

$$f(x) = 0 \simeq f(x_n) + f'(x_n)\Delta x + \frac{1}{2}f''(x_n)(\Delta x)^2. \tag{3.32}$$

This can be rearranged as:

$$\Delta x = \frac{-f(x_n)}{f'(x_n) + f''(x_n)\Delta x / 2}. \tag{3.33}$$

If further the Newton approximation of the correction is used in the denominator term this results in the equation:

$$\Delta x = -\frac{f(x_n)}{f'(x_n)}\left[\frac{1}{1 - f(x_n)f''(x_n)/2f'(x_n)^2}\right]. \tag{3.34}$$

The term in brackets can be considered a correction to the Newton expression due to a finite second derivative term. If the second derivative is zero the expression reduces to the first order Newton equation. The use of this equation is known as Halley's rational formula iteration.

Another approach to the use of Eq. (3.32) is to consider it a second order equation in Δx and this leads to Halley's irrational formula of:

$$\Delta x = \frac{-f'(x_n) + / - \sqrt{f'(x_n)^2 - 2f(x_n)f''(x_n)}}{f''(x_n)}$$

$$= \frac{-1 + \sqrt{1 - 2f(x_n)f''(x_n)/f'(x_n)^2}}{f''(x_n)/f'(x_n)} \tag{3.35}$$

In the second form the proper sign has been retained for obtaining a corrected value closer to the function root. This form has numerical problems when the second derivative is very small and has loss of accuracy as a valid root is approached. These deficiencies can be corrected by multiplying and dividing by:

$$1+\sqrt{1-2f(x_n)f''(x_n)/f'(x_n)^2} \tag{3.36}$$

and making use of the fact that $(1-\sqrt{1-u})(1+\sqrt{1-u})=u$. This leads to the equivalent form of Halley's irrational formula:

$$\Delta x = -\frac{f(x_n)}{f'(x_n)}\left[\frac{2}{1+\sqrt{1-2f(x_n)f''(x_n)/f'(x_n)^2}}\right], \tag{3.37}$$

where the term in brackets is now the correction to Newton's expression and can be compared with the bracket term in Eq. (3.34).

Either Eq. (3.34) or Eq. (3.37) can form the basis for an iterative solution to a nonlinear equation. Eq. (3.37) would appear to be a more exact expression and thus perhaps the favored expression. However, it is relatively easy to find example equations where the quantity within the square root expression is negative and the equation fails even though there are real roots. One such simple expression is $f(x) = \exp(x) - A$ when x is large such that the exponential term dominates. Also the fact that a square root must be calculated means additional computational time. Eq. (3.34) does not have these difficulties and will converge for a wider range of functions. Thus this form of Halley's method will only be considered further.

The use of Halley's method requires an evaluation of the second derivative of a function in addition to the first derivative. If the function is relatively simple in mathematical form an analytical derivative can sometimes be obtained for use in the expression and a customized version of the method can be used for a specific function. For a general root finding method the approach with Newton's method has been to use a numerically evaluated first derivative. Extending this approach to Halley's method requires a numerically evaluated second derivative and this requires an evaluation of the function at three points as opposed to two function evaluations for the first derivative. Thus some (or perhaps most) of the advantage of such a formula will be lost in terms of computational time by the extra function evaluation.

A code segment similar to Newton's method of Listing 3.8 is available in the halley.lua file. This listing is not shown but is very similar to the Newton code with the addition of a numerically evaluated second derivative (using formulas developed in Chapter 5). This can be used as a direct substitute for the newton() function for example in Listing 3.5 by adding near the beginning the code statements: require"halley"; newton=halley. The reader is encouraged to re-execute the code in Listing 3.5 with this change and observe the results. It will be seen that fewer iterations are required with the halley() iterative function as opposed to the newton() function. However, by timing the execution of the code in Listing 3.8 it can be found that the computer execution time is close to the same for either the newton() or halley() function. The advantage of fewer iterative loops is lost in the need for more function evaluations per iterative loop in Halley's method. This method will not be used further in this work as little speed advantage is found for typical examples. Also the method is not extensible to coupled systems of equations as considered in the next chapter.

   This example illustrates one of the guiding principles of this work and this is
the principle of using simple numerical algorithms applied many time to solve
nonlinear problems. While many techniques have over the years been developed
for solving many nonlinear engineering problems, in a great majority of cases the
simplest numerical algorithms are the preferred methods. Iterative techniques are
typically required for solving nonlinear problems. While one can usually develop
more and more complex algorithms that result in improved accuracy in each iterative step, the advantage of such improved equations is in a great majority of cases
lost by the increased complexity of the equations required for each iterative step.
A guiding principle for numerical algorithms should be to first consider the simplest first order solution methods and only consider improved algorithms if such
methods can not for some reason solve a particular problem. In all cases example
solutions should be extensively timed to see if advanced algorithms have any real
computational speed advantage.

3.4 Other Solution Methods

Only a brief discussion will be given of other solution methods in order to direct
the reader to other approaches if the successive substitution or Newton's method is
deemed to not be appropriate for a particular problem.

   Graphical methods have been mentioned before and these are frequently used
in first approaching a problem to find the appropriate range of a possible solution.
They consist of calculating and graphing a given function over a range of values
of the independent variable. From a graphical plot of the data one can get an approximation to one or more roots of the function. This is most appropriate when
one has little idea about the location of roots and as a means of obtaining an initial
guess at a solution for use in more refined techniques such as Newton's method.

   The bisection method starts by assuming that a solution is known to exist between two values x_l and x_u at which points the sign of the function is known with
one value positive and one value negative. A new value is then taken half way between the original points and the function is evaluated at this point. Depending on
the sign of the function at the midpoint, a new interval is now defined which has
the midpoint at one end and one of the initial points at the other end. Each iterative step thus halves the uncertainty in the solution value. The algorithm is repeated decreasing the uncertainty in the function root by one-half at each iterative
step, unless one is very lucky and happen to land on the exact solution value.
While this tends to be a robust technique, the convergence is very slow compared
with Newton's method. If this is used in the early stages of finding a root, one can
also switch to Newton's method as the solution value is approached to obtain the
faster convergence of Newton's method.

   The secant method is similar to Newton's method but uses two consecutive iterative values of the function to approximate the derivative. The basic iterative
algorithm is:

$$x_{n+1} = x_n - f(x_n) \left[\frac{x_n - x_{n-1}}{f(x_n) - f(x_{n-1})} \right]. \tag{3.38}$$

By comparing with Eq. (3.19) it can be seen that two successive evaluations are essentially used to approximate the function derivative. Convergence is slower than with Newton's method which uses a more accurate derivative value. Also two initial points are required before the algorithm can be implemented.

The false-position method is similar to the bisection method but improves on the iterative algorithm by making use of the magnitudes of the function at the upper and lower position values. The iterative algorithm is:

$$x_{new} = x_u - f(x_u) \left[\frac{x_u - x_l}{f(x_u) - f(x_l)} \right]. \tag{3.39}$$

This new position then replaces the old upper or lower value depending on the sign of the function at the new position. Again this is a fairly robust algorithm, but lacks the quadratic convergence of Newton's method and also requires two starting points.

Computer code for these techniques is relatively easy to compose and is left to the interested reader.

3.5 Some Final Considerations for Finding Roots of Functions

The emphasis in this chapter has been on using the Newton method for finding roots of various functions. Newton's method is based upon a local Taylor series expansion of a function and approximating the function locally by the first two terms representing the value of the function and the first derivative as expressed by Eq. (3.17). But what about functions for which the first derivative is zero at the solution value that corresponds to zeros of the function? From Eq. (3.19) one can see that the correction value becomes:

$$x_{n+1} = x_n - \frac{f(x_n)}{f'(x_n)} \rightarrow x_n - \frac{0}{0}, \tag{3.40}$$

if both the function and its derivative approach zero. What happens to Newton's method under such conditions? Also one sees from Eq. (3.23) describing the very desirable quadratic error convergence that the denominator has the first derivative term. So again there is some question as to the convergence to a valid function root if one has a function with zero value of the derivative at the solution point where a zero occurs.

The approach here will be to explore the use of Newton's method with function of this type rather than attempt to theoretically analyze such cases. If the first derivative of a function is zero at the zero solution point, then according to the Taylor series approximation, one can approximate the function near the root by some function such as:

$$f(x) = C(x - x_0)^m,$$ (3.41)

where C and m are constants. If only the first derivative is zero then m will equal 2 while for a general value of m, the first $m-1$ derivatives at the zero point will be zero. Figure 3.8 shows graphs of the function for m values of 10 and 11 and a zero point at 2.0. The $f_1(x)$ function has a minimum at the zero point and the $f_2(x)$ function has what is known as a saddle point at the zero point. It can be seen that evaluating the derivative in the vicinity of the solution point will give a very small slope and Newton's method may potentially have problems with functions of this type. These have been selected as being very extreme examples of cases of 9th and 10th order zeros at the zero point, giving some severe tests for Newton's method.

If Eq. (3.41) is used in Eq, (3.40) one obtains:

$$x_{n+1} = x_n - \frac{(x_n - x_0)}{m}.$$ (3.42)

From this an expression for the relative error can be obtained as:

$$\varepsilon_{n+1} = \left(\frac{m-1}{m}\right)\varepsilon_n.$$ (3.43)

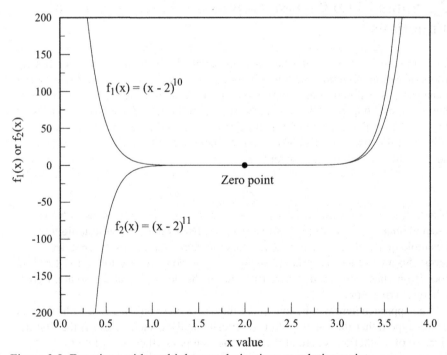

Figure 3.8. Functions with multiple zero derivatives at solution point.

This gives an algebraic reduction factor as opposed to the quadratic reduction as previously seen for the case of a non-zero first derivative. This predicts a much slower convergence rate for this case. If one has only a zero first derivative and a non-zero second derivative ($m = 2$), this would predict a reduction in the relative error by a factor of 2 for each Newton iterative loop. Listing 3.10 shows a code example for applying Newton's method to Eq. (3.41) for $m = 10$ and 11. The nprint variable is set to non-zero on line 4 in order to observe the convergence of the solutions beginning with an initial guess of 0.0. The reader is encouraged to execute the code and observe the results.

```
1 :   --/* File list3_10.lua */ -- equation with multiple roots
2 :
3 : require "newton"
4 : getfenv(newton).nprint = 1 -- Print iterative values
5 :
6 : function f(x) return (x-2)^m end
7 :
8 : m = 10; newton(f,0)
9 : m = 11; newton(f,0)
```

Listing 3.10 Example code of applying Newton's method to function with zero derivatives at solution point.

Some results from the calculation are shown in Figure 3.9 which shows the relative accuracy from the Newton calculations as a function of the iteration number. The open circles are the expected results from Eq. (3.43) and one can see very good agreement except for the last few points where the solution converges to the final zero point. For the case of $m = 10$ the final calculated point is 1.9999999979727 achieved after 115 iterations and for $m = 11$ the final point is 2.0000001071667 achieved after 126 iterations. These give 9 and 8 digits of accuracy respectively while the ERR parameter set in newton() corresponds to 9 digits of accuracy (1.e-9). Thus it is seen that Newton's method can be applied successfully to functions with zero derivatives at the zero point. One just has to accept the fact that convergence is slower. For a more realistic case of a non-zero second derivative, on the order of 30 iterations would be required for 9 digits of accuracy.

Periodic functions require special care in evaluation roots as an infinite number of roots exist (or no roots exist). Figure 3.10 shows an example of a periodic function. Although the generated function is somewhat contrived, it has all the features of any periodic function. From the graph it is seen that the function is periodic with a period of 4.0 and that within each period there are two zero points. The lowest order zeros are seen to be located near 1.5 and 2.5. When this function is used with Newton's method, the result may be a converged value at any of the multiple roots and Newton's method will be perfectly satisfied. However, for many periodic functions one desires the solutions corresponding to the smallest values of the independent variable, in this case the solutions near 1.5 and 2.5. To obtain these solutions one obviously needs to provide initial guesses near these solution points.

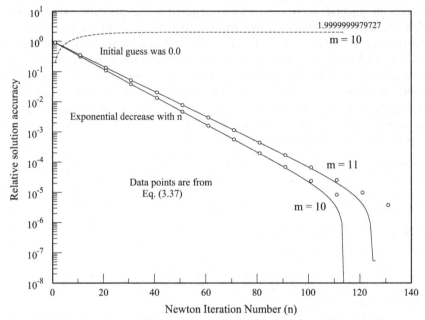

Figure 3.9 Convergence of Newton's method with function of Eq. (3.41).

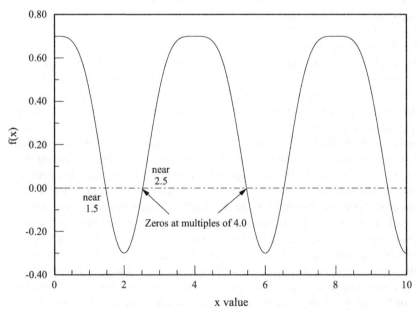

Figure 3.10. Illustration of periodic function for use with Newton's method.

```
 1 :   --/* File list3_11.lua */ -- Example of periodic function
 2 :
 3 : require "newton"
 4 :
 5 : function f(x) return .7 - math.sin(x*math.pi/4)^4 end
 6 :
 7 : i = 1;x,y = {},{}
 8 : for xx=0,10,.002 do
 9 :     x[i],y[i] = xx,f(xx)
10 :     i = i+1
11 : end
12 : plot(x,y);write_data('list3.11.dat',x,y)
13 : print(newton(f,0.1))
14 : print(newton(f,1))
15 : print(newton(f,3))
Output:
462.52973524258      5    -4.2699511299355e-011
1.4702647574209      5     5.0570677651116e-011
2.5297352425791      5    -3.8144709040392e-011
```

Listing 3.11. Code segment for an example of Newton's method with a periodic function.

Listing 3.11 shows a code segment for applying Newton's method to this function defined on line 5 of the listing. Newton's method is called on lines 13, 14 and15 with 3 different initial approximations. For initial guesses of 1 and 3, the solution converges to 1.4702647574209 and 2.5297352425791, both values close to the expected values. However, for the initial guess of 0.1, the returned solution is 462.52973524258. One can envision from the small value of the slope of the function at 0.1 that the linear model constructed at this point intersects the zero axis at some very large value putting the next iterative solution at a very large value. In fact one can see that the solution point can be written as: 462.52973524258 = 115x4 + 2.52973524258 indicating that the point found corresponds to the second root but located 115 cycles of the periodic function from the origin. Once Newton's method obtains a value within some distant cycle it will converge to a solution within that cycle. Thus it is very important with periodic functions to have a good initial approximation to the solution and this can usually be readily obtained by a simple graph of the function being solved. In all cases one should look at a graph before attempting Newton's method or have a good idea of the solution from some other physical knowledge.

3.6 Summary

This chapter has discussed methods for solving for the roots of a single equation. The emphasis has been on two methods, the successive substitution (or zero point iterative) method and Newton's method with the major emphasis on Newton's method. The approach has been different from many textbook treatments of this subject in that the emphasis has been on using Newton's method with a numerical method for evaluating the function derivative required in applying the method. Most authors shy away from the numerical derivative approach and discuss Newton's method in the light of using an analytical expression for the required derivative. This makes Newton's method considerably less attractive in that it is very difficult in many real physical problems to evaluate the derivative analytically. In some cases it is impossible as the function is only defined through an algorithmic approach with computer code. Based upon this writer's experience with applying Newton's method to a wide range of physical problems, one can successfully use numerical techniques to evaluate the derivative required in Newton's method. Limits on taking numerical derivatives are extensively explored in a subsequent chapter. This means that one only has to write code to define a function in order to call Newton's method and it makes possible a more general approach for Newton's method.

Newton's method has been applied to several nonlinear functions as examples. Many of these have been polynomial functions. This has been primarily for convenience in describing functions. A more general method for finding all the roots of polynomial functions is discussed in the next chapter. The highly desirable quadratic convergence nature of Newton's method has been derived and demonstrated with example solutions. Newton's method has also been shown to converge for functions with zero derivatives at the zero point, although with a slower convergence rate. In addition it has been shown that Newton's method can be used for functions with complex roots.

Important code segments developed in this chapter are:

1. ssroot() – Code for finding roots by successive substitutions
2. newton() – Code for finding roots by Newton's method, including limitations on the step size at each iteration

Newton's method and extensions of the technique will form the basis for most of the nonlinear solution approaches used in all the subsequent chapters. The next chapter discusses extending Newton's method to the solution of coupled nonlinear equations in multiple independent variables.

4 Solving Sets of Equations: Linear and Nonlinear

The previous chapter has discussed the solution of a single linear or nonlinear equation to find the roots of the equation or values of a variable for which the equation is zero. This chapter extends that discussion to sets of equations in several variables. Only cases where the number of variables is equal to the number of equations will be considered with equations of the form:

$$
\begin{aligned}
f_1(x_1, x_2, \ldots, x_n) &= 0 \\
f_2(x_1, x_2, \ldots, x_n) &= 0 \\
&\vdots \\
f_n(x_1, x_2, \ldots, x_n) &= 0
\end{aligned}
\tag{4.1}
$$

In general it will be assumed that this is a nonlinear set of equations for which a solution set of $x_1, x_2, \ldots, x_n$ values is desired that satisfy the equations. In the general case there may be (a) no solution values, (b) one set of values or (c) many sets of solution values. The fact that one is trying to find a set of solution values implies that he/she believes that the set of equations has at least one set of solution values. In most physical problems, one has some general idea as to the range of solution values for the variables which can be used as initial guesses at the solution values.

$$
f_1(x_1 + \Delta x_1, \ldots, x_n + \Delta x_n) = f_1(x_1, x_2, \ldots, x_n) + \left(\frac{\partial f_1}{\partial x_1}\right)\Delta x_1 + \cdots \left(\frac{\partial f_1}{\partial x_n}\right)\Delta x_n
$$

$$
f_2(x_1 + \Delta x_1, \ldots, x_n + \Delta x_n) = f_2(x_1, x_2, \ldots, x_n) + \left(\frac{\partial f_2}{\partial x_1}\right)\Delta x_1 + \cdots \left(\frac{\partial f_2}{\partial x_n}\right)\Delta x_n
$$

$$
\vdots
$$
$$
\vdots
\tag{4.2}
$$

$$
f_n(x_1 + \Delta x_1, \ldots, x_n + \Delta x_n) = f_n(x_1, x_2, \ldots, x_n) + \left(\frac{\partial f_n}{\partial x_1}\right)\Delta x_1 + \cdots \left(\frac{\partial f_n}{\partial x_n}\right)\Delta x_n
$$

As with the approach for a single equation, the approach here will be that of "linearize and iterate" (L&I). Assuming a first approximation to the solution values is known, each function can be expanded in a Taylor's series about the approximate solution values and only the first order terms kept. This gives a set of equations as expressed by Eq. (4.2). In these equations the right hand side partial derivatives are to be evaluated at the set of values $x_1, x_2, \ldots, x_n$. Just as Newton's method converts a single nonlinear equation into an approximate linear equation, the generalized Newton's method for systems of equations converts a set of

J.R. Hauser, *Numerical Methods for Nonlinear Engineering Models*, 77–146.
© Springer Science + Business Media B.V. 2009

nonlinear equations into an approximate set of linear equations. This set of equations can be written in matrix form as:

$$
\begin{bmatrix}
\dfrac{\partial f_1}{\partial x_1} & \dfrac{\partial f_1}{\partial x_2} & \cdots & \dfrac{\partial f_1}{\partial x_n} \\[2mm]
\dfrac{\partial f_2}{\partial x_1} & \dfrac{\partial f_2}{\partial x_2} & \cdots & \dfrac{\partial f_2}{\partial x_n} \\[2mm]
\vdots & \vdots & & \vdots \\[2mm]
\dfrac{\partial f_n}{\partial x_1} & \dfrac{\partial f_n}{\partial x_2} & \cdots & \dfrac{\partial f_n}{\partial x_n}
\end{bmatrix}
\begin{bmatrix}
\Delta x_1 \\[2mm]
\Delta x_2 \\[2mm]
\vdots \\[2mm]
\Delta x_n
\end{bmatrix}
=
\begin{bmatrix}
-f_1 \\[2mm]
-f_2 \\[2mm]
\vdots \\[2mm]
-f_n
\end{bmatrix}
\tag{4.3}
$$

where the partial derivatives and the functions are to be evaluated at the set of approximate solution values $x_1, x_2, \ldots, x_n$. The matrix of partial derivatives is known as the Jacobian (or Jacobian matrix) of the system of equations. When this set of equations is solved for the Δx_i values, the updated solution set is obtained by adding these to the original points as follows:

$$
x_i^{k+1} = x_i^k + \Delta x_i^k, \tag{4.4}
$$

where k indicates the iteration index. The algorithm is then iterated until the solution set obtains a desired degree of accuracy. Note that as the solution set is approached the right hand side f_i values in Eq. (4.3) will approach zero and the Δx_i correction terms will approach zero as convergence is achieved. A valid solution set will exist if and only if all the variable values achieve some desired accuracy criteria. As with the single equation, a relative accuracy criteria is much preferred over an absolute accuracy value in order to allow for a wide range of absolute values in the solution variables. Although no attempt at a proof will be given here, the L&I Newton's method presented here also has the highly desirable feature of quadratic convergence for a wide range of physical problems.

4.1 The Solution of Sets of Linear Equations

An approach to solving a set of nonlinear equations is thus established. The solution can be considered as that of two major steps. First the set of nonlinear equations must be expressed in a linear approximation. Then the next step is that of solving a set of linear equations for one step of the L&I approach. Solving such sets of linear equations is thus the topic for consideration in this section. Consider a slightly different notation for Eq. (4.3) as:

$$
\begin{bmatrix}
a_{11} & a_{12} & \cdots & a_{1n} \\
a_{21} & a_{22} & \cdots & a_{2n} \\
& \vdots & & \\
a_{n1} & a_{n2} & \cdots & a_{nn}
\end{bmatrix}
\begin{bmatrix}
x_1 \\
x_2 \\
\vdots \\
x_n
\end{bmatrix}
=
\begin{bmatrix}
b_1 \\
b_2 \\
\vdots \\
b_n
\end{bmatrix}
\tag{4.5}
$$

where the Jacobian matrix terms are represented as the a_{ij} coefficients and the function values are represented as the b_i terms. While matrix operations and matrix inversion can be used to solve sets of linear equations (see Section 4.4), this is not the most efficient method of solving a large set of linear equations. The most frequently used numerical technique for solving simultaneous algebraic equations is the Gauss elimination method.

It is assumed that the reader is somewhat familiar with this technique and only a brief review of the method is given here. In this approach, one uses row operations, such as multiplication by constants and additions of rows to transform the original matrix into a new matrix which has zeros for all the off-diagonal elements below the diagonal elements. This is illustrated below for a 5 by 5 set of equations:

$$
\begin{bmatrix}
a_{11} & a_{12} & a_{13} & a_{14} & a_{15} \\
a_{21} & a_{22} & a_{23} & a_{24} & a_{25} \\
a_{31} & a_{32} & a_{33} & a_{34} & a_{35} \\
a_{41} & a_{42} & a_{43} & a_{44} & a_{45} \\
a_{51} & a_{52} & a_{53} & a_{54} & a_{55}
\end{bmatrix}
\rightarrow
\begin{bmatrix}
a_{11} & a_{12} & a_{13} & a_{14} & a_{15} \\
0 & a_{22} & a_{23} & a_{24} & a_{25} \\
0 & a_{32} & a_{33} & a_{34} & a_{35} \\
0 & a_{42} & a_{43} & a_{44} & a_{45} \\
0 & a_{52} & a_{53} & a_{54} & a_{55}
\end{bmatrix}
$$

$$
\rightarrow
\begin{bmatrix}
a_{11} & a_{12} & a_{13} & a_{14} & a_{15} \\
0 & a_{22} & a_{23} & a_{24} & a_{25} \\
0 & 0 & a_{33} & a_{34} & a_{35} \\
0 & 0 & a_{43} & a_{44} & a_{45} \\
0 & 0 & a_{53} & a_{54} & a_{55}
\end{bmatrix}
\rightarrow
\begin{bmatrix}
a_{11} & a_{12} & a_{13} & a_{14} & a_{15} \\
0 & a_{22} & a_{23} & a_{24} & a_{25} \\
0 & 0 & a_{33} & a_{34} & a_{35} \\
0 & 0 & 0 & a_{44} & a_{45} \\
0 & 0 & 0 & a_{54} & a_{55}
\end{bmatrix}
\qquad (4.6)
$$

$$
\rightarrow
\begin{bmatrix}
a_{11} & a_{12} & a_{13} & a_{14} & a_{15} \\
0 & a_{22} & a_{23} & a_{24} & a_{25} \\
0 & 0 & a_{33} & a_{34} & a_{35} \\
0 & 0 & 0 & a_{44} & a_{45} \\
0 & 0 & 0 & 0 & a_{55}
\end{bmatrix}
$$

At each step, represented by arrows above, a column of the matrix elements below the diagonal is converted to all zero elements. For example in the first step, to eliminate element a_{41}, the first row is multiplied by the factor a_{41}/a_{11} and subtracted from row 4. The general equation for eliminating elements from row k is:

$a_{jk} = a_{jk} - (a_{j1}/a_{11})a_{1k}$ or in general for step n,

$a_{jk} = a_{jk} - (a_{jn}/a_{nn})a_{nk}$ for row j, element k with k $\geq$ j and j>n

$$(4.7)$$

This is the forward elimination step. Although the same notation is used in Eq. (4.6) for the elements following an elimination step as before an elimination step (for example: a_{23}), this is simply for convenience as many of the matrix elements are in fact changed by each elimination step as indicated by Eq. (4.7). For the nth elimination step all the matrix elements below the nth row may be changed in the elimination step. The new matrix at each step is frequently referred to as the augmented matrix. Also at each step the diagonal element, below which one eliminates the matrix elements, is typically called the pivot element.

After the elimination steps, the solutions are readily obtained by a "back substitution" step beginning with the last equation in the matrix where the solution can be readily obtained by $x_n = b_n / a_{nn}$ since the last equation has only a diagonal matrix element. The general back substitution equation is:

$$x_j = \left[b_j - \sum_{k=j+1}^{n} a_{jk} b_k \right] / a_{jj} \qquad (4.8)$$

These equations for forward elimination and backwards substitution are readily programmed in a function for obtaining the solution for a set of linear equation. A slight complication involves the need in a general routine to possible swap rows during the elimination process. For example at any step in the elimination process, the diagonal element at the beginning of an elimination step can potentially be zero. For example at the very beginning the a_{11} coefficient could be zero and the first step would give zero in the denominator of Eq. (4.7). This possibility is typically handled by exchanging the pivot row with the zero diagonal value by another row what is lower in the matrix. The question arises as to which rows to interchange? The most common approach is to interchange the pivot row with the row having the largest value in the column below the pivot element. This approach is called partial pivoting or just pivoting. Another approach at each elimination step is to switch the pivot row with the remaining row with the largest column element under the pivot element even if the pivot element is not zero. Row exchanges are typically fast and do not change the order of the solution variables. Of course if all remaining elements in a column are zero at any step in the elimination process, then one can not solve for the solution values since one has more remaining variables than independent equations. This is a case that should not occur with real world problems and will not be further considered.

The Gauss elimination technique is readily programmed as a computer algorithm and many such codes have been published in many computer languages. An extension of the technique is the "Gauss-Jordan" procedure where all off-diagonal elements at each elimination step are reduced to zero by row operations similar to the Gauss procedure. This procedure eliminates the back substitution step and directly gives the solution values at the end of the elimination process. However, the simpler Gauss elimination method requires fewer mathematical operations and is the preferred solution method. In the elimination process, one frequently normalizes the coefficients for each row so that the largest coefficient is unity. This combined with partial pivoting is known to minimize the round off errors in the many calculations needed for Gauss elimination.

A coded Gauss elimination program is provided for use with this work in the gauss.lua file shown in Listing 4.1. A brief description of the code is given here. First, the number of equations may be omitted from the calling argument and will be evaluated from the number of rows of the input matrix as on line 4 of the code. Lines 6 through 11 handle 1 equation as a special case. Lines 12 through 18 find the largest matrix element on each row and normalize each row so that the largest element value in every row is unity. Lines 20 through 45 implement the elimination steps for each diagonal pivot element. First, the largest element in each column is found with the code from line 22 through 25 with the rows possibly swapped on line 31 so that the row with the largest column element is moved to the diagonal pivot spot. Again all row elements are normalized to a unity diagonal element on lines 33 and 34. Then Eq. (4.7) is implemented from lines 35 through 44 for each remaining column. During this process the diagonal element has already been set to unity and thus does not appear in the denominator of line 40. Finally lines 46 through 50 implement the back substitution of Eq. (4.8).

A few words are perhaps in order regarding the storage method for matrix elements (of a two dimensional matrix for example). Values are stored in a table within a table format as $a = \{\{a_{11}, a_{12}, ...\}, \{a_{21}, a_{22}, ...\}, ... \{a_{n1}, a_{n2}, ...\}\}$ where the outermost table contains a listing of tables and the innermost tables contain the elements of each row of the matrix. A row of the matrix can be accessed by row $= a[i]$ and an individual element can be accessed as $a_{ij} = a[i][j] = \text{row}[j]$ (see lines 13 and 14 of Listing 4.1). The b values of Eq. (4.5) are assumed to be stored in a single table in the form: $b = \{b_1, b_2, ...b_n\}$. On line 51 the solution table (b) is returned along with an integer indicating the status of the solution process. A value of jret different from 1 indicates problems in solving the matrix. This can be checked to insure a proper solution for the set of equations. It should be noted that the solution set is returned in two places – within the argument list and as a result of execution of the function. This is purely for convenience so that one can use the form `sol = gauss(a,b)` as well as the form `gauss(a,b)` with the solution returned in the calling argument b array. Which of these styles is used is a matter of choice.

The coded gauss() function in Listing 4.1 is similar to many published Gauss elimination routines in many different programming languages and will find use in subsequent chapters. However, for many engineering problems, one has a large number of coupled equations, perhaps in the thousands, but the matrix is of a very sparse nature where there are very few non-zero matrix elements in each equation row. Many such examples arise in solving differential equations for engineering problems. For such sparse matrices, the gauss() function in Listing 4.1 becomes impractical because of the large storage requirements and the long times required to solve the set of equations for a fully populated matrix. For example a set of 1000 equations will have a 1000 by 1000 matrix with 1,000,000 elements and for many such practical cases only 10% or fewer of the elements are non-zero. Thus for a vast array of engineering problems it is highly desirable to use sparse storage techniques and one must have a Gauss elimination routine that uses only the non-zero matrix elements. The implementation of such a function called spgauss() (for sparse-Gauss) is now discussed.

The Lua language is ideally suited to implement sparse matrix techniques. A table in Lua is an associative array that is efficiently implemented in the native language in terms of both storage allocation and access speed. A table array can be defined in the language and only the non-zero elements simply be defined. The

```
 1 : -- /* File gauss.lua */
 2 :
 3 : gauss = function (a,b,n)
 4 :   n = n or #a
 5 :   local jret = 1
 6 :   if n < 2 then -- Low dimension matrix -- Special case
 7 :     if n<1 then return b,-1 end
 8 :     row = a[1]
 9 :     if row[1] ~= 0.0 then b[1]= b[1]/row[1] return b,1 end
10 :     return b,-1
11 :   end
12 :   for i=1,n do   -- Find largest value;
13 :     row,ap = a[i], 0.0
14 :     for j=1,n do ap = max(ap,abs(row[j])) end -- large value
15 :     if ap==0.0 then ap=1.0 end
16 :     for j=1,n do row[j]=row[j]/ap end -- Divide by value
17 :     b[i]=b[i]/ap
18 :   end
19 :
20 :   for j=1,n do -- Elimination for each row in matrix
21 :     jp1,ap = j+1, 0.0
22 :     for i=j,n do -- Find largest value in column j n
23 :       am = a[i][j]
24 :       if abs(am) > abs(ap) then ap,imax = am,i end
25 :     end -- At end ap = largest in column j; imax row number
26 :     if abs(ap) <= eps then
27 :       jret=0 -- Probably singular matrix with no solution
28 :       if ap==0.0 then return b,-2 end -- no solution
29 :     end
30 :     if imax~=j then -- Swap rows of a and b
31 :       a[imax],a[j],b[imax],b[j] = a[j],a[imax],b[j],b[imax]
32 :     end
33 :     row,b[j] = a[j],b[j]/ap
34 :     for k=j,n do row[k] = row[k]/ap end
35 :     if j<n then -- Eliminate elements except for last row
36 :       rowj=a[j] -- Selece row j values
37 :       for ix=jp1,n do -- Step rows from j+1 to row n
38 :         row = a[ix]; rowij = row[j] -- Select row ix values
39 :         for jx=jp1,n do -- Step from j+1 to column n
40 :           row[jx]=row[jx]-rowij*rowj[jx]
41 :         end
42 :         b[ix] = b[ix] - b[j]*rowij
43 :       end
44 :     end
45 :   end
46 :   for j=n-1,1,-1 do -- Back substitute from row n-1 to row 1
47 :     for k=n,j+1,-1 do -- Known values from n to j
48 :       b[j] = b[j] - a[j][k]*b[k]
49 :     end
50 :   end
51 :   return b,jret
52 : end
53 : setfenv(gauss,{abs=(Complex or math).abs,max=math.max,
           eps=1.e-12})
```

Listing 4.1. Code segment for implementing the Gauss elimination process for a set of linear equations.

following shows Lua code for defining a two-dimensional matrix as a table of table elements with only some of the elements defined as non-zero:

$$a = \{\} \ \text{-- define matrix}$$
$$a[1] = \{[1] = 2, [2] = -1\}$$
$$a[2] = \{[1] = -1, [2] = 2, [3] = -1\}$$
$$a[3] = \{[2] = -1, a[3] = 2, [4] = -1\} \tag{4.9}$$
$$a[4] = \{[3] = -1, a[4] = 2, [5] = -1\}$$
$$a[5] = \{[4] = -1, [5] = 2\}$$

The undefined table elements are treated as 'nil' or undefined and an attempt to access one of them will return a 'nil' value.

In this example, five rows of a matrix are defined as tables (a[1] through a[5]) with 2 or three non-zero elements defined per row. The total number of defined elements is 13 out of a possible total of 25. Elements are defined along the matrix diagonal and one element on each side of the diagonal. Such a matrix is known as a "tri-diagonal" matrix and such a matrix occurs frequently in engineering problems where the total number of tri-diagonal equations can number in the thousands. In our programming language, nothing special has to be done to accommodate sparse matrices, one just defines the nonzero elements and the language provides an efficient means for storing, managing memory and collecting memory when no longer used. Such is the advantage of modern scripting languages. To implement a Gauss elimination process using sparse matrix techniques, one simply has to implement the code so that the Gauss process steps over the nonzero elements and fills in matrix elements only if the resulting matrix element is nonzero.

A coded spgauss() function is given in Listing 4.2 for such a sparse matrix solver. The code is a little longer than desired for a function but is similar to that of the gauss() function with a few extra features. One should compare the code with that of Listing 4.1. Lines 13 through 18 find the largest element in each row and normalize each row similarly to lines 12 through 18 of Listing 4.1. For a sparse table the means of accessing the element values is through the pairs() function as first seen on line 15 in a do ... end loop. Each call to pairs() returns a pair of values indicating the number of the table index and the value associated with the table index. A null value is returned when no more values are present. The statement "for j,v in pairs(row) do ap = max(ap,abs(v)) end" is the equivalent of the statement "for j=1,n do ap = max(ap,abs(row[j])) end". The latter statement steps over all elements from 1 to n of the row while the former steps only over the defined elements of the row with pairs() returning the table index and value for nonzero elements. With this understanding the spgauss() is almost a line for line replacement for the previous gauss() code. In the column elimination process on lines 36 through 45, matrix elements are set only if a column value has a nonzero value through the check on line 38. Also since the diagonal elements are normalized to unity, the diagonal elements do not need to be stored and are set to 'nil' on line 39. Setting an element to 'nil' essentially eliminates the element from the

```lua
1 : -- /* File spgauss.lua */
2 :
3 : spgauss = function (a,b,n) - sparse Gauss elimination
4 :   local nel = {{},{}} -- arrays for usage statistics
5 :   local jret,jprint,jpold,nct,ug1 = 1,1,1,0,0 - print parms
6 :   n = n or #a
7 :   if n < 2 then -- Low dimension matrix -- Special case
8 :     if n<1 then return b,-1 end
9 :     row = a[1]
10 :     if row[1] ~= 0.0 then b[1]= b[1]/row[1] return b,1 end
11 :     return b,-1
12 :   end
13 :   for i=1,n do  -- Find largest value & divide by value
14 :     ap,row=0.0, a[i]
15 :     for j,v in pairs(row) do ap = max(ap,abs(v)) end
16 :     if ap==0.0 then ap=1.0 end; ap = 1/ap
17 :     for j,v in pairs(row) do row[j] = v*ap end
18 :     b[i] = (b[i] or 0)*ap -- All elements of b[] exist
19 :   end
20 :   for j=1,n do -- Elimination for each row in matrix
21 :     jp1,ap = j+1, 0.0
22 :     for i=j,n do -- Find largest from row j to row n
23 :       am = a[i][j]
24 :       if am and abs(am)>abs(ap) then ap,imax = am,i end
25 :     end -- At end ap=largest in j; imax has row number
26 :     if abs(ap) <= eps then -- Probably singular matrix
27 :       jret=0; if ap==0.0 then return -2 end
28 :     end
29 :     if imax~=j then -- Swap rows of a and b
30 :       a[imax],a[j],b[imax],b[j]=a[j],a[imax],b[j],b[imax]
31 :     end
32 :     row,b[j] = a[j],b[j]/ap; row[j] = nil -- Normalize row
33 :     for k,v in pairs(row) do row[k] = v/ap end
34 :     if j<n then -- Eliminate elements if not last row
35 :       rowj=a[j] -- Selece row j values
36 :       for ix=jp1,n do -- Step rows from j+1 to row n
37 :         row=a[ix]; rowij = row[j] -- Select row ix values
38 :         if rowij then -- Non nil value
39 :           row[j] = nil
40 :           for jx,v in pairs(rowj) do -- columns of row j
41 :             row[jx] = (row[jx] or 0) - rowij*v
42 :           end
43 :           b[ix] = b[ix] - b[j]*rowij
44 :         end
45 :       end
46 :     end
47 :     if usage==2 then -- Collect usage statistics
48 :       ug1= 0
49 :       for _,_ in pairs(a[j]) do nct = nct+1 end -- Row n
50 :       for jj=j+1,n do -- Rows beyond n
51 :         for _,_ in pairs(a[jj]) do ug1 = ug1+1 end
52 :       end
53 :       nel[1][j], nel[2][j] = j, nct + ug1
54 :     end
55 :     if nprint then -- Print at each 100 rows
56 :       jprint = floor(j/100)
57 :       if jprint==jpold then
58 :         jpold=jprint+1; print("Completed row ",j," in
             spgauss");io.flush()
```

```
59 :                if usage==2 then
                       print("Number of matrix elements =",nct+ug1) end
60 :                end
61 :            end
62 :        end
63 :    for j=n-1,1,-1 do -- Back substitute from row n-1 to row 1
64 :        row = a[j]; for k,v in pairs(row) do
                   b[j] = b[j] - v*b[k] end
65 :    end
66 :    if usage==1 then -- Collect usage statistics at end
67 :        nct = 0
68 :        for jj=1,n do
69 :            for _,_ in pairs(a[jj]) do nct = nct+1 end
70 :        end
71 :        nel[1][1],nel[2][1] = n, nct
72 :    end
73 :    return b,jret,nel
74 : end
75 : setfenv(spgauss,{abs=(Complex or math).abs,min=math.min,
            max=math.max,floor=math.floor,
76 :io=io,print=print,pairs=pairs,type=type,nprint=nil,
            usage=nil,eps=1.e-12})
```

Listing 4.2. Code segment for sparse Gauss elimination to solve a set of linear equations.

table thus minimizing the matrix storage requirements. After eliminating elements below the diagonal, the back substitution process is performed on lines 63 through 65, again using only the nonzero elements of each row in the transformed matrix.

Two additional features are added to the code for spgauss() accounting for the additional length of the function. One of these is the collection of "usage" statistics on lines 47 through 54 and lines 66 through 72. A value of 1 or 2 for this parameter causes values to be collected regarding the total number of nonzero matrix elements in the elimination process. If usage equals 1, then the total number of nonzero elements in the a matrix are counted at the end of the elimination process and returned in the nel[] table. If usage equals 2 then the number of nonzero matrix elements is counted for each of the n row elimination steps and the numbers collected in the nel[] table. These values are returned by the function on line 73 as the third returned variable. As the elimination process proceeds for a sparse matrix, the number of nonzero matrix elements tends to grow and this is referred to as matrix "fill-in" or simply "fill". This feature is provided so the user can study the element fill for a given problem. This feature will be used in subsequent chapters for large systems of equations resulting from solving partial differential equations. A second feature is the "nprint" variable which, for a non-nil value, causes printed output after the elimination process has completed each multiple of 100 rows – see lines 55 through 60. This is provided so the user can observe the elimination process for large numbers of equations where the solution time can be somewhat long. Also combined with a usage value, the matrix fill can be observed as the elimination process proceeds. These two features are purely for user convenience and could be eliminated if one desires the maximum in execution speed. However, the two tests for non-nil values of these two variables increase the execution time by an almost insignificant amount.

Listing 4.3 shows a very simple example of using gauss() and spgauss() for solving this simple set of three linear equations:

$$10x_1 - 7x_2 \qquad = 7$$
$$-3x_1 + 2x_2 + 6x_3 = 4 \qquad\qquad (4.10)$$
$$5x_1 - x_2 + 5x_3 \quad = 6$$

```
 1 : --/* File list4_3.lua */
 2 :
 3 : require"gauss"; require"spgauss"
 4 : -- Example of use with gauss()
 5 : A = {
 6 :   {10, -7, 0},
 7 :       {-3, 2, 6},
 8 :       {5, -1, 5}
 9 : }
10 : B = {7, 4, 6}
11 : gauss(A,B) -- or gauss(A,B,3)
12 : table.foreach(B,print)
13 : -- Example of use with spgauss()
14 : A = {{10, -7, 0},{-3, 2, 6},{5, -1, 5}} -- Compact form
15 : B = {7, 4, 6}
16 : sol = spgauss(A,B) -- Optional form of call
17 : table.foreach(sol,print)
Output:
1   1.1102230246252e-016
2   -1
3   1
1   1.1102230246252e-016
2   -1
3   1
```

Listing 4.3 Example of solving linear equations with Gauss elimination program.

For this set of equations, an A matrix is defined on lines 5 through 9 as a table of three row tables and the B table is defined on line 10. The gauss() function is called on line 11 and the results printed on line 12 for the B table values returned by gauss(). This set of equations is known to have the solutions 0, -1, 1 and the solutions can be seen as listed for the Output: lines. For the 0 solution value the returned result is 1.11e-16 and when compared with the 1 value is accurate to about 16 decimal digits. The matrices are redefined on lines 14 and 15 and spgauss() is called on line 16 with the returned solution value set to the sol variable. This simply illustrates the two calling styles as previously discussed and after line 16, the sol variable will be identically the same as the B variable. It is noted that the A and B tables must be redefined between the calls to gauss() and spgauss() because gauss() changes the values of both arrays. Thus if one repeatedly calls gauss() (or spgauss()) within a code loop, one must either save a copy of the tables or repeatedly recalculate the tables before each call to gauss(). It should be noted that one can not copy the elements of a table by a simple code statement such as Acopy = A. This simply makes both variables Acopy and A point to the same table of values – there is still only one table of values. To copy a table requires that a new table be formed (by the {} construct) and that the table values be copied one by one from one table to the other.

Since two coded functions have been presented for solving a set of linear equations, the question arises as to when one would prefer one function over the other (gauss() or spgauss()). If one has a coefficient matrix with all elements non-zero then gauss() is obviously the preferred function. For a full matrix, the spgauss() takes a little longer to execute than gauss() because stepping through a table using pairs() is a little slower than accessing the elements in a loop if all table values are present. As a example, for the 3 by 3 matrix of Listing 4.3 the solution by spgauss() takes approximately 25% longer than the solution by gauss(). However, one must put the solutions in a loop and solve the equations about 1,000,000 times in order to gather reliable statistics on the differences. For solving small sets of equations a small number of times, the differences in execution time are not significant. For typical engineering problems with large numbers of equations, the equation set is typically of a sparse nature and spgauss() is the much preferred function. Thus one can use spgauss() for most all applications, except for small size matrices which must be solved many times. The functions gauss() and spgauss() will be used extensively in subsequent sections of this book for a variety of nonlinear problems. Two other techniques for solving a set of linear equations using matrix inversion and the so called LU decomposition of the coefficient matrix are discussed in a later section of this chapter. But for now we can go back to the main problem of interest in this chapter which is the solution of sets of nonlinear equations and this will use gauss() or spgauss().

4.2 Solution of Sets of Nonlinear Equations

At this point it's appropriate to briefly review the L&I approach to solving a set of nonlinear equations such as represented by Eq. (4.1). First one must have an initial guess at the solution set. Then one forms the linear set of equations given by Eq. (4.2) in terms of corrections to the initial guesses. This linear set of equations is then solved and the solution set is updated with the correction terms according to Eq. (4.4). The algorithm is then repeated until some desired degree of accuracy is obtained in all the solution values.

Code to perform these steps and solve a set of nonlinear equations is shown in Listing 4.4 with the function nsolv(). The function nsolv() accepts four possible arguments of which only the first two are required as input. These are f = name of function describing equations to be solved, and x = table of initial guess at the solution values. The next optional argument is step = limits on step sizes per iteration – one limit for each variable. A final argument may be specified which is the number of equations to be solved. It this is not specified, nsolv() determines

```
 1 : --/* File nsolv.lua */
 2 : require"gauss"; require"spgauss" --load gauss and spgauss
 3 :
 4 : nsolv = function(f,x,step,nx)
 5 :  local nend,wk1,rowptr,b = NMAX,{},{},{} -- work, A, B
 6 :  local c1,c2,dx,cmax,ngauss,stp
 7 :  if full then ngauss=gauss else ngauss=spgauss end
 8 :  if nx==nil then f(b,x); nx = #b end -- Find # of equations
 9 :  for ilp=1,NMAX do -- iterate for at most NMAX steps
10 :    for i=1,nx do rowptr[i] = {} end
11 :    f(b,x) -- Values of equations returned in b[]
12 :    for i=1,nx do -- Set up matrix equations
13 :       c1 = x[i]; dx = FACT*c1
14 :       if dx==0.0 then dx = FACT end
15 :       x[i] = x[i]+dx -- Probe x[i] factor
16 :       f(wk1,x); x[i] = c1 -- Results returned in wk1[]
17 :       for j=1,nx do
18 :          c1 = b[j]-wk1[j]
19 :          if full then rowptr[j][i] = c1/dx
20 :          elseif c1~=0.0 then rowptr[j][i] = c1/dx end
21 :       end
22 :    end
23 :    ngauss(rowptr,b,nx) - Solve equation set using gauss() or
             spgauss() with corrections returned in b[]
24 :    cmax = 0.0
25 :    for i=1,nx do
26 :       if step~=nil and step[i]~=0.0 then --Skip if no step
             Sizes specified
27 :          stp = step[i]
28 :          if stp<0.0 then -- Limits on absolute change
29 :             if abs(b[i])>-stp then
30 :             b[i] = -stp*abs(b[i])/b[i] end
31 :          else -- Limits on percentage change
32 :             if stp<1 then stp=1/stp end
33 :             c1 = 1. + b[i]/x[i]
34 :             if c1>stp then
35 :                b[i] = (stp-1)*x[i]
36 :             elseif c1<1/stp then
37 :                b[i] = (1/stp-1)*x[i]
38 :             end
39 :          end
40 :       end -- End of adjustments, find largest relative error
41 :       c1 = abs(b[i]); c2 = abs(x[i]) + c1
42 :       if c2~=0.0 then c1=c1/c2 end
43 :       if c1>cmax then cmax=c1 end
44 :       x[i] = x[i] + b[i] -- Update solution set
45 :    end
46 :    if linear then nend=ilp; break end --  linear?
47 :    if nprint then -- Print iterative solutions
48 :       fprintf(stderr,'Solutions at iteration %d are:\n',ilp)
49 :       for i=1,nx do fprintf(stderr,'%s   ',tostring(x[i])) end
50 :       fprintf(stderr,'\n'); stderr:flush()
51 :    end
52 :    if cmax < ERROR then nend=ilp; break end
53 :  end
54 :  return nend,cmax -- Return #iterations, error, solution in x
55 : end
56 : setfenv(nsolv,{stderr=io.stderr,gauss=gauss,spgauss=spgauss,
           tostring=tostring,fprintf=fprintf,
```

```
57 :   abs=(Complex or math).abs,ERROR=2.e-6,FACT=1.e-6,NMAX=100,
          nprint=nil,linear=nil,full=nil})
```

Listing 4.4. Code for solving systems of nonlinear equations.

the number of equations from the user specified f() function as on line 8 of the code. Both x and step should be tables of values.

After defining local variables on lines 5 and 6 of the code a major Newton iterative loop is implemented from line 9 through line 55 with a maximum number of possible loops set at NMAX (default value of 100 on line 57). The code to set up the linear equations to be solved is contained in lines 12 through 22 of the listing. On line 11 the user defined function describing the equation set to be solved is called, f(b,x), and this returns the right hand side of Eq. (4.3) through the b calling argument. In nsolv() numerical derivatives are used to have code automatically set up the set of linearized equations for Newton's method. In this manner, the user only has to specify the defining equation set and not the Jacobina matrix of derivative values. The procedure for obtaining the derivatives is essentially the same as that used in the newton() code for linearizing a single equation. This is implemented in lines 12 through 22 which takes each variable in turn, increments the variable by a dx quantity, and calculates a numerical partial derivative to approximate the Jacobian array of partial derivatives in Eq. (4.3). Lines 17 through 20 step through a column of the array, calculating the partial derivatives for each x_i in turn and lines 19 or 20 set the Jacobian matrix values.

If the "full" parameter on line 19 is non-nil every column matrix value is set, otherwise only the nonzero matrix elements are set on line 20 for use with sparse matrix techniques. The default method is to use sparse matrix methods but this can be changed by using a getfenv(nsolv).full = true code statement before calling nsolv(). Then ngauss() is called on line 23 to calculate the corrections and the corrections are finally added to the original values on line 44 of the code. The function ngauss() is either gauss() or spgauss() (see line 7) depending on the value of the "full" parameter with the default value being spgauss().

About half the code in the listing (between lines 26 and 40) is devoted to setting possible limits on the step sizes for each variable. If no step size limitations are set by the input step[] array, this code is bypassed. The step size limitation code is almost identical to that previously discussed in connection with Newton's method for a single variable. A limitation may be specified for each variable independently using either a negative number for a +/- limitation or a positive number for a relative step size limitation. The reader is referred back to Chapter 3 for a more detailed discussion of step size limitation in connection with the newton() function. In addition the convergence criteria (on line 52) is based upon the maximum relative error in any one of the variables and the code tries to obtain a maximum relative error below that specified by the ERROR value (2.e-6 in the code on line 57).

Other features of nsolv() include the possibility of printing the solutions at each iterative step (by setting nprint to non-nil, line 47) and the possibility of specifying that the equations to be solved are linear (by setting linear to non-nil, line 46) in

order to cut the calculation time in half. The nsolv() function returns the number
of iterations required for the solution and the maximum relative error for the least
accurate variable on line 54. This is one of the most useful code blocks that is developed and used in this book. It provides a core routine for solving many nonlinear modeling problems as will be subsequently seen.

As an example of a nonlinear equation set consider the following three equations (in standard math format);

$$f_1 = x_1^2 + 50x_1 + x_2^2 + x_3^2 - 200$$
$$f_2 = x_1^2 + 20x_2 + x_3^2 - 50 \qquad\qquad (4.11)$$
$$f_3 = -x_1^2 - x_2^2 + 40x_3 + 75$$

For use with nsolv() these simply needs to be coded in computer language.

```
1 : -- File list4_5.lua -- Example of use of nsolv()
 2 : require "nsolv"
 3 :
 4 : eqs1 = function(f,x)
 5 :   f[1] = x[1]^2 + 50*x[1] + x[2]^2 + x[3]^2 -200
 6 :   f[2] = x[1]^2 + 20*x[2] + x[3]^2 - 50
 7 :   f[3] = -x[1]^2 - x[2]^2 + 40*x[3] + 75
 8 : end
 9 :
10 : x = {0,0,0} -- Initial guess
11 : nmx,a = nsolv(eqs1,x)
12 : print('For eqs1 : ',x[1],x[2],x[3])
13 : print('Number of iterations, error = ',nmx,a)
14 :
15 : eqs2 = function(f,x)
16 :   f[1] = x[1]*x[1] + math.exp(x[2]) + 1.0/x[3] - 41.
17 :   f[2] = x[1] + 1./x[2] + x[3] - 6.
18 :   f[3] = x[1]*x[1] - 24.0
19 : end
20 : x = {4,4,4}
21 : nmx,a = nsolv(eqs2,x)
22 : print('For eqs2 : ',x[1],x[2],x[3])
23 : print('Number of iterations, error = ',nmx,a)
Output:
For eqs1 :   3.6328279699384    1.7320739297524    -1.4700620210676
Number of iterations, error =    5  5.8261938013078e-012
For eqs2 :   4.8989794855664    2.7500794076873    0.73739465068414
Number of iterations, error =    6  2.6202592146858e-010
```
Listing 4.5. Two examples of using nsolv() for three coupled nonlinear equations.

Listing 4.5 shows an example of the use of nsolv() for this and one other set of
nonlinear equations. This equation set is coded in the eqs1() function on lines 4
through 8. It is seen that the equation subscripts have been converted into table
entry numbers. It should be noted that nsolv() assumes a particular form for the
function defining the equation set and the user must write the defining function in
that particular format. In particular, nsolv() assumes that the table defining the re-

sidual of the equations (the f[] table) is passed to the function as the first argument and the residual values are returned through this table of values. As with many computer language coding decisions there are at least two ways of passing information back to the line of code calling a function. Information can be passed through the argument list or information can be passed through a 'return' code statement (such as 'return f' in this case). However, if the 'return f' method is used and the f[] array is not passed in the argument list, then the user must define the f[] array in his/her function, perhaps with a 'local f = {}' statement. This makes the function definition more prone to errors as the user has to provide more coded statements in the defining equation and may forget to use the 'local' keyword or forget to define the array altogether. Also defining the array within the function requires that a new array be set up each time the equation function is called which uses additional time and memory. Thus the decision made in this work has been to have the table returning the equation residuals defined and supplied by the nsolv() function. One can see in Listing 4.4 this array defined on line 5 (name of b) once and the same table is used for all calls to the used defined equation set. This feature has been discussed in a fair detail because it is a programming style used in this and subsequent coded computer programs. Whenever a user supplied function is required to define tables of residual values, the residual values of the equations will be passed back to the calling program through an entry in the calling argument list. In addition to the function array, nsolv() passes the array of x[] values to the user supplied function as seen on line 4 of Listing 4.5.

A one-to-one correspondence can be readily seen between the terms in the normal algebraic form of Eqs. (4.11) and the computer coded form on lines 5-7 of Listing 4.5. This makes it very easy to code equations for use by nsolv(). As previously noted, the major difference is in the use of bracket [] terms in the coded form in place of subscripts. Each of the coded equations will return the residual value for a particular set of independent variable values (x[] values) and will return zero at the exact solution values. The nsolv() function performs Newton iterative loops forcing the equation residual values toward zero at each iterative step.

For the first set of equations defined on lines 4 through 8, an initial guess is set on line 10 and nsolv() called on line 11 to obtain the solution values. Note that the Newton calculated solution set is returned in the x[] array and not returned by the nsolv() function. This is again a programming decision to have nsolv() return values of the number of iterations and the last achieved error approximation and not to directly return the solution array. Part of this decision is that one already has an x[] array of initial values and one can return the solution set in the same array without having to define a new array. In coding numerical algorithms such decisions of how to pass information back and forth to a called function have to be routinely made and to some extent they are a matter of programming style and preference. However, in order to correctly use coded functions, one must always understand what variables or arrays are passed to a function and how variables and arrays are passed back to the calling program. Hopefully in this work a consistent programming style will be followed.

For each of the two sets of equations in Listing 4.5, a solution set can be obtained for a wide range of initial guesses. For the initial guesses in Listing 4.5, the solutions converged in 5 and 6 iterations with errors well below the convergence limit specified in nsolv(). Although a correct solution set has been obtained with the initial values given in the listing, there may be other equally valid solution sets. For the eqs2 set of equations one readily sees from line 18 that there are two possible solution values for x_1 of $\pm\sqrt{24}$. Thus the eqs2 set of equations has at least one other set of valid solutions with x_1 given by the negative of the answer printed for the eqs2 output in Listing 4.5. This set can be obtained by starting with the initial guess $\{-4, 4, 4\}$. The reader is encouraged to run the code with this initial value and obtain the solution for the other variables. Perhaps there are other valid solution sets for each of the equations, but this will be left as an exercise to see if the reader can find other sets.

```
 1 : -- File list4_6.lua - Using nsolv() with linear equations
 2 : require "nsolv"
 3 :
 4 : lineqs = function(f,x)
 5 :   f[1] = 10*x[1] - 7*x[2] - 7
 6 :   f[2] = -3*x[1] + 2*x[2] + 6*x[3] - 4
 7 :   f[3] = 5*x[1] - x[2] + 5*x[3] - 6
 8 : end
 9 :
10 : x = {0,0,0} -- Initial guess
11 : getfenv(nsolv).linear = true -- Linear set
12 : nmx,a = nsolv(lineqs,x)
13 : print('For lineqs : ', x[1], x[2], x[3])
14 : print('Number of iterations, error = ',nmx,a)
Output:
For lineqs :    4.8706372268725e-011 -0.99999999991752
   0.99999999997769
Number of iterations, error =    1  1
```
Listing 4.6. Example of linear equations solved by nsolv().

Linear equations are just a subset of nonlinear equations so that nsolv() can be used equally readily with sets of linear equations. For this consider the simple set of linear equations previously given in Eq. (4.10). Listing 4.6 shows the use of nsolv() for this set of equations which is also the same as the equations discussed in connection with Listing 4.3. Lines 5 through 7 define this equation set in coded form. By comparing the Listings 4.3 and 4.6, it is seen that the coefficient term in the equations of line 5 through 7 correspond to the A and B matrix table values used in Listing 4.3 with gauss() to solve the same equation set. For this example, the number of iterations is limited to 1 by line 11 of the code. The solution values are seen to be very close to 0,-1,1 with the accuracy being about 10 or 11 digits. The returned error in this case is seen to be 1 which doesn't have much meaning for one iterative loop. Even though one can use nsolv() to solve sets of linear equations, the question arises as to why one would chose this method over simple matrix algebra or the use of the basic gauss() or spgauss() routines. There are at least two reasons. The first is consistency of methods over both linear and nonlinear problems. Only one approach has to be mastered and used. The second is a more natural coding of equations. It is at least the opinion of this author that the coding in lines 5-7 of Listing 4.6 is more natural and less prone to error than setting up the matrices for direct solution by gauss() as in Listing 4.3. Of course nsolv() will not be quite as fast as gauss(), but in many engineering applications, this is not a primary factor since both will execute much faster then the time required to formulate a problem.

We now have a toolbox of code segments which can be used in a variety of nonlinear real world problems. The next section will discuss some applications of the techniques developed thus far.

4.3 Some Examples of Sets of Equations

While there are many examples of sets of nonlinear equations in the real world, several examples will be taken from the field of Electrical and Chemical Engineering. First consider the full wave rectifier circuit shown in Figure 4.1. This uses four nonlinear diodes in a bridge arrangement to convert a sin() voltage waveform to a rectified sin() waveform across the load resistor R_L . The output voltage is the voltage difference $v_1 - v_2$. A brief explanation of the circuit operation is as follows. When the source voltage v_s is positive a low impedance path for current flow exists from v_s through diode D_1, through R_L from v_1 to v_2, and through D_2 back to the source. When the source voltage v_s is negative a low impedance path for current flow exists through diode D_3, through R_L from v_1 to v_2, and through D_4

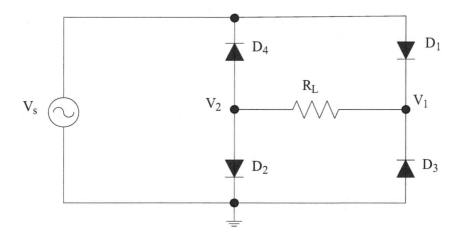

Figure 4.1. Full Wave diode rectifier circuit.

back to the source. In both cases the voltage across the load is positive from the v_1 to the v_2 nodes.

Using the fundamental relationship for current through a diode as expressed back in Chapter 3 (see Eq. (3.27)), two node equations can be written as:

$$\frac{v_1 - v_2}{R_L} - I_s(\exp((v_s - v_1)/v_t) - 1) - I_s(\exp(-v_1/v_t) - 1) = 0$$

$$\frac{v_2 - v_1}{R_L} + I_s(\exp((v_2 - v_s)/v_t) - 1) - I_s(\exp(v_2/v_t) - 1) = 0$$

(4.12)

The first equation sums currents to zero at the v_1 node and the second equation sums currents to zero at the v_2 node. The four exponential terms describe currents through the four diodes. These equations, with parameters, and nsolv() are all that are needed to calculate the voltage response to a sinusoidal source voltage.

```
 1 : -- File list4_7.lua Voltage of a Full Wave Rectifier
 2 : -- Exponential functions present special convergence problems
 3 :
 4 : require"nsolv"
 5 : exp = math.exp; pi2 = math.pi*2 -- Simpler to type
 6 :
 7 : Is,Rl,Vt = 1.e-14,5000,.026 -- Basic parameters
 8 : Vm = 15 -- Magnitude of source voltage -- change as desired
 9 : NMAX = getfenv(nsolv).NMAX -- Get max number of iterations
10 :
11 : fwr = function(I,v)
12 :   I[1] = (v[1]-v[2])/Rl - Is*(exp((vs-v[1])/Vt)-1) -
            Is*(exp(-v[1]/Vt)-1)
13 :   I[2] = (v[2]-v[1])/Rl + Is*(exp((v[2]-vs)/Vt)-1) +
            Is*(exp(v[2]/Vt)-1)
14 : end
15 :
16 : step = {-.8,-.8} -- Not too critical .2 to 1.2 OK
17 : x = {0,0} -- Just need value to get started
18 : xa,v1,v2,v3,vout = {},{},{},{},{} -- Arrays for values
19 :
20 : for i=1,401 do
21 :   xt = (i-1)*pi2/400; xa[i] = xt
22 :   vs = Vm*math.sin(xt)
23 :   nit,err = nsolv(fwr,x,step) -- Use previous value as guess
24 :   if nit==NMAX then
          print('Convergence error at i = ',i,err) end
25 :   v3[i],v1[i],v2[i] = vs, x[1], x[2]
26 :   vout[i] = x[1]-x[2] -- Output voltage
27 : end
28 :
29 : write_data("list4_7.dat",xa,v1,v2,v3,vout)
30 : plot(xa,v3,vout)
```

Listing 4.7. Code for solving full wave rectifier nonlinear equations.

Code for solving for the two unknown node voltages v_1 and v_2 is given in Listing 4.7. The code is fairly straightforward. Basic parameters are defined on lines 7 and 8. The basic equations are defined in the fwr (full wave rectifier) function on lines 11 through 14 and repeat the mathematical expressions of Eqs. (4.12). A computational loop from lines 20 through 27 steps through 401 voltage points along a single cycle of the source voltage sin() wave (see line 22) and tables xa[], v1[], v2[], v3[] and vout[] are used to hold the angle, and voltage points. Calculated values are written to a file on line 30 and presented on a pop-up plot for the user to rapidly see the results. The reader is encouraged to execute the code and observe the output plot.

There are a few features worthy of note in the application of nsolv() to this problem. First, the step size limiting feature of nsolv() is used with a maximum step size of +/- 0.8 volts set on line 16 of the code by defining a table of two limiting values. If some limitation is not used for this problem, the program will not converge due to the highly nonlinear nature of the exponential functions – especially with the v_t factor of only .026 volts as previously discussed in Chapter 3. However the exact value of the step size limitation is not critical as values between 0.2 and 1.2 works just fine with no noticeable difference in execution time. Second on line 23 where the nsolv() function is called, it is seen that the input guess at a solution is either the initially supplied guess or simply the previously solved table of solution values. This is a very useful technique when solving nonlinear problems. If one changes a parameter of the problem set slowly one can bootstrap one's way along the solution by using the results of a previous solution as the initial guess for the next solution point. This technique finds wide application in solving nonlinear problems. In many cases one can arrange to start with some known solution, for example with zero forcing term and slowly increase the forcing term bootstrapping one's way along the nonlinear solution so that at each increment in source voltage, the previous solution is a good initial guess for the next step. For the present solution, the change in source voltage between calls to nsolv() will be small and the solution at the previous voltage should be a good approximation. A check is made at line 24 to insure that the maximum number of iterations has not been reached which would indicate that proper convergence had not been achieved. Note that the maximum number of Newton iterations is obtained on line 9 by use of the getfenv() function. Although the number of iterations is not printed out for each loop, some voltage points take on the order of 30-35 iterations for convergence. If the magnitude of the supply voltage is increased from the 15 value in this example to say 200 volts, then the maximum number of iterations may have to be increased to achieve convergence. As a final comment, note how the changes in vs within the iterative loop on line 22 changes the value of vs back in the previously defined fwr() function on lines 12 and 13. The reader is encouraged to experiment with different voltages and parameter values.

Figure 4.2 shows a plot of the calculated output from the program showing only the source voltage and the voltage across the load resistor. For the positive voltage cycle, the difference between the source voltage and the load resistor voltage is the forward voltage drop across the two diodes. The difference will be about 1.5 volts and can be seen to be fairly constant over the full range of input voltages.

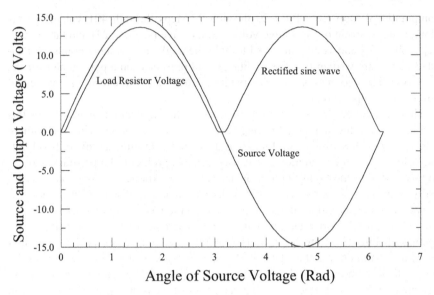

Figure 4.2. Graph of source and load resistor voltages for full wave rectifier example.

A second example is shown in Figure 4.3. This shows a classical BJT biasing circuit with two bias resistors R_1 and R_2 plus an emitter resistor R_S and a load resistor R_L. A typical biasing problem is to pick resistors R_1, R_2 and R_S given a load resistor R_L so as to achieve some objective bias point. For the purposes of this discussion, it will be assumed that a desired R_L, R_S and R_1 have already been selected and that the only remaining task is to select an appropriate value of R_2. The given value of R_L is taken as 2.0 kΩ.

Assuming that all the other resistors are known is not too unrealistic as good design practice normally takes R_S as a fraction of R_L (0.1 of R_L here) and takes R_1 considerably larger than R_L (10 times larger here). There are approximate design techniques which can lead to an appropriate choice of the other bias resistor, R_2. However a more exact analysis will be used here to compute its value making use of the knowledge of solving a nonlinear equation set. The basic equations for this circuit consist of three node voltages identified in the circuit diagram as v_1, v_2 and v_3. The three node current equations are:

$$v_1 / R_2 + (v_1 - V_{cc}) / R_1 + i_b = 0$$
$$v_2 / R_S - i_b - i_c = 0 \qquad\qquad (4.13)$$
$$(v_3 - V_{cc}) / R_L + i_c = 0$$

These must be supplemented with a set of BJT equations giving the device currents in terms of the device terminal voltages and device parameters. For normal forward bias operation of the BJT an appropriate set of device equations is:

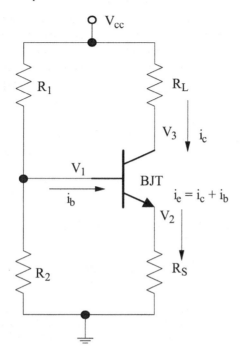

Figure 4.3. Typical BJT biasing circuit.

$$i_b = I_s(\exp((v_1 - v_2)/v_t) - 1)$$
$$i_c = \beta i_b(1 + (v_3 - v_2)/V_a) \tag{4.14}$$

with parameters I_s, β, v_t and V_a. One can obviously combine Eqs. (4.13) and (4.14), eliminating the two currents and giving three resulting equations in the three node voltages alone. However, one can also treat this as a set of five equations and just let nsolv() do the appropriate combining of the equations as it solves for the three voltages plus the two currents. This will be the approach taken here.

Our goal is then to solve the above five coupled equations for a range of values of R_2 and from the computed results pick an appropriate value of the final resistor for a desired bias condition. Code to solve this circuit is given in Listing 4.8. The five defining equations can be seen on lines 10 through 14 in exactly the same form as listed above again with the subscript notation changed to the bracket notation. Also since there is only one name for the table of variables, the i_b and i_c variables are really treated internally to the nsolv(0 function as variables v[4] and v[5]. Line 9 of the code converts these locally to the ib and ic notation to make the equations more readable. The heart of this code is a while loop from line 20 to 27 which steps through values of R_2 from $0.1\,k\Omega$ to $4.0\,k\Omega$. The selected output shows that this range is more than adequate to cover any desired biasing range. Note that this code provides no limits on the steps per iteration and accepts the default number of iterations. The nonlinearities of this problem are much less severe than the previous full wave rectifier problem with four diodes and convergence is much easier to achieve. A pop-up plot (line 29) is used so the user can rapidly observe the results.

```
 1 : -- File list4_8.lua   -- Program to analyze the biasing of BJT
 2 :
 3 : require"nsolv"
 4 :
 5 : Is,vt,B,Va = 1.e-16,.026,100,100 -- Basic trans parameters
 6 : Vcc,R1,R2,Rs,Rl = 15,20.e3,1.e3,200,2.e3 -- circuit params
 7 :
 8 : fbjt = function(i,v)
 9 :   local ib,ic = v[4],v[5] -- Keep local to this function
10 :   i[1] = v[1]/R2 + (v[1] - Vcc)/R1 + ib
11 :   i[2] = v[2]/Rs -   ib - ic
12 :   i[3] = (v[3]-Vcc)/Rl + v[5]
13 :   i[4] = ib - Is*(math.exp((v[1] - v[2])/vt)-1)
14 :   i[5] = ic - B*ib*(1 + (v[3] - v[2])/Va)
15 : end
16 : v = {0,0,0,0,0} -- Initial guesses
17 : Ra,vb,ve,vc,ib,ic = {},{},{},{},{},{} -- Tables for results
18 : R2 = .1e3 -- Initial resistor value, zero causes problems
19 : i = 1
20 : while R2<=4e3 do
21 :     nit,err = nsolv(fbjt,v) -- Use previous solution as guess
22 :     if nit==NXX then
                print('Convergence error at i = ',i,err) end
23 :     print(R2,v[3],v[5])
24 :     vb[i],ve[i],vc[i] = v[1], v[2], v[3]
25 :     ib[i],ic[i],Ra[i] = v[4], v[5], R2
26 :     i = i+1; R2 = R2 + .1e3
27 : end
28 : write_data("list4_8.dat",Ra,vb,ve,vc,ib,ic)
29 : plot(Ra,vc,ve)
Selected Output:
2000   8.8009835135209        0.0030995082432396
2100   8.2834457292202        0.0033582771353899
2200   7.7733705000183        0.0036133147499908
2300   7.270842250979         0.0038645788745105
2400   6.7758965761271        0.0041120517119365
2500   6.2885330740673        0.0043557334629664
2600   5.8087244314725        0.0045956377842637
```
Listing 4.8. BJT biasing example

A graph of the collector voltage (v[3] in the code and v_3 in Figure 4.3) is shown in Figure 4.4. As the value of R_2 increases, the bias voltage applied to the transistor increases and the base and collector currents increase. As the collector current increases, the voltage drop across the load resistor increases and the collector voltage drops as seen in the figure. A typical objective of selecting the biasing resistors is to produce an operating point where the collector voltage is about half of

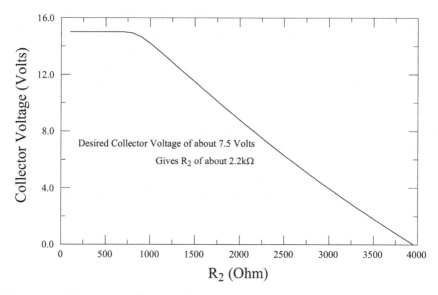

Figure 4.4. Calculated collector voltage as a function of R_2 value.

the supply voltage. For linear operation of the BJT, this allows the collector volt-age to swing above and below this value by about equal values and this is typi-cally a desired feature. Thus an appropriate selection criteria for R_2 will be a value which results in a collector voltage of near 7.5 Volts. From the graph and the selected printed output in Listing 4.8 it is seen that this occurs for an R_2 value of between 2200 and 2300 Ohms. Thus an appropriate value of the bias resistor would be 2.2k Ω as this is a standard resistor value. At this bias point it is seen from the printed output that the collector current will be 3.61mAmp.

One of the major advantages of this type of analysis is that one can easily ex-plore various "what if" cases can be easily explored. What if we change R_L or what if we change R_1 how will the bias point be changed or how must we change R_2 to maintain the same bias point? Or what if the transistor parameters change, how will this change the biasing? Such questions are very easy to explore by a simple change of a parameter and a re-execution of the code.

With a little reflection on this problem it is seen that selecting the value of re-sistor R_2 from Figure 4.4 that gives a collector voltage of 7.5 Volts is equivalent to finding the zero of a function. If the desired 7.5 value is subtracted from the val-ues in Figure 4.4 then the resulting graph passes through zero at the desired value of R_2. Thus finding the desired resistor value can be readily formulated as a prob-lem in finding the root of a function, a task readily implemented with the newton() function of Chapter 3. To do this one simply needs to think of the collector volt-age as a function of R_2 and embed the evaluation of the collector voltage within a function that can then be called by newton(). Code for this modified calculation is shown in Listing 4.9. Line 3 loads both nsolv and newton code segments. The new feature here is the fc() function on lines 18 through 21. The function fc() accepts a value of a resistor which is set to R_2 on line 19. The solver nsolv() is then called to solve the circuit equations, giving a value of v[3] the collector voltage for the input value of resistance. Line 20 then uses this value to return a value that will be zero when the desired v[3] value is achieved. This function is used on line 22 in a call to the newton() function with an initial guess at the resistor value (taken as 1.e3, although one knows that the value is closer to 2e3). At each step in the newton() iteration, the set of nonlinear equations will be solved by nsolv(). The value returned by newton() is then printed on line 23 and the resulting value is seen in the output to be 2254 which is a more precise value than the 2200 value estimated from Figure 4.4. The printed output also shows the final calculated values for all the circuit voltages (output numbers 1, 2 and 3) and transistor currents (output numbers 4 and 5.

```
 1 : -- File list4_9.lua   -- Biasing of a BJT with newton()
 2 :
 3 : require"nsolv"; require"newton"
 4 :
 5 : Is,vt,B,Va = 1.e-16,.026,100,100 -- Basic transistor parameters
 6 : Vcc,R1,R2,Rs,Rl = 15,20.e3,1.e3,200,2.e3 -- Basic parameters
 7 : VC = 7.5 -- Desired collector voltage
 8 :
 9 : fbjt = function(i,v)
10 :   local ib,ic = v[4],v[5] -- Keep definitions local
11 :   i[1] = v[1]/R2 + (v[1] - Vcc)/R1 + ib
12 :   i[2] = v[2]/Rs -  ib - ic
13 :   i[3] = (v[3]-Vcc)/Rl + v[5]
14 :   i[4] = ib - Is*(math.exp((v[1] - v[2])/vt)-1)
15 :   i[5] = ic - B*ib*(1 + (v[3] - v[2])/Va)
16 : end
17 : v = {0,0,0,0,0} -- Initial guesses, try other values?
18 : fc = function(R)   -- Function for newton() call
19 :   R2 = R; nsolv(fbjt,v) -- Use last values for v[]
20 :   return v[3] - VC
21 : end
22 : R2 = newton(fc,R2)
23 : printf('Calculated resistor value = %d\n',R2)
24 : table.foreach(v,print)
Output:
Calculated resistor value = 2254
1   1.4482358030549
2   0.75702736714309
3   7.4999886786965
4   3.5131175063673e-005
5   0.0037500056606518
```

Listing 4.9. Code segment for calculating optimum biasing resistor using newton() function.

The fact that variables are globally defined by default aids in the simplicity of the coding in Listing 4.9. For example since R_2, defined first on line 6, is globally known, the usage of this symbol in the fbjt() function and the fc() function refers to exactly the same memory location and the same variable value. So when R_2 is set in value within the fc() function on line 19, this is reflected in the value used within the fbjt() function. Similarly the use of the fbjt name in the calling argument of nsolv() on line 19 refers to the globally known function defined on lines 9 through 16. While having variables known globally by default can lead to incorrect usage, one simply has to be careful in reusing a variable name to refer to different physical quantities. Within a function any variable names passed through the argument list are known only locally to the defining function code. Thus within the fbjt() function any dummy variable name could be used in place of the v usage and the execution would be unchanged.

While this is a fairly simple example, it illustrates the case of one nonlinear problem embedded within another nonlinear problem. Such problems occur quite frequently within real world engineering problems. In this case the solution of a set of nonlinear equations is embedded in a nonlinear optimization process. Both of the problems are formulated as problems of finding the zeros of an equation or sets of equations. More examples of such problems will be developed during the course of this work. Another important feature of this problem is the observed fact that an analytical function does not exist for the newton() function evaluation. The function for which the zero value is desired in the newton() code is known only through an algorithmic computer solution, i.e. one does not have an analytical expression for $v[3]$ as a function of R_2 in the return of line 20 of the code. Thus it is not possible to take an analytical derivative of the function for use in a Newton's method – the solution must rely on numerical derivatives. This is again characteristic of many real world engineering problems where functions are known only through rather complicated computer algorithmic calculations. Throughout this work one will see more complex engineering problem solutions built upon previously implemented computer algorithms and coded functions.

Nonlinear circuit problems such as those discussed in these two examples are usually solved in Electrical Engineering by use of the SPICE circuit analysis program. For complicated circuits, this should certainly be the means of solving such problems. This program has built-in models for all standard electronic devices and is very advanced in approaches to achieve convergence. However, at the heart of the SPICE program is an approach very similar to that of the much simpler nsolv() program used here. SPICE will automatically set up the equation set to be solved, but uses first-order linearization and iteration to solve the nonlinear equations just as employed in nsolv(). While SPICE is the preferred tool for its domain of application, a tool such an nsolv() can be readily embedded into other computer code for specialized solutions to problems not appropriate for an electronic simulation.

Another example will now be given of the use of nsolv() for a circuits related problem. Calculating the frequency response of a circuit or engineering system is a frequently encountered real world problem. This is really a linear system problem, but is consider here as a set of complex number equations in order to show how the code developed so far can easily be used with such complex number problems. Figure 4.5 shows the schematic of a typical single stage electronic amplifier circuit. The input voltage source is coupled to the amplifier through C_s and coupled to the output through C_c. Three small capacitors, C_p, C_u and C_{ce}, represent internal device capacitances. For the purposes of this work it is not too important to understand all the details of the origin of the circuit. What is desired is an evaluation of the response of the circuit to a sinusoidal voltage source and a calculation of the ratio of output to input voltage in terms of the magnitude and phase angle of the response. This will be done by the use of Phasors where the complex impedance of the capacitors is $1/j\omega C$. The circuit can be analyzed in terms of the three node voltages shown in the figure and the resulting Phasor equations are:

$$(V_1 - V_s)/(R_s + 1/j\omega C_s) + (V_1 - V_2)(1/R_p + j\omega C_p)$$
$$+ (V_1 - V_3)(1/R_u + j\omega C_u) = 0$$
$$(V_2 - V_1)(1/R_p + j\omega C_p) + V_2(1/R_e + j\omega C_e) - g_m(V_1 - V_s)$$
$$+ (V_2 - V_3)(1/R_o + j\omega C_{ce}) = 0 \qquad (4.15)$$
$$(V_3 - V_1)(1/R_u + j\omega C_u) + g_m(V_1 - V_s) + (V_3 - V_2)(1/R_o + j\omega C_{ce})$$
$$+ V_3/R_c + (V_3 - V_4)j\omega C_c = 0$$
$$(V_4 - V_3)j\omega C_c + V_4/R_l = 0$$

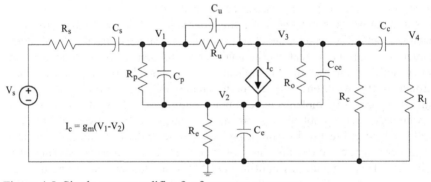

Figure 4.5. Single stage amplifier for frequency response.

In these equations the $V's$ represent voltage Phasors (with amplitude and phase information) for the various nodes and the resulting equation set is linear in the four Phasor voltages but with complex number parameters. By use of the complex algebra enhancement package, previously discussed in Chapter 2, these terms are readily coded with the resulting code shown in Listing 4.10. One only needs to load the Complex code before the nsolv code (see line 4) in order for nsolv() to properly handle complex numbers. Line 5 is used to inform nsolv() that this problem deals with a system of linear equations and that only one iteration is needed. So for this linear problem nsolv() simply sets up the equations and calls gauss() or spgauss() to provide the solutions. The four equations in exactly the same form as Eq. (4.15) are shown on lines 16 through 19 of the listing and one should have no trouble following the equation definitions.

```
 1 : -- /* File list4_10.lua -- amplifier */
 2 : -- Programs for the frequency response of an amplifier
 3 :
 4 : require"Complex"; require"nsolv" -- load complex,nsolv support
 5 : getfenv(nsolv).linear=1 -- Linear equations, one iteration
 6 : -- Circuit Parameters
 7 : Rs,Rp,Re,Ru,Ro,Rc,Rl = 100,20000,200,200000,100000,4000,10000
 8 : Cs,Ce,Cc,Cp,Cu,Cce = 1.e-6,10e-6,2.e-6,5.e-12,1.e-12,1.5e-12
 9 : gm = .01; vs = 1
10 : -- Frequency factors
11 : fact = 10^0.1-- 10 Points per decade in frequency
12 : f = 1.0 -- Begin at frequency of 1Hz
13 : jw = j*2*math.pi*f -- note jw is a variable name
14 :
15 : eqs = function(y,v) -- Phasor equations
16 :   y[1] = (v[1]-vs)/(Rs+1/(jw*Cs)) + (v[1]-v[2])*(1/Rp+jw*Cp) +
             (v[1]-v[3])*(1/Ru+jw*Cu)
17 :   y[2] = (v[2]-v[1])*(1/Rp+jw*Cp)+v[2]*(1/Re+jw*Ce)-
             gm*(v[1]-v[2]) + (v[2]-v[3])*(1/Ro+jw*Cce)
18 :   y[3] = (v[3]-v[1])*(1/Ru+jw*Cu) + gm*(v[1]-v[2]) +
             (v[3]-v[2])*(1/Ro+jw*Cce) + (v[3]-v[4])*jw*Cc + v[3]/Rc
19 :   y[4] = (v[4]-v[3])*jw*Cc + v[4]/Rl
20 : end
21 :
22 : sol = {{},{},{},{},{}} -- Table of 5 empty tables, filled later
23 : local t1=os.clock() -- To check execution speed
24 : nmax = 91; v = {0,0,0,0} -- initial values not critical
25 : for i=1,nmax do
26 :   nsolv(eqs,v)
27 :   v4 = v[4]; v4m = Complex.abs(v4)
28 :   sol[1][i] = f
29 :   sol[2][i] = math.log10(f)
30 :   sol[3][i] = v4m
31 :   sol[4][i] = Complex.angle(-v4)*180/math.pi - 180.
32 :   sol[5][i] = 20*math.log10(v4m)
33 :   f,jw = f*fact, jw*fact
34 : end
35 : print("time taken by nsolv = ",os.clock()-t1) -- Print time
36 : write_data('list4_10.dat',unpack(sol)) -- sol[1],sol[2]--sol[4]
37 : plot(sol[2],sol[5]) -- Magnitude Bode plot
38 : plot(sol[2],sol[4]) -- Angle Bode plot
Output:
time taken by nsolv =    0.125
```

Listing 4.10. Code segment for frequency response of amplifier in Figure 4.5 using nsolv().

To provide the complex quantities, one simply has to set the value of jw as is done on line 13 in terms of the purely imaginary quantity 'j' and the frequency f. The complex quantity 'j' is defined in the Complex package loaded on line 4. The term jw on line 13 is the name of a variable and not j times w. In a typical problem such as this, one is usually interested in the magnitude and phase angle of the frequency response over a broad range of frequencies, typically covering several orders of magnitude in frequency. For this a log frequency scale is typically used with frequency values equally spaced on a log scale. For this each new frequency point is some constant multiplicative factor times the previous frequency point. For N frequency points per decade or order of magnitude in frequency, the proper multiplicative factor is $10^{1/N}$ and line 11 of the code calculates an appropriate factor for 10 points per decade. The core part of the code is a frequency loop from line 25 to line 34 incrementing the frequency over 91 points or over 9 orders of magnitude in frequency from 1 to 1e9 Hz. Within this loop, the equations are solved by calling nsolv() on line 26, the magnitude and phase angle of the output voltage (v[4]) is calculated (lines 27 and 31) and stored for later use and the frequency (f) and jw updated for the next iterative loop calculation (line 33). The new value of jw will automatically be inserted into the defining equations on the next loop through nsolv() because of the globally known nature of variables. Typically the magnitude of the frequency response is plotted in dB units, so line 32 calculates and saves 20*log10 of the voltage magnitude to give the dB values.

By setting the source voltage to unity, the output voltage does not have to be divided by the source voltage to obtain the transfer function which is the ratio of output to input voltage. The resulting magnitude of the amplifier frequency response is shown in Figure 4.6 and is typically referred to as a Bode plot of the magnitude of the transfer function. From such a plot the bandwidth of the amplifier can be determined as the -3dB points which in this case occur at around 200 Hz and 10 MHz. Also shown in the figure is a dotted curve for a separate calculation with a C_e value of 1.0μF. It can be seen that the lower -3dB frequency is almost completely controlled by this value and extending the frequency response to lower frequencies would require a larger value of this capacitor.

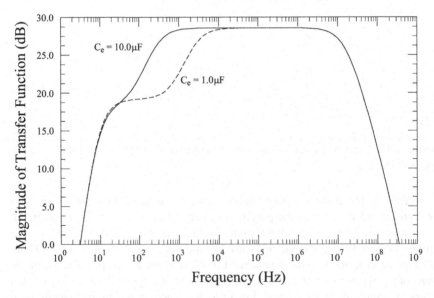

Figure 4.6. Bode plot of amplifier magnitude response

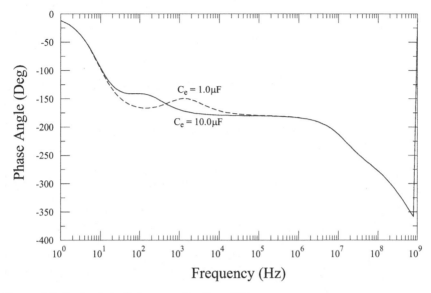

Figure 4.7. Bode plot of phase angle of amplifier response.

This is the type of calculation normally done by Electrical Engineers using SPICE, but is readily calculated from the code developed here. A plot of the angle of the transfer function is shown in Figure 4.7. The spike for the last calculated point is an artifact of the fact that the angle() function returns only the principal value and as the angle goes through -360 deg the curve switches back to 0 degrees.

Since this is a linear coupled equation problem, one can bypass nsolv() and directly solve the equations with only a coupled equation solver such as gauss() or spgauss(). Thus it is perhaps useful to compare the two approaches and similar code to solve this problem directly by gauss() is shown in Listing 4.11. The code is almost identical to that of the previous example. In solving the equations directly using gauss() the matrix of coefficients must be defined each time through the frequency loop and this is done on lines 21 through 25. Note that gmat on line 21 is defined as a series of 4 tables within a table and with 4 entries for each row of the table values. After the matrix coefficients are defined, gauss(gmat,b) is called on line 26 to solve the equations. In order to write the matrix coefficients, the various terms multiplying each voltage term in Eq. (4.15) must be collected together. If this is done, the correspondence between the gmat terms and the terms in Eq. (4.15) are readily identified. However it has been this author's experience that more errors are made in writing the matrix of coefficients terms needed for gauss() than in writing the basic equations needed by nsolv(). This author's preference is thus to use the nsolv() formulation rather than the gauss() approach even for a set of linear equations. However, this does come at some increase in computer time needed for a solution set. From the printed output for the two approaches, it is seen that the direct gauss() solution method in this example is about

```
 1 : -- /* File list4_11.lua */
 2 : -- Programs #2 for the frequency response of an amplifier
 3 :
 4 : require"Complex"; require"gauss" -- load complex math support
 5 : --getfenv(nsolv).full = true -- Try gauss()instead of spgauss()
 6 : -- Circuit Parameters
 7 : Rs,Rp,Re,Ru,Ro,Rc,Rl = 100,20000,200,200000,100000,4000,10000
 8 : Cs,Ce,Cc,Cp,Cu,Cce = 1.e-6,10e-6,2.e-6,5.e-12,1.e-12,1.5e-12
 9 : gm = .01; vs = 1
10 : -- Frequency factors
11 : fact = 10^0.1-- 10 Points per decade in frequency
12 : f = 1.0 -- Begin at frequency of 1Hz
13 : jw = j*2*math.pi*f
14 :
15 : sol = {{},{},{},{},{}} -- Table of 5 empty tables
16 : local c=os.clock()
17 : nmax = 91; v = {0,0,0,0} -- initial values not critical
18 : local t1=os.clock()
19 : -- Second approach -- Set up matrix equations
20 : for i=1,nmax do -- gmat is conductance matrix of coefficients
21 :   gmat={{ (1/(Rs+1/(jw*Cs))+1/Rp+jw*Cp+1/Ru+jw*Cu), (-1/Rp-jw*Cp),
                (-1/Ru-jw*Cu),0},
22 :        { (-1/Rp-jw*Cp-gm), (1/Rp+jw*Cp+1/Re+jw*Ce+gm+1/Ro+jw*Cce),
                (-1/Ro-jw*Cce),0},
23 :        { (-1/Ru-jw*Cu+gm), (-gm-1/Ro-jw*Cce),
                (1/Ru+jw*Cu+1/Ro+jw*Cce+jw*Cc+1/Rc), (-jw*Cc)},
24 :        {0,0, (-jw*Cc), (1/Rl+jw*Cc)}}
25 :   b = {vs/(Rs+1/(jw*Cs)),0,0,0} -- Source terms, currents
26 :   gauss(gmat,b) -- Solutions for v's are returned in the b array
27 :   v4 = b[4]; v4m = Complex.abs(v4)
28 :   sol[1][i] = f
29 :   sol[2][i] = math.log10(f)
30 :   sol[3][i] = v4m
31 :   sol[4][i] = Complex.angle(-v4)*180/math.pi -180.
32 :   sol[5][i] = 20*math.log10(v4m)
33 :   f,jw = f*fact, jw*fact
34 : end
35 : print("time taken by gauss = ",os.clock()-t1)
36 : write_data('list4_11.dat', unpack(sol))
37 : plot(sol[2],sol[5]) -- Magnitude Bode plot
38 : plot(sol[2],sol[4]) -- Angle Bode plot
Output:
time taken by gauss =    0.047
```

Listing 4.11. Code segment for frequency response of amplifier of Figure 4.5 using gauss() or spgauss().

a factor of 2.5 faster than the nsolv() approach. The reader is encouraged to execute the code on one's own computer and observe the differences. However, both sets of calculations are completed in less than 0.13 sec so they both popup a graph of results almost as fast as one can respond to any output. Certainly setting up and writing code for the equations takes much more time than the computer solution, so whichever approach saves time in defining the problem should be preferred. Of course if a problem needed to be solved for such a circuit over and over again for hundreds or thousands of times, the time advantage of programming directly with the gauss() function might be well worth the effort. The reader can execute the code in both listings and compare the output files to verify that they both give the same results as they must if no errors have been made in the coding.

Another variation that one can investigate for the code in Listings 4.10 and 4.11 is to compare the time for the calculations using the gauss() and the spgauss() functions. Since the default case for nsolv() is to use spgauss(), Listing 4.10 already uses this sparse matrix function. One might suspect that some of the additional time might be due to this usage as opposed to gauss(). By un-commenting the code statement on line 5 of Listing 4.10: `getfenv(nsolv).full=true`, the full gauss() function will be used in solving the equation set. The reader is encouraged to re-execute the code with these changes and observe changes in the execution time. However, there will probably be very little difference between using gauss() or spgauss() in either of these code segments.

A set of equations is now explored from the field of Chemical Engineering. A set of nonlinear coupled equations representing a Methane-Oxygen reaction has been given by Carnahan, et.al. (Carnahan, Luther and Wilkes 1969, *Applied Numerical Methods:* Wiley) as the seven coupled equations:

$$x_1/2 + x_2 + x_3/2 - x_6/x_7 = 0$$

$$x_3 + x_4 + 2x_5 - 2/x_7 = 0$$

$$x_1 + x_2 + x_5 - 1/x_7 = 0$$

$$x_1 + x_2 + x_3 + x_4 + x_5 - 1 = 0$$

$$P^2 x_1 x_4^3 - 1.7837 \times 10^5 x_3 x_5 = 0 \tag{4.16}$$

$$x_1 x_3 - 2.6058 x_2 x_4 = 0$$

$$-28837 x_1 - 139009 x_2 - 78213 x_3 + 18927 x_4$$

$$+8427 x_5 + 13492/x_7 - 10690 x_6/x_7 = 0$$

The x_i values represent concentrations of CO, CO_2, H_2O, H_2, CH_4, O_2/CH_4 and TOTAL. A suggested starting value for the solution has been given as x = {.5, 0, 0, .5, 0, .5, 2}. A code segment for this set of equations is given in Listing 4.12. The equation set is defined in a straightforward manner on lines 7 through 14 of the code. The solution then only requires setting an initial guess for the solution set on line 17 and a call to nsolv() on line 19. One can see that the solution took 7 Newton iterative loops with a final expected relative accuracy of about 11 decimal digits. The printed solution set corresponds to the expected values for this problem. A step size array is defined on line 18 but all table elements are set to zero which means that no step size limitations are used. This is included so the reader can experiment with setting step sizes if desired. This example illustrates how simple it is to formulate and solve a set of nonlinear equations with the nsolv() code. One key to such an easy solution is always having a "good' initial guess at the solution set.

At this point we can perhaps go on to another example as the code in Listing 4.12 adequately documents this problem and the obtained solution agrees with the expected result. However, for this solution it was assumed that a good initial

```
 1 :  --/* File list4_12.lua */ -- Equations for chemical reaction
 2 :
 3 :  require"nsolv"
 4 :
 5 :  P = 20 -- Single parameter
 6 :  eqs = function(f,x)
 7 :   f[1] = x[1]/2 + x[2] + x[3]/2 -x[6]/x[7]
 8 :   f[2] = x[3] + x[4] + 2*x[5] - 2/x[7]
 9 :   f[3] = x[1] + x[2] + x[5] -1/x[7]
10 :   f[4] = x[1] + x[2] + x[3] + x[4] + x[5] - 1
11 :   f[5] = P^2*x[1]*x[4]^3 - 1.7837e5*x[3]*x[5]
12 :   f[6] = x[1]*x[3] - 2.6058*x[2]*x[4]
13 :   f[7] = -28837*x[1] - 139009*x[2] - 78213*x[3] + 18927*x[4] +
14 :       8427*x[5] + 13492/x[7] - 10690*x[6]/x[7]
15 :  end
16 :
17 :  x = {.5, 0, 0, .5, 0, .5, 2} -- Initial guesses
18 :  step = {0,0,0,0,0,0,0} -- Try different values
19 :  nmx,a = nsolv(eqs,x,step) -- Solve equations
20 :  print('Number of iterations, error = ',nmx,a)
21 :  table.foreach(x,print)
Output:
Number of iterations, error =    7  3.2334858690716e-012
1  0.32287083947654
2  0.0092235435391875
3  0.046017090960632
4  0.61817167507082
5  0.0037168509528154
6  0.57671539593555
7  2.9778634507911
```

Listing 4.12. Code segment for solving nonlinear equations for Methane-Oxygen reaction.

guess at the solution set is not known. How might one start to pursue the desired solution set in the 7 dimensional space of the unknown variables? When in doubt as to a good guess for the solution set one should experiment with a range of initial values. This can also provide insight into ones physical problem. In many cases an initial guess of zero for the variables gets one started. In this case it is seen that this would not be a good guess as one of the variables ($x[7]$) appears in the denominator of the equations. Another starting point is to assume some constant value for all the variables and observe the solution set. Using this approach the following solutions were obtained:

$$x = \{.1,.1,.1,.1,.1,.1\} \rightarrow \{0.4569,-0.00040,-0.00212,$$
$$0.9151,-.3695,2.610,11.500\}$$
$$x = \{1,1,1,1,1,1,1\} \rightarrow \{-0.1897,0.1897,0.00,$$
$$5.23e-318,1.000,.09487,1.000\}$$

$$(4.17)$$

The first bracket on each line indicates the initial guess at a solution value and the second bracket gives the converged solution set returned by nsolv() using the initial guess. If these new solution sets are inserted into the equations, it is found that these do indeed satisfy the equation set and thus the equations apparently have at least two valid solution sets in addition to the one given in Listing 4.12.

The question then arises as to which solution set corresponds to the real life physical problem that the equation set is supposed to represents? This can not be determine from the pure mathematics of the problem or from the solution algorithm. The relevance of a particular solution set can only be determined from the physical context of the problem being addressed. In this case the solution variables represent concentrations of various gasses in a Methane-Oxygen chemical reaction. For such a physical problem, negative solution values have no physical meaning so these two new solutions with negative values must be rejected as not representative of the physical problem of interest here. However, the mathematical equations or computer algorithm imbedded in nsolv() could care less for physical reality and may return, as in this case, valid mathematical solutions, but not solutions for ones physical reality.

From the above it is obvious that one should seek, for this problem, only solutions for which all the solutions are positive in value. One simple means of forcing the solution values to have a desired sign is through the use of an array of step[] values. If these parameters are positive and positive values are taken for all the initial approximation, then the range of solution values will be restricted to positive values only. Of course a negative initial value can be taken which will restrict a variable to only negative values. This is an important feature of the nsolv() code combined with an array of step[] values. For many such nonlinear coupled equation sets an appropriate set of step values is 2.0 which means that the change in a solution value is limited to a factor of 2 above and below that value in each Newton iterative step. For example if an initial guess for a variable is 1.0 then the value of that variable at the first Newton step would be limited to the range 0.5 to 2.0. In this manner the range of a solution variable can be changed reasonably fast over a limited number of Newton iterative steps, but the variable may change sufficiently slowly to achieve convergence for a wide range of physical problems. Also the sign of the solution value never changes from that of the initial guess.

The code in Listing 4.12 has been re-executed with a range of initial guesses and various step[] array values. With a step array of all 2.0 values (step = {2,2,2,2,2,2,2}), it has been found that nsolv() will converge successfully to the solution set given in Listing 4.12 for sets of constant initial guesses for all variables ranging from {0.1,0.1,...0.1} to {1,1,..1}. In addition, no solution set was found with all positive solution values different from the solution set in Listing 4.12. By probing the sensitivity of a solution set to a range of initial guesses, one can gain confidence that the obtained solution is the real solution desired for a given real physical problem. The reader is encouraged to explore various initial guesses and various step[] values to observe the resulting solution sets – or lack of convergence for some initial guesses.

One lesson from this example is that one must typically have sufficient insight into a physical problem to recognize when a physically valid solution has been obtained or not obtained. Many nonlinear equation sets have multiple solution sets and only one set may have meaning for a particular physical problem.

An example is now given of determining parameters for a voltage waveform with certain specified parameters. Consider the voltage waveform shown in Figure 4.8. Pulses such as this are used in the Electrical Engineering field to characterize voltage surges on power lines and on circuits. A series of definitions for such pulses are specified by IEEE Standards C62.41-2002 and the waveform shown in the figure is known as a 5/50 (t1/t2) waveform. The t1 = 5 parameter is a measure of the rise time of the pulse and the t2 = 50 parameter is a measure of the width of the pulse. The rise time of a waveform is frequently specified in terms of the time taken for the waveform to go from 10% to 90% of its maximum value, the t_{10} and t_{90} times shown in Figure 4.8. The time t1 in the IEEE specification is taken as $1.25(t_{90} - t_{10})$ which is equivalent to the time taken by a linear line passing through the 10% and 90% points to go from 0 to 1. The time t2 is defined as the time taken to the 50% point on the falling waveform from the "virtual origin", where the virtual origin is the time point where a straight line through the 10% and 90% points crosses the 0 function value. This will obviously be near the time origin shown in the figure but will have a slight negative value. The peak of the waveform which occurs at some time t_m in the figure is defined as unity value so any use of the function can be scaled to any desired peak value. The time for the peak is not specified, but is obviously the point where the derivative of the function is zero.

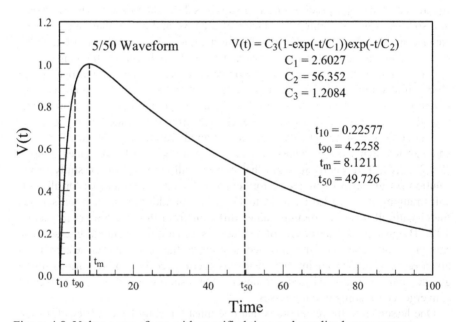

Figure 4.8. Voltage waveform with specified time and amplitude parameters.

One possible mathematical model for such a waveform is the equation shown in Figure 4.8 of:

$$V(t) = C_3(1 - \exp(-t/C_1))\exp(-t/C_2) \tag{4.18}$$

where C_1 and C_2 are some time constants characterizing the initial rise and the longer fall time. The waveform shown in Figure 4.8 is a plot of this function for the parameters shown in the figure and gives a waveform with the specified 5/50 time parameters. The question to be addressed here is how can these parameters be determined from a given set of t1/t2 parameters?

Consider first a slightly simpler set of specifications. Suppose the time parameters t_m and t_{50} are specified. Then the three equations to be satisfied by the $V(t)$ function is:

$$V(t_m) = 1.0$$
$$V(t_{50}) = 0.5 \tag{4.19}$$
$$V'(t_m) = 0.0$$

where the third relationship is that the derivative at the maximum time is zero. This provides three equations from which the three parameters of Eq. (4.18) can in principle be determined (provided a solution is possible). The IEEE specifications as discussed above are slightly more complicated as the time to maximum and the 50% time are not explicitly specified. The specifications on the times are:

$$t_{90} - t_{10} = t_1 \text{ or } t_{90} = t_{10} + t_1$$
$$t_{50} - t_v = t_2 \text{ or } t_{50} = t_v + t_2 = t_{10} - t_1/10 + t_2 \tag{4.20}$$

where t_v is the virtual origin as discussed above and in the second line of this set this has been expressed in terms of t_{10} and t_1 as $t_v = t_{10} - t_1/10$. It is left to the reader to verify this latter expression. With the IEEE specifications the times t_{10} and t_m must then also be considered as unknown parameters to be determined by additional equations. Two additional equations can then be added to those of Eq. (4.19) for a set of five equations as:

$$V(t_{10}) = 0.1$$
$$V(t_{90}) = V(t_{10} + t_1) = 0.9$$
$$V(t_m) = 1.0 \tag{4.21}$$
$$V(t_{50}) = V(t_{10} - t_1/10 + t_2) = 0.5$$
$$V'(t_m) = 0.0$$

This is now a set of 5 equations with the three unknown C values plus the unknown times t_{10} and t_m.

A code segment is shown in Listing 4.13 for formulating and solving this set of equations. Lines 5 through 9 specify a range of pulse parameters that the user can select with values ranging from a 10/1000 long time pulse to a 1.2/50 short pulse. For this example the 5/50 pulse is selected on line 7. The five equations are defined on lines 14 through 18 in essentially the same format as in Eq. (4.21). For the derivative equation, the coded deriv() function from Chapter 5 (details in

```
 1 : --/* File 4_13.lua */ -- Parameters for time pulse
 2 : require "nsolv"; require"deriv"
 3 : exp = math.exp
 4 :
 5 : --t1,t2 = 10,1000 -- Select pulse as desired
 6 : --t1,t2 = 10,350
 7 : t1,t2 = 5,50
 8 : --t1,t2 = 1.2,50
 9 : t3,t4 = t1/1.25,t1/10
10 :
11 : eqs = function(f,c)
12 :  t10,t90 = c[4],c[4]+t3,c[5]
13 :  tm,t50 = c[5],c[4] -t4 + t2
14 :  f[1] = fx(t10,c) -.1 -- 10% point, c[4] = time
15 :  f[2] = fx(t90,c) -.9 -- 90% point x90 = time
16 :  f[3] = fx(t50,c) - .5 -- 50% point x50 = time
17 :  f[4] = deriv(fx,tm,c) -- Derivative at peak = 0
18 :  f[5] = fx(tm,c) -1 -- Peak value = 1 at c[5] = time
19 : end
20 : fx = function(t,c)
21 :  return c[3]*(1-exp(-t/c[1]))*exp(-t/c[2])
22 : end
23 :
24 : c = {t1/2,1.5*t2,1,t1/10,1.5*t1}; nc = #c
25 : nn,err = nsolv(eqs,c,step)
26 : print('at ',t1,t2)
27 : print('Number of iterations, error = ',nn,err)
28 : for i=1,nc do printf('c[%d] = %12.4e\n',i,c[i]) end
29 : i = 1; xv,yv = {},{}
30 : for t = 0,2*t2,t2/1000 do
31 :  xv[i],yv[i] = t, fx(t,c); i = i+1
32 : end
33 : plot(xv,yv); write_data('list4_13.dat',xv,yv)
34 : print('10% point at ',c[4],fx(c[4],c))
35 : print('90% point at',t3+c[4],fx(t3+c[4],c))
36 : print('peak point at',c[5],fx(c[5],c))
37 : print('50% point at',c[4]+t2-t4,fx(c[4]+t2-t4,c))
Selected output:
at    5   50
Number of iterations, error =    5  2.4267509922489e-010
10% point at    0.22577397062501   0.1
90% point at    4.225773970625 0.9
peak point at  8.1210838094714   1
50% point at   49.725773970625   0.5
```

Listing 4.13. Code segment for evaluating parameters of voltage pulse.

next chapter) has been used. Of course, one could, if desired, take the derivative by hand and express it in terms of the coefficients. But why do this when the computer can simply evaluate the derivative? The t_{10} and t_m parameters are set to the $c[4]$ and $c[5]$ coefficients on lines 12 and 13. Initial guesses are set for the unknown times and the parameters of the function on line 24 and in this example very reasonable first guesses can be obtained from the specified function time parameters. For example, the t_{10} parameter must be a small fraction of the t1 parameter, so $c[4]$ is set to t1/10. The reader should be able to understand the other initial guesses. For this example, no step size limitations are needed for convergence and none are used. The five equations are solved by the call to nsolv() on line 25. The remainder of the code prints some values from the solution including the determined coefficients, saves the results and prints the time values for the various critical points. In addition, the value of the function is printed at the critical times to verify that the determined coefficient set does in fact satisfy the desired conditions. For example the value of the function at the evaluated time for the peak (time of 8.12108) is printed as exactly 1 which is the expected value for a correct solution. The obtained function C parameters are not shown in the selected output but are shown in Figure 4.8. The reader is encouraged to select different pulse parameters on lines 5 through 8 and re-execute the code to observe different pulse shapes.

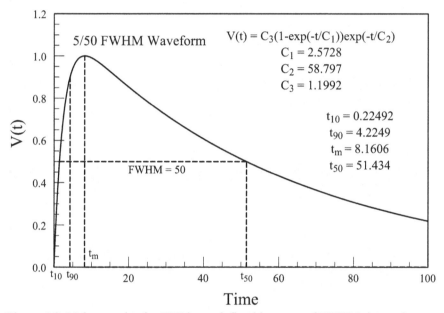

Figure 4.9. Voltage pulse for EFT burst defined in terms of FWHM time value.

Another voltage pulse defined by the IEEE C62.41-2002 standard is an EFT pulse specified as a 5/50 (t_1/t_2) pulse but where the $t_2 = 50$ is now specified as the full width at half maximum (FWTH) time value. This is shown in Figure 4.9 where a waveform satisfying the desired specification is shown. The t_1 parameter is again the same as previously specified, i.e. $t_1 = 1.25(t_{90} - t_{10})$. If t_{51} is the time for the 50% value on the leading edge of the waveform and t_{50} is the time for the 50% value on the trailing edge then a constraint on the times is $t_{50} = t_{51} + t_2$. However, this FWHM specification involves one additional unknown which is the t_{51} value and this can be specified by assigned it to the c[6] parameter value. To solve for the waveform parameters for this set of specifications requires the solution of a set of six equations in six unknowns. A code segment for such a solution is shown in Listing 4.14. A sixth equation has been added to the equations to be

```
 1 : --/* file list4_14.lua */   -- FWHM waveform specifications
 2 :
 3 : require "nsolv"; require"deriv"
 4 : exp = math.exp
 5 :
 6 : t1,t2 = 5,50; t3,t4 = t1/1.25,t1/10
 7 : eqs = function(f,c)
 8 :   t10,t90 = c[4],c[4]+t3
 9 :   tm,t51,t50 = c[5],c[6],c[6] + t2
10 :   f[1] = fx(t10,c) -.1 -- 10% point, c[4] = time
11 :   f[2] = fx(t90,c) -.9 -- 90% point c[4]+x3 = time
12 :   f[3] = fx(t51,c) - .5 -- 50% point x51 = time
13 :   f[4] = deriv(fx,tm,c) -- Derivative at peak = 0
14 :   f[5] = fx(tm,c) -1 -- Peak value = 1 at c[5] = time
15 :   f[6] = fx(t50,c) - .5 -- second 50% point
16 : end
17 : fx = function(t,c)
18 :   return c[3]*(1-exp(-t/c[1]))*exp(-t/c[2])
19 : end
20 :
21 : c = {t1/2,1.5*t2,1,t1/10,1.5*t1,t1/2}; nc = #c
22 : nn,err = nsolv(eqs,c,step)
23 : print('at ',t1,t2); print('Number of iterations,
          error = ',nn,err)
24 : for i=1,nc do printf('c[%d] = %12.4e\n',i,c[i]) end
25 : i = 1; xv,yv = {},{}
26 : for t = 0,100,.1 do
27 :   xv[i],yv[i] = t, fx(t,c);  i = i+1
28 : end
29 : plot(xv,yv); write_data('list4_14.dat',xv,yv)
30 : print('10% point at ',c[4],fx(c[4],c))
31 : print('90% point at',t3+c[4],fx(t3+c[4],c))
32 : print('peak point at',c[5],fx(c[5],c))
33 : print('50% points at',c[6],c[6]+t2,fx(c[6]+t2,c))
Selected output:
at     5   50
Number of iterations, error =      5  9.7422751909592e-008
10% point at    0.22492083410501   0.1
90% point at    4.224920834105 0.9
peak point at  8.1606399675187    1
50% points at  1.4338240647685    51.433824064768    0.5
```

Listing4.14. Code segment for evaluating parameters of EFT voltage pulse with FWHM specification.

satisfied and this is the f[3] equation line 12 for the lower 50% voltage value at the $t_{51} = c[6]$ unknown time. With this addition and an extra initial guess at this parameter value, the code is essentially identical to that of Listing 4.13. Again the printed selected output values verify that the desired points along the function curve have been satisfied by the obtained solution. This can also be verified visually by the resulting curve shown in Figure 4.9. By comparing this figure with Figure 4.8 it can be seen that there is very little difference in the resulting waveform when the 50% point on the trailing edge is defined as the FWHM value or as the time from the virtual origin. This is because, of course, the leading edge of the waveform is very sharply rising and the first 50% crossing occurs at a small value of time. For the FWHM specification it is seen that the trailing edge t_{50} time is slightly larger than the t2 specification and for the previous example it is slightly less then the t2 specification.

This example involves one of the most complicated functional forms for solving a set of nonlinear equations used in this chapter. It also involves a reasonable number of unknown coefficients. However, the nsolv() code is readily able to solve for the set of solution values, and in this case to do so without invoking any limits on the step size for the correction terms of the internal Newton iterations. Part of the good results is the ability to obtain very good initial guesses at an appropriate set of solution values. If one re-executes the coded examples and applies initial guesses far from the values shown in the listings, the nsolv() Newton procedure may not converge. It is certainly easy to obtain non-convergence by changing the initial guesses. It is also possible to obtain convergence over a broader range of initial guesses, by use of an array of step[] parameter values and by restricting the step size changes to something on the order of 20% to 50% of the previously obtained value. The reader can experiment with various initial guesses and step limitations.

A final example in this section is that of a large system of coupled nonlinear equations. For this consider the equation set:

$$f_1 = 3x_1 - 2x_1^2 - 2x_2 + 1 = 0$$
$$f_i = 3x_i - 2x_i^2 - x_{i-1} - 2x_{i+1} + 1 = 0 \text{ for } 1 < i < n-1 \qquad (4.22)$$
$$f_n = 3x_n - 2x_n^2 - x_{n-1} + 1 = 0$$

with the number of equations in the range of 100 to 1000. The coefficient matrix for this equation set is obviously a sparsely populated matrix with only three non-zero elements on each row, with a diagonal element and elements on each side of the diagonal element. Such a matrix is a tri-diagonal matrix and efficient techniques for solving such sets of coupled equations will be presented in later chapters. For the present, however, this will be simply treated as an example of a large number of coupled nonlinear equations appropriate for solution with nsolv() using the default sparse matrix features of the code.

Listing 4.15 shows code for implementing and solving the set of equations for the case of 1000 equations. The equation set is defined on lines 7 through 9 with a loop on line 8 defining all except the first and last equation. For an initial guess, values of -1 are taken for all the variables on line 12. A statement on line 13 then invokes nsolv() to obtain the solution set. The printed output lines indicate that 5 Newton iterative loops were required for an accuracy of about 10 decimal digits. Only samples of the 1000 printed solution values are shown in the listing. After a few varying solution values near the first and last equation, the solution values for the center numbered variables rapidly settles down to the value of $-$ 0.70710678118655 shown for the $500 - 502$ printout lines. One should readily recognize that this value is in fact just $-1/\sqrt{2}$. One can readily verify that this value for all three variables in the equation on line 8 is a solution of the equation set.

```
 1 :   --/* File list4_15.lua */ -- Example of large # of equations
 2 :
 3 : require"nsolv"
 4 :
 5 : N = 1000 -- Number of equations

 6 : eqs = function(f,x)
 7 :   f[1] = 3*x[1] - 2*x[1]^2 -2*x[2] + 1
 8 :   for i=2,N-1 do f[i] = 3*x[i]-2*x[i]^2-x[i-1]-2*x[i+1]+1 end
 9 :   f[N] = 3*x[N] - 2*x[N]^2 - x[N-1] + 1
10 : end
11 :
12 : x = {}; for i=1,N do x[i] = -1 end -- try -0.1 to -10
13 : nmx,a = nsolv(eqs,x)

14 : print(nmx,a); table.foreach(x,print)
Selected Output:
5    4.7958056458624e-010
1   -0.57076119297475
2   -0.68191012886809
3   -0.70248602066765
....
500    -0.70710678118655
501    -0.70710678118655
502    -0.70710678118655
....
998    -0.66579752334218
999    -0.59603531262665
1000   -0.41641230116684
```

Listing 4.15. Code segment for example of 1000 nonlinear coupled equations solved with nsolv()

The initial guess is not very critical for this problem as long as one assumes some initial negative value – positive initial guesses have convergence problems. The reader can experiment with different guesses and values in the range of -.1 to -10 should rapidly lead to convergence in only a few Newton iterative loops. One of the reasons this problem is so readily solved by nsolv() is that sparse matrix techniques are used by default (using spgauss()). If the default case had been selected as using gauss() and assuming a full matrix of coefficients, the solution would take a very long time. The reader can experiment with this by inserting the code line getfenv(nsolv).full=true in Listing 4.15. However be prepared to run the code over night or certainly over lunch.

This is a relatively simple and straightforward example. However, it illustrates the ease with which very large systems of coupled nonlinear equations can be solved – at least in principle and in practice provided the equation set has a sparse coefficient matrix. All of this sparse matrix manipulation is easily handled in our programming language freeing the user from having to worry about memory allocation and freeing memory – such is the beauty of modern scripting languages.

4.4 Polynomial Equations and Roots of Polynomial Equations

Polynomial functions of an independent variable find extensive use in engineering and a rapid and convenient means of performing arithmetic involving polynomials and finding roots of a polynomial equation can be very useful. This section has two goals: (1) to demonstrate a language extension for direct mathematical operations on polynomial equations and (2) to develop a coded function for rapidly finding all the roots of polynomial equations in one variable. Finding roots of a polynomial was briefly considered in Chapter 2 and Newton's method was used to find some of the roots of specific polynomials. For polynomial equations in only one variable, this would seem more appropriate for a Chapter 2 discussion. However, the general approach to be developed for solving for all the roots requires that two simultaneous equations must be solved both of which must be zero. This is just the problem considered in this chapter and the code to be developed for finding all the roots is another example of the use of the nsolv() routine developed here.

Before delving into polynomial root finding, let's consider an extension to the Lua language for easily manipulating polynomials. A general polynomial to be considered is of the form:

$$f(x) = a_1 + a_2 x + a_3 x^2 + \cdots + a_n x^{n-1} \qquad (4.23)$$

Operations that one might desire to perform between such polynomials include the standard mathematical operations of addition, subtraction, multiplication and negation. Division has to be considered a little more carefully. For example multiplying two polynomials one of order 3 and one of order 2 gives as follows:

$$(a_1 + a_2 x + a_3 x^2 + a_4 x^3)(b_1 + b_2 x + b_3 x^2) =$$
$$(a_1 b_1) + (a_1 b_2 + a_2 b_1)x + (a_1 b_3 + a_2 b_2 + a_3 b_1)x^2 \qquad (4.24)$$
$$+(a_2 b_3 + a_3 b_2 + a_4 b_1)x^3 + (a_3 b_3 + a_4 b_2)x^4 + (a_4 b_3)x^5$$

Such operations are easily programmed and a computer program is much more reliable at performing such operations for high order polynomials than are people. A word about notation is in order here. One will note that the coefficients are labeled with subscripts beginning with 1 and not 0. This is in keeping with default table notation in our programming language which uses, by default, the 1 to n notation for a table as opposed to other languages (such as C) that use 0 to n as default labeling. This means that a table labeled from 1 to n will represent a polynomial of order n-1, or one can state that a polynomial of order n has a table representation from 1 to n+1. The 1 to n labeling has advantages in some instances, such as labeling for matrix elements. Also a table of length n has a maximum integer index value of n. The 0 to n labeling has some advantages in other instances. It would make the notation a little easier for the polynomial case considered here. However, with 0 to n labeling, a table with maximum integer index value n has a length of n+1. One can easily be "off by 1" in notation when dealing with tables under either convention.

One simple means of using a table to represent a polynomial is for each entry to represent the coefficient of a power factor term such as:

$$1 + 2x + 3x^2 + 4x^3 + 5x^5 \rightarrow \{1,\ 2,\ 3,\ 4,\ 0,\ 5\} \tag{4.25}$$

There are obviously two choices in the table representation, one can store from the constant value to the highest power or in reverse order from the highest power to the constant value. Storing in increasing order appears more natural (although a matter of choice) so that the nth table value corresponds to the n-1 power of the variable. Note that the fourth power of the polynomial above is missing and is represented by a 0 entry in the fifth table value. From such a representation one can determine the order of the polynomial as being one less than the length of the table.

To implement polynomial objects a coded function is needed that set up a table of values and defines an appropriate metatable of functions for performing standard mathematical operations involving the polynomial objects such as the multiplication shown in Eq. (4.24). To illustrate the ease with which one can set up objects and define operations on the objects, a code segment is shown in Listing 4.16 for defining Polynomial objects and performing simple +, - and * operations. Tables are defined on line 3 to hold the Polynomial and metatable data. A Polynomial.new() function is defined on lines 5 through 7 that simply takes a list of comma separated data values and stores the values in a table and then defines a metatable for the collection of data. Then on lines 8 through 39 metatable functions are defined for performing addition (__add() function), subtraction (__sub() function), multiplication (__mul() function), unitary minus (__unm() function) and converting a Polynomial data structure to a string (__tostring() function) for printing results. Following the definitions, two polynomials are created with the Polynomial.new() calls on lines 42 and 43. These are printed by line 44 and the next two lines test all the 4 mathematical operations so far implemented with printed results. By performing the operations manually the accuracy of the printed output can be verified for the indicated polynomial operations.

This simple code illustrates the principle of a code segment for Polynomial objects with associated operations. However, one can easily recognize that this coding is somewhat incomplete and needs to be supplemented for a more practical object set. For example, the mathematical operations expect both objects to be of Polynomial type, so mathematical operations involving constants will result in an error -- one would have to formulate a constant as a Polynomial table of one constant entry. Also one needs to consider Polynomial division operations and how these are to be interpreted.

A more complete Polynomial object package has been developed and is available by including a `require"Polynomial"` code statement. The use of this set of code functions with some examples is shown in Listing 4.17. Lines 5 through 9 illustrate various forms for setting up Polynomial objects using a list of numbers (line 5) or a table of numbers (line 6) or math operations (lines 7 and 8) on

```
 1 : -- /* File list4_16.lua */ -- examples of polynomial operations
 2 :
 3 : Polynomial = {}; Polynomial_mt = {} -- Poly table and metatable
 4 :
 5 : Polynomial.new = function(...) -- Creates a new Polynomial type
 6 :  return setmetatable({...},Polynomial_mt)
 7 : end

 8 : Polynomial_mt.__add = function(p1,p2) -- p1 + p2
 9 :  local n,sum = math.max(#p1, #p2), {}
10 :  for i=1,n do
11 :     sum[i] = (p1[i] or 0) + (p2[i] or 0)
12 :  end
13 :  return setmetatable(sum,Polynomial_mt)
14 : end

15 : Polynomial_mt.__unm = function(p) -- -p
16 :  local pp = {}
17 :  for i=1,#p do pp[i] = -p[i] end
18 :  return setmetatable(pp,Polynomial_mt)
19 : end
20 : Polynomial_mt.__sub = function(p1,p2) -- p1 - p2
21 :  p2 = -p2; return p1 + p2
22 : end
23 : Polynomial_mt.__tostring = function(p) -- string
24 :  s = tostring(p[1])
25 :  for i=2,#p do
26 :     s = s.." + ("..tostring(p[i])..")*x^"..tostring(i-1)
27 :  end
28 :  return s
29 : end
30 : Polynomial_mt.__mul = function(p1,p2)  -- p1*p2
31 :  local n1,n2,pp,k,fact = #p1, #p2, {}
32 :  for i=1,n1 do
33 :     k,fact = i, p1[i]
34 :     for j=1,n2 do
35 :        pp[k],k = (pp[k] or 0) + fact*p2[j], k+1
36 :     end
37 :  end
38 :  return setmetatable(pp,Polynomial_mt)
39 : end
40 :
41 : -- Now some examples of Polynomial operations
42 : p1 = Polynomial.new(1,2,3,4)
43 : p2 = Polynomial.new(4,3,2,1)
44 : print(p1); print(p2)
45 : p3 = p1*p2; print(p3)
46 : print(p1-p3)
```

Output:
```
1 + (2)*x^1 + (3)*x^2 + (4)*x^3
4 + (3)*x^1 + (2)*x^2 + (1)*x^3
4 + (11)*x^1 + (20)*x^2 + (30)*x^3 + (20)*x^4 + (11)*x^5 + (4)*x^6
-3 + (-9)*x^1 + (-17)*x^2 + (-26)*x^3 + (-20)*x^4 + (-11)*x^5 + (-
4)*x^6
```

Listing 4.16. Code example for simple definition and operations with polynomial objects.

Polynomial objects. Lines 11 through 15 print the results of various Polynomial operations. Also line 14 illustrates the use of object notation to access a function Polynomial.value() that returns the numerical value of a Polynomial at an input value of the polynomial variable (x = 1.5 in this example). This is equivalent to the code statement `Polynomial.value(p3,1.5)`. One can verify the math operations if desired.

```
1 : -- /* File list4_17.lua */ -- Examples of Polynomial operations
2 :
3 : require"Polynomial"
4 :
5 : p1 = Polynomial.new(1,3,5) -- List of numbers
6 : p2 = Polynomial.new{8,6,4,2} -- Table of numbers
7 : p3 = Polynomial.new(4*p2) -- Another Polynomial object
8 : p4 = p1^3 -- Powers of Polynomials
9 : p5 = Polynomial.new(-2,0,0,0,0,0,1) -- x^6 - 2
10 :
11 : print('p3 =',p3);
12 : print('p2*p1 =',p2*p1)
13 : print('p4/5 =',p4/5)
14 : print('Value of p3(x=1.5) =',p3:value(1.5))
15 :
16 : pq,pr = Polynomial.div(p4,p2)
17 : print('Polynomial division of p4 by p2:');
18 : print('Quotient =',pq); print('Remainder =',pr)
Output:
p3 =   32 + (24)*x^1 + (16)*x^2 + (8)*x^3
p2*p1 =   8 + (30)*x^1 + (62)*x^2 + (44)*x^3 + (26)*x^4 + (10)*x^5
p4/5 =    0.2 + (1.8)*x^1 + (8.4)*x^2 + (23.4)*x^3 + (42)*x^4 +
(45)*x^5 + (25)*x^6
Value of p3(x=1.5) = 131
Polynomial division of p4 by p2:
Quotient =   -39 + (-57.5)*x^1 + (-12.5)*x^2 + (62.5)*x^3
Remainder = 313 + (703)*x^1 + (643)*x^2
```

Listing 4.17. Some code examples of using the Polynomial object module.

Finally on line 16 one sees the code term pq,pr = Polynomial.div(p4,p2). This function implements one interpretation of what is meant by dividing one polynomial by another polynomial. This is another function coded in the Polynomial package and implements polynomial long division which results in a quotient and a remainder. For example consider the division:

$$\frac{a_1 + a_2x + a_3x^2 + \cdots a_n x^{n-1}}{b_1 + b_2x + \cdots b_m x^{m-1}} \tag{4.26}$$

Provided $n > m$ this results in a quotient and a remainder. If $m > n$ then only a remainder term exists and this then equals the numerator polynomial. The long division, which one should have learned in high school, is most easily performed for the coefficient of the leading power of x by rearranging the division as:

$$\frac{(a_n/b_m)x^{n-1} + (a_{n-1}/b_m)x^{n-2} + \cdots (a_1/b_m)}{x^{m-1} + (b_{m-1}/b_m)x^{m-2} + \cdots (b_1/b_m)}$$

$$= \frac{a_n' x^{n-1} + a_{n-1}' x^{n-2} + \cdots a_1'}{x^{m-1} + b_{m-1}' x^{m-2} + \cdots b_1'} \tag{4.27}$$

In this reduction, the coefficient of the leading power of x in the denominator has been reduced to unity by dividing by this leading coefficient. A new set of polynomial coefficients result and these are identified as primed coefficients above. In terms of the new primed coefficients, one step in the division can be performed as:

$$\frac{a_n' x^{n-1} + a_{n-1}' x^{n-2} + \cdots a_1'}{x^{m-1} + b_{m-1}' x^{m-2} + \cdots b_1'} \rightarrow a_n' x^{n-m}$$

$$+ \frac{(a_{n-1}' - a_n' b_{m-1}')x^{n-2} + (a_{n-2}' - a_n' b_{m-2}')x^{n-3} + \cdots a_1'}{x^{m-1} + b_{m-1}' x^{m-2} + \cdots b_1'} \tag{4.28}$$

$$\rightarrow a_n' x^{n-m} + \frac{a_{n-1}'' x^{n-2} + a_{n-2}'' x^{n-3} + \cdots a_1''}{x^{m-1} + b_{m-1}' x^{m-2} + \cdots b_1'}$$

In this equation representation, a new set of polynomial coefficients are identified as double primed coefficients. The highest power of the quotient has been determined in such a first step. The reduction process can proceed with a second step determining the coefficient of the next highest power term in x in the quotient as $a_{n-1}'' x^{n-m-1}$. This proceeds until one has a constant term in the quotient or for n-m steps. This assumes of course that $n > m$.

```
  1 : -- /* File Polynomial.lua */
  2 : -- Polynomial Class and associated functions
  3 :
138 : function Polynomial.div(p1,p2)
139 :     n,m = #p1,#p2
140 :     while p1[n]==0.0 do p1[n],n = nil,n-1 end - Zeros?
141 :     while p2[m]==0.0 do p2[m],m = nil,m-1 end
142 :     if m>n then return Polynomial.new{0}, p1 end
143 :     local a,b = {}, {}; fac =  p2[m]
144 :     for i=1,n do a[i] = p1[i]/fac end; for i=1,m do
            b[i] = p2[i]/fac end
145 :     for i=n,m,-1 do
146 :        for j=1,m-1 do a[i-j] = a[i-j] - a[i]*b[m-j] end
147 :     end
148 :     for i=1,m-1 do b[i] = a[i]*fac end; b[m] = nil
149 :     for j=1,n do if j>n-m+1 then a[j] = nil
            else a[j] = a[j+m-1] end end
150 :     while b[m-1]==0.0 do b[m-1],m = nil,m-1 end
151 :     return setmetatable(a,Polynomial_mt),
            setmetatable(b,Polynomial_mt)
152 : end
```

Listing 4.18. Code segment for polynomial long division with quotient and remainder.

Listing 4.18 shows a code segment (from the Polynomial.lua file) implementing this division process. The code implementing the long division algorithm is contained on lines 145 through 147. The outer loop (i=n,m,-1) performs n-m reduction steps as indicated in Eq. (4.28) while the inner loop (j=1,m-1) performs the subtraction process indicated by the first line of Eq. (4.28). The reduction process can be performed within the same table as the original numerator polynomial. However, in order not to change the original polynomials, copies are made of the polynomials on line 144 before the reduction process. Also the tables are checked for possible leading zero coefficients of the input polynomials on lines 140 and 141 so that the function can never divide by zero. The case of the power of the divisor being larger than the numerator is handled by line 142. Finally, the results are transferred from a single table to two tables (a quotient table and a remainder table) on lines 148 and 149 and any leading zero coefficients in the remainder polynomial are eliminated on line 150.

The returned tables are tagged as Polynomial tables on line 151 so further polynomial operations can be performed on the tables if desired. While this is a simple algorithm, it illustrates some of the checking needed for good programming to handle exceptional cases. One can never be too careful about checking ones input data. The reader is encouraged to experiment with the polynomial functions and especially with the div() function. Try experimenting with a divisor that is known to be a factor of the numerator and observe the results returned by the code.

The discussion will now shift to the problem of finding the roots of a general polynomial and the Polynomial objects discussed so far along with the long division code discussed above are very useful for this. Finding all the roots of a polynomial occurs frequently in real world problems. For example, a linear differential equation of the form:

$$a_1 + a_2 \frac{dy}{dt} + a_3 \frac{d^2 y}{dt^2} + \cdots + a_{n+1} \frac{d^n y}{dt^n}, \tag{4.29}$$

has an associated characteristic equation in s given by:

$$a_1 + a_2 s + a_3 s^2 + \cdots + a_{n+1} s^n = 0. \tag{4.30}$$

The transient response for a differential equation has terms of the form $\exp(s_k t)$ where s_k is a root of the characteristic equation. In another application, a linear system transfer function such as that calculated in the previous section and plotted in magnitude and angle in Figures 4.6 and 4.7 is known to be expressible as a ratio of numerator and denominator polynomials in the complex variable, $s = j\omega$. The roots of the denominator polynomial are again related to the transient response and to the stability of a system. These are but two examples of real world applications for finding the roots of a polynomial of some order n.

Newton's method was briefly explored in Chapter 2 as a means for obtaining some of the roots of a polynomial. It is best applied to finding the largest real roots of a polynomial. However, in general even relatively simple polynomials can have complex roots for orders larger than one. What is needed is a general method for finding all the roots of a polynomial including both real and complex roots. The discussion will be restricted to polynomials with "real" coefficients as this is typical of polynomials resulting from real world problems. If the coefficients of the polynomial are real, any complex roots must occur in complex conjugate pairs so that the multiplication of a complex root factor by the conjugate complex root factor gives a second degree equation with real coefficients. This observation forms the basis for a root finding method develop by Bairstow (in 1914). In this algorithm one seeks to obtain quadratic factors of a polynomial reducing the order of the remaining polynomial by 2 for each quadratic factor found. Of course roots of polynomials of order 1 and 2 can be easily found so the method need only be applied to polynomials of order 3 and higher. After each quadratic factor is found, the two roots corresponding to the quadratic factor are obtained by application of the quadratic equation.

To better understand the Bairstow approach, consider a specific fourth order polynomial such as:

$$P(x) = (1-x)(2-x)(3-x)(4-x)$$
$$= 24 - 50x + 35x^2 - 10x^3 + x^4 \quad , \tag{4.31}$$

which has the obvious roots of 1, 2, 3 and 4. Consider dividing $P(x)$ by a second order quadratic term of the form $x^2 + C_1 x + C_2$. Now if the second order term is a factor of $P(x)$, a result polynomial of order 2 less than $P(x)$ will be obtained with a zero remainder term. However, if the term being divided is not a factor of $P(x)$, complete cancellation of all the terms will not occur and a nonzero remainder term will result. Applying long division as discussed above for a quadratic divisor term gives a result of the form:

$$\frac{a_{n+1}x^n + a_n x^{n-1} + \cdots a_2 x + a_1}{x^2 + C_1 x + C_2} = b_{n-1}x^{n-2} + b_{n-2}x^{n-3}$$
$$+ \cdots b_1 + \frac{R_1(C_1, C_2)x + R_2(C_1, C_2)}{x^2 + C_1 x + C_2} \tag{4.32}$$

The remainder as represented by the last fraction above will consist of two terms in x and these terms R_1 and R_2 are shown here as functions of the divisor quadratic coefficients C_1 and C_2. If these terms are of such value that the denominator quadratic is a proper factor, then R_1 and R_2 will both be zero. As a more complete example if one takes $C_1 = -3$ and $C_2 = 2$, then a perfect factor of Eq. (4.31) results. Figure 4.10 shows how the two residuals change as C_1 is held at -3 and C_2 is varied from 1 to 3 for this specific polynomial. It is seen that over this narrow range, the residuals vary almost linearly with the coefficient value and that both residuals go to zero at the value of 2 which is exactly the same point where the quadratic factor is an exact factor of the equation.

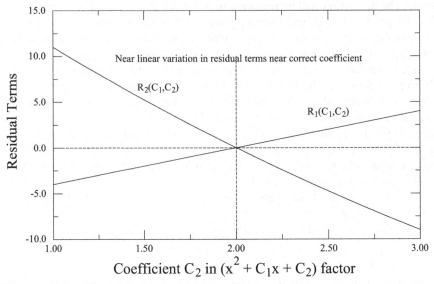

Figure 4.10 Residual variations with C_2 coefficient.

Bairstow's method is based upon the above observations. In this method, one performs division of the polynomial by a quadratic factor with some guess at the coefficients and examines the residual terms. The coefficient values are then changed in some way to obtain an improved solution that makes the residuals closer to zero. This is continued until a solution is obtained with the residuals forced as close as possible to zero. This is a ready made problem for the nsolv() routine. The problem specification has two parameters R_1 and R_2 that are functions of two variables, C_1 and C_2 and values of these are needed which force the function values to zero. As a first step in this approach, an algorithm is needed to divide one polynomial by a quadratic factor and this is what has just been discussed. Code for this is relatively easy as shown in Listing 4.18.

Code for polynomial long division can now be combined with nsolv() to implement Bairstow's method as shown in Listing 4.19. This code is included in the Polynomial package as the function Polynomial.roots() function. Some of the major features of the code will now be discussed. The code for defining the two nonlinear equations to be solved by nsolv() for the coefficients C_1 and C_2 is on lines 175 through 179. For this application there is no explicit equation for generating the equations to be solved. The functional relationship to C_1 and C_2 is generated by a computer algorithm in the Pfactor() function called on lines 176 or 177 with the two Residuals extracted and returned by line 178. The Pfactor() function on lines 168 through 174 performs the polynomial long division given as input a polynomial (p1) and the two coefficients for the denominator polynomial. This code is equivalent to that of the Polynomial.div() function of Listing 4.18 but differs in some aspects as it is customized for only a second order denominator

```
  1 : -- /* File Polynomial.lua */
  2 : -- Polynomial Class and associated functions
153 : function Polynomial.roots(p) -- Solves for roots of p
154 :     if type(p)=='PolyR' then -- Ratios of polynomials
155 :         return {Polynomial.roots(p[1]),Polynomial.roots(p[2])}
                end
156 :     local NMAX = getfenv(nsolv).NMAX -- Maximum #of iterations
157 :     local roots,pr = {}, {} -- Collect roots, local copy of p
158 :     local n, m, f1, pol = #p, 1
159 :     local CMP,REL,typ,nmx,iroot = -1,1,1,1,0
160 :     while p[n]==0 and n>2 do n=n-1 end -- Check for enc zero's
161 :     while p[nmx]==0 and nmx<n-2 do -- Check for initial zeros
162 :         iroot = iroot+1
163 :         roots[iroot], nmx = 0, nmx+1 -- Zero roots
164 :     end
165 :     for i=nmx,n do pr[m],m = p[n-m+1],m+1 end -- Reverse p
166 :     n = n-nmx+1
168 :     local function Pfactor(p1,c1,c2)
169 :         local pr = {} -- leave p1 unchanged
170 :         pr[1],pr[2] = p1[1], p1[2]-c1*p1[1]
171 :         for i=3,n-1 do pr[i]=p1[i]-c1*pr[i-1] - c2*pr[i-2] end
172 :         pr[n] = (p1[n] - c2*pr[n-2])
173 :         return pr -- residuals in n-1 and n terms
174 :     end
175 :     local function eqs(y,c) -- Define function to force to zero
176 :         if typ==REL then pol = Pfactor(pr, c[1]+c[2],c[1]*c[2])
177 :         else pol = Pfactor(pr,2*c[1],c[1]^2+c[2]^2) end--Complex
178 :         y[1],y[2] = pol[n-1], pol[n] -- Force remainder to zero
179 :     end
181 :     while n>3 do -- solve for quadtatic factors
182 :         c = {c0[1],c0[2]}; nmx = nsolv(eqs,c)
183 :         if nmx==NMAX then
184 :             typ,c = -typ, {c0[1],c0[2]}
185 :             nmx = nsolv(eqs,c) -- Reverse real & complex
186 :             if nmx==NMAX then -- Didn't succeed for roots
187 :                 print('Iterations exceed NMAX of ',NMAX, ' in
                        poly_root')
188 :                 print('At root number', iroot+1)
189 :                 print('Try using "getfenv(nsolv).NMAX =
                        ',2*NMAX,'"')
190 :             end
191 :         end
192 :         if typ==REL then pr = Pfactor(pr, c[1]+c[2],c[1]*c[2])
193 :         else pr = Pfactor(pr,2*c[1],c[1]^2+c[2]^2) end--Complex
194 :         n,iroot = n-2, iroot+1
195 :         if typ==CMP then roots[iroot],roots[iroot+1]=
                    -c[1] + j*c[2], -c[1] - j*c[2]
196 :         else roots[iroot],roots[iroot+1] = -c[1], -c[2] end
197 :         iroot = iroot+1; pr[n+2], pr[n+1] = nil, nil -- Reduce 2
198 :     end
199 :     iroot = iroot + 1 -- Final roots, 2 or 1
200 :     if n>2 then -- Quadratic factor left
201 :         f1 = pr[2]^2 - 4*pr[1]*pr[3]
202 :         if f1>=0 then f1=sqrt(f1) else f1 = j*sqrt(-f1) end
203 :         if pr[2]>0 then f1=-0.5*(pr[2] + f1) else
                    f1=-0.5*(pr[2] - f1) end
204 :         roots[iroot] = f1/pr[1]
205 :         iroot = iroot + 1; roots[iroot] = pr[3]/f1
206 :     else roots[iroot] = -pr[2]/pr[1]   end -- Linear factor left
```

```
207 :      return roots
208 : end
209 : setfenv(Polynomial.roots,{sqrt=math.sqrt,nsolv=nsolv,
           j=Complex.j,getfenv=getfenv,table=table,
210 :      Polynomial=Polynomial,print=print,type=type,c0={1,-1}})
```

Listing 4.19. Code for obtaining polynomial roots using Bairstow's method and nsolv().

divisor. For such a second order polynomial divisor, the quotient and remainder coefficients can be evaluated by a single loop as given on line 171. Also no checks on the coefficients are needed as these are done elsewhere in the roots(0 function.

The code also doesn't look exactly like the Bairstow method described above as one of two different Pfactor() function calls are made depending on the value of a "typ" parameter – see lines 176 and 177. While the simple Bairstow method works well with some polynomials, it fails with other polynomials. Cases where it fails involve polynomials with both real and complex roots. To understand some of the shortcomings, consider a polynomial with a real root at $x = r_1$ and a complex root at $x = r_2 + i_2$ where i_2 is small compared with r_2. A quadratic factor formed with these roots will look like:

$$(x - r_1)(x - r_2 + ji_2) = x^2 - (r_1 + r_2 - ji_2)x + (r_1 r_2 - jr_1 i_2)$$
$$= x^2 + c_1 x + c_2$$

(4.33)

When i_2 is small, the quadratic coefficients will be approximately real with values,

$$c_1 \cong -(r_1 + r_2) \text{ and } c_2 \cong r_1 r_2.$$

(4.34)

In such a case, it is possible for Bairstow's method to get close to such a quadratic factor but never be able to extract the factor, since the quadratic coefficients are assumed to always be positive. Thus the algorithm can get into a local minimum in the residuals near a factor consisting of a complex root and a real root and never be able to converge to a proper quadratic factor.

A way around the above difficulty is to force the Bairstow method to search for quadratic factors consisting of either a pair of real roots or a pair of complex conjugate roots. For complex conjugate roots, the form of the quadratic must be

$$(x - r + ji)(x - r + ji) = x^2 - 2rx + (r^2 + i^2)$$
$$= x^2 + 2c_1 x + (c_1^2 + c_2^2)$$

(4.35)

For this case one can take c_1 and c_2 to be the real and imaginary parts of the solution (with negative sign) and take $C_1 = 2c_1$ and $C_2 = c_1^2 + c_2^2$. This is the form coded on line 177 of Listing 4.19. For a pair of real roots, the form of the quadratic equation must be

$$(x - r_1)(x - r_2) = x^2 - (r_1 + r_2)x + r_1 r_2$$
$$= x^2 + (c_1 + c_2)x + c_1 c_2$$

(4.36)

In this case one can take c_1 and c_2 to be the negative of the two real roots and take $C_1 = c_1 + c_2$ and $C_2 = c_1 c_2$. This form is coded on line 176 of the listing. Thus if these forms of the coefficients are used, one can be assured that either a pair of complex roots or a pair of real roots is extracted and that the algorithm never gets into a local minimum near a real and complex root pair. These two forms of the polynomial factors are implemented on lines 176 and 177 of the code depending on the state of the typ variable. The routine nsolv() called on lines 182 or 185 doesn't care if the functional relationship in generated by an explicit mathematical equation or generated by some algorithm as long as it is a smooth, continuous function of the variables.

Line 160 of the code eliminates table entries that are zero for the highest powers of the polynomial. Lines 162 through 164 of the code handle one or more zero roots which mean that the constant term would be missing in the polynomial. The heart of Bairstow's method is on lines 182 through 191. Line 182 calls nsolv() with the two equations to be solved and some initial guess at the roots. If nsolv() returns with the maximum number of iterations, it has not been able to find a solution pair. Then line 184 reverses the sign of the typ variable and switches from looking for pairs of real roots to pairs of complex roots or vice versa. When nsolv() converges at line 182 or 185, it returns with appropriate values of C_1 and C_2 for either two real roots or two complex roots. The remainder of the code again calls Pfactor() on line 192 or 193 with the final coefficients to update the quotient polynomial and then calculates the two found roots using the C_1 and C_2 factors (lines 195 and 196). Then the algorithm repeats the while loop from line 181 to 198 for a new quadratic factor if needed. Finally the program calculates the final root (or roots) when the equation is of order 2 or lower on lines 200 through 206.

The initial guess of C coefficients on lines 182 and 210 needs a little discussion. As this is a general root finding routine, one in general has no knowledge as to what the appropriate coefficients might be in either magnitude or sign. Line 210 simply sets the initial guess to c0 = {1, -1}, the simplest possible means of starting the algorithm. This approach appears to be robust over a wide range of polynomials. So we apply the "use the simplest approach that works" principle. However, a possible improvement is provided for the user, as the user can change this initial value through the c0 coefficient to any desired value by using a code statement of the form `getfenv(Polynomial.roots).c0 = {#1, #2}` with any best guess at the values (#1 and #2).

With the polynomial root finding code, it is very easy to define a polynomial and find its roots. Some example code using the root finding routine is shown in Listing 4.20. The listing shows various examples of these polynomial mathematical operations plus root finding and printing of polynomials. Only selected output is shown from the example code. The reader is encouraged to execute the code and observe the entire output. The roots of a polynomial may be found with the code statement `Polynomial.roots(p)` where p is a table of coefficient values, $p = \{a_1, a_2, \cdots a_{n+1}\}$ or a data structure defined by Polynomial.new(). If p is a Polynomial data structure then one can use the form p:roots(). Examples of polynomial multiplication can be seen on lines 17 and 43.

```
 1 : -- /* File list4_20.lua */
 2 : -- Examples of Roots of Polynomials
 3 : require "Polynomial"
 4 : getfenv(nsolv).NMAX=200
 5 :
 6 : p0 = Polynomial.new{1,1,4} -- 1 + x + 4x^2
 7 : rts = p0:roots() -- Or Polynomial.roots(p0)
 8 : print("Roots of p0 polynomial\n",p0)
 9 : table.foreach(rts,print)
10 :
11 : p1 = Polynomial.new{1,2,3,4,5} -- 1 + 2x + 3x^2 + 4x^3 + 5x^4
12 : rts = Polynomial.roots(p1)
13 : print("\nRoots of p1 polynomial\n",p1)
14 : table.foreach(rts,print)
15 :
16 : p2 = Polynomial.new{-1,1}
17 : p2 = p2*{-2,1}*{-3,1}*{-4,1} -- roots of 1,2,3,4
18 : rts = p2:roots()
19 : print("\nRoots of p2 polynomial\n",p2)
20 : table.foreach(rts,print)
21 :
     .......
31 :
32 :
33 : p2 = Polynomial.scale(p1,100000)
34 : rts = Polynomial.roots(p2)
35 : print("\nPolynomial p1 scaled by factor of 1e5\n",p2)
36 : table.foreach(rts,print)
37 :
38 : p3 = p1:scale(0.001)
39 : rts = p3:roots()
40 : print("\nPolynomial p1 scaled by factor of 1e-3\n",p3)
41 : table.foreach(rts,print)
42 :
43 : p1 = p2*p3
44 : rts = Polynomial.roots(p1)
45 : print("\nProduct of scaled polynomials\n",p1)
46 : table.foreach(rts, print)
Selected Output:
Roots of p1 polynomial
    1 + (2)*x^1 + (3)*x^2 + (4)*x^3 + (5)*x^4
1   (-0.53783227490299) + j(-0.35828468634513)
2   (-0.53783227490299) + j(0.35828468634513)
3   (0.13783227490299) + j(0.67815438910534)
4   (0.13783227490299) + j(-0.67815438910534)
.......
Product of scaled polynomials
    1 + (2000.00002)*x^1 + (3000000.04)*x^2 + (4000000060)*x^3 +
(5000000080000)*x^4 + (100000001.2)*x^5 + (1500.000016)*x^6 +
(0.0200000002)*x^7 + (2.5e-007)*x^8
1   (-0.00053783227490299) + j(-0.00035828468634513)
2   (-0.00053783227490299) + j(0.00035828468634513)
3   (0.00013783227490299) + j(-0.00067815438910534)
4   (0.00013783227490299) + j(0.00067815438910534)
5   (13783.227490243) + j(-67815.438910563)
6   (13783.227490243) + j(67815.438910563)
7   (-53783.227490243) + j(-35828.468634351)
8   (-53783.227490243) + j(35828.468634351)
```

Listing 4.20. Some examples of polynomial algebra and polynomial roots

The first example output is for the four roots of the polynomial:

$$p1 = 1 + 2x + 3x^2 + 4x^3 + 5x^4. \tag{4.37}$$

This equation has four complex roots as shown in the output listing. Using the polynomial scaling feature, this polynomial is x-scaled by factors of 1.e5 and 1.e-3 on lines 33 and 38. Scaling by a factor F means that the x range is changed by the factor F and thus the roots are scaled by 1/F so that the scaled polynomials have roots 1.e5 times smaller and 1.e3 times larger than before scaling. Multiplying the two scaled polynomials as on line 43 then gives an eighth order polynomial with four large roots and four small roots and with the four large roots eight orders of magnitude larger than the four small roots. The resulting eighth order polynomial is:

$$1 + (2000.00002)*x^{\wedge}1 + (3000000.04)*x^{\wedge}2 + (4000000060)*x^{\wedge}3 +$$
$$(5000000080000)*x^{\wedge}4+ (100000001.2)*x^{\wedge}5 + (1500.000016)*x^{\wedge}6 + \tag{4.38}$$
$$(0.0200000002)*x^{\wedge}7 + (2.5e\text{-}007)*x^{\wedge}8$$

This provides a more severe test for the root fining technique since this equation has a large range in magnitude of the coefficients (18 orders of magnitude) plus coefficients needing about 10 digits of precision. The root values printed in the output under the heading "Product of scaled polynomials" are in fact very accurate values as can be seen by comparing with the un-scaled roots. The polynomial root finding routine has been very successful in handling roots which scale over at least eight orders of magnitude.

This provides a reasonably severe test for the root finding technique, since one can see a large range in magnitude of the coefficients (18 orders of magnitude) plus coefficients needing about 10 digits of precision. All of this is obtained with the simple guess of {1,-1} for the quadratic factor coefficients.

The limitations of computer precision can rapidly come into play in seemingly simple polynomial expressions. For example consider a polynomial with roots of 1,2,3,...20 formed in the following manner:

$$p20 = (-1+x)(-2+x)(-3+x) \cdot \cdot (-20+x)$$
$$= 2.4329e018 + (-8.7529e018)x + (1.3803e019)x^2 \cdot \cdot + x^{20} \tag{4.39}$$

The coefficients range over 19 orders of magnitude (20! as the constant value) and exceed the precision of standard double precision floating point numbers. Thus such a simple looking polynomial can not be exactly represented in our double precision computer language. One would thus expect problems with finding the roots of such a polynomial. Listing 4.21 shows the code to generate this 20th order polynomial and the use of roots() to calculate the 20 roots. The answers are surprisingly good and accurate to almost 4 digits which would be good for most engineering applications. The least accurate of the roots is the 13.98589821031 value which is in error by about 0.1%. This is about as far as one can push the accuracy and precision of the root finding function. As with any numerical technique, results are always limited by the precision of the computer programs and machine accuracy used in the numerical modeling. When the range of the polynomial coefficients exceeds about 16 digits for double precision calculations one should be especially aware of possible errors in accuracy.

```
 1 : -- /* File list4_21.lua */
 2 : -- Examples of Roots of 20th order Polynomial
 3 :
 4 : require "Polynomial"
 5 :
 6 : p= Polynomial.new{-1,1}*{-2,1}*{-3,1}*{-4,1}*{-5,1}*{-6,1}
 7 : p = p*{-7,1}*{-8,1}*{-9,1}*{-10,1}*{-11,1}*{-12,1}*{-13,1}
 8 : p = p*{-14,1}*{-15,1,}*{-16,1}*{-17,1}*{-18,1}*{-19,1}*{-20,1}
 9 : print(p)
10 :
11 : rts = p:roots()
12 : for i=1,#rts do
13 :   rtt = math.floor(rts[i]+.5)
14 :   print(i,rtt,rts[i],(rts[i]-rtt)/rtt)
15 : end
Output:
2.4329020081766e+018 + (-8.7529480367616e+018)*x^1 +
(1.3803759753641e+019)*x^2 + (-1.2870931245151e+019)*x^3 +
(8.0378118226451e+018)*x^4 + (-3.5999795179476e+018)*x^5 +
(1.2066478037804e+018)*x^6 + (-3.1133364316139e+017)*x^7 +
(6.3030812099295e+016)*x^8 + (-1.0142299865511e+016)*x^9 +
(1.3075350105404e+015)*x^10 + (-1.3558518289953e+014)*x^11 +
(11310276995381)*x^12 + (-756111184500)*x^13 + (40171771630)*x^14 +
(-1672280820)*x^15 + (53327946)*x^16 + (-1256850)*x^17 + (20615)*x^18
+ (-210)*x^19 + (1)*x^20
1   2   1.9999999999984    -7.8004269710163e-013
2   1   1                   1.2878587085652e-014
3   3   3.0000000001636     5.4540224188789e-011
4   4   3.9999999960608    -9.8480890109442e-010
5   9   9.0001752023672     1.946692969194e-005
6   5   5.0000000486928     9.7385598252231e-009
7   10  9.999239563056     -7.6043694404859e-005
8   6   5.9999995846424    -6.9226261760017e-008
9   7   7.0000036145675     5.163667885658e-007
10  8   7.9999709809809    -3.6273773925677e-006
11  11  11.002427950026     0.00022072272962426
12  16  15.988849566895    -0.00069690206909256
13  12  11.99426531539     -0.00047789038413981
14  20  19.99994665196     -2.6674020043416e-006
15  14  13.985898210313    -0.0010072706919275
16  13  13.010406318604     0.00080048604643556
17  19  19.000512718568     2.6985187807626e-005
18  15  15.014520881093     0.00096805873955645
19  18  17.997727703837    -0.00012623867572654
20  17  17.006055692784     0.00035621722261639
```

Listing 4.21. Roots of 20th order polynomial

This completes the examples of polynomial roots. In summary several observations can be made. A rather robust polynomial root finding technique based upon Bairstrow's technique has been developed and demonstrated by applying the technique to several simple polynomials and to two somewhat difficult polynomial equations. It has been demonstrated that the technique can be used with roots of very different magnitudes, such as eight orders of magnitude in one example. The code also provides a warning in case the specified internal accuracy in root finding is not achieved giving one an opportunity to increase the maximum number of iterations for convergence or to provide a more accurate initial approximation to roots. This is always good programming practice to anticipate problems and alert the user when potential problems arise. This has all been achieved with the use of a single initial guess for the roots which will typically be greatly in error. In specific applications the algorithm could perhaps be improved with a more accurate initial guess. However, as the examples demonstrate the present algorithm is reasonably robust and converges to the desired solutions for a wide range of polynomials. The root finding method builds upon a previously developed function for finding roots of nonlinear functions.

The next section will discuss general numerical operations performed using matrix notation and discuss the development of a set of coded routines for matrix manipulation. These approaches builds upon the solution techniques developed for solving linear equations in Section 4.1.

4.5 Matrix Equations and Matrix Operations

In Section 4.1 the matrix notation was first introduced for writing a set of linear equations. However, the discussion there rapidly moved to a discussion of linear equations and the solution techniques developed were based upon Gauss elimination techniques and did not directly use matrix manipulation techniques. For a number of engineering problems the manipulation of matrix equations is the most convenient approach so this section will discuss and develop some numerical techniques for direct matrix manipulation. Going back to the matrix formulation of a set of linear equations (as Eq. (4.5)) one has:

$$\begin{bmatrix} a_{11} & a_{12} & \ldots & a_{1n} \\ a_{21} & a_{22} & \ldots & a_{2n} \\ & & \vdots & \\ a_{n1} & a_{n2} & \ldots & a_{nn} \end{bmatrix} \begin{bmatrix} x_1 \\ x_2 \\ \vdots \\ x_n \end{bmatrix} = \begin{bmatrix} b_1 \\ b_2 \\ \vdots \\ b_n \end{bmatrix} \qquad (4.40)$$

In simplified matrix form this is $[A][X] = [B]$ or more simply $\mathbf{AX} = \mathbf{B}$. Matrix algebra is a well established branch of mathematics and will not be discussed in detail here. The reader is assumed to be familiar with basic matrix manipulations and with definitions and if not the reader is encouraged to consult any text on linear algebra. One such "formal" solution of a linear system of equations is expressed in terms of the inverse of the $\mathbf{A}$ matrix as:

$$\mathbf{A}^{-1}\mathbf{AX} = \mathbf{X} = \mathbf{A}^{-1}\mathbf{B}. \qquad (4.41)$$

Matrix notation and matrix manipulation is not used extensively in this book. However, there are a few cases where being able to numerically manipulate matrices is very useful. One such case will be encountered in a subsequent chapter on solving coupled differential equations.

```
 1 : --/* File list4_22.lua */
 2 :
 3 : require "Matrix" -- Load matrix manipulation code
 4 :
 5 : --Tests of matrix operations
 6 : ax = Matrix.new({{1,1,1},{2,2,1},{3,3,1}})
 7 : ay = Matrix.new({{3,1,0},{1,-2,1},{-1,2,-1}})
 8 : az = ax+ay -- Matrix addition
 9 : print("Type of az is: ",type(az))
10 :
11 : az = ax*ay -- Matrix multiplication
12 : print('Matrix az is:'); print(az)
13 : a = Matrix.new{ -- New matrix definition
14 :   {1, 0.1, 0, 0, 0},
15 :   {0.1, 1, 0.1, 0, 0},
16 :   {0, 0.1, 1, 0.1, 0},
17 :   {0 ,0, 0.1, 1, 0.1},
18 :   {0, 0, 0, 0.1, 1}}
19 : m,n = Matrix.size(a) -- Obtain size of matrix a
20 : print("Size of matrix a is: ",m,n)
21 : -- Take inverse of matrix
22 : b = a^-1 -- Or b = Matrix.inv(a) or b = a:inv()
23 : print(b,'\n')
24 : print(b*a) -- Should be unity diagonal matrix
Output:
Type of az is:       Matrix
Matrix az is:
{{3, 1, 0},
{7, 0, 1},
{11, -1, 2}}
Size of matrix a is:     5   5
{{1.01020514, -0.102051442, 0.0103092783, -0.00104134124,
0.000104134124},
{-0.102051442, 1.02051442, -0.103092783, 0.0104134124, -
0.00104134124},
{0.0103092783, -0.103092783, 1.0206185, -0.103092783, 0.0103092783},
{-0.00104134124, 0.0104134124, -0.103092783, 1.02051442, -
0.102051442},
{0.000104134124, -0.00104134124, 0.0103092783, -0.102051442,
1.01020514}}
{{1, 3.68628738e-018, -5.01443504e-019, 3.13402190e-019, 2.71050543e-
020},
{-1.38777878e-017, 1, -5.42101086e-018, -2.26327203e-018, -
2.1684043e-019},
{0, -1.38777878e-017, 1, 1.75640751e-017, 1.73472347e-018},
{0, 1.73472347e-018, 1.38777874e-017, 1, 0},
{1.35525271e-020, -2.1684043e-019, -1.73472347e-018, -1.38777878e-
017, 1}}
```

Listing 4.22. Some basic matrix operations and resulting output.

For matrix operations, a package of code segments has been developed and is available for this work in Matrix.lua. Listing 4.22 shows some simple code examples for performing several basic matrix operations. Line 3 (require"Matrix") loads a series of programming code segments that define a set of basic matrix operations. The package of functions provide code for defining new matrix data structures (with Matrix.new()) plus matrix math operations of addition, subtraction, multiplication, negation and calculating integer powers of a matrix. Other matrix operations are provided by the package and these will be subsequently discussed. The code for the basic matrix operations are not discussed here as the operations are well known and relatively straightforward. The interested reader can gain considerable insight into programming in the Lua language by examining the code for the matrix operations. The approach here is to use the matrix operations as needed to solve nonlinear problems.

In Listing 4.22 new matrix objects are created on lines 6, 7 and 13 of the listing by enclosing the elements of each row of the matrix in braces {} and then by enclosing the rows in outside braces {}. This is a natural way of listing a matrix as shown by lines 13 through 18. Note that when a table is passed to a function as the only argument, it does not need to be enclosed in parentheses as exhibited in the Matrix.new() function call on line 13. The example code shows matrix addition (line 8) and multiplication (line 24) as well as printed output (lines 12 and 24) of matrices. These are all made possible by the require"Matrix" statement of line 3 which brings in the basic matrix definitions. When a new matrix is defined, a metatable is associated with the matrix such that the normal operations of addition, multiplication, etc. are properly defined and performed for the associated matrices. Line 22 illustrates taking the inverse of a matrix through the statement a^-1 which is similar to the normal matrix notation of Eq. (4.41). The inverse can also be obtained by the statements Matrix.inv(a) or a:inv(). A check on the inverse operation is provided by line 24 where the result of multiplying the matrix and its inverse is printed. The resulting matrix should be a unity diagonal matrix with all off-diagonal elements zero. The resulting printed output shows unity along the diagonal elements and small residual errors for the off-diagonal elements. The largest off-diagonal elements are of the order of 1.e-17 which represents the inherent error in double precision numerical calculations.

One of the classical methods for solving a set of linear equations such as Eq. (4.40) is to first convert the $\mathbf{A}$ matrix into the product of two special matrices $\mathbf{LU}$ where the $\mathbf{L}$ matrix is a lower triangular matrix and $\mathbf{U}$ is an upper triangular matrix with the forms:

$$\mathbf{L} = \begin{bmatrix} l_{11} & 0 & 0 & \cdots & 0 \\ l_{21} & l_{22} & 0 & \cdots & 0 \\ l_{31} & l_{32} & l_{33} & \cdots & 0 \\ \vdots & \vdots & \vdots & \cdots & \vdots \\ l_{n1} & l_{n2} & l_{n3} & \cdots & l_{nn} \end{bmatrix} \quad \mathbf{U} = \begin{bmatrix} u_{11} & u_2 & u_{13} & \cdots & u_{1n} \\ 0 & u_{22} & u_{23} & \cdots & u_{2n} \\ 0 & 0 & u_{33} & \cdots & u_{3n} \\ \vdots & \vdots & \vdots & \cdots & \vdots \\ 0 & 0 & 0 & \cdots & u_{nn} \end{bmatrix} \quad (4.42)$$

It can be seen that there are $n^2 + n$ terms in these matrices while there are only n^2 terms in the square $\mathbf{A}$ matrix. Thus one is free to select the value of n of the terms in these two matrices. A conventional approach is to set all the diagonal elements of either the $\mathbf{L}$ or $\mathbf{U}$ matrix to unity (known as the Doolittle or Crout method). In this work the $\mathbf{L}$ diagonal elements will be taken as unity ($l_{jj} = 1$).

When multiplying out the matrix terms and setting the results equal to the a_{ij} terms one finds that the elements can be conveniently calculated on a column by column basis beginning with row 1 and proceeding to row n. One also finds that each element of the **A** matrix is used only once in the procedure so that the results of the decomposition can be stored directly back into the **A** matrix. The diagonal elements of the **L** matrix are not stored since they are taken as unity. The procedure for evaluating the elements can be summarized as:

For each j = 1,2,... n do the following steps (on each column)
For each i = 1,2,...j solve for u_{ij} as

$$u_{ij} = a_{ij} - \sum_{k=1}^{i-1} l_{ik} u_{kj} \;\rightarrow\; a_{ij} = a_{ij} - \sum_{k=1}^{i-1} a_{ik} a_{kj} \qquad (4.43)$$

For i = 1 the summation term is taken to mean zero. In the second form, the sum has been shown over the in-place **A** matrix elements.
For each i = j+1,j+2,...n solve for l_{ij} as

$$l_{ij} = \frac{1}{u_{jj}} \left[a_{ij} - \sum_{k=1}^{j-1} l_{ik} u_{kj} \right] \;\rightarrow\; a_{ij} = \frac{1}{a_{jj}} \left[a_{ij} - \sum_{k=1}^{j-1} a_{ik} a_{kj} \right] \qquad (4.44)$$

Again the second form illustrates the fact that all the new elements are stored in the **A** matrix.

Pivoting is needed for a stable decomposition method and again only partial pivoting with the swapping of rows can be efficiently implemented. With pivoting the **LU** decomposition is not exactly on the **A** matrix by on a row-wise permutation of the matrix. One final detail can be implemented with respect to pivoting and the second calculation above. Since all the elements in the second step are divided by the diagonal element, one can perform all the calculations of the elements in a column of the **L** matrix and after the calculations determine which element has the largest value. This row can then be promoted to the diagonal element row and the division by the diagonal element then carried out. This is sometimes called implicit pivoting. With row exchanges, one must keep track of the exchanges if this is to be used to solve a system of equations so that for use with some **B** matrix, this matrix can also have its rows exchanged.

A code segment for implementing the **LU** decomposition is shown in Listing 4.23. This is part of the code in the Matrix package. The first loop from line 146 to 154 collects scaling information by finding the largest element in each row of the matrix. The summations shown in Eqs. (4.43) and (4.44) are readily seen on lines 158 and 164 of the listing. It should be noted that the "for" loops perform as desired when the upper limit is less than the lower limit. For example consider the loop on lines 156 through 160. When j = 1 this loop becomes for i=1,0 ... end with the upper limit less than the beginning value. For such cases, the loop will not be executed as desired in the algorithm of Eq. (4.43). The row with the largest element is identified by lines 165 and 166 and used in lines 168 through 171 for pivoting if needed. Information on row swapping is collected in the indx[] array on line 172. Finally the function returns on line 179 the **LU** decomposition stored in the original **A** matrix along with the index table and a final integer of +

or – unity indicating whether the number of swapped rows in even or odd. This can be used in a routine to calculate the determinant of the matrix.

```lua
  1 : -- /* File Matrix.lua */
  2 : -- Package for Matrix algebra and functions -2D matrices
....
141 : function Matrix.LUdecompose(a,n) -- LU decomposition
142 :     local d,TINY = 1,1.e-100
143 :     local imax,big,dum,sum,temp
144 :     local vv,indx = {},{}
145 :     n = n or #a
146 :     for i=1,n do -- Loop over rows for scaling informatio
147 :         big = 0
148 :         for j=1,n do -- Find largest element in row j
149 :             temp = abs(a[i][j])
150 :             if temp > big then big = temp end
151 :         end
152 :         if big==0 then print("Singular martix in LUdecompose") end
153 :         vv[i] = 1/big -- Scale factor
154 :     end
155 :     for j=1,n do -- This is the main loop for Crout's method
156 :         for i=1,j-1 do
157 :             sum = a[i][j]
158 :             for k=1,i-1 do sum = sum - a[i][k]*a[k][j] end
159 :             a[i][j] = sum
160 :         end
161 :         big = 0
162 :         for i=j,n do
163 :             sum = a[i][j]
164 :             for k=1,j-1 do sum = sum - a[i][k]*a[k][j] end
165 :             a[i][j],dum = sum, vv[i]*abs(sum)
166 :             if dum>= big then big,imax = dum,i end
167 :         end
168 :         if j~=imax then -- Interchange rows
169 :             a[imax],a[j] = a[j],a[imax]
170 :             vv[imax],d = vv[j],-d
171 :         end
172 :         indx[j] = imax
173 :         if a[j][j]==0 then a[j][j] = TINY end - Singular?
174 :         if j~=n then -- Divide by pivot element
175 :             dum = 1/a[j][j]
176 :             for i=j+1,n do a[i][j] = a[i][j]*dum end
177 :         end
178 :     end
179 :     return a,indx,d -- LU matrix, interchange table and sign
180 : end
```

Listing 4.23. Code segment for **LU** matrix decomposition.

Obtaining the **LU** decomposition of a matrix is rarely a goal in itself (except perhaps for pure mathematicians). For one possible application, consider the solution of a set of linear equations as:

$$\mathbf{Ax} = \mathbf{B} \rightarrow \mathbf{LUx} = \mathbf{B}$$

$$\text{Then define: } \mathbf{Ux} = \mathbf{D} \rightarrow \mathbf{LD} = \mathbf{B} \tag{4.45}$$

The problem of finding a solution then reduces to that of solving two equations, first for **D** from **LD = B** and then x from **Ux = D**. Because of the triangular nature of the **L** and **U** matrices these solutions are readily obtained. For the two equations one has the form:

$$\begin{bmatrix} 1 & 0 & 0 & \cdots & 0 \\ l_{21} & 1 & 0 & \cdots & 0 \\ l_{31} & l_{32} & 1 & \cdots & 0 \\ \vdots & \vdots & \vdots & \cdots & \vdots \\ l_{n1} & l_{n2} & l_{n3} & \cdots & 1 \end{bmatrix} \begin{bmatrix} d_1 \\ d_2 \\ d_3 \\ \vdots \\ d_n \end{bmatrix} = \begin{bmatrix} b_1 \\ b_2 \\ b_3 \\ \vdots \\ b_n \end{bmatrix}$$ (4.46)

$$\begin{bmatrix} u_{11} & u_{12} & u_{13} & \cdots & u_{1n} \\ 0 & u_{22} & u_{23} & \cdots & u_{2n} \\ 0 & 0 & u_{33} & \cdots & u_{3n} \\ \vdots & \vdots & \vdots & \cdots & \vdots \\ 0 & 0 & 0 & \cdots & u_{nn} \end{bmatrix} \begin{bmatrix} x_1 \\ x_2 \\ x_3 \\ \vdots \\ x_n \end{bmatrix} = \begin{bmatrix} d_1 \\ d_2 \\ d_3 \\ \vdots \\ d_n \end{bmatrix}$$ (4.47)

Eq. (4.46) is readily solved by forward substitution beginning with the first row and proceeding to the last row:

$$d_j = b_j - \sum_{k=1}^{j-1} l_{jk} b_k$$ (4.48)

Following this, Eq. (4.47) is readily solved by backwards substitution beginning with the last row and proceeding to the first row:

$$x_j = \left(d_j - \sum_{k=j+1}^{n} u_{jk} d_k \right) / u_{jj}$$ (4.49)

Thus **LU** decomposition provides an alternative method to Gauss elimination previously discussed for solving a set of coupled equations. If one has only a single set of equations to be solved once, there is little advantage of the **LU** decomposition. The major advantage of the method occurs when one needs to solve a set of equations for several different right hand side values, i.e. several **B** values in Eq. (4.45). In such cases, the **LU** decomposition, which requires most of the computational effort can be performed only once and for each **B** value one needs to only perform the forward and backwards substitutions to obtain a new solution.

In this work the **LU** decomposition method will not be used extensively since for nonlinear problems, one typically has a different coefficient (or **A**) matrix for each desired solution set. However, one case of multiple solutions is of importance and this is the case of calculating the inverse of a matrix. For this consider one possible defining equation for a matrix inverse as:

$$\mathbf{AD} = \mathbf{I}$$

$$
\begin{bmatrix}
a_{11} & a_{12} & a_{13} & \cdots & a_{1n} \\
a_{21} & a_{22} & a_{23} & \cdots & a_{2n} \\
a_{31} & a_{32} & a_{33} & \cdots & a_{3n} \\
\vdots & \vdots & \vdots & \vdots & \vdots \\
a_{n1} & a_{n2} & a_{n3} & \cdots & a_{nn}
\end{bmatrix}
\begin{bmatrix}
d_{11} & d_{12} & d_{13} & \cdots & d_{1n} \\
d_{21} & d_{22} & d_{23} & \cdots & d_{2n} \\
d_{31} & d_{32} & d_{33} & \cdots & d_{3n} \\
\vdots & \vdots & \vdots & \vdots & \vdots \\
d_{n1} & d_{n2} & d_{n3} & \cdots & d_{nn}
\end{bmatrix}
\qquad (4.50)
$$

$$
=
\begin{bmatrix}
1 & 0 & 0 & \cdots & 0 \\
0 & 1 & 0 & \cdots & 0 \\
0 & 0 & 1 & \cdots & 0 \\
\vdots & \vdots & \vdots & \vdots & \vdots \\
0 & 0 & 0 & \cdots & 1
\end{bmatrix}
$$

where the matrix inverse is written as the matrix **D**. One can solve for the coefficients of the inverse matrix by solving a series of n sets of linear equations the first two of which are:

$$
\begin{bmatrix}
a_{11} & a_{12} & a_{13} & \cdots & a_{1n} \\
a_{21} & a_{22} & a_{23} & \cdots & a_{2n} \\
a_{31} & a_{32} & a_{33} & \cdots & a_{3n} \\
\vdots & \vdots & \vdots & \vdots & \vdots \\
a_{n1} & a_{n2} & a_{n3} & \cdots & a_{nn}
\end{bmatrix}
\begin{bmatrix}
d_{11} \\ d_{21} \\ d_{31} \\ \vdots \\ d_{n1}
\end{bmatrix}
=
\begin{bmatrix}
1 \\ 0 \\ 0 \\ \vdots \\ 0
\end{bmatrix} ;
$$

$$(4.51)$$

$$
\begin{bmatrix}
a_{11} & a_{12} & a_{13} & \cdots & a_{1n} \\
a_{21} & a_{22} & a_{23} & \cdots & a_{2n} \\
a_{31} & a_{32} & a_{33} & \cdots & a_{3n} \\
\vdots & \vdots & \vdots & \vdots & \vdots \\
a_{n1} & a_{n2} & a_{n3} & \cdots & a_{nn}
\end{bmatrix}
\begin{bmatrix}
d_{12} \\ d_{22} \\ d_{32} \\ \vdots \\ d_{n2}
\end{bmatrix}
=
\begin{bmatrix}
0 \\ 1 \\ 0 \\ \vdots \\ 0
\end{bmatrix}
$$

Each solution set provides one column of the inverse matrix and the entire inverse matrix is obtained by solving for n different right hand side values. This is the preferred method for finding a matrix inverse.

A second application of **LU** decomposition is in finding the determinant of a matrix. After decomposition the determinant can be obtained by summing all the diagonal elements of the **U** matrix.

Code segments are shown in Listing 4.24 for using the **LU** decomposition of Listing 4.23 in several further applications. The first function Matrix.LUbsolve() from line 182 through line 199 is designed to take the decomposed matrix from the Matrix.LUdecompose() function along with a b matrix and return the solutions sets. It basically performs the forward and backwards substitutions of Eqs. (4.48) and (4.49) as can be identified on lines 189 and 195. The LUsolve() function from lines 207 through 211 takes as input an **A** and **B** (or a and b) matrices and solves the set of equations using **LU** decomposition and the LUdecompose( ) and

```
  1 : -- /* File Matrix.lua */
  2 : -- Package for Matrix algebra and functions - 2D matrices
....
182 : function Matrix.LUbsolve(a,indx,bb) -- Solve eqs by LU
183 :    local n,ii,ip,sum = #bb,0
184 :    local b = {} -- bb can be a column matrix or a row vector
185 :    if type(bb[1])=='table' then for i=1,n do b[i]=bb[i][1] end
186 :    else b = bb end
187 :    for i=1,n do -- Main loop, Forward substitution
188 :       ip = indx[i];   sum,b[ip] = b[ip],b[i]
189 :       if ii~=0 then for j=ii,i-1 do sum=sum - a[i][j]*b[j] end
190 :       else if sum~=0 then ii = i end end
191 :       b[i] = sum
192 :    end
193 :    for i=n,1,-1 do -- Main loop, Backward substitution
194 :       sum = b[i]
195 :       for j=i+1,n do sum = sum - a[i][j]*b[j] end
196 :       b[i] = sum/a[i][i]
197 :    end
198 :    return b -- Solution set
199 : end
200 :
201 : function Matrix.LUfunction(a,n) -- function to solve equations
202 :    -- by LU decomposition -- stores LU martix locally
203 :    local a,indx = Matrix.LUdecompose(a,n)
204 :    return function(b) return Matrix.LUbsolve(a,indx,b) end
205 : end
206 :
207 : function Matrix.LUsolve(a,b) -- Solve by LU decomposition
208 :    local a,indx = Matrix.LUdecompose(a)
209 :    return Matrix.LUbsolve(a,indx,b)
210 : end
211 :
212 : function Matrix.inv(m)
213 :    local n,b = #m, {}
214 :    local mc,ai = Matrix.new(m),Matrix.new(n,n)
215 :    local fi = Matrix.LUfunction(mc,n)
216 :    for i=1,n do
217 :       for j=1,n do b[j] = 0 end
218 :       b[i] = 1
219 :       b = fi(b)
220 :       for j=1,n do ai[j][i] = b[j] end
221 :    end
222 :    return ai
223 : end
224 :
225 : function Matrix.det(a,n) -- Determinant of matrix.
226 :    n = n or #a
227 :    local a,indx,d = Matrix.LUdecompose(a,n)
228 :    for i=1,n do d = d*a[i][i] end
229 :    return d
230 : end
```

Listing 4.24. Code segments for using **LU** decomposition to solve equations and for matrix inversion.

LUbsolve() functions. The Matrix.LUfunction() on lines 201 through 201 performs **LU** decomposition on a matrix and returns a function that can be later called

```
 1 :   --/* File list4_25.lua */ -- Several LU methods
 2 :
 3 : require"Matrix"; require"gauss"
 4 : -- define some test matrices
 5 : aa=Matrix.new{{2,1,1,3,2},{1,2,2,1,1},{1,2,9,1,5},{3,1,1,7,1},
       {2,1,5,1,8}}
 6 : bb = Matrix.new{{-2},{4},{3},{-5},{1}}
 7 :
 8 : a,b = aa:new(), bb:new() -- make copy of then
 9 : print('Determinant of a = ',a:det(),'\n') -- Modifies a
10 :
11 : a = aa:new() -- Need new copy
12 : ainv = a^-1 -- One form of obtaining inverse
13 : print(ainv*b,'\n') -- #1 solve equation set
14 :
15 : -- a and b are unchanged by above operations
16 : ainv = a:inv() -- Second form for inverse -- changes a matrix
17 : print(ainv*b,'\n') -- #2 solve equation set
18 :
19 : a = aa:new() -- New copy for a, b is still unchanged
20 : fi = a:LUfunction() -- Function for solving with b arrays
21 : sol = fi(b) -- #3 Call function for solution
22 : table.foreach(sol,print); print('\n') -- Solve and print results
23 :
24 : a = aa:new() -- New copy for a, b is still unchanged
25 : sol = Matrix.LUsolve(a,b) -- #4 Another method of solving
26 : table.foreach(sol,print); print('\n')
27 :
28 : a = aa:new() -- New copy, b still unchanged
29 : b = {-2, 4, 3, -5, 1} -- Single table needed for gauss
30 : sol = gauss(a,b) -- #5 Gauss elimination, does not use LU
31 : table.foreach(sol,print)
Selected Output:
Determinant of a =   98

{{-6.4183673469388},
{4.8367346938776},
{-1.0816326530612},
{1.265306122449},
{1.6428571428571}}
......

1  -6.4183673469388
2  4.8367346938776
3  -1.0816326530612
4  1.265306122449
5  1.6428571428571
```

Listing 4.25. Examples of various methods for solving equations using **LU** code.

with a right hand side **B** value to obtain a solution set. It basically hides internally the details of the **LU** decomposition from the user and simplifies the interface. The function Matrix.inv() takes as input a matrix and returns the inverse of the matrix using the technique outlined in Eq. (4.51) above of solving a set of equations n times for each column of the inverse. This makes use of the LUfunction() as can be seen on line 219 of the code. Finally the Matrix.det() function on lines 225 through 230 evaluates the determinant of a matrix by first performing the **LU** decomposition on line 227. This function uses the sign term returned as the third value by the LUdecompose() function. In using these functions it should always be remembered that LUdecompose() performs the decomposition within the original matrix and thus changes the original matrix. If one needs the original matrix for further processing one must copy the matrix before using any function that calls LUdecompose(). Another feature that the user should be aware of is that the LU functions that return the solution to a set of equations returns the solution set in a table and not in the form of a column matrix. This is usually more convenient for further matrix operations, but if one wants the solution set to be a matrix with a single column of values, one can use the form $\mathbf{A}^{-1}*\mathbf{b}$ which will result in a true single column matrix.

This now harkens back to Listing 4.22 where an example of using the Matrix code to obtain the inverse of a matrix was demonstrated. The above discussion has now filled in the details of how this is performed using the **LU** decomposition method. Listing 4.25 further demonstrates some of the matrix operations available with the code supplied in the Matrix.lua file and in Listing 4.24. Several examples are given of alternative formulations for solving a set of linear equations using the matrix notation and the **LU** coded functions. Lines 12 and 16 illustrate two equivalent methods for obtaining the inverse of a matrix. The code illustrates five notational methods for obtaining the solution of a set of equations. All of the approaches that solve the set of equations yield the same solution set so only one set is shown in the printed output. When only a single solution set is needed in subsequent chapters, the simple gauss() function will typically be used. In a few cases the inverse of a matrix will be needed and the **LU** method and functions will then be used.

4.6 Matrix Eigenvalue Problems

Some engineering problems lead to a set of equations to be solved in the form of:
$$[\mathbf{A} - \lambda\mathbf{I}]\mathbf{X} = 0, \tag{4.52}$$
where $\mathbf{A}$ is an n by n matrix, $\mathbf{I}$ is an identity matrix of order n and $\mathbf{X}$ is a column vector of order n. The problem is to solve for the values of λ for which this equation is valid. In many physical problems the values of λ correspond to the natural frequencies of oscillation of some physical structure or problem (Such as an electrical circuit). The valid values of λ are known as the "eigenvalues" and a valid value of $\mathbf{X}$ corresponding to a given eigenvalue is known as an "eigenvector". In some physical problems the formulation leads to a more general eigenvalue problem in the following matrix form:
$$\mathbf{AX} = \lambda\mathbf{BX}, \tag{4.53}$$
where $\mathbf{B}$ is also an n by n matrix. This can be put into the form of Equation (4.23) by use of the following transformation:
$$\mathbf{B}^{-1}\mathbf{AX} = \lambda\mathbf{B}^{-1}\mathbf{BX} = \lambda\mathbf{X}$$
$$\mathbf{DX} = \lambda\mathbf{X} \text{ where } \mathbf{D} = \mathbf{B}^{-1}\mathbf{A} \tag{4.54}$$

Although this is straightforward, and allow one to concentrate on only one form of eigenvalue equation, this is not the recommended procedure when the order of the matrix is very large.

The work presented here is just a brief introduction to the eigenvalue problem and is considered as an additional application of the numerical procedures introduced in the previous sections. Only the simple form of Eq. (4.52) will be considered here and it will be assumed that for problems of small n (n < 20 for example) that the more general problem can be transformed into the simple equation by use of the transformations given in Eq. (4.54). Since the eigenvalue matrix equation of Eq. (4.52) has zero right hand side, it is known that a nontrivial solution can only exist if the determinate of the coefficient matrix of $\mathbf{X}$ is zero, that is,

$$
\begin{vmatrix}
(a_{11} - \lambda) & a_{12} & .. & .. & a_{1n} \\
a_{21} & (a_{22} - \lambda) & .. & .. & a_{2n} \\
.. & .. & .. & .. & .. \\
.. & .. & .. & .. & .. \\
a_{n1} & a_{n2} & .. & .. & (a_{nn} - \lambda)
\end{vmatrix} = 0 \qquad (4.55)
$$

When expanded, this equation leads to an nth order polynomial in λ called the characteristic equation. Thus finding the eigenvalues requires that one find the roots of a polynomial of the form:

$$
\lambda^n - p_1 \lambda^{n-1} - p_2 \lambda^{n-2} \dots - p_{n-1} \lambda - p_n = 0
$$
$$
\text{where } p_1 = a_{11} + a_{22} + \dots a_{nn} \qquad (4.56)
$$

which is exactly the problem discussed in this chapter on finding the roots of a polynomial. Assuming that one can find the n solutions of this equation one has the n eigenvalues as $\lambda_1, \lambda_2, \dots \lambda_n$ and corresponding to each of these eigenvalues, a nontrivial solution of the set of equations can be obtained from the equation:

$$
\mathbf{AX_i} = \lambda_i \mathbf{X_i} \qquad (4.57)
$$

The n solutions of this equation give the eigenvectors corresponding to the eigenvalues. This matrix equation (Eq. (4.57)) can only be solved if one component of the $\mathbf{X_i}$ values is known. The other components can then be evaluated in terms of this component. Another way of stating this is that the eigenvectors can only be determined to within a constant multiplier. One is thus free to normalize the eigenvectors in any convenient manner and one such convenient approach is to take the largest component of the eigenvector as unity.

In order to use the results of the previous section on roots of polynomials, the coefficients in Eq. (4.56) must first be determined. These can be generated most conveniently by use of the Faddeev-Leverrier method. In this method, the polynomial coefficients are generated by successive applications of the following algorithm:

$$\mathbf{P_1} = \mathbf{A}, p_1 = \text{trace}(\mathbf{P_1});$$

$$\mathbf{P_2} = \mathbf{A}(\mathbf{P_1} - p_1\mathbf{I}), p_2 = \frac{1}{2}\text{trace}(\mathbf{P_2});$$

$$\mathbf{P_3} = \mathbf{A}(\mathbf{P_2} - p_2\mathbf{I}), p_3 = \frac{1}{3}\text{trace}(\mathbf{P_3}); \qquad (4.58)$$

$$\cdots$$

$$\mathbf{P_n} = \mathbf{A}(\mathbf{P_{n-1}} - p_{n-1}\mathbf{I}), p_n = \frac{1}{n}\text{trace}(\mathbf{P_n});$$

Computer code for solving for the eigenvalues and eigenvectors is included in the "Matrix" code package. A listing of the most important callable functions associated with eigenvalues and eigenvectors is shown in Listing 4.26. The major features of the above discussion can be readily seen in the code segments. First function Matrix.trace() on lines 240 through 245 calculate the trace of a matrix. The function Martix.eigeneq() from line 247 to 258 sets up the polynomial coefficients using the algorithm of Eq. (4.58). The major eigenvalue code routine is the Matrix.eigenvalues() function on lines 283 through 297. The input arguments may include a **B** matrix in addition to an **A** matrix as discussed in connection with Eq. (4.54). Line 292 of the listing calls Polynomial.roots() to obtain the eigenvalues, after first setting up the polynomial coefficients by calling Matrix.eigeneq() (on line 292). The returned eigenvalues are sorted from largest to smallest on lines 293 throuth 295 and the order of the sorting can be changed by changing the inequality on lines 294 and 295. Finally, the Matrix.eigenvectors() function on lines 260 through 281 evaluates the eigenvectors corresponding to the calculated eigenvalues. The input to this function is the matrix coefficients and possible the eigenvalues. If the eigenvalues are not input, they are calculated within the function on line 263. Thus the eigenvectors() function can be called with just the coefficient matrix and both the eigenvectors and eigenvalues will be returned on line 280. The eigenvectors are evaluated by calling the matrix solution routine gauss()on line 272 which is within a loop over the eigenvalues. The returned eigenvectors are normalized such that the largest X_i value for each eigenvector is set to unity on lines 274 through 278 and each column of the returned matrix corresponds to one of the eigenvalues.

Some applications of the eigenvalues() and eigenvectors() code routines are shown in Listing 4.27. The first example in lines 7 through 20 of the listing is a calculation of the eigenvalues and eigenvectors of a 10 by 10 matrix of the form:

$$[A] = \begin{bmatrix} 2 & -1 & 0 & .. & 0 \\ -1 & 2 & -1 & .. & 0 \\ 0 & -1 & 2 & .. & 0 \\ .. & .. & .. & .. & .. \\ 0 & 0 & 0 & .. & 2 \end{bmatrix} \qquad (4.59)$$

```lua
  1 : -- /* File Matrix.lua */
....
240 : function Matrix.trace(a,n) -- Trace of a matrix
241 :    n = n or #a
242 :    local tx = a[1][1]
243 :    for i=2,n do tx = tx + a[i][i] end
244 :    return tx
245 : end
246 :
247 : function Matrix.eigeneq(a,n) -- Set up eigenvalue equation
248 :    n = n or #a
249 :    local p,pm,px = {}, Matrix.new(a)
250 :    p[n+1] = -1
251 :    for i=1,n do
252 :       px = Matrix.trace(pm)/i
253 :       for j=1,n do pm[j][j] = pm[j][j] - px end
254 :       p[n-i+1] = px
255 :       pm = a*pm -- Matrix multiplication, a and pm matrices
256 :    end
257 :    return p -- form a0 + a1*x + a2*x^2 ...
258 : end
259 :
260 : function Matrix.eigenvectors(ai,eval) -- Calculate Eigenvectors
261 :    local n = #ai
262 :    if type(ai)~='Matrix' then ai = Matrix.new(ai) end
263 :    eval = eval or Matrix.eigenvalues(ai,n) -- Get eigenvalues?
264 :    local evect,b,big,tmp = Matrix.new(n),{}
265 :    b[n] = 1 -- Assume one value for solution
266 :    for j=1,n do -- Loop over eigenvalues
267 :       a = Matrix.new(ai) -- Make copy for use in gauss()
268 :       for i=1,n-1 do
269 :          b[i] = -a[i][n]
270 :          a[i][i] = a[i][i] - eval[j] -- Subtract eigenvalue
271 :       end
272 :       gauss(a,b,n-1) -- eigenvector, a and b changed in gauss()
273 :       big = 0
274 :       for i=1,n do -- Normalize so largest element is +/- 1.000
275 :          tmp = abs(b[i])
276 :          if tmp> big then big = tmp end
277 :       end
278 :       for i=1,n do evect[i][j] = b[i]/big end
279 :    end
280 :    return evect,eval -- Also return eigenvalues possible use
281 : end
282 :
283 : function Matrix.eigenvalues(a,b,n) -- Set up and use roots()
284 :    require"Polynomial" -- Only load if needed for roots()
285 :    local d
286 :    if type(a)=='table' then a = Matrix.new(a) end
287 :    if type(b)=='table' then b = Matrix.new(b) end
288 :    if b==nil then d = a
289 :    elseif type(b)=='number' then d,n = a,b
290 :    elseif type(b)=='Matrix' then d = b^-1*a
291 :    else print('b must be a matrix in eigenvalues') end
292 :    local rts = Polynomial.roots(Matrix.eigeneq(d,n))
293 :    table.sort(rts,function(a,b)
294 :    if type(a)=='Complex' or type(b)=='Complex' then return
                Complex.abs(a)>Complex.abs(b)
295 :    else return a>b end end) -- sort eigenvalues
```

```
296 :     return rts,d -- Sorted most positive to most negative
297 : end
```

Listing 4.26. Matrix code segments for eigenvalue and eigenvector calculations.

This matrix is a tridiagonal matrix with value 2 along the diagonal and -1 on each adjacent element and zeros elsewhere. Such a matrix structure arises in structural analysis problems in mechanical engineering. The two largest eigenvalues and a partial listing of the eigenvalues, are shown in the output listing. The code segments provided are capable of solving this eigenvalue problem for an n by n matrix of about size 20 by 20. This limit is due to the nature of the polynomial solving routine as discussed in the previous section due to the finite number of digits used to represent floating point numbers.

The second example is that of the equation set:

$$
\begin{bmatrix} 300 & -200 & 0 \\ -200 & 500 & -300 \\ 0 & -300 & 300 \end{bmatrix} \begin{bmatrix} X_1 \\ X_2 \\ X_3 \end{bmatrix} = \lambda \begin{bmatrix} 4 & -1 & 1 \\ -1 & 6 & -4 \\ 1 & -4 & 5 \end{bmatrix} \begin{bmatrix} X_1 \\ X_2 \\ X_3 \end{bmatrix} \tag{4.60}
$$

which has both an **A** and B matrix. The corresponding three eigenvalues are shown in the printed output of the listing.

Finally an example of a 3 by 3 matrix with complex coefficients is shown in the listing on lines 28 through 31 with the complex eigenvalues shown in the printed output. The Complex number package is loaded into the code by the Polynomial package and no further action is needed for the routines to operate on complex numbers.

This section has provided a brief introduction to the general field of eigenvectors and eigenvalues. This is only a brief introduction and is used as an example of the combined use of the matrix and root solving code segments developed in previous sections of this chapter. This is an excellent example of how code segments previously developed can be integrated into solution code segments for increasingly more complex physical problems. The general area of eigenvalues has been extensively developed in the numerical methods area and the solution of such problems is highly advanced, especially for problems with matrices of very high order. For such large scale problems, the techniques presented here and developed in the code segments are not the approaches which have been found most effective for handling large size problems. In some cases of large size matrices, only a few of the largest or smallest eigenvalues are of primary importance as they will determine the largest or smallest oscillation frequencies for example of a physical structure. Special fast techniques have been developed for finding such largest and smallest eigenvalues.

The largest size matrix for which the code developed here can be used is probably around a 20 by 20 matrix. This is sufficient to explore eigenvalue and eigenvector problems, but the reader is referred to the extensive literature on eigenvalue problems involving very large matrices.

```
 1 : -- /* File list4_27.lua */
 2 : -- Tests of eigenvalues and eigenvectors
 3 :
 4 : require"Polynomial"
 5 : require"Matrix"
 6 :
 7 : n = 10; A = Matrix.new(n)
 8 : for i=1,n do
 9 :   for j=1,n do
10 :       if j==i-1 then A[i][j] = -1
11 :       elseif j==i then A[i][j] = 2
12 :       elseif j==i+1 then A[i][j] = -1
13 :       else A[i][j] = 0 end
14 :   end
15 : end
16 : print(A)
17 :
18 : ev,rts = Matrix.eigenvectors(A)
19 : print('Eigenvectors \n',ev,'\nEigenvalues')
20 : table.foreach(rts,print)
21 :
22 : A = Matrix.new{{300,-200,0},{-200,500,-300},{0,-300,300}}
23 : B = Matrix.new{{4,-1,1},{-1,6,-4},{1,-4,5}}
24 : rts = Matrix.eigenvalues(A,B)
25 : print('Eigenvalues for A and B matrix problem')
26 : table.foreach(rts,print)
27 :
28 : a = Matrix.new{{2+j,-2-j,3},{2,j,-3+2*j},{j,4-j,3*j}}
29 : rts = Matrix.eigenvalues(a)
30 : print('Eigenvalues for complex matrix')
31 : table.foreach(rts,print)
Selected Output:
{{2, -1, 0, 0, 0, 0, 0, 0, 0, 0},
{-1, 2, -1, 0, 0, 0, 0, 0, 0, 0},
{0, -1, 2, -1, 0, 0, 0, 0, 0, 0},
{0, 0, -1, 2, -1, 0, 0, 0, 0, 0},
{0, 0, 0, -1, 2, -1, 0, 0, 0, 0},
-------
{0, 0, 0, 0, 0, 0, 0, 0, -1, 2}}
Eigenvectors
{{-0.28462967368778, 0.546200349849, -0.76352111819831, ----- },
{0.54620034434861, -0.91898594684424, 1, ----},
------
Eigenvalues
1   3.9189859485549
2   3.6825070637511
-----
Eigenvalues for A and B matrix problem
1   128.76570298984
2   69.994410272701
3   12.560641454441
Eigenvalues for complex matrix
1   (1.1789219308193) + j(5.7960334203025)
2   (-1.3525289229772) + j(-2.5232132599369)
3   (2.1736069921579) + j(1.7271798396344)
```

Listing 4.27. Example applications of eigenvalues() and eigenvectors() code segments.

4.7 Summary

This chapter has discussed methods for solving for the roots of a system of coupled nonlinear equations in the same number of unknown variables as equations.
The technique discussed for solving such systems is the extension of Newton's
method to systems of equations. Using a Taylor series expansion one converts the
nonlinear equations into a set of linear equations which through an iterative process of solving a linear set of equations converges to the solution of the desired
nonlinear equations much as Newton's method is used for a single equation. This
is one of the most fundamental and important approaches for the numerical modeling of nonlinear processes. This is again an example of the L&I approach to
nonlinear engineering problems.

   The development of the roots() algorithm for finding the roots of a polynomial
equation and the associated code has illustrated another feature which will occur
throughout the remainder of the text. This is the reuse of code segments such as
nsolv() to implement nonlinear numerical techniques to attack increasingly more
complex problems. The section on eigenvalues and eigenvectors has also illustrated the reuse of previously developed code segments to attack another type of
engineering problem. While code segments such as those developed in this chapter
can be found elsewhere, such as in MATLAB, the ability to embed the segments
into increasingly more complex algorithms is not so easily done with these packages.

   Important code segments developed in this chapter are:
   1.   gauss() – Code for the solution of coupled linear equations.
   2.   spgauss() – Code for the solution of coupled linear equations using sparse
        matrix storage techniques.
   3.   nsolv() – Code for the solution of coupled nonlinear equations.
   4.   Polynomial.roots() – Code for finding the roots of a polynomial.
   5.   Matrix.eigenvalue() and Matrix.eigenvector() – Code for eigenvalue and
        eigenvector problems.

   In addition the following language extension packages were introduced and
discussed:
   1.   Matrix – Language extensions for standard matrix operations such as ad-
        dition, multiplication, and inversion
   2.   Polynomial – Language extensions for standard manipulation of polyno-
        mial equations such as addition, multiplication and root finding.

The nsolv() approach and this coded function is one of the most important code
segment developed in this chapter and will find extensive use in subsequent chapters.

5 Numerical Derivatives and Numerical Integration

In the previous two chapters numerical derivatives have been used in the implementation of several code segments -- more specifically Newton's method and nsolv(). The most powerful algorithms for handling nonlinear equations are to linearize and iterate on the linearized equations until a valid solution of the nonlinear equations is obtained. To do this one needs to be able to obtain the derivative of the nonlinear equations. Most authors shy away from recommending that this be done by numerical means and stress that if at all possible the derivatives be obtained analytically. This is usually not convenient and in some cases not possible as a set of equations may be generated only by a computer algorithm. The approach taken in this work is that numerical derivatives are acceptable if properly taken and this approach leads to greatly simplified algorithms as one then has to only specify the equations to be solved. This is in keeping with the principle that one should only have to define in a computer program a set of equations and any boundary conditions and the software should then take over and provide solutions.

Simple first order algorithms have been used in the previous chapters for numerical derivatives. These have been adequate for all the examples given in the previous chapters. In this chapter a much more in-depth look is undertaken of numerical derivatives and this provides the background needed to have confidence in the use of numerical derivatives in a wide range of nonlinear problems. In addition code segments are developed that can be used to calculate numerical derivatives with near machine precision for a wide range of functions.

A second major topic for this chapter is the closely related issue of numerically evaluation the value of an integral. Code segments are again developed that can integrate a large range of integrals with near machine precision.

5.1 Fundamentals of Numerical Derivatives

The general problem to be considered is that for some function $f(x)$, one wishes to evaluate the derivative at some point x, i.e.

$$\text{Given } f(x): \text{ Evaluate: } deriv = \frac{df}{dx} \tag{5.1}$$

A starting point for this discussion can again be a Taylor series expansion of $f(x)$ about some point x_o:

J.R. Hauser, *Numerical Methods for Nonlinear Engineering Models*, 147–186.
© Springer Science + Business Media B.V. 2009

$$f(x) = f(x_o) + \frac{df}{dx}\bigg|_{x_o} (x - x_o) + \frac{1}{2}\frac{d^2 f}{dx^2}\bigg|_{x_o} (x - x_o)^2 + \cdots$$

$$\qquad\qquad\qquad\qquad\qquad\qquad\qquad\qquad (5.2)$$

$$f(x + \Delta x) = f(x) + f'(x)(\Delta x) + \frac{1}{2} f''(x)(\Delta x)^2 + \cdots$$

In the second form of Eq. (5.2) the notation has been changed to emphasize that the interest is in the derivative at some point x and Δx is some change in x around that point. If only the first order term in Eq. (5.2) is kept, this leads to the basic single sided derivative definition that is typically learned from Calculus of

$$f'(x) = \frac{f(x + \Delta x) - f(x)}{\Delta x}\bigg|_{\Delta x \to 0} \qquad\qquad (5.3)$$

On the other hand the use of the double sided definition leads to the equation:

$$f'(x) = \frac{f(x + \Delta x) - f(x - \Delta x)}{2\Delta x}\bigg|_{\Delta x \to 0} \qquad\qquad (5.4)$$

While each of these is exact in the limit of Δx going to zero, in numerical computer calculations, this limit can never be achieved since calculations are limited by the precision of the software. However, each of these can provide a basis for calculating a numerical derivative of a function with a certain degree of accuracy. For a finite Δx in Eq. (5.3) terms of order (Δx) in the derivative are being neglected while in Eq. (5.4) terms of order $(\Delta x)^2$ are being neglected. Thus all other things being equal, Eq. (5.4) is expected to give a more accurate estimate of the derivative than Eq. (5.3). Each equation involves two function evaluations, so the computational effort appears to be the same for each function. However, if the function value is needed in addition to the derivative, as in Newton's method, then Eq. (5.4) involves more computational effort.

It appears very straightforward to apply these equations to a given function and obtain an approximate numerical derivative. However, there are subtle problems when developing general computer code for a numerical derivative. The first and most fundamental problem is to determine how big a Δx value should be taken. In searching through a large number of numerical methods books one finds that this problem is either ignored or treated in some superficial way. Most textbook examples appear to assume that x is on the order of magnitude unity and that one can take Δx small compared to unity (perhaps .001 or .0001). However, in real world engineering problems, the magnitude of variables is rarely of order unity. One could require that problems always be scaled appropriately so that variables being used in computer programs are of order unity and certainly the appropriate scaling of variables is a valuable technique to employ. However, not all problems can be scaled so that the range of a variable is always on the order of unity. For example in electrical engineering, current in a semiconductor device may range over many orders of magnitude in a single problem because of the exponential relationship to current to voltage.

The first task in developing a code segment for numerical derivatives is thus to address the question, how does one select a value of Δx for use in either Eq. (5.3) or (5.4)? To address this issue consider two functions and possible problems with selecting a value for each of these:

$$f_1(x) = Ax^n \text{ (with } n \approx 5 \text{ or } 10); \text{ and } f_2(x) = A\exp(x) \qquad (5.5)$$

To explore possible problems let's consider very large values of x, very small values of x and as a special case $x = 0$. These cover some of the most important cases in practice.

First consider the $f_1(x)$ function and x values from some large value such as 1.e10 to some small value such as 1.e-10. It can be seen that no single value of Δx is appropriate for this range of variation in x. If a value appropriate for the smallest range is selected, say 1.e-12, this value will be completely lost in the machine precision when added to the largest range of 1.e10. On the other hand if an appropriate value for the largest range is selected such that it is not lost in the machine precision of say $> 1.e-2$ then this value would completely overwhelm the smallest range values of 1.e-10. Obviously what are needed are values of Δx that scale with the range of x values. This is relatively easy to accomplish if a Δx value is selected that is some relative fraction of x, such as:

$$\Delta x = \varepsilon x \text{ with } \varepsilon \ll 1. \qquad (5.6)$$

As will be seen, appropriate values for ε can vary over a rather wide range. If this value is used in Eq. (5.5) for the $f_1(x)$ function, one obtains:

$$\frac{f_1(x + \Delta x) - f_1(x)}{\Delta x} = \frac{f_1(x + \varepsilon x) - f_1(x)}{\varepsilon x}$$
$$= \frac{A(x + \varepsilon x)^n - Ax^n}{\varepsilon x} = Ax^{n-1}\left[\frac{(1+\varepsilon)^n - 1}{\varepsilon}\right] \qquad (5.7)$$

Now as long as $\varepsilon \ll 1$, $(1+\varepsilon)^n \cong 1 + n\varepsilon$ and the last expression above becomes:

$$\frac{f_1(x + \Delta x) - f_1(x)}{\Delta x} = Ax^{n-1}\left[\frac{(1+\varepsilon)^n - 1}{\varepsilon}\right] \approx Anx^{n-1}; \qquad (5.8)$$

This is the exact answer for the derivative of function f_1 and a correct result is obtained independently of the magnitude of the x variable or rather a result is obtained with a relative accuracy that is independent of the magnitude of the x variable.

Now consider the same definition of Δx used in the f_2 function. In this case one gets:

$$\frac{f_1(x + \Delta x) - f_1(x)}{\Delta x} = \frac{A\exp(x + \varepsilon x) - A\exp(x)}{\varepsilon x}$$
$$= A\exp(x)\left[\frac{\exp(\varepsilon x) - 1}{\varepsilon x}\right] \qquad (5.9)$$

If now it can further be assumed that $|\varepsilon x| \ll 1$ in the exponential term, then the final expression becomes:

$$\frac{f_1(x + \Delta x) - f_1(x)}{\Delta x} = A\exp(x)\left[\frac{\exp(\varepsilon x) - 1}{\varepsilon x}\right] \approx A\exp(x) \qquad (5.10)$$

This again is the exact expression for the derivative. One needs to consider further the requirement that one can take $\varepsilon x \ll 1$. The valid range of x in any exponential term is limited by the maximum and minimum values which can be represented in the computer language. For double precision calculations, this ranges from about 1.e-308 to about 1.e+308. For the exponential function then this gives limits of $-708 < x < 708$. To satisfy $\varepsilon x \ll 1$ then requires that $\varepsilon \ll 0.0014$ which is not a severe restriction on a selection of the relative displacement factor.

By selecting a value of Δx that scales with the magnitude of the independent variable, it is seen in the above examples that one can achieve an appropriate displacement factor for properly evaluating a derivative for these two test functions over a very wide range in values of the function. While these two functions do not cover all possible examples they are representative of highly nonlinear functions and the technique discussed here has been applied to a wide range of functions with much success. It can certainly be appreciated that the value of the displacement must be scaled with the magnitude of variables in a problem if a robust derivative algorithm is to be obtained. There are still potential problems with this choice of Δx, which show up for the special case of $x = 0$. For this special case, the equation $\Delta x = \varepsilon x$ can not be used, as this choice results in zero displacement in the variable and Eq. (5.3) or (5.4) produce the indeterminate form 0/0. A small values of x near zero is not usually the problem, it's the special case of exactly zero. Such cases occur frequently in textbook problems. However, in actual application to real engineering problems, the special case of a variable being exactly zero is much less prevalent. It's rare that a variable is exactly zero in nature. Never the less, one must have some means of handling such a case and for this case there is probably no recourse but to resort to taking $\Delta x = \varepsilon$ and hope that this is a good choice. Accounting for the special case of zero is always a problem in coded algorithms and in this work this will always be treated as a special case. The special case of a zero variable value is perhaps one of the most important limitations in using numerical derivatives.

In the previous discussion, it was possible to place some limits on the selection of an appropriate relative displacement factor ε as being much less than unity for the exponential and power functions. However, how much less than unity is the best choice for this factor? To explore this question, the double sided derivative expression of Eq. (5.4) has been used to evaluate the derivative of the f_1 and f_2 functions previously defined for both large and small values of the function variable. Figure 5.1 shows the relative error in the numerical derivative as a function of ε for the $\exp(x)$ function and for $x = +/-10$ and $+/-20$, where the relative error is defined as the difference between the numerical estimate of the derivative and the true derivative divided by the true derivative.

The relative error is then related approximately inversely to the number of digits of precision in the calculated derivative values. As seen in the figure, the error is slightly larger for $x = +/-20$ as opposed to $x = +/-10$ as expected. As ε decreases from 0.1 the relative error is seen to decrease from somewhat above 0.1 with an approximate dependence of ε^2 as can be seen in the figure from the slope of the

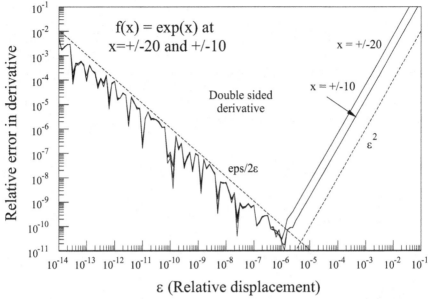

ε (Relative displacement)

Figure 5.1. Relative error in double sided derivative for an exponential function.

dotted line. This is as expected since the error in the double sided derivative is proportional to $(\Delta x)^2$. The relative error is the same for either x = +20 or x = -20. The relative error decreases to less than 1.e-10 at an ε value of about 1.e-6 and then increases as ε is further decreased. This might at first seem unexpected as the relative machine accuracy is about 1.e-16. However the relative error in any calculation such as the numerator of Eq. (5.4) for the derivative is always subject to the machine precision (about 2.2e-16). An analysis of the relative error for the exponential function then gives,

$$RE = (\frac{1}{6}x^2)\varepsilon^2 + \frac{eps}{2\varepsilon}, \qquad (5.11)$$

where eps is the machine precision, as discussed in Chapter 2 (2.2e-16). The figures show that the equation eps / 2ε is a good upper limit to the increasing error due to the machine precision. The above equation has a minimum value of about 7.e-11 at $\varepsilon = 1.2e-6$.

This is in reasonably good agreement with the data in Figure 5.1 especially in consideration of the fact that the error is always some multiple of eps. Just to re-emphasize the range of the function being addressed here, the exponential function has values of 4.85e+8 and 2.1e-9 for the +20 and -20 argument values. Over this range of about 16 orders of magnitude in the value of the function, the relative displacement algorithm gives a very good estimation of the derivative which at the optimum value of ε is accurate to about 10 decimal digits.

Similar relative error data is shown in Figure 5.2 for the x^n function with $n = 5$ and 10 and with large and small values of x of 10^{10} and 10^{-10}. The relative error is again seen to reach a minimum at a value of below 1.e-10 and at about $\varepsilon = 1.e-6$. Also for values above this number, the relative error is the same for the 10^{10} and the 10^{-10} cases. For the n = 10 case, the power function ranges in value from 1.e+100 to 1.e-100, a range of 200 orders of magnitude, and yet the relative displacement algorithm gives a very good estimate of the derivative over this range and the same relative error is achieved for both the large and small values of the function. Again for values of ε below about 1.e-6 the relative error is seen to increase due to machine round off errors.

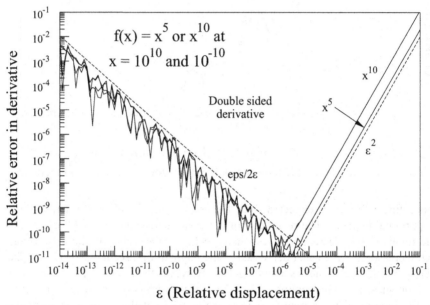

Figure 5.2. Relative error in double sided derivative for a power function at large and small x values.

This discussion and the calculated relative errors have so far considered only the double sided derivative definition of Eq. (5.4). Figure 5.3 shows relative error for the single sided derivative of Eq. (5.3) when evaluated for the power function as used in Figure 5.2. As expected for the larger values of ε the relative error in the derivative now varies directly with the ε value as seen by the slope of the dotted curve in the figure. For small values of ε the relative error increases with the same upper limit due to machine precision as for the double sided derivative. For this example of a power function, the minimum relative error in the derivative is around 10^{-8} and is seen to occur at an ε value of about 10^{-8}. A similar calculation with the exp() function as used in Figure 5.1 would show very similar data and trends.

From the discussion above, it would appear that the best choice of displacement factor would be about 10^{-6} for a double sided numerical derivative calculation and that one can expect to obtain a numerical derivative with an accuracy of

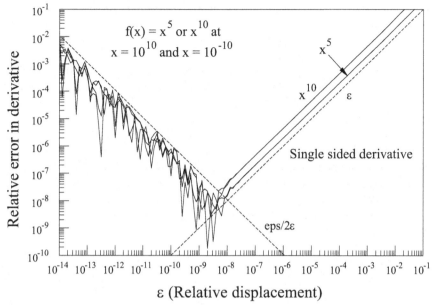

Figure 5.3. Relative error in single sided derivative for a power function at large and small x values.

about 10 to 11 digits. For a single sided derivative the optimum relative displacement would appear to be about 10^{-8} resulting in an accuracy of about 8 digits. However, there is one final factor to be considered and this is very small values of the independent variable for a function that has both a finite value and a finite derivative as x approaches zero. Such a function is the $\exp(x)$ function or a function such as $(1+x)^n$. In numerically evaluating the derivative, the expression $\exp(x(1+\varepsilon))$ must be evaluated and the smallest contribution to the exponent which can be evaluated correctly is eps, the machine epsilon value. Thus the exponent needs to have $\varepsilon x >$ eps. In order to evaluate the derivative correctly for as small an x value as possible, this would suggest that one should take ε as large as possible. Thus this is in conflict with the requirement for a small value of ε. If one takes $\varepsilon = 10^{-6}$ then this would imply that the smallest x value for which an accurate derivative can be computed is about $x = 10^{-10}$. This is the compromise value that has been used in the two previous chapters for evaluating numerical derivatives in the newton() and nsolv() functions.

In the two previous chapters the code segments developed, in particular newton() and nsolv() used a numerical derivative for the first-order Taylor series expansions of various functions. In these routines the simple single sided derivative equation was used with a relative default displacement factor of 10^{-6}. The single sided derivative was used because in all the routines involving Newton's method, the function value is required in addition to the derivative and thus the single sided derivative requires only one additional function evaluation whereas the use of the double sided derivative requires three function evaluations for each step. The relative displacement factor of 10^{-6} is a compromise as discussed above. For the exp() and power functions considered above this value would give a relative error in the range of 10^{-6} for the cases considered in Figures 5.1 through 5.3. This relative error is more than sufficient for Newton's method, as the final accuracy of the technique does not depend on having a very accurate derivative, but depends only on driving the requisite function to zero. The discussion above then provides justification for the approach to a numerical derivative used in the previous two chapters. It also shows that when properly implemented, it is possible to have a robust algorithm for calculating a numerical derivative. The lack of having developed a robust approach to numerical derivatives is perhaps the main reason that most text book authors have recommended against the extensive use of numerical derivatives in computer modeling. As previously stated, numerical derivatives are one of the core techniques used in this book, making the work considerably different from past published texts.

```
 1 : function deriv(f,x)
 2 :    local dx = x*FACT
 3 :    if dx==0 then dx = FACT end
 4 :    return (f(x+dx) - f(x-dx))/(2*dx)
 5 : end
 6 : setfenv(deriv,{FACT=1.e-6})
 7 :
 8 : f1 = function(x) return x^5 end
 9 : f2 = math.exp
10 : print(deriv(f1,1.e8),5*(1.e8)^4)
11 : print(deriv(f2,10),math.exp(10))
output:
5.0000000000421e+032    5e+032
22026.465794261         22026.465794807
```

Listing 5.1. Code segment for simple derivative and its use.

With this as a background it is now straightforward to code a general algorithm for calculating the derivative of a function. Using the double sided derivative for accuracy, such a routine is almost trivial to code as shown in Listing 5.1. Line 2 calculates a displacement value using a relative factor (the FACT term). Line 3 protect against a zero argument value as previously discussed. Two examples are given in the listing of taking the numerical derivative of an exponential function and a power function on lines 8 and 9 corresponding to the test functions previously discussed in this section. If the printed output values are compared with the exact derivative values (also printed), it is seen that the numerical derivatives are accurate to about 11 digits, which is the best expected accuracy based upon the results of Figures 5.1 and5.2. It is left up to the reader to change the value of ε (or FACT) in the code to see its effect on the accuracy of the derivative. Significantly larger or smaller values will result in degraded precision in the calculated derivative. The implementation of a numerical derivative is in fact so simple that code similar to Listing 5.1 is simply embedded within higher level functions in this work that need a numerical derivative. This is the approach taken for newton()

and nsolv() in the two previous chapters and will be the approach used in subsequent chapters.

So far the discussion has explored simple first-order and second-order derivative definitions and discussed intrinsic limits on the accuracy imposed by the relative displacement algorithm and by the internal accuracy of the computer code. A simple algorithm has been implemented that can result in about 10 digits of accuracy for a wide range of functions and function values. For most engineering work this should be of sufficient accuracy. Coupled with the simplicity of the algorithm we could thus move on to other topics. However, there is probably the feeling that one should be able to perhaps achieve a higher accuracy that approaches the intrinsic accuracy of the computer language which for this work is about 16 digits -- 6 orders of magnitude additional accuracy. Thus a little time will be spent on improved algorithms for the derivative.

One approach to a more accurate derivative discussed by most authors is to use higher order derivative equations. For example a central difference equation based upon 4 function evaluations is:

$$f'(x) \cong (-f(x+2\Delta x) + 8f(x+\Delta x) - 8f(x-\Delta x)$$
$$+ f(x-2\Delta x))/(12\Delta x) \tag{5.12}$$

This equation has an error proportional to $(\Delta x)^4$ as opposed to $(\Delta x)^2$ for the second-order central difference of Eq. (5.4). More accurate formulations of a derivative expression such as this are necessarily going to involve function evaluations at more points to gather more information on the function. Thus one approach is to pursue such higher order derivative expressions.

However, there is another approach which uses Richardson's extrapolation (developed by L. F. Richardson in 1910). This technique is useful in a number of numerical modeling applications and will be discussed in general here and applied not only to the derivative problem but in a subsequent section to numerical integration. To understand this approach, consider some numerical algorithm which provides an approximation of a function value but which has an accuracy that depends on some $(\Delta x)^n$. For the derivative equation one has:

$$f' = (f(x+h) - f(x-h)/2h + O(h^2) = F_1(h) + Ch^2. \tag{5.13}$$

In this equation h has been used as the step size in place of Δx for convenience. So in general consider any function which depends on a step size h in the manner of:

$$F = F_1(h) + Ch^n \tag{5.14}$$

In this equation F represents the true value of the function, F_1 represents an approximation to the function, and Ch^n represents the error in the function with n the order of the error dependency on step size. Now consider applying this equation twice with two different step sizes to give two equations as:

$$F = F_1(h_1) + Ch_1^n$$
$$F = F_2(h_2) + Ch_2^n \tag{5.15}$$

Combining these, one can solve for the error term involving C and eliminate it from the answer as:

$$F \cong F_2(h_2) + \frac{F_2(h_2) - F_1(h_1)}{(h_1/h_2)^n - 1}. \tag{5.16}$$

This is basically Richardson's extrapolation which says that if one evaluates a function at two different step sizes, the two approximations can be used to eliminate the largest error term from the result, provided the power dependency of the error term on step size is known. Of course when this is done all errors have not been completely eliminated, because in general there will be an even higher order error term of order h^{2n}. Thus Eq. (5.16) can be thought of as having an additional error term as:

$$F = F_2(h_2) + \frac{F_2(h_2) - F_1(h_1)}{(h_1/h_2)^n - 1} + O(h_2^{2n}). \tag{5.17}$$

If the function is then evaluated for three step sizes, two additional equations can be obtained in the form of Eq. (5.17) with two different step sizes. From these two new equations, one can then in principle eliminate any error of order h^{2n}. In this manner one can get a result accurate, in principle, to an even higher order power of the step size. Each time one evaluates the function at a new step size one can eliminate a higher order error dependency on step size and improve the accuracy of the result.

In the application of Richardson's extrapolation theorem, the step size is typically reduced by a factor of 2 at each iterative step, so that $(h_1/h_2) = 2$ and Richardson's extrapolation becomes:

$$F \cong F_2(h_2) + \frac{F_2(h_2) - F_1(2h_2)}{2^n - 1} = \frac{2^n F_2(h_2) - F_1(2h_2)}{2^n - 1}. \tag{5.18}$$

The value of n is determined by the order of the error term in any numerical algorithm. For the case of the central difference approximation to the derivative, it has the value of 2. For a single sided difference approximation such as Eq. (5.3), it has the value of 1.

Computer code implementing Richardson's extrapolation for the derivative is shown in Listing 5.2. The function deriv() starts out with the basic central difference definition of the derivative on lines 3 through 5 with the result stored in a table (a[]). Then a Richardson improvement loop is coded from lines 7 to 16. Within this loop a test of the accuracy achieved is made on lines 13 through 15 and the loop is exited if sufficient accuracy is achieved. Note that the test is made on line 14 of the difference in value between two iterative values. The loop at lines 10-12 implements a multiple Richardson improvements based upon all previous data which is saved in a table. The deriv() function returns 3 values on line 17: the derivative value, the number of Richardson cycles used and the estimated relative error in the derivative.

In addition to the central difference derivative, code is included for a right-difference formulation rderiv() in lines 23 through 41. This is similar to the central difference formulation but uses only a two point differencing equation and

```
 1 : --/* File deriv.lua */ -- with Richardson's extrapolation
 2 : deriv = function(f,x,...)   -- Central derivative of f(x)
 3 :    local dx = DXX*x -- DXX = .01 to begin
 4 :    if dx==0.0 then dx = DXX end
 5 :    local a = {(f(x+dx,...) - f(x-dx,...))/(2*dx)}
 6 :    local p2, nj
 7 :    for j=1,NMAX do -- Richardson's improvement
 8 :        dx = dx/2; p2 = 4 -- Second order error
 9 :        a[j+1] = (f(x+dx,...) - f(x-dx,...))/(2*dx)
10 :        for k=j,1,-1 do
11 :            a[k],p2 = (p2*a[k+1] - a[k])/(p2-1),  4*p2
12 :        end
13 :        nj = j
14 :        p2 = abs(a[1]-a[2])/abs(a[1]+a[2])
15 :        if p2<err then break end
16 :    end
17 :    return a[1],nj,p2
18 : end
19 : setfenv(deriv,{abs=(Complex or math).abs,NMAX=10,
         DXX=1.e-2,err=1.e-14})
20 : deriv2 = function(f, x,...) -- second derivative of f(x)
21 :    return deriv((function(x,...) return deriv(f, x,...) end),
         x,...)
22 : end
23 : rderiv = function(f,x,...) -- Right derivative of f(x)
24 :    local dx = abs(DXX*x) -- DXX = .01 to begin
25 :    if dx==0.0 then dx = DXX end
26 :    local x0 = x + dx*DXD -- A little right of point
27 :    local a = {(f(x0+dx,...)-f(x0,...))/dx} -- One sided deriv
28 :    local p2, nj
29 :    for j=1,NMAX do -- Richardson's improvement
30 :        dx = dx/2; p2 = 2 -- First order error
31 :        x0 = x + dx*DXD
32 :        a[j+1] = (f(x0+dx,...)-f(x0,...))/dx
33 :        for k=j,1,-1 do
34 :            a[k],p2 = (p2*a[k+1] - a[k])/(p2-1),  2*p2
35 :        end
36 :        nj = j
37 :        p2 = abs(a[1]-a[2])/abs(a[1]+a[2])
38 :        if p2<err then break end
39 :    end
40 :    return a[1], nj, p2
41 : end
42 : setfenv(rderiv,{abs=(Complex or math).abs,NMAX=10,DXX=1.e-2,
         DXD=1.e-6,err=1.e-14})
43 : rderiv2 = function(f, x,...) -- Right second derivative
44 :    return rderiv((function(x,...) return rderiv(f, x,...) end),
         x,...)
45 : end
-----
69 : sderiv = function(f,x,...)
70 :    local dx = abs(DXX*x)
71 :    if dx==0.0 then dx = DXX end
72 :    return (f(x+dx,...)-f(x-dx,...))/(2*dx)
73 : end
74 : setfenv(sderiv,{DXX=1.e-6,abs=(Complex or math).abs})
```

Listing 5.2. Code segments for improved derivative using Richardson extrapolation.

data to the right of the derivative point with a slight offset from the data point by some small value controlled by the DXD variable. Not shown in the listing, but included in the file is a similar lderiv() function from line 46 through line 64 which uses evaluations only to the left of the data point. If one applies the three formulations to the math function abs(x) at x = 0 for example, the derive() function will return 0, the rderiv() function will return +1 and the lderiv() function will return -1. Also included in the listing are functions for evaluating the second derivative of a function by returning the derivative of the derivative as a one line statement. Examples of this are on lines 20 through 22 for the central derivative and on lines 43 through 45 for the right derivative. Finally a simple double sided derivative function sderiv() is coded on lines 69 through 73 without the Richardson enhancement. This is primarily for comparison purposes with the Richardson extrapolations.

Note the use of the three dots (...) as the third argument to the deriv() function. This is included so that additional arguments can be passed along to the function for which the derivative is being computed. This is for optional arguments and is not used in the examples in this chapter, but may be very useful in some applications.

Some discussion is appropriate of the parameters used in the code for the starting step size and the maximum number of Richardson iterations. There are several considerations for these parameters. First, the step size should never become so small that the intrinsic round-off errors are the limiting errors in the basic derivative calculations. If this occurs then the error will no longer be related to step size as assumed in the iterations. As can be seen from Figures 5.1 through 5.3 this means that the step size should not get below about 1.e-5 for the double sided derivative and about 1.e-7 for the single sided derivative. On the other hand, the maximum step size should be such that a reasonably accurate approximation to the derivative is obtained. For the test function data in Figures 5.1 through 5.3 it can be seen that this upper limit is about 1.e-2. Thus if the extrapolation procedure starts with a step size of 1.e-2 and decrease by a factor of 2 for each iteration, then the process can go through at most about 12 iterations. In practice it is found that about 10 Richardson iterations is the maximum number that can be effectively utilized. These are the limits coded into the algorithm by the NMAX parameter used on line 7. For the double sided derivative used with deriv() the error is second order in the displacement so the value of n in Eq. (5.18) is 2 and this is used by setting $2^2 = 4$ on lines 8 and 11 of the code. For the single sided derivative used with rderiv() and lderiv() the error is first order in the displacement leading to a 2 factor on lines 30 and 34. One might be tempted to set the initial step size to a smaller value, such as 1.e-4 and implement the algorithm. However, this typically leads to decreased accuracy as can be verified by changing the constants in the algorithm. Thus while there are some accuracy benefits to be obtained from the use of Richardson's improvements, there are very important limits to the number of cycles and the step sizes that can be used for an improvement in the derivative value.

```
 1 : -- /* File list5_3.lua */ -- Tests of derivative code
 2 :
 3 : require"deriv"
 4 :
 5 : f1 = function(x) return x^5 end
 6 : f2 = math.exp
 7 : val1 = 5*(1.e8)^4
 8 : val2 = math.exp(20)
 9 :
10 : val,n,ex = deriv(f1,1.e8)
11 : print(val,n,ex,(val-val1)/val1)
12 :
13 : val,n,ex = deriv(f2,20)
14 : print(val,n,ex,(val-val2)/val2)
15 :
16 : val,n,ex = rderiv(f2,20)
17 : print(val,n,ex,(val-val2)/val2)
18 :
19 : val,n,ex = lderiv(f2,20)
20 : print(val,n,ex,(val-val2)/val2)
21 :
22 : val,n,ex = deriv2(f1,1.e8)
23 : val2 = 20*(1.e8)^3
24 : print(val,n,ex,(val-val2)/val2)
25 :
26 : f3 = math.sin
27 : angle,vald,err = {},{},{}
28 : for i=1,401 do
29 :    ang = 2*math.pi*(i-1)/400
30 :    val = math.cos(ang)
31 :    angle[i] = ang
32 :    vald[i] = deriv(f3,ang)
33 :    if val==0 then err[i] = 0
34 :    else err[i] = ((vald[i]-val)/val) end
35 : end
36 : write_data("list5_3.dat",angle,vald,err)
Output:
5e+032  3          0           3.3146493257447e-015
485165195.40977  4  6.1427164746483e-017  -5.0615983751099e-014
485165195.40983  6  3.0713582373237e-015   7.9241042522959e-014
485165195.40979  6  4.4227558617466e-015  -4.5456101912395e-015
1.9999999999999e+025  2  3.9728447488003e-015  -7.1940702208e-014
```

Listing 5.3. Examples of numerical derivatives.

Examples of using the improved algorithm for numerical derivatives are shown in Listing 5.3. For each derivative calculation four values are printed in the output: the derivative value, the number of Richardson cycles, the relative error estimate in the derivative and the actual relative error achieved. The first line of output shows the calculated derivative for the x^5 function at x = 1.e8 and the results is the exact value, obtained after 3 Richardson iterations. The second through fourth line of output evaluates the derivative of the exp(x) function at x = 20 using the central difference function and the right and left derivative functions. All of the results are accurate to about the same precision of about 13 decimal digits. This is perhaps surprising since the central differencing expression used in deriv() has a second order error while the one sided differencing expressions used in lderiv() and rderiv() have a first order error in the displacement. However, much of this difference is made up for by the Richardson improvement algorithm. The printed output shows that 6 Richardson extrapolations are used by the lderiv() and rderiv() functions and only 4 are required by the central differencing algorithm. The precision is in the range of 13 to 14 digits, an improvement from the simple derivative expression, as coded in the sderiv() function for example, of 3 to 4 decimal digits in accuracy.

A loop from line 28 to 35 of Listing 5.3 evaluates the numerical derivative and calculates the error in the value for a simple sin() function over a cycle of the function. By repeating the code and substituting the sderiv() function for the derive() function data can be generated to compare the improvement with the Richardson extrapolation. Figure 5.4 shows the results of such a comparison.

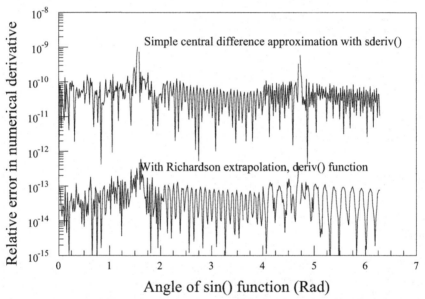

Figure 5.4. Comparison of error in numerical derivatives for sin() function using simple differencing and with Richardson extrapolation.

With the simple differencing of sderiv() a relative error of around 10^{-11} to 10^{-10} is seen in the figure. The Richardson extrapolation improves this by about 3 decimal digits to the 10^{-14} to 10^{-13} range. While this is a very significant improvement in the accuracy of the derivative, it is seen that the relative errors are still somewhat larger than the intrinsic machine accuracy. This should not be a problem for almost all real world problems as the accuracy exceeds what is normally required or expected. The Richardson extrapolation technique is a technique that can be applied to a range of numerical algorithms and will be used again when implementing numerical integration algorithms.

5.2 Maximum and Minimum Problems

In addition to many engineering models that can be formulated as problems requiring finding the zeros or roots of functions, a range of engineering problems can be formulated as finding the maximum or minimum value of a function or finding the parameter values leading to the extreme values of the function. Such problems are generally considered under the heading of "Optimization". An extensive body of research and software exists for treating optimization problems. The discussion here will cover only a very small part of such a broad area. It is included as one application of the derivative techniques developed in the previous section.

At a maximum or minimum of a function the first derivative is zero. Thus if the code segments developed in the previous section are used to evaluate the derivative of a function and then the zeros of the derivative function are calculated one will have evaluated the points where a maximum or minimum value occurs. As an example of this approach consider the three functions shown in Figure 5.5. One function (f_2) has a maximum at some negative x value around -6 and two functions have minimum values at around +6 (f_1) and -1 (f_3).

The equations for the three functions are shown in Listing 5.4 along with code segments for calculating the points of maximum or minimum value. The $f_1()$ function is an exponential function, the $f_2()$ function is a fourth order polynomial and the $f_3()$ function is a polynomial of order 4 that has previously been considered in Chapter 3 as Eq. (3.2). These all have only one extreme value and are rapidly varying around the extreme points.

The $f_3()$ function has a plateau around x = .4 where the derivative is close to zero. The three functions are defined in the listing on lines 6 through 8. The newton() function is called on lines 15, 20 and 25 to obtain the zero points of the first derivatives. This requires that the function passed to newton() be a function of the independent variable that returns the derivative of the function. In turn the derivative function requires as input the name of the function whose derivative is being evaluated. The newton() function can not directly pass along the function name to the derivative function. Thus one has to define in some way a "proxy" function that knows the name of the function but which requires as input only the value at which the derivative is to be determined. While there are different ways of accomplishing this goal, the code on lines 10 through 12 shows one such way that will find many such uses in this work. A function fderiv() is defined on line 10 that returns a function which in turn when called returns the desired numerical derivative value. The name of the function for which the derivative is to be evaluated is passed as an argument to the fderiv() function. This name is embedded within the function returned by fderiv() and the returned function only then needs to be supplied with the value at which the derivative is to be evaluated. For example line 13 then defines a function f1deriv = fderiv(f1) which has embedded the "f1" function name and so when code such as f1deriv(2.0) is executed, the calculated value will be the derivative of the f1 function at the point 2.0. The name of

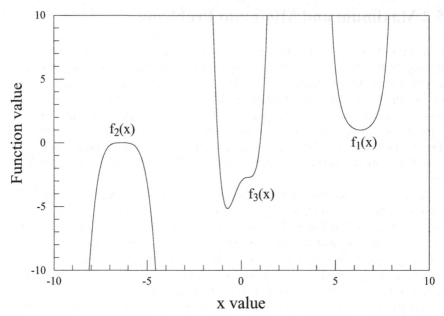

Figure 5.5 Examples of three functions with maximum or minimum values.

this function is then passed to the newton() function on line 15 to evaluate the zero point of the derivative of the f1() function. A second method of passing the proxy function to newton() is shown on line 20 where no intermediate name is used for the function and the calling statement is simply newton(fderiv(f2),0). All this is made possible by a property of the Lua language known as function closure. When a function is defined such as on line 11 of the listing, it has access to all the non-local variables used in defining the function which in this case includes the function name (f) passed to the defining equation. When later the returned function is called with only an x value, it will have correctly remembered the function name and evaluate the derivative for the appropriate function. The reader is encouraged to carefully study this construct and how it operates as this feature will find extensive use in subsequent chapters.

The printed output in Listing 5.4 shows that the extreme points are correctly calculated with an accuracy of about 13 digits for the f1() function and about 7digits for the f2() function. The accuracy for the f3() function is probably about 10 digits. The printed output also indicates that the number of newton() iterative loops for the calculations were 48, 52 and 171 for the f1() to f3() functions. The output also gives the value of the function at the maximum or minimum points which are 1, 0 and -5.1717619591603 for the three functions. In order to determine if the points correspond to maximum or minimum values, a simple test of a nearby point is used as on lines 17, 22 and 27. These are fairly simple functions,

```
 1 : -- /* File list5_4.lua */ -- Finding max and min values
 2 :
 3 : require"deriv"; require"newton"
 4 :
 5 : pos = 6.3324567
 6 : f1 = function(x) return math.exp((x-pos)^2) end
 7 : f2 = function(x) return -(x+pos)^4 end
 8 : f3 = function(x) return 5*x^4-4*x^2+2*x-3 end
 9 :
10 : fderiv = function(f)
11 :    return function(x) return deriv(f,x) end
12 : end
13 : f1deriv = fderiv(f1)
14 :
15 : sol,n,err = newton(f1deriv,0)
16 : print(sol,n,err,f1(sol),(sol-pos)/pos)
17 : if f1(sol)<f1(1.01*sol) then
          print("solution is a minimum at ",sol)
18 : else print("solution is a maximum at ",sol) end
19 :
20 : sol,n,err = newton(fderiv(f2),0)
21 : print(sol,n,err,f2(sol),(sol+pos)/pos)
22 : if f2(sol)<f2(1.01*sol) then
          print("solution is a minimum at ",sol)
23 : else print("solution is a maximum at ",sol) end
24 :
25 : sol,n,err = newton(fderiv(f3),0)
26 : print(sol,n,err,f3(sol))
27 : if f2(sol)<f2(1.01*sol) then
          print("solution is a minimum at ",sol)
28 : else print("solution is a maximum at ",sol) end
Output:
6.3324567       48      1.1351405184306e-014   1        0
solution is a minimum at        6.3324567
-6.332456131 52 -9.963317266e-010 -1.047486802e-025 8.983900040e-008
solution is a maximum at        -6.3324561310984
-0.73247730431733   171   7.1126173076503e-012   -5.1717619591603
solution is a minimum at        -0.73247730431733
```

Listing 5.4. Example of finding a maximum or minimum using derivative function.

but illustrate the principle of combining the numerical derivative calculations with finding the roots of a nonlinear function to determine the maximum and minimum points and values of functions of one variable. The technique can be applied to more complex functions with one caveat: The function for which a maximum or minimum is being calculated must have a smooth derivative at least to as many digits as one desires in the calculated value. For given analytical equations such as in these examples, this is not a problem. However, in more complex problems where the function may be known only thorough some computer numerical algorithm, this can be a potential problem. In the next section this work will be extended to the maximum or minimum of a function of multiple variables.

5.3 Numerical Partial Derivatives and Min/Max Applications

When dealing with a function of several variables, the derivative function becomes a partial derivative where only one variable changes at a time while keeping the other variables constant. For this assume that one has a function of the form $f(x)$, where x is now a table of values of the form $x = \{x_1, x_2, x_3 \ldots x_n\}$. Code for a partial derivative is very similar to that for a single variable function, except that an additional argument to the function is needed that identifies which variable to use in calculating the derivative. Such a pderiv() code segment is shown in Listing 5.5. This uses the simple double sided derivative definition without any Richardson extrapolation enhancement. This should be more then adequate for most applications as it gives on the order of 10 digits of accuracy, as previously shown. The code is straightforward, the only notable feature being that a copy of the $x[i]$ variable must be saved and its value restored its value after changing it in the derivative evaluation. The displacement factor is set at 1.e-6 which is near the value for maximum accuracy as shown in Figures 5.1 and 5.2. This can be changed by the user if desired through the getfenv(pderiv).FACT statement.

```
 1 : --/* File deriv.lua */ -- Derivative
........
75 : function pderiv(f,x,i)
76 :     local xi,fv = x[i]
77 :     local dx = xi*FACT
78 :     if dx==0 then dx = FACT end
79 :     x[i] = xi + dx; fv = f(x)
80 :     x[i] = xi - dx; fv = (fv - f(x))/(2*dx)
81 :     x[i] = xi; return fv
82 : end
83 : setfenv(pderiv,{FACT=1.e-6})
```

Listing 5.5. Code for partial derivative of multivariable function.

An application of partial derivatives is finding the maximum or minimum of a multidimensional function in much the same way as the previous example of finding a maximum or minimum of a single dimension function. A local max/min can be located by requiring that all partial derivatives of the function be zero. For a function such as $f(x)$ of n dimensions, this gives n nonlinear equations to be solved of the form:

$$\frac{\partial f}{\partial x_1} = 0; \frac{\partial f}{\partial x_2} = 0; \ldots \frac{\partial f}{\partial x_n} = 0 . \qquad (5.19)$$

These nonlinear equations can then be solved through use of the nsolv() function.

A textbook-like example of such an application is shown in Listing 5.6. In this code a function of five variables is defined on lines 5 through 10. The function is:

$$f(x) = \exp(\sum_{i=1}^{5}(x_i - x_{io})^2) = \prod_{i=1}^{5}\exp((x_i - x_{io})^2) \qquad (5.20)$$

```
 1 : -- File list5_6.lua -- Minimum problem using pderiv()
 2 :
 3 : require"deriv"; require"nsolv"
 4 :
 5 : xio = {1,2,-3,4,-5} -- minimum points
 6 : fuv = function(x) -- Multi-variable function
 7 :    local fv = 0
 8 :    for i=1,#x do fv = fv + (x[i] - xio[i])^2 end
 9 :    return math.exp(fv)
10 : end
11 :
12 : x = {0,0,0,0,0} -- Initial guess at minimum
13 : print('At initial guess')
14 : print('Partial 1 derivative =',pderiv(fuv,x,1)) -- check on
15 : print('Partial 3 derivative =',pderiv(fuv,x,3))
16 :
17 : eqs = function(y,x) -- Equatuions to be solved
18 :    for i=1,#x do
19 :        y[i] = pderiv(fuv,x,i) -- Force partial deriv to zero
20 :    end
21 : end
22 :
23 : print(nsolv(eqs,x)) -- Now solve them
24 : print('Minimum occurs at')
25 : table.foreach(x,print) -- Print minimum point
26 : print('Minimum value =',fuv(x))
Output:
At initial guess
Partial 1 derivative = -1.5389570545998e+024
Partial 3 derivative = 4.6168711611149e+024
75      6.9389580471016e-011
Minimum occurs at
1       1.0000000000226
2       1.9999999999923
3       -2.9999999999978
4       4
5       -5
Minimum value =        1
```

Listing 5.6. Application of partial derivative to maximum, minimum problem.

The function has one obvious minimum located at $x_i = x_{io}$. An initial guess at the solution point is defined on line 12 as the origin with all values of zero. Lines 14 and 15 simply test the ability to calculate partial derivatives at the initial guess and the output shows that the partial derivatives are on the order of 1.e+24, or rather large values. The five defining equations setting the partial derivatives to zero are defined in the eqs() function on lines 17 – 21. This simply calls the basic partial derivative function, as defined in Listing 5.5 in turn for each variable. The solution of these equations is then a one line statement for calling nsolv() on line 23. The output shows that 75 iterations are required by nsolv() to solve the equation set given the initial guess. The final printed solution values show that the calculated minimum point is accurate to more than 12 digits in all five variables.

This is a fairly straightforward textbook like example and has only one minimum, although the function does vary rapidly away from the minimum. Other more practical problems with multiple minima are likely to be more difficult to solve. However, it does illustrate the calculation of partial derivatives combined with the previously coded nsolv() function and the use of these in evaluation the maximum or minimum of a multivariable function. This is one type of optimization problem that is known in the literature as unconstrained minimization of a function.

Many other optimization problems involve calculating the minimum of a multivariable function subject of certain constraints on the variables. One such example will be shown here to illustrate the use of the partial derivative code with such problems. The problem to be considered is finding the minimum of a function $f(x)$ subject to constraints on the $x = \{x_1, x_2, ..x_n\}$ variables given by equations of the form:

$$g_1(x) = 0; g_2(x) = 0; \ldots g_m(x) = 0, \tag{5.21}$$

where m is the number of such constraint equations. Such a problem can be approached by the use of Lagrange multipliers where m new variables usually designated as $\lambda_1, \lambda_2, \ldots \lambda_m$ are introduced that multiply the constraint equations. The function to be minimized is then taken to be:

$$f(x) + \lambda_1 g_1(x) + \lambda_2 g_2(x) + \ldots \lambda_m g_m(x). \tag{5.22}$$

Partial derivatives of this equation with respect to the independent variables, give n equations and the constraint equations give an additional m equations for a total of n+m equations. These are then solved simultaneously for the n values of x and the m values of λ.

For an application of this constrained optimization approach, consider the following function and constraints:

$$f(x) = \exp(\sum_{i=1}^{5} (x_i - x_{io})^2) = \prod_{i=1}^{5} \exp((x_i - x_{io})^2)$$

$$x_1 - x_2 - 1 = 0 \tag{5.23}$$

$$x_3 + x_4 = 0$$

$$x_4 + x_5 = 0$$

This is the same function as in the previous unconstrained example but now with three constraints on the independent variables. Code for solving for the function minimum with the constraints is shown in Listing 5.7. The function to be minimized is defined on lines 6 through 10 just as before. The three constraints are defined as functions within a table of functions on lines 11 through 15. The eqs() function on lines 21 through 29 then defines the set of equations to be solved corresponding to Eqs. (5.21) and (5.22). This function is defined in such a manner that it can be used unchanged with any desired minimization problem and set of constraints – only the code defining the function and constraints need be changed for another problem. This interface function, eqs(), is then used in the call to nsolv() on line 31 to solve the set of equations with constraints.

```
 1 : -- /* File list5_7.lua */ -- Constrained optimization
 2 :
 3 : require"deriv"; require"nsolv"
 4 :
 5 : xio = {1,2,-3,4,-5} -- minimum points
 6 : f = function(x) -- Multi-variable function
 7 :    local fv = 0
 8 :    for i=1,nx do fv = fv + (x[i] - xio[i])^2 end
 9 :    return math.exp(fv)
10 : end
11 : gs = { -- Table of constraint functions
12 :    function(x) return x[1]-x[2]-1 end, -- #1
13 :    function(x) return x[3]+x[4] end, -- #2
14 :    function(x) return x[4]+x[5] end -- #3
15 : }
16 :
17 : nx,nl = #xio, #gs -- number variables and constraints
18 : x = {0,0,0,0,0,0,0,0} -- Initial guess at minimum
19 : dx = -.2; step = {dx,dx,dx,dx,dx,0,0,0} -- Limit steps
20 :
21 : eqs = function(y,x) -- Equations to be solved
22 :    local yv
23 :    for i=1,nx do
24 :       yv = pderiv(f,x,i)
25 :       for j=1,nl do yv = yv + x[nx+j]*pderiv(gs[j],x,i) end
26 :       y[i] = yv
27 :    end
28 :    for i=1,nl do y[nx+i] = gs[i](x) end
29 : end
30 :
31 : print(nsolv(eqs,x,step)) -- Now solve them
32 : print('Solution values are:')
33 : table.foreach(x,print) -- Print minimum point
34 : print('Value at solution =',f(x))
Output:
48   3.5858001284606e-009
Solution values are:
1    1.9999999999661
2    0.99999999996614
3    -4.0000000000085
4    4.0000000000085
5    -4.0000000000085
6    -109.19630006843
7    109.19630006949
8    -109.19630007244
Value at solution =    54.598150033144
```

Listing 5.7. Example of multi-variable optimization with constraints.

Some features of the code should be noted. First the initial guess of solution values on line 18 has eight values corresponding to the five independent variables and the three Lagrange multiplier terms. All these values are simply taken as zero for the initial guess even though for this simple problem one knows a better set of initial guesses. An array of step parameters is defined on line 19 to limit the maximum change in the solution values between Newton iterations. Without some limitation the nsolv() function will not converge for this problem as the reader can readily verify by removing the step parameter and re-executing the code. The step parameters limit the changes in the independent parameters to +/- 0.2 units between Newton iterations. For the Lagrange multiplier terms, no limits are imposed as expressed by the zero terms for the last three entries in the step table of values. This set of step parameters was chosen through some experimentation with various sets of step parameters. The reader is encouraged to execute the code with various initial values and step parameters and observe the convergence or lack of convergence for some sets of parameters.

The first line of the output listing shows that 48 Newton iterations were required to obtain a solution. From the printed solution values, one can readily verify that the three constraint equations are indeed satisfied. Without the constraints the minimum value occurs at the point $x = \{1,2,-3,4,5\}$ as previously obtained from Listing 5.6. The last three solution values correspond to the three values (-109.19, 109.19 and -109.19) of the Lagrange multipliers. Finally, the minimum value of the function of 54.59815 is printed as the last output value. For this simple problem one can readily see that the obtained solution points are the expected values when the constraints are taken into account.

There is a large volume of literature dealing with optimization problems and the code presented here is certainly not adequate for many such problems. It should be noted that nsolv() takes a numerical derivative of the equations to be solved in order to set up the linearized equations to be solved in each Newton iteration. Thus using nsolv() and the partial derivatives means that the complete code ends up taking a numerical second derivative of the basic equation being solved. Thus for this approach to have a chance of succeeding, the function being minimized must have a smooth second derivative. A large number of optimization problems in engineering practice involve minimizing an equation that is a linear function of the independent variables subject to a set of constraints on the variables. The present approach obviously can not be used for such problems as a second derivative does not exist and the solution of such a minimization problem is found on some boundary of the valid set of variable values imposed by the constraints. Solution methods for such problems are typically identified as linear programming methods and the resulting algorithms that have been developed solve very large problems with thousands of variables and constraints with great efficiency.

Another broad class of optimization problems involves minimizing a function subject to inequality constraints as opposed to the equality constraints in the example used here. Such problems involve the introduction of additional variables and additional solution methods and again such problems are not discussed here. The two examples presented here are simply used as examples of the partial derivative code developed and when combined with Newton's method can be used to solve a limited subset of optimization problems. The reader is referred to the extensive literature on optimization problems if general optimization problems are of major interest.

5.4 Fundamentals of Numerical Integration

The numerical evaluation of integrals is a much studied subject of numerical methods and many diverse algorithms have been developed for such problems. The general mathematical statement is given the integral:

$$I = \int_a^b f(x) , \qquad (5.24)$$

obtain a numerical approximation to the integral accurate to some desired number of decimal digits.

This is equivalent to calculating the area under the curve defined by the function $f(x)$ and the horizontal axis between the points a and b. An example section of such an area is shown in Figure 5.6. A basic approach to the definition of an integral is to subdivide the interval range into panels as shown in the figure of width $h = x_{i+1} - x_i$ and sum the areas of each of the panels as $I \approx I_1 + I_2 + \cdots I_n$.

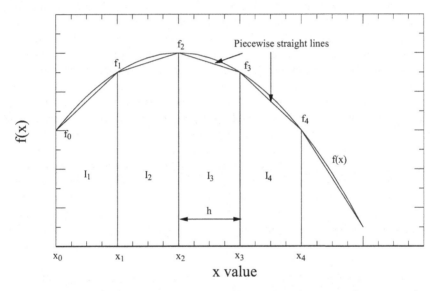

Figure 5.6. Illustration of contributions to integral and straight line approximations.

As $n \to \infty$ the, exact value of the integral can be approached. Of course in practice, an infinite number of intervals can never be achieved and only a numerical approximation to the exact value can be obtained by taking a finite number of panels. Algorithms differ in how the range is subdivided and in how the area of each panel or sub area is approximated. A perusal of most numerical methods texts will find many algorithms for numerical integration discussed under such topics as the rectangular rule, the trapezoidal rule, Simpson's rules, Newton-Cotes formulas, Gauss quadrature, etc. There are usually very limited recommendations on a method and the user is left to ponder which method he/she should use. Generally this author has found that the best approach is usually to use a fairly simple algorithm and repeat it over and over rather than using a very complex algorithm applied a smaller number of times. While there may be exceptions to this rule, it is recommended that one should first think about simple algorithms and only go to complex algorithms if the simpler ones are not adequate.

The discussion here will consider primarily the use of the trapezoidal rule. Several reasons for this preference will be discussed in the course of code development. This algorithm is illustrated in Figure 5.6 and consists of using straight line approximations to each panel. The function is then evaluated at the end points of each panel. The area of each panel is then one-half the sum of the two sides multiplied by the panel width or:

$$I \approx h \sum_{i=1}^{n-1} \left(\frac{f_i + f_{i+1}}{2} \right) = h(\frac{f_1}{2} + f_2 + f_3 + \cdot \cdot f_{n-1} + \frac{f_n}{2}) . \qquad (5.25)$$

One of the virtues of this algorithm is the simplicity of the final expression. All the interior points excluding the two end points are added and then one half of each of the end points is added and the resulting sum multiplied by the panel width. All interior points are treated in the same manner and the number of multiplications is minimized. Another feature of the trapezoidal rule is that the error in the approximation to the area is of order h^2. The accuracy can be increased by decreasing the panel width and increasing the number of panels. A major advantage of this algorithm is the ease with which the number of panels being considered can be doubled. Suppose, for example, that the approximate integral value has been calculated for some h value and it has been concluded that the result does not meet some desired accuracy criteria and that one wants to double the number of panels by dividing each panel into two equal panels. There is now twice the number of interior points to sum over but half of the points are the same as the points included in the previous sum. Thus if one retains the sum of the function values for the interior points, one simply has to add the new interior points to the previous sum and multiply by the new half panel size. As will be seen it is easy to implement an adaptive algorithm for the trapezoidal rule to increase the number of panels to achieve some specified accuracy.

Figure 5.7 shows four test functions that will be used to explore some numerical integration approaches. The humps(x) and $f_y(x)$ functions are typical of smooth functions and should be relatively easy for numerical integration algorithms. The $f_x(x)$ is a little more difficult as it has an infinite first derivative at the x = 0 boundary. Finally the $f_z(x)$ is the most difficult as it has an infinite value at the x = 0 boundary although the area under the curve has a finite value. This is typical of functions with an infinite value at one of the boundaries and requires special attention to for a numerical integration. A straightforward application of Eq. (5.25) to such a function (the f_z function) will obviously result in an infinite value evaluated at the boundary. One advantage of these functions is that the area over the 0 to 4 interval is known exactly so the error in any numerical integration algorithm can be determined exactly for these functions.

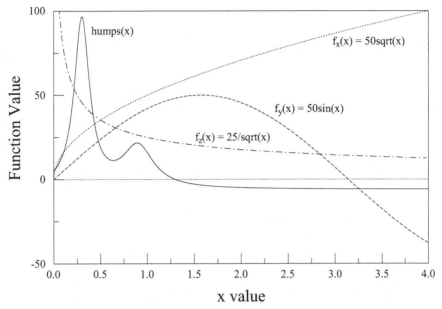

Figure 5.7. Some test functions for exploring numerical integration algorithms.

Listing 5.8 shows code for implementing a simple trapezoidal algorithm based upon Eq. (5.25). The sintg() function on lines 3 through 9 uses a fixed number of panels (4000 here) as shown on line 4. The algorithm is implemented in only 4 lines of code with only two multiplications. The listing shows the area calculated for three of the functions shown in Figure 5.7: $f_x(x) = 25\sqrt{x}$, $f_y(x) = 50\sin(x)$ and $humps(x)$. The $f_z(x)$ function is not shown as it requires special care for the end point. The output given in the listing shows that the integral of the square root function is accurate to about 6 digits while the other two integrals are accurate to about 7 digits. This is with a panel width h of 1.e-3 and many more panels can certainly be used with modern computers.

One question that arises is how many panels should be used in the trapezoidal algorithm for a given accuracy or precision in the computed value? According to theory, the relative error should decrease as approximately the second power of h. So if h is decreased to 1.e-5 or if 400,000 panels are used, an accuracy of about 10 digits should be achieved in the square root function. To test the relationship of relative error to the size of h, the program in Listing 5.8 was executed with varying panel sizes and the results for the 50sin(x) and humps(x) functions are shown in Figure 5.8. For comparison purposes, a dotted line is shown with the h^2 variation. For h sizes above about 1.e-6, the data shows an almost exact variation with the square of the panel size. As h is changed by 5 orders of magnitude from 1.e-1 to 1.e-6, the relative error decreases by approximately 10 orders of magnitude and an accuracy of about 13 digits is achieved for these example functions for

```
1 : -- /* File list5_8.lua */ -- Simple integration
2 :
3 : function sintg(xmin,xmax,f)
4 :    local n = 4000
5 :    local dx = (xmax-xmin)/n
6 :    local sum = 0.5*(f(xmin)+f(xmax))
7 :    for i=1,n-1 do sum = sum + f(xmin+i*dx) end
8 :    return sum*dx
9 : end
10 :
11 : fx = function(x) return 25*x^.5 end
12 : iv = sintg(0,4,fx)
13 : ivv = 25*4^1.5/1.5
14 : print(iv,ivv,(iv-ivv)/ivv)
15 :
16 : fy = function(x) return 50*math.sin(x) end
17 : iv = sintg(0,4,fy)
18 : ivv = 50*(1-math.cos(4))
19 : print(iv,ivv,(iv-ivv)/ivv)
20 :
21 : humps = function(x) --Has 2 maxima and 1 minimum between 0 and 1
22 :    return 1/((x-.3)^2+.01) + 1/((x-.9)^2+.04) -6
23 : end
24 : iv = sintg(0,4,humps)
25 : ivv = 10*(math.atan(37)-math.atan(-3))+5*(math.atan(15.5)-
26 :    math.atan(-4.5))-24
27 : print(iv,ivv,(iv-ivv)/ivv)
Output:
133.33316950568        133.33333333333        -1.2287074353168e-006
82.682174152999        82.682181043181        -8.3333335197028e-008
18.220695625072        18.22070084151 -2.8629185538022e-007
```
Listing 5.8. Simple numerical integration by trapezoidal rule and examples.

an h of about 1.e-6. For panel sizes below about the 1.e-6 value, an intrinsic numerical limit of about 14 decimal digits is seen for the integral of the two functions. This is a considerably better relative accuracy than previously shown to be achievable for the simple numerical derivative. The difference here is that differences between nearly equal quantities are not being takes as with differentiation. From this it would be inferred that for these two typical functions a relative small panel size should be used for h in the range of 1.e-6. With such a small panel size about 4,000,000 function evaluations are required for these examples with an integration range of 0 to 4. This can becomes expensive in terms of computer time, so it is useful to look for ways of increasing the accuracy with a smaller number of panels or with a larger panel size. The key to achieving this goal is through Richardson's extrapolation theorem which was previously used for numerical integration. The trapezoidal integration algorithm again has a theoretical error of $O(h^2)$ so the same Richardson theory applies and if one has approximated the integral by two calculations, an estimate with improved accuracy can be obtained by use of:

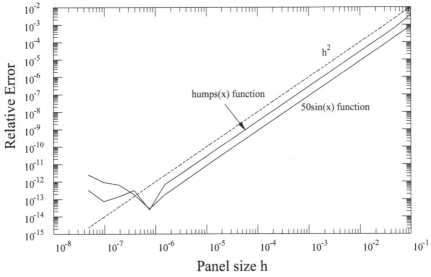

Figure 5.8. Relative error of trapezoidal rule for different panel sizes.

$$I \cong I_2(h_2) + \frac{I_2(h_2) - I_1(h_1)}{2^n - 1} = \frac{2^n I_2(h_2) - I_1(h_1)}{2^n - 1}, \text{ where } h_2 = h_1 / 2. \qquad (5.26)$$

This assumes that the panel width is changed by a factor of 2 in the two calculations. This is easily implemented as previously discussed with the trapezoidal algorithm as all the previously summed function values can be reused as the panel size is reduced at each iterative step by a factor of 2. The use of Richardson's extrapolation when applied to integration is known in the literature as Romberg integration.

Listing 5.9 shows a code segment for the trapezoidal integration algorithm with several enhancements including Richardson's extrapolation. For an initial consideration of the code ignore the code on lines 6 through 16 for special treatment of end points. The main algorithm begins with 100 panels (see n = NINIT on line 3) and calculates the first estimate of the area on lines 17 through 19 just as in the simple trapezoidal algorithm of Listing 5.8. Lines 20 through 31 then implement one iterative step of the Romberg algorithm which doubles the number of panels (line 21) and sums over the additional interior points on line 22 skipping every other integer value. The calculations for Richardson extrapolation are performed on lines 25 through 28 and the error checked on lines 29 and 30. When the convergence criterion is met the area is returned on line 32 along with an estimate of the error in the calculated numerical integral.

Again note the use of the three dots (...) as the fourth argument to the intg() function on line 2. This is for additional arguments to be possibly passed to the function such as seen on line 8 for which the integral is being evaluated. These arguments are optional and the examples in this chapter do not use this feature. However, this can be important in many applications.

```
 1 : --/* File intg.lua */ -- Integration with Romberg enhancement
 2 : intg = function(xmin,xmax,f,...)
 3 :    nend,n = RMAX, NINIT -- Initially set to 100 intervals
 4 :    dx,iend = (xmax - xmin)/n, 0.0 -- Special if infinite
 5 :    fxx = tostring(f(xmin,...)) -- ? infinite value boundaries.
 6 :    if fxx==inf or fxx==und then - Infinite value?, inf = 1/0
 7 :       dx = dx/NDX
 8 :       iend = fend(f(xmin+dx,...),f(xmin+dx/2,...),dx)
 9 :       xmin = xmin+dx; dx = (xmax-xmin)/n
10 :    end
11 :    fxx = tostring(f(xmax,...))
12 :    if fxx==inf or fxx==und then -- Infinite upper limit value?
13 :       dx = dx/NDX
14 :       iend = iend + fend(f(xmax-dx,...),f(xmax-dx/2,...),dx)
15 :       xmax = xmax-dx; dx = (xmax-xmin)/n
16 :    end
17 :    a1 = 0.5*(f(xmin,...)+f(xmax,...)) -- End point for TP rule
18 :    for i=1,n-1 do a1 = a1 + f(xmin+i*dx,...) end -- Sum points
19 :    a = {dx*a1} -- Initial area estimate from trapezoidal rule
20 :    for j=1,RMAX do -- Romberg cycles -- 12 maximum
21 :       n,dx = 2*n, dx/2 -- Double # slices, half spacing
22 :       for i=1,n-1,2 do a1 = a1 + f(xmin+i*dx,...) end -- Sum
23 :       a[j+1] = dx*a1 -- Updated area with additional slices
24 :       p2 = 2; alast=a[1]
25 :       for k=j,1,-1 do -- Romberg convegence accelerations
26 :             a[k] = (p2*a[k+1] - a[k])/(p2-1)
27 :             p2 = 2*p2
28 :       end
29 :       p2 = abs(a[1]-alast)/abs(a[1]+alast) -- Error estimate
30 :       if p2<ERROR then nend=j; break end
31 :    end
32 :    return a[1]+iend,nend,abs(a[1]-alast) -- Add area of ends
33 : end
34 :
35 : local fend = function(f1,f2,dx) - end values for infinite f(x)
36 :    local n = log(f2/f1)/log(2) --Approximates with power law on
x
37 :    if n>=1 then
             print("Value of integral appears to be infinite") end
38 :    return f1*dx/(1-n)
39 : end
40 : setfenv(intg,{abs=math.abs,RMAX=15,NINIT=100,NDX=1000,
          ERROR=1.e-11,tostring=tostring,inf=tostring(math.huge),
          und=tostring(0/0),fend=fend})
41 : setfenv(fend,{log=math.log,print=print})
42 :
43 : intg_inf = function(f,...) -- integral from 0 to infinity
44 :    return intg(0,1,(function(x,...) return f(x,...)+
          f(1/x,...)/x^2 end),...)
45 : end
```

Listing 5.9. Integration with trapezoidal rule and Romberg correction.

One final feature implemented in the code is a technique for approximating end points if the function evaluates to infinity at one or more of the end points. Handling end points where the function evaluates to infinity is not an easy task. In some cases such integrals are in fact infinite, as for example if one tries to integrate $1/x$ from zero to any finite value. On the other hand some functions such as

$1/\sqrt{x}$ have a finite integral value from 0 to some x. The intg() code segment in Listing 5.9 attempts to handle such cases by a separate evaluation of the contribution to the integral of values near the integral boundaries when the function evaluates to infinity (or 0/0) at the boundaries. Code for this is contained in lines 5 through 16 of the listing and in an auxiliary function fend() on lines 35 through 39 following the intg() code. The heart of this correction code is an assumption that the function follows some power law near the boundary with an infinite value such as C/x^n. If n is less than unity, a finite contribution to the area results, and this value is calculated and returned by the fend() function. This technique works with a range of functions which will be demonstrated with some examples.

Finally an intg_inf() is provided to evaluate integrals with infinite limits. This is discussed in a subsequent section of this chapter.

The maximum number of Romberg cycles is set by the RMAX factor (default value of 15). This is based upon the following considerations. Over 15 cycles, the number of panels can increase by a factor of $2^{15} = 32768$. Combining this with the initial 100 panels, the total number of function evaluations is about 3.e6. This begins to take noticeable computer time to execute and this sets an upper limit on the number of cycles and the smallest panel size. The convergence error is set at 1.e-11 and from Figure 5.8 it can be seen that this should be achieved easily for simple test function. It should be noted however, that as the integration interval increases, the required number of panels will also increase and this will increase the number of required Romberg cycles. If only engineering accuracy of 4 to 6 digits is desired, this is easily achieved for most functions. If an accuracy approaching the machine accuracy of 16 digits is desired, this is more difficult to achieve, but can be approached for well behaved functions (no singularities) and non infinite limits.

The code in Listing 5.10 illustrates the use of the intg() function code for integrating four functions three of which are shown in Figure 5.7. The additional test function used here is the fz(x) function on line 23 which is the $x \exp(-x)$ function. These are all fairly simple examples and cases where the functions can be integrated exactly so the accuracy of the numerical integration algorithm can be evaluated. Test cases such as these where the exact results are known are always important trials for any developed software so that the accuracy of the technique can be explored and potential problems with an algorithm can be identified. For each of the four functions in Listing 5.10 there are five columns of printed output giving (a) the number of Romberg iterative cycles used, (b) the numerical value of the integral, (c) the exact value of the integral, (d) the estimated absolute error in the integral from intg() and (e) the exact relative error in the integral from the calculated value and the known exact value.

The output shows that the code took 5 to 15 Romberg cycles in attempting to achieve the specified relative error of 1.e-11. It is seen that the $\sqrt{x}$ function on line 5 took the maximum number of Romberg cycles and thus did not meet the specified error criterion of about 11 decimal digits. However, the achieved accuracy is in fact about 11 decimal digits as can be seen from the achieved relative

```
 1 : -- /* File list5_10.lua */ -- Simple integration
 2 :
 3 : require"intg"
 4 :
 5 : fx = function(x) return 25*x^.5 end
 6 : iv,n,err = intg(0,4,fx)
 7 : ivv = 25*4^1.5/1.5
 8 : print(n,iv,ivv,err,(iv-ivv)/ivv)
 9 :
10 : fy = function(x) return 50*math.sin(x) end
11 : iv,n,err = intg(0,4,fy)
12 : ivv = 50*(1-math.cos(4))
13 : print(n,iv,ivv,err,(iv-ivv)/ivv)
14 :
15 : humps = function(x) --Has 2 maxima and 1 minimum between 0 and 1
16 :     return 1/((x-.3)^2+.01) + 1/((x-.9)^2+.04)
17 : end
18 : iv,n,err = intg(0,4,humps)
19 : ivv = 10*(math.atan(37)-math.atan(-3))+5*(math.atan(15.5)-
20 :     math.atan(-4.5))
21 : print(n,iv,ivv,err,(iv-ivv)/ivv)
22 :
23 : fz = function(x) return x*math.exp(-x) end
24 : iv,n,err = intg(0,4,fz)
25 : ivv = 1 - 5*math.exp(-4)
26 : print(n,iv,ivv,err,(iv-ivv)/ivv)
Output:
15 133.33333333464      133.33333333333 2.4029702672124e-009
9.7820418432093e-012
5 82.68218104318       82.682181043181 8.5265128291212e-013 -
9.2811552010238e-015
7 42.22070084151       42.22070084151  1.4843237750028e-011 -
1.7334127649377e-014
5 0.9084218055563 0.9084218055563 6.883382752676e-015
2.1998607167971e-015
```

Listing 5.10. Some simple examples of numerical integration.

error on the first line of the printed output. This can be seen to have about 5 more decimal digits of accuracy than achieved by the simple trapezoidal algorithm shown in Listing 5.8. For the other three functions, it can be seen that the achieved relative error is always less than the specified relative error for termination of the Romberg cycles. In fact the achieved relative error is in the range of 1.e-14 to 1.e-15. The relative error criterion used in the code to terminate the calculations is that the correction between the two last estimated values be less than the specified relative error (of 1.e-11). This is not the same as the actually achieved error, which could be larger or smaller than the correction at any iteration. However, these examples give us some confidence that the actually achieved accuracy will meet or exceed the termination criterion of at least 11 digits of accuracy.

From these examples, the advantage of using the Romberg or Richardson acceleration technique with numerical integration can be readily seen. If smooth functions such as the sin(x) and humps(x) function are considered, the results of the last column of lines two and four of the output show that the relative accuracy achieved in the integrals is 14 to 15 digits of accuracy. This was achieved with only 5 Romberg cycles which means that the last calculations would have been performed with $(100)2^5 = 3,200$ panels or with an h value of 4/3200 = 1.25e-3. From Figure 5.4 it can be seen that for this step size with just a simple trapezoidal summation a relative accuracy of only about 6 or 7 digits would be expected. To achieve a relative accuracy of 14 to 15 digits would require a step size of less than 1.e-6 or a total number of panels of greater than 1.e6. Since there must be one function evaluation for each panel, it is seen that this would require over 1.e6 function evaluations as opposed to only about 3.2e4 function evaluations with the Richardson acceleration algorithm. The number of function evaluations required with the Romberg technique is thus seen to be reduced for this function by a factor of almost 1000 for the same degree of precision.

When compared with the improvement achieved in taking a numerical derivative, this is a much more effective technique with numerical integration. This is because for differentiation, a much smaller range of step size was found over which the results followed the theoretical decrease in error with step size (see Figures 5.1 and 5.2) as opposed to numerical integration (see Figure 5.4). For numerical differentiation, machine round-off errors are a much more severe problem as step size is decreased and such an increase in error does not occur with numerical integration. It can be concluded from this, that at least for simple, well behaved, function, Richardson acceleration is a very effective algorithm for numerical integration. An exception to this is the square root function or fx() function in Listing 5.10. It is seen that the maximum number of Romberg cycles was employed in an effort to achieve a correction of less than 1.e-11. It is seen that after 15 Romberg cycles, the relative accuracy is in fact about 11 digits, but considerable less accurate than that of the other functions. This is due to the infinite derivative of this function at the lower limit of $x = 0$. This sharp upturn in the function makes contributions to the integral near this limit very important when achieving high precision in the result. Other integrals have special problems with numerical evaluation as are addressed in the next section. In any case the accuracy is certainly better with Richardson acceleration than it would be without it.

The advantage of Richardson acceleration with numerical integration is not so much the increased numerical accuracy that can be achieved but is the reduced computational time required to achieve a given level of accuracy. With the simple trapezoidal algorithm used in Listing 5.8 and without Richardson acceleration the results seen in Figure 5.8 show that an accuracy approaching the machine precision limit can be achieved. However, for the simple functions considered here, this required a very small panel size, a large number of function evaluations and a long computational time. The results in Listing 5.10 show that for the same functions with Richardson acceleration an accuracy near the machine limit can be obtained with orders of magnitude fewer function evaluations and hence orders of magnitude less computational time. This is the real advantage of the intg() code with Richardson acceleration.

Function	MATLAB quad()	MATLAB quadl()	intg()
$\sqrt{x}$	1.36e-6	5.51e-8	9.78e-12
$\cos(x)$	5.85e-9	1.68e-7	9.28e-15
humps(x)	1.22e-7	1.53e-13	1.738e-14
$x\exp(-x)$	9.00e-8	4.84e-7	2.20e-15

Table 5.1. Relative error returned by MATLAB routines compared to intg().

One may wonder how the intg() routine compares with commercially available routines such as available with MATLAB. For a comparison the four above functions were evaluated with two MATLAB integration routines and with the default error tolerance (1.e-6) and the results are shown in Table 5.1 and compared with the results using the default error tolerance of intg(). The results show that the intg() code compares very favorably with the standard MATLAB functions and in all cases gives considerably better accuracy for these MATLAB results with the default parameters.

5.5 Integrals with Singularities and Infinite Limits

Integrals with singularities over the range of integration present special problems for numerical integration techniques. Some examples of such integrals are:

$$I_1 = \int_0^4 \frac{1}{\sqrt{x}} dx$$

$$I_2 = \int_{-2}^2 \frac{1}{(x+1)^{2/3}} dx \tag{5.27}$$

$$I_3 = \int_{-2}^0 \frac{1}{\sqrt{4-x^2}} dx$$

The three integrals have singularities at 0, -1 and +/-2 respectively, all within the specified range of integration. A multiple of the first integral can be seen in Figure 5.7. Of course not all integrals with singularities have a finite value, but the above integrals do have finite values. As previously mentioned, the ingt() function in Listing 5.9 has special code to detect a singularity at one or more of the ends of the integration interval (lines 6 and 12 of the code). If a singularity (inf value or und value) is detected, then a special routine is called to attempt to estimate a value of the integral close to the singularity. The method used it to assume that the function follows some power law near the singularity and estimate the value of the power law exponent and this approximation is then used for the integral value near the singularity. To use this feature, one must know where the singularity occurs and this value must be made the limit of the spatial interval. Thus the second integral in Equation (5.27) can be expressed as:

$$I_2 = I_{21} + I_{22} = \int_{-2}^{-1} \frac{1}{(x+1)^{2/3}} dx + \int_{-1}^{2} \frac{1}{(x+1)^{2/3}} dx . \tag{5.28}$$

In the first integral the singularity is at the upper boundary while it is at the lower boundary for the second integral. Each of these parts can then be used with the intg() code for evaluation of the integral.

```
 1 : -- /* File list5_11.lua */ -- Singular integrals
 2 :
 3 : require"intg"
 4 : getfenv(intg).ERROR=1.e-6
 5 :
 6 : f1 = function(x) return x^-.5 end
 7 : iv,n,err = intg(0,4,f1)
 8 : ivv = 4
 9 : print(n,iv,ivv,err,(iv-ivv)/ivv)
10 :
11 : f2 = function(x) return ((x+1)^2)^(-1/3) end
12 : iv1,n1,err1 =  intg(-2,-1,f2)
13 : iv2,n2,err2 = intg(-1,2,f2)
14 : iv = iv1 + iv2
15 : ivv = 3*(3^(1/3)+1 )
16 : print(n1+n2,iv,ivv,err1+err2,(iv-ivv)/ivv)
17 :
18 : f3 = function(x) return 1/(4 - x^2)^.5 end
19 : iv,n,err = intg(-2,0,f3)
20 : ivv = -math.asin(-1)
21 : print(n,iv,ivv,err,(iv-ivv)/ivv)
Output:
12 4.0000008797496      4                1.510013212247e-006
2.1993740073611e-007
24 7.326759927475    7.3267487109222 9.414428123744e-006
1.5309045329948e-006
12 1.570796629157    1.5707963267949 5.3386886600393e-007
1.9248966782457e-007
```

Listing 5.11. Examples of integrals with singular points at ends of integration range.

Listing 5.11 shows example code for the three integrals of Eq. (5.27) evaluated using the intg() function. The second integral is computed in two parts as explained above on lines 12 and 13 of the code. The requested error is set at 1.e-6 at the beginning, because of the difficulty of achieving highly accurate results for functions with singularities. The output from executing the code shows that for this requested accuracy, each calculation took 12 Romberg iterations, near the maximum allowed by intg(). The calculated values of the integrals are seen to be quite good as indicated by the last entry on each line of the printed output. The values are accurate to 6 or 7 digits, which satisfy the requested relative accuracy of 1.e-6 as set by line 4 of the code. For most engineering work these results would be of sufficient accuracy. The reader is encouraged to comment out line 4 of the code and evaluate the integrals using the internal effort limit and observe the improved accuracy. With the requested increased accuracy, the ingr() function will reach the internal limit on Romberg cycles and the accuracy should improve to about 9 to 11 digits but with an increased computer execution time.

Although the algorithm used to handle the boundary singularities is relatively simple, it can handle the singularities of these test functions quite well. In general the intg() code segment has been found to be a robust function with regard to numerical integration over a wide range of functions.

The final topic in this chapter is how to accurately integrate functions over an infinite range of values. The type of integral to be considered can be represented as:

$$I = \int_a^\infty f(x)dx \text{ or } \int_0^\infty f(x)dx .$$ (5.29)

One of the limits is assumed to be infinity and the other may be 0 or some finite value. The upper limit may be either $+/-\infty$, since a change of variable can always convert a negative limit to a positive limit if so desired. Very few text books address this subject adequately if at all. This is not an easy problem if a very accurate numerical answer is desired. Since a numerical algorithm can not integrate numerically all the way to infinity, a first question is how far the upper limit must be extended, if an accurate answer is desired? This question can not, in general be answered. However, a few examples will provide some insight into the problem. Consider the integral

$$I = \int_0^{x_m} \frac{1}{x^2+1} dx = \tan^{-1}(x_m) \to \frac{\pi}{2} \text{ as } x_m \to \infty .$$ (5.30)

where the value approaches $\pi/2$ as $x_m \to \infty$. The relative error in terminating this integral at a finite upper limit can be investigated by looking at the relative error given by

$$\eta_{RE} = \frac{\pi/2 - \tan^{-1}(x_m)}{\pi/2} .$$ (5.31)

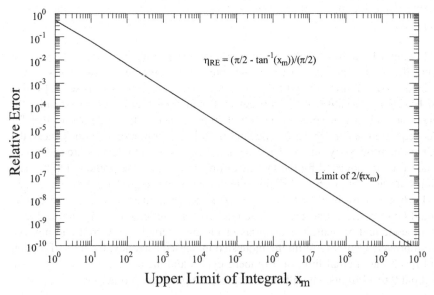

Figure 5.9. Relative error for Eq. (5.30) integral as a function of the upper limit.

A plot of this as a function of x_m is shown in Figure 5.9. As the value of the upper limit increases, the relative error in the integral approaches $2/(\pi x_m)$. It can be seen that in order to achieve an accuracy in numerically evaluating this integral that approaches the machine accuracy of about 15 digits, the integral has to be evaluated over a range of x approaching 1.e15 or 1.e16. Even for a very modest accuracy of 4 digits, the upper limit has to be extended out to about 1.e4. This is not surprising when the integral is examined in more detail. At large values of x the function being integrated approaches $1/x^2$. This means that the integral value will approach $1/x$ for large x and this is exactly the dependence seen in the relative error as the upper limit is increased. This is in fact not the worst possible case as integrals with finite values can have denominator powers of x less than 2.0 and such integrals will converge even more slowly.

For the test integral in Eq. (5.30), one-half of the value comes from the area between 0 and 1, while the remainder comes from 1 to ∞. Thus to achieve accuracy in the 0 to 1 interval a significant number of panels are needed within this range. The problems with this and similar integrals are readily seen if an answer with any reasonable degree of precision is desired. A uniform grid size from 0 out to some x_m can simply not be taken and code like the intg() function applied. Some alternative strategies must be applied to such integrals with infinite limits.

One approach is to use some type of non-uniform panel spacing which could be adjustable within the integration code routine. However, this greatly complicates integration algorithms, especially the Richardson extrapolation approach. Another approach is to integrate by spatial sections using increasing limits for each section. For example integrate from 0 to 1, from 1 to 10, from 10 to 100, from 100 to 1.e3, etc. To obtain 6 digits of accuracy by this approach would require about 6 partial section integrations. This is a doable approach and the use of sub-section integrations is a useful technique for integrals which have peaks over part of the range of an interval. One such function is the humps() function previously shown in Figure 5.7. This type of function is typical of the peaks obtain in spectral data from many types of physical experiments. Such functions are readily handled by the use of multiple integral ranges with a fine mesh used near the peak and a courser mesh away from the peak. However, this function has already been integrated (in Listing (5.10)) and the results show that over the x range shown in the figure, the intg() function works very well without any additional modifications.

A final approach to infinite integration limits is to make a variable transformation such that the upper limit is transformed from infinity to 0. There are a large number of possible transformations which can accomplish this. One of the simplest is to take

$$y = 1/x \text{ or } x = 1/y \rightarrow dx = -dy/y^2. \tag{5.32}$$

If an integration over an infinite range is then broken at some point a, one can write

$$I = \int_0^\infty f(x)dx = \int_0^a f(x)dx - \int_{1/a}^0 \left(f(1/y)/y^2 \right) dy \tag{5.33}$$

If further the value $a = 1$ is taken, the integral can be written as:

$$I = \int_0^\infty f(x)dx = \int_0^1 f_F(x)dx$$

$$\text{where } f_F(x) = f(x) + f(1/x)/x^2 \tag{5.34}$$

In this approach, the integration over an infinite interval is transformed to the range 0 to 1 and the integration is over the original function plus a function folded back from 1 to infinity to the interval 0 to 1. The function $f_F(x)$ will be referred to here as the "folded equivalent" function. One might wonder about the folded part which has a $1/x^2$ factor which goes to infinity at $x = 0$. Since the $f()$ function must be such that it gives 0 at infinity, this will result in a case of 0/0 for the folded part at $x = 0$, which could cause some potential problems. However, the integration algorithm intg() has been designed to handle even the case of an infinite value or undefined value at the integration boundaries. So in practice, this does not cause major problems as will be seen subsequently with example integrals.

Another consideration for this folded equivalent approach to work is that the integration should be dealing with "dimensionless" variables, such that $1/x$ has the same dimensions as x. If this is not the case then it's like adding apples and oranges in Eq. (5.34) which makes no physical sense. Thus before applying this technique, the mathematics must be defined so that "dimensionless" variables are used in the integration. A second consideration is the proper scaling of physical variables. If the break in integration range is to occur at unity, then ideally about half of the integration area should occur for values less than unity and about half of the area should occur for values from unity to infinity. This does not have to be exact, but certainly one would not want almost all of the area to be in one half or the other half of the break. This goal can be achieved by proper scaling of the dimensionless integration variable.

Example code for the numerical integration of four test integrals is shown in listing 5.12. The code calls the intg_inf() function defined in Listing 5.9 (lines 43-45) where the function folding algorithm of Eq. (5.34) is implemented. The calling argument is just the function name as the algorithm assumes the limits are zero to infinity. The evaluated integrals are seen to have good accuracy. The first three functions

$$f_1 = \exp(-x^2), \ f_2 = 2/(1+x^2), \text{ and } f_3 = x\exp(-x) \tag{5.35}$$

are all well behaved functions at $x = 0$ and the folded functions are also well behaved at $x = 0$. The calculated integrals are accurate to more than 12 digits for these cases even with the error termination criterion set to 1.e-8 on line 4 for the intg() function The last two functions:

$$f_4 = \log(x)/(1+100x^2) \text{ and } f_5 = x^{0.2}/(1+10x)^2, \tag{5.36}$$

both have singularity problems, with f_4 having a singularity at $x = 0$ and the f_5 function having a singularity at $x = 0$ for the folded part of the integral. In addition, both of these integrals are not well scaled for folding at $x = 1$ because of the $100x^2$ in the first function and the $10x$ in the second function. Most of the contributions to the integrals come before $x = 1$. Regardless of these problems,

```
 1 : -- /* File list5_12.lua */ Test of integrals with infinite limit
 2 :
 3 : require"intg"; makeglobal(math)
 4 : getfenv(intg).ERROR=1.e-8 -- Try commenting out
 5 :
 6 : f1 = function(x) -- First test integral --> sqrt(pi)/2
 7 :     return exp(-x^2)
 8 : end
 9 : f2 = function(x) -- Second test integral --> pi
10 :     return 2/(1+x^2)
11 : end
12 : f3 = function(x) -- Third test integral --> 2.0
13 :              return x^2*exp(-x)
14 : end
15 : f4 = function(x) -- Fourth test integral --> -pi*ln(10)/20
16 :     return log(x)/(1 + 100*x^2)
17 : end
18 : f5 = function(x) -- Fifth test integral --> .0674467742
19 :     return x^.2/(1 + 10*x)^2
20 : end
21 :
22 : a,b,c,d,e = sqrt(pi)/2,pi,2.0, -pi*log(10)/20,
23 :         -10^(-6/5)*pi/(5*sin(6*pi/5))
23 : aa,ab,ac = intg_inf(f1), intg_inf(f2), intg_inf(f3)
24 : ad,ae = intg_inf(f4), intg_inf(f5)
25 : print('value of f1 = ',aa,(a-aa)/a)
26 : print('value of f2 = ',ab,(b-ab)/b)
27 : print('value of f3 = ',ac,(c-ac)/c)
28 : print('value of f4 = ',ad,(d-ad)/d)
29 : print('value of f5 = ',ae,(e-ae)/e)
Output:
value of f1 =   0.88622692545468      -2.1680135315621e-012
value of f2 =   3.1415926535898       3.8166656177562e-015
value of f3 =   1.9999999999976       1.2022605133666e-012
value of f4 =   -0.36168982109727     -1.6602001384188e-006
value of f5 =   0.067446718930023     8.1879194007103e-007
```

Listing 5.12. Calculation of four test integrals with infinite limits.

the calculated results are seen to be accurate to about 6 digits which would be more than adequate for most engineering applications.

It's interesting to look at a few of the integrals in terms of the function value from 0 to 1 and the value of the folded equivalent function, again from 0 to 1 to see the relative contributions over each part. Figures 5.10 and 5.11 show these contributions for two of the functions, f_1 and f_5. In these figures, we can see the direct contributions for the function from 0 to 1 and the contribution of the function to the total area when the region from 1 to ∞ is folded back into the interval 0 to 1. For the f_1 function, it is seen that the result is a somewhat constant total value of the folded equivalent function over the interval ranging from 1 at $x = 0$ to about 0.8 at $x = 1$. This makes the total area easy to evaluate for the $\exp(-x^2)$ function. For the f_5 function, it is seen that most of the contribution comes from the direct function contribution between 0 and 1. The folded contribution is small over most of the range, although it does have a singularity at $x = 0$ which is difficult to see on the scale of the figure. This integral would be better

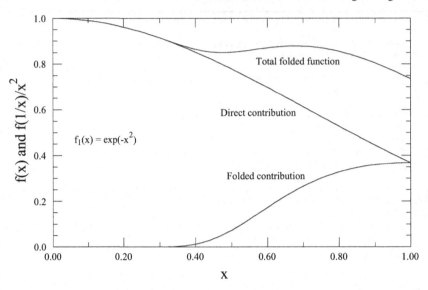

Figure 5.10. Direct and folded contributions to infinite integrals for the $f_1(x)$ function.

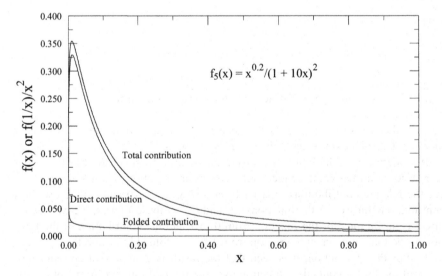

Figure 5.11. Direct and folded contributions to infinite integrals for the $f_5(x)$ function.

balanced if x were scaled such that the folding occurred when $x \simeq 0.1$ on the scale of the figure. This will be left to the reader to reformulate the code with a scaled version of the function and see if the accuracy is increased significantly. In Listing 5.12 the ERROR value for intg() is set at 1.e-8 and it might be expected that improved accuracy could be obtained by decreasing this value. The reader is encouraged to experiment with changing this value as well as with the integration of other functions.

5.6 Summary

The journey through numerical derivatives and numerical integrals is now complete. Along the way several very useful code segments have been developed. Hopefully the reader has also gained a deeper appreciation for some of the fundamental numerical problems with obtaining highly accurate values for both of these quantities with a computer language of limited precision. This is especially significant with numerical derivatives where round-off errors rapidly begin to limit the precision of numerical derivatives. To enhance the accuracy of limited calculations, the concept of Richardson acceleration which is applicable to a wide range of numerical problems has been implemented for both numerical derivatives and numerical integration. Perhaps the most important application of numerical derivatives, at lease for this book is their use in automating such nonlinear algorithms as Newton's method and other methods which depend on a Taylor series expansion of some nonlinear functions. Hopefully the reader has gained an appreciation for the numerical derivative approach used throughout this book in such code segments. Without the ability to implement a robust numerical derivative approach much of the code segment development in this book would not be possible. The application of numerical derivatives to finding the minimum and maximum of a function and to optimization problems has been briefly discussed as applications of numerical derivatives.

For the integration of functions, fundamental limitations on the accuracy that can be achieved on a computer as well as fundamental limits to accuracy of functions with infinite limits have been examined. Some code segments have been developed for numerical integration and for the application to numerical integration. One area not covered is the integration of tabular data that might be obtained in an experimental measurement as opposed to having a mathematical or algorithmic function that can be evaluated at any desired point. This is an important practical subject and will be addressed in a subsequent chapter dealing with approaches to manipulating tabular data.

Important code segments developed in this chapter are:
1. deriv() – Code segments for numerically differentiating a function.
2. rderiv() and lderiv() – Code segments for right and left derivatives or forward and backwards derivatives.
3. pderiv() – Code for the partial derivative of a multi-variable function.
4. intg() – Code for the numerical integration of a function.

5. intg_inf()—Code for numerical integration of a function over the interval
    0 to ∞.

No new language extension packages have been introduced in the chapter. However, the important concept of algorithm improvement through Richardson's extrapolation has been introduced. This can be applied to any computer algorithm where the error in a numerical computation is proportional to some power of an algorithm parameter such as h or Δx. This technique will appear again in subsequent chapters.

6 Interpolation

Several types of functions have been considered in the previous chapters. In all cases it has been assumed that the functions are continuous functions of some independent variable or variables. It has been further assumed that a mathematical expression or a computer algorithm exists that will respond with the value of the function when given a set of values of the independent variables. For a single variable this can be expressed as:

$$\text{Given: } f(x) \text{ for } x_1 < x < x_2 . \tag{6.1}$$

Extensive use has been made of the ability to take a derivative of such a function at any desired point within the allowed range of values of the independent variable. This has been used in such functions as newton() and nsolv() for obtaining the zeros or roots of functions.

In many engineering problems the value of a function is known only at some discrete set of values of the independent variable or variables. This may arise because the function has been calculated or evaluated by the user only over a discrete set of values, or a set of values may be given to the user only for a discrete number of values of the independent variable. A classical numerical problem which then arises for such cases is how to obtain an approximate value of the function at other values of the independent variable.

Interpolation is the process of estimating an intermediate value of a function from a known table of values. For a single independent variable, let's assume that tables of value pairs are given as:

$$y = \{y_1, y_2, \cdot \cdot y_i \cdot \cdot y_n\} \text{ for } x = \{x_1, x_2, \cdot \cdot x_i \cdot \cdot x_n\}$$
$$\text{where } y_i = f(x_i) \tag{6.2}$$

It will be assumed here that the same number of x_i values exist as y_i values. Further assume that the numbering scheme for the data points is from 1 to n as opposed to 0 to n as used in some computer languages. Interpolation seeks to estimate a value of y corresponding to an x value not listed in the table of x values.

One very important physical situation where such tables of values occur is in using experimental data. Such tables of data typically contain not only some expected functional relationship, but experimental data typically contains random measurement errors. The process of estimating function values for tabular data with random errors is very different from the process of estimating function values for tabular data with a smooth functional dependence and negligible random errors. In fact such problems are sufficiently different as to be treated in two separate chapters. This chapter will concentrate on interpolating data known to have zero or negligible random error. The next chapter will concentrate on tabular data

J.R. Hauser, *Numerical Methods for Nonlinear Engineering Models*, 187–226.
© Springer Science + Business Media B.V. 2009

which may also contain random errors. The process of interpolation on experimental data with random errors will be referred to as curve fitting and data plotting as opposed to interpolation. Of course any computer generated data will contain some random errors due to the finite number of digits used in any numerical calculation. However, this is typically sufficiently small to be considered as a negligible error for the algorithms developed in this chapter.

6.1 Introduction to Interpolation – Linear Interpolation

For this chapter it will be assumed that the given tabular data as expressed by Eq. (6.2) represents some underlying smooth functional relationship which may or may not be known. Further assume that the data contains negligible noise or random data, such that a best approximation to the functional relationship is a function passing exactly through each pair of data points (x_i, y_i). The interpolation problem then becomes a problem of how to we describe a function passing through the given set of data points? The simplest form of interpolation is linear interpolation where the approximation is a linear variation in value between the data points. This is illustrated in Figure 6.1 where the dotted curve represents some unknown functional relationship between the y and x values and the 6 data points represent the tabular data pairs. Linear interpolation is represented by the solid straight line segments drawn between the adjacent data points. For a typical linear segment one can write:

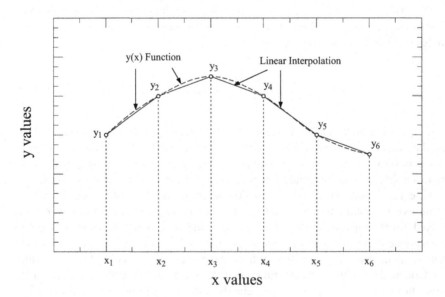

Figure 6.1. Illustration of linear interpolation approach.

$$y_i(x) = y_i \frac{(x - x_{i+1})}{(x_i - x_{i+1})} + y_{i+1} \frac{(x - x_i)}{(x_{i+1} - x_i)} \qquad (6.3)$$

$$\text{for } x_i \le x \le x_{i+1}$$

The mathematical expression as given by Eq. (6.3) can be seen to be a linear function of x and the expression reduces to the two known values of y at $x = x_i$ and $x = x_{i+1}$.

Linear interpolation finds extensive use in one very important arena – that of drawing graphs. Many data plotting programs simply take as input two tables of data values and draws straight line segments between the data points. The plotting program used to generate graphs for this book employs this technique. In fact the dotted curve drawn in Figure 6.1 appears to be a smooth curve simply because a large number of linear segments were used to draw the curve (501 linear segments for this figure). Linear interpolation works best when one has a large number of data points and the functional variation between data points is a small percentage of the overall range of a function.

A typical interpolation problem involves finding the interpolated value for a value of x somewhere within the range of the function given the tabular data. A code segment to accomplish this using linear interpolation is shown in Listing 6.1. While this task is fairly simple, it illustrates several features of implementing interpolation with a computer algorithm. The calling arguments to lintp() in Listing

```
 1 : -- /* File list6_1.lua */ -- Linear interpolation
 2 :
 3 : lintp = function(xd,yd,x,deriv)  -- Linear interpolation
 4 :    local n,y1,y2,x1,x2 = #xd
 5 :    while x>xd[il+1] and il<n-1 do il = il+1 end--Find x interval
 6 :    while x<xd[il] and il>1 do il = il-1 end
 7 :    y1,y2,x1,x2 = yd[il],yd[il+1],xd[il],xd[il+1]
 8 :    y2 = (y2-y1)/(x2-x1)
 9 :    if deriv==1 then return y2 end
10 :    return y1 + y2*(x-x1)
11 : end
12 : setfenv(lintp,{deriv=0,il=1})
13 :
14 : lintpf = function(xd,yd)  -- Function with built-in data table
15 :    local nd,xdd,ydd = #xd, {}, {} -- So user can't change data
16 :    if xd[nd]<xd[1] then for i=1,nd do xdd[i] = xd[nd+1-i] end
17 :    else for i=1,nd do xdd[i] = xd[i] end end
18 :    for i=1,nd do ydd[i] = yd[i] end
19 :    return function(x,deriv) return lintp(xdd,ydd,x,deriv) end
20 : end
21 :
22 : xd = {0,1,2,3,4,5}; yd = {3,4,4.5,4,3,2.5}
23 :
24 : print('value at x = 2.3 is',lintp(xd,yd,2.3))
25 : print('value at x = 5.5 is',lintp(xd,yd,5.5))
Output:
value at x = 2.3 is    4.35
value at x = 5.5 is    2.25
```

Listing 6.1. Code for linear interpolation of a table of data points.

6.1 are the x and y data tables (xd and yd), an x value at which the interpolated value is desired and an optional integer (deriv with default value of 0) specifying that the returned value be either the function value or the derivative estimated from the linear interpolation. The first task of the code is to find an "i" value such that $x_i \le x \le x_{i+1}$. This is done on lines 5 and 6 of the code. Once the proper range of table data is found, the linear interpolation algorithm is easily coded on lines 7 through 10. The function provides for returning either the interpolated value or the derivative from the linear interpolation on lines 9 or 10. To simplify the code one very important assumption is made regarding the tabular data. It is assumed that the x table values are arranged in an increasing order from i=1 to i=n. Hence, before using lintp() the user must ensure that the table date is in increasing order for the independent variable.

A second code segment, lintpf(), is given in the listing in lines 14 through 21. This function solves the x axis ordering by inverting the order of the table entries (line 16) if needed. In addition, the data is stored in tables local to the function (see line 15), so that after the function call, the user can not inadvertently change the data. Finally, the function returns not a value but a function as expressed on line 19 of the code. This returned function takes one or two arguments, (x, deriv), giving an x value and derivative code value. The lintpf() function essentially stores the table data and returns a function which can be called with a single x argument just like any function defined by an equation or by an algorithm.

Examples of the use of lintp() is shown on lines 22 through 25 of the code. After defining tables of data values on line 22, interpolated values for two values of the independent variable between the input data points are printed on lines 24 and 25. An important question arises with interpolation as to what to do if the user requests an interpolated value outside the range of the defined tabular data. This is generally known as function extrapolation. The lintp() code segment makes no restriction on the range of the requested interpolation value and thus allows extrapolation to any desired value. The code uses the straight line approximation through the first two or last two data points to extrapolate outside the table range. It is up to the user to consider whether or not this provides useful information. While extrapolation is sometimes useful for small ranges outside the range of tabular data, going very far outside the table range will obviously not give very accurate results.

A second example of using the code segments in Listing 6.1 is given is Listing 6.2. The linear interpolation code from Listing 6.1 is located in the intp.lua file and is loaded by the statement on line 3. This file contains several interpolation code segments which are discussed and illustrated in this chapter. This simple code segment illustrates the transformation of tabular data to a function on line 5 of the listing using the lintpf() function. The value returned by this function is another function that can then be called as on line 10 to obtain linearly interpolated values from the input data tables. The results of this calculation are shown in Figure 6.2. The data points show the tabular data and the dotted line shows a smooth function (to be discussed later) and the straight line segments are the interpolated data as written to the file on line 13 of Listing 6.2. The calculation shows extrapolated values outside the range of the tabular data. One can judge by eye that these extrapolated values should not be extended very far outside the range of the tabular data.

```
 1 : -- /* File list6_2.lua */ -- interpolation using lintpf()
 2 :
 3 : require'intp'
 4 : xd = {0,1,2,3,4,5}; yd = {3,4,4.5,4,3,2.5}
 5 : ff = lintpf(xd,yd) --- Convert to function
 6 :
 7 : x,y = {},{}
 8 : i,xv =1, -1
 9 : while xv<6.001 do
10 :    x[i],y[i] = xv, ff(xv) -- Call function
11 :    i,xv = i+1, xv+.01
12 : end
13 : write_data('list6_2.dat',x,y); plot(x,y)
```

Listing 6.2. Example of linear interpolation with lintpf().

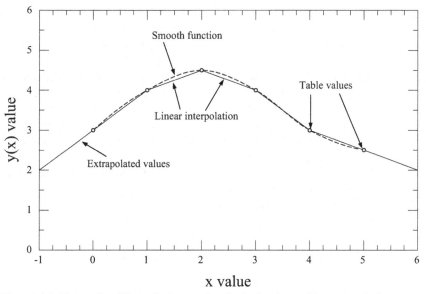

Figure 6.2. Example of linearly interpolated tabular data with extrapolation.

This section has provided a pretty thorough look at linear interpolation of tabular data, the simplest of interpolation techniques. Unless ones tabular data is very closely spaced, one will usually be interested in more advanced interpolation techniques. One such technique is shown by the smooth, dotted curve in Figure 6.2. By eye this looks like a more appropriate fit to the data than the linear interpolation. The next section is devoted to such more advanced techniques. However, the major steps in developing computer algorithms for these more advanced interpolation techniques are almost identical to the steps identified with the linear interpolation technique in this section.

6.2 Interpolation using Local Cubic (LCB) Functions

Interpolation techniques that go beyond linear line segments generally use higher order polynomials to represent the unknown data function. The data in Listing 6.2 and shown in Figure 6.2 has only 6 data points. For n data points a polynomial of order n-1 can be used to pass exactly through all the data points. Thus a 5th order polynomial could be used to fit exactly through the 6 data points. However, if this is done, the resulting curve does not usually get a very satisfactory smooth curve fit to the data. Such high-order polynomials tend to oscillate significantly between the fitted data points and rarely does physical data show such oscillations. Such a fit will not be shown here as this is generally not a useful approach. Another approach is to use low-order polynomials through a limited number of data points and to connect the low-order polynomials in some piecewise manner to move across the data points. The data points are frequently referred to as knots of the fitting polynomials.

The linear interpolation considered in the previous chapter can in fact be considered as a first-order polynomial through the data points (or knots). For improved interpolation accuracy, low-order polynomials can be used such as polynomials of order two and three – quadratic and cubic functions of the independent variable to describe the function between the known data points. Consider the range between two known x values such as x_i and x_{i+1} (or x_1 and x_2) as shown in Figure 6.3. Consider first attempts to fit the data interval to a second degree polynomial or a quadratic function of position. Since a second degree polynomial can pass exactly through 3 data points, the interval from x_1 to x_2 in Figure 6.3 can be approximated by using the points x_0, x_1 and x_2 or by using the points x_1, x_2 and x_3. This gives two slightly different approximating polynomials as shown in the figure. These can be expressed mathematically as:

$$f_1(x) = y_0 \frac{(x-x_1)(x-x_2)}{(x_0-x_1)(x_0-x_2)} + y_1 \frac{(x-x_0)(x-x_2)}{(x_1-x_0)(x_1-x_2)} +$$
$$y_2 \frac{(x-x_0)(x-x_1)}{(x_2-x_0)(x_2-x_1)}$$

$$(6.4)$$

$$f_2(x) = y_1 \frac{(x-x_2)(x-x_3)}{(x_1-x_2)(x_1-x_3)} + y_2 \frac{(x-x_1)(x-x_3)}{(x_2-x_1)(x_2-x_3)} +$$
$$y_3 \frac{(x-x_1)(x-x_2)}{(x_3-x_1)(x_3-x_2)}$$

$$(6.5)$$

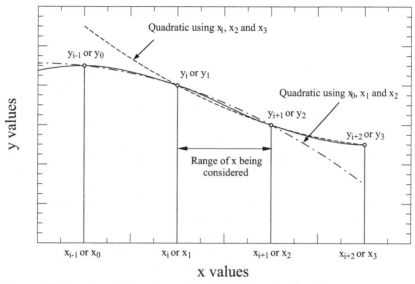

x values

Figure 6.3. Two quadratic polynomials for fitting a function interval.

The form of these equations is the so called second-order Lagrange interpolation polynomial. Each equation is of second order as seen by inspection and reduces to the known data points at the required values of x. In fact Lagrange showed that an interpolation polynomial of any order can be written in the compact form:

$$f(x) = \sum_{i=0}^{n} y_i \prod_{j=0, j \neq i}^{n} \frac{(x - x_j)}{(x_i - x_j)} \qquad (6.6)$$

In this compact form the fitting data point pairs are (x_i, y_i) for $i = 0, 1, \ldots n$ and n is the order of the fitting polynomial.

Since there are two equally valid quadratic polynomials for the interval x_1 to x_2, one can ask as to which is the best approximation to use? It would be expected that each equation would be most appropriate near the midpoint of the three data points used in the fitting polynomial. Thus it would be expected that $f_1(x)$ is most accurate near x_1 and $f_2(x)$ is most accurate near x_2. This expectation is verified from looking at the figure and the two fitting curves. Also near the center of the x_1 to x_2 interval, it appears that an average between the two approximating functions would be best. Thus it is relatively easy to envision that a better concept is to use a "weighted average" of the two local quadratic functions over the interval from x_i and x_{i+1} (or x_1 and x_2). Mathematically this is expressed as:

$$f(x) = f_1(x)\frac{(x_2 - x)}{(x_2 - x_1)} + f_2(x)\frac{(x - x_1)}{(x_2 - x_1)} \quad \text{for } x_1 \le x \le x_2$$

$$= \left[y_0 \frac{(x - x_1)(x - x_2)}{(x_0 - x_1)(x_0 - x_2)} + y_1 \frac{(x - x_0)(x - x_2)}{(x_1 - x_0)(x_1 - x_2)} + y_2 \frac{(x - x_0)(x - x_1)}{(x_2 - x_0)(x_2 - x_1)} \right] \frac{(x_2 - x)}{(x_2 - x_1)} \quad (6.7)$$

$$+ \left[y_1 \frac{(x - x_2)(x - x_3)}{(x_1 - x_2)(x_1 - x_3)} + y_2 \frac{(x - x_1)(x - x_3)}{(x_2 - x_1)(x_2 - x_3)} + y_3 \frac{(x - x_1)(x - x_2)}{(x_3 - x_1)(x_3 - x_2)} \right] \frac{(x - x_1)}{(x_2 - x_1)}$$

It is readily seen from this equation that at $x = x_1$ the weighted function reduces to $f_1(x)$ and at $x = x_2$ the weighted function reduces to $f_2(x)$. Since the starting functions are quadratic equations the resulting Eq. (6.7) is a cubic equation, which depends on four data points, centered about the x_1 to x_2 interval. However, the fitting function uses only local data points and the weighted cubic equation described by Eq. (6.4) through (6.7) will be called a "local cubic function interpolation" or "LCB interpolation". The solid line in Figure 6.3 is in fact an LCB interpolating function fitted to the data points. The LCB function has the property that it uses data from 4 data points symmetrically located about a given spatial interval and provides a cubic equation for describing the function between any two data points.

The local cubic function can apply to interior data points, but how about the first and last intervals where there are not four surrounding data points? For these intervals, the best that can be done is to use a single fitting quadratic function over the first three or last three data points. Thus an LCB algorithm must check for the first and last interval and use a single quadratic there and use the weighted quadratic or local cubic function for interior points. If the data has only two data points, the best that can be done is to fall back on linear interpolation and if there are only three data points, then only a single quadratic function can be used.

An important feature of the LCB algorithm is continuity of the first derivative of the interpolation function. A serious limitation of the linear interpolation algorithm is the abrupt change in slope of the approximating function at each tabular data point. It is certainly highly desirable to have an interpolation function with a continuous derivative (or perhaps derivatives). From Equation (6.7) the derivative can be expressed as:

$$\frac{df(x)}{dx} = \frac{df_1(x)}{dx}\frac{(x_2 - x)}{(x_2 - x_1)} + \frac{df_2(x)}{dx}\frac{(x - x_1)}{(x_2 - x_1)} + \frac{f_2(x) - f_1(x)}{(x_2 - x_1)}. \quad (6.8)$$

If these are evaluated at the two end points of x_1 and x_2, and use is made of the fact that $f_2(x_1) = f_1(x_1)$ and $f_2(x_2) = f_1(x_2)$ the result gives:

$$\left.\frac{df(x)}{dx}\right|_{x_1} = \left.\frac{df_1(x)}{dx}\right|_{x_1} \quad \text{and} \quad \left.\frac{df(x)}{dx}\right|_{x_2} = \left.\frac{df_2(x)}{dx}\right|_{x_2}. \quad (6.9)$$

This means that the first derivative of the LCB polynomial is continuous across the data points, i.e. has the same value on each side of a data point and it is continuous within an interval. Thus the LCB algorithm provides a cubic interpolating polynomial that passes though each data point and has a continuous first derivative. Such is not the case, however, with the second derivative as it is found that this is discontinuous across the data points. In the next section it will be shown that it is possible to obtain a cubic fitting function called a cubic spline function which has continuous first and second derivatives. To achieve this, however, requires that one give up the local nature of the fitting algorithm and the fitting coefficients of the cubic spline depend on all data points in the tabular data file.

Before this discussion, however, it is useful to consider a closely related approach to obtaining an interpolating cubic function known in the literature as piecewise-cubic Hermite interpolation. For this approach, consider a general cubic polynomial used to approximate the function in an interval x_1 to x_2, expressed in the form:

$$f(x) = y_1 + a(x - x_1) + b(x - x_1)^2 + c(x - x_1)^3 \text{ for } x_1 \le x \le x_2 \qquad (6.10)$$

where (x_1, y_1) is the first data point pair and (x_2, y_2) is the second data point pair. In order to evaluate the three coefficients in Eq. (6.10), three additional equations are required. These can be obtained from requiring the function to pass through the (x_2, y_2) data point and from the values of the function derivative at the two end points of the spatial interval. These conditions give the three equations as:

$$y_2 = y_1 + a(x_2 - x_1) + b(x_2 - x_1)^2 + c(x_2 - x_1)^3$$
$$y_1' = a \qquad (6.11)$$
$$y_2' = a + 2b(x_2 - x_1) + 3c(x_2 - x_1)^2$$

The a parameter is simply the derivative of the function at the x_1 point and the b and c coefficients can be readily solved from these equations. The resulting solutions are:

$$b = \frac{1}{(x_2 - x_1)}\left[3\frac{y_2 - y_1}{(x_2 - x_1)} - 2y_1' - y_2'\right]$$

$$c = \frac{1}{(x_2 - x_1)^2}\left[y_1' + y_2' - 2\frac{y_2 - y_1}{(x_2 - x_1)}\right] \qquad (6.12)$$

These are the defining equations for piecewise cubic Hermite interpolation.

The coefficients of the interpolating cubic polynomial are expressed in terms of the function values at the end points of the interval and the derivatives of the function at the end points. One of the points this demonstrates is the fact that an interpolating cubic polynomial over a spatial interval is not a unique function, as Eq. (6.7) and (6.10) are both cubic equations passing through the data points of an interval but they actually may be different cubic polynomials. The next section will discuss a third approach for obtaining a cubic interpolation polynomial.

If the values of the function derivative are known at the data points, then Eq. (6.10) and (6.12) provide an appropriate interpolating polynomial. However, if only a set of data points are known, how can the derivative be determined? One approach is to use a low order polynomial around the end points to obtain an approximate value of the function derivative. Looking back at Figure 6.3 it can be seen that a quadratic polynomial passing through three data points around the end points can provide an appropriate approximate function near the end points. Such equations have been previously gives as Eqs. (6.4) and (6.5) with either $f_1(x)$ or $f_2(x)$ used around the end points. Using these equations, the end point derivatives for use in Eq. (6.12) are obtained as:

$$y_1' = y_0 \frac{x_1 - x_2}{(x_0 - x_1)(x_0 - x_2)} + y_1 \frac{2x_1 - x_0 - x_2}{(x_1 - x_0)(x_1 - x_2)} + y_2 \frac{x_1 - x_0}{(x_2 - x_0)(x_2 - x_1)}$$

$$y_2' = y_1 \frac{x_2 - x_3}{(x_1 - x_2)(x_1 - x_3)} + y_2 \frac{2x_2 - x_1 - x_3}{(x_2 - x_1)(x_2 - x_3)} + y_3 \frac{x_2 - x_1}{(x_3 - x_1)(x_3 - x_2)}$$

(6.13)

Although it is certainly not an obvious conclusion, the use of these derivative evaluations in Eq. (6.10) through (6.12) will give exactly the same cubic interpolation equation as previously given by the LCB approach of Eq. (6.7). How can one know that this is the case? Well they are both cubic polynomials passing through the same data points. Also they have the same identical first derivatives at the end data points. This gives four identical conditions for the cubic polynomials and a cubic polynomial has only 4 parameters. Thus they must be the same cubic polynomial equations. Thus the previous LCB interpolation technique should be more accurately called a "cubic Hermite interpolation with end point derivative values evaluated by a three point numerical derivative". For the sake of simplicity the term "local cubic or LCB" will be used to describe the approach since the cubic equation is determined by the data points at the ends of the interval and the two adjacent data points.

Code for implementing the LCB interpolation algorithm is shown in Listing 6.3. In addition the code implements first and second derivative calculations based upon the local cubic function approximation to the data. This is controlled by the deriv parameter in the calling arguments (the default value is 0 on line 5). Lines 4-11 of the code define local variables, implement parameter checks and print an error message if the number of data points is less than three. With two data points only linear interpolation can be performed. Lines 17 and 18 implement a search for the appropriate interval for the requested x value. The largest section of code evaluates the end point derivatives. Lines 20 through 32 calculate the derivative at the lower x value, or left side, of the spatial region. The first data point must be considered as a special case (see line 21) as there is no left side data point to use in evaluating the derivative. The upper or right side derivative is evaluated on lines 36 through 50 with again the end data point (il = n-1 on line 37) treated as a special case. The general equations for the derivatives can readily be followed from Eq. (6.13). In each case a table (ypd in the code) is tested for a possible stored value of the data point derivative. Such a table of derivatives can be supplied to the interpolation function through a table in the calling argument (the ypd parameter). Alternatively, an empty table may be supplied in the calling argument and the code will store the calculated derivative values as they are calculated in the function (see line 33 and 49). If a subsequent call is made to the function requesting another interpolated value within the same spatial interval, the stored table value will then be used and the derivative value not recalculated. Providing the possibility of a table of derivative values serves two purposes. First

```
 1 : --/*  File intp.lua */ Interpolation code segments
 2 :
 3 : intp = function(xp,yp,x,deriv,ypd) -- Local cubic interp
 4 :    if type(deriv)=='table' then deriv,ypd = ypd, deriv end
 5 :    deriv = deriv or 0
 6 :    local n = #xp
 7 :    local x0,x1,x2,x3,y0,y1,y2,y3,c1,c2,c3,yp1,yp2
 8 :    if il>n-2 then il=n-2 end
 9 :    if n<3 then fprintf(stderr,
10 :       'Number of points must be > 2 in intp.\n'); return nil
11 :    end
12 :    if xp[n]<xp[1] then
13 :       fprintf(stderr,"Array of x values in intp must be monotonic
              increasing\n")
14 :       fprintf(stderr,"Change x values to negative values and try
              again\n")
15 :    end
16 :    -- Find appropriate interval for requested interpolation
17 :    while x>xp[il] and il<n-1 do il = il+1 end
18 :    while x<xp[il] and il>1 do il = il-1 end
19 :    if ypd and ypd[il] then yp1 = ypd[il]
20 :    else -- must evaluate derivative at lower point
21 :       if il==1 then
22 :          x0,x1,x2 = xp[1],xp[2],xp[3]
23 :          y0,y1,y2 = yp[1],yp[2],yp[3]
24 :          c1,c2,c3 = x1-x0,x2-x1,x2-x0
25 :          yp1 = -y0*(c1+c3)/(c3*c1)+y1*c3/(c1*c2)-y2*c1/(c2*c3)
26 :          x1,x2,y1,y2 = x0,x1,y0,y1
27 :       else -- general point
28 :          x0,x1,x2 = xp[il-1],xp[il],xp[il+1]
29 :          y0,y1,y2 = yp[il-1],yp[il],yp[il+1]
30 :          c1,c2,c3 = x1-x0,x2-x1,x2-x0
31 :          yp1 = -y0*c2/(c1*c3)+y1*(c2-c1)/(c1*c2)+y2*c1/(c2*c3)
32 :       end
33 :       if ypd then ypd[il] = yp1 end -- Save if table available
34 :    end
35 :    if ydp and ypd[il+1] then yp2 = ypd[il+1]
36 :    else -- must evaluate derivative at upper point
37 :       if il==n-1 then
38 :          x1,x2,x3 = xp[n-2],xp[n-1],xp[n]
39 :          y1,y2,y3 = yp[n-2],yp[n-1],yp[n]
40 :          c1,c2,c3 = x3-x2,x2-x1,x3-x1
41 :          yp2 = y1*c1/(c2*c3)-y2*c3/(c1*c2)+y3*(c1+c3)/(c1*c3)
42 :                x1,x2,y1,y2 = x2,x3,y2,y3
43 :       else -- general point
44 :          x1,x2,x3 = xp[il],xp[il+1],xp[il+2]
45 :          y1,y2,y3 = yp[il],yp[il+1],yp[il+2]
46 :          c1,c2,c3 = x3-x2,x2-x1,x3-x1
47 :          yp2 = -y1*c1/(c2*c3)+y2*(c1-c2)/(c1*c2)+y3*c2/(c1*c3)
48 :       end
49 :       if ypd then ypd[il+1] = yp2 end
50 :    end
51 :    c1,c2 = x2-x1,x-x1
52 :    b = (3*(y2-y1)/c1 - 2*yp1 - yp2)/c1
53 :    c = (yp2 + yp1 - 2*(y2-y1)/c1)/(c1*c1)
54 :    if deriv==0 then return y1+((c*c2+b)*c2+yp1)*c2 end
55 :    if deriv==1 then return yp1+(3*c*c2+2*b)*c2 end
56 :    if deriv==2 then return 2*b+6*c*c2 end
57 : end
```

```
58 : setfenv(intp,{il=1,fprintf=fprintf,type=type})
```
Listing 6.3. Code for Local Cubic Function interpolation (LCB).

the user may provide such a table as input data in which the LCB algorithm becomes essentially the cubic Hermite interpolation algorithm. Second, for applications where the interpolation code is called repeatedly for many x values, the code will execute faster as the data derivatives only have to be computed once. Following the derivative evaluations, the code on lines 51 through 54 implement the interpolation algorithm of Eq. (6.10) through (6.12). The reader should be able to easily follow the coded equations in the listing as lines 52 and 53 correspond to Eq. (6.12). The interpolated value is returned on line 54 of the listing or the first or second derivative is returned on lines 55 and 56 if requested. The LCB code is contained in the intp.lua code file.

Another function coded for the user is an intpf() function as shown in Listing 6.4. This function has as input data the user supplied arrays of data points (xd and yd). The function returns a function on line 67 than is subsequently callable with just an x value (and possibly a derivative code) as the argument. The data points in the input tables are essentially embedded within the returned callable function so the user can treat the function just as any known function of a general independent variable. In addition, the code stores the supplied data in an internal table available only to the intpf() function so the user can not change the data after the function executes and returns. In addition, the stored data table will be reversed in x direction, on lines 63 and 64 if the data is not in an increasing direction. This works the same as the linear interpolation function return discussed in the previous section. In most applications, the intpf() function is more appropriate to call than the intp() function since one does not have to worry about the direction of the tabular data. However, the function does require that the data x points be in monotonic increasing or decreasing order.

```
 1 : --/*  File intp.lua */
 2 :
.......
60 : intpf = function(xd,yd)  -- Embed data and convert to f(x)
61 :    local xdd,ydd,ydrv = {},{},{} -- Store data locally
62 :    nd = #xd
63 :    if xd[nd]<xd[1] then for i=1,nd do -- reverse data
64 :       xdd[i],ydd[i] = xd[nd+1-i],yd[nd+1-i] end
65 :    else for i=1,nd do
66 :       xdd[i],ydd[i] = xd[i],yd[i] end end
67 :    return (function(x,drv) return intp(xdd,ydd,x,drv,ydrv) end)
68 : end
```
Listing 6.4. Code segment for embedding the interpolation algorithm within a callable function.

Examples of the use of the LCB interpolation function will be given following the development of the cubic spline approach in the next section. This is a second general approach to a cubic data interpolation polynomial.

6.3 Interpolation using Cubic Spline Functions (CSP)

A cubic function can be passed through any given four data points. Then with reference back to the interval x_1 to x_2 in Figure 6.3 it can be seen that a cubic equation can be constructed so as to pass exactly through the data points at x_0, x_1, x_2 and x_3. From the Lagrange formulation such an equation is:

$$f(x) = y_0 \frac{(x-x_1)(x-x_2)(x-x_3)}{(x_0-x_1)(x_0-x_2)(x_0-x_3)} + y_1 \frac{(x-x_0)(x-x_2)(x-x_3)}{(x_1-x_0)(x_1-x_2)(x_1-x_3)} +$$
$$y_2 \frac{(x-x_0)(x-x_1)(x-x_3)}{(x_2-x_0)(x_2-x_1)(x_2-x_3)} + y_3 \frac{(x-x_0)(x-x_1)(x-x_2)}{(x_3-x_0)(x_3-x_1)(x_3-x_2)} \tag{6.14}$$

While this looks like a straightforward approach to increasing the accuracy of an approximation, this has potential problems. There is no assurance that the first or second derivative will be continuous across the data points, and in fact the derivatives will not in general be continuous. If a function is to appear to the eye as a good approximation it must have at least a continuous first derivative. Such a function was derived in the previous section with a continuous first derivative. If an approach is to improve on that algorithm, a cubic function must be considered with both a continuous first and second derivative. However, is spite of these potential difficulties, this algorithm is coded and supplied for reference purposes in the intp file with the intp4(xd,yd,deriv) function name and calling arguments. Code segments for this implementation are not presented here, but the reader can explore the coding of this algorithm in the supplied file.

The discussion will now consider the possibility of using a cubic equation for interpolation within an interval bounded by two data points such as x_1 and x_2 in Figure 6.3 plus the additional constraint of having continuous first and second derivatives at all the data points. Of course the coefficients of the interpolating cubic polynomials must change as one moves from interval to interval in the table data. Starting with the assumption that a continuous second derivative is desired in the cubic equations, the simplest way to achieve this is for the second derivative to vary linearly with distance between the given data points. Then for the interval from x_i to x_{i+1} this can be expressed as:

$$f_i''(x) = f''(x_i)\frac{(x_{i+1}-x)}{(x_{i+1}-x_i)} + f''(x_{i+1})\frac{(x-x_i)}{(x_{i+1}-x_i)} \quad \text{for } x_i \le x \le x_{i+1}, \tag{6.15}$$

where the second derivatives at the data points are taken as unknown constants. If this equation is now integrated twice one obtains a cubic term from the above terms plus an additional linear function of x and a constant. The integrated function can be written as:

$$f_i(x) = f''(x_i)\frac{(x_{i+1}-x)^3}{6(x_{i+1}-x_i)} + f''(x_{i+1})\frac{(x-x_i)^3}{6(x_{i+1}-x_i)} +$$
$$c(x-x_i) + d(x_{i+1}-x) \tag{6.16}$$

where c and d are the constants of integration. The subscript on the function is used to specifically indicate that a different cubic equation is needed for each interval between data points. The equation constants can be evaluated by requiring that the function pass through the two data points (x_i, y_i) and (x_{i+1}, y_{i+1}). Using these two constraints and evaluating the c and d constants gives the final equation:

$$f_i(x) = f''(x_i) \frac{(x_{i+1} - x)^3}{6(x_{i+1} - x_i)} + f''(x_{i+1}) \frac{(x - x_i)^3}{6(x_{i+1} - x_i)} +$$

$$\left\{ \frac{y_{i+1}}{(x_{i+1} - x_i)} - f''(x_{i+1}) \frac{(x_{i+1} - x_i)}{6} \right\} (x - x_i) + \qquad (6.17)$$

$$\left\{ \frac{y_i}{(x_{i+1} - x_i)} - f''(x_i) \frac{(x_{i+1} - x_i)}{6} \right\} (x_{i+1} - x)$$

This equation still has two unknown parameters, the second derivatives at the two end points of the interval. To evaluate these parameters, the additional requirement that the first derivative be continuous at each end point of the interval can be imposed. This leads to the further requirement that

$$f_i'(x_i) = f_{i-1}'(x_i) \text{ for } i = 2, 3, \ldots n-1. \qquad (6.18)$$

This leads to the important recursion relationship:

$$f''(x_{i-1})(x_i - x_{i-1}) + 2f''(x_i)(x_{i+1} - x_{i-1}) + f''(x_{i+1})(x_{i+1} - x_i)$$

$$= 6 \left\{ \frac{y_{i+1} - y_i}{x_{i+1} - x_i} - \frac{y_i - y_{i-1}}{x_i - x_{i-1}} \right\} \qquad (6.19)$$

This equation applies to all except the first and last points of the data intervals. In this equation the second derivative terms evaluated at the data points are the values that must be determined from the set of coupled equations.

These equations now provide the formulation of the equations for a cubic spline interpolation. The set of relationships of Eq. (6.19) provide n-2 equations while for n data points there are n unknown second derivative terms. Thus two additional equations are needed in order to obtain a set of valid solutions. These are the two second derivatives at the two end points of the data set. In most applications of cubic splines the second derivatives are set equal to zero at the boundaries and this results in what are known as a "natural cubic spline". A second choice is to set the derivatives at the two neighbor end points equal. These two choices are expressed mathematically as:

$$\text{(a) } f''(x_1) = 0; f''(x_n) = 0$$
$$\text{(b) } f''(x_1) = f''(x_2); f''(x_n) = f''(x_{n-1}) \qquad (6.20)$$

Another choice, not pursued here, is to take the end point second derivatives as a linear extrapolation of the two next closest data points to the two end points. Eqs. (6.19) and (6.20) now form a set of n equations in the n unknown second derivatives of the piecewise cubic interpolation functions. The set of n equations forms a tri-diagonal matrix with only the diagonal element and the two closest off-diagonal elements non-zero. The form of the matrix equations is:

$$
\begin{bmatrix}
b_1 & c_1 & 0 & 0 & \cdots & 0 \\
a_2 & b_2 & c_2 & 0 & \cdots & 0 \\
0 & a_3 & b_3 & c_3 & \cdots & 0 \\
0 & 0 & a_4 & b_4 & \cdots & 0 \\
\vdots & \vdots & \vdots & \vdots & \vdots & \vdots \\
0 & 0 & 0 & 0 & \cdots & b_n
\end{bmatrix}
\begin{bmatrix}
f_1'' \\ f_2'' \\ f_3'' \\ f_4'' \\ \vdots \\ f_n''
\end{bmatrix}
=
\begin{bmatrix}
d_1 \\ d_2 \\ d_3 \\ d_4 \\ \vdots \\ d_n
\end{bmatrix},
\tag{6.21}
$$

where

$$
a_i = (x_i - x_{i-1}); \quad b_i = 2(x_{i+1} - x_{i-1}); \quad c_i = (x_{i+1} - x_i) \text{ and}
$$
$$
d_i = 6(y_{i+1} - y_i)/(x_{i+1} - x_i) + 6(y_{i-1} - y_i)/(x_i - x_{i-1})
\tag{6.22}
$$

A tri-diagonal matrix of this form is easily solved by eliminating the a_i elements under the diagonal. The recursion relationships for this reduction are:

$$
a_i = a_i - (a_i / b_{i-1}) b_{i-1} = 0
$$
$$
b_i = b_i - (a_i / b_{i-1}) c_{i-1}
\tag{6.23}
$$
$$
d_i = d_i - (a_i / b_{i-1}) d_{i-1}
$$

The second derivative terms are then found by back substitution as

$$
f_i'' = (d_i - c_i f_{i-1}'')/b_i .
\tag{6.24}
$$

These recursive relationships can be performed on simple table values of the parameters and a matrix or matrix techniques do not have to be invoked.

A code segment for the cubic spline algorithm is shown in Listing 6.5. The calling argument list for the spline() function is identical to that previously discussed for the intp() function in Listing 6.3. In addition to the input data points (xp and yp), it includes two optional parameters: a code (deriv with value 0, 1 or 2) for possibly returning a derivative value and a possible table (d2x) of spline second derivative coefficients. This last argument is provided so the function can be repeatedly called for various x values and the spline coefficients may be calculated only once. For this, a test is made on lines 7 and 8 to see if a supplied table has the same number of entries as the number of data points. If so, the evaluation of the second derivative parameters from line 14 through 32 is bypassed and the supplied table of coefficients is used.

If the spline coefficient set needs to be evaluated, the a, b and c coefficients are defined on lines 15 through 22 just as written above. The forward elimination and back substitution is on lines lines 24 through 31. The data point interval corresponding to the x value at which the interpolated value is desired, is evaluated on lines 33 and 34 again assuming a monotonically increasing array of x values. If the interval differs from the last used interval (see line 35) then the equation coefficients are updated on lines 36 through 39 and the interpolated value or derivative value is returned on lines 41 through 44.

```
 1 : -- /* File spline.lua */  Spline function for interpolation

 2 : spline = function(xp,yp,x,deriv,d2x)
 3 :    if type(deriv)=='table' then deriv,d2x = d2x,deriv end
 4 :    deriv = deriv or 0
 5 :    local n,nd2x = #xp, true
 6 :    local a,b,c,d = {},{},{},{}
 7 :    if d2x==nil then d2x = {} -- Table of coefficients input?
 8 :    else if #d2x==n then nd2x = false end end
 9 :    if n<3 then
10 :          print('Insufficient number of data points in spline
                  function')
11 :          return nil
12 :    end
13 :    -- Set up coefficients,
14 :    if nd2x then -- Need coeficients?
15 :        a[1],b[1],d[1] = 0,1,0
16 :        b[n],c[n],d[n] = 1,0,0
17 :        if stype==2 then c[1],a[n] = -1, -1
18 :        else c[1],a[n] = 0, 0 end
19 :        for i=2,n-1 do
20 :            a[i],b[i],c[i] = xp[i]-xp[i-1],
                   2*(xp[i+1]-xp[i-1]),xp[i+1]-xp[i]
21 :            d[i]=6/(xp[i+1]-xp[i])*(yp[i+1]-yp[i])+
                   6/(xp[i]-xp[i-1])*(yp[i-1]-yp[i])
22 :        end
23 :        -- Solve tridiagonal system of equations
24 :        for i=2,n do -- Forward reduction of elements
25 :            ei = a[i]/b[i-1]
26 :            b[i],d[i] = b[i] - ei*c[i-1], d[i] - ei*d[i-1]
27 :        end
28 :        d2x[n] = d[n]/b[n] -- Last coefficient now calculated
29 :        for i=n-1,1,-1 do -- Back substitution for final solution
30 :            d2x[i] = (d[i]-c[i]*d2x[i+1])/(6*b[i]) -- 6 factor
31 :        end
32 :    end -- End of coefficients, if needed
33 :    while x>xp[il] and il<n do il = il+1 end
34 :    while x<xp[il-1] and il>2 do il = il-1 end
35 :    if il~=ilold then
36 :        c0 = xp[il]-xp[il-1]
37 :        c1,c2 = d2x[il-1]/c0, d2x[il]/c0
38 :        c3,c4 = yp[il-1]/c0-d2x[il-1]*c0, yp[il]/c0-d2x[il]*c0
39 :        ilold = il
40 :    end
41 :    if deriv==0 then return c1*(xp[il]-x)^3+c2*(x-xp[il-1])^3+
42 :        c3*(xp[il]-x)+c4*(x-xp[il-1]) end
43 :    if deriv==1 then return 3*(-c1*(xp[il]-x)^2+
                c2*(x-xp[il-1])^2)-c3+c4 end
44 :    if deriv==2 then return 6*(c1*(xp[il]-x)+c2*(x-xp[il-1])) end
45 : end
46 : setfenv(spline, {type=type,print=print,il=2,ilold=0,stype=1})
47 :
48 : splinef = function(xd,yd,nd) -- Embed data and convert to f(x)
49 :    local xdd,ydd,d2x = {},{},{} -- Store data locally
50 :    nd = nd or #xd
51 :    if xd[nd]<xd[1] then for i=1,nd do
52 :        xdd[i],ydd[i] = xd[nd+1-i],yd[nd+1-i] end
53 :    else for i=1,nd do
```

```
54 :       xdd[i],ydd[i] = xd[i],yd[i] end end
55 :     return (function(x,drv) return spline(xdd,ydd,x,drv,d2x) end)
56 : end
```

Listing 6.5. Code segment for cubic spline interpolation.

The spline interpolation routine can also be accessed through the splinef() code segment also shown in Listing 6.5 on lines 48 through 56. This function checks for input data values with increasing values of the x variable and reverses the data table if needed, since the spline() function assumes monotonically increasing values of the independent variable. It also saves a copy of the data points so the user can not change the data and returns a function which can be called to evaluate any interpolated data point. In addition it provides an empty named table (d2x) to the spline() function for storing the spline coefficients. In most cases the splinef() function is probably the preferred function to call first before interpolating over a large number of data points. Finally if the spline() function is called repeatedly with the same data points and without a named table for storing the coefficients, it will respond with the correct interpolated values, but the set of spline coefficients will be recalculated each time the function is called. By using the splinef() function the spline coefficients are calculated only once for a given data set.

As a final feature of the code, an integer parameter stype (default value of 1) is defined on line 46 and used on line 17 to implement either the natural boundary conditions or the equality of second derivatives at two adjacent boundary points – see Eq. (6.20). These two types of cubic spline will be identified as CSP1 and CSP2 in the next section.

6.4 Interpolation Examples with Known Functions

The most important interpolations techniques have now been discussed and code has been developed for linear interpolation (LIN) with lintp() and three forms of cubic polynomial interpolation, the local cubic technique (LCB) with intp(), the cubic spline technique (CSP1 and CSP2) with spline() and the 4 point cubic technique (LCB4) with intp4(). This section discusses some applications of the techniques and compares the various interpolation approaches. When using interpolation, one normally has a set of tabular data for which a mathematical formula is not known for the data either because the data was obtained from some complex computer algorithm or because the data was given to the user in some manner. However, in understanding and comparing the interpolation techniques, it is useful to use data generated from known mathematical equations. In this manner the accuracy of the techniques with representative types of tabular data can be evaluated. This will be done first in this section and then applications to unknown tabular data will be discussed in the next section.

As test functions typical of physical data these five functions will be considered:

$$f_1(x) = \exp(-x^2); \text{ for } -2 < x < 2$$
$$f_2(x) = x\exp(-x); \text{ for } 0 < x < 5$$
$$f_3(x) = 1/(1 + 25x^2); \text{ for } -1 < x < 1 \qquad (6.25)$$
$$f_4(x) = \log(x); \text{ for } 0.5 < x < 4$$
$$f_5(x) = \sin(x); \text{ for } 0 < x < 2\pi$$

Code to test these functions is shown in Listing 6.5. The five functions are defined on lines 7 through 11 of the code. Derivative functions for the test functions
are defined on lines 12 through 16 of the code. The function to be tested and some
initial conditions are defined on lines 18 through 22 with one of these lines selected each time the file is executed. The tabular data to be used in the interpolation is generated on lines 24 through 27 using the selected function. The call to
the interpolation code segments is a single line statement given on either line 28 or
29 with the statement:

```
fintp,fspline = intpf(xd,yd), splinef(xd,yd)
```
                                or
```
fintp,fspline = intp4f(xd,yd), splinef(xd,yd); get-
                  fenv(spline).stype=2
```

If the first form is used, the generated data will be for the LCB and CSP1 interpolation approaches while the second form generates data for the LCB4 and CSP2
interpolation approaches. In all cases the calls convert the tabular data to functions with the data embedded within the returned functions so the functions can be
called with a single argument. These interpolation functions are then used on lines
37 and 38 of the code where interpolated values are calculated for a finely spaced
grid of points over the specified range of the functions (from the defined xmin to
xmax values). Finally, RMS error values are calculated on lines 43 through 50 of
the code between the exact mathematical function and the interpolated functions
for the function value plus the first and second derivatives evaluated from the
function and from the interpolating functions. The code then prints the RMS error
values and writes the calculated data to a data file for further use in plotting and
analysis.

Comparisons of the interpolation techniques for the $f_1(x) = \exp(-x^2)$ function
with the exact function values are shown in Figures 6.4 and 6.5. The nine data
points used for the table of values for the interpolation techniques are shown as
open data points. For each of the four interpolation functions the interpolated values are seen to be very close to the exact function values as given by the solid
curves throughout the range of values plotted in the figure. Although it is difficult
to see in Figure 6.4, the cubic spline (CSP1) function provides a slightly better fit
to the original function than does the local cubic function (LCB). Also in Figure
6.5, the cubic spline CSP2 function is a better fit to the function than is the 4 point
cubic LCB4 function. Because of the close fits, the printed RMS errors give a better indication of the accuracy of the interpolations. The reader is encouraged to
execute the code and observe the printed RMS errors. These values are summarized in a table later in this section.

```
 1 : -- /* File list6_5.lua */ -- Tests of interpolation functions
 2 :
 3 : require"intp"; require"spline"; require"deriv"
 4 : makeglobal(math) -- So we can use exp() in place of math.exp()
 5 :
 6 : xd,yd = {},{} -- Arrays for tabular data
 7 : f1 = function(x) return  exp(-x^2) end -- Five test functions
 8 : f2 = function(x) return x*exp(-x) end
 9 : f3 = function(x) return 1/(1+25*x^2) end
10 : f4 = function(x) return log(x) end
11 : f5 = function(x) return sin(x) end
12 : f1d = function(x) return -2*x*exp(-x^2) end
13 : f2d = function(x) return (1-x)*exp(-x) end
14 : f3d = function(x) return -50*x/(1+25*x^2)^2 end
15 : f4d = function(x) return 1/x end
16 : f5d = function(x) return cos(x) end
17 : -- Select function to test -- Change for different functions
18 : funct,functd,xmin,xmax,dx1,dx2 = f1,f1d,-2,2,0.5,0.01-- f1 tests
19 : --funct,functd,xmin,xmax,dx1,dx2 = f2,f2d,0,5,.5,.01 -- f2 tests
20 : --funct,functd,xmin,xmax,dx1,dx2 = f3,f3d,-1,1,.25,.01--f3 tests
21 : --funct,functd,xmin,xmax,dx1,dx2 = f4,f4d,0.5,5,.5,.01--f4 tests
22 : --funct,functd,xmin,xmax,dx1,dx2 = f5,f5d,0,2*pi,.5,.01 -- f5
23 : i,x = 1,xmin
24 : while x<=xmax+.001 do -- Define tabular data
25 :    xd[i],yd[i] = x, funct(x)
26 :    x,i = x + dx1, i+1
27 : end -- Select one of following two lines
28 : fintp,fspline = intpf(xd,yd), splinef(xd,yd)
29 : --fintp,fspline = intp4f(xd,yd),splinef(xd,yd);
           getfenv(spline).stype = 2
30 :
31 : xi,y_lcb,y_csp,y_fun={},{},{},{} -- Tables for function and data
32 : y1_lcb,y1_csp,y1_fun,y2_lcb,y2_csp,y2_fun = {},{},{},{},{},{}
33 : x = xmin; i = 1
34 : while x<=xmax+.001 do -- Generate interpolated data values
35 :    xi[i] = x
36 :    y_fun[i],y1_fun[i],y2_fun[i] = funct(x), functd(x),
           deriv(functd,x)
37 :    y_lcb[i],y1_lcb[i],y2_lcb[i] = fintp(x), fintp(x,1),
           fintp(x,2)
38 :    y_csp[i],y1_csp[i],y2_csp[i] = fspline(x), fspline(x,1),
           fspline(x,2)
39 :    x = x+dx2; i = i+1
40 : end
41 : ndat = #xi
42 : er0lcb,er0csp,er1lcb,er1csp,er2lcb,er2csp = 0,0,0,0,0,0
43 : for i=1,ndat do -- Calculate average errors
44 :    er0lcb = er0lcb + (y_fun[i] - y_lcb[i])^2
45 :    er0csp = er0csp + (y_fun[i]-y_csp[i])^2
46 :    er1lcb = er1lcb + (y1_fun[i]-y1_lcb[i])^2
47 :    er1csp = er1csp + (y1_fun[i]-y1_csp[i])^2
48 :    er2lcb = er2lcb + (y2_fun[i]-y2_lcb[i])^2
49 :    er2csp = er2csp + (y2_fun[i]-y2_csp[i])^2
50 : end
51 : print("Average mean square errors for "..fname(funct).." are:")
52 : print("Functions, LCB, CSP = ",sqrt(er0lcb/ndat),
sqrt(er0csp/ndat))
53 : print("First derivatives, LCB, CSP = ",
       sqrt(er1lcb/ndat),sqrt(er1csp/ndat))
```

```
54 : print("Second derivatives, LCB, CSP = ",
         sqrt(er2lcb/ndat),sqrt(er2csp/ndat))
55 : --write_data("intp_org.dat",xd,yd)
56 : write_data("list6_5.dat",xi,y_fun,y_lcb,y_csp,y1_fun,y1_lcb,
         y1_csp,
57 :     y2_fun,y2_lcb,y2_csp,xd,yd)
58 : plot(xi,y_fun,y_lcb,y_csp)
```

Listing 6.5. Code segment to test interpolation techniques for five test functions.

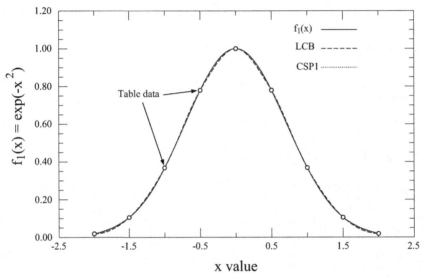

Figure 6.4. Comparison of the $f_1(x)$ function with LCB and CSP1 interpolated values.

The exact first derivative of the function is compared with the first derivatives of the interpolated functions in Figures 6.6 and 6.7. In general the first derivatives agree reasonably closely with the exact first derivative. Again it is seen that the CSP1 and CSP2 interpolation algorithms agrees more closely with the function derivative than do the local cubic functions (LCB and LCB4). It is also seen that the first derivative from the interpolation functions is continuous for all except the LCB4 interpolation function of Figure 6.7. It was previously noted that the 4 point interpolation cubic does not insure a continuous first derivative while a continuous first derivative is inherent in the basic formulations of the other algorithms.

Larger differences are expected to be seen in the second derivative of the interpolating functions and this is shown in Figures 6.8 and 6.9. It is seen that the second derivative of the LCB and LCB4 functions shows abrupt changes at the table data points while the CSP1 and CSP2 functions shows a continuous second derivative. All of the interpolation functions provide a reasonably good fit to the

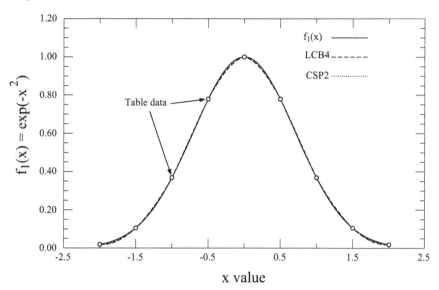

Figure 6.5. Comparison of the $f_1(x)$ function with LCB4 and CSP2 interpolated values.

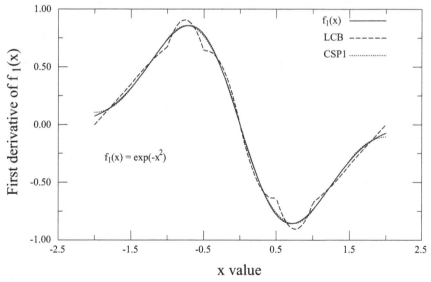

Figure 6.6. First derivative of $f_1(x)$ compared with LCB and CSP1 interpolated derivative values.

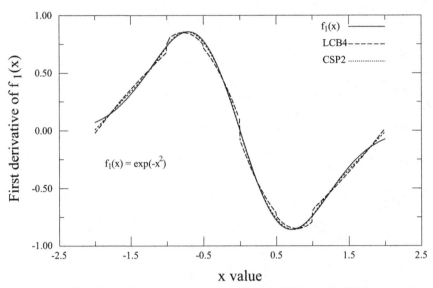

Figure 6.7. First derivative of $f_1(x)$ compared with LCB4 and CSP2 interpolated derivative values.

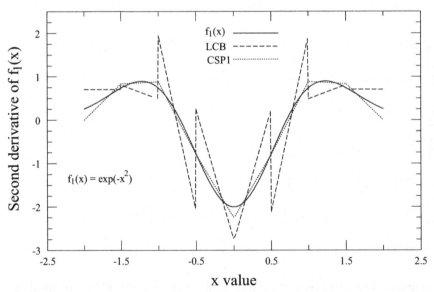

Figure 6.8. Second derivative of $f_1(x)$ compared with values from the LCB and CSP1 interpolated functions.

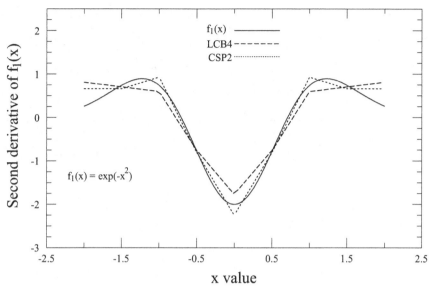

Figure 6.9. Second derivative of $f_1(x)$ compared with values from the LCB4 and CSP2 interpolated functions.

function value as Figures 6.4 and 6.5 indicate. For this particular function, the best to worst interpolation function appears to rank the interpolation functions in the order of CSP1, CSP2, LCB4 and LCB. This is based upon looking at the error in the function value and the first and second derivatives. However, this order may not be the same for all the functions used in Listing 6.5.

Table 6.1 shows a comparison of the RMS errors for the two LCB and the CSP1 interpolation techniques when applied to the five test functions over the range of independent variable values shown in Listing 6.5. As with the above example, all of the interpolation functions give good agreement with the test data for all five of the test functions. The lowest RMS errors for each condition are shown in bold in the table. In all five cases the interpolation function with the lowest RMS error fit to the function also gives the lowest RMS error for both the first and second derivative approximations. This is probably to be expected although one might expect that the LCB techniques with a discontinuous second derivative might give the largest error for the second derivative. However, this is not always the case. For four of the test functions, the local cubic (LCB or LCB4) functions give a lower RMS error than the CSP1 cubic spline function.

Data is not shown for the CSP2 function which uses equal second derivatives at the boundary points. The errors for this fitting function are similar to the data in the table. However, this function did not give the smallest fitting error for any of the five functions tested. The reader is encouraged to experiment with using these interpolation functions for other data sets.

Function	Function Value (X100)			First Derivative			Second Derivative		
	LCB	CSP1	LCB4	LCB	CSP1	LCB4	LCB	CSP1	LCB4
$f_1(x)$	.600	**.136**	.589	.045	**.009**	.037	.505	**.091**	.263
$f_2(x)$	.322	.531	**.122**	.022	.037	**.008**	.212	.333	**.075**
$f_3(x)$	**.980**	2.07	1.34	**.149**	.282	.277	**4.24**	5.23	7.87
$f_4(x)$	.600	.865	**.347**	.043	.062	**.025**	.434	.608	**.284**
$f_5(x)$	.465	.279	**.180**	.034	.020	**.013**	.222	**.090**	.064

Table 6.1. Comparison of RMS errors for LCB and CSP interpolation techniques.

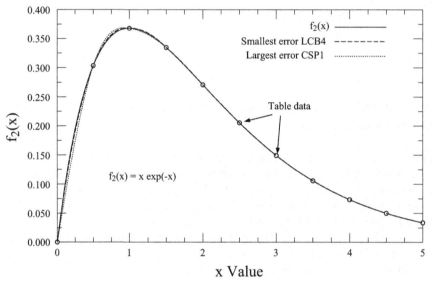

Figure 6.10. Comparison of $f_2(x)$ function with interpolated functions with smallest and largest errors.

Shown in Figures 6.10 through 6.13 are comparisons of the other four test functions in Listing 6.5 with the most accurate and least accurate interpolation functions. In almost all cases the interpolated values overlap with the actual function curves, so it is difficult to see any differences between the test function and any of the interpolation functions. The largest differences are perhaps seen in the $f_3(x)$ shown in Figure 6.11. In this case the CSP1 or CSP2 algorithms show some tendency to oscillate between the data points. This is an especially difficult function to interpolate from a few data points because of the rapid changes is slope of the function. The data in Table 6.1 shows that the LCB technique provides the best fit to the function. Since this function has the smallest deviation from the original function, a comparison of the first and second derivatives is shown in

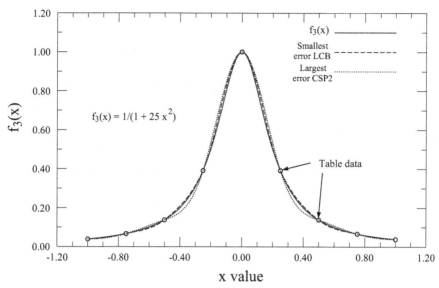

Figure 6.11. Comparison of $f_3(x)$ function with interpolated functions with smallest and largest errors.

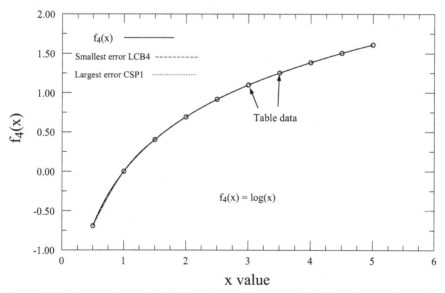

Figure 6.12. Comparison of $f_4(x)$ function with interpolated functions with smallest and largest errors.

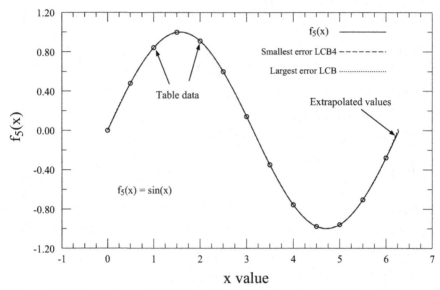

Figure 6.13. Comparison of $f_5(x)$ function with interpolated functions with smallest and largest errors.

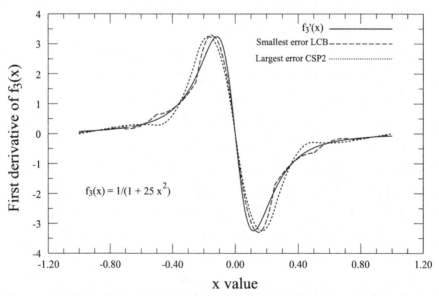

Figure 6.14. First derivative of $f_3(x)$ compared with derivatives of the interpolation functions.

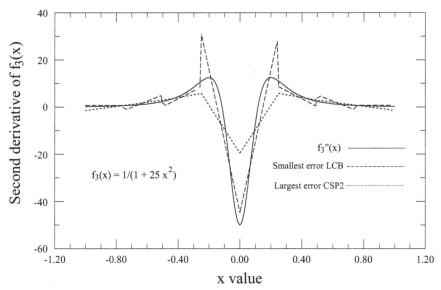

Figure 6.15. Second derivative of $f_3(x)$ compared with derivatives of the interpolation functions.

Figures 6.14 and 6.15. In both figures the derivative of the original function is shown as the solid line and the interpolation function derivatives are shown as dotted lines. For the first derivative, it is seen that the LCB interpolating function gives a slightly better value on average for the derivative than the CSP2 function. Both of the functions also have continuous first derivative values which may be important in some engineering applications. From Figure 6.15 it can be seen that even though the LCB interpolation function has discontinuous jumps in the second derivative at the table points, it still gives a smaller average deviation from the second derivative than does the CSP2 function with its continuous second derivative.

First and second derivative comparisons have only been shown for two of the five functions. However, the comparisons shown are representative of the other functions as well. For all four interpolation techniques, the second derivative shows a continuous variation between the table points but the local cubic functions show a jump in value at the table points while the cubic spline functions have no such jump in the second derivative at the table points. This was of course one of the criteria used in the derivation of the cubic spline functions.

From this brief look at interpolating data based upon five known functions some general conclusions can be drawn – at least for data representative of these types of functions. The first conclusion is that both the local cubic techniques and the cubic spline techniques provide good approximations to the functions. The number of simulated data points ranged from 9 to 13 for the five test cases. Obviously as the number of points decrease, the accuracy will decrease and the accuracy will increase for a larger number of data points. For a good approximation, a few data points are needed in regions of the function where the function is changing rapidly. For the test functions and simulated data points the largest interpolation error occurred for the $f_3(x)$ function shown in Figure 6.11 and the largest fitting RMS error was around 1% of the maximum function value.

The normal distribution function or standard error function occurs frequently in engineering statistical problems. It is defined by the equation

$$f_{nd}(x) = \frac{1}{\sqrt{2\pi}} \exp(-x^2/2) . \tag{6.26}$$

The cumulative distribution function which is the integral of this function from $-\infty$ to some value is also a very important function and is given by:

$$f_{cd}(x) = \frac{1}{\sqrt{2\pi}} \int_{-\infty}^{x} \exp(-z^2/2)dz . \tag{6.27}$$

This function can not be expressed in terms of elementary math functions and is normally tabulated in the form of standard normal tables in most books on statistics and probability. This function is a suitable candidate for interpolation as the value can be computed numerically (or table values used) at several suitable values of x and some interpolation technique can be used between the table values. It is also know that most of the change in the integral comes from values of x between -4 and +4, but one might need to evaluate the function for larger values of x. The integrand is also symmetric about $x = 0$ so all information about the integral can be obtained by considering the range $x = 0$ to $x = \infty$ and tabulating the integral only over this range. This is normally done is statistical tables.

Listing 6.6 shows a code segment for using interpolation for defining a normal cumulative error distribution function. The code uses several techniques discussed in previous sections of this text. Two tables of values are calculated – one for the interval 0 to 1 on lines 8 through 13 and one for the interval 1 to ∞ on lines 14 through 19. For the later interval the technique of function folding is used to transform the infinite interval into the range 0 to 1. Lines 7 and 8 of the code define the direct and folded function. Eleven table values are used over each interval for the table of values. The code then returns a function on lines 24 through 33 that uses the previously defined tables in the intp() function to interpolate the table values for any desired value in the cumulative error function. Note that outside the range -1 to +1 the interpolation is on the $1/x$ value as this is the way the table values are generated. This is done so the interval 1 to ∞ can be conveniently covered. The table values are generated using numerical integration with the intg() function defined in the previous chapter. Alternatively, the table values could just be entered directly into the tables using values published in many reference books. They are generated here as an example of the use of the intg() function and as an example of how table values might be generated for more complicated problems.

An example of using this code segment is included on lines 35 through 41 of the listing. The statement errfcn = ncerr() on line 36 returns a function which is then called as errfcn(x) on line 38 as one would invoke any function such as the

```
 1 : -- /* File list6_6.lua */ -- Normal error function definition
 2 : require"intp"; require"intg"
 3 :
 4 : ncerr = function() -- Error function using 22 data points
 5 :     local x1,y1,x2,y2,x2t,y2t = {},{},{},{},{},{}
 6 :     local f1 = function(x) return math.exp(-x^2/2) end
 7 :     local f2 = function(x) return f1(1/x)/x^2 end -- Folded curve
 8 :     local i = 2
 9 :     x1[1],y1[1] = 0,0
10 :     for xi=.1,1.01,.1 do -- Integral from 0 to 1
11 :         x1[i],y1[i] = xi, intg(xi-.1,xi,f1)/math.sqrt(2*math.pi)+
                y1[i-1]
12 :         i = i+1
13 :     end
14 :     i = 2
15 :     x2[1],y2[1] = 1,y1[#x1]
16 :     for xi=.9,-.01,-.1 do -- Integral from 1 to infinity
17 :         x2[i],y2[i] = xi, intg(xi,xi+.1,f2)/math.sqrt(2*math.pi)+
                y2[i-1]
18 :         i = i+1
19 :     end
20 :     local n = #x2 -- Need to reverse table entries
21 :     for i=1,n do y1[i],y2[i]=.5*y1[i]/y2[n],.5*y2[i]/y2[n] end
22 :     for i=1,n do x2t[i],y2t[i] = x2[n+1-i],y2[n+1-i] end
23 :     x2,y2 = x2t,y2t -- Now have increasing x2 values
24 :     return function(x) -- Return Normal cumulative error function
25 :         if x>=0 then
26 :             if x<=1 then return intp(x1,y1,x)+0.5
27 :             else return intp(x2,y2,1/x)+0.5 end
28 :         else
29 :             if x<-1 then return 0.5-intp(x2,y2,-1/x)
30 :             else return 0.5-intp(x1,y1,-x) end
31 :         end
32 :     end
33 : end
34 : -- Now test function
35 : i,x2t,y2t = 1,{},{} -- Now test function
36 : errfcn = ncerr()
37 : for x=-3,3,.5 do
38 :     x2t[i],y2t[i],i = x, errfcn(x), i+1
39 :     print(x,y2t[i-1])
40 : end
41 : write_data("list6_6.dat",x2t,y2t)
```

Listing 6.6 Code segment to define Normal Cumulative Error Function.

built-in sin(x) function. With 11 values stored in the base tables for the interpolation, the accuracy of the returned function is about 4 decimal digits. This is sufficient for most applications but the accuracy can be increased by using more stored table values. The reader can examine the values from the calculation and stored in the data table. This example illustrates several techniques developed in this text: (a) function folding to more readily handle an infinite range of values, (b) numerical integration and (c) local cubic interpolation. Other numerical techniques for approximating an error function will be discussed in a subsequent chapter.

6.5 Interpolation Examples with Unknown Functions

While the examples in the previous section give some confidence in the use of interpolation techniques for typical functions encountered in practical engineering problems, many applications of interpolation involve tabular data where the underlying physical equations are not known or are too complex to express in closed form. In many cases the tabular data in generated by some complex computer algorithm which can not be explicitly expressed in an equation such as the cumulative error function at the end of the previous section. Another possible example is data generated by a complex equation (or set of equations) where the variable of interest can not be explicitly solved for. For example in the equation

$$y = \frac{\log(1+x^2)}{1+0.2\tan^{-1}(x)}, \tag{6.28}$$

there is an explicit relationship giving y in terms of x. However, suppose the real interest is in the inverse relationship, i.e. a representation of x in terms of y is desired. Such a functional relationship exists even though an explicit equation can not be written for that relationship. One approach is to use Newton's method to calculate an x value for a given y value. However, the use of function interpolation now gives a second possible approach.

If a range of values exists over which the inverse relationship is of interest, it is relatively easy to define an inverse function through the use of an interpolation function. Code for defining such an inverse relationship is shown in Listing 6.7. In this listing it is assumed that the range of x values of interest is from 0 to 18. Table values over this range spaced 0.5 x units apart are generated by the code on lines 4 through 10 in the listing. The xd[] and yd[] tables of values are then used on line 11 to generate an interpolation function fx. Since the first variable in the intpf(yd,xd) calling argument is treated as the independent variable, the fx function defined in line 11 will return x values for any given y argument value. Three test values for y = 1, 2 and 3 are printed by the listing and shown as output. On lines 14 through 17 the interpolation function is used to generate tables of x values as a function of the independent y parameter values. The data written to the output file is shown in Figure 6.16. The three printed output values can be compared with the graphical curve to verify that in deed the interpolation function describes the inverse relationship of Eq. (6.28). Once an interpolated function is defined from tabular data as in line 11 of the listing, it is as easy to use in any subsequent calculations in the same manner as any other mathematically defined and known function. It's just defined by an algorithm from the tabular data. This is a function illustrating a point made in Chapter 5 regarding numerical derivatives. In numerical computer work a function is frequently defined by some computer algorithm as in this example and not by an explicit equation so it may not be possible to obtain an analytical equation for the derivative of a function. However, a numerical derivative can be readily obtained for the functional relationship shown in Figure 6.16.

```
 1 : --/* File list6_7.lua */ Function inversion by Interpolation
 2 : require"intp"
 3 :
 4 : x,i = 0,1
 5 : xd,yd = {},{}
 6 : while x<18. do -- Generate table of values
 7 :     xd[i] = x
 8 :     yd[i] = math.log(1+x^2)/(1+.2*math.atan(x))
 9 :     i,x = i+1, x+.5
10 : end
11 : fx = intpf(yd,xd) -- Define interpolation function
12 : xf,yv = {},{}
13 : i=1
14 : for y=0,4,.01 do
15 :     yv[i] = y
16 :     xf[i] = fx(y) -- Use the new inverse function
17 :     i = i+1
18 : end
19 : print(fx(1),fx(2),fx(3)) -- Print selected values
20 : write_data('test1.dat',xd,yd)
21 : write_data('test2.dat',yv,xf)
22 : plot(yv,xf)
Output:
1.5210860078657    3.3673682372369    6.7986352233613
```

Listing 6.7. An example of function inversion by interpolation.

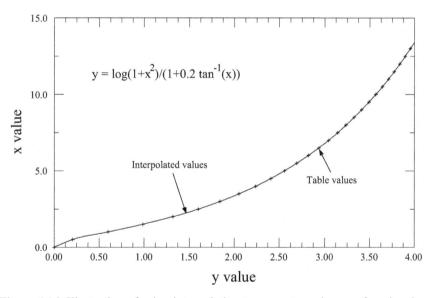

Figure 6.16. Illustration of using interpolation to generate an inverse functional relationship.

This is the first example of using the interpolation functions where the independent variable is not defined on a uniformly spaced grid of data points. While directly generated table data is frequently generated on a uniform spacing of the independent variable, when the functional data is inverted, the table data is certainly not uniformly spaced and the need to handle non-uniformly spaced data is readily seen in the interpolation routines for such applications.

A second example of using interpolation for function inversion is afforded by the frequency response of the voltage amplifier electronic circuit analyzed in Chapter 4 with the code shown in Listing 4.10. As a review, in that example the magnitude and phase response of an amplifier as a function of frequency was calculated and the magnitude of the amplifier voltage gain (in dB) was plotted and shown in Figure 4.6. The data from that calculation and figure was stored in a file (named list4_10.dat) for future use. An expanded graph of the data saved is shown here in Figure 6.17. The data points in this figure are the actual calculated values saved in Listing 4.10. Some expanded information is shown on this figure including the maximum gain of 28.58 dB. In analyzing the frequency response of an amplifier, some important parameters are the Bandwidth and the Upper and Lower "half power points". These are identified on the figure as f_L and f_H. In terms of the gain these points are identified as the points where the gain has been reduced from its maximum in dB by the value $10\log_{10}(2) = 3.01$ dB, the so called -3dB frequency points. These are identified on the graph as the points where the gain is 25.574 dB.

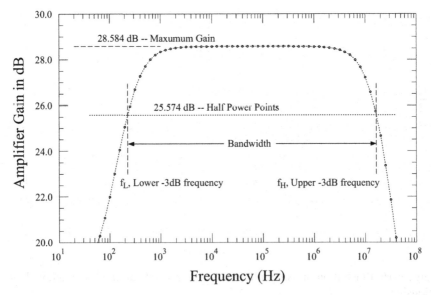

Figure 6.17 Expanded graph of Chapter 4 example of Amplifier Gain (from Listing 4.10).

For the purpose of this discussion, this will be considered as a problem of finding the points on the graph where the gain is 25.574. These frequency points
could, of course, be estimated from the graph using the vertical dashed lines on the
graph. However, with interpolation, this process can be automated and the upper
and lower -3dB frequencies can be automatically obtained with computer code
with high precision. This is again a type of function inversion, because the calculated data is of gain in terms of the independent variable frequency and now what
is desired is to determine the frequencies corresponding to a specified gain. For
this the independent variable is the gain and the dependent variable is the frequency. This can be viewed by considering the graph in Figure 6.17 as rotated by
90 degrees. When this rotation is performed, it is seen that the frequency as a
function of gain is a multi-valued function, i.e. for any value of gain there are two
frequencies an upper and a lower frequency. For the purpose of using interpolation, the functional relationship must be broken into two single valued functions –
one for the low frequency range and one for the high frequency range. The break
between the two functions can occur at the maximum gain point.

```
 1 : -- /* File list6_8.lua */ -- Interp example with amplifier data
 2 : require"intp"
 3 :
 4 : xd,yd,_ = {},{},{} -- dummy table _ = {}
 5 : read_data('list4_10.dat',yd,_,_,_,xd)
 6 :
 7 : npt,xm,im = #xd, 0, 0
 8 : -- Search for largest value in gain -- xd here
 9 : for i=1,npt do if xd[i]>xm then xm,im = xd[i],i end end
10 :
11 : h3db = xm-10*math.log10(2) -- -3Db gain value
12 : fl = intp(xd,yd,h3db) -- Lower -3dB frequency
13 :
14 : i=1
15 : for j= npt,im,-1 do -- Define table for upper -3dB range
16 :     xd[i],yd[i],i = xd[j], yd[j], i+1
17 :   end -- Reverse x data for increasing table values
18 :
19 : fh = intp(xd,yd,h3db) -- Upper -3dB frequency
20 : printf("Maxumim gain in dB =   %12.4e dB\n",xm)
21 : printf("Lower 3dB Frequency = %12.4e Hz\n",fl)
22 : printf("Upper 3dB Frequenty = %12.4e Hz\n",fh)
23 : printf("Amplifier Bandwidth  = %12.4e Hz\n",fh-fl)
Output:
Maxumim gain in dB =   2.8584e+001 dB
Lower 3dB Frequency =  2.2620e+002 Hz
Upper 3dB Frequenty =  1.6487e+007 Hz
Amplifier Bandwidth =  1.6487e+007 Hz
```

Listing 6.8. Code for calculating critical frequencies of the amplifier example.

With this as background, a code segment to evaluate the two -3dB frequencies
is shown in Listing 6.8. Line 5 reads in the amplifier gain data stored by the code
in Listing 4.10 in a file named list4_10.dat. Note that when the data is read, the xd
variable is set to the gain data and yd variable is set to the frequency data. Note

also that a dummy table variable named " _ = {}" is defined on line 4 and used on line 9 for the tabular input data that is not used in this example. Lines 7 through 9 search the xd table for the point of maximum gain so the frequency range and the gain function can be separated into two single valued functions. The lower -3dB frequency in calculated using the intp() function (LCB algorithm) on line 12 using the input data and the lower frequency range of values. For the upper frequency range, the table function values are redefined on lines 15 through 17 of the code. In this redefined table, the independent variable (the gain) values are reversed in order so that the gain function will be a monotonically increasing function as required by the intp() function. Thus line 19 of the code then calls intp() to obtain the upper -3dB frequency value. The code then prints the maximum gain, the upper and lower -3dB frequencies and the bandwidth. The printed frequencies have been used to draw the f_L and f_H values shown in Figure 6.17 so it can be seen that the interpolated values do in fact correspond to the desired frequencies. The code segment in Listing 6.8 can easily be added to the original calculation of the data in Listing 4.10 to automatically calculate these parameters as the data is generated.

In this example the data points are sufficiently close that linear interpolation also gives very good results. To recalculate the critical frequencies using linear interpolation, one needs only to change the calls in lines 12 and 19 from intp() to lintp() to use linear interpolation. This is left to the reader, but the f_L and f_H values calculated using linear interpolation are: 228.99 Hz and 1.6487e7 Hz, both very close to the values with the local cubic interpolation. Also cubic spline interpolation can be used if desired by using the spline() function for the interpolation.

These two examples have illustrated two fairly typical engineering applications of interpolation with tabular data to define the inverse of a function or to extract specific additional data from a series of computer calculations. The second example has also illustrated how interpolation can be performed with multi-valued functional data. Many times a single valued function in terms of one variable will result in a multi-valued function when the role of the independent and dependent variable are reversed. To use the interpolation code segments, the data must always represent a single valued function of the independent variable. Also for the simple interpolation functions (such as intp() or spline()) the independent variable must have a monotonically increasing range of values. The functions that return a function such as intpf() and splinef() will reverse the direction of the data if needed but the data must still represent a single valued function. In the two examples used here the inverse functional relationship needed for the calculations is too complicated to derive an implicit mathematical expression and consequently the desired inverse functional calculation must rely upon tabular data.

A few final examples will be used to illustrate the use of interpolation with tabular data. Several such tables of data are shown in Listing 6.9 from lines 5 through 25. Each data set is listed on three lines with x_d, y_d table values defined for each set of data along with minimum and maximum ranges of the data. To explore each data set, the user must comment out all data sets except one and define appropriate files to store the calculated data. The current statements in the listing are appropriate to explore the table of type-T thermocouple data. Figures 6.18

```
 1 : -- /* File list6_9.lua */ -- Interpolation for several tables
 2 : require"intp"
 3 : require"spline"
 4 :
 5 : --xd = {-2,-1.5,-1,-.5,0,.5,1,1.5,2} -- pulse-like function
 6 : --yd = {0,0,0,.87,1,.87,0,0,0} -- example a
 7 : --xmin,xmax,dx = -2,2,.01
 8 : --xd = {1,2,3,4,5,6,7,8,9,10} -- linear line with step
 9 : --yd = {3.5,3,2.5,2,1.5,-2.4,-2.8,-3.2,-3.6,-4} -- example b
10 : --xmin,xmax,dx = 1,10,.02
11 : --xd = {-4,-3,-2,-1,0,1,2,3,4} -- Impulse like function
12 : --yd = {0,0,0,0,1,0,0,0,0} -- example c
13 : --xmin,xmax,dx = -4,4,.02
14 : --xd = {-1,-.5,0,.5,1} --Second pulse like
15 : --yd = {0.0385,0.1379,1,0.1379,0.0385} -- example d
16 : --xmin,xmax,dx = -1,1,.005
17 : --xd = {0,1,2,3} -- Sparse data table
18 : --yd = {0,1,4,3} -- example e
19 : --xmin,xmax,dx = 0,3,.05
20 : --xd = {40,48,56,64,72} -- vapor pressure of water
21 : --yd = {55.3,83.7,123.8,179.2,254.5} -- example f
22 : --xmin,xmax,dx = 40,72,.05
23 : xd = {-200,-100,0,100,200,300,400,500} -- Type T thermocouple
24 : yd = {-4.111,-2.559,-.67,1.517,3.967,6.647,9.525,12.575} -- g
25 : xmin,xmax,dx = -200,500,1
26 :
27 : flcb = intpf(xd,yd) -- Define local cubic interpolation fcn
28 : fcsp = splinef(xd,yd) -- Define cubic spline interpolation fcn
29 :
30 : x,ylcb,ycsp = {},{},{}
31 : xv,i=xmin,1-- Calculate interpolated values
32 : for xv=xmin,xmax+.001,dx do
33 :    x[i], ylcb[i], ycsp[i] = xv, flcb(xv), fcsp(xv)
34 :    i = i+1
35 : end
36 : write_data("list6_9table.dat",xd,yd)
37 : write_data("list6_9interp.dat",x,ylcb,ycsp)
```

Listing 6.9. Several interpolation examples using data from tables.

through 6.22 show several examples of data interpolation from this listing. Each of these examples has been selected to illustrate one or more features of the interpolation approaches. In each figure, the table data is shown as open circle data points along with intermediate interpolated values using both the local cubic interpolation (LCB) with a solid line and the natural cubic spline interpolation (CSP1) with a dotted line. Taken together these five examples illustrate important similarities and differences between the two interpolation techniques. Both of the techniques demonstrated here provide interpolation function with continuous first derivative values and are thus the two most useful techniques.

Figure 6.18 illustrates the interpolation techniques with a pulse-like table of values. The pulse is not well defined because of the few data points used in the table. The interpolation techniques cause a smoothing to occur around the data points and some undershoot and oscillation as the function transitions to zero values on both the negative and positive sides of the x axis. One noticeable difference is seen between the LCB and CSP techniques with regard to these oscillations. For the LCB technique, such oscillations die out after one data point away from any step change in the value. This is because the local cubic interpolation function depends on only 4 local data points. However, the spline cubic fitting coefficients are globally defined and depend to some extent on data over the entire range of the function. This causes damped oscillations to extend to many data points following an abrupt change in function value. These differences are most noticeable in the -2 to -1.5 and 1.5 to 2 x axis range in Figure 6.18.

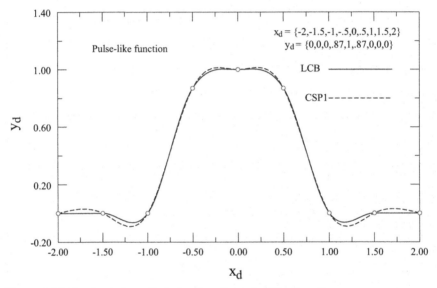

Figure 6.18. Interpolation of tabular data for pulse-like function.

These oscillations in interpolated values are also seen in Figure 6.19 which shows tabular data describing a linear line with a step in the line occurring between $x = 5$ and 6. Again in order to obtain a continuous derivative, the interpolation functions must overshoot or undershoot the line adjacent to the step. Again the LCB function shows less overshoot and undershoot and the oscillation dies out after one data point whereas the CSP1 technique shows a damped oscillation extending through several data points around the step. This local evaluation of cubic fitting parameters for the LCB technique vs. the global evaluation of spline fitting parameters is the major difference between the two interpolation techniques. Which technique is best for these particular tables of data is perhaps debatable. For this particular data set a simple linear interpolation as shown by the dotted line in the figure might be preferred over either of the cubic interpolation functions.

Figure 6.19 shows a sparse set of data points with a fairly wide variation in value between the data points. Both the LCB and CSP techniques give reasonable

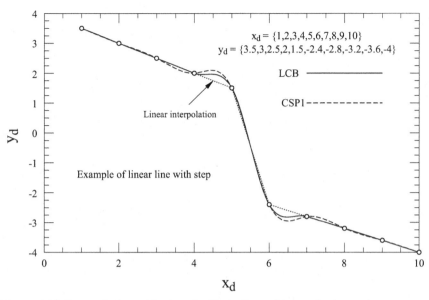

Figure 6.19. Interpolation table data for linear line with step in data.

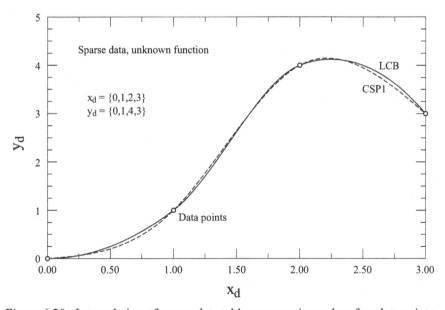

Figure 6.20. Interpolation of sparse data table representing only a few data points.

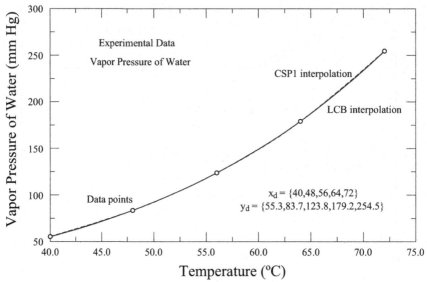

Figure 6.21. Interpolation of vapor pressure of water vs. temperature experimental data.

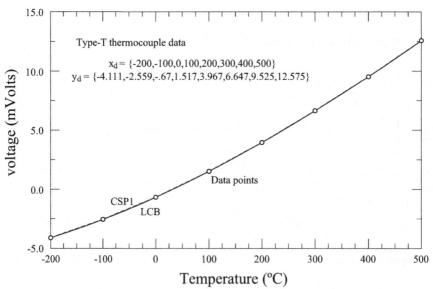

Figure 6.22 Interpolation of experimental thermocouple data for Type-T thermocouple.

smooth flowing function through the data points. Without more data to define a function, either of these interpolation functions would be a reasonable interpolation of the table data. They both agree fairly closely as to the functional relationship defined by the data points. As the number of data points decreases, larger differences are to be expected between the interpolated values provided by the various interpolation techniques. For various sets of data the user would probably try different interpolation techniques and use the one providing the most reliable results.

Finally, Figures 6.21 and 6.22 show interpolation functions applied to two sets of experimental data, one for the vapor pressure of water vs. temperature (lines 20 and 21 of listing) and the other for thermocouple voltage vs. temperatures (lines 23 and 24 of listing). Both sets of data show very smooth variations and the CSP1 and LCB interpolation functions fall essentially on top of each other. Either technique could be used equally successfully with these sets of experimental data. Interpolating experimental data such as these sets is one of the most important applications of interpolation.

In spite of the good results shown for these two sets of experimental data, the reader should be very cautious about using any simple interpolation technique with experimentally generated data sets. In these examples, the data points are obviously defined with a high degree of precision (however the degree of precision is unknown). For data sets such as these with very little random measurement error, simple interpolation works well. However, with many sets of experimental data, the data values include significant random measurement errors. For such data sets, the techniques discussed in the next chapter should be employed rather than the simple interpolation techniques of this chapter.

6.6. Summary

This chapter has explored several data interpolation techniques. This is the art and process of estimating an intermediate value of a function when only a limited number of values of the function are known from a table of values. Two major interpolation techniques were discussed and code segments developed for these techniques. There are the local cubic approximation technique (LCB and LCB4) and the cubic spline technique (CSP1 and CSP2). For each of the two major approaches two variations of the techniques have been presented. Both approaches are based upon using a cubic function of an independent variable to describe the unknown function over a range of values between the table data points. The difference arises in how the cubic coefficients are determined. With the LCB approach, the coefficients are determined by four adjacent table values. With the CSP approach, the coefficients are globally determined through solving a set of coupled equations involving all the data table values. Both approaches give a function which passes through the table data points. All of the interpolation techniques except the LCB4 approach provide an interpolated function with a continuous first derivative. Major differences occur in the second derivative of the cubic approximating functions. For the CSP approach, the second derivative is also continuous across the table data points, while this is not the case for the LCB fitting cubic. The tradeoff in having a cubic with a continuous second derivative is that the cubic fitting parameters are not locally determined, but globally determined.

For most applications, both techniques give very good interpolation results. When fitting the interpolations to known functions, sometimes the LCB technique produces smaller RMS errors and sometimes the CSP technique produces smaller RMS errors. In terms of computer resources, for a single point interpolation, the LCB technique is faster since it does not require the solution of a set of coupled equations. For calculations at many data points, both techniques are somewhat comparable, since the coupled equations need only be solved once for a given table of data and the equation coefficients can be stored for subsequent calculations.

Important code segments developed in this chapter are:

1. lintp() – Code for linear interpolation
2. lintpf() – Code for embedding a table of values into a linear interpolation function
3. intp() – Code for the local cubic interpolation (LCB) technique
4. intpf() – Code for embedding a table of values into the LCB technique
5. intp4() – Code for the 4 point local cubic interpolation (LCB4) technique
6. intp4f() – Code for embedding a table of values into the LCB4 technique
7. spline() – Code for the cubic spline interpolation (CSP1 and CSP2) techniques
8. splinef() – Code for embedding a table of values into the CSP techniques

These code segments will be used in building up more complicated applications in subsequent chapters. In particular the intp() code will be used in the next chapter as an important core technique for one type of data plotting with experimental data sets.

7 Curve Fitting and Data Plotting

The previous chapter addressed the problem of fitting a mathematical function through a given set of data points and approximating the value of a smooth curve between the given data points by an interpolation function. The techniques work well for data points which have been evaluated from some known function or computer algorithm or for data points with zero or negligible random errors in the data values. For experimental data and thus for a wide range of engineering problems, a given set of data has some uncertainty in the data points or values, i.e. the data with some randomness. There are several things that one might want to do with such data. First, one might just want to plot the data and draw a "smooth" curve through the data to aid the user in visualizing the functional dependency. Second, one might want to fit a particular mathematical model to the data and analyze the "goodness" of the fit of the model to the data. Third, one might want to fit a particular physical model with unknown parameters to the data and determine the "best" values of some set of physical parameters when fitted to the data. All of these approaches to data analysis and plotting are considered in this chapter under the heading of Curve Fitting and Data Plotting. The general difference from the previous chapter is that in this chapter the data is assumed to have some randomness so that one does not necessarily want to fit a function "exactly" through the given set of data points, but a curve or function is desired that "best" represents the data. The terms in quotes in the above discussion are not well defined at this point, but hopefully they will become better defined as this chapter progresses.

7.1 Introduction

The general problem to be addressed in this chapter is perhaps best illustrated by an example such as the data set shown in Figure 7.1. This shows a set of data points as open circles and some type of smooth curve drawn through the data points. As opposed to the previous chapter on interpolation where techniques were developed for obtaining a smooth curve passing exactly through each data point, in this case a smooth curve is desired that, in some way, "averages" out the local variations in the data and follows the major trends of the data. Generating such a smooth curve is a much more difficult problem than the interpolation problem because questions involving the "goodness" of fit of a smooth curve and how one measures such a parameter must be addressed. However for typical experimental data there is no choice but to address these issues. It would be foolish to argue for the data in Figure 7.1 that the functional relationship represented by the data points should pass exactly through each of the data points which would be the

J.R. Hauser, *Numerical Methods for Nonlinear Engineering Models*, 227–311.

relationship generated by an interpolation function as discussed in the previous chapter. The more logical conclusion from data such as that shown is that the experimental data has some degree of uncontrollable randomness associated with the data and that the functional relationship between the two variables more closely approximates the solid curve in the figure. In many cases of experimental data, the variations are not completely due to random processes, but to processes which appear random to the particular measurement process. For example, 60 Hz electromagnetic signals are present everywhere due to the electrical distribution system. Electrical instruments are susceptible to such fields and it is almost impossible to completely shield instruments from all such noise. Since measurement times are rarely synchronized to the 60 Hz frequency, such pickup can appear random due to the random phase angle at which measurements are made. This is just one example of how random variations occur in experimental data.

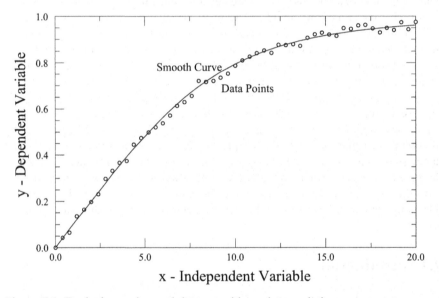

Figure 7.1. Typical experimental data set with random variations.

There are several reasons for wanting to include a smooth curve with such data and not just present a set of data points. First, the curve aids the reader in following the data trend and seeing the general functional relationship as represented by the data. In the old days before computers, such solid curves where drawn by hand using a set of special curves known as "French curves". Drawing a "good" curve required some skill and experience in selecting the proper portion of a set of French curves for various regions of the data such that the resulting curve had a continuous derivative and passed through some mean value of the data points. With the present use of computers to do almost all drafting work, not only is the art of using a French curve lost, but it is also highly desirable to have the computer automatically perform the same task. This turns out to be a difficult task as anyone who has attempted this with computer software can verify. A second reason for a smooth curve in many cases is to compare some mathematical or algorithmic model to the experimental data. In this case the intent and purpose may be to find a set of parameters in the theoretical model which "best fits" in some sense the experimental data. In this case one often speaks of performing "parameter estimation" or "parameter extraction" from the data. Comparing theory and experiment is one of the prime reasons for fitting or drawing a smooth curve through experimental data.

In drawing the solid line curve of Figure 7.1 the horizontal axis has been identified as the independent variable and the vertical axis as the dependent axis. One generally likes to think of one variable being dependent on the value of some other variable. In the real physical world this is usually the case and one typically knos which variable depends on which. However, in terms of the pure analysis of the data, either variable can be considered as being the independent one and the other the dependent one. The analysis mathematics depends little on a choice of dependent or independent variable. In some cases it is convenient to reverse the role of the variables in curve fitting techniques.

These are some of the topics to be covered in this chapter. The key feature of all the techniques is the need to fit a set of data points with noise to some smooth functional relationship which may be an explicit mathematical equation or may be some functional relationship which can be expressed only algorithmically by some computer code. Since the major emphasis of this book is on nonlinear problems, the approach will be to fairly rapidly get to a general approach to curve fitting which is applicable to data analysis with any form of nonlinear function. The case of fitting with linear functions can then be considered as a special case of the general nonlinear fitting approach. However, to build up to this general approach, the next section will review some concepts associated with linear least squares data fitting. This will help introduce many of the concepts needed for the more general case.

7.2 Linear Least Squares Data Fitting

One of the commonly encountered or proposed functional relationships for a data set is a linear relationship between the dependent variable and the independent variable. Figure 7.2 shows a data set where such a postulated relationship appears to be valid. The solid line shows a "linear least squares" line fit to the data of the form:

$$y = ax + b \tag{7.1}$$

with the appropriate values of a and b given on the figure. The topic of this section is how one obtains such "best" straight line fits to a set of data. In the statistics literature this problem is known as "linear regression". In general the term "regression" is used to denote methods for estimating the value of one variable which is assumed to be a function of one or more other variables.

In fitting a smooth curve to a set of data it is obvious that the fitting curve should pass as closely as possible to all the data points as seen in Figure 7.2. At each value of the independent variable the difference in value between the data point and the assumed functional dependency can be expressed as:

$$\varepsilon_i = y_i - f(x_i) \tag{7.2}$$

where $f(x)$ is some assumed functional relationship and y_i is the actual data value at that value of the independent variable. In the present case of fitting with a linear function, the assumed functional relationship is: $f(x) = ax + b$. Eq.(7.2) will continue to be an appropriate definition of fitting error for more complicated forms of fitting function. For a "best fit" one would like to minimize in some way the total error associated with the fitting function. The first thought might be to minimize the sum of the errors from all the data points. However it can be quickly realized that there is no unique solution which minimizes the sum of the errors. Next one might try the sum of the magnitudes of the errors. Again it can be concluded that this does not provide a unique answer and in fact does not necessarily even give a good fit to the data.

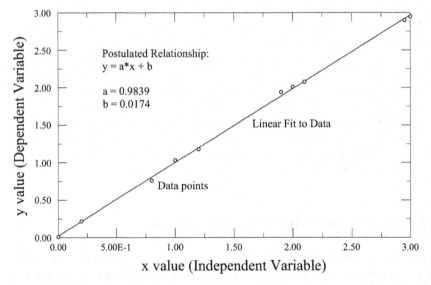

Figure 7.2. Illustration of linear functional dependence.

The almost universal technique for determining the "best" data fit is to *minimize the sum of the squares of the error* the so called "least squares" technique. Stated mathematically the procedure is to minimize the quantity:

$$\varepsilon^2 = \sum_{i=1}^{n}(y_i - f_i)^2, \text{ where } f_i = f(x_i). \tag{7.3}$$

subject to the assumed functional relationship. In the case of a linear functional approximation this becomes:

$$\varepsilon^2 = \sum_{i=1}^{n}(y_i - ax_i - b)^2 \tag{7.4}$$

For a given data set this gives an equation in two unknowns a and b. The standard solution method to find a minimum of the function is to take partial derivatives with respect to the two variables and set each equation to zero giving two equations in the two unknown variables: a and b. This leads to the two equations:

$$a\sum_{i=1}^{n}x_i^2 + b\sum_{i=1}^{n}x_i = \sum_{i=1}^{n}x_i y_i.$$

$$a\sum_{i=1}^{n}x_i + b\sum_{i=1}^{n}1 = \sum_{i=1}^{n}y_i. \tag{7.5}$$

These are readily solved to give the a and b coefficients as:

$$a = \frac{n\sum_{i=1}^{n}x_i y_i - \sum_{i=1}^{n}x_i \sum_{i=1}^{n}y_i}{n\sum_{i=1}^{n}x_i^2 - \left(\sum_{i=1}^{n}x_i\right)^2}, \quad b = \frac{\sum_{i=1}^{n}x_i^2 \sum_{i=1}^{n}y_i - \sum_{i=1}^{n}x_i y_i \sum_{i=1}^{n}x_i}{n\sum_{i=1}^{n}x_i^2 - \left(\sum_{i=1}^{n}x_i\right)^2} \tag{7.6}$$

These are the a and b coefficients given in Figure 7.2 for the linear data fit shown there. It is seen that a is the slope of the fitting line and b is the intercept on the dependent axis.

It is seen that the residual error squared as given by Eq. (7.4) represents a measure of the "goodness" of fit to the model. If the data points all fall exactly on a straight line, then the sum of the residual errors squared will be zero and the closer this comes to zero the better the fit. A measure of the quality of the fit for linear regression is normally taken as the correlation coefficient, r, defined by the set of equations:

$$r = \left(\frac{\varepsilon_o^2 - \varepsilon^2}{\varepsilon_o^2}\right)^{1/2}, \quad \text{where } \varepsilon_o^2 = \sum_{i=1}^{n}(y_i - \bar{y})^2, \quad \bar{y} = \frac{1}{n}\sum_{i=1}^{n}y_i \tag{7.7}$$

In this case a good fit corresponds to an r value close to unity. In all cases one should plot the data and the straight line fit to assess visually the goodness of the fit of the line to the data. If there is considerable scatter in the data, the r value may not be very close to unity, but there may be no obviously better relationship between the variables. The correlation coefficient depends on the degree to which the assumed functional relationship (in this case linear) agrees with the data and on the degree of randomness associated with the data values.

Listing 7.1 gives an example of evaluating the linear least squares parameters for two sets of data. The code segment for implementing Eqs. (7.6) and (7.7) is very straightforward and is not included in the listing but is contained in the prob.lua file and is loaded into the code on line 2. The interested reader can examine this file for the computer code. The listing simply shows the use of the clinear() function which returns the a,b and r coefficients in that order. The first set of data (x7_1,y7_1) is that shown in Figure 7.1 and the coefficients shown in Figure 7.1 agree with the values of the output in Listing 7.1. The r value for the first data set is 0.9996 indicating the very good fit between the data and linear line. The second data set (x7_2,y7_2) is shown in Figure 7.3 along with the fitted linear line. In this case the data is more scattered, but the correlation coefficient, r, is still good and has the value 0.987.

```
 1 : -- /* File listing7_1.lua */ -- Some statistical functions
 2 : require"prob"
 3 :
 4 : x7_1 = {0.0,0.2,0.8,1,1.2,1.9,2,2.1,2.95,3}
 5 : y7_1 = {0.01,0.22,0.76,1.03,1.18,1.94,2.01,2.08,2.9,2.95}
 6 : x7_2 = {1,2,3,4,5,6,7,8,9,10}
 7 : y7_2 = {6,9.2,13,14.7,19.7,21.8,22.8,29.1,30.2,32.2}
 8 :
 9 : a1,b1,rsq1 = clinear(x7_1,y7_1) -- Linear least squares routine
10 : print(a1,b1,rsq1)
11 : a2,b2,rsq2 = clinear(x7_2,y7_2)
12 : print(a2,b2,rsq2)
13 : yy7_1 = {}
14 : for i=1,#x7_1 do yy7_1[i] = a1*x7_1[i] + b1 end
15 : write_data('list7_1a.dat',x7_1,y7_1,yy7_1)
16 : yy7_2 = {}
17 : for i=1,#x7_2 do yy7_2[i] = a2*x7_2[i] + b2 end
18 : write_data('list7_1b.dat',x7_2,y7_2,yy7_2)
19 : a3,b3,rsq3 = clinear(y7_2,x7_2)
20 : print(a3,b3,rsq3)
21 : xx7_3 = {}
22 : for i=1,#y7_2 do xx7_3[i] = a3*y7_2[i] + b3 end
23 : write_data('list7_1c.dat',x7_2,y7_2,xx7_3)
Selected Output:
0.98387806172746        0.017424736482901       0.99922355939632
2.9678787878788         3.5466666666667         0.9867794661099
0.3324864445745        -1.1065056536954         0.9867794661099
```
Listing 7.1. Example of data fitting with linear least squares formulas.

In addition to the two calculations described above, Listing 7.1 shows on lines 19 and 20 a linear fit to the data with the y and x inputs reversed as the calling arguments to the clinear() function. This means that the y data is being considered as the independent variable and the x data is considered as the dependent variable. This minimizes the error in the fitting in the horizontal direction of the line and data as opposed to the vertical direction when x is considered as the independent variable. The third line of output lists the fitting coefficients for this case. The coefficients now describe x in terms of y. The linear regression line obtained by reversing the role of the dependent and independent axis is not exactly the same regression line as shown in Figure 7.3. It is slightly below the solid line in Figure 7.3 for the smallest values of x and slightly above the solid line for the largest values of x. It is not shown on the figure however, because it almost overlaps with the solid line and in practice there is little difference in the fitted linear line when either one or the other of the axis is taken as the independent variable. This is not true for data fitting with nonlinear functions where there can be considerable differences, in some cases, between the fitted curve depending on which axis is considered the independent variable. This will be considered again later.

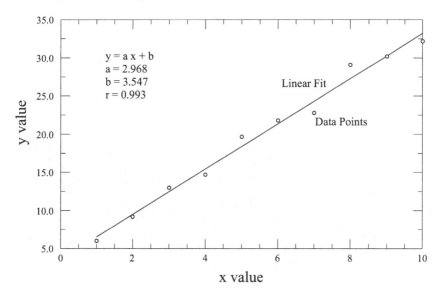

Figure 7.3. Another linear data fit with more data scatter.

As another example of linear data fitting consider the data for Wolf's monthly sunspot activity (http://sidc.oma.be/index.php3) as shown in Figure 7.4 for years 1750 to 2007. It's important to know when peaks in sunspot activity occur because of the potential for interruption of earth based communications systems. The data shows a periodicity of somewhere around 11 years as readily seen from the graph. For each cycle there is a period of very low sunspot activity followed by a period of increasing activity with a peak in activity about equally spaced between the years of minimum activity. Also shown in the figure are data points indicating years of minimum activity (open circle points) and years of maximum activity (open triangle points). If these points are plotted as a function of an integer associated with each solar period the results is the data shown in Figure 7.5. As seen in the figure a linear model fits almost exactly to the data points. From this the solar period can be identified as 11.05 years from the maximum data or 11.11 years from the minimum data or 11.08 years for the average of the two curves. This is frequently stated as simply an 11 year solar cycle.

A code segment for analyzing the solar data is shown in Listing 7.2. The data is input on line 5 from the data file monthssn.dat. The code on lines 11 through 30 searches first for a local peak in the data and then searches for a local minimum in the data. The years of maximum and minimum values are stored in two data files (pks[] and mns[]). The printed output indicates that 24 peak values and 24 minimum values are identified from the data and this can be verified visually from the graph in Figure 7.4. These solar cycles are typically identified as cycles 0 to 23 with cycle 24 as the next cycle which will occur for the years of approximately 2008 through 2019. A linear data fit is obtained on line 32 to the data and the

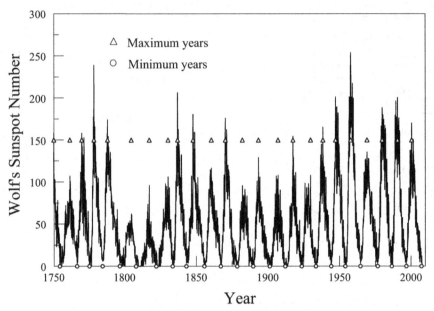

Figure 7.4. Monthly sunspot activity from year 1750 to 2007.

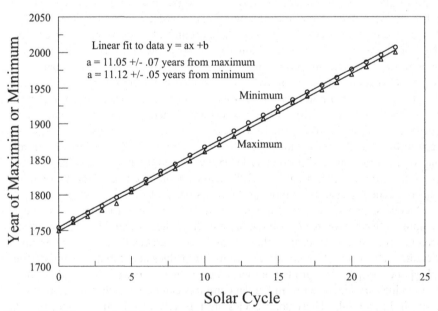

Figure 7.5. Linear data fit to years of solar maximum or minimum activity.

```
 1 : --/* File list7_2.lua */  Example of solar data for linear fit
 2 : require'prob'
 3 :
 4 : jk,yyd,ssd,_ = {},{},{},{}
 5 : read_data('monthssn.dat',jk,yyd,ssd) -- read data
 6 : nmths = #jk -- Number of months
 7 :
 8 : prd = 60 -- Approximate period in months
 9 : pn,pks,mns,ssp = {},{},{},{}
10 : i,k,ixx = 1,1,1
11 : while i<nmths do
12 :    mx = 0
13 :    for j=ixx,prd+ixx do -- search for peak
14 :            if i>=nmths then break end
15 :            if ssd[i]>mx then mx,mtn,ixm = ssd[i],yyd[i],i end
16 :            i = i+1
17 :    end -- Now have peak
18 :    i = ixm; prd = 120
19 :    pks[k] = mtn; ssp[k] = mx
20 :    mx,ixx = 1.e10,ixx+ixm
21 :    for j=ixx,prd+ixx do -- search for minimum
22 :            if i>=nmths then break end
23 :            if ssd[i]<mx then mx,mtn,ixm = ssd[i],yyd[i],i end
24 :            i = i+1
25 :    end -- Now have minimum
26 :    pn[k],mns[k],k = k,mtn,k+1
27 :    ixx = ixx+ixm
28 :    if i>=nmths then break end
29 :    i = ixm
30 : end
31 : print('Number peaks and minima =',#pks,#mns)
32 : a1,a2 = clinear(pn,pks), clinear(pn,mns)
33 : print('solar cycle in years =',a1,a2)
34 : write_data('list7_2.dat',pks,mns,ssp)
Output:
Number peaks and minima =    24      24
solar cycle in years = 11.05   11.11
```

Listing 7.2. Code segment for analyzing solar sunspot activity with a linear model for years of maximum and minimum sunspot activity.

slope of the line is printed on the output (by line 33) giving the number of years per cycle of solar sunspot activity. The data shown in Figure 7.5 is plotted from the data saved by Listing 7.2. This is a good example of experimental data that shows a very excellent fit to a linear data model.

Linear regression is very simple and in some cases nonlinear fitting functions and data can be transformed to a linear relationship so that this simple case can be applied. One such example is the model:

$$y = a\exp(bx), \text{ where } a \text{ and } b \text{ are constants.} \qquad (7.8)$$

This can be linearized by taking natural logs of both sides as:

$$\ln(y) = \ln(a) + bx . \qquad (7.9)$$

Thus the $\ln(y)$ of the fitting function is a linear function of x, and a linear function of x can be fit to the natural log of the dependent variable data (the y data).

Another function which can be linerized is the relationship:

$$y = a\frac{x}{b+x}.$$
(7.10)

This function can be transformed to:

$$\frac{1}{y} = \frac{b}{a}\frac{1}{x} + \frac{1}{a}$$
(7.11)

To consider this a linear equation, one needs to make a change of variables to consider x^{-1} as the independent variable and to consider y^{-1} as the dependent variable. There are several such nonlinear equations which can be linearized by appropriate changes of variables. However, in a subsequent section a general nonlinear least squares technique will be developed that will make all this variable manipulation unnecessary as it will then be possible to fit any general nonlinear function directly to a set of data.

7.3 General Least Squares Fitting with Linear Coefficients

A step up in complexity from the linear least squares technique is the consideration of using various fitting functions with unknown coefficients included in a linear manner. The most general such fitting function can be expressed as:

$$y = C_1 f_1(x) + C_2 f_2(x) + C_3 f_3(x) + \ldots C_n f_n(x),$$
(7.12)

where the C_i's are fitting constants and the $f_i(x)$'s are any desired general functions of the independent variable. In practice, several specific forms are popular. Included in these are polynomial regression using a power series in x and sinusoidal functions of ever increasing frequency. For these two cases the general equation takes on the specific forms:

Polynomial: $y = C_1 + C_2 x + C_3 x^2 + \cdots C_n x^n$

Sinusoidal: $y = C_1 \cos(\omega_1 t) + C_2 \cos(\omega_2 t) + \cdots C_m \cos(\omega_m t) +$ (7.13)
$\qquad C_{m+1} \sin(\omega_{m+1} t) + C_{m+2} \sin(\omega_{m+2} t) + \cdots C_n \sin(\omega_n t)$

For the sinusoidal case the function has been written as a series of m general cos() terms plus n-m general sin() terms each with a corresponding frequency ω_i. In addition the independent variable has been written as t instead of x which is the common independent variable in such problems. In many cases the sinusoidal frequencies are taken as some multiple of a fundamental frequency ω_o (as in a Fourier series approximation). However, there is no such requirement in terms of a general fitting procedure using sinusoids and, if desired, any combinations of sinusoidal frequencies can be taken in the general case as represented by Eq. (7.13).

In terms of the general notation of Eq. (7.12) the mean square error from Eq. (7.4) can be written as:

$$\varepsilon^2 = \sum_{k=1}^{nd}(y_k - C_1 f_1(x_k) - C_2 f_2(x_k) - \dots C_n f_n(x_k))^2$$

$$= \sum_{k=1}^{nd}\left(y_k - \sum_{i=1}^{n}C_i f_i(x_k)\right)^2$$

(7.14)

where nd is the number of data points and n is the number of assumed functions in the representation. Taking the partial derivative of this equation with respect to each of the C_i terms now gives n equations in the C_i coefficients. This then results in the same number of equations as fitting coefficients, so one can hopefully solve for the coefficients. This process leads to the set of equations shown in matrix form in Eq. (7.15), where the notation f_{ik} is used for the function f_i evaluated at independent variable, x_i, i.e. $f_{ik} = f_i(x_k)$. Each of the sums in the equation is over the number of data points.

$$\begin{bmatrix} \sum_{k=1}^{nd} f_{1k}f_{1k} & \sum_{k=1}^{nd} f_{1k}f_{2k} & \cdots & \sum_{k=1}^{nd} f_{1k}f_{nk} \\ \sum_{k=1}^{nd} f_{2k}f_{1k} & \sum_{k=1}^{nd} f_{2k}f_{2k} & \cdots & \sum_{k=1}^{nd} f_{2k}f_{nk} \\ \vdots & \vdots & \vdots & \vdots \\ \sum_{k=1}^{nd} f_{nk}f_{1k} & \sum_{k=1}^{nd} f_{nk}f_{2k} & \cdots & \sum_{k=1}^{nd} f_{nk}f_{nk} \end{bmatrix} \begin{bmatrix} C_1 \\ C_2 \\ \vdots \\ C_n \end{bmatrix} = \begin{bmatrix} \sum_{k=1}^{nd} y_k f_{1k} \\ \sum_{k=1}^{nd} y_k f_{2k} \\ \vdots \\ \sum_{k=1}^{nd} y_k f_{nk} \end{bmatrix}$$

(7.15)

A popular special case of this general formulation is the polynomial power series case as previous shown in Eq. (7.13). For this special case the matrix elements are sums over power series in x_i as shown in Eq. (7.16). In this formulation, the numbering of the coefficients is started at 1, whereas in some texts the numbering

$$\begin{bmatrix} \sum_{k=1}^{nd} 1 & \sum_{k=1}^{nd} x_k & \cdots & \sum_{k=1}^{nd} x_k^{n-1} \\ \sum_{k=1}^{nd} x_k & \sum_{k=1}^{nd} x_k^2 & \cdots & \sum_{k=1}^{nd} x_k^n \\ \vdots & \vdots & \vdots & \vdots \\ \sum_{k=1}^{nd} x_k^{n-1} & \sum_{k=1}^{nd} x_k^n & \cdots & \sum_{k=1}^{nd} x_k^{2(n-1)} \end{bmatrix} \begin{bmatrix} C_1 \\ C_2 \\ \vdots \\ C_n \end{bmatrix} = \begin{bmatrix} \sum_{k=1}^{nd} y_i \\ \sum_{k=1}^{nd} y_k x_k \\ \vdots \\ \sum_{k=1}^{nd} y_k x_k^{n-1} \end{bmatrix}$$

(7.16)

starts at 0. This numbering scheme is in keeping with the presently used software where the natural numbering scheme for arrays begins with 1 and not 0.

Code segments could now be written to handle this special case or other special cases as is the approach of many textbooks. However, a more important goal in this work is to develop software for nonlinear least squares data fitting where the functional form of the fitting equation can be of any arbitrary form. This will be done in a subsequent section and after that development such cases as the polynomial equation or the more general linear coefficient case of Eq. (7.12) will just become special cases of the general analysis program. Thus the linear coefficient case will not be developed further here. In developing the general least squares fitting approach, it will be necessary to iterate over a set of fitting parameters and to do this will require a solution at each iterative step of a matrix equation with essentially the same form as Eq. (7.15). So the development leading up to this equation provides a good background for a more general approach to fitting with nonlinear equations.

An interesting special case of the general formulation is the case of fitting with a series of sinusoids as represented by the second line of Eq. (7.13). If an infinite number of fitting terms are taken, then one has the Fourier series representation of a data set. This is a linear data fitting problem and as such is not in the main thrust of this book which is to develop approaches for nonlinear problems. However, the Fourier series is so important in many engineering disciplines, that a more detailed discussion and development of some code segments for this special case appears warranted. This is done in the next section devoted to the Fourier series method. Since this is somewhat of a diversion from the main thrust of this chapter, the reader can skip the next section, if desired, and come back later if she/he has an interest in the Fourier method of data fitting.

7.4 The Fourier Series Method

Before beginning the Fourier series method, it is interesting to explore further the use of sinusoidal signals in the least squares method as discussed in connection with Eq. (7.12). Suppose a series of sin() and cos() functions are taken as the approximating functions in the form:

$$y = \sum_{i=1}^{m} C_i \cos(\omega_i t) + \sum_{i=m+1}^{n} C_i \sin(\omega_i t) \tag{7.17}$$

This leads to terms in the matrix Eq. (7.15) of three general types given by:

$$\text{(a)} \sum_{k=1}^{nd} \cos(\omega_i t_k) \cos(\omega_j t_k)$$

$$\text{(b)} \sum_{k=1}^{nd} \cos(\omega_i t_k) \sin(\omega_j t_k) \tag{7.18}$$

$$\text{(c)} \sum_{k=1}^{nd} \sin(\omega_i t_k) \sin(\omega_j t_k)$$

Each a_{ij} factor of the matrix will have a term which has one of these forms. The diagonal terms will have $i = j$ and these terms are of the form (a) or (c) while the off diagonal terms will have $i \neq j$ and can be of any of the three forms. For the diagonal terms, it is obvious that the terms will be nonzero regardless of how the frequency terms are selected since these terms are the square of a sin() or cos() function. In general the off-diagonal terms will also be nonzero unless special conditions are met. If the matrix has zero off-diagonal terms it is easy to solve for the coefficients as the matrix is a diagonal matrix. This will be the case if all of the frequencies are selected as some multiple of a fundamental frequency, ω_o. In this special case, Eq. (7.18) takes on the form:

$$(a) \sum_{k=1}^{nd} \cos(i\omega_o t_k) \cos(j\omega_o t_k)$$

$$(b) \sum_{k=1}^{nd} \cos(i\omega_o t_k) \sin(j\omega_o t_k) \quad\quad (7.19)$$

$$(c) \sum_{k=1}^{nd} \sin(i\omega_o t_k) \sin(j\omega_o t_k)$$

A further restriction is that the fundamental frequency must be selected so that the maximum value of t_k in the equations corresponds to one fundamental period. In other words if the maximum time of the data set is T, then ω_o must be selected as $\omega_o = 2\pi / T$. Under these conditions, the sums in the above equations will be zero for $i \neq j$. This does require one final condition and that is that the number of data points be more than twice the highest harmonic value included in the approximating sinusoids. This is in fact the *Nyquist* sampling criteria which requires that the number of data points exceed twice the highest harmonic or frequency in the sinusoidal representation in a signal.

Under the above conditions for a least squares fit, the diagonal terms of the matrix will be of one of the two forms:

$$(a) \sum_{k=1}^{nd} \cos^2(i 2\pi (k-1)/n_d) = \frac{n_d}{2}$$

$$(b) \sum_{k=1}^{nd} \sin^2(i 2\pi (k-1)/n_d) = \frac{n_d}{2} \quad\quad (7.20)$$

which has the same value independently of the order of the harmonic and the off-diagonal elements in the matrix of Eq. (7.15) will be zero. The linear fitting coefficients can then be evaluated as:

$$C_i = \frac{2}{n_d} \sum_{k=1}^{n_d} y_k \cos(i\omega_o t_k) \text{ or } C_{m+i} = \frac{2}{n_d} \sum_{k=1}^{n_d} y_k \sin(i\omega_o t_k) \quad\quad (7.21)$$

The cos() or sin() term correspond to the two sets of terms in Eq. (7.17). The above equations are in fact just the equations for the normal Fourier series coefficients as used to approximate a function if the trapezoidal rule is used as the numerical technique for evaluation the Fourier integrals. These Fourier coefficients have been obtained through the minimum least squares technique under special selection criteria for the approximating sinusoidal function set. The general least squares technique has no such limitations and for minimum simplicity in approximating a function, one might want to take sub-harmonica of the fundamental time period in an approximating set. This subject will be approached again after a general nonlinear fitting technique has been developed in the next section.

The standard approach to the Fourier series is based upon the 1807 work of Fourier who showed that an arbitrary function could be expressed as a linear combination of sin() and cos() functions with the series in general having an infinite number of terms. The series are normally expressed in terms of a set of a and b coefficients or coefficients with the defining equations being:

$$f(t) = \frac{a_o}{2} + \sum_{i=1}^{n} a_i \cos(i\omega_o t) + \sum_{i=1}^{n} b_i \sin(i\omega_o t)$$

$$f(t) = \frac{c_o}{2} + \sum_{i=1}^{n} c_i \cos(i\omega_o t + \theta_i)$$

(7.22)

These two forms are equivalent since a sin() plus cos() term at the same frequency can always be combined into a single cos() term at some phase angle (θ_i). A final form frequently used in Electrical Engineering is the complex exponential series expressed as:

$$f(t) = \sum_{i=-n}^{n} \alpha_i \exp(ji\omega_o t)$$

(7.23)

This is again equivalent to the sin() and cos() form through the application of Euler's theorem.

The Fourier coefficients are known to be given by the equations:

$$a_i = \frac{2}{T} \int_0^T f(t) \cos(i\omega_o t) dt$$

$$b_i = \frac{2}{T} \int_0^T f(t) \sin(i\omega_o t) dt$$

(7.24)

The Fourier series is periodic in time with period T with the fundamental frequency in the series related to the period by:

$$\omega_o = 2\pi / T$$

(7.25)

The a_o value of Eq. (7.24) for $i = 0$ is twice the DC value of the function over the period T and the corresponding value of b_o is zero. The term for $i = 1$ corresponds to the fundamental frequency associated with the signal and the terms for increasing i are known as the "harmonics" associated with the signal with the second and third harmonic frequencies being associated with the $i = 2$ and 3 terms for example.

For many given analytical functions of time, the integrals in Eq. (7.24) can be calculated exactly and the corresponding coefficients evaluated exactly. For this work the interest is more in applying the theory to experimental data or to evaluating the integrals numerically. If the time function is known at some equally spaced time intervals *over a period* then the trapezoidal rule for integration can be used to approximate the integrals. It is important that the function evaluations be over a period and it is further assumed that there are n_d intervals over the period. Thus there will be $n_d + 1$ function values if the function has been evaluated at both ends of the period. Let's identify the function data points as $y_1, y_2, \cdots y_{nd+1}$. The equation for a_i then becomes:

$$a_i = \frac{2}{T} \sum_{k=1}^{n_d} y_k \cos(i2\pi t_k / T)(T / n_d)$$

(7.26)

The trapezoidal rule says that one should take ½ of the first and last function value while the above equation uses the full value of the first point and uses no contribution from the last point. This provides the same answer since the function is assumed to be periodic with equal first and last value. In addition it can be seen that this is in fact identical to Eq. (7.21) which was obtained from considering the least squares criteria for a similar sinusoidal approximation. It can thus be concluded that both the Fourier approach and the least squares approach lead to the same coefficient values, which must be the case in the limit of a large number of terms.

The above equation can be further simplified by expressing the t_k / T term as $(1 - k) / n_d$, which is valid for uniformly sampled data. The coefficient equations can then be expressed as:

$$
\begin{aligned}
a_i &= \frac{2}{n_d} \sum_{k=1}^{n_d} y_k \cos(i2\pi(k-1)/n_d) \\
b_i &= \frac{2}{n_d} \sum_{k=1}^{n_d} y_k \sin(i2\pi(k-1)/n_d)
\end{aligned}
\tag{7.27}
$$

It is noted that these expressions no longer contain any reference to time. They simply involve the data points, and the total number of data points. The implied assumption is of course that the data points are evaluated at equally spaced intervals over the fundamental period of the signal. The Fourier coefficients can then be evaluated simply from one string of data giving the signal values. The harmonic content of a signal is determined by the shape of the signal over a fundamental period. Of course the actual frequency values depend on the actual time of a period of the function.

Code segments are provided to perform both forward Fourier analysis and inverse Fourier analysis to reconstruct a signal from its Fourier components and Listing 7.3 shows the major functions provided. The code consist of three callable functions, fourier(), Fourier() and iFourier(). First the fourier() function takes 1 to 3 arguments with one being required. The first argument must be a table of function values, corresponding to the y_i values in Eq. (7.27). As previously noted this is the only input needed to evaluate the Fourier components and this function returns the Fourier coefficients in three different formats (fca, fexp or fcs) on line 22. The returned order is Cos-Angle, Exponential and Cos-Sin in that order. Other arguments which can be supplied to the fourier() routine are the desired number of Fourier components (the nc parameter) with 20 as the default value, and a time shift parameter (the thmin parameter) which is the fraction of a period to time shift the function before evaluating the Fourier components. This is useful to explore the effects of a time shift on the Fourier components without having to re-evaluate the data values. (You can give it a try.) The heart of the routine is the code on lines 10 through 18 which implement the basic relationships of Eq. (7.27) in a very straightforward manner as the reader should be able to readily verify.

```
 1 : -- /* Fourier Series Analysis */
 2 :
 3 : require"Complex"
 4 :
 5 : fourier = function(fk,nc,thmin) -- Evaluate Fourier Components
 6 :    nc = nc or 20
 7 :    thmin = thmin or 0
 8 :    local nd,fcs,fca,fexp = #fk,{},{},{}
 9 :    local pi2,sum,th = 2*math.pi/nd
10 :    for i=0,nc do
11 :       fac,th = pi2*i,i*thmin
12 :       sum1,sum2 = 0,0
13 :       for k=1,nd do
14 :             sum1 = sum1 + fk[k]*math.cos(fac*(k-1)+th)
15 :             sum2 = sum2 + fk[k]*math.sin(fac*(k-1)+th)
16 :       end
17 :       fcs[i+1] = {2*sum1/nd, 2*sum2/nd} -- Cos-Sin Components
18 :    end -- Next line does Cos-Ang components
19 :    fcs = fourier_trunc(fcs)
20 :    for i=1,nc+1 do fca[i] = {math.sqrt(fcs[i][1]^2+fcs[i][2]^2),
            math.atan2(-fcs[i][2],fcs[i][1])} end
21 :    for i=1,nc+1 do fexp[i] = {fca[i][1]/2,fca[i][2]} end -- Exp
22 :    return setmetatable(fca,Fourierca_mt),setmetatable(fexp,
            Fourierexp_mt),setmetatable(fcs,Fouriercs_mt),nc
23 : end
24 :
25 : -- Assumed forms are f(t) = a[1] + sum(over n>1)
26 : -- Or for exponential form f(t) = sum(over all -/+ n)
27 : Fourier = function(f,tmin,tmax,nc) -- Evaluat nc terms of Series
28 :    local nd,fd,t = 1024,{}
29 :    nc,tmax,tmin = nc or 20,tmax or 1,tmin or 0
30 :    dt = (tmax-tmin)/nd
31 :    for i=1,nd do fd[i] = f(tmin + (i-1)*dt) end -- 1024 samples
32 :    return fourier(fd,nc,tmin/(tmax-tmin)*2*math.pi)
33 :    --return fca,fexp,fcs
34 : end -- Only positive harmonics returned for Exp form
35 :
36 : iFourier = function(fcs,nt,nc) -- Inverse Fourier Series
37 :    nt,nc = nt or 512, nc or 1.e20 -- Number of time components
38 :    local nm,typ,fl = #fcs, type(fcs), {}
39 :    if nc>nm then nc = nm end -- Can't use more than available
40 :    if typ=='CosSin' then for i=1,nc do fl[i] = fcs[i] end
41 :    elseif typ=='CosAng' then for i=1,nc do fl[i] =
            {fcs[i][1]*math.cos(fcs[i][2]),
42 :             -fcs[i][1]*math.sin(fcs[i][2])} end
43 :    elseif typ=='ExpAng' then for i=1,nc do fl[i] =
            {2*fcs[i][1]*math.cos(fcs[i][2]),
44 :                -2*fcs[i][1]*math.sin(fcs[i][2])} end
45 :    end
46 :    local ft,tv = {},{}
47 :    local pi2,sum,erun,efac = 2*math.pi/nt
48 :    local fdc = fl[1][1]/2 -- a[0]/2 is DC term
49 :    for i=1,nt do
50 :       fac,sum = (i-1)*pi2,0
51 :       for k=2,nc do -- Cos-Sin form is faster than Exp form
52 :             sum=sum+fl[k][1]*math.cos((k-1)*fac)+
                    fl[k][2]*math.sin((k-1)*fac)
53 :       end
54 :       ft[i], tv[i] = sum + fdc,  (i-1)/nt
```

```
55 :    end
56 :    return tv,ft -- Time, between 0 and 1 and Function values
57 : end
```

Listing 7.3. Code segments for Fourier analysis and inverse Fourier calculations.

The Fourier() routine accepts a function name, a tmin and tmax value along with an optional number of Fourier components value. It evaluates 1024 data values on line 31 from the supplied function over the tmin to tmax time interval. It then calls fourier() on line 32 to return the three forms of Fourier coefficients. This author's personal preference for physical interpretation is the Cos-Angle format from which the magnitude of each harmonic component is given simply by the magnitude component. A "type" tag is associated with each of the formats ('CosSin', 'CosAng', or 'ExpAng' using the setmetatable() statements on line 22 of the code) in order to make printing of the formats easy and so computer code can determine which format is supplied as input to various functions.

The final function iFourier() takes a Fourier series and returns the corresponding time function. It requires as input a table of Fourier coefficients and optionally a number of time points for the time evaluation (the default is 512 time values as set on line 37) and the number of Fourier components to be used in the evaluation (the default is the number in the input table). By specifying the number of Fourier components the user can experiment with the effects of varying number of components on the inverse waveform without redoing the Fourier series. The heart of iFourier() is lines 49 through 55 which implements the time series form the Cos-Sin coefficient forms. The routine will accept any of the 3 types of Fourier coefficient formats so lines 41 through 44 transform other formats into the Cos-Sin format. One might think that the Exp format would be faster to evaluate the coefficients. However, experience has shown that the sin() and cos() functions evaluate much faster than an exp() function with complex argument. The time values returned by iFourier range from 0 to 1.0 and represent fractions of a period of the signal. These values can be scaled and time shifted to any desired time scale. As previously discussed, the Fourier components do not depend on the exact time values. A function (not shown in Listing 7.3) called expandFt(time_function, tmin,tmax,np) is supplied to convert a returned time function to any desired time interval over tmin to tmax and in addition to expand the function to np periods of the periodic function.

The Fourier code segments are supplied in the fourier.lua package. In addition to the functions listed in Listing 7.3, another code segment, plotFourier() will give a quick pop-up plot of the Fourier coefficients in the various formats when supplied with the appropriate table of Fourier coefficients in any one of the three formats. This is very useful for rapid looks at the Fourier components. The function determines the coefficient format from the type tag associated with the three formats.

A few examples are now given of the Fourier analysis with the code in Listing 7.4. Four simple functions are defined on lines 4 through 20 and one of these is selected for analysis on line 21 through 24. The code on lines 29 through 32

```
 1 : --/* File list7_4.lua */ -- Test of Fourier functions
 2 :
 3 : require"Fourier"
 4 : fsqw = function(t) -- Square wave
 5 :    if t<.5 then return 1
 6 :    else return 0 end
 7 : end
 8 : frsw = function(t) -- Rectified sin() wave
 9 :    if t<.5 then return natg,sub(2*math.pi*t)
10 :    else return 0 end
11 : end
12 : fstooth = function(t) -- Saw-tooth wave
13 :    if t<.5 then return math.pi*2*t
14 :    else return 0 end
15 : end
16 : fstooth2 = function(t) -- Full period saw-tooth wave
17 :    if t<.5 then return 2*t
18 :    else return 2*(t-1) end
19 : end
20 :
21 : ft = fsqw
22 : --ft = frsw
23 : --ft = fstooth
24 : --ft = fstooth2
25 :
26 : dt,nt = 1/1024, 1024
27 : fd,ta = {}, {}
28 : for i=1,nt do
29 :    fd[i] = ft(dt*(i-1))
30 :    ta[i] = (i-1)/512
31 : end
32 :
33 : plot(ta,fd,{'Function of time','Time values','Function Values'})
34 : fx,fy,fz = Fourier(ft,0,1,40) -- Get 40 Fourier Coefficients
35 : print(fx) -- Print Cos-Ang coefficients
36 : plotFourier(fx) -- Plot Cos-Ang coefficients
37 : fcm,fca = {},{} -- Collect Fourier coefficients into tables
38 : for i=1,#fx do fcm[i],fca[i] = fx[i][1],fx[i][2] end
39 : write_data('list7_4a.dat',fcm,fca)
40 :
41 : tt,fft = iFourier(fy,nt,10) -- Inverse Fourier transform
42 : plot(tt,fft,fd) -- Plot original signal and inverse transform
43 : write_data('list7_4b.dat',tt,fft,fd)
44 : xx,yy = expandFt(fft,-1e-4,2.e-4) -- Expand to 3 cycles
45 : plot(xx,yy) -- plot 3 cycles of wave
46 : write_data('list7_4c.dat',xx,yy)
```
Selected Output:
```
((1)Cos(0 + (0 Rad))
((0.63662077105409)Cos(1 + (-1.5677283652191 Rad))
((0)Cos(2 + (0 Rad))
((0.21220958687503)Cos(3 + (-1.5615924420676 Rad))
((0)Cos(4 + (0 Rad))
```

Listing 7.4. Some examples of Fourier Analysis with known waveforms.

calculates nt sample points of the selected time waveform. The Fourier() function is called on line 34 with a request for 40 Fourier harmonic coefficients and iFourier() is called on line 41 to reconstruct the waveform using 10 Fourier harmonic components. The reader can change the function analyzed to look at a square wave (line 21), a rectified sin() wave (line 22), or two forms or triangular ramps (lines 23 and 24). Selected output from the analysis for the square wave is shown in Figures 7.6 through 7.9. First in Figure 7.6, the square wave is shown with the Fourier approximation using only 5 harmonic components. It is obvious that many more components are necessary for a good approximation to the waveform. The use of a simple known waveform allows one to verify the proper operation of the Fourier code segments. Figure 7.7 and 7.8 show the magnitude and angle of the first 20 Fourier components when expressed in the Cos-Ang format of Eq. (7.22). For a square wave of unit amplitude, the Fourier components are known to have the magnitude:

$$c_0 = 1; \ c_n = \frac{2}{n\pi} \text{ for odd } n \text{ and } c_n = 0 \text{ for even } n . \tag{7.28}$$

The plotted values in Figure 7.7 are seen to agree closely with these theoretical values.

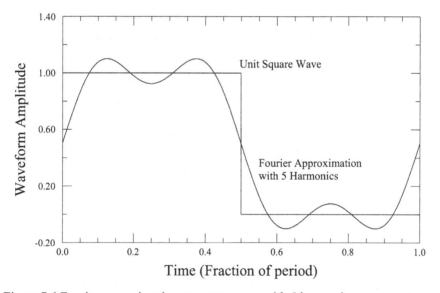

Figure 7.6 Fourier approximation to square wave with 5 harmonic components.

Also the given square wave is known to have a sin() representation so the phase angle when expressed in terms of cos() functions should be +/- 180 degrees. This is seen to be the case in Figure 7.8 which shows the angle of the Cos-Ang representation.

Finally Figure 7.9 shows an expanded view of several cycles of a square wave with two approximations using 20 and 40 Fourier components to approximate the waveform. This was obtained from Listing 7.4 by executing the code with different values of the requested number of Fourier components in iFourier() on line 41.

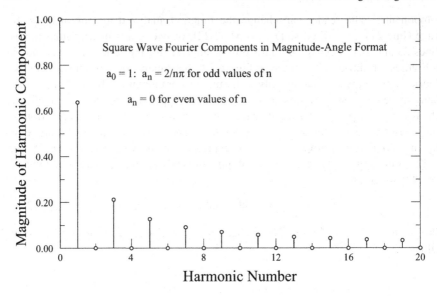

Figure 7.7. Magnitude of the first 20 Fourier Components for square wave.

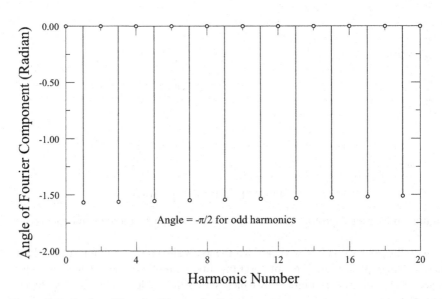

Figure 7.8. Angle of Fourier Harmonic Components.

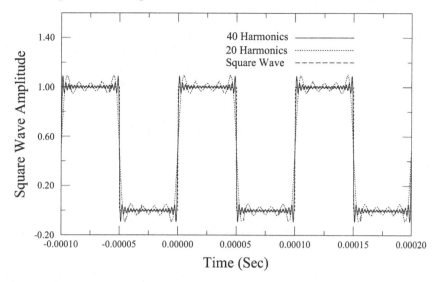

Figure 7.9. Fourier approximation to Square Wave using 20 and 40 harmonic components.

This plot has also made use of the expandFt() function on line 44 to expand a periodic function to several cycles and at the same time to time scale the function to a desired time range, in this case the default 3 cycles on a time scale of 1.e-4 to 2.e-4 is used. This is another example of the functions available in the Fourier.lua code package.

The Fourier analysis method is not a very good general purpose data fitting approach. In many cases, a large number of Fourier components are required for a good fit to a periodic function. Its major use is in allowing engineers to transform their thinking and analysis methods from the time domain to the frequency domain. As another example consider again the data for Wolf's monthly sunspot activity as previously shown in Figure 7.4. The data shows a periodicity of somewhere around 11 years as readily seen from the graph and this should show up in a Fourier analysis of the data. Listing 7.5 shows code for the Fourier analysis of this monthly solar sunspots data. The listing calls the routine fourier() on line 9 since the data already exists in a data file and no sampling of a function is needed. The fundamental period for the Fourier analysis will be the total time interval from 1749 to 2004. The code segment on line 11 through 14 converts the harmonic number into a frequency component in order to better interpret the data.

Figure 7.10 shows the resulting Fourier amplitude vs. time periods for the data with time periods up to 30 years. The figure shows a strong peak in the Fourier components at a time interval of around 11 years as expected. The identification of frequency components composing a signal is one of the most important applications of the Fourier series method. However in this example the identification of the fundamental period, is not nearly as clear as the data in Figure 7.5 showing a simple linear line fitted to the time of both the solar peaks and solar minimum activity. So a variety of analysis techniques should always be used with experimental data to determine which analysis technique provides the most reliable estimate of some desired parameters. The next chapter will be devoted to a discussion of various data analysis techniques.

```
 1 : --/* File list7_5.lua */ Fourier analysis of sunspot activity
 2 :
 3 : require"Fourier"
 4 :
 5 : jk,yyd,ssd = {},{},{}
 6 : read_data('monthssn.dat',jk,yyd,ssd) -- read data
 7 : nmths = #yyd -- Number of months
 8 : plot(yyd,ssd) -- Look at data
 9 : fx,fy,fz = fourier(ssd,100) -- Get Fourier components
10 : freq,fmag = {},{} -- Tables of magnitude and angle

11 : for i=1,#fx do
12 :    freq[i] = nmths/(12*i) --Convert from harmonic # to frequency
13 :    fmag[i] = fx[i][1]
14 : end
15 :
16 : plotFourier(fx) -- View Fourier components
17 : plot(iFourier(fx)); plot(freq,fmag)
18 : write_data('list7_5.dat',freq,fmag) -- Save analysis
```

Listing 7.5. Code for Fourier analysis of sunspot activity.

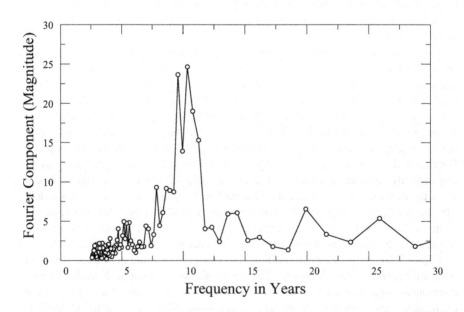

Figure 7.10. Magnitude of Fourier component vs. time period in years.

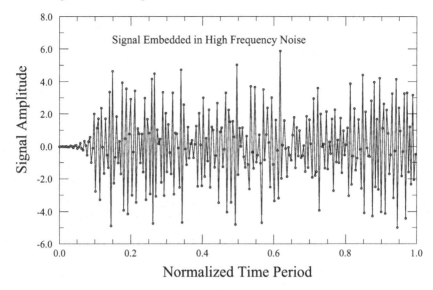

Figure 7.11 Example of signal embedded in high frequency noise.

Data for another Fourier example is shown in Figure 7.11. This shows what looks like a very noisy signal as a function of time. The time signal looks rather random and there is no apparent evidence of any structured signal present. However when one looks at the Fourier coefficient spectra as a function of harmonic number as seen in Figure 7.12 it is seen that the spectra has two rather different components. From DC out to about the 60'th harmonic, there is evidence of a signal spectra with a structured amplitude spectra and beyond this region there is an amplitude spectra with rather random amplitudes. The low frequency signal component can be recovered by eliminating the high frequency noise and by reconstruction the signal using only the first 60 harmonics. This is shown in the recovered signal in Figure 7.13. The reconstructed signal shows that there is indeed a structured pulse embedded in the signal of Figure 7.11. The peak of the pulse occurs at a normalized time of about 0.55 and there is no evidence in Figure 7.11 of any signal around that time. The signal amplitude at its peak is seen to be only 1.5 units in height while the signal with noise has peak amplitudes of 5 to 6, so it is not surprising that the signal is not readily seen in the original signal.

The code segment to perform this analysis using the Fourier code routines is shown in Listing 7.6. The time signal is read in from a data file on line 6 along with the embedded signal waveform on line 7. Line 8 performs the Fourier analysis for 128 Fourier components (this is the maximum unique value for a signal with 256 data points). The reconstructed signal is generated on line 16 using iFourier() for 256 time intervals and using 60 Fourier components. The remainder of the code generates pop-up plots and writes data to files for further plotting. A plot of the embedded signal is not shown but the reader can compare Figure 7.13

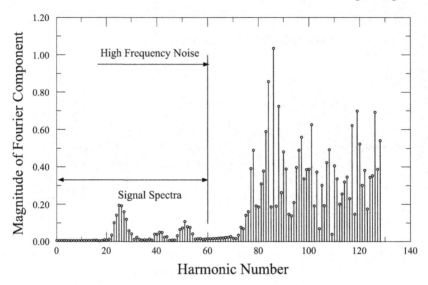

Figure 7.12 Fourier magnitude spectra for signal in Figure 7.10.

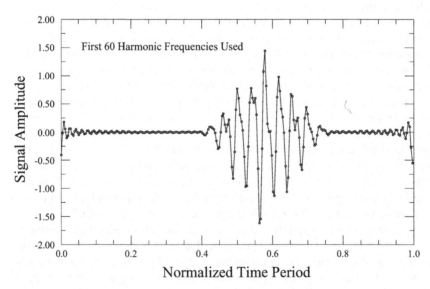

Figure 7.13 Recovered signal from inverse Fourier transform.

with the data stored in the target_signal.txt file and see that the recovered signal closely matches the embedded signal waveform. This example has illustrated how Fourier analysis can be used to extract a low frequency signal from high frequency noise.

```
 1 : -- /* File list7_6.lua */ -- Signal in noise example
 2 :
 3 : require"fourier"
 4 :
 5 : tpn,tsn = {},{} -- Read signal_noise and signal, 256 points
 6 : read_data('target_plus_noise.txt',tpn)
 7 : read_data('target_signal.txt',tsn)
 8 : fx,fy,fz = fourier(tpn,128) -- Fourier analysize
 9 : fm,fang = {},{} -- Magnitude and angle of Fourier comp
10 : for i=1,#fx do fm[i],fang[i] = fx[i][1],fx[i][2] end
11 : plotFourier(fx) -- Plot coefficients
12 : ttn = {} -- Generate time points
13 : for i=1,#tpn do ttn[i] = i-1 end
14 : plot(ttn,tpn) -- Plot signal plus noise
15 : plot(ttn,tsn) -- Plot signal
16 : tt,sig = iFourier(fx,256,60) -- Recover signal
17 : write_data('list7_6a.dat',fm,fang)
18 : plot(tt,sig) -- Plot recovered signal
19 : write_data('list7_6b.dat',tt,sig,tpn,tsn)
```
Listing 7.6. Code for signal in noise example.

Calculations such as these are useful in designing and understanding physical electronic systems. In the real engineering world, such signal extraction might be performed in near real time by passing the signal plus noise through a low pass filter to eliminate the high frequency components and keep only the lower frequency components However, Fourier analysis is a key analysis tool to aid in the design of such filters. By knowing the frequency spectra of signal and noise, the engineer can properly design a real time filter (either analog or digital) to perform the signal extraction.

A Fourier series expansion is not good at approximating a curve such as that of Figure 7.1 if one assumes that the data presented represents a full cycle of some periodic signal. The difficulty arises from the large step in the function at the boundary required to make the function periodic as required by the Fourier analysis. In some cases such as this it is most convenient to think in terms of the data representing only a partial segment of some periodic signal – for example a half or quarter of a periodic cycle. Since the interest is only in a given range of data, one is free to extrapolate in any useful way so that a cycle of data can be completed. Figure 7.14 shows three possible interpretations of the data in Figure 7.1 in terms of a periodic function. In (a) the given data period (0 to 20) is assumed to be a complete period, while in (b) and (c) respectively, the given data period is assumed to represent only one-half and one-fourth of a complete period respectively. Since the interest is only in the range 0 to 20 any one of these data extension techniques can be used as a possible curve for a Fourier expansion. Obviously if the original data period is selected as a complete cycle of the function then many terms of a Fourier series are needed because of the discontinuity in the function in (a) at the assumed period boundaries. Using a half period for the data in (b) requires fewer Fourier terms, but the slope discontinuity at 40 in the figure still requires many Fourier series terms. The selection of a quarter of a period for the data interval as in (c) looks close to a single sinusoidal function and should be easily represented by only a few Fourier series components. This is the obvious choice in this case. The Fourier series representation will then be in terms of subharmonics of the data period, since the fundamental frequency will have a period of 80 while the data of interest has a period of only 20 x valued units. Also because of the symmetry of the quarter period function, it can be seen that this representation will have only odd sin() harmonic terms in the Fourier series with zero DC component.

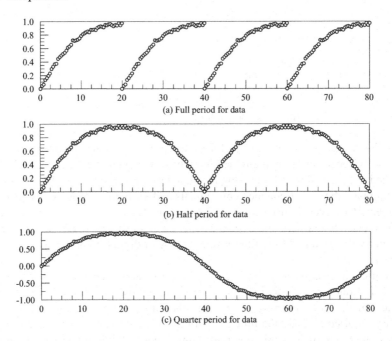

Figure 7.14. Three possible extensions of data in Figure 7.1 for a periodic function.

Listing 7.7 shows a code segment for the quarter period Fourier analysis of the data shown in Figure 7.14(c). Lines 8 through 19 expand the input data to a full period from the quarter period input data. The Fourier analysis is performed on line 20. The pop-up plot of the Fourier coefficients verifies that only the fundamental and third harmonic components have significant amplitude. These amplitudes are extracted on line 23 and printed. The angles of the Fourier components

are not needed since it is known from the symmetry of the waveform that only
sin() components should exist.

```
 1 : -- /* File list7_7.lua */ -- data fitting for Figure 7.1
 2 :
 3 : require"Fourier"
 4 :
 5 : xd,yd = {},{}
 6 : read_data('list7_1.dat',xd,yd) -- Data for Figure 7.1
 7 :
 8 : n1 = #xd
 9 : i = n1+1
10 : for j=n1-1,1,-1 do -- Expand to half cycle
11 :     xd[i],yd[i] = xd[j+1]-xd[j]+xd[i-1], yd[j]
12 :     i = i+1
13 : end
14 : n1 = #xd
15 : for j=2,n1 do -- Expand to full cycle
16 :     xd[i] = xd[i-1]+xd[j+1]-xd[j]
17 :     yd[i] = -yd[j]
18 :     i = i+1
19 : end
20 : fx,fy,fz = fourier(yd,40) -- Fourier coefficients -- 40 compo-
nents
21 : plotFourier(fx) -- plot Fourier coefficients
22 :
23 : c1,c3 = fx[2][1],fx[4][1]
24 : print('c1,c3 = ',c1,c3)
25 : ycalc = {}
26 : for i=1,n1 do
27 :     ycalc[i] = c1*math.sin(2*math.pi*xd[i]/80) +
            c3*math.sin(6*math.pi*xd[i]/80)
28 : end
29 : write_data('list7_7.dat',xd,yd,ycalc)
Output:
c1,c3 =          1.0325212059091              0.078882627688746
```

Listing 7.7 Code for Sub-Harmonic analysis of data in Figure 7.1.

Figure 7.15 shows a plot of the data compared with only two of the sin() Fourier components, which are the fundamental of the quarter cycle analysis and the
third harmonic. The figure shows that these two sin() components give a good fit
to the data. The use of sub-harmonics or half- and quarter-range Fourier expansions can be a very useful data fitting technique for cases where one is interested
only in a limited range of data and where such an extension of the data provides a
much smoother function for Fourier analysis.

7.5 Nonlinear Least Squares Curve Fitting

This section continues with the discussion at the end of Section 7.3 dealing with
the general fitting of a function to data which contains some random noise. In this
section code segments are developed that allow one to fit a general nonlinear

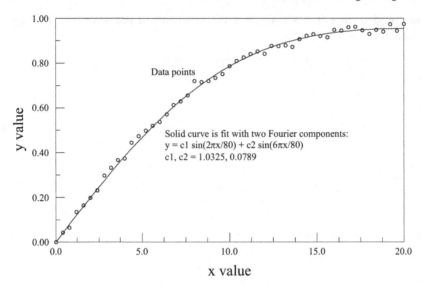

Figure 7.15. Quarter-cycle, Fourier analysis of Figure 7.1 data. Only two Fourier components give a good fit to the data.

equation with one or more unknown coefficients to a set of experimental data and minimize the square error between the fitting equation and the experimental data.

The fitting function may have a general nonlinear relationship between the dependent and the independent variable and the fitting coefficients may be included in very general nonlinear terms. In addition the fitting relationship may be an implicit equation where the dependent variable may not be expressable in an explicit manner in terms of the dependent variable. Finally, the concept of an independent variable will be extended to include sets of independent variables on which a dependent variable may depend. This is a rather ambitious task for this chapter and as such in one of the most important section in this book. The code segments developed in this section can find very important applications in a variety of physical modeling applications. The author has used the code as the basis for several computer modeling and parameter estimation applications.

For the present development consider y as some dependent variable and x as some independent variable. For the purposes of this section it will be assumed that some functional relationship exists between the variables involving coefficients $C_1, C_2, \cdots C_n$, which in functional form can be expressed as:

$$F(y, x, C_1, C_2, \cdots C_n) = F(X, C) = 0,$$
$$\text{where } X = \{y, x\} \text{ and } C = \{C_1, C_2, \cdots C_n\}$$

(7.29)

In the second form the dependent and independent variables have been combined into an X array with the dependent variable as the first element and the coefficients have been combined into a linear C array containing all the fitting coefficients. In order to simplify the equations, the array notation has been omitted and it will be understood that the X (capital) values include the dependent variable as the first element and the C notation without a subscript is used to denote an array of coefficient values.

An example or two will help clarify the notation:

$$F_1(X,C) = C_1 \exp\left[-((C_2 - x)/C_3)^2\right] - y = 0$$
$$F_2(X,C) = C_1 \left[\exp((x - C_2 y)/C_3) - 1\right] - y = 0$$

(7.30)

In F_1 the dependent variable y can be solved for in terms of the independent variable, but in the F_2 function the variable y can not be explicitly solved for since it occurs both as a linear term and in the exp() function. Thus in F_2 there is only an implicit relationship for the dependent variable in terms of the independent variable. In both equations the constants enter into the relationships in nonlinear manners. In the most general case then a functional relationship can exist between the independent and dependent variables such that one can not solve explicitly for either variable. In fact the relationship between the variables might be expressed only in terms of some computer algorithm. In order to handle the most general case it will be required only that the functional relationship be expressible in the general form of Eq. (7.29).

The general direction this discussion is headed is to develop a nonlinear least squares fitting approach which can handle any general functional relationships as shown above. However, in order to develop the process in incremental steps, first consider the case where an explicit expression is available for the dependent variable in terms of the independent variables. This will handle cases such as F_1 above but not F_2. Let's assume that the model for fitting to a set of data is of the form:

$$y = f(X,C) \tag{7.31}$$

where the notation is again that X may be an array of independent variables and C is an array of fitting parameters. The sum mean square error between data and model can then be expressed as:

$$\varepsilon^2 = \sum_{k=1}^{n_d}(y_k - f(X_k,C))^2 = \sum_{k=1}^{n_d}(y_k - f(X_k,C_1,C_2,\cdots C_n))^2 \tag{7.32}$$

where the sum is over the n_d data points. In the second form of the equation the functional dependence on all the fitting parameters has been explicitly shown. The goal is thus to determine a set of fitting parameters which minimizes the square error.

The primary technique for approaching the solution of any nonlinear problem in any general fashion is as has been stated before the L&I or "linearize and iterate" method. To begin such a process it is assumed that some initial guess at the C coefficients is available and that the fitting function can be expanded in terms of a set of corrections to these coefficients. In functional form then the f function can be expanded in a Taylor series and only first order derivative terms kept as in:

$$f(X_k, C) + \frac{\partial f}{\partial C_1} \delta C_1 + \frac{\partial f}{\partial C_2} \delta C_2 + \cdots + \frac{\partial f}{\partial C_n} \delta C_n$$

$$= f_k + f_{1k} \delta C_1 + f_{2k} \delta C_2 + \cdots + f_{nk} \delta C_n \qquad (7.33)$$

$$\text{with } f_k = f(X_k, C) \text{ and } f_{ik} = \frac{\partial f}{\partial C_i}$$

In this it is understand that the partial derivatives are to be evaluated at the original values of the C coefficients and the δC_i's represent correction terms. The second form of the equation simplifies the notation with the simpler double subscript definitions of the partial derivatives.

With this expansion, Eq. (7.32) becomes:

$$\varepsilon^2 = \sum_{k=1}^{n_d} (y_k - f_k - f_{1k} \delta C_1 - f_{2k} \delta C_2 - \cdots - f_{nk} \delta C_n)^2 \qquad (7.34)$$

This is now a linear equation in terms of the corrections to the coefficients and is in much the same form as Eq. (7.14) previously considered in connection with linear least squares regression. The difference here is that the solution for the corrections will have to be added to the coefficients and the process iterated until a consistent set of coefficients is obtained.

$$\begin{bmatrix} \sum_{k=1}^{n_d} f_{1k} f_{1k} & \sum_{k=1}^{n_d} f_{1k} f_{2k} & \cdots & \sum_{k=1}^{n_d} f_{1k} f_{nk} \\ \sum_{k=1}^{n_d} f_{2k} f_{1k} & \sum_{k=1}^{n_d} f_{2k} f_{2k} & \cdots & \sum_{k=1}^{n_d} f_{2k} f_{nk} \\ \vdots & \vdots & \vdots & \vdots \\ \sum_{k=1}^{n_d} f_{nk} f_{1k} & \sum_{k=1}^{n_d} f_{nk} f_{2k} & \cdots & \sum_{k=1}^{n_d} f_{nk} f_{nk} \end{bmatrix} \begin{bmatrix} \delta C_1 \\ \delta C_2 \\ \vdots \\ \delta C_n \end{bmatrix} = \begin{bmatrix} \sum_{k=1}^{n_d} (y_k - f_k) f_{1k} \\ \sum_{k=1}^{n_d} (y_k - f_k) f_{2k} \\ \vdots \\ \sum_{k=1}^{n_d} (y_k - f_k) f_{nk} \end{bmatrix} \qquad (7.35)$$

To obtain a set of equations for the corrections, the usual procedure of taking partial derivatives with respect to the variables, here the δC_i's, need to be taken with the resulting terms set to zero. The resulting sets of equations when expressed in matrix form are then as shown in Eq. (7.35).

This set of equations is very similar to Eq. (7.15), the set developed for the general least squares case with linear coefficients. This is not surprising since the fitting function was linerized before formulating this set of equations. However, the meaning of the f_{ik} terms is different for this nonlinear case and the right hand side of the equation involves different terms as can be seen by referring back to Eq. (7.15).

Looking back at the basic least squares equation (Eq. (7.32)) one can take a derivative with respect to the constants and see that the condition for a minimum can be expressed as:

$$\sum_{k=1}^{n_d} (y_i - f_k) \frac{\partial f_k}{\partial C_i} = \sum_{k=1}^{n_d} (y_i - f_k) f_{ik} = 0. \qquad (7.36)$$

This is seen to be the right hand side terms of each of the rows of the matrix in Eq. (7.35). Thus it can be seen that when a set of coefficients that minimize the square error has been obtained, the forcing terms on the right of Eq. (7.35) will approach zero and the correction terms in the solution of the matrix will go to zero. This is similar to what happens in the Newton method for solving a set of nonlinear equations and the solution will be approached in a quadratic manner. So it can be conclude that if this iterative approach converges with the correction terms approaching zero, a coefficient set will be obtained which gives minimum least square error and the convergence is expected to be very rapid as convergence is approached.

An approach to the nonlinear least squares data fitting problem can now be formulated. First an initial approximation to the coefficient values is needed. Second the fitting function is linearized around the coefficient values with linear correction terms to the coefficients. Third a set of linear equations is formulated for the corrections to the coefficients. Fourth the correction values are obtained from the linear equations. Fifth the corrections to the coefficients are added to the coefficients and a check made for convergence. Last, the process is repeated until the corrections are smaller than some termination criteria. The entire process is very similar to that developed in Chapter 4 for solving sets of nonlinear equations, only the details of the matrix coefficients are different.

One final detail remains before presenting computer code for this problem. In the original discussing it was stated that a method was desired to fit any general nonlinear function of the form of Eq. (7.29) where the function returns zero when satisfied by the dependent and independent variables. In the above formulation a function is needed that returns the value of the dependent variable when called with the coefficients and independent variables, such as represented by Eq. (7.31). How are these two ideas connected and how is this formulated for use in Eq. (7.32) where it was assumed that the dependent variable was known in terms of the independent variable? The key is Newton's method for solving a nonlinear equation. Back in Chapter 3 an approach was developed that takes an equation formulated in terms of returning zero when satisfied and returns the value of the dependent variable. Thus the connection here is that one needs to interpose a Newton's method solution routine between the equation formulation of Eq. (7.29) and the least squares approach. Formally stated this means that the $f(X,C)$ function should really be expressed as:

$$f(X,C) = newtonfc(F,X,C) \qquad (7.37)$$

where *newtonfc*() represents computer code for a Newton iterative method taking an $F(X,C) = 0$ function and returning the solved value of the dependent variable. The major difference here as opposed to Chapter 3 for Newton's method is that the Newton's method to be used here must pass the coefficient array along to the function, whereas in Chapter 3 the discussion was not concerned with a coefficient matrix. This will hopefully become clearer when the code for implementing the method is discussed.

A final issue involves evaluation the multitude of partial derivatives needed in the formulation. Partial derivative must be evaluated with respect to each coefficient and these partial derivatives must be evaluated at each data point. So if there are n_d data points and n_c coefficients, there will be $n_d n_c$ partial derivative values. The majority of authors advise against trying to evaluate these derivatives numerically and insist that one should provide mathematical equations from which these derivatives can be evaluated. However, this is very inconvenient, very tedious, and in cases of implicit equation formulations this can not be accomplished. In Chapter 5 a very robust approach to numerical derivatives was developed and this will be extensively used in this chapter to evaluate the required partial derivatives. This is essential if the formulation of Eq. (7.37) is to be used to handle nonlinear problems, because the function evaluations and derivatives are only available through a Newton's method computer algorithm and not directly expressible in term of a mathematical equation. As will be seen the developed algorithm using numerical derivatives for least squares fitting is rather robust over a wide variety of physical problems.

Computer code for the nonlinear least squares algorithm is shown in Listing 7.8. This is the longest and most involved code segment presented so far and only the main features will be discussed. Before discussing the code it is important to consider the calling arguments to the function as given below:

$$\text{nlstsq} = \text{function}(x, fw, f, c, actv, step) \qquad (7.38)$$

These calling arguments are:

1. x – this is an array of input data values of dimension nx by nd where nx is the number of data variables and nd is the number of data points. The number of data variables counts the dependent variable as well as the independent variable so nx $>= 2$. If yd and xd represent arrays of input variables of length nd then the proper format for the input x array would be x = {yd, xd}, with the dependent variable always being the first table

2. fw – is an array of length nd of weighting factors for the data points. Normally these weighting factors will simply be taken as equal to unity. However, this is not required. If one has "outlier" data that should not be included in the analysis that particular weighting factor can be set to zero. A null array ({}) or undefined variable may be passed as the value of this argument and the code will assume unity value for all the weighting factors.

3. f – the nonlinear function to be fit to the data, coded so that f(x,c) = 0 is the desired functional relationship.

4. c – array of C coefficients, of dimension nc.
    actv – this is an array of dimension nc which specifies whether or not to let the C coefficients vary in the analysis. An actv[i] value of nonzero specifies that the i'th value of the C's will be optimized in the analysis. There are important cases where one wishes to keep a particular C value constant and optimize on the other C values. Setting the actv[i] value associated with that particular C to zero accomplishes this. This parameter

```
 1 :  --/* File nlstsq.lua */
 2 :
 3 :  require "gauss"; require"Minv" -- Matrix inversion needed
 4 :  nlstsq = function(x, fw, f, c, actv, step)
 5 :     actv,step,fw = actv or {},step or {},fw or {}
 6 :     local nx,nd,nc = #x,#x[1],#c
 7 :     local iend,nend,cv,cmax,err = 0, NMAX -- Max # of iterations
 8 :     local del,p,wk1,wk2,rowptr,b,fx ={},{},{},{},{},{},{}--Work
 9 :     for i=1,nc do
10 :        rowptr[i],step[i],actv[i] = {},step[i] or 0,actv[i] or 1
11 :        if step[i]>1.0 then step[i] = 1./step[i] end
12 :     end
13 :     for ilp=1,NMAX do -- Main loop for Newton iterations
14 :        ix = 1
15 :        for i=1,nc do -- Loop over C coefficients
16 :           if actv[i]~=0 then -- An active coefficient
17 :              b[ix],row,wk2[ix] = 0.0, rowptr[ix], {}
18 :              for j=1,nc do row[j] = 0.0 end--Initialize to zero
19 :                 dx = FACT*c[i] -- Set up dC factors
20 :                 if dx==0.0 then dx = FACT end
21 :                 wk1[ix], ix = dx, ix+1 -- Store dC factors
22 :              end
23 :           end
24 :           for k=1,nd do -- Evaluate function at data points
25 :              for i=1,nx do p[i] = x[i][k] end
26 :              fx[k] = newtonfc(f, p, c)
27 :           end -- fx[k] has function values at data points
28 :           ix, jx = 1, 1
29 :           for i=1,nc do -- Loop over active coefficients
30 :              if actv[i]~=0 then
31 :                 c1, dx = c[i], wk1[ix]
32 :                 c[i] = c1+dx
33 :                 pwk2 = wk2[ix] -- Pointer to work array
34 :                 for k=1,nd do -- Loop over data points
35 :                    for i=1,nx do p[i] = x[i][k] end
36 :                    pwk2[k] = (newtonfc(f, p, c) -fx[k])/dx
37 :                 end
38 :                 c[i], ix = c1, ix+1
39 :              end
40 :           end -- Partial derivatives stored in wk2[i][k] arrays
41 :           for j=1,nc do -- Loop over active rows in matrix
42 :              if actv[j]~=0 then
43 :              row, pwk2 = rowptr[jx], wk2[jx]-- ptr matrix & wk2
44 :              for k=1,nd do -- Loop over data points
45 :                 b[jx]=b[jx]+(x[1][k]-fx[k])*pwk2[k]*(fw[k] or 1)
46 :              end
47 :              ix = 1
48 :              for i=1,nc do -- Accumulate row of a[][] matrix
49 :                 if actv[i]~=0 then
50 :                    pwk3 = wk2[ix] -- Another pointer
51 :                    for k=1,nd do -- Loop over data points
52 :                       row[ix]=row[ix]+pwk3[k]*pwk2[k]*(fw[k] or 1)
53 :                    end
54 :                    ix = ix+1
55 :                 end
56 :              end
57 :                                  jx = jx+1
58 :        end
```

```
59 :       end -- Back for another coefficient

60 :       if iend==0 then -- Update values except for last time through
61 :          gauss(rowptr, b) -- Now solve for Coefficient corrections
62 :          ix, cmax = 1, 0.0
63 :          for i=1,nc do -- Loop over active coeff, with limitations
64 :             if actv[i]~=0 then -- and update active coefficients
65 :                if step[i]~=0.0 then
66 :                   if step[i]<0.0 then -- +/- step change limits
67 :                      if abs(b[ix])>-step[i] then
68 :                         if b[ix]>0.0 then b[ix] = -step[i]
69 :                         else b[ix] = step[i] end
70 :                      end
71 :                   else -- percentage change limits
72 :                      c1 = 1. + b[ix]/c[i]
73 :                      if c1<step[i] then b[ix] =
                               (step[i] - 1.)*c[i] end
74 :                      if c1>1./step[i] then b[ix] =
                               (1./step[i]-1.)*c[i] end
75 :                   end
76 :                end
77 :                c1 = abs(b[ix]) -- Check on convergence
78 :                c2 = abs(c[i]) + c1
79 :                if c2~=0.0 then c1 = c1/c2 end
80 :                if c1>cmax then cmax = c1 end
81 :                c[i], ix = c[i] + b[ix], ix+1 -- Updated C's here
82 :             end
83 :          end
84 :          if nprint~=0 then -- Print iteration values if desired
85 :             printf('Coefficients at iteration %d are:\n',ilp)
86 :             for i=1,nc do printf(' %e ',c[i]) end;
                      printf('\n'); flush()
87 :          end
88 :       end
89 :       if iend==1 then nend=ilp; break end -- End now
90 :       if cmax <ERROR then iend = 1 end -- once more for matrix
91 :    end
92 :    local cv = minv(rowptr,jx-1) --Covar matrix, bypass gauss
93 :    ix, err = 1, 0.0 -- Accumulate error
94 :    for k=1,nd do -- Sum over data points
95 :       for i=1,nx do p[i] = x[i][k] end
96 :       c1 = x[1][k] - newtonfc(f, p, c) -- Error evaluation
97 :       err = err + c1*c1 -- Square and accumulate
98 :    end
99 :    for j=1,nc do -- Calculate standard deviations of coeff
100 :      if actv[j]==0 then del[j] = 0
101 :      else del[j],ix = sqrt(err*((type(cv)=='table' and
                cv[ix][ix]) or 0)/(nd-nc)),ix+1 end
102 :   end
103 :   return del, sqrt(err/nd), nend-1 -- error and # of iterations
104 : end
```

Listing 7.8 Code for nonlinear least squares data fitting.

may be omitted or a null array ({}) may be passed and the code will assume all unity values for the actv[i] elements.

5.   step – an array of dimension nc, specifying the maximum change to be allowed in each iteration step. This can be specified with a negative sign

for a +/- change limitation or with a positive sign for a percentage change limitation. This parameter may be omitted or a null array ({}) may be passed and the code will assume all zero values for the step[i] elements which will impose no limits on the step sizes.

This function has a rather long and involved parameter list. However, this is one of the most versatile and most important functions developed in this entire book. It is thus important that the reader understand fairly completely the input variables if the reader is to effectively use this routine.

A brief discussion is now given for the code in Listing 7.8. The first few lines, 5 through 9, set up arrays needed in the program and check input data. The main Newton iterative loop goes from line 13 to 91 and will be executed for a maximum of NMAX times (with a default value of 50). Lines 16 through 23 sets up delta C's to be used in the numerical derivatives and the code is very similar to derivative factors used in previous functions where some relative change is as the displacement factor (the dx term on line 19). Lines 24 through 59 set up the matrix equations in three steps.

To understand the steps in the code it is useful to again consider the matrix terms as given in Eq. (7.35). Each matrix term involves a sum over the data points of the product of two partial derivatives of the data fitting function with respect to the fitting coefficients. Thus at each data point there are nc of these paratial derivatives. The simplest approach to evaluating the matrix terms would appear to be to select each data point in turn, evaluate the nc partial derivatives, multiply out the terms and add the result to each matrix coefficient. This approach requires the storage on nc derivative values before forming the derivative products for each of the nc by nc matrix elements. However, this requires that each c[] coefficient be changed once for each data point for a total of nd times. For most applications of this nlstsq() function this is not an important factor as the data fitting function can be readily evaluated for either different coefficients or different data values. However, there are some potential applications, such as fitting a differential equation to a set of data, where each change in the c[] coefficients requires a lengthy calculation within the data fitting function. To readily handle such applications, what is desired is to change a c[] coefficient once and evaluate the paratial derivatives for all data points before considering another c[] coefficient. This requires that nd partial derivative values be stored for each c[] coefficient and a total of (nc + 1)*nd values stored before the matrix elements can be calculated. While this requires considerably more data storage, the requirement is easily met for several thousands of data points and a reasonable number of fitting coefficients. This approach of evaluating all partial derivatives and then calculating the matrix elements is used in the nlstsq() function of Listing 7.8. This approach also executes slightly faster than considering each data point in turn.

Actually when first coded, the nlstsq() function used the simpler approach discussed above of considering each data point in turn and evaluation all the partial derivatives for a single data point before moving on to the next data point. The intermost loop was then on the c[] coefficients and not on the data points. It was only after the material of Chapter 11 was developed that it was realized that the code segment would be more generally useful if the c[] coefficients were changed as few times as possible in the nlstsq() function. The code setment for nlstsq() was then rewritten in the form shown in Listing 7.9. This is a good example of how a slight change in the numerical evaluation of an algorithm can result in a more general and more useful code segment (see Section 11.7 for usage of nlstsq() with differential equations).

With the above discussion, the three major steps in setting up the matrix elements can be considered. First lines 24 through 27 calculates and stores values returned by the fitting equation for all data points in the fx[] table corresponding to the f_k terms in Eq. (7.35). Second, lines 29 through 40 evaluate and store the nc by nd partial derivatives with a loop over the c[] coefficients and an intermost loop over the data points with the values stored in the wk2[][] nc by nd tables. Finally, lines 41 through 59 loops over the active coefficient rows of the matrix evaluating the matrix elements on line 52 within the sum over the data points on lines 51 through 53. After calculating the matrix elements, the corrections to the c[] coefficients are obtained on line 61 by use of the gauss() function.

Lines 63 through 83 implement any specified maximum change limits on the coefficients. These limits are similar to the limits implemented in Chapter 3 involving Newton's method with coupled nonlinear equations and will not be discussed in detail here. The coefficient values are updated on line 81 and if the convergence criterion is met the Newton loop is terminated on line 89 and 90 and the solution process is complete. The remainder of the code, lines 92 through 102 calculates the actual value of mean square error achieved and estimates the achieved uncertainty in the coefficient values. These lines will not be discussed in detail here. The estimate of the uncertainty in the del[] array is returned by the function on line 103 along with the average RMS error between the data and the fitting function and the number of Newton cycles needed for convergence of the algorithm.

The calls to the function to be fitted to the data in the code on lines 26 and 36 are through another function newtonfc() as previously discussed. The code for this function is shown in Listing 7.9. This is very similar to code previously presented for Newton's method in Chapter 3, Listing 3.4. The major differences are that x is now treated as an array of values with x[1] being the dependent variable and the other x values the independent variables. The newtonfc() function uses the '...' variable argument convention (on line 105) to pass any number of additional calling arguments through to the f() function as f(x, ...) on lines 119 and 121. When newtonfc() is called from Listing 7.8 as newtonfc(f, p, c) the c argument will simply be passed through to the function unchanged within the newtonfc() function.

Since Newton's method is not needed if one is fitting a function where the dependent variable occurs only as a linear term, a tag (ylinear = 1) can be set to bypass Newton's method and save computer time for such cases. Lines 108 through 113 handle this special case. Execution time can be considerably reduced for problems where the fitting equation is linear in the dependent vatiable by using the statement getfenv(nlstsq).ylinear = 1. Newton's method requires an initial guess for the solution value and in the case of fitting a data set a good initial guess which is just the dependent data point is readily available and this is used on line 119 of the code. All of the nonlinear least squares code is stored in a single file nlstsq.lua and can be called into other programs by the require"nlstsq" code statement.

As previously stated this nonlinear least squares routine is one of the most important and versatile programs developed in this book. This author has used this as the basis for several real world engineering applications of data fitting and parameter extraction. The remaining sections of this chapter will illustrate the use of this program is different applications. This function is also used in several subsequent chapters in important nonlinear data modeling applications, especially in the area of estimating the parameters of a nonlinear model describing experimental data.

```
  1 : --/* File nlstsq.lua */
  2 :

105 : newtonfc = function(f, x, ...) -- Newton's with parameters
106 :    -- Form of f must return f(x,c) = 0, x an array of values
107 :    local y,dy,fy,cy,ysave
108 :    if ylinear~=0 then -- Bypass Newton if linear function
109 :        y,x[1] = x[1],0.0 -- First variable is being solved for
110 :        fy = f(x, ...) -- f must return calculated_val - x[1]
111 :        x[1] = y
112 :        return fy -- Return value of function
113 :    else
114 :        y = x[1] -- First variable solved for, x[1] initial guess
115 :        ysave = y -- Save initial guess
116 :        for i=1,NT do -- Main Newton loop
117 :            dy = y*FACT
118 :            if dy==0.0 then dy = FACT end -- Increment variable
119 :            fy = f(x, ...)
120 :            x[1] = x[1] + dy
121 :            cy = fy - f(x, ...) -- Difference for derivative
122 :            if cy~=0.0 then cy = fy*dy/cy end -- Correction term
123 :            y = y + cy -- New value of variable being solved for
124 :            x[1] = y; dy = abs(cy) + abs(y)
125 :            if dy==0.0 or abs(cy/dy)<ERR then break end – Return?
126 :        end
127 :        x[1] = ysave -- Restore initial guess, valid data point
128 :        return y -- Rerurn solved value
129 :      end
130 : end

131 : setfenv(nlstsq,{ERROR=5.e-4,FACT=1.e-8,NMAX=50,ERR=1.e-3,
        NT=50,
132 :    nprint=1,ylinear=0,abs=math.abs,sqrt=math.sqrt,
        flush=io.flush,
133 :    printf=printf,gauss=gauss,newtonfc=newtonfc,type=type,
134 :    minv=minv}) -- Set parameters for nlstsq()
135 : setfenv(newtonfc,getfenv(nlstsq)) -- Same params as nlstsq()
```

Listing 7.9 Code listing for Newton's method with passed coefficient array.

7.6 Data Fitting and Plotting with Known Functional Forms

This section will consider least squares data fitting and plotting of data using known or assumed functional forms. In order to get a feel for the use of the nlstsq() code, a simple example is first shown in Listing 7.10. This is a very simple example where data is generated by lines 5 through 9 from the equation:

$$y = 2(1 - \exp(-.8x)) \tag{7.39}$$

```
1 :    -- /* File list7_10.lua */ -- First example of data fitting
2 :
3 :    require"nlstsq"
4 :
5 :    xd,yd,i = {},{},1
6 :    for t=0,5,.1 do -- Generate some data
7 :     xd[i],yd[i] = t, 2*(1-math.exp(-.8*t))
8 :     i = i+1
9 :    end
10 :
11 :    ft = function(x,c) -- Define function to fit data
12 :     return c[1]*(1 - c[2]*math.exp(c[3]*x[2])) -x[1]
13 :    end
14 :
15 :    c = {4,.5,-1} -- Initial guess at coefficients
16 :    actv,step = {1,1,1},{0,0,0} -- Set up arrays
17 :    nd,nx,nc = #xd,2,3
18 :    fw = {} -- Weighting factors
19 :    for i=1,nd do fw[i] = 1 end
20 :    x = {yd,xd} -- Data values
21 :    del,err,nmax = nlstsq(x,fw,ft,c,actv,step) -- Call fiting
22 :    print(err,nmax) -- print results
23 :    for i=1,nc do printf('c[%d] = %12.4e +/- %12.4e\n',i,c[i],
           del[i]) end
Output:
Coefficients at iteration 1 are:
 1.985264e+000  7.458772e-001  -7.822460e-001
Coefficients at iteration 2 are:
 1.999812e+000  1.001820e+000  -8.060560e-001
Coefficients at iteration 3 are:
 1.999979e+000  9.999951e-001  -7.999956e-001
Coefficients at iteration 4 are:
 2.000000e+000  1.000000e+000  -8.000000e-001
2.4748817195222e-011   5
c[1] =   2.0000e+000 +/-   8.7463e-012
c[2] =   1.0000e+000 +/-   7.2571e-012
c[3] =  -8.0000e-001 +/-   1.4085e-011
```

Listing 7.10. Simple example of data fitting with nlstsq() code.

The function used to fit the generated data is defined in function ft() on lines 11 through 13 as:

$$c[1](1 - c[2] * \text{math.exp}(c[3] * x[2])) - x[1] \tag{7.40}$$

In a conventional non-computer math notation this would be:

$$C_1(1 - C_2 \exp(-C_3 x)) - y = 0 , \qquad\qquad (7.41)$$

where three coefficients (c[1], c[2] and c[3]) have been used in the assumed fitting function. The example illustrates the fact that the first variable in the x array passed to the fitting function is the dependent variable, in this case y (x[1]), and the second variable is the independent variable, in this case x. This order of definition can be seen on line 20 of the listing. Lines 15 through 20 set up all the input variables needed by the fitting routine and similar definitions will be required each time one uses the nlstsq() function. Line 15 provides an initial guess at the C coefficients, in this case taken as 4, .5, -1. It will be noted that the actual values used in generating the data are 2, 2, and -.8 as seen in Eq. (7.39) and as used on lines 7 in generating the data. This illustrates that the routine is fairly robust with respect to initial estimations. The most critical initial guess in this case is the exponential factor. The closer the initial guess comes to the actual best fitting parameters the faster the code will converge. Also as with any nonlinear problem, it is always possible to be so far off on the initial approximations that the iterative algorithm will not converge to a physically correct solution. The reader can execute the program with different initial conditions to explore the range of values for which the function will properly converge to the known values. Line 16 defines all coefficients as being active (by setting all the actv[] array parameters to 1) and imposes no limits on the step size per iteration (by setting the step[] array parameters to 0). The weighting factors are all set to unity on line 19 and the data is setup in the x array on line 20 with the dependent variable data listed first followed by the independent variable. Note that this matches the order of the data supplied to the fitting function as used on line 12.

After setting up the required inputs, the nlstsq() fitting program is called on line 21 returning the estimated uncertainty in the parameters as a table (del[]) the actual RMS error achieved in the fit and the number of nonlinear iterations required by nlstsq() for the fit. The listing shows output generated by this program. The default option is for nlstsq() to print achieved values of the coefficients at each iteration. This is useful in insuring proper convergence of the program as 50 iterations would indicate that the program did not reach a converged result. In this case it is seen that 4 iterations were required and the RMS error achieved was below 2.5e-11. The calculated three parameters are 2, 1, and -.8, exactly as expected for a properly working data fitting routine. In this case an exact fit to the data is achieved with the obtained parameters equal to the values used in generating the data set and the returned del[] values, representing estimated uncertainties in the coefficient values are in the 1.e-11 range. Such small values can only be achieved if an exact equation is available for the data model and if the data has no random errors associated with the data values. This example illustrates the manner in which the program is setup, called and the manner in which a fitting function with unknown coefficients is written. Recall that there is some freedom in writing the fitting function as long as the function returns 0 when the proper value of independent variable and dependent variable appears in the equation. While this is a very simple example, it is always essential in developing numerical algorithms to have examples where the expected result is known exactly in order to check the accuracy of any numerical algorithms.

```
 1 : -- /* File list7_11.lua */ -- Data fitting for Figure 7.1
 2 :
 3 : require"nlstsq_new"
 4 :
 5 : xd,yd = {},{}
 6 : read_data('list7_1.dat',xd,yd)
 7 :
 8 : ft = function(x,c) -- Define function to fit data
 9 :     return c[1]*(1 - c[2]*math.exp(c[3]*x[2])) - x[1]
10 : end
11 :
12 : c = {1,1,-.2} -- Initial guess at coefficients
13 : actv = {1,1,1}
14 : del,err,nmax = nlstsq({yd,xd},fw,ft,c,actv) -- Call fiting
15 : print(err,nmax) -- print results
16 : for i=1,#c do printf('c[%d] = %12.4e +/- %12.4e\n',i,c[i],
          del[i]) end
17 : ycalc = {}
18 : for i=1,#xd do
19 :     x = {0,xd[i]}
20 :     ycalc[i] = ft(x,c)
21 : end
22 : write_data('list7_11.dat',xd,yd,ycalc)
23 : plot(xd,yd,ycalc)
Output
0.019294406815624        5
c[1] =   1.0678e+000 +/-  1.2429e-002
c[2] =   1.0375e+000 +/-  9.8042e-003
c[3] =  -1.3176e-001 +/-  4.2635e-003
```

Listing 7.11. Fitting of function to data in Figure 7.1.

Let's now look at fitting a known functional form to some data with noise such as the data in Figure 7.1. This looks somewhat like a function which rises with an exponential time constant so an appropriate trial function might be the same exponential form as used in the previous example. Listing 7.11 shows code for such an analysis. The data is input from a file on line 6 and the fitting function is defined on line 9. After defining c[] and actv[] arrays, the nlstsq function is called on line 14. With an initial guess of 1, 1, -.2 for the three parameters (line 12), the code takes 5 iterations to converge to the parameters shown in the output. In this case the RMS error achieved is about 0.019 and the uncertainty in the fitting parameters is in the range of 1.e-2 or about 1% of the calculated parameter values. This example has two features worthy of mention. First it can be noted that the fw[] table parameter in the calling argument list to nlstsq() is not defined anywhere in the code listing. This is the table of data weighting factors. Being undefined, a value of nil will be passed to the nlstsq() function where the code will detect this value and use default values of unity for all the weighting factors. Thus weighting factors only need to be defined if they need to be unequal for some data points. Second, the argument list to nlstsq() is missing the last parameter of step[] which is a table of step limitations for Newton's method. Again this will result in a nil value detected in the function and zero assumed for all the step parameters (implying no limits on the step sizes). Again this parameter only needs to be specified if limits are to be used.

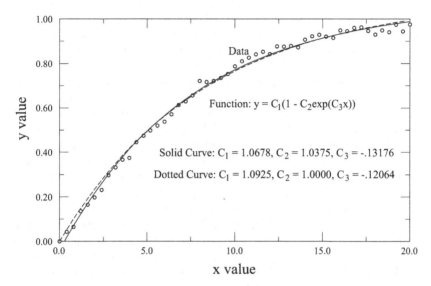

Figure 7.16 Least squares data fit of Figure 1 data with exponential function.

A graph of the original data and the fitted function with the calculated parameters is shown in Figure 7.16 as the solid curve. In general this function has a pretty good fit to the data. However, it has one slightly annoying feature and that is the fact that the best fitted function does not give $y = 0$ at $x = 0$. This may in deed be acceptable under some circumstances but the data appears to the eye to be such that the real relationship should probably be zero at $x = 0$. It is readily seen that the function will pass through the origin if the C_2 parameter is identically 1. This can be accomplished if the fitting function is defined in terms of only two unknown parameters. This can be done by retyping the fitting function definition, but can be accomplished more easily by use of the actv[] feature of the software. To fix the value of C_2 at 1 in the fitting function requires only that the initial guess be set to 1 as is already done in Listing 7.11 and in addition that the appropriate actv[] value is set to 0 (actv[2] = 0). The only change needed in the listing is to change line 13 to actv = {1,0,1}. The 0 entry corresponding to the second coefficient causes the nlstsq() routine to use the input value for this coefficient parameter in all the calculations and not include this factor in the updated coefficient matrix parameters. When the program is re-executed, the second set of parameters shown in Figure 7.16 is obtained and the function values now give the dotted line shown in the figure. This set of parameters gives an RMS error of 0.0223 as opposed to the value of 0.0193 obtained for the solid curve. However, it is this author's opinion that the dotted curve is a more pleasing fit to the data than the solid curve even though the solid curve is "better" according to the strict definition of minimum least square error. This illustrates the fact that one should never blindly accept the results of a computer analysis. Engineering judgment should always be used in accepting the "best" computer generated results. Either curve in the figure is a reasonably good fit to the data and would certainly be acceptable as a publishable curve for describing the data trend. However if it is known from physical principles that the data should pass exactly through the origin then this information should be utilized in fitting the data to a functional form.

At this point it is probably important to discuss a little more the significance of the del[] values returned by the nlstsq() code routine – one value for each fitting coefficient. As previously stated these represent a measure of the uncertainty with which one can trust the returned values of the coefficients. Such values are frequently called 'standard errors' or 'standard deviations'. In the case of Listing 7.11 the second line of the printed output shows $C_1 = 1.0678$ +/- .0124. This implies that the nlstsq() fitting routine can only determine this coefficient to an accuracy of +/- .0124 and the C_1 value should only be quoted to about 3 digits of accuracy and not the 5 digits as shown in the printed output. The question arises as to how these uncertainties are obtained. Basically a very simple explanation is that they represent the amount of change needed in the coefficients to double the sum square error of the fitting. The smaller the sum square error of the fitting equation, the smaller will be these values. Another simple explanation is that the coefficient has approximately 68% probability of lying within the fitted value and +/- the one-standard deviations and about a 95% probability of lying within two standard deviations of the fitted values. The accuracy of the fitting parameters is very important when one is performing parameter estimation for a particular physical model. The situation is actually considerably more complicated than the simple explanations stated above, because of possible correlations between the fitting parameters. This topic will be discussed in much greater detail in a subsequent chapter on parameter estimation. For the present, it is sufficient to think in terms of the simple explanations stated above and to realize that the smaller the del[] error terms, the more accurately the fit of the curve to the data. The value depends not only on the scatter in the data, but also on how well the model function approximates the general trends of the data.

As a final variation on the data fit to the Figure 7.16 data, one can minimize not only the vertical error in the fitted function, but one can also minimize the horizontal error in the fitted function. This can be achieved by considering the y-values in Figure 7.16 as the independent values and the x-values as the dependent values. This is sometimes a useful technique in obtaining a "better" fit to data. This is easily done with the code developed here since the data describing function is of the form $F(X, C) = 0$ with no real distinction made between the dependent and independent variable. The only requirement is that the first array value passed to the function must correspond to the independent variable.

```
 1 : -- /* File list7_12.lua */ -- Reversed indep and dep variables
 2 :
 3 : require"nlstsq"
 4 : xd,yd,ycalc = {},{},{}
 5 : read_data('list7_1.dat',xd,yd)
 6 :
 7 : ft = function(x,c) -- Define function to fit data
 8 :    return c[1]*(1 - c[2]*math.exp(c[3]*x[2])) - x[1]
 9 : end
10 : frev = function(x,c) -- Reverse variables to fitted function
11 :    return ft({x[2],x[1]},c)
12 : end
13 :
14 : c = {1,1,-.2} -- Initial guess at coefficients
15 : del,err,nmax = nlstsq({xd,yd},fw,frev,c) -- Call fiting
16 : print(err,nmax) -- print results
17 : for i=1,#c do printf('c[%d] = %12.4e +/- %12.4e\n',i,c[i],
          del[i]) end
18 : for i=1,#xd do ycalc[i] = ft({0,xd[i]},c) end
19 : write_data('list7_12.dat',xd,yd,ycalc)
20 : plot(xd,yd,ycalc)
Selected Output:
0.77510681135336         10
c[1] =   1.0400e+000 +/-  1.2177e-002
c[2] =   1.0757e+000 +/-  4.5745e-002
c[3] =  -1.4554e-001 +/-  8.5612e-003
```

Listing 7.12. Data fitting with reversed independent and dependent variables.

Listing 7.12 shows a code segment for reversing the role of the dependent and independent axes in the data fitting. The nlstsq() function calls a frev() function defined on lines 10 through 12 which simply reverses the calling arguments to the original function definition ft() on lines 7 through 9. This change could also have been made by reversing the x[1] and x[2] values in the original ft() function. However, this approach works regardless of the complexity of the defining function and allow one to think in terms of the original variables in the defining function. The other significant change is on line 15 in the first argument to nlstsq() where the array is specified as {xd,yd} which treats the x data as the dependent variable and the y data as the independent variable. It can also be noted that the actv[] and step[] arrays are both omitted from the argument list so that nlstsq() will use the default values for these parameters (evaluate all coefficients and use no step size limitations). The printed output shows that 10 Newton iterations are required for this evaluation as opposed to only 5 iterations shown in Listing 7.11. The larger number of Newton iterations is required because in this case the assumed dependent variable occurs in the defining equation in a nonlinear manner.

Figure 7.17 compares the results of the two function fits to the data. In both cases all three coefficients are allowed to vary to optimize the data fit. As can be seen from the two curves and from the best fit values of the parameters, somewhat different curves and parameters are obtained depending upon whether the least squares technique minimizes the vertical distance between the theory and the data or minimizes the horizontal distance between the model and the data. This is a difference of nonlinear curve fitting as opposed to linear curve fitting. For a linear line fit to data, the difference between such two fits is typically very small. Which curve "best" fits the data is somewhat a matter of preference. One might think that some fitting approach which minimized the perpendicular distance from the data points to the curve might be a better choice than minimizing either the vertical or horizontal distances. However, such an algorithm is not easily implemented because the two axes in most cases have very different dimensions and very different scales as in Figure 7.17. Perhaps the best that can be done is to look at both possibilities and select the one that seems to give the best fit to all the data points, including any known initial and final value for the curve. The example does illustrate the fact that there is not a "single" unique least squares data fit between a given model and a set of data unless it is know that the error in one variable is much smaller than the error in the other data variable so that one variable is indeed the independent variable with negligible error and the other variable the dependent variable with random error.

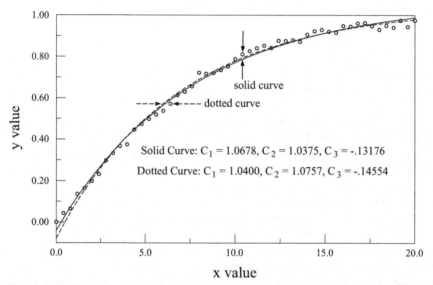

Figure 7.17 Comparison of minimizing vertical or horizontal least squares error.

A second example of fitting experimental data to a given mathematical function is shown in Figure 7.18. This shows data believed to be two Gaussian peaks on an exponential background and described by the model equation:

$$y = C_1 \exp(-C_2 x) + C_3 \exp(-(x - C_4)^2 / C_5^2) + C_6 \exp(-(x - C_7)^2 / C_8^2). \qquad (7.42)$$

This is an equation with eight fitting coefficients. The two Gaussian peaks are located at $x = C_4$ and $x = C_7$ with half-widths of C_5 and C_8. Even though there are a large number of fitting coefficients, good initial estimates can be obtained from the experimental data since the two peaks are seen to occur at x values of about 110 and about 150. Also the peak widths are easily estimated to be in the range

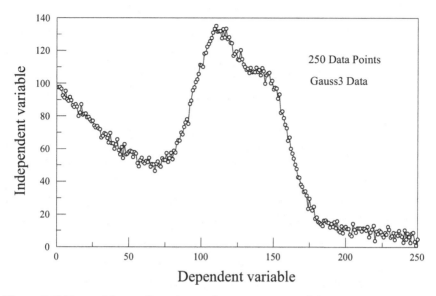

Figure 7.18 Data with two Gaussian peaks on an exponential background. (From: www.itl.nist.gov/div898/strd/nls/data/gauss3.shtml).

of about 15 units. Finally the exponential background can be estimated to have a value of about 100 at x = 0 and decreasing to about 10 at x = 200. All these can be used to give good initial estimates of the c[] coefficients needed for the data fit. The code segment shown in Listing 7.13 illustrates the use of nlstsq() for fitting this data. The data is read in from a data file on line 6, the fitting function is defined on lines 9 through 13 and the initial C values are defined on line 15. The call to nlstsq() on line 16 calculates the parameters using default values for the actv[] and step[] parameters. When executed, the selected output in the listing indicates that the fitting obtains convergence in 5 iterations, with the parameters and standard deviations shown in the listing.

An advantage of this example is that the NIST site referenced in Figure 7.18 provides expected converged parameter values and certified standard deviation values of the estimates. All of the values obtained here agree exactly with the expected values. This provides a cross check on the accuracy of the nlstsq() code and other routines used in the nonlinear least squares data fitting such as the newtonfc() function. As a final point with this example, it should be noted that the input data set (the gauss3.dat file) has the y data stored in the first column and the x data stored in the second column (see the order of x1 and g1 on line 6).

A large part of the code in Listing 7.13 (lines 20 through 29) is concerned with evaluating the model for each of the individual components of the total model curve. These tables are written to a data file on line 31and Figure 7.19 shows the resulting Gaussian peak fits to the original data and plots of the individual components of the fit. The three components are seen to provide a very good fit to the

```
 1 : -- /* File list7_13.lua */ -- Fit to two gaussian peaks
 2 :
 3 : require"nlstsq"
 4 :
 5 : g1,x1 = {},{}
 6 : read_data('gauss3.dat',g1,x1) -- Read data from file
 7 : plot(x1,g1)
 8 :
 9 : gfit = function(xa,c) -- Define function for two peaks
10 :    local x = xa[2]
11 :    return c[1]*math.exp(-c[2]*x) +
           c[3]*math.exp(-(x-c[4])^2/c[5]^2) +
12 :            c[6]*math.exp(-(x-c[7])^2/c[8]^2) - xa[1]
13 : end
14 :
15 : c = {100,.01,90,110,25,80,150,25} -- c[7] ~ 150, c[4] ~110
16 : del,err,nmax = nlstsq({g1,x1},fw,gfit,c) -- Call fitting program
17 :
18 : print('RMS error =',err,'#Iter =',nmax)
19 : nd = #x1
20 : for i=1,#c do printf('c[%d] = %12.4e +/- %12.4e\n',i,c[i],
          del[i]) end
21 : gcalc,gp1,gp2,gb = {},{},{},{} -- arrays for total and peaks
22 : for i=1,nd do gcalc[i] = gfit({0,x1[i]},c) end -- Total function
23 : c1,c3,c6 = c[1],c[3],c[6]
24 : c[3],c[6] = 0,0 -- Set peaks to zero
25 : for i=1,nd do gb[i] = gfit({0,x1[i]},c) end -- Background only
26 : c[1],c[3] = 0,c3
27 : for i=1,nd do gp1[i] = gfit({0,x1[i]},c) end -- Peak one only
28 : c[3],c[6] = 0,c6
29 : for i=1,nd do gp2[i] = gfit({0,x1[i]},c) end -- Peak two only
30 : plot(x1,g1,gcalc,gp1,gp2,gb) -- plot individual components
31 : write_data('list7_13.dat',x1,g1,gcalc,gp1,gp2,gb) -- Save data
Selected Output:
RMS error =      2.2311294323847         #Iter = 5
c[1] =  9.8940e+001 +/-  5.3005e-001
c[2] =  1.0946e-002 +/-  1.2554e-004
c[3] =  1.0070e+002 +/-  8.1257e-001
c[4] =  1.1164e+002 +/-  3.5318e-001
c[5] =  2.3301e+001 +/-  3.6585e-001
c[6] =  7.3705e+001 +/-  1.2091e+000
c[7] =  1.4776e+002 +/-  4.0488e-001
c[8] =  1.9668e+001 +/-  3.7807e-001
```

Listing 7.13 Code segment for fitting two Gaussian peaks to data of Figure 7.19.

data in this example. The reader is encouraged to change the initial guesses for the parameters on line 15 and execute the code to explore the sensitivity of the final results to the initial guesses.

In Section 7.4 on Fourier series, an example was given of fitting such a Fourier series to the data of Figure 7.1. It was found that only two term of a quarter-range Fourier series provided a good fit to the data. The general nonlinear least squares fitting approach provides a second avenue for such a data fit. For this assume that a two term sin() series of the form:

$$y = C_1 \sin(C_2 x) + C_3 \sin(C_4 x) \tag{7.43}$$

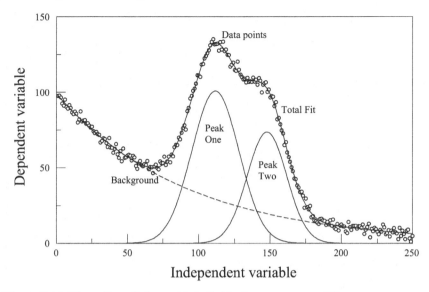

Figure 7.19. Fit to Gauss3 data with individual components of fit.

can be used as a good representation of the data. From the previous results it is expected that one of the sin() periods should be approximately 4 times the data period and the other sin() term should have a frequency approximately 3 times the first. The advantage of the nlstsq() approach is that both the amplitude and the periods of the sin() functions can be optimized to achieve the best fit while with the Fourier analysis, one needs to specify the fundamental period and all other periods are some multiple of that fundamental frequency. Listing 7.14 shows the code for this data fitting analysis. A two component sin() series is defined on lines 8-11 in ft() with both amplitude and frequency to be determined by the best least squares fit criteria. The fitted frequency values are converted to sin() periods and printed in the output as 72.04 and 22.17. The period of 72.04 indicates that slightly more than one fourth of a period occurs during the data interval of 0 to 20. Also the 22.17 period indicates that slightly less than one period of the second sin() term occurs during the data interval.

The resulting data plot and the two sin() component terms are shown in Figure 7.20. The two components show that adjusting both the amplitudes and frequencies of the two component sin() waves provides the best least squares fit to the data points. This occurs not exactly at the quarter-cycle and the third harmonic as assumed in the Fourier analysis, but at slightly different frequencies for both terms. The nlstsq() technique is more general than a Fourier analysis in that each component can be adjusted individually to achieve the best fit. On the other hand

```
 1 : -- /* File list7_14.lua */ -- Figure 7.1 with nlstsq()
 2 :
 3 : require"nlstsq"
 4 :
 5 : xd,yd = {},{}
 6 : read_data('list7_1.dat',xd,yd)
 7 :
 8 : ft = function(x,c) -- Define function to fit data
 9 :    local xx = x[2]
10 :    return c[1]*math.sin(c[2]*xx) +c[3]*math.sin(c[4]*xx)- x[1]
11 : end
12 :
13 : c = {1,2*math.pi/80,.1,6*math.pi/80} -- Initial guess
14 : del,err,nmax = nlstsq({yd,xd},fw,ft,c) -- Call fiting
15 : print('RMS error =',err,'#Iter =',nmax) -- print results

16 : for i=1,#c do printf('c[%d] = %12.4e +/- %12.4e\n',i,c[i],
         del[i]) end

17 : print('periods = ',2*math.pi/c[2],2*math.pi/c[4])
18 : ycalc,y1,y2 = {},{},{}
19 : for i=1,#xd do
20 :    y1[i] = c[1]*math.sin(c[2]*xd[i])
21 :    y2[i] = c[3]*math.sin(c[4]*xd[i])
22 :    ycalc[i] = y1[i] + y2[i]
23 : end
24 : write_data('list7_14.dat',xd,yd,ycalc,y1,y2)
25 : plot(xd,yd,ycalc)
Selected Output:
RMS error =    0.012928899982909      #Iter = 6
c[1] =   1.0050e+000 +/-  2.1891e-002
c[2] =   8.7211e-002 +/-  5.1510e-003
c[3] =   5.4680e-002 +/-  1.4322e-002
c[4] =   2.8341e-001 +/-  2.7950e-002
periods =       72.045763667524         22.169677577089
```

Listing 7.14. Data fitting for Figure 7.1 with two general sinusoidal terms.

it would be very difficult to fit a large number of frequency components in both amplitude and frequency using nlstsq() as can be done in the Fourier analysis.

These example data files provide a useful introduction to using the nlstsq() code for fitting a nonlinear function to a data set. Listings 7.10 through 7.14 are typical of the user generated code needed for fitting a model to any data set. The features needed are a definition of the data arrays, a fitting function with adjustable parameters, a set of initial guesses for the parameters or C[] coefficients. In addition a set of actv[] coefficients, step[] coefficients and weighting factors wf[]may be defined if desired. A simple call to the nlstsq()function then performs the nonlinear least squares fit returning the best set of equation parameters and estimates of the uncertainty in the determined parameter values.

These examples have not so far exercised the full range of capabilities using the step limits to improve convergence and the weighting factors. Some examples of using these parameters are shown later in this chapter with other examples.

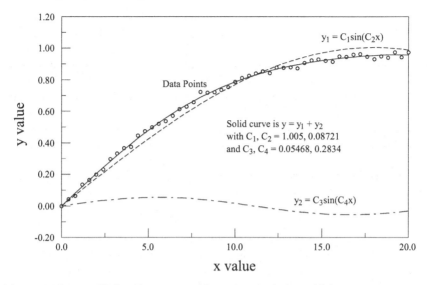

Figure 7.20. Data fit for Figure 7.1 with two general sinusoidal terms.

7.7 General Data Fitting and Plotting

In the examples of the previous section, it was assumed that a mathematical model was known to which the data was to be fitted. This is one broad general class of data fitting and plotting applications. Another general case is one where a good mathematical functional form representing the data is not known. Or one may be primarily interested in drawing a smooth curve through the data for publication without postulating any explicit functional relationship between the variables. For example considering the data in Figure 7.1, one might simply want to draw a smooth curve through the data as shown in the figure without having a known functional form of the curve. Even though a general nonlinear least squares data fitting technique has been developed, it can only be as good as a function used to model the data. If the fitting function does not adequately represent the data, the resulting least squares curve will be a "best fit" in the sense of the least squares technique, but may not provide a good or a pleasing fit to the actual data. In the case of the data again in Figure 7.1 and the fit of the data to an exponential function in Figure 7.17 the fit is not extremely good and a good draftsman with a French curve would draw a smooth curve through the data that would give a more pleasing fit to the data than that obtained with either of these exponential functions. The emphasis in this section is on data fitting approaches which more closely approximate the function of an electronic French curve and which don't depend on having a good mathematical function for fitting the data.

How should one proceed if a good mathematical function which has the proper functional dependency is not known? One could of course search for additional mathematical functions and one would probably eventually find some combination of functions that provides a good fit to a given set of data. However, this is not always an easy task. In many texts on data fitting, the authors stress the use of polynomials for general data fitting. This possibility was previously discussed and in fact the matrix equations for this approach was set up in Eq. (7.16) for this case as an example of fitting functions with linear coefficients. However, this is just a special case of fitting with a general nonlinear function and is easily handled by the nlstsq() code segment. All one has to do is set up a general function to evaluate the assumed fitting polynomial. The code segment in Listing 7.15 shows such an example of a coded function. A general polynomial function is defined on lines 10 through 14 which can be used for a polynomial fitting function of any desired order. The order of the polynomial is determined in the function by looking at the number of terms in the c[] coefficient array on line 11. Since a polynomial fit has linear fitting coefficients, the initial guess on the coefficient values can be any value including zero and the calculation will converge in two iterations. Thus the only parameter which needs to be specified for the data fit is the number of coefficients, as on line 17 of the code. All coefficients can be active and no limits are needed on the iteration steps as the problem becomes a linear matrix problem. Listing 7.15 can be used as a prototype code for fitting any function with a polynomial function. The only changes that need to be made are in the input file, for example selecting line 6 to 8, and in the calculation of the fitted curve as on lines 22 through 26.

Listing 7.15 with the input file selected on line 6 applies the polynomial fitting to the data in Figure 7.21. This shows data on the number of PhD graduates per year as a function of Research Expenditures in millions of dollars per year for US ECE university graduates in a recent year. As might be expected there is a strong correlation between research expenditures and the number of graduates per year. The largest scatter is in the largest departments with a tendency for the data to saturate at 60 to 70 graduates per year. At the low end of the graph there is a linear relationship between graduates per year and research expenditures per year with a slope of very close to 1 graduate per $500,000 of research expenditures. From this data it can thus be concluded that for US university ECE Departments it costs approximately $500,000 to graduate one PhD student. This data would also indicate that smaller ECE departments are more efficient at producing PhD's than are the larger ECE departments in terms of dollars spent per graduate student.

For fitting a smooth curve to the data, three polynomial curves are shown in the figure ranging from second degree to fourth degree polynomials. The fitting and curve for the dot-dash 4th degree polynomial fitting is what will be generated by executing the code in Listing 7.15 (with nc = 5). To generate the other fitted curves, the code in Listing 7.15 needs to only be changed on line 16 with nc values of 3 and 4 for the 2nd and 3rd order polynomials. The reader is encouraged to make such changes and execute the code. The selected output indicates that the nlstsq() routine required only 2 Newton iteration for convergence. Actually since

```
 1 : -- /* File list7_15.lua */ -- fitting with polynomial functions
 2 :
 3 : require"nlstsq"
 4 :
 5 : xd,yd = {},{}
 6 : read_data('PhDgrads.dat',xd,yd)
 7 : --read_data('gauss3.dat',yd,xd) -- Another data set
 8 : -- read_data('list7_1.dat,xd,yd) -- Yet another data set
 9 :
10 : polyft = function(x,c) -- Define polynomial function to fit data
11 :    local nc,xx,sum = #c, x[2], 0
12 :    for i=nc,1,-1 do sum = sum*xx + c[i] end
13 :    return sum - x[1]
14 : end
15 :
16 : nc = 5 -- Specify number of polynomial coefficients
17 : c = {}; for i=1,nc do c[i] = 0 end -- zero initial guesses
18 : del,err,nmax = nlstsq({yd,xd},fw,polyft,c) -- Call fiting
19 : print('RMS error =',err,'#Iter =',nmax) -- print results
20 : for i=1,#c do printf('c[%d] = %12.4e +/- %12.4e\n',i,c[i],
         del[i]) end
21 : xcalc,ycalc = {},{}
22 : x1,x2,i = 0,60, 1; dx = (x2-x1)/100
23 : for x=x1,x2,dx do
24 :    xcalc[i],ycalc[i] = x, polyft({0,x},c)
25 :    i = i+1
26 : end
27 : write_data('list7_15.dat',xcalc,ycalc,xd,yd)
28 : plot(xcalc,ycalc);plot(xd,yd)
```
Selected output:
```
RMS error =    4.2971861190456       #Iter = 2
c[1] =  9.0021e+000 +/-  6.8731e+000
c[2] = -2.9510e-001 +/-  1.7623e+000
c[3] =  2.0752e-001 +/-  1.3821e-001
c[4] = -6.7829e-003 +/-  4.0251e-003
c[5] =  6.3657e-005 +/-  3.8612e-005
```

Listing 7.15. Data fitting with general polynomials. Applied to data in Figure 7.21.

all the coefficients enter into the defining equations in a linear manner only one iterative step is needed to solve for the coefficients. The second iterative step is needed so that the code can detect that the error criteria has been satisfied in the nonlinear solution loop. If desired a code statement of "getfenv(nlstsq).NMAX = 1" can be used to limit the execution to one nonlinear loop. This will cut the execution time by close to 1/2 but this is probably only important if one has a very large data set.

One can certainly debate as to which of the curves in Figure 7.21 provides the most appropriate approximate fit to the data. However, the general conclusion would probably be either the second or third order polynomial fit with this author's preference being the second degree polynomial, although a curve which saturated rather than decreases at large values would probably be more appropriate for the given data. The fourth degree polynomial obviously has too many turns to provide a "good fit" to the data although it will certainly have a smaller average RMS error to the data.

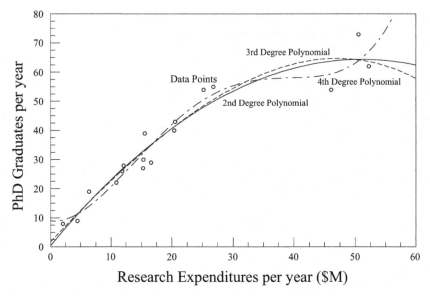

Figure 7.21. PhD graduates per year as a function of Research Expenditures per year for US ECE Departments.

Figure 7.21 illustrates the type of data fits obtained with polynomials for many types of data sets. In general, only low order polynomials provide any type of pleasing fit to the data. If the data has any significant scatter in the data or any significant number of slope changes, it is difficult to obtain a good data fit with global polynomials of very high order. A second example is shown by Figure 7.22 which shows a 9th order polynomial fit to the Gaussian Peaks data previously shown in Figure 7.18. Complete code for this analysis is not shown but requires only a few changes to Listing 7.15. The input data file needs to be that shown as a comment on line 7. In addition the nc on line 16 is set to 10 for a 9th order polynomial. Finally the x2 value on line 22 needs to be changed to 250. The reader is encouraged to make these changes and re-execute the code. As Figure 7.22 shows, the global polynomial has the same general shape as the data, but does not provide any fit to the data that would be appropriate to use in a publication or any representation of the data. Also an oscillating-like behavior can be seen outside the peaks region which is characteristic of any attempt to use high order polynomials for a global data fit. To reproduce the structure near the peaks would require even higher order polynomials and there would be considerably more oscillations about the data set away from the peaks.

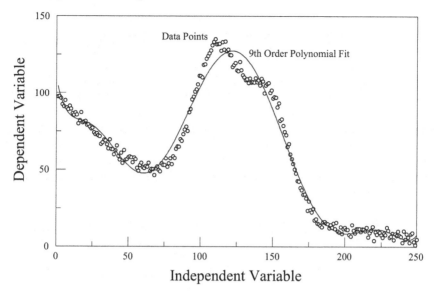

Figure 7.22 Polynomial fit to Gaussian Peaks data of Figure 7.18.

The above examples illustrate that global polynomials are of very limited use in any general data fitting approach. However it was found in Chapter 6 on Interpolation that low order polynomials could be used very effectively in fitting a wide range of functions over limited data ranges. In fact local pieces of cubic polynomials provide the basis for both the Spline and Local Cubic interpolation techniques developed in that chapter. In summary, the technique discussed was to connect local pieces of cubic polynomials in such a manner that first derivatives (and second derivatives for the Spline) are continuous from local region to local region. With interpolation, the desired value of the function is known at a selected set of points and an appropriate set of equations can then be formulated to determine the coefficients of the local low order approximating polynomials. For the present work, the exact placement of any point along a curve is not known because of assumed noise or error in the curve and points along a fitting curve can only be determined from minimizing the least square error between some fitting function and the data. However, the possibility of combining local cubic functions to approximate the data over a local region with the least squares minimization technique appears to be a promising approach. This is the technique to be explored in the remainder of this section.

The combination of cubic function interpolation with least squares data fitting can be illustrated with reference to Figure 7.23. Shown in the figure as open circles is the data set from Figure 7.1. The solid points represent a set of interpolation data points through which the solid line has been drawn. The solid points are uniformly spaced along the independent variable axis at 0, 4, ... 12 and have dependent values labeled as $C_1, C_2, \ldots C_6$. In this case there are six interpolation points. How these are determined will become clear a little later. For the present, assume that such a data set is known which in the present case appear to give a good fit to the data and which provides a small least square error, if not the minimum least square error. The dotted curve shows how the local cubic interpolation function would change as the value of the C_3 coefficient is changed but keeping all the other coefficients fixed. As this coefficient is increased or decreased, the change in the fitting curve occurs primarily within the two adjacent intervals. Actually with the Local Cubic Function (LCB) interpolation, some change occurs out through at most two intervals on each side of the point in question. It's readily seen from the two dotted curves, that with the other points fixed, there is an optimum value of the C_3 interpolation data point which will minimize the least square error of the interpolation function.

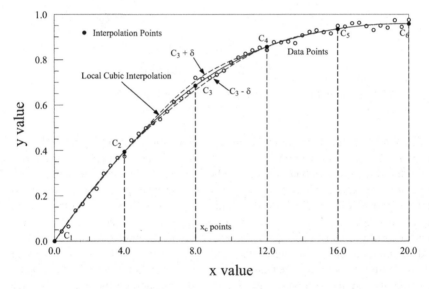

Figure 7.23. Illustration of least squares fitting with Local Cubic Function (LCB) interpolation.

The same argument can be made for any of the interpolation points, so it is expected that there exists a unique set of interpolation points which will give a minimum least square error between the data and the interpolation curve. Of course the minimum lease square error will depend to some extent on the exact location and number of the interpolation data points. If the number of interpolation data points is increased to the point where they equal the number of data points, then the interpolation curve would pass exactly through each data point and one would have the case of data interpolation as considered in the previous chapter. On the other hand too few interpolation data points and a piecewise set of local cubic functions may not provide a good fit to the data. In using this approach some judgment must be exercised in the selection of the "best" number of interpo-

lation data points and the "best" location of these points which does not have to be
a uniform spacing as shown in Figure 7.23. As a starting point, good results have
been achieved for many situations with around 10 data points per interpolation
point and with a uniform spacing along the independent variable axis. These will
be used as an initial starting point and can be adjusted as needed in particular
situations.

In interpolation terminology, the interpolation points are frequently referred to
as knots. The approach to be developed here can be thought of as interpolation
with "floating" knot values to be determined by a least squares criteria. To implement this in computer code and use the previously developed nlstsq() function,
a coded function is needed that returns the difference between the fitting function
and the data points. In Section 6.2 on the local cubic or LCB interpolation technique, a function intp(xp,yp,x) was introduced in Listing 6.3 which takes a set of
interpolation data points, xp and yp and returns the value of the interpolation function at some value, x, of the independent variable. In the present application, the
values of the yp array are simply the data fitting parameters or the c, coefficients.
A little thought indicates that setting up the fitting function to be used in the least
squares routine is in fact pretty simple. Listing 7.16 shows code for the fitting
function as well as a general data fitting code segment called datafit(). The data
fitting function called fdf() is defined in terms of intp() on lines 12 through 14. To
be compatible with the nlstsq() function this function must accept as arguments a
pair of data points to be fitted in the form yx = {yd,xd} and the array of c coefficients to be fitted to the data. This provides all the parameters for nlstsq(). However, the intp() function requires an additional set of parameters that includes the
interpolation data points, identified as xc on line 8 of the code. Since these parameters are previously defined on line 5 of the code they will be properly added
to the argument list for intp() on line 8 of the code. This is another example of an
interface or proxy function that adds additional calling arguments between a defining argument list and the arguments needed for a calling function internal to the
function being used. The only requirement is that the xc array be properly defined
before the fdf() function is invoked.

To simplify the use of the LCB least squares data fitting for the user, the function datafit() in Listing 7.16 is provided with a simplified calling argument list. In
it's simplest use, this function takes only a listing of the experimental data to be
fitted, performs the evaluation of the interpolation points using nlstsq() and the
LCB interpolation and returns a function, which when called with x values, will
evaluate points along the lease squares interpolation function. This provides a
very simple interface to the curve fitting procedure as will be seen in subsequent
applications. In order to use such a simple interface, the datafit() function must
determine an appropriate number of interpolation intervals and set up all the necessary arrays to call the nlstsq() fitting program. With no guidance from the user,
the datafit() routine sets up somewhere between 3 and 15 interpolation points or
knots equally spaced along the independent variable axis. It attempts to use at
least 10 data points per interpolation point unless that gives a number outside the
range 3 to 15. Code for setting up the default interpolation points is contained on

```
 1 : -- /* File Data_Fit.lua */
 2 : -- Program to fit data with noise to a smooth curve for plotting
 3 : -- Number of knots specified by user or determined by program <=
10
 4 : require"nlstsq"; require"intp"
 5 : local checkorder
 6 :
 7 : datafit = function(yx,nc)
 8 :    local nd,c,xc = #yx[1] -- yx array contains {yd, xd}
 9 :    yx = checkorder(yx,nd) -- increasing x array values?
10 :    local xmin,xmax = yx[2][1], yx[2][nd] -- x min and max values
11 :    local actv,dx,del = {}
12 :    local fdf = function(yx,c)--datafit Interpolation function
13 :       return intp(xc,c,yx[2])-yx[1] -- function with knots
14 :    end
15 :    if type(nc)=='table' then -- arrays for xc and fixed knots?
16 :       xc,c = nc[1] or {}, nc[2] or {} -- floating knots and c?
17 :       nc = #xc -- All xc values must be passed
18 :    else
19 :       if (nc==nil) or (nc==0) then -- Use default # of knots
20 :          nc = math.min(nd/10,15); nc = math.max(nc,3)
21 :       end -- End result should be between 3 and 15
22 :       nc,xc,c = math.ceil(nc), {}, {}
23 :    end
24 :    dx = (xmax-xmin)/(nc-1) -- Use equal spacing for knots
25 :    for i=1,nc do -- Set initial x and y values of floating knots
26 :       xc[i] = xc[i] or (xmin + (i-1)*dx) - calculate?
27 :       if c[i] ~=nil then actv[i] = 0 else actv[i] = 1 end--use?
28 :       c[i] = c[i] or 0 -- linear problem, guesses can be zero
29 :    end
30 :    local nlenv = getfenv(nlstsq);
          local lcp,lyl = nlenv.nprint, nlenv.ylinear
31 :    getfenv(nlstsq).nprint = 0; getfenv(nlstsq).ylinear = 1
32 :    del,err = nlstsq(yx, fw, fdf, c, actv) -- call data fit
33 :    nlenv.nprint, nlenv.ylinear = lcp, lyl -- restore values
34 :    return function(x,df) return intp(xc,c,x,df) end,
          {xc, c, del}, err
35 : end
```

Listing 7.16. Code segments for simple data fitting with Local Cubic Function interpolation.

lines 19 through 24. The user can also specify the number of interpolation points by specifying the number as a second argument to the datafit() routine. A more general capability exists of providing both the number and location of the interpolation points by making the nc argument a table of interpolation points and initial values. Code for implementing this case is on lines 15 through 17. The use of this feature will become clearer with a subsequent example. Lines 25 through 30 set up the arrays needed by nlstsq() and finally the fitting routine is called on line 32. Note that line 26 uses either the input array of specified knot locations or generates an array of uniformly spaced values. Finally line 34 returns several items with the first returned value being a function which when called will evaluate points along the least squares interpolating function. This returned function can simply be called as f(x) in the same manner as any computer defined function.

The second item returned by the datafit() function is a table containing three items: a table of the x locations of the interpolation knots (the xc values), a table of the y values associated with the knots (the c values) and the uncertainties associated with the determination of the know y values from nlstsq() (the del values). In most used of this function this additional information will not be needed as the proof of the technique will simply reside in an observation of how well the returned function provides an acceptable smooth curve representation of the input data points.

The use of datafit() is best illustrated by an example as shown in Listing 7.17. The use of datafit() is extremely simple as the code shows. The data fitting code of Listing 7.16 is input on line 3 by the require"DataFit" statement. A data set is read into the program on line 6, the fitting function is called with the data on line 7 which returns a function. The returned function fxc() is then used on line 8 to generate a set of fitted values at the data points and the results are plotted and saved on lines 9 and 10. The data used in this example is in fact the data set in Figures 7.1 and 7.16. The least squares generated interpolation function from the data generated by Listing 7.17 is the dotted line curve shown in Figure 7.24.

```
 1 : -- /* File list7_17.lua */
 2 :
 3 : require "DataFit"
 4 :
 5 : xd,yd,ycalc={},{},{}
 6 : read_data('list7_1.dat',xd,yd) -- Read data
 7 : fxc = datafit{yd,xd} -- Now do fit with floating knots
 8 : for j=1,#xd do ycalc[j] = fxc(xd[j]) end -- Calculate fit values
 9 : plot(xd,yd,ycalc) -- Plot data and fit
10 : write_data("list7_17.dat",xd,yd,ycalc) -- Save data
```

Listing 7.17. Illustration of fitting data with datafit().

It can be seen that the dotted curve provides a very good smooth curve representation of the data points. This datafit() function is a computer generated equivalent to the draftsman's French Curve. Although it is not shown in Listing 7.17, the analysis actually generates 6 interpolation points or knots uniformly spaced along the data with a knot at each end of the data range and with 4 internal knots.

A close examination of the dotted curve and the saved output file shows that the fitted curve does not go exactly through the origin (the 0,0) point. The value is close to zero so the fit might be acceptable and no further improvement would be needed. However, in cases such as this where it is know that the curve should go through a particular point, the datafit() function provides additional flexibility. As seen in Listing 7.16 a second parameter can bc passed to the datafit() function in addition to the {y,x} data. This second parameter can have one of two forms. The first and simplest form is simply a number which will be used as the number of interpolation points used in the data fit. This is useful when the standard multiple of 10 data points per interpolation point does not produce a good curve. With this parameter the user can experiment with varying numbers of interpolation data points. The second form, of the second parameter is a table of two entries which are also tables. The first entry is a table that explicitly specifies the desired location along the independent axis of the interpolation points. The second entry is a table of possible c values. In the c[] array any desired fixed interpolation points can be specified, such as for example the (0,0) point.

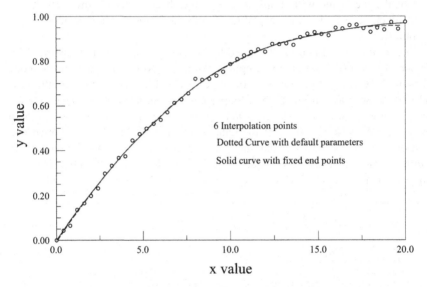

Figure 7.24. Example of data fitting with datafit() both with default parameters and with fixed end points and variable interpolation spacing.

```
 1 : -- /* File list7_18.lua */  example with specified knots
 2 :
 3 : require "DataFit"
 4 :
 5 : xd,yd,ycalc={},{},{}
 6 : read_data('list7_1.dat',xd,yd) -- Read data
 7 : xc,c = {0,6,9,11,16,20},{} -- Set up location of knots
 8 : c[1],c[6] = 0,0.97 -- Specify two end points
 9 : fxc = datafit({yd,xd},{xc,c}) -- Now do fit with floating knots
10 : for j=1,#xd do ycalc[j] = fxc(xd[j]) end -- Calculate fit values
11 : plot(xd,yd,ycalc) -- Plot data and fit
12 : write_data("list7_18.dat",xd,yd,ycalc) -- Save data
```
Listing 7.18. Example of datafit() with fixed end points.

The use of this feature is perhaps best demonstrated with an example such as that shown in Listing 7.18. The calling arguments to datafit() on line 9 now consist of two tables, each with two sub-tables. The first table consist of the {y,x} data which is always required. The second table consists of the interpolation points in the form {xc,c} where xc is an array of the location of the interpolation points and c is an array of fitting coefficients. The xc locations are defined on line 7 of the listing. It will be noted that the selected points are not uniformly spaced along the independent axis, but may be of any desired spacing as illustrated in the code. On line 8 of the listing only the first and last $c[1]$ and $c[6]$ values of the coefficients are defined while on line 7 all of the xc location values are specified. This second table of values provides a mechanism for (a) specifying the exact location of the interpolation knots along the independent variable axis and (b) specifying a fixed value for any desired points along the fitting curve. In a typically use one might want to specify one or both end points as is done in this example. If this feature of datafit() is used, the location of all the interpolation points along the independent axis must be specified but only fixed values of the C coefficients are specified. The result of fixing the end points and using the non-uniform spacing is shown as the solid curve in Figure 7.24. The difference near the origin is difficult to see in the figure, because the dotted curve with the default parameters passes close to zero. However, there are slight differences in the solid and dotted curves in the figure especially near the two end points. This example illustrates the flexibility of the datafit() function to handle fixed end points or other fixed points and non-equally spaced interpolation points.

Some additional features are worth noting about this LCB least squares fitting approach. First, the data fitting function using intp() does not provide an explicit expression for the fitting function in terms of the fitting parameters (the c's). The intp() routine only provides a computer algorithm for the error in terms of the parameters. This is perfectly acceptable for the nlstsq() routine, since it determines through numerical partial derivatives the relationship between the coefficients and the error function. Without the numerical evaluation of the partial derivatives, the formulation of this approach would be much more difficult. Thus the importance of the emphasis on accurate numerical derivatives in Chapter 5 is seen in this example. While an explicit relationship can in principle be derived for the error values in terms of the interpolation points, it would be very difficult. However, the relationship is in fact linear, so the LCB least squares fitting is a case of fitting with a set of linear coefficients. This means that the nlstsq() routine converges after two iterations. In fact the first calculated values are correct, but the routine takes another iteration to determine that the coefficients are correct. Since the equations are linear, the initial guess on the c values is not important, and the datafit() routine simply uses all zeros for the initial guess.

Since data fitting and plotting of data is such an important topic, several examples of using datafit() with a variety of data will now be given. Figure 7.25 shows the graph of PhD graduates vs. research expenditures previously shown in Figure 7.19. The code listing for this example is not shown but is essentially the same as Listing 7.17 except for use of a different data file. The data file used here is the PhDgrads.dat file. This fitting is with the default parameters and in this case the generated fitting curve is close to the global second degree polynomial fit previously shown as the solid curve in Figure 7.21. The default number of interpolation points in this example is only three, two end points and a mid-point. The reader can generate this data and fitted curve by re-executing the code in Listing 7.17 with line 6 changed to: read_data('PhDgrads.dat',xd,yd).

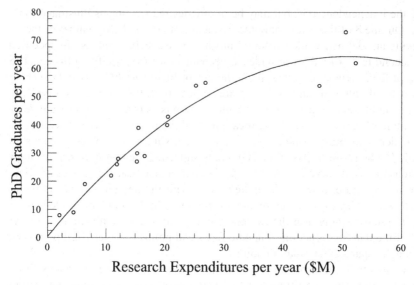

Figure 7.25. Example with limited number of data points and large scatter.

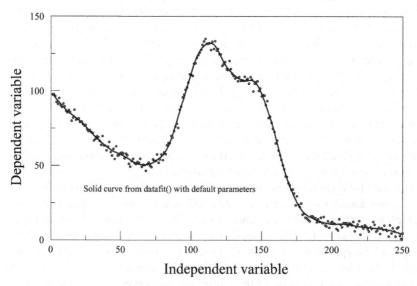

Figure 7.26. Fit of Gauss3 data with default parameters in datafit().

Next Figure 7.26 shows the datafit() with default parameters for the two Gaussian peaks data previously shown in Figure 7.18. This is again achieved by the code in listing 7.17 but with line 6 changed to: read_data('gauss3.dat',xd,yd). The fit is very reasonable considering the scatter in the data. This curve using local cubic segments can be compared with the attempt in Figure 7.22 to use a 9th order global polynomial to fit the same data. On a local scale, low order polynomials provide a good fit to functions, but global polynomials have very limited use in fitting most sets of data. With 250 data points in the file, the default parameters in datafit() will lead to 25 interpolation points equally spaced along the x-axis. Such a fine spacing is needed near the peaks and along the rapidly changing slopes of the curve. However, the finely spaced interpolation points lead to some oscillations in the solid curve in the slowly changing regions beyond the peaks which is probably not justified by the data. A good draftsman with a set of French curves would probably not show such oscillations in the 0 to 70 and 200 to 250 regions of the independent variable.

```
 1 : -- /* File list7_19.lua */  gauss3.dat with selected knots
 2 :
 3 : require "DataFit"
 4 :
 5 : xd,yd,xcalc,ycalc={},{},{},{}
 6 : read_data('Gauss3.dat',yd,xd) -- Read data
 7 : xc = {0,35,70,75,85,95,103,112,117,125,135,140,148,
 8 : 162,180,190,210,250} -- define knots
 9 : c = {}
10 : fxc = datafit({yd,xd},{xc,c})
11 : for i=1,#xd do ycalc[i] = fxc(xd[i]) end
12 : plot(xd,yd,ycalc)
13 : write_data("list7_19.dat",xd,yd,ycalc) -- Save data
```

Listing 7.19. Datafit() for two Gaussian peaks with hand selected interpolation points.

The fitting can be improved a little by using a set of hand selected interpolation points. The code for such a fitting is shown in Listing 7.19. In this case 18 interpolation points have been hand selected and listed on lines 7 and 8 of the code. To select this set, several runs were made of the fitting program with the points adjusted after each run. This is rapidly done since the fitting routine runs fast and the pop-up plot can be used to rapidly view the output. The fitted curve with dotted lines showing the selected interpolation knots is shown in Figure 7.27. Although the solid fitted curve is not very different from that shown in Figure 7.26 using the default parameters, the small oscillations are no longer present outside the peak regions and the solid curve is much closer to what one would expect the real function to look like. The key to improving the fitting is to use fewer interpolation points outside the peak regions and to use finely spaced points around the peak regions and where the function is changing rapidly. The solid curve in this author's opinion would be an acceptable solid curve for use in publishing such data and thus it can be argued that this computer generated solid curve is about as good as a draftsman would produce with a set of French curves. Thus the datafit()

function can be labeled as a software replacement of a draftsman with a set of French curves. This was one of the previously stated goals of this section.

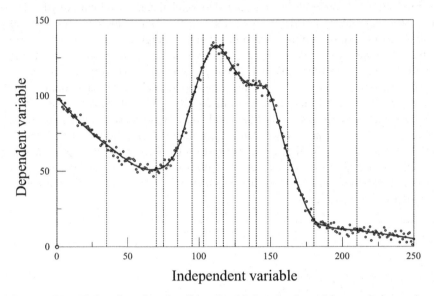

Figure 7.27. Fitting function for Gauss3 with hand selected interpolation points. The vertical dashed lines show the selected locations of the interpolation points.

With some data sets such as the gauss3 data the default parameters of datafit() will not give a good fit to the data because of the uniformly spaced data points. Some data sets need only a few extra interpolation points at strategically placed points to greatly improve the fitting. To partially automate this process, some additional functions are contained in the DataFit package. Listing 7.20 shows these additional code segments. The first function refitdata() adds two additional knots to the interpolation points. The question arises as how to have the computer code automatically recognize where within the range of the independent variable additional interpolation knots would be most useful. For this it should be recalled that the LCB interpolation technique provides continuous first derivatives but gives discontinuous second derivatives at the interpolation points.

It is thus reasonable to assume that the region most in need of a finer spaced interpolation grid is around the interpolation point where the second derivative exhibits the largest discontinuity. This point can be readily determined and this is the point selected by the refitdata() function. This point is located by the search on lines 44 through 47 using the intpf() function passed in through the calling argument. The second argument of 2 in this intpf() function on line 45 indicates that the interpolation function is to return the second derivative. When used, the intpf() function passed in the argument list is intended to be a previously generated LCB interpolation function. After identifying the appropriate node, additional knots are inserted into the table of interpolation points on lines 48 and 49. Following this nlstsq() is again called on line 52 to generate the set of new least squares coefficients and the results are returned as a callable function on line 53 along with the same table returned by the datafit() function.

```
 1 : -- /* File Data_Fit.lua */
------
35 : refitdata = function(yx,intpf,xcc) -- Add two additional knots
36 :    local j,nd = 0,#yx[1]
37 :    local xmin,xmax = 0,0
38 :    local c,nc = xcc[2],#xcc[2]
39 :    local xc,del -- Needed for interpolation function
40 :    local fdf = function(yx,c)   -- Interpolation function
41 :       return intp(xc,c,yx[2])-yx[1] -- function with knots
42 :    end
43 :    xc = xcc[1] -- Global value for fdf
44 :    for i=2,nc-1 do -- Find point of max second derivative change
45 :       xmin = math.abs(intpf(xc[i]*(1.001),2) -
                 intpf(xc[i]*(.9999),2))
46 :       if xmin>xmax then j = i; xmax = xmin end
47 :    end
48 :    table.insert(xc,j+1,(xc[j+1] + xc[j])*.5) -- Add first point
49 :    table.insert(xc,j,(xc[j] + xc[j-1])*.5) --added points in xc
50 :    table.insert(c,j+1,0); table.insert(c,j,0)
51 :    nc = nc + 2
52 :    del,err = nlstsq(yx,fw,fdf,c) -- call with 2 extra knots
53 :    return function(x,df) return intp(xc,c,x,df) end,
             {xc, c, del}, err
54 : end
55 : datafitn = function(yx,na,nc)
56 :    local intpf,xcc,err = datafit(yx,nc)
57 :    for i=1,na do
58 :            intpf,xcc,err = refitdata(yx,intpf,xcc)
59 :    end
60 :    return intpf,xcc,err
61 : end
```

Listing 7.20 Additional helper functions for data plotting. These add additional knots to the interpolation data sets.

The refitdata() function is useful when only a small number of additional points are needed at some critical point along the curve. The argument list to the function is selected so that it can be used directly with the results returned by datafit() in the form refitdata(yx,datafit(yx)). A higher level function datafitn() is shown on lines 55 through 61 using this function. The calling argument for this function is similar to datafit() with the addition of an na parameter that specifies the number of additional times to call the refitdata() function. The iterative loop on lines 57 through 59 repeatedly calls the refitdata() function na times adding two additional interpolation knots for each call.

An example of the use of this function is shown in Listing 7.21. The call of the data fitting routine is on line 7 and uses the datafitn() function. The argument list passes the data arrays and requests an initial data fit with 5 interpolation points. This will be passed along to datafit() by datafitn() and then the redatafit() function will add two additional data points (from the second argument of 1 to datafitn())

and reevaluate the fitting function. Finally, the return from datafitn() will provide a function for the data fit just as the datafit() function does.

```
 1 : -- /* File list7_21.lua */ -- Example of datafitn() function
 2 :
 3 : require "DataFit"
 4 :
 5 : xd,yd,xcalc,ycalc={},{},{},{}
 6 : read_data('thurber.dat',yd,xd) -- Read data
 7 : fxc = datafitn({yd,xd},1,5)
 8 : nd = #xd
 9 : for i=1,nd do ycalc[i] = fxc(xd[i]) end
10 : xx,yy = {},{}
11 : xmin,xmax,nx = xd[1],xd[nd],1000
12 : dx = (xmax-xmin)/nx
13 : for i=1,nx do
14 :   xx[i] = xmin+dx*(i-1)
15 :   yy[i] = fxc(xx[i])
16 : end
17 : plot(xx,yy)
18 : write_data("list7_21.dat",xx,yy,xd,yd) -- Save data
```
Listing 7.21. Data fitting using datafitn() to give additional critical knots.

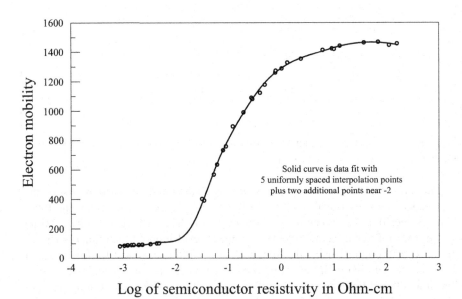

Figure 7.28. Data fit of electron mobility as function of log10 of semiconductor resistivity.

The resulting data and the generated curve fit are shown in Figure 7.28. The data in this case is electron mobility for a semiconductor as a function of the log10 of the resistivity in Ohm-cm. The curve is a relatively smooth function except near -2 where there is a rather sharp turn in the curve to a relatively flat region of the curve. This is the type of data fitting curve where the use of a few additional data points makes a tremendous difference in the fitting. Starting with 5 fitting points, the interpolation points are a little over 1 unit apart from -3 to 2. Because of the large curvature near -2, the refitdata() routine will insert additional interpolation points on either side of the point near -2. This is sufficient to give a good fit to the data. The reader can execute the code in Listing 7.21 with modified parameters and without the additional points and see the resulting fits. The solid curve in Figure 7.28 provides a good fit to the experimental data although the number of data points is somewhat limited. In this case the default parameters will not provide sufficient interpolation points for a good fit – hence the imposed initial condition of 5 points in the call to datafitn().

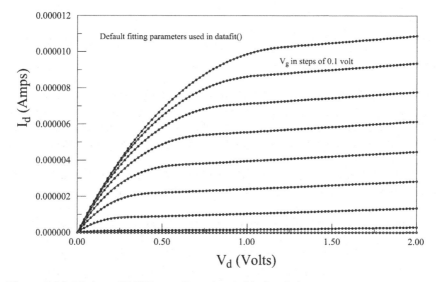

Figure 7.29. Fitting of MOS transistor data with datafit().

A final example is shown in Figure 7.29 of fitting MOS transistor I_d data with datafit(). This is experimental data but is very smooth and relatively free of noise. For this case the datafit() function is perhaps not needed for plotting a smooth curve, since the data is so free of noise. However, the general data fitting approach easily handles this case also with a small number of interpolation points. Since the data file has 100 data points for each curve, the default number of interpolation points used will be only 10 for each curve. The advantage of datafit() is of course more obvious for data with noise. Since this example has multiple curves, it is perhaps useful to also include the computer code used to analyze this data as an example of handling data files with multiple sets of data. This is shown in Listing 7.22. Nine sets of data are read in from a data file on line 5 of the listing. These sets of data are then fit one by one with datafit() on line 12 of the code with each curve generating a new function that is stored in a table of functions, the fxc[] table. A series of 1000 data points is then evaluated along each curve on line

```
 1 : -- /* File list7_22.lua */ -- datafit() to transistor data
 2 :
 3 : require "DataFit"
 4 : xd,yd={},{{},{},{},{},{},{},{},{},{}}
 5 : nd = read_data('mosidd1.5.0.6.dat',xd,yd)
 6 : xx,yy = {},{{},{},{},{},{},{},{},{},{}}
 7 : xmin,xmax,nx = xd[1],xd[nd],1000
 8 : dx = (xmax-xmin)/nx
 9 : for i=1,nx do xx[i] = xmin+dx*(i-1) end
10 : ncrv,fxc = #yd, {}
11 : for j=1,ncrv do
12 :    fxc[j] = datafit{yd[j],xd}
13 : end
14 : for j=1,ncrv do
15 :    for i=1,nx do yy[j][i] = fxc[j](xx[i]) end
16 : end
17 : plot(xx,unpack(yy))
18 : write_data("list7_22.dat",xx,yy)  -- Save data
```

Listing 7.22. Code segment for fitting multiple curve from a data file. MOS transistor data.

15 and collected into an array of calculated curves (the yy[] tables) with each set of data generated using a separate function from the table of functions (called by fxc[j]()). Since each function returned by the datafit() function has embedded within it the evaluated array of interpolation points and values, each function returned (each fxc[j]) will calculate one of the series of transistor curves. Finally, all the calculated data is saved in a file for plotting on line 16. Note the use of the unpack(yy) statement on line 17 of the listing. This extracts the elements of a table and is equivalent to yy[1], yy[2], ... yy[n]. This is not needed in the read_data() or write_data() functions as these properly handle tables within tables.

Although it is difficult to see in Figure 7.29, the solid line fitting function returned by datafit() essentially overlays all the experimental data points and does a very excellent job of fitting the experimental data with its approach of using local cubic functions to describe a much more complicated relationship on the global scale. The interpolation functions have been saved in this example in a table to illustrate how a series of functions can be generated each with a different set of embedded interpolation parameters for use in subsequent calculations.

This section has considered the general problem of generating a smooth curve for representing a set of experimental data with error or noise in the data. This is not an easy problem and most books on numerical techniques provide little guidance on this if a good mathematical function is not known which has a proper mathematical functional form to fit to the data. The code developed and discussed in this section is primarily oriented toward cases where the user simply wants to generate a smooth least squares curve for fitting the data for example for publication without fitting any particular physical model for the data. The approach has been based upon use of the Local Cubic (LCB) interpolation method of Chapter 7. In that chapter the Cubic Spline (or CSP) interpolation method was also developed and discussed. One might think about using the CSP method with a least squares approach for data fitting similar to that developed here. However using this interpolation technique is somewhat more complicated than the LCB method. For example in the fdf() function used in datafit() for the function called by nlstsq() – see Listing 7.16, line 8 – an interpolation function is needed that returns the difference between the predicted value and the data points. For a cubic spline function, each time one of the knots, or in this case the C values, changes, a set of coupled equations must be solved to determine the fitting functions within each interval of the interpolation region. In the nlstsq() least squares determination of the c values or the interpolation points, the c values are changed many times to determine the optimum values. This means that using CSP interpolation as the basis of such a data fitting technique, while possible in theory, is somewhat more difficult and involved than using the LCB interpolation. The implementation of a data fitting approach based upon cubic splines will be left as an exercise for the reader if she/he feels that the CSP interpolation is sufficiently better than the LCB technique. In Chapter 6 it was shown that there is little difference in the fitting ability of either the LCB or CSP technique and each approach was better for some particular functional forms of data.

The datafit() code segment is another useful illustration of how previously developed code segments and functions can be combined in new and very useful ways. The datafit() function combines in a very short code segment the interpolation function of Chapter 6 with the nonlinear least squares data fitting routine nlstsq() of the previous section of this chapter. The result is another very useful and practical software application. In subsequent chapters many other applications of the code segments already developed will be seen.

7.8 Rational Function Approximations to Implicit and Transcendental Functions

Many engineering applications of curve fitting involve representing an implicit function by some continuous function. This may be for data plotting and publication of results or for ease in calculating approximate values of the function at selected values of the independent variable. This section further considers such cases and the use of the previously developed nonlinear least squares approximation approach for finding such approximate fitting functions.

As a first example, consider the electron mobility data shown in Figure 7.28. In the previous section the datafit() code was used to obtain an LCB interpolation function for the data and this is shown as the solid line in the figure. In some cases, this might be an acceptable function to approximate the data. In other cases one might desire an explicit mathematical equation for the physical function representing the data. In this case an explicit exact mathematical equation is not available to fit to the data. The datafit() returns only a function that is evaluated by a computer algorithm. Even if one had a detailed theory for the relationship provided by this algorithm, it might be in implicit form so one would not necessarily have a simple relationship. There are many possible explicit mathematical functional forms that might be used to approximate data such as the mobility data.

Polynomials are frequently suggested, but as discussed in a previous section, global polynomials are usually not very good functions for describing data such as shown in the figure. However, the quotient of two polynomials, which is usually called a rational function, is much more useful for function approximations. Thus this section will consider fitting the mobility data and other data sets to a rational function of the form:

$$f(x) = \frac{a_1 + a_2 x + a_3 x^2 + \cdots a_n x^n}{1 + b_2 x + b_3 x^2 + \cdots b_m x^m} .$$ (7.44)

Such approximating forms are so frequently used that it is useful to develop an approach for fitting data to such equations. If the order of the denominator polynomial is equal to or greater than the order of the numerator polynomial, the function remains finite for large and small values of x. This is a useful feature of rational functions and for data such as seen in Figure 7.28 it would be expected that taking $m = n$ would be the best choice as the data appears to saturate for large values. The selected order of the polynomials for fitting a set of data is subject to some guess and experimentation. For the mobility data in Figure 7.28, a third order numerator and denominator polynomial will be selected for fitting. Then in terms of a set of adjustable c coefficients the proposed fitting function can be written as:

$$f(x) = \frac{C_1 + C_2 x + C_3 x^2 + C_4 x^3}{1 + C_5 x + C_6 x^2 + C_7 x^3}$$ (7.45)

where there are seven coefficients to be determined. These can be determined by the nonlinear least squares technique for example.

Before listing the code segment for fitting the data with this function, it is useful to discuss some features and problems with obtaining convergence for such a rational function. Convergence to the minimum least squares error requires some special care in selecting initial approximations for the coefficients. Also the range of x in Figure 7.28 goes from about -3 to about +3. If the denominator value ever approaches zero, the function will become infinitely large, so it can be concluded that the denominator should always have a positive value over the range of the independent variable. This is generally true for using rational polynomials for all types of approximations. Obtaining good initial guesses for all the coefficients is not easy. If the denominator coefficients (C_5, C_6, C_7) were all zero, the fitting function would be linear in the $C_1 \cdots C_4$ coefficients and the nlstsq() routine would converge in two iterations independently of any initial guesses for the coefficients. Thus if very little knowledge is known of the proper coefficients, some strategy is needed to properly converge to a valid final set of coefficients.

One possible convergence strategy will be briefly outlined and is implemented in the code segment in Listing 7.23. The rational polynomial is defined on lines 5 through 8. As noted above, if the denominator coefficients are set to zero and a least squares analysis is made using only the numerator coefficients, a best set of parameters can be obtained for the numerator constants independently of any initial guess. After obtaining values for the numerator polynomial, one can then try to evaluate the denominator polynomial. This will of course change the best numerator polynomials. However, if the denominator coefficients can be made to change slowly, then perhaps the nlstsq() function can adjust the numerator polynomials as the denominator coefficients are determined. One additional strategy is to turn on each of the denominator coefficients one-by-one and reevaluate the coefficients each time. This is the strategy used in Listing 7.22 through a combination of the actv[] and step[] parameters. By setting a table entry in actv[] to 0 for any desired coefficient, the nlstsq() fitting with respect to that parameter will be bypassed. By setting step[] to some positive value near 1.0, the nlstsq() function will be forced to slowly adjust that particular coefficient and hopefully the other coefficients can also slowly adapt at each step to match the changing value. This strategy is implemented in Listing 7.23.

Line 12 in the listing sets all the initial coefficient values to 0 while lines 15 and 16 make coefficients 1 through 4 active with no limit on the step size. The step 1 call to nlstsq() on line 19 then determines values of C_1 to C_4 with C_5 to C_7 set to zero. The first line of the printed output then shows that the RMS error obtained with just the numerator is 68.46 and was obtained after 2 iterations. Line 22 then makes C_5 active with an initial value of 0.3 and a step[5] value of 1.2. The initial value is selected so that the denominator will be positive at all values of x over the range of -3 to 2, i.e. so that $1+C_5*x>0$. The step selection value will allow nlstsq() to change C_5 by approximately 20% at each iteration step. Line 23 then re-executes nlstsq() to evaluate the first denominator coefficient. For this step, the previously determined coefficients C_1-C_4 are also again re-evaluated as the C_5 value is adjusted. The second line of the output then shows that the achieved RMS error is 51.99 and was achieved after 14 Newton iterative steps.

The above procedure is repeated two more times on lines 26 through 31 making the other two denominator coefficients active. In each case the previously determined coefficients are updated as each denominator coefficient is added to the calculation. In each case the initial guess of the coefficient is selected so that the denominator remains positive at the smallest value of independent variable, i.e. for C_7, the initial guess is such that $1+C_7*(-3)^3>0$. This procedure allows nlstsq() to slowly converge on the correct coefficients, without having good initial guesses for the coefficients. The reader can experiment with fitting the data in one step by making all coefficients active and selecting various initial guesses. Initial guesses close to those in the selected output listing will lead directly to convergence of the fitting program. However, if initial values very far from the final values are used, the program will not converge to a proper set of coefficients as will be obvious in the fitted function and in nlstsq() using the maximum number of iterations for the calculations. For the code in Listing 7.23, the maximum number of iterations required at any step is 24 for the last call to nlstsq() on line 31. It is also noted that the RMS error decreases as each additional term is added to the denominator polynomial.

The calculated values of the final coefficients are listed in the selected output with the standard deviation in the calculated values. The standard deviations

```
 1 : -- /* File list7_23.lua */
 3 : require"nlstsq"
 5 : fft = function(yx,c)
 6 :    x = yx[2]
 7 :    return (c[1]+ c[2]*x+c[3]*x^2+c[4]*x^3)/(1+c[5]*x+c[6]*x^2+
          c[7]*x^3) - yx[1]
 8 : end
 9 :
10 : xd,yd={},{}
11 : nd = read_data('thurber.dat',yd,xd) -- Read data
12 : c = {0,0,0,0,0,0,0}
13 : nc = #c
14 : actv,step = {},{}
15 : for i=1,nc do actv[i],step[i] = 1,0 end
16 : actv[5],actv[6],actv[7] = 0,0,0
17 : yx = {yd,xd} -- Data values
18 :
19 : del,err,nmax = nlstsq(yx,fw,fft,c,actv,step) -- Call fiting #1
20 : print(err,nmax,'-- Numerator only') -- print results
21 : for i=1,nc do printf('c[%d] = %12.4e +/- %12.4e\n',i,c[i],
          del[i]) end
22 : c[5],actv[5],step[5] = .3,1,1.2 -- Include c[5]
23 : del,err,nmax = nlstsq(yx,fw,fft,c,actv,step) -- Call fiting #2
24 : print(err,nmax,'-- Add c[5]') -- print results
25 : for i=1,nc do printf('c[%d] = %12.4e +/- %12.4e\n',i,c[i],
          del[i]) end
26 : c[6],actv[6],step[6] = .1,1,1.2 --Include c[6]
27 : del,err,nmax = nlstsq(yx,fw,fft,c,actv,step) -- Call fiting #3
28 : print(err,nmax,'-- Add c[6]') -- print results
29 : for i=1,nc do printf('c[%d] = %12.4e +/- %12.4e\n',i,c[i],
          del[i]) end
30 : c[7],actv[7],step[7] = .03,1,1.2 -- Include c[7]
31 : del,err,nmax = nlstsq(yx,fw,fft,c,actv,step) -- Call fiting #4
32 : print(err,nmax,'-- Add c[7]') -- print results
33 : for i=1,nc do printf('c[%d] = %12.4e +/- %12.4e\n',i,c[i],
          del[i]) end
34 : xmin,xmax,nx = xd[1],xd[nd],1000
35 : dx = (xmax-xmin)/nx
36 : xx,yy={},{}
37 : for i=1,nx do
38 :    xx[i] = xmin+dx*(i-1)
39 :    yy[i] = fft({0,xx[i]},c)
40 : end
41 : plot(xx,yy)
42 : write_data("list7_23.dat",xx,yy) -- Save data
Selected Output:
68.46576011431 2         -- Numerator only
51.993230789667       14      -- Add c[5]
15.308573619913       11      -- Add c[6]
12.34931747043 24       -- Add c[7]
c[1] =  1.2881e+003 +/-  4.6648e+000
c[2] =  1.4911e+003 +/-  3.9543e+001
c[3] =  5.8327e+002 +/-  2.8679e+001
c[4] =  7.5422e+001 +/-  5.5635e+000
c[5] =  9.6632e-001 +/-  3.1314e-002
c[6] =  3.9799e-001 +/-  1.4976e-002
c[7] =  4.9735e-002 +/-  6.5795e-003
```

Listing 7.23. Code for fitting of rational function to electron mobility data.

range from 0.36% for C_1 to 7.3% for C_4 and 13.2% for C_7. The coefficients after the intermediate steps are not shown in the listing but are printed when the code is executed. The reader is encouraged to execute the code and observe how the "best fit" coefficients change as the number of denominator terms is increased. The data and fitted rational function are shown in Figure 7.30. It can be seen that the solid curve provides a good, smooth approximation to the experimental data over the entire range of the function. The average RMS error in the fitted function is 12.3 out of a typical value of 100 to 1400. The rational function approximation is close to the LCB interpolation function as can be seen by comparing Figures 7.28 and 7.30. However there are important differences around the -2 independent variable value where the curvature changes rapidly. The rational polynomial has the advantage of being a mathematical expression which can be more easily included in other computer code.

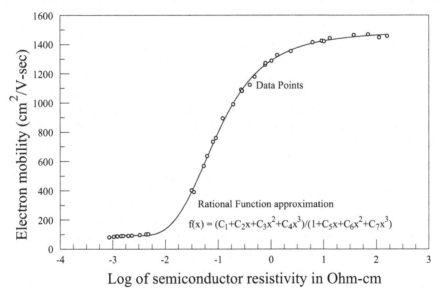

Figure 7.30. Rational function approximation to electron mobility data.

There is one other approach worthy of exploring for obtaining the coefficients of a fitting rational function such as Eq. (7.45). If the equation is rewritten by multiplying through by the denominator, the following equation results:

$$f(x) = C_1 + C_2 x + C_3 x^2 + C_4 x^3 - f(x)(C_5 x + C_6 x^2 + C_7 x^3) \qquad (7.46)$$

This could also be the function used for a call to nlstsq() since it contains the same functional content as the original rational equation. This equation would then be called for each data point. If the equation were a "perfect" fit to the data points with no error, the $f(x)$ value would always equal one of the data point 'y' values and the $f(x)$ value on the right hand side of the equation could be replaced by the data point values. Thus if the fitted curve closely approaches the data points as it does in Figure 7.30, a good approximation is:

$$C_1 + C_2 x_k + C_3 x_k^2 + C_4 x_k^3 - y_k (C_5 x_k + C_6 x_k^2 + C_7 x_k^3) - f(x_k) = 0 . \qquad (7.47)$$

This equation is now a linear fitting equation in terms of all the C coefficients. Fitting to this equation will then give an initial approximation to all the coefficients in one iterative step. Using these initial approximations to the coefficients, one can then execute nlstsq() with the original fitting function for the final determination of the coefficients. This approach is shown in Listing 7.24. Two fitting functions are defined in the code. Lines 9 through 12 define the first linear equation approximation to obtain initial approximations to the coefficients. This function is used on line 21 to call nlstsq() for the initial approximations. Note in this call, the data array passed to nlstsq() is defined as a three element array on line 19 as yx = {yd,xd,yd}. This may appear as unnecessary, since the first element in the data array is yd. However, the value passed to the fitting function at each data point for the first element of the data array, is not necessarily the data point value. In nlstsq() the value of the independent variable (the first array element) is modified as required to probe the linear nature of the fitting function and the first value of the yx array passed to the fitting functions on line 5 or 9 can not be counted on to be the exact data value point. Passing the y value as the third element in the array of data values ensures that this element will be available when each data point is passed to the fitting function. This code segment is a useful example of how additional parameters can be passed to the fitting function. In subsequent chapters the nlstsq() function will be used to analyze data where the independent variable is a function of multiple variables.

The selected output shows the C values (first set of printed values) obtained from the first call to nlstsq() which uses the approximate linear equation. The final set of C values is also shown in the selected output listing, so they can be compared with the initial values. It can be seen that the linear approximate equation does provide a good initial approximation to the fitting parameters. After the first call to nlstsq() the second call is made on line 26 of the listing, this time using the complete nonlinear fitting function as defined on lines 5 through 8. The only change between the calls to nlstsq() is to set limits on the allowed steps of the denominator coefficients (C_5, C_6 and C_7) to 20% changes per iteration on line 25 of the listing. The second call to nlstsq() for final evaluation of the coefficients converges in 11 iterations to the same final values as shown in Listing 7.23. It is noted that the resulting values are almost identical with difference occurring only in the fifth decimal place which is beyond the accuracy of the obtained coefficients. A graph of the results would be identical to that shown in Figure 7.30. This technique of obtaining a first approximation from a linearized approximation of the fitting equation is a very useful technique when fitting data to a rational polynomial.

This example illustrates the fact that several valid approaches may typically be used to find a set of nonlinear least squares parameters for fitting a function. All nonlinear data fitting approaches require some type of iterative approach and a

```
 1 : -- /* File list7_24.lua */ --Two step fit to rational polynomial
 3 : require"nlstsq"
 4 :
 5 : fft = function(yx,c) -- fitting function, Nonlinear in C's
 6 :    x = yx[2]
 7 :    return (c[1]+c[2]*x+c[3]*x^2+c[4]*x^3)/(1+c[5]*x+
            c[6]*x^2+c[7]*x^3) - yx[1]
 8 : end
 9 : fft1 = function(yx,c) -- First approximation, Linear in C's
10 :    x,xy = yx[2],yx[2]*yx[3]
11 :    return (c[1]+ c[2]*x+c[3]*x^2+c[4]*x^3) - xy*(c[5]+
            c[6]*x+c[7]*x^2) - yx[1]
12 : end
13 :
14 : xd,yd={},{}
15 : nd = read_data('thurber.dat',yd,xd) -- Read data
16 : c = {0,0,0,0,0,0,0}; nc = #c
17 : actv,step = {},{}18 : for i=1,nc do actv[i],step[i] = 1,0 end
19 : yx = {yd,xd,yd} -- Data values, pass y values as third argument
20 :
21 : del,err,nmax = nlstsq(yx,fw,fft1,c,actv,step) --#1 Linear in C's
22 : print(err,nmax,'-- First Approximation') -- print results
23 : for i=1,nc do printf('c[%d] = %12.4e +/- %12.4e\n',i,c[i],
            del[i]) end
24 :
25 : for i=5,nc do step[i] = 1.2 end -- limits on steps for coeff
26 : del,err,nmax = nlstsq(yx,fw,fft,c,actv,step) -- #2, Nonlinear
27 : print(err,nmax,'-- Final Calculation') -- print results
28 : for i=1,nc do printf('c[%d] = %12.4e +/- %12.4e\n',i,c[i],
            del[i]) end
29 :
30 : xmin,xmax,nx = xd[1],xd[nd],1000
31 : dx = (xmax-xmin)/nx; xx,yy={},{}
32 : for i=1,nx do
33 :    xx[i] = xmin+dx*(i-1)
34 :    yy[i] = fft({0,xx[i]},c)
35 : end
36 : plot(xx,yy)
37 : write_data("list7_24.dat",xx,yy) -- Save data
Selected output:
15.227249490757       2        -- First Approximation
c[1] =   1.2874e+003 +/-  6.1925e+000
c[2] =   1.2692e+003 +/-  1.3904e+002
c[3] =   4.1438e+002 +/-  1.0063e+002
c[4] =   4.3093e+001 +/-  1.9401e+001
c[5] =   7.7345e-001 +/-  1.0955e-001
c[6] =   2.9674e-001 +/-  5.2114e-002
c[7] =   3.2930e-002 +/-  1.8856e-002
------
12.349317631229       11       -- Final Calculation
c[1] =   1.2881e+003 +/-  4.6648e+000
c[2] =   1.4911e+003 +/-  3.9540e+001
c[3] =   5.8327e+002 +/-  2.8676e+001
c[4] =   7.5423e+001 +/-  5.5630e+000
c[5] =   9.6633e-001 +/-  3.1312e-002
c[6] =   3.9799e-001 +/-  1.4975e-002
c[7] =   4.9736e-002 +/-  6.5789e-003
```

Listing 7.24. Two step fitting to rational function with linear first approximation.

"good" initial set of parameters is required for such iterative approaches to converge. In many cases experimentation is required with more than one approach to arrive at a technique which rapidly converges. Unfortunately this is a fact of life and one should not be discouraged if a first approach does not lead immediately to a properly converged set of parameters. In using nlstsq() one key parameter is to always look at the number of iterations returned by the function. If it equals the maximum number set in the function (50 by default) then a properly converged set of parameters has not been obtained. In most applications the fitting routine will converge in considerable fewer iterative steps than the default limit of 50. In all cases, the final proof of a "good" fit must be ascertained by evaluating the fitting function and comparing the calculated results with the original data.

Rational polynomials or variations of such polynomials are frequently used to approximate implicit functions occurring in engineering problems. For example the probability function defined as:

$$P(x) = \frac{1}{\sqrt{2\pi}} \int_{-\infty}^{x} \exp(-t^2/2)dt \tag{7.48}$$

has no closed form expression in terms of more elemental functions. The evaluation of this function by use of numerical integration and interpolation was considered in Section 6.4 where a code segment was presented which can give about 4 digits of accuracy for this function. While that approach is satisfactory in some cases, a more straightforward approach is to use some rational function to represent implicit functions such as this. A more general functional form for fitting functions with polynomials than Eq. (7.44) is an equation of the form:

$$f(x) = \frac{(a_1 + a_2 x + a_3 x^2 + \cdots a_n x^n)^p}{(1 + b_2 x + b_3 x^2 + \cdots b_m x^m)^q}. \tag{7.49}$$

In this form powers of p and q can be used to provide a more rapid change with the variable for a limited number of fitting coefficients. This form is particularly useful for functions that rapidly approach zero for large x values and which requires that $q > p$.

For one such example consider the probability function above and conside only positive x values. The function then starts at 0.5 for x = 0 and asymptotically approaches 1.0 for large x values. A possible fitting function for positive x values is then:

$$P(x) \simeq f_p(x) = 1.0 - 0.5/(1 + C_1 x + C_2 x^2 + \cdots C_m x^m)^q. \tag{7.50}$$

For negative x values, the value $1.0 - f_p(-x)$ can be evaluated due to the symmetry of the probability function. There is some choice in the selection of values of m and q to use in this equation. One would expect that the fitting accuracy increases as m increases and perhaps also as q increases, within some limits. From various experiments, it can be found that good results can be achieved with combinations such as m = 4, q = 4 and m = 6, q=16. In order to experiment with approximating the probability function with equations of this form one must first have a means of accurately evaluating the integral for some finite number of points. For this work, this will be done by use of the numerical integration techniques discussed in Chapter 5 where an intg() function was developed which was found to give an accuracy of about 8 digits for typical integrals.

```
 1 : -- /* File list7_25.lua */ -- Rational approximation to prob fct
 2 : require"intg"; require"nlstsq"
 4 : x1,y1,x2,y2 = {},{},{},{}
 5 : f1 = function(x) return math.exp(-x^2/2) end -- Normal curve
 6 : f2 = function(x) return f1(1/x)/x^2 end -- Folded normal curve
 7 : x1[1],y1[1],i = 0,0,2
 8 : for xi=.05,1.01,.05 do -- Integral from 0 to 1
 9 :    x1[i],y1[i] = xi, intg(xi-.05,xi,f1)/math.sqrt(2*math.pi)+
            y1[i-1]
10 :    i = i+1
11 : end
12 : x2[1],y2[1],i = 1,y1[#x1],2
13 : for xi=.95,-.01,-.05 do -- Integral from 1 to infinity
14 :    x2[i],y2[i] = xi, intg(xi,xi+.05,f2)/math.sqrt(2*math.pi)+
            y2[i-1]
15 :    i = i+1
16 : end
17 : n = #x2; j = n+1
18 : for i=1,n do y1[i],y2[i]=.5*y1[i]/y2[n]+.5,.5*y2[i]/y2[n]+.5 end
19 : for i=2,n-1 do -- Combine results into table from x=0 to x=20
20 :    x1[j],y1[j] = 1/x2[i],y2[i]
21 :    j = j+1
22 : end -- x1,y1 now contains P(x) values
23 :
24 : fft = function(yx,c) -- Function to fit data
25 :    x = yx[2]
26 :    return 1. - 0.5/(1+c[1]*x+c[2]*x^2+c[3]*x^3+c[4]*x^4+
27 :            c[5]*x^5+c[6]*x^6)^16 - yx[1]
28 : end
29 :
30 : c = {0,0,0,0,0,0} -- Six possible coefficients
31 : yx = {y1,x1} -- Data values
32 :
33 : del,err,nmax = nlstsq(yx,fw,fft,c) -- Call fiting
34 : print(err,nmax) -- print results
35 : for i=1,#c do printf('c[%d] = %14.6e +/-
%12.4e\n',i,c[i],del[i]) end
Selected output:
5.423778962316e-008    21
c[1] =  4.986768e-002 +/-  7.2804e-008
c[2] =  2.113827e-002 +/-  4.6359e-007
c[3] =  3.284452e-003 +/-  1.0329e-006
c[4] =  3.100412e-005 +/-  1.0166e-006
c[5] =  5.197688e-005 +/-  4.4925e-007
c[6] =  4.900997e-006 +/-  7.2404e-008
```

Listing 7.25. Approximation to probability function with Eq. (7.50).

The code in Listing 7.25 shows code for fitting a test function of the form of Eq. (7.50) to the probability function. Lines 5 through 22 define the integral function and use intg() to evaluate the integral at selected points from 0 to 20 (for a total of 40 points). The approximating polynomial function is defined on lines 24 through 28 of the code and uses six fitting coefficients in the form of Eq. (7.50) with the power q in the denominator taken in the code to be 16. This is based upon experimenting with several values and selecting the one with the smallest fitting error. The printed output shows that the RMS error in fitting the data is 5.4 e-8, which is also about the accuracy with which the integral values is calculated by intg(). Thus it might be expected that the overall error in approximating the probability function with the evaluated coefficients to be of the order of 1.e-7. This is a fairly rough estimate and should not be taken as an exact answer. The C defining array on line 30 and the function definition on line 31 allow for up to 6 coefficients. The reader is encouraged to experiment with different numbers of fitting parameters and with different powers on the denominator polynomial. The nlstsq() routine converges for this function with zero initial guesses over a fairly wide range of coefficient numbers and power values. In the present example nlstsq() function converges after 21 Newton iterations from the zero input coefficient values and with no step limits on the changes in the parameter values. The code executes rapidly so it is easy to experiment with different combinations of powers and numbers of parameters. This is a good example to experiment with different approximations and explore the effect on RMS error. The resulting six model parameters are shown in the listing. The largest reported uncertainty in the coefficients is for the c[4] parameter with an uncertainty of about 3.4%.

The values of m and q used in this example were actually not selected at random, but were based upon a recommended approximating function given in the "Handbook of Mathematical Functions", National Bureau of Standards, US Dept. of Commerce, 1964. This publication recommends as one approximation to the probability function, an equation of the form of Eq. (7.50) with six coefficients and with $q = 16$. The published coefficients for such an approximation are given as:

$$C_1 = 0.0498673470 \quad C_2 = 0.0211410061 \quad C_3 = 0.0032776263$$
$$C_4 = 0.0000380036 \quad C_5 = 0.0000488906 \quad C_6 = 0.0000053830$$

(7.51)

This set of parameters is said to give an approximation with an error of less than 1.5e-7 over the complete range of positive x values. A comparison of these parameters with the set produced by Listing 7.25 shows that the two sets are fairly close in value but not identical. A comparison of the functions produced by the two sets of parameters, shows that the difference in values calculated for the probability function between the two sets is always less than 1.5e-7 so without more effort it can be said that the nlstsq() fitting has produces a set of polynomial parameters that gives an accuracy of better than about 3.0 e-7. Because of the effort used in developing the approximating functions listed in the "Handbook of Mathematical Functions", the set of coefficients in Eq. (7.51) is probably slightly better than the set obtained with Listing 7.25. However, the code provides a useful example of how the nlstsq() function can be used to obtain closed form rational polynomial approximations to implicit functions for which an exact closed form representations can not be obtained.

Much literature exists on obtaining finite polynomial approximations to various transcendental functions and many examples are given in the "Handbook of

Mathematical Functions". Several of these have been implemented in code and are included in the software supplied with this work. A selection of these can be loaded with the require"elemfunc" statement. Included in this computer code are the following functions:

1. $P(x)$ – The probability function as discussed above.
2. $Q(x) - 1.0 - P(x)$.
3. $\mathrm{erf}(x)$ – The error function $= \dfrac{2}{\sqrt{\pi}} \int_0^x \exp(-t^2)dt$.
4. $\mathrm{erfc}(x)$ – Complementary error function $= 1 - \mathrm{erf}(x)$.
5. $\mathrm{gamma}(x)$ – Gamma function $= \int_0^\infty t^{x-1} \exp(-t)dt$.
6. $\mathrm{beta}(x,w)$ – Beta function $= \int_0^1 t^{z-1}(1-t)^{w-1}\, dt$
7. $E1(x)$ – Exponential integral $= \int_x^\infty t^{-1} \exp(-t)dt$.
8. $En(x)$ –General Exponential integral of order n $= \int_1^\infty t^{-n} \exp(-xt)dt$.
9. $Si(x)$ – Sine Integral $= \int_0^x t^{-1} \sin(t)dt$.
10. $Ci(x)$ – Cosine Integral $= \gamma + \ln(x) + \int_0^x t^{-1}(\cos(t)-1)dt$.
11. $K(x)$ – Complete Elliptic Integral of First Kind
    $= \int_0^1 [(1-t^2)(1-xt^2)]^{-1/2}\, dt$.
12. $E(x)$ – Complete Elliptic Integral of Second Kind
    $= \int_0^1 (1-t^2)^{-1/2}(1-xt^2)^{1/2}\, dt$.
13. $J0(x), J1(x), Y0(x), Y1(x)$ – Bessel function (from solution of differential equation).
14. $Jn(x,n), Yn(x,n)$ – Bessel functions of order n (form solution of differential equation).

These functions are available for use in subsequent computer code. The probability function will be used extensively in the next chapter. These coded functions use some form of rational polynomials for approximating the functions. In many cases the rational polynomials are combined with other functions such as the exp() function which is assumed to be a known function.

Listing 7.26 shows some example code segments for some of these supplied functions. The listing uses a mixture of limited polynomials and rational polynomials to approximate the indicated functions over the complete range of function arguments. The numerical coefficients have been taken from the Handbook of Mathematical Functions. The accuracy of the numerical approximations is reported to be about 7 decimal digits for the implementations shown in the listing. The approach here of using polynomials to approximate implicit functions over a limited range of values appears similar to the interpolation approach discussed in the previous chapter. However, there is a fundamental difference. In interpolation the polynomial is designed to pass exactly through a set of assumed known data points. In the use here of polynomials, the desire is to approximate the function in such a manner that the maximum error between the function and the approximating polynomial is always less than some minimum value. The approximating polynomial may or may not be exactly equal to the exact function value at some particular value of the independent variable. The code in Listing 7.26 is given to illustrate the flavor of the function definitions in the elemfunc.lua file. The reader is encouraged to look at the complete file for more details on the various functions provided.

```
 1 : -- /* File elemfunc.lua */ -- Some elementary functions
 2 : -- Probability, Error, Complementary Error, Gamma function
 3 :
 4 : elemfunc = {} -- Table for elementary functions
 5 : local pp = function(x) -- Function used by Pn,Qn
 6 :    return 0.5*((((((.0000053830*x + .0000488906)*x +
         .0000380036)*x +
 7 :       .0032776263)*x + .0211410061)*x + .0498673470)*x + 1)^-16)
 8 : end
 9 : elemfunc.Pn = function(x) -- Probability (~7 digits of accuracy)
10 :    if x>=0 then return 1.0 - pp(x)
11 :    else return pp(-x) end
12 : end
13 : elemfunc.Qn = function(x) -- 1-P(x) (~7 digits of accuracy)
14 :    if x>=0 then return pp(x)
15 :    else return 1.0-pp(-x) end
16 : end
17 :
18 : elemfunc.erf = function(x) -- Error function (~7 digits )
19 :    local sgn = 1; if x<0 then x,sgn = math.abs(x), -1 end
20 :    return sgn*(1-((((((.0000430638*x + .0002765672)*x +
         .0001520143)*x +
21 :       .0092705272)*x + .0422820123)*x + .0705230784)*x + 1)^-16)
22 : end
23 : elemfunc.erfc = function(x) return 1-elemfunc.erf(x) end
24 :
25 : elemfunc.gamma = function(x) -- Gamma (x-1)! (~7 digits)
26 :    local ans,sign,fac=1,1,1
27 :    if x<0 then sign,fac,x = -1,
             math.pi/(x*math.sin(-math.pi*x)), -x end
28 :    if x<1 then ans,x = ans/x, x+1 end
29 :    x = x-1
30 :    while x>1 do ans,x = ans*x, x-1 end
31 :    ans = ans*(((((((.035868343*x - .193527818)*x +
         .482199394)*x - .756704078)*x +
32 :       .918206857)*x - .897056937)*x + .988205891)*x -
         .577191652)*x + 1)
33 :    if sign>0 then return ans*fac
34 :    else return fac/ans end
35 : end
```

Listing 7.26. Example code segments for calculating some elementary mathematical functions using limited polynomials and rational polynomials.

```
 1 : -- /* File list7_27.lua */ -- tests of elementary functions
 2 :
 3 : require"elemfunc"
 4 : gamma = elemfunc.gamma -- Gamma function
 5 : prob = elemfunc.Pn -- Probability function
 6 : J0 = elemfunc.J0 -- Bessel function
 7 : i,xt,y1t,y2t,y3t = 1,{},{},{},{} -- Now test function
 8 : for x=-5,5.0001,.01 do
 9 :     xt[i] = x
10 :     y1t[i] = gamma(x)
11 :     y2t[i] = prob(x)
12 :     y3t[i] = J0(x)
13 :     i = i+1
14 : end
15 : plot(xt,y1t,{'gamma plot','x','y','set yrange [-5:5]'})
16 : plot(xt,y2t); plot(xt,y3t)
17 : write_data("list7_27.dat",xt,y1t,y2t,y3t)
```

Listing 7.27. Example of the use of some of the elementary functions in elemfunc.

An example of the use of some of these functions is shown in Listing 7.27. In this case the gamma function, the probability function and the zero order Bessel function are evaluated on lines 9 through 14 over the interval -5 to 5. Pop up plots of the resulting functions are then coded on lines 15 and 16 so the user can rapidly see the results of the calculations. The data is also saved into a file on line 17 for plotting if desired. This listing is provided simply to indicate how the functions can be invoked in subsequent calculations. The reader should also take note of the calling arguments for the plot() function on line 15 that produces the pop-up plot for the gamma function. In addition to the data array arguments (xt and y1t) a final table is included with the terms {'gamma plot','x','y','set yrange [-5:5]'}. This optional plot() function parameter can be used to give title information (gamma plot), x axis title (x), y axis title (y) and optional input to the plotting code (set yrange [-5:5]). In this example, the optional input is used to indicate to the plotting function to plot the y data range over the values of -5 to +5 so the user can readily see the generated data. This is a new parameter for the pop-up plots that has not been used before. The user may also restrict the x range by use of a set xrange statement.

For such optional information, the user must include data for the title and axes labels as in this example. Finally for this section, Figure 7.31 shows the type of pop-up plot that should be obtained from the code for the gamma function graph on line 15. For those familiar with the properties of the gamma function this will be a familiar graph.

This section has discussed several examples of using the nonlinear least squares data fitting technique to approximate various functions using rational polynomials. Examples have included functions known only through a set of experimental data and examples known only through various integral definitions.

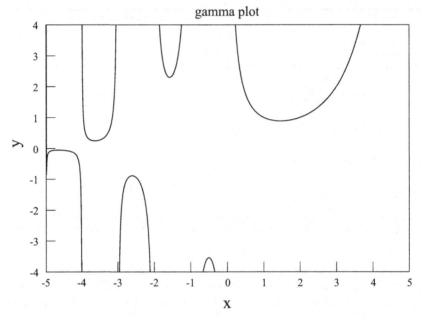

Figure 7.31. Gamma function plot obtained from the code in Listing 7.27.

7.9 Weighting Factors

The nlstsq() program includes a user supplied table of weighting factors with a possible value specified for each data point. So far all the examples have simply used the default value of unity for all data points. This treats all data points with equal emphasis and is typically used in nonlinear least squares data fitting. One of the implicit assumptions behind using equal weighting is the assumption that the random error associated with each data point is the same for each data point regardless of the magnitude of the data values. For some data sets this is not a good assumption and some different form of weighting is appropriate. For example if the "relative" error is approximately the same for each data point then the use of a constant weighting factor, overly emphasizes the largest data values. For such a case, a more appropriate factor to consider would be not the absolute error squared, $(y_i - f(x))^2$, but the relative error squared:

$$\varepsilon_r^2 = \left[\frac{(y_i - f(x))}{f(x)} \right]^2 \quad \text{or} \quad \left[\frac{(y_i - f(x))}{y_i} \right]^2 \tag{7.52}$$

Two forms are given for the relative error – one with the data value in the denominator (second form) and one with the predicted data value in the denominator (first form). Ideally the first form with the predicted value would probably be preferred, since it hopefully eliminates some of the random error that might be associated with the experimental data term. However, this form of the weighting factor is more complicated to implement, since it requires an additional set of iterations as the predicted values are not known before the data fitting and a set of weighting factors is required to determine the fitting function. In actual practice, there is typically little difference in using either of these forms and the second form with the data value in the denominator is typically used. This leads to a weighting factor of:

$$fw_i = \frac{1}{y_i^2} \qquad (7.53)$$

and this is frequently referred to as $1/y^2$ data weighting.

Another form of weighting frequently used is $1/y$ where the weighting is inversely proportional to the data value. This provides a weighting somewhat between that of a unity weighting factor and the relative error weighting factor. One must be careful with either of these inverse weighting factors to protect against a zero data value which would give an infinite weighting factor. This can be done by establishing a maximum value for the weighting factor if any data point is near zero.

One example will be used to illustrate the use of a non-unity weighting factor. Consider again Figure 7.30 for the rational function fitting of the experimental mobility data. The solid line fit in Figure 7.30 looks relatively good in the plotted figure. The same data and the previously fitted curve is shown again in Figure 7.32 with a semilog scale now used for the vertical mobility axis. The dotted curve in the figure is the previously fitted rational polynomial from Listing 7.23 and is the same curve as shown in Figure 7.30. The semilog plotting expands the smaller values of the mobility and shows that the dotted curve does not provide a very good fit at the small values of mobility. In fact the dotted curve shows a local peak in value at an x value of about -2.7 while the experimental data shows no such peak.

When data is plotted on such a semilog scale, relative errors show up as equal increments on the vertical scale. It is thus seen that the data fit previously obtained has much larger relative errors at the small values of mobility as opposed to the relative errors at the largest values of mobility. This is to be expected as discussed in the previous paragraphs when a unity weighting factor is used in the least squares data fitting.

The solid curve shows a fit to the experimental data to the same rational polynomial function but with a weighting factor as shown in Eq. (7.53), i.e. $1/y^2$ weighting. The code for this fitting is shown in Listing 7.28. This code is essentially identical to that of Listing 7.25 except for line 20 where the weighting factors are set to(1000/yd[i])^2. The use of the 1000 factor is simply for convenience to make the weighting factors closer to unity. The resulting rational polynomial for the fitted function has only small changes in the coefficients, but the resulting curve shows significant changes for the small mobility values. Visually it can be seen that the relative errors between the data points and the solid curve are much smaller for the low mobility values as compared with the dotted line in the figure.

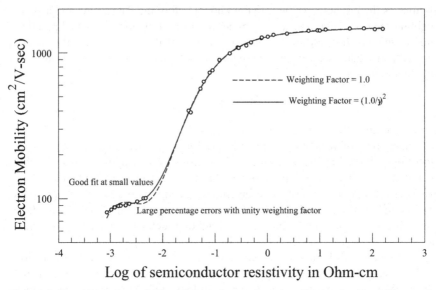

Figure 7.32. Electron mobility data and two function fits on log mobility scale. Dotted curve is for unity weighting factor. Solid curve is for relative error weighting factor.

It is also seen that the difference between the solid curve and the data points is approximately the same for all the data points, whereas the differences for the dotted curve are much larger at low mobility values. With the relative weighting factors, the nlstsq() routine also converged in only 5 iterations as opposed to requiring 12 iterations with equal weighting. The actual coefficient values are close for the two cases as can be seen by comparing the printed values in Listings 7.25 and 7.28. The coefficient with the largest change is the C_7 coefficient which from Listing 7.25 has the value 4.9736e-002 +/- 6.5789e-003 and from Listing 7.28 the value 3.2702e-002 +/- 4.5874e-003. It is interesting to compare the RMS error achieved by the unity weighting factor with that achieved with a weighting factor. From Listing 7.24 the RMS error is seen to be 12.35 while from Listing 7.28 the RMS error is 14.05. As might be expected, the unity weighting factor gives a slightly smaller RMS error which is heavily weighted by the largest values in the data. However, if the RMS value of the relative error were calculated, it would be expected that the fitting of Listing 7.28 would provide the smallest value. The use of a weighting factor in the least squares fitting approach is most important when values for a variable extend over a wide range of values.

This example illustrates several important features with polynomial data fitting. As previously mentioned polynomials tend to oscillate and have undesirable peaks and valleys. This is especially true with global polynomials but is also true with rational polynomials, although it is not as much a problem with rational polynomials. Whenever one tries to fit a function with a rapidly changing derivative, polynomials, even rational polynomials, tend to give peaks and valleys in the

```
 1 : -- /* File list7_28.lua */ Example of 1/y^2 weithting factor
 2 :
 3 : require"nlstsq"
 4 :
 5 : fft = function(yx,c) -- Final fitting function, Nonlinear in C's
 6 :     x = yx[2]
 7 :     return (c[1]+ c[2]*x+c[3]*x^2+c[4]*x^3)/(1+c[5]*x+
            c[6]*x^2+c[7]*x^3) - yx[1]
 8 : end
 9 : fft1 = function(yx,c) -- First approximation, Linear in C's
10 :     x,xy = yx[2],yx[3]*yx[2]
11 :     return (c[1]+ c[2]*x+c[3]*x^2+c[4]*x^3) - xy*(c[5]+
            c[6]*x+c[7]*x^2) - yx[1]
12 : end
13 :
14 : xd,yd={},{}
15 : nd = read_data('thurber.dat',yd,xd) -- Read data
16 : c = {0,0,0,0,0,0,0}; nc = #c
17 : actv,step = {},{}
18 : for i=1,nc do actv[i],step[i] = 1,0 end -- No limits on steps
19 : yx = {yd,xd,yd} -- Data values, pass y values as third argument
20 : fw = {}; for i=1,nd do fw[i] = (1000/yd[i])^2 end
21 :
22 : del,err,nmax = nlstsq(yx,fw,fft1,c,actv,step) -- #1 Linear
23 : print(err,nmax,'First Approximation') -- print results
24 : for i=1,nc do printf('c[%d] = %12.4e +/- %12.4e\n',i,c[i],
            del[i]) end
25 :
26 : for i=5,nc do step[i] = 1.2 end -- Set limits on steps
27 : del,err,nmax = nlstsq(yx,fw,fft,c,actv,step) -- #2 Nonlinear C's
28 : print(err,nmax,'Final Calculation') -- print results
29 : for i=1,nc do printf('c[%d] = %12.4e +/- %12.4e\n',i,c[i],
            del[i]) end
30 : xmin,xmax,nx = xd[1],xd[nd],1000
31 : dx = (xmax-xmin)/nx
32 : xx,yy={},{}
33 : x0 = 600 -- Initial guess at mobility
34 : for i=1,nx do
35 :     xx[i] = xmin+dx*(i-1)
36 :     yy[i] = fft({0,xx[i]},c) -- Required for nonlinear function
37 : end
38 : plot(xx,yy)
39 : write_data("list7_28.dat",xx,yy) -- Save data
Selected output:
14.050500172239         5         Final Calculation
c[1] =   1.2867e+003 +/-   6.6956e+000
c[2] =   1.4319e+003 +/-   1.4516e+001
c[3] =   5.4386e+002 +/-   9.6790e+000
c[4] =   6.7485e+001 +/-   1.8360e+000
c[5] =   9.3632e-001 +/-   1.2972e-002
c[6] =   3.8369e-001 +/-   6.9373e-003
c[7] =   3.2702e-002 +/-   4.5874e-003
```

Listing 7.28. Fitting of mobility data with non-unity weighting factor.

function as seen in this example. In this present example it is found that these oscillations can be eliminated using a relative error weighting factor. While the dotted line fit appears to provide a good approximate equation when viewed on a linear scale, the log scale in Figure 7.32 shows that this is not an adequate equation to describe the experimental data for small mobility values. If the dotted line fitting were used in subsequent calculations, it could easily lead to incorrect conclusions. If, for example, some physical effect was being explored for low mobility values, incorrect conclusions could possible be drawn because of the physically unreal negative slope of the solid curve for low mobility values. There is no physical reason for expecting such a negative slope as a function of the x parameter and in fact the data does not show such a slope change. One lesson to be learned from this example is that one should always examine very carefully an approximation to experimental data in all regions of the data to see if there are any unreal physical effects introduced as a result of the data fitting.

There are at least two other important cases where the use of a non-unity weighting factor is important. First, suppose one desires to force the fitting function to go through or very close to one (or more) data points. This can occur for example when one has a data set beginning at x,y = 0,0 and from physical reasons it is known that the data should be exactly 0 at x = 0. A straightforward least squares data fit may not pass exactly through the origin. A forced fit through the origin (or extremely close to the origin) can be implemented by setting a large weighting factor for the data point at the origin. Exactly how large a weighting factor to be used must be determined from some experimentation.

A second use of weighting factors is in handling outlier data points. These are data points which are obviously so far removed from other data points as to be considered non-valid data. Such points can occur from misreading instruments or from a large noise spike which just happened to occur as a data point was taken. For data with a large scatter, care must be used in removing such data points. However, for data points that are many standard deviations from a fitted curve, one sometimes wishes to eliminate such points from consideration. Such outliers can simply be removed from the data set. However, another technique is to simply retain the outlier data points and simply set the weighting factor to zero for such data points. This allows one to retain the data for plotting purposes, but to remove them from the data fitting.

7.10 Summary

This chapter has explored the general topic of fitting a smooth curve to data and with the general topic of plotting of data for use in publications. As opposed to the previous chapter on data interpolation, in the fitting of a curve to such data a curve is not desired to pass through each data point, because of random errors in measurement data, but a smooth curve is desired that passes as closely as possible to every data point and give a good representation to the data. The fitted function may come from a known mathematical model or the exact functional form of the data being represented may not be known. The general approach to modeling such data is through the use of the "least squares" data fitting technique. The major computer code developed in this chapter is a general nonlinear least squares fitting routine, known as nlstsq(). This is a very versatile code segment which only requires the user to define a fitting function in terms of a set of unknown coefficients and supply some initial guess at the fitting coefficients. Linear curve fitting equations are handled as special cases of the general nonlinear fitting approach.

A second major code segment developed is a simple datafit() routine for drawing a smooth curve through experimental data which is intended primarily for publishing graphs. This general routine requires very little input from the user, except for the y-x data to be fit to a smooth function. This approach builds upon the general interpolation approach developed in the previous chapter. This comes reasonably close to replacing the function of a draftsman with a set of French curves by a software routine.

Important code segments developed in this chapter are:
1. clinear() – code for fitting data with a linear least squares line.
2. fourier() – code for evaluating Fourier coefficients for a periodic function
3. ifourier() – code for inverse Fourier series analysis – converting form spectrum to time function.
4. nlstsq() – code for nonlinear least squares data fitting – major routine.
5. datafit() – code for simple data fitting with minimum user input.
6. elemfunc – Collection of transcendental functions defined using various polynomials.

Of these code segments, the nlstsq() routine is one of the most important and versatile in this book and the datafit() routine is one of the most useful for preparing data for publishing graphs.

The fitting of experimental data to a smooth function is an extremely important application area for computer based numerical methods. This chapter has concentrated primarily on obtaining a good fit to data either for data plotting or for subsequent use of some approximating function for further data analysis. Another very broad range of applications is estimating a set of parameters from a data model fitted to a set of experimental data. This is generally called "parameter estimation" or in some circles "parameter extraction". Chapter 9 is devoted to this closely related and very important subject. This is discussed after the next chapter develops some important statistical methods and statistical functions needed in the parameter estimation chapter.

8 Statistical Methods and Basic Statistical Functions

The previous chapter has addressed the general problem of fitting a set of data with a smooth function or curve where the data set has either some random component or is generated from some transcendental function that can not be expressed in a closed mathematical form. The major objective of the chapter was to obtain a continuous functional representation of the data set which could be used either for publication of the data or for further analysis of the data set. This chapter and the next chapter continues some of the discussions of Chapter 7 with an emphasis on the analysis of the statistical properties of data sets and on the fitting of data sets to physical models for the purpose of estimating various parameters of the model – sometimes called "parameter estimation". There are a broad range of such applications to data analysis and data fitting. In most data fitting applications, the interest in not only in obtaining a smooth curve which gives a good visual representation of the data, but also in determining how accurately various model parameters can be estimated from the experimental data. This is where the statistical properties of the data become important. This emphasis on confidence limits of the parameters associated with data fitting is what distinguishes the examples in this chapter and the next chapter from the previous chapter. The present chapter concentrates on developing various statistical methods that are useful in data analysis and the next chapter concentrates on parameter estimation for data in the presence of random noise using many of the statistical properties developed in this chapter

8.1 Introduction

The emphasis of this chapter can perhaps best be discussed with reference to some examples. Figure 8.1 shows two hypothetical data sets with accompanying noise. Also shown are two possible approximations to the data sets obtained for example by use of the least-squares data fitting of the previous chapter. Data Set 1 appears to be essentially constant with random variations about some mean value as represented by the solid line while Data Set 2 has some nonlinear relationship to the independent variable and is fitted by some smooth nonlinear solid line curve such as shown in the figure. Several important physical situations lead to data sets of type 1. This includes cases where the independent variable is time and the dependent variable might represent some physical parameter in a manufacturing process which must be held between some limits. As another example, it might represent various types of reliability measurements of a physical parameter where the inde-

J.R. Hauser, *Numerical Methods for Nonlinear Engineering Models*, 313–368.
© Springer Science + Business Media B.V. 2009

pendent variable is measurement number and the dependent variable is some measure of failure, such as failure time or yield strength. Type 2 data is more general in that there is apparently some underlying physical relationship between the dependent variable and the independent variable. The independent variable might be time and the dependent variable any of a wide range of time dependent parameters from drug response to force to voltage, etc. The independent variable might be temperature and the dependent variable resistance or thermocouple voltage or chemical reaction rate, etc.

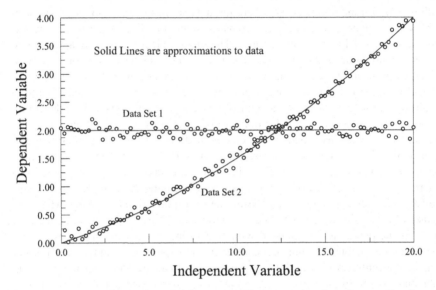

Figure 8.1. Types of experimental data sets of interest in engineering applications.

Regardless of the type of data and origin, some questions of usual interest are: (a) How well does some physical or mathematical model fit the data?, (b) How accurately can any parameters in some model be known?, (c) Does a particular model agree with the data more accurately than some another model?, and (d) What are the properties of the random variations about the model values? If the solid line values in Figure 8.1 are subtracted from the data values, the random variations around the model values will remain and this random noise will appear similar for both types of data sets. Thus it would be expected that methods to study the properties of the random variations would be similar for the two data sets. However the type 1 data set is simpler to analyze since there is only one model parameter and the present chapter will concentrate primarily on this type of data set. A large literature exists associated with the analysis of the mean and standard deviation of such data sets. This will also provide an opportunity to develop some statistical concepts and develop several statistical functions that will be of use in the more general case represented by type 2 data. The following chapter (Chapter 9) will be devoted primarily to an analysis of type 2 data sets.

The general area of parameter estimation and statistical analysis is very broad and many texts have been written devoted just to these topics. The purpose here is not to review or cover all these areas, but to illustrate the use of previously developed code segments to this important area of engineering. Very good reviews of statistical terms and parameter estimation techniques can be found on the web at the following locations:

(a) www.itl.nist.gov/div898/handbook; *NIST/SEMATECH e-Handbook of Statistical Methods*, and (b) http://www.curvefit.com; *Graphpad.com* home.

The reader is referred to these references for reviews of statistical topics and more general discussions of techniques for the analysis of the mean and standard deviation of data sets.

In order to discuss and evaluate the uncertainty of the parameters evaluated by data fitting, use will be made of some properties of statistics and some statistical functions. The next section provides some general statistical background and discusses code segments for some statistical functions to be used in the remainder of this chapter.

8.2 Basic Statistical Properties and Functions

Random or uncontrollable variables frequently contribute to engineering data taken to represent various physical parameters. Such random variables may be a natural consequence of uncontrollable initial conditions for an experiment or they may come from various physical sources such as random induced electromagnetic pickup in wires used in making electrical measurements. Space is alive with various forms of electromagnetic energy, some of it naturally occurring and much of it man-made from various radiation sources. Although some of the induced "noise" from such sources such as 60Hz electrical systems is in principle predicable, as far as most electrical measurements are concerned the noise appears random because of the random phase at which measurements are made. One must always be aware of the possibility of non-random variations in data and make tests for randomness. The next chapter will discuss some techniques for accessing the randomness of data sets.

It is frequently valuable to have software simulations of random events and random errors in data, so some means of generating random data variables is very useful. Computers and computer programs are deterministic machines so some special care must be exercised to generate data numbers from computers that appear to have the properties of random numbers. Much theory and thought has been given over many years to how a sequence of numbers that appear to be random can be generated by a deterministic program. Such computer generated sequences are frequently termed "quasi-random" as all such sequences must have a finite number of discrete values since all computer generated numbers have a finite number of digits of accuracy (about 14 digits for the presently used software).

Perhaps the most fundamental random number sequence is a series of uniform deviates which are just a series of random numbers which lie within a specified range (typically 0 to 1) with any one number within the range just as likely to occur as another number. Most computer languages supply such random number generators that typically generate random numbers between 0 and 1 with a uniform probability density (see math.random() for the present language). Such system supplied random number generators typically generate the sequence from the recurrence relationship (also known as a linear congruential generator):

$$n_{i+1} = an_i + c \bmod(m) = (an_i + c)\%m = \bmod(an_i + c, m) \tag{8.1}$$

where m is the modulus and a and c are positive integers called the multiplier and increment. This recursive relationship will generate numbers between 0 and m so the sequence will eventually repeat. With properly selected values the sequence will generate all possible integers between 0 and $m-1$. A sequence between 0 and 1 can then be obtained by dividing the results of Eq. (8.1) by the modulus value. For a good random number generator a large value of m is desired. However, many C language (and Lua) have implementations that use a rather small value of modulus, with the ANSI C standard requiring only that the modulus be at least 32767. Such a random number generator is available here with the call math.random() and will at best generate 32767 separate random values.

In addition to the built-in language random function, four other uniform random number generation code segments are provided in the prob.lua code segments supplied with this work. These are the ran0(), ran1(), ran2() and ran3() functions in the prob.lua file. These random number generator functions are based upon recommendations in the book *Numerical Recipes in C*, Cambridge Press, 1988. These are said to be better uniform random number generators than those normally provided in computer languages. It is usually good to have several different random number generators available so one can make sure no features of a problem are due to a specific random number generator function. The ran0() function is a portable language random number generator based upon Eq. (8.1) and code for this is shown in Listing 8.1. It is seen that the modulus is taken as 1771875 which provides over 1e6 unique values for the resulting random numbers. The returned sequence is normalized to the interval 0 to 1 by line 322 of the code.

```
  1 : -- /* File prob.lua */ -- basic probability and statistical
functions
-------

-------
320 : ran0 = function() -- Very simple random number generator
321 :    jran = mod(jran*ia+ic,im)
322 :    return jran/im
323 : end
324 : setfenv(ran0,{jran=885937,im=1771875,ia=2416,ic=374441,
         mod=math.fmod})
```

Listing 8.1. Portable random number generator ran0() as implemented in prob.lua.

The reader can explore the other random number generators supplied in the prob.lua file. The ran1() function is based upon three linear congruental generators of the type in Eq. (8.1). One generator is used for the most significant part of the output number, the second for the least significant part and the third for a shuffling routine that saves a series of generated numbers and randomly selects a third random number from the saved series of numbers. The repeat period for ran1() is for all practical purposes infinite and is thought to have no sensible sequential correlations between the random numbers.

The ran2() function uses a table of generated numbers from only one linear congruental generator but uses a table of shuffled values similar to the table of ran1(). For the programmed constants in the code, the maximum number of discrete values returned is 714025 and the repeat period is again essentially infinite. Finally ran3() is not based upon the linear congruental technique but uses a subtractive method for generating random numbers. The reader is referred to the above mentioned book for further details on this method. It's importance here is that it provides a different approach so this method can be substituted for the others if there is any suspected difficulty with the other techniques.

In general the user of random number generators should be very cautious about results of using such generators. What is random enough for one application may not be random enough for another application. However, by having several random number generators more than one can be tried to see if it changes any conclusions. Also most random number generators provide a means of changing an initial seed number so different deterministic sequences can be generated again to explore any effects on generated results. For the internal math.random() function the appropriate statement is math.randomseed(mseed). For the other supplied functions, the seed can be set by supplying a seed number to the functions as an argument value (ran2(mseed) for example).

Listing 8.2 illustrates the use of the random number functions for generating a series of uniformly distributed numbers between 0 and 1. The listing also illustrates the use of several additional code segments also present in the prob library of probability functions. The functions used in the listing aid in displaying the properties of a sequence of variable values with some random component – in this case a purely random quantity. Line 5 defines a random number generator to be used in the code. Options are shown for five uniformly distributed generators (math.random, ran0, ran1, ran2 and ran3).

The reader is encouraged to re-execute the code with the various random number generators and with various number of generated random variables (such as 1000 or 10000 or 100000) as indicated on line 6. The random numbers are generated on line 8 and stored in an x[] table of values.

The remainder of the code illustrates some important statistical functions and plots which will be useful in this chapter. The mean, standard deviation, variance and skewness coefficient of the data are calculated and printed on line 9 using the stats() function. These quantities are defined in the standard way by:

```
1  :  -- /* File list8_2.lua */ Test of random number generators
2  :
3  :  require"prob"
4  :
5  :  rand = ran0 -- or math.random, ran1, ran2, ran3 or gnormal
6  :  nmb = 400 -- or 1000 or 10000 or 100000
7  :
8  :  x = {}; for i=1,nmb do x[i] = rand() end
9  :  print('mean,std,variance,skew = ',stats(x))
10 :
11 :  xx,yy = makeODF(x)
12 :  write_data('list8_2a.dat',xx,yy); plot(xx,yy)
13 :
14 :  xx,yy = lag(x)
15 :  write_data('list8_2b.dat',xx,yy); scatterplot(xx,yy)
16 :
17 :  xx,yy = makeCDF(x)
18 :  write_data('list8_2c.dat',xx,yy); plot(xx,yy)
Output:
mean,std,variance,skew =        0.51594  0.28674  0.08222  -
0.001264000367204
```

Listing 8.2. Examples of uniform random number generation and generation of Distribution Function for a random variable.

$$\text{mean} = \overline{x} = \frac{1}{n}\sum_{i=1}^{n} x_i$$

$$\text{variance} = \sigma^2 = \frac{1}{n-1}\sum_{i=1}^{n}(x_i - \overline{x})^2$$

$$\text{standard deviation} = \sigma = \sqrt{\frac{1}{n-1}\sum_{i=1}^{n}(x_i - \overline{x})^2}$$

$$\text{Skewness coefficient} = \frac{1}{\sigma^3(n-1)}\sum_{i=1}^{n}(x_i - \overline{x})^3$$

(8.2)

These are useful for a first-order look at the characteristics of sets of random data. The mean is also the result one would get by performing a least squares fit to the data set with a constant value.

Line 11 of the listing uses the function makeODF() to produce a 'rank order function" or an "order distribution function" (ODF) from the random data. This function returns an independent variable (xx[] on line 11) containing an ordered array of the variable values and a dependent variable (yy[] on line 11) consisting of an integer value ranging from 1 to n giving the order rank of the variable values. Figure 8.2 shows a plot of the value returned on line 11 of the code from makeODF(). Ideally for a uniformly distributed variable, the ODF curve values should all fall on a straight line, such as the solid line running from 1 to n as the variable value ranges from 0 to 1. For 400 generated values the data points are seen to be close to the ideal line. The reader can re-execute the program with larger numbers of points, such as 10000 and observe that the data points become much closer to the ideal straight line. More data points were not included in the figure simply because it becomes impossible to see the individual data points with a large number of points.

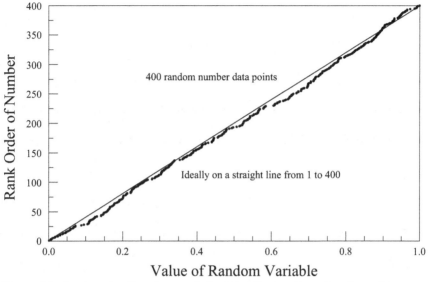

Figure 8.2. Rank order function or Order Distribution Function for uniform random variable. Y variable ranges from 1 to 400 (or 1 to n) for $0 < x < 1$.

The ODF function in Figure 8.2 is closely related to the "Cumulative Distribution Function" (CDF) (or just "Distribution Function") which gives the probability that a variable will be observed with a value less than or equal to a given value. If the data set considered, is a complete data set which means that it includes all possible values of the random variable, then simply dividing the ODF by the total number of samples gives the CDF. When the data set represents a sampling from a larger data set, such as in this example, the ODF curve can be used to approximate the CDF curve. The CDF function always varies between 0 and 1 for the possible range of values of the random variable. The ODF curve divided by the number of samples provides an estimate of the CDF for a given random variable. This holds in all cases regardless of the shape of the ODF curve. A simple way to obtain a CDF is to simply scale the order number of the ODF so that it ranges from 0 to 1 instead of from 1 to n. The simplest way to achieve this is to divide each value on the vertical axis in Figure 8.2 by the maximum value n. However, this is not quite correct when the sample set is not a complete set. If the data set is recalculated with a larger number of samples, there is a high probability of achieving some numbers less than the smallest number in an original sample data set and achieving some numbers larger than the largest number in an original sample data set. Thus one would not expect the sample random variables of a finite data set to correspond exactly to a probability of 0 on the low side or to a probability of exactly 1 on the high side. The values should be offset from 0 on the low side and 1 on the high side. There appears to be three major ways to convert a rank number into a CDF number and these are:

$$P(x_i) = \frac{i}{n+1} \qquad \text{Mean Rank}$$

$$P(x_i) = \frac{i-0.3}{n+0.4} \qquad \text{Median Rank} \qquad (8.3)$$

$$P(x_i) = \frac{i-0.5}{n} \qquad \text{Symmetrical Rank}$$

In all of these equations, i is the order variable going from 1 to n and n is the total number of variable values. All of these equations approach the same curve as n becomes large and there is little difference in the curves for n > 100.

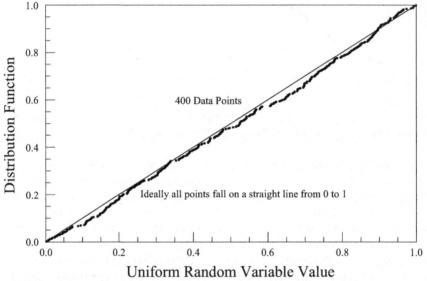

Figure 8.3. Distribution function for uniform random variable data.

With any one of these transformations an order distribution can be converted into a probability distribution and this is done on line 17 of the code using the makeCDF() function. The default method for this function is the median rank, but the other options can be selected by passing a second parameter to makeCDF() of 'mean' or 'sym'. Figure 8.3 shows the resulting distribution function generated by line 17 of Listing 8.2 for the 400 data points as shown in Figure 8.2. The curve looks essentially identical to Figure 8.2 except for the range of 0 to 1 on the vertical axis which now represents the probability that a value less than or equal to the horizontal axis value will be observed in the random data set. Ideally the curve is a straight line from 0 to one on both axes and the data points should all fall on the line. The reader can re-execute the code with larger numbers of data points and verify that the points fall closer and closer to the ideal line as the number of data points increase. The reader can also run the other random number generator functions and verify that all the functions give very good results when viewed on the gross scale of 0 to 1.

In probability theory, the probability density function is another important functional quantity. In theory for a continuous random variable this is related to the distribution function by the equation:

$$P(x) = \int_{-\infty}^{x} p(x')dx'$$
$$p(x) = \frac{dP}{dx} \tag{8.4}$$

where $P(x)$ is a distribution function and $p(x)$ is the corresponding probability density function. Thus the probability density function is the derivative of the distribution function and can be seen to be approximately a constant for the data shown in Figure 8.3. For a data set which takes on only discrete values, the distribution function, $P(x)$, values are a sum over the probability density, $p(x)$, values instead of the integration. In this work only random variables that are assumed to take on continuous values will be considered and for such cases, Eq. (8.4) is assumed to be valid.

The previous discussion has shown that the random number generators produce a set of random values between 0 and 1 with approximately uniform density. In many applications a random variable is desired that ranges from some minimum value x_{min} to some maximum value x_{max}. The appropriate scaling for such a range is then:

$$x = (x_{max} - x_{min})rand() + x_{min} \tag{8.5}$$

where $rand()$ is assumed to return numbers over the range 0 to 1.

Another highly desirable feature of a random number generator is that each call to the number generator should produce a value that is independent of the previously generated random number, i.e. there should be no correlation between numbers generated by two calls to the random number function. Such independence is frequently tested by use of a "lag" plot. This is a plot on the vertical axis of the y_i data value as a function of the y_{i-1} value on the horizontal axis. Such a lag plot will reveal any nearest neighbor correlations in data points. An example of this is the call to the lag() function on line 14 of the code. This is a very simple function that returns data for such a plot and the results of such a plot are generated by a pop-up scatter plot on line 15. A scatter plot simply shows the data points with no line drawn between the data points. Such a plot is shown in Figure 8.4 from the code in Listing 8.2.. This data was generated using 10000 random numbers to improve the statistics. Ideally the points should be uniformly scattered throughout the graph plane in the figure. The data for this example does in fact appear to have such a uniform density indicating that there is no visual correlation between successive calls to the random number generator. The reader is encouraged to re-execute the code with various combinations of data point numbers and various random number generators to verify that all four random number generators produce essentially equivalent results for large numbers of random variables.

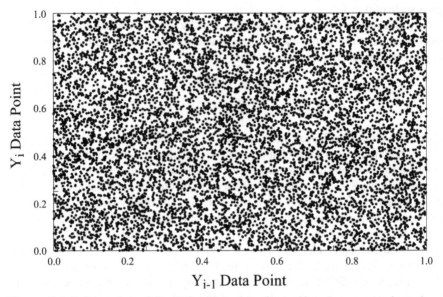

Figure 8.4 Scatter plot of lag() function data for uniformly generated random numbers. For this plot 10000 points were used.

In addition to uniformly distributed random numbers, the Gaussian distribution is also of classical and practical importance. The function gnormal() (also in the prob.lua file) can be used to generate such random numbers by setting rand = gnormal as on line 5 of Listing 8.2. This function returns a Gaussian distribution with zero mean and a unity standard deviation. For other values of these parameters, the equation:

$$x = std(gnormal() + mean) \qquad (8.6)$$

can be used where *mean* and *std* are the desired mean and standard deviation respectively. The gnormal() function generates Gaussian random numbers by first obtaining two uniformly distributed random numbers within a unit circle and then transforming this uniform distribution to the Gaussian distribution over all of space. For the details of the method, the reader is referred to the book <u>Numerical Recipes in C</u> and to the coding of the function in the prob.lua file.

Listing 8.3 illustrates the use of gnormal() to generate a series of Gaussian random numbers, in this case 10,000 random numbers as defined on line 6 of the listing. This also illustrates several other functions in the prob.lua file for use with random numbers. The random numbers are converted into a CDF by the makeCDF() call on line 11. Following this a theoretical Gaussian comparison CDF is generated by the normalCDF() call on line 12 of the listing. The experimental data is passed to the function, so that the normalCDF() function can generate a function with the same mean and standard deviation as the data set. In this way, a comparison theoretical Gaussian CDF can be generated for any desired random data set. Such a comparison to a Gaussian data set is typically done for many random data sets. The CDF data is saved on line 13 and the results are shown in Figure 8.5. In order to see some of the data points on the graph only every 100th data point is saved to the data file and plotted on the graph (see the use of the 100 first calling argument to the write_data() function on line 13). Visually it is seen that the generated random numbers for 10000 values closely approaches the ideal Gaussian CDF curve. A comparison like this can be made for any random data set by use of the normalCDF(x), where x is any random data set. The normalCDF() function will match the mean and standard deviation of the passed data set (x). The printed values of mean (0.01 instead of 0.0) and standard deviation (1.003 instead of 1.0) are close to what is expected.

```
1 : --/* File list8_3.lua */Test of gaussian random number generator
2 :
3 : require"prob"
4 :
5 : rand = gnormal -- Normal density function
6 : nmb = 10000 -- or 1000
7 : x = {}
8 : for i=1,nmb do x[i] = rand() end -- Generate random numbers
9 : print('mean,std,variance,skew = ',stats(x))
10 :
11 : xx,yy = makeCDF(x) -- Make CDF from random numbers
12 : xn,yn = normalCDF(x) -- Make normal CDF for comparison
13 : write_data(100,'list8_3a.dat',xx,yy,yn) -- Save every 100 point
14 : plot(xx,yy,yn)
15 :
16 : xl,yl = lag(x) -- Generate lag data
17 : write_data('list8_3b.dat',xl,yl) -- Save lag data
18 : scatterplot(xl,yl) -- Plot lag data
19 :
20 : xd,yd = hist(x,50,-4,4) -- histogram, 50 bins from -4 to +4
21 : xp,yp = histnorm(x,50,-4,4) -- Normal histogram for comparison
22 : write_data('list8_3c.dat',xd,yd,yp) -- Save histograms
23 : plot(xd,yd,yp) -- Plot histograms
Output:
mean,std,variance,skew =        0.01627 1.00311 1.00623 -0.02270
```

Listing 8.3. Generation of Gaussian random numbers, and histogram plots.

The code in Listing 8.3 also illustrates a lag plot on lines 16 through 18 for the Gaussian data. The lag plot for this data set is shown in Figures 8.6. It is readily seen that the lag data is clustered around a zero mean value with no evidence of any visual pattern in the data and this is consistent with an ideal Gaussian random variable.

Another important visual display of random data is by means of a histogram. For such a plot, the range of random variable is divided into a number of bins usually of equal width and the number of data values falling into each bin is then determined. The collected numbers of data points within each bin are then plotted as

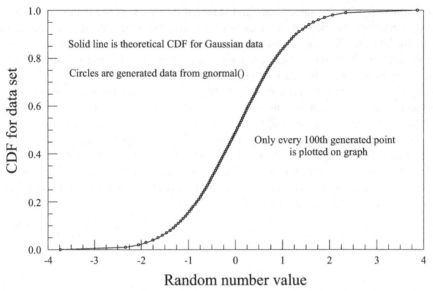

Figure 8.5. Comparison of CDF for gnormal() data with ideal Gaussian random data.

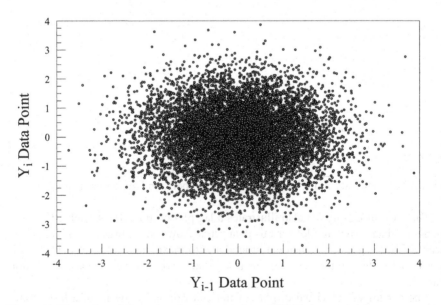

Figure 8.6 Lag plot of Gaussian random numbers (10000 data points).

a function of the mean value of the random variable within the bin. As the number of data points becomes very large and the size of the bin becomes small, the resulting histogram approaches the shape of the probability density function for the variable. Histogram data is calculated on lines 20 and 21 of the code using the hist() function and the results are plotted in Figure 8.7. The calling arguments for the hist() function on line 20 are the data table, the number of bins to use in the histogram and the min and max values of the variable range to consider. For the example, 50 bins are used in collecting the data between -4 and +4, with a bin size of 8/50 = 0.16.

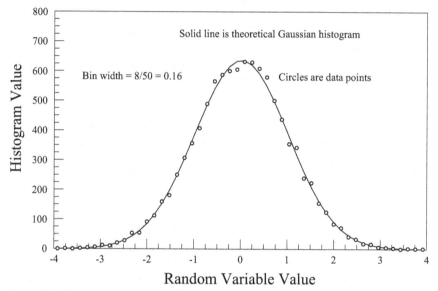

Figure 8.7 Histogram of gnormal() generated random numbers with comparison to theoretical Gaussian curve.

The generated random data, shown as data points in Figure 8.7, are seen to closely follow a theoretical Gaussian distribution for the total number of data points and for the bin size. The theoretical histnorm() function on line 21 of the code takes a random data set, (x) as input and generates a theoretical Gaussian histogram data set with the same mean and standard deviation as the input data set. In this manner, the histnorm() function can be used for generating a comparison histogram for any desired experimental random variable data set. The experimental histogram data set points are seen to agree closely with the theoretical expectation for the Gaussian random variable generated in Listing 8.3.

One of the problems with a histogram plot is that many data points and many bins are needed to get an accurate representation of the probability density function. For a limited number of data points – several hundred or fewer – it is probably better to concentrate on the CDF plot since all the data points can be meaningfully displayed on such a plot. The reader is encouraged to re-execute the code in Listing 8.3 with smaller numbers of data points such as 1000 and 100 and see the increased variation between the data and the exact Gaussian model especially in the histogram plots. Also the code can be executed with much larger data sets to more closely approach the theoretical Gaussian curve.

8.3 Distributions and More Distributions

The uniform distribution and the Gaussian distribution for random variables have been introduced in the previous section. Engineers and scientists have developed many additional probability distributions for specialized applications. Some of the major ones are presented in this section. This is by no means a complete listing of published distribution functions. The objective is to give a flavor for other distributions and probability density functions and to provide some of the functions that will be used to fit experimental data in subsequent sections of this chapter. Code segments for computing these distributions are included in the prob.lua file. Each of the distributions will now be defined and a brief description given of the primary uses of the function. For more complete information on the functions and uses, the reader is referred to the web. Simply type the distribution name and many references to each distribution will be obtained.

8.3.1. Normal Distribution for $-\infty \leq x \leq \infty$

$$Pn(x) = \frac{1}{\sqrt{2\pi}} \int_{-\infty}^{x} \exp(-t^2 / 2)dt \qquad (8.7)$$

The normal distribution is perhaps the most widely used distribution because of the central limit theorem which basically states that as the sample size (n) for a random variable becomes large, the sampling distribution of the mean becomes approximately a normal distribution regardless of the distribution of the original variable.

8.3.2. Uniform Distribution for $0 \leq x \leq 1$

$$Punif(x) = \begin{cases} x \text{ for } 0 \leq x \leq 1 \\ 0 \text{ Otherwise} \end{cases} \qquad (8.8)$$

One of the most important applications is the generation of uniformly distributed random numbers. Other random number distributions can be generated by appropriate transformations.

8.3.3. Cauchy Distribution for $-\infty \leq x \leq \infty$

$$Fcauchy(x) = 0.5 + \tan^{-1}(x) / \pi \qquad (8.9)$$

The Cauchy distribution looks very similar to a Gaussian distribution near the origin, but has much larger tails in the probability density for large and small values of the variable. In the physics community this distribution is frequently known as the Lorentz distribution.

8.3.4. t Distribution for $-\infty \leq x \leq \infty$

$$Ptdist(x,n) = 1 - 0.5Ibeta(n/(n+x^2), n/2, 1/2) \qquad (8.10)$$

where $Ibeta(y, a, b)$ is the incomplete beta function. The t distribution is used in many applications of finding the critical regions for hypothesis tests and in determining confidence intervals. The most common example is testing to see if data are consistent with the assumed process mean.

8.3.5. F Distribution for $0 \leq x \leq \infty$

$$Pfdist(x, n_1, n_2) = 1 - Ibeta(n_2 / (n_2 + n_1 x), n_2 / 2, n_1 / 2) \qquad (8.11)$$

where $Ibeta(y, a, b)$ is the incomplete beta function. The F distribution is used in many cases for the critical regions for hypothesis tests and in determining confidence intervals. Two common examples are the analysis of variance and the F test to determine if the variances of two populations are equal.

8.3.6. Chi-squared Distribution for $0 \leq x \leq \infty$

$$Pchisq(x, n) = Igamma(x / 2, n / 2) \qquad (8.12)$$

where $Igamma(y, a)$ is the incomplete gamma function. The chi-square distribution results when n independent variables with standard normal distributions are squared and summed. One example is to determine if the standard deviation of a population is equal to a pre-specified value.

8.3.7. Weibull Distribution for $0 \leq x \leq \infty$

$$Pweibull(x, \gamma) = 1 - \exp(-x^\gamma) \qquad (8.13)$$

The Weibull distribution is used extensively in reliability applications to model failure times.

8.3.8. Lognormal Distribution for $0 \leq x \leq \infty$

$$P \log normal(x, \sigma) = Pn(\log(x) / \sigma) \qquad (8.14)$$

where $Pn(x)$ is the normal distribution function defined by Eq. (8.7) above. The lognormal distribution is used extensively in reliability applications to model failure times. The lognormal and Weibull distributions are probably the most commonly used distributions in reliability applications.

8.3.9. Gamma Distribution for $0 \leq x \leq \infty$

$$Pgamma(x, \gamma) = Igamma(x, \gamma) \qquad (8.15)$$

where $Igamma(y, a)$ is the incomplete gamma function. This distribution can take on many shapes depending on the value of γ ranging from an exponential like probability density to a Gaussian like probability density.

8.3.10. Beta Distribution for $0 \leq x \leq 1$

$$Pbeta(x, p, q) = Ibeta(x, p, q) \qquad (8.16)$$

where $Ibeta(y, a, b)$ is the incomplete Beta function. This is another general distribution function that can take on many different shapes depending on the p and q parameters.

The names of the above distribution functions are given without the use of subscripts so that the names can match computer supplied subroutines for the functions. Several of the above distributions have been defined in terms of more elementary transcendental functions such as the incomplete Beta function and the incomplete Gamma function. These are defined as follows:

$$Ibeta(x,a,b) = \frac{1}{B(a,b)} \int_0^x t^{a-1}(1-t)^{b-1}\,dt$$

$$\text{for } (a,b > 0), (0 \le x \le 1)$$

$$B(a,b) = \int_0^1 t^{a-1}(1-t)^{b-1}\,dt \tag{8.17}$$

$$Igamma(x,a) = \frac{1}{\Gamma(a)} \int_0^x e^{-t} t^{a-1}\,dt$$

$$\Gamma(a) = \int_0^\infty e^{-t} t^{a-1}\,dt$$

Computer code for evaluating these math functions is discussed in Section 7.8.

The distribution functions have also been defined in terms of so-called "standard form" with either zero mean value or with the minimum value at zero. All of the functions can be shifted in value along the horizontal axis and scaled in value. For most of the functions this is achieved by the substitution:

$$x \to \frac{(x-\mu)}{\sigma} \tag{8.18}$$

where μ is a location parameter and σ is a scale parameter. For those distributions centered about zero such as the normal and Student's t distribution, this shifts the center point to μ and for those distributions beginning at zero such as the uniform, Weibull and F distributions, this shifts the origin to μ. In all cases, the range of the variable is changed by σ. For the two distributions ranging from 0 to 1, Punif() and Pbeta() the appropriate substitution is:

$$x \to \frac{(x-a)}{(b-a)} \tag{8.19}$$

in order to switch the limits to the range $a \le x \le b$.

Figure 8.8 shows three of the zero centered distributions in standard form and Figure 8.9 shows the corresponding probability density functions. It is seen that the Student's t function is very similar to the normal distribution function and the Cauchy function as previously mentioned is similar to the normal distribution but with a longer tail region.

Plots are shown in Figures 8.10 and 8.11 of the standard form of the other seven distributions for some selected values of the distribution parameters. It can be seen that this gallery of distribution functions provides a wide range of functional forms. The shifting and scaling of the x axis variable provides further variation as well as the flexibility provided by the various parameters of the distribution functions.

As an example of the type of functions possible with just one of the functions, Figure 8.12 shows various Weibull probability density plots for a range of Weibull shape parameters. The ability of this rather simple function to take on many different functional forms is one of the reasons it is frequently used for modeling of reliability data. The shape can range from an exponential like decrease with x to approximate a Gaussian for a shape parameter around 4 and an even more sharply

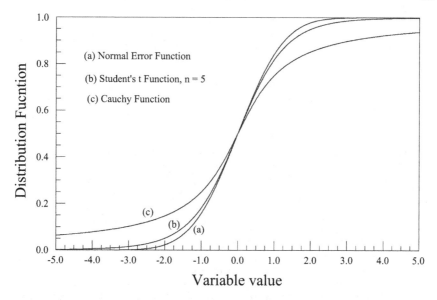

Figure 8.8 Three zero centered distribution functions.

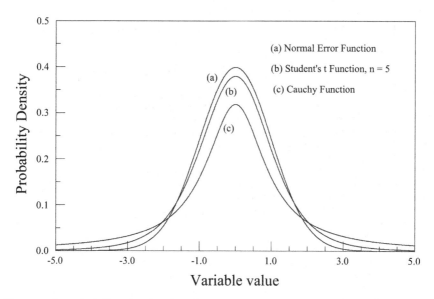

Figure 8.9 Probability density functions corresponding to Figure 8.8 distributions functions.

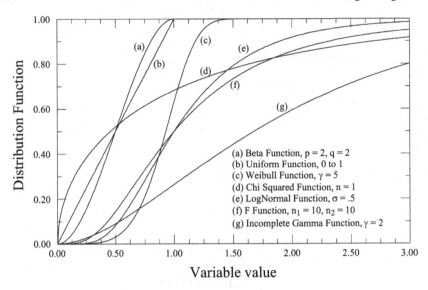

Figure 8.10. Illustration of various distribution functions with selected parameters. Values have been selected so functions have similar variable ranges.

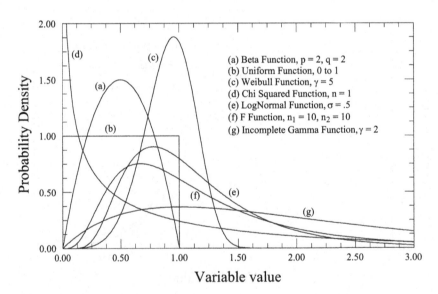

Figure 8.11. Probability density functions for distributions in Figure 8.10. Range is 0 to ∞, except of curves (a) and (b) which range over 0 to 1.

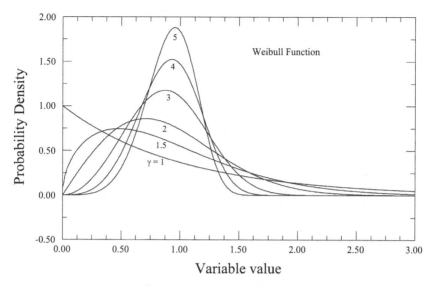

Figure 8.12. Various Weibull probability densities with a range of shape parameters. Shifting and scaling provides additional flexibility.

peaked function for larger shape parameters. The use of shifting and scaling parameters provides additional flexibility in the shape of the probability function or distribution function which would be the integral of the curves shown. The file prob.lua provides code segments for evaluating all the distribution functions discussed above. Functions are not provided for the probability density functions such as shown in Figure 8.12. However these can be evaluated from the distribution functions by use of the deriv() function as discussed in Chapter 5. Although it is useful to visualize the probability density functions, the distribution functions are more useful in the fitting of experimental data to probability functions.

An example of the use of the derivative function to evaluate the probability density function is shown in Listing 8.4 for the Weibull function. This code in fact generates and saves the data used in plotting Figure 8.12. This code also gives pop-up plots of both the distribution function and the density function over a range of function parameters which in this case is the shape parameter of the Weibull function. The reader can view families of other distributions and probability densities by modifying the code on lines 8 and 16 for any of the other distribution function defined in this section.

This section has presented an array of probability distribution functions. Several of these will be used in subsequent sections as examples of the statistical analysis for several types of data. For a more complete discussion of these and other functions the reader is referred to books on probability, to various web sites and to numerical texts such as *Numerical Recipes in C*, Cambridge Press, 1988.

```
 1 : -- /* File list8_4.lua */ -- generating prob density data
 2 :
 3 : require"prob"; require"deriv" -- rderiv for derivative at x = 0
 4 :
 5 : x,y,yd = {},{},{}
 6 : v = {1,1.5,2,3,4,5,6}; nv = #v -- v values for Weibull function
 7 : -- Probability density calculated from derivative
 8 : f = function(x) return Pweibull(x,vv) end - F() for derivative
 9 :
10 : for i=1,nv do y[i],yd[i] = {},{} end
11 : i = 1
12 : for xv=0,5,.01 do -- Step over x values
13 :    x[i] = xv
14 :    for j=1,nv do -- Step over v values
15 :        vv = v[j]
16 :        y[j][i] = Pweibull(xv,vv) -- distribution values
17 :        yd[j][i] = rderiv(f,xv) -- derivative values
18 :    end
19 :    i = i+1
20 : end
21 : plot(x,unpack(y))
22 : plot(x,unpack(yd))
23 : write_data('list8_4.dat',x,yd)
```

Listing 8.4. Illustration of evaluating the probability density function from derivative of a distribution function. Code also gives plots of Weibull distribution and
density functions.

8.4 Analysis of Mean and Variance

If some constant value is fitted to a data set such as in Figure 8.1 for Data Set 1, an
application of the least-squares technique will lead to the conclusion that the "best
fit" parameter is the mean of the data set. Thus the two most fundamental parameters of a set of measurements with noise or random variations are the sample mean
and the sample variance. Various tests and issues regarding these parameters are
discussed in this section. As examples of applications, several typical engineering
type data sets will be considered.

The first data set to be considered is a set of 95 measurements of the thickness
of pages of paper. The data set is contained in paper.txt and the Listing in 8.5
shows a basic analysis of the statistical properties of the data as well as the generation of four plots (4-Plot) for the measurement data. The 4-Plot approach provides
a detailed look at the data set and tests some of the assumptions that underlie most
statistical data analysis approaches. A 4-Plot consists of:

      1. Run sequence plot

      2. Lag plot

      3. Histogram

      4. Distribution plot

The 4-Plots are shown in Figures 8.13 through 8.16. The run sequence plot
(line 9) shows the order of the measurement set, i.e. the measured value vs. the

```
 1 : -- /* File list8_5.lua */  Analysis of data for paper thickness
 2 :
 3 : require"prob"
 4 :
 5 : x,y = {},{}
 6 : read_data('paper.txt',y)
 7 : nd = #y
 8 : for i=1,nd do x[i] = i end
 9 : plot(x,y)
10 : m,std,var,skew = stats(y)
11 : print('mean,std,variance,skew = ',m,std,var,skew)
12 : xx,yy = makeCDF(y)
13 : xn,yn = normalCDF(y)
14 : plot(xx,yy,yn)
15 : xl,yl = lag(y)
16 : scatterplot(xl,yl)
17 : xh,yh = hist(y,20)
18 : xph,yph = histnorm(y,20)
19 : plot(xh,yh,yph)
20 : write_data('list8_5a.dat',x,y,xx,yy,yn)
21 : write_data('list8_5b.dat',xl,yl)
22 : write_data('list8_5c.dat',xh,yh,yph)
Output:
mean,std,variance,skew = 0.07709  0.00377 1.42225e-005 -1.72885e-008
```

Listing 8.5. Generation of 4-Plot data for random data set of paper thickness measurements.

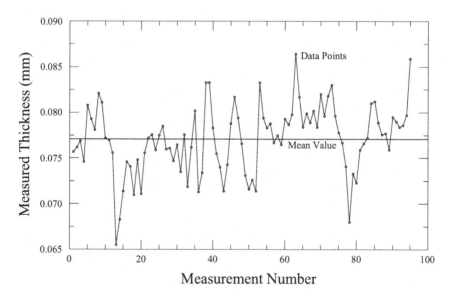

Figure 8.13. Run sequence plot. Measured value vs. Measurement number.

number of the measurement. This is used to indicate any drift in the measured data or any dependency on measurement number or on time if the data set is taken over time. For random data the run sequence should show random scatter about

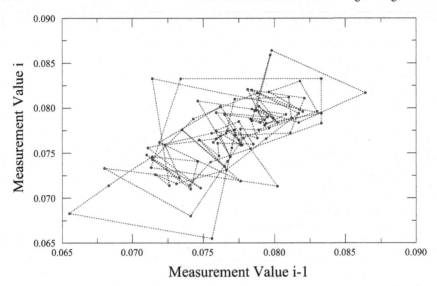

Figure 8.14. Lag Plot of Measurement value i vs. Measurement value i-1.

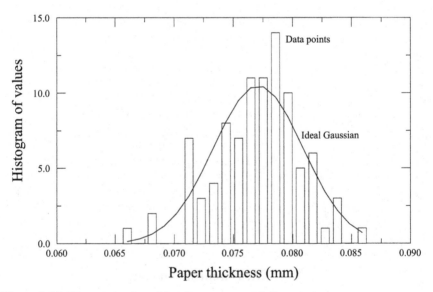

Figure 8.15. Histogram plot with reference normal Gaussian curve.

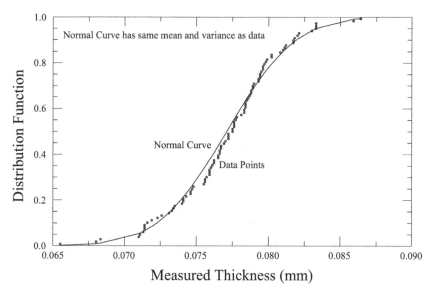

Figure 8.16. Distribution function for paper thickness data compared with normal Gaussian CDF.

the mean and the vertical spread in values should be approximately the same over the entire horizontal axis. The present data set has some questionable regions such the 10 to 30 measurement number region where almost all values are below the mean and the 55 to 75 region where almost all the measurements are above the mean. This may indicate some problems in the sequence of measurement values.

The data for the lag plot is generated on line 15 of the listing with the lag() function execution. This plot should be structure-less and completely random for a random data set. For the data in Figure 8.14 there appears to be some tendency for clustering along a 45 degree line indicating some correlation between adjacent data points. This is consistent with the long runs of negative and positive values in the run plot. This plot is primarily for observing any data correlations between adjacent measurement values.

A histogram plot is shown in Figure 8.15 and compared with a Normal curve with the same mean and standard deviation. This data is generated with two function calls to hist() and histnorm() on lines 17 and 18 of the code listing. A histogram plot is always a tradeoff between the desire to have a large number of bins and the need to have a large number of points per bin for good statistics. In the present example, a total of 20 bins were selected based upon some trial and error with several values. The comparison in Figure 8.15 with a normal Gaussian plot shows some skewness in the data with larger bin numbers above the ideal Gaussian curve and smaller bin numbers below the ideal Gaussian curve but with a longer tail on the low side. From the random nature of such data, one can not make too much of the differences in the histogram. Also the maximum number of samples in a bin near the peak is around 10 which does not provide good statistics for the data in Figure 8.15.

Finally the cumulative distribution data is shown in Figure 8.16 compared again with a normal Gaussian distribution. This data is generated with the two calls makeCDF() and normalCDF() on lines 12 and 13 of the code. This figure is consistent with the histogram curve with data points below the curve for thicknesses below the mean value and data points above the curve for thicknesses above the mean value. The agreement between the data and an ideal Gaussian is not especially good and the data appears to have a larger slope near the mean than does the reference Gaussian.

The 4-Plot as illustrated above can be used to examine the time independence of data and many of the assumptions about data independence and randomness assumed in most statistical analysis. In spite of some concerns about this data set, it will be further analyzed with conventional statistical techniques.

After looking at the character of the data set with the 4-Plot, the mean, and standard deviation (std) of this data set can be seen from the printed output of Listing 8.5 as:

sample mean = $\bar{x} = 0.07709$

sample standard deviation = s = 0.00377

These are called the sample mean and sample standard deviation, as they are estimates of the true population mean and true population standard deviation. It is assumed that the measurement set is not a complete set of possible measurements. The true mean and standard deviation will be denoted by μ and σ respectively.

Although these are printed with many digits from the computer calculation, the real accuracy is expected to be considerably less. The question to be addressed is what can be said about the accuracy of these computed parameters relative to the true mean and true standard deviation? One measure of the uncertainty of the mean is the so called "standard error" given by

$$stderr = s/\sqrt{n} \qquad (8.20)$$

where n is the number of data points. For this data set of 95 data points this is calculated as 0.00039. This estimate of the true mean with uncertainty would then be:

$$\mu = 0.07709 +/- 0.00039 \text{ or } 0.07670 < \mu < 0.07748. \qquad (8.21)$$

This simple standard error is rarely used now in practice, because there are more accurate ways of specifying one's knowledge of the uncertainty in the mean. This problem was first addressed by William Sealy Gosset in 1908 publishing under the pen name of "Student". He derived the statistics describing the mean when estimating the mean from a set of data and derived the t-distribution function or the more commonly called Student's t distribution. This is listed in the previous section as the t distribution (see Section 8.3.4). Rather than the t distribution, a more commonly used function is the following At() function:

$$At(t,n) = \frac{1}{n^{1/2}B(n/2,1/2)} \int_{-t}^{t} (1+x^2/n)^{-\frac{n+1}{2}} dx$$ (8.22)

$$= 1 - Ibeta(n/(n+t^2), n/2, 1/2)$$

where $B(a,b)$ is the Beta function as defined in Equation 8.17. As opposed to the t distribution, the At() function represents the area under the t probability function between $-t$ and $+t$. It represents the probability of observing a value of the variable described by the statistic, which in the present case is the mean, between the values of $-t$ and t. In this function a parameter n appears which might appear to be the number of samples. However, this "n" value, known as the "degrees of freedom" is to be interpreted as the number of samples minus the number of parameters to be estimated. For the statistics of the mean this degrees of freedom is the sample size $- 1$, i.e. n-1. When the number of degrees of freedom is large, the t distribution approaches the normal distribution. A comparison can be seen in Figures 8.8 and 8.9 between the normal Gaussian and the t distribution for 5 degrees of freedom. For n approaching 100 these is very little difference between the two distributions.

The parameter t in Student's distribution is:

$$t_s = \frac{(\bar{x} - \mu)}{(s/\sqrt{n})}$$ (8.23)

The t distribution function gives the probability that in a given data set with n degrees of freedom the observed mean value will deviate from the true mean by a certain amount. Since the distribution is symmetric, Eq. (8.23) can be written as:

$$\mu = \bar{x} \pm t_s \frac{s}{\sqrt{n}}$$ (8.24)

Now it remains to be determined as to what value of t_s is to be used in the equation? Well this depends on the degree of certainty one wish to place on the limits. If one wish to have a 99% probability that the mean falls within a given range, then a value of $t = t_s$ in the At() function of Eq. (8.22) must be found such that the resulting value of At() is 0.99. If a 90% probability of enclosing the true mean is desired then a value of t_s such that the integral equals 0.9 must be found.

In the application of Student's t distribution to sample means what is needed is not the direct integral of Eq. (8.22) but the inverse function – a table of t values for given values of the integral and number of degrees of freedom. Such tables of t values have been published in many probability books and reference books. A short listing of values is given here in Table 8.1 where probability is the desired value of the integral and the table values are the t_s values to achieve that level of probability. A table of values is fine for hand calculations but for computer calculations it is very desirable to have a software version of the table of t values. Such a function is provided in the prob.lua code segments as ttable(n, prob) where the returned value corresponds to the values in Table 8.1 for the specified number of degrees of freedom and the desired probability.

Degrees of Freedom	Probability			
	0.50	0.90	0.95	0.99
1	1.000	6.314	12.706	63.657
2	0.816	2.920	4.303	9.925
4	0.741	2.132	2.776	4.604
6	0.718	1.943	2.447	3.707
10	0.700	1.812	2.228	3.169
20	0.687	1.725	2.086	2.845
40	0.681	1.684	2.021	2.704
120	0.677	1.658	1.980	2.617
∞	0.674	1.645	1.960	2.576

Table 8.1 Short list of Student's t values.

Listing 8.6 shows a code segment for calculating the Student's t values such as in Table 8.1 using the previously defined t distribution function in Eq. (8.22). The code makes use of newton() and the Atstud() function to solve for the integral limit corresponding to a value of the integral. The reader can examine how this is done in the listing on lines 16 through 22 of the code. Executing the code in Listing 8.6 will generate the table values of Table 8.1 as the reader can verify. Actually the values for infinitely large degrees of freedom is approximated by n = 10000 which gives values very close to the infinity value. Student's t values can now be used to improve an estimate of the uncertainty in the mean value of the previous paper thickness data at various levels of confidence. The code on line 16 through 21 of Listing 8.6 is reproduced in the prob.lua file and the ttable() function can be called directly in subsequently coded examples.

```
 1 : -- /* File list8_6.lua */ -- Example of Student's t values
 2 :
 3 : require"prob"
 4 :
 5 : pbv = {.5,.9,.95,.99,.999} -- Define probabilities
 6 : n = {1,2,4,6,10,20,40,120,10000} -- Define DOF's
 7 : ts = {}
 8 : for i=1,#n do
 9 :    for j=1,#pbv do
10 :          ts[j] = ttable(n[i],pbv[j]) -- Get t values for table
11 :    end
12 :    print(ts[1],ts[2],ts[3],ts[4])
13 : end
14 :
15 :   -- Following code is actually in prob.lua
16 : ttable = function(n,pbv)
17 :    local Atf = function(x)
18 :          return Atstud(x,n) - pbv
19 :    end
20 :    return newton(Atf,0)
21 : end
```

Listing 8.6. Example of Student's t Table calculations.

```
 1 : -- /* File list8_7.lua */
 2 :
 3 : require"prob"
 4 :
 5 : x,y = {},{}
 6 : f = 1.e-3
 7 : ds = 'm'
 8 : read_data('paper.txt',y)
 9 : nd = #y
10 : for i=1,nd do y[i] = f*y[i] end
11 : m,std = stats(y)
12 : print('mean,std = ',engr(m,ds),engr(std,ds))
13 : stderr = std/math.sqrt(nd-1)
14 :
15 : pbv = {.6827,.9545,.9,.95,.99,.999}
16 : for i=1,#pbv do
17 :     ts = ttable(nd-1,pbv[i])
18 :     dm = ts*stderr
19 :     print(engr((m-dm),ds)..' < mean <'..engr((m+dm),ds)..' at
             '..100*pbv[i]..' percent confidence level')
20 : end
21 : print(engr((m-stderr),ds)..' < mean <'..engr((m+stderr),ds)..
             ' for standard error bounds ')
22 : print(engr((m-2*stderr),ds)..' < mean <'..engr((m+2*stderr),
             ds)..' for 2 standard error bounds ')
Selected Output:
76.70 µm < mean < 77.48 µm at 68.27 percent confidence level
76.31 µm < mean < 77.88 µm at 95.45 percent confidence level
```

Listing 8.7. Example of calculating bounds on sample mean at difference confidence values.

Lower Bound	Measured mean	Upper Bound	Confidence Level
76.70 µm	77.09 µm	77.48 µm	68.27%
76.31 µm	77.09 µm	77.88 µm	95.45%
76.45 µm	77.09 µm	77.74 µm	90.00%
76.32 µm	77.09 µm	77.87 µm	95.00%
76.07 µm	77.09 µm	78.12 µm	99.00%
75.77 µm	77.09 µm	78.42 µm	99.90%
76.70 µm	77.09 µm	77.48 µm	Standard Error
76.32 µm	77.09 µm	77.87 µm	2 * Standard Error

Table 8.2 Calculated bounds on the mean for several confidence levels. The paper thickness are expressed in engineering units – µm in this case.

The code segment in Listing 8.7 illustrates such a set of calculations and the results of the calculations are summarized in Table 8.2. In the listing bounds on the mean are calculated for confidence levels of 68.27%, 95.45%, 90%, 95% 99% and 99.9% as defined in the listing on line 15. The code then makes use of the ttable() function on line 17 to generate the corresponding Student's t values which are then used on line 19 to calculate the corresponding mean uncertainties. Two lines of output are shown with the remaining results shown in Table 8.2.

Several features can be seen from the results. First the bounds are fairly tight with the 99.9% confidence level representing about a 1.7% uncertainty in the sample mean. This means that out of 1000 such collections of sample data one would expect the sample mean to be outside the range of 76.07 μm to 78.12 μm only 1 time. This tight bound is due to the small deviation around the mean and the large sample size. From the values in Table 8.1 it can be seen that the bounds go up quite rapidly for small sample sizes. Another feature of the results can be seen by comparing the 68.27% confidence level with the "standard error" bounds. It is seen that the bounds for these two cases differ only in the fourth digit. Classically the standard error corresponds to a 1 sigma variation about the mean and would correspond to the 68.27% probability level. Also the 2 sigma variation classically corresponds to 95.45% of the values for a Gaussian distribution and the last line in Table 8.2 can be compared with the second line for the 95.45% confidence level from the t values. Again it is seen that these results agree to at least 3 digits. The Student's t approach is essential when the sample size is small but will give essentially the same results as the classical standard error results when the sample size is large as in this example.

The output from Listing 8.7 makes use of the engr() function (on lines 19, 21 and 22) to format the output for printing. This function returns a string for printing with the number represented by 4 digits of accuracy – generally considered engineering accuracy – and with the number expressed in powers of 1000. The powers of 1000 are expressed by standard engineering symbols of 'y', 'z',' a', 'f', 'p ,'n ,'μ ','m', ' ', 'k', 'M', 'G', 'T', 'P', 'E' ,'Z', 'Y' for numbers ranging from 1.e-24, 1.e-21, 1.e-18, … 1.e18, 1.e21,1.e24 respectively. This limiting of the number of digits and use of symbols for powers of 1000 makes the results in the table more readable. In addition, a string passed to the engr() function as a second argument can be used to print the dimensions of the number. In the present case the numbers are in meters so the symbol 'm' is passed to the engr() function as can be seen on line 19 for example. The use of engineering units causes the mean value of paper thickness to be printed as 77.09 μm for example.

Estimating the uncertainty in the sample variance or standard deviation is now considered. The statistics of the sample variance is described by the Chi-squared distribution – see Section 8.3.6. The Chi-squared cumulative distribution is given by the equation:

$$Pchisq(t,n) = \frac{1}{2^{n/2}\Gamma(n/2)} \int_0^t e^{-x/2} x^{n/2-1} dx \qquad (8.25)$$

where $\Gamma(a)$ is the normal gamma function. The parameter described by the Chi-squared distribution is:

$$t = \chi^2 = \frac{(n-1)s^2}{\sigma^2} \qquad (8.26)$$

where s^2 is the measurement sample variance and σ^2 is the variance of the total population. The Chi-squared distribution gives the probability of observing a sample value of the variable t given a sample with n-1 degrees of freedom. The equation provides a corresponding distribution function for the variance similar to the t distribution for the sample mean discussed above. In this case the distribution is asymmetric and always positive as the variance must be positive.

Bounds on the variance can then be obtained by finding appropriate values of t in Eq. (8.25). If the confidence level is expressed as $1 - \alpha$ for example, then one needs to find the lower bound by setting the integral equal to $\alpha / 2$ and to find an upper bound by setting the integral equal to $1 - \alpha / 2$. For a confidence level of 99%, the value of α would be .01 and the value of $\alpha / 2$ would be .005. From these two values, the bounds on the standard deviation can be expressed as:

$$\frac{\sqrt{(n-1)}}{\sqrt{t_{Chi-H}}} s < \sigma < \frac{\sqrt{(n-1)}}{\sqrt{t_{Chi-L}}} s \tag{8.27}$$

where t_{Chi-L} and t_{Chi-H} are the lower and upper limits found from Eq. (8.25) for the desired confidence level. Probability and reference books normally supply Chi-Square distribution tables with entries for typical confidence levels of 99, 95 or 99.9 percent and a range of degrees of freedom values. It is noted that the lower limit corresponds to the largest t value and the upper limit corresponds to the smallest t value.

A code listing for the calculation of upper and lower bounds on the standard deviation is shown in Listing 8.8. This is very similar to Listing 8.7 except for the call to tchisq() on line 17. This call returns two values corresponding to the high and low value of the Chi-Square distribution as discussed above. Code for the tchisq() function is not shown but can be viewed in the prob.lua file. It is similar to the code for the ttable() function in Listing 8.6 except for using the Chi-Square distribution and calling the function two times for the two limit values. The reader is encouraged to study the code in prob.lua for details. Calls to the tchisq() routine will generate Chi-Square table values similar to those published in many texts and reference books.

The calculated bounds on σ the population standard deviation at various confidence levels are shown in Table 8.3. Again the increased range that must be specified as the confidence level is increased can be readily seen in the table results. In this case a comparison of the 68.27% bounds from the Chi-Sq analysis shows a reasonable agreement with the standard error bounds. However the 95.45% confidence bounds show a larger difference when compared with the 2 standard error bounds. As the sample size is reduce the standard error bounds provides a less useful estimate of the bounds. For reference, the standard error for the standard deviation is evaluated as:

$$\text{stderr} = s / \sqrt{2n} \tag{8.28}$$

```
 1 : -- /* File list8_8.lua */ Example of prob limits on std
 2 :
 3 : require"prob"
 4 :
 5 : x,y = {},{}
 6 : f = 1.e-3
 7 : ds = 'm'
 8 : read_data('paper.txt',y)
 9 : nd = #y
10 : for i=1,nd do y[i] = f*y[i] end
11 : m,std = stats(y)
12 : print('mean,std = '..engr(m,ds)..engr(std,ds))
13 : stdf,fdfc = std*math.sqrt(nd-1), 1/math.sqrt(2*nd)
14 :
15 : pbv = {.6827,.9545,.9,.95,.99,.999}
16 : for i=1,#pbv do
17 :    tchL,tchH = tchisq(nd-1,pbv[i])
18 :    varL,varH = stdf/math.sqrt(tchL),stdf/math.sqrt(tchH)
19 :    print(engr(varH,ds)..' < std <'..engr(varL,ds)..
            ' at '..100*pbv[i]..' percent confidence level')
20 : end
21 : print(engr(std*(1-fdfc),ds)..' < std <'..engr(std*(1+fdfc),ds)..
            ' for standard error bounds ')
22 : print(engr(std*(1-2*fdfc),ds)..' < std <'..
            engr(std*(1+2*fdfc),ds)..' for 2 standard error bounds ')
Selected Output:
mean,std =   77.09 µm 3.771 µm
 3.524 µm < std < 4.079 µm at 68.27 percent confidence level
 3.292 µm < std < 4.414 µm at 95.45 percent confidence level
```

Listing 8.8. Code for calculating bounds on the sample standard deviation (std) for the paper thickness data. Note that the results are multiplied by 1000 to aid in viewing.

Lower Bound	Measured standard deviation	Upper Bound	Confidence Level
3.524 µm	3.771 µm	4.079 µm	68.27%
3.292 µm	3.771 µm	4.414 µm	95.45%
3.371 µm	3.771 µm	4.290 µm	90.00%
3.301 µm	3.771 µm	4.400 µm	95.00%
3.170 µm	3.771 µm	4.627 µm	99.00%
3.029 µm	3.771 µm	4.916 µm	99.90%
3.498 µm	3.771 µm	4.045 µm	Standard Error
3.224 µm	3.771 µm	4.318 µm	2*Standard Error

Table 8.3. Calculated bounds on the standard deviation for several confidence levels. The paper standard deviations is output in engineering units of µm.

Calculating bounds on both the mean and standard deviation require essentially that a distribution function which is known only as an integral value be inverted – in one case the t distribution and in the other case the Chi-squared distribution.

Since these two functions are so important, tables of these inverse functions are readily available. However, for other distributions, tables are not readily available, but it may still be important to know the inverse function value from a distribution in order to find the range of a variable over which the probability has a certain value. The inverse function for any probability distribution can be obtained in the same manner as demonstrated in the previous code segments for the inverse of the t and Chi-squared distributions. Figure 8.17 shows the CDF for the Incomplete Gamma function of Section 8.3.9 with a γ parameter of 10. This can represent a general distribution function for which one wishes to find a value of the x-axis random variable corresponding to a particular probability. For example to find the range of 90% probability of the variable about the mean value, one needs to find the values along the x-axis where the probability is 0.05 and 0.95. These limits are shown as horizontal dotted lines on the figure. For the 99% probability range one needs the values corresponding to the .01 and .99 values on the vertical axis. Typically the CDF value is known as a function of the x-axis value and in most cases only as an integral function that has been fitted to some experimental data.

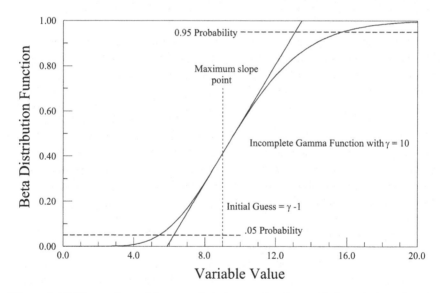

Figure 8.17 Illustration of inverse function evaluation with Incomplete Gamma Function.

Finding the x-values for given CDF values is an ideal application for Newton's method with a nonlinear function as has previously been shown for the t distribution and the chi-squared distribution. Listing 8.9 shows a general subroutine that can be used for such purpose. In this listing the example is the Incomplete Gamma distribution as shown in Figure 8.17. Lines 5 through 9 of the listing

```
 1 : -- /* File list8_9.lua */
 2 :
 3 : require"prob"
 4 : local Pigamma = elemfunc.Pigamma
 5 : x,y = {},{} -- Display Incomplete Gamma function
 6 : j,gam = 1,10
 7 : for i=0,22,.1 do x[j],y[j],j = i,Pigamma(i,gam),j+1 end
 8 : plot(x,y); write_data('list8_9.dat',x,y) -- plot &save data e
 9 :
10 : iCDF = function(CDF,pbv,xinit,...) -- General inverse CDF func-
tion
11 :    local arg = {...}
12 :    local iCDFx =  function(x) -- Function for Newton's method
13 :            return CDF(x,unpack(arg)) - pbv
14 :    end
15 :    return newton(iCDFx,xinit) -- Calculate inverse value
16 : end
17 :
18 : xinit = gam-1 -- Peak slope ocurs here
19 : print(iCDF(Pigamma,.05,xinit,gam))
20 : print(iCDF(Pigamma,.95,xinit,gam))
Output:
5.4254057744594          6        -1.923087660837e-011
15.705216429026          7        -6.7662855785605e-013
```

Listing 8.9 Code segment for an inverse CDF evaluation using Newton's method.

generate the data for Figure 8.17 and are not part of the inverse function which is iCDF() and is given on lines 10 through 16. The function requires three inputs, CDF the name of the distribution function, pbv the value of the probability for which the x-axis value is to be calculated and xinit an initial guess at the solution value. Any number of optional parameters can be passed to the distribution function through the use of the '...' argument in iCDF() which is converted into an arg[] table on line 11 and then the corresponding unpack(arg) term in the call to the CDF() function. All the previously discussed distributions require at most two parameters but the function is just as easily written for any arbitrary number of CDF parameters. Examples of using the inverse CDF function are on lines 19 and 20 where the 0.05 and 0.95 probability points are evaluated and printed for the example Incomplete Gamma function. The printed output gives values of 5.43 and 15.70 which agree with what can be estimate from Figure 8.17. It is also seen that 6 and 7 Newton iterations were required for the two solution values.

The only real difficulty in using the iCDF() function is in specifying a good initial guess for the newton() function. This is always a problem with a Newton's method of solution. From the graph in Figure 8.17 it can be seen that if an initial guess is too small or too large a linearization of the CDF function will give a small slope and the first iteration of Newton's method will put the next iterative solution value very far from the final solution value. In fact an initial guess of below about 4 or above about 15 will possibly lead to serious trouble with Newton's method. An ideal initial guess is the vertical dotted line in the figure labeled as the maximum slope point. The first application of Newton's method will then give the intersection of the maximum slope line with one of the horizontal dotted lines and convergence will occur very rapidly. Fortunately the point of maximum slope is relatively easy to obtain – at least for the standard distribution functions. The maximum slope point occurs at the peak in the density function and this in turn is typically called the "mode" of the distribution. This is obtained from setting the derivative of the density function to zero and is known for all the previously discussed distributions. For the incomplete gamma function this value is given by $\gamma - 1$ as indicated on the graph and as used in the code listing. For the t and the Chi-Squared distributions, the mode value is 0 and n-2 respectively and these values are used as initial guesses in the ttable() and tchisq() functions previously used for the inverse t and Chi-Squared functions. The iCDF() code segment on lines 10 through 16 is included in the prob.lua file so it can be used in other applications.

A fairly detailed analysis of the mean and variance of the paper thickness data set has been presented. A second example will now be presented on mean and variance evaluation. The second data set stored in file steel_yield.txt is a set of 50 measurements of the yield strength of steel. The data values are in kpsi units.

Listing 8.10 shows an analysis of this data set and serves as a prototype analysis program for any similar data set. This code segment is a composite of the previous Listings 8.5, 8.7 and 8.8. The code performs the 4-Plot analysis with 4 popup plots on lines 9 through 16. This is followed by the bounds analysis on the mean for several confidence levels on lines 18 through 27. Finally the bounds analysis of the standard deviation is done on lines 29 through 37 of the code listing. Such an analysis provides a fairly complete look at a random data set. No files are saved in the listing but the reader can add such statements to the code to make a more complete analysis tool. These are omitted in the interest of shortened the code. An additional more general feature would be to probe the user to input a file name for use in the analysis. The read_data() statement on line 6 can be changed to input any desired data file. The reader is encouraged to apply the program and analysis to a variety of data sets.

Figure 8.18 shows one of the 4-Plots, that compares the experimental CDF with that of a normal Gaussian with the same mean and same standard deviation. The data points in the graph show reasonably good agreement with a Gaussian curve. The other 3 plots are not shown but the reader can generate these by executing the code in Listing 8.10. Although not shown, the run data shows no indication of measurement-to-measurement correlations and this is further verified by a very random scatter of data points in the lag plot. Finally, there are insufficient data points to obtain a good histogram, but the histogram data is consistent at first look with a Gaussian distribution of data. The 99.9% confidence limits represent about a 2.25% variation from the mean value and with 99.9% confidence it would be stated that the mean yield strength is between 29.45 kPSI and 30.81 kPSI as indicated by the selected output in Listing 8.10.

While the analysis presented here can be performed by hand using Student's t tables and Chi-Squared tables, there is some advantage to being able to perform the analysis completely by computer. First, one is less likely to make errors from reading values from a table and second one can analyze many more data sets and

```
 1 : -- /* File list8_10.lua */
 2 :
 3 : require"prob"
 4 :
 5 : x,y = {},{}
 6 : read_data('steel_yield.txt',y)    -- Change for any desired file
 7 : nd = #y; ds = 'kpsi'

 8 : -- Do 4-Plots
 9 : for i=1,nd do x[i] = i end -- Generate run number
10 : plot(x,y) -- Run plot
11 : xx,yy = makeCDF(y) -- Generate distribution data
12 : xn,yn = normalCDF(y)
13 : plot(xx,yy,yn); write_data('list8_10.dat',xx,yy,yn) - Dist
14 : scatterplot(lag(y)) -- Lag plot
15 : xh,yh = hist(y,15); xph,yph = histnorm(y,15)
16 : plot(xh,yh,yph) -- Histogram plot

17 : -- Do mean analysis with bounds
18 : m,std = stats(y)
19 : print('number of data points = ',nd); print('mean, std = ',
         engr(m,ds),engr(std,ds))
20 : stderr = std/math.sqrt(nd-1)
21 : probv = {.6827,.9545,.9,.95,.99,.999}
22 : for i=1,#probv do
23 :     ts = ttable(nd-1,probv[i]); dm = ts*stderr
24 :     print(engr(m-dm,ds)..' < mean <'..engr(m+dm,ds)..' at'..
            engr(100*probv[i])..'percent confidence level')
25 : end
26 : print(engr(m-stderr,ds)..' < mean <'..engr(m+stderr,ds)..
        ' for standard error bounds ')
27 : print(engr(m-2*stderr,ds)..' < mean <'..engr(m+2*stderr,ds)..
        ' for 2 standard error bounds ')

28 : -- Do std analysis with bounds
29 : stdf = std*math.sqrt(nd-1)
30 : for i=1,#probv do
31 :     tchL,tchH = tchisq(nd-1,probv[i])
32 :     varL,varH = stdf/math.sqrt(tchL),stdf/math.sqrt(tchH)
33 :     print(engr(varH,ds)..' <   std <'..engr(varL,ds)..' at'..
            engr(100*probv[i])..'percent confidence level')
34 : end
35 : fdf = 1/math.sqrt(2*nd)
36 : print(engr(std*(1-fdf),ds)..' <   std <'..
        engr(std*(1+fdf),ds)..' for standard error bounds ')
37 : print(engr(std*(1-2*fdf),ds)..' <   std <'..
        engr(std*(1+2*fdf),ds)..' for 2 standard error bounds ')
Selected Output:
mean, variance, std, skew =  30.13 kpsi 1.844 1.358 kpsi     24.59 m
29.81  kpsi < mean < 30.46 kpsi at 90.00  percent confidence level
29.45  kpsi < mean < 30.81 kpsi at 99.90  percent confidence level
29.94  kpsi < mean < 30.33 kpsi for standard error bounds
1.167  kpsi < std  < 1.632 kpsi at 90.00  percent confidence level
1.012  kpsi < std  < 1.991 kpsi at 99.90  percent confidence level
```

Listing 8.10. Composite 4-Plot and bounds analysis for example of steel yield strength data set.

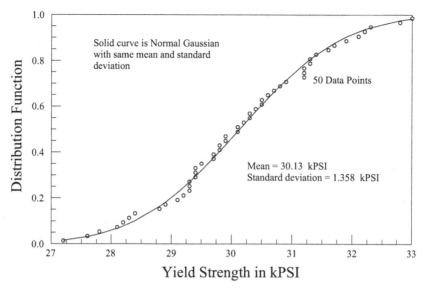

Figure 8.18 Comparison of CDF for Steel yield strength with Gaussian distribution.

do it generally in less time. With tabular data the needed inverse values are only known in a functional way from the specified limits on integrals or it is only specified in a transcendental way. This is where Newton's method can be brought to bear very effectively. The code needed to generate equivalent data to the normally used tables from references can be put together very easily and readily as the examples illustrate from code segments developed in previous chapters. This again emphasizes the need to develop modular code segments and functions which can be integrated into a variety of nonlinear problems.

8.5 Comparing Distribution Functions – The Kolmogorov-Smirnov Test

A comparison has been shown in Figures 8.16 and 8.18 of an experimental cumulative distribution function (CDF) with a theoretical CDF for a Gaussian distribution with the same mean and standard deviation. The question to be addressed in this section is with what confidence level can one assert the fact that the two distributions are the same? Put another way, can the observed difference between the two distributions be expected to occur by chance for the number of sample observations or does the difference imply a real difference in the two distributions? The paper thickness distribution of Figure 8.16 is shown again as an example set of

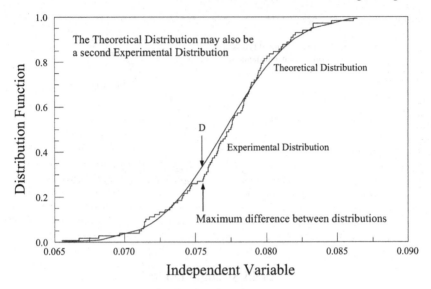

Figure 8.19 Comparing two distribution functions.

distributions in Figure 8.19 in order to illustrate the comparison between an experimental distribution and a theoretical distribution. However, the same comparison could just as easily be made between two experimental distributions.

There are many statistics that could be used to characterize the difference between two distributions. One could take the area between the distributions for example, or the sum of the difference squared as characterizing the difference. The Kolmogorov-Smirnov (K-S) technique to be discussed here takes a simple measure which is the maximum value of the difference between the two distributions. This is identified as D in Figure 8.19 and is defined as:

$$D = \max \left| CDF_1(x) - CDF_2(x) \right|_{-\infty < x < \infty} \tag{8.29}$$

where CDF_1 and CDF_2 are the two distribution functions. What makes the K-S statistic very useful is that its distribution in the case of the null hypothesis can be calculated, at least to an approximation. The significance of any observed nonzero value of D can thus be calculated. The important function entering into the calculation of the significance is the Kolmogorov-Smirnov function expressed as:

$$Q_{KS}(x) = 2\sum_{i=1}^{\infty} (-1)^{i-1} e^{-2i^2 x^2} . \tag{8.30}$$

This is a monotonic function with the limiting values:

$$Q_{KS}(0) = 1 ; \quad Q_{KS}(\infty) = 0 . \tag{8.31}$$

In terms of this function, the significance level of an observed value of D (as a disproof of the null hypothesis that the distributions are the same) is given by the formula:

$$\text{Probability}(D > D_O) = Q_{KS}(\sqrt{N}D_O) \tag{8.32}$$

where D_O is the observed value of D and N is the number of data points. This applies when comparing a distribution to a theoretical distribution. When comparing two experimental distributions the appropriate equation is the same expect for the replacement of N by a weighted average of the two samples as follows:

$$N \rightarrow \frac{N_1 N_2}{N_1 + N_2} \tag{8.33}$$

where N_1 and N_2 are the number of samples in each distribution.

Figure 8.20 shows a plot of the K-S probability function calculated with the Qks() function in prob.lua. The important transition region occurs over the range of x values from about 0.5 to 1.5. In applying this theory to the question of whether two distributions are equal or different, values in the tail of the distribution for x > 1 are usually of most importance.

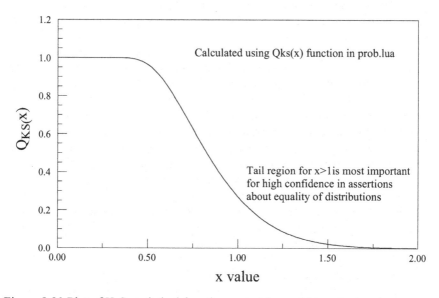

Figure 8.20 Plot of K-S statistical function. x > 1 is most important region.

The K-S statistic is used in proving or disproving an assertion about the equality of the two distributions with a certain probability. For example consider the assertion that a certain theoretical distribution fits the experimental data. The K-S test statistic as given by $\sqrt{N}D$ is calculated. This number is then used in the

Qks() function to calculate the probability that such a large value could occur purely by chance if the two distributions are in fact equal. If this Qks() number is very small – say .0.01 – then there is only a 1% probability that the difference occurs by chance. One can then assert that the two distributions are different at the 99% confidence level. When the K-S function predicts a high probability that the difference could occur by chance, one would then accept the hypothesis that the distributions are the same as far as the given experimental data can determine.

Listing 8.11 shows some code segments for the analysis of two distributions by the K-S algorithm. The analysis is packaged in a code routine called KS_test() beginning on line 11 and extending to line 49 of the code. The routine accepts as input either (a) two distributions with the values in arrays or (b) one distribution as a table of values for the first argument and the name of a theoretical distribution for comparison as the second argument. If a function name is passed as the second argument, a series of parameters for the theoretical distribution can also be passed in the argument list with any number of additional arguments as needed to be passed to the function describing the theoretical distribution. An example of this use is given on line 52 where the KS_test() function is called with the paper thickness data file and with a Gaussian theoretical distribution function for comparison. The Gaussian is defined on lines 8 through 10 of the code and includes provisions for a nonzero mean and non-unity standard deviation. This is the reason the Pn() function can not be passed directly to the KS_test() function but is modified in the Pnms() function to allow for any desired mean and standard deviation (std).

When a function is passed to the routine, the D parameter is evaluated on lines 16 through 21 of the code and when two distributions are passed D is evaluated in lines 22 through 34 by stepping through both arrays and finding the largest difference in the arrays. When the second distribution is a table of values, the D parameter is evaluated by lines 22 through 33 of the code. Common code is then used on lines 35 through 48 to perform the KS analysis of the data sets. The KS_test() code segment is also included in the prob.lua file so it can be easily called by other programs when the prob file is included in any code segment. Note on line 44 the use of the iCDF() function to find the inverse value of the K-S function (the Qks function) as represented by Figure 8.20.

On lines 52, 53 and 54 three calls are made to the KS_test() function. The first of these compares the paper thickness data to a Gaussian distribution with the same mean and std as the data set. The second call compares the data to a Gaussian with a 5% larger mean and the third call compares the data with a Gaussian with a 50% larger std. Executing the code in Listing 8.11 produces the output shown in Listing 8.12. The question addressed by the K-S test is the null hypothesis listed in the output that the "Distribution Fits the Data". For two data sets the equivalent hypothesis is "The Two Data Sets are Equivalent". The alternative hypothesis is that the "Distribution does not Fit the Data". The printed output shows the KS parameters D and $\sqrt{ND}$. This is followed by a table of confidence levels

```
 1 : -- /* File list8_11.lua */ -- KS test
 3 : require"prob"; require"elemfunc"
 4 : local Pn = elemfunc.Pn
 6 : paper = {}; read_data('paper.txt',paper) -- Data set
 7 :
 8 : Pnms = function(x,mean,std) -- Gaussian definition for tests
 9 :    return Pn((x-mean)/std)
10 : end
11 : KS_test = function(data1,data2,...) -- with possible arguments
12 :    local arg,nd1,nd2,imax = {...}
13 :    local x1,x2,d1,d2
14 :    local D,dt,dt1,dt2=0,0,0,0
15 :    x1,d1 = makeCDF(data1) -- Make distribuition function
16 :    if type(data2)=='function' then -- Compar data to function()
17 :       nd = #data1
18 :       for i=1,nd do -- Step over data set
19 :          dt = math.abs(d1[i] - data2(x1[i],unpack(arg)))
20 :          if dt>D then D,imax = dt,i end
21 :       end
22 :    else -- Comparison of two distributions in table form
23 :       nd1,nd2 = #data1,#data2
24 :       nd = nd1*nd2/(nd1+nd2)
25 :       x2,d2 = makeCDF(data2)
26 :       local j1,j2,xv1,xv2 = 1,1
27 :       while j1<=nd1 and j2<=nd2 do -- Step over both data sets
28 :          xv1,xv2 = x1[j1],x2[j2]
29 :          if xv1<=xv2 then dt1,j1 = d1[j1],j1+1 end -- next step
30 :          if xv2<=xv1 then dt2,j2 = d2[j2],j2+1 end -- data2
31 :          dt = math.abs(dt1-dt2)
32 :          if dt>D then D,imax = dt,j1 end
33 :       end
34 :    end
35 :    local kstest = D*math.sqrt(nd)
36 :    print('\n\t\t Kolmogorov-Smirnov Goodness-of-Fit Test\n')
37 :    print('Null Hypothesis HO:      Distribution Fits the Data')
38 :    print('Alternate Hypothesis HA:     Distribution does not
         Fit the Data')
39 :    print('Number of observations = ',engr(nd))
40 :    print('\nK-S test statistic: D, sqrt(N)*D = '..
         engr(D)..','..engr(kstest)..'(Test value)')
41 :    print('\nConfidence Level\tCutoff\t\tConclusion')
42 :    prob = {.1,.05,.01,.001}
43 :    for j=1,#prob do
44 :       lv = iCDF(Qks,prob[j],1)
45 :       if lv>kstest then print('',engr((1-prob[j])*100,'%'),
            '',engr(lv),'\t Accept HO')
46 :       else print('',engr((1-prob[j])*100,'%'),'',engr(lv),
            '\t Reject HO') end
47 :    end
48 :    print('Accept Above Confidence Level of ',
         engr((1-Qks(kstest))*100,'%\n'))
49 : end
50 :
51 : mean,std = stats(paper)
52 : KS_test(paper,Pnms,mean,std) -- Call with comparison equation
53 : KS_test(paper,Pnms,1.05*mean,std)
54 : KS_test(paper,Pnms,mean,1.5*std)
```

Listing 8.11. Code segments for Kolmogorov-Smirnov Test of two distributions.

and a conclusion regarding accepting the HO hypothesis or rejecting the HO hypothesis.

```
                 Kolmogorov-Smirnov Goodness-of-Fit Test

Null Hypothesis HO:          Distribution Fits the Data
Alternate Hypothesis HA:     Distribution does not Fit the Data
Number of observations =     95.00

K-S test statistic: D, sqrt(N)*D =  66.91 m,  652.1 m(Test value)

Confidence Level      Cutoff           Conclusion
       90.00  %              1.224                      Accept HO
       95.00  %              1.358                      Accept HO
       99.00  %              1.628                      Accept HO
       99.90  %              1.949                      Accept HO
Accept Above Confidence Level of        21.13  %

                 Kolmogorov-Smirnov Goodness-of-Fit Test

Null Hypothesis HO:          Distribution Fits the Data
Alternate Hypothesis HA:     Distribution does not Fit the Data
Number of observations =     95.00

K-S test statistic: D, sqrt(N)*D =  424.0 m,  4.132   (Test value)

Confidence Level      Cutoff           Conclusion
       90.00  %              1.224                      Reject HO
       95.00  %              1.358                      Reject HO
       99.00  %              1.628                      Reject HO
       99.90  %              1.949                      Reject HO
Accept Above Confidence Level of       100.00  %

                 Kolmogorov-Smirnov Goodness-of-Fit Test

Null Hypothesis HO:          Distribution Fits the Data
Alternate Hypothesis HA:     Distribution does not Fit the Data
Number of observations =     95.00

K-S test statistic: D, sqrt(N)*D =  126.9 m,  1.237   (Test value)

Confidence Level      Cutoff           Conclusion
       90.00  %              1.224                      Reject HO
       95.00  %              1.358                      Accept HO
       99.00  %              1.628                      Accept HO
       99.90  %              1.949                      Accept HO
Accept Above Confidence Level of        90.62  %
```

Listing 8.12. Example output from execution the code in Listing 8.12.

For the first comparison, the two distributions being compared can be seen in Figure 8.19. For this comparison, the conclusion of the KS analysis is "Accept HO" i.e. that one has no reason not to accept the Gaussian fits to the data set at all the listed confidence levels. The last line reading "Accept Above Confidence Level of 21.13% means that there is a 78.87% probability (100%-21.13%) of observing a difference between the theory and the model as large or larger than that actually observed. Thus assuming the Gaussian is the true distribution one would expect to observe the measured difference in distributions or a larger difference about 79% of the time. Since this is a large percentage, the postulated Gaussian distribution can not be eliminated.

When the mean of the theoretical distribution is increased by about 5% on line 53 and the test re-executed on line 53 of the code, the results are very different. In this case the conclusion is to reject the HO hypothesis that the Gaussian distribution fits the data. This conclusion is rejected at all the confidence levels specified as high as 99.9% confidence. This means that less than 1 time out of 1000 would a difference in the distributions as large as that actually observed be expected. This can be readily seen by taking the value of $\sqrt{N}D = 4.132$ for this case and looking at the KS distribution in Figure 8.20 for this value of x. For an x-axis value this large it is seen that the function is essentially zero indicating an essentially zero probability of observing such a large KS parameter at random. Thus we can be 99.9% sure that the Gaussian distribution with an increased mean of 5% does not fit the data. This also agrees with our previous Student's t analysis and the results in Table 8.2 which gives the 99.9% confidence limits on the mean as about 0.5% from the mean.

The third comparison with an increased std of 50% shows mixed results as the third series of outputs shows in Listing 8.12. The HO hypothesis would be rejected at the 90% confidence level, but accepted at essentially all higher confidence levels. This is due to the fact that a difference as large as the obtained D value would be expected by chance about 10% of the time and a smaller value would be observed about 90% of the time. Thus one can not be sure with above 90% confidence that the two distributions are different so one must accept the possibility that the model actually describes the underlying distribution from which the measured data was obtained.

Listing 8.13 shows code for comparing two experimental distributions and the accompanying output produced by the code. The distributions being compared are shown in Figure 8.21. The reader is referred to this web site for the physical background and meaning of the data sets:

www.physics.csbsju.edu/stats/KS-test.html.

For the purpose of this discussion they are just two typical data distributions which one wishes to compare. Visually some significant differences can be seen in the two distributions; however one can sometimes be fooled by just a visual look. The output from the K-S test in Listing 8.13 says that at the 95% confidence level the hypothesis that the two distributions are the same would be rejected. However, at the 99% confidence level the hypothesis that the two distributions are the same can not be eliminated. Thus the intuition conclusion that the two distributions are probably not the same is somewhat confirmed, but the K-S test allows one to make more precise statements regarding the two distributions and the certainty with which these statements can be made. Because of the nature of statistics absolute statements are always very difficult to make. Note that the output states that the number of observations is 39.75. This fraction is because of the weighted equivalent number used with the two distributions and as given by Eq. (8.33).

```
 1 : -- /* File list8_13.lua */
 2 :
 3 : require"prob"
 4 :
 5 : redwell,whitney = {},{} -- Two data sets for comparison
 6 : read_data('redwell.txt',redwell)
 7 : read_data('whitney.txt',whitney)
 8 : plot({makeCDF(whitney)},{makeCDF(redwell)})
 9 :
10 : KS_test(redwell,whitney) -- K-S Distribution Test
Output:
              Kolmogorov-Smirnov Goodness-of-Fit Test

Null Hypothesis HO:          Two Distributions are the Same
Alternate Hypothesis HA:     Two Distribution are not the Same
Number of observations =     39.75

K-S test statistic: D, sqrt(N)*D =  219.4 m 1.383   (Test value)

Confidence Level       Cutoff         Conclusion
          90.00  %              1.224              Reject HO
          95.00  %              1.358              Reject HO
          99.00  %              1.628              Accept HO
          99.90  %              1.949              Accept HO
Accept Above Confidence Level of       95.64  %
```

Listing 8.13. Sample code for comparing two experimental distributions.

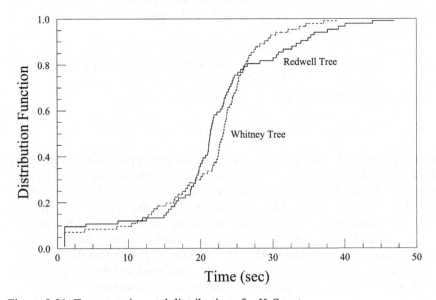

Figure 8.21. Two experimental distributions for K-S test.

This section has illustrated one of the most important techniques to compare two probability distributions and illustrated some uses of previously developed code segments for this application.

8.6 Monte Carlo Simulations and Confidence Limits

Limits to the mean and standard deviation have been discussed in the previous sections based upon Student's t function and the Chi-squared function. While the theory of confidence intervals for these two quantities is well developed, such is not the case for the general nonlinear fitting of parameters to a data set. This will be discussed in the next chapter on general parameter estimation. For such cases, about the only approach to confidence intervals is through the Monte Carlo simulation of a number of test data sets. This approach is also applicable to limits on the mean and standard deviation and will be discussed here partly as background for the next chapter and as another approach to obtaining confidence intervals on statistical quantities.

In beginning this discussion, it is useful to think of an experimental data set as a sampling of a much larger data set with essentially an infinite number of possible measurement values – at least for continuous variables – with some characteristic background random noise. The measured data set provides only estimates of the true parameters of the underlying physical phenomena being sampled. If it were possible to draw additional data set of the same size from the universe of possible data sets it would then be possible to get a distribution of measured parameters such as the mean and standard deviation. The more replicas of the observed data set that were obtained, the closer to the true value of the measured parameters would the obtained estimates become. In the Monte Carlo (MC) approach, one attempts to computationally generate additional data sets as substitutes for actual physical measurements which in most cases is not possible. In the simulations some fundamental assumptions must be made regarding the relationship of parameters in the simulated sample data set and the parameters in the larger real world data set. One of the fundamental assumptions is that the manner in which random errors enter into the measured data set is the same as the manner in which random errors affect the entire data set. For a more complete discussion of MC assumptions the reader is encouraged to read the discussion in Chapter 14 of *Numerical Recipes in C*, Cambridge Press, 1988.

In the MC approach the analysis starts with a set of sample parameters, in this case the sample mean and standard deviation, and constructs synthetic data sets with the same number of data points and with a random distribution of errors that mimic the best understanding of the error distribution in the measured data set. These synthetic data sets are then used to construct a distribution of measured parameters such as mean and standard deviation. This distribution of parameters is then used to make arguments about confidence intervals on the measured parameters.

```
 1 : -- /* File list8_14.lua */ MC analysis of mean and std limits
 2 :
 3 : require"prob"
 4 : local Pn = elemfunc.Pn
 5 :
 6 : yield = {}
 7 : read_data('steel_yield.txt',yield) --Read data change as desired
 8 : nd = #yield; snd = math.sqrt(nd)
 9 : mean,std = stats(yield) -- Mean and std
10 : xd,yd = makeCDF(yield) -- CDF data
11 : print('Number of data, mean, std = ',nd,engr(mean),engr(std))
12 : nmc = 10000 -- Number of Monte Carlo runs
13 :
14 : Pnms = function(x,m,s) return Pn((x-m)/s) end -- Gaussian
15 :
16 : ymn,xmn = {},{}
17 : mmc,mstd,Dn = {},{},{} -- collect mean and std for MC runs
18 : for j=1,nmc do -- Begin MC runs
19 :    for i=1,nd do -- Generate MC data set
20 :            ymn[i] = std*gnormal()+mean
21 :    end
22 :    mmc[j],mstd[j] = stats(ymn) -- Now evaluate mean and std
23 :    xmn,ymn = makeCDF(ymn) -- CDF for Dn evaluation
24 :    Dn[j] = 0
25 :    for i=1,nd do -- Step over data set, collect Dn data
26 :            dt = math.abs(ymn[i] - Pnms(xmn[i],mean,std))
27 :            if dt>Dn[j] then Dn[j] = dt end
28 :    end
29 : end
30 : pbv = {68.27,95.45,90,95,99,99.9}
31 : np = #pbv
32 : for j=1,nmc do Dn[j] = snd*Dn[j] end -- Final K-S statistic
33 : x,y = makeCDF(mmc) -- CDF of MC mean values
34 : for i=1,np do print('Limits for mean at '..engr(pbv[i],'%')..
        ' = ',climits(x,pbv[i])) end
35 : write_data('list8_14a.dat',x,y)
36 : x,y = makeCDF(mstd) -- CDF of MC std values
37 : for i=1,np do print('Limits for std at      '..engr(pbv[i],'%')..
        ' = ',climits(x,pbv[i])) end
38 : write_data('list8_14b.dat',x,y)
39 : x,y = makeCDF(Dn) -- CDF of MC D values
40 : for j=1,nmc do y[j] = 1 - y[j] end
41 : write_data('list8_14c.dat',x,y)
```

Listing 8.14. Monte Carlo analysis of confidence intervals for mean and standard deviation.

All of this is perhaps best illustrated with an example and some code for executing the MC approach. Listing 8.14 shows code for applying the MC approach to a typical data set which for this example is the steel failure data previously used in Listing 8.10 and Figure 8.18. Lines 5 through 11 import the data set, calculate the sample mean and standard deviation and print the values. Line 14 defines a Gaussian function with specified mean and std using the normal Gaussian function Pn() which is defined with zero mean and unity std. Line 12 defines the number of MC synthetic data sets to be generated as 10000 and the MC simulation loop runs from line 18 to line 29 of the code. Lines 19 through 21 generate a synthetic data set (ymn[i]) of nmc length with the same mean and std as the measured data set but with random Gaussian noise. The random noise is assumed to have the same mean and std as the measured data set. For this simulation step, use should be made of the best estimate of the statistical properties of the measured data set. In MC calculations it is typical to use a Gaussian random distribution for generating the synthetic data sets. However, this is not required if a better distribution is evidenced by the measured data set. For the steel yield data as seen in Figure 8.18 the CDF curve closely follows a Gaussian distribution and this can be further verified by performing a DK-S test as in the previous section.

The mean and std of the generated data set is collected into arrays on line 22. An additional feature of the code on lines 25 through 28 is the collection of K-S D data values. It will be recalled that this is the amount by which the distribution differs from a test Gaussian distribution. The D values are finally converted to the K-S statistic value -on line 32 by multiplying by the square root of the number of data points. Finally results of the analysis are printed and saved on lines 33 through 41. The printed values include confidence intervals evaluated from the generated CDF functions using the climits() function. This function takes a CDF table of sorted x-y values and returns the x-value corresponding to a given probability range. For example for a 90% confidence interval it returns the 5% and 95% points in the CDF table. The MC analysis of Listing 8.13 performs 10,000 MC evaluations, but takes only a few seconds on a modern desktop computer.

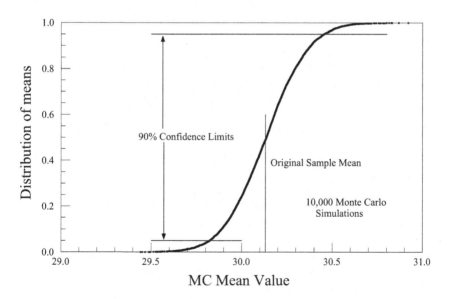

Figure 8.22 Monte Carlo simulations of the distribution of mean values.

Figures 8.22, 8.23 and 8.24 show the important results from the analysis. First Figure 8.22 shows the distribution of simulated mean values from the MC simulations. The original sample mean is at the center of the distribution as would be expected since this was used for the mean of the synthetic data sets. However, there is a range of calculated values due to the limited sample sizes and the statistical nature of the noise associated with the data. What looks like a continuous black curve in the figure is really just a collection of the 10,000 points evaluated by the MC calculations. It can be seen that 10,000 points is sufficient to generate a rather smooth distribution of simulated values along the curve. The 90% confidence limits are shown on the figure by neglecting the lower and upper 5% of the points. With 10,000 points, this corresponds to neglecting 500 points on each end of the CDF table. The 99.9% confidence limits correspond to ignoring the lower and upper 5 points in the CDF table.

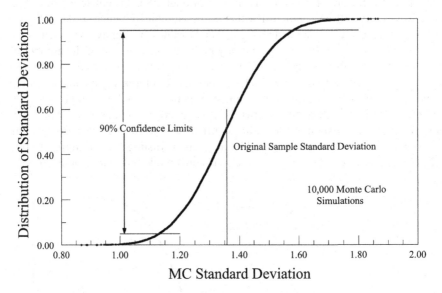

Figure 8.23 Monte Carlo simulations of the distribution of standard deviations.

Similar calculated values for the distribution of standard deviation values are shown in Figure 8.23. Again 90% confidence limits are shown on the figure for reference. Even though the synthetic data sets have values drawn from a Gaussian data set with the same std, the random nature of the limited sample size causes individual synthetic data sets to have a std significantly different from the underlying Gaussian data set.

Finally, Figure 8.24 show the MC generated distribution of the K-S parameter which it will be recalled is $\sqrt{N}D$ where D is the maximum difference in the sample distribution from a test distribution which in this case is the assumed Gaussian distribution. The original measured data set had a very small K-S value as identified on the figure. The MC simulations show that almost all the synthetic data sets had larger K-S parameter values. In that sense, the original data set was a somewhat unique data set with a much closer fit to a Gaussian distribution than would normally be expected from the given number of sample points and the given std. However, a little thought will reveal that the comparison Gaussian curve was selectd as a curve with exactly the same mean and standard deviation as the original sample data. Therefore it is not surprising that the comparison Gaussian has a closer fit to the original sample data than a typical simulated data set which will have a different mean and std.

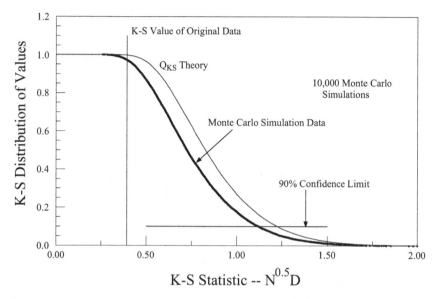

Figure 8.24. Monte Carlo simulation of the distribution of the K-S parameter.

Also shown in the figure is the theoretical K-S distribution as given by the $Q_{KS}()$ function of Eq. (8.30). The Monte Carlo simulations have a general shape similar to the theory, but have a somewhat smaller value at any confidence value. In this case a 90% confidence limit ignores the upper 10% of the points in the K-S parameter array.

Figure 8.25 shows a comparison of the calculated bounds on the mean value at several confidence levels obtained from the standard error, from the Student's t table approach and from the MC approach. It can be seen that there is relatively good agreement between Student's t values and the MC simulated values. The MC generated bounds tend to be slightly smaller than the classical t table values for this particular example. In this example the 2 standard error results agree very well with the t table results and MC 95% confidence bounds. The MC results are from the values printed by Listing 8.14. The reader can execute the code to observe the calculated results.

A comparison of the confidence limits on the standard deviation by the classical standard error approach, the Chi-Squared error analysis and the MC approach (dotted lines) is shown in Figure 8.26. In this case there are larger differences

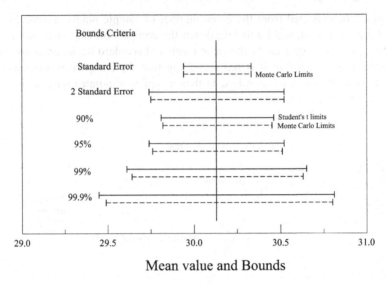

Mean value and Bounds

Figure 8.25. Comparison of bounds on mean value as calculated by Student's t table (solid lines) and by Monte Carlo simulations (dotted lines).

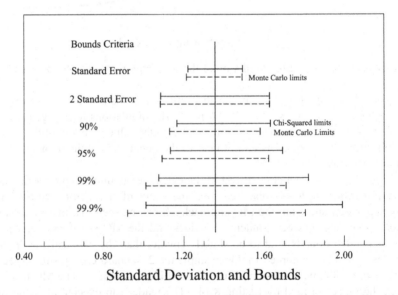

Standard Deviation and Bounds

Figure 8.26. Comparison of bounds on standard deviation value as calculated by Chi-Squared table (solid lines) and by Monte Carlo simulations (dotted lines).

between the Chi-Squared table results and the MC simulated results. Also the Chi-Squared results tend to be considerably more asymmetrical about the sample std value than the MC simulated results which for this example are fairly symmetrical about the sample std value. Again for this example, the classical 2 standard error results tend to be reasonably close to the 95% MC bounds.

Monte Carlo simulations provide a valuable alternative approach to studying many of the random properties of measured data. For the mean and standard deviation of a data set much theoretical work has been done on classical approaches to the confidence limits. In the next chapter, the general problem of fitting a nonlinear function to a set of data will be addressed. For such cases the MC approach is about the only available means for studying some of the important properties of parameter estimation and for obtaining bounds on the accuracy of parameter estimation. This section provides some of the background of the MC approach that will be useful in extending the MC approach to the general problem of parameter estimation in the next chapter.

8.7 Non-Gaussian Distributions and Reliability Modeling

The Gaussian distribution is an important classical distribution function and is applicable to many physical measurements. However, not all distributions follow such a model and some of the other important distributions have been discussed previously in Section 8.3 and computer code is provided for evaluation many such distributions. Distributions other than the Gaussian typically require additional parameters beyond the mean and standard deviation to describe the function. Going beyond the Gaussian distribution begins to get into the realm of general parameter estimation for a model equation which is the major subject of the next chapter. So in a sense this section provides an introduction to the next chapter. It has in common with the present chapter the fact that the interest is purely in the statistical properties of some physical variable whereas in the next chapter the interest will be in measured data where there is one or more independent parameters and one or more dependent parameters. So this section partly belongs in this chapter and partly belongs in the next chapter.

The general subject of non-Gaussian distribution functions is very broad and this section will only discuss one small, but very important application to the area or reliability modeling. In such applications the interest is typically in how some physical parameter or some part of a physical system fails – i.e. becomes non-usable in some important application. Examples include metal fatigue, stress failure of concrete or steel beams, voltage or current stress failure of electrical components, etc. Failures may be rather benign in such thing as one's cell phone stopping operating or very catastrophic is such things as a plane crash due to metal fatigue. There is a vast literature and practice associated with reliability modeling and reliability measurements and this section will only touch briefly on one aspect of this field and that is the use of non-Gaussian distribution functions to characterize measured reliability data. One of the difficulties of reliability measurements and modeling is the problem of getting meaningful data. One is generally trying to characterize phenomena that occur very rarely – at least one usually hopes that failures are a very rare event. Thus some type of accelerated testing is usually required – such as at elevated stress levels or elevated temperatures and accurate models are needed to extrapolate from test conditions to normal operating conditions. This extrapolation is one of the very important applications of reliability models.

Two of the most important distributions used in reliability modeling are the Weibull distribution and the log-normal distribution – see Section 8.3. The Weibull distribution finds wide application because of its simplicity and because it can describe a wide ranging shape of failure distribution. The basic Weibull model is given by the equation:

$$F = 1 - \exp(-((x - \mu)/s)^m) \text{ with } x \geq \mu \text{ and } s, \mu > 0 \qquad (8.34)$$

where m is a shape parameter, s is a scale parameter and μ is a location parameter. In many applications the location parameter is taken to be zero. For small values of m the distribution looks exponential like and for m near 4 the distribution looks similar to a Gaussian as has been shown in Figure 8.12. One of the attractive features of the Weibull distribution is the simplicity of the model which does not involve an integral as many of the classical distributions do. If one manipulates the equations in a particular manner one can arrive at the equation:

$$m(\ln(x - \mu)) - m\ln(s) = \ln(-\ln(1 - F))$$
$$\text{or } m\ln(x) - m\ln(s) = \ln(-\ln(1 - F)) \quad \text{if } \mu = 0 \qquad (8.35)$$

When the location parameter, μ, is zero the second form of Eq. (8.35) has a particularly simple form. If one plots not F but $\ln(-\ln(1 - F))$ on the vertical axis vs. $\ln(x)$ on the horizontal axis, a straight line is obtained with slope m and intercept $m\ln(s)$. Under these conditions, the parameters of the Weibull distribution can easily be obtained and one can easily see how accurately a data set conforms to the model.

Figure 8.27 shows a Weibull plot with a selected set of parameters as identified in the figure. The solid straight line is for a zero value of the location parameter, while the dotted curves are for both negative and positive location parameters which give the function some curvature on this graph at small parameter values. The right hand scale on the figure shows values of the distribution function F corresponding to the Weibull parameter shown on the left hand scale. An important feature of this type of Weibull plot is that it greatly compresses the high probability values (near 1) and greatly expands the low probability values (near 0). For small F values the vertical axis corresponds essentially to the log of the probability values. This is very important in reliability modeling since it is the low probability values or early failures that are generally of most importance. Some slight curvature of a data set on a Weibull plot can be accounted for by a non-zero position parameter.

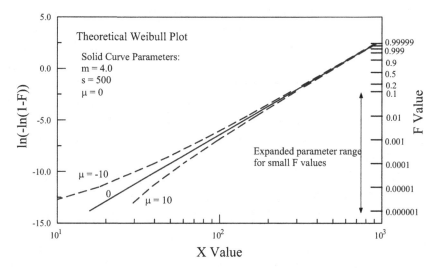

Figure 8.27. Weibull probability plot with selected set of parameters. For zero location parameter, a straight line results.

Listing 8.15 shows code for calculating the Weibull parameters for a data set and for plotting the resulting data on a Weibull plot. Data is imported from a file on line 6. A Weibull data fitting function Fwloglog() is defined on lines 9 through 13 corresponding to Eq. (8.35) with three unknown constants corresponding to:

$$c[1] = m, \ c[2] = m\ln(s), \ c[3] = \mu \tag{8.36}$$

The input data in terms of the distribution function F is converted to the Weibull log-log parameter on lines 16 through 19 of the code. Parameters for a call to the nonlinear data fitting function nlstsq() on line 2 are set up on line 21. Without the c[3] parameter, the fitting equation would be linear in the coefficients and convergence obtained in one iteration with any initial guesses for the parameters. The initial parameters are selected with reasonable values, but are not critical as the program will converge even for zero initial parameters as the reader can verify by re-executing the program with different initial guesses on line 21. The final part of the code on lines 23 through 33 calculates the theoretical Weibull curve with the evaluated parameters and plots and saves the data to a file on line 33.

Figure 8.28 shows the data being analyzed by the code and the resulting solid line fit of the data to the Weibull distribution model. The data analyzed is from: E. Wu et al., 1999 *IEDM Technical Digest*, p. 441. The data represents the amount of charge in Coulomb per cm^2 flowing through an MOS structure with an oxide thickness of 2.5nm before destructive breakdown occurs. This value, known as the Charge-to-Breakdown value (Q_{bd}), is an important measure of the reliability and lifetime of semiconductor MOS devices.

```
 1 : -- /* File list8_15.lua */ -- Weibull distribution analysis
 2 :
 3 : require"prob"; require"nlstsq"
 4 :
 5 : qbd,F = {},{}
 6 : read_data('qbd.dat.txt',qbd,F) -- Data in form of F already
 7 : nd = #qbd
 8 :
 9 : Fwloglog = function(x,c) -- log-log Weibull function
10 :    local xx = x[2] - c[3]
11 :    if xx<0 then return -55 -x[1] end
12 :    return c[1]*math.log(xx) - c[2] - x[1]
13 : end
14 :
15 : xx,yy = {},{}
16 : for i=1,nd do -- Convert F values to log-log Weibull data
17 :    xx[i] = qbd[i]
18 :    yy[i] = math.log(-math.log(1-F[i]))
19 : end
20 :
21 : c = {4,xx[nd]/2,xx[1]/2} -- Initial guesses
22 : del,err,nn = nlstsq({yy,qbd}, fw, Fwloglog, c);
           print('Number of Newton steps =',nn)
23 : for i=1,#c do printf('c[%d] = %12.4e +/-
%12.4e\n',i,c[i],del[i]) end
24 : print('m,s,mu = ',c[1],math.exp(c[2]/c[1]),c[3])
25 :
26 : xp,yp = {},{}
27 : xs = {0,0}
28 : for i=1,nd do -- Theoretical Weibull function
29 :    xp[i] = math.log(qbd[i])
30 :    yp[i] = Fwloglog({0,qbd[i]},c)
31 : end
32 : scatterplot(xp,yp,yy)
33 : write_data('list.8_15.dat',qbd,yy,yp)
```

Listing 8.15. Code for plotting Weibull data and calculating Weibull parameters.

Although there is some scatter in the data there is generally good agreement between the data and the Weibull model, except perhaps at the largest Q_{bd} values where the data appears to fall significantly below the model curve. This illustrates the importance of performing the data fitting with the log-log Weibull function values on the vertical axis and not directly using the F probability values. If one attempts to fit the Weibull model directly to the F probability data, the fitting will be dominated by the largest F values and a good fit to the low probability data values would not be obtained. By performing the data fit on the log-log Weibull data, the low probability values are weighted heavily in the fitting and these low probability values are the most important for reliability studies.

It is noted that the data set appears to have a slight curvature and the best fit parameters occur with a non-zero value of the position parameter of 0.8426 C/cm^2. In reliability modeling such a non-zero position parameter is sometimes thought of as a threshold value below which no failures will occur, since F is zero in the model at this value. Whether this is a good interpretation is left to speculation and

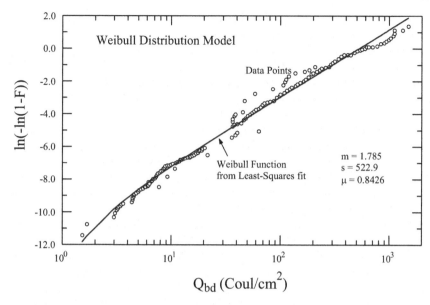

Figure 8.28. Weibull distribution analysis of Charge-to-Breakdown data.

interpretation. If a zero position parameter is used, the nlstsq() function is not required to fit the data since any linear least squares data fitting routine will give the appropriate values of parameters. However, the nlstsq() routine is only slightly more complicated to employ and allows for a more general fit to the Weibull distribution function. Many references in the reliability literature recommend plotting curves with several values of the position parameter and selecting the best value from the curve that fits the data most closely, but with the nlstsq() function this is not needed as the best fit is obtained as easily as with a zero position parameter.

From the vertical axis values of the log-log Weibull function in Figure 8.28 it can be seen that the log-log values go down to values of almost -12. From Figure 8.27, it can be seen that such values correspond to F values of around 1×10^{-5}. Traditionally such small experimental values would require that one collect data from about 1×10^{5} samples since it represents 1 failure in 1×10^{5} – a very difficult task. So how are such small values of F obtained for the Q_{bd} data set? This was done for this data set by making measurements on devices with varying areas and using a known scaling principle for devices of different areas. The data were scaled to an equivalent MOS structure area of 8.4×10^{-7}. This is just one of the tricks of the trade used in reliability analysis.

Results for a second data set analyzed by the Weibull technique are shown in Figure 8.29. The computer code for this analysis is very similar to Listing 8.15, but is not shown here. It is included in the software as listing8.16.lua for the reader to use. The data being analyzed here is the yield strength of steel previously shown in Figure 8.18. In that figure the distribution function was compared to a Gaussian distribution and it was found to agree quite well with a standard Gaussian distribution. In Figure 8.29 the data is plotted on a Weibull plot and compared with two theoretical Weibull distributions. First the dotted curve shows the best fit Weibull function when the position parameter to set to zero. This is basically the best fit straight line to the data. The horizontal axis in Figure 8.29 is linear in the yield data and not a log scale as predicted for a straight line. However, because of the limited range of the data, the difference between a log scale and a linear scale is very small. The experimental data points have considerable curvature on the Weibull plot and the straight line fit is seen to give a poor fit to the data. Also the slope parameter has a very large value of 26.9, again raising a flag with regard to the fitting. This dotted line would obviously not be a good representation of the statistical properties of the data.

The solid line curve is the least squares best fit Weibull distribution including a position parameter which in this case is 25.7 kPSI. The solid line is visually seen to provide a very good fit to the data. Whether it is a better fit then the Gaussian fit seen in Figure 8.18 is open to debate. The fit also has a very reasonable value of the slope parameter of 3.56. This example illustrates the importance of being able to include a position parameter in the use of the Weibull function for reliability modeling. This is where the nlstsq() code segment becomes the appropriate tool for the model analysis.

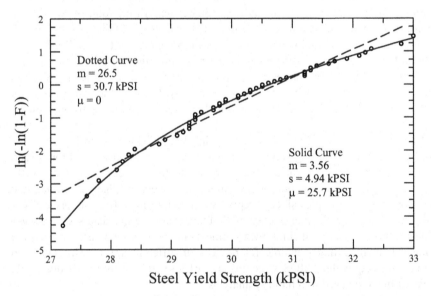

Figure 8.29. Analysis of Steel Yield Strength by Weibull distribution function.

One final note on the results for the yield strength of steel will be made for consideration. If a positive position parameter is interpreted as a threshold for the reliability data, then for the steel yield strength data essentially zero failures are expected below a stress level of about 25.7 kPSI. This is what the Weibull reliability analysis would suggest which is fundamentally different from what one would predict from Gaussian statistics where a finite probability would still be found at 0 kPSI. One would also probably want to know the confidence levels and corresponding bounds on this predicted value. The nlstsq() function returns a standard error estimate of 0.15 kPSI which is a very tight limit. The decision of whether these are correct implications, will be left for others to decide.

This section has shown how previously developed code segments such as the nlstsq() function can be brought to bear on random data fitted to non-Gaussian distribution functions. The frequently used Weibull reliability analysis has provided some examples of this use.

8.8 Summary

This chapter has focused on the development of code segments and numerical approaches for studying the statistical nature of measured data that has associated random variations. The emphasis has been on single variable data where the mean value and deviations around the mean value are of prime importance. Many of the functions developed here will be useful in the next chapter which focuses on the more general problem of fitting a function to a set of data when the measured data set containing some random errors but where the data is also a function of one or more independent parameters. This is the general problem of estimating parameters given a model to be fitted in some way to a set of experimental data.

Several new code segments have been developed in this chapter. These include random number generators ran0(), ran1(), ran2() and ran3() for uniformly distributed random numbers and gnormal() for Gaussian random numbers. Code segments for the following probability distribution functions have also been discussed with developed code segments:

1. Normal Distribution
2. Uniform Distribution
3. Cauchy Distribution
4. Student's t Distribution
5. F Distribution
6. Chi-Squared Distribution
7. Weibull Distribution
8. Lognormal Distribution
9. Gamma Distribution
10. Beta Distribution

In addition code segments were developed and demonstrated for manipulating random data. The most important of these are the functions:

1. stats() for evaluating mean and std of a data set
2. makeCDF() for converting random data into a cumulative distribution function
3. normalCDF() for generating a comparison Gaussian CDF

4. hist() for generating histogram data
5. lag() for generating data for a lag plot
6. At() for Student's t table evaluation
7. ttable() for obtaining the inverse of Student's t table results and bounds on mean values
8. tchisq() for obtaining the inverse of Chi-Squared table results and bounds on std
9. iCDF() for obtaining the inverse of any distribution function
10. KS_test() for the Kolmogorov-Smirnov test of equality of two distributions
11. climits() for obtaining confidence limits from a table of distribution values

Finally some of the concepts were brought together in a section on the modeling of reliability data using non-Gaussian statistics. The next chapter will continue with the analysis of data with random errors but will expand the concepts in this chapter to examples where measured data sets depend on one or more independent variables.

9 Data Models and Parameter Estimation

A frequently encountered engineering problem is that of having a set of data which one wishes to describe by a mathematical model and determine a set of parameters that characterize the model. This is similar to the topic addressed in Chapter 7 of curve fitting and data plotting. However, in that chapter the emphasis was only on obtaining a smooth curve representation of a set of data with noise. The intent of the curve fitting there was to obtain a function that could be used in subsequent manipulations of the data or in visually presenting the data in graphical form. The particular form of the fitting function was secondary to the goal of obtaining a good smooth approximation to the data. In the present chapter, the emphasis will still be on obtaining a "good" approximation to the data set; however, the major emphasis will shift to the fitting parameters of a data model which will be assumed to have some particular physical or mathematical significance. For example the fitting parameters could be some time constant of a physical process or some saturated value of a physical process or some time period for a physical process. The emphasis will thus be on estimating the value of the parameters used in describing a data set. This will naturally lead to question about how accurately the values of the parameters can be obtained and questions associated with confidence limits on the estimated parameters. Thus this chapter continues the discussion begun in Chapter 7 on curve fitting and brings in many of the statistical concepts developed in Chapter 8 to address the general area of modeling of data and the estimation of parameter values from the data for a given model of the data. In some engineering fields, this process is also called parameter extraction.

9.1 Introduction

Figure 9.1 shows a simple example of a data set (which has previously been considered) to illustrate some of the factors to be considered in this chapter. Shown is a data set with random errors but which has some obvious nonlinear functional variation of the Y variable with the value of an assumed independent parameter, X in this case. The solid line curve represents a simple proposed functional relationship between the independent variable and the dependent variable as indicated in the figure. In this chapter the emphasis will be primarily on estimating values of a set of parameters in the fitting model equation and secondarily on the characteristics of the random noise in the data. In the previous chapter, the primary emphasis was on the characteristics of the random variations about some mean value. For the data set in Figure 9.1 it can be seen that the model curve is about half the time above the data points and about half the time below the data points as would be

369

J.R. Hauser, *Numerical Methods for Nonlinear Engineering Models*, 369–460.
© Springer Science + Business Media B.V. 2009

expected for a least-squares fit to the data. If one looks at the difference between the fitted curve and the data points, one would expect to see a random data set very similar to the data sets analyzed in the previous chapter. Thus if the solid line represents a good model equation for the data, it would be expected that all the techniques discussed in the previous chapter can be brought to bear to study the "difference between the model curve and the data set".

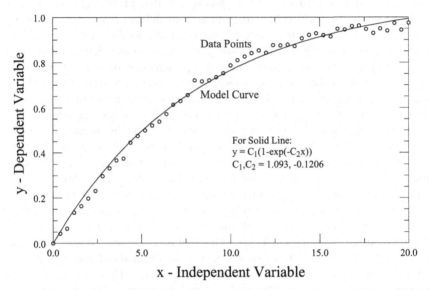

Figure 9.1. Example of data set for parameter estimation.

The data set in Figure 9.1 is a function of only one independent parameter. In general data sets may be a function of several independent parameters and such cases will be considered in later sections of this chapter. The general approach to fitting a model equation to a data set will be through the least-squares minimization technique developed in Chapter 7. As a review, in the least squares approach one seeks to minimize the square of the error between a fitting function and the data set as expressed in the equation:

$$esq = \sum_{i=1}^{n_d}(y_i - f(x_i, C_1, C_2, \cdots C_m))^2 = \sum_{i=1}^{n_d}\varepsilon_i^2 \tag{9.1}$$

It will be assumed that there is some exact functional relationship between the independent variable and the dependent variable that can be represented as $y_i = F(x_i)$ and it is this functional relationship that one seeks to approximate by some mathematical $f(x_i, C_1, C_2, \cdots C_m)$ function. Sources of error in the measured value of the dependent variable can come about because of fundamental measurement errors in both the value of the independent and/or dependent variable. Thus for any particular measured value of y one can write:

$$y_i = F(x_i + \delta_i) + \varepsilon_{yi} \simeq F(x_i) + F'(x_i)\delta_i + \varepsilon_{yi} \tag{9.2}$$

where δ_i is the error in the independent variable and ε_{yi} is the error in the dependent variable. When this is combined with Eq. (9.1) the error to be minimized can be written as:

$$\varepsilon_i = [F(x_i) - f(x_i, C_1, C_2, \cdots C_m)] + F'(x_i)\delta_i + \varepsilon_{yi} = \varepsilon_{mi} + \varepsilon_{xi} + \varepsilon_{yi} \qquad (9.3)$$

The three terms on the right of Eq. (9.3) represent three sources of error in the model equation fitting procedure: 1) an error term because the model equation doesn't exactly correspond to the underlying physics, 2) an error due to the measurement of the independent variable and 3) an error due to the measurement of the dependent variable. In some cases, the first two sources of error can be neglected and only the third source, the error in measuring the dependent variable considered as dominant. However, it is difficult to separate out contributions from errors in the independent variable from errors in the dependent variable. In any case one would hope and expect that these two sources of error are basically random in nature. This is in contrast to errors due to the model equation, the first right hand term in Eq. (9.3). This tends to not be random in nature and arises because of an inability in many cases to accurately model all the physical effects contribution to the relationship between independent and dependent variables. This term is by no means negligible in many cases and the more accurately measurements can be made and any random variables in a measurement process eliminated, the more important becomes this term. Examples will be shown where the model errors are the dominant source of error in some of the subsequent least squares fit to experimental data. Of course the larger the number of fitting parameters in a proposed model equation, the smaller this error term would be expected to become. Some important tests will be developed to expose such error contributions due to the model equation from the other random errors.

Listing 9.1 illustrates some code for a detailed look at fitting a model equation to a data set. In this case the set of data points is that shown in Figure 9.1 and the data model is the exponential model defined by the equation:

$$y = C_1(1 - \exp(C_2 x)) \qquad (9.4)$$

which has two constants to be determined in fitting the model to the data. Line 3 loads previously defined code segments, nlstsq (from Chapter 7), prob (from Chapter 8) and DataFit (from Chapter 7). The data set as shown in Figure 9.1 is input on line 6 of the code. The model equation for the data is defined on lines 8 through 10. The code up through calling the nlstsq() function for fitting the data on line 14 is very similar to several previous listings. The model is defined, an initial guess is defined for the model equations (line 12) and then the nlstsq() function is called. The resulting model parameters and the fitted equation (solid line curve) are shown in Figure 9.1 as the solid line curve.

The overall agreement between model equation and data is as would be expected in Figure 9.1 from a least-squared fitting to a model. The model falls about equally above and below the data points. However does the model provide a "good fit" to the data and how does one define a good fit? This is a question addressed by the code in Listing 9.1 between lines 17 and 35. This is also the topic addressed by the next section of this chapter.

```
 1 : -- /* File list9_1.lua */ --Data fitting and 6 plots for Fig9.1
 2 :
 3 : requires("nlstsq", "prob", "DataFit")
 4 :
 5 : xd,yd = {},{} -- Data arrays
 6 : read_data('list7_1.dat',xd,yd); nd = #xd -- data set
 7 :
 8 : ft = function(x,c) -- Define function to fit data
 9 :    return c[1]*(1 - math.exp(c[2]*x[2])) - x[1]
10 : end
11 :
12 : c = {1,-.2}; nc = #c -- Initial guess at coefficients
13 : del,err,nmax = nlstsq({yd,xd},fw,ft,c) -- Call fiting
14 : print('err, nmax =',err,nmax) -- print results
15 : for i=1,nc do printf('c[%d] = %12.4e +/- %12.4e\n',i,c[i],
           del[i]) end
16 : -- Generate 6-Plot data
17 : ycalc,res,yres1,yres2 = {},{},{},{}
18 : for i=1,nd do -- Calculate fitted values and residuals
19 :    ycalc[i] = ft({0,xd[i]},c)
20 :    res[i] = yd[i]-ycalc[i]
21 : end
22 : fres1,fres2 = datafit({res,xd}), datafit({res,ycalc})--residuals
23 : for i=1,nd do yres1[i],yres2[i]=fres1(xd[i]),fres2(ycalc[i]) end
24 : xp,yp = makeCDF(res); xpn,ypn = normalCDF(res) -- CDF residuals
25 : xh,yh = hist(res,15); xhn,yhn = histnorm(res,15) -- Histogram
26 : xl,yl = lag(res) -- lag data for residuals
27 : write_data('list9_1.dat',xd,yd,ycalc,res,yres1,yres2,xp,yp,
           ypn,xl,yl,xh,yh,yhn)
28 : -- Now show 6 plots
29 : plot(xd,yd,ycalc) -- Plot 1 -- Predicted values and data points
30 : plot(xd,res,yres1) -- Plot 2 -- Residuals vs x
31 : plot(ycalc,res,yres2) -- Plot 3 -- Residuals vs y
32 : scatterplot(xl,yl) -- Plot 4 -- Lag plot of residuals
33 : plot(xh,yh,yhn) -- Plot 5 -- Histogram plot of residuals
34 : plot(xp,yp,ypn) -- Plot 6 -- Distribution plot of residuals
Selected Output:
err, nmax =     0.022294069882552        5
c[1] =   1.0925e+000 +/-  1.4103e-002
c[2] = -1.2064e-001 +/-  3.5264e-003
```

Listing 9.1. Example code for fitting a model equation to data and analyzing quality of data fit.

9.2 Goodness of Data Fit and the 6-Plot Approach

The process of fitting a model equation to a set of data using a program such as nlstsq() is just a first step in the process of modeling data and of parameter estimation. While it sometimes takes special care to formulate the model equations appropriately and to obtain a good initial guess at the parameters, a robust fitting program makes the process of obtaining a least squares set of model parameters relatively easy as the example in Listing 9.1 shows. However, the following questions should always be asked of any model fitted to a set of data:

1. Is there a good fit between the model and the data?
2. How do we know when there is a good model fit?
3. How can the quality of fit between the model and the data be measured?

These are difficult questions to answer in an absolute manner. However, there are certain factors which should be considered in addressing these questions and in accepting any least squares fit of a model equation to a set of data.

Listing 9.1 provides a set of 6 graphical plots to aid in determining the "goodness of data fit" for a model equation to a data set. (The reader is encouraged to see www.itl.nist.gov/div898/handbook/ for a more complete discussion of this technique and the graphical plots.) This will be referred to as the 6-plot approach and consists of the following plots:

1. Plot of the predicted values and data values versus the independent variable.
2. Plot of the residuals versus the independent variable.
3. Plot of the residuals versus the dependent variable.
4. Lag plot of the residuals.
5. Histogram plot of the residuals.
6. Cumulative distribution plot of the residuals.

The code in Listing 9.1 from line 18 through line 28 generates data for these 6 plots. The data is written to a permanent file on line 28 and then the 6 plots are produced as pop-up plots by lines 30 through 35 of the code. The reader is encouraged to execute the code and view the 6-plots.

The first plot is the data and model comparison plot of Figure 9.1. A comparison of model equation and data should show close agreement and the data points should be scattered about the model curve with equal numbers of points above and below the model curve. Also the calculated model parameters should have reasonable numerical values and in most engineering examples they will have some physical significance. For most engineering problems, reasonable ranges of model parameters are also known before a data fitting analysis. If the model curve deviates greatly from the data points then it should be concluded that the model equation is not a valid model or that an invalid set of model parameters has been obtained. If the model deviates appreciably from the data, the number of Newton iterations returned by the nlstsq() function should be examined to see if convergence was achieved. For this example the selected output line shows that 5 Newton iterations were required for convergence on the model parameters, so a valid set of parameters is expected to have been computed.

The second and third plots explore more closely the obtained agreement between the model equation and the data set. These plots are shown in Figures 9.2 and 9.3 for the Figure 9.1 data. The figures show the residual values which are the differences between the data points and the model predictions as functions of both the independent variable and the dependent variable. The solid lines in the figures are obtained from the DataFit() function discussed in Chapter 7 and provide a fast means of visualizing an averaged fit through the set of residual values. These solid lines are generated by the Listing 9.1 code on line 22. (This is just another example of the progressive use of code segments as the discussion progresses through this book.) These solid line curves have no particular physical significance but are simply included to aid the reader in following the general trends of the residual data plots.

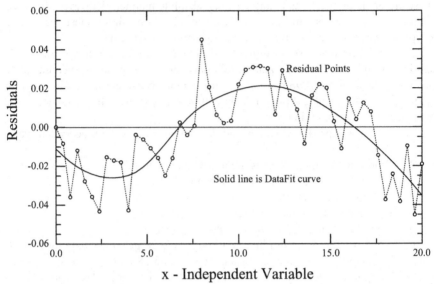

x - Independent Variable

Figure 9.2. Plot of residuals versus independent variable for Figure 9.1 data and Eq. (9.4) model.

The residual plots provide very useful information about the goodness of fit of the model equation to the data set. Ideally one would hope to see essentially random data points as a function of either the independent or dependent variable value. However for this example, this is not the case as the figures show. As the independent variable increases in Figure 9.2 a set of residuals with negative values are seen to be followed by a set of residuals with predominately positive values and finally a set of residuals with predominately negative values. The same trends are clearly seen in Figure 9.3 when the residuals are plotted as a function of the dependent variable. These trends can also be seen by a careful look at Figure 9.1, but are more clearly identified by the residual plots.

While on average the residuals tend to occur equally above and below the model equation as they should for a least squares fit, clearly they are not uniformly distributed with respect to the model equation. Such trends in the residuals clearly indicate that the theoretical model used for representing the data is not very accurately describing the underlying relationship between the independent and dependent variable for this data set. In terms of the three error sources in Eq. (9.3) one can clearly see a major contribution from the first model dependent term which depends on the independent or dependent variable value. The solid line in Figure 9.2 represents an approximate value of how much the model equation differs from the actual underlying relationship between the dependent and independent variables assuming that the data points should be randomly distributed about the model equation. The peak value of this difference is about 0.025 which in some cases might be an adequate approximate to the real relationship – in other cases it might lead one to seek a more accurate model equation.

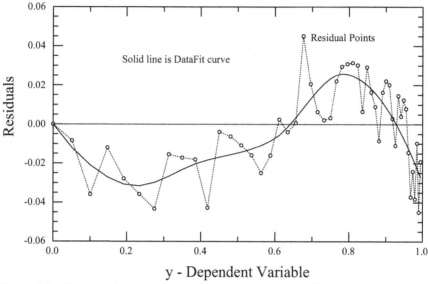

Figure 9.3. Plot of residuals versus dependent variable for Figure 9.1 data and Eq. (9.4) model.

The residual lag plot shown in Figure 9.4 provides somewhat similar information regarding the randomness of the residual data. In this case the plot shows any correlation between adjacent residual values. The lag data points are aligned along a generally upward slope and not uniformly scattered over the plane of residual values. This indicates a generally positive correlation of residuals – a positive (or negative) residual tends to be followed by a positive (or negative) residual. This provides further confirmation of the trends seen in Figures 9.2 and 9.3. For a good model fit to a data set, the lag plot should have essentially random scatter over the entire range of the residuals.

The final two plots, the residual histogram and the residual cumulative distribution, provide additional information about the characteristics of the data. These two plots are shown as Figures 9.5 and 9.6. Because of the limited number of data points, it is difficult to gleam very much information from the histogram plot. The distribution plot in Figure 9.6 shows that the residuals follow reasonably well a Gaussian distribution. However, it must be recalled that the residual values include not only a random component but also a non-random model dependent component as demonstrated by earlier figures. It is not clear how this might affect the distribution plots if it were removed by use of a more accurate model. These two plots are perhaps most useful when one has residuals with a smaller model

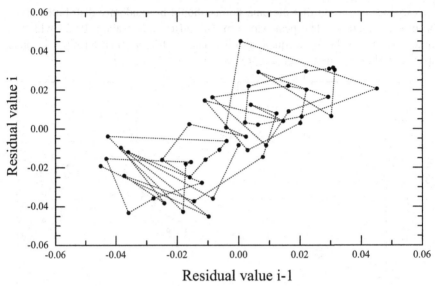

Figure 9.4 Lag plot of residual values for Figure 9.1 data and Eq. (9.4) model

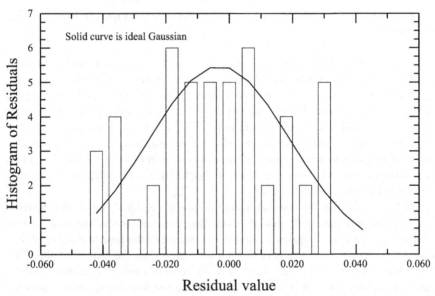

Figure 9.5. Histogram of residual values compared with Gaussian histogram for Figure 9.1 data and Eq. (9.4) model.

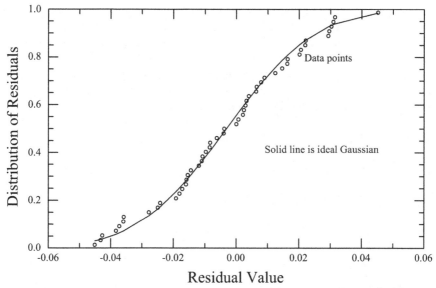

Figure 9.6. Distribution of residual values compared with Gaussian distribution for Figure 9.1 data and Eq. (9.4) model.

dependent contribution. However, it illustrates the fact that having a Gaussian distribution of residuals is no guarantee that a modeled equation is an accurate model for describing a measured relationship between a set of physical variables.

The 6-Plot allows one to draw some qualitative conclusions regarding the goodness of fit of a model equation to a set of data by looking at the residuals between the model and the data. For the data in Figure 9.1 and the exponential model equation with two parameters, it can be concluded that the model equation does not describe in detail the shape of some underlying relationship between the independent and dependent variables. Depending on the intended use of the model and the estimated parameters, the fit and the model might be acceptable. However, in other applications, the model might not be accepted as adequately describing the data set. The latter approach will now be explored with an attempt to improve on the model equation and reapply the 6-Plot approach.

There are many mathematical functional forms that have the general shape of the data set in Figure 9.1 with the exponential being one of the simplest. However, it can be seen that this function increases too fast at low values of x and doesn't saturate fast enough at large values of x. Another simple function which has an adjustable shape and a saturation value is the function:

$$y = C_1 x / (x^{C_2} + C_3)^{1/C_2} \tag{9.5}$$

involving three parameters. At large values of x this saturates at C_1 which should be close to 1.0 for the data in Figure 9.1. The function has additional flexibility with respect to the shape of the function relative to a simple exponential function.

Listing 9.2 shows the changes needed to re-execute the code in Listing 9.1 and
generate new 6-Plot results for this data model. Only the code on lines 8 through
12 which defines the fitting function and the initial parameters for the fitting coefficients is changed and only these changes are shown in Listing 9.2. In execution
the code it was found that convergence can be obtained with a range of initial
guesses for the C_3 parameter from about 40 to 200. The selected output shows
that convergence was obtained with 8 Newton iterations. The evaluated model parameters with their standard error values are also shown in the selected output section of Listing 9.2.

```
 1 : -- /* File list9_2.lua */ --Data fitting and 6 plots for Fig9.1
--------
 8 : ft = function(x,c) -- Define function to fit data
 9 :    return c[1]*x[2]/(x[2]^c[2]+c[3])^(1/c[2]) - x[1]
10 : end
11 :
12 : c = {1,2,100}; nc = #c -- Initial guess at coefficients
---------- Same as Listing 9_1
35 : nr,prb = runs(res)
36 : print('Prob of observed '..nr..' runs is '..engr(100*prb,'%'))
Selected output:
err, nmax =     0.013430311472409        8
c[1] =   1.0247e+000 +/-  1.3828e-002
c[2] =   2.7098e+000 +/-  1.7173e-001
c[3] =   5.3607e+002 +/-  2.1605e+002
Prob of observed 23 runs is   32.22  %
```

Listing 9.2. Code for recalculated 6-Plot analysis for Figure 9.1 data using Eq.
(9.5). Only changes to Listing 9.1 are shown.

Any improvement in the model fit can now be accessed by looking at some of the
6-Plot results with Figure 9.7 showing the agreement between the model equation
and the data points. A close look at the figure and comparison with Figure 9.1
shows that this function does in fact provide an improved functional fit to the experimental data with the data points more randomly distributed around the curve
for both small and large values of the independent variable. The residual plots
more clearly show this improved agreement with the data and the two residual
plots are shown in Figures 9.8 and 9.9. The solid lines are again fits to the residual data from the DataFit() routine which provides a somewhat averaged fit to the
residuals. The plots of the residuals show much more random data values about
the zero line than that previously shown in Figures 9.2 and 9.3 as can be readily
seen by comparing the figures. A quantitative measure of the randomness of the
residuals can be obtained by looking at the so called "runs" associated with the
plots. A "run" is a series of values above or below the mean value – zero in this
case. This also equals the number of zero crossings of the residual curves. With
the improved model there are 21 runs in Figures 9.8 or 9.9 while there are only 9
runs in Figure 9.2 or 9.3. A theory of runs indicates that for ideally random data
the mean and std of the number of runs should be as follows:

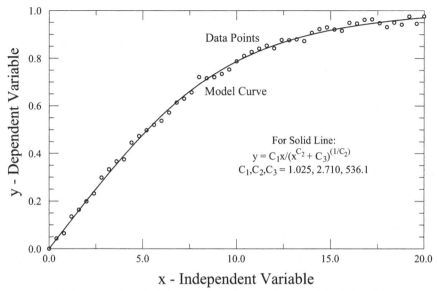

Figure 9.7. Fitting of data for Figure 9.1 with improved model of Eq. (9.5).

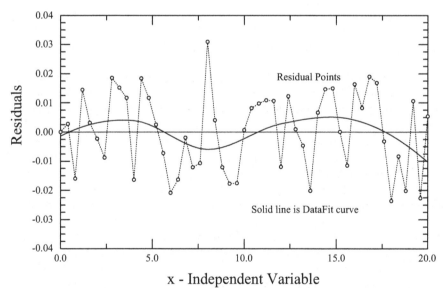

Figure 9.8 Residuals versus independent variable for Figure 9.1 data and Eq. (9.5) improved model.

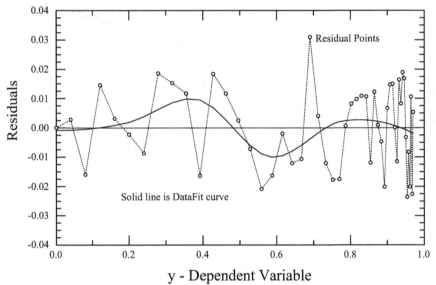

Figure 9.9 Residuals versus dependent variable for Figure 9.1 data and Eq. (9.5) improved model.

$$\text{mean of runs} = \frac{n_d}{2} + 1$$

$$\text{std of runs} = \sqrt{\frac{(n_d-1)^2-1}{4(n_d-1)}} \simeq \frac{\sqrt{(n_d-1)}}{2} \tag{9.6}$$

with approximately a Gaussian distribution for a large number of data points. For computer analysis a runs() function is included in the prob.lua file that takes a data set as input and returns the number of observed runs and a probability of that number being observed for purely random data. This function is shown on lines number 35 and 36 of Listing 9.2 as added calculations. In the present case with 51 data points the expected mean is 26.5 and the std is 3.53 from Eq. (9.6). The very small number of runs, 9, observed with the exponential model of Eq. (9.4) has a very low probability of occurring by chance while the number observed with the improved model, 23, as given in the output of Listing 9.2 is much more consistent with the residuals being a purely random data set. The probability of observing a value this far from the mean purely by chance is calculated as 32.22%. Thus apparently 1/3 of the time this number of runs would be expected with purely random residuals.

While the improved model may still not completely describe the underlying relationship between the variables, it comes much closer as the residuals data plots show and especially as the runs analysis indicates. The solid line curves in the residuals plots are much smaller in value than for the exponential model and considerably smaller in magnitude than the residual values. Thus it can be concluded that the improved model adequately captures the underlying relationship between the independent and dependent variables. In this case a meaningful physical interpretation of the parameters may be more difficult. This will not be attempted at this point but Eq. (9.5) will simply be considered as a mathematical model which provides a good fit to the data leaving only an essentially random set of residuals. The reader is encouraged to execute the code in Listing 9.2 and file list9_2.lua and observe the remaining pop-up plots of the 6-Plot assessment. As would be expected the lag plot shows a much improved random scatter, indicative of a random set of residuals.

This section has provided a set of functions and code segments for rapidly assessing to some extent the degree by which a given model equation provides a "good" fit to a set of data. The hope and desire of using a model equation is that it captures the underlying physical relationship between the independent and dependent variables, leaving only a random data set for the residuals. The 6-Plot approach can provide useful data to aid in determining how successfully a given model equation comes to achieving that goal. The code segment given in Listing 9.1 provides a prototype for fitting any desired function to a set of data read in from a file. The only changes one needs to make are in the data file, the defining function and initial guesses to the fitting parameters. The model equations in the example in this section were generated by an explicitly defined mathematical function but this is not necessary. The model equation can be a relationship generated purely by some computer algorithm or the model may be defined as a transcendental relationship between the independent and dependent variables

9.3 Confidence Limits on Estimated Parameters and MC Analysis

The previous section has provided a 6-Plot approach for accessing the goodness of fit between a model equation and a data set. After an appropriate model equation with unknown parameters has been developed for describing a data set, one of the next questions one asks is how accurate are the estimated parameters obtained from the least squares data fitting? In many cases the entire point of fitting a function to a data set is to obtain estimates of the model equation parameters. The nlstsq() function provides a first estimate on the accuracy of the parameters by returning a set of del() parameters, known as "standard errors" or standard deviations. These parameters represent the equivalent of the standard deviation values obtained from a Gaussian distribution of random variables. If the number of data points is very large, (100 or more) these values represent the 68.3% confidence limits and 2 standard error values represent the 95.4% confidence limits. This means that if one did a large number of experiments collecting equivalent data sets and went through the same data fitting one would expect to find 95% of the time the extracted parameter values to lie between the calculated value and +/- 2 standard errors. For small numbers of data sets, use must be made of Student's t table values to evaluate the correct multiple for the standard errors as discussed in the previous chapter. The correct equation is:

$$C_i \rightarrow C_i \pm t_s \Delta_i \qquad\qquad (9.7)$$

where t_s is the Student's t table value for a desired confidence level and Δ_i is the standard error value returned by nlstsq() (as the first returned value) for the given parameter. The value of t_s is very close to 2 for a large number of data points and a 95% confidence limit. For other confidence values and for limited data sets, the appropriate value of t_s can be evaluated by the ttable() function (see previous chapter).

The standard error values provide only approximate ranges of confidence intervals for general non-linear least squares data fitting. They are estimated from the so called covariance matrix. In this approach one parameter at a time is allowed to vary with the other parameters allowed to change so as to minimize the square error criteria with only the one parameter changed. This is the only easy calculation one can make in the general non-linear case with an unknown distribution of errors. For a detailed discussion of standard errors the reader is referred to the literature (such as *Numerical Recipes in C*, Cambridge Press, 1988).

To move beyond the standard error evaluation and obtain information about the expected distribution of parameter values is very difficult in theory for general non-linear problems. There is really only one avenue for making such estimations and this is through the idea of Monte Carlo simulations. In this approach additional hypothetical data sets are simulated similar to the measured data, the simulated data sets are analyzed and conclusions drawn about the accuracy of the estimated parameters based upon the MC simulations. This is somewhat similar to the simulations of means and standard deviations discussed in the previous chapter in Section 8.6. The major assumption of the MC approach is that the distribution of parameter values about the "true" value of any parameter in the real world is essentially the same as the distribution of parameter values about the parameter values extracted from the measured data. The MC approach does not assume that the extracted parameter values equal the true values, but only that variations around the true values will be similar to variations around the extracted values. While this is extremely difficult (or impossible) to prove in general, it does appear to be a good assumption in a large range of real world physical problems. In any case it is essentially the only method available for the general case of least squares data fitting with nonlinear coefficients.

Thus in the MC approach to confidence limits, a large number of synthetic data sets are generated with randomly selected errors around model values that have been fitted to the measured data set. The synthetic data sets are then analyzed in exactly the same manner as the real data set and distributions of model parameter values are then obtained. Arguments are then made about the confidence intervals of the fitting parameters based upon the distribution of parameter values obtained from the synthetic data sets. The synthetic data sets are constructed to simulate as closely as possible the random nature of the measured data set and are typically generated as follows. First the model equation with the extracted parameter values is taken as a good approximation to the real world relationship between the independent and dependent variables. Thus it is important that the model equation accurately represent the real world relationship so that most of the error comes from random errors and not from model deficiencies as discussed in the previous section. The synthetic data sets are then generated by stepping through the independent variable values and calculating a model equation value to which is added some random variable with a random distribution as closely matching the random variables determined from the actual data set as possible. The random values to be added to the model values are typically taken from a Gaussian distribution with zero mean and the same standard deviation as the residuals extracted from the measured data set. If sufficient information can be gleamed from the distribution of the residuals, to suggest a distribution different from a Gaussian, any other distribution can be used in place of the Gaussian to generate the synthetic data sets. This is the reason a study of the residuals is so important before a MC analysis and this can be done by the 6-Plot approach of the previous section. Following the generation and analysis of the synthetic data sets, the distribution of parameter values is used to generate confidence limits and to explore correlations among the parameter values. Before proceeding with a MC analysis it is important to perform the 6-Plot analysis as discussed in the proceeding section.

Listing 9.3 shows code segments for performing a MC analysis of the parameter distributions as outlined in the above discussion. The code begins as in Listing 9.1 with a fitting of the data to the model equation (defined on lines 8 through 10). This part of the analysis is completed on line 15 with a printing of the evaluated parameter set, just as in Listing 9.1. The Monte Carlo analysis is performed by a call to the function MCpar() on line 19 of the listing. This function takes exactly the same parameter list as the previous call to nlstsq()for the data fitting with an optional additional parameter at the end of the list which in Listing 9.3 is set to 1000, for the generation and analysis of 1000 synthetic sets of data by the MC approach. This parameter may be omitted and a default value of 1000 will be used. The MCpar() function returns a single table containing the results of the MC analysis for all the parameters fitted to the data sets, in this case the three "C" parameters in the model equation of line 9 plus a second integer parameter indicating the maximum number of Newton iterations required in calls within MCpar() to the nlstsq() function to re-evaluate the model coefficients. The remainder of the code in Listing 9.3 from line 22 through 36 prints results and displays the results in various pop-up plots as well as saving the coefficient arrays and the generated coefficient distribution functions to a file on line 24.

The selected output in the listing shows that the maximum number of Newton iterations for the nlstsq() analysis of all the synthetic data sets was 29. This number should be below the value set in nlstsq() (which has a default value of 50). The MC generated parameter limits at various confidence limits are shown for the $c[1]$ parameter. The last line at 99.9% confidence is somewhat suspect as it corresponds to using all the generated data points. At the 99% confidence level, the limit corresponds to neglecting only 5 of the largest and smallest generated data points. This is probably as far as one should push the calculations for 1000 synthetic data sets. Of course more data sets can be used at an increase in execution time. Confidence limits are printed for all three parameters, but results for $c[2]$

```
 1 : -- /* File list9.3.lua */ -- MC analysis for Figure 9.1
 2 :
 3 : requires("nlstsq","mcpar")
 4 :
 5 : xd,yd = {},{} -- Data arrays
 6 : read_data('list7_1.dat',xd,yd) -- Read in data set to be fitted
 7 :
 8 : ft = function(x,c) -- Define function to fit data
 9 :    return c[1]*x[2]/(x[2]^c[2]+c[3])^(1/c[2]) - x[1]
10 : end
11 :
12 : c = {1,2,100}; nc = #c -- Initial guess at coefficients
13 : del,err,nmax = nlstsq({yd,xd},fw,ft,c,actv,step) -- Call fiting
14 : print(err,nmax); io.flush() -- print results
15 : for i=1,nc do printf('c[%d] = %12.4e +/- %12.4e\n',i,c[i],
           del[i]) end
16 : -- Now perform MC simulations
17 : getfenv(MCpar).nprint=1 -- Follow MC development
18 : step = {1.1,1.1,1.2}
19 : cvar,nm = MCpar({yd,xd},fw,ft,c,actv,step,1000) -- 1000 MC loops
20 : print('Maximum Newton iterations =',nm)
21 : -- Analyze results
22 : xcdf,ycdf = {},{}
23 : for i=1,nc do xcdf[i],ycdf[i] = makeCDF(cvar[i]) end
24 : write_data('list9_3.dat',cvar[1],cvar[2],cvar[3],xcdf[1],
           ycdf[1],xcdf[2],ycdf[2],xcdf[3],ycdf[3])
25 : for i=1,nc do plot(xcdf[i],ycdf[i]) end
26 : for i=1,nc do stem(hist(cvar[i],40)) end
27 : for i=1,nc do print(stats(cvar[i])) end
28 : -- Correlation plots
29 : scatterplot(cvar[1],cvar[2]); scatterplot(cvar[1],cvar[3]);
           scatterplot(cvar[2],cvar[3])
30 : prob = {68.27,95.45,90,95,99,99.9}
31 : np = #prob
32 : for j=1,nc do
33 :    print('') -- spacer line
34 :    for i=1,np do printf('Limits for c[%d] at %s = %12.4e  to
           %12.4e\n',
35 :    j,engr(prob[i],'%'),climits(xcdf[j],prob[i])) end
36 : end
```
Selected Output:
```
Maximum Newton iterations =   29
Limits for c[1] at  68.27 % =  1.0119e+000  to   1.0376e+000
Limits for c[1] at  95.45 % =  9.9882e-001  to   1.0531e+000
Limits for c[1] at  90.00 % =  1.0037e+000  to   1.0475e+000
Limits for c[1] at  95.00 % =  9.9913e-001  to   1.0516e+000
Limits for c[1] at  99.00 % =  9.9297e-001  to   1.0605e+000
Limits for c[1] at  99.90 % =  9.9013e-001  to   1.0786e+000
```

Listing 9.3. Code for MC analysis of confidence limits for estimated fitting parameters.

and c[3] are not shown in the output. The reader is encouraged to execute the code and observe all outputs and the generated plots.

Figures 9.10 through 9.12 show the MC generated distribution plots for the three model parameters. The generated distributions for c[1] and c[2] are fairly symmetrical about the default model values obtained directly from the measured

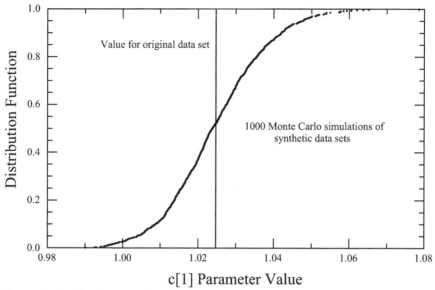

Figure 9.10. Distribution of c[1] parameter values from MC analysis of 1000 synthetic data sets for data in Figure 9.1.

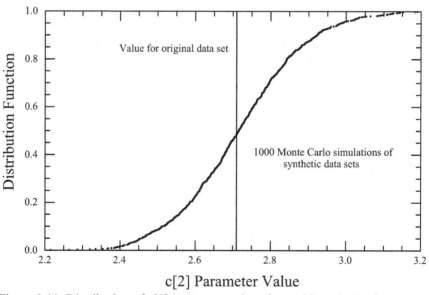

Figure 9.11. Distribution of c[2] parameter values from MC analysis of 1000 synthetic data sets for data in Figure 9.1.

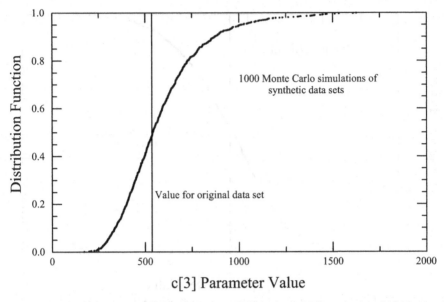

Figure 9.12. Distribution of c[3] parameter values from MC analysis of 1000 synthetic data sets for data in Figure 9.1.

data set as shown by the vertical solid line in the figures. However, the c[3] distribution shows a very asymmetrical distribution with some MC generated values extending to very large values. In all cases the original values of the parameters correspond to about the 0.5 point on the vertical axis of the distribution function as would be expected for Gaussian generated MC parameter values. The distributions for c[1] and c[2] appear to be Gaussian in nature while the c[3] distribution has a considerably longer tail region at large values than would be expected for a Gaussian distribution.

Code for the MCpar() function used in the MC analysis is shown in Listing 9.4. The algorithm is fairly straightforward. First on lines 12 through 15, the fitted function defined by the best-fit set of c parameters is used to generate an array of standard function values on line 13. The newtonfc() function is used to generate the model data set in order to handle functions that are defined in a transcendental manner. Line 14 then calculates the random deviation of the measured data set from the fitted function. All this assumes that the data set has been fitted to a function defined by the function name "ft" in the calling argument list before the call to MCpar() so that the values of the c[] parameters in the calling argument to the MCpar() function are the set of "best-fit" model parameters. The standard deviation between the best-fit function and the measured data set is then evaluated on line 16 and then subsequently used with the gnormal() function on line 21 to generate synthetic random data sets for analysis. The heart of the MC analysis is the while loop from line 19 through line 35 that repeatedly generates new sets of

```
 1 : --/* File mcpar.lua */
 2 :
 3 : require"prob"; require"nlstsq"
 4 :
 5 : MCpar = function(yx, fw, ft, c, actv, step, nmx)
 6 :     local cvar,cc = {},{}
 7 :     local xd = yx[2]
 8 :     local nc,nd,y,yv = #c,#xd,{},{}
 9 :     local k,kp,nnmx,nnm = 1,1,0
10 :     local NTM = getfenv(nlstsq).NMAX-1 -- Max Newton iterations
11 :     for i=1,nc do cvar[i] = {} end
12 :     for i=1,nd do
13 :             y[i] = newtonfc(ft,{yx[1][i],xd[i]},c)
14 :             yv[i] = y[i] - yx[1][i]
15 :     end
16 :     local _,sdev = stats(yv)
17 :     getfenv(nlstsq).nprint=0 -- Don't print iteration values
18 :     nmx = nmx or NMAX -- Default of 1000 iterations
19 :     while k<=nmx do -- Now MC loop
20 :         for i=1,nc do cc[i] = c[i] end -- Initial guess at C's
21 :         for i=1,nd do yv[i] = y[i] + sdev*gnormal() end --MC
22 :         _,_,nnm = nlstsq({yv,xd},fw,ft,cc,actv,step) -- C's
23 :         if nnm>=NTM then
24 :             print('Convergence not achieved for MC #',k);io.flush()
25 :         else
26 :             for i=1,nc do cvar[i][k] = cc[i] end -- Save C values
27 :             k = k+1
28 :         end
29 :         if nnm>nnmx then nnmx = nnm end
30 :         if nprint~=nil then
31 :             if kp==NP then print('Completed MC simulation #',k-1)
32 :             io.flush(); kp = 0 end
33 :         end
34 :         kp = kp+1
35 :     end
36 :     return cvar, nnmx --Return as {{c[1]'s},{c[2]'s},...{c[n]'s}}
37 : end
38 : setfenv(MCpar,{NMAX=1000,NP=100,stats=stats,print=print,io=io,
        newtonfc=newtonfc,
39 :     getfenv=getfenv,nlstsq=nlstsq,gnormal=gnormal,nprint=nil})
```

Listing 9.4. Code listing for the Monte Carlo analysis of the distribution of model equation data fitting parameters.

synthetic data on line 21, calls nlstsq() on line 22 to evaluate new c[] parameters and saves the c[] parameters on line 26 of the code. Note that before entering the MC loop the printing of iteration values in nlstsq() is eliminated by setting the nprint parameter on line 17 to zero. Otherwise the execution would generate many lines of iteration parameter values. Finally on line 36 of the MCpar() function, the array of MC generated c value is passed back to the calling function. A check is made on line 23 to see if the nlstsq() evaluation converged properly. If not a message is printed and the invalid solution set is not accepted (by not incrementing the k loop counter). If the nlstsq() function converged properly, the set of parameters is saved on line 26 and the loop counter incremented on line 27.

It should be noted in Listing 9.3 on line 18 before calling the MCpar() function, a set of step[] parameters are set to non-zero values. This limits the rate at which the nlstsq() function will change the model parameters in searching for a new set of coefficients with each new synthetic data set. Since the generated synthetic data sets are expected to be very similar in nature to the original data set, the model parameters should not have to be changed very much from the original set of values. Setting the step[] parameters aids in achieving convergence for some of the data sets that exhibit significant differences to the original data sets. The reader can experiment with different values for the step[] parameters. Setting the step[] parameters to zero and using no limits on the iterative steps in nlstsq() will result in the lack of convergence for a few of the MC generated data sets as the reader can verify. However, this will make little difference in the overall conclusions from the MC analysis as the results for such cases will simply be ignored in the MCpar() function.

From the MC generated cumulative distributions of the c[] parameters as shown in Figures 9.10, 9.11 and 9.12 estimated limits on the parameters at various confidence levels can be readily obtained by neglecting appropriate numbers of lower and upper values from the parameter distributions. The function climits() as used on line 35 of Listing 9.3 can be used to automatically generate such limits. Various confidence level bounds on the parameter values are shown for the three parameters in Figures 9.13 through 9.15. The solid lines show the bounds as estimated by the MC analysis while the dotted lines show the one standard error

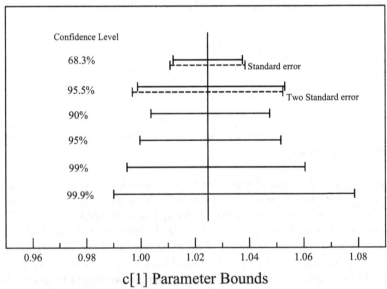

c[1] Parameter Bounds

Figure 9.13. Bounds on c[1] parameter at various confidence levels from MC analysis of data in Figure 9.1 with Eq. (9.5) model. Dotted curves show conventional standard error bounds.

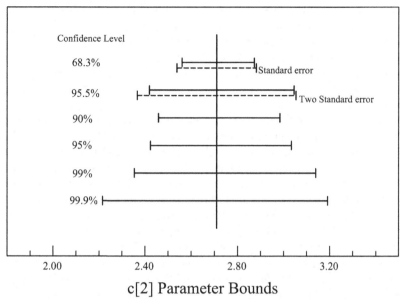

c[2] Parameter Bounds

Figure 9.14. Bounds on c[2] parameter at various confidence levels from MC analysis for data in Figure 9.1 with Eq. (9.5) model. Dotted curves show conventional standard error bounds.

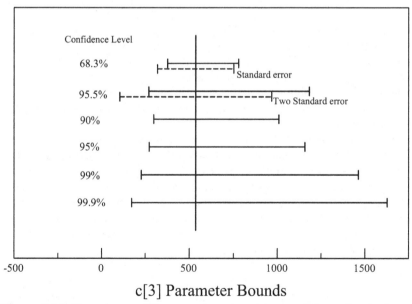

c[3] Parameter Bounds

Figure 9.15. Bounds on c[3] parameter at various confidence levels from MC analysis for data in Figure 9.1 with Eq. (9.5) model. Dotted curves show conventional standard error bounds.

bounds and the two standard error bounds as determined by the values returned by the nlstsq() analysis. For the $c[1]$ and $c[2]$ parameters, the results of the MC simulations and the standard errors agree reasonably closely at the 68.3% and 95.45% confidence levels. For the $c[3]$ parameter, the agreement on the width of the confidence band is also reasonably good. However, the MC results for the $c[3]$ parameter are very asymmetrical as previously noted and the bounds are offset toward larger values.

Looking at the 95% confidence level bounds for the three parameters it will be noted that the range from maximum-to-minimum value is about 5.1% for $c[1]$, about 22.5% for $c[2]$ and about 212% for $c[3]$. This is somewhat disturbing since the measured data points as seen in Figure 9.7 that are being fitted to the model equation shows relatively small percentage deviations from the model curve. One would expect to be able to estimate a set of parameters with much higher precision from the data. When such wide percentage ranges in the bounds of estimated parameters are observed, further investigation of the parameter set is needed.

The MC analysis provides a very important means of looking at possible correlations among the model parameters. This has already been included in Listing 9.3 through the pop-up scatterplots() on line 29. These are shown in Figures 9.16 through 9.18 for all combinations of parameter pairs. For the $c[1]$ vs. $c[2]$ comparison in Figure 9.16 some negative correlation is seen with larger values of $c[1]$ corresponding to smaller values of the $c[2]$ parameter. There is a general scatter of the fitting parameters about some negative sloping line with the width of the scatter around the negative sloping line being about the same for the various $c[1]$ values.

For the $c[1]$ vs. $c[3]$ correlation in Figure 9.17, again a negative correlation is seen. However in this case the scatter in $c[3]$ values is considerable larger in width at small values of $c[1]$ than at large values of $c[1]$. Also a best-fit line through the data points would show considerable curvature as opposed to the approximate linear relationship of Figure 9.16. The large uncertainties in the $c[3]$ parameter are obviously related to the large scatter in the $c[3]$ values fitted to the curve at small $c[1]$ values.

Finally, Figure 9.18 shows the correlation between the $c[2]$ and $c[3]$ parameter values. This curve is most disturbing as it shows that there is an extremely close correlation between the $c[2]$ and $c[3]$ parameters in the proposed data model for fitting sets of experimental data similar to the curve of Figure 9.7. With nonlinear models, it is not unusual to see some correlation between extracted parameter values such as seen in the previous two figures. However, this correlation is about as close as one can get to an exact relationship between the model parameters. The solid line in the figure is a power law fit to the MC data points with the shown parameter value.

One of the most valuable results of the MC analysis is the ability to look at parameter correlations such as those in the above three figures. If the interest is only in obtaining a good fit of a function to a set of data, these correlations may not be of prime interest. However, if one wishes to attach some physical significance to

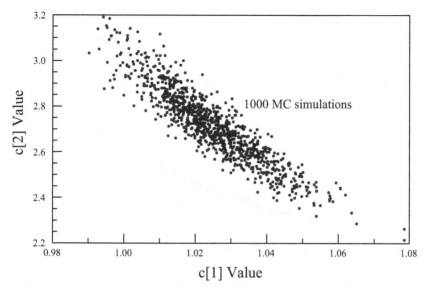

Figure 9.16 Correlation plot between extracted c[1] and c[2] parameter values for MC simulations of Figure 9.1 data with Eq. (9.5) model.

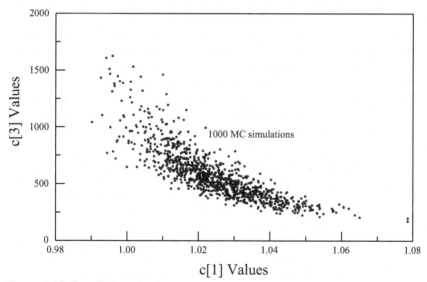

Figure 9.17 Correlation plot between extracted c[1] and c[3] parameter values for MC simulations of Figure 9.1 data with Eq. (9.5) model.

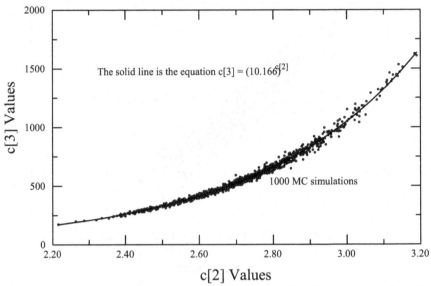

Figure 9.18 Correlation plot between extracted c[2] and c[3] parameter values for MC simulations of Figure 9.1 data with Eq. (9.5) model.

a set of model parameters, one is very interested in how accurately a set of parameters can be determined and in possible correlations between the parameters. The above correlations help in understanding the large uncertainty, percentage wise, in the model parameters such as the 212% range in the c[3] parameter at the 95% confidence level. The strong correlation shown in Figure 9.18 indicates that approximately the same level of data fitting or least squares values can be obtained over a wide range of c[2] and c[3] values as long as the parameters are changed according to the relationship seen in Figure 9.18. In cases such as this it is frequently difficult to obtain convergence of a nonlinear data fitting technique such as nlstsq() because coupled changes in the model parameters have very little affect on the total mean square error of the fitting technique.

The strong correlation between c[2] and c[3] is a strong signal that reliably values can not be determined separately for these two parameters in the proposed data model. If at all possible it indicates that one needs to fix one or the other of these parameters, possibly from other physical considerations, and re-evaluate the data with only a two parameter model. For the purpose of discussion here, this will be done by assuming that the c[2] parameter is fixed at the 2.7 value which is about the mid-range of the MC determined values in Figure 9.18. A physical justification for this choice will not be attempted, since the source of the data set is unknown. This change can be readily accomplished with a small change to the code in Listing 9.3. When this change is made a few of the results will be presented here. The reader is encouraged to go back to Listing 9.3 and replace lines 12 with:

```
12 : c,actv = {1,2.7,100},{1,0,1}; nc = #c -- Initial guess
     at coefficients
```

This fixes the c[2] parameter at a value of 2.7, although any value between about 2.4 and 3.2 would work about as effectively, if any additional information were available on which to select a value. With these changes, the previous analysis in Listing 9.3 and the MC analysis can be preformed with only c[1] and c[3] treated as model parameters. The reader is encouraged to perform the replacement and re-execute the code.

With a fixed c[2] value the re-execution of Listing 9.3 produces the following estimated parameter values and standard errors:

> c[1] = 1.0254e+000 +/- 5.3448e-003
> c[2] = 2.7000e+000 +/- 0.0000e+000
> c[3] = 5.2391e+002 +/- 1.7203e+001

The standard error on c[3] has now dropped from 216 in the previous analysis to 17.2. This now represents about 3.3% of the estimated value whereas the previous value (letting c[2] also vary) represented about 40% of the extracted value. The 3.3% standard error in a parameter value is much more what would be expected from the random error in the data set. The MC analysis now produces the following 95.45% confidence limits bounds on c[1] and c[3]:

> Maximum Newton iterations =   3
> Limits for c[1] at 95.45 % = 1.0147e+000 to   1.0357e+000
> Limits for c[3] at 95.45 % = 4.9083e+002 to   5.5589e+002

These are very close to the 2 standard error bounds as can be seen from the values for c[1] and c[3] in the previous listing. In executing Listing 9.3 with fixed c[2], the pop-up graphs will also verify that the cumulative distributions for the c[1] and c[3] parameters appear to be much closer to Gaussian distributions than the corresponding distributions shown in Figures 9.10 through 9.12. Also the maximum number of Newton iterations needed for converging with the synthetic data sets is only 3 for this case, which indicates that the least squares fitting occurs much easier with a fixed c[2] value.

The correlation plot between the c[1] and c[3] parameters for fixed c[2] is shown in Figure 9.19. In this case the range of both parameters is much less than previously shown in Figure 9.17. There is positive correlation between the two parameters as seen by the general upward trend in the data and as shown by the solid line in the figure. However the simulated values are relatively uniformly distributed above and below the solid line with about equal spread in the random values along the line. Again some general linear trend in parameter correlations is not unusual with nonlinear models, but the MC data should be relatively uniformly distributed around the trend in values. What should be seen in such correlation plots is a generally oval distribution of parameter values around the best fit values for the data set and this correlation plot has these desired features.

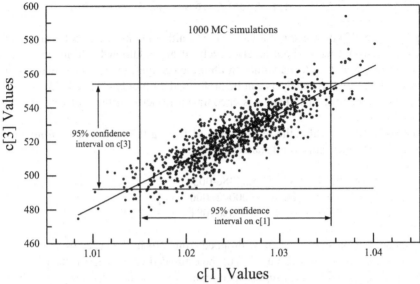

Figure 9.19. Correlation plot between extracted c[1] and c[3] parameter values for MC simulations with c[2] fixed at 2.7and for Figure 9.1 data with Eq. (9.5) model.

The parameter bounds and confidence values so far discussed are for the single parameters individually. These limits are shown on Figure 9.19 as horizontal or vertical lines for the 95% confidence intervals. The 95% confidence intervals contain 950 of the 1000 MC simulation points, so for the c[3] intervals, there are 25 excluded points above the upper horizontal solid line and 25 excluded points below the lower horizontal line. Similarly 50 points are excluded for the 95% confidence intervals for the c[1] variable with 25 points to the left of the leftmost vertical solid line and 25 points to the right of the rightmost vertical solid line.

Sometimes the 95% confidence region is desired considering both parameters jointly. In this case one needs to form in some way an elliptical region around the mean values that contains the closest 95% of the points in Figure 9.19 to the mean values. This must be done taking into account the different standard deviations of the two parameters and subtraction out any general trend line in the correlation betweem the parameters. More specifically, after subtracting out any bias dependence of the data, such as the solid line in Figure 9.19, the two sets of random values can be put on an equal footing by transforming to a set of χ^2 variables which are obtained by further subtracting the mean and dividing by the standard deviation. The largest combined χ^2 values are then excluded with the number depending on the desired confidence level of the evaluation. A function for performing this joint confidence bounds analysis for two joint parameters is provided as cltwodim() (for confidence limits in two dimensions).

```
 1 : -- /* File list9_5.lua */ Analysis of joint confidence limits
 2 :

 3 : require"cltwodim"

 4 :
 5 : c = {{},{},{}} -- C[] arrays
 6 : read_data('list9_3a.dat',c) -- Read first 3 columns of file
 7 : clim,xyb,cxx = cltwodim(c[1],c[3],90) -- 90% confidence limits
 8 : clim2,xyb2,cxx2 = cltwodim(c[1],c[3],95) -- 95% confidence lims
 9 : for i=1,2 do
10 :    for j=1,2 do
11 :       print('c['..i..'], 95% limits = ',clim2[i][j],
12 :             '90% limits = ',clim[i][j])
13 :    end
14 : end

15 : scatterplot(c[1],c[3],xyb,cxx,xyb2,cxx2)
16 : write_data('list9_5.dat',c,xyb,cxx,xyb2,cxx2) -- Save for plots
Output:
c[1], 95% limits = 1.0125579413551    90% limits = 1.0141857243023
c[1], 95% limits = 1.0377875039991    90% limits = 1.0361597210519
c[2], 95% limits = 483.51889261519    90% limits = 488.42457318949
c[2], 95% limits = 564.73740487182    90% limits = 559.14677206047
```

Listing 9.5. Code segment for calculating joint confidence bounds between two parameters using data saved from Listing 9.3.

Use of the joint confidence limit function is shown in Listing 9.5 and the data used for the analysis is that shown in Figure 9.19 which is read from a stored file produced by Listing 9.3. The arguments to the cltwodim() function are the two tables of MC generated coefficient values and a desired confidence level. The cltwodim() function returns three tables: (a) a table with the upper and lower joint probability bounds on the two passed coefficients (the clim on line 7), (b) a table of x,y values describing the two dimensional confidence level boundary in the coefficient space (the xyb on line 7), and (c) a table listing all the coefficient value pairs lying outside the desired confidence boundary (the cxx on line 7). These returned values are perhaps best understood by considering a plot of the returned values as in Figure 9.20. This shows all the MC generated coefficient values with two elliptical curves labeled the 90 % and 95 % confidence boundaries. These elliptical curves are plots of the second table values returned by the function (the xyb and xyb2 tables). Outside the 95 % confidence region there are 5 % or 50 of the MC generated points with the largest χ^2 values (with values returned in the cxx2 table). For the 90% confidence region the excluded region contains 10% of the MC generated points (with values returned in the cxx table). For a good analysis, the excluded points should be distributed rather uniformly outside the various confidence boundaries. This is seen to be the case in Figure 9.20.

The two horizontal solid lines and the two solid vertical lines, touching the 95% confidence boundary, provide the confidence bounds for c[1] and c[2] considered jointly. These values are from the first table of values returned by the cltwodim() function (the clim and clim2 tables) and are the values printed in the output section of Listing 9.5. By comparing Figure 20 with the results in Figure 9.19 it can be seen that the joint confidence bounds are always somewhat larger than the bounds considering each parameter individually. In considering only one variable as in Figure 9.19, 5% of the points lie either above or below the two horizontal lines identifying the 95% confidence bounds. In Figure 9.20, on the other hand, 5% of the points lie outside the two dimensional elliptical region with the highest and lowest values identified by the solid horizontal lines. Thus there will of necessity be fewer points below and above the joint boundary lines and the joint confidence interval must be wider than the bounds for each parameter considered individually. Computer code for the cltwodim() function is not shown but the reader can view the code in the cltwodim.lua file.

With the aid of Figure 9.20 some of the assumptions underlying the MC analysis of confidence limits on fitted model parameters can be reconsidered. With any particular data set, a least squares analysis will evaluate a set of model parameters that are not exactly the real world values assuming that an infinite set of parameter values could be collected. Hopefully a set of data will be a good representative data set for the complete real world data set and the values of the obtained model parameters by least squares fitting to the data will be close to the actual real world parameters. When a MC analysis is performed, sets of fitted model parameters are obtained that are distributed around the measured data set values as illustrated in Figure 9.20. The underlying assumption of the MC analysis is not that the analysis evaluates the distribution around the set of exact real world parameters, but that if the starting parameter set were moved around in the two dimensional space of

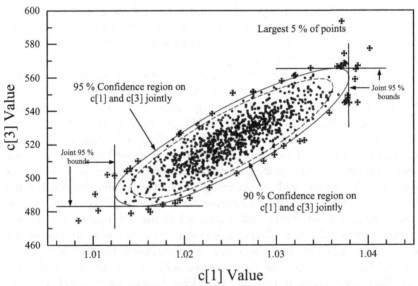

Figure 9.20. Joint confidence intervals for c[1] and c[2] as generated by Listing 9.5. 95 % of points lie within solid line ellipse and 90 % of points within the dotted ellipse.

Figure 9.20 and the MC simulations were re-evaluated, a similar set of elliptical confidence regions would be obtained centered around any starting set of parameters, provided one does not move too far from the measured set of parameters. For example, if the starting set of parameter values were moved to perhaps some point on the 95% boundary of Figure 9.20 and the MC simulations were re-evaluated, a similar elliptical region would be expected now centered around the new starting set of parameters and the elliptical center in Figure 9.20 would simply shift to the 95% boundary of the new distribution. This assumes of course that the random variables observed in the measured data set are characteristic of the underlying random variables of the physical process. This is why it is very important to remove as much as possible any model dependent errors in the data fitting as discussed with respect to Figure 9.2 and the initial model used to fit this data set. It is also important to remove any major correlation between the model parameters (as done in the MCpar() function) so that the underlying random errors can be modeled.

This section has discussed in length the evaluation of parameter bounds and confidence levels of error bounds for one particular data set. Several important factors that must be considered in any data fitting problem have been illustrated by this example. The parameter correlation plots generated from the MC simulations can provide joint confidence levels considering more than one parameter. However, the most useful application of these correlation studies is perhaps the information that they supply about interdependencies of the parameters of any proposed nonlinear data models. This is information not readily obtainable by any other means and is essential if one is to understand a data fitting model and the meaning of estimated limits on parameters. The model considered here is relatively simple and in real world situations, much more complicated models are typically involved making the MC simulations even more important. The next section will apply these developed techniques to several examples of data models and parameter estimation.

9.4 Examples of Single Variable Data Fitting and Parameter Estimation

This section illustrates the fitting of data and parameter estimation with several examples where there is a single independent variable and a single dependent variable. Also the examples use only a constant weighting factor with data distributed relatively uniformly along the independent variable axis. Many practical examples are of this type. More complex examples are shown in subsequent sections.

9.4.1. Energy from Lamp vs. Temperature

This first example is the experimental measurement of energy from a lamp vs. the temperature of the lamp filament as seen in Figure 9.21. (For a description of the

DanWood data set see: www.itl.nist.gov/div898/strd/nls/nls_main.shtml) The proposed model equation for the data is a two parameter model of the form:

$$y = C_1 x^{C_2} \qquad (9.8)$$

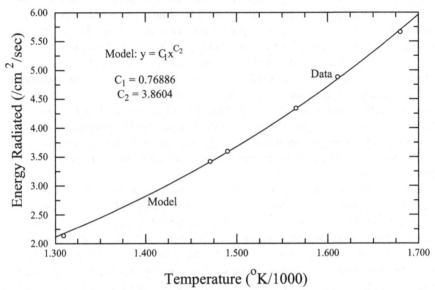

Figure 9.21 Plot of lamp energy data and fitted equation.

The code segment is Listing 9.6 illustrates the analysis of this data set. The code is arranged into 5 sections as shown in the listing. The major tasks of each section of the code are:

1. Section 1(lines 5 through 15): Reads input data file and defines model with coefficients to be fitted to the data.
2. Section 2 (lines 16 through 20): Performs the least squares fitting of the model to the data using nlstsq().
3. Section 3 (lines 21 through 32): Generates the data for the 6 plots discussed in the previous section and plots 4 of the most important of the 6 plots.
4. Section 4 (lines 33 through 46): This code performs a Monte Carlo analysis of the fitting parameters and plots scatter plots of the fitting parameters as well as printing MC generated confidence bounds on the fitting parameters.
5. Section 5 (lines 47 through 54): Generates joint confidence bounds for the parameters and plots a joint scatterplot of the resulting joint confidence bounds for the first two estimated parameters.

The code segment of Listing 9.6 is a prototype of a general purpose routine that can be used with minor modifications to analyze a wide range of data, generating a set of estimated function parameters and looking at the general properties of the

```
 1 : -- /* File list9_6.lua */ -- General file for Param estimation
 2 :
 3 : requires("nlstsq","mcpar","DataFit","cltwodim")
 4 :
 5 : -- Section 1. Input data file and define fitting function
 6 : -- Change lines 6 through 14 for different data sets
 7 : infile = 'DanWood'
 8 : xd,yd = {},{}; read_data(infile..'.txt',yd,xd) -- y stored first
 9 : nd = #xd
10 :
11 : ft = function(x,c) -- Define function to fit data
12 :    return c[1]*x[2]^c[2] - x[1]
13 : end
14 : c = {1,1}; nc = #c -- Initial approximations. End for data sets
15 :
16 : -- Section 2. Perform data fit and print fitted parameters
17 : del,err,nm = nlstsq({yd,xd},fw,ft,c) -- print max iterations
18 : print('Max iteration number =',nm)
19 : for i=1,nc do printf('c[%d] = %12.4e +/- %12.4e\n',i,c[i],
         del[i]) end
20 :
21 : -- Section 3. Generate 6-Plots, Only 1,2,3 and 6 are used here
22 : ycalc,res = {},{} -- Calculate fitted values and residuals
23 : for i=1,nd do ycalc[i] = ft({0,xd[i]},c);
         res[i] = yd[i] - ycalc[i] end
24 : fres1,fres2 = datafit({res,xd}), datafit({res,yd}) -- residuals
25 : yres1,yres2,xyp,xypn = {},{},{},{}
26 : for i=1,nd do yres1[i],yres2[i] = fres1(xd[i]),
         fres2((ycalc[i])) end
27 : xyp[1],xyp[2]=makeCDF(res);xypn[1],xypn[2]=normalCDF(res) --CDF
28 : plot('test2.gfx',{xd,yd,ycalc},{"Data plot"});
29 : scatterplot(xd,yres2,res,{"Residual vs X","X","Res"})
30 : scatterplot(ycalc,yres2,res,{"Residual vs Y","Y","Res"});
         plot(xyp,xypn,{"CDF plot","Res","CDF"})
31 : write_data(infile..'a.dat',xd,yd,ycalc,res,yres1,yres2,xyp,xypn)
32 :
33 : -- Section 4. Generate MC Data
34 : cvar = MCpar({yd,xd},fw,ft,c) -- 1000 MC loops -- default value
35 : xcdf,ycdf = {},{}
36 : for i=1,nc do xcdf[i],ycdf[i] = makeCDF(cvar[i]) end
37 : scatterplot(cvar[1],cvar[2],{"c[2] vs c[1]","c[1]","c[2]"});
38 : if nc>2 then scatterplot(cvar[1],cvar[3],
         {"c[3] vs c[1]","c[1]","c[3]"})
39 :    scatterplot(cvar[2],cvar[3],{"c[3] vs c[2]","c[2]",
         "c[3]"}) end
40 : prob = {68.27,95.45,90,95,99,99.9}; np = #prob
41 : print('Individual confidence limits')
42 : for j=1,nc do
43 :    for i=1,np do printf('Limits for c[%d] at %s =
         %12.4e  to  %12.4e\n',j,engr(prob[i],'%'),
44 :             climits(xcdf[j],prob[i])) end
45 : end
46 :
47 : -- Section 5. Generate data for joint confidence bounds
48 : cl,xyb,cx = cltwodim(cvar[1],cvar[2],90); cl2,xyb2,cx2 =
         cltwodim(cvar[1],cvar[2],95)
49 : print('Joint confidence limits')
50 : for i=1,2 do
51 :    for j=1,2 do print('c['..i..']', 95% limits = ',
```

```
             c12[i][j],'90% limits = ',cl[i][j]) end
52 : end
53 : scatterplot(cvar[1],cvar[2],cx,cx2,'with lines',xyb,xyb2,
         {"c[2] vs c[1] with limits","c[1]","c[2]"}) -- Joint limits
54 : write_data(infile..'b.dat',cvar,xyb,cx,xyb2,cx2)
```

Listing 9.6. General code segment for data fitting and parameter estimation with various plots to view the degree of fit of data to model.

estimated model parameters. This code segment forms the basis of the analysis for all the data sets and model equations in this section.

A plot of the data and the model equation with the estimated parameters for the first set of data is shown in Figure 9.21. For this data set there are only 6 data points so this is an example of parameter estimation with a limited number of data points. Shown in Figure 9.22 is a plot of the fitted residuals as a function of the independent data variable (or temperature). As seen in the figure, the residuals do not appear to be random but show a basic second degree type of dependency with the end point residuals being negative and the mid range residuals being positive. This is perhaps an indication that the basic assumed functional relationship is not adequate to describe the underlying physics of the data and has too rapid a dependency on temperature. However, because of the limited data set, it is difficult to draw too many conclusions from the residual plot. It does raise a flag with regard to the assumed data model which would need further exploration, perhaps with a more extensive data set covering a wider temperature range. A plot of the residuals vs. the dependent variable (not shown her) shows a similar nonrandom variation. However, the residuals are in fact only a very small percentage of the data point values as the model equation closely matches the data points.

The MC generated model parameter correlations are shown in Figure 9.23 based upon 1000 Monte Carlo generated data sets. The figure shows a very strong correlation between the two model parameters indicating that it is difficult to obtain very accurate independent values of the two parameters from the limited range of the experimental data. The printed output when executing the code listing of Listing 9.6 provides the following 95% joint confidence limits for the two model parameters from the joint plot in Figure 9.23:

$$C_1 = 0.729 \text{ to } 0.809$$
$$C_2 = 3.749 \text{ to } 3.975$$

$$(9.9)$$

If one of the parameters, for example the exponent in the temperature model (C_2), could be fixed at some value, the other parameter could be determined much more accurately. However this example of a strong correlation between estimated parameters is typical of many parameter estimation examples where there is a limited range of experimental data.

Executing the code in Listing 9.6 generates six pop-up data plots with only three of these shown here (Figures 9.21, 9.22 and 9.23). The plots not shown are for the residuals vs. the y variable and the scatter plot of C_2 vs. C_1 which is similar to Figure 9.23 but without the boundary lines. Also the printed output is not shown in Listing 9.6. This consists of printed confidence limits for the individual

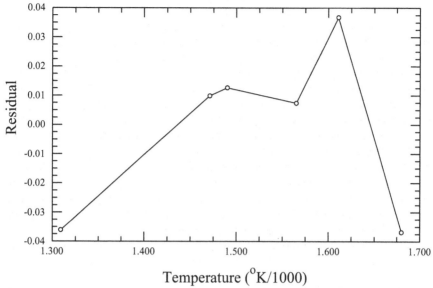

Figure 9.22. Plot of fitted residuals as a function of the independent variable for the data set of Figure 9.21.

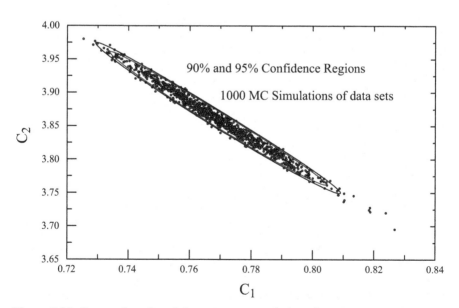

Figure 9.23. Scatterplot of model parameter correlations from Monte Carlo generated data for the data of Figure 9.21.

coefficients from lines 42 through 45 of the listing and joint confidence limits from lines 50 through 52 of the code. The reader is encouraged to execute the code and observe all plots and the printed output as Listing 9.6 with modifications is used to analyze several data sets in this section.

A word of caution is in order with respect to the use of Listing 9.6 and general data sets. The code uses the routine read_data() on line 8 to read in data values from a file. This routine assumes that the data is stored on a line-by-line basis with the data values separated by tabs, commas or spaces and each line terminated by a line return character. For data sets stored in other manners, the routine will not properly read in the data sets. The routine will also ignore text lines in the data file following a data set provided there is a blank line separating the data from any text lines. If data files are downloaded from the nist.gov data site referenced above, the downloaded data files do not satisfy these requirements and the files must be changed into the proper form for use with the read_data() routine. This data mismatch with the read_data() function is also possible with other data sites, so one must be careful to match the data format to the format assumed in the read_data() routine.

9.4.2 Biochemical Oxygen Demand vs. Incubation Time

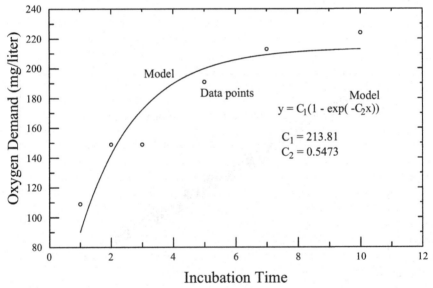

Figure 9.24. Oxygen demand data vs. incubation time and fitted two parameter model equation.

This is a second example of a two parameter model used to fit a set of experimental data, in this case the oxygen demand in mg/liter vs. incubation time for a biochemical experiment. (See www.itl.nist.gov/div898/strd/nls/nls_main.shtml for a description of the BoxBOD data set.) This data set again has a limited number of only 6 data points and the data set also exhibit considerable random variations.

The data and proposed model are shown in Figure 9.24. The model equation is plotted in the figure along with values for the best fit model parameters. As noted, the figure shows a large scatter between the model and the experimental data points. Also the residuals show a definite trend depending on the independent parameter value. The data points can readily be seen to be above the model equation for the two lowest data points, below the model for the two mid points and again above the model equation for the two largest data points. This dependence of the residuals on the incubation time perhaps indicates too sharp a saturation effect for the proposed mode. However, again the limited number of data points makes definitive conclusions difficult.

```
 1 : -- /* File list9_7.lua */ -- file for Parameter estimation
 2 :
 3 : requires("nlstsq","mcpar","DataFit","cltwodim")
 4 :
 5 : -- Section 1. Input data file and define fitting function
 6 : -- Change lines 6 through 14 for different data sets

 7 : infile = 'BoxBOD'
 8 : xd,yd = {},{}; read_data(infile..'.txt',yd,xd) -- y stored first
 9 : nd = #xd
10 :
11 : ft = function(x,c) -- Define function to fit data
12 :    return c[1]*(1 - math.exp(-c[2]*x[2])) - x[1]
13 : end
14 : c = {100,1}; nc = #c -- Initial approximations. End of changes

. . . Same as Listing 9.6
Selected Output:
Max iteration number = 8
c[1] =   2.1381e+002 +/-  1.2354e+001
c[2] =   5.4727e-001 +/-  1.0457e-001
Individual confidence limits
Limits for c[1] at  68.27  % =   2.0398e+002  to   2.2663e+002
Limits for c[1] at  95.45  % =   1.9400e+002  to   2.3787e+002
Limits for c[1] at  90.00  % =   1.9708e+002  to   2.3295e+002
Limits for c[1] at  95.00  % =   1.9430e+002  to   2.3681e+002
Limits for c[1] at  99.00  % =   1.9045e+002  to   2.4207e+002
Limits for c[1] at  99.90  % =   1.8814e+002  to   2.4617e+002
Limits for c[2] at  68.27  % =   4.6222e-001  to   6.5416e-001
Limits for c[2] at  95.45  % =   3.9194e-001  to   7.9147e-001
Limits for c[2] at  90.00  % =   4.1276e-001  to   7.2840e-001
Limits for c[2] at  95.00  % =   3.9265e-001  to   7.7671e-001
Limits for c[2] at  99.00  % =   3.6523e-001  to   8.5648e-001
Limits for c[2] at  99.90  % =   3.4963e-001  to   1.0549e+000
Joint confidence limits
c[1], 95% limits = 188.37790297693    90% limits = 192.01967801776
c[1], 95% limits = 241.54243438149    90% limits = 237.90065934066
c[2], 95% limits = 0.32184067945722   90% limits = 0.35193429551886
c[2], 95% limits = 0.8144608655263    90% limits = 0.77629440168876
```

Listing 9.7. Changes to Listing 9.6 for Oxygen Demand example as shown in Figure 9.24.

Changes to the code of Listing 9.6 needed to analyze this data set with the proposed model are shown in Listing 9.7. The only changes are in the name of the input file on line 7, the data fitting model on line 11 through 13 and the initial guess at the parameters on line 14. From the maximum data values and the model equation, it is seen that the value of the $c[1]$ parameter must be in the range of 200, although the initial guess used was only 100. Also shown in Listing 9.7 is the printed output produced by execution the code in Listing 9.7. Similar data is generated by Listing 9.6, although the results were not presented for the previous example. The printed output shows standard error results for various probability limits of the individual model parameters. Also the final 4 output lines show joint bounds for the variables at 90% and 95% probabilities. As previously discussed, these always cover a wider range of values than the single variable limits.

Figure 9.25 shows the results of the Monte Carlo simulation for 1000 generated random data sets for the distribution of model parameter values. Also shown in the figure are the 90% and 95% probability bounds on the parameter values. While there is some correlation between the two model parameters, the correlation is much weaker than in the previous example. The data points in the figure show a generally downward trend of the C_2 value with increasing C_1 value. The data points within the 90% boundary appear to be relatively randomly distributes. However the points outside the 95% boundary do not appear to be randomly distributed but concentrate at the largest C_1 and C_2 values. The uncertainty in the predicted model parameters is relatively large percentage wise because of the relatively large scatter in the experimental data.

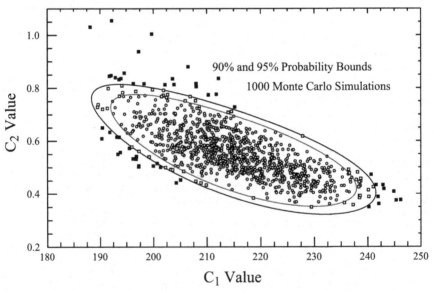

Figure 9.25. Joint probability bounds on model parameters for Oxygen Demand data.

9.4.3 Ultrasonic Sensor Response vs. Metal Distance

This example is the experimental response of an ultrasonic sensor as a function of metal distance. Such a sensor can be used to measure distance or thickness of a metal. (See www.itl.nist.gov/div898/strd/nls/nls_main.shtml for a description of the Chwirut1 data set.) The experimental set data is shown in Figure 9.26 along with the model used to fit the data and the best-fit parameter values. The model equation is a three parameter model as given by the equation shown within the figure. For this example, the number of data points is 214 with several repeated data points at the same value of metal distance.

There is a significant amount of scatter in the data points, but it can be seen that the points scatter rather uniformly above and below the solid line given by the model equation. This can also be seen in Figure 9.27 which shows a plot of the residuals as a function of the independent parameter. The residuals are distributed rather uniformly about zero with no long runs of positive or negative values. The data does show some tendency for larger residual values for the smaller values of metal distance. From this data, it would be concluded that the proposed model provides a good functional relationship to the underlying physics at least as evidenced by the experimental data.

Listing 9.8 shows the modifications to Listing 9.6 used to analyze this data set with 3 unknown parameters. The changes are again only on lines 7 through 14 which define the fitting function and the initial guesses for the model parameters on line 14. In addition, Section 5 of the code that generates the joint confidence

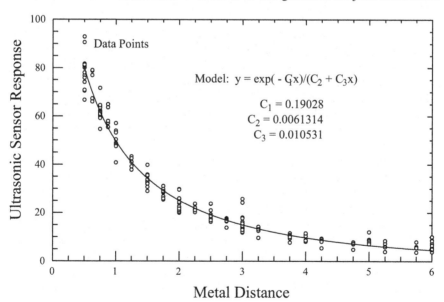

Figure 9.26. Ultrasonic sensor response vs. metal distance, data and model equation.

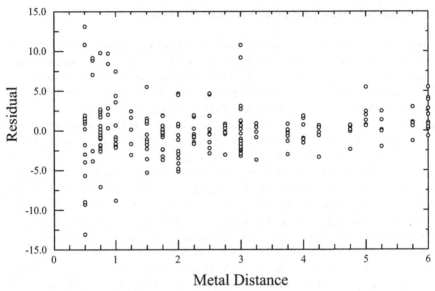

Figure 9.27. Residuals vs. independent parameter of model for Ultrasonic Sensor Response from Figure 9.26.

bounds will now use the code on lines 38 and 39 of Listing 9.6. This requires 3 plots and 3 evaluations for each possible pair of parameters. These three joint probability bounds plots are shown in Figures 9.28, 9.29 and 9.30. The data shows a positive correlation between the C_1 and C_2 parameters, and a negative correlation between C_1 and C_3 and C_2 and C_3. The C_3 parameter has a rather strong correlation to both C_1 and C_2 although the correlation is not as strong as that seen in the two parameter model used in Section 9.4.1. Since this data set has over 200 data points, the MC evaluations will take more time than the similar calculations for the earlier examples in this section. Again 1000 MC data sets are used for the MC simulations.

In addition to the generated cross correlation plots, Figure 9.31 shows the CDF of the measured residuals compared with a Gaussian model of the residuals with a mean and standard deviation equal to that of the measured residuals. As opposed to most of the previous data sets, the residuals here show considerable differences from a Gaussian model. The data points have considerably longer tail regions for residuals far removed from the mean. For this data a better representation for the residuals might be a Cauchy or Student's t distribution as previously shown in Figure 8.8. Because of these differences, the use of a Gaussian distribution for the MC analysis can certainly be questioned. If a more representative distribution of the residuals were used, it would probably be expected that the parameter correlation plots of Figures 9.28 through 9.30 would have the data points clustered more

```
 1 : -- /* File list9_8.lua */-- file for Parameter estimation
 2 :
 3 : requires("nlstsq","mcpar","DataFit","cltwodim")
 4 :
 5 : -- Section 1. Input data file and define fitting function
 6 : -- Change lines 6 through 14 for different data sets
 7 : infile = 'Chwirut1'
 8 : xd,yd = {},{}; read_data(infile..'.txt',yd,xd) -- y stored first
 9 : nd = #xd
10 :
11 : ft = function(x,c) -- Define function to fit data
12 :    return math.exp(-c[1]*x[2])/(c[2] + c[3]*x[2]) - x[1]
13 : end
14 : c = {.1,.01,.02}; nc = #c -- Initial approximations. End of
changes
-------- same as Listing 9.6
Selected output:
Max iteration number = 7
c[1] =   1.9028e-001 +/-   2.1939e-002
c[2] =   6.1314e-003 +/-   3.4500e-004
c[3] =   1.0531e-002 +/-   7.9282e-004
Joint confidence limits
c[1], 95% limits = 0.1363899395656       90% limits = 0.14215675423834
c[1], 95% limits = 0.2453598343529       90% limits = 0.23959301968016
c[2], 95% limits = 0.0052410900597       90% limits = 0.0053393292620691
c[2], 95% limits = 0.0069625900415       90% limits = 0.0068781698992996
```

Listing 9.8. Example code segment for data analysis with three fitting parameters for data of Figure 9.26.

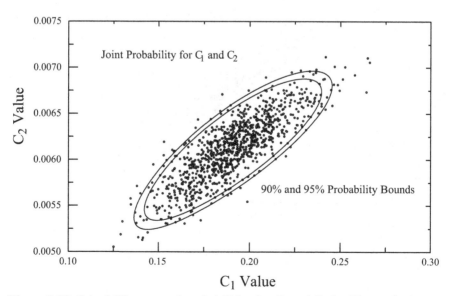

Figure 9.28. Joint MC generated probabilities for C_1 and C_2 for Ultrasonic Sensor Response.

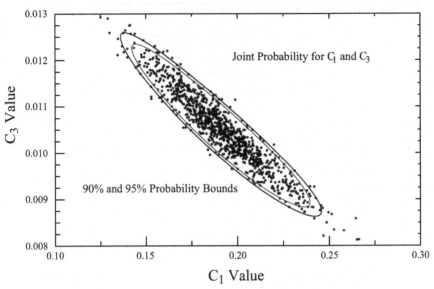

Figure 9.29. Joint MC generated probabilities for C_1 and C_3 for ultrasonic sensor response.

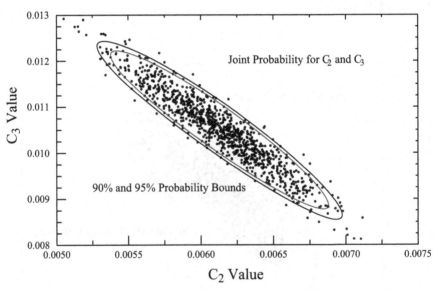

Figure 9.30. Joint MC generated probabilities for C_2 and C_3 for ultrasonic sensor response.

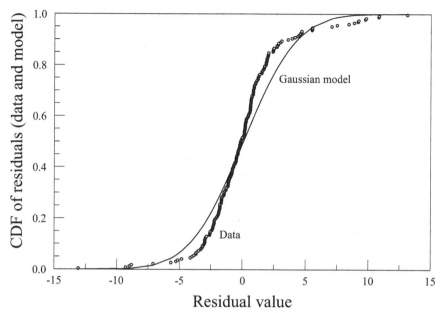

Figure 9.31. CDF plot of experimental and Gaussian model residuals for ultrasonic sensor response.

toward the centers of the ovals in the figures. However, a more realistic distribution might not make much difference for the values in the 90% and 95% regions as these probably arise from values clustered in the tails of the residual distributions. It will be left for the reader to implement a more appropriate distribution for the MC analysis and see if there are any appreciable differences in the MC predicted values. Note that such a change requires modifications to the MCpar() function in listing 9.4 as this function uses the gnormal() Gaussian random number function to generate the MC data sets. As a general rule the CDF plot should always be observed in order to access how much confidence one can place in the MC analysis.

9.4.4 Pasture Yield vs. Growing Time

Many sets of experimental data in the natural sciences tend to follow a general sigmoidal curve of the form (for 3 parameters):

$$y = C_1 / (1 + \exp(C_2 - C_3 x)) \tag{9.10}$$

One such data set is that of pasture yield vs. growing time as shown in Figure 9.32. (See www.itl.nist.gov/div898/strd/nls/nls_main.shtml for a description of the Rat42 data set.) Data of the sigmoidal type have a low saturation value and high saturation value as the independent variable changes from low to high values and exhibit a smooth almost symmetrical transition between the two saturation limits. For the data in Figure 9.32, the two limiting values are only approached approximately, but some value near zero and some value slightly above 70 appears to be the saturation values. Listing 9.9 shows the required code changes to analyze this data set. Changes again are only in the function definition and the initial guesses at the parameter values.

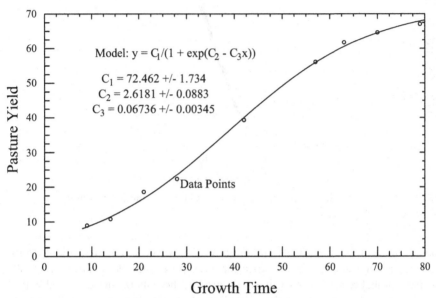

Model: $y = C_1/(1 + \exp(C_2 - C_3 x))$

$C_1 = 72.462$ +/- 1.734
$C_2 = 2.6181$ +/- 0.0883
$C_3 = 0.06736$ +/- 0.00345

Figure 9.32 Pasture yield vs. growth time with sigmoidal model.

A few words about selecting initial values are in order for this function since convergence of the data fitting is not as readily obtained as with the previous data sets in this section. In fact line 14 of the code listing shows that step size limitations are imposed on the nlstsq() function by setting all the step size parameters to 0.5. It will be recalled that a positive value of the step parameter limits the change in a parameter value to the factor set in the step[] array. This means that the value of a fitting parameter can change by a factor within the range of 0.5 to 1/0.5 (or 2) between each iterative step of the data fitting procedure. This is the first example in this section where a step size limitation has been used in order to obtain convergence of the fitting procedure. The exact value of the step size limitation is not critical as the reader can verify by running the code with different limiting values. In fact the factor of 2 limitation used is not a very severe limitation but prevents the procedure from straying too far from the initial guesses.

Now for a discussion of the initial guesses of the fitting parameters. From the form of the equation, it is seen that the C_1 value should be the upper saturation value of the data which in this case appears to be a little over 70. Thus a good initial guess for this parameter would be around 70. The value on line 13 of the code listing is 100 and this value was deliberately selected somewhat larger than expected to explore convergence issues with the algorithm. The midpoint of the data range occurs when the argument of the exponent in Eq. (9.10) equals zero and this is seen to occur at an x value of around 40. Thus we expect that $40C_3 \simeq C_2$. Finally it can be seen that about 90% of the change in the function occurs over a time scale of magnitude about 80, so it would be expectd that $\exp(80C_3) \simeq 10$. Combining this with the above relationship between the fitting parameters, the following estimates can be obtained for the three fitting parameters: $C_1, C_2, C_3 \simeq 70$, 1.2, 0.03. These are much closer to the optimum values than the initial guesses shown on line 13 of Listing 9.9. In fact if the code is re-executed with these initial guesses, there is no need to impose any limitations on the step size of the parameter changes between iterations and the fitting procedure converges in 6 iterations. The reader is encouraged to make these changes and observe the convergence to the same estimated parameters. The values in Listing 9.9 show that the fitting

```
 1 : -- /* File list9_9.lua */ -- file for Parameter estimation
 2 :
 3 : requires("nlstsq","mcpar","DataFit","cltwodim")
 4 :
 5 : -- Section 1. Input data file and define fitting function
 6 : -- Change lines 6 through 14 for different data sets

 7 : infile = 'Rat42'
 8 : xd,yd = {},{}; read_data(infile..'.txt',yd,xd) -- y stored first
 9 : nd = #xd
10 :
11 : ft = function(x,c) -- Define function to fit data
12 :    return c[1]/(1 + math.exp(c[2] - c[3]*x[2])) - x[1]
13 : end
14 : c = {100,1,.1}; nc = #c; step ={.5,.5,.5}

----- Same as Listing 9.6

Selected output:
Max iteration number = 7
c[1] =  7.2462e+001 +/-  1.7340e+000
c[2] =  2.6181e+000 +/-  8.8297e-002
c[3] =  6.7360e-002 +/-  3.4466e-003
Individual confidence limits
Limits for c[1] at  68.27  % =  7.1055e+001  to  7.3957e+001
Limits for c[1] at  95.45  % =  6.9681e+001  to  7.5738e+001
Limits for c[2] at  68.27  % =  2.5435e+000  to  2.6950e+000
Limits for c[2] at  95.45  % =  2.4791e+000  to  2.7683e+000
Limits for c[3] at  68.27  % =  6.4619e-002  to  7.0382e-002
Limits for c[3] at  95.45  % =  6.1705e-002  to  7.3183e-002
Joint confidence limits
c[1], 95% limits = 68.924142645163   90% limits = 69.368102014539
c[1], 95% limits = 76.144089776572   90% limits = 75.700130407197
c[2], 95% limits = 2.4384561173302   90% limits = 2.4609482460916
c[2], 95% limits = 2.8028153528519   90% limits = 2.7804954455003
```

Listing 9.9. Code segment showing changes for analyzing Pasture Yield data set of Figure 9.32.

procedure is rather robust with respect to selecting the initial guesses provided a step size limitation is set on the iterative procedure.

The above discussion has illustrated how simple features of the data can be used to obtain good initial estimates of the fitting parameters and improve convergence and also how a step size limitation parameter can be used to improve convergence of the Newton iterative algorithm in the nlstsq() function if the initial guesses are far from the correct values. The values of the model parameters given in Figure 9.32 are the values returned by nlstsq() along with the one sigma estimates of the uncertainty in the parameter values.

The joint probability bounds on the three fitting parameters taken in pairs are shown in Figures 9.33 through 9.35. The MC data shows some correlation among the model parameters but also exhibit more elliptical like areas than in the previous example. Perfect independence of the parameters would result in an ellipse with the major axis parallel to one of the variable axes and not tilted ellipses as shown in the figures. One can estimate the joint bounds on the parameter values from the figures or from the printed values as shown in the selected output of Listing 9.9. Again the MC generated limits for the individual parameters are close to the corresponding one sigma values returned by the nlstsq() function and the joint probability limits are slightly larger as expected. Because the data points in Figure 9.32 agree rather closely with the model curve, the uncertainties in the obtained model parameters are relatively small in percentage terms. From this analysis one would conclude that the proposed data model provides a good model for the experimental data.

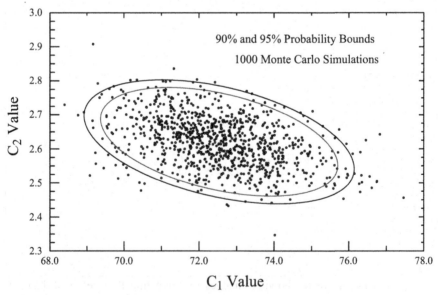

Figure 9.33. Joint MC generated probabilities for C_1 and C_2 for pasture yield data.

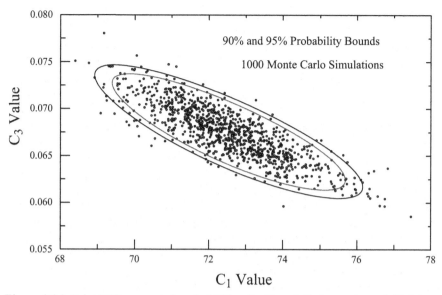

Figure 9.34. Joint MC generated probabilities for C_1 and C_3 for pasture yield data.

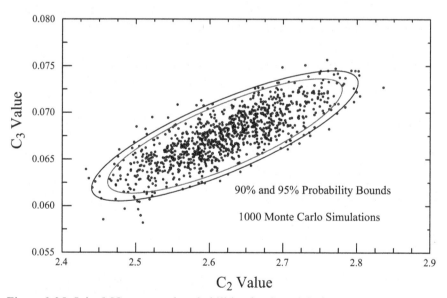

Figure 9.35. Joint MC generated probabilities for C_2 and C_3 for pasture yield data.

9.4.5 Two Gaussian Curves with Exponential Base Line

This example illustrates parameter estimation with a large number of data points and a large number of model parameters. This is illustrative of many types of material analysis where one observes a Gaussian-like response of some variable on a baseline with noise. In this particular data set two Gaussian curves are located close to each other with a baseline response. (for a description of the Gauss3 data set see www.itl.nist.gov/div898/strd/nls/nls_main.shtml.) The data and model function are shown in Figure 9.36. The two peaks partially overlap with no clear peak from the second Gaussian. The model function has a total of 8 parameters to be determined from the least squares fitting so this is a more severe test of parameter estimation than the previous examples in this section. Each Gaussian peak is described by 3 parameters related to peak height, peak location and peak width (C_1, C_2 and C_3 for first peak). In terms of the least squares fitting approach and obtaining initial estimations of the parameters, good initial estimations are relatively easily obtained from the data. The most important parameter is the location of the maximum value of the peaks which can be estimated as occurring at about 115 and 140 units along the x axis. Estimated peak heights can be seen to be about 100 and 80 units along the y axis. The constants C_3 and C_6 then can be estimated from the half width of the peaks to be about 20 squared or 400.

Code for the data fitting and parameter estimation is partially shown in Listing 9.10. Only shown are the lines changed from Listing 9.6 in order to analyze the Gauss3.txt data set. Again the changes are only in the function definition of lines 11 through 15 and the initial approximations on line 16. No step limitations are

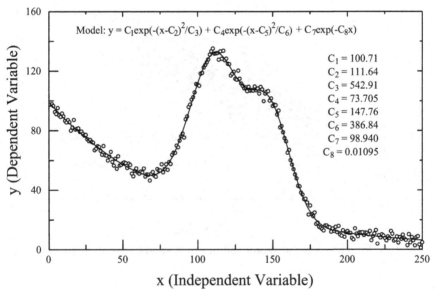

Figure 9.36. Two Gaussian Peaks data with model fit to data.

```
 1 : -- /* File list9_10.lua */ -- General file for Parameter estima-
tion
 2 :
 3 : requires("nlstsq","mcpar","DataFit","cltwodim")
 4 :
 5 : -- Section 1. Input data file and define fitting function
 6 : -- Change lines 6 through 14 for different data sets
 7 : infile = 'Gauss3'
 8 : xd,yd = {},{}; read_data(infile..'.txt',yd,xd) -- x stored first
 9 : nd = #xd
10 :
11 : ft = function(x,c) -- Define function to fit data
12 :    xx = x[2]
13 :    return  c[1]*math.exp(-(xx - c[2])^2/c[3]) +
           c[4]*math.exp(-(xx - c[5])^2/c[6]) +
14 :    c[7]*math.exp(-c[8]*xx) - x[1]
15 : end

16 : c = {100,115,400,80,140,400,100,.01}; nc = #c -- Initial c's
```

Listing 9.10. Partial code segment for analysis of the two Gaussian peaks data shown in Figure 9.36.

needed with the initial guesses shown for the model parameters. Executing the code in Listing 9.10 produces convergence to the model parameter values in 7 iterations and Figure 9.36 shows that there is good agreement between the experimental data and the proposed model where the solid line curve is the model equation with the parameter values given in the figure.

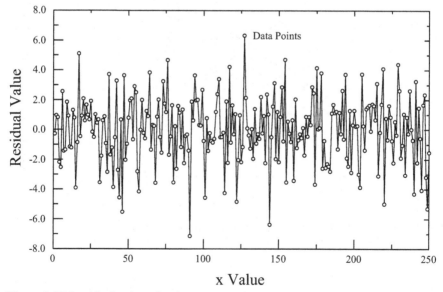

Figure 9.37 Residual values for the two Gaussian curve fit shown in Figure 9.36.

Figure 9.37 shows a plot of the fitted curve residuals as a function of the x variable. A relatively uniform scatter of the data points about the model equation values can be seen, indicating that the proposed equation is a good model for describing the data points. This is about as good a distribution of random residual values as would ever be expected from a set of experimental data. Also a graph of the CDF for the residuals shows a very good fit to a Gaussian distribution. Such a curve can be seen as a pop-up plot when the code in Listing 9.10 is executed. The reader is encouraged to execute the cods and observe the generated graphs. Because of the large number of data points, the MC analysis will require more computer time than any of the previous examples in this section.

As in previous examples, executing Listing 9.10 also produces a Monte Carlo analysis of the distribution of parameter values. With 8 parameters, there are a large number of joint parameter distributions that could be explored and studied. In many such types of data fitting what is most desired is an accurate estimate of the location of the peak heights (C_2 and C_5 in the model equation). Joint probability distributions for the parameters associated with one of the peaks, C_1, C_2 and C_3 are shown in Figures 9.38, 9.39 and 9.40. While there is some correlation among the three parameters, the three values can be seen to be relatively independent of each other.

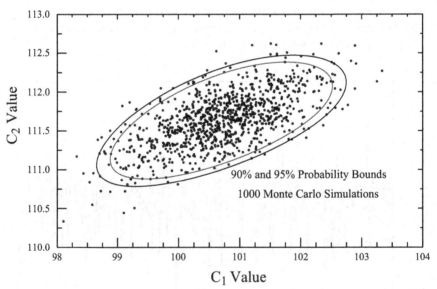

Figure 9.38. Joint MC generated probabilities for C_1 and C_2 for two Gaussian peaks data.

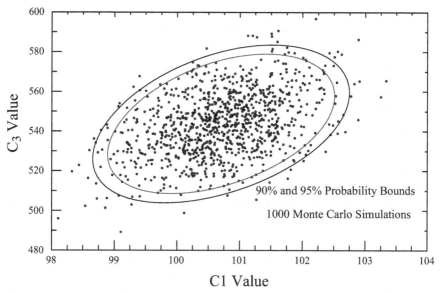

Figure 9.39. Joint MC generated probabilities for C_1 and C_3 for two Gaussian peaks data.

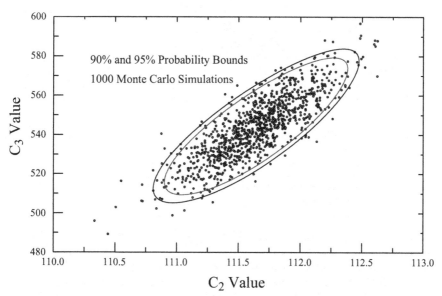

Figure 9.40. Joint MC generated probabilities for C_2 and C_3 for two Gaussian peaks data.

The 95.5% probability bounds for the first peak height location can be summarized as follows from the calculations resulting from Listing 9.10:

(a) Standard error (2 sigma):   110.93 to 112.35
(b) Monte Carlo (One dimension): 110.93 to 112.32
(c) Monte Carlo (Joint C_1 and C_2): 110.78 to 112.47
(d) Monte Carlo (Joint C_2 and C_3): 110.81 to 112.47

These values indicate that the Monte Carlo generated one dimensional probability bounds are almost identical to the 2-sigma standard error bounds. In addition, the joint bounds cover only slightly larger range for the C_1-C_2 and C_2-C_3 joint bounds. For this particular case, the standard error bounds provide a good estimate of the uncertainty with which the peak location can be determined from the experimental data and the model location.

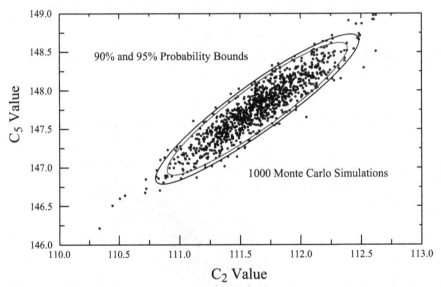

Figure 9.41. Joint MC generated probabilities for C_2 and C_5, the two peak locations for the Gaussian peaks data.

Figure 9.41 shows the MC generated joint probabilities for the two peak height parameters, C_2 and C_5. Ideally one would like to determine the two peak height locations independently of each other. However, the figure shows that for the model equation and the experimental data there is a significant degree of correlation between the best fit values of the two peak locations. If the lower peak near 111 happens to have a large fitting value, the higher peak near 148 also tends to have a large fitting value. This is a feature of the model fitting equation and the random errors in the measured data. In obtaining an estimate of the peak locations, the correlation shown in Figure 9.41 shows that completely independent values of the peak height locations can not be obtained, but the values have some degree of correlation.

It might be expected that the difference between the two peaks could be determined more accurately on an absolute scale than either peak individually. This is in fact the case. If one looks at the MC generated distribution of values for the peaks and the peak difference, the following values can be obtained:

$$C_5 = 147.76 \pm 0.405$$

$$C_2 = 111.34 \pm 0.353 \tag{9.11}$$

$$C_5 - C_2 = 36.131 \pm 0.155$$

where the $\pm$ values represents the one sigma standard errors of the various terms. The error for the difference in peak locations is less than half the standard error of either peak individually. Thus the peak differences can be estimated with approximately half the error of the individual peaks. This might be useful information in some applications of the fitting and parameter estimation process and this is a useful result that can be obtained from the Monte Carlo analysis.

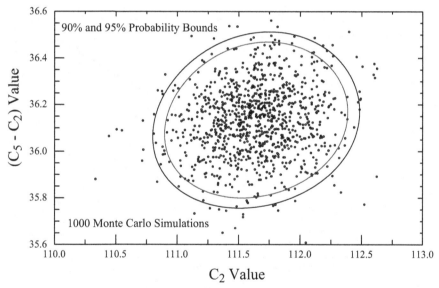

Figure 9.42. Joint MC generated probabilities for C_2 and $(C_5 - C_2)$, the lower peak and the peak differences for the Gaussian peaks data.

Figure 9.42 shows the joint probability distribution for the peak difference and the lower peak value. The probability bounds are very close to circles in this joint plane, indicating the near complete independence of the difference in peak values from one of the peak values. This again indicates that the estimate of the peak difference can be determined essentially independently of the location of the first peak. Again this knowledge may be important in a particular application. Such information about correlations between various model parameters is some of the useful information that can be gleamed from a MC analysis of data fitting parameters.

9.4.6 Superconductivity Magnetic Field Data

In the previous examples in this section, the spacing of data on the independent variable axis has been relatively uniform. This example and the next example illustrate data with a very non-uniform spacing of independent parameter values. As the examples show this makes little difference to the data fitting and parameter estimation approach. The data shown in Figure 9.43 is for magnetic field strength taken in a NIST study involving superconductivity magnetization as a function of time in seconds. (See www.itl.nist.gov/div898/strd/nls/nls_main.shtml for a description of the Bennett5 data set). The figure shows that the data points when expressed in terms of log Time are very non-uniform. In fact the independent variable points are uniformly spaced in terms of linear time but are very non-uniform on the log time scale. The model used to fit the data is a simple two parameter model as shown in the figure in terms of log time as the independent parameter.

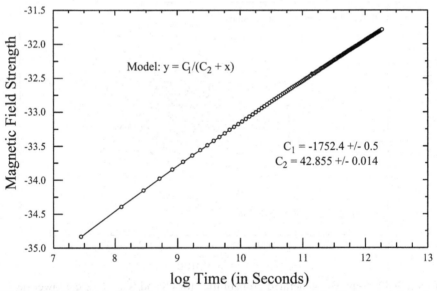

Figure 9.43 Magnetic field strength vs. log time for superconductivity experiment.

Listing 9.11 shows the code changes to Listing 9.6 for this example. The best fit values of the parameters as shown in the selected output of the listing and have a very small standard error. This arises because the model fits very accurately the data and there is very little random error associated with the data points. In fact almost all the fitting error occurs in the two data points with the smallest time value. The RMS error in the fitting is only 0.0019 units along the dependent variable axis. With a model that provides an excellent functional fit to the data and very little random error in a data set, the least squares procedure will return a set of model parameters with very small estimated one sigma relative error.

```
 1 : -- /* File list9_11.lua */ -- file for Parameter estimation
 2 :
 3 : requires("nlstsq","mcpar","DataFit","cltwodim")
 4 :
 5 : -- Section 1. Input data file and define fitting function
 6 : -- Change lines 6 through 14 for different data sets
 7 : infile = 'Bennett5'
 8 : xd,yd = {},{}; read_data(infile..'.txt',yd,xd) -- x stored first
 9 : nd = #xd
10 :
11 : ft = function(x,c) -- Define function to fit data
12 :    return  c[1]/(c[2] + x[2]) - x[1]
13 : end
14 : c = {-3000,50}; nc = #c -- Initial approximations
Selected output:
c[1] = -1.7524e+003 +/-  4.6698e-001
c[2] =  4.2855e+001 +/-  1.4417e-002
```

Listing 9.11. Model and code changes for Magnetic field strength example.

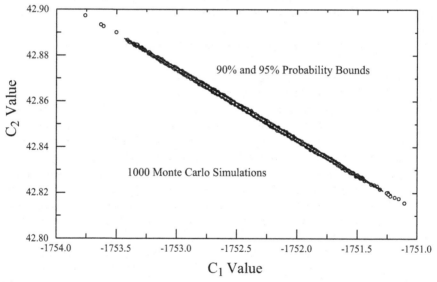

Figure 9.44. Joint MC generated probabilities for C_1 and C_2 for the magnetic field data as seen in Figure 9.43.

Figure 9.44 shows the joint probabilities between the C_1 and C_2 model parameters as determined in the Monte Carlo analysis of the model and data. While the analysis accurately determines the best fit values of the two model parameters, the joint probability figure shows that the two parameters are very closely correlated. In fact the data points collapse to essentially a straight line with negative slope indicting that one of the parameters can essentially be predicted if the other parameter is known. This strong correlation should be a warning about placing much confidence in knowing accurate values for both of the model parameters. The range of values of the independent variable is not large enough to accurately determine both parameters. The joint probability plot gives a very different view of the accuracy of the parameters from that gleamed just from the standard error of the parameters.

Part of the purpose of this example has been to show that the use of non-uniform independent variable spacing has little effect on the parameter estimation approach. It does however mean that not all ranges of the independent variable (or dependent variable) have equal weighting in the data fitting. For example the number of data points between 8 and 9 on the x-axis in Figure 9.43 is 4 while there are 73 data points in the interval 11 to 12. This means that the x-axis interval 11 to 12 has about 18 times as much weighting on the least-squares fitting error as the interval from 8 to 9. If equal weighting over any given x interval is desired, then a weighting factor should be used with the data points. The reader can re-execute the code in Listing 9.11 with a weighting factor such as fw[i] = math.exp(-xd[i]) and verify that such a weighting factor makes very little difference in the resulting best fit model parameters. Models with weighting factors are further explored with examples in Section 9.5.

9.4.7 Volumetric Water Content vs. Matric Potential Data

Many parameters in the modeling of soil physics exhibit nonlinear relationships. One such relationship is the water retention rate as a function of matric potential, which defines for water flow a term similar to electric potential for current flow. (See *J. Nat. Resour. Life Sci. Educ.*, Vol. 27, 1998, pp. 13-19 for a description of the data set for this example.) Figure 9.45 shows experimental data for soil volume water content vs. matric potential. The data points are approximately equally spaced on a log scale of the independent variable. From the data points plotted on a log scale it seems apparent that the data points should all be weighted equally and this is the default used in all the previous data fittings. This will occur naturally for the data fitting even when the points are non-uniformly spaced along the x-axis.

The proposed model function shown in Figure 9.44 is the model described in the referenced source for the data set. The function definitions and initial parameter estimations are shown in Listing 9.12 along with the evaluated model parameters with standard errors printed in the selected output listing. The model has four parameters, but the data range covers a sufficient range of the functional dependency to obtain good estimates of all the parameters.

In viewing the joint probabilities from the Monte Carlo analysis, it is found that all the parameters are relatively independent of each other. The strongest correlation is between the C_3 and C_4 parameters, so only this joint probability is shown here in Figure 9.46. To calculate the correlation between the C_3 and C_4 parameters requires additional modifications to the code of Listing 9.6 from the changes

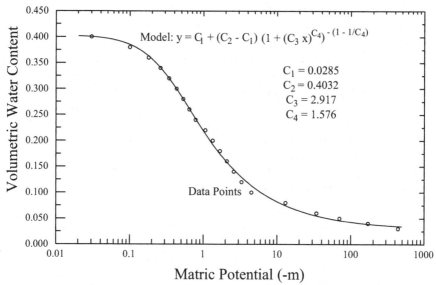

Figure 9.45. Volumetric Water Content vs. Matric Potential

```
1 : -- /* File list9_12.lua */ -- file for Parameter estimation
2 :
3 : requires("nlstsq","mcpar","DataFit","cltwodim")
4 :
5 : -- Section 1. Input data file and define fitting function
6 : -- Change lines 6 through 14 for different data sets
7 : infile = 'matricpot'
8 : xd,yd = {},{}; read_data(infile..'.txt',xd,yd) -- x stored first
9 : nd = #xd
10 :
11 : ft = function(x,c) -- Define function to fit data
12 :     return  c[1]+ (c[2]-c[1])*(1 +
            (c[3]*x[2])^c[4])^-(1-1/c[4]) - x[1]
13 : end
14 : c = {0,.4,1,2}; nc = #c -- Initial approximations.
------- Same as Listing 9.6
Selected Output:
c[1] =   2.8512e-002 +/-   4.2426e-003
c[2] =   4.0324e-001 +/-   3.6013e-003
c[3] =   2.9174e+000 +/-   1.7985e-001
c[4] =   1.5756e+000 +/-   2.8459e-002
```

Listing 9.12. Code segment showing changes from Listing 9.6 for Volumetric Water Content Data.

shown in Listing 9.12. Changes are required in Sections 4 to obtain a pop-up plot similar to Figure 9.46 on lines 38 or 39 to generate plots for cvar[3] vs. cvar[4]. For joint probability limits for these variables, changes are needed on line 48 of

Section 5 to use the cvar[3] and cvar[4] arrays. While there is some correlation among the model parameters in this example, the correlation is much less than that seen in most of the examples in this section. Since the strongest correlation among any of the parameters is that seen in Figure 9.46, it can be concluded for this example that the four model parameters can be estimated with relative small relative uncertainty and that the estimated values are relatively independent of each other. To achieve this accuracy in the estimation of parameter, one needs experimental data which covers a broad range of the functional dependency of the model as can be seen for this example in Figure 9.45 where the data range is sufficient to observe almost complete saturation of the response at both large and small values of the independent parameter. The experimental data points also show little scatter about the model curve as seen in Figure 9.45.

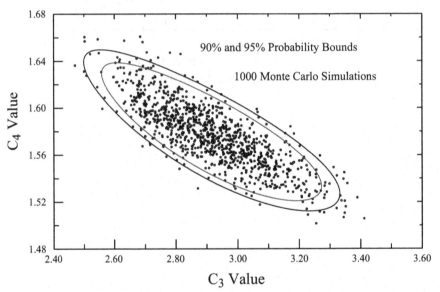

Figure 9.46. . Joint MC generated probabilities for C_3 and C_4 for the volumetric water potential data shown in Figure 9.45.

9.4.8 Poles and Zeros of Amplifier Transfer Function

The frequency response of a single stage electronic amplifier was calculated in Section 4.3 as an example of solving several simultaneous complex equations. The result of this was a Bode plot of the magnitude and phase angle of the response as shown in Figure 4.7. This is repeated here for reference as Figure 9.47. Such a network transfer function is known to be the ratio of numerator and denominator polynomials in powers of the frequency f, the independent variable, i.e.

$$H(f) = K \frac{1 + a_1(jf) + a_2(jf)^2 + \cdots a_n(jf)^n}{1 + b_1(jf) + b_2(jf)^2 + \cdots b_m(jf)^m} \qquad (9.12)$$

where j is the complex variable operator and always multiplies the frequency factor. From looking at the Bode plot in Figure 9.47 many properties of the form of Eq. (9.12) can be inferred. The form of the falloff of the response at low frequencies and at high frequencies is determined by the order of the numerator and denominator polynomials and by the poles and zeros of the transfer function. For example since the response falls off at about 40 dB per decade at high frequencies, the order of the denominator must be larger than the order of the numerator and in fact be of order 2 larger than the order of the numerator. The points where the slope of the Bode plot changes (always by 20 dB) are known as break frequencies. The response thus indicates a break frequency in the denominator at around 3e8 Hz and another at around 1e8 Hz. At low frequencies the response indicates 2 break frequencies in the denominator at around 7 Hz and around 300 Hz (for the 10 µF data) and a numerator break frequency at around 70 Hz. For someone that is skilled in interpretation of Bode plots, Figure 4.46 indicates that an expression for the transfer function should be of the form:

$$H(f) = K \frac{(1 + f_1 / jf)}{(1 + f_2 / jf)(1 + f_3 / jf)(1 + f_4 / jf)} \frac{(1 + jf / f_7)}{(1 + jf / f_5)(1 + jf / f_6)}$$

$$= K \frac{f_1}{f_2 f_3 f_4} \frac{f^2(1 + jf / f_1)}{(1 + jf / f_2)(1 + jf / f_3)(1 + jf / f_4)} \frac{(1 + jf / f_7)}{(1 + jf / f_5)(1 + jf / f_6)} \qquad (9.13)$$

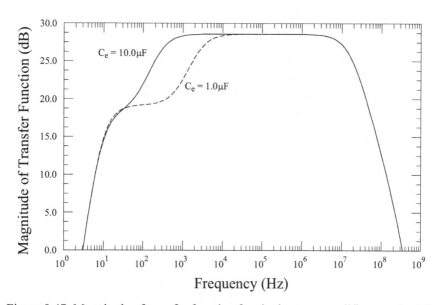

Figure 9.47. Magnitude of transfer function for single stage amplifier as solved for in Section 4.4.

where the first terms involving break frequencies f_1 through f_4 account for the falloff at low frequencies and the second terms involving f_5 through f_7 account for the falloff at high frequencies. Admittedly the presence of the zero expressed as f_7 is difficult to discern from the plot in Figure 4.46 since the frequency range is not high enough to clearly distinguish a high frequency zero in the function. However, such a zero is known to exist from other sources such as electronic books.

To describe a transfer function to be fitted to the data then requires seven break frequencies and a constant term giving a total of eight fitting parameters. A proposed data model equation can then be expressed in a form more consistent with the other discussions in this section as:

$$H = C_1 \frac{x^2(1+C_2 x)}{(1+C_3 x+C_4 x^2 +C_5 x^3)} \frac{(1+C_8 x)}{(1+C_6 x+C_7 x^2)} \tag{9.14}$$

$$y = 20\log 10(|H|), \text{ with } x = jf$$

where x is now a complex independent variable and y is the dependent variable. In this formulation, the three low frequency break terms and two high frequency break terms have been multiplied out into cubic and quadratic terms to possibly account for any interactions among the break frequencies or for possible complex poles. The high frequency terms have been kept separate from the low frequency terms because they are seen from the data to be well separated in this example. If desired, the numerator and denominator terms could be combined into single polynomials in the numerator and denominator with a polynomial of order 5 in the denominator and of order 4 in the numerator.

Listing 9.13 shows a code segment for implementing this data fitting procedure on the amplifier transfer function data that is stored in the amplifier2.dat(or amplifier3.dat) file. Estimates of the various numerator and denominator break frequencies are selected on line 9 of the code. The Polynomial package of code routines are used on lines 11 and 12 as a simple means of multiplying the various polynomials to obtain the initial approximations to the model equation coefficients on line 13. Code for complex arithmetic is also loaded by the Polynomial.lua package on line 3. The model fitting function is implemented on lines 15 through 20 in essentially the same form as given by Eq. (9.14). The evaluation of the model coefficients is performed on line 24 by calling the nlstsq() function.

The selected output shows that the nlstsq() routine takes 5 iterations to achieve an average error of 4.5e-10. Also the evaluated coefficients are seen to be accurate to about 10 digits. This gives valid evidence that the functional form selected for the approximation function is indeed an exact match to the underlying mathematical structure of the problem. Such agreement is only possible for computer generated data and when one has an exact match of the data set to a model equation. If the high frequency zero term represented by the f_7 term for example had not been included in the model the fitting would have been very good, but only in the range of 3 or 4 digits and not the achieved 10 digits of accuracy. A comparison plot is not shown for the model equation and the initial data as the model

```
 1 : -- /* File list9_13.lua */ -- estimation for transfer function
 2 :
 3 : requires("nlstsq","Polynomial")
 4 :
 5 : -- Section 1. Input data file and define fitting function
 6 : infile = 'amplifier2' -- or 'amplifier3'
 7 : xd,yd,_ = {},{},{}; read_data(infile..'.dat',xd,_,_,yd)
 8 :
 9 : c1,f1,f2,f3,f4,f5,f6,f7 = 22,70,7,10,300,3e7,1e8,1e9 -- Initial
10 : c1 = c1*f1/(f2*f3*f4)

11 : p1 = Polynomial.new{1,1/f2}*{1,1/f3}*{1,1/f4}
12 : p2 = Polynomial.new{1,1/f5}*{1,1/f6}
13 : c = {c1,1/f1,p1[2],p1[3],p1[4],p2[2],p2[3],1/f7}; nc = #c
14 :
15 : ft = function(x,c) -- Define function to fit data
16 :    w = j*x[2]
17 :    h = c[1]*w^2*(1+c[2]*w)/(1+c[3]*w+c[4]*w^2+c[5]*w^3)*
18 :          (1+c[8]*w)/(1+c[6]*w+c[7]*w^2)
19 :    return  20*math.log10(Complex.abs(h)) - x[1]
20 : end
21 :
22 : --- Section 2. Perform data fit and print model parameters
23 : del,err,nx =nlstsq({yd,xd},fw,ft,c) -- Call fiting
24 : print('Number of iterations,err =',nx,err)
25 : for i=1,nc do printf('c[%d] = %12.4e +/- %12.4e\n',i,c[i],
          del[i]) end
26 : print('Zeros of Transfer Function at (in Hz):')
27 : print(-1/c[2]); print(-1/c[6])
28 : roots = Polynomial.roots{1,c[3],c[4],c[5]}
29 : print('Poles of Transfer Function at (in Hz):')
30 : print(roots[1]); print(roots[2]); print(roots[3])
31 : roots = Polynomial.roots{1,c[6],c[7]}
32 : print(roots[1]); print(roots[2])
```

Selected output:
```
Number of iterations,err =     5        4.5037680659612e-010
c[1] = 1.1751e-001 +/-  3.0235e-012
c[2] = 1.2579e-002 +/-  1.0208e-012
c[3] = 2.2347e-001 +/-  9.0879e-012
c[4] = 1.4758e-002 +/-  5.9299e-013
c[5] = 5.5018e-005 +/-  4.4728e-015
c[6] = 6.3028e-008 +/-  1.2643e-018
c[7] = 1.4857e-016 +/-  1.5845e-026
c[8] = 6.2863e-010 +/-  2.7093e-019
Zeros of Transfer Function at (in Hz):
-79.496588520183
-15865935.499628
Poles of Transfer Function at (in Hz):
(-7.9025036360333) + j(-3.0908959234148)
(-7.9025036360333) + j(3.0908959234148)
-252.4333218307
-407726433.29397
-16508327.143326
```

Listing 9.13. Code segment for data fitting to amplifier transfer function with data as shown in Figure 9.47.

equation exactly matches the data shown in Figure 9.47 to the degree that can be shown on a graph.

An important application of such an analysis is to determine the pole and zero locations perhaps as a function of some circuit parameters. For example in this case the value of the C_e capacitor is seen from Figure 9.47 to change the low frequency poles and zeros. By executing the program using both of the data sets (amplifier2.dat and amplifier3.dat), these parameters can be evaluated for the two values of capacitance. The results of such an analysis are given in Table 9.1. The table results are close to the values that can be estimated from the graph of the magnitude of the frequency response.

One perhaps unexpected result is the existence of the complex conjugate pair of poles at low frequencies. Simple first-order circuit calculations do not lead to such a complex conjugate pair of poles but to simple real poles. However, this fitting of the response curve to an exact functional form shows the correctness of the complex poles. This complex conjugate pair of poles would not have been recovered if the denominator polynomial at low frequencies had not been expressed as a cubic factor in the denominator of Eq. (9.14).

The results in Table 9.1 show that changing C_e the emitter capacitor value changes predominately two of the table terms, one zero and one pole. This is as expected from first order models which relate one of the low frequency poles to one of the low frequency capacitors. As the C_e value changes by a factor of 10 two of the terms, a zero and a pole, change in value by approximately a factor of 10. From these changes it can be concluded that these two frequencies are determined predominately by this circuit parameter. There are also some smaller changes in the location of the complex pair of poles, indicating a coupling of low frequency effects among the pole frequencies. There are no changes in the high frequency poles and zeros as would be expected.

Amplifier Frequency Response Example	$C_e(\mu F)$	
	10.0	1.0
Zeros (Hz)	-79.496	-794.96
	-1.586E7	-1.586E7
Poles (Hz)	$-7.903 - j3.090$	$-8.225 - j3.143$
	$-7.903 + j3.090$	$-8.225 + j3.143$
	-252.43	-2344.1
	-1.651E7	-1.651E7
	-4.077E8	-4.077E8

Table 9.1 Evaluated Pole and Zero locations for Amplifier Frequency Response Example of Figure 9.47.

This example has shown an interesting double example of a series of computer analysis methods for an engineering problem. First one computer analysis was used (in Chapter 4) to evaluate the frequency response of an electronic circuit.

Then a second computer analysis has been used in this chapter to determine a very accurate analytical model for the frequency response of the circuit in terms of the general frequency parameter f. Such an analysis is frequently performed by hand where one must solve a set of coupled equations, keeping all important frequency dependent terms. This becomes very difficult and beyond our ability to correctly perform the math for more than a simple circuit. However, with the approach used here, considerably more complicated problems can be addressed. By performing similar analyses with different circuit parameters, one could map out the dominant dependences of the transfer function poles and zeros on the values of all the circuit parameters.

9.5 Data Fitting and Parameter Estimation with Weighting Factors

Although weighting factors have been mentioned in the previous examples, all the examples so far in this chapter have used equal weighting of the data points in the least squares data fitting and parameter estimation. In fact for the previous calls to the nlstsq() function as on line 23 of Listing 9.13 the fw parameter in the argument list has simply been an undefined variable. When the nlstsq() function detects this as being undefined, the function simply assumes that unity weighting factors are to be used as a default case. When a set of data covers a large range of dependent variable values, it is frequently convenient, and necessary, to use an appropriate set of data weighting factors to obtain a good model fit over the entire range of the dependent variable. This is not required, as the previous example shows when the independent variable covers a large range of values or when the independent variable has a log scale.

In Electrical Engineering, the semiconductor diode provides one of the most fundamental nonlinear devices. The current in the forward or conduction region is approximately an exponential function of diode voltage. Data for such a semiconductor diode is shown in Figure 9.48, along with a typical model equation of the form:

$$i = I_s(\exp(v/nV_t) - 1)$$
$$\text{or } y = C_1(\exp(x/C_2V_t) - 1) \tag{9.15}$$

where $V_t = kT/q = 0.0259$ Volts at room temperature. The other parameters (I_s and n) are known as the saturation current and the diode factor. Typically n is between 1 and 2, although the best fit value listed in Figure 9.46 for the large current range is 4.755. The model parameters were obtained from the code in Listing 9.14. The model definition is shown on line 12 and the initial approximations to the model parameters are given on line 15. A nonzero step parameter is specified on line 16 for the C_2 parameter to limit the step change between Newton iterations to about 20% (a 1.2 factor). The remainder of the listing is similar to previous data fitting routines in this section. The code in Listing 9.14 uses the standard fitting procedure with unity weighting factor on the data points as the fw parameter is defined with all unity values on line 16. As seen in Figure 9.48, the model provides a reasonable fit at the high current values. However, some deviation between the data and the model can be seen in Figure 9.48 at low currents.

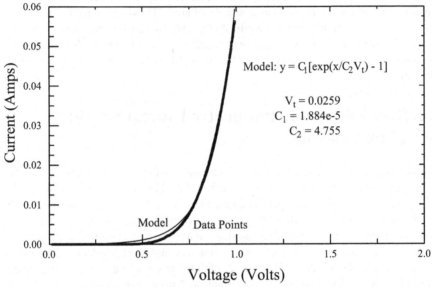

Figure 9.48. I-V data for semiconductor diode and model fit at high current.

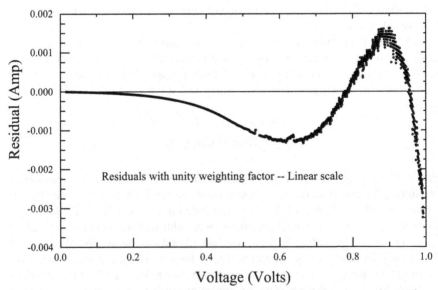

Figure 9.49 Residuals for diode model with unity weighting factors on data points.

A plot of the fitting residuals as a function of diode voltage is shown in Figure 9.49. The figure shows that the residuals are anything but randomly distributed around the fitted model. This is typical of residuals due to model inadequacies where the residuals are not due to any random measurement error but are almost entirely due to the model not having a functional form that closely matches the experimental data. This can be seen even more dramatically in Figure 9.50 which shows a plot of the experimental data and the fitted model equation with a semilog plot on the current axis. The figure shows that the fitted model is relatively accurate at large voltages (above about 0.8 Volts) but fails badly at low voltage or current values, where the model and data differ by many orders of magnitude in value. This model fit and set of model parameters is certainly not adequate if an accurate evaluation of the current is needed for voltage values below about 0.6 Volts.

```
1 : -- /* File list9_14.lua */ -- Parameters for diode equation
2 :
3 : requires("nlstsq","mcpar","DataFit")
4 :
5 : -- Section 1. Input data file and define fitting function
6 : -- Change lines 6 through 12 for different data sets
7 : infile = 'ivdata'
8 : xd,yd = {},{}; read_data(infile..'.txt',xd,yd) -- x stored first
9 : nd = #xd
10 :
11 : ft = function(x,c) -- Define function to fit data
12 :    return  c[1]*(math.exp(x[2]/(0.0259*c[2])) - 1) - x[1]
13 : end
14 :
15 : c = {1.e-9,2}; nc = #c; step ={0,1.2} -- Initial approximations
16 : fw = {}; for i=1,nd do fw[i] = 1.0 end
17 :
18 : -- Section 2. Perform data fit and print fitted parameters
19 : del,err,nn = nlstsq({yd,xd},fw,ft,c,actv,step) -- Call fiting,
20 : print("Number of iterations, err =",nn,err)
21 : for i=1,nc do printf('c[%d] = %12.4e +/- %12.4e\n',i,c[i],
         del[i]) end
22 :
23 : -- Section 3. Generate 6-Plot Data, Only 1,2,3 and 6 used here
24 : ycalc,res = {},{} -- Calculate fitted values and residuals
25 : for i=1,nd do ycalc[i]=newtonfc(ft,{yd[i],xd[i]},c);
         res[i]=yd[i]-ycalc[i] end
26 : fres1,fres2 = datafit({res,xd}), datafit({res,yd}) -- residuals
27 : yres1,yres2,xyp,xypn = {},{},{},{}

28 : for i=1,nd do yres1[i],yres2[i] = fres1(xd[i]),
         fres2((ycalc[i])) end
29 : xyp[1],xyp[2] = makeCDF(res); xypn[1],xypn[2] = normalCDF(res)

30 : plot(xd,yd,ycalc); plot(xd,yres2,res); plot(ycalc,yres2,res);
         plot(xyp,xypn)
31 : write_data(infile..'a.dat',xd,yd,ycalc,res,yres1,yres2,xyp,xypn)
```

Listing 9.14 Code segment for first analysis of the diode I-V data seen in Figure 9.48.

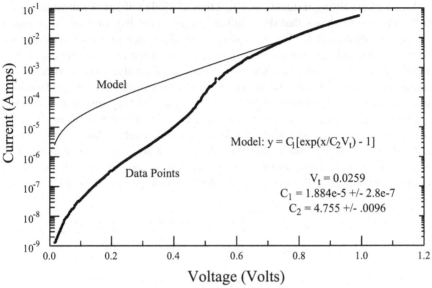

Figure 9.50. Plot of fitted model equation and experimental data on log current scale.

It will be noted that the code in Listing 9.14 omits any Monte Carlo analysis. This is because such an analysis would not give meaningful results. The MC analysis assumes that the model errors are due entirely to a random variable and this is certainly not the case in this example. One must be careful in using the MC analysis to ensure that the conditions appropriate to the analysis are valid or otherwise the results will be meaningless as they would in the present example. A more meaningful avenue is to pursue a more accurate model for the diode – as done in the following paragraphs and in the next section of this chapter.

There are two avenues for obtaining a better model representation over the entire independent variable range when one has data such as this were the dependent variable changes by many orders of magnitude – in this case the dependent variable changes by about 8 orders of magnitude. The first approach is to use an appropriate data weighting factor and the second approach is to take the log of the dependent variable data and perform a least squares data fitting using the log of the variable as the fitting variable. Both of these approaches are now considered and compared.

First for the use of a weighting factor, as discussed in Section 7.9, a $1/y^2$ weighting factor provides equal relative errors for each data point. This is implemented in the code of Listing 9.14 by changing line 16 to that shown in Listing 9.15. Executing the code in Listing 9.14 with this simple change then produces a new set of model parameters. These are shown in Figure 9.51 along with a comparison of the data with the newly obtained model equation. The diode factor (C_2)

```
15 : ---- Same as Listing 9.14
16 : fw = {}; for i=1,nd do fw[i] = 1/y[i]^2 end -- 1/y^2 Weighting
17 : ---- Same as Listing 9.14
```

Listing 9.15. Code changes to Listing 9.14 for non-unity weighting factors and diode I-V data.

for this fit is 2.33 which is much closer to the expected range of 1 to 2 than is the value obtained for the data fit with unity weighting which heavily weights the large current values. The figure shows that the model equation now provides a better "overall" fit to the experimental data, although the model and data differ by almost an order of magnitude in the range of 0.6 to 0.8 volts. A little thought will indicate that there will be no set of model parameters that will very closely fit the data points over the entire range of data because the model equation will give essentially a straight line fit when plotted on a log scale at least for voltages larger than about 0.1 Volt. This is a fundamental limitation of the data model and not a limitation of the least squares data fitting technique. It can be seen from the figure that this set of model parameters will do a poorer job of describing the high current data than the parameters in Figure 9.49. In fact a plot of the data on a linear current scale such as shown in Figure 9.48 shows very poor agreement between the model and data at high currents.

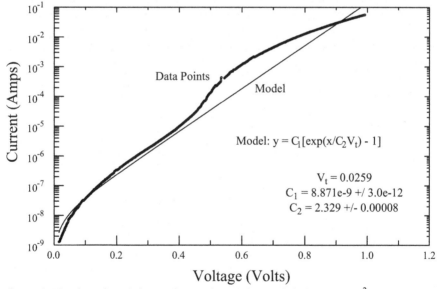

Figure 9.51. Plot of model equation and experimental data using $1/y^2$ data weighting factors.

The second approach to fitting the data over the entire range is to first take the log of the experimental dependent variable values and then fit the log values. This will convert the y values of the diode data from the range of about 1.e-9 to 1.e-1 to the range of about -21 to -2, providing a more reasonable range to obtain a meaningful error term for each data point. This is done by making the code changes shown in Listing 9.16 which shows only the important changes from Listing 9.14. This fitting uses a unity data weighting factor as indicated on line 16 of the code for the log converted data values. A loop over the input dependent variable values has been added on line 9 to change to the log() of the data values. In addition the model equation definition on line 12 takes the log() of the previously defined model equation in order to match the dependent variable modified data.

Figure 9.52 shows the resulting data fit using this approach and the resulting model equation values. The parameter values and data fit are close to that of the $1/y^2$ weighting, but the model now gives a slightly better fit for the lower current values at the expense of a slightly poorer fit at the very high current values. With this data fit, the model predictions are within about a factor of 4 for all voltage values below about 0.8 Volt. Since this is the range normally encountered with a diode, this is about the best that can be done for this data with the simple exponential diode model used in the examples.

Again no Monte Carlo data is shown as the errors are almost entirely model related errors and such an analysis will not lead to meaningful error bounds on the parameters. The basic nlstsq() fitting routine does return standard error values and these are also shown in Figure 9.52 for this fitting. The "best fit" slope parameter (the c[2] value) has a very small standard error, about 0.3%, while the saturation current value shows about a 3% standard error. Before a better agreement between the data and model can be achieved, an improved diode model must be used to describe the data.

```
 1 : -- /* File list9_16.lua */ -- Parameters for diode equation
 2 :
 3 : requires("nlstsq","mcpar","DataFit")
 4 :
 5 : -- Section 1. Input data file and define fitting function
 6 : -- Change lines 6 through 12 for different data sets

 7 : infile = 'ivdata'
 8 : xd,yd = {},{}; read_data(infile..'.txt',xd,yd) -- x stored first
 9 : nd = #xd; for i=1,nd do yd[i] = math.log(yd[i]) end
10 :
11 : ft = function(x,c) -- Define function to fit data
12 :     return  math.log(c[1]*(math.exp(x[2]/(0.0259*c[2])) - 1)) -
            x[1]
13 : end
14 :
15 : c = {1.e-9,2}; nc = #c; step ={0,1.2} -- Initial approximations
16 : fw = {}; for i=1,nd do fw[i] = 1 end

------ Same as Listing 9.14
```

Listing 9.16. Code changes for fitting data using a log y scale with the diode I-V data.

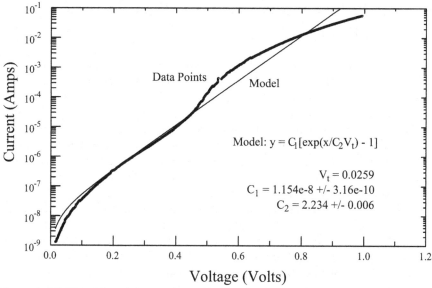

Figure 9.52. Plot of model equation and experimental data using log scale for dependent variable.

Before that however, some of the results seen so far can be summarize. First, the use of an appropriate weighting factor has been seen to greatly improve the agreement between a model equation and a data set that has a large range in the independent parameter values. The use of $1/y^2$ data weighting is an easy process to implement and gives essentially equal weighting to the percentage error of each data point. This has been seen to be very similar to first taking the log of the dependent variable data set and fitting an equation to the log of the data although the two approaches are not identical. A final approach that gives similar results to the use of a log scale is to reverse the role of the independent and dependent variable and treat the y axis or current as the independent variable and the x axis or voltage as the dependent variable. Since the voltage values are uniformly spaced on a linear scale this will appropriately weight the error associated with the data points, although the minimization will be on the horizontal difference in model and data as opposed to the vertical difference. This is easily implemented in the code examples, by simply reversing the role of the axis and reversing the [1] and [2] notation in the defining equation for the data fit. This is left to the reader, but performing this fit leads to the set of parameter values: $C_1 = 1.995e-8$ and $C_2 = 2.137$. The resulting fit between the data and model is not very different from that shown in Figure 9.52. The reader is encouraged to just give it a try.

9.6 Data Fitting and Parameter Estimation with Transcendental Functions

Many physical models can only be expressed in terms of some transcendental equation, where an explicitly solution for the dependent variable can not be obtained in terms of the independent variable. In some cases one can not solve explicitly for either variable in terms of the other variable. Such cases are readily handled by the code segments so far developed. This is one of the features of the nlstsq() data fitting code segment that the defining equation can be any general function of the independent and dependent variables as long as the equation returns zero when satisfied by the two variable values. This section considers one such example, building on the diode I-V data example of the previous section.

An improved diode equation model that is especially appropriate for large voltages and currents is to consider a physical diode as composed of an ideal diode with an I-V relationship as given in Eq. (9.15) but with the ideal diode in series with a resistance resulting from the bulk materials of the semiconductor and the contacts. This means that the voltage appearing in the diode equation will be the real terminal diode voltage minus the voltage drop across the internal resistance and in turn the internal voltage drop is proportional to the current. This leads to the modified diode model of:

$$i = I_s(\exp((v - iR_s)/nV_t) - 1)$$
$$\text{or } y = C_1(\exp((x - C_3 y)/C_2 V_t) - 1) \tag{9.16}$$

where a new parameter R_s represents the diode series resistance. This is a transcendental equation for y in terms of x. It is true that in this case x can be solved for explicitly in terms of y, but not the other way around. So if a data fitting is performed with the diode current values as the dependent parameter the resulting equation is a transcendental equation for the dependent variable in terms of the independent variable.

Listing 9.17 shows the code changes for defining this data fitting function including the series resistance on line 12. At first the code appears strange since the added term is of the form: (x[2] - c[3]*math.exp(x[1])), i.e. the term multiplying the added series resistance (c[3]) is the exp() of the dependent variable term. This arises because the data is converted to a log scale on line 9 of the code so the dependent variable becomes the log of current and the exponential function is needed to convert back to current in the fitting equation. One other item to note in Listing 9.18 is the use of the function newtonfc() on line 24 to calculate the fitted value of the dependent variable. This function is required since the fitting equation is a transcendental equation. This uses Newton's method to solve an equation similar to the normal newton() method. However, it also allows one to pass a coefficient array of values to the function being solved as is needed in this example.

```
 1 : -- /* File list9_17.lua */ -- Transcendental diode model
 2 :
 3 : requires("nlstsq","mcpar","DataFit")
 4 :
 5 : -- Section 1. Input data file and define fitting function
 6 : -- Change lines 6 through 12 for different data sets
 7 : infile = 'ivdata'
 8 : xd,yd = {},{}; read_data(infile..'.txt',xd,yd) -- x stored first
 9 : nd = #xd; for i=1,nd do yd[i] = math.log(yd[i]) end
10 :
11 : ft = function(x,c) -- Define function to fit data
12 :    return  math.log(c[1]*(math.exp((x[2]-c[3]*math.exp(x[1]))/
           (0.0259*c[2])) - 1)) - x[1]
13 : end
14 :
15 : c = {1.e-9,2,1}; nc = #c -- Initial approximations.
16 :
17 : -- Section 2. Perform data fit and print fitted parameters
18 : del,err,nn = nlstsq({yd,xd},fw,ft,c) -- Call fiting
19 : print('Newton iterations =',nn)
20 : for i=1,nc do printf('c[%d] = %12.4e +/- %12.4e\n',i,c[i],
           del[i]) end
21 :
22 : -- Section 3. Generate 6-Plot Data, Only 1,2,3 and 6 used here
23 : ycalc,res = {},{} -- Calculate fitted values and residuals
24 : for i=1,nd do ycalc[i] = newtonfc(ft,{yd[i],xd[i]},c);
           res[i] = yd[i] - ycalc[i] end
25 : fres1,fres2 = datafit({res,xd}), datafit({res,yd}) -- residuals
26 : yres1,yres2,xyp,xypn = {},{},{},{}
27 : for i=1,nd do yres1[i],yres2[i] = fres1(xd[i]),
           fres2((ycalc[i])) end
28 : xyp[1],xyp[2] = makeCDF(res); xypn[1],xypn[2] = normalCDF(res)
29 : plot(xd,yd,ycalc); plot(xd,yres2,res); plot(ycalc,yres2,res);
           plot(xyp,xypn)
30 : write_data(infile..'d.dat',xd,yd,ycalc,res,yres1,yres2,xyp,xypn)
```

Listing 9.17. Code changes for including series resistance and fitting using log current scale.

The resulting data fit and evaluated parameters are shown in Figure 9.53. This now provides a much better data fit over the entire data range from low current to high current values. The maximum error in the model is now less than a factor of two over the entire data range. Thus great improvement in the functional dependency of the model equation can be seen when the small corrective resistance term is included. The major effect on the model values is in the high current regime. However, it provides a better overall fit to the data throughout the entire data range. For many applications, this model may now provide an adequate match to the experimental data.

While this model now provides a much better fit to the data, the errors between the model and the data are still seen to not be due to random errors but represent systematic deviations from the model equation indicating that the basic model does not incorporate some level of physics present in the experimental data. The model parameter $c[3]$ is now the diode series resistance which is evaluated as 4.394 +/- 0.067 Ohms. The diode factor ($c[2]$) is now evaluated as 1.898 which is closer to any expected value and in the range of 1 to 2 as expected. However, such a model is to some extent an empirical model as no semiconductor theory predicts a diode factor of this value. The next section goes further in discussing an improved physical model for a semiconductor diode.

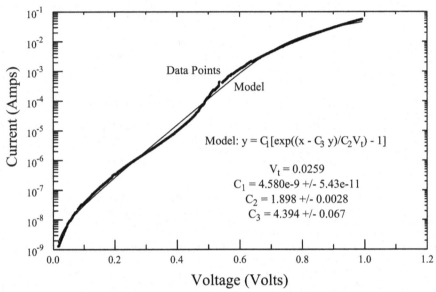

Figure 9.53. Diode I-V data and model including series resistance and using log data range for current.

Before that however, it is useful to summarize some of the results of this section on the use of a transcendental model for data fitting. The procedure for data fitting and parameter estimation is essentially the same whether the fitting model can be solved explicitly for the dependent variable equation or if it can only be expressed only in terms of some transcendental relationship. One simply needs to express the functional relationship to be satisfied by an equation (or some algorithm) and the nlstsq() fitting routine is sufficiently general to handle any desired transcendental relationship. Of course as the model becomes more nonlinear, more care may be needed in selecting the initial guesses at the parameters and/or limits may be required on the parameter step changes from iteration to iteration to obtain convergence. For this example, no limits have been imposed on the step size in Listing 9.17 and the listing converges to correct parameter values for a rather wide range of initial estimates, as the reader may verify by re-executing the code with different initial values.

9.7 Data Fitting and Parameter Estimation with Piecewise Model Equations

Many physical models can only be expressed in terms of several equations that provide good estimates of a set of measurements over different ranges of the independent variable – a piecewise definition of model equations. The semiconductor diode is one such example. In terms of theoretical operation of a diode, it is known that there are three functional forms for the diode current depending on the range of the voltage. These can be expressed in equation form as:

$$i = \begin{cases} i_1 = C_1(\exp(x/2V_t) - 1) & \text{for low voltages} \\ i_2 = C_2 \exp(x/V_t) & \text{for intermediate voltages} \\ i_3 = C_3 \exp(x/2V_t) & \text{for high voltages} \\ \text{where } x = v - R_d i = v - C_4 i \text{ and } V_t = kT/q \end{cases} \tag{9.17}$$

In the second and third expressions the -1 term has been omitted as the exponential term will be many orders of magnitude larger than unity for the region of operation. There three forms that can be derived from basic semiconductor theory correspond essentially to the fundamental diode equation (Eq.(9.16)) with diode factors of either 1 or 2. For each exponential factor the voltage is corrected by an internal resistance term from any measured terminal voltage value. The diode factor fitted in Figure 9.52 of 1.898 is understood as being some composite value from a combination of the fundamental terms expressed in Eq. (9.17). Figure 9.54 shows the diode experimental data with individual plots of the three terms given above and with coefficients selected to match the data in the three regions of operation. As the dotted lines in the figure indicate, different models can provide an improved approximation to the experimental data over different regions of operation and there are theoretical justifications for the functional form of the different models and approximations.

With such a piecewise definition of model equations, the question arises as to how can the three approximations be combined into one overall diode model? For the low to intermediate transition region, it is seen than one needs to retain in some way the "largest" of the values between the two approximations (between i_1 and i_2). However, for the intermediate to high transition one needs to retain in some way the "smallest" of the values between the two approximations (between i_2 and i_3). These represent the most common transition types between approximate regionally applicable models – either combine two models so that the largest approximation is used or combine then so that the smallest approximation is used. One would also like to achieve such a transition in equation form with a continuous first derivative. While there are many possible ways to combine terms to obtain this feature, the following is one of the simplest:

$$\text{Largest Term Dominant: } y = (y_1^n + y_2^n)^{1/n}$$

$$\text{Smallest Term Dominant: } y = (y_1^{-m} + y_2^{-m})^{-1/m} \tag{9.18}$$

where y is the combined model for two piecewise dominant model terms y_1 and y_2. For n or m near unity in value, there is a gradual transition between the two terms. As n or m becomes much larger than unity the transition becomes more abrupt. For m=1, the second form will be recognized as the same as the formula for combining parallel resistors. The n term here obviously has no relationship to the diode n factor as given in Eq. (9.16). As n or m becomes large compared to 1, the transition between the two models becomes more abrupt, but always with a continuous derivative.

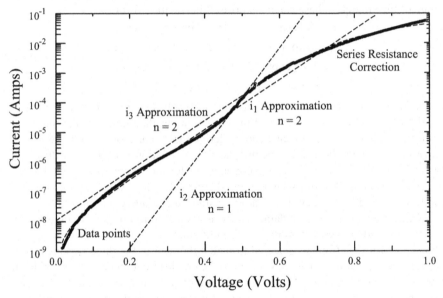

Figure 9.54. Regional approximations to diode I-V characteristic with three model equations.

For the present example, an i_1 and i_2 term from Eq. (9.17) can be combined making the largest term dominant and then the resulting equation combined with an i_3 term making the smallest term dominant. The code for this piecewise regional model is shown in Listing 9.18. The major difference from previous listings is the model equation definition on lines 11 through 15. Values of the exponents used in the regional combining expressions are shown as 4 and 4 on line 10 of the code. These were selected after running the program with trial values of 1 to 5. It can be seen from Figure 9.54 that a rather abrupt transition between the three regions is probably going to be needed for the best fit to the data. These values give rather abrupt transitions between the regional approximations as can be seen in Figure 9.55 which shows a comparison between the overall model and the experimental data using these regional approximations. This fit of the thin solid line is seen to be considerably better than any of the data fits in the previous sections (Sections 9.5 and 9.6). This section could in fact be considered an extension

```
 1 : -- /* File list9.18.lua */ -- Piecewise diode model
 3 : requires("nlstsq","mcpar","DataFit","cltwodim")
 4 :
 5 : infile = 'ivdata'
 6 : xd,yd = {},{}; read_data(infile..'.txt',xd,yd) -- x stored first
 7 : nd = #xd
 8 : for i=1,nd do yd[i] = math.log(yd[i]) end
 9 :
10 : n,m,Vt = 4,4,.0259 -- Experiment with values for best fit
11 : ft = function(x,c) -- Define function to fit data
12 :    v = (x[2] - c[4]*math.exp(x[1]))/Vt
13 :    i12 = ((c[1]*(math.exp(v/2)-1))^n +
             (c[2]*math.exp(v))^n)^(1/n)
14 :    return math.log((((i12)^-m + (c[3]*math.exp(v/2))^-m)^-(1/m))-
             x[1]
15 : end
16 :
17 : c = {5e-9,5.6e-13,8e-9,4}; nc = #c -- Initial approximations
18 :
19 : del,err,nn = nlstsq({yd,xd},fw,ft,c) -- Call fiting
20 : print('Newton iterations, err =',nn,err)
21 : for i=1,nc do printf('c[%d] = %12.4e +/- %12.4e\n',i,c[i],
             del[i]) end
22 :
23 : ycalc,res = {},{} -- Calculate fitted values and residuals
24 : for i=1,nd do ycalc[i] = newtonfc(ft,{yd[i],xd[i]},c);
             res[i] = yd[i] - ycalc[i] end
25 : fres1,fres2 = datafit({res,xd}), datafit({res,yd}) -- residuals
26 : yres1,yres2,xyp,xypn = {},{},{},{}
27 : for i=1,nd do yres1[i],yres2[i] = fres1(xd[i]),
             fres2((ycalc[i])) end
28 : xyp[1],xyp[2] = makeCDF(res); xypn[1],xypn[2] = normalCDF(res)
29 : plot(xd,yd,ycalc);plot(xd,yres2,res);plot(ycalc,yres2,res);
             plot(xyp,xypn)
30 : write_data(infile..'e.dat',xd,yd,ycalc,res,yres1,yres2,xyp,xypn)
Selected Output:
Newton iterations, err =  9    0.16691286719695
c[1] =   5.5406e-009 +/-  2.7063e-011
c[2] =   6.0134e-013 +/-  1.5657e-014
c[3] =   1.1076e-008 +/-  1.0043e-010
c[4] =   4.8044e+000 +/-  5.0791e-002
```

Listing 9.18. Listing for I-V data fit using piecewise regional models.

of the previous section on transcendental functions since the resulting model equation as described in lines 11 through 15 is certainly a transcendental function where neither the dependent variable nor the independent variable can be solved for explicitly in terms of the other variable. A separate section has been used however, because the concept of regional approximations to an overall physically based model occurs very frequently in engineering problems and this provides a good example of this approach.

The four model parameters evaluated in this fitting have the closest connection to the underlying physics of diode operation and can be further correlated with the physical properties of a semiconductor diode such as doping density and physical size. To pursue this further is beyond the scope of this work. However, the point

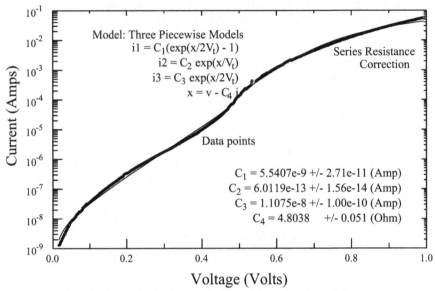

Figure 9.55 Comparison of I-V data and piecewise regional model.

is that the closer one can come to a model derivable from underlying physical concepts, the closer one is likely to obtain a good fit to experimental data, even if the models are described by regionally valid equations. Even though the model now fits relatively well over the entire current range of about eight orders of magnitude, a plot of the residuals (not included here) shows that the errors are still dominated by model errors and not by random errors. Executing the code produces a pop-up plot of the residuals. Thus it still does not make any physical sense to attempt a Monte Carlo analysis of the model and data set to explore joint probabilities of the model parameters. The best we can do is to use standard error uncertainties and these are still due almost entirely to inaccuracies between the model equations and the data points.

9.8 Data Fitting and Parameter Estimation with Multiple Independent Parameters

All of the examples so far considered in this chapter involve one independent variable and one dependent variable. While this is a very common engineering problem, many measurement problems and parameter estimation problems involve physical situations with more than one independent variable. This type of problem was anticipated when the nlstsq() routine was developed and such problems are easily handled in essentially the same manner as the previous examples. To illustrate such problems, two data fitting examples and parameter estimation examples are given in this section. Both of these involve only two independent parameters, but the examples are easily extendable to any number of independent parameters.

The first example is that of dielectric breakdown as a function of time and temperature. (See www.itl.nist.gov/div898/strd/nls/nls_main.shtml for a description of the Nelson data set.) Figure 9.56 shows selected parts of the data set which consists of measurements taken at times (T_m) of 1, 2, 4, 8, 16, 32, 48 and 64 sec. and at temperatures (T) of 180, 225, 250 and 275 °C. The other data curves lie between those shown in Figure 9.56 but are not included so that the data shown can be clearly identified. As the figure shows the data has considerable scatter at any temperature and time. A proposed model for the observed data contains three parameters as shown by the equation given in the figure. To aid in fitting the data to a model, the log() of the data is taken and fitted to a model. Without the log() the proposed model would have a double exponential function of time.

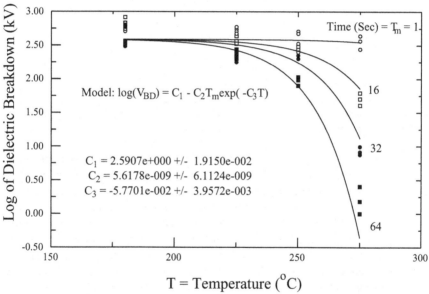

Figure 9.56. Example of data fitting with two independent parameters – Dielectric breakdown as a function of time and temperature.

The computer code for data fitting to the breakdown data is shown in Listing 9.19. As can be seen very little change in the code is needed to handle multiple independent variables. Note that on line 9 the raw input data is converted to a log() function of the data before the data fitting is accomplished. In the read_data() call on line 8 three variables are read from the data file, the dependent variable (yd) followed by the two independent variable values in column format (x1d and x2d). In the fitting function definition on lines 11 through 13, it is seen that three x[] variables are used. The code always expects x[1] to be the dependent variable on which the error is being taken. Any number of other variables can

```
 1 : -- /* File list9_19.lua */ -- Data with multiple parameters
 3 : require"nlstsq"
 4 :
 5 : -- Section 1. Input data file and define fitting function
 6 : infile = 'Nelson'
 7 : yd,x1d,x2d,yy,ycalc = {},{},{},{},{}
 8 : read_data(infile..'.txt',yd,x1d,x2d)
 9 : nd = #yd; for i=1,nd do yy[i] = math.log(yd[i]) end
10 :
11 : ft = function(x,c) -- Define function to fit data
12 :    return c[1] - c[2]*x[2]*math.exp(-c[3]*x[3]) - x[1]
13 : end
14 :
15 : c = {2,5e-9,-5e-2}; nc = #c
16 : step={1.2,1.2,1.2}
17 : --- Section 2. Perform data fit and print fitted parameters
18 : del,err,nn = nlstsq({yy,x1d,x2d},fw,ft,c,actv,step) -- fitting,
19 : print('Newton iterations, err =',nn,err)
20 : for i=1,nc do printf('c[%d] = %12.4e +/- %12.4e\n',i,c[i],
         del[i]) end
21 : for i=1,nd do ycalc[i]=newtonfc(ft,{yy[i],x1d[i],x2d[i]},c) end
22 : write_data(infile..'.dat',x1d,x2d,yd,ycalc,yy)
Selected Output:
Newton iterations, err =        5        0.17224807958031
c[1] =   2.5907e+000 +/-  1.9150e-002
c[2] =   5.6178e-009 +/-  6.1124e-009
c[3] =  -5.7701e-002 +/-  3.9572e-003
```

Listing 9.19. Code segment for breakdown voltage data fitting example with two independent variables.

then be specified as $x[2]$, $x[3]$, … $x[n]$ for n independent variables. The only additional change in the code is in the calling argument to the nlstsq() function. The first calling argument contains arrays for the dependent variable values followed by the corresponding arrays for any independent variables. In the present case this is the {yy, x1d,x2d} term on line 18. The number of data variables is automatically determined in nlstsq() by the size of the first calling argument. Finally the defining function for fitting the data then uses the dependent variable ($x[1]$) and the two independent variables ($x[2]$ and $x[3]$) as seen on line 12. That's all there is to data fitting with multiple independent variables. One just adds additional arrays in the nlstsq() argument and uses the corresponding values of dependent and independent variables in the function definition code. All the hard work is simply done in nlstsq() for handling any number of dependent variables and for handling any possible transcendental relationships between the independent and dependent variables. One of the goals in writing any reusable code segments should always be to make them as general as possible and to treat the most common cases as simply special cases. For the nlstsq() function the common case of a single independent variable is just a special case of multiple independent values.

A few words are in order about this specific example and the results shown in Figure 9.56. The data has considerable scatter as can be seen in the figure. This causes some difficulty in the convergence of the nlstsq() routine and Listing 9.19 on line 16 shows that step limits of about 20% have been set on the changes in fitting parameters at each Newton iterative step. For the initial guesses shown some limits are required to achieve a valid converged result. The reader is encouraged to re-execute the code changing the initial guesses and/or step limits and explore the robustness of this fitting. It will be found that the code is much less robust with respect to initial guesses than the typical data fittings of this chapter. Sometimes trial and error is needed to obtain properly converged and fitted solutions especially when the data contains large random variations.

The best-fit coefficients as seen in Figure 9.56 also have large standard error. For example the C_2 coefficient has a standard error that is larger than the best estimated value. It would thus be very difficult to make many arguments about the exact value of this coefficient or of the other coefficients. This might lead one to search for an improved model for the breakdown voltage. However that is beyond the scope of this work.

The next example of two independent variables is almost the opposite of the above in that the data to be analyzed has very little random error – at least it appears that way. The data set shown by the thick dark lines is that of the drain current (i_d) of an MOS transistor as a function of two voltages, the gate voltage (v_g) and the drain voltage (v_d) as shown in Figure 9.57. The data set consists of 301 data points along the drain voltage axis at each set of gate voltages for a total of 2107 data points. As can be seen from the figure, there is very little random error associated with the measured data points.

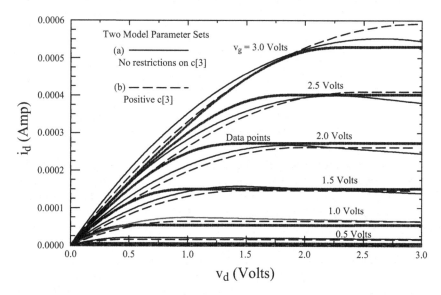

Figure 9.57. MOS drain current as a function of two independent variables, drain voltage and gate voltage. Fitted curves are shown for two first-order MOS device models.

Also shown in the figure are least-squares fitted curves for two sets of model parameters using a simple first-order MOS device model. Any elementary text book on electronics will give a first-order MOS device model as represented by the following set of equations:

$$i_d = \begin{cases} 0 \text{ for } v_g < V_t \\ K[(v_g - V_t)v_d - 0.5v_d^2](1 + \lambda v_d) \text{ for } v_d \leq v_g - V_t \\ \dfrac{K}{2}(v_g - V_t)^2(1 + \lambda v_d) \text{ for } v_d \geq v_g - V_t \end{cases} \qquad (9.19)$$

This is a 3 parameter model with fitting parameters K (in Amps/V^2), V_t (in Volts) and λ (in 1/Volts). In the nlstsq() formulism these will be identified as c[1], c[2] and c[3] respectively. This is another example of a piece-wise defined model (as discussed in Section 9.7) as three defining equations are needed for various operating regions of the device. The three defining equations have the important feature that the first derivatives of the model with respect to the two independent variables are continuous across the boundaries of the operating regions. Having such continuous derivatives is usually a very important feature of a model representing some physical phenomena as nature typically gives data with continuous values and first derivatives.

The last two terms in Eq. (9.19) can be combined into a single equation as follows:

$$i_d = \begin{cases} 0 \text{ for } v_g < V_t \\ K[(v_g - V_t)v_{dx} - 0.5v_{dx}^2](1 + \lambda v_d) \text{ where} \\ v_{dx} = \begin{cases} v_d \text{ for } v_d \leq v_g - V_t \\ v_g - V_t \text{ for } v_d \geq v_g - V \end{cases} \end{cases} \qquad (9.20)$$

The defining equation for v_{dx} implies a saturation limit. As v_d increases, v_{dx} follows the increasing value until v_d equals $v_g - V_t$ and at this point the value of v_{dx} stays fixed, regardless of the increasing value of v_d. In MOS device terminology this is known as the "drain saturation voltage", because the current at this point becomes almost constant as can be seen from Figure 9.57 at large drain voltages.

Such an upper or lower saturation limit value is very common in modeling of physical phenomena with piecewise models. Because of this common occurrence, the discussion will digress slightly to discuss how such saturation limits can be mathematically modeled. One straight forward way of modeling such hard saturation limits is obviously with two piecewise equations, one valid below the saturation limit and one above the saturation limit. In coding such a limit use can be made of the traditional "if" language statement. However by using a root function, it is possible to model such saturation limits by a single mathematical expression. Consider for example the following equation:

$$y = -0.5[\sqrt{(x-a)^2} - (x+a)]$$
$$= \begin{cases} x \text{ for } x \leq a \\ a \text{ for } x \geq a \end{cases} \tag{9.21}$$

This simple equation has exactly the desired properties of providing a saturation limit at the value $x = a$. An added feature of this single equation formulation is that it is very easy to extend the equation slightly to provide not only a continuous value of the function at the saturation point but to provide a continuous derivative at the transition point by modifying the equation to the form:

$$y = -0.5[\sqrt{(x-a)^2 + (2\varepsilon)^2} - (x+a)]$$
$$= \begin{cases} x \text{ for } x \ll a \\ a - \varepsilon \text{ for } x = a \\ a \text{ for } x \gg a \end{cases} \tag{9.22}$$

This provides a smooth transition from the linear regime to the saturation regime and an appropriate value of ε can be selected to give as abrupt a transition as desired between the linear and saturation regimes.

Variations of this basic equation can be used to provide upper and lower saturation limits to increasing or decreasing functions of some variable. Figure 9.58 shows four possible limiting functions using variations of this formulism. The curve in the first quadrant of the figure shows a function that limits an increasing function at some upper saturation limit. The curve in the second quadrant shows a function with an upper saturation limit but with a decreasing value for large positive values of x. The other two quadrants show functions with lower saturation limits for increasing or decreasing parameter values. These four cases cover all the various types of upper or lower saturation limits that might be encountered in the piecewise specification of a model equation. The Figure also shows the slight rounding of the saturation effect that can be achieved by use of a small ε value.

Now using this concept, the MOS device model can be written as:

$$i_d = \begin{cases} 0 \text{ for } v_g < V_t \\ K[(v_g - V_t)v_{dx} - 0.5v_{dx}^2](1 + \lambda v_d) \text{ where} \\ v_{dx} = -0.5[\sqrt{(v_d - v_g + V_t)^2} - (v_d + v_g - V_t)] \end{cases} \tag{9.23}$$

In this equation no rounding of the function at the saturation point has been used.

```
 1 : -- /* File list9_20.lua */ -- Data fitting, Parameter estimation
 2 :
 3 : require"nlstsq"
 4 :
 5 : -- Section 1. Input data file and define fitting function
 6 : infile = 'mosiv9.20'
 7 : id,vd,vg,vgd,ycalc = {},{},{},{},{}
 8 : read_data(infile..'.txt',id,vd,vg); nd = #id
 9 :
10 : ft = function(x,c) -- Define function to fit data
11 :    local vgt,vd = x[3] - c[2], x[2]
12 :    local vdx = 0.5*(vd+vgt - math.sqrt((vd-vgt)^2))
13 :    if vgt<0 then return -x[1] end
14 :    return c[1]*(vgt*vdx - 0.5*vdx^2)*(1 + c[3]*vd) - x[1]
15 : end
16 :
17 : c = {1.e-4,.20,0.01};nc = #c
18 : --step = {2,2,5}
19 :
20 : --- Section 2. Perform data fit and print fitted parameters
21 : del,err,nn = nlstsq({id,vd,vg},fw,ft,c) -- Call fitting
22 : for i=1,nc do printf('c[%d] = %12.4e +/- %12.4e\n',i,c[i],
        del[i]) end
23 : for i=1,nd do ycalc[i] = newtonfc(ft,{id[i],vd[i],vg[i]},c) end
24 : write_data(infile..'a.dat',id,vd,vg,ycalc)
25 : plot(vd,id,vg,ycalc)
Selected output:
c[1] =   1.4920e-004 +/-   7.2001e-007
c[2] =  -3.3838e-002 +/-   5.8287e-003
c[3] =  -6.8843e-002 +/-   1.7154e-003
```

Listing 9.20. Code for fitting the first-order MOS model to experimental data.

The computer code for fitting this model to the MOS data is shown in Listing 9.20. The MOS data is stored in a file in three column format with the current, drain voltage and gate voltage in the three columns. This is read into the program on line 8 of the listing. The model defining equations are given on lines 10 through 15 of the code in essentially the same form as Eq. (9.23) with the use of the $c[1] - c[3]$ notation. An "if" statement could just as easily have been used to give identical results in the code as formulated in Eq. (9.19). Initial approximations to the model coefficients as given on line 17 are not too critical and can be easily estimated from the data curves in Figure 9.57. No limits are placed on the size of the coefficient changes in each nlstsq() Newton iterative step as no step[] table is used (definition on line 18 is commented out).

The model coefficient values resulting from the data fitting are shown as output in Listing 9.20 and the resulting fit of the model to the data is shown in Figure 9.57 as the set of thin solid line curves. In performing a least squares fit of a model equation to a set of data one is frequently surprised by the results and this is one such case. The thin solid curves in Figure 9.57 are not seen to provide a very good fit to the experimental data. The errors are obviously model deficiency errors as the general functional form of the experimental data is not well modeled

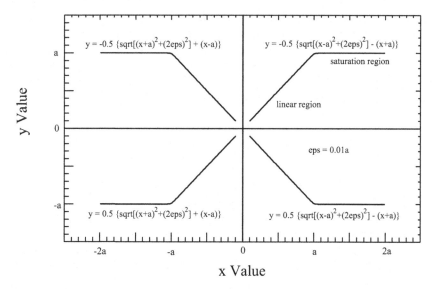

Figure 9.58. Illustration of upper and lower saturation limits for increasing or decreasing functions.

by the text-book MOS device equations with only 3 parameters. Because these equations are so commonly used in Electrical Engineering text books, it might be expected that the model would show better agreement with typical data for modern day MOS devices, but such is not the case for this example set of data.

Perhaps even more surprising, is the result that the λ (or $c[3]$) coefficient for the best fit to the data has a negative value. This can be seen by the negative slopes on the solid curves of Figure 9.57 in the saturation region or at large positive drain voltages. Such negative values are physically unreal as it is known that the slope of the curves in the saturation region has a near zero or positive slope. Thus a straight forward least squares fitting of the MOS data to the proposed model equation leads to physically unreal values of one of the model parameters. This again is not too uncommon in an unconstrained fitting of a model to experimental data. For the present case it would appear more appropriate to fit the data with a constraint on $c[3]$ such that the fitted value must be positive or zero. Such a desire to force a particular sign on a fitting parameter is not an uncommon occurrence in fitting a model to experimental data and such a constraint is easily included in the nlstsq() evaluation. To restrict a given coefficient to a desired sign one simply has to give the desired parameter an initial value with the desired sign and use a positive step size constraint. Recall that a positive step[] parameter for any coefficient limits the relative change in the parameter to the specified value. So if a step[3] = 2.0 is specified, for example, the $c[3]$ parameter will change at most by a factor of 2 at each iterative step in the fitting process. With such a step size limitation, the sign of the parameter can never change, only the magnitude can change to larger or smaller multiples at each iterative step. Line 18 of Listing 9.20 shows the changes needed to force the λ coefficient to remain positive in the data fitting procedure. In this case a factor of 5 change is specified in the step[3] value. The reader can simply uncomment this line and re-execute the code to obtain a new set of model fitting parameters. Such a re-execution of the code produces the following output set of model parameters:

$$
\begin{aligned}
c[1] &= 1.3309\text{e-}004 \\
c[2] &= 1.9239\text{e-}002 \\
c[3] &= 4.0960\text{e-}011
\end{aligned}
\tag{9.24}
$$

This set of parameters produces the dotted set of curves in Figure 9.57. The resulting value of c[3] is very small and approaching zero as expected since the optimum negative value can never be reached with the positive step size constraint. This set of model parameters, gives more reasonable physical values, but can not provide as good an overall fit to the data as the unconstrained set of fitted parameters.

The overall fit of the simple MOS model to the experimental data as seen in Figure 9.57 is not very good with the simple proposed model. It can thus be concluded that the first order text book model equations are not very good at representing the MOS data shown in Figure 9.57 and an improved physical model is needed. One such correction to the basic model is to include a denominator factor in the K parameter that effectively decreases the mobility of the MOS carriers as a function of the gate voltage. Such an improved set of MOS device equations can be formulated as:

$$
i_d = \begin{cases}
0 \text{ for } v_g < V_t \\
K\dfrac{[(v_g - V_t)v_{dx} - 0.5v_{dx}^2]}{[1 + \theta(v_g - V_t)]}(1 + \lambda v_d) \text{ where} \\
v_{dx} = -0.5[\sqrt{(v_d - v_g + V_t)^2} - (v_d + v_g - V_t)]
\end{cases}
\tag{9.25}
$$

This introduces a fourth parameter, θ, in the denominator of the equation. The code for fitting this equation to the MOS data is shown in Listing 9.21. The improved model equations are given on lines 10 through 15 of the code now with 4 c[] table parameters. The initial value of the extra parameter is set at 0 on line 17. Again a step limit of 5 is used for the λ coefficient to force it to have a positive value, but no step size limits are imposed on the other model parameters (see line 18 of the code). The resulting best-fit set of parameters is shown in the output lines in the listing along with standard error values. For this improved model, the value of the λ coefficient (the c[3] value) does in fact have a physically real positive value as expected.

The comparison between the model equation with the evaluated parameters and the experimental data is shown in Figure 9.59. As seen in this figure, the agreement between the data and the model is much improved over the simple three parameter model. The agreement is especially improved in the saturation region. There is still some inadequacy of the model in the low drain voltage and large gate voltage regime. However, a more complex MOS model would take the present discussion too far away from the main theme of this work. For the purpose of a modeling example as used here, this will be considered a sufficiently good fit to the experimental data. However, the errors between the data and the model are still seen to be due to an inadequate model equation and not due to random effects, so a Monte Carlo analysis of the accuracy of the coefficients would not be warranted for this example.

```
 1 : -- /* File list9_21.lua */ -- MOS transistor, improved model
 2 :
 3 : require"nlstsq"
 4 :
 5 : -- Section 1. Input data file and define fitting function
 6 : infile = 'mosiv9.20'

 7 : id,vd,vg,vgd,ycalc = {},{},{},{},{}
 8 : read_data(infile..'.txt',id,vd,vg); nd = #id
 9 :
10 : ft = function(x,c) -- Define function to fit data
11 :    local vgt,vd = x[3] - c[2], x[2]
12 :    local vdx = 0.5*(vd+vgt - math.sqrt((vd-vgt)^2))
13 :    if vgt<0 then return -x[1] end
14 :    return c[1]*(vgt*vdx - 0.5*vdx^2)*(1 + c[3]*vd)/(1+vgt*c[4])
          - x[1]
15 : end
16 :
17 : c = {1.e-4,.20,0.01,0}; nc = #c
18 : step = {0,0,5,0}
19 :
20 : --- Section 2. Perform data fit and print fitted parameters
21 : del,err,nn = nlstsq({id,vd,vg},fw,ft,c,actv,step) -- fitting

22 : for i=1,nc do printf('c[%d] = %12.4e +/- %12.4e\n',i,c[i],
          del[i]) end
23 : for i=1,nd do ycalc[i] = newtonfc(ft,{id[i],vd[i],vg[i]},c) end

24 : write_data(infile..'c.dat',id,vd,vg,ycalc)
25 : plot(vd,id,ycalc)

Selected output:
c[1] =  7.0213e-004 +/-  1.4175e-005
c[2] =  5.1464e-001 +/-  4.5525e-003
c[3] =  1.7904e-002 +/-  1.1947e-003
c[4] =  1.3078e+000 +/-  3.5076e-002
```

Listing 9.21. Code for MOS data fitting with improved four parameter model.

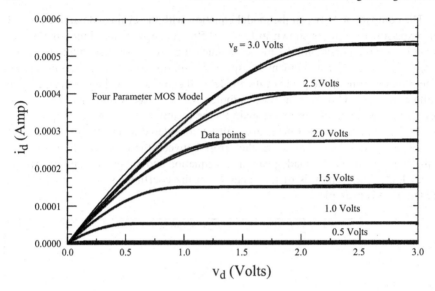

Figure 9.59. MOS data fitted with improved four parameter model.

9.9 Response Surface Modeling and Parameter Estimation

An important analysis tool for use with the design of experiments is that of response surface modeling. In a typical engineering experiment, the interest is in determining the response of some manufacturing system to changes in a set of manufacturing parameters. As an example of such an experiment, a set of data involving the chemical vapor deposition of a thin film as given in Table 9.2 will be considered. (See www.itl.nist.gov/div898/handbook/ for a description of the data set). The data consists of 6 data runs with two independent parameters, Pressure of the CVD system and gas flow ratio H_2/WF_6. Two parameters are measured as characteristics of the deposited film, the film uniformity (in %) and stress in the film. The center point values for pressure and gas flow are the Run Number 2 entry with values of 42 and 6. Around these center point values, the two independent parameters are varied and the values of the two response variables recorded. In the present case low values of both the uniformity parameter and the stress parameter are desired. There is considerable theory about the so called "design of experiments" which can be used to determine how to select a minimum number of such experiments to obtain maximum knowledge about a physical process. The reader is encouraged to consult the literature on the design of experiments for more knowledge and background on such experiments if desired.

Run Number	Pressure	H_2/WF_6	Uniform-ity	Stress
1	80	6	4.6	8.04
2	42	6	6.2	7.78
3	68.87	3.17	3.4	7.58
4	15.13	8.83	6.9	7.27
5	4	6	7.3	6.49
6	42	6	6.4	7.69
7	15.13	3.17	8.6	6.66
8	42	2	6.3	7.16
9	68.86	8.83	5.1	8.33
10	42	10	5.4	8.19
11	42	6	5.0	7.90

Table 9.2. Experimental data for response surface example.

For this work the interest is in modeling and characterizing the relationship between the two response variables (uniformity and stress) and the two independent variables (pressure and flow ratio). In response surface modeling, the usual assumption is to take the simplest possible forms for the model relationship, consisting of linear and quadratic functions of the independent variables. Thus for this two parameter experiment, a proposed response surface model is:

$$R = C_1 + C_2 x_1 + C_3 x_2 + C_4 x_1^2 + C_5 x_2^2 + C_6 x_1 x_2 \tag{9.26}$$

where R is a measured response, in the present case either uniformity or stress and x_1 and x_2 are the independent parameters. This equation describes a surface in the two dimensional space of the independent variables – hence the term response surface analysis. As the number of independent parameters grows, the number of model parameters grows very fast and the technique becomes difficult to apply for more than 3 or 4 variables.

What is desired is a model to predict the response as a function of the dependent variables and some measures of how good the model is in prediction a response. The response surface can then be used to hopefully find an optimum operating point that gives the best combination of response variables. The work here will concentrate only on modeling the response surface and not on how such a surface might be used to determine the optimum operating point. The previous nlstsq() function can readily determines an optimum set of model parameters for Eq. (9.26). In fact the model is linear in the model coefficients and the fitting procedure will converge in two iterations (One to determine the values and a second to realize that the exact values have been calculated).

Computer code for evaluating the model coefficients is shown in Listing 9.22. The model definition on lines 12 through 15 is readily identified from Eq. (9.26). Either line 8 or 9 can be executed to input either the uniformity data or the stress

```
 1 : -- /* File list9_22.lua */ -- Response surface analysis
 2 :
 3 : requires("nlstsq","prob")
 4 :
 5 : -- Section 1. Input data file and define fitting function
 6 : infile = 'rspsurf'
 7 : p,gr,y,ycalc,ydata = {},{},{},{},{}
 8 : read_data(infile..'.txt',{},p,gr,y) -- For uniformity analysis
 9 : --read_data(infile..'.txt',{},p,gr,{},y) -- For stress analysis
10 : nd = #y
11 :
12 : ft = function(x,c) -- Define function to fit data
13 :    local x1,x2 = x[2],x[3]
14 :    return c[1]+c[2]*x1+c[3]*x2+c[4]*x1^2+c[5]*x2^2+c[6]*x1*x2 -
          x[1]
15 : end
16 :
17 : c = {0,0,0,0,0,0}; nc = #c
18 : --actv = {1,1,1,0,0,1} -- Uncomment to eliminate c[4] and c[5]
19 :
20 : --- Section 2. Perform data fit and print fitted parameters
21 : del,err,nn = nlstsq({y,p,gr},fw,ft,c,actv)
22 : for i=1,nc do printf('c[%d] = %12.4e +/- %12.4e \t (+/-
          ratio)%9.4f\n',
23 :          i,c[i],del[i],del[i]/math.abs(c[i])) end
24 : x1,x2,i = {},{},1
25 : for xx1=4,80.1,1 do -- Points for surface plot
26 :    for xx2=2,9,.1 do
27 :          x1[i],x2[i],ycalc[i],i = xx1,xx2,ft({0,xx1,xx2},c),i+1
28 :    end
29 : end
30 : for i=1,nd do ydata[i] = newtonfc(ft,{y[i],p[i],gr[i]},c) end
31 : _,_,rsq = clinear(ydata,y)
32 : print('Rsquared for model = ',rsq)
33 : write_data(infile..'a.dat',y,p,gr,ydata,{ycalc,x1,x2})
```
Selected Output:
```
c[1] =   1.1373e+001 +/-  1.9798e+000   (+/-ratio)    0.1741
c[2] =  -1.2517e-001 +/-  4.6737e-002   (+/-ratio)    0.3734
c[3] =  -5.5065e-001 +/-  5.0868e-001   (+/-ratio)    0.9238
c[4] =   9.2777e-005 +/-  4.2350e-004   (+/-ratio)    4.5647
c[5] =   2.0847e-003 +/-  3.8190e-002   (+/-ratio)   18.3192
c[6] =   1.1174e-002 +/-  4.7824e-003   (+/-ratio)    0.4280
Rsquared for model =    0.87068809771198
```

Listing 9.22. Code segment for response surface analysis of film uniformity.

data. The initial values of the coefficients are all taken as zero which is appropriate for a model with linear coefficients. About the only unique features of the code is the addition of two loops on lines 25 through 29 to generate a table of values for subsequent plotting of the resulting response surface. Finally line 31 performs a linear model fit between the measured uniformity values and the model predicted values. The resulting R^2 value is one measure of the agreement of the model and the experimental data. A graph of this is subsequently shown, but the obtained value of 0.8707 is reasonable although not exceptionally good.

The evaluated fitting parameters provide important information about the proposed data model. To aid in viewing the parameters, the ratio of the standard error to the value of the coefficient is calculated on line 23 and included in the printed output. A small ratio means that the given parameter is significant in the model while a large ratio means that the given parameter cannot be determined from the data. The printed output values show that the standard error for the c[4] parameter is 4.5 times larger than the value itself and that the standard error for the c[5] parameter is 18.3 times larger than the value. Thus the data does not support the proposed model dependence on these two parameters which are the squared terms on the pressure and the gas ratio. However, the c[6] standard error value indicates a significant dependence on the product of pressure and gas ratio.

The large uncertainty in the c[4] and c[5] coefficients suggests that the data should be re-evaluated with a response surface model which takes these coefficients as zero. This could certainly be done by recoding the basic equation on line 14 of Listing 9.22. However, this can also be done by setting the entries in an actv[] table to zero for these parameters. The code line to make this change is shown as a comment on line 18 of Listing 9.22. The reader is encouraged to un-comment this line and re-execute the code. The resulting executed code gives the following values for the new coefficients:

$$c[1] = 1.1194e+001 +/- 1.4048e+000 \ (+/\text{-ratio}) \ 0.1255$$

$$c[2] = -1.1736e-001 +/- 3.0367e-002 \ (+/\text{-ratio}) \ 0.2588$$

$$c[3] = -5.2553e-001 +/- 2.2135e-001 \ (+/\text{-ratio}) \ 0.4212 \qquad (9.27)$$

$$c[6] = 1.1170e-002 +/- 4.8053e-003 \ (+/\text{-ratio}) \ 0.4302$$

$$\text{Rsquared for model} = 0.86944407962076$$

The values all now have a standard error less than the fitted value. The best fit values for these coefficients differ only slightly from the values shown in Listing 9.22. The R^2 value for the fitting is essentially unchanged.

Figure 9.60 shows a plot of the measured uniformity value vs. the model predicted uniformity value. Ideally all the points would fall along the solid curve in the figure. Points are shown for the model with all 6 coefficients evaluated and with only 4 coefficients evaluated ($C_4 = C_5 = 0$). There is very little difference in the two sets of points again illustrating the point that no dependence on the square of pressure or square of gas ratio is justified in the model by the experimental data.

Another measure of the appropriateness of the proposed model is the randomness of the residuals. A graph of this is shown in Figure 9.61 as a function of run number and the data points are seen to be rather uniformly distributed about the ideal prediction line. While the limited number of data points makes definitive conclusions impossible, there is no obvious dependency of the observed residual values on the run number. Such possible time trends should always be checked to be sure that the observed residuals are truly random in nature.

An analysis of the stress values for the data can be obtained by substituting line 9 in Listing 9.22 for line 8. The reader is encouraged to execute the modified code and observe the following evaluated coefficients for the stress data:

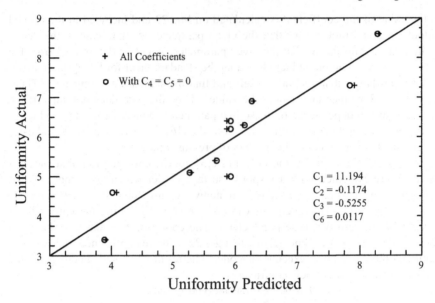

Figure 9.60. Predicted vs. modeled uniformity. $R^2 = 0.869$ for linear fit.

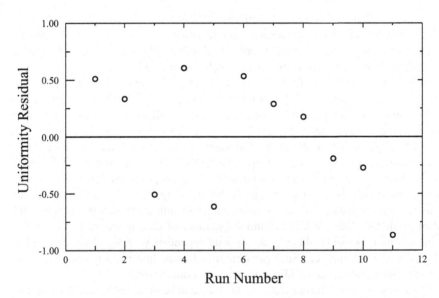

Figure 9.61. Uniformity residual value vs. run number for thin film deposition data.

$c[1] = 5.4275e+000 +/- 2.0834e-001$ (+/-ratio)   0.0384
$c[2] = 4.7454e-002 +/- 4.9182e-003$ (+/-ratio)   0.1036
$c[3] = 1.9496e-001 +/- 5.3529e-002$ (+/-ratio)   0.2746
$c[4] = -3.6685e-004 +/- 4.4565e-005$ (+/-ratio)   0.1215
$c[5] = -7.4928e-003 +/- 4.0187e-003$ (+/-ratio)   0.5363
$c[6] = 4.6196e-004 +/- 5.0325e-004$ (+/-ratio)   1.0894
Rsquared for model =   0.99189303685573

For this data, the $c[6]$ coefficient has a standard error larger than the evaluated value, so can be neglected. This means that there is no significant variation of film stress with the product of pressure and gas ratio. Re-executing the code setting $actv[6] = 0$, results in the fitting parameters shown in Figure 9.62 along with a comparison of the predicted stress and measured stress values. The relationship between predicted stress and measured stress is seen to have considerably less random variation than the modeled uniformity data in Figure 9.60. The R^2 value of 0.992 indicates a very good linear fit of the data to the model. Although not shown, the residuals show good random scatter about the predicted values. It can thus be concluded that the proposed response surface model provides a very good model for the stress data and an acceptable model for the uniformity data.

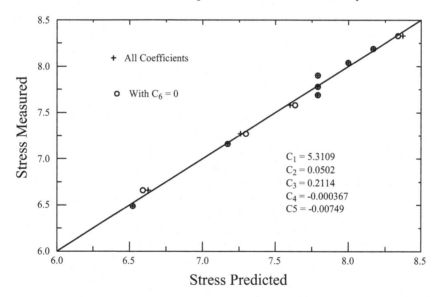

Figure 9.62. Predicted vs. modeled film stress. $R^2 = 0.992$ for linear fit.

So far the means of generating a response surface and some evaluation of the goodness of fit of the resulting response surface have been discussed. The object of such an analysis is typically to determine an optimum operating point – in this case the objective might be to determine an optimum operating point that simultaneously produces low stress and low uniformity values. To achieve this objective, one needs to visualize the surface response data and one way of displaying the response surface is a 3-D plot such as that shown in Figure 9.63 showing the modeled stress value as a function of both the system pressure and gas ratio. Also shown are the data points from which the response surface has been generated. One could also display such a response surface for the uniformity values.

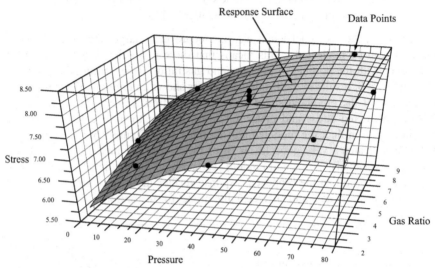

Figure 9.63. Response surface of stress as a function of pressure and gas ratio.

For low stress values, it can be seen from Figure 9.63, that low pressure and low gas ratios are desired. However, low uniformity values are found for large pressure values and low gas ratios. Thus one has a case where tradeoffs are required between the desire for low stress and low uniformity values. For better viewing of the tradeoffs a contour plot provides one such analysis tool. This is shown in Figure 9.64 where curves of constant uniformity and constant stress are both shown in the pressure-gas ratio plane. For the lowest values of stress, operation should occur in the lower left corner of the plane while for lowest – or best uniformity, operation should operate in the lower right corner. This is typically of such response surface analyses where some compromise must typically be made in the operating point to achieve a set of desired objectives. The optimum operating point will depend upon the relative weight one places on the importance of low stress and on the importance of a highly uniform film. If both of these objectives are of relative equal importance, then a relatively broad operating point can be seen in the plane with a stress value of around 7 and a uniformity value of around 7. The response surface analysis provides the data needed for making such decisions, but the importance of the relative response factors must be specified before an optimum operating point can be specified.

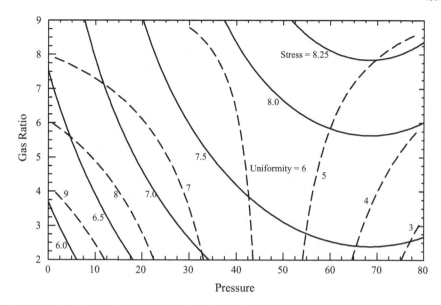

Figure 9.64. Plots of constant uniformity and constant stress contours in the pressure-gas ratio plane. Solid curves are for stress and dotted curves are for uniformity.

The literature on the response surface methodology and the design of experiments is quite extensive and the reader is referred to the literature for a more in-depth discussion. The intent here is to demonstrate how the previously developed software can be used in this type of analysis. This example has also illustrated the use of the actv[] table as input to nlstsq() for eliminating the use of particular parameters from a least squares data fitting analysis.

9.10 Summary

This chapter has taken an extended look at the modeling of experimental data and the estimation of a set of parameters associated with some mathematical model representing the experimental data. Very little new code or theory has been developed in this chapter, but rather the material has provided an extensive application of the code and analysis methods developed in the two previous chapters. In particular, extensive use has been made of the nonlinear least-squares data fitting routine nlstsq(). The Monte Carlo analysis and statistical code developed in the previous chapter has also been extensively employed to evaluate important statistical bounds on estimated model parameters.

Two code segments that have been developed as extensions of previous code segments are:

1. mcpar() – Code segment for performing a Monte Carlo analysis of the residuals associated with the least squares parameter estimation method.
2. cltwodim() – Code segment for evaluating the two dimensional confidence limits from the results of a Monte Carlo analysis of residual errors.
3. runs() – Code segment for a "runs" analysis of random data.

The concept of a 6-plot and its use to explore the "goodness-of-fit" of a model to experimental data was developed and illustrated with several examples. This was combined with the Monte Carlo analysis to explore parameter correlations for a number of examples. It was found from the examples, that significant parameter correlations are quite common in fitting nonlinear equations to experimental data and such correlations need to be understood in any given application.

Examples were given of data fitting with general transcendental functions of the dependent and independent variables as well as examples with several independent variables. All of these use the same basis nlstsq() data fitting routine. Finally, data fitting with piecewise functions was discussed and demonstrated with examples. With the examples in this chapter, the reader should be able to apply the code segments and analysis approach to a wide range of data fitting and parameter estimation problem.

10 Differential Equations: Initial Value Problems

A large number of engineering problems are formulated in terms of a differential equation or a system of differential equations. A typical engineering problem requires finding the solution of a set of differential equations subject to some set of initial values or subject to some set of boundary values. In this chapter only the initial value type of problem will be considered. A subset of general differential equations is the set of linear differential equations with constant coefficients. For such systems, closed form solutions can always be found as the solutions are always sums of exponential functions. For general differential equations and especially for non-linear differential equations, closed form solutions can not in general be found and one must resort to numerical solutions. What is meant by a numerical solution is a set of tabular values giving the value of the dependent variable (or variables) as a function of the independent variable (or variables) at a finite number of values of the dependent variable.

This chapter begins by discussing the simple case of a single first-order differential equation. The independent variable will be assumed to be time (t), but the discussing is independent of whether the independent variable is time or some spatial coordinate. Some fundamental properties of all numerical solutions to differential equations will be developed for some simple cases. The discussion will then be expanded to systems of first-order differential equations and then to systems of second and higher order differential equations. Some general computer code segments will be developed for use in solving general non-linear differential equations. Finally several examples will be given to illustrate the application of the code segments to typical problems.

10.1 Introduction to the Numerical Solution of Differential Equations

Consider first a single first-order differential equation written in the form:

$$\frac{dy}{dt} = y' = f(t, y) \tag{10.1}$$

where the function specifying the derivative may be any linear or nonlinear function of y and t. Consider a function of time such as shown in Figure 10.1. For this discussion, assume that the solid curve is the solution of some differential equation and one desires to obtain a numerical approximation to the solid curve at some finite set of data points identified in the figure as $t_{n-1}, t_n, t_{n+1}, t_{n+2}$. The corresponding solution points are identified as $y_{n-1}, y_n, y_{n+1}, y_{n+2}$ where n is some gen-

461

J.R. Hauser, *Numerical Methods for Nonlinear Engineering Models*, 461–574.
© Springer Science + Business Media B.V. 2009

eral time point. Further assume that the solution at some point n and t_n is known and one desires to calculate a solution value at the next time point of $n+1$ and t_{n+1}. The spacing between time points is assumed to be uniform with value

$$h = t_{n+1} - t_n \tag{10.2}$$

The simplest algorithms make use of the solution at the nth time step and the derivative at the n'th and possibly the (n+1)th time step.

Perhaps the simplest algorithm is Euler's formula of:

$$y_{n+1} = y_n + hy_n' = y_n + hf(t_n, y_n) \tag{10.3}$$

which uses a future value of the solution variable to approximate the derivative at the nth time point. The second form of the equation expresses the dependence of the derivative on the function value and the time value at point n. From Figure 10.1, it can be seen that this will give a solution point along the derivative line passing through the nth point. For a solution with downward curvature, it can be seen that this will give a solution point that is somewhat too large and for a curve with an upward curvature, this will give a solution point that is somewhat too small. However, if the step size, h, is taken small enough, the evaluated point may be sufficiently accurate for a given purpose. This simple equation is also known as the "forward difference" approximation since it uses the forward point only in the derivative approximation. The expression of Eq. (10.3) provides an explicit equation for evaluating the solution at the next time step using only values evaluated at the nth time step.

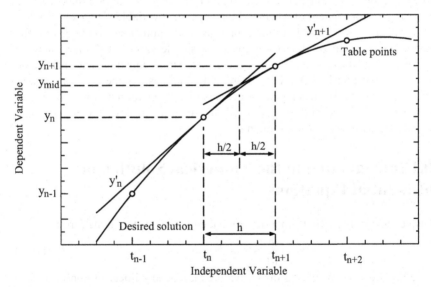

Figure 10.1. Illustration of function and derivatives at various time points.

A second approximation to the relationships shown in Figure 10.1 is the "backwards difference" approximation that is expressed as:

$$y_{n+1} = y_n + hy'_{n+1} = y_n + hy'_{n+1}(t_{n+1}, y_{n+1})$$
$$= y_n + hy'_{n+1}(t_n + h, y_{n+1}) \qquad (10.4)$$

This uses a backwards value of the solution variable to approximate the derivative at the t_{n+1} point. The second form of the expression in Eq. (10.4) expresses the fact that the derivative function at point n+1 depends on the solution value and time value at that point. The third form emphasizes the fact that the time is known from the previous time and the time step value. From Figure 10.1 it can be seen that this will give a solution along a line passing through the nth solution value but with a slope that is evaluated at the (n+1)th point. For a curve with downward curvature, it can readily be seen that this will give a solution point that is somewhat too small and for a curve with an upward curvature, this will give a solution point that is somewhat too large. However, again if the step size is taken sufficiently small, a reasonably accurate approximation to the solution can be obtained.

Either of the above algorithms can be used as the basis for a numerical solution algorithm. The forward difference (FD) algorithm gives an explicit expression for the solution at time step n+1 in terms of values that can be explicitly evaluated at time step n. The backwards difference (BD) algorithm gives an implicit expression for the solution at time step n+1 in terms of values at time step n+1 since y_{n+1} appears on the left side of Eq. (10.4) and on the right side in the derivative of the $f()$ term. Thus in general when the derivative function is a nonlinear function of the dependent variable, the BD algorithm must employ some iterative approach to obtain a consistent solution. One approach frequently used is to use the FD approximation to obtain an estimate of the next solution point and then use that approximation in the BD equation to obtain a more accurate value of the solution. The process can be iterated if desired to obtain a more accurate solution. However, as subsequently shown, it is possible to employ the previously developed nonlinear equation solver to obtain a solution using Newton's method, which typically converges much faster than a successive substitution approach.

By looking at the derivative curve in Figure 10.1 and the FD and BD algorithms, it can readily be seen that regardless of the curvature of a function, one of the algorithms tends to overestimate the solution at the next time step and the other tends to underestimate the solution. Thus it might be expected that some combination of these algorithms would give a better algorithm. This can be achieved by use of a weighted average of the two derivatives in an algorithm for the derivative as:

$$(y_{n+1} - y_n)/h = (y'_n + y'_{n+1})/2 \qquad (10.5)$$

This can be rearranged into the form:

$$y_{n+1} = [y_n + (h/2)y'_n] + (h/2)y'_{n+1} = y_{mid} + (h/2)y'_{n+1} \qquad (10.6)$$

This is also known as the "trapezoidal" (TP) rule because of its relationship to the trapezoidal integration rule as discussed in Chapter 5. The final form of Eq. (10.6) provides an implementation of this algorithm which illustrates that this algorithm is equivalent to taking a half step (h/2) from the y_n point using the slope at the nth point to obtain the y_{mid} point shown in Figure 10.1 then taking another half step from this mid point using the slope at the n+1 point. It can be seen graphically from the figure that this should give a much better estimate of the function at the next time step than the individual FD or BD algorithms. However, this also gives an implicit equation for the next time step value since the derivative at point n+1 in Eq. (10.6) also depends on the value of the function which is to be determined. However, this is similar to the functional dependency of Eq. (10.4) and there is little difference in the complexity of either the TP or the BD algorithms as most of the effort is involved with solving the resulting implicit equations for the new solution point.

It is expected that the TP algorithm will give better accuracy than the FD and BD algorithms. However, in addition to accuracy, stability is also very important. A stable algorithm means that the solution value will remain finite as the number of solution steps increases indefinitely. Stability can be usefully studied by looking at the simple linear differential equation:

$$y' = \lambda y \tag{10.7}$$

with the exact solution:

$$y = y(0)\exp(\lambda t) \tag{10.8}$$

The constant λ may take on any real or complex value but for a stable system at $t \to \infty$ one must have the requirement $\mathrm{Re}(\lambda) < 0$. It can be noted that the solution of any linear differential equation with constant coefficients or any system of linear differential equations with constant coefficients is the sum of a series of exponential terms with each term similar to that of Eq. (10.8). Thus the study of Eq. (10.7) is equivalent to studying the stability of any one component of the solution of a general linear differential equation.

Applying the forward difference equation to Eq. (10.7) gives the recursion relationship

$$y_{n+1} = (1 + \lambda h)y_n \tag{10.9}$$

By considering the application of this algorithm to n steps beyond the initial value one gets

$$y_n = (1 + \lambda h)^n y_0 \tag{10.10}$$

If this solution value is to remain finite as the number of steps approaches infinity, then one must have

$$\left|1 + h\lambda\right| \leq 1 \tag{10.11}$$

This is the condition of stability for the forward difference equation. Since λ can be complex, it can be written as:

$$h\lambda = h\alpha + jh\beta \tag{10.12}$$

where j is the purely imaginary unity value. Putting this result into Eq. (10.11) leads to the requirement that one must have

$$(1 + h\alpha)^2 + (h\beta)^2 \leq 1 \tag{10.13}$$

In the $h\alpha - h\beta$ plane this describes the inside of a circle centered at point (-1,0) and of radius 1 passing through the origin at (0,0) and the point (-2,0). For a solution with a real decaying exponential, λ will have only a real negative part and a stable solution will require that: $-2/\mathrm{Re}(\lambda) = -2/\alpha \leq h$. For a rapidly decaying solution with a large value of λ this will require a very small time step, h. This would not be too bad if one had only a single exponential in a solution. However, when many exponentials are present with a wide range of time constants, the stability is determined by the most rapidly changing exponential term.

Now looking at the backwards difference equation and Eq. (10.7) it is found that the recursion relationship after n steps gives:

$$y_n = \frac{1}{(1-\lambda h)^n} y_0 \tag{10.14}$$

If this is to remain finite after an infinitely large number of steps then one must have

$$|1-h\lambda| \geq 1 \tag{10.15}$$

Again since λ can be complex, this leads to the requirement

$$(1-h\alpha)^2 + (h\beta)^2 \geq 1 \tag{10.16}$$

In the $h\alpha - h\beta$ plane this describes the region outside a circle centered at point (1,0) and of radius 1 passing through the origin at (0,0) and the point (2,0). For a solution with a real decaying exponential (negative real λ), any point in the negative half plane will result in stability and thus the backwards difference algorithm will be stable for any desired value of the time step. This is a very important feature of the backwards difference algorithm. However, in addition to stability, accuracy is also of importance.

The third approximation to the derivative discussed here is the trapezoidal rule as expressed in Eq. (10.5). Applying this to the test differential of Eq. (10.7) leads to the recursion relationship after n steps of

$$y_n = \frac{(1+\lambda h/2)^n}{(1-\lambda h/2)^n} y_0 \tag{10.17}$$

A comparison of this equation with Eqs. (10.10) and (10.14) shows that the trapezoidal algorithm has features of both the forward difference and backwards difference equations. For stability after an infinite number of steps this expression requires that

$$\frac{|1+\lambda h/2|}{|1-\lambda h/2|} \leq 1 \tag{10.18}$$

In the $h\alpha - h\beta$ plane this becomes

$$\frac{(1+h\alpha/2)^2 + (h\beta/2)^2}{(1-h\alpha/2)^2 + (h\beta/2)^2} \leq 1 \tag{10.19}$$

This will be satisfied by any value in the negative half plane, i.e. by any value of $h\alpha \leq 0$. Thus for an exponentially decaying solution, the trapezoidal algorithm will be stable for any desired step size, similar to the backwards difference algorithm.

Figure 10.2 illustrates the stability regions of the three algorithms in the $h\alpha - h\beta$ plane. For most engineering problems one deals with differential equations that have stable solutions as time approaches infinity. This means that typically differential equations have a negative real part of the exponential term

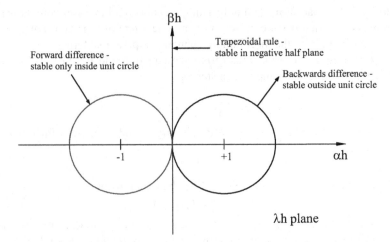

Figure 10.2. Illustration of stability regions of three differencing approximations in the complex plane.

– or α is typically negative. For such problems, the forward difference formula will only give stable solutions if a step size h is selected such that $h\lambda$ is inside the unit circle centered at (-1,0). For the backwards difference or trapezoidal formulas, any value of $h\lambda$ within the entire negative half plane will result in a stable solution. Examples will be given in subsequent sections of the stability and accuracy of the three formulas. However, it can be seen that from a stability viewpoint, that either the backwards difference or the trapezoidal formulas are much to be favored. This is especially true for so-called "stiff" systems of differential equations. A stiff system is one in which there are two or more solution term with greatly differing time constants, i.e. one solution term decays very rapidly with respect to one or more other solution terms. If the forward difference formulation is used, a step size small enough to bring the most rapidly decaying solution term into the circle of stability must be selected, regardless or whether or not the term is on major importance on the time scale of interest. An attempt to use large time steps to observe the longer time constant terms will result in instability in the solution. The backwards or trapezoidal rules have no such limitations. However, both of these stable formulations generate an implicit expression for the solution variable at each time step and for a general nonlinear differential equation require some type of iterative solution at each time step. Thus the property of stability can be achieved but at the cost of added complexity in the solution technique.

Stability and accuracy are both very important in solving any differential equation. The above discussion has focused on the three simplest algorithms for numerically approximating the derivative term in a differential equation. The algorithms allow one to bootstrap a solution along the solution path moving from one solution point to the next solution point. Since the numerical solution of differential equations is so important, much effort and many publications have been devoted to the topic of the best algorithms for solving such equations. Many more complicated algorithms have been developed and studied. One such class of algorithms goes by the term "predictor-corrector" techniques. In this approach, a first equation is used to "predict" a new solution along the curve and then a second equation is used to generate a "corrected" solution point. The equations may use any number of previously evaluated solution points, such as the solution points at times $t_n, t_{n-1}, t_{n-2}, \cdots t_{n-k}$ to predict and correct the solution at time t_{n+1}. However, absolutely stable multi-step solution techniques are known to occur only for low order formulas and it is known that the trapezoidal rule provides the lowest possible truncation error for an absolutely stable formula. Thus such higher order formulas will not be considered here but the reader is encouraged to explore other numerical books and references for more general discussions of predictor-corrector techniques for solving differential equations. The Runge-Kutta method is a technique very frequently used for solving differential equations and is easily programmed if one desires to compare this technique with the methods discussed in this text.

10.2 Systems of Differential Equations

Many problems of interest involve systems of coupled differential equations. Thus before looking at coded solution techniques, approaches to handling systems of possibly coupled differential equations will be considered. As one example, it is possible to formulate a higher order differential equation in terms of a coupled system of first order differential equations. As an example consider a simple second-order differential equation of the form

$$f_1(t)\frac{d^2 y}{dt^2} + f_2(t)\frac{dy}{dt} + f_3(t)y + f_4(t) = 0 \qquad (10.20)$$

Consider the following definitions:

$$y_1 = y \text{ and } y_2 = \frac{dy}{dt} \qquad (10.21)$$

The original second order equation can now be written in terms of the two equations

(1) $\quad \dfrac{dy_1}{dt} - y_2 = 0$

(2) $\quad f_1(t)\dfrac{dy_2}{dt} + f_2(t)y_2 + f_3(t)y_1 + f_4(t) = 0$

$\qquad (10.22)$

An nth order differential equation can in a similar manner be written in terms of an equivalent set of n first order differential equations. In the above set of equations, each equation has only one derivative term. While many coupled sets of equations are of this form, other sets of equations may have mixed derivative terms in any given equation. Such sets of equations result from formulating the differential equations of an electronic circuit for example.

Thus the most general form of a set of k coupled first order differential equations can be expressed in the notation

$$eq1 = F_1(t, y_1, y_2, \ldots, y_1', y_2', \ldots) = 0$$
$$eq2 = F_2(t, y_1, y_2, \ldots, y_1', y_2', \ldots) = 0$$
$$\ldots \qquad \ldots \tag{10.23}$$
$$eqk = F_k(t, y_1, y_2, \ldots, y_1', y_2', \ldots) = 0$$

In this set of equations a prime has been used to denote the derivative to simplify the notation. Also no assumptions have been made with regard to the form of the equations. In each equation it is assumed that all the function values and derivatives may possible be present. Also it is assumed that any of the terms may appear in any nonlinear or implicit form. It is only assumed that the equations are written in a form that gives zero as the answer when satisfied by the set of variables and derivatives and at each time point. To update this set of coupled differential equations at some time point using the known value of the function at some previous time point simply requires that the derivative terms be replaced by one of the previously discussed three approximations. For example, if the backwards difference operator of Eq. (10.4) is used the replacement is:

$$\frac{dy_i}{dt} \rightarrow \frac{(y_{i,n+1} - y_{i,n})}{h} \tag{10.24}$$

for each derivative operator. This gives the set of equations

$$eq1 = F_1(t, y_1, y_2, \ldots, \frac{(y_1 - y_{1,n})}{h}, \frac{(y_2 - y_{2,n})}{h}, \ldots) = f_1(t, y_1, y_2, \ldots) = 0$$

$$eq2 = F_2(t, y_1, y_2, \ldots, \frac{(y_1 - y_{1,n})}{h}, \frac{(y_2 - y_{2,n})}{h}, \ldots) = f_2(t, y_1, y_2, \ldots) = 0 \tag{10.25}$$

$$\ldots \qquad \ldots$$

$$eqk = F_k(t, y_1, y_2, \ldots, \frac{(y_1 - y_{1,n})}{h}, \frac{(y_2 - y_{2,n})}{h}, \ldots) = f_k(t, y_1, y_2, \ldots) = 0$$

In this set of equations, the $n+1$ notation as a subscript has been dropped on each variable ($y_1, y_2 \ldots$ etc.) with the understanding that the evaluated parameters are at the next time point. As the right hand side of each term in Eq. (10.25) indicates, the resulting equations are a coupled set of k equations in k unknowns which are the updated solutions of the differential equation variables at the next time increment. If the trapezoidal integration algorithm is selected instead of the backwards difference algorithm, the form of the equations is similar except for the following replacements:

$$h \rightarrow h/2 \text{ and } y_{i,n} \rightarrow y_{mid,i,n}$$

where $y_{mid,i,n} = [y_{i,n} + (h/2)y'_{i,n}]$ is a function of the known time step point parameters of function values and derivatives. In either the BD or TP cases the resulting set of equations is of the same form in terms of the unknown solution points.

A brief discussion is perhaps in order on the selected form for representing the equation set as in Eq. (10.25). Some numerical packages used to solve differential equations require that a set of functions be defined that return the derivative values of the differential equations when evaluated. This has several disadvantages. First such a formulation does not allow one to have a set of equations expressed in terms of combinations of the derivative terms. Second, the general case of nonlinear derivative terms or transcendental functions precludes such a simple formulation. Lastly, some important sets of equations, as discussed later, are formulated in terms of a combined set of differential and algebraic equations where some equations do not have a derivative term. These important cases can not be handled if the defining equation formulism simply requires the return of the derivative value. The form selected for representation here has no such limitations and is in keeping with other problem formulations in this work where equations were coded such that the function returns zero when satisfied by a set of solution variables.

Now the important question of how to solve the resulting set of equations can be addressed. Fortunately, this has already been addressed in Chapter 4. The resulting form of Eq. (10.25) is exactly of the form of Eq. (4.1) discussed in that chapter. The reader is encouraged to review the Chapter 4 material and the code developed there for the solution of sets of simultaneous equations. The code developed in that chapter, especially the nsolv() routine that uses Newton's method to solve a nonlinear or linear set of such coupled equations can be directly used in this application. Most of the hard work in solving set of differential equations has thus already been done in Chapter 4. This is another important example of the concept of reusable computer code.

Before developing some general functions for solving systems of differential equation, the next section will consider some stability and accuracy issues with simple differential equations with known solutions. This will provide some important insight into the features needed for solving differential equations.

10.3 Exploring Stability and Accuracy Issues with Simple Examples

The background needed to implement a simple numerical integration algorithm has now been developed in the preceding sections. This section explores the stability and accuracy issues associated with the forward difference, backwards difference and trapezoidal rule approximations to the derivative values. A simple code listing is shown in Listing 10.1 for this application. A typ parameter is used on lines 5 through 7 to specify the type of derivative approximation used in the code. The differential equations to be solved are specified by the feqs() function on lines 9 through 13. In the example shown the equations being solved are three uncoupled simple differential equations with the mathematical form

$$eq1 = y_1' + y_1 = 0$$
$$eq2 = y_2' + 10y_2 = 0 \qquad\qquad (10.26)$$
$$eq3 = y_3' + 100y_3 = 0$$

The listing shows the equations coded in essentially this form except for the equal zero part on lines 10 through 12. The notation yp[] is used for specifying the first derivative terms in the equations. The solutions are known exponential functions of time with decaying time constants of 1, 0.1 and 0.01 for the three equations respectively. The initial values of the solutions are set on line 18 and 19 to be 1.0 at zero time. The major part of the solution technique is the time loop from line 36 through line 45. For each time step the code performs the following steps:

(a) Evaluates the present derivative values on line 37.
(b) Sets up the yn values as defined in Eq. (10.25) and used in the derivatives on lines 38 and 39, depending on the BD or FD algorithm being used.
(c) Calculates the forward difference approximation on line 40.
(d) Updates time on line 41.
(e) Calls nsolv() to obtain the new solution value on line 42 if the algorithm being used is the BD or TP equation. For the FD algorithm the predicted value is the final solution value.
(f) Saves the time and calculated solution values on lines 43 and 44.

All the detailed computer work is taken care of by the previously developed nsolv() code.

The transformation of the defining differential equation form on line 9 into a form appropriate for use in the nsolv() code needs some brief explanation. Looking back at Chapter 4 will reveal that the nsolv() routine has the following calling arguments: nsolv(f, x, step) where f is a function defining the equations to be solved, x is the solution values and step is an optional array specifying any limits on the solution steps at each iteration in the solution (no limit is used in the present case). However, the function defining the set of differential equations on line 9 can not be directly supplied to the nsolv() function. The reason for this can be understood in conjunction with Eq. (10.25). The function supplied to nsolv() must be defined only in terms of the unknown variables, i.e. it corresponds to the lower case f functions on the right hand side of Eq. (10.25). The nsolv() function knows nothing about derivatives only function values. On the other hand the differential equations defined on line 9 are in terms of derivative values and corresponds to the capital F functions on the left side of Eq. (10.25). What is needed is an intermediary function, or proxy function, that translates the differential equation representation into the representation needed in the nsolv() routine. This is provided by the function defined on lines 25 through 32 of the code. This function that is passed to the nsolv() code accepts as arguments a set of equations and the solution values. In turn this proxy function takes the solution values and uses the desired algorithm to approximate the time derivative, on lines 27 through 30 using

```
 1 : -- /* File list10_1.lua */
 2 : -- Programs to explore stability for differential equations
 3 :
 4 : require"nsolv"; exp = math.exp
 5 : typ = 'TP' -- Trapeziodal rule -- Select as desired
 6 : --typ = 'BD' -- Backwards difference
 7 : --typ = 'FD' -- Forward difference
 8 :
 9 : feqs = function(eq, t, y, yp) -- Define differential equations
10 :    eq[1] = yp[1] + y[1]
11 :    eq[2] = yp[2] + 10*y[2]
12 :    eq[3] = yp[3] +100*y[3]
13 : end
14 :
15 : fac = 1-- Set h increment
16 : h,kmax = .01*fac, 40/fac; neq,t = 3,0
17 :
18 : y = {}; sol = {{t},{1},{1},{1},{},{},{}} -- Initial values
19 : for i=1,neq do y[i] = sol[i+1][1] end -- initial y value array
20 :
21 : fderiv = function(eqs,yp) -- Function for derivatives - yp
22 :    feqs(eqs,t,y,yp) -- Add t and y to arguments
23 : end
24 :
25 : fnext = function(eqs,y) -- Function for next y values
26 :    local yp,h2 = {}
27 :    if typ=='TP' then h2=h/2 else h2=h end
28 :    for i=1,neq do -- TP or BD algorithms
29 :       yp[i] = (y[i] - yn[i])/h2 -- trapezoidal rule
30 :    end
31 :    feqs(eqs,t,y,yp) -- Add t and yp to arguments
32 : end
33 :
34 : yp,yn = {},{}; for i=1,neq do yp[i] = 0 end -- Deriv and step
35 :
36 : for k=1,kmax do -- Main time loop
37 :    nsolv(fderiv,yp) -- Update y' values
38 :    if typ=='TP' then for i=1,neq do yn[i]=y[i] + 0.5*h*yp[i] end
39 :    else for i=1,neq do yn[i] = y[i] end end -- BD
40 :    for i=1,neq do y[i] = y[i] + h*yp[i] end --predict, for FD
41 :    t = t+h -- Update time
42 :    if typ~='FD' then nsolv(fnext,y) end -- Calculate new values
43 :    sol[1][k+1] = t -- Save calculated time values
44 :    for i=1,neq do sol[i+1][k+1] = y[i] end
45 : end

46 : for i=1,#sol[1] do
47 :    sol[neq+2][i] = exp(-(i-1)*h)
48 :    sol[neq+3][i] = exp(-(i-1)*h*10)
49 :    sol[neq+4][i] = exp(-(i-1)*h*100)
50 : end
51 : write_data('list10_1'..typ..'.dat',sol); plot(sol)
```

Listing 10.1. Code segment for exploring stability and accuracy for different derivative approximations – forward difference, backwards difference and trapezoidal rule.

the appropriate formula for the TP or BD algorithms. The function then has the appropriate argument values to call the function defining the differential equations on line 31. This intermediate function needs additional information such as h, t and the tn[] values and these are obtained from the calculated values in the main loop of the code as previously discussed in connection with lines 38 through 41. These values can be known by the proxy function because they are "global" values defined in the code listing. The only requirement is that values must be defined before they are used in the fnext() function. The name of this intermediate proxy function is arbitrary and the only requirement is that the name be passed as an input parameter to the nsolv() routine.

A second proxy function is shown on lines 21 through 23. This is used on line 37 of the code in a call to nsolv() to evaluate the values of the derivative terms. This function (called fderiv()) adds the additional terms beyond the derivative needed to call the differential equations. For the simple form of the differential equations in this example, other means could be used to obtain the derivative values. The function feqs() could simply be called with the yp[] array values set to zero and the returned values in the eq[] array would be the derivative terms. However, this only works if the equation set is defined as linear in the derivative terms with unity coefficients.

Execution of the code as shown in Listing 10.1 (using line 5 for TP) will calculate the solution of the three differential equations using the trapezoidal rule for 40 time values over the time interval 0 to 0.4. For comparison purposes, the theoretical solution values for the three differential equations are calculated on lines 46 through 50 and saved as part of the output file on line 51. Different derivative algorithms can be selected by selecting the desired type on lines 5 through 7. Different values of the time step, while keeping the same total time, can be selected by changing the "fac" parameter on line 15. Finally the differential equations being explored can be changed by changing the functions defined on lines 10 through 12, or additional equations can simply be added as desired. This code makes a convenient template for rapidly exploring the different derivative formulations and different equations. The reader is encouraged to experiment with a range of values and functions. The following discussion will illustrate some of the important properties of the various derivative formulations with the three simple linear differential equations.

When the code in Listing 10.1 is executed with the three possible derivative formulations, the solutions shown in Figure 10.3 are generated. For the selected time step of 0.01shown in the listing, the time step corresponds exactly to the time constant of the fastest changing function and to 0.1 and 0.01 times the time constant of the other two functions. There are several features that can be gleamed from the solutions. First, the BD solution tends to always be above the exact solution while the FD solution tends to always be below the exact solution. This was previously noted in connection with the discussion of Figure 10.1. The TP solution tends to be between the other two solutions and much closer in all these cases to the true solution. For the longest time constant, all three algorithms provide a reasonably good solution, although one would expect the TP solution to be considerable more accurate. The three individual curves can not be distinguished in the figure for the upper curves (the y_1 solution). It is also noted that all the numerical solutions are stable in the sense that they all approach zero for a large number of time steps. In fact the value of $h\alpha$ for the three solutions is -1, -.1 and -.01 for the three curves. All of these values are along the negative real axis and within the region of stability as shown in Figure 10.2. The FD algorithm is the only one that will become unstable if the step size is increased to above 0.2.

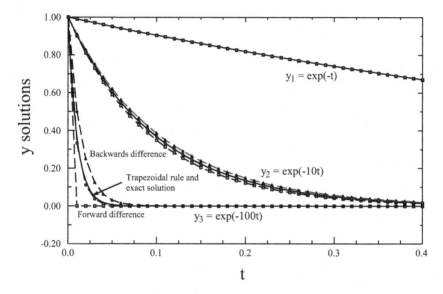

Figure 10.3. Illustration of the accuracy of the three formulations for the parameters shown in Listing 10.1.

A stiff set of differential equations is a much more severe test of a solution algorithm than the simple exponential solutions in the above example. Consider the following two coupled differential equations:

$$y_1' = -1001y_1 + 999y_2$$
$$y_2' = 999y_1 - 1001y_2 \tag{10.27}$$

For the initial conditions of $y_1(0) = 2, y_2(0) = 0$ these two equations have the solution

$$y_1 = \exp(-2t) + \exp(-2000t)$$
$$y_2 = \exp(-2t) - \exp(-2000t) \tag{10.28}$$

The solution involves two decaying exponential terms with time constants that differ by a factor of 1000. On a time scale to observe the details of the second term in the equations, the first term is almost constant (at value 1) and on a time scale to observe the details of the first term, the second term is very close to zero.

Listing 10.2 shows the code changes needed in Listing 10.1 to solve this set of differential equations using the three differencing algorithms. Only the changes in the code are shown and lines 9 through 12 define the two coupled equations as defined in Eq. (10.27). The time step size is specified on line 15 and is shown as 1.e-5 in the listing. To observe the entire time solution the time step needs to be changed from some small value such as that shown in the listing to larger values such as the optional value of 1.e-3 shown on line 14. The listing shows 1000 time steps per calculation.

```
 1 : -- /* File list10_2.lua */
 2 : -- Programs to explore stability for differential equations
 3 :
 4 : require"nsolv"; exp = math.exp
 5 : typ = 'TP' -- Trapeziodal rule -- Sekect as desured
 6 : --typ = 'BD' -- Backwards difference
 7 : --typ = 'FD' -- Forward difference
 8 :
 9 : feqs = function(eq, t, y, yp) -- Define differential equations
10 :     eq[1] = yp[1] + 1001*y[1] - 999*y[2]
11 :     eq[2] = yp[2] - 999*y[1] + 1001*y[2]
12 : end
13 :
14 : --h,kmax = 1.e-3, 1000 -- Select desired time scale
15 : h,kmax = 1.e-5, 1000
16 : neq,t = 2,0
17 :
18 : y = {}; sol = {{t},{1},{1},{},{}} -- arrays with Initial values
19 : for i=1,neq do y[i] = sol[i+1][1] end -- initial y value array
20 :
     ---------- Same as Listing 10.1
46 : for i=1,#sol[1] do
47 :     f1,f2 = exp(-2*(i-1)*h), exp(-2*(i-1)*h*1000)
48 :     sol[neq+2][i],sol[neq+3][i] = f1 + f2, f1 - f2
49 : end
50 : write_data('list10.2'..typ..'.dat',sol); plot(sol)
```

Listing 10.2. Code changes for exploring two coupled differential equations with vastly differing time constants.

Typical solutions for the trapezoidal rule are shown in Figures 10.4 and 10.5. Two figures are needed to illustrate the full range of time for the solution. Curves for the exact solutions are not shown as they would be indistinguishable in the figures from the results calculated by the trapezoidal rule for the cases shown. Figure 10.4 shows the solution at short times where the $\exp(-2000t)$ term dominates. At $t = 0$ one solution starts at 0 and the other solution starts at 2.0 as seen in the exact solution of Eq. (10.28). The fast time constant term rapidly dies out and both solutions become essentially equal at the value of approximately 1.0 as seen in the figure after about a time of 0.004. The two solutions then decay to zero on a much longer time scale as can be seen in Figure 10.5. On this long time scale, the solutions appear to jump almost instantaneously from their initial values of 0 and 2 to 1.0 and then decay slowly to zero with a time constant of value 0.5. The figure shows the solution out to 2 time constants of the slowly varying component.

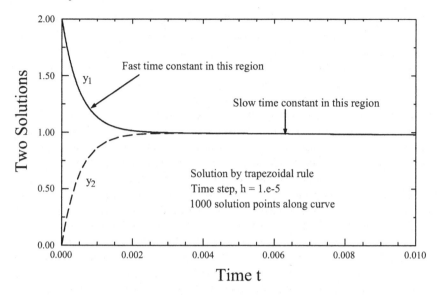

Figure 10.4. Solution for stiff differential equation example of Listing 10.2 at small times.

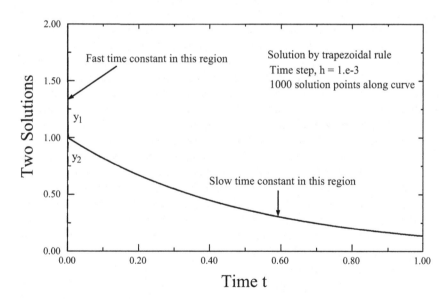

Figure 10.5. Solution for stiff differential equation example of Listing 10.2 at long times.

While the solution is shown only for the trapezoidal case, the backwards difference rule produces almost the same curves, at least on the scale of the solution shown in the figures. However, this is not the case for the forward difference or FD rule. Stable solutions are only observed for this case if the time step is less than 0.001 and accurate solutions are only found for much smaller values of the time step. Solutions such as those shown in Figure 10.5 can not be obtained by the FD rule with the shown time step. The reader is encouraged to explore the full range of algorithm options and time steps by executing the code in Listing 10.2 for various time steps and various "typ" option values as set on lines 5 through 7.

One important conclusion can be drawn from this simple example. The forward difference technique has severe problems with differential equations that have differing time constants. Also as shown by these examples, the accuracy does not approach that of the trapezoidal or backwards differencing rule. Thus the forward differencing algorithm will be eliminated from further consideration as a general purpose technique for the numerical solution of differential equations.

To explore the accuracy of the TP and BD solution methods requires a more detailed look at the difference between the exact solutions values of Eq. (10.28) and the calculated table values. Plots of these differences in absolute value are shown in Figures 10.6 and 10.7 for the BD and TP algorithms. Curves are shown for the $y_1(t)$ solution and for time steps varying from 1.e-6 to 1.e-2. For all the curves, the calculations were performed using a total of 1000 time steps. The 1.e-5 and 1.e-3 time step cases in Figure 10.7 correspond to the actual solution curves shown in Figures 10.4 and 10.5 for the TP algorithm. The first major trend to be observed from Figures 10.6 and 10.7 is that the accuracy of the TP algorithm is considerably better than the BD algorithm for all the time step cases considered. In the region of time below about 1.e-2 in Figure 10.6 it can be seen that the accuracy varies approximately linearly with the step size h for the BD algorithm. For the TP algorithm, the absolute accuracy in the same region varies approximately with h^2 as can be seen in Figure 10.7. From the discussion in Chapter 5 on numerical integration, it will be recalled that the trapezoidal rule is accurate to order h^2 so this dependency is expected. This same dependence is seen in the error in the time region from 1.e-2 to about 1. The dotted curve is each figure is the true solution and it can be readily seen that the relative accuracy at any given time depends on the time step value. The best accuracy is obtained with the smallest time steps. However, to calculate out to a time of 1.0 with a time step of 1.e-6 would require 1.e6 iterative time step calculations. A better approach for such vastly varying time constants is probably to use varying time steps and this approach will be discussed in a subsequent section.

From an accuracy viewpoint, the TP algorithm appears very attractive. However, for this example the TP algorithm has some problems when the step size becomes too large as in fact occurs for the time step of 1.e-2 shown in Figure 10.7. The curves show that the accuracy for small values of time is essentially equal to the solution value. Even more problematic is the fact that the TP solution tends to oscillate about the true solution for the 1.e-2 case as seen near a time of about 6.

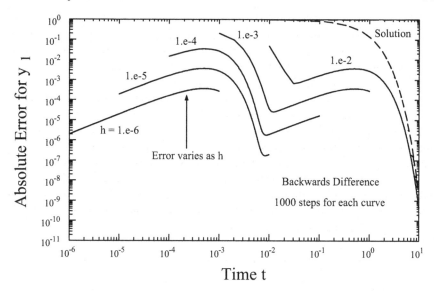

Figure 10.6. Accuracy of backwards difference solution calculated with Listing 10.2 for various time steps.

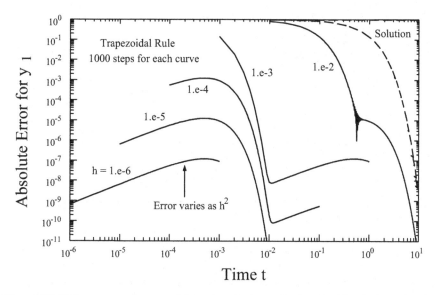

Figure 10.7. Accuracy of trapezoidal rule solution calculated with Listing 10.2 for various time steps.

The case of a time step of 0.001 shows no oscillations and decays normally as expected. However, increasing the time step to 0.01 causes oscillations to occur in the solution for several time steps as the solution starts and rapidly makes a jump from the value of 2.0 at t=0 to the value of 1.0. This can be more clearly seen in Figure 10.8 which shows the generated numerical solution using a time step of 1.e-3 and 1.e-2 on a linear scale. For the time step of 0.001, it can be seen from Figure 10.4 that a couple of time points occur during the fast transient. This appears sufficient to generate an acceptable solution (with no oscillations) for longer times as seen in Figure 10.8. However, for the time step of 0.01, the fast transient is essentially over by the first time point calculation. In this case it can be seen in Figure 10.8 that the TP solution shows several cycles of oscillation in the solution before this artifact of the algorithm is damped out. In fact the large error shown in Figure 10.7 for the 0.01 time step in the time region from 0.01 to beyond 0.1 is due to the oscillations shown in Figure 10.8.

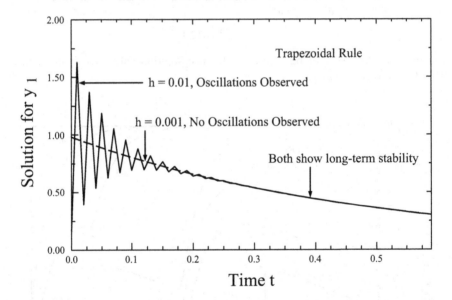

Figure 10.8. Oscillations observed in trapezoidal rule solution for example in Listing 10.2 with large time steps.

From these observations it can be concluded that the TP algorithm will show oscillation problems near the starting point of a solution anytime the solution has a fast transient part and several calculated time points are not taken within the range of the fast transient change. The TP algorithm is unconditionally stable in that the oscillations will always die out as seen in Figure 10.8 for long times. However, these oscillations present somewhat of a potential problem for the accuracy of the TP algorithm when beginning a solution. The BD algorithm does not have such an oscillation problem even for large step sizes at least in the present example.

Even for a step size of 0.01 the accuracy of the BD algorithm is on the order of 0.01 to 0.001 as can be seen from Figure 10.6 even when no time step points are taken within the range of the fast transient. From this example at least two important conclusions can be drawn. First, the TP algorithm has an inherent accuracy advantage, with the error proportional to h^2 as opposed to h for the BD algorithm. Second, the BD algorithm is able to much more readily handle initial fast steps in a solution than is the TP algorithm which can show oscillatory artifacts. These conclusions are important in developing a general algorithm for integrating differential equations.

As a final simple example consider a second order differential equation which has sin(t) or cos(t) solution. Such an equation (or equations) can be formulated as:

$$y'' + y = 0$$
$$\text{Let } y_1 = y \text{ and } y_2 = y' \qquad (10.29)$$
$$\text{Then } \begin{cases} y_1' - y_2 = 0 \\ y_2' + y_1 = 0 \end{cases}$$

The final form of the equations represents the second order equation as two first order equations. Listing 10.3 shows the code changes needed for this set of equations with initial conditions such that $y_1(t) = \cos(t)$ and $y_2(t) = -\sin(t)$. The total time for the solution has been selected so that about 5.5 cycles of the sinusoidal waveform are simulated. The total number of points calculated along the 5.5 cycles is 3000.

The calculated solutions using the three differencing algorithms are shown in Figure 10.9. The data points shown in the figure are the exact solution for the differential equation with points shown only for every 20th calculated point along the curve. The solid curve passing through all the data points is the solution obtained using the TP algorithm. In the figure it is difficult to distinguish between the exact solution values and the curve for the TP numerical solution. Looking back at Eq. (10.18) for the stability of the TP algorithm it can be seen that for a purely imaginary value of λ in that equation and in Eq. (10.7) the magnitude of the solution at each time step should remain constant regardless of the time step. This is indeed found for the TP algorithm where the magnitudes of the peaks of the sinusoidal solution in Figure 10.9 all appear to be the same. Even if the time step is increased so that only a few points per cycle are calculated, the TP algorithm will continue to give a sinusoidal solution with constant amplitude. This is in sharp contrast to the two other algorithms. For the BD algorithm it is seen in Figure 10.9 that the amplitude of the solution continues to decrease with each cycle of the sinusoid. The decay is even faster if the time step size is increased. For the FD algorithm the amplitude increases with each cycle. This is consistent with the results shown in Figure 10.2 where it can seen that any value along the imaginary axis, representing a sinusoid, is outside the stability region for the FD algorithm. Again in this example, the superiority of the trapezoidal algorithm in terms of accuracy is readily seen. It is the only algorithm that will show constant amplitude

for a purely sinusoidal solution. The absolute accuracy of the TP solution shown
in Figure 10.9 is about 1.e-3 or better over the cycles shown in the figure.

```
 1 : -- /* File list10_3.lua */
 2 : -- Programs to explore stability for differential equations
 3 :
 4 : require"nsolv"; exp = math.exp
 5 : typ = 'TP' -- Trapeziodal rule -- Sekect as desured
 6 : --typ = 'BD' -- Backwards difference
 7 : --typ = 'FD' -- Forward difference
 8 :
 9 : feqs = function(eq, t, y, yp) -- Define differential equations
10 :     eq[1] = yp[1] - y[2]
11 :     eq[2] = yp[2] + y[1]
12 : end
13 :
14 : h = .01
15 : kmax = 3000
16 : neq,t = 2,0
17 :
18 : y = {}; sol = {{t},{1},{0},{2}} -- arrays with Initial values
------Same as Listing 10.1
46 : for i=1,#sol[1] do sol[neq+2][i] = math.cos(-(i-1)*h) end
47 : write_data('list10_3'..typ..'.dat',sol); plot(sol)
```

Listing 10.3. Code segment for second order equation with sin(t) or cos(t) solutions.

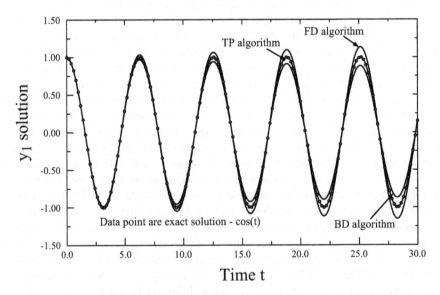

Figure 10.9. Oscillatory differential equation solution with different derivative algorithms.

Several examples have now been explored evaluating the stability and accuracy of the three basic algorithms considered for approximating the derivative in a differential equation. While this in not an exhaustive exploration of these issues, the examples do provide sufficient information to guide the development of a robust integration approach. No nonlinear problems have been considered. However, one must infer knowledge of accuracy and stability from linear problems or at least from simple nonlinear problems for which exact solutions can be obtained. Accuracy will continue to be addressed after the development of a more general routine for solving differential equations in the next section and the accuracy of solutions for some nonlinear problems will be addressed.

10.4 Development of a General Differential Equation Algorithm

Based upon the examples in the previous section, it can readily be concluded that the trapezoidal rule is much preferred over the forward difference and backwards difference algorithms for accuracy in a numerical solution. The only potential problem identified with the examples and the TP algorithm is the possibility of an oscillation in the early time solution points when a solution value changes very rapidly in the first few time steps such as shown in Figure 10.8. This leads to the concept of using the TP algorithm in general but handling in some special way the first few time steps. The BD algorithm becomes the best candidate for handling a few initial time steps as it does not exhibit such oscillatory properties.

Computer code for a general purpose differential equation solution routine named odebiv() (for ordinary differential equations – basic initial value) is shown in Listing 10.4. The calling arguments shown in the listing on line 5 are identified as feqs, tvals, u, up and upp where feqs is the name of a function defining the differential equation set to be solved, tvals is an array listing the desired time points at which to evaluate the solution, u is an array of the initial values for the variables and up is a possible arrays containing values for the first derivatives. The inclusion of the derivatives in the argument list is at first perhaps surprising and questionable since the discussion has so far dealt only with first order differential equations and only initial values can be specified for such a problem. This argument is included so that the routine may be called several times in succession with varying time steps for the same set of differential equations and with a saved set of initial conditions from some previous calculation. In such cases, the derivative values returned by a previous call are used to pick up a calculation at some later time point. For a simple call to the odebiv() routine, these derivative variables may be omitted from the argument list. This will become clearer in the next section where multiple time steps are discussed. For the present section and discussion, the derivative arguments can be ignored.

The tvals parameter is expected to be of a simple table form {tmin, tmax, ntval}, where tmin and tmax specify the minimum and maximum desired time

```
 1 :  -- /* File odeiv.lua */
 2 :  -- Programs to integrate first order differential equations

 3 :  require"nsolv"; local atend
 4 :
 5 :  odebiv = function(feqs,tvals,u,up,upp)
 6 :     local j, neq, t, h, h2,hs,hx -- Local variables for function
 7 :     local sol, yn, jfirst = {}, {}, 0
 8 :     local nit,nitt,ntx = 0,0,0 -- Number of iterations
 9 :     local neq = #u
10 :     local tmin,tmax,ntval = tvals[1],tvals[2],tvals[3]
11 :     -- Function to calculate next time values using nsolv()
12 :     local fnext = function(eqs,u)
13 :        for m=1,neq do
14 :           up[m] = (u[m] - yn[m])/h2 -- h2 = h/2 for TP , h for BD
15 :        end
16 :        feqs(eqs,t,u,up) -- Now call user defined diff equations
17 :     end
18 :     -- Use input value for # of intervals or set at default value
19 :     up = up or {}
20 :     for m=1,neq+1 do -- Array to return values with t value first
21 :        sol[m] = {}
22 :        if m==1 then sol[m][1] = tmin else sol[m][1] = u[m-1] end
23 :     end
24 :     t = tmin -- Initial t value
25 :     hs = (tmax - t)/ntval -- Equal increments in t used
26 :     -- If init deriv not specified, use BD for first 4 points
27 :     if #up~=neq then -- up array not input
28 :        for m=1,neq do up[m] = 0 end
29 :        jfirst,h,h2,hx = 0,hs/4,hs/4,0 -- Set to BD parameters
30 :     else jfirst,h,h2,hx = 4,hs,hs/2,hs/2 end -- TP parameters
31 :     for k=1,ntval do -- Main loop for independent variable
32 :        repeat -- Use BD with 4 sub intervals, size h/4
33 :           jfirst = jfirst+1
34 :           -- Set up yn array and for next solution value
35 :           for m=1,neq do
36 :              yn[m] = u[m] + hx*up[m] -- hx = 0 or h/2
37 :              u[m] = u[m] + h*up[m] -- Predicted value of u array
38 :           end
39 :           t = t + h -- Now increment t to next t value
40 :           -- Calculate new u values at next time,u returns values
41 :           nitt = nsolv(fnext,u,step)
42 :           if nitt>nit then nit,ntx = nitt,ntx end --Monitor #its
43 :           -- New Derivative values, same function as in fnext
44 :           for m=1,neq do up[m] = (u[m] - yn[m])/h2 end
45 :        until jfirst>=4 -- End of first interval using BD
46 :        if k==1 then h,h2,hx = hs,hs/2,hs/2 end -- Set to TP
47 :        sol[1][k+1] = t -- Save calculated values
48 :        for m=1,neq do sol[m+1][k+1] = u[m] end
49 :     end -- End of main loop on t, now return solution array
50 :     sol[1][ntval+1] = tmax; return sol,nit,ntx -- Solutions
51 :  end -- End of odebiv
52 :  setfenv(odebiv,{nsolv=nsolv,step=nil})
```

Listing 10.4. Code segment for basic general purpose integration of sets of first order differential equations with given initial values.

values and ntval specifies the number of time intervals to be used in the calculation. It should be noted that ntval is specified as the number of time intervals. As
calculations are made at the beginning and end time points, the returned solution
array will consist of ntval+1 values. This is the "off by one" problem that always
occurs in computer problems.

As this routine is intended to be called primarily by other routines, no error
checking is made of the input variables. If this routine is called directly, the user
must make sure the time variables are in the proper format. Much of the code between lines 6 and lines 31 is taken up with declaring initial values of variables and
arrays. Lines 12 through 17 of the code define the proxy function that translates
from the form of the user defined differential equations into the functional form
needed in nsolv(). This function fnext() on line 12 is the same as the previously
discussed translation function in Listing 10.1 and has essentially the same form.
As will be recalled it performs the functional translation between the capital F and
lower case f functions in Eq. (10.25) adding time and derivative values between
the fnext() and feqs() functions. The only difference here is that the BD and TP
differencing approximations are combined into a single equation through the use
of a common parameter and table (yn[]) defined on line 36 of the code. The heart
of the solution algorithm is very similar to the simple solution algorithm previously discussed in connection with Listing 10.1 and is contained between lines 31
and 49 of the listing. This loop steps through the time points one by one and
evaluates the solutions at each time point as time is incremented on line 39.

The algorithm uses the TP algorithm except for the first time interval which is
subdivided into 4 sub intervals and the BD rule is used for each of the four sub intervals. These four sub intervals are handled by the repeat-until loop from lines 32
to 45. The first time this loop is encountered, it is repeated four times as jfirst is
initially set to zero on line 7. Subsequent encounters of this loop cause only one
execution of the loop. Before entering the main loop and the repeat-until loop for
the first time, three parameters h, h2 and hx are set on line 29 to values appropriate
to the BD algorithm. After the repeat-until loop is executed four times, these parameters are set to the appropriate TP parameters on line 46 for all subsequent
time points. Even though the first time interval is subdivided into 4 intervals, only
the solution values for the last of the four sub intervals is saved in the output files
on lines 47 and 48. The calculation of the solution values for the differential equations requires only four basic steps for each desired time interval. On line 36 the
yn parameter needed in the derivative approximation is evaluated. On line 37 a
predicted value of the solution variables is calculated. On line 39, time is incremented to the next time point. On line 41 nsolv() is called to solve the set of implicit equations for the new solution values. On line 44 new values of the derivatives are evaluated so they can be used in the next iterative step on line 36 in the
evaluation of the yn parameters.

It is seen in this code that the derivative values are calculated without resorting
to a second call to nsolv() as in Listing 10.1. This is possible by using the same
equation on line 44 for the derivative as used on line 14 in solving the equation
set. If a proper solution set has been returned by nsolv() then this relationship

must have been satisfied. Finally the new solution values are saved on lines 47 and 48 and the loop is repeated for the specified number of time points. The algorithm is so simple because the fnext() function performs the replacement of the derivative by the desired algorithm in terms of the function values and time increment and the nsolv() routine solves the resulting set of implicit equations. As previously stated the hard work was already done in developing the nsolv() code in Chapter 4. As a final comment, the maximum number of Newton iterations taken by nsolv() is monitored on lines 42 and returned by the function so convergence problems can be identified. Also no step size limitations are used with nsolv() as the solution values are assumed to change by small percentages between solution times and the values at one time step are used as initial approximations for the next time step.

```
1 : -- /* File list10_5.lua */
2 : -- Programs to integrate diff. equation using odebiv()
3 :
4 : require"odeiv"
5 :
6 : feqs = function(eqs,t,y,yp) -- test of stiff differential eqns
7 :     eqs[1] = yp[1] + 1001*y[1] - 999*y[2]
8 :     eqs[2] = yp[2] - 999*y[1] + 1001*y[2]
9 : end
10 :
11 : yin = {0,2} -- Initial values, 0 and 2
12 : sol,nit = odebiv(feqs,{0,1,1000},yin)
13 : print(nit); plot(sol)
14 : write_data("list10_5.dat",sol)
```

Listing 10.5. Example of solving stiff differential equations with the basic integration routine, odebiv().

Now let's look at some simple examples of the use of this routine. Listing 10.5 shows a simple example of the use of this routine for solving the set of stiff differential equations previously defined in Eq. (10.27) and with the solution plotted in Figures 10.4 and 10.5. The code is very minimal and simple. The basic differential equations are defined in a function on lines 6 through 9 in essentially the same form as one would write them mathematically. The set of initial values are defined in a table on line 11 as 0 and 2 for the two solutions. Then the routine odebiv() is called on line 12 to solve the differential equation set and return the solution set and a number giving the maximum number of iterations of nsolv() taken to solve the equation set. Since this is a linear set of equations, the number of iterations returned will be 2 and could be omitted for a linear set of equations. For a nonlinear set of equations, the number of iterations should always be checked to see if it equals the maximum value set in nsolv() (default value of 100). If this number is reached, the solution is not likely to be very accurate and further work is needed to obtain an accurate solution. The output from this listing is saved in a file for plotting on line 14. The results are not shown here, since the plot will be identical to that shown in Figure 10.5. The reader is encouraged to verify this as well as verifying that the subroutine gives identical results to the previous code in Listing 10.1.

The ability to solve nonlinear differential equations as readily as linear equations is one of the major advantages of the numerical solution of differential equations. For one such example, the Van der Pol equation is a classical nonlinear equation that has been extensively studied in the literature. It is defined in second order form and first order differential equation form as:

$$y'' - \mu(1 - y^2)y' + y = 0$$

$$\text{or in equivalent form} \begin{cases} y_1' - y_2 = 0 \\ y_2' - \mu(1 - y_1^2)y_2 + y_1 = 0 \end{cases} \tag{10.30}$$

The second form is of two coupled first order differential equations. The parameter μ determines the degree of nonlinearity with large positive values giving a highly nonlinear solution. A code segment for solving this differential equation is shown in Listing 10.6. For this example a value of $\mu = 20$ has been selected as shown on line 6 of the listing. The call to odebiv() indicates a desired solution from 0 to 100 using 16000 time points. The user is encouraged to experiment with different values of these parameters. These values were selected to give about 2.5 cycles of the solution with reasonable accuracy. The resulting solution for the y_1 variable is shown as the solid line curve in Figure 10.10. The solution shows rather sharp vertical transitions at certain time points along the time axes. Because of these sharp transitions, for a small number of solution points, there are very few solution points along the solution curve in these rapidly changing regions. Even with the sharp transitions for the case of 16000 time points, there are many calculated points along the sharp vertical transitions shown in the solution.

```
 1 : -- /* File list10_6.lua */

 2 : -- Programs to integrate diff. equation using odebiv()
 3 :
 4 : require"odeiv"
 5 :
 6 : mu = 20
 7 : feqs = function(eqs,t,y,yp) -- test of stiff differential eqns
 8 :     eqs[1] = yp[1] - y[2]
 9 :     eqs[2] = yp[2] - mu*(1 - y[1]^2)*y[2] + y[1]
10 : end
11 :
12 : yin = {1,0} -- Initial values, 0 and 2
13 : sol,nit = odebiv(feqs,{0,100,16000},yin)
14 : print('Number of iterations =',nit); plot(sol[1],sol[2])
15 : write_data("list10_6.dat",sol)
Output:
Number of iterations = 3
```

Listing 10.6. Code segment for solving the Van der Pol equation.

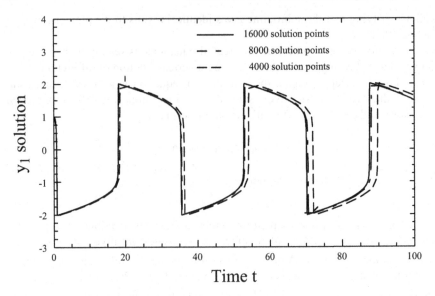

Figure 10.10. Example solution of nonlinear Van der Pol equation with a $\mu = 20$ parameter and with varying numbers of time steps.

Since this differential equation is a nonlinear differential equation and no explicit equation is known for the solution, the question naturally arises as to how accurate is any generated numerical solution? One way to explore the accuracy of a solution is to perform the calculation on some given time interval with varying numbers of time steps and observe any changes in the calculated solution. Three such solutions are shown in Figure 10.10 with 4000, 8000 and 16000 total time points along each solution curve. The solid curve with the largest number of solution points is expected to be the most accurate as the TP algorithm has an error proportional to h^2. It can also be observed that there is a small change in the solution in going from 8000 to 16000 total time points. Although not shown, one would expect that an increase to 32000 points would give an even more accurate solution, but not much difference would be expected between such a solution and the solid curve shown. It is left to the reader to re-execute the code in Listing 10.6 with such an increased number of solution points to verify that increasing the number of points beyond 16000 produces very small differences in the solution. This illustrates one very important means of exploring the accuracy of a numerical solution, especially for nonlinear equations where an exact solution can not be obtained. In fact this is probably the only practical way of verifying the accuracy of a numerical solution. In a subsequent section a formal method will be developed to obtain an estimate of the error in a numerical solution based upon calculations at two different time intervals.

10.5 Variable Time Step Solutions

The basic code and algorithm developed in the previous section uses a uniform time step throughout the solution space for obtaining a numerical solution of a set of linear or nonlinear differential equations. From the example of the "stiff" differential equations considered and the Van der Pol equation example, it is readily seen that such a uniform step distribution is not always the best choice for a calculation with some total number of time point. To achieve an accurate solution over the entire range of time values when greatly varying time constants are involved as shown in Figures 10.4 and 10.5, small step sizes are needed at small values of time. However, it is impractical to employ such small time steps at large values of time because of the large number of calculations that would be involved and the computer time required for such a calculation. Also such small time steps are not required for accuracy at large times in most cases as illustrated by the results shown in Figure 10.7. Also in the Van der Pol example of Figure 10.10 it can be seen that it would be very desirable to have small step spacing for regions of time where the function is changing rapidly and larger step spacing for region where the function is changing slowly. Thus it is very useful for a general solution method to handle a series of variable time step regions.

There are many ways to go about developing a code segment for varying the size of the time steps. In the most general case a program that automatically adapts the time steps to the problem to obtain the "best" solution for some specified criteria such as a given total number of calculation points might be the ideal solution. However, this is not an easy problem. So to build up to such more general cases, it is useful to take a few less ambitious steps. In the simplest of cases, a given time interval can be subdivided into subintervals and a constant time step can be used within each interval. A simple modification of the previous calling argument for this case is to replace the tvals parameter in the calling argument with a more general table representation of desired time values in the form: tvals = {tinitial, {t1, t2, t3, ... tk}, {n1, n2, n3, ... nk}}, where tinitial specifies the starting time (typically 0), the second array specifies times at which the step increment is to change and the third array specifies the number of steps to use for each time increment. To be more specific, the above specification would use n1 steps between time tinitial and t1, n2 steps between t1 and t2, n3 steps between t2 and t3, etc. The number of specified time increments could be less than the number of time steps with the last number specified being used for any remaining time steps. A single specified step number would then perform the calculation with the same number of time steps for all time intervals.

The code for implementing such a multi-time step algorithm is shown in Listing 10.7. This code is actually part of the odeiv.lua file and is loaded whenever this file is included in a code segment. The multi time step function is odeiv() and is intended to be the basic interface to the trapezoidal differential equation solving routine as defined in Listing 10.4. If called with a single set of time parameters it performs the same calculation as the odebiv() routine but also allows for a more

```
  1 : -- /* File odeiv.lua */
  2 : -- Programs to integrate first order differential equations

  3 : require"nsolv"; local atend
  4 :
 65 : odeiv = function(feqs,tvals,u,up,upp) -- Multi Time Step Solver
 66 :   local sa,sb,tl,ns = {}
 67 :   local ntp,ni,nit,ntx
 68 :   local NMAX,ND = getfenv(nsolv).NMAX,1000 -- 1000 intervals?
 69 :   up,upp = up or {},{}
 70 :   if type(tvals)=='number' then tvals = {0,tvals,0} end
 71 :   local j = #tvals
 72 :   if j==1 then tvals = {0,{tvals[1]},2*ND} end
 73 :   if j==2 then tvals[3] = 2*ND end
 74 :   if type(tvals[2])=='number' then tvals[2] = {tvals[2]} end
 75 :   ntp = #tvals[2]
 76 :   if type(tvals[3])=='table' then ns=tvals[3][1]
          else ns = tvals[3] end
 77 :   nit = 0
 78 :   for i=1,ntp do
 79 :      if i>1 then tl = tvals[2][i-1] else tl = tvals[1] end
 80 :      if type(tvals[3])=='table' then ns = tvals[3][i] or ns end
 81 :      sb,ni,ntx = odebiv(feqs,{tl,tvals[2][i],ns},u,up,upp)
 82 :      if ni==NMAX and ntx~=ns then
 83 :          print("Error: Maximum number of iterations exceeded
                  in nsolv")
 84 :          print("Results may not be accurate!");
                  print(tl,tvals[2][i],ns)
 85 :      end
 86 :      if ni>nit then nit = ni end
 87 :      if i>1 then sa = atend(sa,sb) else sa = sb end
 88 :   end
 89 :   return sa,nit
 90 : end
 91 :
 92 : odeivqs = function(feqs,tvals,u,up) -- ODE Quick Scan function
 93 :   local NPTS,NPS = 20,2
 94 :   local ttvals,nl,nu,fact = {},10
 95 :   local nt,j = #tvals,1
 96 :   if nt<2 then print('Error, must specify two times
          in obeivqs'); return end
 97 :   NPTS = floor((((tvals[3] or NPTS)+1)/2)
 98 :   nl = 10^(floor(log10(tvals[2][1])))
 99 :   nu = 10^(ceil(log10(tvals[2][2])))*1.000001
100 :   fact = 10^(1/NPTS)
101 :   ttvals[1],ttvals[2],ttvals[3],nl = nl,{},NPS,nl*fact
102 :   while nl<= nu do ttvals[2][j],nl,j = nl,nl*fact,j+1 end
103 :   ttvals[2][#ttvals[2]] = 10^(ceil(log10(tvals[2][2])))--at end
104 :   u,up = u or {},up or {} -- Linear steqs for first interval
105 :   odeiv(feqs,{tvals[1],ttvals[1],NPTS},u,up) -- NPTS points
106 :   return odeiv(feqs,ttvals,u,up)
107 : end
108 : setfenv(odeivqs,{floor=math.floor,ceil=math.ceil,
          log10=math.log10,odeiv=odeiv})
```

Listing 10.7. Code segment for multi step trapezoidal differential equation solving routines.

general specification of the time parameters for integrating an equation with the trapezoidal algorithm. The code between lines 66 and 77 basically handles various possible formats for the time interval specifications as discussed above. The heart of the routine is the loop from line 78 through line 88 that steps through the various time segments calling the basic odebiv() code to integrate the equations over the various specified time segments with the call to the previously defined odebiv() function on line 81. Other features of the loop are the monitoring of the maximum number of iterations in the nsolv() routine as returned by the odebiv() routine on lines 82 through 87. If the maximum number of nsolv() iterations is exceeded, the code prints an error message on line 83, but continues with the calculations. The solution results for the various time segments are collected together on lines 87 using an atend() function and returned as a single table of values on line 89. In order to change the step size with the trapezoidal algorithm and not have to use a "starter" segment that uses the backwards differencing technique for the initial time step, both the value of the solution and the derivative must be known. This is seen in the basic algorithm of Eq. (10.6) for the trapezoidal rule. With this it can now be seen why the odebiv() routine has the derivative parameter, as an argument. This is returned by the odebiv() routine so that it may be input back into the routine on subsequent calls and the trapezoidal routine can pick up its calculation with a new time step just as if it had never left the odebiv() routine. (The return of the second derivative upp still remains to be clarified in a subsequent section, but it is for a similar reason when the code is generalized to handle equations with a second derivative.)

Listing 10.7 also contains code for a function named odeivsq(), called a quick scan function. This illustrates one application of the multiple time step routine. This function is intended to provide the user with a "quick scan" look at the general properties of a system of differential equations. In this case the tvals parameter is specified as a list of parameters in the form tvals = {tinitial, {tllimit, tulimit}, ndecade} where tinitial is some initial time such as zero at which the initial conditions are specified, tllimit and tulimit are the times over which a logarithmic step size is to be used and ndecade is the number of time points per decade in time to solve the set of differential equations. This function is intended to rapidly scan the solution over many orders of magnitude in time using logarithmically spaced time values and get a quick first look at the solution. For example using tvals = {0,{1.e-6, 1.e6}} the code will integrate the set of differential equations from 0 to 1.e-6 using a linear step spacing and then from 1.e-6 to 1.e6 taking logarithmic time steps and using a default number of time steps per decade in time (the default value is 20 points per decade in time as defined on line 93). The user may change the number of time steps per decade by specifying a third parameter in the tvals parameter. As an example, for a range of 12 orders of magnitude as above and a value of 20 points per decade, this is only a total of 240 time points for the calculation. This can be rapidly performed and one can rapidly get an indication of the time ranges needed for a more accurate solution. This can provide a quick look at the solution of a set of equations with minimal calculation effort and provide very valuable information for more detailed subsequent calculations. The default calculation actually uses 10 logarithmically spaced intervals per decade with two equally spaced subintervals within each major time interval – see code on lines 101 through 105, where the NPS parameter is set to 2. The reason for the two equally spaced increments is for ease in automatically obtaining an estimate of the error in the solution as will be subsequently discussed. The odeivqs() function calls the odeiv() function on line 105 after defining a table of logarithmically spaced time points in the tvals[] and ttvals[] tables.

```
 1 : -- /* File list10_8.lua */
 2 : -- Programs to integrate first order diff. equation
 3 : require"odeiv"
 4 :
 5 : f1 = function(eqs,t,y,yp) -- test of stiff differential eqns
 6 :     eqs[1] = yp[1] + 1001*y[1] - 999*y[2]
 7 :     eqs[2] = yp[2] - 999*y[1] + 1001*y[2]
 8 : end
 9 : f2 = function(eqs,t,y,yp) -- test of sinusoidal equation
10 :     eqs[1] = yp[1] - y[2]
11 :     eqs[2] = yp[2] + y[1]
12 : end
13 : mu = 20
14 : f3 = function(eqs,t,y,yp) -- test of Van der Pol equation
15 :     eqs[1] = yp[1] - y[2]
16 :     eqs[2] = yp[2] - mu*(1 - y[1]^2)*y[2] + y[1]
17 : end
18 :
19 : s1 = odeivqs(f1,{0,{1.e-5,1e3}},{0,2}); plot(s1)
20 : s2 = odeivqs(f2,{0,{1.e-5,1e3}},{1,0}); plot(s2)
21 : s3 = odeivqs(f3,{0,{1.e-5,1e3}},{1,0}); plot(s3)
22 :
23 : write_data("list10_8.dat",s1,s2,s3)
Output:
Error: Maximum number of iterations exceeded in nsolv
      Results are probably not accurate!
```

Listing 10.8 Illustration of the use of multiple time steps to obtain a quick time scan of various second order differential equations.

Listing 10.8 illustrates how the quick scan function and the variable time step functions can be used to rapidly obtain valuable information about the nature of the solution to three differential equations. The code in the listing scans the time solution space for three sets of two first order differential equations, one with vastly different time constants, one for a periodic sinusoidal solution and one for the nonlinear Van der Pol equation. All three of these sets of equations have previously been discussed in this section and good solutions are shown in Figures 10.5, 10.9 and 10.10. The code in Listing 10.8 is very simple, after defining the three sets of differential equations with three f1,f2 and f3 functions the call to the odeivqs() function is made on lines 19 through 21 for the three function with the desired initial values for the solutions. All three calculations use the time span from 1.e-6 to 1.e3 (9 orders of magnitude in time) with the default value of 20 calculations per order of magnitude in time. Also shown in the listing is the output indicating that the maximum number of iterations was exceeded in nsolv(). Although this can not be identified from the output, this actually occurred in the calculations for the Van der Pol equations set. The other two equation sets are linear and will only require 2 iterations of nsolv(). The results of the quick scans are saved in an output file on line 23.

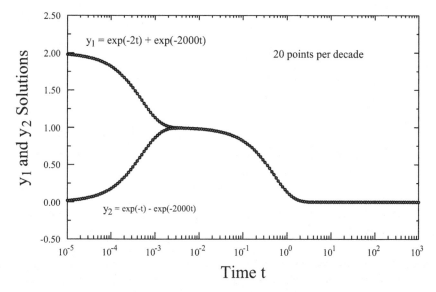

Figure 10.11. Quick scan results for the solution with vastly different time constants.

Figure 10.11 shows the quick scan results for the first function which is the stiff differential equation with two vastly different time constants. The data points shown are the calculated solution values. Both solutions are evaluated quite nicely over the entire range of time. It can be seen that the solutions begin to change from the initial values at times of about 1.e-5, rapidly change in the 1.e-4 to 1.e-3 range, reach a plateau in the mid 1.e-3 time range and then rapidly change in the 0.1 to 1 time range. By about a time of 10, the solutions are near zero. Thus the quick scan function can be used to rapidly envision the entire solution of this set of equations and rapidly determine a strategy for obtaining a more accurate solution by a more careful placement of additional time points if desired for the solution. It can also be readily seen that the solution time must cover in some way the time range from at least about 1.e-5 to about 10 to obtain the complete solution and at the low end of time, steps of about 1.e-6 are needed while at the high end of time much larger time steps can be used.

From the relatively smooth function obtained by the quick scan solution, it might be expected that the solution is already in fact relatively accurate even with this rather course time grid. This is in fact the case as can be seen in Figure 10.12 which shows the error in either the y1 or y2 solution values for this quick scan

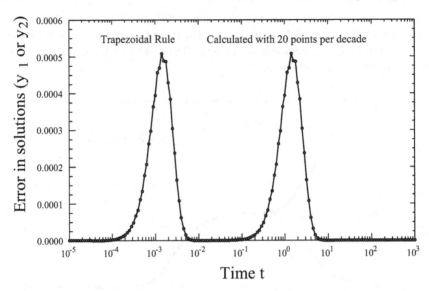

Figure 10.12. Error in the y1 or y2 solution of Figure 10.11 for the stiff equation test case and with the quick scan function.

when compared with the theoretically known solutions. The error is actually identical for either of the two solutions. As the figure indicates, the solution with this rather course time grid is accurate to better than three decimal places at all time points and the maximum error is about 0.0005 and occurs at approximately two time points. By looking at the time solution in Figure 10.11 it can be seen that the maximum error points occur in regions where the time function has a large curvature and where a large number of solution points are needed.

It would be especially useful if a solution such as that obtained by the quick scan function could be used to estimate the accuracy for numerical solutions where an exact solution is not known. As a general rule one would expect that a good solution would be obtained when there is a relatively small percentage change in the function between each time step, but a more quantitative assessment of the accuracy of a solution is desired. This subject will be further addressed in the next section where an algorithm will be developed for estimating the accuracy of a numerical solution for a differential equation.

The results of the quick scan calculation for the sinusoidal functions are shown in Figure 10.13. In this case it is seen that the results are rather scattered for large times. The figure does not show the results for the 1.e-6 to 1.e-2 time range as the solutions are essentially constant for these time regions. What can be learned from this type of quick scan results? First it is readily seen that the solutions do not begin to change until times on the order of 1.e-2, so this determines a minimum time scale for further exploration. Second it is seen that the solutions exhibit some type of oscillations with a period of something less than a time value of 10.

This period can be estimated as 6.3 from the first peak in the y_1 solution. The quick scan thus shows that the solution needs to be more carefully examined out to times corresponding to several periods of about 6.3 and that a minimum time value of around 1.e-2 is needed. So a next trial solution for this function might be a time range of zero to 40 with 4000 time points which could be used to cover the time range with a minimum time step of 1.e-2 over the entire range. Based upon the results of that calculation, one might further modify a solution strategy. The reader is encouraged to run this calculation and verify the results which will be very similar to the center curve previously shown in Figure 10.9. For that calculation 3000 time steps were used over the time range of 0 to 30.

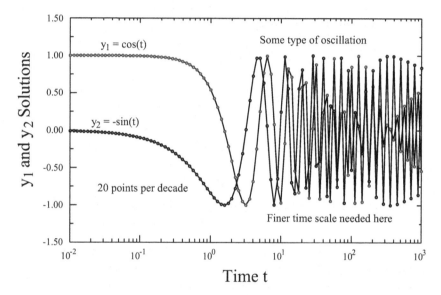

Figure 10.13. Results of quick scan solution for two equations with sinusoidal solutions.

The results of the quick scan solution for the nonlinear Van der Pol equation are shown in Figure 10.14. In this case only the results for the y_1 solution are shown. Again the results are not shown below a time of 1.e-2 as essentially nothing happens there. In fact the solution only begins to change for times on the order of about 0.5 and then rapidly changes to about the -2 range at about a time of 1. The quick scan solution has relatively few time points during the first downward transition which is rather quickly followed by an upward transition between times of 1 and 2. This is followed by another downward transition between times of 10 and 20 and further oscillations. If this quick scan solution is compared with the much more accurate results in Figure 10.10 it is seen that this calculation lacks much of the details of the accurate solution. However, it does indicate some type of periodic oscillation with a period of 20 to 40 time units and it indicates the need for a much more accurate time grid beginning before a time of 1.0. Armed with this information, one can then develop a new strategy for an improved time grid for subsequent calculations. An appropriate next solution step then might be to evaluate the solution over the time interval from 0 to 100 with a minimum time step of around 0.1.

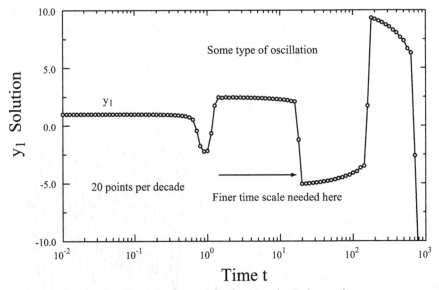

Figure 10.14. Results of quick scan results for Van der Pol equation.

While the quick scan function odeivqs() is one important application of the multi time step code routine odeiv() any desired time step sequence can be devised for use by this function. In fact a separate time step interval can be used for each solution point by generating a table of desired time steps for the tvals[2] parameter used in the calling argument and specifying a unity value for the tvals[3] parameter. While this is certainly not the optimum way of handling individual time steps, it can be used if desired. A test of performing the numerical integration for the si-nusoidal equations of Listing 10.3 using first a single call to the odeiv() function from 0 to 30 with 1000 time points vs. using 1000 calls to odeiv() with 1 time point each has shown that the overhead in using individual points vs. a single range with the same number of time points is around 60%. It takes about 1.6 times longer for the code to execute with the 1000 individual time steps vs. the single range of 1000 points. Thus the odeiv() code sequence can be considered a general purpose routine adaptable to any desired strategy for selecting appropriate time points for the solution of a differential equation.

A stiff differential equation set is a somewhat ideal equation for use with mul-tiple ranges of time steps or in fact with an enhanced logarithmically spacing. For example consider a piecewise multi time step approach for the solution shown in Figure 10.11 consisting of the following tvals parameter: tvals = {0,{1.e-4, 1.e-3, 1.e-2, 0.1, 1, 10}, 400}. This uses a time step increment of 2.5e-7 in the interval 0 to 1.e-4, up to a time step interval of 0.025 in the interval from 1 to 10. The total number of time points is 2400, which is a very reasonable number.

Listing 10.9 shows code for incorporating this type of multi step time specification into the call to the odeiv() routine. The defining table for the multi time steps is seen on line 10 of the code as the second calling argument to the odeiv() function. In this example, the same numbers of steps are used for each time interval as indicated by the third argument of 400. However, different time step numbers can be specified by making the third entry in the tvals array an array of step numbers. The only other addition to the code is the calculation of the exact solutions on lines 15 through 23 of the code. This is included so that a printout of the maximum error between the numerical calculation and the exact solution can be obtained. The printed output shown in the listing indicates that this maximum error is 8.55e-6 and occurs at a time point of about 1.5. The solution is thus accurate to better than 5 decimal digits of the maximum solution value. In terms of relative accuracy, the solution at this point is accurate to about 4 digits. Such a solution would be adequate for most engineering work.

```
 1 : -- /* File list10_9.lua */
 2 : -- Programs for first order diff. equation using odeiv()

 3 : require"odeiv"
 5 : f1 = function(eqs,t,y,yp) -- test of stiff differential eqns
 6 :    eqs[1] = yp[1] + 1001*y[1] - 999*y[2]
 7 :    eqs[2] = yp[2] - 999*y[1] + 1001*y[2]
 8 : end
 9 :
10 : s1 = odeiv(f1,{0,{1.e-4,1.e-3,1.e-2,0.1,1,10},400},{0,2});
        plot(s1)
11 : --s1 = odeivqs(f1,{0,{1.e-6,1.e3},400},{0,2}); plot(s1)-- Use?
12 : st1,st2 = {},{}
13 : nd = #s1[1]
14 : err,terr = 0,0
15 : exp,abs = math.exp,math.abs
16 : for i=1,nd do
17 :    time = s1[1][i]
18 :    st1[i] = exp(-2*time)-exp(-2000*time)
19 :    st2[i] = exp(-2*time)+exp(-2000*time)
20 :    err1 = abs(st1[i] - s1[2][i])
21 :    if err1 > err then err,terr = err1,time end
22 :    err1 = abs(st2[i] - s1[3][i])
23 :    if err1 > err then err,terr = err1,time end
24 : end
25 : print('Maximum error is ',err,' and occurs at time ',terr)
26 : write_data("list10.9_dat",s1,st1,st2)
Output:
Maximum error is 8.5558711942213e-006  and occurs at time 1.495
```

Listing 10.9. Example of user selected multiple time intervals and time steps using odeiv() routine.

The solid line in Figure 10.15 shows the error profile in the solution using the time step specifications in Listing 10.9. Also shown in the figure is the error obtained by using a logarithmic spacing of step sizes throughout the entire time interval. This was obtained by re-execution the code of Listing 10.8 with line 11 replacing line 10 and using the odeivqs() function for the solution. This change executes the quick scan code with 400 time intervals per decade instead of the default value of 20. It is seen from the results in Figure 10.15, that the solution errors are very similar whether uniformly spaced time points on a logarithmic time scale are used or linearly spacing time points within each decade of time are used. The general trends of the two curves are very similar and the maximum errors are very similar, although the peak error for the logarithmically spaced time points is slightly smaller.

By looking back at Figure 10.12 it can be seen that the solution with 400 time points per decade is almost 100 times more accurate than the solution with 20 time points per decade. By comparing these results it can easily be seen that a further increase in the number of time steps in the regions of 0.001 to 0.1 and 1 to 10 would further improve the accuracy. In order to properly select time steps and control solution accuracy, it is important to better understand how the accuracy of the TP algorithm varies with time step size. Now that a more general algorithm for varying time steps has been developed, this subject is explored in more depth in the next section. This will lead to some important concepts regarding accuracy

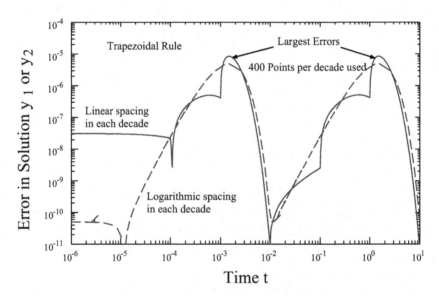

Figure 10.15 Error profile for stiff differential equation set with 400 solution points per decade in time using uniform logarithmically spaced or uniformly spaced points.

of numerical solutions to differential equations and some methods for determining the accuracy of numerical solutions for nonlinear differential equations where exact known solutions are not available.

10.6 A More Detailed Look at Accuracy Issues with the TP Algorithm

Accuracy issues were addressed a little in Section 10.3 where it was shown that the TP algorithm was significantly more accurate than the FD or BD algorithms by looking at some examples of the algorithms for simple equations. With the code segments developed in the previous section sufficient background and code segments have been developed to explore in more depth issues of accuracy. On a theoretical note, one can think of expanding some function in the interval from a value at point n to point n+1 about the midpoint of the interval with a Taylor series expansion of the form

$$y = y_{mid} + (y'_{mid})x + \frac{1}{2}(y''_{mid})x^2 + \frac{1}{6}(y'''_{mid})x^3 + \cdots \tag{10.31}$$

where terms to third order have been shown. The derivatives are to be evaluated at the mid point and x is the distance of the independent variable from the mid point. For the y_n and y_{n+1} values, this expression can be evaluated at points $x = -h/2$ and $+h/2$ respectively (where x is now used instead of t). Two equations can be obtained for the value of the solution at point n+1: one from Eq. (10.31) and one from the trapezoidal formula of Eq. (10.6) where the above equation can be used to evaluate all the terms on the right hand side. This results in the two equations:

$$y_{n+1} = y_{mid} + (y'_{mid})\left(\frac{h}{2}\right) + \frac{1}{2}(y''_{mid})\left(\frac{h}{2}\right)^2 + \frac{1}{6}(y'''_{mid})\left(\frac{h}{2}\right)^3$$

$$y_{n+1}^{TP} = y_{mid} + (y'_{mid})\left(\frac{h}{2}\right) + \frac{1}{2}(y''_{mid})\left(\frac{h}{2}\right)^2 + \frac{5}{6}(y'''_{mid})\left(\frac{h}{2}\right)^3 \tag{10.32}$$

where the second expression is the TP approximation to the solution at point n+1 evaluated to third order in the step size h. As readily seen, the difference is in the last term involving the third derivative and the third power of the step size h. Thus the difference between these expressions provides an estimate of the error of a TP integration algorithm as;

$$y_{n+1} - y_{n+1}^{TP} = Error_{TP} = -\frac{1}{12}h^3\left(y'''_{mid}\right) \tag{10.33}$$

where the third derivative is evaluated at the mid point as indicated by the subscript in this equation.

Equation (10.33) is suggestive that the error in the TP algorithm is proportional to the product of the third derivative of the function being calculated and h^3 and that by decreasing the step size by a factor of 10 the solution accuracy would increase by three orders of magnitude. However care must be exercised in using this expression as it represents the error associated with a single time step. In a numerical solution of some differential equations one is typically interested in many time steps and errors may accumulate over many time steps. In fact if the third derivative of some function is constant over some region then after n time steps the cumulative error will not be Eq. (10.33) but will be:

$$Error_{TP} = -\frac{1}{12}nh^3 y_{mid}''' = -\frac{1}{12}th^2 y_{mid}''' \tag{10.34}$$

where $t = nh$ is the total time of a given solution point. This gives a somewhat different picture of the error associated with the TP algorithm. So for a solution with constant third derivative and at some time t, an error is obtained that varies as the second power of the time step and not the third power and the error increases linearly with the time of a calculation. Thus if a constant time step is maintained, the error in a numerical solution by the TP algorithm will tend to increase with the total time for which a solution is obtained.

In solving a particular set of equations, the situation is likely to be considerably more complicated with the third derivative changing sign many times during the time of a solution. In regions of positive third derivative, the error will build up in a negative magnitude direction and in regions of negative third derivative the error will build up in a positive magnitude direction. Thus one might expect to observe regions of error in one direction followed by regions of error in the other direction as a solution progresses and with the error passing through zero between these regions. The error in a numerical solution is thus a global function of the properties of the solution and the time step values within various regions of the solution. Many authors discuss developing a time step algorithm based upon the local error criteria of Eq. (10.33). However, this can lead to incorrect conclusions because, as will be seen from examples, the error tends to depend more closely on the second power of step size instead of the third power of step size when a complete solution over a range of times is considered.

Let's now look at how the above theory holds with some simple examples and the TP algorithm. The code of the odeiv() algorithm is not completely a TP algorithm as it uses the BD algorithm for the first time interval, so for a completely TP algorithm one needs to go back to the code listed in Listing 10.1. Consider first a simple single exponential differential equation of the form:

$$y' + \lambda y = 0 \tag{10.35}$$
$$\text{with solution } y = C\exp(-\lambda t)$$

This function has a third derivative of $y''' = -\lambda^3 y = -C\lambda^3 \exp(-\lambda t)$. According to Eq. (10.33) the absolute error should be given by:

$$Error_{ATP} = \frac{1}{12}th^2 \lambda^3 C\exp(-\lambda t) \tag{10.36}$$

If one is more interested in the relative error then this should be divided by the function value to give:

$$Error_{RTP} = \frac{1}{12}th^2 \lambda^3 \tag{10.37}$$

How do these compare with actual calculations? The results of a numerical calculation comparing the relative and absolute errors are shown in Figure 10.16. The solid lines are the numerical results achieved with Listing 10.1 and with a step size of 0.001 and for a single exponential function with a $\lambda = 10$ value. The data points are theoretical values computed from Eqs. (10.36) and (10.37). The results are in excellent agreement with the theoretical equations. The dotted curve is the absolute error for a step size of 0.0005 or a factor of 2 smaller than the solid curve. As can be seen the error is reduced by a factor of exactly 4 across the entire range of the solution. At any given value of time the error is seen to vary by the second power of h according to Eq. (10.36) and not the third power of h as might be inferred from Eq. (10.33).

From Eq(10.36) it can be seen that the peak in the absolute error will occur for a simple exponential when $t \exp(-\lambda t)$ has a maximum value and this occurs at $t = 1/\lambda$. For the data in Figure 10.16 this occurs at a time of 0.1 as can be verified from the graph. With this as a background, the previous two error peaks seen in Figure 10.15 for the stiff differential equation case can be much better understood. The solution for that example has two exponential solutions with exponential factors of 2 and 2000, so two peaks in the solution error should be expected with times of 5e-4 and 0.5. The actual peaks observed in Figure 10.15 occur somewhat later (about a factor of 2 later), because of the non-uniform grid spacing and the global nature of errors.

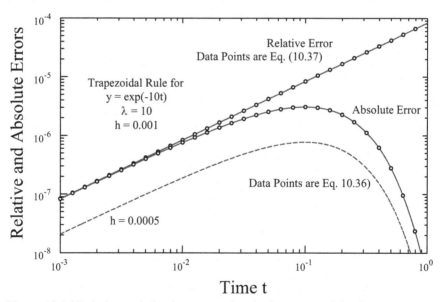

Figure 10.16 Relative and absolute errors for single exponential using TP integration algorithm. Data points are theoretical values from Equations (10.36 and (10.37).

A second easy case to look at is that of a linear second order differential equation with a sinusoidal solution such as that in Eq. (10.29) and used in Listing 10.3. When this set of equations is solved as in Listing 10.3 and the error calculated between the numerical solution and the exact cos(t) solution the resulting error is shown in Figure 10.17. The solid curve in the figure is the absolute error between the known solution and the numerical solution using the TP algorithm for a step size of 0.02 while the open data points are the expected error results calculated from Eq. (10.34) using the fact that the third derivative is simply the sin(t) function. Again there is excellent agreement between the actually observed error with the numerical integration and the predictions of Eq. (10.34). The general trend is for the error to grow linearly with time but to show zero crossings at points where the sin() equals zero. The error also alternates between positive and negative values. The relative error approaches large values where the cos() function approaches zero so the most useful result is to think in terms of the absolute error or the error normalized to the peak of the periodic function. For this solution the peak of the function was taken as 1.0, so the normalized error is the same as the absolute error. In general the absolute error would scale with the amplitude of the sinusoidal solution. From this it can be seen that integrating a periodic function over many cycles can lead to some rather large errors.

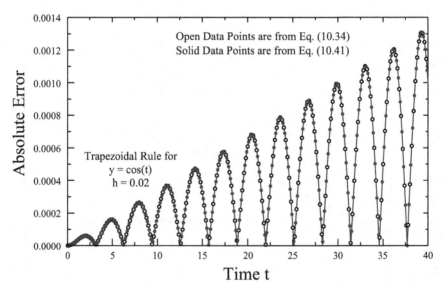

Figure 10.17 Absolute errors in the numerical solution of second order equation with sinusoidal solution. Solid curve is numerical error from numerical integration with TP algorithm. Open and solid points are theoretical errors.

It is interesting to note that the error in the case of the pure sinusoidal solution can be considered not as a result of an error in amplitude of the numerical solution but an error in the frequency of the numerically calculated sinusoidal solution. As previously discussed, the amplitude of the numerical solution will not change over a very large number of cycles using the TP numerical solution. This was noted earlier as a unique feature of the TP algorithm. It is thus somewhat surprising at first thought that the error in the numerical solution for a pure sinusoidal function continues to increase with time. The increasing error is not due to an incorrect time dependent amplitude of the numerical solution. Then why does the numerical error continue to increase with each cycle of solution? To understand this, consider again an imaginary value of λ in Eq. (10.17) and the TP algorithm. This gives after n time iterations:

$$y_n = \frac{(1 - j\beta h/2)^n}{(1 + j\beta h/2)^n} y_0 = y_0 \angle [2n \tan^{-1}(\beta h/2)] = y_0 \angle \theta$$

$$y_n = y_0(\cos(\theta) + j\sin(\theta))$$

$\qquad(10.38)$

This equation will rotate through a complete cycle when

$$2n \tan^{-1}(\beta h/2) = 2\pi = (2T/h)\tan^{-1}(\beta h/2)$$

$$\text{or } T = 2\pi h/\tan^{-1}(\beta h/2)$$

$\qquad(10.39)$

The last form of the equation gives the frequency of the resulting numerical solution. When $\beta h/2 \to 0$ this becomes $T = 2\pi/\beta$, agreeing with the expected time period of the differential equation. However, for a finite step size the time period of the numerical solution will be increased and the frequency (to first order in h^2) of the numerical solution will be decreased to:

$$F/F_0 = 1 - (\beta h/2)^2/3$$

$\qquad(10.40)$

where F_0 is the frequency of the differential equation and F is the frequency at which the numerical solution repeats. If this is correct, then the numerical error in solving the differential equation for a pure sinusoid should also be expressible as

$$Error_{Cos} = \cos(\beta t) - \cos(\beta t(1 - (\beta h)^2/12))$$

$\qquad(10.41)$

The solid data points in Figure 10.17 are points calculated from this equation, illustrating the correctness of this model for the error. In fact after a very large number of cycles, this equation will continue to give correct values of the error while Eq. (10.34) will become inaccurate as the phase difference between the numerical solution and the true solution becomes large.

From this discussion, it becomes apparent that the error in the numerical integration of a purely periodic function can be interpreted as an error in the frequency of the numerical solution. An error is still an error regardless of the cause. However, one might feel a little differently about the error when viewed as a frequency difference than when viewed as a magnitude error. When integrating a purely sinusoidal function by the TP algorithm, the magnitude of the function is maintained through many cycles of the function, but the numerical algorithm will result in a solution with a slightly lower frequency than inherent in the differential equation.

From the theory presented above it is expected that at any given time for a solution, the error in the TP algorithm varies with the square of the step size. These concepts can now be tested in more depth for the TP integration code and the actual error obtained by the computer code so far developed can be explored. For this test the selected equations will first be the set of stiff differential equations as previously used in Listing 10.9. The error associated with the TP algorithm and a multi step solution with 400 linear points per decade has previously been shown in Figure 10.15. Some code segments for multiple step solutions are now shown in Listing 10.10. The stiff differential equation set is defined on lines 5 through 8. This is followed by the solution of the equations on lines 10 and 11 using the odeiv() function with multiple time step intervals (taken as orders of magnitude in time) and with 200 and 400 points per time interval for the two calculations.

```
 1 : -- /* File list10.10.lua */
 2 : -- Programs for first order diff. equation using odeiv()
 3 : require"odeiv"; exp = math.exp; abs = math.abs
 4 :
 5 : f1 = function(eqs,t,y,yp) -- test of stiff differential eqns
 6 :    eqs[1] = yp[1] + 1001*y[1] - 999*y[2]
 7 :    eqs[2] = yp[2] - 999*y[1] + 1001*y[2]
 8 : end
 9 : -- TP Solution with step sizes of 2h and h
10 : s1 = odeiv(f1,{0,{1.e-4,1.e-3,1.e-2,0.1,1,10},200},{0,2})
11 : s2 = odeiv(f1,{0,{1.e-4,1.e-3,1.e-2,0.1,1,10},400},{0,2})
12 :
13 : -- Evaluate estimated error given h and 2h step sizes
14 : erra = odeerror(s1,s2) -- Evaluation of estimated error
15 : print(errstat(erra[2])) -- statistics of error estimate
16 : print(errstat(erra[3])) -- returns statistics of errors
17 :
18 : -- Calculate exact and corrected solution and errors obtained
19 : yexact,ycorrected,err1,err2,nd = {},{},{},{},#s2[1]
20 : for i=1,nd do yexact[i]=exp(-2*s2[1][i])+exp(-2000*s2[1][i]) end
21 : for i=1,nd do ycorrected[i] = s2[3][i] - erra[3][i] end
22 : for i=1,nd do
23 :    err1[i] = abs(yexact[i] - s2[3][i]) -- Error in TP solution
24 :    err2[i] = abs(yexact[i] - ycorrected[i]) -- Error in solution
25 : end
26 : print(errstat(err1))
27 : print(errstat(err2))
28 : plot(s1[1],err1)
29 :
30 : -- Solve ODE by TP and return solution and error estimate
31 : s3,err,n2 = odeive(f1,{0,{1.e-4,1.e-3,1.e-2,0.1,1,10},400},
        {0,2})
32 : print(errstat(err[2]))
33 : print(errstat(err[3]))
34 : write_data("list10_10.dat",s2,s1,s3,erra,err,err1,err2)
35 : write_data(20,"list10_101.dat",s2,s1,s3,erra,err,err1,err2)
Selected Output:
1.6015988433583e-006 8.5631809141456e-006 2023 -6.5693035762722e-009
1.4997301268234e-006 8.5631809141456e-006 2023 -5.6199664602728e-007
1.4960853042557e-006 8.555871193458e-006  2023  5.7083136619233e-007
2.2153039509691e-009 4.1749225854204e-008 802   3.618703841064e-010
```

Listing 10.10. Code segments for exploring TP solution accuracy with multi step size solutions.

Before continuing with a description of the code in Listing 10.10 it is instructive at this point to look at the two solutions with time steps differing by a factor of 2 as generated by the calls to odeiv() on lines 10 and 11. Both solutions will accurately reproduce the solution previously shown in Figure 10.11 and the reader is referred back there to see the wide range of the two time constants involved in the solutions. Here the emphasis will just be on the errors in the two solution sets as calculated from the known exact solution. These errors are shown in Figure 10.18. The dotted curve is the error for the case of 200 time points per decade in time going from 1.e-4 to 1.e2 or 8 orders of magnitude in time with an additional 200 points from 0 to 1.e-4; a total of 1201 data points. The solid curve is similarly the error in the numerical solution with an increase by a factor of 2 in the number of data points for each time interval. As previously shown for more simple cases, the error is seen to vary to a high degree of accuracy with the square of the step size with the solid curve being a factor of 4 smaller than the dotted curve and agreeing with the previous theory of Eq. (10.34). This uniform dependence on step size on the global scale is the major factor that makes possible a computer generated estimate of the accuracy of numerical solutions. Points are also noted where the error becomes very small and these are points were the error changes sign.

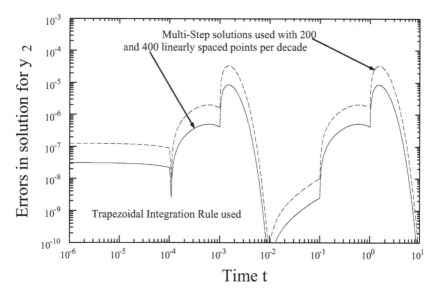

Figure 10.18. Errors in solution of stiff differential equation using TP rule with different time step sizes obtained form the solutions in Listing 10.10.

Based upon the background discussed so far, it should be possible to use a step size squared model for the solution error at any point in time and thus to write the true, y, value of a solution in terms of the numerical solution, y_{TP}, and as error term, $K_{ER}h^2$, as:

$$y_{TP} = y + K_{ER}h^2 \qquad (10.42)$$

If this relationship is applied to two numerical calculations at two different step sizes one obtains

$$y_{1,TP} = y + K_{ER}h_1^2$$
$$y_{2,TP} = y + K_{ER}h_2^2 = y + K_{ER}h_1^2 r^2 \qquad (10.43)$$
$$\text{where } r = h_2 / h_1$$

Let's further assume that $r = h_2 / h_1$ is less than unity and in this example is equal to 1/2, since the second solution set uses twice the number of steps as the first. By taking the difference in these two equations, the first order error term in the solution can be calculated as:

$$h^2 K_{ER} = (y_1 - y_2)/(1 - r^2)$$
$$y = y_{2,ER} - (y_1 - y_2)/(r^{-2} - 1) = y_{2,ER} - Error \qquad (10.44)$$
$$Error = (y_1 - y_2)/3$$

The last line above shows the estimated error to the solution obtained with the smallest step size as being simply $(y_1 - y_2)/3$. In terms of Figure 10.18 the first order estimate of the solution error is then simply 1/3 of the difference between the dotted curve and the solid curve. This is the error estimate of the most accurate solution with the largest number of data points. This technique of numerically estimating the global error of a solution will be referred to as the "h-2h" error estimation technique.

A special function odeerror() as shown on line 14 of Listing 10.10 is provided in the "odeiv" package loaded on line 3 to perform this calculation. The reader can consult the code file to see the details of the calculation but it is rather straightforward. The function will accept two solutions in any order and return the estimated error "under the assumption that the step sizes of the two files are related by a factor of 2". If any other multiple of step size is used, the results will not be correct and it is up to the user to supply appropriate files to the odeerror() routine. The only snag in evaluating the error estimate of Eq. (10.44) is the fact that the function with the largest step size is known only at one-half the data points of the function with the smallest step size. In order to obtain an estimate at all the data points of the most accurate function, interpolation must be performed on the y_1 function for the extra data points. For ease in plotting results, the array returned by the odeerror() function repeats the array of time values as the first entry. The reader can view how this is done in the code for odeerror() contained in the odeiv package.

Now let's see how this evaluation performs with the known solution for the stiff differential equation of Listing 10.10. The comparison is shown in Figure 10.19. Three sets of data and curves are shown in the figure. First the solid curve is the actual error between the TP numerical solution of the differential equation for the y_2 variable and the exact known solution. The data points shown are the estimated errors from performing the calculation indicated in Eq. (10.44) and are 1/3 of the difference between the two curves shown in Figure 10.18. The data points are also the table values returned by the odeerror() function on line 14 of the listing. Only a few of the calculated points are shown in order to see the solid curve behind the data points. The results show excellent agreement between the theory and the actual errors verifying the dependence of the global errors on the square of the step size for the TP algorithm.

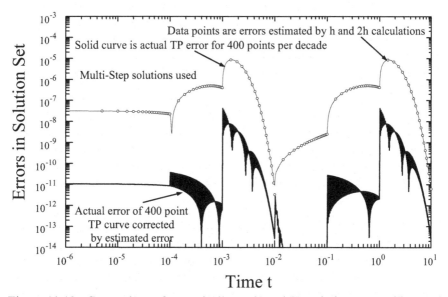

Figure 10.19. Comparison of numerically evaluated TP solution error with actual error obtained from the difference in the TP solution and the exact solution for the stiff differential equation of Listing 10.10.

Figure 10.19 has one final lower curve that needs an explanation. By looking back at Eq. (10.44) it can be seen that a "better" estimate of the true solution can be obtained by subtraction the error estimate from the most accurate numerical solution as indicated on the middle line of Eq. (10.44). This calculation is performed on line 21 of Listing 10.10. The difference between this corrected solution and the exact solution is the curve that is plotted as the lower solid curve in Figure 10.19. While the error associated with this corrected solution shows some oscillations in magnitude, the results indicate that this corrected curve is on average about 2 orders of magnitude more accurate than the uncorrected solution and the maximum error is improved by about one order of magnitude by using the correction factor. At each even factor of 10 on the time axis, the step size used in the numerical solution was changed by one orders of magnitude. It can be seen that the corrected solution shows some jump in error value at these points and that the oscillations tend to die out between the jumps in time step. This provides some indication that it would perhaps be better to change the step size continuously rather than abruptly. This is addressed further in the subsequent discussion.

In order to rapidly evaluate the statistical properties of the error terms, a special function errstat() is provided and it's usage is shown on lines 15, 16, 26 and 27. The reader can examine the code in the odeiv package, but basically this returns the RMS value of the error, the maximum error, the point number of the maximum error and the average error. The selected output shows examples of the returned values. The first two output lines give statistical properties of the y_1, y_2 solutions as obtained from the odeerror() evaluation. For example, for the upper solid curve in Figure 10.19 the second line of output shows values 1.496e-6, 8.58e-6, 2022, -5.62e-7 for the RMS error, the maximum error, the point of maximum error and the average error respectively. The most important value is probably the maximum error of 8.58e-6 which one can see agrees with the peak value shown for the upper curve in Figure 10.19. The third output line shows corresponding values for the difference between the exact solution and the TP numerical solution again indicating the accuracy of the numerically evaluated error estimate. The fourth line of output shows corresponding values of the TP numerical solution after correcting for the evaluated error at each point. These values correspond to the lower solid curve shown in Figure 10.19. The maximum error for this calculation is seen to be 8.63e-7 which is about an order of magnitude smaller than the uncorrected solution error value.

The remaining code in Listing 10.10 can now be discussed. The code from lines 18 through 29 evaluates the exact solution values, calculates exact error values in the TP solutions and evaluates some statistics of the solutions. Finally line 31 shows a second way of solving a set of differential equations and generate error statistics using the supplied function odeive() function. This has identical calling arguments to the basic odeiv() function but essentially combines the odeiv() function with the odeerror() function. Basically the odeive() function integrates the set of given differential equations with the supplied time parameters. It then performs a second evaluation of the differential equation with a second step size arrangement essentially doubling the step size of the first calculation. It then uses these two h-2h solutions to generate an estimate of the error in the solution. As indicated on line 31, the odeive() function returns two arrays with the first an array of solutions to the differential equation and the second an array of estimated errors for the solutions at each time point. Finally an integer is returned indicating the maximum number of iterations taken in the calculations, similar to that returned by odeiv(). The odeive() function is then basically two calls to odeiv() with step sizes a multiple of 2 apart in the second call and this is then followed by a call to odeerror(). This odeive() function should be used if one desires not only a numerical TP solution but also an estimate of the accuracy of the obtained solution. Of course nothing comes completely free. In this case the cost is approximately twice the execution time of a single call to odeiv(). The obvious advantage is that an estimate of the accuracy of the numerical solution is obtained with the odeive() function. If one feels confident in the accuracy of a given numerical solution, than only the single solution obtained with odeiv() can be used. The reader is encouraged to view the code for the odeive() and odeerror() functions which are not listed in the text as they are reasonably simple code segments.

```
1 : -- /* File list10_11.lua */
2 : -- Programs for first order diff. equation using odeiv()
3 : require"odeiv"; exp = math.exp; abs = math.abs
4 :
5 : f1 = function(eqs,t,y,yp) -- test of stiff differential eqns
6 :    eqs[1] = yp[1] + 1001*y[1] - 999*y[2]
7 :    eqs[2] = yp[2] - 999*y[1] + 1001*y[2]
8 : end
9 : -- TP solution with Quick Scan function, also error evaluation
10 : s1,err = odeivqse(f1,{0,{1.e-6,10},400},{0,2})
11 : print(errstat(err[2]));print(errstat(err[3]))
12 : -- Calculate exact and corrected solution and errors obtained
13 : yexact,ycorrected,err1,err2,nd = {},{},{},{},#s1[1]
14 : for i=1,nd do yexact[i]=exp(-2*s1[1][i])+exp(-2000*s1[1][i]) end
15 : for i=1,nd do ycorrected[i] = s1[3][i] - err[3][i] end
16 : for i=1,nd do
17 :    err1[i] = abs(yexact[i] - s1[3][i]) -- Error in TP solution
18 :    err2[i] = abs(yexact[i] - ycorrected[i]) -- Error in solution
19 : end
20 : print(errstat(err1));print(errstat(err2))
21 : write_data("list10_11.dat",s1,err,err1,err2)
22 : write_data(20,"list10_111.dat",s1,err,err1,err2)
```

Listing 10.11. TP integration with error estimation using Quick Scan function for the stiff differential equation.

The evaluation of estimated error can also be obtained with a non-uniform distribution of step sizes such as the logarithmic distribution used in the quick scan function of Listing 10.7. An example of this is shown in Listing 10.11. The odeivqse() function used on line 10 combines the quick scan function with the error estimation based upon a reduction in step size of a factor of 2. The error results of using 400 points per decade and scanning the stiff differential equation from 1.e-6 to 10 are shown in Figure 10.20. The odeivqse() function adds the h-2h calculation to the quick scan odeivqs() function and returns the calculated function plus an estimate of the solution error similar to the odeive() function. In Listing 10.11 this function is called on line 10 with the same set of time parameters previously used with the odeivqs() function in Listing 10.9. In Figure 10.20 the upper solid curve is the actual error in the numerical TP solution as compared with the known solution while the data points are the error terms estimated by the use of the two different time sizes and the h-2h algorithm. Again excellent agreement is seen between the actual error and the estimate obtained by the h-2h algorithm with the peak error being around 1.e-6. The lower curve is the error obtained when the 400 point solution is corrected by the numerically evaluated error term. In this case a reduction in the maximum error by a factor of about 10000 is seen, with a peak error in the range of 1.e-10. This is a very excellent result, but the equation is a special case of a linear differential equation. However, the results do suggest that the TP algorithm with h-2h error correction is capable of excellent accuracy in solving such equations. It is also useful to compare the solution errors in

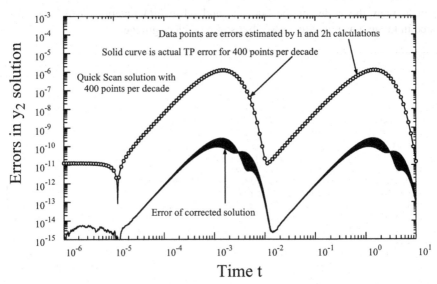

Figure 10.20. Errors in the solution for the stiff differential equation of Listing 10.11 using a logarithmically spaced time grid.

Figure 10.20 with those shown in Figure 10.19. The difference in the calculations is the use of abruptly changing time step increments for Figure 10.19 and smoothly changing time step increments for Figure 10.20. The solution accuracy is considerably improved for the case of Figure 10.20 and again suggests that the best approach for problems with very different time constants (or stiff differential equations) is to use smoothly varying time steps in the solution so a wide range of time values can be observed.

As a third test of the TP integration error, the sinusoidal differential equation pair of Listing 10.8 will be considered. The typical error from a TP numerical solution was illustrated in Figure 10.17 where it was shown that the absolute error tends to grow linearly with time. Listing 10.12 shows a code segment for a new integration of this equation using the odeive() function evaluating the solution and estimating the solution error. The solution and error is obtained on line 10 of the code using a two segment stepping sequence. In order to obtain good initial values, 100 points are evaluated in the interval 0 to 0.01. This is followed by 2000 points over the interval of 0.01 to 40. The first interval has a time step of 0.0001 and the second interval has a time step of approximately 0.02. The technique of using a small initial time step over some finite interval is very useful in a range of problems in order to start the integration process with the BD algorithm and then continue with the TP algorithm. Smaller step sizes are needed with the BD algorithm in order to achieve the same accuracy as with the TP algorithm. The computed results of the calculations for the resulting errors in the solution are shown in Figure 10.21. Again a log scale is used on the errors in order to cover a large

```
 1 : -- /* File list10_12.lua */
 2 : -- Programs for first order diff. equation using odeive()
 3 : require"odeiv"
 4 :
 5 : f1 = function(eqs,t,y,yp) -- test of sinusoidal diff eqns
 6 :     eqs[1] = yp[1] - y[2]
 7 :     eqs[2] = yp[2] + y[1]
 8 : end
 9 : -- TP solution with two step intervals, also error evaluation
10 : s1,err = odeive(f1,{0,{.01,40},{100,2000}},{1,0})
11 : print(errstat(err[2]));print(errstat(err[3]))
12 : -- Calculate exact and corrected solution and errors obtained
13 : yexact,ycorrected,err1,err2,nd = {},{},{},{},#s1[1]
14 : for i=1,nd do yexact[i] = math.cos(s1[1][i]) end
15 : for i=1,nd do ycorrected[i] = s1[2][i] - err[2][i] end
16 : for i=1,nd do
17 :     err1[i] = math.abs(yexact[i] - s1[2][i]) -- Error in TP
18 :     err2[i] = math.abs(yexact[i] - ycorrected[i]) -- Error
19 : end
20 : print(errstat(err1));print(errstat(err2))
21 : write_data("list10_12.dat",s1,ycorrected,err,err1,err2)
22 : write_data(20,"list10_121.dat",s1,ycorrected,err,err1,err2)
23 : plot(err)
```

Listing 10.12. Code segments for numerical solution of sinusoidal differential equation with error calculation.

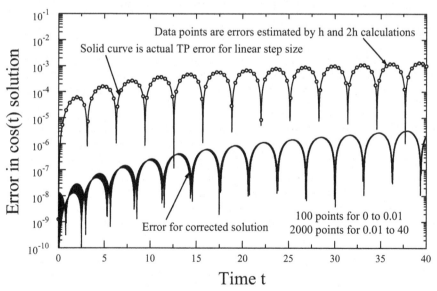

Figure 10.21. Errors in a sinusoidal solution with TP algorithm obtained in Listing 10.12. Both the uncorrected and corrected solution errors are shown.

range of error values. The upper solid curve and the data points are the numerically evaluated errors from the h-2h algorithm and the actually achieved error in the numerical solution. If plotted on a linear vertical scale, the curves would be essentially identical to the curves shown previously in Figure 10.17 with the peaks of the curve increasing linearly with time. The resulting peak error is in the range of 1.e-3 which would be sufficient for most engineering applications. However, since it increases linearly with time, it would be in the range of 1.e-2 if calculations were carried out to about 60 cycles of the solution. In most applications this would be more than adequate. Also shown in Figure 10.21 is the accuracy of the numerical solution if it is corrected by the numerically evaluated error terms. Although the error calculation is an approximate value based upon the h-2h algorithm, it can be seen that the corrected solution has a greatly improved accuracy. The corrected solution is some 500 times more accurate than the solution obtained by direct integration. This corrected solution is quite accurate with a peak error of only 3.2e-6, but it is still not as good as achieved in the previous example of the solution with two exponential terms.

The examples of computational accuracy so far have been for linear differential equations. This is reasonable since for such equations the exact solution of such linear equations is known. It would also be instructive to explore similar results for nonlinear equations. In general, nonlinear differential equations do not have closed form solutions so one must extrapolate to such equations. There are, however, a small set of nonlinear differential equations for which an exact solution can be obtained and one such equation is the second order equation:

$$\frac{d^2 y}{dx^2} = \exp(y) \tag{10.45}$$

This equation arises in approximating the electric potential in an accumulation (or inversion) layer in a semiconductor and the above is a normalized form of the equation in normalized potential and distance. The solution of this equation is:

$$y = \log(C_1^2 \sec^2(C_1(x + C_2)/\sqrt{2})) \tag{10.46}$$

where C_1 and C_2 are the two arbitrary constants that must accompany any second order differential equation. The method of closed form solution will not be given here, but the reader can verify that this is an exact solution by substituting into the equation and taking two derivatives. For a simple set of boundary conditions of $y(0) = 0, y'(0) = 0$ the constants can be evaluated as $C_1 = 1, C_2 = 0$.

Converting the second order differential equation into a set of two first order differential equations with initial conditions gives:

$$y_1' - y_2 = 0; \qquad y_1(0) = 0$$
$$y_2' - \exp(y_1) = 0; \qquad y_2(0) = 0 \tag{10.47}$$

with solution: $y_1 = \log(\sec^2(x/\sqrt{2}))); \; y_2 = \sqrt{2}\tan(x/\sqrt{2})$

The exact solution becomes infinitely large as $x \to \pi/\sqrt{2} = 2.22144....$ Thus a critical test of the integration algorithm (or any solution algorithm for that matter) can be obtained by numerically integrating this equation from zero to a point close to this divergent value.

```
 1 : -- /* File list10_13.lua */
 2 : -- Programs to integrate first order diff. equation using
odeive()
 3 : require"odeiv"
 4 :
 5 : --odebiv = odebrk
 6 : f1 = function(eqs,t,y,yp) -- test of nonlinear differential eqns
 7 :    eqs[1] = yp[1] - y[2]
 8 :    eqs[2] = yp[2]  - math.exp(y[1])
 9 : end
10 : sr2 =math.sqrt(2)
11 : xmax = 2.221 -- must be less than 2.22144
12 : -- TP solution with two step intervals, also error evaluation
13 : s1,err1,n1 = odeive(f1,{0,{xmax*.99,xmax},1000},{0,0})
14 : print(errstat(err1[2]));print(errstat(err1[3]));print('n = ',n1)
15 : s2,err2,n2 = odeive(f1,{0,{xmax*.99,xmax},2000},{0,0})
16 : print(errstat(err2[2]));print(errstat(err2[3]));
         print('n2 = ',n2)
17 :
18 : y1,y2 = {},{} -- Evaluate exact solutions
19 : for i=1,#s2[1] do
20 :    y1[i] = math.log(1/math.cos(s2[1][i]/sr2)^2)
21 :    y2[i] = sr2*math.tan(s2[1][i]/sr2)
22 : end
23 : write_data("list10_13.dat",s1,s2,y1,y2,err1,err2)
24 : write_data(50,"list10_131.dat",s1,s2,y1,y2,err1,err2)
Output:
0.032678694317672   0.43817107269992    2001    0.011970957193953
77.028720978766     1638.1181172715     2001    10.147628459349
n =      3
0.0065167420545363  0.074447874229755   4001    0.0027035028254573
10.238791715513     185.07693880659     4001    1.6695693431679
n2 =     2
```

Listing 10.13. Code for TP analysis of highly nonlinear second order differential equation.

Sample code for the integration of this set of equations is shown in Listing 10.13. The equations are defined on lines 6 through 9 as expressed above. Two sets of calculations are made first on line 13 with odeive() and two independent variable intervals of 1000 steps each. The maximum x value is set on line 11 to 2.221 which is just below the point of infinite values at 2.22144... The step distribution in the listing on line 13 uses 1000 points from 0 to 99% of the maximum value and then uses 1000 additional points in the last 1% of the spatial interval. This is an attempt to use considerably more spatial points in the region where the functions are known to change rapidly. Actually, this step selection resulted from a couple of initial trial calculations with a uniform step distribution. As can be seen this step distribution is by no means optimal. A second integration is performed on line 15 with the same two spatial intervals but with 2000 points in each interval. In both cases the odeive() function is used to return not only the solution but an estimate of the errors in the solutions.

The exact solutions are evaluated on lines 20 and 21 and the RMS and average errors are printed as output and the results are included in the listing. The last three lines of the listing apply to the calculation with a total of 4001 points. It is seen that the maximum estimated errors are 0.0744 and 185.0 for y_1 and y_2 respectively as printed on the last three lines of the listing. One can not know if these are good or bad errors until the actual magnitudes of the solutions are known. Figure 10.22 shows the calculated values of the two solution variables as the two upper curves labeled y_1 and y_2. Shown as a dotted vertical line on the right of the figure is the boundary at which the exact analytical solution approaches infinity for both variables. The very sharp vertical solution values near the boundary can be seen even on a log scale. Values for the exact analytical solution are not shown, but they would be essential identical to the solid curves on the scale shown.

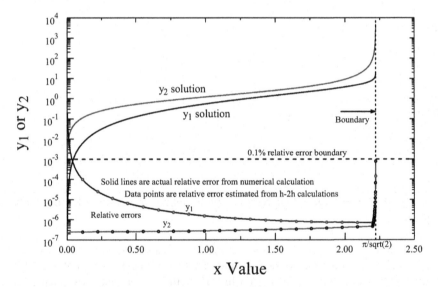

Figure 10.22. TP integration results for highly nonlinear problem obtained from code in Listing 10.13.

It is seen that the y_2 solution which is the derivative has a much sharper increase than does the potential which is the y_1 variable. Since what is important is the relative error in the solution and not the exact value of the error, shown in Figure 10.22 as the lower two figures are relative error values as function of distance along the solution. For the solid relative error curves, the values were evaluated by taking the difference between the exact analytical solution and the numerical integration values and dividing by the value of each variable at each point along the solution. For the data points shown, the values were calculated by taking the estimated errors returned by the odeive() function obtained from an h-2h evaluation and again dividing by the value of the solution. Several conclusions can be drawn from the curves in this figure. First, the h-2h technique of estimating errors is in excellent agreement with the actually achieved errors even for such a highly nonlinear problem. Second, the achieved relative errors are very good even with the rather rough spatial step selection. A sharp increase in the relative error is readily seen near the singularity boundary. The accuracy in that region could be improved, if desired, by using a finer spacing of grid points there. However, the relative accuracy is seen to be better than the 0.1% value line shown on the figure and this would be sufficient for most engineering work. Thus additional effort on an improved step distribution is probably not needed.

The reader is encouraged to re-execute the code in Listing 10.13 with additional grid points near the right boundary and observe the effects on accuracy of the solution. Also an improved solution can be obtained by again subtracting the error estimate returned by the h-2h algorithm from the generated numerical solution. Although not shown in the figure, the corrected solution is several orders of magnitude more accurate over most of the spatial range. However, the corrected solution still exhibits about the same error in the region near the singularity boundary where the function is changing very rapidly. The only technique to improve the accuracy near the boundary is to use a finer spatial grid in the numerical solution.

```
 1 : -- /* File list10.14_lua */
 2 : -- Programs for first order diff. equation using odeive()
 3 : require"odeiv"
 4 :
 5 : mu = 10
 6 : f1 = function(eqs,t,y,yp) -- Van der Pol equation
 7 :     eqs[1] = yp[1] - y[2]
 8 :     eqs[2] = yp[2] - mu*(1 - y[1]^2)*y[2] + y[1]
 9 : end
10 : -- TP solution with one step interval, also error evaluation
11 : s1,err = odeive(f1,{0,{100},8000},{2,0})
12 : print(errstat(err[2]));print(errstat(err[3]))
13 :
14 : write_data("list10_14.dat",s1,err)
15 : plot(s1[1],s1[2],err[2])
```

Listing 10.14 Illustration of one use of the h-2h error estimation for the nonlinear Van der Pol equation.

Listing 10.14 shows one excellent way the h-2h error estimate can be used. The code in this listing solves the nonlinear Van der Pol equation and is very similar to the code in Listing 10.6 with the results shown in Figure 10.10. The addition here is the use of the TP equation solver odeive() on line 11 with the h-2h error estimation obtained. Figure 10.23 shows the pop-up plot that the code produces from line 15 showing an overlay of the estimated error on the solution variable. The solution variable is shown with a light dotted curve so that the solid curve representing the error estimate can be more clearly seen in the figure. This gives a very clear graphical indication of the accuracy of the numerical solution. The near linear growth in the error peaks with time can also be readily seen indicating that the first few cycles are more accurate than the latter cycles. However the error is primarily an error in the frequency of the periodic signal as was previously shown for a linear equation with a sinusoidal solution. This can also be seen from the previous calculations shown in Figure 10.10 for the same differential equation but with a slightly different nonlinear parameter and with different numbers of grid points. By plotting both the solution and the estimated error on the same plot the accuracy of a solution can readily be accessed and also a clear indication can be obtained of where a finer mesh of grid point is needed in order to improve the accuracy.

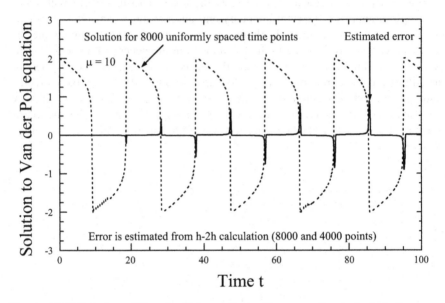

Figure 10.23 Solution of Van der Pol equation with error estimate included.

The results obtained from Listing 10.14 are for a total of 8000 time points. If this number is increased by a factor of 4 to 32000 time points the h-2h estimated error would be expected to decrease by a factor of 16 which would make it quite small on the scale of the solution. This is left as an exercise for the reader to execute the code in Listing 10.14 with various numbers of time steps and observe the effect on the estimated error of the solution. For a linear step distribution, little improvement in the solution would be expected after a total number of 32000 steps. With modern desk top computers, this calculation can be performed in a few seconds even with the interpretative language used in this work. Interestingly in research for this work, a reference was found to such a calculation in 1974 using a UNIVAC 1108 computer and performing three simulations from 0 to 10 with a step size of 0.001 (total of 10,000 points per calculation) that was performed in a little less than 30 min! Just shows how fortunate we are today with modern desk top computers and how our few seconds of simulation time will seem so long 30 years or so from now.

This section has presented a rather detailed discussion of the accuracy of the trapezoidal algorithm for integrating first order differential equations. It has been shown that one must consider the global aspects of solution errors and not just the local truncation error in a single integration step to understand the accuracy of an integration algorithm. Although the discussion has been in terms of several specific example equations, the results can be extrapolated to other differential equations and the TP integration technique is expected to produce solutions with very acceptable engineering accuracy. It is also expected that the technique of estimating the solution error by examining two solutions with only the step size reduced by a factor of 2 will also be applicable to other linear and nonlinear differential equations. It has also been found in all the examples studied that the solution accuracy can be greatly improved by correcting the numerical solution with the estimated h-2h first-order error in the solution. The accuracy improvement ranged from about a factor of 10 to about a factor of 10000 in the best case. Although it is difficult to make general statements about the improvement in accuracy because it depends on higher order derivatives of a solution, it would be expected that the corrected solution will be considerably more accurate than the uncorrected solution. However, it is only for the uncorrected solution that definite statements can be made regarding the expected accuracy of the numerical solutions.

10.7 Runge-Kutta Algorithms

The trapezoidal algorithm so far discussed in this chapter provides a very excellent algorithm for the numerical solution of nonlinear and linear differential equations because of its excellent stability properties and because the local and global errors associated with the algorithm are readily understood. The one possible shortcoming of the algorithm is that the accuracy is only second order in the step size and global errors in a numerical integration are proportional to solution time and the square of the step size as demonstrated by many examples in the previous section. Many numerical algorithms exist in the literature that provides higher order accuracy with regard to local step size – such as accuracy to the fourth power of the step size. Such algorithms can potentially provide increased accuracy with the same total number of time steps in certain applications. In general such algorithms require more numerical evaluations of the differential equations per step so some the potential advantages may be lost by the increased complexity of the higher order algorithms. In many cases the best approach to numerical computer work is to employ simple algorithms and apply them many times in iterative loops as opposed to more complicated algorithms applied fewer times. This is the principle of keep it simple, if possible, at work. Also stability can be a potential problem with many higher-order algorithms. Higher order algorithms do not always mean higher accuracy, but in some cases it does particularly for easily solved differential equations where stability is not a problem.

Of all the possible higher order methods studied, the Runge-Kutta (RK) algorithms have probably found the most application. Many commercial package, such as the solvers supplied by MATLAB, use variations of these algorithms so it is useful to have some knowledge of these algorithms and how they compare with the TP algorithm that is emphasized in this work. For this discussion let's consider first a single differential equation written in the form of Eq. (10.1), i.e. $y' = f(y,t)$ where $f(y,t)$ represents any general function describing the first order derivative. For such a derivative description, the general class of Runge-Kutta integration methods expresses the single step integration algorithm as

$$y_{i+1} = y_i + h(C_1 f(y_i + \alpha_1, t_i + \beta_1) + C_2 f(y_i + \alpha_2, t_i + \beta_2) +$$
$$\cdots C_n f(y_i + \alpha_n, t_i + \beta_n)) \tag{10.48}$$

For this algorithm one sees on the right hand side n terms involving the evaluation of the derivative at different time points and with different additional terms, identified as α_i, added to the solution variable (for example $y_i + \alpha_1$). Typically for n such derivative terms one has an nth order RK algorithm although for some RK algorithms some of the terms may be equal. In terms of the times at which the derivative is evaluated, the times all lie between t_i and t_{i+1} so that $0 <= \beta_j <= h$ for all the values added to the time point. For the value selected to be added to the solution variable in the derivative evaluations it is perhaps best to consider several specific examples of the algorithm.

There are a large number of possible second order RK algorithms which are accurate to the second power in step size. Three possible implementations are given below:

a. Heun Method

$$m_1 = f(y_i, t_i)$$
$$m_2 = f(y_i + m_1 h, t_i + h) \tag{10.49}$$
$$y_{i+1} = y_i + h(m_1 + m_2)/2$$

b. The Midpoint Method

$$m_1 = f(y_i, t_i)$$
$$m_2 = f(y_i + m_1 h/2, t_i + h/2) \tag{10.50}$$
$$y_{i+1} = y_i + hm_2$$

c. Ralston's Method

$$m_1 = f(y_i, t_i)$$
$$m_2 = f(y_i + 3m_1 h/4, t_i + 3h/4) \tag{10.51}$$
$$y_{i+1} = y_i + h(m_1 + 2m_2)/3$$

Each of these methods requires two evaluations of the derivative term. The first method can be viewed as a "poor man's" trapezoidal method since the two derivatives are evaluated at the t_i and t_{i+1} time points. Instead of using the unknown value of the solution variable at the second time point, a first order estimation of the solution value is used based upon the pervious time point and the slope at the previous time point. In this manner an estimate of the trapezoidal rule derivatives is obtained without involving an implicit equation as in the true trapezoidal rule. The second method similarly seeks to approximate the derivative at the mid point of the interval by using the mid point time and a first order estimate of the solution value at the mid point based upon the pervious solution value and the previous slope. Finally the third equation attempts to evaluate a second value for the derivative term from a time 3/4 of the way across the time interval using a similarly extrapolated value of the solution value. All of these algorithms are accurate to second order in the step size and each has different advantages.

One possible third order RK algorithm is based upon this set of equations

$$m_1 = f(y_i, t_i)$$
$$m_2 = f(y_i + m_1 h/2, t_i + h/2)$$
$$m_3 = f(y_i - m_1 h + 2m_2 h, t_i + h)$$
$$y_{i+1} = y_i + h(m_1 + 4m_2 + m_3)/6$$

(10.52)

This algorithm requires three evaluations of the derivative function at different times corresponding to the initial time, the mid point and the final time points with approximations to the solution values used at the three points. In all of these algorithms the derivative terms are evaluated in the order presented in the equation set.

The most popular RK methods are fourth order. Again there are an infinite number of possible forms, but the most common is the classical fourth order RK method given by this set of equations

$$m_1 = f(y_i, t_i)$$
$$m_2 = f(y_i + m_1 h/2, t_i + h/2)$$
$$m_3 = f(y_i + m_2 h/2, t_i + h/2)$$
$$m_4 = f(y_i + m_3 h, t_i + h)$$
$$y_{i+1} = y_i + h(m_1 + 2m_2 + 2m_3 + m_4)/6$$

(10.53)

Not surprisingly this requires four evaluations of the derivative function performed in the order presented as each derivative beyond the first depends on the previous evaluation. This algorithm forms the basis for many successful computer programs used to solve differential equations.

Although higher order RK algorithms are possible, rarely are RK methods beyond the fifth order used. Beyond the fourth order the gain in accuracy is offset by the added computational effort and complexity. The discussion here will concentrate on the fourth order RK method of Eq, (10.53) which is the most common RK algorithm employed. The algorithm is readily extended to a coupled set of first order equations. The major difference being that each derivative may depend upon a number of solution values and a number of derivative values as the y_i values in the equation form an array or table of solution values. At each step each solution value is incremented by the previously evaluated slope value and inserted into the next formula for the derivative evaluation. The equations will not be rewritten in terms of an array of solution values as the exercise is very straightforward.

```
  1 : -- /* File odeiv.lua */
  2 : -- Programs to integrate first order differential equations
  3 : require"nsolv"; local atend
----------
110 : odebrk = function(feqs,tvals,u,up,upp)-- Basic Runge-Kutta code
111 :    local j, neq, t, h, h2 -- Local variables for function
112 :    local sol, m1,m2,m3,m4,u1 = {}, {}, {}, {}, {}, {}
113 :    local nit,nitt = 0,0 -- Number of iterations
114 :    local neq = #u -- Number of equations
115 :    local tmin,tmax,ntval = tvals[1],tvals[2],tvals[3]
116 :
117 :    fderiv = function(eqs,up) -- Function for derivatives
118 :             feqs(eqs,t,u1,up)
119 :    end
120 :    -- Use input value for number of intervals or default value
121 :    if type(up)~='table' then up = {} end
122 :    for m=1,neq+1 do -- return solution values with t value first
123 :       sol[m] = {}
124 :       if m==1 then sol[m][1] = tmin else sol[m][1] = u[m-1]   end
125 :    end
126 :    t = tmin; hs = (tmax - t)/ntval -- Equal increments in t used
127 :    for m=1,neq do up[m] = 0 end
128 :    h,h2 = hs,hs/2 -- Set to RK parameters
129 :    for k=1,ntval do -- Main loop for incrementing variable
130 :       for m=1,neq do u1[m] = u[m] end
131 :       nitt = nsolv(fderiv,up,step) -- Update up values
132 :       for m=1,neq do m1[m] = h*up[m]; u1[m] = u[m] + m1[m]/2 end
133 :       t = t+h2; nitt = nsolv(fderiv,up,step)+nitt -- next value
134 :       for m=1,neq do m2[m] = h*up[m]; u1[m] = u[m] + m2[m]/2 end
135 :       nitt = nsolv(fderiv,up,step)+nitt -- next up value
136 :       for m=1,neq do m3[m] = h*up[m]; u1[m] = u[m] +m3[m] end
137 :       t = t+h2; nitt = nsolv(fderiv,up,step)+nitt -- next value
138 :       for m=1,neq do m4[m] = h*up[m] end
139 :       for m=1,neq do u[m] = u[m] + (m1[m]+2*m2[m]+2*m3[m]+
             m4[m])/6 end
140 :       sol[1][k+1] = t -- Save calculated values
141 :       for m=1,neq do sol[m+1][k+1] = u[m] end
142 :       if nitt>nit then nit = nitt end -- Monitor # of iterations
143 :    end -- End of main loop on t, now return solution array
144 :    return sol,nit/4 -- Solution and maximim number of iterations
145 : end -- End of odebrk
146 : setfenv(odebrk,{type=type,nsolv=nsolv,step=nil})
```

Listing 10.15. Code for implementing a fourth order Runge-Kutta integration algorithm.

Code for implementing a basic step of the RK algorithm in a function odebrk() is shown in Listing 10.15. The code between lines 110 and 129 is almost identical to the corresponding code for the basic TP algorithm. This section of code checks input values, defines needed arrays for the algorithm and defines a proxy function on lines 117 through 119 to translate between the nsolv() calling arguments and the differential equation calling arguments. The major loop of the RK algorithm is the main loop from line 129 through 143. Four calls are made to nsolv() on lines 131, 133, 135 and 137 to evaluate the four derivatives as defined in Eq. (10.53) above. The implementation of the fourth order RK algorithm should be readily correlated between the equations in the code and the equation set of Eq. (10.53).

The only additional step needed is a temporary array u1[] used on lines 130, 132, 134 and 136 to hold the intermediate values of the solution variables.

The odebrk() RK code in Listing 10.15 has been written in the same format as the odebiv() TP code in Listing 10.4. Each routine implements one or more steps of integration using either the TP or RK algorithms. Each can be called repeatedly to implement a multi step integration algorithm and the possibility of repeated calls is the reason for the up argument in the argument list to the odebrk() function. The RK algorithm can be simply substituted for the TP algorithm in other integration schemes such as odeiv() or odeive() using multiply defined time intervals.

```
 1 : -- /* File list10_16.lua */
 2 : -- Programs for first order diff. equation using odebrk()
 3 : require"odeiv"
 4 : odebiv = odebrk
 5 :
 6 : f1 = function(eqs,t,y,yp) -- equation for cos(t), sin(t)
 7 :    eqs[1] = yp[1] - y[2]
 8 :    eqs[2] = yp[2] + y[1]
 9 : end
10 : -- RK solution with one step interval, also error evaluation
11 : s1,err1 = odeive(f1,{0,100,2000},{1,0})
12 : print(errstat(err1[2]))
13 : s2,err2 = odeive(f1,{0,100,1000},{1,0})
14 : print(errstat(err2[2]))
15 : plot(s1[1],s1[2],err1[2])
16 :
17 : write_data("list10_16.dat",s1,err1,s2,err2)
Output:
2.1398828157025e-006  5.153786947278e-006  1983  -4.3427701467537e-008
3.4190865621653e-005  8.2458320143231e-005   993  -6.6325647173493e-007
```

Listing 10.16. Example code for replacing TP algorithm with fourth order RK algorithm.

Example code is shown in Listing 10.16 for substituting the RK algorithm for the previously developed TP algorithm. The change in the basic integration algorithm is made by the substitution on line 4 of the code with the statement odebiv = odebrk. This simply makes any subsequent calls to the odebiv() function call the RK function odebrk(). This can be used in any desired program in order to use the RK algorithm. The resulting numerical solution for the sin()-cos() differential equation using the RK algorithm is shown in Figure 10.24. The solid line numerical solution is compared with the data points which are the expected exact solution. The results agree so closely that it is hardly worth the use of a figure to show the results. However, this will be compared with some results from MATLAB in a subsequent section. One observation is that the amplitude of the oscillation remains fixed at essentially the same value of 1.0 at least through the 15 cycles of oscillation shown. This is not true of all integration algorithms as previously found with the BD and FD algorithms and it is not true with regard to some commonly used commercial code. A better indication of the accuracy of the RK

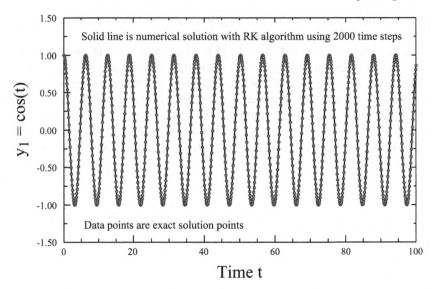

Figure 10.24. Numerical solution for cos(t) using RK algorithm. Data points are exact solution compared with numerical solution as given by the solid line.

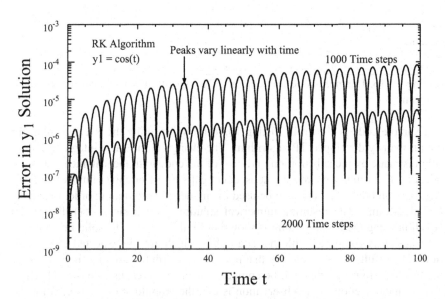

Figure 10.25. Error in solving differential equation for cos(t) solved by numerical integration using RK algorithm with different step sizes.

algorithm can be seen from Figure 10.25 showing only the difference between the numerical solution and the true solution of the differential equation set. The two error curves correspond to time steps or h values of 0.01 and 0.005 for the 1000 and 2000 time steps respectively. The accuracy shown in the figure is quite good being less that1.e-4 or 1.e-5 for the two curves. However it will be recalled that the RK algorithm makes 4 nsolv() evaluations so it is equivalent in terms of function evaluations to a TP algorithm of 4 times as many time points. When plotted on a linear vertical scale the peaks in the error vary linearly with time in the same manner as seen for the TP algorithm and as plotted in Figure 10.17. The code in Listing 10.16 evaluates the solution for 1000 and 2000 time steps and the two curves in Figure 10.25 show that a doubling of the number of time steps reduces the error by almost exactly a factor of 16. This is in keeping with the expected error being proportional to h^4. Since this is a fourth order RK technique, the error in a single step is proportional to h^5 but the error varies as the fourth power for accumulated errors over a fixed time interval. One can go back to the discussion in connection with Eqs. (10.42) through (10.44) and derive an expected error for the fourth order RK algorithm from two integrations with two step sizes varying by a factor of 2. The result will be the equation:

$$Error_{RK} = (y_1 - y_2)/(2^4 - 1) = (y_1 - y_2)/15 \tag{10.54}$$

In turn this should replace the result of Eq. (10.44) which has a factor of 3 in the denominator. Thus the RK error using two different step sizes should be a factor of 5 more accurate than the same expression for the TP algorithm. The odeerr() routine used to return the error values with the odeive() routine contains statements to determine if the RK basic integration routine is being used and to apply the appropriate factor to return the correct error estimate. The printed output values shown in Listing 10.16 give the RMS, mean and peak errors as estimated by the h-2h algorithm and it can be seen that the peak error values are 5.15e-6 and 8..25e-5 for the two step sizes respectively. The ratio of these values is almost exactly 16 as theory would predict. It can also be seen from Figure 10.25 that this ratio of error values agree closely with the maximum values actually achieved near the right boundary of the figure. Thus by knowing the first order dependence of the error term on step size an estimate the expected error from the RK algorithm can be obtained as easily as for the TP algorithm.

The Runge-Kutta algorithm provides an alternative integration technique to the trapezoidal algorithm. For some sets of differential equations, it provides improved accuracy over the TP algorithm. However, because it involves 4 calls to nsolv() for each time step as opposed to 1 call to nsolv() for the TP algorithm, the accuracy can not be compared on the basis of equal time steps. If it is assumed that most of the computational time will be taken in nsolv(), then one should compare the RK algorithm with the TP algorithm on the basis of taking a time step of 4 times as large for the RK algorithm as for the TP algorithm. Even with this difference, for many problems the RK algorithm will show improved accuracy. However, the RK algorithm with increased step size will in some cases not give valid results when the TP algorithm will. Also the RK algorithm is not as general as the TP algorithm. Many engineering problems involve sets of differential and algebraic equations where not every equation involves a derivative term or other types of equations where the equations may involve coupled derivatives of more than one variable. Such equation sets are not readily solvable by the RK algorithm but can be readily handled by the TP algorithm. Such sets of equations are considered in a subsequent section. The reader is encouraged to use the RK algorithm to solve the Van der Pol equation as shown in Listing 10.14 with varying step sizes and compare with the TP algorithm.

Several approaches have now been developed for the numerical solution of differential equations. From the examples of solving several rather simple differential equations, it is obvious that many equations of interest require solutions over values of the independent variable covering many orders of magnitude in value. For such problems step sizes of varying magnitude must be used over different intervals of the independent variable. Other equations such as the nonlinear Van der Pol equation exhibit solutions that change very rapidly over certain time intervals and very slowly over other time intervals. The step size used for such problems can be manually adjusted, but an algorithm that automatically adjusts the step size as the solution progresses would be highly desirable for such problems. This task is addressed in the next section.

10.8 An Adaptive Step Size Algorithm

The ability to automatically adjust the step size in the numerical integration is very convenient in solving many types of differential equations. One example is the previous stiff differential equation with exponential time constants differing by a factor of 1000. A small step size is needed for the fast transient and a much longer step size is needed to calculate in a reasonable time the response for the long time constant term. It was found that such cases can be easily handled with a step size variation approach such as the use of logarithmically spaced time steps. In many other cases it would be very advantageous to be able to adapt during the solution to a very rapid change in the solution and to transition from large step sizes to small step sizes as the solution progresses. This is especially the case for highly nonlinear problems such as shown in Figure 10.22 and for problems with a periodic solution such as the Van der Pol solution shown in Figure 10.23. Such cases are much more difficult to handle because it is not known beforehand exactly when in the solution small step sizes are needed. For such problems an adaptable step size algorithm can be of considerable value. This section will discuss some automatic step size adjustment algorithms and develop one function for solving differential equations implementing such an algorithm.

One approach to an adaptive step size is simply to first assume some best guess at a good step distribution and then calculate the solution and obtain an estimated error as has been done in the previous section. Based upon the resulting global error a new step size distribution could be developed and the calculation could be repeated, hopefully until a given accuracy had been achieved. This would probably require at least several iterations of the solution depending upon how good an initial guess was used as the first step distribution. However, such an approach is probably not a good general approach because a poor initial guess for the step distribution, may lead one astray as to where extra steps are needed in the solution. For example if the quick scan function is used as a first approximation, it will work well for solutions with exponential decays, but fails badly for periodic functions as previously demonstrated in Figures 10.13 and 10.14.

Most approaches to an adaptive step size algorithm works on a local basis to determine an appropriate step size and not on a global basis. This has the disadvantage that the accumulation of errors across a solution can not be accounted for but has the advantage that a reasonably good step distribution can be generated on a point-by-point basis as the solution steps along in time (or distance). This will be the approach taken here. A general approach to step size adjustment can now be outlined. Let's assume a good solution of a set of differential equations has been obtained up to some value of the independent variable with an appropriate step size already known at the solution point and a new step size is desired to advance the solution to the next time (or spatial) point.

One approach is to use the previous step size as a trial step size and generate a solution at the new time point based upon the TP (or RK) algorithm. In order to obtain an estimate of the local error in the solution, a second solution is also needed with a different step size. The previously verified equations for accuracy based upon step size can then be used to estimate the local error achieved with the two trial step sizes. If some established error criteria has been met, the solution so generated can then be accepted. If the estimated error exceeds some established criteria, the step size can be adjusted to a smaller value and another try attempted to generate an acceptable solution value at a new time point. Also if the estimated error is sufficiently below some established criteria, the step size can also be increased in the next increment along the solution. In this manner an integration algorithm can step along the solution curves adjusting the step size as the solution progresses. This technique is not unique here but has been well established in the literature. The limitation of this approach is that meeting some local error criteria does not necessarily mean that some desired global error criteria will also be met. The global errors that accumulate over a complete solution depend on how the local errors add and subtract in the solution. This difference will be demonstrated in some examples with the step adjustment algorithm developed here.

It is important to determine some good local criteria for measuring the accuracy of a solution on a local basis. Assuming that a good estimate of the local error from an h-2h step calculation can be obtained (this has been demonstrated with several previous examples), with what value should this estimated error be compared? One criterion would be to require that the magnitude of the local error always be below some value, such as 1.e-6. However this is very difficult to implement in a meaningful way for a general solution routine. As a solution is started (t = 0) no reliable idea can be known, in general, as to how large the solution variables will become in the course of the solution. In particular problems, a solution is sometimes known to die out with time and has its largest value at the

origin of the solution. In other problems the solution has some oscillatory properties with constant amplitude. In general it is not known if the solution variable
will be in the range of 1.e-20 or 1.e+20 or somewhere in between. The generated
solution may be such as that in Listing 10.13 and Figure 10.22 where the solution
increases without bound as a function of the independent variable. In fact for this
set of two equations, the initial values of both variables are zero so there is no initial value with which to compare any error as a solution is started. Another approach would be to always try to normalize the differential equations so that the
range of the solution variables is on the order of unity, but a general differential
equation solver should not have to make any such assumptions about the solution
and the user should not be required to supply such information.

The only other simple local error criterion that can be applied is to require that
the local "relative error" remain within some bound. The solution variables will
always have some local values and the important local criterion with respect to the
accuracy of a solution is how the error in the solution compares with the local values of the solution variables. But what about a problem such as that mentioned
above where the initial boundary condition on all variables is zero? It appears that
there are no finite solution values with which to compare the error. However, after having taken the first step in a solution the variables will have some finite values even if the values may be very small. Thus a comparison can be made of an
estimated error with the average of the solution variables over a possible time step.

To make this more concrete, consider the following notation for a single variable:

$$y_0 = y_n = \text{solution at time step } n$$

$$y_1 = y_{n+1}\,|_{\text{evaluated with 1 step of size h}} = \text{first estimate at } n+1 \qquad (10.55)$$

$$y_2 = y_{n+1}\,|_{\text{evaluated with 2 steps of size h/2}} = \text{second estimate at } n+1$$

An estimate of the error in the solution and the relative error can be obtained from
these values as (see Eq. (10.44)):

$$ERR = (y_1 - y_2)/3$$

$$ERR_{REL} = \frac{(y_1 - y_2)/3}{(|y_1| + |y_2|)/2} \approx \frac{(y_1 - y_2)}{(|y_1| + |y_2|)} \qquad (10.56)$$

Taking this relative error as proportional to the square of the step size, this can be
used with some desired relative error to obtain a new step size as

$$ERR_{REL} = \frac{(y_1 - y_2)}{(|y_1| + |y_2|)} = Kh_{old}^2$$

$$ERR_{MAX} = \text{Maximum desired relative error} = Kh_{new}^2 \qquad (10.57)$$

$$h_{new} = \left(\frac{ERR_{MAX}}{ERR_{REL}}\right)^{0.5} h_{old} = \left(\frac{ERR_{MAX}(|y_1| + |y_2|)}{(y_1 - y_2)}\right)^{0.5} h_{old}$$

In the above ERR_{MAX} is some desired maximum relative error in the solution (such as 1.e-5), h_{old} is the old step size used in evaluation the relative error and h_{new} is a new estimated step size needed to give a relative error equal to the desired maximum error of ERR_{MAX}. In the last equation it can be seen that if the estimated relative error is larger than the desired maximum error the new step size will be appropriately smaller than the old step size. In this manner the local step size can be adjusted as the solution progresses to achieve a desired relative error criterion. It should be noted that the power of 2 and 0.5 in the above equations is based upon the TP algorithm. For use with the RK algorithm the appropriate values would be 4 and 0.25 respectively since the error varies as the fourth power of the step size. For multiple variables with several differential equations, a relative error can be associated with each variable in the same form as the above equations. In such a case the criterion should use the largest calculated relative error among all the variables at each time step so that the relative error of all variables will satisfy some desired criterion.

Let's now summarize the major features of an adaptive step size algorithm that will be implemented in computer code. At each time step including the first time point, two trial values of the solution vector will be obtained, one using a single time step of size h and the other using two time steps of size h/2. This will move the solution along two time steps of size h/2 at each iteration loop. Based upon the two trial solution values an estimate of the relative error associated with each solution variable will be evaluated. The maximum relative error will be compared with a predetermined desired relative error value. If the maximum relative error exceeds the desired value, the step size will be reduced appropriately according to Eq. (10.57) and the process repeated until the obtained maximum relative error is below the desired value. The two solution points will then be accepted as valid solution points and the time increased by the final value h used in the calculation and the procedure repeated for a new advanced time point. The use of two time steps of equal size has the added advantage that one can come back if desired and use a global h-2h evaluation of the global errors to calculate more accurately the real error on a global basis for the obtained solution.

A similar approach can be used to increase the step size in regions where the functions are changing slowly and the estimated relative error in less than the desired specified value. If the relative error in Eq. (10.57) is less than the maximum specified value the new calculated step size will be larger than the old step size and in principle the step size can be increased. However, this will have to be careful considered and an algorithm should not be too impatient to increase the step size or a situation can potentially arise where the step size simply has to be reduced back in the next iterative step. Also a few steps in the solution with a smaller relative error than that specified will improve the overall error at the expense of a few more solution points. Because of this the step size will not be increased unless the estimated error is some specified fraction (default value of 0.2) of the desired relative error. Also the increase in step size will be taken as something less than a factor of 2. This is from experience with a number of nonlinear equations where it is observed that convergence difficulties are less likely to occur if the step size increases slowly. An increase in step size by a factor of 1.2 for example is somewhat arbitrary, but it allow for the step size to increase gradually and yet, if needed, for the step size to increase by a factor of 9000 in only 50 steps. Smaller step sizes are no problem with accuracy, but step sizes that are too large must be reduce immediately to meet the error criteria.

One prominent question is always how to get started with such an iterative algorithm. Once an appropriate step size is found that meets the relative error criteria, it would be expected that the step size would only need to change gradually. However, at the beginning of a solution an appropriate step size is not known. Somewhat related to the selection of an initial step size is the need for a minimum step size value. Since numerical values have some maximum precision (around 15 decimal digits) a step size that is too small will be lost in numerical precision for the larger values of the independent parameter. Thus a minimum step size must be selected as some fraction of the total range of the independent variable – perhaps 1.e-12 times the maximum range of the independent variable. For simplicity in the algorithm developed here, the initial step size will be selected at some larger fraction of the range of the independent variable – perhaps 1.e-6 times the range. Many differential equations need a small step size to start the solution as has been previously shown with the example of the differential equation with very different time constants. If a smaller initial step size is needed it will be achieved by the iterative algorithm. If such a small size is not needed, an adaptive step size algorithm can increase the size after a number of steps to a more appropriate value.

Code for implementing an adaptive step size algorithm is shown in Listing 10.17. The code function odeivs() makes use of the basic single step integration routines already developed and can be used with either the TP algorithm of odebiv() or the RK algorithm of odebrk(). The calling argument to this routine is similar to that of previous codes and is of the form: s, n = odeivs(feqs, tvals, u, up) where feqs is again the set of differential equations to be solved, u is the table of initial values for the variables and up can be ignored for now and in fact omitted in the calling argument. The tvals parameter specifies the time values in the form: tvals = {tinitial, {tfinal, hmin, hmax},errmax} or {tinitial, tfinal} where only the initial and final time values are required inputs. The other parameters are optional and include errmax, the desired goal for the maximum relative error as discussed above, hmax, a maximum value for step size and hmin, a minimum value for the step size. Default values (1.e-2*(tfinal – tinitial) and 1.e-12*(tfinal – tinitial)) of maximum and minimum step sizes are specified in the code on line 160 if omitted from the calling arguments.

The code in listing 10.17 is very similar in structure to the previously discussed multi step integration routines odeiv() and odeivqs(). Code between lines 149 and 164 define default parameters, handle the optional input forms and set initial values of various parameters. One notable feature is on line 154 where default parameters are changed if the RK routine odebrk() has been substituted for the TP routine odebiv(). The heart of the code is the major itterative loop from line 165

```
  1 : -- /* File odeiv.lua */
  2 : -- Programs to integrate first order differential equations
  3 : require"nsolv"; local atend
----------
148 : odeivs = function(feqs,tvals,u,up) -- Adaptive Step Size Solver
149 :   local ttvals,sa,upp,t,s1,s2 = {},{},{}
150 :   local k,ni,nit,err,fac,relerr,relmin,t1,t2,t3,h,
          h2,hmax,hmin,tmax
151 :   local neq,NMAX,abs = #u, getfenv(nsolv).NMAX, math.abs
152 :   local u1,u2,up1,up2,upp1,upp2 ={},{},{},{},{},{} -- arrays
153 :   local NTMIN,NTMAX,TFAC,HFAC,FAC,RELERR,fe,fd,tt,nhmax =
          1000,25000,1.e-6,1.e-12,.8,1.e-5,.5,1,1+1.e-12,0
154 :   if odebiv==odebrk then fe,fd = 0.25,0.2 end -- Set for RK
155 :   up = up or {} -- Below is for different tvals formats
156 :   if type(tvals)=='number' then tvals = {0,tvals} end
157 :   if #tvals==1 then tvals = {0,tvals[1]} end
158 :   if type(tvals[2])=='number' then tvals[2] = {tvals[2]} end
159 :   t,tmax = tvals[1],tvals[2][1]
160 :   hmin,hmax = tvals[2][2] or (tmax-t)*HFAC,tvals[2][3]
          or (tmax-t)/NTMIN
161 :   relerr = tvals[3] or RELERR*neq; relmin = relerr/5
162 :   nit,k,h = 0,1,(tmax-t)*TFAC; h2 = h/2 -- Use TFAC initially
163 :   for i=1,neq+1 do sa[i] = {} end; sa[1][1] = t -- Set initial
164 :   for i=1,neq do t1,t2 = u[i],up[i];
          sa[i+1][1],u1[i],u2[i],up1[i],up2[i] = t1,t1,t1,t2,t2 end
165 :   while 1 do -- Major time step loop
166 :     while 1 do -- Adjust step size until relative error met
167 :         ttvals[1],ttvals[2],ttvals[3] = t,t+h,1
168 :         s1,nx = odebiv(feqs,ttvals,u1,up1,upp1) -- One step
169 :         ttvals[3] = 2; s2,ni = odebiv(feqs,ttvals,u2,up2,upp2)
170 :         err = 0 -- Evaluate maximum relative error
171 :         for i=1,neq do
172 :           fac =  fd*abs(u2[i]-u1[i])/(abs(u1[i]) + abs(u2[i]))
173 :           if fac>err then err = fac end
174 :         end
175 :         if h==hmin then break end -- Just accept,
176 :         if err<relerr then break end -- Accept error met
177 :         if nx==NMAX then -- Didn't converge try half step size
178 :             if h==hmax then hmax = hmax/2 end
179 :             h,h2 = h/2,h2/2
180 :         elseif err==1 then h,h2 = h/2,h2/2 -- Try half step
181 :         else h = (relerr/err)^fe*FAC*h; h2 = h/2 end --
182 :         if abs(h)<abs(hmin) then h,h2 = hmin,hmin/2 end
183 :         for i=1,neq do t1,t2,t3 = u[i],up[i],upp[i]
184 :             u1[i],u2[i],up1[i],up2[i],upp1[i],upp2[i] =
                  t1,t1,t2,t2,t3,t3 end
185 :     end -- loop back if relerr criteria not met
186 :     if ni==NMAX and err>relerr then -- Print warning message
187 :         print("Error at t =" ,t," : Maximum number of
                iterations exceeded in nsolv")
188 :         print("     Results are probably not accurate!")
189 :     end
190 :     if ni>nit then nit = ni end
191 :     for i=1,2 do -- Save 2 time points
192 :         k,t = k+1,t+h2; sa[1][k] = t; for j=2,neq+1 do
                sa[j][k] = s2[j][i+1] end
193 :     end
194 :     if k>NTMAX then -- Limit solution to NTMAX data points
195 :         print("Number of adaptive data points exceeds ",NTMAX)
```

```
196 :            print("       Best effort at solution returned!"); break
197 :       end
198 :       for i=1,neq do t1,t2,t3 = u2[i],up2[i],upp2[i];
 u[i],u1[i],up[i],up1[i],upp[i],upp1[i] = t1,t1,t2,t2,t3,t3 end
199 :       if h>0 then if t>=tmax then break end -- Exit if finished
200 :       elseif t<=tmax then break end -- Negative time step
201 :       if err<relmin then h = h*1.4 end -- Adjust factor of 1.4
202 :       if abs(h)>abs(hmax) then
203 :           h = hmax; nhmax = nhmax+1
204 :           if nhmax>10 then nhmax,hmax = 0,hmax*1.4 end
205 :       else nhmax = 0 end
206 :       if h>0 then if t+h+h>tmax then h = tmax*tt - t   end
207 :       elseif t+h+h<tmax then h = tmax*tt - t end
208 :       h2 = h/2 -- Finalize h and h/2
209 :   end -- loop back for next time step
210 :   sa[1][#sa[1]] = tmax; return sa,nit -- Set limit to tmax
211 : end
```

Listing 10.17. Code segment for odeivs() an adaptive step size integration algorithm.

through 209. A sub loop from line 166 to line 185 handles the task of downward step size adjustments. The two trial solutions are obtained using odebiv() calls with one step of size h (on line 168) and two steps of size h/2 (on line 169). The only tricky part of the coding is that temporary variables must be used to hold the solution values before the calls because the functions change the values of the solution values in the calling arguments − thus the use of u1 and u2 variables (similarly for the derivative values). The maximum relative error is evaluated on lines 171 through 174 and used on line 181 to adjust the new step size. The only difference here is that the theoretical value for step adjustment of Eq. (10.57) on line 181 is multiplied by a FAC term with a default value of 0.8. This is so the reduced step size will be sufficiently small that hopefully a reduced step size does not have to occur every time through the loop. If the value set by Eq. (10.57) is used the error limit would just barely be met and the next iteration might put the next value slightly over the limit. Again smaller step sizes are good for the overall error.

Three conditions are checked for possible step size reductions. First on line 177 the maximum number of Newton iterations in nsolv() is checked and if at the limit set in nsolv(), the step size is reduced by a factor of 2. Next if the obtained relative error is 1, the step size is reduced by a factor of 2 on line 180. This is perhaps a surprising test. However, nsolv() can return with some variable set at the maximum numerical limit due to convergence not being achieved in nsolv(). When nsolv() terminates with such a maximum value, the error calculation on line 172 gives a unity value. Finally the desired relative error limit is used on line 181 to reduce the step size. When the step size is adjusted downward, the initial parameters to the odebiv() function must be reset to initial values and this is accomplished on lines 183 and 184.

The downward step size adjustment loop from lines 166 to 185 exits under two possible conditions − either the relative error meets the desired specification (see line 176) or the step size is at the specified minimum value (see line 175). After obtaining an acceptable step size, the two solution values are saved in arrays on lines 191 through 193 using the last obtained odebiv() solution for the two steps of h/2 which is the most accurate solution. Line 199 through 200 exits the time loop when finished. Two tests are made on time depending on whether the time variable is increasing (h>0) of decreasing (h<0). Finally, line 201 increases the step size by a factor of 1.2 if the estimated relative error is below a specified value of relmin which is set on line 161 to 0.2 times the relative error limit. In principle the step size could be doubled when the relative error was 0.25 times the relative error limit assuming a perfect dependence of error on the second power of h. The use of the 0.2 factor on the relative error and the 1.2 factor on h is again to more gradually increase the step size and to better ensure that increasing the step size will not push the next evaluation into needing to reduce the step size. If an increase in step size is pushed too aggressively, it can be expected that many step size increases would then have to be followed by a decrease in the step size. One other increase is implemented on lines 202 through 205. If 10 iterative steps are taken in a row at the maximum step size, then the maximum allowed step size is increased by a factor of 1.4 on line 204. This is included so that the algorithm can continually increase the allowed step size if a solution is found which becomes essentially constant as time progresses. Finally the test on the number of solution points on lines 194 through 197 prevents an infinite time loop where the relative error criteria can not be met in an acceptable number of time steps. Solutions that go off to infinity such as the example of Figure 10.22 can cause such problems.

One feature to note in the code is on line 153 where a default local error limit is set to 1.e-5. This is adjusted on line 161 to account for the number of equations being solved. For example if there are 10 simultaneous equations, the local error is adjusted upward to 1.e-4. This is a compromise between accuracy and computational time. As the number of equations increases the computational time increases roughly as the square of the number of coupled equations. By reducing the required local accuracy, some of this increase in time can be offset with a reduced accuracy. Also with the error criteria searching for the maximum relative error among all the solutions, the algorithm naturally tends to require more solutions points as the number of equations increases. Thus it is naturally to adjust the required local error upward as the number of equations increases. In all cases the user can specify a required error by inputting the desired value into the tvals[] input array.

The adaptive step selection code in Listing 10.17 has of necessity several constants that have been selected for the adaptive step size algorithm. The majority of these are defined on line 153 of the code. For most of these the resulting algorithm will work satisfactorily over some range of values for these parameters. The present values have been selected based upon considerable exercise of the code for a range of differential equations. However, these parameters are perhaps not optimal for many differential equations. In using this function, the reader is encouraged to experiment with a range of these values to observe the effects on the accuracy of solutions and on the number of solution points generated by the code.

As a final discussion point, it can be argued that Eq. (10.57) does not have the correct dependency of the local error on the local step size, but from Eq. (10.33) the dependency should be on the third power of the step size. If this is taken to be the relationship, then the exponent in Eq. (10.57) would be 1/3 instead of 1/2. This can certainly be implemented if desired into the code for Listing 10.17. The reader would perhaps like to try this modification. However, this will make very little difference in any of the example solutions to be subsequently shown in this chapter. The change will result in some minor differences in the total number of solution points and in the maximum achieved error. However, the algorithmic approach of Listing 10.17 is not very sensitive to the details of how the step sizes are reduced as long as a reasonable reduction is made when the evaluated relative error exceeds the desired relative error. If the step size reduction is more than adequate, it simply adds a few extra calculated points, if it is not adequate, the size will be further reduced in an added iterative step.

A second function called odeivse() with the same calling arguments is also provided in the odeiv package of programs. The reader can view the code if desired. This routine is essentially the odeivs() program plus a second call to odeiv() for using the h-2h technique for estimating the global errors in a numerical integration solution. Thus this routine is called as follows:

```
s, err, n = odeivse(feqs, tvals, u, up)
```

where the extra returned array of values, err, is the estimated global error associated with the adaptive solution obtained by the h-2h technique. The evaluation of the associated error is facilitated by the use of the fact that the solution values occur in pairs of equal time steps within the odeivs() routine. Because this odeivse() function requires an additional call to the odeiv() routine, this function will take somewhat longer to execute, but provides extremely valuable information on the accuracy of a numerical solution. Most of the examples to follow use this odeivse() function so the accuracy of the numerical solutions can be accessed.

Listing 10.18 illustrates the use of the adaptive step size integration algorithm for solving three systems of differential equations. The selected equations are those previously discussed: (a) a stiff differential equation with exponential terms differing by a factor of 1000 (the f1 function), (b) two equations with pure sinusoidal solutions (the f2 function) and (3) the Van der Pol nonlinear equation set (the f3 function). These provide a good test of the algorithm because of the wide time range to be covered in the first set, the accumulated error that occurs with a sinusoidal solution and the very abrupt changes of the nonlinear equation. The code is essentially identical to previous listings where these equations have been solved by different calling functions. The solutions along with an error estimation is made for the three equation sets by a call to odeivse() on lines 19, 21 and 23. A similar call to odeivs() with the same arguments would produce the same solutions but without the h-2h error estimates of the shown solutions. Summaries of the estimated errors are printed with calls to the errstat() function on lines 20, 22 and 24 and the results are shown in the printed output. The four columns of printed output give the RMS error, the maximum error, the point in the array where the maximum error occurs and the average error. These values are from the h-2h algorithm and of course represent estimated errors for the solutions. The maximum error values will be discussed in connection with some plots of the results.

```
 1 : -- /* File list10_18.lua */
 2 : -- Programs for diff. equation using adaptive step algorithm
 3 : require"odeiv"
 4 :
 5 : mu = 20
 6 : f1 = function(eqs,t,y,yp) -- Stiff differential equation
 7 :     eqs[1] = yp[1] + 1001*y[1] - 999*y[2]
 8 :     eqs[2] = yp[2] - 999*y[1] + 1001*y[2]
 9 : end
10 : f2 = function(eqs,t,y,yp) -- Sinusoidal differential equation
11 :     eqs[1] = yp[1] -y[2]
12 :     eqs[2] = yp[2] +y[1]
13 : end
14 : f3 = function(eqs,t,y,yp) -- Van der Pol equation
15 :     eqs[1] = yp[1] - y[2]
16 :     eqs[2] = yp[2] - mu*(1 - y[1]^2)*y[2] + y[1]
17 : end
18 : -- Now solve three equations
19 : s1,err1 = odeivse(f1,{0,10},{0,2})
20 : print(errstat(err1[2]));print(errstat(err1[3]))
21 : s2,err2 = odeivse(f2,{0,100},{2,0})
22 : print(errstat(err2[2]));print(errstat(err2[3]))
23 : s3,err3 = odeivse(f3,{0,100},{1,0})
24 : print(errstat(err3[2]));print(errstat(err3[3]))
25 : write_data("list10_18.dat",s1,err1,s2,err2,s3,err3)
Output:
1.9873418571013e-006  7.9283881346074e-006  92   2.1375902061827e-007
1.7782619844538e-006  7.9283882562879e-006  92   -9.1222105798481e-007
0.00046318254873528   0.0011183075852802    5202 5.0045451298107e-006
0.00047455218887791   0.001137410276563     5282 -8.1517253588828e-006
0.0059478323311414    0.033335071261276     3808 0.00045378119610977
0.1196459444609       0.71699473556522      3838 -0.00054189417845999
```

Listing 10.18. Example of use of adaptive step size algorithm for solving three differential equations.

Figure 10.26 shows the two solution values for the first differential equation (the stiff differential equation) as the two upper solid curves and the returned error estimates as the lower solid curve in the figure. The h-2h error estimates are essentially identical for the two solutions. It can be seen that the estimated errors are in the range of 1.e-5 and the first two printed lines of output indicate RMS errors of about 2.e-6 and peak errors of about 0.8e-5. It is readily seen that these agree with the graph of the errors. Also plotted in the figure is the actual error achieved by the adaptive integration algorithm as the dashed line. This curve is in excellent agreement with the h-2h estimation over the entire time range and in fact is so close to the h-2h estimate that the dotted curve can barely be seen in the figure. It will be recalled that the adaptive step size algorithm attempts to maintain a certain relative error (below 1.e-5). The dotted curve shows the relative error achieved for the y_2 solution. This agrees well with the absolute error except in the long time regime where the relative error increases to about 3.e-4 at a time of 10 where the actual solution value is about 2.e-9. The desired relative error is seen to be achieved, except for the long time values. This illustrates the difficulty of any numerical integration algorithm in obtaining a small relative error when the magnitude of the solution becomes very small. Nevertheless these values are quite good as will be seen in the next section where a comparison is made to commercial code as represented by MATLAB routines. The odeivse() adaptive routine actually used only 917 solution points along the curve in trying to maintain the local relative error at the default limit of 1.e-5 as set in the routine. The data points in the figure show the error for every 20th actual solution points. These points show relatively few solution points in regions where the solution is slowly changing and much closer spaced time points in regions where the relative error is the largest. It is thus seen that the adaptive routine was able to maintain a relative error below the desired value over much of the solution range, but the relative error tended to grow in the long time, but even in this range the absolute error decreased very appropriately.

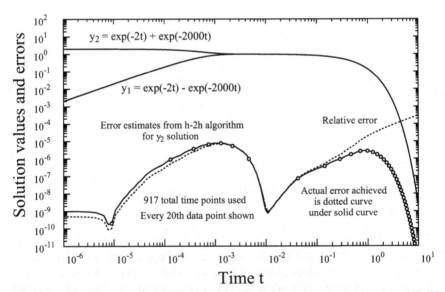

Figure 10.26. Comparison of solution errors and solution values for the stiff differential equation example using adaptive step size algorithm. Solution from Listing 10.18.

A final view of the adaptive step algorithm can be seen in Figure 10.27 which shows the accumulation time step distribution for this numerical solution. Plotted on a log time scale it is seen that the adaptive step algorithm generated about 100 time points in the region between about 1.e-4 and 4.e-3, but generated relative few time points in the region around 1.e-2 where the functions are changing slowly.

This is after the fast time constant has died out and before the slow time constant comes into play. It is also seen that about 3/4 of the total number of time steps were generated for the region between 1 and 10 where the solutions are decaying rapidly. This is the region noted above where the adaptive step size algorithm was having difficulty in maintaining a relative error in the range of 1.e-5. This example illustrates the ability of such an adaptive step size algorithm to adjust to a rapidly changing solution and a slowly changing solution as the solution progresses in time.

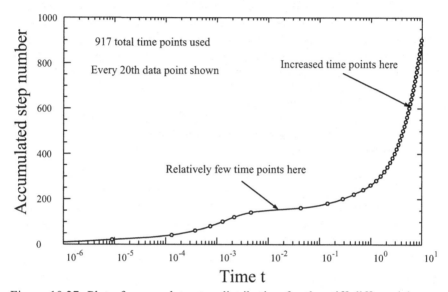

Figure 10.27. Plot of accumulate step distribution for the stiff differential equation, solved by odeivse() in Listing 10.18.

Detailed results from the solution of the second set of equations – the sinusoidal equations – in Listing 10.18 will not be shown. The reader can plot the results if desired. The printed peak errors for this example shown on the third and fourth lines of output are about 1.1e-3. This might be larger than one would first expect since the local relative error criteria used in the adaptive step algorithm is 1.e-5. However, as previously noted in connection with Figures 10.17 and 10.21 errors tend to accumulate for periodic solutions and vary linearly with time as a solution progresses. The total time of the simulation is 100 corresponds to about 16 cycles of a sinusoid and the magnitude of the error is consistent with the results previously shown in Figures 10.17 and 10.21. For the sinusoidal solution the adaptive step algorithm generates an almost uniform distribution of step sizes as might be expected and the total number of time steps generated was 5333 time points or about 333 points per cycle of the sinusoidal signal. A few hundred points per cycle of a sinusoidal wave is usually sufficient for a good numerical representation.

More interesting is the ability of the adaptive step algorithm to handle the nonlinearities of the Van de Pol equation as generated by the f3() function in Listing 10.18 and the third set of numerical calculations. Results for the Van der Pol equation are shown in Figure 10.28. Only the solution variable is shown and not the derivative solution. The data points are plots of every 20th data point since plotting ever data point makes it impossible to see any distribution of points. As can be clearly seen, the adaptive step technique has placed more data points in regions where the function is changing rapidly and fewer data points in regions where the function changes slowly. The data points also show that the algorithm is able to rapidly adjust to the changing conditions of the solution. The dotted curve along the zero axis is the estimated error in the solution from the h-2h estimation technique of the odeivse() code. The maximum error is about 0.03 as can be seen from the next to last line in the output line in Listing 10.18. This is essentially zero on the scale of the plot in Figure 10.28. Since an exact solution for such a nonlinear equation can not be generated, this is the only estimate of the error in the solution that can be obtained. Without this estimate, one would be guessing at the accuracy of the solution. By now, however, considerable confidence has hopefully been generated in the h-2h error estimation technique with the TP integration algorithm so one can be confident in the accuracy of the solution and the estimated solution error.

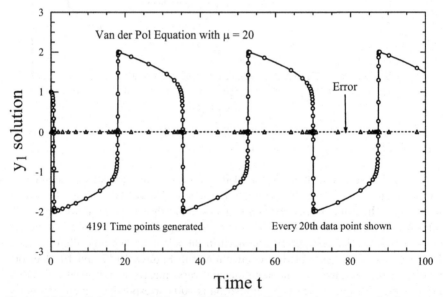

Figure 10.28. Solution variable for the Van der Pol equation using adaptive step generation algorithm.

It is also interesting to look back at the solution of this differential equation obtained in Listing 10.14 and the results shown there in Figure 10.23. The solution there was obtained with a uniform distribution of 8000 time points. The solution error shown there is much larger than that obtained in this calculation even though almost twice as many solution time points were used. This again emphasizes the advantage of an adaptive step distribution algorithm to place solution points at critical regions in the solution.

Finally Figure 10.29 shows the accumulated step distribution from the ode-iose() code for this present solution. It can be seen that each pulse in the solution generates essentially the same distribution of time steps but within each pulse the number of time points automatically adjusts to put more points in the rapidly changing regions of the solution. As the solution begins to rapidly change, the algorithm rapidly adds additional time points and when the solution begins to slowly change, the algorithm rapidly decreases the number of solution points.

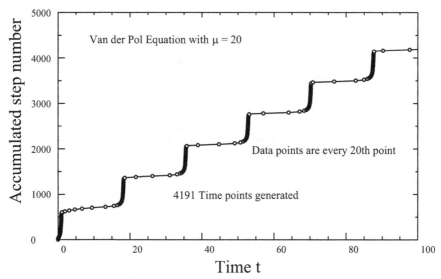

Figure 10.29. Distribution of time steps generated by adaptive step algorithm for Van der Pol example of Listing 10.18.

With this example, the maximum error of about 0.033 might be larger than one would expect from the desired local error criteria of 1.e-5 set in Listing 10.17. However as with the periodic sinusoidal solution, the error in this example accumulates with time. Although the error is too small to be seen in Figure 10.28, if one plots just the error, it can be readily seen that the maximum error occurs at the abrupt transitions in the function and these maximum values tend to increase linearly with the total time of the numerical integration. This type of error accumulation occurs not only with the integration routines in this work but also occur with commercial numerical integration packages such as those with the commercial package of MATLAB. A set of general purpose numerical integration routines have now been developed and this is probably a very good time to compare with other available options. This will be done in the next section by comparing with MATLAB integration code routines.

10.9 Comparison with MATLAB Differential Equation Routines

So far in this work numerical integration approaches have been developed based primarily on the TP algorithm and the fourth order RK algorithm. These algorithms have been used to explore the solution of some simple differential equations for which known solutions exist. Considerable attention has been paid to the numerical accuracy of the resulting algorithms and this can only be done with differential equations with known solutions. Compared with other text on numerical methods for engineering, considerably more attention has been paid to the numerical accuracy of the algorithms than is typically present in texts. This is in keeping with the general theme of this text where considerable attention has been paid to accuracy of numerical differentiation, integration and parameter estimation. Without good measures of numerical accuracy, one is just guessing that numerical results have any valid physical meaning. The old adage that garbage-in produces garbage-out is certainly valid in any numerical algorithm. At this point, it appears appropriate to compare the algorithms developed here with other commercially available code for the solution of differential equations. MATLAB is one such package with a large following in the engineering and scientific community so this seems like a logical comparison basis.

   MATLAB has a variety of numerical routines for solving differential equations. A listing of the code names and a very brief description of some of these is given below:

> ode15s() – Solve stiff differential equations with variable order method.
> ode23() – Solve non-stiff differential equations with low order method.
> ode23s() – Solve stiff differential equations with low order method.
> ode23t() – Solve moderately stiff ODEs with trapezoidal rule.
> ode23tb() – Solve stiff differential equations with low order method.
> ode45() – Solve non-stiff differential equations with medium order method.
> ode113() – Solve non-stiff differential equations with variable order method.

The ode23() functions use second and third order Runge-Kutta algorithms with adaptive step size algorithms known as Runge-Kutta-Fehlberg methods. The ode45() function uses forth and fifth order RK techniques with adaptive step size algorithms. The adaptive step size algorithms attempt to obtain an absolute error of 1.e-6 and a relative error of 1.e-3 according to the MATLAB documentation. The simplest way to call the routines is without a step size specification and let the internal algorithms compute a step size and step distribution that in general will not be a uniform step distribution.

   The differential equation set for cos(t) has proven valuable in understanding numerical errors in code developed here, so Listing 10.19 shows example

MATLAB code and present Lua code for solving the same differential equation set. For MATLAB the numerical integration function ode15s() is shown, although any of the above functions can be used with simply a different function name.

MATLAB Code	Present Code

```function f = dfuc(t,y)```
``` f = zeros(2,1);```
``` f(1) = y(2);```
``` f(2) = -y(1);```
```return```
```y0 = [0 1];```
```[t,y]=ode15s('dfuc',[0 100],```
```y0);```
```plot(t, y(:,1))```
```save ode15s.dat -ASCII -TABS y```
```x```
	```require"odeiv"```
	```f = function(eqs, t, y, yp)```
	``` eqs[1] = yp[1] + y[2]```
	``` eqs[2] = yp[2] - y[1]```
	```end```
	```y0 = {0, 1}```
	```sol = odeivs(f, {0, 100}, y0)```
	
	```plot(sol[1], sol[2])```
	```write_data("odeiv.dat", sol)```

Listing 10.19 Code example for MATLAB numerical solution of DE for y = cos(t). A comparison with present coding style is also shown.

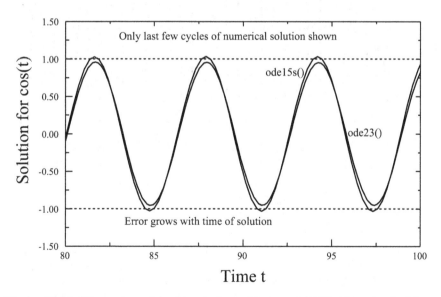

Figure 10.30. Illustration of cos(t) solution with some MATLAB numerical integration routines.

It is seen that there is a one-to-one correspondence between the problem formulated in the present code or in MATLAB code. The solution interval is specified as 0 to 100 with no step size specification. Figure 10.30 shows some of the re-

sults of the MATLAB solutions for the sinusoidal solution vs. time. Only the solution for the ode15s() and the ode23() functions are shown as these give the largest deviations from the exact solution in absolute magnitude. Only the last few
cycles of the solution from t = 80 to 100 are shown. The ode15s() solution shows a
tendency to increase in amplitude as the number of cycles increases, while the
ode23() function shows a tendency to decrease in amplitude as the number of cycles increases. For all the other MATLABfunctions, the amplitude appears to remain closely at 1.0, the expected exact amplitude value. The causes of amplitude
growth or decay have been previously discussed with respect to the backwards and
forward differencing techniques. It is thus seen that the ode15s() and ode23() algorithms appear to have problems with maintaining a constant amplitude when
solving a differential equations set with sinusoidal solutions. This might lead one
to be somewhat skeptical of using these functions for highly accuracy solutions.

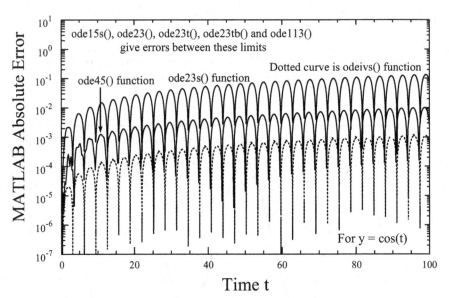

Figure 10.31. Absolute error in MATLAB numerical solutions for the y = cos(t)
solution.

A somewhat better comparison of the MATLAB numerical integration algorithms can be obtained by taking the MATLAB solutions and comparing with the
exact solution of the simple differential equations or comparing to the cos(t) function. This comparison in terms of absolute error in the solution is shown in Figure
10.31. Only results for the ode23s() and ode45() functions are shown as these represent the worst and best obtained errors. All the other MATLAB functions gave
errors within the order of magnitude range of the results shown in Figure 10.31 for
these two functions. Also shown in the figure is the error obtained from solving
the same problem with the presently developed odeivs() function discussed in the

previous section. The dotted curve shows the results generated by Listing 10.18. It can be seen that the errors from all the numerical integration techniques have similar behavior. Basically the peak errors of the curves in the figure grow linearly with time. There is no magic with the MATLAB functions with regard to the basic growth of numerical integration errors with time. The ode45() error using a fourth and fifth order numerical algorithm is more accurate than ode23s() using a second and third order numerical algorithm as expected. However, the increased accuracy of about a factor of 10 is not as large as might have been expected.

The data shows that the odeivs() adaptive step size and TP algorithm developed here performed somewhat better than all the MATLAB routines. However, it must be admitted that this is with the default local error criteria for both sets of routines. For the MATLAB routines, the default value is set at 1.e-3 while for the odeivs() routine a default value of 1.e-6 times the number of equations has been used. However, if the data for odeivs() is executed with an error limit of 1.e-3 an error of a little better than the ode23s() data in Figure 10.31 will be obtained. Thus it can be concluded that the odeivs() numerical integration code is very comparable to the MATLAB functions, and for the default error tolerance will give more accurate results than the default MATLAB routines. It can also be readily seen that the actually achieved error for all the routines after about 16 cycles of the sinusoids is considerably larger than the local error goals. This just verifies once again the fact that global errors are quite different from local errors achieved in a single step of integration and that global errors tend to accumulate. One must be very careful and not assume that a local error objective of 1.e-3 or 1.e-6 means that a numerical solution of that accuracy after many integration steps will be achieved.

The stiff differential equation set of Listing 10.8 and with the solution shown in Figure 10.11 provides a second good test for the MATLAB functions. The two solutions are $y_1, y_2 = \exp(-2t) \pm \exp(-2000t)$ which have exponential terms with time constants differing by a factor of 1000. This tests the ability of an integration algorithm to handle widely differing time constants. The MATLAB code for solving this set of differential equations will not be presented here; only the results will be shown. All the MATLAB functions are able to get a reasonable solution for the differential equation. However, the interest here is primarily in the error achieved by the solutions. Figure 10.32 shows the error actually achieved in the solution for the range of time from 0 to 1 with five of the MATLAB functions. Even though some of the integration functions are supposedly better suited for stiff differential equations, the actually achieved error in the early part of the solution, for t < 1.e-3, is essentially the same for all five of the functions with the ode15s() function being a little better than the other functions. Also the errors achieved at the largest values of t, around 1.0 are similar. The largest difference in the algorithms occurs in the region around 1.e-2 where the solution is making a transition between the two exponential terms. In this region the ode23tb() function which uses low order integration methods giving the best accuracy. Interestingly, the ode15s() function which is advertised as especially suited for stiff

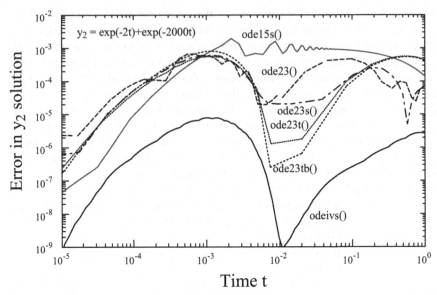

Figure 10.32. Error achieved by MATLAB integration functions with a stiff differential equation example.

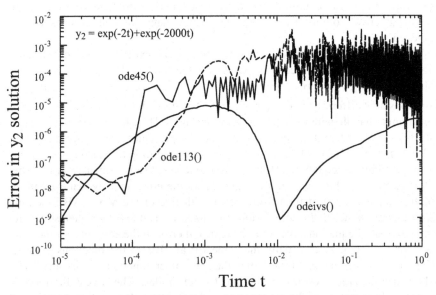

Figure 10.33. Error achieved by additional MATLAB integration functions with a stiff differential equation example.

differential equations using a variable order method gives the poorest accuracy in the transition region.

The errors achieved with the remaining two MATLAB functions are shown in Figure 10.33. These two functions are recommended for use with non-stiff differential equations and the results indicate that they do not provide a good solution for this example. However, the actual error achieved is not very different from the functions shown in Figure 10.32 which are supposed to be better able to handle stiff differential equations. Shown in each figure is the error achieved by the adaptive step size coupled with the TP algorithm function odeivs() developed in this work. It is seen that for the default parameters, the achieved error of odeivs() is better than that achieved by any of the MATLAB functions.

Based upon this look at the errors achieved by the MATLAB functions for these two test cases, considerable confidence can be gained in the differential equation solvers developed in this chapter. The code has been developed with the intent of easily estimating the accuracy of a solution using the h-2h technique. Because of the internal routines used in the MATLAB routines, it is not possible to readily evaluate the accuracy of the MATLAB codes for general nonlinear differential equations. From the comparison in this section it can be expected that the codes developed in this work are comparable in accuracy to the MATLAB integration routines and with care in the selection of step distributions or with the use of the automatic adaptive selection algorithm to be more accurate than the MATLAB routines with default parameters.

10.10 Direct Solution of Second Order Differential Equations

Second order differential equations occur very frequently in practical engineering type problems, for example in f = ma type problems. Such second order equations can certainly be converted into two first order differential equations which have only been so far considered in this chapter. This applies to either linear or nonlinear equations. However, another approach is to formulate the basic numerical integration algorithm to directly handle second order differential equations. Why do this as opposed to converting to two second order equations? Well, there are several valid reasons. First the formulation of the equations is simpler and less prone to error since one can just write an equation set in the natural format given. Second, the extra numerical effort in extending the TP algorithm to second order equations is very small. Finally, the numerical solution can actually be obtained faster since the order of the resulting matrix equation set to be solved by nsolv() is reduced by one. This is perhaps the most important reason for reformulation the equations in terms of a second order differential equation. For this development then consider a set of equations formulated as:

$$eq1 = F_1(t, y_1, y_2, \ldots, y_1', y_2', \ldots, y_1'' y_2'', \ldots) = 0$$

$$eq2 = F_2(t, y_1, y_2, \ldots, y_1', y_2', \ldots, y_1'' y_2'', \ldots) = 0$$

$$\ldots \qquad \qquad \ldots$$

$$eqk = F_k(t, y_1, y_2, \ldots, y_1', y_2', \ldots, y_1'' y_2'', \ldots) = 0$$

(10.58)

This is a generalization of Eq. (10.23) to include second order derivatives. To use an approach similar to that developed so far, a means is needed for replacing not only the first derivative but the second derivative in these equations by some function of the solution values. Using the trapezoidal rule for both the first and second derivative terms a similar procedure as used for Eq. (10.5) can be used to generate the averaged derivative equations:

$$\frac{y_{n+1}' + y_n'}{2} = \frac{y_{n+1} - y_n}{h}$$

$$\frac{y_{n+1}'' + y_n''}{2} = \frac{y_{n+1}' - y_n'}{h} = \frac{(2/h)(y_{n+1} - y_n) - 2y_n'}{h}$$

(10.59)

For the TP algorithm one needs to replace the first and second derivative terms by quantities already known at solution point n and by the solution value at point $n+1$. This can be doe if these equations are written in the form:

$$y_{n+1}' = (y_{n+1} - yn)/(h/2)$$

$$y_{n+1}'' = (y_{n+1} - ynn)/(h/2)^2$$

where

(10.60)

$$yn = y_n + (h/2)y_n' = y_n + h_x y_n'$$

$$ynn = y_n + (h)y_n' + (h/2)^2 y_n'' = y_n + h_y y_n' + h_z y_n''$$

where the terms yn and ynn are functions of quantities involving only values at the point n in the solution. To solve the system of equations as represented by Eq. (10.58) with the TP algorithm then simply requires replacing the first and second derivative by the expressions given in Eq. (10.60). This results in a coupled set of equations in terms of the solution values at point $n+1$ and in terms of the solution values and derivatives (both first and second) at the previous solution point. For a set of first order equations, the procedure only needs to keep up with the solution value and first derivative while for a set of second order equations the procedure needs to keep up with the second derivative in addition as can by seen by the ynn function in Eq. (10.60). Basically in using this replacement of the second derivative in terms of the variable value, one is doing in algebraic form exactly what the computer code does in numerical form when a second order differential equation is replaced by two first order differential equations.

Listing 10.19 shows code for implementing the TP algorithm directly for systems of second order differential equations. This is very similar to the TP code in Listing 10.4 for integration of systems of first order differential equations. The major difference occurs in setting up the call to the user defined set of equations on lines 220 through 226 where replacements are generated for both the first and second derivatives according to Eq. (10.60) The call of the user supplied functions

```lua
  1 : -- /* File odeiv.lua */
  2 : -- Programs to integrate first order differential equations
  3 : require"nsolv"; local atend
213 : odeb12 = function(feqs,tvals,u,up,upp)
214 :   local j, neq, t, h, h2,h2sq,hs,hx,hy,hz -- Local variables
215 :   local sol,ynn,yn,jfirst = {},{},{},0
216 :   local nit,nitt = 0,0 -- Number of iterations
217 :   local neq = #u
218 :   local tmin,tmax,ntval = tvals[1],tvals[2],tvals[3]
219 :   -- Function to calculate next time values using nsolv()
220 :   local fnext = function(eqs,u)
221 :     for m=1,neq do
222 :         up[m] = (u[m] - yn[m])/h2 --h/2 for TP rule, h for BD
223 :         upp[m] = (u[m] - ynn[m])/h2sq -- (h/2)^2 or, (h)^2
224 :     end
225 :     feqs(eqs,t,u,up,upp) -- Now call user defined equations
226 :   end
227 :   -- Use input value for #intervals or set at default value
228 :   up,upp = up or {}, upp or {}
229 :   for m=1,neq+1 do -- Array for solution values with t first
230 :     sol[m] = {}
231 :     if m==1 then sol[m][1] = tmin else sol[m][1] = u[m-1]  end
232 :   end
233 :   t = tmin -- Initial t value
234 :   hs = (tmax - t)/ntval -- Equal increments in t used
235 :   -- If initial deriv not specified, use BD for first 4 points
236 :   if #up~=neq then for m=1,neq do up[m] = up[m] or 0 end end
237 :   if #upp~=neq then for m=1,neq do upp[m] = 0 end
238 :     jfirst,h = 0,0.25*hs; h2,h2sq,hy,hx,hz = h,h*h,h,0,0
239 :   else jfirst,h = 4,hs; h2,h2sq,hy,hx,hz =
          hs/2,h*h/4,h,h/2,h*h/4 end
240 :   for k=1,ntval do -- Main loop for independent variable
241 :     repeat -- Use BD for first with 4 sub intervals of h/4
242 :         jfirst = jfirst+1
243 :         -- Set up yn, and ynn arrays to solve equations
244 :         for m=1,neq do
245 :             yn[m] = u[m] + hx*up[m] -- hx = 0 or h/2
246 :             ynn[m] = u[m] + hy*up[m] + hz*upp[m] --
247 :             u[m] = u[m] + h*(up[m] + 0.5*h*upp[m]) --Predicted
248 :         end
249 :         t = t + h -- Now increment t to next t value
250 :         -- Calculate new u values returned in u array
251 :         nitt = nsolv(fnext,u,step)
252 :         if nitt>nit then nit = nitt end -- Monitor maximun #
253 :         -- New derivative values, same function as in fnext
254 :         for m=1,neq do up[m],upp[m] =
                (u[m] - yn[m])/h2, (u[m] - ynn[m])/h2sq end
255 :     until jfirst>=4 -- End of first interval repeat using BD
256 :     if k==1 then h = hs;
          h2,h2sq,hy,hx,hz = h/2,h*h/4,h,h/2,h*h/4 end
257 :     sol[1][k+1] = t; for m=1,neq do sol[m+1][k+1] =
          u[m] end -- Save
258 :   end -- End of main loop on t, now return solution array
259 :   sol[1][ntval+1] = tmax; return sol,nit - Solution, #iter
260 : end -- End of odeb12
261 : setfenv(odeb12,{nsolv=nsolv,step=nil})
```

Listing 10.19. Code segment for direct solution of systems of second order differential equations by TP algorithm

with replaced first and second derivatives is on line 225. In this case the user supplied function is assumed to include a second derivative argument as the last value
in the list of arguments. The other major difference is on lines 245 and 246 where
the new set of *yn* and *ynn* values are calculated after the solution values at a new
time point have been evaluated. The same equations are used in evaluation these
functions as are used in the derivative replacements before the application of the
TP algorithm. Other than these changes along with definitions for the hx, hy and
hz parameters used in Eq. (10.60) the code is almost identical to that for solving
systems of first order differential equations. A final difference is the need to return both the first and second order derivative values at the end of a series of calculations if one is to be able to recall the function and resume the calculation with
a new step distribution. Thus the calling argument to this odeb12() function contains the second derivative as well as the first derivative. This is also the reason
the earlier code development for such functions as odeiv(), odeivs(), etc. included
a place holder for a possible second derivative term. By including this possibility
in the previous multi step integration code segments, these functions can be used
with the present odeb12() basic integration routine as well as the previously developed integration routines. The odeb12() function is a direct replacement code
segment for either the odebiv() routine or the odebrk() routine.

```
 1 : ---[[ /* File list10_20.lua */
 2 : -- Programs for first or second order differential equations
 3 : require"odeiv"
 4 : odebiv = odeb12 -- Substitute 2nd order solver
 5 : -- mu = 20 Try different values
 6 : mu = 0 -- Gives pure sinusoidal solutions
 7 : -- Example of second order formulation
 8 : f1 = function(eqs,x,y,yp,ypp) -- Van der Pol equation
 9 :    eqs[1] = ypp[1] -mu*(1-y[1]^2)*yp[1] + y[1]
10 : end
11 : -- Example of equivalent first order formulation
12 : f2 = function(eqs,t,y,yp)
13 :    eqs[1] = yp[1] - y[2]
14 :    eqs[2] = yp[2] - mu*(1 - y[1]^2)*y[2] + y[1]
15 : end
16 : t1 = os.clock()
17 : s1,err1 = odeive(f1,{0,100,8000},{1},{0})
18 : print('first time =',os.clock()-t1); plot(s1)
19 : t1 = os.clock()
20 : s2,err2 = odeive(f2,{0,100,8000},{1,0})
21 : print('second time =',os.clock()-t1); plot(s2[1],s2[2])
22 :
23 : write_data("list10_20,dat",s1,s2,err1,err2)
```

Listing 10.20. Illustration of code segments for solving either first or second order
differential equations with TP algorithm

Listing 10.20 shows an example of using this basic function for solving the
second order Van der Pol equation as both a second order equation and as two first
order equations. On line 4 of the code the original odebiv() routine definition is
replaced by the odeb12() routine. When the solutions are generated on lines 17

and 20 by calls to odeivse(), the odeb12() code will be used in place of odebiv() to integrate the equation sets. The f1() function is used to define the single second order Van der Pol differential equation on lines 8 through 10 using y, yp and ypp values which correspond to the solution plus first and second derivatives. An equivalent formulation in terms of two first order equations is shown with the f2() function beginning on line 12. This is the same formulation in terms of two first order differential equations as previously used for the Van der Pol equation. Note that in the f2() argument list there is no ypp argument corresponding to the second derivative since this is not used in the formulation. If an argument is not used at the end of an argument list it is acceptable to omit the name from the argument list and the software does not complain about a missing argument. In this manner the odeb12() routine can be used in exactly the same manner for only first order equations as in previous code listing in this chapter. The only additional items of note is in the calling argument to the odeive() routine. For the second order equation the calling arguments must have a third and fourth entry which is a table of initial values and initial first derivative values. For this example these are the tables {1} and {0} shown on line 17 of the listing. For the two coupled first order differential equations, equivalent initial data is shown on line 20 as the third argument with the table values {1, 0}. For an initial value problem with a second order differential equation both the initial values and the initial first derivative values must be specified in two separate tables. When calling the function with only a set of first order differential equations as in the call using the f2 function on line 20, the initial list of first derivative values may simply be omitted. It should be understood that both of the calls to odeive() on lines 17 and 20 use the new odeb12() function which handles both first order equations and second order equations. What about the case of a mixed set of first and second order differential equations? Such a set is perfectly fine for use with the ode12() function. For the first derivative terms, values are only used for the equations involving a second derivative. For example if one had a system of three equations with the first two being first order differential equations and the third equation being a second order equation with first derivative value of -1, the array {_,_,-1} could be used to pass the derivative values to the integration routine. The first two entries are undefined, but that is OK since they would not be needed by the function used to define the differential equations.

Results will not be shown for the code in Listing 10.20 as the results are essentially identical to previous results. For the value of mu shown on line 6, the solution will be a pure sinusoidal solution. This is used in the example because it is easier to compare the results for such a simple case. The reader is encouraged to execute the code and probe the solutions using either the single second order equation or the two first order equations. Such a comparison of results will show that the two generated solutions for a pure sinusoidal solution differ by less that 1.e-12 at all the generated solution points. The results from the two solutions will not be exactly identical as slightly different equations are being solved by nsolv() and the relative accuracy of this routine is specified as 5.e-6. In one case two coupled equations are being solved and in the other only one equation is being solved. As a result of the finite accuracy of this routine and of numerical calculatios, the error in the two solutions will not be identical after many steps of integration. However, the results of this simple example can be used to verify that the two approaches generate the same solution to within the accuracy specified by the various code segments.

If the adaptive step generation routines odeivs() or odeivse() are used to generate the solutions, additional differences will be observed. Because the second order equation has only one relative error to consider while the two first order equations have two relative errors to consider, the adaptive step generation routine used in the solution will not generate the same time grid for solving the equations. For this example, the odeivse() routine with default parameters generates 4733 time points for the single second order differential equation with an estimated maximum error of 0.0018 and 5367 time points for the two first order differential equations with an estimated maximum error of 0.0012. When used with systems of only first order differential equations, the odeb12() routine will generate identical results to the odebiv() routine as they both use the same TP algorithm. The question then is why not always use the odeb12() routine? The answer is that the odebiv() routine will execute about 10% faster if one has only first order equations since the additional calculations for possible second derivatives do not have to be made. However, in most cases this additional speed increase is hardly noticeable, so the odeb12() routine can be used as a general replacement for the odebiv() routine if desired.

The code in Listing 10.20 also prints the time taken to solve the equations for both the single second order equation and the two first order equations. As might be expected the single second order equation is solved considerable faster than the two first order equations. In this example the two first order equations take about 3 times longer to obtain a solution as the reader can verify by executing the code. This time difference is also one of the major reasons for developing the second order TP algorithm.

10.11 Differential-Algebraic Systems of Equations

Many engineering problems consist not just of systems of differential equations, but many times coupled systems of some differential equations and some algebraic equations. In such cases, some of the equations involve one or more derivatives of the solution variables and some of the equations simply involve an algebraic relationship between the variables with no derivative involved in the equation. It would be very nice if the code so far developed for the numerical solution of differential equations could accommodate such systems of differential-algebraic equations. As will be seen much of the work in this chapter is directly applicable to such physical problems.

As an example of such systems of equations, consider the electrical circuit shown in Figure 10.34 consisting of five resistors, two capacitors and one inductor plus a voltage source. It will be assumed that the circuit has been unexcited for a

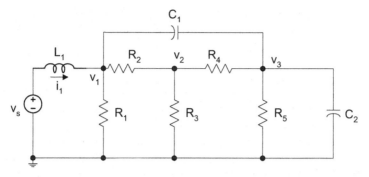

Figure 10.34. Electrical circuit for example of differential-algebraic systems of equations.

long time before t = 0 and at t = 0 the voltage source represented by v_s is switched into the circuit. It will further be assumed that the voltage source has a constant DC value independent of time after being switched into the circuit. Any electrical engineer who has had an introductory course in electric circuits can write a set of so called node voltage equations describing the voltages present at the three nodes, v_1, v_2, and v_3 as the following equations:

$$(1) \quad -i_1 + \frac{v_1}{R_1} + \frac{(v_1 - v_2)}{R_2} + C_1 \frac{d(v_1 - v_3)}{dt} = 0$$

$$(2) \quad \frac{(v_2 - v_1)}{R_2} + \frac{v_2}{R_3} + \frac{(v_2 - v_3)}{R_4} = 0$$

$$(3) \quad \frac{(v_3 - v_2)}{R_4} + \frac{v_3}{R_5} + C_1 \frac{d(v_3 - v_1)}{dt} + C_2 \frac{dv_3}{dt} = 0$$

$$(4) \quad v_s - v_1 - L_1 \frac{di_1}{dt} = 0$$

$$(10.61)$$

The first three equations are node voltage equations, one for each node and the fourth equation expresses the relationship between voltage and current for the inductor where a fourth variable which is the current in the inductor must be introduced. Thus the four variables in the equation set are v_1, v_2, v_3 and i_1, giving four equations in terms of four unknown variables. It is noted that three of the equations (1), (3) and (4) involve derivatives of the variables while one equation (2) does not involve any derivatives of any of the variables. This is what makes the system a case of coupled differential-algebraic equations. It is certainly true for such simple cases that one could solve for one of the variables such as v_2 in the second of the set , for example, and substitute into the other equations leaving three equations in three unknowns. The resulting equations would then all have a derivative term and be a system of three first order differential equations. However, as the complexity of a electrical circuit increases, this becomes more difficult and for a general set of nonlinear mixed differential-algebraic system of equations is not always possible. Also such an approach is prone to errors in algebra if one has to do the work. Computers are much better adapt at manipulating equations and people are best adapt at formulating problems in the simplest, most basic way which in this example is in the form of separate equations for each node of an electrical problem.

The circuit analysis program SPICE which is familiar to every electrical engineer takes the formulation of a circuit's problem one step further and allows the user to skip the formulation in terms of node voltage equations and allows the user to only specify the nodal connections. If one makes a list of the elements and the nodes they are connected to, a so-called nodal list, then it is a straightforward process of generating the node voltage equations from such a nodal element list. Such an approach will not be pursued here, although it is not particularly difficult to develop such a program for the case of only sources, resistors, capacitors and inductors as in this example. This will be left up to the reader if desired. For a circuit such as that shown, SPICE would internally formulate a set of equations such as those above and apply numerical techniques to solve the resulting set of differential-algebraic equations much as will now be done here with the code segments already developed.

The basic TP integration code routines developed in the previous sections of this chapter have no difficulty in handling such a set of coupled differential-algebraic equations. The TP algorithm does not require that a derivative term be present in the equations, it only permits such a term to be present. However, this is not the case with the fourth order Runge Kutta integration algorithm developed. As will be recalled, this algorithm evaluates the derivative at several points within the time step being considered and this is how the solution values are updated at the new solution point. If no derivative term is present, a variable will not have its value changed in an integration step with the RK algorithm. This feature is also present with many commercial integration packages, such as those supplied in MATLAB. These can handle systems with mixed derivatives, such as occur in (1) and (3) above, but not the second equation (2) where no derivative is present. With the TP algorithm and the implicit approach used here to solve a set of differential equations, an updated solution value can be found even if any or all derivative terms are missing. In fact the integration routines will execute perfectly well if all derivative terms are missing and one is simply solving a set of coupled algebraic equations. Of course the resulting solutions would be constant with no time dependence and a simple call to nsolv() would be more appropriate.

Listing 10.21 shows code to solve the set of equations for the circuit example above. The equation set is defined in the f() function on lines 8 through 13. The reader should be readily able to correlate the code formulation with the basic equation set of Eq.(10.61). The major differences are the use of the v[4] variable for the inductor current in the equation set and the notation vp[] for the various derivatives. A selected set of element values are defined on lines 5 and 6 to complete the equation definition. The set of equations is solved on line 15 by a call to the function odeivse() which is the adaptive step size algorithm with a returned error estimate. Other choices for functions to generate the solution include the odeiv() routine with a specified range of times and number of solution time points

```
 1 : --/* File list10_21.lua */
 2 : -- Programs to solve electric circuit equations using odeivse()
 3 : require"odeiv"
 4 :
 5 : R1,R2,R3,R4,R5 = 4e3,2e3,4e3,2e3,2e3
 6 : C1,C2 = 1e-6,4e-6; L1 = 2.e-3; vs = 5
 7 :
 8 : f = function(eqs,t,v,vp)
 9 :     eqs[1] = -v[4] + v[1]/R1 + (v[1]-v[2])/R2 + C1*(vp[1]-vp[3])
10 :     eqs[2] = (v[2]-v[1])/R2 + v[2]/R3 + (v[2]-v[3])/R4
11 :     eqs[3] = (v[3]-v[2])/R4 + v[3]/R5 + C1*(vp[3]-vp[1]) +
C2*vp[3]
12 :     eqs[4] = vs - v[1] - L1*vp[4]
13 : end
14 :
15 : s1,err1 = odeivse(f,{0,2e-2},{0,0,0,0})
16 : print('#points=',#s1[1])
17 :
18 : print(unpack(maxvalue(s1))); print(unpack(maxvalue(err1)))
19 : plot(s1)
20 :
21 : write_data("list10_21.dat",s1,err1)
Output:
#points=         11955
0.02 9.8066804162957 4.709170457191 1.9662463185984 0.10052933970712
0.02 0.0047183933258 0.002264843965 0.0009437165874 9.4870456882e-005
```

Listing 10.21. Example of solution of coupled differential-algebraic system of equations.

or several ranges of time values or the odeivqs() quick scan function. With a range of possible functions, how does one know which of these functions to use and how does one select the time range of the solution such as the 0 to 2.e-2 values used on line 15 of the code? Well, this requires some approximate knowledge of the solution or some trial and error. From prior knowledge of the time constants of transients in electrical circuits, it is known that the use of resistors in the kOhm range, capacitors in the μF range and inductors in the mH range typically give transient response times in the mSec range – hence a first guess at the time scale. However, the time range parameters in Listing 10.21 were not selected without some preliminary calculations with other values. In fact the first execution of the code was with the odeivqs() function using a time range from 1.e-4 to 1 to get a general feel for the solution. Following this the time interval of interest was refined to the value shown in the listing.

Figure 10.35 shows two of the calculated voltages as saved into a file on line 20 of Listing 10.21. The v_2 response is not shown but has a similar behavior and with a steady state value of 2.5. The resistor values were selected so that a quick check could be made of the solution since the steady state voltage values for the three nodes are 0.5 0.25 and 0.125 times the source voltage. These values are being approached for long times as can be seen in Figure 10.35 for the two node voltages shown. The response shows a damped sinusoidal response characteristic of a fairly high Q resonant circuit with many cycles of oscillation before the transient dies out. This makes the full transient solution somewhat difficult to obtain since for high accuracy in the solution, many calculated points are needed per cycle of oscillation, but many cycles must be followed to observe the complete transient. The results illustrate again that the adaptive step selection technique provides a good solution at short times and out to many cycles of the response. In this example it generates 11955 time point values (see first line of output in Listing 10.21).

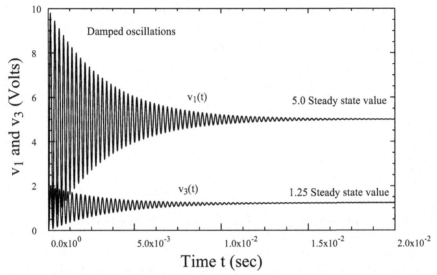

Figure 10.35. Calculated response of circuit of Figure 10.34 with parameters of Listing 10.21.

The calculation on line 15 of the listing includes an error estimate. The maxvalue() function shown on line 16 shows a new way to use the error estimate data. The maxvalue() function takes a series of data arrays such as the solution array or error array and returns the maximum values found for each input data array. For example the maxvalue(s1) call will return the maximum values of all the solution values as a single array of values. When these are printed by the print(unpack(maxvalue(s1))) statement the printed output as shown in the first line of output gives in order the maximum value of the time variable, followed by the maximum values of the four solution variables. Thus from the second line of output the maximum calculated values of v_1, v_2, v_3 and i_1 are 9.806, 4.709, 1.966 and 0.100 respectively. The third line of output similarly list the maximum estimated errors in the corresponding variables as 0.0047, 0.0022, 0.00094 and 0.000094. From these values it can be seen that the estimated maximum errors are about 0.048% of the peak variable values for all the voltages and about 0.094% of the peak value of the inductor current. If a more detailed analysis of the estimated errors and their location in time is desired then the error values can be plotted as a function of time. In many cases an estimate of the maximum errors will be sufficient to determine if a more detained analysis is needed.

This section has shown that the TP integration routines developed in this chapter can be used not only with coupled differential equations, but with a more general class of coupled differential-algebraic systems of equations. A fairly complete set of code segments have now been developed for solving systems of differential equations with boundary conditions of the initial value type. As opposed to MATLAB which supplies a number of possible routines for solving differential equations, the present work concentrates on one general approach based upon the TP algorithm. The next section will consider several examples of using these routines for a broad range of engineering type problems.

10.12 Examples of Initial Value Problems

In addition to the examples in the previous sections, several new examples of initial value differential equations will be presented in this section. The emphasis will be on nonlinear first and second order differential equations.

10.9.1 Second Order Mechanical Systems

Many types of mechanical systems lead to second order differential equations of the initial value type. For example from Newton's second law the net force acting on a body of mass m, leads to the equation (in one dimension)

$$ma = m\frac{d^2x}{dt^2} = F(\frac{dx}{dt}, x, t) \qquad (10.62)$$

where the net force is some function of velocity, position and time. A typical initial value problem has known values of the initial velocity and position at some starting time (usually taken as 0). Application to a generic spring-mass-damper system with spring constant k, mass m and damping coefficient c leads to the specific equation:

$$\frac{d^2x}{dt^2} + 2\varsigma\omega\frac{dx}{dt} + \omega^2 x = f(t) \qquad (10.63)$$

where $\omega = \sqrt{k/m}, \varsigma = c/(2\sqrt{km})$ and $f(t)$ is some external force applied to the spring-mass-damper system. This is a linear second order differential equation that is easily solved by use of the previous software packages and the solution is left to the reader.

Nonlinear equations result for mechanical systems such as the above when the damping and/or the restoring force is nonlinear in the velocity or position. A classical example is that of one dimensional sliding friction where the equation becomes:

$$\ddot{x} + \mu g\frac{\dot{x}}{|\dot{x}|} + \omega^2 x = f(t) \qquad (10.64)$$

where μ is the coefficient of friction and g is the acceleration due to gravity. The damping term is positive or negative depending on the direction of the velocity but does not depend on the magnitude of the velocity.

Listing 10.22 shows the code for solving this differential equation with selected parameter values. The $\dot{x}/|\dot{x}|$ function is implemented with a slight modification of the equation to the form $\dot{x}/\sqrt{\dot{x}^2 + eps}$ in order to have a continuous derivative at zero velocity. The value of eps has been selected to be sufficiently small as to not have a significant effect on the solution. The listing employs the second order version of equation solver by use of the odebiv = odeb12 statement on line 5 of the listing.

```
 1 : --- /* File list10_22.lua */
 2 : -- Programs for differential equation with sliding friction
 3 :
 4 : require"odeiv"
 5 : odebiv = odeb12
 6 :
 7 : g = 983.21
 8 : w = 100
 9 : u = 0.02
10 : eps = 1.e-2
11 : ug = u*g
12 :
13 : f1 = function(eqs,t,x,xp,xpp)
14 :    eqs[1] = xpp[1] + ug*xp[1]/math.sqrt(xp[1]^2+eps) + w*x[1]
15 : end
16 :
17 : s1,err1 = odeivse(f1,{0,9},{10},{0})
18 : plot(s1)
19 : write_data("list10_22.dat",s1)
```

Listing 10.22. Code segment for solution of Harmonic oscillator with sliding friction.

A plot of the obtained solution is shown in Figure 10.36. The amplitude of oscillation is seen to decrease linearly with time until the mass becomes fixed at some location due to the friction force and does not oscillate any further. This behavior is decidedly different from that of the Harmonic oscillator with damping proportional to the first derivative where the amplitude of oscillation decays exponentially with time toward zero. The reader is encouraged to change the parameters in Listing 10.22, especially the w, u and eps value and re-execute the code observing any changes in the calculated response.

A second example from classical mechanics is that of a classical pendulum of length L and mass m which gives the equation:

$$\frac{d^2\theta}{dt^2} + \frac{c}{mL}\frac{d\theta}{dt} + \frac{g}{L}\sin(\theta) = f(t) \qquad (10.65)$$

where θ is the displacement of the pendulum from the vertical position. A classical damping term has been included in the equation and the nonlinear term arises from the restoring force being proportional to $\sin(\theta)$.

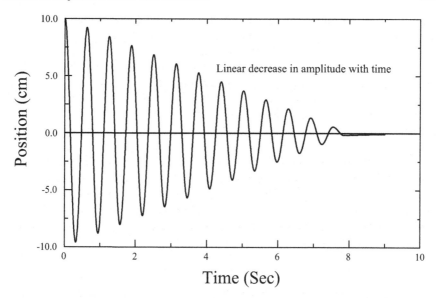

Figure 10.36. Damped oscillation of Harmonic oscillator with sliding friction damping term.

```
 1 : --- /* File list10_23.lua */
 2 : -- Programs to integrate differential equation for pendulum
 3 : require"odeiv"
 4 : odebiv = odeb12
 5 :
 6 : g,L,c,m = 983.21, 100, 400, 20
 7 : dc,w2 = c/(m*L), g/L
 8 : th0 = 0.95*math.pi
 9 :
10 : f1 = function(eqs,t,x,xp,xpp)
11 :    eqs[1] = xpp[1] + dc*xp[1] + w2*math.sin(x[1])
12 :    eqs[2] = xpp[2] + dc*xp[2] + w2*x[2]
13 :    eqs[3] = xpp[3] + w2*math.sin(x[3]) -- No Damping
14 :    eqs[4] = xpp[4] + w2*x[4] -- No Damping
15 : end
16 :
17 : s1,err1 = odeivse(f1,{0,20},{th0,th0,th0,th0},{0,0,0,0})
18 : plot(s1)
19 : write_data("list10_23.dat",s1)
```

Listing 10.23. Code segment for nonlinear pendulum problem with and without damping.

Listing 10.23 shows the code for solving this second order differential equation for a set of selected parameters. The listing shows four second order equations being solved at once. Since there is no coupling between the parameters in defining the differential equations this is equivalent to solving four separate differential equations. In this example the equations are for the pendulum problem with damping (lines 11 and 12) and without damping (lines 13 and 14) and with the nonlinear form of Eq. (10.65) (lines 11 and 13) and with the classical linear differential equation that replaces the $\sin(\theta)$ function with the linear term θ (lines 12 and 14). The simultaneous solution of multiple equations is a convenient way of comparing solutions with a single execution of the code.

Figure 10.37 shows the solution results for the case of no damping (solutions from lines 13 and 14). The solid curve is the nonlinear more exact solution while the dotted curve is the solution of the classical linear equation. In each case the initial angular displacement has been taken as 0.9*pi which is a rather large angle of about 171 degrees. The solid curve is decidedly non-sinusoidal with much more time spent at the large angles than would be the case for a sinusoidal waveform. The period of oscillation is also significantly increased and more than twice the classical period for this large an angular oscillation.

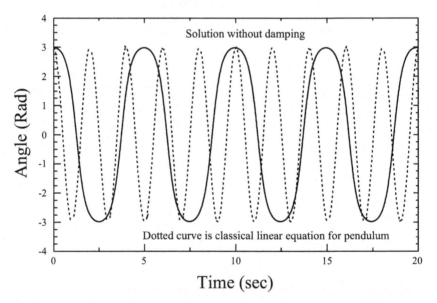

Figure 10.37. Solution of pendulum equation without damping obtained from Listing 10.23.

The corresponding solutions for the pendulum with one selected level of damping are shown in Figure 10.38. In this case the damping has been taken as a very significant factor in order to see the decay over only a few cycles of oscillation. As would be expected, the exact solution and the classical linearized solution tend to approach the same oscillation period and decay time constant as the pendulum angle of displacement   becomes small and approaches zero. For the initial few cycles, a decidedly nonlinear oscillatory behavior of the solution can be seen resulting from the sin() term in Eq. (10.65).

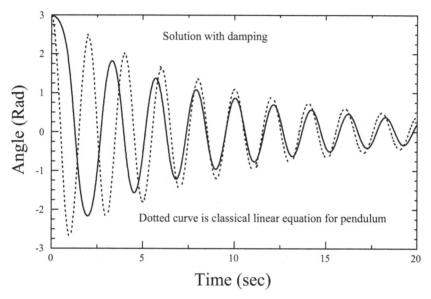

Figure 10.38 Solution of pendulum equation with significant level of damping obtained from Listing 10.23.

A third example from classical mechanics is that of a planet or comet orbiting the sun in the x-y plant which can be described by the set of second order differential equations:

$$\frac{d^2x}{dt^2} + Kx/r^3 = 0, \quad \frac{d^2y}{dt^2} + Ky/r^3 = 0, \quad \text{where } r = \sqrt{x^2 + y^2} \qquad (10.66)$$

When distance is measured in AU units (1AU = 1.496 e13 m) and time is measured in years the value of K is approximately 40. This example illustrates two coupled second order differential equations or they may be considered a system of four first order coupled differential equations. Many solutions are possible depending on the set of initial conditions for the position and velocities.

Listing 10.24 shows code for solving this set of equations with three different initial velocity conditions. The initial locations of the planet are taken as (1.0, 0.0) in AU units and the three initial velocities in AU/year units are taken as (0.2, 0.0), (1.0, 0.0) and (1.2, 0.0) as can be seen in the listing on lines 11 through 13. The solutions are obtained for times of 0 to 20 (in years). The resulting calculated planetary orbits are shown in Figure 10.39. The calculated orbits range from a circular orbit for the center curve to elliptical orbits with the orbit passing very close to the central mass (the sun) for the low initial velocity. The data points on the calculated orbits are at one year intervals. One can visualize the equal angles swept out by the motion for equal time intervals as expressed by Kepler's laws. One could also move to three-dimensional motion by including a third equation and move to the interaction of multiple bodies by expanding the equations to account for gravitational interactions between multiple bodies. Such an expansion is straightforward and will be left to the reader to implement.

```
 1 : --- /* File list10_24.lua */
 2 : -- Programs to integrate second order F=ma equations
 3 : require"odeiv"
 4 : odebiv = odeb12
 5 :
 6 : f = function(eqs,t,f,fp,fpp)
 7 :    eqs[1] = fpp[1] + f[1]/(f[1]^2 + f[2]^2)^1.5
 8 :    eqs[2] = fpp[2] + f[2]/(f[1]^2 + f[2]^2)^1.5
 9 : end
10 :
11 : s1 = odeivs(f,{0,20},{0,1},{1,0})
12 : s2 = odeivs(f,{0,20},{0,1},{.2,0})
13 : s3 = odeivs(f,{0,20},{0,1},{1.2,0})
14 :
15 : plot({s1[2],s1[3]},{s2[2],s2[3]},{s3[2],s3[3]})
16 : write_data("list10_24.dat",s1,s2,s3)
```

Listing 10.24. Classical mechanics for the motion of a planet orbiting the sun.

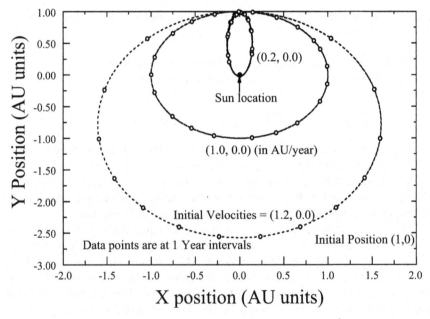

Figure 10.39. Typical solutions of planetary motion equations for three initial velocities.

10.12.2 Predator-Prey Equations

This example is from a biological system consisting of a model for the time dynamics of two interdependent species of animals. One species, the prey, is the primary food source for the other species, the predator. This might apply for example to populations of wolves and rabbits. The set of differential equations describing the population dynamics is given by the two coupled equations:

$$\frac{dp_1}{dt} = \alpha_1 p_1 - \delta_1 p_1 p_2$$
$$\frac{dp_2}{dt} = \alpha_2 p_1 p_2 - \delta_2 p_2$$

(10.67)

where p_1 and p_2 are the populations of prey and predators, α_1 and α_2 are growth rate coefficients and δ_1 and δ_2 are death rate coefficients. In this model the prey (p1) is assumed to have no shortage of food. Example code for solving this set of equations is shown in Listing 10.25. A set of growth rate and death rate parameters have been selected for the model on line 5. How these are selected is left to another discussion. The initial population of prey is set at 5000 for all the calculations and three solutions are obtained for initial predator populations of 100, 200 and 300 species.

```
 1 : --- /* File list10_25.lua */
 2 : -- Programs to integrate predator prey equations
 3 : require"odeiv"
 4 :
 5 : alf1,alf2,del1,del2 = 2.0, 0.0002, 0.02, .8
 6 :
 7 : f = function(eqs,t,p,pp)
 8 :     eqs[1] = pp[1] - alf1*p[1] + del1*p[1]*p[2]
 9 :     eqs[2] = pp[2] - alf2*p[1]*p[2] +del2*p[2]
10 : end
11 :
12 : s1 = odeivs(f,{0,30},{5000,100})
13 : s2 = odeivs(f,{0,30},{5000,200})
14 : s3 = odeivs(f,{0,30},{5000,300})
15 : plot(s1);plot(s2);plot(s3)
16 : write_data("list10_25.dat",s1,s2,s3)
```

Listing 10.25. Code for integration of predator prey differential equations.

Solutions are shown in Figure 10.40 for the three cases. Several interesting features of the solutions are illustrated by these examples. First for the 5000, 100 initial conditions, the solutions oscillate slightly with time but the populations are close to an equilibrium condition, matching the two populations. In fact a time independent solutions exists for the selected parameters if the initial conditions are selected with 4000 prey and 100 predators. With 5000 initial prey and 100 predators, the population of predators will initially grow slightly with time as the population of prey decreases. Then as the population of prey decreases, the predator population decreases due to the shortage of food and a cycle repeats. For an assumed fixed prey initial population, as the initial assumed predator population increases, the amplitude of the population cycles increases. A large initial predator population rapidly gives rise to a decrease in the prey population and this is followed by a rapid decay in the predator population as the lack of food leads to increased deaths. For 300 initial predators, the population of prey is almost completely wiped out in each cycle and the number of prey is almost completely wiped out by the lack of food. In fact if one increases the initial predator population to 400, the minimum predicted predator population will drop below 1, which is not a recoverable population.

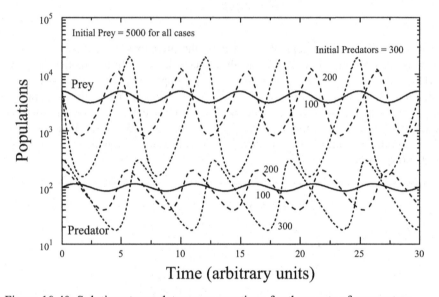

Figure 10.40. Solutions to predator prey equations for three sets of parameters.

Although the predator prey problem has a fairly simple set of nonlinear differential equations, it is typical of the nonlinear coupling found in real world problems involving many sets of physical parameters.

10.12.3 Electrical Engineering Problems

This section will discuss three examples from the field of Electrical Engineering, (a) an AC voltage multiplier circuit, (b) the transient response of a CMOS circuit and (c) the pulse response of a linear amplifier. Almost all electronic circuits provide nonlinear differential equations as the basic active circuit elements of bipolar or MOS transistors are nonlinear devices.

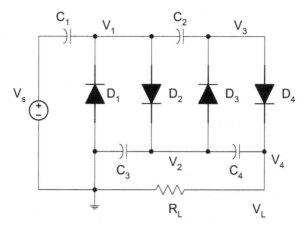

Figure 10.41. Voltage multiplier circuit with 4X multiplication factor.

The circuit diagram of an AC voltage multiplication circuit is shown in Figure 10.41. In this case the circuit is used to produce a DC voltage across the load resistance that is approximately 4 times the magnitude of the sinusoidal voltage source. The voltage at the node between C_3 and C_4 will be approximately 2 times the magnitude of the voltage source. The diodes are all assumed to be equal with an ideal diode I-V characteristic.

The circuit is described by 4 node voltage equations of :

$$C_1 \frac{d(v_1 - v_s)}{dt} + C_2 \frac{d(v_1 - v_3)}{dt} - i_D(-v_1) + i_D(v_1 - v_2) = 0$$

$$C_3 \frac{dv_2}{dt} + C_4 \frac{d(v_2 - v_4)}{dt} + i_D(v_2 - v_3) - i_D(v_1 - v_2) = 0$$

$$C_2 \frac{d(v_3 - v_1)}{dt} - i_D(v_2 - v_3) + i_D(v_3 - v_4) = 0 \qquad (10.68)$$

$$C_4 \frac{d(v_4 - v_2)}{dt} - i_D(v_3 - v_4) + v_4 / R_L = 0$$

with $i_D(v) = I_S(\exp(v / v_{th}) - 1)$

where the last equation is the assumed ideal diode equation. Computer code for the time integration of this initial value problem is shown in Listing 10.26. The definition of the differential equation set on lines 13 through 17 follows the same form as the above set of equations. An extra equation on line 17 is used to define the source voltage (v[5]) as a sinusoidal time dependent voltage. A separate function (id) on lines 9-11 is used to define the diode I-V relationship.

The calculated time dependence of two of the node voltages is shown in Figure 10.42. The voltage across the load resistor or $V_4(t)$ is the desired output voltage. Ideally this rectified voltage after a long time should be 4 times the peak of the sinusoidal voltage or 40 Volts in this example. It can be seen from the figure that the voltage peaks at about 37.5 Volts or about 2.5 Volts below the ideal value.

```
 1 : --- /* File list10_26.lua */
 2 : -- Programs to solve AC voltage multiplier circuit
 3 : require"odeiv"
 4 :
 5 : C1,C2,C3,C4,Rl = 100e-6, 100e-6, 100e-6, 100e-6, 100E3
 6 : Vm,w,tmax = 10.0, 2*math.pi*60, 0.7
 7 : Is,vth = 1.e-12, 0.026
 8 :
 9 : id = function(v)
10 :    return Is*(math.exp(v/vth) - 1)
11 : end
12 : f = function(eqs,t,v,vp)
13 :    eqs[1] = C1*(vp[1]-vp[5])+C2*(vp[1]-vp[3])-id(-v[1])+
             id(v[1]-v[2])
14 :    eqs[2] = C3*vp[2] + C4*(vp[2]-vp[4]) + id(v[2]-v[3]) -
             id(v[1]-v[2])
15 :    eqs[3] = C2*(vp[3]-vp[1]) - id(v[2]-v[3]) + id(v[3]-v[4])
16 :    eqs[4] = C4*(vp[4]-vp[2]) - id(v[3]-v[4]) + v[4]/Rl
17 :    eqs[5] = v[5] - Vm*math.sin(w*t)
18 : end
19 :
20 : s1 = odeivs(f,{0,tmax},{0,0,0,0,0})
21 : plot(s1)
22 : write_data("list10_26.dat",s1)
```
Listing 10.26 Code for time solution of AC voltage multiplier circuit.

This is due to the forward voltage drops across the four diodes in the circuit. The 2.5 Volts distributed among the four diodes, represents a voltage drop of about 0.625 Volts per diode which is very reasonable. It can also be seen that the circuit takes about 25 cycles of the AC voltage source to reach approximately the steady state value. The small ripple on the output voltage after reaching steady state is due to the current supplied to the load resistor and the ripple increases as the value of the load resistor is decreased.

This example illustrates the time dependent solution of 5 coupled nonlinear equations with the automatic step selection algorithm used by the odeivs() integration routine. In this example, the automatic step selection routine is found to generate a solution at 8105 time points. However, about 1000 of these time steps are used in the first 0.01 time interval to really get the solution started with acceptable accuracy. Following this initial interval the odeibse() routine generates an average of about 285 time steps per cycle to achieve the accuracy specified in the code.

An alternative means of execution the code is with the odeiv() function using a fixed number of time intervals in the range of 5000 to 10000. The solution obtained with this alternative will be essentially the same as that shown in Figure 10.42. The reader is encouraged to experiment with this code segment for obtaining the solution. Also the estimated error for the solution is generated by Listing 10.26 and shown in a pop-up plot. The reader is encouraged to examine this data and resulting graph.

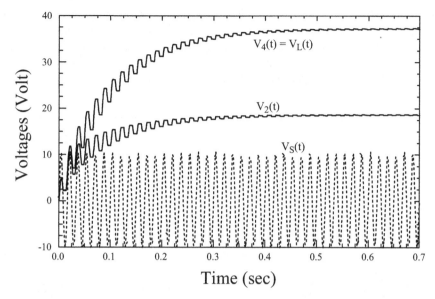

Figure 10.42. Time solution for voltage multiplier DC supply voltages obtained from Listing 10.26.

A differential equation arising in the study of charge layers at the surface of a MOS semiconductor can be expressed as

$$\frac{df}{dx} = a(\frac{1+f}{f}) - 1 \tag{10.69}$$

with $f = 0$ at $x = 0$

where a is a constant value. The initial condition presents some difficulty in evaluation the derivative for the initial time point. However, the differential equation can be rearranged by multiplying through by f to give the equivalent form

$$f\frac{df}{dx} + f(1-a) - a = 0 \tag{10.70}$$

with $f = 0$ at $x = 0$

This formulation now has no difficulty with a zero denominator form for the initial condition. Many available solution packages, such as those available with MATLAB, require one to express the equations in the form of Eq. (10.69) where the derivative is equal to some function. Such solution packages can have difficulty with differential equations such as this. The present formulation simply requires that the defining equation equate to zero and is very general in requirements on specifying the differential equation.

Code for solving this differential equation is illustrated in Listing 10.27. The code solves the same differential equation for 7 values of the "a" parameter as defined on line 6 of the listing. This is accomplished in the code by line 10 where a "for" loop is used to step over the same differential equation with the array of a coefficient values. The f() function is essentially returning the results for 7 differential equations with the same functional form but with different coefficients so seven solutions can be obtained by one call to the differential equation solver which is taken to be the odeivse() function on line 13.

```
 1 : --- /* File list10_27.lua */
 2 : -- Programs to integrate surface charge differential equation
 3 :
 4 : require"odeiv"
 5 :
 6 : a,a0 = {.1,.2,.5,1,2,5,10}, {0,0,0,0,0,0,0}
 7 : na = #a
 8 :
 9 : f = function(eqs,t,f,fp) -- Define equations, multiple a values
10 :    for i=1,na do eqs[i] = f[i]*fp[i] + f[i]*(1-a[i]) - a[i] end
11 : end
12 :
13 : s1,err1 = odeivse(f,{0,100},a0)
14 : print(unpack(maxvalue(s1))) ; print(unpack(maxvalue(err1)))
15 : plot(s1); plot(err1)
16 :
17 : write_data("list10_27.dat",s1,err1)
```

Listing 10.27. Code for solving differential equation involving surface charge.

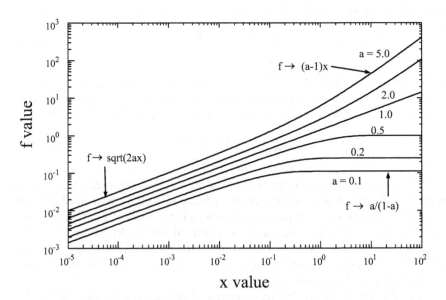

Figure 10.43. Typical solutions to differential equation for surface charge layers as obtained from Listing 10.27.

The resulting solutions for the differential equation are shown in Figure 10.43. The selected values of the "a" parameter cover the interesting range of the parameter. For all values of the parameter the solution at small values of x follows a square root of x dependence as indicated on the figure. For large values of x the solution changes nature with the results depending on the value of "a". For a < 1, the solution approaches a constant value for large x as indicated on the figure. For a > 1, the solution approaches a linear dependence on x as also indicated on the figure. The case of a = 1 is a special case where the solution has a sqrt() functional dependence for all values of x.

Again in this problem the automatic step selection integration function is used to obtain a solution. This provides an accurate solution for both small and large values of x as one can see from a detailed examination of the output file. For an accurate solution at small x values, the odeivse() code used step sizes in the range of 1e-10. The spatial range up to 100 is then covered in only 647 total spatial points as the step size automatically adjusts to larger increments as the solution progresses to larger spatial values. The resulting solution is a very efficient use of computer computational effort to cover a wide range of spatial values. The example also illustrates the use of the code for solving a differential equation with a nonlinear specification of the derivative term.

This next example is for the transient response of an electronic amplifier stage including both low frequency and high frequency capacitors. The circuit to be analyzed is shown in Figure 10.44. This is the same electronic circuit as presented in Chapter 4 as Figure 4.5 where the frequency dependent transfer function was calculated as an example of solving sets of coupled complex equations. While this is a linear circuit and does not exercise the ability of the differential equation solvers to handle nonlinear problems, it does have widely separated time constants and is another example of an ill-conditioned set of differential equations. The voltage source is taken to be a pulse going from 0 to some fixed value (5 mV) and then back to 0 at some later time.

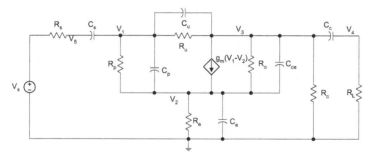

Figure 10.44. Example of amplifier stage for transient analysis.

This time dependent problem can be formulated as a set of five coupled differential equations as:

$$C_s \frac{d(V_1 - V_5)}{dt} + C_u \frac{d(V_1 - V_3)}{dt} + C_p \frac{d(V_1 - V_2)}{dt} + \frac{V_1 - V_2}{R_p} +$$

$$\frac{V_1 - V_3}{R_u} = 0$$

$$C_e \frac{dV_2}{dt} + C_p \frac{d(V_2 - V_1)}{dt} + C_{ce} \frac{d(V_2 - V_3)}{dt} + \frac{V_2 - V_1}{R_p} + \frac{V_2}{R_e} +$$

$$\frac{V_2 - V_3}{R_o} - g_m(V_1 - V_2) = 0$$

$$C_u \frac{d(V_3 - V_1)}{dt} + C_{ce} \frac{d(V_3 - V_2)}{dt} + C_c \frac{d(V_3 - V_4)}{dt} + \frac{V_3 - V_1}{R_u} +$$

$$\frac{V_3 - V_2}{R_o} + \frac{V_3}{R_c} + g_m(V_1 - V_2) = 0$$

$$C_c \frac{d(V_4 - V_3)}{dt} + \frac{V_4}{R_l} = 0$$

$$C_s \frac{d(V_5 - V_1)}{dt} + \frac{V_5 - V_s}{R_s} = 0$$

(10.71)

```
1 : --- /* File list10_28.lua */
2 : -- Programs to solve for transient response of amplifier
3 : require"odeiv"
5 : --Circuit Parameters
6 : Rs,Rp,Re,Ru,Ro,Rc,Rl = 100, 20e3, 200, 200e3, 100e3, 4e3, 10e3
7 : Cs,Ce,Cc,Cp,Cu,Cce = 1e-6, 10e-6, 2e-6, 5e-12, 1e-12, 1.5e-12
8 : gm = 0.01
9 : Vm,t1,t2 = -.1, 2.e-7, 4.e-7 -- Short pulse (200, 400 nsec)
10 :
11 : vs = function(t)
12 :     if t<t1 then return Vm else return 0 end
13 : end
14 : f = function(eqs,t,v,vp)
15 :     eqs[1] = Cs*(vp[1]-vp[5])+Cu*(vp[1]-vp[3])+Cp*(vp[1]-
            vp[2])+(v[1]-v[2])/Rp+(v[1]-v[3])/Ru
16 :     eqs[2] = Ce*vp[2]+Cp*(vp[2]-vp[1])+Cce*(vp[2]-vp[3])+
            (v[2]-v[1])/Rp+v[2]/Re+(v[2]-v[3])/Ro-gm*(v[1]-v[2])
17 :     eqs[3] = Cu*(vp[3]-vp[1])+Cce*(vp[3]-vp[2])+Cc*(vp[3]-vp[4])+
            (v[3]-      v[1])/Ru+(v[3]-v[2])/Ro+v[3]/Rc+gm*(v[1]-v[2])
18 :     eqs[4] = Cc*(vp[4]-vp[3])+v[4]/Rl
19 :     eqs[5] = Cs*(vp[5]-vp[1])+(v[5]-vs(t))/Rs
20 : end
21 :
22 : s1 = odeivs(f,{0,t2},{0,0,0,0,0})
23 : plot(s1)
24 : Vm,t1,t2 = -.1, 5.e-3, 10.e-3 -- Long pulse (5, 10 msec)
25 : s2 = odeivs(f,{0,t2},{0,0,0,0,0})
26 : plot(s2)
27 : write_data("list10_28.dat",s1,s2)
```

Listing 10.28. Code for the transient response of single stage transistor amplifier.

Code for solving for the transient response of this circuit is given in Listing 10.28. The parameters are the same as those used in Listing 4.10. The defining set of equations on lines 15 through 19 of the code can be readily compared with the set of Eq. (10.71). A transient solution is obtained on lines 22 and 25 for two different pulse conditions. For the first solution the source pulse is of short duration, being 200 nsec in duration, while the second solution is for a long pulse of 5 msec duration. The circuit has both very small time constants and long time constants and these two pulses demonstrate these features.

Figure 10.45 shows the output voltage for the short time source pulse. It can be seen that the output rises rapidly to a value of about 2.66 and essentially remains constant for the remainder of the time of the pulse. When the input pulse goes to zero at 2×10^{-7} sec, the output rapidly goes back to zero. For such a short pulse, the time constants are determined essentially by the high frequency device capacitances of C_p, C_u and C_{ce} of Figure 10.44. In Chapter 9, in Listing 9.13 the frequency response of this circuit was analyzed in terms of the poles and zeros of the system transfer function. The two last entries in Table 9.1 show the high frequency poles occurring at -1.651×10^7 and -4.077×10^8 Hz. If these are converted to time constants, transient time constants of $(1/(2\pi f))$ 9.60×10^{-9} and 3.90×10^{-10} sec are obtained. The longest of these, corresponds closely to the time constant observed in Figure 10.44 for the rise and fall times of the pulse.

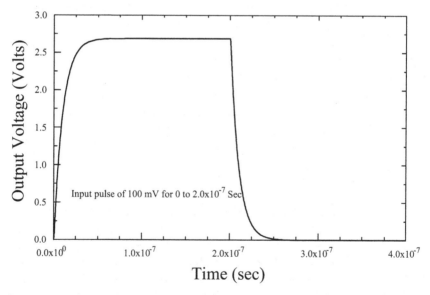

Figure 10.45. Response of amplifier to short time pulse as obtained from Listing 10.28. Output is taken as V_4.

For a long time pulse, the low frequency capacitors, consisting of C_s, C_e and C_c begin to limit the voltage response and the output voltage begins to sag with time as illustrated in Figure 10.46. The rapid rise in the response at the leading edge of the pulse still shows the fast time constant as seen in Figure 10.45. This can be seen by taking the saved output file and plotting the resulting solution on a greatly expanded time scale near t = 0. The fall or sag in the output is controlled by the low frequency poles of the transfer function. Table 9.1 shows that there are 3 low frequency poles with real parts of the poles at -7.903 and -252.4 Hz. These correspond to time constants of 6.30×10^{-4} and 2.01×10^{-2} sec. The fastest fall in the response seen in Figure 10.46 corresponds to the smallest of these values while the slower decrease toward zero corresponds to the largest of these values. This larger time constant of about 0.02 sec can be observed by changing the pulse length to 5×10^{-2} in Listing 10.28 and re-executing the example. This is left to the reader.

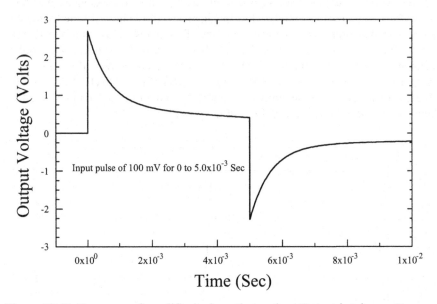

Figure 10.46. Response of amplifier to long time pulse. Output is taken as V_4.

An important feature of this example is the large range of time constants encountered – from about 2×10^{-2} to about 1×10^{-10} sec. This is a good test of the automatic step selection algorithm used in the odeivs() code for solving differential equations. For the long time pulse the algorithm has to implement small time steps to resolve the fast changes in the variables when the input abruptly changes, but then has to implement long time steps in order to resolve the long time constants. The success of the time step algorithm can be seen from Figure 10.47 which plots the time step size used in solving the differential equations vs. the number of the time step. This step distribution corresponds to the time solution shown in Figure 10.46. The results show that near t = 0, corresponding to the first step number and near t = 5 msec, the evaluated step size is very small, being on the order of 1×10^{-13}. However the step size grows in a controlled manner to about 1×10^{-4} in regions where the response changes slowly. The adaptive step algorithm is seen to rapidly transition to very small step sizes in the region just before the abrupt change is input and output that occurs at the 5 msec time. During this rather abrupt decrease in step size, one can see several attempts by the algorithm to increase the step size, but in each case after a small increase in step size, a further decrease is seen until a minimum step size is reached at the 5 msec time. The decrease in step size just before the abrupt change in input at a time of 5 msec is seen to occur in a very controlled manner. The total number of time steps resulting from the automatic step adjustment algorithm for this problem is a convenient number of 1067. This is a very respectable number considering the large range of step sizes used which covers a range of almost 10 orders of magnitude. This example illustrates the ability of the code segment developed to handle transient problems with widely varying time constants and the ability of the adaptive algorithm to rapidly adjust to both small step sizes and large step sizes.

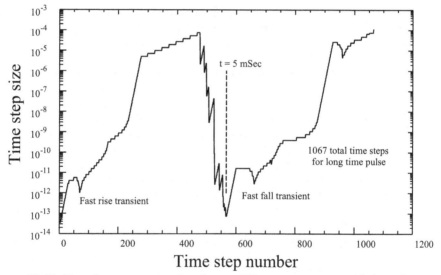

Figure 10.47. Step size vs. step number for amplifier step response with long time pulse.

10.12.4 Two Additional Examples

The shape of a mirror that focuses light to a single point (in two dimensions) is determined by the differential equation

$$x\left(\frac{dy}{dx}\right)^2 - 2y\frac{dy}{dx} - x = 0 \tag{10.72}$$

where $y(x)$ describes the relative shape of the mirror. Both x and y may be scaled in this equation by the same factor and the resulting differential equation is unchanged. In order to express the derivative term as an explicit quantity, as required by some differential equation packages, this quadratic equation for the derivative would have to be solved for an explicit expression for the first derivative. However, the software developed here requires only that an expression be written involving some function of the derivative and its variables and the result set equal to zero as in the above equation.

```
1 : -- /* File list10_29.lua */
2 : --Shape of lamp reflector
3 :
4 : require"ode12"
5 :
6 : f = function(eqs,x,y,yp,ypp)
7 :    eqs[1] = x*yp[1]^2 - 2*y[1]*yp[1] - x
8 : end
9 :
10 : s1 = odeivs(f,{-2,2},{0})
11 : plot(s1); write_data("list10_29.dat",s1)
```
Listing 10.29. Code for solving for the shape of a reflector mirror.

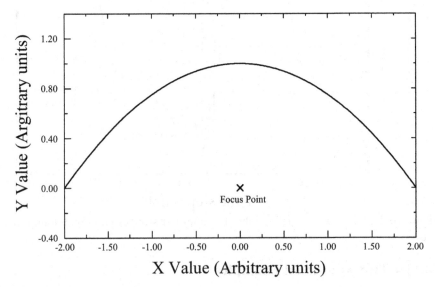

Figure 10.48. Shape of optical reflector for mommon focus point obtained from Listing 10.29.

Listing 10.29 shows the simple code for solving this differential equation over the range of -2 to 2. This was selected as a convenient range as the y shape value varies from 0 to 1 over this range. The resulting calculated shape is shown in Figure 10.48 and is in fact a simple parabolic shape as might be expected. The shape may be extended to larger x values or scaled in both dimensions by any convenient multiplicative factor.

While this is a fairly simple example it again emphasizes the general nature with which the differential equation may be specified for use with the developed software code. Again the automatic step size algorithm software is used so the user does not have to worry about step size selection.

Many coupled time dependent differential equations exhibit types of solutions known as "chaotic behavior". In principle, the solution of an initial value differential equation problem is completely determined by the differential equation and the set of initial conditions. However, for some types of coupled nonlinear differential equations an extremely small change in the initial conditions produces a very distinguishably different time behavior. Such systems are said to exhibit chaotic behavior. One such system of equations is the Lorentz equations defined by:

$$\frac{dx}{dt} + a(x - y) = 0$$

$$\frac{dy}{dt} + y + xz - bx = 0 \tag{10.73}$$

$$\frac{dz}{dt} + cz - xy = 0$$

where a, b and c are constants

```
 1 : -- /* File list10_30.lua */
 2 : -- Lorentz equations
 3 : require"odeiv"
 4 :
 5 : a,b,c = 10,28,8/3
 6 : f = function(eqs,t,y,yp)
 7 :    eqs[1] = yp[1] + a*(y[1] - y[2])
 8 :    eqs[2] = yp[2] + y[2] + y[1]*y[3] - b*y[1]
 9 :    eqs[3] = yp[3] + c*y[3] - y[1]*y[2]
10 : end
11 :
12 : s1 = odeiv(f,{0,40,10000},{5,5,5})
13 : s2 = odeiv(f,{0,40,10000},{5.0001,5,5})
14 : s3 = odeiv(f,{0,40,10000},{5.000001,5,5})
15 : plot(s1,s2)
16 : write_data("list10_30.dat",s1,s2,s3)
```

Listing 10.30. Code for solving the Lorentz equation set.

Listing 10.30 shows the code for solving this set of equations with one particular set of constant parameters. Three solutions are generated as S1, S2 and S3 using slightly different initial conditions for the x variable. The initial conditions are 5.0, 5.0001 and 5.000001, representing a change in one of the initial conditions by 2×10^{-3} % and 2×10^{-5} %. The solutions are generated for times from 0 to 40 as seen on lines 12-14 of the code. The initial conditions for the other two variables are kept fixed at 5.0. In order to compare the results for the same step distribution, the odeiv() function is used to generate solutions at 10000 uniformly spaced time intervals.

Solutions generated by the code are shown in Figure 10.49 for the x variable and for times from 0 to 20. As the curves show for initial times all three solutions are essentially identical for the slightly different initial conditions. However, at a little beyond t = 15, the S_2 solution begins to deviate from the S_1 solution and exhibits very different behavior for larger times. For the S_3 solution the differences begin to show up after about t = 18 on the graph. It can thus be seen that the smaller the difference in initial condition, the longer the solutions essentially agree, but the solutions eventually begin to deviate from each other. Any slight difference in initial conditions within the numerical accuracy of the software will eventually result in significantly different time solutions. This is what is identified as chaotic behavior of the solutions.

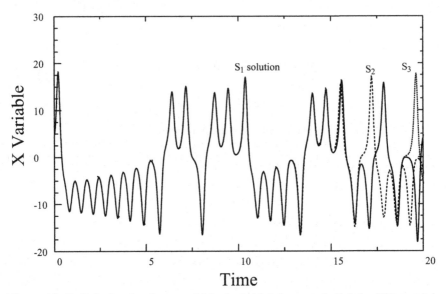

Figure 10.49. Solution for the x variable for initial times and slightly different initial conditions from Listing 10.30.

The chaotic behavior is more clearly seen in the three solutions for longer time intervals such as the t = 30 to 40 time range shown in Figure 10.50. While the three solutions exhibit some similar general features such as the maximum and minimum ranges over which the solutions exist, the detailed time development is very different for the three solutions.

The same type of behavior is also exhibited by the other variables in the solutions. Figure 10.51 shows the z variable in the solution for the longer time intervals. Again very different time solutions are seen for any given time but similar overall limits in the maximum and minimum values of the variable are observed.

Different long term solutions will also be generated if the code in Listing 10.30 is re-executed with different numbers of time intervals instead of the 10000 used here. The reader is encouraged to experiment with different values and with the results using the odeivs() function with an adaptive step generation.

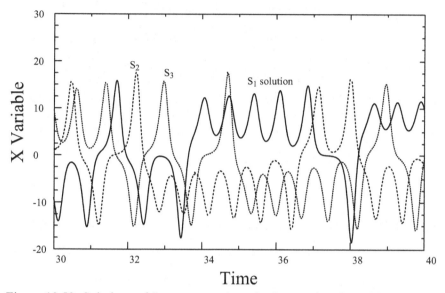

Figure 10.50. Solutions of Lorentz equations for longer time intervals and for x variable from Listing 10.30.

A somewhat different view of the solutions is provided by the state-space projection views of the solutions shown in Figures 10.52 and 10.53. These show solution points traced out in the xy and xz planes as functions of time for the three solutions. The same dotted and dashed curves are shown in these figures for the three solutions as shown in the previous Figures 10.50 and 10.51. Another state-space plot for the yz plane could also be drawn but is not shown here as the general features are similar to the figures shown. The three solutions are seen to trace out similar trajectories in the state-space plots. The solutions are seen to form orbits around two critical points in each plane. Such points are called strange attractors in the language of those who study such chaotic systems of differential equations.

The study of nonlinear systems of chaotic differential equations is of considerable research interest. However, the subject will not be pursued further here but the reader is referred to the literature (or Web sites) for more details. For the purpose here, one of the major observations is the very sensitive nature of the solution to the initial conditions which is one of the distinguishing features of such chaotic systems of nonlinear equations. This is sometimes referred to as the "butterfly" effect where it has been observed that if the weather equations for our world are a set of chaotic equations then the flap of a butterfly's wings in one part of the world

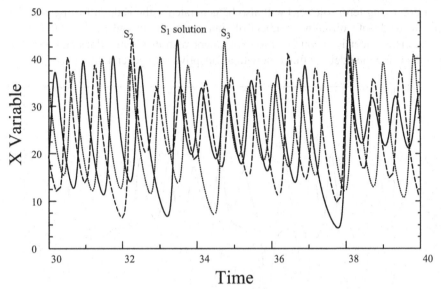

Figure 10.51. Solution of Lorentz equations for the z variable and long times from Listing 10.30.

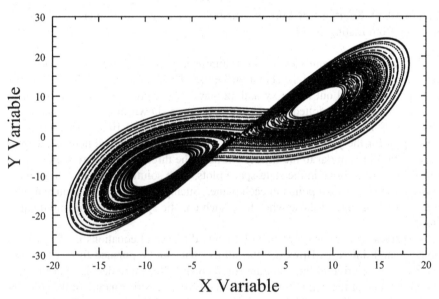

Figure 10.52. State-space plot for Lorentz equation – xy projection from Listing 10.30.

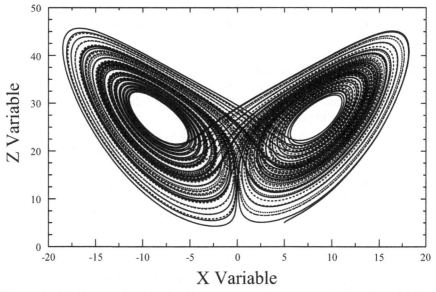

Figure 10.53. State-space plot for Lorentz equation – xz projection from Listing 10.30.

could set off severe weather conditions such as a tornado in another part of the world. This assumes of course that the world's weather patterns exhibit such sensitivity to initial conditions as some sets of coupled nonlinear differential equations. Hopefully this is not the case.

10.13 Summary

This chapter has concentrated on the numerical solution of systems of differential equations in one independent variable with specified initial values of the variables. A general discussion of numerical techniques applicable to the numerical solution of such equations has been presented with a special emphasis on the accuracy of such solution techniques. Following this the trapezoidal rule of time step integration was selected as providing a majority of desirable attributes for implementing into software code. The accumulative errors associated with the trapezoidal rule when solving differential equations was extensively explored and a technique was developed for estimating the accuracy of a numerical solution using the trapezoidal rule. Finally the development of an automatic step selection approach was discussed and such a first order technique was combined with the trapezoidal rule for solving sets of differential equations.

Several important code segments were developed in this chapter that can be readily used to integrate systems of differential equations. The most important of these are:

1. odebiv() – basic code segment for systems of first order differential equations solved at specified time points using the trapezoidal rule.
2. odeb12() – basic code segment for systems of second order differential equations solved at specified time points using the trapezoidal rule. This may replace the odebiv() function for solving equations.
3. odeiv() – code for multiple time step solutions of systems of first order or second order differential equations. This uses odebiv() (or odebrk() or odeb12()) for the individual time steps.
4. odeive() – code for an odeiv() calculation plus a returned estimate of the error in the solution using the h-2h algorithm.
5. odeivqs() – code for solving sets of differential equations using a log based time scale – called a "quick scan" solution.
6. odeivqse() – code for an odeivqs() calculation plus a returned estimate of the error in the solution.
7. odeivs() – code for an adaptive step size solver. Combines the trapezoidal rule of odebiv() with an adaptive step size algorithm.
8. odeivse() – code for an adaptive step size solver combined with a return of the estimated error in the numerical solution using the h-2h algorithm.
9. odebrk() – code for implementing the Runge-Kutta algorithm as the basic integration step. This may replace odebiv() in all the other algorithms above.

The developed code segments provide a very useful set of functions for the solutions of sets of first order or second order differential equations with specified initial values.

The developed code was used to solve a number of typical sets of differential equations. With the examples the reader should be able to apply the developed code to a wide range of real-world problems. The next chapter will concentrate on sets of second order differential equations when applied to two point boundary value problems as opposed to initial value problems considered in this chapter.

11 Differential Equations: Boundary Value Problems

The previous chapter has discussed the solution of differential equations of the "initial value" type, where all the values needed to specify a specific solution are given at one specific initial value of the independent variable. Many time dependent differential equations in engineering are of this type where some dependent variable is governed by a differential equation in time and the initial conditions are specified at some initial time that can usually be taken as t = 0. For such problems, the differential equation can then be integrated into the future and in principle to any desired value of time. The previous chapter has developed several general computer algorithms and software packages for addressing such problems. The developed code can be applied to nonlinear differential equations just as easily as linear differential equations although as with all nonlinear problems iterative approaches must be used in obtaining a solution. One of the features of the numerical solution of such problems, either linear or nonlinear, is that the relative error in the solution tends to increase as such equations are integrated further into the future or further from the initial starting point.

This chapter addresses a different type of differential equation problem, the so called "boundary value" problem. For this class of problems, specific values of the dependent variable (can be either values or derivatives) are not specified at one particular point but are specified at two different values of the independent variable. Engineering problems of this type usually involve some spatial independent variable as opposed to time as the independent variable. For example for a second-order differential equation, the values of the independent variable may be specified at two values of x such as x = 0 and x = L. Again this chapter addresses differential equations of only one independent variable. Partial differential equations involving two or more independent variables are discussed in the next two chapters.

11.1 Introduction to Boundary Value Problems in One Independent Variable

A typical boundary value problem arising in Engineering is of the following type:

$$a\frac{d^2U}{dx^2} + b\frac{dU}{dx} + cU = f(x)$$

$$\text{with } U(0) = U_1 \text{ and } U(L) = U_2$$

(11.1)

575

J.R. Hauser, *Numerical Methods for Nonlinear Engineering Models*, 575–703.
© Springer Science + Business Media B.V. 2009

In this equation, U is some variable that depends on the independent variable x. While this example is a linear second order differential equation with constant coefficients, the equation could just as easily be some nonlinear function of the solution variable, or first derivative or second derivative of the variable. In addition the a, b and c coefficients could be functions of the independent variable. More general equations will be considered as the chapter develops. As with any second order differential equation, two arbitrary constants can be specified and these are specified as the value of the independent variable at two different spatial points – these are the "boundary" values. The specified values could just as well be values of the first derivative or some combination of dependent value and first derivative at the boundary points. All of these are possibilities.

The code and approaches developed in the previous chapter can not be used to directly solve such a problem, since all initial conditions are not known at either boundary point. Solutions to such boundary value problems can be obtained by (a) modifying the initial value problem approach of the previous chapter or by (b) developing a new approach to solving such problems. Both of these avenues will be pursued in this chapter.

Before developing numerical approaches to solving two point boundary value problems, it is useful to review the solution to a linear two point boundary value problem such as represented by Eq. (11.1) where complete closed form solutions can be obtained for many types of $f(x)$ functions. A solution to Equation (11.1) is known to be of the form:

$$U(x) = C_1 e^{s_1 x} + C_2 e^{s_2 x} + U_p(x)$$

$$\text{where } s_1 = (-b + \sqrt{b^2 - 4ac})/2a \text{ and} \qquad (11.2)$$

$$s_2 = (-b - \sqrt{b^2 - 4ac})/2a$$

The first two terms constitute the complementary solution and the function $U_p(x)$ is a particular solution of the equation that depends on the functional form of the $f(x)$ term. The complementary solution containing the two arbitrary constants comes from solving the differential with the forcing function $f(x)$ set to zero. The functional form of the complementary solution depends on the sign of the term under the square root in the s_1 and s_2 definitions and is typically classified as:

$$\text{(a) Underdamped for } b^2 < 4ac$$

$$\text{(b) Critically damped for } b^2 = 4ac \qquad (11.3)$$

$$\text{and (c) Overdamped for } b^2 > 4ac$$

For the underdamped case the complementary solution consist of damped sinusoidal terms and for the overdamped case it consists of two real exponential terms. In either case two equations can be formulated for the boundary conditions as:

$$C_1 + C_2 = U(0) - U_p(0)$$

$$C_1 e^{s_1 L} + C_2 e^{s_2 L} = U(L) - U_p(L) \qquad (11.4)$$

Provided this set of equations can be solved for the two constants, a complete solution is then obtained for the two point boundary value problem. Solutions do not always exist and especially for an important class of problems known as eigenvalue problems where $f(x) = 0$. Such problems will be considered in a special subsequent section of this chapter.

So far the boundary value problem has been considered in terms of a second order differential equation. It is know however from the previous chapter that such a second order equation can also be written as two first order differential equations in two dependent variables by introducing the first derivative as a second variable. In addition one can have a single differential equation of higher order than two with a set of conditions for the variable and derivatives specified at two points and this again constitutes a boundary value problem. Thus the general boundary value problem can be specified as a set of first order differential equations:

$$F_1(U_1', U_1, U_2 \cdots U_N) = 0$$
$$F_2(U_2', U_1, U_2 \cdots U_N) = 0$$
$$\cdots$$
$$F_N(U_N', U_1, U_2 \cdots U_N) = 0$$

$$(11.5)$$

with N values of $U_1, U_2 \cdots U_N$ specified at either of two boundary values of the independent variable. One can also have any mixture of first order and second order equations as long as the overall order of the system of equations is second order or above. The most common type of engineering problem is a second order differential equation or a coupled system of two or more second order differential equations. For some problems such as beam deflection in mechanical engineering, a two point boundary value problem with a fourth order differential equation must be solved.

11.2 Shooting(ST) Methods and Boundary Value Problems (BVP)

Shooting methods seek to convert the two point boundary value problem into an iterative approach to solving an initial value problem. The approach can be summarized by the following steps:

(a) A guess is used for the unspecified initial conditions at one of the two boundaries.

(b) The differential equation is then solved as an initial value problem by integrating the solution to the other boundary point. The direction of integration may be forward or backwards in the independent variable.

(c) The boundary values of the variables at the second boundary are computed and compared with the required boundary values.

(d) Differences between the computed boundary values and the desired boundary values are then used to correct the guessed initial conditions at the starting boundary.

(e) The new guessed initial conditions are then used to repeat the solution which hopefully gives boundary conditions closer to the desired condition.

(f) The iterative process is repeated until the guessed initial conditions result in a set of final boundary conditions satisfying the required second boundary conditions.

The term "shooting method" comes from the use of the method when values of the solution are specified at both boundaries and the unknown initial condition is the value of the derivative. Selecting a guessed value of the missing derivative at one boundary one can essentially shoots the solution from one of the boundaries and observe the solution value at the other boundary. If the value exceeds the desired value, the initial derivative was probably too large and a smaller value is selected to shoot a second solution to the second boundary. By using some algorithm to correct the guessed initial derivative value, the procedure can be iterated until a correct boundary value is achieved at both boundaries. The convergence of the process and speed of convergence depends on the algorithm used to correct the initial guessed value of the missing derivative. As will be seen this can be done using a Newton's method such that quadratic convergence can be achieved in the iterative process.

To illustrate the shooting method consider the deflection of a beam supported at two ends and subject to uniform transverse load w and with tension T. This is described by the BVP:

$$\frac{d^2y}{dx^2} - \frac{T}{EI}y - \frac{wx(x-L)}{2EI} = 0, \ 0 \le x \le L \tag{11.6}$$

$$y(0) = y(L) = 0$$

Although not essential to understanding the numerical method, the physical parameters are the modulus of elasticity, E, and the central moment of inertia, I. The two point boundary conditions are specified at the end points of the beam. This second order equation can be formulated as two first order equations as:

$$\frac{du_1}{dx} - u_2 = 0$$

$$\frac{du_2}{dx} - \frac{T}{EI}u_1 - \frac{wx(x-L)}{2EI} = 0 \tag{11.7}$$

$$u_1(0) = u_1(L) = 0$$

Listing 11.1 shows computer code for solving this BVP using the initial value odeiv() code developed in the previous chapter. Lines 11 through 14 define the differential equation in the same form as used in the previous chapter. The addition for the boundary value problem is to embed the odeiv() differential equation solver within a boundary evaluation function called bvalue() defined on lines 16 through 19. This function accepts a value of the unknown derivative of the

```
 1 : -- /* File list11_1.lua */
 2 : -- Shooting method for boundary value problem
 3 :
 4 : require"odeiv"; require"newton"
 5 :
 6 : -- Parameters
 7 : T,E,I,w,L = 500, 1.e7, 500, 100, 100
 8 : EI = E*I; print('w,EI =',w,EI)
 9 : y0, yL,nx = 0, 0, 2000 -- Boundary values, #x values
10 :
11 : f = function(eq,x,u,up) -- Differntial equations
12 :    eq[1] = up[1] - u[2]
13 :    eq[2] = up[2] - T*u[1]/EI - w*x*(x-L)/(2*EI)
14 : end
15 :
16 : bvalue = function(up) -- Boundary function, up is derivative
 7 :    s1 = odeiv(f,{0,L,nx},{0,up}) -- Solve initial value problem
18 :    return s1[2][nx+1] - yL -- difference in boundary value
19 : end
20 :
21 : yp = 0 -- Initial guess at derivative
22 : yp,nm,err = newton(bvalue,yp) -- Newton's method for BV
23 : print('Initial derivative, #iterations, errors =',yp,nm,
         err,bvalue(yp))
24 : s2 = odeiv(f,{0,L,nx},{0,yp*1.1}) -- Larger derivative
25 : s3 = odeiv(f,{0,L,nx},{0,yp/1.1}) -- Smaller derivative
26 : plot(s1[1],s1[2],s2[2],s3[2])
27 : write_data("list11_1.dat",s1,s2,s3)
Output:
w,EI = 100     5000000000
Initial derivative, #iterations, errors =
0.00083325011258534    2  2.3293312412038e-012  1.1410858383562e-015
```

Listing 11.1. Illustration of shooting method for solving two point boundary value differential equation.

equation at the starting point of the solution ($x = 0$) and returns the difference between the desired value of the solution at the second boundary point ($x = L$) and the actual value obtained from a solution of the initial value differential equation. This requires only two lines of code, line 17 which call the odeiv() differential equation solver and line 18 that returns the difference in the solution value at the L boundary and the desired value at this boundary (of 0 in this example). In calling the odeiv() solver on line 17, the two initial conditions for the equation variables are taken as {0,up} where the second value is the unknown initial derivative condition that is being calculated by the newton() function and passed to the bvalue() function. To complete the solution, this function and an initial guess at the unknown first derivative or second variable value only needs to be supplied to the newton() function as shown on line 22. The newton() function developed in Chapter 3 does all the work in calling the bvalue() function and adjusting the initial value until the desired two point boundary value problem is satisfied. Note again the use of previously developed code segments to implement a higher level numerical algorithm.

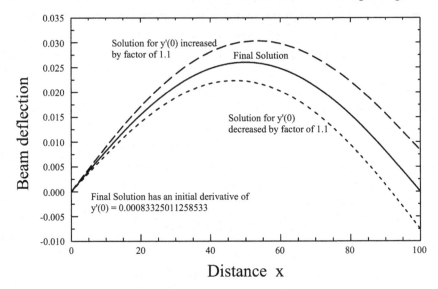

Figure 11.1. Illustration of shooting method for solving two point boundary value problem. Example of uniformly loaded beam deflection between two rigid points.

The code in the listing calculates three solution curves and saves the three solutions on line 27. These are plotted in Figure 11.1 to further illustrate the shooting method with this BVP. The solid line in the figure is the final converged solution that corresponds to an initial derivative value of 0.00083325 which is returned as the yp value on line 22 from the newton() function. Two other curves are shown for the differential equation, one with the initial slope increased by a factor of 1.1 and the other with the initial slope decreased by a factor of 1.1 as evaluated on lines 24 and 25. The three solutions illustrate in a graphical way how the initial slope can be adjusted to obtain a final solution point passing through any desired value such as the specified value of 0.0. The output line printed in Listing 11.1 shows that the actual computed value of the final solution at x = L differs from exactly 0.0 by about 1.1e-15 which is about the accuracy of the computer language and well within the specified accuracy of the newton() method. The fact that this value is so small is somewhat fortuitous and in general such high precision can not always be expected.

A few other points are worthy of note in Listing 11.1 and this example. First, the initial guess for the unknown derivative value is simply 0.0 as seen on line 21 of the code. This is the simplest guess that can be made although it is known that there must be some finite derivative value. Even with this, the printed output shows that only 2 Newton iterations are required to achieve excellent accuracy. This is due to the fact that the equation being solved is a linear differential equation with a linear boundary condition. Each Newton iteration requires two calls to the function being solved, so to achieve the final solution requires 4 calls to the bvalue() function on line 16. This can be verified by putting a print statement within the function. Since each call to this function requires an integration of the differential equation using the initial value solution routine odeiv() this shooting method will require essentially 4 times the computational time for a solution which is a pure initial value problem where all the initial conditions are known. Additional iterations might be required if the initial guess is extremely far from the solution value. This factor of 4 (or more) will apply to any linear BVP problem with a linear differential equation and linear boundary conditions. For nonlinear equations or nonlinear boundary conditions, more Newton iterations will in general be required.

In this example, the two boundary conditions specify the value of the function on each of the boundaries. While this type of boundary condition is frequently encountered, other types of boundary conditions may exist, such as the mixed boundary type where the boundary condition involves some combination of the value and derivative at the two boundaries. One such test problem is the following boundary value problem:

$$y'' - \frac{2x}{x^2+1} y' + \frac{2}{x^2+1} y - x^2 - 1 = 0,$$

$$\text{with } y'(0) + y(0) = 0 \text{ and } y'(1) - y(1) = 3$$

(11.8)

The second line gives mixed boundary conditions at both boundaries involving in this case a linear combination of the function value and the first derivative. A more general case would be some nonlinear combination of the function value and first derivative. Expressing this in terms of two first order differential equations gives:

$$\frac{du_1}{dx} - u_2 = 0$$

$$\frac{du_2}{dx} - \frac{2x}{x^2+1} u_2 + \frac{2}{x^2+1} u_1 - x^2 - 1 = 0$$

$$\text{with } u_2(0) + u_1(0) = 0 \text{ and } u_2(1) - u_1(1) = 3$$

(11.9)

Since a boundary value problem is completely specified by the differential equation and a set of boundary conditions, it would be very nice to have computer programs that required only this set of specifications. The details of using Newton's method to satisfy the boundary conditions could hopefully be a common feature to all types of boundary value problems using the shooting method. The code segment in Listing 11.2 provides such an example. This code defines the above differential equations through a function f() written on lines 8 through 11 and a set of left and right boundary conditions defined by the function fbound() on lines 14 through 17. The differential equation formulation is the same as previously used. The boundary condition function receives the values of dependent variables at the boundaries and specifies each boundary condition in terms of an array value to be forced to zero when the boundary value is satisfied. The boundary conditions for the above example are specified in a slightly more general form of:

```
 1 : -- /* File list11_2.lua */
 2 : -- Shooting method for BV problem with mixed conditions
 3 : require"odebv" -- Boundary value by shooting method solver
 4 :
 5 : nx,xmax = 2000,1 -- Independent variable parameters
 6 : a1,b1,c1 = 1,1,0; a2,b2,c2 = -1,1,-3 -- BV parameters
 7 :
 8 : f = function(eqs,x,u,up) -- Differntial equation
 9 :     eqs[1] = up[1] - u[2]
10 :     eqs[2] = up[2]-(2*x/(x^2+1))*u[2]+(2/(x^2+1))*u[1] - x^2 - 1
11 : end
12 :
13 : -- Define Left and Right boundary equations
14 : fbound = function(bv,uL,uR) -- uL,uR, Left, Right values
15 :     bv[1] = a1*uL[1] + b1*uL[2] + c1 -- Left, x = 0 condition
16 :     bv[2] = a2*uR[1] + b2*uR[2] + c2  -- Right, x = L condition
17 : end
18 :
19 : s,ns,nm,err = odebvst({f,fbound},{0,xmax,nx},{0,0})
20 : print(ns,nm,err); plot(s[1],s[2])
21 : write_data("list11_2.dat",s)
Output:
2      2      2.909779220633e-009
```

Listing 11.2. Code for general solution of boundary value problem by shooting method.

$$a_1 u_1(L) + b_1 u_2(L) + c_1 = 0$$
$$a_2 u_1(R) + b_2 u_2(R) + c_2 = 0 \qquad (11.10)$$

where the L and R notation indicates the left and right boundary values, or the $x = 0$ and $x = 1$ values of the variable for this case. In the computer notation of Listing 11.2 these values are written as uL[1], uL[2], uR[1] and uR[2]. After specifying the differential equation and boundary conditions, the listing calls a function odebvst() to solve the boundary value problem on line 19. The calling arguments of the function are almost the same as the argument list used for a purely initial value problem with the odeiv() function in the previous chapter. The additional information needed to solve the boundary value problem is the name of the function specifying the boundary values, fbound() in this example. This is included by changing the first calling argument from a pure function name to a table listing the two function names f and fbound (see line 19). This choice was made in order to collect all the function information in one place and to keep the same number of calling arguments as in the simple initial value problem. The second entry specifies the time interval for solution (0, xmax, nx) and the third argument specifies the initial conditions on function values just as in the initial value problem (0, 0). In this case these initial conditions are simply initial guesses, since the final values are determined by the values specified in the boundary condition function. Some initial conditions are needed to begin the shooting method and these are the input values. In this example the initial guesses are simply specified as zero values which are perfectly acceptable values for linear differential equations with linear boundary conditions.

For nonlinear problems the initial guesses will need to be specified with more care. The two boundaries are called the "left" and "right" boundaries. However, they could probably be more accurately called the "initial" and "final" boundaries, since the final boundary value of the independent variable can be less than the initial value (or one can integrate in the negative direction).

```lua
 1 : -- /* File odebv.lua */
 2 : -- Shooting(ST) method for boundary value problems
 3 :
 4 : require"odeiv"; require"nsolv"
 5 : local loc_odebiv = odebiv
 6 :
 7 : odebvst = function(feqs,tvals,u,up) -- Basic shooting BV solver
 8 :    local s,ns,neq,nu,ni,fbound,err -- If no BV fct, return IV
 9 :    if type(feqs)=='table' then feqs,fbound = feqs[1],feqs[2]
10 :    else return odeiv(feqs,tvals,u,up) end
11 :    local inbvs,uL,upL = {},{},{}
12 :    nu = #u
13 :    -- Define local function for Newton's method
14 :    local fbeval = function(beqs,bvs) -- bvs[] is array of BVs
15 :       for i=1,nu do u[i] = bvs[i]; uL[i] = bvs[i] end -- IV
16 :       for i=nu+1,neq do up[i-nu] = bvs[i]; upL[i-nu]=bvs[i] end
-- 
17 :       odeiv(feqs,tvals,u,up) -- Solve IV problem
18 :       fbound(beqs,uL,u,upL,up) -- Evaluate errors in BV eqs
19 :    end
20 :    -- End of function for Newton's method
21 :    fbound(inbvs,u,u,up,up) -- get # of boundary equations
22 :    neq = #inbvs -- Number of boundary value equations
23 :    if neq~=nu then odebiv = odeb12 end -- Second degree eqn?
24 :    for i=1,nu do inbvs[i] = u[i] end -- Initial values
25 :    for i=nu+1,neq do inbvs[i] = up[i-nu] end -- derivative IVs
26 :    ni,err = nsolv(fbeval,inbvs) -- Solve BV problem
27 :    for i=1,nu do u[i] = inbvs[i] end -- Final starting values
28 :    for i=nu+1,neq do up[i-nu] = inbvs[i] end -- Final deriv
29 :    s,ns = odeiv(feqs,tvals,u,up) -- final solution
30 :    odebiv = loc_odebiv -- Always leave with odebiv() reset
31 :    return s,ni,ns,err
32 : end
```

Listing 11.3. Code segment for odebvst() function that solves boundary value problems by use to the shooting method.

The work in solving the boundary value problem with the listed code by the shooting method is all involved in the new function odebvst() and the code for this function is shown in Listing 11.3. The code implements in a general way the procedure illustrated in Listing 11.1 using nsolv() for forcing the specified boundary values to be satisfied. Looking first at the code on lines 21 through 26, the boundary value function is called on line 21 in order to determine the number of returned values which is the number of boundary value equations to be solved. For a system of first order equations, this should be the same as the number of equations. However this call to the function is included so the code segment can be extended to handle second order differential equations where two boundary values are needed for a single second order differential equation. The code assumes that the user has specified a sufficient set of boundary values to solve the problem.

If the number of specified boundary values exceeds the number of differential equations, then the formulation must be in terms of one or more second order differential equations. This information is used on line 23 of the code to switch to the second-degree equation solver odeb12() if required. Lines 24 and 25 then set up an array of initial values and derivatives from the values input to the function in the u and up arrays. Again the derivative values only apply when the code is extended to second order equations. This set of initial values is then passed to the nsolv() routine on line 26 along with the function specifying the set of equations to be solved and named fbeval() in the code. This function is not directly the function specifying the equations to be solved but rather a proxy function defined on lines 14 through 19 of the code. The user supplied differential equation function, feqs(), can not be directly passed to the nsolv() function because it is not in the explicit form required by nsolv() and an intermediate or proxy function is required to properly interface between the nsolv() function and the user specified differential equations. The nsolv() function expects only a single array of solution values while the differential equation specifies two array of values – one for the solution values and one for the derivative values (the u[] and up[] arrays). The code on lines 24 and 25 place these values into a single table of values in the inbvs[] array.

The fbeval() function reverses this process by extracting out the sets the initial values of the functions and derivatives on lines 15 and 16 from the single array of input values as well as saving the left boundary values and derivatives in the uL[] and upL[] arrays. The initial value differential equation solver odeiv() is then called with the proper argument arrays on line 17 to solve the differential equation as an initial value problem. The reader is referred to the previous chapter for a discussion of this function which normally returns a set of solution values. However, in this case, the only important values as far as the nsolv() routine is concerned are the solution values and derivatives at the last spatial point which are returned by the odeiv() function in the u[] and up[] arrays. These are the right hand boundary values computed by the differential equation solver. The two sets of solution values are then passed to the user defined boundary value function on line 18 in order to evaluate the error values between the actual values and the specified values of the boundary conditions. These errors are returned from the user supplied fbound() function in the beqs[] array which is in turn passed back to the nsolv() function so the initial values can be appropriately updated.

After the nsolv() function on line 26 has converged to a set of initial function values and initial derivatives satisfying the user defined boundary values, a final solution is calculated on line 29 after first setting the initial parameters to the converged values on lines 27 and 28. The solution is then returned on line 31 along with the number of iterations required by nsolv(), the number of iterations required in the final solution, and an estimate of the maximum relative error in satisfying the boundary values.

The code in Listing 11.3 is an interesting example of the reuse of previously developed code segments in new ways to solve a new numerical problem. Although not demonstrated so far, all the code routines work not only with linear differential equations and boundary conditions, but also work with nonlinear differential equations and boundary conditions. Note that by defining fbeval() as a local function internally to the odebvst() function all the calling arguments to the odebvst() function are available to be used as needed in the fbeval() function, for example in calling the odeiv() function. In this manner the additional arguments required in the fbeval() function can be added to only the arguments required by the nsolv() function. This use of proxy functions with added parameters is one of the keys to being able to reuse many of the previously developed code segments in many other applications.

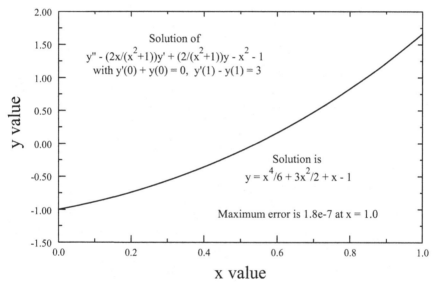

Figure 11.2. Example of mixed boundary value conditions with a second order differential equation.

The solution obtained by Listing 11.2 is shown in Figure 11.2. The solution is a very smooth function and is not a very difficult boundary value problem. It does, however, illustrate the shooting technique with other than fixed boundary conditions on either the value or the derivative. A second advantage of this example is that an exact solution as given in the figure is known. The accuracy of the code in solving this boundary value problem with 2000 spatial intervals specified can be readily determined. As indicated in the figure, the maximum error occurs at $x = 1.0$ and is 1.8e-7 as compared to a maximum solution value of about 1.5. The accuracy of the shooting method for solving boundary value problems will be discussed in considerable more detail in the next section.

Now that a general approach and computer routines have been developed for the solution of boundary value problems with the shooting method, several examples, such as higher order equations with specified boundary values will now be explored. One such classical problem is the transverse deflection of a uniform beam subject to a distributed load w(x). The basic differential equation is a fourth-order equation:

$$\frac{d^4 y}{dx^4} = \frac{w(x)}{EI} \qquad (11.11)$$

where E and I are the elasticity and moment of inertia. For a constant w, this can be integrated twice, giving the second order boundary value problem as previously illustrated by Eq. (11.6). This problem can be formulated as two second-order differential equations or as four first-order differential equations. These two formulations are summarized below:

    (a)   Four first-order equations

$$\frac{du_1}{dx} = u_2 \quad (u_1 = y)$$

$$\frac{du_2}{dx} = u_3$$

$$\frac{du_3}{dx} = u_4 \qquad (11.12)$$

$$\frac{du_4}{dx} = \frac{w(x)}{EI}$$

    (b)   Two second-order equations:

$$\frac{d^2 u_1}{dx^2} = u_2 \quad (u_1 = y)$$

$$\frac{d^2 u_2}{dx^2} = \frac{w(x)}{EI} \qquad (11.13)$$

Either of these formulations can be used with the shooting method to solve the boundary value problem. To complete the formulation appropriate boundary conditions must be specified. For a simply supported beam, the displacement and the second derivative are both zero on both boundaries. For a fixed-fixed beam at both boundaries, the boundary conditions are zero for the displacement and the first derivative at both boundaries. As an example, this second set of boundary conditions will be used and the boundary conditions can be specified for the two formulations as:

    (a)   Four first-order equations:

$$u_1(0) = 0, \quad u_2(0) = 0$$

$$u_1(L) = 0, \quad u_2(L) = 0 \qquad (11.14)$$

(b)   Two second-order equations:

$$u_1(0) = 0, \quad u_1'(0) = 0$$
$$u_1(L) = 0, \quad u_1'(L) = 0$$

(11.15)

In each case there are four boundary conditions that must be satisfied in combination with the differential equations. In all such boundary value problems, the number of boundary value equations must match the total order of derivatives in the coupled differential equation set – two second order derivatives in one case and four first order derivatives in the other case.

```
 1 : -- /* File list11_4.lua */
 2 : -- Shooting method for fourth order boundary value problem
 3 : require"odebv" -- Boundary value by shooting method solver
 4 :
 5 : E,I,w,nx,xmax = 1.e7,500,100,1000,100 -- parameters
 6 : EI = E*I
 7 :
 8 : f2 = function(eqs,x,u,up,upp) -- Two second-order Diff equations
 9 :     eqs[1] = upp[1] - u[2]
10 :     eqs[2] = upp[2] - w/EI
11 : end
12 : fb2 = function(bv,uL,uR,upL,upR) -- Left, Right values
13 :     bv[1] = uL[1]; bv[2] = upL[1] --u[1](Left)=u'[1](Left) = 0
14 :     bv[3] = uR[1]; bv[4] = upR[1] --u[1](Right)=u'[1](Right) = 0
15 : end
16 :
17 : f4 = function(eqs,x,u,up) -- Four first-order equations
18 :     eqs[1] = up[1] - u[2]
19 :     eqs[2] = up[2] - u[3]
20 :     eqs[3] = up[3] - u[4]
21 :     eqs[4] = up[4] - w/EI
22 : end
23 : fb4 = function(bv,uL,uR) -- Left, Right values, no derivatives
24 :     bv[1] = uL[1];  bv[2] = uR[1]  -- u[1](Left) = u[2](Left) = 0
25 :     bv[3] = uL[2]; bv[4] = uR[2]  -- u[1](Right) = u[2](Right) = 0
26 : end
27 :
28 : s2 = odebvst({f2,fb2},{0,xmax,nx},{0,0},{0,0})
29 : s4 = odebvst({f4,fb4},{0,xmax,nx},{0,0,0,0})
30 :
31 : plot(s2[1],s2[2]); plot(s4[1],s4[2])
32 :
33 : nx,dmax = #s2[1], 0
34 : for i=1,nx do
35 :     x = math.abs(s2[2][i] - s4[2][i])
36 :     if x>dmax then dmax = x end
37 : end
38 : print('maximum difference = ',dmax)
39 : write_data("list11_4.dat",s2,s4)
Output:
maximum difference = 1.0708448017205e-014
```

Listing 11.4. Example of solving fourth order equation by two coupled second-order equations or by four coupled first-order equations.

Listing 11.4 shows code for solving this fourth-order beam deflection problem by the use of both formulations described above. For simplicity, a constant load is assumed in the problem, although any given distribution of load can be solved just as easily using a specified $w(x)$. For the two second-order equations, the differential equations are defined by the f2() function on lines 8 through 11 and by the four boundary values with the fb2() function on lines 12 through 15. For the four first-order equations, the corresponding functions are f4() on lines 17 through 22 and fb4() on lines 23 through 26. For the second order equations, the defining boundary conditions are on the boundary values and first derivatives. For the first order equations, the defining boundary conditions are on the four function values. For the second order formulation it is seen that there are more boundary conditions (4) than there are differential equations (2) and this is how the odebvst() function determines whether to use the odeb12() or odebiv() initial value equation solver. After defining the differential equation sets, the solutions are obtained in Listing 11.4 by calls to the odebvst() function on lines 28 and 29. The second derivative is not needed for the f4() function so the upp term is not included in the calling argument list for the defining differential equations as seen on line 17.

Similarly, for the fb4() function, the derivatives are not needed in the boundary conditions and only the boundary values are listed in the calling argument list to this function. Similarly, no initial conditions are needed for the derivatives of the f4() formulation, so the initial derivative array is not included in the calling argument list to the odebvst() function on line 29. As these examples demonstrate, only values that are needed by the functions need to be included in the calling arguments. For comparison purposes, the maximum difference in the solution values for the two formulations is calculated on line 34 through 37 and printed out on line 38. The resulting output shown in the listing indicates that the solutions for the u[1] variable with the two formulations differ at most by 1.07×10^{-14} which is close to the numerical accuracy of the computer.

The solution of the boundary value problem for beam deflection is shown in Figure 11.3. It can easily be seen that the zero boundary values on both the function and the first derivatives are satisfied by the resulting solution. The plotted solution could be the results from either of the formulations as the maximum difference in the two formulations is very small.

A few final points are appropriate in using the supplied odebvst() code for solving boundary value problems. First for a set of first-order equations, the Runge-Kutta integration routine may also be evoked by using the statement odebiv = odebrk before calling the odebvst() function. Finally one can mix first-order and second-order differential equations provided one defines all second-order differential equations before defining any first-order differential equations. In all cases the number of boundary conditions must match the total number of derivatives in all the defined differential equations. For example a third-order differential equation with boundary values could be specified as three first-order differential equations or one second-order differential equation followed by one first-order differential equation. In either case an appropriate set of three boundary values equations would need to be specified.

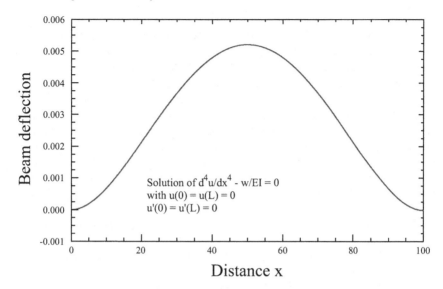

Figure 11.3. Soultion to fourth-order boundary value problem of beam deflection.

The shooting method can be employed with slightly nonlinear boundary value problems. The method is not reliable for many highly nonlinear boundary value problems and another technique more appropriate for such problems is developed in a later section of this chapter. One convenient test nonlinear boundary value problem is the following equation with boundary conditions:

$$\frac{d^2u}{dx^2} + 4\left(\frac{du}{dx}\right)^2 = 0; \text{ with } u(0) = 0,\ u(1) = 2 \tag{11.16}$$

This is reasonable nonlinear near x = 0 and has the closed form solution:

$$u(x) = \frac{1}{4}\ln[x(e^{4u(1)} - e^{4u(0)}) + e^{4u(0)}], \text{ and } u'(0) = [e^{4(u(1)-u(0))} - 1]/4 \tag{11.17}$$

The two solution constants have been expressed in terms of the value of the function at the upper and lower boundary values or $u(0)$ and $u(1)$. The first derivative of the solution at x = 0 is also given above and it can be seen that the solution has a large first derivative at this lower boundary even for a rather small value of the variable at the upper boundary.

Listing 11.5 shows a code segment for solving this two point boundary value problem using the shooting method with the odebvst() function. Based upon the previous listings this should be pretty self explanatory. The software routines handle nonlinear differential equations in exactly the same manner as linear equations, with line 8 in this case defining the nonlinear differential equation. The left and right boundary value equations on line 12 and 13 are for fixed boundary values, but the procedure would be the same if the boundary conditions were some nonlinear functions of the boundary values and derivatives. The initial value input to the odebvst() function for the function value and first derivative at 0 is seen to be simply zero as shown on line 16 where the odebvst() function is called (last two table values in the argument list). These simple values are used although it is known from the exact solution of Eq. (11.17) that the initial derivative has a rather large value at the left boundary. A simple initial value is input to the boundary value solver and it is left up to nsolv() to sort out the correct first derivative. For some problems one might have to have a closer initial guess to the derivative in order to obtain convergence. However the simple approach leads to a converged solution in this case.

```
 1 : -- /* File list11_5.lua */
 2 : -- Shooting method for boundary value problem with nonlinear DE
 3 : require"odebv" -- Boundary value by shooting method solver
 4 :
 5 : nx,ubL,ubR = 2000, 0, 2 -- #points, Left, Right Boundary values
 6 :
 7 : f = function(eqs,x,u,up,upp) -- Differntial equation
 8 :    eqs[1] = upp[1] + 4*up[1]^2
 9 : end
10 : -- Define Left and Right boundary equations
11 : fb = function(bv,uL,uR)
12 :    bv[1] = uL[1] - ubL -- Left boundary
13 :    bv[2] = uR[1] - ubR -- Right boundary
14 : end
15 : -- Use multiple intervals, from 0 to .1 and from .1 to 1
16 : s1,n1,n2,err = odebvst({f,fb},{0,{.1,1},nx},{0},{0}) --Solve BV
17 : print(n1,n2,err)
18 : plot(s1[1],s1[2])
19 : write_data("list11_5.dat",s1)
Output:
9     3  8.8377682964176e-007
```

Listing 11.5 Example code for solving a second-order nonlinear boundary value problem with fixed boundary values.

The solution for this BVP is shown in Figure 11.4. The large derivative at the left hand boundary can be seen as expected from the exact solution of Eq. (11.17). The solution looks like what might be expected from the exact solution and if one plots the exact solution the two curves will fall on top of each other, indicating a reasonably accurate solution. However, an obvious question for any solution is how one knows if the solution is an accurate solution. A more basic first question might be how does one check on convergence of the solution? To aid in evaluating convergence, the odebvst() function returns two integers shown as n1 and n2 on line 16 and an err value. The n1 value is the number of nsolv() iterations used to obtain convergence of the solution and n2 is the maximum number of iterations needed by the initial value solver for convergence of the numerical integration routine at any point along the solution. Both of these have maximum allowed values of 100 since they both use nsolv() for the required nonlinear iterations. The printed output in Listing 11.5 shows that these values are 9 and 3 respectively, well below the maximum number specified in nsolv(). Also the printed value of err indicates that the relative error in achieving the upper boundary value as specified in the boundary conditions is 8.8 x 10^{-7}. A look at the output file shows that the upper value achieved is actually 1.9999999999974 which is accurate to about 12 decimal digits. The accuracy of the shooting method will be explored in considerable depth in the next section.

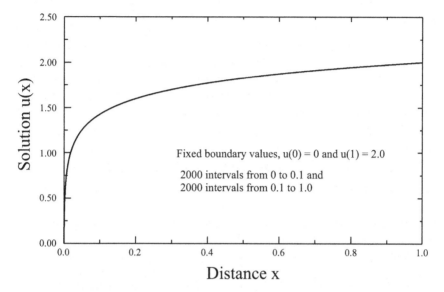

Figure 11.4. Solution values for the example nonlinear boundary value problem of Listing 11.5.

A new feature of the odebvst() function is also illustrated in this example. This is the ability to specify multiple spatial intervals for the solution. The second argument to the odebvst() function is specified on line 16 as the table {0, {.1,1},nx} with an array of values as the second entry in the list of values. The first entry (0) specifies the starting value of the independent variable and the last entry (nx) specifies the number of spatial intervals. The second entry can be either a single number specifying the upper limit to the spatial interval or a table with a list of increasing spatial values. For a table of values, the solution will be calculated with nx spatial intervals between each listed set of values. In this case the solution will be obtained with nx intervals between 0 and 0.1 and with nx intervals between 0.1 and 1.0. More intervals can be included as desired. This is used in this example to obtain more spatial resolution near x = 0 where the function is changing very rapidly as can be seen from the solution in Figure 11.4. This use of multiple intervals is identical to the discussion in the previous chapter on the variable time step solutions and the reader is referred to Section 10.3 for a review. Since the present boundary value approach is built upon the code developed there, specifically the odeiv() function, the variable time step solution method can also be used here with no required changes. In fact the adaptive step size algorithm odeivs() developed in Section 10.5 can also be used in solving boundary value problems by the shooting method. To use this adaptive step size algorithm one simple needs to include the statement odeiv = odeivs in the code somewhere before calling the BV solver. This will force the use of the variable step size solver, odeivs(), for the basic initial value solver in place of the odeiv() solver.

The shooting method of solving boundary value problems is a relatively straightforward extension of the initial value solution technique of the previous chapter. This combined with the nsolv() code for iterating over the set of initial conditions in order to satisfy the two point boundary values constitute the heart of this boundary value approach. Some typical examples have been given of the method applied to both linear and nonlinear BV problems. The next section takes a more in depth look at the accuracy possible with this method and develops a means by which the accuracy of this method can be estimated for both linear and nonlinear problems.

11.3 Accuracy of the Shooting Method for Boundary Value Problems

One must always be concerned about the accuracy of any numerical technique and especially with numerical techniques for nonlinear problems. The previous chapter contained considerable discussion of the inherent accuracy of various numerical approaches to the solution of differential equations of the initial value type. Since the heart of the shooting method for boundary value problems is the initial value problem, much of the discussion in the previous chapter is relevant here. For the trapezoidal integration technique used in odebiv(), the previous chapter has shown that the relative solution accuracy tends to decrease as the length of the integration period increases and the accuracy increases with a decrease in the step size taken in the numerical integration. The additional factor to be considered here is any effect of the two point boundary value specification on the numerical accuracy. Consider first a single second order differential equation of the classical two point boundary value type where the function value is specified on both boundaries. If a numerical integration algorithm begins with exactly the correct initial conditions on both the function value and the derivative at the starting boundary, the solution will have some inaccuracy by the time the numerical solution reaches the second boundary due to inherent inaccuracies of any finite numerical technique. It can thus be concluded that in order to obtain an exact value at both boundaries, an initial value problem must start with some slight inaccuracy in the derivative value. The accuracy with which the solution matches the second boundary value depends on the accuracy of the boundary iterative technique used in the solution. For a linear differential equation the accuracy with which such a second boundary condition can be determined is limited only by the accuracy with which two linear equations can be solved and this is essentially the relative accuracy of the programming language (about 15 decimal digits for this work). The example in Listing 11.4 provides such an example. If one looks at the output from executing this program, it can be seen that the error in the upper boundary value of zero is in the range of 1×10^{-16} for that example.

For a nonlinear differential equation or for a nonlinear boundary value specification, the accuracy of matching the boundary condition will be limited by the accuracy specified in the nsolv() function where the default relative accuracy is set at 2×10^{-6}. It can thus be expected that the accuracy of a BV solution at the spatial boundaries will be somewhere between this value and the machine accuracy. The achieved error of a solution for a boundary value problem by the shooting method is thus not expected to show a general increase from the starting boundary to the final boundary as in a pure initial value problem. Forcing the solution to match a second boundary value will give a solution with about the same relative accuracy near both boundaries. One can think of the tendency of a solution to become more inaccurate with distance as being forced back toward zero error at the second boundary by the accurately known boundary value. Thus for fixed end point values, the error in the shooting method to expected to show a peak somewhere along the solution and perhaps near the center of the interval of the solution.

The three examples in the previous section with results shown in Figures 11.2, 11.3 and 11.4 provide excellent examples for looking at the accuracy of the shooting method. One provides mixed boundary conditions, one is a high-order equation and one is a nonlinear equation and in all cases exact solutions are known with which to compare the numerical solutions. Consider first the fourth order differential equation for beam deflection with fixed boundaries as defined in Listing 11.4 and with the solution shown in Figure 11.3. By executing the code in Listing 11.4 for different numbers of solution points and comparing with the theoretical solution, the error results shown in Figure 11.5 can be generated. Three curves are shown in the figure for three different integration step sizes corresponding to 500, 1000 and 2000 spatial intervals taken over the 0 to 100 range. These errors are for the formulation in terms of two second-order equations. However, the errors for the four first-order equations are essentially identical. The center solid curve corresponds to the values shown in Listing 11.4 from which the solution was previously plotted. The maximum error is about 2×10^{-8} while the maximum solution value seen in Figure 11.3 is 5×10^{-3} so the maximum error relative to the peak solution value is about 4×10^{-6}. As discussed in a previous paragraph, it can be seen that the error near the right boundary becomes very small as the numerical solution value is forced to match the known solution value. Somewhat surprising is the fact that the left boundary value does not also become very small. In fact the error at x = 0 is zero as the solution value matches exactly the initial condition. The errors near the left boundary are most likely related to the first-order algorithm used to start the solution in the initial spatial interval.

In Figure 11.5 the error varies essentially as h^2 as can be seen by the factor of 16 difference between the upper and lower curves in the figure. This is to be expected from the extensive discussion in Chapter 10 on the accuracy of the TP algorithm for initial value problems. This means that a good estimate of the error in the solution can be obtained by use of the h-2h algorithm discussed in the previous

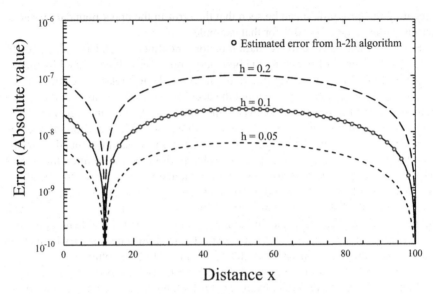

Figure 11.5 Error in solution for fourth-order differential equation for beam deflection. Solution for beam deflection is seen in Figure 11.3.

chapter. One essentially solves the problem for two values of the step size differing by a factor of 2 and then uses the algorithm of Eq. (10.44) to estimate the error. The results of such an evaluation using step sizes of 0.2 and 0.1 are shown as the circular data points in Figure 11.5. These points essentially match the actual error for the 0.1 curve verifying the accuracy of this error estimation technique. Since error estimation is very important, a special function called odebvste() has been coded and included in the odebv listing of available functions. This is a replacement for the odebvst() boundary value solver and returns both the solution for the BV problem plus an array containing the estimated errors in the solutions. This is similar to the odeive() function in the previous chapter which can be used in place of odeiv() to return a solution plus an error estimate. Code for the odebvste() function will not be shown, but is a straightforward implementation of one call to odebvst() with the input spatial parameters and a second call to odebvst() with the spatial steps reduced by a factor of 2. The results are used in the h-2h algorithm to then calculate an error estimate. The code shown in Listing 11.6 illustrates the use of this function to generate data for the center curve and data points shown in Figure 11.5. The code is similar to that previously presented in Listing 11.4 with the replacement of the odebvst() function by the odebvste() function on line 18. This replacement function can be used whenever an estimate of the error in the solution is desired. Of course this will require some additional execution time as another solution has to be obtained with half the data points, so the execution time will increase by approximately a 1.5 factor.

```
 1 : -- /* File list11_6.lua */
 2 : -- Shooting method for BV problem with mixed conditions

 3 : require"odebv" -- Boundary value by shooting method solver
 4 :
 5 : E,I,w,nx,xmax = 1.e7,500,100,1000,100 -- Independent parameters
 6 : EI = E*I
 7 :
 8 : f2 = function(eqs,x,u,up,upp) -- 2 second-order Diff equations
 9 :     eqs[1] = upp[1] - u[2]
10 :     eqs[2] = upp[2] - w/EI
11 : end

12 : fb2 = function(bv,uL,uR,upL,upR) -- Left, Right values
13 :     bv[1] = uL[1]; bv[2] = upL[1] -- u[1](Left) = u'[1](Left) = 0
14 :     bv[3] = uR[1]; bv[4] = upR[1] -- u[1](Right) = u'[1](Right) =
0
15 : end
16 :
17 : -- Solution with estimated error
18 : s2,err = odebvste({f2,fb2},{0,xmax,nx},{0,0},{0,0})
19 : write_data(20,"list11_6.dat",s2,err)
```

Listing 11.6. Example code for calculating boundary value solution plus error estimate using odebvste() function.

The nonlinear BV problem of Listing 11.5 provides a more severe test of the accuracy of the present shooting solution methodology. Figure 11.6 shows the accuracy of the solution as calculated by the code in Listing 11.5 which uses two intervals for spatial points with an equal number of grid points from 0 to 0.1 and from 0.1 to 1.0. This was an attempt to get more accuracy near x = 0 because of a large derivative in the function near that boundary. In Figure 11.6 ignore for the moment the dot-dash curve labeled "Adaptive step size algorithm" and concentrate on the three curves with associated N values. The actually achieved error in the solution is seen to not be as good for this case as the previous case with absolute errors near the origin being in the range of 1×10^{-4}. Also it can be seen that the error near the terminal or right boundary again approaches zero as the shooting method forces the final value to agree with the exact solution value.

Even though the absolute error is somewhat larger in this case, it can be seen that the error still depends essentially on the square of the spatial interval taken in the solution. As N goes from 1000 to 4000, the error decreases by essentially the 16 factor expected from the h-2h algorithm. The estimated error using the h-2h algorithm for the N = 2000 case is also shown as the dotted line in the figure. The agreement is reasonably good with the actual error achieved in the solution. This estimated error was obtained by using the odebvste() function as previously indicated in Listing 11.6. Although not a general proof of the h-2h algorithm for any nonlinear BV problem, the agreement in this example gives credence to the use of this algorithm for estimating the error for other nonlinear BV problems solved by the shooting method.

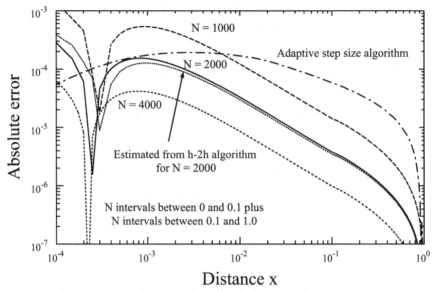

Figure 11.6. Error in solution of nonlinear boundary value problem with formulation of Eq. (11.16). See Figure 11.4 for solution and Listing 11.5.

In the previous chapter an adaptable step size algorithm was developed for cases where the solution of an initial value problem changed rapidly at a boundary or somewhere along the solution. The question arises as to how this could also be combined with the shooting method to perhaps improve the accuracy of problems such as this nonlinear BV example. This is in fact relatively easy to do. To make use of the adjustable step size solver, odeivs() in place of the odeiv() solver simply requires a a code statement of 'odeiv = odeivs' placed before calling the BV solver. Listing 11.7 shows code for such a replacement when used for solving the present nonlinear BV problem using the adaptive step size initial value solver. The only code change needed is seen on line 4 which redefines the odeiv() function as indicated above. This is again an interesting example of the ability to reuse developed code segments in important new ways with a minimum of additional effort. The error in the solution obtained with this adaptive solver is shown as the dot-dash curve in Figure 11.6. The error is seen to be better than the other multiple interval technique at small distances, but is somewhat less accurate at larger distances. It can also be seen that the adaptive step algorithm attempts to maintain a somewhat constant error throughout the entire solution interval, but again the error drops to zero near the right boundary where the solution is forced to give the exact value. The total number of spatial steps generated in the adaptive solution is only 449 as opposed to 2000, 4000 and 8000 for the other cases shown in Figure 11.6. The achieved error for the adaptive algorithm is thus seen to be very good for the number of step sizes used in the calculation when compared with the uniformly spaced step size calculations. Which of the solutions would be better in any particular engineering application would depend on whether the solution was needed with greatest accuracy at small or large values of the independent variable. The adaptive step size solver can be used with any of the examples in this section.

```
 1 : -- /* File list11_7.lua */
 2 : -- Shooting method for boundary value problem with nonlinear DE

 3 : require"odebv" -- Boundary value by shooting method solver
 4 : odeiv = odeivs -- Use adaptive step IV solver
 5 :
 6 : nx,ubL,ubR = 2000, 0, 2 -- #points, Left, Right Boundary values
 7 :
 8 : f = function(eqs,x,u,up,upp) -- Differntial equation
 9 :     eqs[1] = upp[1] + 4*up[1]^2
10 : end

11 : -- Define Left and Right boundary equations
12 : fb = function(bv,uL,uR)
13 :     bv[1] = uL[1] - ubL -- Left boundary
14 :     bv[2] = uR[1] - ubR -- Right boundary
15 : end
16 :
17 : s1,n1,n2,err = odebvst({f,fb},{0,1},{0},{0}) -- Solve BV problem
18 : print(n1,n2,err)

19 : plot(s1[1],s1[2])
20 : write_data("list11_7.dat",s1)
```

Listing 11.7. Code segment example for the use of an adaptive step size IV solver with the BV solver. Compare with Listing 11.5.

As a final example of solution accuracy, the mixed boundary value problem of Listing 11.2 with the solution given in Figure 11.2 will be considered. This is of some interest because the BV algorithm does not force the solution to a particular value at the boundaries, but only some combination of the solution value and the boundary derivative is forced to zero. For such a case it might be expected that the solution error will behave differently at the two boundaries, whereas in the other examples, the error was forced to be zero at the boundaries of the solution interval. Figure 11.7 shows the error achieved in the solution for three different spatial step sizes corresponding to 1000, 2000 and 4000 spatial intervals in the solution. This data was generated by re-executing Listing 11.2 with different numbers of spatial intervals. Again it is seen that the fundamental h^2 variation in the solution error is observed. In this case it can also be seen that the error continues to increase with an increasing value of the independent variable. This is the type of growth in error seen in the previous chapter associated with the basic initial value solution algorithms. In this example with the mixed boundary condition, there is no fixed boundary value that forces the solution error to go to zero on the right or final boundary as in the previous BV problems, so the fundamental growth in error with solution distance shows up in the BV solution. In any case the error in the solution is small and within the range expected from the TP numerical integration algorithm. Also shown in the figure are circular points of estimated error based upon the h-2h algorithm and the results show excellent agreement with the actually achieved error.

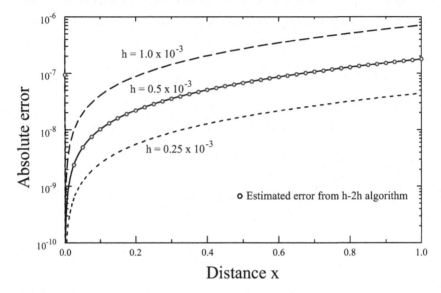

Figure 11.7. Error in solution of example BV problem with mixed boundary conditions. See Figure 11.2 for solution value.

In summary, the inherent accuracy of the shooting method for solving a boundary value problem is limited by the accuracy of the basic initial value integration algorithm used to implement the procedure. This is what one would expect form first principles. However, for fixed boundary conditions, the error does not always grow from the starting boundary to the final boundary as in a pure initial value problem. The two fixed boundary conditions can force the solution error to approach zero at both boundaries with the maximum solution error occurring somewhere in the middle of the range of the independent parameter. For the integration algorithms used in this work, the h-2h algorithm provides an excellent means of estimating the error in the solution and when accuracy of a solution is of great importance, this algorithm should always be used to estimate the accuracy of a solution.

The shooting method can be applied to a wide range of BVPs involving both linear and nonlinear differential equations and boundary conditions. However, this approach is not always the best method for solving such problems, especially for highly nonlinear differential equations which are the major emphasis of this book. For many such BVPs the method of finite difference equations is the most appropriate solution technique as this approach tends to solve for all solution points simultaneously. This technique is developed in Section 11.5. However, before going to that approach, the next section discusses a type of engineering boundary value problem closely related to that discussed in Section 11.2 and for which a shooting solution type method is very appropriate.

It's also appropriate to consider solution times and any factors that can be used to decrease the time needed to obtain a solution. If one has only a few solutions to be obtained, the time needed for a solution is probably not the most important factor as the time spent in coding the equations and in looking at any solution will typically exceed the time needed for a numerical calculation. However when many solutions are desired, solution time can become a significant factor. In the solution of the fourth order equation, two formulations have been used as shown in Listing 11.4 one with four first-order equations and one with two second-order equations. The two formulations give essentially identical results, but the solution time for the two second-order equations is significantly shorter than for the four first-order equations. This is easily understood by the fact that at each spatial point, a matrix needs to be solved to increment the solution along the spatial dimension. In one case a 2x2 matrix is involved and in the other a 4x4 matrix is involved. In general a 4x4 matrix will take about 4 times as long to solve as a 2x2 matrix. Since not all the computational time is spent in solving the matrix equations, the additional solution time should be somewhat less than the 4x factor, but will be significantly longer as the number of equations is increased. Thus for the fastest solution times, it can be said that one should always formulate the solution in terms of second-order equations as much as possible. A third-order equation could be formulated as one second-order and one first-order equation.

If one has a linear set of differential equations and linear boundary conditions, an additional speed-up can be obtained in the solution. The general formulation is such that nsolv() is used to solve for the boundary values as well as to solve for the solution at each updated spatial point. If the equation is linear, only one iteration is needed by Newton's method in nsolv() to obtain an accurate solution. For two coupled equations, a single iteration of Newton's method requires 3 calls to the function defining the differential equations. In order to determine that the first iteration gives the correct solution an additional 3 calls are required to the differential equation for a total of 6 calls. If nsolv() can assume that the equation set is linear, then the additional calls are not necessary and essentially the solution can proceed at twice the speed. One can so inform nsolv() that a linear equation set is involved by the code statement: getfenv(nsolv).linear=1. In this manner, one can speed up the calculation by a factor of about 2x. In fact the speed up is even larger because the decreased calls to nsolv() apply not only to the evaluation of each spatial point, but also to the evaluation of the unknown initial boundary value. The use of this single statement can give a very significant reduction is execution time if one has a linear differential equation and linear boundary conditions.

In order to experimentally explore these effects on execution speed, the fourth-order, linear differential equation for beam displacement as given in Listing 11.4 was executed under several different conditions and the time required for a solution evaluated. Below are some of the results, which the reader is encouraged to verify:

Solution Method	Normalized time
a. 2 Second-order formulation	1.0
b. 2 Second-order formulation + linear equations	0.4
c. 4 First-order formulation	3.2
d. 4 First-order formulation + linear equations	1.1

The results are normalized to the time taken for case a, 2 second-order equations with no specification of linear equations. The case of 4 first-order equations with no specification of linear equations is seen to take about 3.2 times longer than the two second-order equations. Finally informing nsolv() that the equations are linear, showed in each case more than a factor of 2 improvement is execution speed. An even greater improvement in speed is seen by using the four second-order equations as opposed to four first-order equations. This can be employed with either linear or nonlinear sets of differential equation.

11.4 Eigenvalue and Eigenfunction Problems in Differential Equations

A special type of two point boundary value problem arises in many areas of engineering. Such problems are frequently referred to as the Sturm-Liouville problem after the two mathematicians who made the first extensive study of the problem and published results in 1836. A typical formulation of the problem is the following second order differential equation with associated boundary conditions:

$$\frac{d}{dx}\left[r(x)\frac{du}{dx}\right]+[q(x)+\lambda p(x)]u = 0$$

$$a_1 u(a)+b_1 u'(a) = 0; \quad a_2 u(b)+b_2 u'(b) = 0$$

(11.18)

The boundary conditions are in general of the mixed type involving a combination of the function value and derivative at the two boundaries taken here to occur at $x = a$ and $x = b$. Special cases of this equation lead to many classical functions such as Bessel functions, Legendre polynomials, Hemite polynomials, Laguerre polynomials and Chebyshev polynomials. In addition the Schrodinger time independent wave equation is a form of the Sturm-Liouville problem.

Under rather general conditions on the functions r, q and p it is known that a non-zero solution exists only for a discrete set of values of the λ parameter which can be denoted by $\lambda_1, \lambda_2 \cdots \lambda_n$. These values are called the eigenvalues and the solutions corresponding to these values are called the eigenfunctions and can be denoted by $u_1, u_2 \cdots u_n$. Other known properties are the fact that over the interval a to b the eigenfunctions are orthogonal with respect to the weight function $p(x)$. Another feature of the formulation of Eq. (11.18) is the fact that the solution is specified only to within an arbitrary scale factor. This is due to the fact that the differential equation is linear in terms of the solution variable. Thus if $u(x)$ is a solution of the equation so is $Au(x)$ where A is any arbitrary constant. To obtain a unique eigenfunction some additional constraint must be imposed on the magnitude of the solution. For Schrodinger's wave equation for example, the constraint is that the magnitude squared of the solution integrated over the solution interval must equal unity.

A numerical solution of the Sturm-Liouville problem is similar to the boundary value problem considered in Section 11.2, but with some important differences. In the previous sections, the major problem was to find a set of initial parameters, value and derivative, such that the second boundary condition could be satisfied after numerically integrating the equation across the spatial interval. In the eigen-value-eigenfunction problem, one must determine the value of the eigenvalue such that the differential equation and boundary condition can be satisfied. Some thought on this problem leads to the possibility that with an appropriate formulation the shooting method combined with some type of Newton's method could perhaps be used to evaluate the eigenvalue. To understand some of the factors in such a formulation, consider one of the simplest examples of such a problem, that of Schrodinger's wave equation for a one-dimensional box with infinitely high boundary walls. The basic equation is then:

$$-\frac{\hbar^2}{2m}\frac{d^2\psi}{dx^2}+V(x)\psi = E\psi \tag{11.19}$$

$$V(x) = 0 \text{ for } 0 < x < a, \text{ and } \psi(0) = \psi(a) = 0$$

This can be put into dimensionless form with the substitution $x \to x/a$ and written as:

$$\frac{d^2u}{dx^2}+\lambda u = 0 \text{ with } \lambda = 2mEa^2/\hbar^2 \tag{11.20}$$

$$u(0) = u(1) = 0$$

The spatial interval is now 0 to 1 and λ, the eigenvalue is related to energy of the particle as in the above equation. Closed form solutions of this equation are easily obtained as:

$$u(x) = A\sin(\sqrt{\lambda}x) + B\cos(\sqrt{\lambda}x) \tag{11.21}$$

To satisfy the $x = 0$ boundary condition requires that $B = 0$ and the $x = 1$ boundary condition can only be satisfied for a nontrivial solution by the eigenvalue requirement:

$$\sqrt{\lambda_n} = n\pi \text{ or } \lambda_n = n^2\pi^2 = 2mEa^2/\hbar^2 \text{ with } n = 1, 2, 3 \cdots \tag{11.22}$$

The corresponding eigenfunctions are:

$$u_n(x) = A\sin(n\pi x) \text{ for } 0 < x < 1 \tag{11.23}$$

The first three eigenfunctions are shown in Figure 11.8 where the constant has simply been taken as $A = 1$.

These results are of course well known and are reproduced here for discussion purposes and to aid in understanding how a numerical approach might be developed to obtain the same results. After developing a numerical approach the formulation can then be applied to more complicated eigenvalue problems for which closed form solutions can not be so readily obtained. It can be noted that as the eigenvalue increases, the solution has more zero crossings over the spatial interval. If the end points are neglected, it is seen that the number of internal zeros is equal to the order of the eigenvalue less one. For example the $n = 3$ solution has 2 internal zero crossings. This general trend of course occurs for the ever increasing eigenvalues. For more complicated eigenvalue problems, it is also possible to classify the solutions by the number of internal zero crossings.

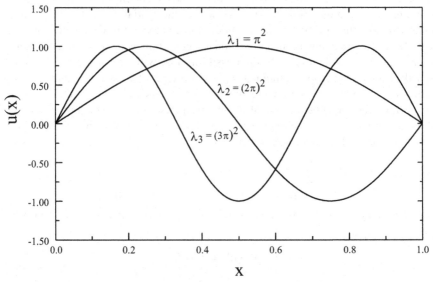

Figure 11.8. First three eigenfunctions for infinite potential well problem.

Also it can be seen that while the end point values of the solution are fixed at a 0 value, the end point derivatives are not determined by the eigenvalue or the eigenfunction. If the derivative of Eq. (11.23) is evaluated, it is seen that the end point derivatives depend on the constant A or the scale of the solution which as previously noted is not determined by the eigenvalue. Thus thinking in terms of applying the shooting method for integrating the differential equation for this problem, an approach would be to start at the left boundary with known values of the function and derivative. But what value of the starting derivative should be used? Well it doesn't matter what value is selected as long as some finite value is selected, since this only affects the scale of the solution and not the eigenvalue. For example some convenient value such as unity value can be selected for the derivative. The previous shooting method formulation for integrating an initial value problem can then be used and Newton's method applied to evaluate the appropriate eigenvalue that will satisfy the second boundary condition. With this approach a solution can be started at the left boundary (or right boundary) and integrated to the right boundary (or left boundary).

A second approach would be to start an initial value problem at the center of the spatial interval as seen in Figure 11.8. At the center point, the solutions are of two types. For odd integer values, the solution has a maximum or minimum at the center line and zero derivative value. For even integer values, the solution has zero value at the center line, but has a finite derivative value. In one case there is a symmetrical solution and in the other case there is an antisymmetrical solution. In either case if an initial value integration starts at the center point, the values of both the function and first derivative which would be needed for the initial value problem are known. However, this approach depends on the symmetry of the differential equation about the center line. For many problems of interest, one does not have a symmetrical function about some center line, so this approach is not applicable to all types of eigenvalue problems. For this reason an approach will be developed that integrates the differential equation from one boundary of the problem to the other boundary.

```
 1 : -- /* File list11.8.lua */
 2 : -- Shooting method for boundary value plus eigenvalue problems
 3 : require"odebv"
 4 :
 5 : f = function(eqs,E,x,u,up,upp) -- Differntial equation
 6 :    eqs[1] = upp[1] + E[1]*u[1]
 7 : end
 8 :
 9 : nx,xmin,xmax = 2000,0,1
10 : Ei = 0; E = {Ei} -- Guess at an eigenvalue
11 : -- Set initial value to zero and derivative to 1.0
12 : s,ns,nm,err = odebvev({f,E},{xmin,xmax,nx},{0},{1})
13 :
14 : plot(s); print(E[1],ns,nm,err)
15 : print('Eigenvalue error =',E[1]-math.pi^2)
16 : write_data('list11.8.dat',s)
17 : print('number of zeros = ',nzeros(s))
Output:
9.869608458314 2        6        2.7058808255092e-010
Eigenvalue error =      4.0572246629011e-006
number of zeros =       0
```

Listing 11.8. Code example of minimum definitions needed to solve an eigenvalue boundary value problem.

With this discussion as a background, computer code for the numerical solution of eigenvalue differential equation problems can be developed. Listing 11.8 is example code of the minimum definitions needed to solve an eigenvalue problem. The differential equation to be solved is defined on lines 5 through 7 in the same manner as previously used for differential equations. The difference is the need here to include one additional parameter in the argument list, the eigenvalue which is the second parameter on line 5 and is labeled E. Actually the eigenvalue is taken as an array of eigenvalues which is in keeping with the fact that our basic differential equation solvers can handle multiple second order differential equations, each of which could be associated with a different eigenvalue. For the present discussion only one differential equation will be considered with one eigenvalue. The additional argument for the function is included in the argument list as the second parameter before the listing of the independent variable, but the location is somewhat a matter of choice. The differential equation plus an initial guess on the eigenvalue is all that one should have to specify and the software should perform the remainder of the initial value equation integration as well as optimizing the eigenvalue to satisfy the boundary value problem. On line 12 a function odebvev() is called which is assumed to perform this function. This function differs from the BV solver odebvst() in Section 11.2 only by the additional energy eigenvalue parameter. In order to keep the same number of total arguments to the solver as before, the eigenvalue (E here) is included in the first array of values to the function in the form of a table of function names and eigenvalues as: {f,E}. This is in much the same way that the boundary value array was included in the calling argument for the odebvst() function. Looking back at Listing 11.7 for example one see the form {f,fb} in the same argument position. Here, the unknown eigenvalues are the unknowns whereas in the previous examples the initial conditions for the differential equations were the unknowns.

The appropriate code for implementing the eigenvalue algorithm is called into the program through the odebv listing of programs on line 3 of the code. Before presenting results of running the code in Listing 11.8 it is appropriate to discuss further the implementation of the working part of the code, the odebvev() function. The code for this function is shown in Listing 11.9. The code is similar to Listing 11.3 for the basic shooting boundary value problem so it is assumed that the reader is familiar with that code. The core problem being addressed is finding a value of the eigenvalue such that the boundary conditions for the differential equation can be satisfied. The initial boundary conditions are satisfied by inputting the correct boundary conditions for the starting boundary, for example in Listing 11.8 the {0} initial value and {1} initial derivative on line 12. The final initial condition is satisfied by using the solver nsolv() to adjust the energy eigenvalue. This is on line 67 of Listing 11.9 where the function being called by nsolv() is the internally defined function fbeval() defined on lines 59 through 66. This in turn is a function that returns any error in matching the boundary conditions after the solution of the differential equation. So inherent in this function is the integration of the differential equation treating it as an initial value problem. The solution of this initial value problem is obtained by the call to the odeiv() function on line 64 after the initial values are set on line 63.

The only additional difficulty is that the normal calling sequence for the initial value code does not expect the additional eigenvalue parameter (E) that is now required in defining the differential equation. This in turn requires that a proxy function be used which can insert this additional parameter between the function needed by the IV solver and the eigenvalue defining differential equation. This proxy function is defined internally on lines 60 through 62 of the code and it can be seen that this function takes the arguments needed by odeiv() and adds the eigenvalue argument before calling the feqs() function on line 61. This in turn is the user supplied function defining the differential equation set with the unknown eigenvalue. Because this proxy function needs the eigenvalue, it must be defined internally to the fbeval() which in turn is being called by the nsolv() function which is in turn varying the eigenvalue to satisfy the boundary conditions. This can all be a little confusing to new users of the code, but can become clear with a little study of the calling arguments and calling sequence. As a final comment, the odebvev() function returns (see line 72) the solution values in the s array, the maximum number of iterations required in the odeiv() function on line 70, plus the number of newton iterations required on line 69 for obtaining the eigenvalue and the estimated error in the obtained eigenvalue.

```
 1 : -- /* File odebv.lua */
 2 : -- Shooting(ST) method for boundary value problems
    -----
46 : -- Shooting method for boundary value plus eigenvalue problems
47 : odebvev = function(feqs,tvals,ui,upi) -- Basic EV solver
48 :    odebiv = odeb12 -- Second order equations
49 :    local nr,s,ns,ni,fbound,E,feq,feqse,err = 'false'
50 :    feq,E,fbound = feqs[1],feqs[2],feqs[3]
51 :    if type(E)=='function' then nr,E,fbound = 'true',fbound,E end
52 :    local u,up,nu = {},{},#ui
53 :    -- Define local default function for boundary values
54 :    local defb = function(beqs,uF) -- Default function for fbound
55 :       for i=1,ns do beqs[i] = uF[i] end -- BV conditions
56 :    end
57 :    fbound = fbound or defb; ns = #E
58 :    -- Define local function for Newton's method
59 :    local fbeval = function(beqs,E) -- E[] is array of EVs
60 :       feqse = function(eqs,x,du,dup,dupp) -- Local function
61 :             feq(eqs,E,x,du,dup,dupp) -- Adds E to calling argument
62 :       end -- Now use in calling equation solver
63 :       for i=1,nu do u[i],up[i] = ui[i],upi[i] end -- Set IVs
64 :       odeiv(feqse,tvals,u,up) -- Solve initial value problem
65 :       fbound(beqs,u,up) -- Update boundary errors
66 :    end -- End of function for Newton's method
67 :    ni,err = nsolv(fbeval,E) -- Solve BVP subject to fbeval()
68 :    for i=1,nu do u[i],up[i] = ui[i],upi[i] end
69 :    s,ns = odeiv(feqse,tvals,u,up) -- Update with final values
70 :    odebiv = loc_odebiv -- Always leave with odebiv() reset
71 :    if nr then nr,feqs[2],feqs[3] = 'false',fbound,E end --
72 :    return s,ns,ni,err,E
73 : end
```

Listing 11.9. Code for implementing the shooting method with an eigenvalue differential equation solution.

A few other remarks can clear up some of the code on lines 48 through 58. Line 48 simply ensures that the basic IV solver is the second order equation solver, the odeb12() function. A local boundary value function defb() is defined on lines 54 through 56. This function simply defines the final boundary values such that the function value is forced to zero on the final boundary. Most boundary value problems are of this type and can use this internal function. However, how about the possibility of mixed boundary conditions as given by the general formulation in Eq. (11.18)? To handle such cases, the code is arranged so the user can supply an additional function defining the final boundary conditions. This is done by adding a third argument to the first calling argument table to odebvev() in the form of {f, fbound, E} where fbound() is a function defining the desired boundary conditions and is a replacement for the default defb() function on lines 54 through 56. This would replace the simpler calling sequence of {f, E} with the table {f, fbound, E}. The code for handling an additional argument and setting the appropriate boundary function is on lines 50, 51 and 57.

Now that the eigenvalue solution code has been discussed, the simple example in Listing 11.8 can be reconsidered. The initial guess at an eigenvalue is 0 and the output when running the code gives a found eigenvalue of 9.869608458314. The printed output also shows that 2 iterations were required in solving the linear differential equation (as expected) and 6 newton iterations were required in obtaining the final eigenvalue. A separate function available in the odebv code is the nzeros() function called on line 16 of the code in Listing 11.8. This evaluates the number of zero crossings for a function and shows that the obtained solution has 0 zero crossings and thus corresponds to the n = 1 curve in Figure 11.8. This is also verified by the popup plot of the obtained solution on line 14. The results are not shown here since they are identical to the λ_1 curve in Figure 11.8. The n = 1 numerical eigenvalue from the number of spatial points and the code used here then differs from the exact value of π^2 by the value 4.06×10^{-6} so that the numerical solution for the eigenvalue is accurate to about 5 decimal digits. There are two possible factors that might limit the accuracy of the numerical eigenvalue. First is some inaccuracy in matching exactly the second boundary condition. Second because of the finite spatial step size there are inherent inaccuracies in the numerical integration algorithm for solving the differential equation.

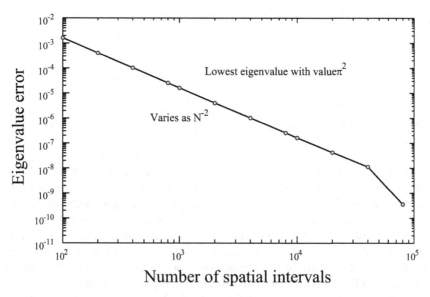

Figure 11.9. Variation of error in the lowest energy eigenvalue with number of spatial intervals in solving differential equation.

In order to ascertain which of these is the limiting factor, the code in Listing 11.8 was executed with varying numbers of spatial steps ranging from 100 to 80000 and the results are shown in Figure 11.9. The eigenvalue error is seen to essentially vary as the inverse square of the number of spatial intervals, which leads to the expectation that the accuracy is being limited by the numerical integration and not by the accuracy with which the upper boundary condition is satisfied. This also leads to a simple method of increasing the accuracy of the eigenvalues using a Richardson type extrapolation (see Eq. 5.18). For example if E_1 and E_2 are the eigenvalues calculated at 1000 and 2000 spatial intervals, then the extrapolated value of the eigenvalue would be given by $E = (4E_2 - E_1)/3$. Applying this to the above data at 1000 and 2000 spatial intervals, gives the following:

$E_1 = 9.8696206237853$

$E_2 = 9.869608458314$

$E = 9.8696044031569$ (exact value is 9.8696044010894)

This extrapolated value differs from the exact value by only 2.07×10^{-10} which is accurate to about 10 decimal digits. The use of Richardson type extrapolation in this example decreases the error in the eigenvalue by about a factor of 1000.

The h-2h calculation needed for the Richardson type extrapolation is also the calculation needed to estimate the error in the solution of a boundary value problem as discussed in the previous section. Thus it seem appropriate to combine these into an eigenvalue solver and such a function has been coded as the function odebveve() which is also available with the require"obebv" statement. An example of using this function for a higher order eigenvalue and function of the same constant potential problem is shown in Listing 11.10. The difference from Listing 11.8 is the call to odebveve() on line 12. The returned values by this function are the eigenfunction array (s here), the estimate of error in the eigenfunction (err here) and the extrapolated eigenvalue array (Ex here). The code also illustrates another function sqnorm() that is useful with Schrodinger's equation or many other eigenfunction problems. This function takes a solution and returns two functions as shown on line 14 of the code. The first function (s on line 14) is the same as the input function but has been scaled so that the square integral over the spatial region equals some value specified as the second argument to the sqnorm() function (1.0 here). The second returned function is the square of the eigenfunction. For Schrodinger's equation this would represent a probability density function. In many eigenfunction problems, this is the function of most direct physical interest. The code shows popup plots of both the original function and estimated error (on line 13) and the square normalized function (on line 14).

The printed output shows the order of the eigenfunction evaluated by the initial guess of an eigenvalue (order 5 in this case with 4 internal zeros for an initial guess of 210 on line 10). The printed output also shows the eigenvalue (EV) evaluated by the nsolv() function, the extrapolated eigenvalue, the theoretical eigenvalue and errors between the evaluated and extrapolated eigenvalue. Perhaps the most important information is contained in the error between the evaluated eigenvalue or the extrapolated eigenvalue and the theoretical value. The last two printed output values indicate that the directly calculated eigenvalue is accurate to about 5 decimal digits while the extrapolated eigenvalue is accurate to about 8 decimal digits. This again illustrates the advantage to be gained by use of the h-2h algorithm for error estimation and for the extrapolated eigenvalue. The reader is encouraged to execute the code in Listing 11.10 with different numbers of spatial intervals and observe the effect on accuracy of the eigenvalue. For 20000 spatial intervals, the accuracy increases to about 12 decimal digits, but at the expense of a somewhat longer execution time.

```
 1 : -- /* File list11_10.lua */
 2 : -- Shooting method for boundary value plus eigenvalue problems
 3 : require"odebv"
 4 :
 5 : f = function(eqs,E,x,u,up,upp) -- Differntial equation
 6 :    eqs[1] = upp[1] + E[1]*u[1]
 7 : end
 8 :
 9 : nx,xmin,xmax = 2000,0,1
10 : Ei = 210; E = {Ei} -- Guess at an eigenvalue
11 : -- Set initial value to zero and derivative to 1.0
12 : s,err,Ex = odebveve({f,E},{xmin,xmax,nx},{0},{1})
13 : plot(s,err) -- Solution and estimated error
14 : s,sq = sqnorm(s,1.0); plot(sq) -- Normalize square integral to
1.0
15 : nz = nzeros(s)+1; Et = (nz*math.pi)^2
16 : write_data('list11_10.dat',s,err)
17 : print('Eigenvalue Order = ',nz)
18 : print('EV, Extrapolated EV, exact EV =',E[1],Ex[1],Et)
19 : print('EV error, Extrapolated EV error =',E[1]-Et,Ex[1]-Et)
Output:
Eigenvalue Order =     5
EV, Extrapolated EV, exact EV = 246.74264579321 246.74011120713
246.74011002723
EV error, Extrapolated EV error = 0.0025357659738461
        1.1799008348135e-006
```

Listing 11.10. Illustration of code for solving eigenvalue problem with extrapolated eigenvalue and with estimated error for the solution, using the odebveve() function.

This simple example has been used to explore several factors in the numerical solution of eigenvalue boundary value problems. However, the spatial range for the infinite potential well problem has definite finite boundaries while many eigenvalue problems exist over an infinite spatial range, for example from 0 to ∞ or from −∞ to +∞. Additional considerations are necessary for such open ended problems. Again a relatively simple example with known eigenvalues and eigenfunctions can prove useful in exploring such problems. Therefore consider the application of Schrodinger's equation to a simple harmonic oscillator problem defined by the equations:

$$-\frac{\hbar^2}{2m}\frac{d^2\psi}{dx^2} + \frac{1}{2}Kx^2\psi = E\psi$$

$$\text{with } \psi(-\infty) = \psi(\infty) = 0$$

(11.24)

This can be put into dimensionless form by an appropriate change of variables:

$$x \to \alpha x \text{ and } E \to \lambda$$

where $\alpha^4 = mK / \hbar^2$ and $\lambda = \dfrac{2E}{\hbar}\sqrt{\dfrac{m}{K}} = \dfrac{2E}{\hbar\omega_o}$ (11.25)

The eigenvalue problem then becomes:

$$\frac{d^2u}{dx^2} + (\lambda - x^2)u = 0$$

    (11.26)

with $u(-\infty) = u(\infty) = 0$

In this equation λ is a dimensionless eigenvalue.

Solutions of this eigenvalue problem are known to result in the Hermite polynomials for the eigenfunctions and the dimensionless eigenvalues are known to be given by:

$$\lambda_n = 2n + 1 \text{ where } n = 0, 1, 2\cdots$$

    (11.27)

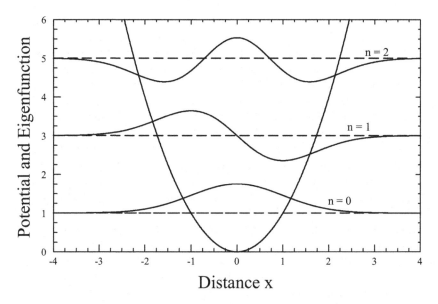

Figure 11.10. First three eigenfunctions for harmonic oscillator Schrodinger equation.

The first three eigenfunctions for this eigenvalue problem are shown in Figure 11.10. Also shown is the quadratic potential energy and each eigenfunction has been displaced vertically by the corresponding value of the eigenvalue. Again one sees the characteristic oscillatory behavior inside the potential well where the second order differential equation has an underdamped characteristic and one sees a rapid decrease to zero in the eigenfunction outside the potential well where the equation has an overdamped characteristic. Again it is noted that the order of the eigenfunctions can be correlated with the number of zero crossings of the eigen-function.

The question to be pursued here is how to develop a numerical algorithm to obtain the eigenfunctions and eigenvalues based upon the use of a shooting method for solving the basic differential equation coupled with an algorithm for adjusting the eigenvalue to match the required boundary conditions. The boundary conditions are really specified at $+/-\infty$ which cannot be reached in a numerical algorithm, but a spatial distance can hopefully be taken sufficiently large to accurately approach the requisite boundary conditions. From Figure 11.10 it can be seen that the required spatial distance to achieve a given small value of the eigenfunction depends on the order of the solution and larger spatial distances are required as the energy eigenvalue increases. Thus one must be careful in selecting a spatial range over which to apply any numerical algorithm. For the results in Figure 11.10 it appears that the range -4 to +4 would probably be sufficient for the first three eigenfunctions.

So how can a numerical solution begin at some spatial point with known boundary conditions, integrate the solution by the shooting method and then try to match the required final boundary conditions at some final spatial point? The three solutions shown in the figure suggest one approach. It is observed that at $x = 0$ the solutions are of two types: (a) the eigenfunction is a maximum (or minimum) with zero derivative value or (b) the eigenfunction is zero with a positive or negative derivative value. The even eigenvalues correspond to case (a) while the odd eigenvalues correspond to case (b). This is a known property of any eigenvalue problem with a symmetrical potential. Thus in this case an obvious approach might be to start at the origin with known values of the function and derivative and integrate to a sufficiently large spatial value where the eigenfunction can be taken as essentially zero with zero derivative value. But how would the starting value of the function and the derivative value be selected? A little thought will convince one that it doesn't matter what starting value is selected. Since the differential equation is a linear differential equation, the scale of the solution is independent of the differential equation and must be determined by some external factors. In fact the scales of the functions shown in Figure 10.11 have been determined by the requirement of unity value for the square of the eigenfunction integrated over the range of the function. Thus for the even symmetry functions one can select unity as the starting value and zero as the derivative value, while for the odd functions one can select zero for the starting value and unity for the starting derivative value. The eigenvalue should be independent of the choice of starting values.

The above procedure works as can be verified by execution the procedure. However, it depends on the basic differential equation being symmetrical with respect to the spatial variable. Not all eigenvalue problems have this property, for example Schrodinger's equation for the Hydrogen atom. It would thus be much more convenient if an algorithm could be developed that worked just as well for any potential. Thus it is important to further explore a solution procedure. Assume for the moment that the procedure outlined above of starting at the origin and shooting a solution to some large spatial value has been implemented with ad-justment of the eigenvalue to satisfy the zero boundary condition at the upper boundary and thus a valid solution and eigenvalue has been obtained. One could then think of reversing the integration process by starting at the large spatial value with a known value of the function and derivative and integrating back to the ori-gin. If one has a good integration algorithm, the value obtained back at the origin should be exactly the initial starting values or the exact solution values. However, in this case the integration could be continued through the origin and over the en-tire spatial interval since one would have good starting values for the negative spa-tial interval. Thus the integration procedure can be conceptually reversed where one first integrates from the origin to some negative spatial value and then follows this by integrating over the entire spatial range from negative to positive values.

From the above thought process it can be concluded that it should be possible to apply the shooting method over the entire spatial range provided one has a good set of starting parameters for the integration at some large negative (or positive) spatial value. But how can a "good" set of starting parameters be obtained for large spatial values where both the function and its derivative value approach zero? In starting the shooting method at the origin and going to some large spatial value one would apply a zero function value at the upper range. This would result in some finite non-zero value for the function derivative value at the upper range. Thus if this non-zero derivative value were known, the integration process could be reversed by starting at the upper boundary and integrating to the origin. But wait, the eigenvalue can't depend upon some initial derivative value as it was pre-viously argued when starting from the origin. Thus it can be similarly argued that the eigenvalue can't depend upon some assumption about the derivative value if the shooting method starts at some large positive or negative spatial value. For example suppose the method starts at the origin with a function value of 1.0 (and zero derivative value) and evaluate a converged solution to some large spatial point such as 3.0 in Figure 11.10. Suppose further that the evaluated solution de-rivative at this point is some small value such as 1×10^{-6}. If the process is then re-versed starting at the large spatial value with an assumed derivative value of unity, it would be expected that the solution value then obtained at the origin would be $1 \times 10^{+6}$. However, the obtained eigenvalue should be the same, since the scale of the solution can not be determined from the differential equation and the eigen-value is independent of the scale of the solution.

From the thought process discussed above it can be concluded that the shooting method can be applied over the entire spatial range of the solution just as easily as starting at the origin. With what starting boundary conditions one then asks? Well the answer is it doesn't matter since the scale of the solution is not deter-mined by the differential equation. One is free to choose a convenient value such as unity value for the starting derivative since this choice only determines the scale factor of the eigenfunction and not the eigenvalue. While this sounds almost too good to be true, does it really work in practice, or do some numerical errors invalidate the arguments. The answer is that the procedure works surprisingly well as long as the range of spatial values is taken sufficient large so as to ensure that the value of the eigenfunction is very small at the initial and final starting values.

```
 1 : -- /* File list11_11.lua */
 2 : -- Shooting method for boundary value plus eigenvalue problems
 3 :
 4 : require"odebv"
 5 :
 6 : f = function(eqs,E,x,u,up,upp) -- Quantum Harmonic oscillator
 7 :    eqs[1] = upp[1] + (E[1] - x^2)*u[1]
 8 : end
 9 :
10 : nx,xmax,Ei = 2000, 5.0, 5.1 -- Try different ranges
11 : E = {Ei}
12 : -- Try different initial conditions for shooting method
13 : s,err,Ex = odebveve({f,E},{-xmax,xmax,nx},{0},{1})
14 : --s,err,Ex = odebveve({f,E},{-xmax,xmax,nx},{1},{0})
15 :
16 : plot(s,err)
17 : sn,sqn,fac = sqnorm(s,1.0)
18 : if fac>1e-5 then
19 :    print('Sparial range is not large enough for high accuracy')
20 : end
21 : nz = nzeros(s,300)
22 : print(xmax,nx,E[1],Ex[1],E[1]-2*nz-1,Ex[1]-2*nz-1)
23 :
24 : plot(sn); plot(sqn)
25 : print('number of zeros = ',nz,fac)
26 : write_data('list11_11.dat',s,sn,sqn)
```

Listing 11.11. Code to implement shooting method for eigenvalue problem of harmonic oscillator.

Listing 11.11 shows code for solving the harmonic oscillator eigenvalue problem by the shooting method using this approach. The differential equation is defined on lines 6 through 8 in the usual manner. After defining the number of spatial points and the integration end points on line 10 the eigenvalue solver odebveve() function is called on line 13 with an initial value of zero for the initial function value and unity for the initial derivative both taken at -xmax which for the listed values is -5.0. It will be recalled that the odebveve() function solves the boundary value problem as well as returning an estimate of the error in the solution and an extrapolated energy eigenvalue, set to Ex on line 13. An optional call is shown on line 14 which used an initial set of solution conditions of unity for the eigenfunction value and zero for the derivative value({1},{0} on line 14). The reader can execute the code with both options and verify that there is little difference in the accuracy of the solution using either set of boundary conditions. Because of the different scale factors with different initial conditions, a normalized solution is obtained from the sqnorm() function on line 17. A printout on line 22 compares the evaluated eigenvalues with the theoretical values based upon the number of function zeros obtained on line 21. The test for spatial range on lines 18 through 20 will be subsequently discussed after looking at some results.

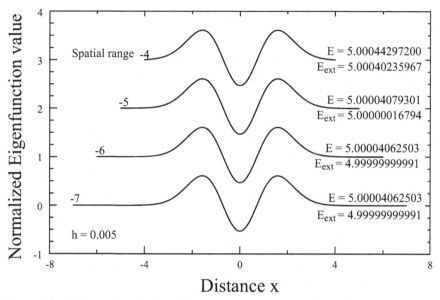

Figure 11.11. Example solutions for n = 2 eigenvalue obtained with different total spatial ranges. The exact eigenvalue is 5.0.

Figure 11.11 shows the normalized output from running the code in Listing 11.11 for several different total spatial ranges and for the n = 2 eigenvalue obtained with the initial guess of 5.1 on line 10. The solutions in the figure for the top three curves have been shifted upward to simply separate the curves in the figure. Before actually normalizing the various solutions, the actual calculated eigenfunctions will have vastly different magnitudes. For example the solution for the -4 to +4 range has a peak amplitude of about 16, while the solution for the -7 to +7 range has a peak amplitude of about 3.8×10^7, about 6 orders of magnitude larger. One can see that the different scale factor on the solution has little effect on the normalized eigenfunction or on the eigenvalue. The different spatial ranges provide important information on what is needed to obtain an accurate eigenfunction and eigenvalue when one has an infinite spatial range. The solutions demonstrate that there is an optimum spatial range over which the shooting method should be applied for the most accurate evaluation of the eigenfunction and eigenvalue and this is somewhere around -6 to +6 for this particular example. The directly determined eigenvalue is not very different for all the cases shown in the figure, having about 4 to 5 correct decimal digits. However, the extrapolated eigenvalue (E_{ext} in figure) is much more accurate for the -6 to +6 range and is accurate to about 10 decimal digits.

If the spatial range is increased to too large a value, such as -8 to +8 beyond the values shown in the figure, one begins to see problems in the solution. First, the accuracy of the extrapolated eigenvalue begins to decrease. More importantly, the final boundary value becomes very difficult to satisfy. This trend is very evident if the code in Listing 11.11 is executed with a spatial range of -8 to +8. This is a known problem with shooting methods and differential equations. For large spatial values (positive or negative) the Schrodinger differential equation takes on the characteristic of an overdamped case where the solution has a rapidly decreasing solution and a rapidly increasing solution as a function of distance. The obvious solution that satisfies the boundary condition is only the rapidly decreasing solution. However, numerical errors can easily cause a numerical solution to gain a small component of the rapidly increasing solution and deviate in an exponential-like manner toward large positive or negative values. The further one attempts to extend the solution into such a region, the more difficulty is encountered in retaining only the rapidly decreasing solution branch.

Thus when an eigenvalue problem has an infinite spatial range, one must be careful in selecting an appropriate range over which to employ the shooting method to obtain an accurate solution. Is there any characteristic of the solution that can be used to quantify the spatial range needed for an accurate solution? The answer is yes, there is such a factor and this is the scale factor by which the solution must be multiplied in order to obtain a normalized solution. This factor is returned as the third argument to the sqnorm() function as shown by the fac parameter on line 17 of Listing 11.11. If this factor, for example, is 1×10^{-6} it means that the solution obtained with the unity initial derivative value has to be scaled by this factor and thus the initial derivative value for the normalized solution will not be unity but 1×10^{-6}. This scale factor can be used as an indicator of when the spatial range of the solution should be increased. This test is shown on line 18 through 20 of the code and prints a message if the factor is not sufficiently small. For this example a desired scale factor is set at 1×10^{-5} on line 18 of Listing 11.11. The code does not include an automatic correction for too small a spatial range, but the user can re-execute the code with an increased spatial range if this condition is not met. The reader is encouraged to re-execute the code in listing 11.11 with different spatial ranges. It will be observed that calculations for the spatial ranges of -4 to 4 and -5 to 5 will both trigger the printing stating that the spatial range should be increased.

There is very little difference in starting the shooting method with a unity function value and zero derivative value as indicated by the optional code on line 14 of the listing. Again if the scale factor needed to normalize the solution is less than 1×10^{-5}, then the starting value for a normalized solution will be less than 1×10^{-5}, a value that should be sufficiently close to zero to obtain eigenvalues accurate to more than 5 decimal digits. The reader is encouraged to execute the code with different spatial ranges, different starting parameters and with different starting values for the eigenvalues. By increasing the Ei value in the code, one can obtain solutions for increasing eigenvalues. Figure 11.12 shows solutions for the square amplitude of the eigenfunctions for a range of normalized eigenvalues ranging up to 25. Only solutions for even n values are shown and all the solutions were obtained with a spatial range of -6.5 to +6.5. The reader is encouraged to execute the code in Listing 11.11 to reproduce some or all of these results.

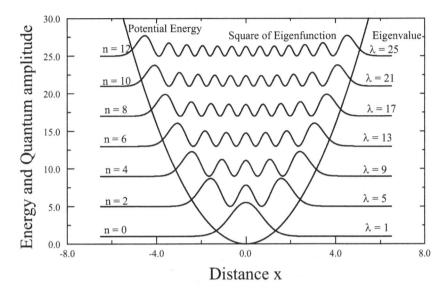

Figure 11.12. Square of eigenfunctions for harmonic oscillator and for normalized eigenvalues up to 25. Only even integer eigenfunctions are shown.

Based upon the results for the two previous examples, code segments and an approach have now been developed that can be used to solve a wide range of eigenvalue, boundary value problems. Problems for which analytical solutions are not known can be solved numerically. Before concluding this section a couple of such examples will be given.

The first example is that of a fourth order potential well in the Schrodinger equation. Code for solving this eigenvalue problem is shown in Listing 11.12. The coefficient for the potential well as defined on line 7 is selected such that the potential has the same value as a second order harmonic oscillator at a distance of $x = \sqrt{10}$ or at the point where the potential is equal to 10. The code is straight-forward and the initial guess (of 0) at an eigenvalue on line 10 is just an example (for the lowest energy state). The reader is encouraged to execute the code for varying initial guesses at the eigenvalues and observe from the popup plots the normalized eigenfunction solutions. If an increased spatial interval (beyond the 5.5 value on line 10) is needed, a statement is printed on line 15 of the code. With the code in Listing 11.12 it is easy to experiment with different parameters and rapidly observe the effects on the eigenfunctions and eigenvalues. Based upon the previous examples, the extrapolated eigenvalues (from Ex on line 12) are expected to be accurate to better than 6 decimal digits.

```lua
 1 : -- /* File list11_12.lua */
 2 : -- Shooting method for fourth power potential well
 3 :
 4 : require"odebv"
 5 :
 6 : f = function(eqs,E,x,u,up,upp) -- Fourth order potential
 7 :    eqs[1] = upp[1] + (E[1] - x^4/10)*u[1]
 8 : end
 9 :
10 : nx,xmax,Ei = 2400, 5.5,0 -- Typical parameters
11 : E = {Ei}
12 : s,err,Ex = odebveve({f,E},{-xmax,xmax,nx},{0},{1})
13 : sn,sp,fac = sqnorm(s,1.0)
14 : if fac>1.e-5 then -- Test for spatial interval increase
15 :    print('Need to increase spatial interval')
16 : end
17 :
18 : print(nx,xmax,Ex[1],E[1],nzeros(s))
19 : plot(sn); plot(sp)
20 : write_data('list11_12.dat',s,sn,sp,err)
```

Listing 11.12. Code for eigenvalue problem with fourth order potential well.

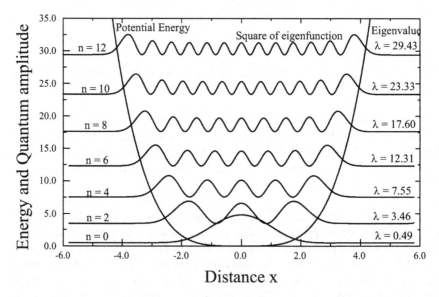

Figure 11.13. Square of eigenfunctions for fourth order potential well and for the first 12 even integer eigenvalues.

Collected results from executing the code for a range of eigenvalues are shown in Figure 11.13 for the even integer eigenvalues. This figure should be directly compared with the second order potential well shown in Figure 11.12. It can be seen that the fourth order potential well has a shape that is somewhat between that of the second order potential well and an infinite barrier square well potential.

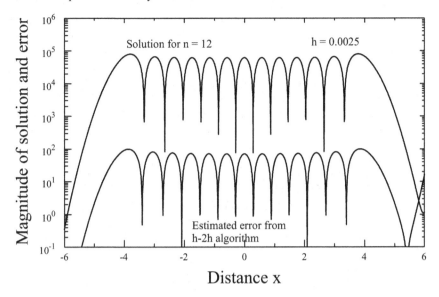

Figure 11.14. Plot of magnitude of eigenfunction and estimated error for the n = 12 case.

The eigenfunctions and eigenvalues are similar in the two cases with the order of the eigenfunction being characterized in each case by the number of zero crossings of the eigenfunction with the lowest order eigenfunction having no zero crossings. For the second order potential well, the energy eigenvalues are equally spaced while for the fourth order well, the spacing between eigenvalues increases with the order of the eigenvalue, but increases more slowly than for the square well with infinite barrier potential well case. To obtain all the data for Figure 11.13 requires a little experimenting with various guesses at the eigenvalues, since the values are not known from some closed form equation for the fourth order potential well. The reader is encouraged to experiment with the code observing the eigenfunctions for the even eigenvalues which always have a zero crossing at the origin.

The code in Listing 11.12 employs the odebveve() solver which returns not only the eigenvalue solution but also an estimate of the error in the solution and the extrapolated eigenvalue. One of the major reasons for using this is to obtain the much more accurate eigenvalue. However, an estimate of the solution error is also obtained as a byproduct of the calculation (returned as err on line 12 of the code). In a problem such as this one should examine some of the error results to obtain confidence in the accuracy of the solutions. Such results are shown in Figure 11.14 for the case of the n = 12 solution shown in Figure 11.13. This plot is for the "raw" solution before the normalization of the solution. It can be seen that the estimated error is relatively constant over the interval of the solution and approaches small values near both end points as expected. Also the estimated error is about a factor of 1000 below the solution value, so one would estimate that the solution values are accurate to about 3 decimal digits. The accuracy could be increased by using more spatial points in the solution, but this accuracy would probably be adequate for most engineering problems. Because of the increased accuracy of the extrapolated eigenvalue, this value can be considerably more accurate than the accuracy of the eigenfunction.

This section will close with a final eigenfucntion problem. The Schrodinger equation for the radial part (R) of the wavefunction of an electron in a central potential is:

$$-\frac{\hbar^2}{2m}\frac{1}{r^2}\frac{d}{dr}\left(r^2\frac{dR}{dr}\right)+\left[\frac{Zq^2}{4\pi r}+E-\frac{l(l+1)\hbar^2}{2mr^2}\right]R=0 \qquad (11.28)$$

where the third term in brackets involving l arises from the angular momentum of the electron and is zero for the lowest energy state. This equation can be simplified by use of the substitution $R = U / r$ and by the use of dimensionless variables with the replacements:

$$r \to r / a_o \text{ where } a_o = \frac{\varepsilon h^2}{\pi m Z q^2}$$

$$E \to E / E_o \text{ where } E_o = \frac{m Z^2 q^4}{8\varepsilon^2 h^2} \qquad (11.29)$$

With these substitutions the differential equation becomes:

$$\frac{d^2U}{dr^2}+\left(\frac{2}{r}+E-\frac{l(l+1)}{r^2}\right)U=0 \qquad (11.30)$$

with $U(0) = U(\infty) = 0$

This equation has of course been extensively studied and the eigenfunctions and eigenvalues are well known. The lowest normalized eigenvalue for $l = 0$ is known to occur at the normalized energy value of $E = -1.0$ in Eq. (11.30). Although it is interesting to solve this numerically, a slightly modified equation will be considered here and this is the solution for a "screened Coulomb potential". Such a modified potential is important in heavily doped semiconductors for determining the binding energy of doping impurities. For such a case the potential energy term has an exponential decay term multiplying the usual $1/r$ dependency. With this modification, the BV problem to be considered here is:

$$\frac{d^2U}{dr^2}+\left(\frac{2}{r}e^{-r/L_D}+E-\frac{l(l+1)}{r^2}\right)U=0 \qquad (11.31)$$

with $U(0) = U(\infty) = 0$

In this equation the L_D term is the normal Debye screening distance divided by the distance normalization constant a_o. For the purposes here it is not too important to know the exact meaning of this term but to know that it can range from very large values (1000 or more) when screening is relatively unimportant to small values (near 1) when screening becomes very important. Solutions of this eigenvalue problem are not known exactly, so one must resort to numerical solutions.

A code segment for solving for the eigenvalues of the screened Coulomb potential is shown in Listing 11.13. The code is very straightforward with the differential equation defined in the usual manner on line 6. The definition includes the angular momentum term although it is set to zero in the example here on line 9. The reader can re-execute the example code with different L values to see the effects of angular momentum on the solutions. The particular example is for a fairly severe screening effect where the normalized energy eigenvalue has been changed from -1.0 with no screening to -0.296. This degree of screening has reduced the lowest energy state to about 30% of the unscreened binding energy. As L_D is decreased further, the binding energy is decreased even further.

```
 1 : -- /* File list11_13.lua */
 2 : -- Eigenvalue problem for screened Coulomb potential
 3 : require"odebv";require"intp"
 4 :
 5 : f1 = function(eqs,E,x,u,up,upp) -- Hydrogen atom
 6 :    eqs[1] = upp[1] + (2*math.exp(-x/Ld)/x + E[1] -
            L*(L+1)/x^2)*u[1]
 7 : end
 8 :
 9 : L = 0; Ld = 2
10 : nx,xmax,Ei = 2000, 20, -1
11 : E = {Ei}
12 :
13 : s = odebvev({f1,E},{0,xmax,nx},{0},{1})
14 : print('Energy eigenvalue = ', E[1])
15 : sn,sqn = sqnorm(s,1.0)
16 : plot(sn); plot(sqn)
17 : write_data('list11_13.dat',s,sn,sqn)
Output:
Energy eigenvalue = -0.29619979319569
```

Listing 11.13. Code for eigenvalue problem with screened Coulomb potential.

Typical solutions for three values of the screening distance are shown in Figure 11.15 which shows the screened Coulomb potential along with the resulting electron probability densities (proportional to $|U(r)|^2$) as a function of distance from the potential center. The base lines for the vertical positions of the density functions are taken as being at the value of the eigenvalue. The screening is seen to sharpen the Coulomb potential as well as reducing the magnitude of the potential. The result is a decreasing eigenvalue or binding energy for an electron in the screened Coulomb potential. For sufficiently large screening, the potential is no longer able to support any bound states and this occurs around the normalized screening distance of unity. The reader can re-execute the program with progressively smaller L_D values and try to determine the value where the potential can no longer support a single bound state. However, for smaller screening distances the range of the solution variable must increase, since the lower energy state has an increasingly larger mean radius. For a screening distance of 1.0, the code will return an energy eigenvalue of about -0.02, indicating that this is close to the value for which a bound energy state can no longer be obtained.

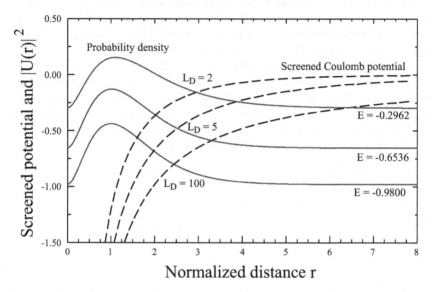

Figure 11.15. Plots of typical numerical solutions for the screened Coulomb Schrodinger wave equation eigenvalue problem.

This section has concentrated on a special type of two point boundary value problem associated with linear second order differential equations. Such systems only have nontrivial solutions for a discrete set of value of one equation parameter called the eigenvalue of the equation. This is a very important subset of boundary value differential equations. It has been shown that the shooting method of solving boundary value problems can be extended to such eigenfunction problems and accurate numerical solutions obtained. A set of callable code functions that can be used for these problems have been developed and discussed. The next section will return to the general problem of solving two point boundary value problems and attack the problem with a different approach.

11.5 Finite Difference Methods and Boundary Value Problems

The finite difference method of formulating a boundary value problem starts by approximating the differential equation by a set of equations obtained by approximating any derivative terms in the equation (or equations) by finite difference expressions. The reader is assumed to be somewhat familiar with approximating derivatives by finite differences as discussed in Chapter 5 as well as the material in Chapter 6 on fitting polynomials to sets of data points.

As a starting point with a simple equation consider the simple second order differential equation as given below:

$$a(x)\frac{d^2u}{dx^2} + b(x)\frac{du}{dx} + c(x)u + f(x) = 0 \tag{11.32}$$

The coefficients a, b and c may be functions of the independent variable. A more general form of nonlinear second order differential equation will be subsequently considered after some fundamental concepts are developed with this simpler equation. For any numerical solution, a solution is sought of the differential equation at some finite set of spatial points as the solution can not be obtained for the infinite range of values as contained in a continuous variable. Let's assume that the solution is desired at some set of spatial points as represented by the values in Figure 11.16 where the solid line represents the true solution of the boundary value problem. In the simplest case the spatial points may be equally spaced but

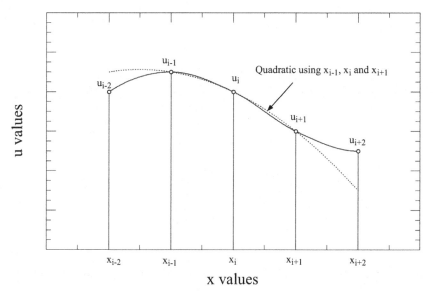

Figure 11.16. Representative spatial points for which solution of differential equation is desired.

in the more general case a nonuniform grid spacing or spacing of spatial points may be more appropriate.

From the discussion in Chapter 6 it is known that any three adjacent spatial points can be used to approximate the function around the point x_i by the equation (see Eq. (6.4)):

$$u = u_{i-1}\frac{(x-x_i)(x-x_{i+1})}{(x_{i-1}-x_i)(x_{i-1}-x_{i+1})} + u_i\frac{(x-x_{i-1})(x-x_{i+1})}{(x_i-x_{i-1})(x_i-x_{i+1})} +$$
$$u_{i+1}\frac{(x-x_{i-1})(x-x_i)}{(x_{i+1}-x_{i-1})(x_{i+1}-x_i)} \tag{11.33}$$

Using this approximation the first and second derivative at point x_i can be approximated by the expressions:

$$u_i' = \frac{u_{i+1}+u_i(\alpha_i^2-1)-u_{i-1}\alpha_i^2}{\alpha_i(s_i+s_{i-1})}$$

$$u_i'' = 2\frac{u_{i+1}-u_i(\alpha_i+1)+u_{i-1}\alpha_i}{s_i(s_i+s_{i-1})} \tag{11.34}$$

where $s_i = (x_{i+1}-x_i)$ and $\alpha_i = s_i/s_{i-1}$

When the spatial points have equal spacing, these equations simplify greatly to:

$$u_i' = \frac{u_{i+1}-u_{i-1}}{2h}$$

$$u_i'' = \frac{u_{i+1}-2u_i+u_{i-1}}{h^2} \tag{11.35}$$

where $s_i = s_{i-1} = h$ and $\alpha_i = 1$

If the finite difference approximations to the derivatives of Eq. (11.34) are substituted into Eq. (11.32) the result is a set of difference equations in the form:

$$u_{i-1}\left(\frac{2a_i\alpha_i/s_i-\alpha_ib_i}{(s_i+s_{i-1})}\right) + u_i\left(c_i+\frac{b_i(\alpha_i^2-1)/\alpha_i-2a_i(\alpha_i+1)/s_i}{(s_i+s_{i-1})}\right)$$
$$+u_{i+1}\left(\frac{2a_i/s_i+b_i/\alpha_i}{(s_i+s_{i-1})}\right) + f_i = 0 \tag{11.36}$$

where $a_i = a(x_i)$, $b_i = b(x_i)$, $c_i = c(x_i)$ and $f_i = f(x_i)$. This equation couples the solution at each spatial point to the preceding and subsequent spatial point and is an equation of the form:

$$A_iu_{i-1} + B_iu_i + C_iu_{i+1} + D_i = 0 \tag{11.37}$$

The complete set of coupled equations for all spatial points forms a tri-diagonal matrix of the form:

$$
\begin{bmatrix}
B_1 & C_1 & 0 & 0 & \cdots & 0 & 0 \\
A_2 & B_2 & C_2 & 0 & \cdots & 0 & 0 \\
0 & A_3 & B_3 & C_3 & \cdots & 0 & 0 \\
0 & 0 & \vdots & \vdots & \cdots & 0 & 0 \\
0 & 0 & 0 & \cdots & \cdots & \cdots & 0 \\
0 & 0 & 0 & \cdots & A_{N-1} & B_{N-1} & C_{N-1} \\
0 & 0 & 0 & \cdots & 0 & A_N & B_N
\end{bmatrix}
\begin{bmatrix}
u_1 \\ u_2 \\ u_3 \\ \vdots \\ \vdots \\ u_{N-1} \\ u_N
\end{bmatrix}
+
\begin{bmatrix}
D_1 \\ D_2 \\ D_3 \\ \vdots \\ \vdots \\ D_{N-1} \\ D_N
\end{bmatrix}
= 0 \tag{11.38}
$$

where N is the total number of spatial points for which the solution is desired. Such tri-diagonal matrices have been extensively studied and can be solved relatively easily. However before discussing the solution and considering boundary conditions, let's consider how to extend this finite difference approach to the solution of nonlinear second order differential equations.

A general second order differential equation can be expressed in the functional form:

$$F(x,U,U',U'') = 0 \qquad (11.39)$$

where no assumptions are made about the functional form of the dependence on the derivatives or on the function itself. An iterative approach to solving such a differential equation can be formulated by applying the fundamental principle for nonlinear problems which is that of "linearize and iterate". To linearize the equation it is assumed that an approximate solution to the equation is known and that a Newton-Raphson expansion is functional space can be utilized to linearize the equation keeping only first order terms in a correction to the approximate solution. In mathematical form one assumes that the solution U in the equation can be replaced by $U(x)+u(x)$ where the upper case function is an approximation to the solution and the lower case function is a correction (hopefully small) to the approximation. The nonlinear function can then be expanded about the approximate solution keeping only the first order correction terms in the lower case function. When this procedure is performed the resulting equation in functional notation is:

$$\left(\frac{\partial F}{\partial U''}\right)\frac{d^2u}{dx^2}+\left(\frac{\partial F}{\partial U'}\right)\frac{du}{dx}+\left(\frac{\partial F}{\partial U}\right)u+F(x,U,U',U'') = 0 \qquad (11.40)$$

This is now a linear differential equation in terms of the lower case correction function. Assuming this linear equation can now be solved, the correction function (lower case u) can then be added to the approximation function (upper case U) and an improved solution obtained. The procedure can then be repeated as many times as necessary to achieve a desired degree of accuracy in the solution. It can be seen that if a valid solution is obtained, the F function in Eq. (11.40) approaches zero so the correction term will approach zero. As with other Newton like methods, the solution is expected to converge rapidly as the exact solution is approached. This technique is frequently referred to as quasilinerization and it has been shown that quadratic convergence occurs if the procedure converges.

The functional form of Eq. (11.40) is exactly the same as Eq. (11.32) previously considered. To step through one iterative loop of a solution for a nonlinear equation, the coefficients of Eq. (11.32) are evaluated in terms of the partial derivative terms as given in Eq. (11.40). For this nonlinear equation, the equivalent of Eq. (11.36) is:

$$u_{i-1}\left(\frac{2(F_{U'})\alpha_i / s_i - \alpha_i(F_{U'})}{(s_i + s_{i-1})}\right)$$

$$+ u_i\left((F_U) + \frac{(F_{U'})(\alpha_i^2 - 1)/\alpha_i - 2(F_{U'})(\alpha_i + 1)/s_i}{(s_i + s_{i-1})}\right) \qquad (11.41)$$

$$+ u_{i+1}\left(\frac{2(F_{U'})/s_i + (F_{U'})/\alpha_i}{(s_i + s_{i-1})}\right) + F = 0$$

In each case the F function and its partial derivatives are to be evaluated using the first and second derivatives of the approximate solution. As with any Newton iterative method, convergence may not be achieved if the initial guess is too far from the final solution, but more on this later.

Now consider the solution of the resulting tridiagonal system of equations as given by Eq. (11.37) or the matrix form as in Eq. (11.38). One of the simplest methods of solution is to assume that there exist some functions e_i and f_i such that:

$$u_{i-1} = e_i u_i + f_i \qquad (11.42)$$

If this form is put into Eq. (11.37) one gets after some rearrangement:

$$u_i = -\left(\frac{C_i}{A_i e_i + B_i}\right) u_{i+1} - \left(\frac{A_i f_i + D_i}{A_i e_i + B_i}\right) \qquad (11.43)$$

From this relationship and the original equation it is readily seen that the relationship of Eq. (11.42) requires the recursive expressions:

$$e_i = -\left(\frac{C_{i-1}}{A_{i-1} e_{i-1} + B_{i-1}}\right)$$

$$\qquad (11.44)$$

$$f_i = -\left(\frac{A_{i-1} f_{i-1} + D_{i-1}}{A_{i-1} e_{i-1} + B_{i-1}}\right)$$

This provides a concise method of solving the system of tridiagonal equations. The approach is as follows. Beginning at one boundary of the solution region, such as the left boundary, with initial values of e and f, values of e and f are computed across the spatial region using the recursive expressions of Eq. (11.44). Then beginning at the other boundary (the right boundary), and the known solution value at that boundary, solution values are calculated for all the spatial points using the recursive relationship of Eq. (11.42). This procedure is equivalent to reducing the below diagonal elements of the matrix to zero in the first pass and then doing a back substitution on the matrix equations to obtain the solution values.

The remaining problem is how to begin the solution with some initial e and f value and this is related to how to include boundary conditions in the set of equations. The key to a starting relationship is the basic recursive relationship of Eq. (11.42). If this is applied to the first data point ($i = 1$), one obtains:

$$u_1 = e_2 u_2 + f_2 \qquad (11.45)$$

If one has a fixed boundary condition such that u_1 is some known value, then the equation can be satisfied by the beginning values $e_2 = 0$, $f_2 = u_1$ and this is all that is needed to begin the recursive definition process. In general however, mixed boundary conditions may be specified or perhaps some nonlinear boundary condition is required involving the function value and derivative on the boundary. This can be expressed in general for a left (L) and right (R) boundary condition as:

$$FL(U_1, U_1') = 0 \text{ and } FR(U_N, U_N') = 0 \tag{11.46}$$

where the subscripts are used to indicate the first and last spatial point or the values of the function and derivative on the boundaries. In keeping with a first order expansion of the solution about some approximate solution and correction term these can be linearized to give:

$$FL + \left(\frac{\partial FL}{\partial U}\right)u_1 + \left(\frac{\partial FL}{\partial U'}\right)u_1' = 0$$

$$FR + \left(\frac{\partial FR}{\partial U}\right)u_N + \left(\frac{\partial FR}{\partial U'}\right)u_N' = 0 \tag{11.47}$$

A first order approximation can now be used to approximate the initial and final derivative. Working first with the initial boundary and using $u_1' = (u_2 - u_1)/s_1$, the following expressions are readily obtained:

$$u_1 = -\frac{(\partial FL/\partial U')u_2 + (FL)s_1}{s_1(\partial FL/\partial U) - (\partial FL/\partial U')} \tag{11.48}$$

$$u_1 = e_2 u_2 + f_2$$

From this set of equations an expression for the initial values of e_2 and f_2 which satisfy some set of mixed boundary conditions can be obtained as:

$$e_2 = -\frac{(\partial FL/\partial U')}{s_1(\partial FL/\partial U) - (\partial FL/\partial U')}$$

$$f_2 = -\frac{(FL)s_1}{s_1(\partial FL/\partial U) - (\partial FL/\partial U')} \tag{11.49}$$

A similar consideration at the final boundary using a first order approximation to the upper derivative value gives the set of equations:

$$u_{N-1} = \frac{[(\partial FR/\partial U) + (\partial FR/\partial U')/s_{N-1}]u_N + (FR)}{(\partial FR/\partial U')/s_{N-1}} \tag{11.50}$$

$$u_{N-1} = e_N u_N + f_N$$

The question now is what is to be done with this set of equations at the upper boundary? They do not determine the e_N and f_N values as these are determined by the recursive relationships as the iterative procedure is applied across the spatial region. However, they do determine the upper boundary value as u_N can be expressed as:

$$u_N = \frac{f_N(\partial FR/\partial U')/s_{N-1} - FR}{(\partial FR/\partial U) + (1 - e_N)(\partial FR/\partial U')/s_{N-1}} \tag{11.51}$$

If the upper boundary condition is some fixed value this will force the upper boundary value to match that fixed value as one can easily verify from this equation. With a mixed boundary value or nonlinear condition, this will iterate to the correct boundary value. The mathematical formulism has now been developed for coding a finite difference approach to solve two point boundary value problems.

A code listing for such a finite difference routine is shown in Listing 11.14. This is one of the longer computer routines so far developed in this work. The heart of the code is contained between lines 18 and 77 where a major loop (j = 1,NMAX) steps through an appropriate number of Newton iterations to solve the set of differential equations (NMAX is the maximum permitted number of Newton iterations). A brief description of the various parts of the code will now be given. First lines 20 through 26 evaluate numerically the first (up) and second (upp) derivatives of the solution value for a given approximation to the solution value using Eqs. (11.34). The end points require separate equations on lines 27 through 32. At the same time average values of the derivatives are accumulated in variables fctupp and fctup (see lines 23, 25). These are used on line 33 through 35 to compute factors to be used in calculating numerical derivatives required in the linearization of the differential equations. The initial boundary condition is evaluated on lines 36 through 38 using a locally defined boundary value function bound() which is in turn defined on lines 11 through 16. This function simply calls the user supplied boundary functions, to obtain the Newton linearized version of the boundary condition (see lines 12 through 15) corresponding to Eq. (11.47). The spatial loop for each Newton iteration is contained between lines 39 and 52 of the code. At each spatial point, the user supplied equation is linearized by calculating the partial derivatives with respect to the function value (line 44), its first derivative (line 43) and second derivative (line 42). These terms correspond to the various partial derivative of Eq. (11.40).

A word or two is perhaps in order about the use of numerical derivatives here as opposed to actually implementing equations for the partial derivatives. The use of numerical derivatives is extremely convenient as the user of the code has to only define the differential equation and not supply expressions for the various partial derivatives. On the other hand it is known that one must be careful in using numerical partial derivatives as discussed in Chapter 5. In fact many authors on numerical methods recommend against the use of numerical derivatives and always recommend that explicit equations be supplied to software routines for the partial derivatives. However, as discussed in Chapter 5 numerical derivatives can be very successfully used if appropriate care is taken in their evaluation. One of the keys is in selecting an appropriate increment in the value for which the derivative is being evaluated, i.e. an appropriate value of the increment in function value, in first derivative and in second derivative, the fctu, fctup and fctupp values on lines 42 through 44 of the code. Too large or too small a value for these can result in inaccurate partial derivative values as discussed in Chapter 5. For the present code, an appropriate value for these factors is obtained by first calculating average values for the function and its first and second derivative over the spatial points. Then some small factor (FACT = 1.e-3) of these average values is used as the increment for evaluating the partial derivatives (see line 33). A significant

```lua
 1 : -- /* File odebvfd.lua */
 2 :
 3 : odebv1fd = function(eqs,x,u)
 4 :    local s,up,upp,e,f,du = {},{},{},{},{},{}
 5 :    local fval,fu,fup,fupp,uppi,upi,ui,xi,duu
 6 :    local fctupp,fctup,fctu = FACT,FACT,FACT
 7 :    local feq,nx = eqs[1], #x
 8 :    local nxm1,nend = nx-1, NMAX
 9 :    for i=1,nx-1 do s[i] = x[i+1] - x[i] end
10 :
11 :    bound = function(nb,nxb) -- Function to evaluate BVs
12 :       upi,ui,xi = up[nxb],u[nxb],x[nxb] -- boundary values
13 :       fval = eqs[nb](ui,upi)
14 :       fup = (eqs[nb](ui,upi+fctup) - fval)/fctup
15 :       fu = (eqs[nb](ui+fctu,upi) - fval)/fctu
16 :    end
17 :
18 :    for j=1,NMAX do -- Major loop for iterative solution
19 :       fctupp,fctup = 0,0
20 :       for i=2,nx-1 do -- Calculate second and first derivative
21 :          si,sisi,alfi = s[i],(s[i]+s[i-1]),s[i]/s[i-1]
22 :          duu = 2*(u[i+1]-(alfi+1)*u[i]+alfi*u[i-1])/(si*sisi)
23 :          upp[i],fctupp = duu, fctupp+abs(duu)
24 :          duu = (u[i+1]+(alfi^2-1)*u[i]-alfi^2*u[i-1])/
                 (alfi*sisi)
25 :          up[i],fctup = duu, fctup+abs(duu)
26 :       end
27 :       alfi = s[2]/s[1] -- Handle end points, lower boundary
28 :       upp[1] = upp[2] - (upp[3]-upp[2])/alfi
29 :       up[1] = (-u[3]+u[2]*(1+alfi)^2-u[1]*(2/alfi+1))/
                 (alfi*(s[2]+s[1]))
30 :       alfi = s[nxm1]/s[nx-2] -- Upper boundary
31 :       upp[nx] = upp[nxm1] + (upp[nxm1]-upp[nxm1-1])*alfi
32 :       up[nx] = (u[nx]*(1+2*alfi) - u[nxm1]*(1+alfi)^2 +
                 u[nx-2]*alfi^2)/(alfi*(s[nxm1]+s[nx-2]))
33 :       fctupp,fctup = FACT*fctupp/nx, FACT*fctup/nx
34 :       if fctupp==0 then fctupp = FACT end -- protect against 0?
35 :       if fctup==0 then fctup = FACT end
36 :       bound(2,1) -- Evaluate lower boundary conditions
37 :       duu = fup - fu*s[1] -- Determines first values of e and f
38 :       e[2],f[2] = fup/duu, fval*s[1]/duu
39 :       for i=2,nx-1 do -- Set up a,b,c,d arrays, save e and f
40 :          uppi,upi,ui,xi = upp[i],up[i],u[i],x[i]
41 :          fval = feq(xi,ui,upi,uppi)
42 :          fupp = (feq(xi,ui,upi,uppi+fctupp) - fval)/fctupp
43 :          fup = (feq(xi,ui,upi+fctup,uppi) - fval)/fctup
44 :          fu = (feq(xi,ui+fctu,upi,uppi) - fval)/fctu
45 :          si,sisi,alfi = s[i],(s[i]+s[i-1]),s[i]/s[i-1]
46 :          ai = (2*fupp/si - fup)*alfi/sisi
47 :          bi = fu - (2*(alfi+1)*fupp/si - (alfi^2-1)*
                 fup/alfi)/sisi
48 :          ci,di = (2*fupp/s[i] + fup/alfi)/sisi, fval
49 :          -- Forward reduction of tridiagonal system
50 :          gi = 1/(ai*e[i] + bi)
51 :          e[i+1],f[i+1] = -gi*ci, -gi*(di + ai*f[i])
52 :       end
53 :
54 :       bound(3,nx) -- Evaluate upper boundary conditions
55 :       -- Now back substitute for correction values
```

```
56 :          du[nx] =-(fval*s[nxm1]-f[nx]*fup)/(fu*s[nxm1]+
                  fup*(1-e[nx]))
57 :          for i=nx,2,-1 do  -- Calculate correction values
58 :              du[i-1] = e[i]*du[i] + f[i]
59 :          end
60 :          -- Now update solution and check for desired accuracy
61 :          cmax,umax,imax,fctu = 0,0,0,0
62 :          for i=1,nx do
63 :              c1 = abs(du[i]); if c1>umax then umax = c1 end
64 :              c2 = abs(u[i]) + c1; if c2~=0.0 then c1=c1/c2 end
65 :              if c1>cmax then cmax=c1; imax=i end
66 :              u[i] = u[i] + du[i] -- Add corection to previous solu-
tion
67 :              fctu = fctu+abs(u[i])
68 :          end
69 :          fctu = fctu/nx; if fctu==0 then fctu = FACT end
70 :          if nprint~=0 then
71 :              printf("Iteration number %i, Maximum  relative,
                  absolute correction = %e, %e at %i\n",j,cmax,umax,imax)
72 :              io.flush()
73 :          end
74 :          if cmax<ERROR then nend=j; break end
75 :          if FABS~=0 then if umax<fctu*ERROR then nend=j;
                  break end end
76 :          fctu = FACT*fctu
77 :      end
78 :      return nend, cmax, umax -- Solution values returned in u
79 : end
80 : setfenv(odebv1fd,{abs=math.abs,ERROR=1.e-6,FABS=1,FACT=1.e-3,
         EPXX=1.e-14,NMAX=50,nprint=0,io=io,printf=printf,
81 : table=table})
```

Listing 11.14. Code for solving second order boundary value problem using finite difference approach.

fraction of the code in the listing is devoted to obtaining such values in an attempt to properly evaluate the partial derivatives. While this procedure may not be optimum, it seems to work for a large number of differential equations for which the code has been applied.

Lines 46 through 48 set up the a, b, c and d coefficients at each spatial point and essentially follow the formulation of Eq. (11.41). This is followed by an evaluation of the e and f coefficients on line 51. It is noted the values of the a, b, c and d coefficients are not saved and only the e and f coefficients are saved in arrays. After stepping across the spatial region in one direction, the upper boundary condition is used on lines 54 through 56 to evaluate the upper solution value. Then the corrections throughout the spatial region are calculated in the du[] array on lines 57 through 59. The remainder of the code on lines 60 through 76 is devoted to evaluation the accuracy of the solution and determining when to terminate the Newton iterations with an accurate solution. This is always a difficult task and a significant fraction of the code is devoted to this task. The code evaluates both the maximum relative error (line 65) in the solution and the maximum absolute error (line 63) throughout the spatial region. The termination criteria can occur either from a sufficiently small relative error (default value of 1.e-6 on line

74) or from an absolute error (default of 1.e-6 times the average solution value on line 75). For a wide variety of equations these termination criteria have been found to work well.

Perhaps a few words are appropriate as to why the code simply does not use the relative error criteria instead of a more complicated combination of criteria. For some problems, the solution value can be very small (approximately zero) over much of the solution space. For such problems, a small correction to the solution can represent a large relative error, even though it might be an insignificant correction with respect to the overall range of the solution. For such problems where the solution value in near zero, the relative error criteria does not work very well while an absolute correction criteria can work appropriately. This is again a case of problems in numerical algorithms when a function value is near zero. One must always try to protect the code against zero values and resulting problems. The absolute error criteria can be turned off by setting FABS = 0 and the relative error criteria can be changed by setting the ERROR term to any desired value. For the default values, the code terminates the Newton iterative loop when the relative error or absolute error indicates an accuracy of about 6 decimal digits which should be adequate for most engineering problems. The actually achieved accuracy will be discussed later with specific examples as this depends on the number of spatial intervals.

Listing 11.15 shows sample code for using the odebv1fd() routine to solve a very simple two point boundary value problem. The differential equation used here is a linear second order differential equation for beam deflection as previously used in Listing 11.1. While this is a simple example, some important facts can be learned from the results about the finite difference approach. The function and boundary conditions are defined in the standard way on lines 9 through 13. Lines 15 through 19 set up an appropriate spatial array of uniformly spaced points and computes the exact solution y1[] array on line 18 for comparison purposes. The boundary value solver is called on line 20 and the returned results printed on the output. Note that the functions defining the differential equation and the two boundary functions are passed as names in a table supplied as the first argument to the odebv1fd() function. The two other arguments to the function are the table of x values and an initial guess at the solution which is taken here as simply zero. In addition the nprint parameter of odebv1fd() is set on line 4 to a non-zero value so that the maximum and relative errors at each Newton iteration will be printed. The output shows that two iterations are required for accurate solutions with the maximum relative and absolute corrections being 3.7×10^{-13} and 6.8×10^{-15} respectively at the second iteration. Since this is a linear differential equation, only one iteration is in fact needed to obtain an accurate solution. However, the code does not know that the equation is linear and requires a second Newton iteration to determine that an accurate solution has been obtained in the first Newton step. For linear equations with the finite difference code, the Newton iterative loop could be eliminated, but the cost of having a general routine appropriate to nonlinear equations, is the additional Newton iteration for the special case of a linear equation. This is usually not a problem as the code executes relatively fast for linear equations. The output also shows that the values returned by the odebv1fd() function is the number of Newton iterations as well as the final relative and absolute corrections, as shown as the printed output on the last line of Listing 11.15.

```
 1 : -- /* File list11_15.lua */
 2 : -- Use of boundary value solver for second order equation
 3 :
 4 : require"odebvfd"; getfenv(odebv1fd).nprint=1
 5 :
 6 : -- Parameters
 7 : E,I,w,L = 1.e7, 500, 100, 100; EI = E*I
 8 :
 9 : f = function(x,y,yp,ypp) -- Differntial equation
10 :    return  ypp - w*x*(x-L)/(2*EI)
11 : end

12 : fl = function(y,yp) return y end -- Left boundary, y=0
13 : fr = function(y,yp) return y end -- Right boundary, y=0
14 :
15 : x,y,y1,nx = {},{},{}, 1001; dx = L/(nx-1)
16 : for i=1,nx do
17 :    x[i], y[i] = (i-1)*dx, 0
18 :    y1[i] = w/(24*EI)*(x[i]^4-2*L*x[i]^3+L^3*x[i])
19 : end
20 : print(odebv1fd({f,fl,fr},x,y))
21 : plot(x,y,y1); write_data('list11_15.dat',x,y,y1)
```

Output:
```
Iteration number 1, Maximum  relative, absolute correction =
1.000000e+000,  2.604175e-002 at 2
Iteration number 2, Maximum  relative, absolute correction =
3.760693e-013,  6.814268e-015 at 101
2        3.7606934972851e-013   6.8142682748889e-015
```

Listing 11.15. Simple example of finite difference boundary value routine used to solve a linear differential equation.

Now consider some of the results for this simple case. The solution results are shown in Figure 11.17. In this case the results are plotted on a log scale, but the results are essentially the same as shown on the linear scale in Figure 11.1 where a very similar problem was solved by the shooting method. The log scale is used here in order to also show the error in the solution for several different spatial step sizes, corresponding to 500, 1000 and 2000 spatial intervals across the solution space. For 500 intervals, the maximum absolute error is in the range of 1×10^{-7} for a relative solution accuracy of about 5 decimal digits. This is considerable less than the errors reported back by the odebv1fd() routine and printed in listing 11.15. These errors are however two different things. The error reported by the odebv1fd() routine is the error in solving the set of finite difference equations. One may have a very accurate solution of the finite difference equations, but the solution is still left with the inherent numerical inaccuracies of the finite difference approximation to the differential equation. These inherent inaccuracies are seen to again be essentially proportional to the square of the step size as it can be seen that the error reduces by approximately a factor of 4 for each of the error curves in Figure 11.17. While the inherent errors of the finite difference method have not so far been discussed, the errors are expected to vary as the square of the step size for a three point finite difference method. If this relationship holds for more general nonlinear finite difference solutions, the previously discussed h-2h algorithm can again be used to obtain good estimates of the error in various solutions to boundary value problems with the finite difference method.

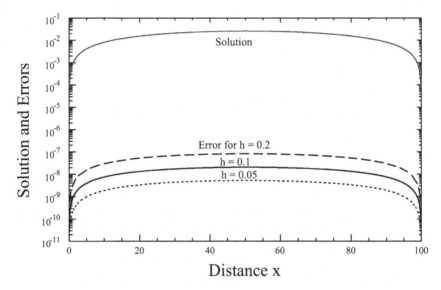

Figure 11.17. Solution and error for simple beam deflection boundary value problem of Listing 11.15.

Let's now explore a slightly more difficult nonlinear boundary value problem. The equation selected is that of Eq. (11.16) with known solution as given by Eq. (11.17). While this is a fairly simple nonlinear equation, the advantage of exploring this differential equation is that the exact solution is known so the accuracy of the finite difference method can be more easily explored. A code segment for solving this equation is shown in Listing 11.16. The differential equation is defined on lines 7 through 9 in the usual manner. However, the function used to solve the equation is ode1fd() and not the previously defined odebv1fd() function. Also it is noted that no boundary value functions are defined in the listing or passed to the solving function. This illustrates a simpler function, the ode1fd(), function for solving certain types of boundary value problems. Many boundary values problems specify constant values for the function on the two boundaries and are not of the mixed boundary value type. Also many times a uniformly spatial grid is used for the solution. For such cases the important parameters for the spatial range are the starting value, the final value and the desired number of spatial points. These are defined on line 11 for the x array. Also line 11 defines a ta-

ble of the two desired boundary values for the solution as the "left or lower" and
"right or upper" values (the {ubL, ubR} table values).

```
 1 : -- /* File list11_16.lua */
 2 : -- Finite difference method for BVP with nonlinear DE
 3 : require"odebvfd" -- Use ode1fd() function
 4 :
 5 : nx,L,ubL,ubR = 1001, 1,0, 2 -- #points, and Boundary values
 6 :
 7 : f = function(x,u,up,upp) -- Differntial equation
 8 :    return upp + 4*up^2
 9 : end
10 : -- x range values, boundary values specified
11 : x,u = {0,L,nx},{ubL,ubR}
12 : u,nn,err1,err2 = ode1fd(f,x,u) -- Solve BV problem
13 :
14 : print(nn,err1,err2); plot(u)
15 : C2,uex = math.exp(8),{} -- Exact solution parameters
16 : for i=1,nx do uex[i]=math.abs(0.25*math.log(u[1][i]*(C2-1)+1) -
           U[2][i]) end
17 : write_data(2,"list11_16.dat",u,uex)
Output:
7        2.3135109752643e-006   8.9418164949335e-007
```

Listing 11.16 Code segment for solving simple nonlinear differential equation.

The ode1fd() function simply takes this information and generates a uniform
spatial grid and generates an initial guess at the solution which is simply taken to
vary linearly between the two fixed boundary values. The reader can view the details of the ofe1fd() function in the odebvfd.lua file. In addition the function defines default boundary functions that fix the boundary values at the specified end
points. Since this type of BV problem is so common, a simpler calling function is
supplied to have the software set up the initial arrays and boundary functions. The
user can then simply concentrate on supplying the minimum needed information
for the solution. This is simply what the ode1fd() function provides by employing
a simpler interface for BV problems with fixed solution values on the boundary
and for uniformly spaced solution grid points. The call to the solution function of
line 12 returns as the first argument (u on line 12) a table of spatial values (u[1])
and a table of solution values (u[2]) in addition to the number of Newton iterations
and the achieved maximum errors in the solution values.

The printed output shows that 7 Newton iterations are required to achieve correction errors in the range of 1×10^{-6} for both the relative and absolute errors. It is
noted again that these numbers specify the accuracy of solving the finite difference equations and not the accuracy of the solution. The achieved solution accuracy is shown by the data in Figure 11.18. The solution value is essentially the
same as that in Figure 11.4 shown on a linear scale. The log scale is used here in
order to show both the solution and the error in the solution. As would be expected because of the rapid change in the solution value near the left boundary, the
error is considerably larger than for the previous linear boundary value problem
and in fact considerably larger than the accuracy with which the finite difference

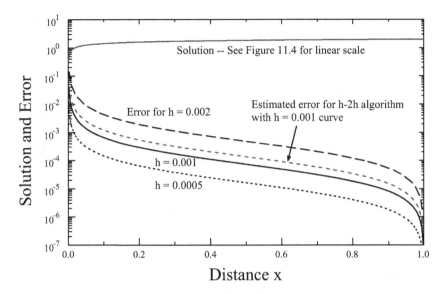

Figure 11.18. Illustration of solution and error for simple nonlinear equation. Code for solution is in Listing 11.16.

equations are solved. For the case of 1000 spatial intervals, the accuracy is in the range of 1×10^{-3} to 1×10^{-5} over most of the range with the largest values occurring near the left boundary. The range of the solution value is 0 to 2 so the relative error becomes quite large near the left boundary. It can also be seen that the error decreases as the spatial step size decreases or the number of spatial points increases as expected. The error is roughly proportional to step size squared as can be seen from the three error curves for three different step sizes. If it is assumed that the error is exactly proportional to the step size squared then the h-2h algorithm of Eq. (10.44) can be used to estimate the error from solutions at two different step sizes. When this is applied to the h sizes of 0.001 and 0.002, the resulting estimated error is the dotted curve shown in the figure. This can be compared with the actually achieved error given by the solid curve labeled with the h = 0.001 value. While the h-2h algorithm does not exactly predict the error it provides a reasonable estimate of the achieved error. This give some good indication that the h-2h algorithm can be used to obtain a good estimate of the error in the finite difference code even for nonlinear problems. This will be subsequently used in new callable functions that return not only the BV solution but also error estimates.

This nonlinear BV problem looks like an excellent case for employing a non-uniform spatial grid, as the solution changes very rapidly at one boundary. This was one of the reasons for formulating the equation set and developing the software codes to handle non-uniform grid spacings. If only uniform spatial grids were of interest, the code could be made much simpler as the finite difference equations are greatly simplified for the case of a uniform grid. This is thus an appropriate equation to explore advantages of non-uniform spatial points for such nonlinear BV problems. A slight digression from the main theme is thus needed to discuss the generation of an array of non-uniform spatial points.

A desirable non-uniform set of grid points starts with small step sizes in some region, the left boundary in the present case, and gradually increases the step size as the spatial variable moves away from the boundary. A very gradual change in step size is desired in order to retain accuracy in the finite difference equations. There are many different ways such a non-uniform grid can be achieved. One such approach has already been discussed in Chapter 10 and especially in Section 10.5 where an adaptive step size algorithm was used. The concept of a logarithmically spaced time grid was introduced and shown to significantly improve the solution accuracy for initial value problems with both fast and slow time constants (see for example Figure 10.20). Logarithmically spaced grid points are thus one approach that can be used.

Another approach comes from the concept of a geometrical progression where one has a first term of size 'a', a ratio 'r' between terms and a sum of 'n' terms. The appropriate equations are:

$$a = \text{ first term}; \quad l = \text{ last term}; \quad s = \text{ sum of } n \text{ terms}$$

$$l = ar^{n-1}; \quad s = a(r^n - 1)/(r-1) \tag{11.52}$$

For a given initial step size and a desired number of spatial points, a ratio factor can be calculated which will satisfy the requirements. The resulting equation for the ratio is nonlinear, but by now that in not a determent as nsolv() or newton() can easily solve the resulting equation. A final approach is to equally space the solution points along a log position vs log step number graph, a log-log distribution of spatial points. These approaches give three possible methods for generating a non-uniform spatial step distribution for problems such as the one under consideration. Details of the code for generating these step distributions will not be discussed here as it is relatively straightforward but routines are included in the odebvfd package. They can be called as xlg(), xgp() or xll() for the log distribution, the geometric progression or the log-log distribution respectively. In each case the calling arguments are the initial value and final value of the spatial interval followed by the minimum step size and the number of desired spatial intervals.

Example code segment for solving the nonlinear BV problem with a nonlinear spatial grid is shown in Listing 11.17. The code shown the use of the log distribution function xlg() on line 5. Alternative step distributions are shown commented out on lines 6 and 7. The reader is encouraged to re-execute the code with each of these distributions. The code is a model of simplicity. After defining the spatial array and the function on lines 9 through 11, the solution is obtained with a call to ode1fd() on line 13. The calling arguments illustrate another feature of the ode1fd() routine and this is the use of only the two boundary value points {0, 2} as the initial solution values. The function internally generates an array of initial solution values to match the spatial array as a convenience to the user. Another feature of ode1fd() is that it returns as the first argument a table containing the x values followed by the solution values (u = {x, y} on line 13). This again is for convenience in using the results. For the code shown 2000 spatial intervals are specified by each of the x array generation functions with the spatial values ranging from 0 to 1 with a minimum step size of 1.e-6. The number of spatial points including the two end points will be 2001 in each case.

```
 1 : -- /* list11_17.lua */
 2 : -- Solution of BV problem with non-uniform spatial grid
 3 : require"odebvfd"
 4 :
 5 : x = xlg(0,1,1.e-6,2000) -- Log distribution -- Try others below
 6 : --x = xgp(0,1,1.e-6,2000) -- Geometric distribution
 7 : --x = xll(0,1,1.e-6,2000) -- Log-log distribution
 8 :
 9 : f = function(x,y,yp,ypp) -- Differntial equation
10 :    return ypp + 4*yp^2
11 : end
12 :
13 : u,nn,err1,err2 = ode1fd(f,x,{0,2}) -- Solve equation
14 :
15 : print(nn,err1,err2) -- # Newton iterations and errors
16 : plot(u); write_data('list11_17.dat',u)
Output:
   11              2.8132557956183e-007   7.3209456067983e-008
```

Listing 11.17. Example of using non-uniform spatial grid with nonlinear BV problem.

The printed output from the program indicates that 11 Newton iterations are required to achieve errors in the range of 1×10^{-7} in solving the finite difference equations. The geometric and log-log distributions are found to require 5 and 6 Newton iterations. The important question is does the use of a non-uniform grid improve the accuracy of the numerical solution? Figure 11.19 shows the important results with respect to the solution error. The top curve shows the solution to the equation on a log scale for the x values as well as solution values, so the accuracy for small distances can be easily seen. The lower three curves in the figure show the achieved errors with the three non-uniform spatial distributions. The most accurate is the geometric distribution which has a maximum error of about 1×10^{-5} and for small values of x gives a relative accuracy of about 4 decimal digits. The logarithmic distribution of spacing gives almost as good an accuracy and the log-log spacing is somewhat less accurate in the small x region due to the more rapid increase in step size for this distribution. All three of the non-uniform distributions give considerably better accuracy than the uniform spacing which is also shown for comparison. The minimum step size for the uniform grid is 1.e-3. In all cases the total number of spatial intervals has been taken as 2000.

It is interesting to relate the achieved accuracy in the solution to the three distributions of grid points used in the non-uniform grid calculations. Figure 11.20 shows the x position variable as a function of the spatial grid point number for the three non-uniform grid generation functions presented here. All of the distributions create the first internal grid point at the same specified point which in this

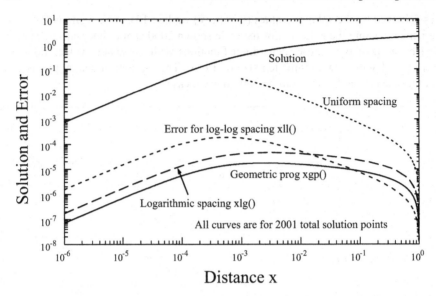

Figure 11.19. Illustration of solution error for nonlinear BV problem of Listing 11.17 with different non-uniform spatial grids.

2000.

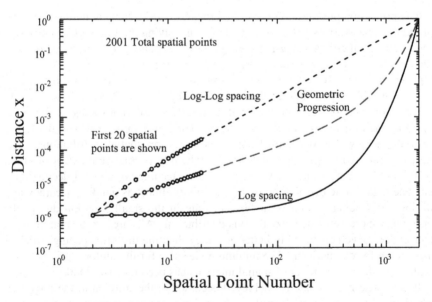

Figure 11.20 Distribution of grid points for three non-uniform functions for generating spatial points.

case is $x = 1.0\text{x}10^{-6}$. The plots are on a log-log graph so the log-log generating function produces spatial grid points varying almost linearly on this graph between the first internal grid point and the final grid point at $x = 1.0$. If the graph were a log y-axis and linear x-axis, the log spacing curve would be approximately linear on such a graph, while for the present graph, it has a more exponential like appearance. The first 20 generated points are shown as data points on the three curves. Very significant differences can be seen in the spatial grid point distributions for small distances. The log-log spacing results in the first 20 grid points going from $1\text{x}10^{-6}$ to about $3\text{x}10^{-4}$, while the log spacing function puts about 1000 grid points within this same interval.

The geometric progression spacing is about mid way between the log-log and log spacing distributions when plotted on this log-log scale. When comparing this with the error curves in Figure 11.19 we can see that the smallest error actually occurs with the geometric progression which is somewhat of a compromise between the other two curves, providing small initial step sizes but also providing more grid points for larger x values. The log spacing curve appears to put an unnecessarily large number of grid points at small x values leaving an insufficient number at larger values for achieving a minimum error. The log-log distribution produces many more spatial points at large distances and a slightly improved accuracy for the larger x values. However all things considered, the geometric progression seems to produce the best spatial distribution for this case. It is difficult to generalize these results to all nonlinear BV problems as the best distribution of spatial points will be very dependent on the exact equation being solved and how rapidly the solution changes within the spatial region being considered. These three functions provide simple spatial distributions that may be of value, as in this case, in generating a non-uniform spatial distribution of grid points.

For the finite difference equations using three grid points, the solution error is expected to be proportional to the second power of step size as previously shown for some simple cases. This again leads to the concept of using an h-2h calculation to obtain an estimate of the error in a BV problem. Code for performing such an evaluation is included in the functions provided as the ode1fde() function with identical calling arguments to the ode1fd() function. The routine simple evaluates the solution using the ode1fd() function and then sets up a second call to the same function using a grid spacing with half the number of original grid points, obtained by eliminating every second grid point. The difference between the solutions is then used with the h-2h algorithm of Eq. (10.44) to obtain an error estimate at each grid point.

Listing 11.18 shows an example of the use of this function for the same nonlinear BV problem. This is about as simple as one can get for defining a differential equation, boundary conditions and obtaining a solution with estimated error. Note that in the xgp() call the number of desired spatial points is omitted as the function has a default value of 2000 spatial intervals. The printed output shows that 5 Newton iterations are required for this solution. This is smaller than the 11 iterations required if the log spatial distribution is used as seen on the output line of Listing 11.17. This smaller number of Newton iterations also indicates a more appropriate spatial distribution for this case. The user is encouraged to re-execute the code and explore the estimated error returned by the function (err in Listing 11.18). The results will not be shown here, but the estimated error obtained by the h-2h algorithm is almost identical to the exact error in the solution. The exact error has been previously shown for this case in Figure 11.19 as the lower solid curve. The err curve obtained by Listing 11.18 and saved in file list11.18.dat if plotted will fall essentially on top of this exact error curve in Figure 11.19. This is one of the reasons it is not shown here.

```
 1 : -- /* list11_18.lua */
 2 : -- Solution of BV problem with non-uniform spatial grid
 3 : require"odebvfd"
 4 :
 5 : f = function(x,y,yp,ypp) -- Differntial equation
 6 :     return ypp + 4*yp^2
 7 : end
 8 :
 9 : x = xgp(0,1,1.e-6) -- Geometric distribution
10 : s,err,nn = odelfde(f,x,{0,2}) -- Solve equation
11 : print('Number of Newton iterations = ',nn) -- # Newton itera-
tions
12 : plot(s); write_data('list11_18.dat',s,err)
Output:
Number of Newton iterations = 5
```

Listing 11.18. Sample code segment for solving boundary value problem with non-uniform spatial grid and with estimated error in solution.

Before leaving the topic of non-uniform grids, there is one other technique for obtaining such grids that should be discussed. This is based upon the following considerations. A "good" step distribution should be one in which the dependent variable changes very little from step to step along the independent variable. One might consider the ideal step distribution as one in which the dependent variable changes by equal increments as the solution progresses from step to step. For example if a certain number of spatial intervals is desired, say 2000 spatial increments along the solution and the solution varies from 0 to 2 as in the previous example, then an ideal step distribution might be considered as one where the solution varies by $2/2000 = 0.001$ between each spatial step. This leads to the concept of using an approximate solution to generate a good step distribution. For example if an approximate solution of U(x) is available, then interpolation can be used to invert the functional dependency and step along values of U and calculate corresponding values of x for which uniform changes in U occur. In other words function inversion can be used to obtain a step distribution.

The code in Listing 11.19 illustrates the use of this concept for the previous nonlinear equation. On line 10 the geometric progression grid generator is called for only 100 spatial intervals. This is used on line 11 to generate an approximate solution with only these few grid points. The approximate solution is then used on lines 13 through 16 with the intp() function to generate a new spatial grid with 2001 grid points using the principle that the new grid will correspond to equal steps in the approximate solution variable. This new grid is then used on line 17 to obtain a new, and hopefully more accurate, solution to the BV problem. As shown in the output, the programs require 8 Newton iterations for the first solution and 5 for the second solution. This example can be used to explore any improvement in the spatial grid generation by use of the function inversion approach.

```
 1 : -- /* list11_19.lua */
 2 : -- Solution of BV problem with non-uniform spatial grid
 3 : require"odebvfd"
 4 : require"intp"
 5 :
 6 : f = function(x,y,yp,ypp) -- Differntial equation
 7 :    return ypp + 4*yp^2
 8 : end
 9 :
10 : xx = xgp(0,1,1.e-6,100) --approximate solution, few grid points
11 : s1,nn = odelfd(f,xx,{0,2}) -- Solve equation
12 : plot(s1); print('Number of Newton iterations = ',nn)
13 : x = {0} -- New spatial array, from approximate solution
14 : for i=2,2001 do -- Expand to 2001 grid points
15 :    y = 2*(i-1)/2000; x[i] = intp(s1[2],s1[1],y)
16 : end
17 : s,err,nn = odelfde(f,x,{0,2}) -- More accurate solution
18 : print('Number of Newton iterations = ',nn)
19 : plot(s); write_data('list11_19.dat',x,s,err,xx)
Output:
Number of Newton iterations =        8
Number of Newton iterations =        5
```

Listing 11.19. Example code for generating non-uniform grid spacing from approximate solution using function inversion.

Some results for this example are shown in Figure 11.21. The lower dotted curve shown the spatial point distribution of distance vs. point number for the grid generated directly by the geometric progression function xgp(). The upper solid curve shown the grid distribution generated by Listing 11.19 from the approximate solution. The curves are amazingly similar and in fact almost identical. No, it was not known beforehand that this would occur. This close agreement for this example is a fortuitous result and such agreement is not to be expected in other cases. In fact if the xgp() function is used to generate another grid distribution beginning at 1×10^{-5} or any value other than the 1×10^{-6} value the agreement will also not be as close as shown in the figure. Even the use of more or fewer grid points for the trial solution on line 10 will not result in the same agreement. It can be seen however from this example why the xgp() distribution beginning at 1×10^{-6} gives excellent results, as it corresponds almost exactly to equal changes in the solution variable between grid points.

The generation of a grid point distribution based upon an approximate solution provides another tool for generating non-uniform spatial distributions for BV problems. It is useful in some problems, but must be used with caution. If in Listing 11.19 an initial uniformly spaced grid distribution is attempted in place of the xgp() distribution, the solution will not converge unless a fairly large number of

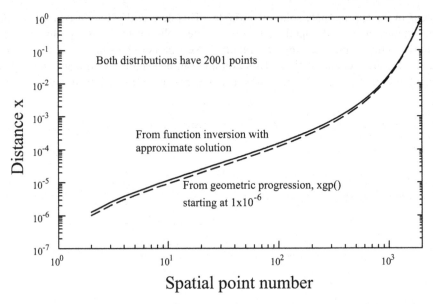

Figure 11.21 Illustration of spatial grid distribution generated by xgp() function and by function inversion from an approximate solution. Equation solve is in Listing 11.19.

uniformly spaced gird points are used (such as 1000). In some BV problems, the solution variables will be approximate constant over some regions of the spatial range. One such example for Poisson's equation in a semiconductor will be given later. For such cases, generating the grid points using the changes in the solution variable can result in very large step sizes and or large changes in step sizes between adjacent grid points. Both of which are undesirable. A good grid point distribution is one for which the change in solution variable is small between grid points but also one in which the spatial step size changes very little between adjacent grid points. General algorithms for automatically generating such grid distributions are not easy to compose and program. In any particular case the user must use some engineering judgment in selecting an appropriate non-uniform grid distribution for any particular problem.

This nonlinear BV problem has illustrated several important facts. One is that the solution accuracy can be greatly improved by using non-uniformly spaced grid points for many problems. This could also be inferred from the improved accuracy of using an adjustable time step for the initial value problems considered in the previous chapter. Second it has been demonstrated that the h-2h algorithm can be effectively used with the finite difference method to obtain an estimate of the error in a BV solution. One final example will be given of using the code segments developed so far for a highly nonlinear BV problem.

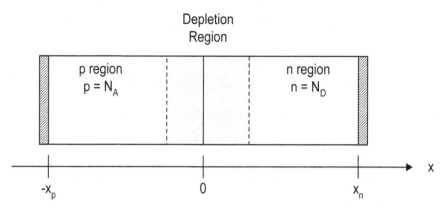

Figure 11.22. Sketch of one dimensional p-n junction

An important nonlinear BV problem in semiconductor physics is the solution of Poisson's equation around a p-n junction. A sketch of such a one dimensional p-n junction is shown in Figure 11.22. The structure consists of two doped semiconductor regions as identified as the p region and n region consisting of primarily holes on the left side and electrons on the right side of the interface which occurs at x = 0. In equilibrium a so-called depletion region exists around the center interface where the electron and hole densities are greatly reduced below the doping densities in the respective regions. The electric potential, V, along the structure is described by Poisson's equation which for this case is:

$$\frac{d^2V}{dx^2} = \frac{q}{\varepsilon}(p - n + N_{Net}), \text{ with } N_{Net} = \begin{cases} N_D \text{ for } x > 0 \\ -N_A \text{ for } x < 0 \end{cases} \quad (11.53)$$

$$p = n_i \exp(-qV/kT) \text{ and } n = n_i \exp(qV/kT)$$

In this formulation the zero of potential is taken at the point in the structure where $n = p = n_i$. This results in the boundary values for the potential as:

$$V(-x_p) = -(kT/q)\log(N_A/n_i)$$
$$V(x_n) = (kT/q)\log(N_D/n_i) \quad (11.54)$$

This set of equations then constitute a nonlinear two point BV problem. The solution of this problem is reasonably difficult because of the exponential relationship of the carrier densities to potential and because of the small value of kT/q (0.026 Volts at room temperature). The finite difference method is appropriate for this problem but the problem is essentially impossible to solve with the shooting method since the first derivative is essentially zero on both boundaries.

Listing 11.20 shows code for solving this BV problem for a p-n junction that is doped with 1×10^{18} impurities on each side of the junction. Lines 5 through 13 define the parameters of the problem and the differential equation. Lines 15 through 19 set up initial arrays for a uniformly spaced grid of points and an initial approximation to the voltage across the junction. Note that on line 18, a very simple

```
 1 : -- /* list11_20.lua */
 2 : -- Solution of semiconductor depletion layer as a BV problem
 3 : require"odebvfd"; exp = math.exp
 4 :
 5 : L,Na,Nd = 2.e-5, 1e18, 1e18 -- Size and doping densities
 6 : q,eps,vt,ni = 1.6e-19, 11.9*8.854e-14, .026, 1.45e10
 7 : qdep = q/eps
 8 : v1,v2 = -vt*math.log(Na/ni), vt*math.log(Nd/ni) -- BVs
 9 :
10 : f = function(x,v,vp,vpp) -- Poisson's equation
11 :    if x<=0 then Nnet = -Na else Nnet = Nd end
12 :    return vpp + qdep*(ni*exp(-v/vt) - ni*exp(v/vt) + Nnet)
13 : end
14 :
15 : nx = 2001; dx = L/(nx-1); x,v = {},{}
16 : for i=1,nx do -- Initial voltage approximation
17 :    x[i] = (i-1)*dx - L/2
18 :    if x[i]<0 then v[i] = v1 else v[i] = v2 end
19 : end
20 : s,err,nn = odelfde(f,x,v) -- Solve equations, fixed BVs
21 : p,n = {},{} -- To calculate holes and electrons
22 : for i=1,nx do -- Now calculate then
23 :    p[i],n[i] = ni*exp(-s[2][i]/vt),ni*exp(s[2][i]/vt) end
24 :
25 : print('Number of Newton iterations = ',nn)
26 : plot(s); plot(s[1],n);plot(s[1],p)
27 : plot(err); write_data('list11_20.dat',s,err,n,p)
Output:
Number of Newton iterations =        7
```

Listing 11.20. Code segment for solving Poisson's equation as applied to a semi-conductor p-n junction.

approximation is used for the initial potential and this is an abrupt step in the potential from the left boundary value to the right boundary value at the p-n junction interface. Other simple approximations that one might try are to use a linear function for the voltage varying from the two boundary values. The reader is encouraged to experiment with this initial value. However, it will be found that the program will not converge in 50 Newton iterations with a simple linear variation for the initial approximation. It is simply too far from the exact solution, especially far from the junction. One might also expect that a better approximation would be needed than the abrupt step at the junction. However, this is not the case with the present physical parameters. Other values of length and/or doping densities might require a more accurate initial guess. Perhaps the approach used here can be stated as: "use a simple initial guess at the solution and see if it works". But one must be sure that it can be determined if some initial approximation does in fact-work. This is the point of returning the number of Newton iterations and with printing the value on line 25. It the value is the default limit (50 used here) then it will be know that the initial guess didn't work. This might be because of an initial guess that is too far from the solution or possibly a mistake could have been made in defining the differential equation or the boundary values. All avenues for possible problems should be explored when the solution of a BV problem does not converge. In Listing 11.20, no boundary value functions are defined. This is acceptable as the boundary values are fixed and the initial approximation has the correct end point boundary values. The ode1fde() function will detect the missing boundary evaluation functions and supply default ones that are appropriate for fixed boundary values. Finally the electron and hole densities are evaluated from the voltage on line 23 and the results are then plotted and saved.

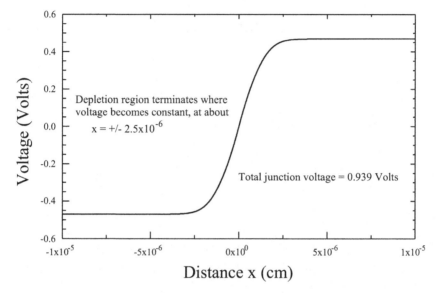

Figure 11.23 Voltage across symmetrical p-n junction. Parameters are given in Listing 11.20.

The reader is encouraged to execute the program and rapidly see the result which for the voltage is presented in Figure 11.23. As seen the voltage exhibits a rather abrupt transition around the p-n junction. From this it can be seen that an initial approximation that makes an abrupt step change at the origin is probably a better approximation than a linearly varying voltage between the two end points. Since it is known that the voltage exhibits this behavior and the computer code converged in only 7 Newton iterations, one can be reasonably sure that an accurate solution to this nonlinear problem has been obtained. A look at the number of Newton iterations should always be the first step in accepting a solution returned by the code. An estimate of the error in voltage is also obtained in this example from the ode1fde() function and saved in a file. The reader is encouraged to look at the estimated error, but the results will not be presented here. The reader can verify that the estimated error indicates that the results for the potential are accurate to several decimal digits. One should always be very skeptical of numerical solutions and always try to evaluate the accuracy of any such solution. With the present code the very useful h-2h algorithm is available to give an estimate of accuracy achieved in any solution.

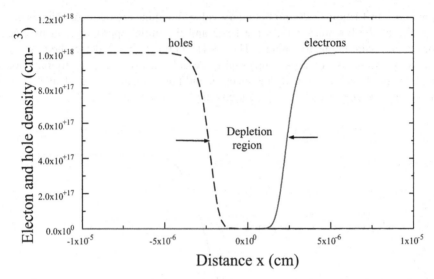

Figure 11.24 Electron and hole densities for symmetrical p-n junction on linear scale. Parameters are given in Listing 11.20.

Two separate graphs of the electron and hole densities are shown in Figures 11.24 and 11.25. The linear graph of Figure 11.24 clearly indicates a region near the origin that is void or depleted of either electrons or holes. The boundaries of this region are not completely abrupt, but can be approximated by the points where the densities are approximately half of their maximum values. Such points are identified on the figure. A quick calculation with first order semiconductor device equations predicts that the depletion region boundaries should be at approximately 2.49×10^{-6} cm on each side of the junction. This is very close to what would be estimated from the figure at the point of the arrows. Finally Figure 11.25 shows the carrier densities on a log scale so that the continuous nature of the transition can be seen across the junction. Far away form the junction interface the densities are controlled by the doping densities with one type of carrier being identified as a majority carrier density and the other type as a minority carrier density. For many semiconductor device, it is the minority carrier density that is of interest (such as diodes and bipolar transistors) so the accurate evaluation of these small values is not only of academic interest but also of much practical importance. The need to accurately evaluate minority carrier densities is one of the reasons that semiconductor BV problems are especially difficult nonlinear BV problems. Although calculated points are not shown in Figure 11.25, it can be seen that the small densities of minority carriers on each side of the junction are accurately calculated.

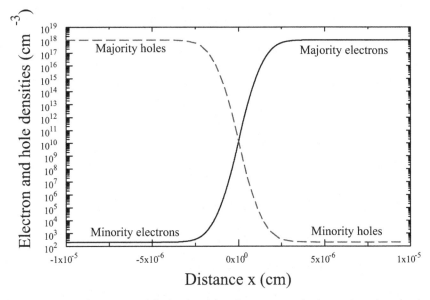

Figure 11.25. Electron and hole densities for symmetrical p-n junction on log scale. Parameters are given in Listing 11.20.

One additional example of Poisson's equation with a p-n junction is given with Listing 11.21. There are two major changes from Listing 11.20. First the doping density on the p-side is reduced by two orders of magnitude to 1×10^{16}. For such a case most of the depletion region is on the lightly doped side of the junction ($x < 0$ region). Also since the doping density is reduced by a factor of 100, the depletion layer width will be increased by about a factor of 10 so the spatial distance for the solution is increased on line 5 to 4×10^{-4} cm. The other major difference is the use of a non-uniform spatial grid near the p-n junction boundary. This is implemented as an example on lines 15 through 18 of the code. First a non-uniform grid is generated for the spatial range 0 to L/2 using the geometrical progression xgp() on line 15. A mirror image of this is taken on line 17 for negative x values and added to the positive values on line 18. The result is a spatial distribution that uses small steps on both sides of the $x = 0$ value and larger spatial steps near the negative and positive x boundaries. This is provided as one example of how more complicated non-uniform step distributions can be generated from the basic xgp() function.

The minimum step size specified for xgp() is 1.e-9 and this is about a factor of 200 smaller than would be generated by a uniform step distribution with the same total number of points. It would be expected that this would increase the accuracy of the solution, and in fact it does decrease the maximum estimated error in the solution by about two orders of magnitude. To verify this the reader is encouraged to execute the code in Listing 11.21 and then re-execute the code with a uniform step distribution and compare the resulting estimated errors from the two sets of calculations. This is left up to the reader.

```
 1 : -- /* list11_21.lua */
 2 : -- Solution of semiconductor depletion layer as a BV problem
 3 : require"odebvfd"; exp = math.exp
 4 :
 5 : L,Na,Nd = 2.e-4, 1e16, 1e18 -- One sided parameters
 6 : q,eps,vt,ni = 1.6e-19, 11.9*8.854e-14, .026, 1.45e10
 7 : qdep = q/eps
 8 : v1,v2 = -vt*math.log(Na/ni), vt*math.log(Nd/ni) -- BVs
 9 :
10 : f = function(x,v,vp,vpp) -- Poisson's equation
11 :    if x<=0 then Nnet = -Na else Nnet = Nd end
12 :    return vpp + qdep*(ni*exp(-v/vt) - ni*exp(v/vt) + Nnet)
13 : end
14 :
15 : nx1 = 1000; xx = xgp(0,L/2,1.e-9,nx1) -- grid on 0 to L/2
16 : x,v = {}, {}; i=1 -- Now mirror to negative values
17 : for j=nx1+1,1,-1 do x[i] = -xx[j]; i = i+1 end
18 : for j=2,nx1+1 do x[i] = xx[j]; i = i+1 end
19 :
20 : nx = #x
21 : for i=1,nx do -- Initial voltage approximation
22 :    if x[i]<0 then v[i] = v1 else v[i] = v2 end
23 : end
24 : s,err,nn = odelfde(f,x,v) -- Solve equations, fixed BVs
25 : p,n = {},{} -- To calculate holes and electrons
26 : for i=1,nx do -- Now calculate then
27 :    p[i],n[i] = ni*exp(-s[2][i]/vt),ni*exp(s[2][i]/vt) end
28 :
29 : print('Number of Newton iterations = ',nn)
30 : plot(s); plot(s[1],n);plot(s[1],p)
31 : plot(err); write_data('list11_21.dat',s,err,n,p)
```

Listing 11.21. Second example of p-n junction with asymmetrical doping.

Results for this listing are illustrated by Figures 11.26 and 11.27. The voltage curve shows that almost all the voltage drop is on the lightly doped side of the junction (for negative x values). Also the total voltage across the junction is decreased by the lighter doping on the p-side. The carrier density plots in Figure 11.27 also show that almost all the changes in the densities occur on the lightly doped side of the junction. On the lightly doped side, the majority carrier density is reduced by two orders of magnitude while the minority carrier density is increased by two orders of magnitude from the values in the previous solution. From the estimated errors in the potential, it can be assured that the solutions for both the voltage and carrier densities are quite accurate solutions of the differential equation and BV problem.

This application of the finite difference method to Poisson's equation as applied to semiconductors provides only a glimpse of how the code routines can be applied to nonlinear boundary value problems. However, not all problems of interest involve only a single differential equation. Many problems of interest involve equations of higher order than two or involve systems of coupled second order differential equations. Such problems are the subject of the next section where algorithms and code segments are discussed and developed for these more general

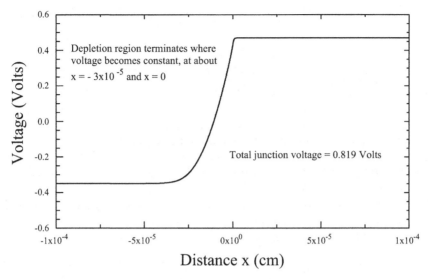

Figure 11.26. Voltage for asymmetric p-n junction with parameters of Listing 11.21.

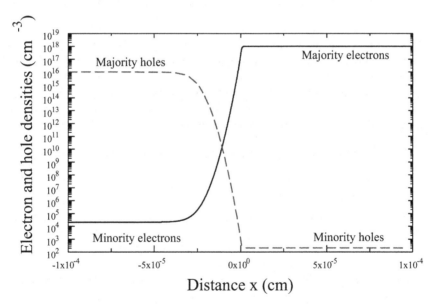

Figure 11.27. Electron and hole densities for asymmetric p-n junction with parameters of Listing 11.21.

problems. The solution methods will be extensions to the finite difference methods developed in this section.

11.6 Boundary Value Problems with Coupled Second Order Differential Equations

A large number of boundary value problems of interest in engineering involve coupled systems of second order differential equations and this is the topic of discussion for this section. For discussion purposes it will be assumed that one has a system of three coupled second order differential equations, although hopefully it will be obvious how the discussion could be extended to a larger number of coupled equations. One formulation of such a system of equations is of the form of Eq. (11.55) where F_1, F_2 and F_3 are any arbitrary functions of the three function values and the three first derivatives.

$$\frac{d^2 U_1}{dx^2} + F_1(x, U_1, U_2, U_3, U_1', U_2', U_3') = 0$$

$$\frac{d^2 U_2}{dx^2} + F_2(x, U_1, U_2, U_3, U_1', U_2', U_3') = 0 \qquad (11.55)$$

$$\frac{d^2 U_3}{dx^2} + F_3(x, U_1, U_2, U_3, U_1', U_2', U_3') = 0$$

This assumes that the equations can be written in terms of linear uncoupled second derivative terms. This formulation describes many physical problems. However, the most general formulation would allow nonlinear second derivative terms as well as providing for the possibility of coupled nonlinear second derivative terms. Since this is the most general case and is almost as easy to handle as the above formulation, the form assumed here is the more general form of:

$$F_1(x, U_1, U_2, U_3, U_1', U_2', U_3', U_1'', U_2'', U_3'') = F_1(x, \mathbf{U}, \mathbf{U}', \mathbf{U}'') = 0$$
$$F_2(x, U_1, U_2, U_3, U_1', U_2', U_3', U_1'', U_2'', U_3'') = F_2(x, \mathbf{U}, \mathbf{U}', \mathbf{U}'') = 0 \qquad (11.56)$$
$$F_3(x, U_1, U_2, U_3, U_1', U_2', U_3', U_1'', U_2'', U_3'') = F_3(x, \mathbf{U}, \mathbf{U}', \mathbf{U}'') = 0$$

In these equations no specific form is assumed for any of the dependences on the function values, or on the first derivatives or second derivatives. The bold symbols are used to indicate a vector or table of values, i.e. $\mathbf{U} = \{U_1, U_2, U_3\}$ and similarly for the bold prime and double prime quantities indicating tables of first and second derivatives. In addition to the differential equations, a set of three boundary value equations must exist (in perhaps nonlinear form) relating the functions and first derivatives on both boundaries.

To solve this potentially nonlinear set of equations the fundamental principle of linearize and iterate can be applied. Functional linearization similar to that of Eq. (11.40) for one variable gives the set of equations as expressed in Eq. (11.57). Since no assumptions are made regarding the form of the equations, each of the three quasilinearized equations can possible have second and first derivative terms for the corrections on each variable. While the algebra gets a little involved, the concept is straightforward to extend to any number of coupled equations.

$$
\left(\frac{\partial F_1}{\partial U_1''}\right)\frac{d^2 u_1}{dx^2}+\left(\frac{\partial F_1}{\partial U_1'}\right)\frac{du_1}{dx}+\left(\frac{\partial F_1}{\partial U_1}\right)u_1+\left(\frac{\partial F_1}{\partial U_2''}\right)\frac{d^2 u_2}{dx^2}+\left(\frac{\partial F_1}{\partial U_2'}\right)\frac{du_2}{dx}+\left(\frac{\partial F_1}{\partial U_2}\right)u_2
$$

$$
+\left(\frac{\partial F_1}{\partial U_3''}\right)\frac{d^2 u_3}{dx^2}+\left(\frac{\partial F_1}{\partial U_3'}\right)\frac{du_3}{dx}+\left(\frac{\partial F_1}{\partial U_3}\right)u_3+F_1(x,\mathbf{U},\mathbf{U}',\mathbf{U}'')=0
$$

$$
\left(\frac{\partial F_2}{\partial U_1''}\right)\frac{d^2 u_1}{dx^2}+\left(\frac{\partial F_2}{\partial U_1'}\right)\frac{du_1}{dx}+\left(\frac{\partial F_2}{\partial U_1}\right)u_1+\left(\frac{\partial F_2}{\partial U_2''}\right)\frac{d^2 u_2}{dx^2}+\left(\frac{\partial F_2}{\partial U_2'}\right)\frac{du_2}{dx}+\left(\frac{\partial F_2}{\partial U_2}\right)u_2
$$

$$
+\left(\frac{\partial F_2}{\partial U_3''}\right)\frac{d^2 u_3}{dx^2}+\left(\frac{\partial F_2}{\partial U_3'}\right)\frac{du_3}{dx}+\left(\frac{\partial F_2}{\partial U_3}\right)u_3+F_2(x,\mathbf{U},\mathbf{U}',\mathbf{U}'')=0
$$

$$
\left(\frac{\partial F_3}{\partial U_1''}\right)\frac{d^2 u_1}{dx^2}+\left(\frac{\partial F_3}{\partial U_1'}\right)\frac{du_1}{dx}+\left(\frac{\partial F_3}{\partial U_1}\right)u_1+\left(\frac{\partial F_3}{\partial U_2''}\right)\frac{d^2 u_2}{dx^2}+\left(\frac{\partial F_3}{\partial U_2'}\right)\frac{du_2}{dx}+\left(\frac{\partial F_3}{\partial U_2}\right)u_2
$$

$$
+\left(\frac{\partial F_3}{\partial U_3''}\right)\frac{d^2 u_3}{dx^2}+\left(\frac{\partial F_3}{\partial U_3'}\right)\frac{du_3}{dx}+\left(\frac{\partial F_3}{\partial U_3}\right)u_3+F_1(x,\mathbf{U},\mathbf{U}',\mathbf{U}'')=0
$$

(11.57)

The next step in the development of a solution approach is to use finite differences to approximate the derivatives in Eq. (11.57). Looking at these equations and the development for a single equation, it can be readily seen that each of the three equations above at some spatial point i will involve the solution variables at points i-1, i and i+1 for each of the three variables. The resulting three equations will be of the form:

$$
\begin{bmatrix} A_{11} & A_{12} & A_{13} \\ A_{21} & A_{22} & A_{23} \\ A_{31} & A_{32} & A_{33} \end{bmatrix}\begin{bmatrix} u_{1,i-1} \\ u_{2,i-1} \\ u_{3,i-1} \end{bmatrix}+\begin{bmatrix} B_{11} & B_{12} & B_{13} \\ B_{21} & B_{22} & B_{23} \\ B_{31} & B_{32} & B_{33} \end{bmatrix}\begin{bmatrix} u_{1,i} \\ u_{2,i} \\ u_{3,i} \end{bmatrix}+
$$

$$
\begin{bmatrix} C_{11} & C_{12} & C_{13} \\ C_{21} & C_{22} & C_{23} \\ C_{31} & C_{32} & C_{33} \end{bmatrix}\begin{bmatrix} u_{1,i+1} \\ u_{2,i+1} \\ u_{3,i+1} \end{bmatrix}+\begin{bmatrix} D_1 \\ D_2 \\ D_3 \end{bmatrix}=0
$$

(11.58)

This is the generalization of Eq. (11.37) for a single variable and in the simplified form of a matrix equation becomes:

$$\mathbf{A}\mathbf{u}_{i-1}+\mathbf{B}\mathbf{u}_i+\mathbf{C}\mathbf{u}_{i+1}+\mathbf{D}=0 \qquad (11.59)$$

Except for the presence of 3x3 matrices and vectors of solution values, this is the same as the corresponding equation for a single variable. Increasing the number of coupled equations will just increase the size of the matrices. It is perhaps useful to give expressions once for the A_{jk}, B_{jk} and C_{jk} coefficients:

$$A_{jk} = \left(2\left(\frac{\partial F_j}{\partial U_k''} \right) \alpha_i / s_i - \alpha_i \left(\frac{\partial F_j}{\partial U_k'} \right) \right) / \left(s_i + s_{i-1} \right)$$

$$B_{jk} = \frac{\partial F_j}{\partial U_k} + \left(\left(\frac{\partial F_j}{\partial U_k'} \right) (\alpha_i^2 - 1) / \alpha_i - 2\left(\frac{\partial F_j}{\partial U_k''} \right) (\alpha_i + 1) / s_i \right) / \left(s_i + s_{i-1} \right) \qquad (11.60)$$

$$C_{jk} = \left(2\left(\frac{\partial F_j}{\partial U_k''} \right) / s_i + \left(\frac{\partial F_j}{\partial U_k'} \right) / \alpha_i \right) / \left(s_i + s_{i-1} \right)$$

$$D_j = F_j$$

In these equations, a non-uniform spatial grid has been assumed and the subscript j refers to the row of the matrix element and the subscript k refers to the column of the matrix. The subscript i is used to refer to the spatial point around which the finite difference equation is being written. If this formulation of the coefficients is used for three equations with no coupling between the equations, only the diagonal elements would have non-zero terms. In many practical problems, several of the partial derivatives may be zero; for example the second derivative may occur uncoupled in the form of Eq. (11.55) in which case the A matrix would be of diagonal form. However, to retain as much flexibility as possible with the equation formulation, all terms will be kept in the implementation here.

To evaluate a given row of the A, B or C matrix it can readily be seen what has to be done from the equation set. The partial derivatives of one of the functions must be evaluated with respect to each solution variable, each variable first derivative and each variable second derivative. These partial derivatives allow the elements in Eq. (11.60) to be evaluated row by row, or column by column if more convenient. In fact it is more efficient to evaluate the terms on a column by column basis as one then only has to select an increment value once for each variable and derivative and then apply this incremented value to all the functions. In terms of a single variable BV problem, there is roughly N^2 as much computational effort in setting up the finite difference equations, where N is the number of coupled equations. Then there is additional time required to solve the set of coupled matrix equations as represented by Eq. (11.58).

Now consider the solution of the finite difference equations as written in matrix form in Eq. (11.59). Looking back at the single variable problem and the corresponding equation (Eq. (11.37)) it is seen that the equations have the same functional form it's just that one is a single variable equation and the other is a matrix equation. This leads to the same approach for solution of assuming that one can write:

$$\mathbf{u}_{i-1} = \mathbf{e}_i \mathbf{u}_i + \mathbf{f}_i \qquad (11.61)$$

where $\mathbf{e}_i$ is now an NxN matrix and $\mathbf{f}_i$ is a column vector of length N. Substituting this into Eq. (11.59) and rearranging gives

$$(\mathbf{A}_i \mathbf{e}_i + \mathbf{B}_i)\mathbf{u}_i = -\mathbf{C}_i \mathbf{u}_{i-1} - (\mathbf{A}_i \mathbf{f}_i + \mathbf{D}_i) \qquad (11.62)$$

These are now matrix operations so it is important to maintain the order of the matrix operations. Formally one can multiply by the matrix inverse of the first factor in parentheses and replace i by i-1 to give the equation

$$\mathbf{u}_{i-1} = -(\mathbf{A}_{i-1}\mathbf{e}_{i-1} + \mathbf{B}_{i-1})^{-1}(\mathbf{C}_{i-1}\mathbf{u}_i + \mathbf{A}_{i-1}\mathbf{f}_{i-1} + \mathbf{D}_{i-1})$$ (11.63)

Comparing this with Eq. (11.61) if is seen that the recursive relationship for the factors in Eq. (11.61) must satisfy the expressions:

$$\mathbf{e}_{i+1} = -(\mathbf{A}_i\mathbf{e}_i + \mathbf{B}_i)^{-1}\mathbf{C}_i$$
$$\mathbf{f}_{i+1} = -(\mathbf{A}_i\mathbf{e}_i + \mathbf{B}_i)^{-1}(\mathbf{A}_i\mathbf{f}_i + \mathbf{D}_i)$$ (11.64)

If these are compared with the corresponding equations for the single variable case (see Eq. (11.44)) it is readily seen that the equations are essentially the same except that for the single variable case the negative power operation can be replaced by ordinary division by the first factor in parentheses whereas for this case the operation must be a matrix inverse operation followed by matrix multiplication.

As a solution technique then the same fundamental procedure can be used with a system of coupled equations as with the single variable problem. Starting at one boundary and assuming know initial starting values for the $\mathbf{e}$ and $\mathbf{f}$ factors the calculation can step across the independent variable range using the recursive relationships of Eq. (11.64) at each grid point. Then a back substitution using Eq. (11.61) will give the solution values for all variables as the solution is stepped back across the independent variable space. In principle it's the same as for the single variable case; the difference is working with matrices instead of single values. As noted for the single variable case, values for the $\mathbf{A}, \mathbf{B}, \mathbf{C}$ and $\mathbf{D}$ matrices do not have to be stored at the spatial points. They can be computed once for each spatial point, then used in the recursive equations and then discarded. However one does have to store at each spatial grid point the $\mathbf{e}$ and $\mathbf{f}$ matrix factors. This corresponds to one NxN matrix and one column vector of length N at each spatial point. This is a storage requirement of Nx(N+1) at each spatial grid as compared with only 2 stored values for the single variable problem. For three coupled equations with 2001 grid points this is a storage requirement of 4,006,002 values. Fortunately with the code here the software manages the memory storage so the work can concentrate on numerical concepts.

Computer code for implementing the above finite difference approach for coupled second order differential equations is shown in Listing 11.22. This is a rather long listing and represents the most complicated code example so far presented. However the code is very similar to Listing 11.14 with appropriate additions to handle multiple differential equations. A fairly detailed discussion of the code for the single differential equation of Listing 11.14 has preciously been given. For those interested in the details of the expanded implementation for multiple equations, a comparison of the major sections of the code in both listings is given in Table 11.1 below. Because of the multiple equations, most sections of the one-dimensional code are expanded to implement various sums over the number of equations being solved. In general the terms like "for m=1,neq do" in Listing 11.22 represen t sums over the differential equations. The reader's attention is

```
 1 : -- /* File odefd.lua */
 2 :
 3 : require"Matrix" -- Needed for matrix algebra
 4 : ode2bvfd = function(eqs,x,u)
 5 :   local neq,s,upp,up,uppi,upi,ui = #u, {}, {}, {}, {}, {}, {}
 6 :   a,b,c,d = Matrix.new(neq,neq),Matrix.new(neq,neq),
            Matrix.new(neq,neq),Matrix.new(neq,1)
 7 :   local alfi,e,f = 0, {}, {}
 8 :   local feq ,nend = eqs[1], NMAX
 9 :   local fval,fu,fctupp,fctup,fctu = {}, {}, {}, {}, {}
10 :   local uppi,upi,ui,cmax,imax,umax = {}, {}, {}, {}, {}, {}
11 :   for m=1,neq do fctupp[m],fctup[m],fctu[m]=FACT,FACT,FACT end
12 :
13 :   nx = #x; nxm1 = nx-1 -- #x values associated with index 'i'
14 :   for i=1,nxm1 do s[i] = x[i+1] - x[i] end
15 :
16 :   bound = function(nb,nxb)  -- Function to evaluate BVs
17 :     xi = x[nxb]
18 :     for m=1,neq do upi[m],ui[m] = up[m][nxb],u[m][nxb] end
19 :     eqs[nb](fval,ui,upi)
20 :     if nb==2 then c1,c2 = -1/s[1],1/s[1]
           else c1,c2 = 1/s[nx-1],-1/s[nx-1] end
21 :     for m=1,neq do
22 :       d[m][1] = fval[m]
23 :       upi[m]=upi[m]+fctup[m];eqs[nb](fu,ui,upi);
              upi[m]=upi[m]-fctup[m]
24 :       for n=1,neq do -- Probe up factor
25 :         fjk = (fu[n] - fval[n])/fctup[m]
26 :         b[n][m] = c1*fjk
27 :         if nb==2 then c[n][m]=c2*fjk
             else a[n][m]=c2*fjk end
28 :       end -- Then probe u factor below
29 :       ui[m] = ui[m]+fctu[m]; eqs[nb](fu,ui,upi,ui);
              ui[m]=ui[m]-fctu[m]
30 :       for n=1,neq do b[n][m] = b[n][m] + (fu[n] -
              fval[n])/fctu[m] end
31 :     end
32 :   end -- Return boundary equations involving a,b,c,d
33 :
34 :   for m=1,neq do upp[m],up[m] = {},{} end -- Derivative arrays
35 :   for ni=1,NMAX do -- ni = iteration number count
36 :     for m=1,neq do fctupp[m],fctup[m],fctu[m] = 0,0,0 end
37 :     for i=2,nx-1 do -- Calculate second and first derivatives
38 :       si,sisi,alfi = s[i],(s[i]+s[i-1]),s[i]/s[i-1]
39 :       c1,c2,c3 = 2/(si*sisi),-2*(alfi+1)/(si*sisi),
              2*alfi/(si*sisi)
40 :       for m=1,neq do
41 :         fctu[m] = fctu[m] + abs(u[m][i]*si)
42 :         duu = c1*u[m][i+1] + c2*u[m][i] + c3*u[m][i-1]
43 :         upp[m][i],fctupp[m] = duu,fctupp[m]+abs(duu*si)
44 :       end
45 :       c1,c2,c3 = 1/(alfi*sisi),(alfi-1)/si,-alfi/sisi
46 :       for m=1,neq do
47 :         duu = c1*u[m][i+1] + c2*u[m][i] + c3*u[m][i-1]
48 :         up[m][i],fctup[m] = duu,fctup[m]+abs(duu*si)
49 :       end
50 :     end
51 :     alfi = s[2]/s[1]
52 :     for m=1,neq do -- Special treatment for lower end point
```

```
53 :            upp[m][1] = upp[m][2]
54 :            up[m][1]=(-u[m][3]+u[m][2]*(1+alfi)^2-
                    u[m][1]*alfi*(2+alfi))/(alfi*(s[2]+s[1]))
55 :        end
56 :        alfi = s[nxm1]/s[nx-2]
57 :        for m=1,neq do -- Special treatment for upper end point
58 :            upp[m][nx] = upp[m][nxm1]
59 :            up[m][nx]=(u[m][nx]*(1+2*alfi)-u[m][nxm1]*(1+alfi)^2 +
60 :                u[m][nx-2]*alfi^2)/(alfi*(s[nxm1]+s[nx-2]))
61 :        end
62 :        for m=1,neq do -- protect against large values
63 :            fctupp[m]=fctupp[m] + abs(upp[m][1]) + abs(upp[m][nx])
64 :            fctup[m] = fctup[m] + abs(up[m][1]) + abs(up[m][nx])
65 :            fctu[m] = fctu[m] + abs(u[m][1]) + abs(u[m][nx])
66 :        end
67 :        for m=1,neq do --Average values, variables and derivatives
68 :            fctupp[m],fctup[m],fctu[m]=FACT*fctupp[m],
                        FACT*fctup[m],FACT*fctu[m]
69 :            if fctupp[m]==0 then fctupp[m] = FACT end -- zero?
70 :            if fctup[m]==0 then fctup[m] = FACT end
71 :            if fctu[m]==0 then fctu[m] = FACT end
72 :        end
73 :        if umin[1]~=nil then -- limit fctu values
74 :            for m=1,neq do fctu[m] = max(fctu[m],FACT*umin[m]) end
75 :        end
76 :
77 :        bound(2,1) -- Evaluate lower boundary conditions
78 :        gi = b^-1 -- Matrix algebra for e[] and f[]
79 :        if type(gi)=='number' then
80 :            printf('Error in left boundary values\nCheck boundary
                        equations\n')
81 :            return end
82 :        e[2],f[2] = -gi*c,-gi*d
83 :
84 :        for i=2,nx-1 do -- Set up a,b,c,d and e and h as arrays
85 :            xi,si,sisi,alfi = x[i],s[i],(s[i]+s[i-1]),s[i]/s[i-1]
86 :            for m=1,neq do -- Set up arrays for derivatives
87 :                uppi[m],upi[m],ui[m] = upp[m][i],up[m][i],u[m][i]
88 :                for n=1,neq do a[n][m],b[n][m],c[n][m] = 0,0,0 end
89 :            end
90 :            feq(fval,xi,ui,upi,uppi,i) -- Evaluate equations
91 :            for m=1,neq do -- increment each variable in order
92 :                d[m][1] = fval[m] -- Set d[] array value
93 :                c1,c2,c3=2*alfi/(si*sisi), -2*(alfi+1)/(si*sisi),
                        2/(si*sisi)
94 :                uppi[m] = uppi[m] + fctupp[m]
95 :                feq(fu,xi,ui,upi,uppi,i) -- Probe upp factor
96 :                for n=1,neq do -- Now collect changes
97 :                    fjk = (fu[n] - fval[n])/fctupp[m] -- Update
98 :                    a[n][m],b[n][m],c[n][m]=
                            a[n][m]+c1*fjk,b[n][m]+c2*fjk,c[n][m]+c3*fjk
99 :                end
100 :               c1,c2,c3 = -alfi/sisi, (alfi-1)/si, 1/(alfi*sisi)
101 :               uppi[m],upi[m]=uppi[m]-fctupp[m], upi[m]+fctup[m]
102 :               feq(fu,xi,ui,upi,uppi,i) -- probe up factor
103 :               for n=1,neq do
104 :                   fjk = (fu[n] - fval[n])/fctup[m] --Update a,b,c
105 :                   a[n][m],b[n][m],c[n][m] =
                            a[n][m]+c1*fjk,b[n][m]+c2*fjk,c[n][m]+c3*fjk
```

```
106 :             end
107 :             upi[m],ui[m] = upi[m] - fctup[m], ui[m] + fctu[m]
108 :             feq(fu,xi,ui,upi,uppi,i) -- Probe u factor
109 :             for n=1,neq do b[n][m] = b[n][m] + (fu[n]-
                        fval[n])/fctu[m] end
110 :             ui[m] = ui[m] - fctu[m]
111 :           end
112 :           -- Solve tridagonal matrix equations
113 :           gi = (a*e[i] + b)^-1-- Matrix algebra to e[] and f[]
114 :           e[i+1] = -gi*c; f[i+1] = -gi*(d + a*f[i])
115 :         end -- Now have [i] array of e[] and f[] Matrix factors
116 :
117 :         bound(3,nx) -- Evaluate upper boundary condition,
118 :         gi = (a*e[nx] + b)^-1
119 :         if type(gi)=='number' then
120 :             printf('Error in right boundary values\nCheck boundary
                        equations\n')
121 :             return end
122 :         d = -gi*(a*f[nx] + d)
123 :
124 :         for m=1,neq do -- Zero error factors
125 :             cmax[m],imax[m],umax[m]  = 0,0,0
126 :         end
127 :         for i=nx,1,-1 do -- Now go from nx to 1 for corrections
128 :             for m=1,neq do
129 :                 du = d[m][1] -- Each correction is taken in turn
130 :                 c1 = abs(du); if c1>umax[m] then umax[m] = c1 end
131 :                 c2 = abs(u[m][i])+c1;if c2~=0.0 then c1=c1/c2 end
132 :                 if c1>cmax[m] then cmax[m]=c1; imax[m]=i end
133 :                 u[m][i] = u[m][i] + du -- Update solutions,
134 :             end
135 :             if i==1 then break end
136 :             d = e[i]*d + f[i] -- Matrix algebra for correction
137 :         end -- Back for next point, i-1
138 :
139 :         if nprint~=0 then -- Print iteration data
140 :             printf("-- %i-- Iteration number, Maxumum relative,
                        absolute corrections are: \n",ni)
141 :             for m=1,neq do printf("(%i) %e, %e, at %i ; ",
                        m, cmax[m], umax[m],imax[m]) end
142 :             printf("\n"); io.flush()
143 :         end
144 :
145 :         c1 = 1-- Now see if solution meets accuracy criteria
146 :         for m=1,neq do      if cmax[m]>ERROR then c1 = 0 end end
147 :         if c1==1 then nend=ni; break end -- Relative error met
148 :         if umin[1]~=nil then -- Absolute accuracy limits specified
149 :             c1 = 1
150 :             for m=1,neq do
151 :                 if umin[m]~=nil then -- Limit for variable m
152 :                     if umin[n]~=0 then if umax[m]>umin[m] then c1=0 end
153 :                     else if umax[m]>fctu[m]*ERROR/FACT then c1 = 0 end
154 :                 end end
155 :             end
156 :             if c1==1 then nend=ni; break end -- Absolute error  met
157 :         end
158 :     end
159 :     return nend, cmax, umax, upp, up -- Solution returned in u
160 :     -- Derivatives returned in case user needs values,
```

```
161 : end
162 : setfenv(ode2bvfd,{type=type,abs=math.abs,max=math.max,
          Matrix=Matrix,
163 :   ERROR=1.e-5,umin={},FACT=1.e-6,NMAX=50,nprint=0,
          printf=printf,io=io,
164 :   math=math,table=table,unpack=unpack,ode2bvfd=ode2bvfd})
```

Listing 11.22 Code for coupled second order boundary value problems by the finite difference algorithm.

Major Sections of Code	Listing 11.14 Single DEQ	Listing 11.22 Multiple DEQ
1. Basic definitions	Lines: 1-9	Lines: 1-14
2. Function to evaluate boundary conditions	Lines: 11-16	Lines: 16-32
3. Major iterative loop	Lines: 18-79	Lines: 35-158
4. Evaluate derivatives at present iteration	Lines: 20-26	Lines: 37-61
5. Evaluate increments for partial derivatives	Lines: 33-35	Lines: 67-72
6. Evaluate lower boundary conditions	Lines: 36-37	Lines: 77-82
7. Calculate a,b,c and d values across structure	Lines: 46-48	Lines: 91-111
8. Calculate e and f arrays	Lines: 51	Lines: 113-114
9. Evaluate upper boundary conditions	Lines: 55-56	Lines: 117-122
10. Back calculations for corrections to solutions	Lines: 57-59	Lines: 127-137
11. Calculate relative and absolute corrections	Lines: 61-68	Lines: 128-134
12. Print iterative results, if desired	Lines: 70-73	Lines: 139-143
13. Check for desired accuracy	Lines: 74-76	Lines: 145-157
14. Return results	Line: 78	Line: 159

Table 11.1. Comparison summary of major sections of code for finite difference method with single differential equation and multiple coupled differential equations.

called to lines 92, 98, 105 and 109 of the code where the values of the a, b, c and d matrix values are calculated using equations corresponding to Eq. (11.60). The corresponding values of the e and f matrices are evaluated on lines 113 and 114. These calculations are performed using matrix algebra because the a, b, c and d parameters are defined as matrices on line 9 of the code following the loading of the matrix code routines on line 3. The use of the Matrix package discussed in Chapter 4 greatly simplifies the development of the code here.

The code in Listing 11.22 implements the basic finite difference method for coupled equations. However, as with the case of a single equation, the use of an interface function can provide additional flexibility in the use of the software. For example, with many problems, the use of a uniform spatial grid is sufficient for solving the BV problem. In such a case computer code can easily generate the spatial grid from a simple specification of the starting and ending points and the desired number of grid points. In addition for problems with fixed boundary values, standard boundary functions can be used to specify the boundary conditions freeing the user from possible errors in writing such functions. To handle such simpler cases an interface function odefd() has been coded and is available in the odefd package of routines. The code will not be presented here but the user can look at the code if desired in the odefd.lua file.

```
 1 : -- /* list11_23.lua */
 2 : -- Solution of nonlinear BV problem with coupled equation code
 3 :
 4 : require"odefd"
 5 :
 6 : f = function(eqs,x,y,yp,ypp)  -- Differntial equation
 7 :    eqs[1] = ypp[1] + 4*yp[1]^2
 8 : end
 9 :
10 : u = {0,2}; x = {0,1,2000}
11 : s,nn,err1 = odefd(f,x,u)
12 : print(nn,err1[1])
13 : plot(s[1],s[2]); write_data('list11_23.dat',s)
Output:
7        3.3944896910937e-007
```

Listing 11.23. Illustration of using coupled equation FD code for solving single differential equation.

Listing 11.23 shows an example of using the ode2bvfd() function of Listing 11.22 through the odefd() calling function. The BV problem being solved is the nonlinear equation previously solved in Listings 11.16 through 11.18 with several spatial grids. The example does not illustrate the multiple equation feature of the code, but is used to illustrate the basic calling sequence and to illustrate that the code can also be used for a single second order BV problem. The basic differential equation is defined on lines 6 through 8 in the same format as used in other solvers in this work. In terms of the calling arguments for the odefd() function, the first argument is just the function defining the differential equation. A more general calling argument is the form {f, fbl, fbr} where fbl() and fbr() are functions defining the left and right boundary conditions. With no specification for these functions, the odefd() function uses default functions that fix the boundary values at the first and last value input by the u[] array. In this case this array specifies only two values 0 and 2. A more general form acceptable to odefd() is an array specifying an initial guess at the solution for all spatial points across the solution region. In the present case, the odefd() function generates an initial approximation to the solution which varies linearly between the two boundary values. For the spatial grid, the simple form of initial value, final value and number of spatial intervals is used (x = {0,1,2000}). The odefd() function in this case generates a uniform array of spatial points based upon this specification.

The reader is encouraged to execute the code in Listing 11.23 and see from the popup plot that the solution is indeed what has been previously obtained for this nonlinear BV problem. Results are not shown here as the solution does not differ significantly from that previously shown in Figure 11.4 and 11.19. The printed output shows that 7 iterations were required to reach a maximum relative correction term of 3.39e-7. An accuracy of about 7 decimal digits would thus be expected. Since this code uses the same basic finite differencing algorithm as the single differential code of the previous section, it is expected that the error in the solution will be proportional to the step size squared or to the square of the number of spatial points. This leads to the proposed use of the h-2h algorithm for obtaining an estimate of the error in such solutions. As in the previous section a separate function odefde() is provided to complement odefd() which returns not only the solution for coupled differential equations but also an error estimate in the solution. This function can by used in place of the one used on line 11. Subsequent examples will use this function as a substitute for the odefd() function. The reader is encouraged to re-execute the code using this function which is also contained in the odefd.lua listing of functions.

This example illustrates that the multi-equation form of the FD code can be used to solve a single second order BV problem. In the interest of having a unified approach to BV problems one might wish to use this for all single differential equations. However, if execution time is important, the single equation form as in Listing 11.16 might still be preferred since there is a definite increase in execution time when using the multi-equation form because of the extra loops and the use of matrix algebra in the solution. The reader is encouraged to compare the execution speed of the solution in Listing 11.23 with the solution using the code for a single differential equation in Listing 11.16. It will be found that the single BV code executes about 15 times faster than the more general BV code for this example of a single differential equation. However, for multiple coupled differential equations this more general code must be used as in the next example.

An example from the field of Chemical Engineering will now be given. In the book, *Quasilinearization and Invariant Imbedding* by E. S. Lee (Academic Press, 1968), the following nonlinear boundary value problem is presented:

$$\frac{1}{M}\frac{d^2c}{dx^2} - \frac{dc}{dx} - \beta c^2 \exp(-E/RT) = 0$$

$$\frac{1}{M}\frac{d^2T}{dx^2} - \frac{dT}{dx} + Q\beta c^2 \exp(-E/RT) = 0$$

$$0 < x < 48 = x_f$$

With Boundary Conditions: (11.65)

$$c(0) - \frac{1}{M}\frac{dc(0)}{dx} = c_e; \quad T(0) - \frac{1}{M}\frac{dT(0)}{dx} = T_e$$

$$\frac{dc(x_f)}{dx} = 0; \quad \frac{dT(x_f)}{dx} = 0$$

This set of equations models an adiabatic tubular chemical reactor with axial mixing where c is the concentration of a particular reactant, x is a dimensionless reactor length, and T is the temperature along the reactor. The exponential term involving T arises from an Arrhenius reaction rate which occurs frequently in engineering problems with E being the activation energy and R the gas constant. The other parameters in the equations (M, Q and β) are constants characteristic of the reactor and the reaction process. The exact definition and nature of these constants are not essential to the problem being considered here. The boundary values specified at the ends of the reactor are of the mixed type involving both the function values and the derivatives. The constants c_e and T_e are initial values of reactant concentration and temperature before entering the reactor. As with many engineering problems the presence of an exponential term involving one of the solution variables makes the problem highly nonlinear and increases the difficulty of solution.

Listing 11.24 shows a code segment for solving these two coupled boundary value equations with the odefde() function which returns both the solution and an estimate of the error in the solution using the h-2h algorithm. As the listing illustrates, the use of the BV code is relatively straightforward. The two differential equations are defined by the f() function on lines 8 through 11 with any difference between the values of the equation and the ideal zero value being returned in the eqs[] arrays. Two boundary value functions are defined on lines 12 through 19 in the fl() and fr() functions. These implement the mixed boundary conditions as given by Eq. (11.65). Constants for the problem as given by Lee are defined on line 6. The first argument to the odefde() function is a list of the three defining functions {f, fl, fr}. The x calling argument is a table listing of the initial and final spatial values along with the number of desired spatial intervals (1000). The odefde() function will set up the full array of spatial points. In the same manner a simplified form of initial guesses at the solution values are specified as just the initial (x = 0) and final (x = 48) solution values in the u[] tables. A uniform initial guess of 0 is taken for the concentration variable and a uniform value of 1250 taken for the temperature. The odefde() function will take the input end point values and generate a table of values matching the x[] table values. The reader is encouraged to re-execute the code with various initial values for the two variables. It is relatively easy to find sets of initial guesses for which a converged solution can not be obtained. For example changing the initial concentration to 0.1 will result in a divergence of the code from the valid solution. However, a range of initial guesses around those given in the listing can be used to obtain a properly converged solution.

This naturally brings up the question of how does one know that a properly converged solution has been obtained? One way is to print out the maximum corrections found at each iteration step. This is done by including the statement on line 4 which sets the nprint value of the basic equation solver, the odefd() function to a non-zero value. This causes the selected output in the listing to be printed for each Newton iterative step in the solution. This particular example required 6 iterations to achieve a maximum relative correction of less than 1.e-5 for both

```
 1 : -- /* list11_24.lua */
 2 : -- Solution of nonlinear Chemical Engineering BV problem
 3 : require"odefd"
 4 : getfenv(ode2bvfd).nprint = 1
 5 :
 6 : M,EdR,b,Q,ce,Te = 2, 22000, .5e8, 1000, 0.07, 1250
 7 :
 8 : f = function(eqs,x,u,up,upp) -- Differntial equation
 9 :    eqs[1] = upp[1]/M - up[1] - b*u[1]^2*math.exp(-EdR/u[2])
10 :    eqs[2] = upp[2]/M - up[2] + Q*b*u[1]^2*math.exp(-EdR/u[2])
11 : end
12 : fl = function(eqs,u,up) -- Left boundary conditions
13 :    eqs[1] = u[1] - up[1]/M - ce
14 :    eqs[2] = u[2] -up[2]/M - Te
15 : end
16 : fr = function(eqs,u,up) -- Right boundary conditions
17 :    eqs[1] = up[1] -- Zero slopes
18 :    eqs[2] = up[2]
19 : end
20 :
21 : u = {{0,0},{1250,1250}} -- initial guesses, end points only
22 : x = {0,48,1000} -- Linear grid spacing
23 : s,se,nn = odefde({f,fl,fr},x,u)
24 : plot(s[1],s[2]); plot(s[1],s[3])
25 : write_data('list11_24.dat',s,se)
Selected Output
 -- 1-- Iteration number, Maxumum relative, absolute corrections are:
(1) 1.000000e+000, 6.999996e-002, at 1001 ;
(2) 3.020200e-006, 3.775262e-003, at 1001 ;
 -- 2-- Iteration number, Maxumum relative, absolute corrections are:
(1) 4.903990e-001, 6.735869e-002, at 1001 ;
(2) 5.113137e-002, 6.735854e+001, at 1001 ;
----
 -- 6-- Iteration number, Maxumum relative, absolute corrections are:
(1) 4.435235e-006, 3.853136e-008, at 1001 ;
(2) 2.938385e-008, 3.853141e-005, at 1001 ;
----
```

Listing 11.24. Code segment for solving two coupled boundary value equations using finite difference odefde() routine.

variables. The output also shows where the maximum correction was occurring in terms of the spatial point number. For the results printed in Listing 11.24 it is seen that this is at point number 1001 or at the right side boundary. In many problems, this can provide information as to where additional spatial points should be placed to improve the accuracy of the solution or the speed of convergence. It is highly recommended that these iteration values be printed out when one is exploring the solution of a new nonlinear BV problem. There is no such thing as a foolproof method of solving nonlinear BV problems. Usually some type of Newton's method is the best method at one's disposal. When such problems converge, convergence is typically of the second order type which means that the error rapidly approaches zero. For example if the error is 1.e-2 at a particular iteration, it is 1.e-4 at the next iteration and 1.e-8 at the next. One very important check on a solution is to verify that the reported maximum errors are rapidly approaching zero in the printed iteration steps. After one has familiarity with valid solutions for a particular problem, one can in many cases verify that a solution is correct simply by looking at the solution and visually verifying that the solution has the features expected for a valid solution. Only if one has confidence in the solution space should the user not look at the iteration steps and the rate of convergence. When calling the odefde() function, there will be two series of calculations, the first with the specified step distribution and a second with the step sizes reduced by one half in order to implement the h-2h algorithm for the estimated error in the solution. Although not shown in the output, this second solution required 3 iteration steps for convergence.

The resulting solutions for the two independent variables are shown in Figures 11.28 and 11.29. It can be seen in Figure 11.28 that a gradual loss in the concentration of the entering solution occurs as a chemical reaction occurs within the reactor. Along with this is a gradual increase in temperature along the reactor as energy is given up in the chemical reaction in the adiabatic chemical reactor as seen in Figure 11.29. In reality nothing very unusual occurs in the solution and the results are qualitatively what would be expected from the physics and chemistry of the problem. The solutions agree with the numerical solutions obtained in the Lee reference book cited at the beginning of this problem.

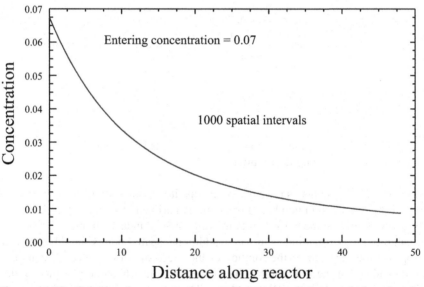

Figure 11.28. Calculated concentration profile along adiabatic tubular chemical reactor.

Perhaps of more interest here is the accuracy with which the solutions are calculated by the finite difference algorithms. This can be estimated by the arrays returned by the odefede() routine on line 23 of the listing. Results for the estimated accuracy are shown in Figure 11.30. A log scale is used in the figure to present

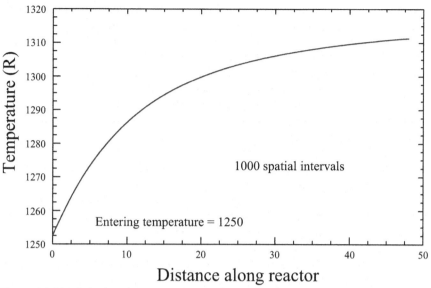

Figure 11.29. Calculated temperature profile along adiabatic tubular chemical reactor.

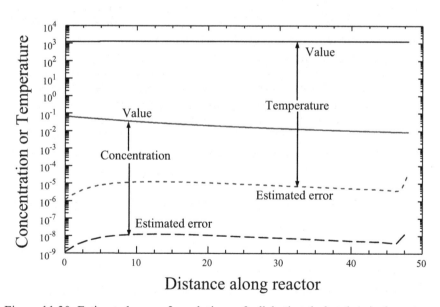

Figure 11.30. Estimated errors for solutions of adiabatic tubular chemical reactor.

both solution variables and the estimated errors. For the concentration variable, the estimated error is about a factor of 1×10^{-6} smaller than the concentration over the range of the reactor while for the temperature the estimated error is about a factor of 1×10^{-8} smaller than the temperature values. From this it would be estimated that the concentration values should be accurate to about 6 decimal digits and the temperature values to about 8 decimal digits. This is a very good accuracy and indicates that fewer spatial grid points could be used in the solution and a very acceptable engineering accuracy still obtained in the solutions. This would speed up the solution if one had many such calculations to perform. It is also of interest to examine the cost in terms of computer resources to obtain the error estimate as opposed to just obtaining the solution. This can be explored by re-executing the code in Listing 11.24 with the odefde() call replaced by odefd() and by including a calculation of the time for the execution using the os.clock() function. This will not be shown here, but performing such a study shows that the additional time for obtaining the error estimate for this problem is an increase in computer execution time of about 29%. This is a reasonable penalty to pay in order to obtain the increased confidence in the accuracy of a numerical solution. This is especially true when starting to explore a new BV problem.

Obtaining the error estimate may not be necessary if similar problems have been solved many times and one has confidence in the accuracy with which solutions are being obtained. Such an error calculation can also indicate spatial regions where a finer spatial grid may be required to obtain an accurate solution. For the present problem, the estimated error is relatively uniform across the spatial region, so a uniformly spaced grid appears to be appropriate.

In Electrical Engineering a common source of coupled nonlinear BV problems is the study of semiconductor devices. A set of fundamental device equations consist of Poisson's equation, two current density equations and two continuity equations. For the time independent case and one-dimensional geometry as previously shown in Figure 11.22, a formulation of the appropriate equations is:

$$\frac{d^2V}{dx^2} = \frac{q}{\varepsilon}(p - n + N_{Net}), \text{ with } N_{Net} = \begin{cases} N_D \text{ for } x > 0 \\ -N_A \text{ for } x < 0 \end{cases}$$

$$J_n = q\mu_n nE + qD_n \frac{dn}{dx}; \quad \frac{1}{q}\frac{dJ_n}{dx} - U = 0 \tag{11.66}$$

$$J_p = q\mu_p pE - qD_p \frac{dp}{dx}; \quad -\frac{1}{q}\frac{dJ_p}{dx} - U = 0$$

where J_n and J_p are the electron and hole current densities, μ_n and μ_p are the carrier mobility, D_n and D_p are diffusion coefficients and U is the net carrier recombination-generation rate. The second and third lines of Eq. (11.66) include both the current density equations and the continuity equations. A relationship for U, the net recombination-generation rate is needed and this is usually taken as:

$$U = \frac{(pn - n_i^2)}{\tau_{no}(p + p_1) + \tau_{po}(n + n_1)} \tag{11.67}$$

where τ_{no} and τ_{po} are lifetime parameters and p_1 and n_1 are parameters of some recombination level (frequently taken as $p_1 = n_1 = n_i$).

A direct application of the continuity equations in Eq. (11.66) to the current density equations leads to second-order differential equations in the potential and the two carrier densities. This is one method of formulating the equations. A second method is to use a Quasi-Fermi potential representation where the new variables are exponentially related to the carrier densities by the relationships:

$$n = (Kn_i)\exp((V - \phi_n)/V_T)$$
$$p = (n_i / K)\exp((\phi_p - V)/V_T) \tag{11.68}$$

where the constant K depends on the choice of zero for the potential. Since only potential differences have direct physical meaning one is free to select a zero of potential at any desired point. In a previous application of these equations, the zero of potential was taken where the carrier densities were both equal to n_i which is equivalent to taking $K = 1$. For the present problem a different reference will be taken where the zero of potential for the equilibrium case is taken at the ohmic contact to the p-region at $x = -x_p$ in Figure 11.22. This is equivalent to taking $K = n_i / N_A$ and in this case Eqs. (11.68) become:

$$n = (n_i^2 / N_A)\exp((V - \phi_n)/V_T)$$
$$p = N_A \exp((\phi_p - V)/V_T) \tag{11.69}$$
$$\text{with } V = \phi_n = \phi_p = 0 \text{ at } x = -x_p$$

In terms of electric potential and Quasi-Fermi potentials, the three semiconductor device equations then become:

$$\frac{d^2V}{dx^2} - \frac{q}{\varepsilon}(p - n + N_{Net}) = 0, \text{ with } N_{Net} = \begin{cases} N_D & \text{for } x > 0 \\ -N_A & \text{for } x < 0 \end{cases}$$

$$\frac{d^2\phi_p}{dx^2} + \frac{1}{V_T}\frac{d\phi_p}{dx}\left(\frac{d\phi_p}{dx} - \frac{dV}{dx}\right) - \frac{U}{p\mu_p} = 0 \tag{11.70}$$

$$\frac{d^2\phi_n}{dx^2} + \frac{1}{V_T}\frac{d\phi_n}{dx}\left(\frac{dV}{dx} - \frac{d\phi_n}{dx}\right) + \frac{U}{n\mu_n} = 0$$

To complete the formulation a set of boundary conditions are needed for the three solution variables. The reader is referred back to Figure 11.22 which shows the physical geometry of a p-n junction to which this set of equations is to be applied. In equilibrium a built-in voltage exists across the junction with the n-side positive with respect to the p-side. This static equilibrium potential has previously been demonstrated by solving just Poisson's equation and typical solutions are shown in Figures 11.23 or Figure 11.26. For the equilibrium case, the current densities are zero and the two Quasi-Fermi potentials are zero at all points along the device.

The problem to be addressed here is the application of a forward bias to the p-n junction such that the applied voltage tends to reduce the built-in potential and cause current to flow in the device. To forward bias the junction, the voltage of the p-side is made positive relative to the n-side or the n-side is made negative relative to the p-side. To complete the mathematical formulation of the BV problem, the following sets of boundary conditions are specified at the two boundaries:

$$V(-x_p) = \phi_n(-x_p) = \phi_p(-x_p) = V_a$$

$$V(x_n) = V_T \ln(N_A N_D / n_i^2), \quad \phi_n(x_n) = \phi_p(x_n) = 0 \tag{11.71}$$

This now constitutes a complete mathematical formulation of a BV problem applied to a semiconductor p-n junction. Even if the reader doesn't follow all the physics associated with the problem, the equations defined by Eq. (11.70) with the subsidiary relationships of Eqs. (11.67) and (11.69) constitute three coupled second-order BV equations. Solutions of this set of equations with the BV solvers developed in this chapter are now explored.

Listing 11.25 shows a code segment for implementing this BV problem. Material and device parameters are defined on lines 7 through 15. The doping densities are taken as lightly doped (1×10^{16}/cm^3) on the p-side and heavily doped (1×10^{19}/cm^3) on the n-side. Also of note is the definition of different mobility values for electrons and holes depending on the sides of the junction (lines 12 and 13). These values are switched in the function eq() defining the equation set on lines 18 and 19. The three differential equations are defined in a straightforward manner on lines 22 through 24. Left and right boundary value functions, efl() and efr() are defined to implement the boundary values. The default boundary functions are not used because different applied voltages are of interest (different va values). An x array of uniformly spaced grid points are defined on lines 34 through 40. For an initial approximation to the solution variables, the potential function is set at a step at x = 0 as was done previously for only Poisson's equation. The initial Quasi-Fermi levels are set at 0 on line 39.

After setting up the equations, boundary functions and initial arrays, the BV equation solver is called on line 32 after defining an applied voltage of 0 on line 41. This is done to verify a correct solution with no applied voltage and the expected solution for the potential should match that obtained from simply solving a single Poisson equation. Also the correct solution for the two Quasi-Fermi levels should be zero. To monitor the progress of the solution, the nprint parameter is set to 1 on line 4, so the iterative steps will be printed. The reader is encouraged to execute the code in Listing 11.25 and observe the printed iterative values as the odefd() routine attempts to find a solution. This simple execution results in some unexpected results. One would expect the solver to go through a few (less than 10) iterations and converge to a solution which is the same as obtained from a single BV problem. However this is not what actually occurs. The BV solver goes through 50 iterations without terminating with what it decides is an acceptable solution, i.e. one that meets the convergence criteria specified in the routine. Selected output is shown in the above listing at Newton iterative loops 1, 2, 8 and 50. The odefd() function has a default limit of 50 Newton iterations.

```
 1 : -- /* File list11_25.lua */ -- semiconductor equations
 2 :
 3 : require"odefd"
 4 : getfenv(ode2bvfd).nprint = 1 -- Print iterations if desired
 5 :
 6 : -- Material and device parameters
 7 : q = 1.6e-19; eps = 11.9*8.854e-14; L = 2.e-4
 8 : vt = .026; ni = 1.45e10
 9 : Na = 1e16; Nd = 1.e19 -- Doping densities
10 : tno = 1.e-12; tpo = 2.e-12 -- Lifetimes
11 : n1 = ni; p1 = ni
12 : unp,unn = 1020, 100 -- Electron mobilities
13 : upp,upn = 420, 50 -- Hole mobilities
14 : qdep = q/eps; vj = vt*math.log(Na*Nd/ni^2); no = ni^2/Na
15 : va = 0.0
16 :
17 : eq = function(fv,x,v,vp,vpp)
18 :    if x<=0 then Nnet, un, up = -Na, unp, upp
19 :    else Nnet, un, up = Nd, unn, upn end
20 :    p, n  = Na*math.exp((v[2]-v[1])/vt),
                 no*math.exp((v[1]-v[3])/vt)
21 :    U = (p*n-ni^2)/(tpo*(n+n1)+tno*(p+p1))
22 :    fv[1] = vpp[1] + qdep*(p - n + Nnet) -- V term
23 :    fv[2] = vpp[2] + vp[2]*(vp[2]-vp[1])/vt - U/(p*up)-- Up term
24 :    fv[3] = vpp[3] + vp[3]*(vp[1]-vp[3])/vt + U/(n*un)-- Un term
25 : end
26 :
27 : efl = function(fv,v,vp) -- Left boundary values, all zero
28 :     fv[1], fv[2], fv[3] = v[1]-va, v[2]-va, v[3]-va
29 : end
30 : efr = function(fv,v,vp) -- Righr boundary values
31 :     fv[1], fv[2], fv[3] = v[1] - vj, v[2], v[3]
32 : end
33 :
34 : x, v, nx = {}, {}, 2001; dx = L/(nx-1)
35 : for j=1,3 do v[j] = {} end
36 : for i=1,nx do -- Set initial values -- step in v[1][]
37 :     x[i] = (i-1)*dx - 2*L/3
38 :     if x[i]<0 then v[1][i] = 0 else v[1][i] = vj end -- Step in V
39 :     for j=2,3 do v[j][i] = 0 end -- Zero Quasi-Fermi levels
40 : end
41 : va = 0.0
42 : s,nn = odefd({eq,efl,efr},x,v)
43 : print(nn); plot(s)
44 : write_data('list11_25.dat',s) -- Save solution
Selected Output:
-- 1-- Iteration number, Maxumum relative, absolute corrections are:
(1) 1.000000e+000, 8.394363e-001, at 1334 ;
(2) 1.000000e+000, 1.984028e-014, at 2000 ;
(3) 1.000000e+000, 1.800656e-014, at 2000 ;
-- 2-- Iteration number, Maxumum relative, absolute corrections are:
(1) 9.988042e-001, 3.082875e-001, at 69 ;
(2) 8.803050e-001, 1.425088e-013, at 1306 ;
(3) 8.134392e-001, 6.464864e-014, at 939 ;
----------
-- 8-- Iteration number, Maxumum relative, absolute corrections are:
(1) 2.533992e-006, 2.769842e-014, at 2 ;
(2) 9.982036e-001, 1.489475e-006, at 1509 ;
(3) 9.999228e-001, 4.869938e-009, at 1131 ;
```

```
----------
-- 50-- Iteration number, Maxumum relative, absolute corrections are:
(1) 4.336538e-007, 2.857878e-014, at 30 ;
(2) 9.999202e-001, 1.177044e-006, at 1765 ;
(3) 9.946253e-001, 1.047554e-008, at 630 ;
```

Listing 11.25. Initial code for semiconductor BV problem

Assuming an error has not been made in coding the problem, the cause of the solver not terminating needs to be explored. Actually if the corrections to the solution are examined, it can be seen that a good solution has in fact been obtained. In fact the printed output shows that the maximum corrections obtained at step 8, for example (see Listing 11.25) are in fact 2.7×10^{-14}, 1.5×10^{-6} and 4.9×10^{-9} for the three solution variables. These should be perfectly acceptable values as the solutions are accurate to about 6 decimal digits. However, the BV solver does not recognize this as an acceptable solution. The problem here is the old "zero value" problem. In numerical work protecting algorithms against the possibility of a "zero" value for the solution causes as much, or perhaps more, grief as any other problem. The problem here is the fact that the exact solution for the Quasi-Fermi levels is 0. At iteration number 8, for example, it is seen that although the absolute corrections are small, the relative correction for variables 2 and 3 printed on the output is almost 1.0. This is due to the fact that any small correction to 0 will be a large relative percent correction. The termination criteria programmed into the BV solver is that the relative correction should be below some specified value (default of 1.e-5 in Listing 11.22). The other "zero value" problem relates to the increment in solution value used to numerically evaluate the partial derivatives. The size of this increment value is based upon the average value of the solution variable across the solution space. For a near identical zero solution, this makes the increment used in obtaining the partial derivatives exceedingly small leading to increased errors in the numerical partial derivative. This problem was discussed in Chapter 5 when discussing numerical derivatives. For the present problem, this use of a very small increment for the partial derivative evaluations is probably the more important of the "zero value" problems. The net result of these problems is that a solution variable identically equal to zero for all solution points causes many numerical problems.

Now that we know that the problem is not one of obtaining a valid solution but various problems due to the "zero value" of two of the solution variables, the question is what can be done about this problem in an automatic manner? One could simply ignore the problem, accept the fact that 50 iterations will occur but that a good solution has been obtained and move on to an applied voltage where the correct solution will not result in zero values for the Quasi-Fermi potentials. However, this approach requires considerable computer time and is not very elegant or universally valid. A second approach is to simply skip the zero applied voltage calculation, using the results from a single equation and apply some small non-zero voltage such that the Quasi-Fermi potentials are not identically zero. The reader is encouraged to explore this approach by setting va on line 41 to some small value and execute the code. However a value of about 1.e-2 is required before the code will recognize an acceptable solution with the default testing on the relative accuracy. A final solution is to modify the basic BV solver, the ode2bvfd() function to test the absolute accuracy of the solution variables as well as the relative accuracy and use this as a secondary termination criteria. This is a somewhat more satisfying solution. However, the problem with this is than an acceptable absolute correction criterion is not known for general BV problems where the range of a variable is not known. If the range of solution variables is on the order of unity as in this problem then a maximum absolute correction in the range of 1×10^{-5} or 1×10^{-6} would probably be acceptable. However, if the code is to be used for a wide range of engineering problems, an acceptable absolute correction can not be known for general problems. This is the reason for relying so heavily on the relative error criterion, since this can ensure that results are accurate to a specified number of decimal digits. However, this problem has been anticipated and the code shown in Listing 11.22 for the ode2bvfd() function has provisions for a user specified absolute error criterion for terminating the iterations. This is provided through the use of a umin[] table and the testing of these values on lines 148 through 157 of the ode2bvfd() code. This table is set in the listing on line 163 to a table with no values. The user can modify this table entry to specify any acceptable values for the minimum corrections on the solution variables. For example to specify that the iteration should stop when the maximum absolute corrections are below 1×10^{-5} one only needs to include the statement getfenv(ode2bvfd).umin = {1.e-5, 1.e-5, 1.e-5} before calling the BV solver software. Different values may be used for each solution variable depending on the range of the solution variable.

Listing 11.26 shows how this is implemented in the code for solving the semiconductor device equations. The listing is similar to Listing 11.25 but with a few changes. First, as noted, line 4 sets limits on the absolute values of the corrections for the three variables. The other change is a voltage loop from line 41 to 46 that steps through an applied voltage range from 0 to 1.1 Volts, incrementing the voltage by 0.1 Volt per step. The BV solver is called for each voltage and the resulting solution obtained on line 43 is saved to various output files on line 45. The iterative steps are not printed as experience has shown that this code has no problems in convergence. The number of iterative steps required at each voltage value is printed and although only a few printed values are shown, the number is between 7 and 5 with the largest number of Newton iterations required at the smallest applied voltages. The reader is encouraged to execute the code and observe the results.

Several features of this code are particularly noteworthy. It can be seen that within the voltage loop incrementing the voltage, there is no apparent evaluation of initial parameters for use in calling the odefd() function. However the value returned by the calling argument list (x and v) will be the actual solution values found from the converged calculation. This means that when the voltage is incremented and the odefd() function is called for the new voltage step, the initial approximation passed to the odefd() solver will be the converged solution set from the previous voltage step. For example the initial approximation passed to the

```
 1 : -- /* File list11_26.lua */ -- semiconductor equations
 2 :
 3 : require"odefd"
 4 : getfenv(ode2bvfd).umin = {1.e-6, 1.e-6, 1.e-6}
 5 :
 6 : -- Material and device parameters
 7 : q = 1.6e-19; eps = 11.9*8.854e-14; L = 2.e-4
 8 : vt = .026; ni = 1.45e10
 9 : Na = 1e16; Nd = 1.e19 -- Doping densities
10 : tno = 1.e-12; tpo = 2.e-12 -- Lifetimes
11 : n1 = ni; p1 = ni
12 : unp,unn = 1020, 100 -- Electron mobilities
13 : upp,upn = 420, 50 -- Hole mobilities
14 : qdep = q/eps; vj = vt*math.log(Na*Nd/ni^2); no = ni^2/Na
15 : va = 0.0
16 :
17 : eq = function(fv,x,v,vp,vpp)
18 :    if x<=0 then Nnet, un, up = -Na, unp, upp
19 :    else Nnet, un, up = Nd, unn, upn end
20 :    p, n  = Na*math.exp((v[2]-v[1])/vt),
              no*math.exp((v[1]-v[3])/vt)
21 :    U = (p*n-ni^2)/(tpo*(n+n1)+tno*(p+p1))
22 :    fv[1] = vpp[1] + qdep*(p - n + Nnet) -- V term
23 :    fv[2] = vpp[2] + vp[2]*(vp[2]-vp[1])/vt - U/(p*up)-- Up term
24 :    fv[3] = vpp[3] + vp[3]*(vp[1]-vp[3])/vt + U/(n*un)-- Un term
25 : end
26 :
27 : efl = function(fv,v,vp) -- Left boundary values, all zero
28 :    fv[1], fv[2], fv[3] = v[1]-va, v[2]-va, v[3]-va
29 : end

30 : efr = function(fv,v,vp) -- Righr boundary values
31 :    fv[1], fv[2], fv[3] = v[1] - vj, v[2], v[3]
32 : end
33 :
34 : x, v, nx = {}, {}, 2001; dx = L/(nx-1)
35 : for j=1,3 do v[j] = {} end

36 : for i=1,nx do -- Set initial values -- step in v[1][]
37 :    x[i] = (i-1)*dx - 3*L/4
38 :    if x[i]<0 then v[1][i] = 0 else v[1][i] = vj end -- Step in V
39 :    for j=2,3 do v[j][i] = 0 end -- Zero Quasi-Fermi levels
40 : end
41 : for i=0,11 do
42 :    va = i/10
43 :    s,nn = odefd({eq,efl,efr},x,v)
44 :    print('Number of iterations is ',nn,' at va = ',va);
              io.flush()
45 :    write_data('list11_26'..i..'.dat',s) -- Save solution
46 : end
```
Selected output:
```
Number of iterations is       7       at va =       0
Number of iterations is       6       at va =       0.1
------
Number of iterations is       5       at va =       1.1
```

Listing 11.26. Code segment for solving semiconductor equations for forward biased pn junction.

function for the 0.8 voltage step will be the converged solution set for the 0.7 voltage step. This is done automatically by returning the converged solution values within the calling argument table of values. This provides a good initial approximation for the next voltage step. In this manner the solution is able to progress up the voltage range from 0 to 0.1 to 0.2 ... to finally 1.1 Volt. If a solution is attempted directly at 1.1 Volts without the intermediate calculations, the solver will have great difficulty in obtaining convergence because the initial approximation will be so far from an accurate solution.

This technique of slowly increasing a parameter of a nonlinear equation is a very useful technique in obtaining the solution to a highly nonlinear set of BV problems. While the final interest may be in a solution for some large value of a physical parameter, by slowly incrementing the parameter from some known initial solution, one can bootstrap the way to a final solution that is far from any known initial approximation. For the present problem, the voltage can probably be incremented is steps of 0.2 Volt and still obtain convergence if a solution at large voltages is desired. For the present problem, however, the solutions at low voltages are as important and perhaps more important than the high voltage solutions, although the high voltage solutions are more difficult to obtain numerically.

They actually contain more physical effects such as high injection and ohmic resistance affects than do the low voltage solutions. For Electrical Engineers, the total current would be just as important, or perhaps more important than the potential profile and Quasi-Fermi potentials. The current can be evaluated from the three solution variables in a straightforward manner but will not be done here since the emphasis is on solving coupled nonlinear BV equations and this can be done without evaluating the total current. A calculation of the current is left to the interested Electrical Engineers.

Plots of the three solution variables are shown in Figure 11.31 for the p-n junction with the parameters of Listing 11.26. The selected boundary values fix the potentials at the right boundary at their equilibrium values (the $V_a = 0.0$ values) and result in left boundary values for all three of the variables which are incremented by the applied voltage value (0.1 Volt steps in the solution). For the lower voltage steps, below about 0.6 Volts applied voltage, it can be seen that almost all the voltage step (the $V(x)$ variable) is taken up by a change in the electric potential across the depletion region, as evidenced by the essentially zero slope on the $V(x)$ curves for $x < -5.0E-5$. For larger applied voltages, an increasing fraction of the applied voltage is dropped outside the depletion region as evidenced by the negative slope on the $V(x)$ curves for $x < -.5.0E-5$ and for applied voltages larger than about 0.7 volts. This is a well known phenomena referred to as "high injection" in device terminology. The electron Quasi-Fermi potential closely follows the electric potential in the heavily n-type region while the hole Quasi-Fermi potential closely follows the electric potential in the heavily p-type region. Both of these are well known effects and expected from the solutions.

The electron and hole densities are also of important to Electrical Engineers and plots of these resulting from the solutions are shown in Figure 11.32. These have been evaluated from the three solution variables and Eq. (11.69). Code for

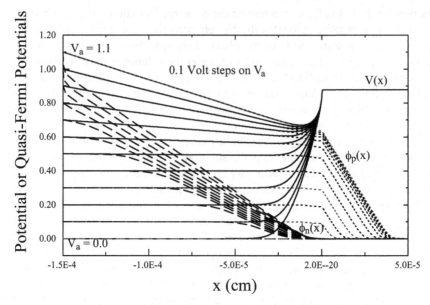

Figure 11.31. Solutions of potential and Quasi-Fermi potentials for Listing 11.26.

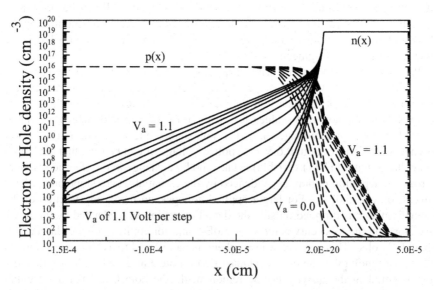

Figure 11.32 Electron and hole densities around p-n junction using output from Listing 11.26.

this evaluation is not shown but is straightforward to implement. It can be seen that the carrier densities vary over a range of about 18 orders of magnitude which requires accurate solutions for the three differential equation variables. The smooth curves for the carrier densities support the expected accuracy in the three solution variables. Again the results are as expected and the straight line plots on the log scale indicate essentially an exponential variation of injected minority carrier density as electrons or holes diffuse away from the p-n junction. Again this exponential behavior is a known physical effect.

The above example illustrates the use of the finite difference method with a set of three highly nonlinear equations with fixed boundary values. However, a reader knowledgeable in the physics of semiconductor devices will recognize that at least one set of parameters is not characteristic of typical parameters for Silicon devices. This is the set of lifetime parameters which are taken in this example as 1.e-12 and 2.e-12 seconds on line 10 of the code. These values are much smaller than typically present for Si devices -- more reasonable values might be in the 1.e-8 second range or even larger. In semiconductor device theory, a parameter called the diffusion length is determined by the diffusion coefficient and the lifetime as

$$L_{Dn} = \sqrt{D_n \tau_n} \qquad (11.72)$$

for electron diffusion length and a similar expression for the hole diffusion length. This represents the mean distance a minority carrier will diffuse away from the junction interface before recombining with a majority carrier and represents the characteristic exponential decay lengths of the curves in Figure 11.32 where the densities are straight lines on the log plot. For the lifetimes used in the example, these values are about 5.2e-6 cm and 1.6e-6 cm for electrons and holes. These are both much less than the lengths of the device dimensions used in solving the equations in Listing 11.26. These agree with the curves in Figure 11.32 where the carrier densities decay very rapidly away from the junction interface. If the lifetimes are increased to 1.e-8 and 2.e-8 the corresponding diffusion lengths will increase to about 5.2e-4 cm and 1.6e-4 cm which are now much larger than the device dimensions (of 1.5e-4cm and 0.5e-4 cm). This changes the nature of the solutions and for such cases it is known that the injected minority carriers should vary approximately linearly with distance away from the junction interface. This actually makes the solution more difficult to obtain numerically.

To explore this case the reader is encouraged to change the lifetime values in Listing 11.27 to 1.e-8 and 2.e-8 and re-execute the code. The solution will almost immediately run into problems when the code tries to obtain valid solutions. At a forward voltage of 0.1, 11 Newton iterations are required for convergence and for a forward voltage of 0.3, convergence is not obtained after 50 iterations. For an attempt at a solution at 0.4 the code fails with a printed message of "Singular matrix". This is a sure sign of failure to find a converged solution and typically results from some solution value diverging to some very large value. The major problem with obtaining a valid solution for this set of parameters can be seen from the converged solution for a forward voltage of 0.2 Volts as shown in Figure 11.33. It can readily be seen that the solution for the Quasi-Fermi potentials,

ϕ_n or ϕ_p exhibits very large derivative values at the boundaries of the x dimension. This results from the boundary conditions which fix the Quasi-Fermi potential values at the applied voltage on the left boundary and at 0 on the right boundary. Put another way, the carrier densities are fixed at these boundaries at the equilibrium values. This is a case where it is obvious that more spatial points are needed near the boundaries to resolve the rapidly changing solution values in the boundary regions. This provides another example to pursue where a non-uniform spatial grid of points is needed for obtaining a converged solution.

Several techniques have already been introduced with a single BV problem to generate non-uniform spatial grids such as the log distribution xlg(), the geometric distribution xgp() and the log-log distribution xll() discussed in connection with Listing 11.17. In addition the concept of generating a spatial distribution based upon function inversion was introduced in Listing 11.19 which attempted to generate a spatial distribution such that the change is the solution variable occurs in essentially equal steps between the spatial points. This latter approach will be further developed here since it is an adaptive technique which can be used to update a spatial grid as a solution develops. For example, with the present problem, a solution at one voltage can be used to generate a spatial grid that can then be used to solve the equations at an increasing device voltage.

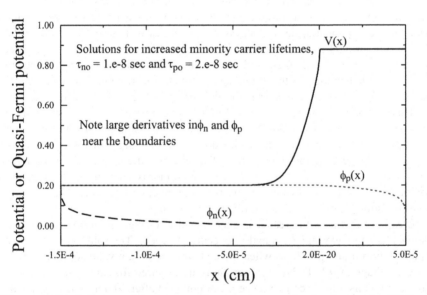

Figure 11.33. Solution of device equations for increased minority carrier lifetimes and for an applied voltage of 0.2 Volts.

Let's explore conceptually how the solution at one voltage such as shown in Figure 11.33 could be used to obtain a non-uniform spatial grid. A fine spatial grid is needed in regions where any one of the solution variables is changing rapidly. From Figure 11.33 it is seen that such a fine grid is needed around the junction interface where the potential changes rapidly and near the left and right boundaries where the Quasi-Fermi potentials change rapidly. A function which accumulates the changes in the variables across the device will then have the general shape needed for obtaining an appropriate spatial grid. However, if two or more variables change rapidly in the same spatial region, it is not necessary to sum the functions but only include the most rapidly changing function. Also in order to accommodate solution variables with different orders of magnitude in value, one should work with the percentage changes along the solution variables.

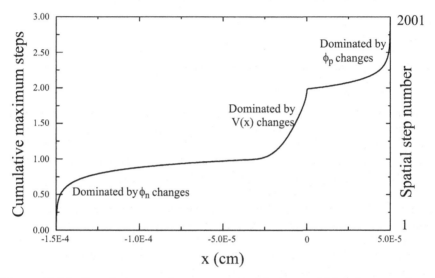

Figure 11.34. Cumulative maximum steps in solution variables from results of Figure 11.33.

Figure 11.34 shows such a cumulative sum of the maximum step size of the three solution variables as shown in Figure 11.33. If this function is inverted so that the vertical axis becomes step number going from step 1 to the maximum step size, the result will be a non-uniform step distribution for which the maximum change in any of the three solution variables is essentially the same between each spatial step. This is illustrated in Figure 11.34 on the right hand side of the figure with a spatial step number going from 1 to 2001 for 2000 spatial intervals uniformly spaced along the vertical axis. A function to generate such an adaptive spatial step size distribution is provided as a function xad(x,u) where two calling arguments are used, the original spatial step distribution, x, and the array of solution variables, u. The function returns a new spatial grid using the cumulative maximum solution relative changes as discussed above along with the solution values evaluated at the new set of spatial points. The function is used as: x, u = xad(x, u) and can be called after an appropriate initial solution has been generated with some initial spatial grid. The coding for the function is relatively straight-forward and is not shown here. It is included in the code listing odefd.lua, where the interested reader can examine the code.

```
 1 : -- /* File list11_27.lua */ -- semiconductor equations
-----
 4 : getfenv(odefd).umin = {1.e-6, 1.e-6, 1.e-6}
 5 : getfenv(odefd).nprint = 1 -- Added statement
 6 :
-----
10 : Na = 1e16; Nd = 1.e19 -- Doping densities
11 : tno = 1.e-8; tpo = 2.e-8 -- Lifetimes -- Changed statement
12 : n1 = ni; p1 = ni
-----
46 :    write_data('list11_27'..i..'.dat',s) -- Save solution
47 :    x,v = xad(x,v) -- Added statement
48 : end
```

Listing 11.27. Code changes needed for using adaptive step size adjustment with the p-n junction problem. See Listing 11.26 for other lines of code.

Listing 11.27 shows code changes needed for using this adaptive step genera-tion function with the p-n junction problem and with the increased lifetime pa-rameters. The listing is the same as Listing 11.26 but with changes in three lines of code. Only these changes and surrounding lines are shown in Listing 11.27. Line 5 has been added to view the individual iterative steps. Line 11 has changed for the longer lifetimes and line 47 has been added to give an improved spatial step size distribution based upon the cumulative solution step size function inverse technique as discussed above. The xad() step size function is placed within the loop on device voltage so the step size is updated at each voltage based upon the solution of the BV problem at that voltage value. When this code is executed the solution converges for voltages up to 0.3 Volts but then fails to converge at 0.4 Volts, which is only slightly better than the case of a uniform step size distribu-tion. This is somewhat discouraging – but read on.

To explore further the difficulty with the solution, Figure 11.35 shows the three solution variables for 0.3 Volts applied voltage and Figure 11.36 shows the step distribution generated from this converged solution. The solution for the Quasi-Fermi potentials has a very large derivative near the two boundaries as seen in Figure 11.35. The resulting generated step size distribution results in very small step sizes near the boundaries as can be seen in Figure 11.36. The insert in the figure showing the step size as a function of step index shows that the step sizes near the boundaries are in the 1.e-10 cm range. Since this is smaller than the inter atom spacing in silicon, one would expect that this should be a sufficiently small step size. However, even with this small step size the solution will not converge with a modestly large applied voltage.

Other step distributions could be pursued in hopes of obtaining a sufficient variation for a converged solution. However, when one encounters a problem where it becomes so difficult to obtain a solution to a mathematically defined physical BV problem, one should also step back and explore the possibility that a

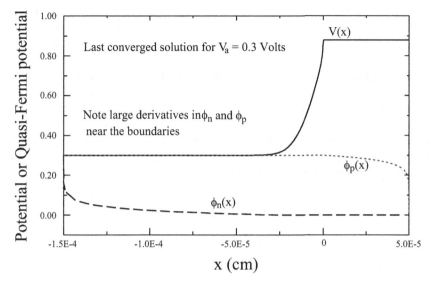

Listing 11.35. Converged solution for $V_a = 0.3$ Volts with adaptive spatial step size.

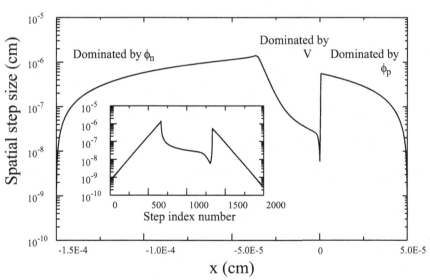

Figure 11.36. Adaptive step size evaluated from converged solution at $V_a = 0.3$ Volts.

physically impossible boundary condition is being imposed. Nature does not have great difficulty in satisfying the differential equation set and the boundary values and numerical techniques should not have such difficulty in approximating nature. In the present case it is instructive to look at the value of the derivative of the Quasi-Fermi potentials in the regions near the boundaries. Such a graph is shown in Figure 11.37 for the 0.3 Volt converged solution. The derivatives have been plotted as a function of the spatial step index in order to expand the spatial range near the boundaries. It is seen that the derivatives exhibit extremely rapid increases near the boundaries with values above 1.e7 Volts/cm. The question is are such large values physically meaningful and if so what they would imply about semiconductor physics near the boundaries?

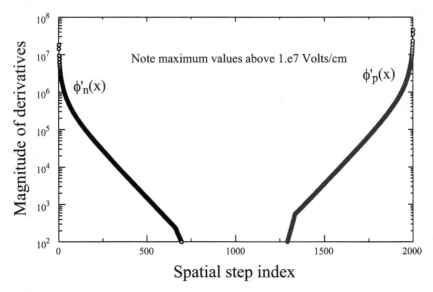

Figure 11.37. Magnitudes of Quasi-Fermi potential derivatives near the two boundaries of the p-n junction problem. For an applied voltage of 0.3 Volts.

To explore this question, consider the electron current density in terms of the electron Quasi-Fermi potential as:

$$J_n = -q\mu_n n \frac{d\phi_n}{dx}; \quad J_p = -q\mu_p p \frac{d\phi_p}{dx} \tag{11.73}$$

The current densities can also be written in terms of the average drift velocities as:

$$J_n = -qn\overline{v}_n; \quad J_p = qp\overline{v}_p \tag{11.74}$$

From these it can be seen that the carrier drift velocities can be directly related to the derivatives of the Quasi-Fermi potentials as:

$$\overline{v}_n = \mu_n \frac{d\phi_n}{dx}; \quad \overline{v}_p = -\mu_p \frac{d\phi_p}{dx} \tag{11.75}$$

From the peak derivative values seen in Figure 11.37 and the mobility values in the model, one can readily see that the resulting magnitudes of average carrier velocities would exceed 1.e9 cm/sec. Physically this is impossible in Si as it is known that at high fields the carrier velocity saturates at around 1.e7 cm/sec and in a low field region such as an ohmic contact, the velocity can not exceed that due to the thermal motion of the carriers when all carriers of positive (or negative) velocity are extracted by an ohmic contact (again this is around 1.e7 cm/sec). The reason for this unphysical high carrier velocity is the boundary condition which fixes the boundary value of the minority carrier density at the equilibrium value. While this is frequently used as a boundary condition it leads to physically unrealistic values of average carrier velocity at the contacts in problems such as the present. To correct this the boundary condition must be relaxed so that the minority carrier density can increase at the contact and then the maximum carrier velocity will not exceed any physically possible value in a given semiconductor.

From this discussion it can be concluded that the route to obtaining converged solutions to the set of semiconductor device equations in this BV problem is not to be found in searching for some improved numerical techniques that can handle extremely large solution derivatives at the boundaries, but in modifying the boundary conditions so that known semiconductor physical limits are not exceeded. A more realistic set of boundary conditions on the minority carriers at the boundaries of this problem is needed. This means the boundary condition for ϕ_n on the left boundary and ϕ_p on the right boundary. Without much discussion an acceptable set of boundary conditions can be formulated as:

$$\frac{n(-x_p)}{n_o(-x_p)}\left[1-\frac{\mu_n}{v_{max}}\left|\frac{d\phi_n(-x_p)}{dx}\right|\right]-1=0$$

$$\frac{p(x_n)}{p_o(x_n)}\left[1-\frac{\mu_p}{v_{max}}\left|\frac{d\phi_p(x_n)}{dx}\right|\right]-1=0 \qquad (11.76)$$

where the maximum carrier velocity has been taken for simplicity as the same for electrons and holes (for more accuracy different values might be used). For small values of the derivatives, these equations reduce to the requirement that the carrier densities equal the equilibrium densities. They provide limits to the derivatives when the carrier densities at the boundaries exceed the equilibrium values. Also the carrier velocities at the boundaries will never exceed the maximum values given by the v_{max} parameter.

Listing 11.28 shows code for solving the coupled semiconductor equations with these modified boundary conditions. The code is similar to Listing 11.26 with modified boundary conditions on lines 31 and 35 and with the maximum velocity defined on line 15. Since the derivatives are negative for the polarity of device dimension selected here, the magnitude operator is not included in the defining equations for the boundary conditions. The other addition to the code is the use of the adaptive step size routine xad() on line 52 to update the spatial steps for each applied voltage based upon the three solutions at the previous voltage step. This is not actually needed with the more physical boundary conditions, but is included as

```
 1 : -- /* File list11_28.lua */ -- semiconductor equations
 2 :
 3 : require"odefd"
 4 : getfenv(odefd).umin = {1.e-6, 1.e-6, 1.e-6}
 5 : getfenv(odefd).nprint=1
 6 :
 7 : -- Material and device parameters
 8 : q = 1.6e-19; eps = 11.9*8.854e-14; L = 2.e-4
 9 : vt = .026; ni = 1.45e10
10 : Na = 1e16; Nd = 1.e19 -- Doping densities
11 : tno = 1.e-8; tpo = 2.e-8 -- Lifetimes
12 : n1 = ni; p1 = ni
13 : unp,unn = 1020, 100 -- Electron mobilities
14 : upp,upn = 420, 50 -- Hole mobilities
15 : vsat = 1.e7 -- Saturated drift velocity
16 : qdep = q/eps; vj = vt*math.log(Na*Nd/ni^2); no = ni^2/Na
17 : va = 0.0
18 :
19 : eq = function(fv,x,v,vp,vpp)
20 :    if x<=0 then Nnet, un, up = -Na, unp, upp
21 :    else Nnet, un, up = Nd, unn, upn end
22 :    p, n  = Na*math.exp((v[2]-v[1])/vt), no*math.exp((v[1]-
          v[3])/vt)
23 :    U = (p*n-ni^2)/(tpo*(n+n1)+tno*(p+p1))
24 :    fv[1] = vpp[1] + qdep*(p - n + Nnet) -- V term
25 :    fv[2] = vpp[2] + vp[2]*(vp[2]-vp[1])/vt - U/(p*up)-- Up term
26 :    fv[3] = vpp[3] + vp[3]*(vp[1]-vp[3])/vt + U/(n*un)-- Un term
27 : end
28 :
29 : efl = function(fv,v,vp) -- Left boundary values
30 :    fv[1], fv[2] = v[1]-va, v[2]-va
31 :    fv[3] = math.exp((v[1]-v[3])/vt)*(1 + unp/vsat*vp[3]) - 1
32 : end
33 : efr = function(fv,v,vp) -- Right boundary values
34 :    fv[1], fv[3] = v[1] - vj, v[3]
35 :    fv[2] = math.exp(v[2]/vt)*(1 + upn/vsat*vp[2]) - 1
36 : end
37 :
38 : x,xi, v, nx = {},{}, {}, 2001; dx = L/(nx-1)
39 : for j=1,3 do v[j] = {} end
40 : for i=1,nx do -- Set initial values -- step in v[1][]
41 :    x[i] = (i-1)*dx - 3*L/4
42 :    xi[i] = i-1
43 :    if x[i]<0 then v[1][i] = 0 else v[1][i] = vj end -- Step in V
44 :    for j=2,3 do v[j][i] = 0 end -- Zero Quasi-Fermi levels
45 : end
46 : for i=0,11 do
47 :    va = i/10
48 :    s,nn = odefd({eq,efl,efr},x,v)
49 :    print('Number of iterations is ',nn,' at va = ',va);
          io.flush()
50 :    write_data('list11_28'..i..'.dat',s) -- Save solution
51 :    plot(x,unpack(v))
52 :    x,v = xad(x,v)
53 : end
```

Listing 11.28. Code for solving semiconductor junction equations with modified boundary conditions.

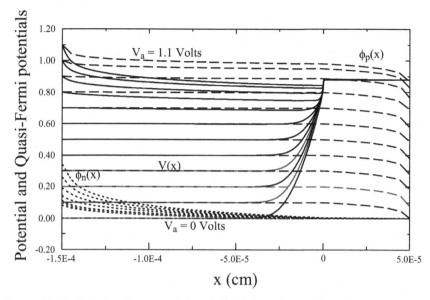

Figure 11.38. Solution for potential and Quasi-Fermi potentials for semiconductor p-n junction with modified boundary conditions and long lifetime. Compare with Figure 11.31 for short lifetime case.

an example of using the adaptive step selection approach. With the modified boundary conditions, the code is able to solve relatively easily the set of semiconductor equations. In Listing 11.28 the odefd() function requires from 7 to 4 Newton iterations to obtain converged solutions for the range of voltages used from 0 to 1.1 Volts. The upper voltage value is well above the built-in junction voltage and into the regime where high-injection and ohmic resistance effects tend to dominate the forward bias behavior.

Solutions for the three device equation variables are shown in Figure 11.38 for applied voltages between 0 and 1.1 Volts. The solutions can be compared with Figure 11.31 for the very short lifetime case. For the present parameters, the carrier diffusion length is much larger than the device dimensions and most of the injected carriers reach the boundary regions rather than recombine with majority carriers within the device. It can be seen that the modified boundary conditions have greatly relaxed the boundary value derivative of the Quasi-Fermi potentials at both boundaries and made the task of obtaining a converged solution much easier than the original boundary value equations.

Device engineers probably think more readily in terms of the electron and hole densities rather than the Quasi-Fermi levels and these can be calculated from the standard equations which are given on line 22 of Listing 11.28. Figure 11.39 shows the results for these densities plotted on a log scale in order to show the vast

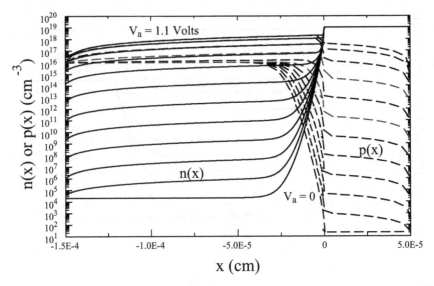

Figure 11.39. Electron and hole densities around p-n junction for the potentials shown in Figure 11.38. Compare with Figure 11.32 for the case of a very short lifetime.

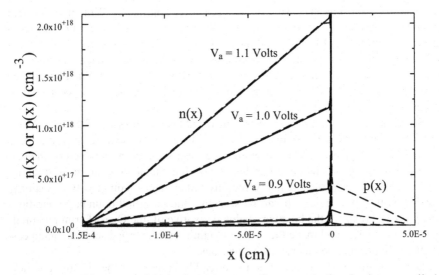

Figure 11.40. Electron and hole densities on a linear scale for the large applied voltages.

range of values needed to describe the operation over the specified voltage range. For negative x values away from the depletion region, electrons are minority carriers in the p-type region and for positive x values, holes are minority carriers in the n-type region. The logarithmic type shapes of the density curves on the log scales, actually indicates that the densities vary approximately linearly with distance away from the junction depletion region. This is more clearly seen in Figure 11.40 which shows the carrier densities for the largest applied voltages on a linear scale. As can be seen a linear approximation is a very good representation of the carrier densities. It can also be seen for the negative x values that the electron and hole densities are essentially equal implying charge neutrality. The reader is encouraged to re-execute the code in Listing 11.28 and change any desired parameters to observe the effects on the solutions as well as on convergence of the FD BV algorithms.

There is an important lesson to be learned from this example and this is one of the reasons it has been pursued so far. The lesson is that when great difficulty is encountered in the numerical solution of a physical problem, many times the answer is not in pursuing the difficult solution but in a careful look at the mathematical formulation of the problem. Many times the problem is not in the numerical algorithms used in solving the problem but in the physical formulation of the problem either in terms of the basic differential equation or in the specified boundary conditions. One must be very careful in formulating not only a set of differential equations but any boundary conditions. Any set of mathematical equations is only an approximation to nature and any set of equations is also an approximation when applied to a physical problem. Nature in general seems to solve nonlinear problems relatively easily. If great difficulty is encountered in solving such a problem it may very well be that a proper mathematical formulation of the problem has not been given. For the present problem, using a more realistic set of boundary conditions relaxes a boundary condition that was found to be extremely difficult to satisfy and one that imposes physical unrealistic values on an important physical parameter of electrons and holes (the carrier velocity).

This section has demonstrated how the finite difference method can be applied to the solution of several coupled second order differential equations. The code routines developed are rather robust and can be applied to many such engineering problems. The next section expands some on the application range of the code routines developed in this section.

11.7 Other Selected Examples of Boundary Value Problems

The code segments developed in the previous section can be applied to a wide variety of boundary value problems. However, there are some BV problems that are not amendable to the FD approach developed in that section. For example in Listing 11.4 the following BV problem was considered:

$$\frac{d^2 u_1}{dx^2} = u_2; \quad \frac{d^2 u_2}{dx^2} = \frac{w(x)}{EI}$$

with boundary conditions: (11.77)

$$u_1(0) = 0, \quad u_1'(0) = 0$$

$$u_1(L) = 0, \quad u_1'(L) = 0$$

This BV problem was solved in Listing 11.4 by use of the shooting method for the case of a constant $w(x)$. While we can certainly formulate two second order coupled equations for use with the FD approach, there is a major problem in the boundary conditions. The code developed in the previous section requires that one be able to obtain values of the solution variables on the boundaries from the given boundary conditions. With the above boundary conditions, there is no dependency on the u_2 variable as the four boundary conditions are all specified on the u_1 variable. This is certainly sufficient to solve the BV problem as was previously demonstrated by use of the shooting method. However, applying these boundary conditions with the FD code of the previous section will result in a "singular matrix" problem as the code formulates the boundary conditions and tries to solve for updated boundary values. Thus there are certain BV problems that are not directly amendable to solution by the FD method. For such problems one can resort to the shooting method for a solution.

Also as formulated, the FD method appears to be applicable only to differential equations (or sets of differential equations) of even order (second, fourth, sixth, etc.). If one has a third order differential equation for example, this could be expressed as one first order equation and one second order equation. However, what does one do with the first order equation, as there will be only one boundary condition? A prototype of this problem is the question of whether the FD method developed here can be used with a single first order differential equation with only one boundary condition. Such a problem is in fact an initial value problem and can be solved by the techniques of Chapter 10. However, the question remains as to whether the formulism of the FD method can be used for such a problem? Several authors have suggested that the way to handle such a problem is to simply take another derivative of the given first order differential equation and convert it into a second order differential equation. For the second boundary value one then uses the original first order differential equation at the second boundary.

For a specific example of the above discussion consider the following first order non-linear equation considered in Section 10.9.3:

$$f \frac{df}{dx} + f(1-a) - a = 0$$ (11.78)

with $f(0) = 0$ and $0 < x < 10$

By taking another derivative, this equation can be converted into the following second order BV problem:

$$f\frac{d^2f}{dx^2} + \left(\frac{df}{dx}\right)^2 + (1-a)\frac{df}{dx} = 0 \tag{11.79}$$

with $f(0) = 0,\ f(L)f'(L) + (1-a)f(L) - a = 0,\ x < 0 < L$

This is now in an appropriate form for use with the FD code developed in the previous section.

However, this approach should not just blindly be accepted. We can ask the question of do we really need to take the extra derivative? After all the only difference in the solution of Eq. (11.78) and (11.79) is that the solution of Eq. (11.79) can possibly have an additional constant in the solution. But any constants will be evaluated from the boundary conditions. Why can't one simply use the original first order differential equation with the same boundary conditions or simply use the following formulism:

$$f\frac{df}{dx} + f(1-a) - a = 0 \tag{11.80}$$

with $f(0) = 0,\ f(L)f'(L) + (1-a)f(L) - a = 0,\ x < 0 < L$

For this formulism the upper boundary condition is simply the same as the differential equation. But the question is will this work? Rather than attempt a theoretical answer, the proof will be simply to see if the FD code generates a valid solution with either or both of these cases.

Listing 11.29 shows code for solving both of these BV problems using the FD routine odefde(). This function returns an estimated error as well as the solution. The specific value of "a = 1" shown in the listing is only one of many possible values. The reader is encouraged to explore the solution with other possible values of this parameter and compare the solutions with those shown in Figure 10.43 obtained by treating the problem as an initial value problem with the solution using an adaptive spatial step. In Listing 11.29, two functions are used in solving the equation, the first order equation on lines 8-10 and the second order equation on lines 11-13. For each solution on lines 22 or 23, the same boundary value equations are used. A simple linear guess at the initial solution varying from 1 to 500 is used on lines 21 and 23 to begin the iterative solution. This is not a very accurate initial guess, but is sufficient for convergence of the FD algorithms. The major conclusion from running the code in Listing 11.29 is that the FD method provides a valid solution using either the first order equation of line 9 or the second order equation of line 12. Both solutions converge in 9 Newton iterations as the printed output shows. Plots of the resulting solutions are not shown here as they would be essentially identical to the results plotted in Figure 10.43. The reader can execute the code and observe the popup solutions for any desired value of the "a" parameter.

From this example it can be concluded that it is not necessary to convert a first order equation to a second order equation in order to use the FD solution method. One simply needs to supplement the one boundary condition with a second boundary condition that reproduces the first order differential equation. However, it may be of some interest to look at the estimated error in the solution using either

```
 1 : -- /* list11_29.lua */
 2 : -- Solution of first order diff. eqn. with FD method
 3 : require"odefd"
 4 : getfenv(odefd).nprint=1
 5 : getfenv(odefd).umin = {1.e-5,1.e-5}
 6 :
 7 : a = 1 -- or 0.5 or 1 or 4 or any number
 8 : f1 = function(eqs,x,u,up,upp)
 9 :     eqs[1] = u[1]*up[1] + (1-a)*u[1] - a
10 : end
11 : f2 = function(eqs,x,u,up,upp)
12 :     eqs[1] = u[1]*upp[1] +up[1]^2 + (1-a)*up[1]
13 : end
14 : fbl = function(eql,u,up)
15 :     eql[1] = u[1]
16 : end
17 : fbr = function(eqr,u,up)
18 :     eqr[1] = u[1]*up[1] +(1-a)*u[1] - a
19 : end
20 :
21 : x = {0,100,2000}; u = {{1,500}}
22 : s1,er1,n1 = odefde({f1,fbl,fbr},x,u)
23 : x = {0,100,2000}; u = {{1,500}}
24 : s2,er2,n2 = odefde({f2,fbl,fbr},x,u)
25 : plot(s1,er1); plot(s2,er2)
26 : print('Number of iterations =',n1,n2)
27 : write_data('list11_29.dat',s1,s2,er1,er2)
Selected output:
Number of iterations = 9        9
```

Listing 11.29. Code segment for solving first order differential equation using FD techniques.

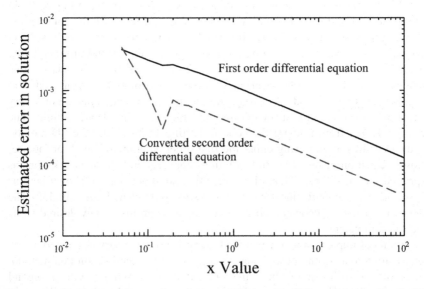

Figure 11.41. Estimated error in solution of first order differential equation in Listing 11.29 using FD code.

the first order or second order differential equation. This error for the example of Listing 11.29 is shown in Figure 11.41. The estimated error in the solution is slightly smaller over most of the solution interval using the converted second order differential equation as opposed to using the original first order differential equation. However, the maximum error at small x values is approximately the same and occurs at the very first spatial point for this example. For this particular solution, the maximum value of the solution is about 14.0. The accuracy could obviously be improved by using a non-uniform spatial grid with small spatial steps for small x values.

There is probably little advantage in considering a single first order differential equation as a BV problem as opposed to treating it simply as an initial value problem. However, the above discussion has application to the solution of BV problems with differential equations of "odd" order. One such example is the Blasius equation which arises in the study of boundary layer problems in fluid mechanics. For the discussion here, the details of the physical parameters are not important and in normalized form the Blasius equation as usually discussed can be presented as the following BV problem:

$$2\frac{d^3 f}{d\eta^3} + f\frac{d^2 f}{d\eta^2} = 0$$ (11.81)

with $f(0) = f'(0) = 0,\ f'(\infty) = 1$

In keeping with a third order equation, three boundary values are specified with two specified at $x = 0$ and one at $x = \infty$. Of course in any numerical approach the infinite limit will have to be approximated by some large finite value. This problem could be solved by the shooting method discussed in Section 11.2. The discussion here will be to see if the FD approach is also applicable. The original equation can be converted into three first order differential equations or one first order equation and one second order equation. The latter approach seems most appropriate for the FD solution method. If the first variable in a formulation becomes the function and the second variable becomes the first derivative, the following two equations are obtained:

$$\text{(a)}\quad \frac{du_1}{dx} - u_2 = 0\ \text{ or }\ \frac{d^2 u_1}{dx^2} - \frac{du_2}{dx} = 0$$

$$\text{(b)}\quad 2\frac{d^2 u_2}{dx^2} + u_1\frac{du_2}{dx} = 0$$ (11.82)

The appropriate set of boundary conditions then becomes:

$$u_1(0) = 0;\quad u_2(0) = 0$$
$$u_1'(L) - u_2(L) = 0;\quad u_2(L) - 1 = 0$$ (11.83)

In the above it is assumed that some finite distance is used for the infinite spatial boundary. Two possible forms are shown for the first equation (a) using either the simple first order differential equation or converting to a second order differential equation. The extra boundary condition needed at the upper boundary is simply taken as identical to the first order differential equation. This is seen as the first entry on the second line of Eq. (11.83) and matches the first order differential equation on the first line of Eq. (11.82).

Listing 11.30 shows code for solving this BV problem by two methods -- the shooting method and the FD method. The two differential equations are defined for the FD method on lines 8-11 with the formulation being given in terms of two second order differential equations. Boundary conditions for the FD method are given on lines 12-19. For the shooting method the equations are formulated as three first order equations on lines 21-15 with the three boundary conditions given on lines 26 through 29. An advantage of the shooting method is that no additional boundary conditions must be specified. The same spatial grid and initial conditions are used on lines 31-34 to call the two solution methods. This provides the reader with a good example of the similarities and differences in solving such a BV problem with the shooting method vs. the FD method.

```
1 : -- /* list11_30.lua */
2 : -- Solution of nonlinear third order BV problem by two methods
3 : require "odebv"; require"odefd"
4 : getfenv(odefd).nprint=1
5 : getfenv(odefd).umin = {1.e-3,1.e-3}
6 :
7 : L = 10
8 : f = function(eqs,x,u,up,upp) -- Equations for FD approach
9 :     eqs[1] = upp[1] - up[2] -- Or use up[1] - u[2]
10 :    eqs[2] = 2*upp[2] + u[1]*up[2]
11 : end
12 : fbl = function(eql,u,up) -- Left boundary conditions for FD
13 :     eql[1] = u[1]
14 :     eql[2] = u[2]
15 : end
16 : fbr = function(eqr,u,up) -- Right boundary conditions for FD
17 :     eqr[1] = up[1] - u[2]
18 :     eqr[2] = u[2] - 1
19 : end
20 :
21 : fst = function(eqs,x,u,up) -- Equations for shooting approach
22 :     eqs[1] = up[1] - u[2]
23 :     eqs[2] = up[2] - u[3]
24 :     eqs[3] = 2*up[3] + u[1]*u[3]
25 : end
26 : fb2 = function(bv,uL,uR) -- Boundary values for shooting method
27 :     bv[1] = uL[1]; bv[2] = uL[2] -- Left boundary values
28 :     bv[3] = uR[2] -1 -- Right boundary values
29 : end
30 :
31 : x = {0,L,2000}; u = {{0,10},{0,1}} -- FD method
32 : s1,n1 = odefd({f,fbl,fbr},x,u); plot(s1)
33 : x = {0,L,2000}; u = {0,0,1} -- Shooting method
34 : s2,n2 = odebvst({fst,fb2},x,u); plot(s2)
35 : write_data('list11_30.dat',s1,s2)
```

Listing 11.30. Solution of Blasius third order boundary value equation by two methods.

The solutions for the Blasius BV problem are shown in Figure 11.42. Plotted are the Blasius function and its first derivative. It is readily seen that the derivative approaches zero in the limit of large distances and in fact the limiting value appears to be approached closely for values beyond a distance of about 5 units on this graph. Thus the use of an upper limit of 10 should provide a sufficiently large limiting value. Also shown in the figure are data points taken from a web posted numerical solution to the same problem. The presently calculated values and curves are in excellent agreement with these previously published values. The reader may also wish to re-execute the code using the odefde() function on line 32 in place of the odefd() function to obtain an estimated error in the solutions. The results should indicate that the solutions are estimated to be accurate to about 7 decimal digits.

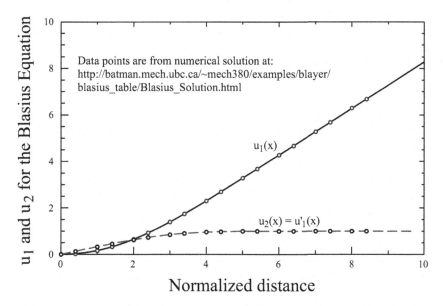

Figure 11.42. Plots of solution variables for the Blasius problem in fluid flow.

The reader is encouraged to re-execute the code in Listing 11.30 replacing the second order differential equation on line 9 with the equivalent first order equation. The results when plotted on the scale of Figure 11.42 should be essentially the identical solution. However, if this is done and one makes a careful comparison between the FD solutions and the shooting method solution, one will find that the solution using the second order differential equation formulation is somewhat more accurate that the solution using the first order differential equation. This is a second example where converting the first order differential equation into an equivalent second order equation for use with the FD method provides better accuracy than simply leaving the equation in first order form. It will be left up to the interested reader to further explore why this occurs and if a simple explanation can be developed of why this occurs. It will just be recommended here that for using the FD solution method with an odd order of differential equations that one convert any resulting first order differential equation into a second order equation and use the first order differential equation as the appropriate missing boundary condition. Although this is not necessary for a converged solution, it appears to give somewhat improved accuracy in the resulting solutions.

11.8 Estimating Parameters of Differential Equations

A final type of problem to be considered in this chapter is that of fitting the solution to a differential equation to a set of experimental data and estimating a set of parameters associated with the differential equation. The differential equation can be of either the initial value type or the boundary value type. A discussion of this general engineering problem has been postponed until both types of differential equations have been considered. One simple example of this comes from the field of Chemical Engineering where a model of the form:

$$\frac{dc}{dt} = -kc^n \tag{11.84}$$

is frequently used to describe the rate of change of some concentration, c, in a chemical reaction. The parameter k is the reaction rate and n is the order of the reaction. Experimental data taken of concentration as a function of time can then be used to obtain the reaction rate and rate constant by fitting the data to the differential equation and determining the best fit value of the parameters. This is a problem combining the results of this chapter on solving differential equations with the results of Chapter 9 on parameter estimation. Thus providing another example of reusable code segments developed throughout this book.

A set of data on concentration vs. time is shown in Table 11.2 and a graph of this data is shown in Figure 11.43. Also shown in the figure is a least squares best fit solution of the differential equation to the data. The best fit parameters are shown in the figure. Developing a procedure for determining such a best fit curve is the objective to be explored in this section. It can be seen in Figure 11.43 that with an appropriate set of parameters the differential equation provides a very good fit to the data. It can also be seen that three parameters are needed to specify the solution, the two parameters of the differential equation plus the constant of integration which in this case is the initial concentration c(0). This will be the case of fitting any differential equation to a set of experimental data. Fitting parameters will consist of parameters associated with the differential equation plus initial parameters or boundary values associated with the solution of the differential equation or equations.

t (min)	0	1	3	5	10	15	20
c (ml)	18	16	13.5	10.6	6.9	3.8	2.7

Table 11.2. Data for concentration vs. time.

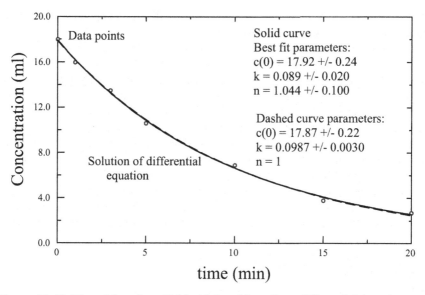

Figure 11.43. Plot of data from Table 11.2 and best fit to differential equation.

To begin the discussion of fitting a differential equation to a set of data, let's review the set of equations that must be solved to determine an update to a set of coefficients by the least squares technique. The reader is referred back to Section 7.5 for a complete discussion of the least squares technique. The major equation (Eq. (7.35)) is reproduces below:

$$
\begin{bmatrix}
\sum_{k=1}^{n_d} f_{1k} f_{1k} & \sum_{k=1}^{n_d} f_{1k} f_{2k} & \cdots & \sum_{k=1}^{n_d} f_{1k} f_{nk} \\
\sum_{k=1}^{n_d} f_{2k} f_{1k} & \sum_{k=1}^{n_d} f_{2k} f_{2k} & \cdots & \sum_{k=1}^{n_d} f_{2k} f_{nk} \\
\vdots & \vdots & \vdots & \vdots \\
\sum_{k=1}^{n_d} f_{nk} f_{1k} & \sum_{k=1}^{n_d} f_{nk} f_{2k} & \cdots & \sum_{k=1}^{n_d} f_{nk} f_{nk}
\end{bmatrix}
\begin{bmatrix}
\delta C_1 \\
\delta C_2 \\
\vdots \\
\delta C_n
\end{bmatrix}
=
\begin{bmatrix}
\sum_{k=1}^{n_d} (y_k - f_k) f_{1k} \\
\sum_{k=1}^{n_d} (y_k - f_k) f_{2k} \\
\vdots \\
\sum_{k=1}^{n_d} (y_k - f_k) f_{nk}
\end{bmatrix}
\tag{11.85}
$$

In this the f_{ik} terms are partial derivatives of the fitting function with respect to the coefficients, i.e.

$$
f_{ik} = \frac{\partial f(x_k)}{\partial C_i} \tag{11.86}
$$

In the present application the $f(x)$ function is the numerical solution of a differential equation known at a finite set of x values. In Chapter 7 a code function nlstsq() was developed to formulate and solve this set of equations for a given fitting function with a set of fitting coefficients. For the present example, this means that Eq. (11.85) would give a set of 3 equations to determine updates to the three fitting coefficients. However, each time a coefficient is changed in developing the equation set the differential equation must be solved or the $f(x)$ function must be recalculated. It is important in setting up the solution to minimize the number of times the differential equation must be solved. This minimization of the number of changes in the fitting parameters was anticipated in Chapter 7 and has been discussed there. The approach to coding the nlstsq() function was to make the intermost evaluation loop over the data points with each c[] parameter changed only once for all the data points. The trade off is that all the partial derivative for all the coefficients and all the data points must be stored before the matrix coefficients in Eq. (11.85) can be evaluated. For this example with 7 data points and three coefficients, 35 values need to be stored before evaluating the matrix elements. The previously coded nlstsq() function can thus be used directly for determining the parameters of a differential equation fitted to a set of experimental data.

Now expressing the fitting differential equation and initial condition in terms of a coefficient set gives:

$$\frac{du}{dx} + c[1]u^{c[2]} = 0$$

$$u(0) = c[3], \quad 0 < x < 20$$

(11.87)

where u is now the concentration and x is the time variable. Code for solving this differential equation and interfacing to the least squares data fitting code is shown in Listing 11.31. The code imports the required routines on line 3 and inputs the data set on line 6. Initial guesses at the coefficients are specified on line 8 and arrays and constants are defined on lines 9-13. The differential equation to be solved is defined on lines 15-17 in a manner that should be familiar to the reader. The only difference here is the use of the coefficients c[1] and c[2] in the definition. These values are not passed to the function through the argument list, but are simply specified as globally known table values defined outside the function, for example on line 8. Note that since this is an initial value problem, the odeiv() differential equation solver is used on line 22 to solve the differential equation. For a boundary value problem the odefd() solver can be used.

The second function that must be defined is a function to be passed to nlstsq() for returning the difference between the model values at the data points and the experimental values at the data points. This is the feq() function defined on lines 19-26. The name of this function is passed to the nlstsq() function on line 27 as the model function used to fit the data. The reader should consult Chapter 7 for a review of this nlstsq() function.

In order to better understand the code for fitting a differential equation, a brief description of the operation will be given. The main data fitting function nlstsq() calls the supplied data fitting function (feq()) in this case) to obtain values of the difference between the experimental data points and the model values. When feq() is called the first argument in the argument list is an array of data points (yx on line 18) with yx[1] being the dependent value of the data (u in this case) and yx[2] being the independent value (x in this case). The second argument (c) is the array of fitting coefficients.

```
 1 : -- /* list11_31.lua */
 2 : -- Estimation of parameters for first order DEQ

 3 : require"odeiv"; require"nlstsq"; require"intp"
 4 : getfenv(nlstsq).ylinear = 1 -- Not required
 5 :
 6 : t,ct = {},{}; read_data('conc.txt',t,ct) -- Data points
 7 :
 8 : c = {.3, 1.0, 24} -- Guess at Differential eqn. coefficients
 9 : nt = 200 -- Number of time intervals for integration
10 : xmin,xmax,nd = 0.0, 20, #t -- number of data points
11 : xx = {xmin, xmax, nt} -- nt time intervals
12 : step = {2,1.1,0} -- nlstsqi parameters
13 : --actv = {1, 0, 1} -- Fix exponent at 1.0 -- Try it !!!
14 :
15 : ft = function(eqt,x,u,up)  -- Differential equation
16 :    eqt[1] = up[1] + c[1]*u[1]^c[2]
17 : end
18 :
19 : feq = function(yx,c,id) -- Equation for nlstsq()
20 :    xv = yx[2]
21 :    if xv==xmin then -- Some coefficient has changed
22 :             s = odeiv(ft,xx,{c[3]}) -- Solve differential equation
23 :             xt,con = s[1],s[2]
24 :    end -- New solution now available
25 :    return intp(xt,con,xv) - yx[1] -- Return solved DE value
26 : end

27 : del,err,nmax = nlstsq({ct,t},fw,feq,c,actv,step)
28 : for i=1,#c do printf('c[%d] = %12.4e +/- %12.3e\n',i,c[i],
          del[i]) end

29 : s = odeiv(ft,xx,{c[3]}) -- Final solution of DE
30 : write_data('list11_31.dat',s,t,ct)
31 : plot(s,{t,ct})
```

Selected output:
```
Coefficients at iteration 1 are:
 1.500000e-001  1.100000e+000  1.834401e+001
--------
Coefficients at iteration 6 are:
 8.950985e-002 1.043644e+000 1.791829e+001

c[1] =  8.9510e-002 +/-   2.037e-002
c[2] =  1.0436e+000 +/-   1.002e-001
c[3] =  1.7918e+001 +/-   2.438e-001
```

Listing 11.31. Code for determining coefficients of differential equation by least squares fit to experimental data.

When the feq() function is called for the first time, no solution of the differential equation exists. However the first call will be for the first data point which has an x value of xmin = 0.0 (or t value of 0.0). This data point value is obtained on line 20 from the passed yx[] table values. If this value equals xmin, lines 21 through 23 of the code are executed and this solves the differential equation on line 22 using the initial guesses at the c[] parameters. Note that the differential equation is solved for 200 uniformly spaced time intervals over the 0 to 20 range using the definitions from lines 9 through 11 to define the xx[] table values. The numerical solution to the differential equation is then used on line 25 to obtain an interpolated model value for each data point. Interpolation is used because the linear array of points used in solving the differential equation may not match the time points of the data values. On subsequent calls to the feq() function for other data points, the code on line 21 through 24 will be skipped.

When the nlstsq() function makes a subsequent change in a c[] parameter, the next call to the feq() function will be for the first data point. This will trigger a new solution of the differential equation by line 22 using the new values of the c[] parameters. Since there are three parameters for the differential equation, nlstsq() will call the feq() function four times for each data point and the differential equation will be solved four times for each Newton iterative loop in nlstsq(). The reader can verify by print statements in the feq() function that the code requires only 4 solutions of the differential equation for each update of the unknown coefficients – one solution for the present value of the coefficients and one solution for each partial derivative evaluation. No check of the solution convergence is made, so it is being assumed in this example that the differential equation solver has no problems with integrating the differential equation. In other cases, the user might want to check on convergence of the differential equation solver.

One note about the c coefficients and the value passed to the feq() function is worth discussing. First it is noted that the c[1] and c[2] coefficients are parameters of the differential equation definition while c[3] is the value of the initial concentration. So the coefficients play different roles in the overall solution. It is seen that only the c[3] coefficient seems to be passed along to the differential equation solver on line 22 while the other two coefficients are required in the function to be integrated on line 16. So the question arises as to how the correct c values get incorporated into the differential equation on line 16. Normally a calling argument such as c in the feq() function on line 19 would not be known outside the function. However, the c argument is an array of values and arrays are passed as addresses. Additionally, the nlstsq() function does not change the address of the c coefficient values. In turn the address of the c array is determined by the definition on line 8 which occurs before the definition of the ft() function on line 15. Thus the c values on line 16 will reference the address established on line 8 and this in turn will be the address passed as the argument to the feq() function. In this manner all values of c[] throughout the program code reference the same address and the same set of values. This is again used on line 29 for a final solution of the differential equation with the converged set of fitting parameters.

The selected output in Listing 11.31 shows that 6 iterations of the c[] parameters were required for converged values. The evaluated coefficient values are shown in the listing as well as in Figure 11.43. It can be seen that the resulting differential equation solution provides a very good fit to the data points – see Figure 11.43. From the fitted coefficients it can be seen that a value of c[2] = 1 is well within the estimated error of the coefficients, indicating a first order reaction process. The reader is encouraged to re-execute the code in Listing 11.31 fixing the c[2] parameter and observe the output. This can be easily done by un-commenting the statement actv = {1, 0, 1} on line 13 which will then skip over the second coefficient in the least squares fitting. The results of this are also shown in Figure 11.43 with the resulting parameters given and with the plotted dashed line. It can be seen that the fit to the data is essentially identical and well within the un-certainty of the data.

A second somewhat more complex differential equation fitting example will now be considered. Earlier in this chapter in Section 11.6 two coupled second or-der differential equations were considered as model equations for an adiabatic tu-bular chemical reactor with axial mixing – see Eq. (11.65). This model was used to describe chemical concentration of a reacting species and temperature along the length of a chemical reactor and for a specific set of parameters a solution of the coupled set of differential equations was given in Figures 11.28 and 11.29. For the present example this solution has been taken as a basis for generating a model set of experimental data by taking the solutions in these figures and applying some random noise on the data. This technique has been used to generate a set of data shown in Figures 11.44 and 11.45. Exactly how the data points were randomized from the original solution is not important for the problem at hand. For this ex-ample it will be assumed that measurements of chemical concentration and tem-perature have been taken at 11 points along a tubular chemical reactor and the data points in the two figures have been obtained. For the moment simply consider the solid curves as just smooth fits to the experimental data in order to better visualize the data points. It is seen that there is some measurement error associated with the data points.

To complete the formulation of the parameter estimation problem and the dif-ferential equations, it is assumed that this set of data is described by the set of dif-ferential equations below (same as Eq, (11.65))

$$\frac{1}{M}\frac{d^2c}{dx^2} - \frac{dc}{dx} - \beta c^2 \exp(-E/RT) = 0$$

$$\frac{1}{M}\frac{d^2T}{dx^2} - \frac{dT}{dx} + Q\beta c^2 \exp(-E/RT) = 0$$

$$0 < x < 50 = x_f$$

With Boundary Conditions:

$$c(0) - \frac{1}{M}\frac{dc(0)}{dx} = c_e; \ T(0) - \frac{1}{M}\frac{dT(0)}{dx} = T_e$$

$$\frac{dc(x_f)}{dx} = 0; \ \frac{dT(x_f)}{dx} = 0$$

(11.88)

In these equations, c is the concentration as given by Figure 11.44 and T is the temperature as given by Figure 11.45. In the differential equation set there are 4 parameters that may be considered as adjustable constants -- M, β, Q and E/R.

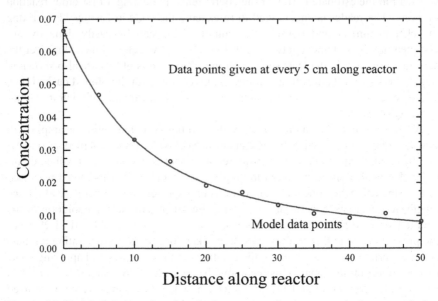

Figure 11.44. Model data for chemical concentration measured along a tubular chemical reactor.

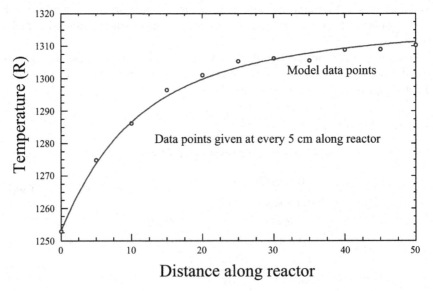

Figure 11.45. Model data for temperature measured along a tubular chemical reactor.

In addition, the boundary values of c_e and T_e might be considered as adjustable constants. However, for the purpose here it will be assumed that the initial concentration and temperature (c_e and T_e) in the boundary conditions are known with exact precision and are not adjustable parameters in modeling the data.

The problem can thus be reformulated in terms of unknown constants as:

$$c[1]\frac{d^2u_1}{dx^2} - \frac{du_1}{dx} - c[2]u_1^2 \exp(-c[3]/u_2) = 0$$

$$c[1]\frac{d^2u_2}{dx^2} - \frac{du_2}{dx} + c[4]u_1^2 \exp(-c[3]/u_2) = 0$$

$$0 < x < 50 = x_f \tag{11.89}$$

With Boundary Conditions:

$$u_1(0) - c[1]\frac{du_1(0)}{dx} = c_e; \quad u_2(0) - c[1]\frac{du_2(0)}{dx} = T_e$$

$$\frac{du_1(x_f)}{dx} = 0; \quad \frac{du_2(x_f)}{dx} = 0$$

In this set of equations the concentration and temperature are represented by u_1 and u_2 respectively and the constant array values represent the original constants as:

$$c[1] = 1/M$$
$$c[2] = \beta$$
$$c[3] = E/R \tag{11.90}$$
$$c[4] = Q\beta$$

It can be noted that the boundary conditions, as well as the differential equations, involve one of the fitting parameters.

Code for solving this set of differential equations and simultaneously fitting the data set is shown in Listing 11.32. The major sections of the code will now be discussed. Starting with the function definitions, code for the differential equations and the boundary conditions are shown on lines 34-43. The equations are in essential identical form to Eq. (11.89) and should be readily understood. The equation passed to the nonlinear least squares fitting routine is given on lines 44-52. The call to the least squares fitting routine can be noted on line 52 using the feq function in the argument list as specifying the fitting function along with other parameters as standard input. Finally the evaluated coefficients are printed on line 54 and 55 and a final integration of the differential equation is obtained on line 56 and written to a saved file on line 57 for comparison with the experimental data. The first half of Listing 11.32 defines the input data, sets up appropriate data arrays and initial conditions. The data for concentration and temperature is read from a data file on line 8 into the x, u1 and u2 arrays.

First consider the calls to the datafit() functions on line 7. For this discussion consider using the least squares error squared minimization technique applied simultaneously to the concentration data and the temperature data in Figures 11.44 and 11.45. One would like to minimize the error between the model and the data simultaneously for both sets of data – concentration and temperature. However the data values differ by about 5 orders of magnitude. Because of this large difference in magnitudes of the two variables, some type of weighting factor is probably essential. If equal (or unity) weighting factors are used for both data sets, the mean square error will be completely dominated by the temperature data with virtually no contribution from the concentration data because of the much larger data values for the temperature. Thus with unity weighting a good fitting is expected to the temperature data but a poor fitting is expected to the concentration data. For such data sets that varying greatly in magnitude it is clear that different weighting factors must be used with the concentration data and the temperature data, but how can appropriate weighting factors be determined? In Section 7.9 different weighting factors were discussed, but the discussion did not consider cases such as this where one desires to use least squares data fitting for two different physical variables with vastly different magnitudes.

For a general data set a frequently used measure of the deviation from some smooth curve is the standard deviation of the data set. If a smooth fit to the data as illustrated by the solid curves in Figures 11.44 and 11.45 were available, the standard deviation from the solid curve could be used as a measure of the average error. However, a valid smooth solution such as these solid curves is not known when given simply a data set. Hence enters the datafit() function developed and discussed in Chapter 7, Section 7.7. The reader is referred back to this section for a discussion where the code was developed as a computer aid to plotting data. However, for use here, this function evaluates a smooth fitting function to the data set which can in turn provide an estimate of the standard deviation of the data. The third variable returned by this datafit() function is in fact the standard deviation between the data points and the smooth data fitting function. These returned values are used on line 8 to determine two values std1 and std2 which are estimates of the standard deviations associated with the data. These values are shown as the first line of the output in Listing 11.32. The reciprocals of these values squared are used as weighting factors for the data on lines 24 and 25.

Other possible weighting factors are known as $1/y_i$ and $1/y_i^2$ weighting. If all the data points have the same relative error than the $1/y_i^2$ weighting would be an appropriate factor. In the present example, this would weight the errors in the concentration data with a factor of about $(6.e4)^2$ times the errors in the temperature data. This will cause the concentration data to completely dominate the fitting and result in a good fit to the concentration data but a poor fit to the temperature data. It might be expected that using the $1/y_i^2$ weighting would be an appropriate weighting factor. However, when a physical problem has two or more physically different types of data such as in this example, there is no inherent reason to expect the relative error associated with the different measurements to be the same.

```
 1 : -- /* list11_32.lua */
 2 : -- Fitting of data to BV problem for parameter estimation
 3 : require"odefd"; require"nlstsq"; require"intp"; require"DataFit"
 4 : exp = math.exp
 5 : x,u1,u2 = {},{},{}
 6 : read_data('list11_32_in.dat',x,u1,u2)
 7 : _,_,std1 = datafit{u1,x}; _,_,std2 = datafit{u2,x} -- Est. STD
 8 : print('std1, std2 = ',std1, std2)
 9 : nd = #x -- number of data points
10 : utp = {1}; for i=2,nd do utp[i] = 2 end -- flag for u data
11 : u = {}; for i=1,nd do u[i] = u1[i] end
12 : for i=1,nd do
13 :    x[i+nd] = x[i] -- repeat x values
14 :    u[i+nd] = u2[i] -- make single array
15 :    utp[i+nd] = 3 -- flag for u2 data
16 : end -- 3 data values, one of which is data flag
17 : yx = {u,x,utp} -- data arrays, dependent values first
18 : c = {0.5, 0.5e8, 22000, .5e11} -- Guess at coefficients
19 : nt = 200 -- Number of points for integration
20 : nx,nc,xx = #yx, #c, {0,50,nt}
21 : nd = #x -- 2*number of data points
22 : actv,step,del,fw = {},{},{},{} -- nlstsq parameters
23 : for i=1,nd do
24 :    if i<= nd/2 then fw[i] = 1/std1^2
25 :    else fw[i] = 1/std2^2 end -- 1/std^2 weighting
26 :    --fw[i] = 1 -- Unity weighting -- Try these
27 :    --fw[i] = 1/u[i] -- 1/Y weighting
28 :    --fw[i] = 1/u[i]^2 -- 1/y^2 weighting
29 : end
30 : ce, te = 0.07, 1250
31 : ui = {{0,0},{1250,1250}} -- Initial approximations to solutions
32 : actv = {0,1,1,0} -- Tru any two combinations
33 : ft = function(eqt,x,u,up,upp)  -- Differential equation
34 :    eqt[1] = c[1]*upp[1] - up[1] - c[2]*u[1]^2*exp(-c[3]/u[2])
35 :    eqt[2] = c[1]*upp[2] - up[2] + c[4]*u[1]^2*exp(-c[3]/u[2])
36 : end
37 : fbl = function(eqb,u,up) -- Left boundary conditions
38 :    eqb[1] = u[1] - c[1]*up[1] - ce
39 :    eqb[2] = u[2] - c[1]*up[2] - te
40 : end
41 : fbr = function(eqb,u,up) -- Right boundary conditions
42 :    eqb[1] = up[1]; eqb[2] = up[2] -- Zero derivatives
43 : end
44 : feq = function(yx,c,new) -- Equation for nlstsq()
45 :    flg = yx[3]
46 :    if flg==1 then
47 :       s = odefd({ft,fbl,fbr},xx,ui) -- Solve DEQ
48 :       xt, flg = s[1], 2
49 :       ff = {intpf(s[1],s[2]), intpf(s[1],s[3])}
50 :    end -- New solution now available
51 :    return ff[flg-1](yx[2]) - yx[1]
52 : end
53 : del, err, nmax = nlstsq(yx,fw,feq,c,actv,step)
54 : print('RMS error =',err)
55 : for i=1,nc do printf('c[%d] = %12.4e +/- %12.3e\n',i,c[i],
           del[i]) end
56 : s = odefd({ft,fbl,fbr},xx,ui) -- Solve differential equation
57 : write_data('list11_32.dat',s,x,u); plot({s[1],s[2]},{x,u1});
```

```
                    plot({s[1],s[3]},{x,u2})
Selected output:
std1, std2 =     0.0026765410105568        3.3962850026022
```

Listing 11.32. Code for solving data fitting for tubular reactor differential equations.

For example temperature is a variable that can be measured with high precision while concentration of some chemical species is a variable that is much more difficult to measure with high precision. Thus in the present example it is natural that the average relative error in the temperature data is considerably less than the average relative error in the concentration data. The reader can verify this by using the computed standard deviations printed in Listing 11.32 along with the average data values to estimate the average relative error of the two data sets. This large difference in the average errors is the reason the $1/y_i^2$ weighting gives poor fitting results in this example. The $1/y_i$ weighting is somewhat between that of unity weighting and the $1/y_i^2$ weighting. For the present example, as will be seen, the use of the $std1^{-2}$ and $std2^{-2}$ weighting gives a good fit of the model equations to both sets of data.

A final requirement on the data set used in the nlstsq() code is the need to supply the data set to be fitted as a single column array of dependent variables and a single column array of independent variables. This is set up by the code on lines 11 through 16. For the loop over the data points, the code on line 16 repeats the x array values and for the u array values are defined so that the first nd array values contains the concentration data and the second nd array values contains the temperature data. If only these two arrays are passed to the nlstsq() function each call to the feq() function will pass one x value and one u value in the argument list. From these two values the feq() function can not easily determine if the values correspond to the concentration data or to the temperature data. For this an additional variable is needed and this is supplied by the utp array defined on lines 10 and 15. The first nd values of this variable are set to 2 and the last nd values are set to 3. Actually the very first value is set to 1 as subsequently discussed. This variable provides a unique identifier for the concentration data and the temperature data and is included in the set of input data to nlstsq() as the third entry in the definition of the yx array on line 17. With this third variable, the set of values passed by nlstsq() to feq() at each call will be one x value, one u value and one utp value and the utp value will indicate whether the data is for concentration (value of 1 or 2) or temperature (value of 3).

Now consider the details of the feq() function on line 44 that is used by the nlstsq() least squares fitting function. When this function is first called for a model value at the first data point, the flg parameter at line 45 equals 1 and the code between lines 46 and 50 is executed to solve the differential equation as a boundary value problem using the odefd() function on line 47. The returned solution array will have a table of three values with s[1] the array of position values (the independent variable), s[2] the concentration values and s[3] the temperature values. Note that 200 spatial intervals are defined on line 19 for use in solving the differential equation. This value can be changed over a rather wide range with little effect on the best fit model parameters.

After solving the differential equation, two interpolation functions are defined on line 49 for the concentration solution and the temperature solution. The feq() function then returns on line 51 the difference between an interpolated value of the solution to the differential equation and a data point value. The flg parameter, with value of 2 or 3 is used on line 51 (as ff[flg-1]) to select either the concentration solution or the temperature solution. From this the usefulness of the third parameter value can be readily seen. For each Newton iterative loop in nlstsq() the differential equation must be solved 5 times for this example so for the 6 iterations seen in Listing 11.32 the differential equation will be solved a total of 30 times.

The only remaining code in Listing 11.32 not discussed is lines 18-32 which should be pretty self explanatory as it sets up needed arrays and initial conditions. The initial approximations to the solutions of the differential equation are defined simply as constants for the two end points on line 31.

On line 32 of the code it is seen that only the $c[2]$ and $c[3]$ coefficients are allowed to vary in the data fitting and the $c[1]$ and $c[4]$ coefficients are held fixed (using actv = {0, 1, 1, 0}). This choice was selected after first experimenting with having all four coefficients variable in attempting a fit to the data. This was found to result in non-convergence of the nlstsq() function or in large uncertainties in the evaluated values. After some thought this is perhaps not too surprising. The differential equation has four variable parameters $c[1] - c[4]$. However, in integrating the differential equation four additional parameters are varied which are the initial and final concentration and temperature values since these are not fixed by the boundary conditions. Thus there are eight adjustable parameters being used to fit the data and only a total of 22 data points with considerable random error. The net result is that a wide range of possible parameter values can give a "good" fit to the experimental data with measurement errors. This is not atypical of sets of experimental data and one must then attempt to fix some of the model parameters. In this example it was simply decided to fix the $c[1]$ and $c[4]$ parameters (or the M and $Q\beta$ parameters) and determine values of the $c[2]$ and $c[3]$ parameters (or the β and E/R parameters).

As shown in Listing 11.32 the code under these conditions converges after 3 iterations of the nlstsq() function and the best fit solution values for the four differential equation parameters are:

$$c[1] = 5.0000\text{e-}001 +/- 0.000\text{e+}000$$
$$c[2] = 5.1992\text{e+}007 +/- 1.117\text{e+}006$$
$$c[3] = 2.1974\text{e+}004 +/- 5.225\text{e+}001$$
$$c[4] = 5.0000\text{e+}010 +/- 0.000\text{e+}000$$

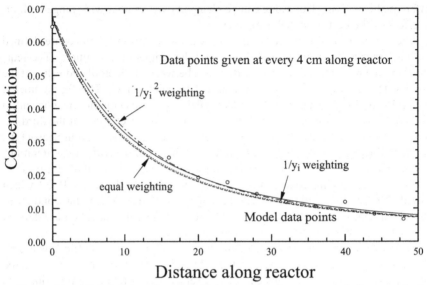

Figure 11.46. Model data for chemical concentration measured along a tubular chemical reactor fitted with different weighting factors.

The uncertainty in the model parameters is seen to be about 2% for $c[2]$ and 0.2% for $c[3]$. The resulting solutions to the differential equation give a good fit to the data points. This can be seen from the solid curves in Figures 11.46 and 11.47 which are the results of solving the differential equation set with the above coefficient set. Three other curves are shown in the figures obtained by the use of equal, $1/y_i$ and $1/y_i^2$ data weighting. These curves can be obtained by un-commenting appropriate lines of code in the listing (lines 26 – 28) and re-executing the code. The equal and $1/y_i$ weighting curves are seen to too heavily weight the temperature data in the fitting and not provide a good fit to the concentration data. On the other hand the $1/y_i^2$ weighting gives a good fit to the concentration data but a poor fit to the temperature data. These curves are consistent with the previous discussion and the $1/std^2$ fitting is seen to provide the best compromise to fitting both sets of data. The reader is encouraged to experiment with the various weighting factors and with other combinations of model parameters. One should attempt to fit the data to all 4 model parameters and observe the results. The data fitting is most sensitive to the $c[3]$ parameter value as it occurs in an exponential factor in the differential equation.

The two examples in this section have demonstrated how the nonlinear least squares parameter estimation technique can be combined with the numerical solution of sets of differential equations to estimate a set of parameters associated with one or more differential equations. A general computer routine will not be developed to handle such problems, but the code in Listing 11.32 can be used as a template for solving other such problems. These examples have also provided excellent examples of the reuse of previously developed code segments such as the datafit() and intp() routines.

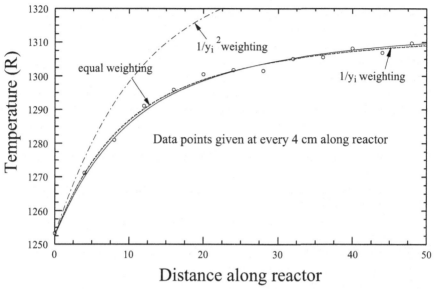

Figure 11.47. Model data for temperature measured along a tubular chemical reactor fitted with different weighting factors.

11.9 Summary

This chapter has concentrated on the numerical problem of solving general nonlinear differential equations of the boundary value type. These are typically of second order or higher order differential equations or of systems of coupled second order differential equations. Boundary values are typically specified at two ends of a range of values of the independent variable. Many engineering problems are of this type. Code segments and routines have been developed for solving single differential equations or coupled systems of nonlinear differential equations. In keeping with the emphasis of this book the code developed can be applied to both nonlinear and linear differential equations.

Basically two general techniques for solving BV problems have been discussed and developed. These are known as shooting methods and finite difference methods. The shooting method is an extension of solving initial value differential equations while finite difference methods provide in many cases more robust BV solvers. A section was devoted to using the shooting method for solving eigenvalue and eigenfunction differential equation problems which are not amenable to the FD approach. Such equations are of special engineering importance as Schrodinger's equation from quantum mechanics falls into this type of problem.

Considerable attention has been given to accessing the accuracy of various methods of solving BV problems. This has been done by looking at selected linear and nonlinear problems for which exact solutions can be obtained. From this a general approach was developed for estimating the accuracy of the solution of any BV problem based upon solving the problem for two different spatial step sizes – the h-2h algorithm approach. In most cases general code routines were developed to provide both solutions to BV problems plus estimates of the accuracy of the solutions.

Several code segments and important routines for solving BV problems have been developed in this chapter. A brief summary of the most important ones is given below:

1. odebvst() – Code for solving boundary value problems by the shooting method.
2. odebvste() – Code for solving boundary value problems by the shooting method and with an error estimate.
3. odebvev() – Code for solving eigenvalue differential equations by the shooting method.
4. odebveve() – Code for solving eigenvalue differential equations by the shooting method with an error extimate.
5. sqnorm() – Code for normalizing the solution of an eigenvalue problem
6. nzeros() – Code for counting the number of internal zeros of the solution for an eigenvalue problem.
7. odebv1fd() – Basic code for solving a single BV equation with the FD approach.
8. ode1fd() – More user friendly code for solving single BV problem with the FD approach.
9. ode1fde() – Code for solving single BV problem with the FD approach and with error estimate.
10. ode2bvfd() – Basic code for solving coupled BV equations with the FD approach.
11. odefd() – More user friendly code for solving coupled BV equations with the FD approach.
12. odefde() – Code for solving coupled BV equations with the FD approach and with error estimate.
13. xlg() – Code for generating nonuniform spatial grid with logarithmic spacing.
14. xgp() – Code for generating nonuniform spatial grid with geometric spacing.
15. xll() – Code for generating nonuniform spatial grid with log-log spacing.
16. xad() – Code for adaptive spatial grid generation based upon solution values.

These code segments provide a very useful set of functions for solving a variety of BV problems with nonlinear coupled differential equations. The code developed was used to solve a number of typical sets of nonlinear BV equations. These should provide sufficient detail for the reader to apply the code to many more interesting engineering problems.

The next chapter will explore some examples of extending the techniques of this chapter to partial differential equations and systems of partial differential equations. This extends the numerical approaches developed in this chapter to functions of several independent variables.

12 Partial Differential Equations: Finite Difference Approaches

The two previous chapters have considered differential equations which involve only one independent variable – typically a spatial variable or a time variable. The present chapter expands the range of equations to one or more differential equation in two or more independent variables which may be spatial variables or spatial variables and a time variable. The simplest partial differential equations involve only two variables while some physically interesting engineering problems involve as many as four independent variables (three spatial variables and a time variable). Even more general problems with more independent variables can of course be considered.

For the numerical solution of any physical problem, the solution is typically obtained for the physical variables at some discrete set of solution points. For the one-dimensional problems in the previous two chapters, typical solution values have been obtained for several hundred or a few thousand values of the independent variable. For similar accuracy with partial differential equations the number of solution points increases by the number of solution points in each dimension raised to the power of the number of independent variables. For example to keep 1000 solution points in each dimension, one needs a total of $(1000)^4 = 10^{12}$ solution points for a partial differential equation in four independent variables. Such a large number of solution points is usually not practical from either a memory storage viewpoint or from a computational time viewpoint. On the other hand, a total of 10,000 total solution points only gives 10 solution points for each dimension in a problem in four independent variables. Thus one of the major problems with the numerical solution of partial differential equations is the selection of an appropriate set of solution points – the so called discretization problem.

The discussion in this chapter will concentrate primarily on partial differential equations in only two independent variables. This is due in part to the exponential growth in needed solution points as the number of dimensions increases. Also for a large number of engineering problems the symmetry of the problem makes it possible to reduce the number of primary variables to two. Finally, much of the insight into real engineering problems can be obtained from a study of partial differential equations in only two dimensions. The discussion in this chapter also concentrates on finite difference approaches to solving partial differential equations. This approach is most appropriate to problems where time is one variable and a spatial variable is the second variable. For many engineering problems involving two or more spatial variables, the finite element approach discussed in the next chapter is a more appropriate technique, because the discretization of spatial

J.R. Hauser, *Numerical Methods for Nonlinear Engineering Models*, 705–881.
© Springer Science + Business Media B.V. 2009

points can more easily be make to conform to appropriate spatial boundaries of a
given problem.

12.1 Introduction to Single Partial Differential Equations

To begin the discussion of partial differential equations (PDE) consider a linear
second-order partial differential equation with constant coefficients of the form:

$$A\frac{\partial^2 U}{\partial x^2} + B\frac{\partial^2 U}{\partial x \partial y} + C\frac{\partial^2 U}{\partial y^2} + D\frac{\partial U}{\partial x} + E\frac{\partial U}{\partial y} + GU + H = 0 \qquad (12.1)$$

The two variables $(x$ and $y)$ may be two spatial variables or one spatial and one
time variable. In general, the solution of the differential equation is desired over
some region R in the independent variables. The equation can be classified as follows:

$$B^2 - 4AC < 0 \quad \text{Elliptic equation}$$

$$B^2 - 4AC = 0 \quad \text{Parabolic equation} \qquad (12.2)$$

$$B^2 - 4AC > 0 \quad \text{Hyperbolic equation}$$

Such a classification can also be applied to higher order equations involving more
than two independent variables. Typically elliptic equations are associated with
physical systems involving equilibrium states, parabolic equations are associated
with diffusion type problems and hyperbolic equations are associated with oscil-
lating or vibrating physical systems. Analytical closed form solutions are known
for some linear partial differential equations. However, numerical solutions must
be obtained for most partial differential equations and for almost all nonlinear
equations.

   Prototypical examples of the three forms of partial differential equations are:
   A. Elliptic equation – Poisson equation (in two dimensions)

$$\frac{\partial^2 U}{\partial x^2} + \frac{\partial^2 U}{\partial y^2} = \rho(x.y) \qquad (12.3)$$

   B. Parabolic equation – Diffusion equation (in one spatial dimension)

$$\frac{\partial U}{\partial t} - \frac{\partial}{\partial x}(D\frac{\partial U}{\partial x}) = 0 \qquad (12.4)$$

   C. Hyperbolic equation – Wave equation (in one spatial dimension)

$$\frac{\partial^2 U}{\partial t^2} - v^2 \frac{\partial^2 U}{\partial x^2} = 0 \qquad (12.5)$$

In each of these examples, the partial differential equation is linear in the partial
derivative terms.

   For a general nonlinear partial differential equation in two dimensions, the following general equation will be considered:

$$F(x,y,\frac{\partial^2 U}{\partial x^2},\frac{\partial U}{\partial x},U,\frac{\partial U}{\partial y},\frac{\partial^2 U}{\partial y^2},\frac{\partial^2 U}{\partial x \partial y}) = 0 \tag{12.6}$$

where the function may be any arbitrary function of the variable list. A more general problem is that of two or more coupled partial differential equations involving a set of solution variables and the partial derivatives of each of the solution variables. This case will be considered later in the chapter. Although a vast majority of partial differential equations of engineering interest are linear in the partial derivatives, in keeping with the major emphasis of this work, the example code developed in this chapter will consider the more general case of nonlinear partial differential equations.

The general approach to nonlinear equations is still the "linearize and iterate" approach. In this case consider that some initial approximation is known to the solution and an improved approximation to the solution is desired. In the L&I approach the solution is considered to be composed of the form: $U \rightarrow U + u$, where u is a correction term for the approximate solution. If the differential equation is then expanded in function space and only first order terms in the correction variable are kept, this leads to a linear partial differential equation of the form of Eq. (12.1) with

$$A = \frac{\partial^2 F}{\partial U_{xx}^2}, B = \frac{\partial^2 F}{\partial U_{xy}^2}, C = \frac{\partial^2 F}{\partial U_{yy}^2}, D = \frac{\partial F}{\partial U_x}, E = \frac{\partial F}{\partial U_y}, G = \frac{\partial F}{\partial U}$$

$$H = F(x,y,\frac{\partial^2 U}{\partial x^2},\frac{\partial U}{\partial x},U,\frac{\partial U}{\partial x},\frac{\partial^2 U}{\partial y^2},\frac{\partial^2 U}{\partial x \partial y}) \tag{12.7}$$

$$\text{and } U_{xx} = \frac{\partial^2 U}{\partial x^2}, U_{xy} = \frac{\partial^2 U}{\partial x \partial y}, U_{yy} = \frac{\partial^2 U}{\partial y^2}, U_x = \frac{\partial U}{\partial x}, U_y = \frac{\partial U}{\partial y}$$

Given a nonlinear equation of the form of Eq. (12.6), a linearized version of the equation to be iterated to a converged solution is obtained by this technique. In keeping with the approaches in previous chapters, the linearization can be accomplished automatically by computer code using the approach of numerical partial derivatives developed in Chapter 5. If the equation being considered is in deed a linear equation, then only one iterative solution of the linerized equation is required for a valid numerical solution. If the iterative series converges to a solution, then the H term in Eq. (12.1) and (12.7) approaches zero and the correction term will in turn approach zero.

12.2 Introduction to Boundary Conditions

A partial differential equation is typically specified over some multi-dimensional spatial region or some spatial region and some time interval. In addition to the differential equations, the values of the function on the boundaries of the solution space are of prime importance. While mathematically there is little difference in the forms of the equations between a spatial variable and a time variable, boundary conditions are typically different depending on whether the variable is a spatial or time variable. If one of the variables is time, the boundary condition on time is typically of the initial value type, where the variable and possibly time derivatives are known at some specific time. The time evolution of the solution is then typically desired beginning with this initially known time solution. For spatial variables, the boundary conditions are more typically of the type where the solution is known over some boundary of the spatial region – the boundary value type. Thus boundary conditions may be of the initial value type or the boundary value type, or of mixed initial value, boundary value type for a problem involving both time and spatial dimensions. Boundary conditions where the function is known on a set of boundaries is typically referred to as *Dirichlet conditions* while boundary conditions specifying the normal gradients on the boundary are referred to as *Neumann conditions*. More general boundary conditions involving some combination of normal gradients and boundary values are referred to as *mixed conditions*.

Perhaps the simplest partial differential equation and set of boundary conditions is that of an equation which is first order in time and second order in a single spatial dimension, such as that of Eq. (12.4). The boundary conditions for such a problem are typically initial conditions in the time variable, i.e. the solution value at some initial time is known for the range of spatial variable and the solution then develops as a function of time and position. For the spatial variable, the boundary conditions are typically of the boundary value type where the solution value is known as a function of time on the boundaries of the spatial region. The spatial region of interest for such a problem may be either finite or of infinite extent. Equations that are second order in time such as Eq. (12.5) are also typically of the initial value type in time and of boundary value type in the spatial dimensions.

The solution methods discussed in this chapter are based upon finite difference approximations to the partial derivatives. Such an approach is most appropriate for solutions that exist in a rectangular region of solution space. For example one might desire a solution over the time interval $0 \leq t \leq T$ and the spatial interval $0 \leq x \leq L$. For such a rectangular solution space, the finite difference method matches very conveniently to the boundaries of the solution space. For the vast majority of engineering problems involving two or more spatial dimensions, the spatial region of interest is not rectangular in dimensions, but involves complicated spatial boundaries. Such problems do not match naturally to easily generated finite difference spatial grids. For such problems, the *finite element* approach to be discussed in the next chapter provides a more natural method of matching the numerical solution to non-rectangular boundary regions.

12.3 Introduction to the Finite Difference Method

The method of approximating derivatives by finite differences has been extensively covered in Chapter 10 with regard to single variable differential equations. The reader is assumed to be familiar with that material so that the extension to partial derivatives can be easily made in this chapter. In order to have a specific example to discuss, the model diffusion equation of Eq. (12.4) will be considered with constant diffusion coefficient:

$$\frac{\partial U}{\partial t} - D\frac{\partial^2 U}{\partial x^2} = 0 \tag{12.8}$$

It is further assumed that this is to be applied to a problem with a linear grid of time points (separated by Δt) and a linear grid of spatial points (separated by Δx). The time points will be identified by the integer index n while the spatial points will be identified by the integer i so that each solution value in the two dimensional space of time and distance can be identified as U_i^n . Approximating the spatial derivative by finite differences for a uniform spatial grid gives:

$$\frac{\partial^2 U}{\partial x^2} \rightarrow \frac{U_{i+1}^n - 2U_i^n + U_{i-1}^n}{(\Delta x)^2} \tag{12.9}$$

For the time derivative there are several possible approximations. Three of these have been discussed in detail in Chapter 10 and are known as the explicit forward differencing (FD) method, the implicit backwards differencing (BD) method and the trapezoidal rule (TP) which averages the time derivative between two successive time points. From the discussion of these methods in Section 10.1, one would expect different long term stability results for each of these methods and this is certainly the case for partial differential equations as well as single variable differential equations. The forward and backwards time differencing methods leads to the set of equations:

$$\text{FD Method:} \quad \frac{U_i^{n+1} - U_i^n}{\Delta t} = D\left[\frac{U_{i+1}^n - 2U_i^n + U_{i-1}^n}{(\Delta x)^2}\right]$$

$$\text{BD Method:} \quad \frac{U_i^{n+1} - U_i^n}{\Delta t} = D\left[\frac{U_{i+1}^{n+1} - 2U_i^{n+1} + U_{i-1}^{n+1}}{(\Delta x)^2}\right] \tag{12.10}$$

These equations are valid at each interior point of the spatial variable while separate equations must be specified at the boundaries of the spatial region. The FD method gives an equation for the solution at time step $n+1$ explicitly in terms of the solution at time step n since only one term in the first of Eq. (12.10) involves the $n+1$ time point. On the other hand the BD method gives a set of coupled equations that must be solved at each time step for the solution variable since the equation involves spatial points $i, i-1$ and $i+1$ at time step $n+1$. The resulting set of equations for the BD method forms a set of tri-diagonal matrix equations. The solution of such a set of equations has previously been covered in Section 11.5 and the techniques discussed there are applicable to solving the set of BD equations in this application.

The third method to be discussed here is similar to the trapezoidal method for single variable equations and can be considered as averaging between the FD and BD methods to give an improved differencing technique which for partial differential equations is known as the *Crank-Nicholson* (CN) method:

CN Method: $\dfrac{U_i^{n+1}-U_i^n}{\Delta t}$

$$= \frac{D}{2}\left[\frac{(U_{i+1}^{n+1}-2U_i^{n+1}+U_{i-1}^{n+1})+(U_{i+1}^n-2U_i^n+U_{i-1}^n)}{(\Delta x)^2}\right] \qquad (12.11)$$

This again leads to a set of coupled equations that are tri-diagonal in nature. Since this method used a centered time difference approximation, it is accurate to second order in both time and space while the FD and BD methods are only accurate to first order in the time variable. Thus one would expect an improved accuracy with the Crank-Nicholson method. However, numerical stability of the three techniques is perhaps more important than accuracy.

The von Neumann stability analysis can be used to examine the long term time stability of the different finite difference approximations. This technique only applies to linear partial differential equations with constant coefficients, but much can be learned from such simple cases. This analysis begins by assuming that the solution of the finite difference system can be expressed as a superposition of Fourier modes having the form

$$U_i^n \rightarrow \xi^n e^{jk(i\Delta x)} \qquad (12.12)$$

where k is a Fourier wave number and $j=\sqrt{-1}$ (i and n are the spatial and time integers). The factor ξ is a complex quantity depending on k and is known as the *amplification factor*. The von Neumann stability analysis then consists of substituting Eq. (12.12) into the difference equation and imposing the requirement of $\left|\xi^{n+1}/\xi^n\right|=|\xi|<1$. Putting the functional form of Eq. (12.12) into the difference equations for the three methods produces the following equations for the amplification factor:

$$\text{FD Method: } \xi = 1 - \frac{4D\Delta t}{(\Delta x)^2}\sin^2(k\Delta x/2)$$

$$\text{BD Method: } \xi = \frac{1}{1+\dfrac{4D\Delta t}{(\Delta x)^2}\sin^2(k\Delta x/2)} \qquad (12.13)$$

$$\text{CN Method: } \xi = \frac{1-\dfrac{2D\Delta t}{(\Delta x)^2}\sin^2(k\Delta x/2)}{1+\dfrac{2D\Delta t}{(\Delta x)^2}\sin^2(k\Delta x/2)}$$

For a large number of time steps, a stable solution requires that the magnitude of the above amplification factors remain less than or equal to unity. For the FD method this requirement is:

$$\left| 1 - \frac{4D\Delta t}{(\Delta x)^2} \sin^2(k\Delta x/2) \right| \le 1 \quad \Rightarrow \quad \frac{2D\Delta t}{(\Delta x)^2} \le 1$$

$$\text{or } \Delta t \le \frac{2D}{(\Delta x)^2} \tag{12.14}$$

A physical interpretation of the above equation is that the maximum step in time is up to a numerical factor the diffusion time across a spatial cell of width Δx. The restriction on the allowable time step for stability is very severe in most practical problems as times of interest are typically much larger than the maximum allowable time step. Note that as the spatial resolution increases, the requirement on the time steps become very small. As previously discussed with regard to single variable differential equations, the FD method is of little practical use in solving partial differential equations.

For the BD and CN methods, the amplification factor is unconditionally less than or equal to unity for all time steps, so both of these methods exhibit the desired stability and are possible differencing approaches. However, it is know from the previous work with single variable differential equations, that the CN method, which is essentially the trapezoidal time method, provides higher accuracy than the BD method. The numerical complexity of the CN method is only slightly greater than the BD method and thus the CN differencing method will be emphasized in this work and this chapter. It should of course be noted that stability is not the same as accuracy. While the methods are stable with regard to large time steps, time steps for a desired accuracy may have to be very small. Also just as with single variable problems, a particular partial differential equation may be ill conditioned with time constants of vastly differing magnitudes, requiring small time steps to see part of a transient solution and much larger time steps to see the remainder of a transient solution. Much of the consideration of selecting appropriate time steps for single differential equations from Chapter 10 carries over to this chapter.

The stability discussion has focused on one particular type of partial differential equations and has even assumed that the equation is linear in the derivatives with constant coefficients. This is obviously a restrictive type of analysis when one may be interested in nonlinear partial differential equations. However, such a stability analysis can only be done analytically for very simple cases. No general analytical method of studying stability of differencing methods exists for more complicated cases such as nonlinear equations. One can only extrapolate some limited knowledge from these simple cases and hope that the conclusions also hold for more complicated cases. It would not be expected that long term stability would improve with more complicated equations, so the FD method should be eliminated from consideration for any reasonable analysis, except perhaps to get a solution started for some very small initial time steps. Also since the CN algorithm appears to be the best for simple cases, this will be the method of choice for developing general algorithms for any type of equation in this chapter.

A procedure can now be outlined for numerically solving a partial differential equation that is first order in time and second order in a spatial variable. Assume that the solution is known at some initial time point and that the solution is to be advanced by some time interval. The Crank-Nicholson approach gives a modified equation for the solution variable at the next time step. The equation to then be solved at each time point is very similar to that of a single boundary value problem in a spatial variable as discussed in detail in Chapter 11. In fact as will subsequent be developed, the boundary value code developed in that chapter will be used to solve the present problem at each time step. An additional control loop is needed to increment the time variable to cover a desired time range. This is very similar to the initial value problem of a single variable covered in detail in Chapter 10. A complete algorithm is then very similar to a combination of an initial value algorithm (developed in Chapter 10) to increment the time variable and a boundary value algorithm (developed in Chapter 11) to solve for the spatial solution at each time step.

If the BD algorithm was used at each time step, it would be simple to replace the time derivative in any given equation by:

$$\frac{\partial U_i}{\partial t} \rightarrow \frac{U_i^{n+1} - U_i^n}{h} \tag{12.15}$$

where h is the time step value and solution values at time step n are assumed to be know. For the CN algorithm the replacement is slightly more complicated as it involves an average of the time derivative at two time points in the form:

$$\frac{U_i^{n+1} - U_i^n}{h} = \frac{1}{2}\left[(U')_i^{n+1} + (U')_i^n\right] \tag{12.16}$$

In the right hand side of the equation, primes have been used to indicate time derivatives. This equation can be rearranged to give the form:

$$\frac{\partial U}{\partial t} \rightarrow (U')_i^{n+1} = \frac{U_i^{n+1} - U_i^{mid}}{(h/2)}; \text{ with } U_i^{mid} = U_i^n + (h/2)(U')_i^n \tag{12.17}$$

From this it can be seen that the CN algorithm is equivalent to two half-step calculations. First one estimates the mid-point solution value from taking the value and derivative at point n as given by the U_i^{mid} expression. Then the derivative at the new time point $n+1$ is estimated by a half-step difference expression using the final function value and the mid-time value. The advantage of this formulation is that the only replacement in the fundamental differential equation involves the time derivative term so that Eq. (12.8) for example can be expressed as:

$$\frac{U_i^{n+1} - U_i^{mid}}{(h/2)} - D\frac{\partial^2 U_i^{n+1}}{\partial x^2} = 0 \tag{12.18}$$

This is entirely equivalent to the conventional CN formulation of Eq. (12.11) if one were to substitute back into the equation for the mid point value in terms of the second derivative of the function at the initial time point. A major advantage of this formulation is that it provides an easy route to extend the CN approach to more complicated equations such as nonlinear equations in either the time or spatial derivatives. One simply has to replace the time derivative by the appropriate expression as given by Eq. (12.17) and solve the resulting nonlinear boundary value problem. It should be noted that the mid-time value in the equation is a known function of position, since it is assumed that one has obtained a valid solution of the partial differential equation at time point n so that the function value and first derivative are known at this previous time point. This formulation is exactly the same as previously used in Chapter 10 for solving initial value differential equations in one time variable.

Before implementing the CN algorithm is computer code, it is appropriate to consider how a second derivative with respect to time can be implemented in an initial value problem on the time variable. This discussion is very similar to the equivalent discussion in Section 10.7 for a single variable differential equation. In keeping with the same level of approximation for the second derivative with respect to time an equation equivalence to Eq. (12.16) can be written as:

$$\frac{(U')_i^{n+1} - (U')_i^n}{h} = \frac{1}{2}\left[(U'')_i^{n+1} + (U'')_i^n\right] \tag{12.19}$$

The finite difference expression for the second derivative in terms of the first derivative on the left hand side is equated to the average of the second derivative at the initial and final time values on the right hand side. This can be further expressed as

$$\frac{\partial^2 U}{\partial t^2} \rightarrow (U'')_i^{n+1} = \frac{(U')_i^{n+1} - (U')_i^{mid}}{(h/2)}; \tag{12.20}$$

$$\text{with } (U')_i^{mid} = (U')_i^n + (h/2)(U'')_i^n$$

Finally this can be expressed in terms of the function value at the final time point by combining with Eq. (12.17) to give

$$\frac{\partial^2 U}{\partial t^2} \rightarrow (U'')_i^{n+1} = \frac{U_i^{n+1} - U_i^{ext}}{(h/2)^2}; $$

$$\text{with } U_i^{ext} = U_i^{mid} + (h/2)(U')_i^{mid} \tag{12.21}$$

$$\text{or } U_i^{ext} = U_i^n + h(U')_i^n + (h/2)^2(U'')_i^n$$

In these equations U_i^{ext} can be physically interpreted as the linearly extrapolated final solution value obtained from the mid-point value and the derivative at the mid-point value.

This equation provides an appropriate replacement for a second-order time derivative with an explicit expression of the function value at the final time point and a function (U_i^{ext}) that depends on the function value, first derivative and second derivative at the initial time point all of which are assumed to be known from a valid solution previously obtained. This is again identical to the second order time differencing method discussed in Section 10.7 with respect to single variable differential equations.

With the above formulism a method is now defined for forming a finite difference set of equations for a partial differential equation of the initial value type in time and of the boundary value type in a spatial variable. The method can be applied to both linear and nonlinear partial differential equations. The result is an implicit equation which must be solved for the spatial variation of the solution variable at each desired time point. To begin the solution an initial value of the solution variable (for all spatial points) must be known for an equation with only a first-order time derivative and in addition an initial value of the first derivative with respect to time must be known for an equation with a second-order time derivative. Examples will be subsequently given of typical engineering problems involving such equations.

12.4 Coupled Systems of Partial Differential Equations

Many practical engineering problems involve not just a single variable and a single partial differential equation but a set of coupled partial differential equations. The extension of the finite differencing technique to coupled partial differential equations of the initial value, boundary value type is relatively straightforward and will be briefly discussed before developing code for solving such systems of equations.

Consider a set of partial differential equations in a time variable, t, and one spatial variable, x, with $U_1, U_2, \cdots U_M$ being the physical quantities coupled in the differential equations. A general set of coupled partial differential equations can be expressed as:

$$F_1(t, x, U_1, \cdots U_M, U_{1,x}, \cdots U_{M,x}, U_{1,xx}, \cdots U_{M,xx}, U_{1,tt}, \cdots U_{M,tt}) = 0$$

$$F_2(t, x, U_1, \cdots U_M, U_{1,x}, \cdots U_{M,x}, U_{1,xx}, \cdots U_{M,xx}, U_{1,tt}, \cdots U_{M,tt}) = 0$$

$$\vdots \qquad\qquad\qquad\qquad\qquad\qquad\qquad\qquad\qquad\qquad\qquad (12.22)$$

$$F_M(t, x, U_1, \cdots U_M, U_{1,x}, \cdots U_{M,x}, U_{1,xx}, \cdots U_{M,xx}, U_{1,tt}, \cdots U_{M,tt}) = 0$$

In these equations the partial derivatives are expressed as:

$$U_{k,xx} = \frac{\partial^2 U_k}{\partial x^2}, U_{k,x} = \frac{\partial U_k}{\partial x}, U_{k,t} = \frac{\partial U_k}{\partial t}, U_{k,tt} = \frac{\partial^2 U_k}{\partial t^2} \qquad (12.23)$$

The first subscript in each case now refers to the physical variable in the set of equations and not to the spatial point as in Eq. (12.10) through (12.21).

Using the CN algorithm of the previous section at each time interval, all the partial derivatives with respect to time (both first order and second order) can be converted into values involving the solution variables and known functions as expressed in Eq. (12.17) and (12.21). The set of coupled equations is then converted into a coupled set of equations of the form:

$$G_1(x, U_1, \cdots U_M, U_{1,x}, \cdots U_{M,x}, U_{1,xx}, \cdots U_{M,xx}) = 0$$

$$G_2(x, U_1, \cdots U_M, U_{1,x}, \cdots U_{M,x}, U_{1,xx}, \cdots U_{M,xx}) = 0$$

$$\vdots \qquad\qquad\qquad\qquad\qquad\qquad\qquad\qquad\qquad\qquad\qquad (12.24)$$

$$G_M(x, U_1, \cdots U_M, U_{1,x}, \cdots U_{M,x}, U_{1,xx}, \cdots U_{M,xx}) = 0$$

This set of coupled second order differential equations for each time increment involves only a boundary value problem essentially identical to that previously discussed in Section (11.6). The reader is referred back to that section for a discussion of such a set of equations and the development of computer code for solving such a set of equations. The code developed there can be reused in this chapter for solving a set of partial differential equations at each time increment.

Before writing computer code for this problem, it is useful to consider how the code should be arranged. To define such an initial value, boundary value problem, three functions are needed for each solution variable in the equation set. One function to define the partial differential equation and left and right boundary functions to define the boundary values on the spatial variable. As in previous chapters, it is convenient to define all the partial differential equations and boundary values in single code functions. It is also convenient to define a code segment that performs a calculation of the solution for one desired increment in the time variable. This basic code segment can then be used in higher level code segments to implement various time integration schemes. This approach was previously used in Section 10.4 and Listing 10.4 to solve initial value differential equations in one variable. However, with a partial differential equation one generally has at least two time steps to consider. First there is the basic time step needed to increment the solution and this must be chosen appropriately small to obtain some desired accuracy in the solution. Generally, for an accurate solution, this fundamental time step must be selected so that the solution variables change by small factors between time steps.

Because of the small time steps, a code segment can generate considerable data for the solution variables and if all this data is saved for a particular problem, it can easily become an excessive amount of data. More typically one would be interested in saving data on the solution set with a larger time increment than is used in incrementing the basic differential equation. For example one might save only every 10th calculated solution set rather than the solution set at each fundamental time increment. This could still give sufficient resolution in the solution to obtain an excellent view of the time development of the solution set. Thus one needs to consider a basic integration time step and a basic time step for saving the solution values.

In terms of writing a basic time step integration code segment, the code could be implemented to return a solution set after each fundamental time step or it could be implemented to return with a solution set only after incrementing the time by an amount set by the time desired for saving a data set. This later approach will be used here to implement a fundamental time step algorithm since the intermediate time step solutions are to be eventually discarded anyway. The fundamental time integration routine will then need time information on 1) the initial time, 2) the final time and 3) the number of fundamental time increments between the initial and final time.

In addition to functions and time information, a code segment will need as input an array of spatial values on which the solution is desired (the spatial grid may be uniform or nonuniform), plus initial values of the solution variables at the initial time value and initial derivative values if the equation is second order in the time derivative. A possible calling sequence for such a fundamental time step integration routine is then pdebivbv(eqsub, tvals, x, u, ut, utt), where eqsub is a table containing the names of the functions (defining equations and boundary functions), tvals is a table of desired time values, x is the spatial array of points, u is a multi-dimensional array of initial solution values and ut is a multi-dimensional array of initial derivative values. A final variable in the calling argument list is utt, a multi-dimensional array of second derivative values. This set of values is to be returned by the code segment so that the fundamental code segment can be recalled and the code can pick up the solution with an increasing time value and have the solution algorithm continue with a new time increment. This will perhaps become clearer when the code segment is shown in detail. Solution values at the end of the time integration increment can be returned in the u array, and first time derivatives can be returned in the ut array.

Computer code for implementing one basic time step with a set of partial differential equations is now shown in Listing 12.1. The heart of the code segment is the time loop from line 37 to line 61 which increments the time variable from some tmin to tmax value with ntval increments. Within each loop the factors needed in replaceing the time derivatives at each time step (xt and xtt in the code) are evaluated on lines 42 and 43 for each solution variable. These factors correspond to U_i^{mid} and U_i^{ext} in Eqs. (12.17) and (12.21). The reader should be able to readily identify the terms in these equations. The call to function ode2bvfd() on line 47 then solves the set of boundary value problems using the finite difference boundary value code segment developed in Chapter 11. A key to this call is the function eqtr() defined on lines 8 through 13 of the code. This provides a translation of the defined PDE equation set in terms of the time derivatives into a set of equations involving at each time step a set of equations in only the spatial variable. The translations of the first and second time derivatives into finite difference equations is performed on lines 10 and 11 of the code where the equations implement the replacements as expressed in Eqs. (12.17) and (12.21) using the xt and xtt values calculated on lines 42 and 43. A second key to the solution is the two new boundary condition functions, eqbr() and eqbl() defined on lines 14 through 23. The boundary conditions required by the ode2bvfd() function can only be functions of the solution variables and the spatial derivatives at the boundary points. However, for the partial differential equation, the boundary conditions can be specified not only in terms of the function values and spatial derivatives, but in terms of time and the first time derivative of the solution values. The new boundary functions thus add the time variable and translate the first time derivatives into the function values. This is done on lines 15 through 17 and lines 20 through 22. After adding these values, the boundary functions are called on lines 17 and 22. After the return of the new solution set from the pde2bvfd() call on line 47, new values of the first and second order time derivatives are evaluated on lines 55 and 56 using the same basic equations as used on lines 10 and 11 in replacing the time derivatives in obtaining the solution set. These values are needed in the next time step for the partial time derivative replacements.

A problem in implementing the time loop is in starting the set of calculations since the time derivative replacements require known values of the first and second time derivative at the initial time point. When beginning the solution of a partial differential equation set these are not generally known. If the equation set is

```
 1 : -- /* File pdeivbv.lua */
 2 : -- Programs to solve partial differential equations of IV,BV
 3 : require"odefd"; require"intp"
 4 : pdebivbv = function(eqsub,tvals, x,u,ut,utt) --Basic PDE solver
 5 :     local neq,nx,nut,nutt,t,h,h2,h2sq,hs,att,btt,at =
            #u, #x, #ut, #utt
 6 :     local nit,nitt,eqsu = 0,0, eqsub[1] -- Number of iterations
 7 :     local fu,xt,xtt,uti,utti = {},{},{},{},{}
 8 :     eqtr = function(fu,x,u,ux,uxx,i) -- Map time to spatial vals
 9 :         for m=1,neq do
10 :             uti[m] = (u[m] - xt[m][i])/h2 -- h/2 or h
11 :             utti[m] = (u[m] - xtt[m][i])/h2sq -- second deriv
12 :         end; eqsu(fu,x,t,u,ux,uxx,uti,utti) -- Call equation set
13 :     end
14 :     eqbl = function(fu,u,ux) -- map left boundary condition
15 :         for m=1,neq do
16 :             uti[m] = (u[m] - xt[m][1])/h2 -- Left time derivative
17 :         end; eqsub[2](fu,u,ux,t,uti)
18 :     end
19 :     eqbr = function(fu,u,ux) -- map right boundary condition
20 :         for m=1,neq do
21 :             uti[m] = (u[m] - xt[m][nx])/h2 -- Left time derivative
22 :         end; eqsub[3](fu,u,ux,t,uti)
23 :     end
24 :     local eqsx = {eqtr,eqbl,eqbr}
25 :     local tmin,tmax,ntval = tvals[1],tvals[2],tvals[3]
26 :     t,hs = tmin, (tmax-tmin)/ntval -- Initial t value
27 :     if nutt~=neq then for m=1,neq do utt[m] = {} end end
28 :     if nut~=neq then for m=1,neq do ut[m] = {} end end
29 :     for m=1,neq do
30 :         if #utt[m]~=nx then nutt = 0;
                for k=1,nx do utt[m][k] = 0 end end
31 :         if #ut[m]~=nx then nut=0;for k=1,nx do ut[m][k]=0 end end
32 :     end
33 :     if nutt~=neq then -- utt aray not input, use BD initially
34 :         jfirst,h,h2,h2sq,att,at,btt=0,hs/4,hs/4,
                (hs/4)^2,hs/4,0,0
35 :     else jfirst,h,h2,h2sq,att,at,btt=4,hs,hs/2,
                (hs/2)^2,hs,hs/2,(hs/2)^2
        end
36 :     for m=1,neq do xt[m],xtt[m],fu[m] = {},{},0 end
37 :     for k=1,ntval do -- major time loop -- Heart of solution
38 :         repeat -- Use BD for first interval with 4 sub intervals
39 :             jfirst = jfirst+1
40 :             for i=1,nx do -- Set up xx arrays, time derivatives
41 :                 for m=1,neq do
42 :                     xt[m][i] = u[m][i] + at*ut[m][i]
43 :                     xtt[m][i] = u[m][i]+att*ut[m][i]+btt*utt[m][i]
44 :                 end
45 :             end -- These are used by eqtr() function when called
46 :             t = t + h -- Now increment t to next t value
47 :             nitt = ode2bvfd(eqsx,x,u) -- Calculate new u values
48 :             if nitt>nit then nit = nitt end -- Monitor maximum #
49 :             if nprint~=0 then
50 :                 printf('Time = %e Number of iterations in
                        pdebivbv = %d \n\n',t,nitt)
51 :                 io.flush()
52 :             end
53 :             for i=1,nx do -- values at new time point
```

```
54 :                 for m=1,neq do
55 :                     ut[m][i] = (u[m][i] - xt[m][i])/h2
56 :                     utt[m][i] = (u[m][i] - xtt[m][i])/h2sq
57 :                 end
58 :             end
59 :         until jfirst>=4 -- End of first interval with BD
60 :         if k==1 then h,h2,h2sq,att,at,btt=hs,hs/2,(hs/2)^2,
                 hs,hs/2,(hs/2)^2
            end
61 :     end -- End of major loop, Go back and do for another time
62 :     return nit, tmax, tmin, ntval
63 : end          -- End of function
64 : setfenv(pdebivbv,{ode2bvfd=ode2bvfd,printf=printf,type=type,
                 nprint=0,io=io,plot=plot})
```

Listing 12.1. Code segment for single time step solution of a set of coupled partial differential equations of the initial value, boundary value type.

first order in the time derivative, then only the initial value of the solution set is known. If the equation set is second order in the time derivative then the initial value problem requires that one also know the initial value of the first derivative, but in no case is the second derivative known for the initial starting time point. Thus to begin a solution the code in the time loop uses an initial sub time interval of 4 time points with the backwards difference approach used to solve the set of equations. The BD approach requires only knowledge of known initial values. These first 4 sub time intervals are performed in a repeat-until loop on lines 38 through 59 of the code. The switches between the various parameters of the BD and CN algorithms are made on lines 33 through 35 before the time loop and on lines 60 after the first 4 sub time intervals. The basic solution function, pde-bivbv() defined on line 4 contains the first and second time derivatives (ut and utt on line 4) as arguments so that these can be returned by a call to the solution routine. These can then be used in a subsequent call to the code segment so that the CN algorithm can pick up with a new time interval and not have to use an initial BD algorithm for the first time interval.

The use of an initial set of BD intervals has another very useful feature in addition to allowing the solution to begin with a set of known data. If the equation set is an ill conditioned set of equations with time constants of vastly different magnitudes, then the BD algorithm can rapidly damp out an initial very short time constant solution and allow one to observe a longer time constant response. The CN algorithm can under such conditions show oscillatory behavior. This was discussed in Section 10.3 and the reader is encouraged to review the discussion there since the CN algorithm is essentially the same as the trapezoidal algorithm in two dimensions as opposed to the one dimension discussed there.

The sections of Listing 12.1 not so far discussed are basically housekeeping code such as defining required local variables. In addition, the code on lines 27 through 32 ensures that proper arrays are set up for the first and second order time derivatives. This is done so that one does not have to supply known data arrays to the calling arguments. For example, initial ut and utt arrays are not required in the calling argument list if one has a set of equations with only a first order time derivative. The pdebivbv() function will automatically set up the arrays if not user supplied. The function returns on line 62 the maximum number of iterations required in solving the boundary value problem so that one can monitor the progression of the solution set. In summary for Listing 12.1, the basic time step integration routine takes as input a set of partial differential equations and set of boundary value equations plus an initial time, a final time and a number of time steps and integrates the PDE set using the CN algorithm over the specified time interval. Once the time derivatives are replaced by difference equations, the boundary value code from Chapter 11 is used to solve this coupled set of equations. The function does not save any of the calculated solution values, but returns with the final solution set obtained, plus values of the first and second time derivatives at the final time point. The basic time step routine is intended to be used by a higher level solution routine that saves the solution sets and increments time over a more complete time interval. For example if one wishes to save calculated values for every 10th calculated time interval, then this routine can be called to perform calculations over 10 basic time intervals and return the final calculated value which would then be saved and the function recalled with a new increased time interval. The code segment can pick up where left off in time again calculation over 10 time intervals and returning another solution set to be saved.

A code segment to implement a more complete solution using this basic time step routine is shown in Listing 12.2. Although this function, pdeivbv() has the same set of calling arguments as the pdebivbv() function, the arguments have slightly different meanings.

This function is intended to be the commonly called routine for solving a set of PDEs. The first calling argument (eqsub on line 67) is the same as for pdebivbv() in Listing 12.1 which is a table of functions in the order of a function defining the PDE set, followed by two functions defining the two sets of boundary values for the spatial interval. The second argument is a set of time values with the basic form: tvals = {t0, {t1, t2, --- tm}, {nt1, nt2, --- ntm }} where t0 is the initial time for the solution and {t1, t2, --- } are a set of time values at which one desired a saved set of solution values and {nt1, nt2, ---} are a number of sub-time intervals to be used within the corresponding t1, t2, -- time intervals. Solution sets are not saved at the intermediate time points. If all the sub time intervals are to be the same, the table can be replaced by one sub time interval number, ntval. This is perhaps best illustrated with an example such as tvals = {0, {1.e-5, 1.e-4, 1.e-3, 1.e-2}, 10} which would use 10 sub time intervals between 0 and 1.e-5, between 1.e-5 and 1.e-4, between 1.e-4 and 1.e-3 and between 1.e-3 and 1.e-2 for the solution of the equation set but would save only the calculated values at times 0, 1.e-5, 1.e-4, 1.e-3 and 1.e-2. Thus for 4 saved solution sets, the code segment would calculate solutions for a total of 4X10 = 40 time intervals and save values at 5 time values. If equal time increments are desired, then a simpler form of the time input can be used in the form: tvals = {tmin, tmax, nkeep, ntval} where nkeep solutions would be kept, linearly spaced over the time interval tmin to tmax and ntval sub time intervals would be used between each saved time solution set.

```
  1 : -- /* File pdeivbv.lua */
-------
 66 : local usave,utsave
 67 : pdeivbv = function(eqsub,tvals,x,u,ut,utt) -- Simple IV,BV
 68 :    local tmin,ntkeep,ntval,dtkeep,nit
 69 :    local usol = {x} -- Array to hold solutions
 70 :    if type(u[1])~='table' then u = {u} end -- For one D case
 71 :    local neq,nx = #u, #x -- Number of solution variables
 72 :    ntkeep,ntval = tvals[3] or NKEEP, tvals[4] or NTVAL -- Set #
 73 :    if type(tvals[2])=='table' then ntkeep,ntval = #tvals[2],
              tvals[3] or NTVAL end
 74 :    if type(tvals[2])=='number' then -- Simple input format
 75 :        if type(tvals[3])~='table' then tvals[3] = {} end --table?
 76 :        dtkeep = (tvals[2] - tvals[1])/ntkeep; tvals[2] = {}
 77 :        for j=1,ntkeep do -- Set up time arrays for simple input
 78 :              tvals[2][j] = tvals[1] + j*dtkeep
 79 :              tvals[3][j] = tvals[3][j] or ntval
 80 :        end
 81 :    elseif type(tvals[3])=='number' then
 82 :        tvals[3] = {};
 83 :        for j=1,ntkeep do tvals[3][j] = ntval end
 84 :    end
 85 :    if type(utt)~='table' then utt = {} end -- Ensure tables
 86 :    if type(ut)~='table' then ut = {} end
 87 :    local nutt,nut = #utt, #ut
 88 :    if nut~=neq then for m=1,neq do ut[m] = {} end end
 89 :    for m=1,neq do
 90 :        if #ut[m]~=nx then for k=1,nx do ut[m][k] = 0 end end
 91 :    end
 92 :    tmin = tvals[1]
 93 :    for i=1,ntkeep do -- Step over time increments
 94 :        usol = usave(usol,u)  -- Save solution values
 95 :        nit = pdebivbv(eqsub,{tmin,tvals[2][i],tvals[3][i]},
              x,u,ut,utt)
 96 :        tmin = tvals[2][i] -- Increment initial time value
 97 :        if nprint~=0 then
 98 :            printf('Time = %e Number of iterations in pdeivbv =
                  %d \n\n' ,tmin,nit)
 99 :            io.flush()
100 :        end
101 :        if nprint==2 then for j=1,neq do plot(x,u[j]) end end
102 :    end
103 :    usol = usave(usol,u) -- Save final values
104 :    return usol
105 : end
106 : setfenv(pdeivbv, {pdebivbv=pdebivbv,printf=printf,
             plot=plot,type=type,
107 : nprint=0,io=io,NKEEP=10,NTVAL=10,usave=usave})
```

Listing 12.2. Code segment for calculating set of solutions for PDE of the initial value and boundary value type.

The code in Listing 12.2 from line 72 through 84 is simply to handle the different possible time specification options and to ensure that appropriate time derivative arrays are present. For a first-order time derivative PDE the ut and utt arrays are not required as input to the pdeivbv() function and for a second-order time derivative PDE the utt array is not required as input data. If these are not supplied as input, arrays are set up on lines 88 through 91.

The heart of Listing 12.2 is the time loop from line 93 to 102 which steps over the time intervals for which a saved solution is desired and calls the pdebivbv() function on line 95 passing the initial, final and number of sub time steps to this function. Lines 94 and 103 saves the calculated values and accumulates them in a single array of solution sets using the usave() function. The first entry to the saved array is set to the x array on line 69. The usave() function simply adds solution sets to an existing set of solutions and the reader can review the code in the pdeivbv.lua file.

This function pdeivbv() provides a versatile multi-time step function for solving sets of PDEs of the IV, BV type. As a simple example let's consider a linear diffusion problem into a slab of finite thickness defined by the following equation set:

$$\frac{\partial U}{\partial t} - D\frac{\partial^2 U}{\partial x^2} = 0 \text{ for } 0 \le x \le L$$

$$U(x,0^-) = 0; \text{ and } U(0,t) = U_0; \quad U(L,t) = 0 \text{ for } t \ge 0^+$$

(12.25)

The initial concentration within the slab of material is assumed to be 0 before the diffusion process begins and at $t = 0$ the concentration of the diffusant is abruptly increased to U_0 at $x = 0$ and held fixed at that value for all subsequent time. The boundary at $x = L$ is taken as an ideal absorbing boundary where the diffusant concentration is always 0. Note the use of the $t = 0^-$ and $t = 0^+$ engineering concepts to distinguish conditions just before the beginning of the transient and just after the beginning of the transient diffusion process. Of course this is an idealization of a physical process as it requires an infinite derivative of the diffusant at the origin as the process begins. This problem can be transformed into a dimensionless variable problem by use of the following substitutions:

$$u \to U/U_0; \quad x \to x/L; \quad t \to Dt/L^2$$

(12.26)

The formulation of the problem then becomes:

$$\frac{\partial u}{\partial t} - \frac{\partial^2 u}{\partial x^2} = 0 \text{ for } 0 \le x \le 1$$

$$u(x,0^-) = 0; \text{ and } u(0,t) = 1; \quad u(1,t) = 0 \text{ for } t \ge 0^+$$

(12.27)

A code segment for solving this PDE IV-BV problem is shown in Listing 12.3. The diffusion equation and the boundary conditions are defined by the eq(), efl() and efr() functions on lines 9 through 17. These are set up in a form that should be very familiar by now as the notation is the same as used in previous chapters. The only addition in the defining equations is the presence of a possible time derivative identified in the eq() calling argument list as ut. The spatial derivatives are identified as up and upp for the first and second order derivatives. Only one equation is used in this example but each variable and derivative must be identified with the [1] array notation as the solution routines are coded to handle multiple sets of equations. It can be noted that since the second time derivative is not present in the defining PDE, the utt argument is omitted in the calling list to the eq() function. Similarly, since time and the first time derivative are not needed for the boundary functions, these arguments are not included in the argument lists to the efl() and efr() functions.

```
 1 : -- /* File list12_3.lua */
 2 : -- Programs to integrate diffusion eqn in one spatial variable
 3 :
 4 : require"pdeivbv"
 5 : getfenv(odefd).nprint = 1; getfenv(pdeivbv).nprint = 1
 6 : -- Model equations to be solved
 7 : L,D = 1,1; u = {}
 8 :
 9 : eq = function(fu,x,t,u,up,upp,ut) -- PDEs
10 :     fu[1] = ut[1] - D*upp[1] -- Diffusion equation
11 : end
12 : efl = function(fu,u,up) -- Left boundary fixed at 1.0
13 :     fu[1] = u[1] - 1.0
14 : end
15 : efr = function(fu,u,up) -- Right boundary, fixed at 0.0
16 :     fu[1] = u[1] - 0.0
17 : end
18 :
19 : x = {}; Nx = 400; tmax = 0.01
20 : for i=1,Nx+1 do x[i],u[i] = (i-1)/Nx, 0.0 end -- Set up x and u
21 :
22 : sol = pdeivbv({eq,efl,efr},{0,tmax,5,5},x,u)
23 : plot(sol[1],sol[2],sol[3],sol[#sol])
24 : write_data('list12_3.dat',sol)
Selected Output:
-----
-- 1-- Iteration number, Maxumum relative, absolute corrections are:
(1) 6.788155e-001, 1.308514e-002, at 400, 50 ;
-- 2-- Iteration number, Maxumum relative, absolute corrections are:
(1) 7.938991e-013, 2.429211e-013, at 53, 49 ;
-- 1-- Iteration number, Maxumum relative, absolute corrections are:
(1) 6.578206e-001, 1.241403e-002, at 400, 51 ;
-- 2-- Iteration number, Maxumum relative, absolute corrections are:
(1) 3.223653e-013, 5.467929e-014, at 107, 59 ;
Time = 8.000000e-003 Number of iterations in pdeivbv = 2
------
```

Listing 12.3. Example code segment for solving Diffusion Equation, PDE

The initial conditions and the spatial array of x points are defined on lines 19 and 20 of the listing. The call to the PDE solver on line 22 then returns a set of solutions from 0 to tmax (0.01 in this example) with 5 time solutions returned and with 5 sub time intervals used in solving the equations. Finally line 23 plots three of the solutions and line 24 saves all the calculated results. In order to monitor the progress of the solution set, the nprint = 1 parameters are set on line 5 for both the odefd() and pdeivbv() functions. This produces printed output such as that shown in the selected output listing. It can be noted in the selected output that the solutions at each time increment are requiring 2 iterations of the basic Newton method used to solve the sets of equations. It is noted that the corrections on the second iteration are always within the machine accuracy limits. This is because the equation set is linear and does not require Newton iterations for a solution. However, the coded solution algorithms can only determine that by performing a second iterative step at which negligible correction terms are obtained. Approximately half of the execution time could be saved by not performing this extra iterative step. An easy means of accomplishing this will be shown in a following example.

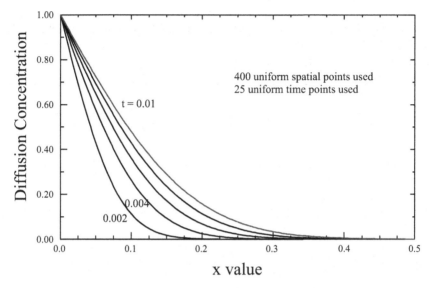

Figure 12.1. Solution set for diffusion problem from Listing 12.3. Uniform spatial grid and uniform time grid used in the solution

Figure 12.1 shown results for the solutions obtained from Listing 12.3. For the time interval used in the calculation, it is seen that diffusion is confined to approximately half of the defined spatial region and only the 0 to 0.5 region is shown in the figure. Also it is seen that at the earliest saved solution set at $t = 0.002$, the diffusant front is already considerably into the spatial region. While this is a useful start and illustrates the use of the PDE code, it provides a solution set only over a limited time and spatial range. While the code execution saves 5 solution sets, 5 additional time solutions have been calculated for each of the saved sets. So between $t = 0$ and the first saved set at $t = 0.002$, 5 additional solutions were calculated in order to achieve a small time step between the solution sets and improve the accuracy of the saved solutions. From this set of solutions several features can be deduced. First, it can be seen that significant diffusion is occurring on a much shorter time scale than used in the obtained solution set. This is to be expected since diffusant is entering the spatial region from the $x = 0$ boundary and can rapidly change the concentrations near the left boundary while long times are required for the diffusant to reach deep into the spatial region. Thus a uniform time grid is perhaps not the best choice for obtaining an accurate solution to the overall behavior of the diffusion process. On the other hand, if calculations are made for shorter times, the solution behavior will be very confined to small spatial values and a fine spatial grid near the left boundary will be required to obtain an accurate solution set. Thus a uniform spatial grid is perhaps not the best choice either. Fortunately, both non-uniform spatial grids and time grids are relatively easy to accommodate with no further changes to the solution code. The FD spatial solver used by this new code is the same as developed in Chapter 11 where a number of examples of non-uniform spatial grids were demonstrated. The use of non-uniform time grids has also been covered in Chapter 10 in connection with single variable initial value problems and the multi-step routine pdeivbv() has been de-signed to handle multiple time intervals.

```
 1 : -- /* File list12_4.lua */
 2 : -- Programs to integrate diffusion equation in one x variable
 3 :
 4 : require"pdeivbv"; require'odebvfd'
 5 : getfenv(odefd).nprint = 1; getfenv(pdeivbv).nprint = 1
 6 : getfenv(ode2bvfd).NMAX = 1 -- Only one iteration required
 7 : -- Model equations to be solved
 8 : L,D = 1,1; u = {}
 9 :
10 : eq = function(fu,x,t,u,up,upp,ut) -- Equations
11 :     fu[1] = ut[1] - D*upp[1] -- Diffusion equation
12 : end

13 : efl = function(fu,u,up) -- Left boundary fixed at 1.0
14 :     fu[1] = u[1] - 1.0
15 : end

16 : efr = function(fu,u,up) -- Right boundary, fixed at 0.0
17 :     fu[1] = u[1] - 0.0
18 : end
19 :
20 : Nx = 400; x = xlg(0,1,1.e-6,Nx) -- Set up x and u arrays
21 : for i=1,Nx+1 do u[i] = 0.0 end
22 :
23 : sol = pdeivbv({eq,efl,efr},{0,{1.e-6,1.e-5,1.e-4,1.e-3,1.e-2,
        1.e-1},5},x,u)
24 : plot(sol[1],sol[2],sol[3],sol[#sol])
25 : write_data(10,'list12_4.dat',sol)
Selected Output:
------
-- 1-- Iteration number, Maxumum relative, absolute corrections are:
(1) 9.731521e-001, 7.771088e-002, at 400 ;
-- 1-- Iteration number, Maxumum relative, absolute corrections are:
(1) 9.594194e-001, 5.945364e-002, at 400 ;
-- 1-- Iteration number, Maxumum relative, absolute corrections are:
(1) 9.454885e-001, 4.821901e-002, at 400 ;
Time = 1.000000e-003 Number of iterations in pdeivbv = 1
-------
```

Listing 12.4. Code segment for solving diffusion equation with non-uniform spa-tial and time grids.

Listing 12.4 shows the same problem formulated in terms of non-uniform spa-tial and time grids. The code is very similar to Listing 12.3 with two major changes. First on line 20 a non-uniform spatial grid is set up using the xlg() function introduced in Chapter 10. For the parameters on line 20, the call generates a logarithmically spaced array of 400 spatial points between 0 and 1 with a minimum step size of 1.e-6. Second the call to the PDE solver, pdeivb() on line 23 has a tvals[] argument of {0, {1.e-6, 1.e-5, 1.e-4, 1.e-3, 1.e-2, 1.e-1},5} which results in 5 time solutions for each of the time intervals 0 to 1.e-6, 1.e-6 to 1.e-5, 1.e-5 to 1.e-4, 1.e-4 to 1.e-3, 1.e-3 to 1.e-2, and 1.e-2 to 1.e-1. Solutions are saved at only the specified beginnings and ends of the time intervals. Thus the returned array in sol[] on line 23 will contain solutions of the PDE for times of 0, 1.e-6, 1.e-5, 1.e-4, ... 1.e-2 or every decade in time. A final change in Listing 12.4 is on line 6 which sets the NMAX variable in function ode2bvfd() to 1. This change results in only one iteration of the Newton solution loop in that routine and cuts the solution time in half. This is possible only for a problem that is known to be a linear partial differential equation where Newton iterations are not required. One can see the difference by looking at the selected output in Listing 12.4 as compared with the selected output in Listing 12.3. In this case only one iteration is shown for each time step.

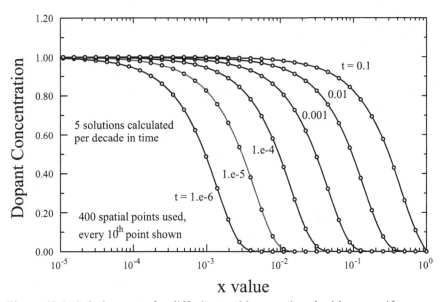

Figure 12.2. Solutions sets for diffusion problem produced with non-uniform spatial and time grids of Listing 12.4.

Figure 12.2 shows graphs of the solution set produced by Listing 12.4. The solution sets are shown on a log spatial scale so that the short time solutions may be clearly seen. The data points indicate every 10th spatial point used in the solution and one can see the equally spaced spatial points on the log scale. The solutions are shown, as saved, for each decade in time from 1.e-6 to 1.e-1 and are seen to be essentially equally spaced on the log distance scale. The curves in Figure 12.2 show basically a diffusion front that advances through the spatial variable as time increases with little change in shape when plotted on the log distance scale. For diffusion with constant diffusion into a semi-infinite medium, the solution is known to have a closed form solution of:

$$u(x) = u(0)erfc(\frac{x}{2\sqrt{Dt}})$$ (12.28)

Expect for the longest time solutions, the curves in Figure 12.2 should follow this solution. It is seen that the x value for a constant solution value should vary as the square root of time, so that time changes of two orders of magnitude causes the diffusion front to advance by one order of magnitude in distance. This is seen to be the case in the solution set of Figure 12.2. Equal values along the diffusion front advance in x by an order of magnitude for each two orders of magnitude in time. Accuracy of the numerical solution set is explored in the next section.

The ability to rapidly cover a large range of time values is useful not only in this example but in many PDEs of the initial value type. This is similar to time dependent initial value problems discussed in Chapter 10 for single variable differential equations. A function odeivqs() has been developed in that chapter for easily integrating an equation using log based steps in time. Listing 12.5 shows code for such a function named pdeivbvqs() for use with partial differential equaitons. To use this function to reproduce the same results as Listing 12.4 one simply needs to replace the call on line 23 with the statement sol = pdeivbvqs({eq,efl,efr},{0,{1.e-6, 1.e-1},1,10},x,u). In the time specification of this calling argument, the 1.e-6 and 1.e-1 terms specify the beginning and ending time intervals for keeping saved solutions, the 1, 10 parameters specify 1 saved solution per decade in time and the 10 specifies an additional 10 logarithmically spaced sub time intervals for actually solving the equation set. These parameters may be omitted and default values of 1, 10 will be used as defined on line 176 of Listing 12.5. The reader is encouraged to modify Listing 12.4 with this change and re-execute the code. The results are not shown here as they are essentially identical to those shown in Figure 12.2. This code segment simply provides a simple interface to rapidly scanning a time solution over a wide range of values of the time variable with logarithmically spaced time intervals. There is one significant difference between using the multi-time step approach illustrated in Listing 12.4 and using the quick scan function in Listing 12.5 and this is the treatment of the sub-time intervals between the saved solution values. With the multi-time interval call as used in Listing 12.4, the sub time intervals are linearly spaced between the saved time points while with the quick scan function of Listing 12.5, the sub-time intervals are logarithmically spaced.

Now that numerical algorithms and code have been developed for solving PEDs of the IV BV type it is useful to explore the accuracy that can be achieved in solving such equations with the finite difference method and with the coded routines so far developed. This is explored in the next section.

```
  1 : -- /* File pdeivbv.lua */
------
------
166 : pdeivbvqs = function(eqsub,tvals,x,u,ut,utt) -- log time
167 :   local nps,npts -- save per decade, additional steps per save
168 :   local ttvals,nl,nu,fact = {}
169 :   local nt,neq,j = #tvals, #u, 0
170 :   local nitt,nit,sol = 0
171 :   local nprint = nprint or getfenv(pdeivbv).nprint
172 :   ut = ut or {}; utt = utt or {}
173 :   if type(u[1])~='table' then u = {u} end -- For single EQN
174 :   if nt<2 then print('Error, must specify two times in
                pdeivbvqs')
175 :     return end
176 :   nps,npts = math.floor(tvals[3] or NPS),
                math.floor(tvals[4] or NPTS)
177 :   fact = 10^(1/(nps*npts)) -- Factor between steps
178 :   nl = 10^(math.floor(math.log10(tvals[2][1])))
179 :   nu = 10^(math.ceil(math.log10(tvals[2][2])))*
                1.00000000001/fact
180 :   sol = pdeivbv(eqsub,{tvals[1],nl,1,nps*npts},x,u,ut,utt)
181 :   while nl<=nu do -- Step over log time spacings
182 :     nit = pdebivbv(eqsub,{nl,nl*fact,1},x,u,ut,utt)
183 :     if nit>nitt then nitt = nit end
184 :     nl = nl*fact; j = j+1
185 :     if j==npts then
186 :       sol = usave(sol,u); j = 0 -- Save solutions
187 :       if nprint~=0 then
188 :         printf('Time = %e Number of iterations in
                   pdeivbvqs = %d \n\n' ,nl,nitt)
189 :         io.flush(); nitt = 0
190 :       end
191 :     end
192 :   end
193 :   return sol
194 : end
195 : setfenv(pdeivbvqs,{pdebivbv=pdebivbv,pdeivbv=pdeivbv,
                printf=printf,type=type,
196 : math=math,nprint=nprint,io=io,NPS=1,NPTS=10,usave=usave,
                getfenv=getfenv})
```

Listing 12.5. Code segment to solve PDEs with saved solutions at logarithmically spaced time intervals.

12.5 Exploring the Accuracy of the Finite Difference Method for Partial Differential Equations

The question of numerical stability of various time differencing methods as applied to PDEs was discussed in the previous sections. The method selected to pursue, known as the Crank-Nicholson (or CN) method was shown to have good numerical stability. However, the equally important issue of numerical accuracy was not addressed. It was noted that the developed method was accurate to second order in both the time and space numerical differencing techniques and from previous differencing approaches as applied to single variable differential equations, this was assumed to be a good implementation. In this section, the accuracy of the numerical approach and developed code segments will be explored.

It would be very nice to have a general theory of numerical accuracy for solving PDEs. However, this is beyond our reach especially for non-linear equations. No theory exists even for the existence of solutions to general non-linear problems much less a theory of the accuracy of solution methods. The best one can do is to explore the solution accuracy of some known solutions and then attempt to draw some conclusions from such particular examples. For this approach some PDE problems of the IV-BV type are needed where an exact solution is known. One of the simplest such cases is that of the diffusion equation as applied to diffusion from a fixed boundary concentration into a semi-infinite one dimensional medium. The known solution for such a problem has been previously given in Eq. (12.28). Also the solutions shown in Figures 12.1 and 12.2 approximate such a case except for the longest time interval solution shown in Figure 12.2. From this figure it can be seen that if the normalized time is limited to less than about 0.01, then the finite boundary condition at the right side of the finite spatial region has little influence on the solution. A finite spatial region can then be used to approximate a semi-infinite region provided the time variable is kept sufficiently small (less than about 0.01 in dimensionless value).

The simplest case to look at solution accuracy corresponds to using a uniform time and spatial grid in the solution as was used in generating the data for the solution shown in Figure 12.1 (see Listing 12.3 for program). These curves can be compared with the theoretical results of Eq. (12.28) to explore the error in the numerical solutions. The resulting errors are shown in Figures 12.3 and 12.4 for the solutions at times of 0.002 and 0.01 which are the earliest and latest solutions in Figure 12.1. There are several conclusions that can be drawn from the figures. First the error depends on the time at which the solution is observed. This is not too unexpected, but it is seen that the error for $t = 0.01$ is less than the error for the earlier time of $t = 0.002$ which might not be expected.

Let's first concentrate on the solution errors for the earlier time solution as shown in Figure 12.3. Taking curve a as the reference error, curve b shows the results for doubling the number of spatial increments used in the solution while curve c shows the results for doubling the number of time increments. The results show that increasing the number of spatial intervals from 400 to 800 has very little effect on the solution error while doubling the number of time increments has a very significant effect. From the discussion of the finite difference method for one-dimensional problems in Chapter 11, one would expect an error proportional to the square of the step size used in the finite difference method. However, for partial differential equations there are two (or more) variables. There is obviously a limit to the accuracy that can be achieved by increasing the number of grid points in any given dimension. Even if an infinite number of grid points is used in one dimension, there will still be solution errors due to the differencing scheme associated with the other dimensions. For the present example, it can readily be seen that the solution error is due much more to the limited number of time steps used as opposed to the number of spatial points used in the solution. Doubling the

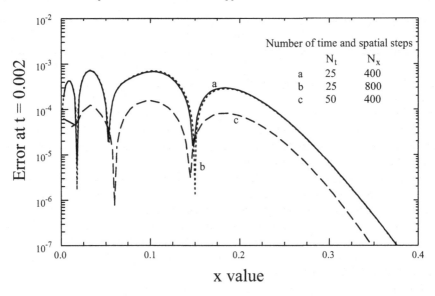

Figure 12.3. Solution error at t = 0.002 with different numbers of linearly spaced time and spatial grid points.

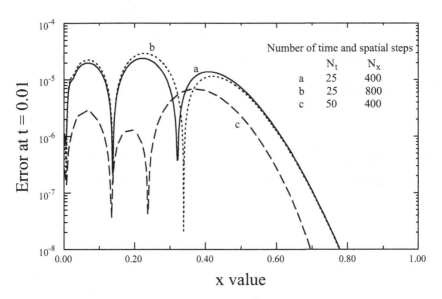

Figure 12.4. Solution error at t = 0.01 with different numbers of linearly spaced time and spatial grid points.

number of spatial intervals increases the computational time by essentially the same factor as doubling the number of time increments so it can easily be concluded for this example that adding additional time steps is the route to take to improve the accuracy. It can also be seen that curve c is approximately a factor of 4 lower than curve a so that the solution error is approximately proportional to the square of the time step used in the solution as would be expected for a differencing method that is accurate to second order in the step size. Of course if one continues to decrease the step size a point will eventually be reached where the error will be limited by the spatial step size and further reductions in the time steps will not further decrease the solution error.

For the solution at the longer time shown in Figure 12.4 it can again be seen that decreasing the time step size results in improved solution accuracy while increasing the number of spatial grid points actually causes a slight decrease in the accuracy of the solution. Again the maximum solution error is reduced by about a factor of 4 for a doubling of the number of time increments used in the solution. It is also seen in the figures that the maximum error occurs at different spatial point as the number of time increments is doubled.

For curves c in the figures, the maximum solution error is about 1×10^{-4} for the earliest time and about 1×10^{-5} for the later time. The improved accuracy for longer times results from the very rapid diffusion occurring at short times. This is not accurately reproduced by the time grid used in the solution set so the accuracy at short times is actually poorer than the accuracy at longer times. For this example it is obvious that the route to improved accuracy in the solution is the use of a finer time grid or perhaps the use of a non-uniform time grid so that the rapid changes at short times can be much more accurately calculated.

The quick scan function of Listing 12.5 provides a simple interface for obtaining solutions at short time intervals and at longer time intervals. Listing 12.6 shows a code segment for using this function for solving the diffusion equation for both short times and long times. The functions and boundary condition definitions are on lines 11 through 19 in the same format as in Listing 12.4. The logarithmically spatial steps are set up on line 23 and the call to the logarithmical time solver is made on line 26 with an initial time step of 1×10^{-6} and time extending to 1×10^{-2} or over 4 orders of magnitude in time with 20 time steps/decade in time. The solutions are saved only for each magnitude in time, i.e. at 1×10^{-6}, 1×10^{-5}, 1×10^{-4}, 1×10^{-3} and 1×10^{-2}. Longer times are not used as the finite boundary begins to influence the solutions for longer times as noted in connection with Figure 12.2. The resulting solutions for this calculation are not shown as they are essentially identical to those shown in Figure 12.2. The interest here is in the error in the solutions as the number of spatial steps and time steps are varied. By uncommenting lines 22 and 27 the reader can calculate corresponding solutions for half the number of spatial intervals (200 vs. 400) and half the time steps/decade (10 vs. 20). Finally the code on lines 29 through 44 calculate theoretical values of the solutions for times of 1.e-6 and 1.e-2 and prints out the maximum error between the numerical solutions and the exact solutions.

```
 1 : -- /* File list12_6.lua */
 2 : -- Programs to integrate diff equation in one spatial variable
 3 :
 4 : require"pdeivbv"; require'odebvfd'
 5 : require"elemfunc"; erfc = elemfunc.erfc
 6 : getfenv(pdeivbv).nprint = 1; getfenv(pdeivbvqs).nprint=1
 7 : getfenv(ode2bvfd).NMAX = 1 -- Only one iteration required
 8 : -- Model equations to be solved
 9 : L,D = 1,1; u = {}
10 :
11 : eq = function(fu,x,t,u,up,upp,ut) -- Equations
12 :    fu[1] = ut[1] - D*upp[1] -- Diffusion equation
13 : end
14 : efl = function(fu,u,up) -- Left boundary fixed at 1.0
15 :    fu[1] = u[1] - 1.0
16 : end
17 : efr = function(fu,u,up) -- Right boundary, fixed at 0.0
18 :    fu[1] = u[1] - 0.0
19 : end
20 :
21 : Nx = 400
22 : --Nx = 200 -- Half spatial steps
23 : x = xlg(0,1,1.e-6,Nx) -- Set up x and u arrays
24 : for i=1,Nx+1 do u[i] = 0.0 end
25 :
26 : sol = pdeivbvqs({eq,efl,efr},{0,{1.e-6,1.e-2},1,20},x,u)
27 : --sol = pdeivbvqs({eq,efl,efr},{0,{1.e-6,1.e-2},1,10},x,u)
28 : u1,u2 = {},{} -- Theoretical values
29 : t = 1.e-6
30 : for i=1,Nx+1 do u1[i] = erfc(x[i]/(2*math.sqrt(t))) end
31 : t = 1.e-2
32 : for i=1,Nx+1 do u2[i] = erfc(x[i]/(2*math.sqrt(t))) end
33 : emax = 0.0
34 : for i=1,Nx+1 do
35 :    ex = math.abs(sol[3][i] - u1[i])
36 :     if ex>emax then emax = ex end
37 : end
38 : print('Max error at t = 1.e-6 is ',emax)
39 : emax = 0.0
40 : for i=1,Nx+1 do
41 :    ex = math.abs(sol[7][i] - u2[i])
42 :     if ex>emax then emax = ex end
43 : end
44 : print('Max error at t = 1.e-2 is ',emax)
45 : plot(sol[1],sol[2],sol[3],sol[#sol])
46 : write_data('list12_6.dat',sol,u1,u2)
Selected output:
Max error at t = 1.e-6 is     0.00012301494673156
Max error at t = 1.e-2 is     0.00017154152885468
```

Listing 12.6. Code segment for solving diffusion equation with logarithmically spaced time and spatial steps.

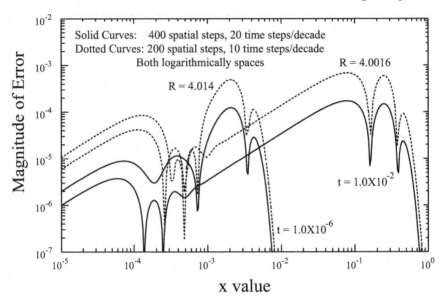

Figure 12.5. Solution errors using logarithmically spaced spatial and time points for two time values. The R values are the ratios of the peak of the dotted curves to the peak of the solid curves for both time solutions.

The magnitudes of the errors in the numerical solutions are shown in Figure 12.5 for the shortest and longest time solutions. The solid curves are for 400 spatial steps and 20 time steps/decade while the dotted curves are for one-half of each of these values. Since both the spatial steps and time steps are reduced for the dotted curves, the computation time is about one-quarter of that of the solid curves. There are several features that can be seen in the curves of Figure 12.5. First, it can be seen that the logarithmically spaced time and spatial points provide good non-uniform parameters for both short and long time solutions since the maximum errors are approximately the same for both types of solution. The many dips in the curves in the figure indicate spatial points where the sign of the error is reversing value from positive to negative values. Second, it is seen that the errors are reduced as the number of spatial and time points is doubled. The R values in the figure give the ratio of the maximum solution errors in the short time and long time solutions for the two sets of spatial and time points. From the discussion in Chapters 10 and 11 on the accuracy of the differencing techniques, one would expect that a doubling of the number of time and spatial points would reduce the error by approximately a factor of 4 since the differencing techniques are expected to be accurate to second order in both variables. This is essentially the case as seen in the figure where the ratio of the peak errors is approximately 4 for each set of data (actually 4.014 and 4.0016). This provides some confidence in the numerical implementation techniques for both the time and spatial differencing techniques.

The accuracy for the solid curves in the figure is just slightly larger than 1×10^{-4} as seen from the figure and the printed values in Listing 12.6. To improve the accuracy, the calculations would have to be made with smaller spatial and/or time steps. Since both the number of spatial and time points were changed for the data in the figure, one does not know if it is best to change the number of spatial steps or time steps or if both need to be changed. To determine that fact, one would need to re-execute the code in Listing 12.5 changing only the number of spatial steps and only the number of time steps and observe the changes in solution accuracy. The reader is encouraged to perform these additional calculations. The result should be that "for this case", one needs to increase both the number of spatial points and the number of time points as both of these make significant contributions to the solution error. For other problem and other choices of parameters, one of the parameters may dominate the error as previously discussed in connection with the use of uniformly spaced grid points in connection with Figures 12.3 and 12.4. To increase the accuracy by a factor of 10 would require an increase in the number of time and spatial steps by about a factor of 3.2 each and a corresponding increase in computer execution time by about a factor of 10.

Since the solution error is seen to vary approximately as the square of the number of steps, the question arises as to whether the h-2h technique can be used to obtain an estimate of the error in the numerical solutions. This technique discussed in depth in Chapters 10 and 11 compares solutions with some standard step size and with twice the step size and estimates the solution error for the most accurate of the two solutions (step size h) by taking a linear factor times the difference in the h and 2h solutions. Listing 12.7 shows a code segment to perform such a calculation for the present linear diffusion equation. The code on lines 21 through 24 is similar to Listing 12.6 and generates a solution for 400 spatial intervals and 20 time intervals/decade. The code on lines 26 through 29 generates a solution for 200 spatial intervals and 10 time intervals/decade. Finally lines 31 and 32 calls the odeerror() function discussed in Chapter 10 with the two solutions at $t = 1 \times 10^{-6}$ and 1×10^{-2} to obtain error estimates based upon the h-2h algorithm. Figure 12.6 shows the important results of this comparison. The solid curves in the figure are the actual errors in the numerical solutions obtained from the numerical solution and the theoretical erfc() function while the dotted curves are the estimated errors as calculated by the h-2h technique of Listing 12.7. The curves show that the maximum values of the errors are approximated very accurately by the h-2h algorithm. Only at very small spatial values ($x < .001$) do the actual errors differ significantly from the h-2h estimated errors.

The numerical error in the PDE initial value, boundary value solution by finite differences has been extensively explored for one PDE which is first order in the time derivative. In order to have an exactly known solution with which to compare the numerical techniques, a simple PDE boundary value problem must be considered. While it is sometimes dangerous to extrapolate too far from a known result, the results obtained here can provide some guidance with regard to the expected accuracy of solving PDEs by using the code segments developed in the previous section. First, it is expected that the algorithm is accurate to second order in both the spatial and time step variables. This provides some confidence in the ability to use the h-2h algorithm for more complex problems where an exact solution is not known or for nonlinear problems where an exact solution is not possible.

```
 1 : -- /* File list12_7.lua */
 2 : -- Programs to integrate diff equation in one spatial variable
 4 : require"pdeivbv"; require'odebvfd'
 5 : require"elemfunc"; erfc = elemfunc.erfc
 6 : getfenv(pdeivbv).nprint = 1; getfenv(pdeivbvqs).nprint=1
 7 : getfenv(ode2bvfd).NMAX = 1 -- Only one iteration required
 8 : -- Model equations to be solved
 9 : L,D = 1,1; u = {}
10 :
11 : eq = function(fu,x,t,u,up,upp,ut) -- Equations
12 :    fu[1] = ut[1] - D*upp[1] -- Diffusion equation
13 : end
14 : efl = function(fu,u,up) -- Left boundary fixed at 1.0
15 :    fu[1] = u[1] - 1.0
16 : end
17 : efr = function(fu,u,up) -- Right boundary, fixed at 0.0
18 :    fu[1] = u[1] - 0.0
19 : end
20 :
21 : Nx = 400; Nt = 20
22 : x1 = xlg(0,1,1.e-6,Nx) -- Set up x and u arrays
23 : for i=1,Nx+1 do u[i] = 0.0 end
24 : sol1 = pdeivbvqs({eq,efl,efr},{0,{1.e-6,1.e-2},1,Nt},x1,u)
25 :
26 : Nx = 200; Nt = 10 -- Half spatial steps, half time steps
27 : x2 = xlg(0,1,1.e-6,Nx) -- Set up x and u arrays
28 : for i=1,Nx+1 do u[i] = 0.0 end
29 : sol2 = pdeivbvqs({eq,efl,efr},{0,{1.e-6,1.e-2},1,Nt},x2,u)
30 :
31 : err1 = odeerror({x1,sol1[3]},{x2,sol2[3]}) -- t = 1.e-6
32 : err2 = odeerror({x1,sol1[7]},{x2,sol2[7]}) -- t = 1.e-2
33 : write_data('list12_7.dat',err1,err2)
```

Listing 12.7. Code segments to estimate PDE solution error using h-2h technique.

One technique that is always available when solving numerical PDEs is to vary the spatial and time grids and compare the computed solutions. If significant differences are observed in numerical solutions with different spatial and time grids, then little confidence can be placed in the solutions obtained. The technique of varying grid sizes and numerically experimenting with the obtained solutions is perhaps the only resort one has for nonlinear equations. One can only feel confident in a solution when different spatial and time grid sizes give essentially the same solution (at least to some accuracy level). However, one can always go to extreme values where numerical accuracy is lost due to extremely small grid sizes. Further the use of non-uniform spatial and time grids can be very useful for many problems as exhibited in this diffusion problem.

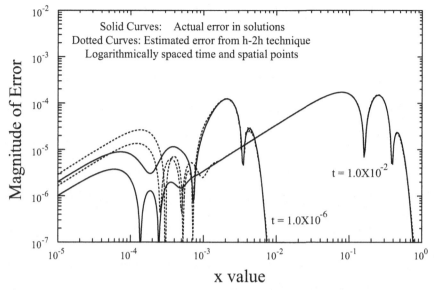

Figure 12.6. Comparison of exact error for diffusion equation with error estimate based upon the h-2h algorithm. Curves are very close near the peak errors for each solution.

12.6 Some Examples of Partial Differential Equations of the Initial Value, Boundary Value Type

This section explores some examples of IV, BV PDEs using the computer code developed in the previous sections.

12.6.1 The Nonlinear Diffusion Equation

A step up in complexity from the linear diffusion equation is a nonlinear version of the diffusion equation where the diffusion coefficient depends on the concentration of diffusing species. The equation to be considered is then

$$\frac{\partial U}{\partial t} - \frac{\partial}{\partial x}\left(D(U)\frac{\partial U}{\partial x}\right) = 0 \tag{12.29}$$

For impurity diffusion in semiconductors a frequently used model for the diffusion coefficient is:

$$D(U) = D_0 + D_1(U/n_i) + D_2(U/n_i)^2 \tag{12.30}$$

where n_i is the intrinsic carrier density in the semiconductor. All of these terms in turn are functions of the temperature at which the diffusion process occurs and are typically described by an Arrhenius form of equation. For this form of diffusion coefficient, the diffusion equation can be expressed as:

$$\frac{\partial U}{\partial t} - \left[D_0 + D_1\left(\frac{U}{n_i}\right) + D_2\left(\frac{U}{n_i}\right)^2 \right]\frac{\partial^2 U}{\partial x^2} - \left[\frac{D_1}{n_i} + 2\frac{D_2}{n_i}\left(\frac{U}{n_i}\right) \right]\left(\frac{\partial U}{\partial x}\right)^2 = 0 \quad (12.31)$$

```
 1 : -- /* File list12_8.lua */
 2 : -- Programs to integrate nonlinear diffusion equation
 4 : require"pdeivbv"; require'odebvfd'
 5 : getfenv(pdeivbv).nprint = 1; getfenv(pdeivbvqs).nprint=1
 6 : --getfenv(ode2bvfd).nprint=1 -- See detailed convergence
 7 :
 8 : -- Model equations to be solved
 9 : D00,D10,E0,E1 = 0.05, 0.95, 3.5, 3.5 -- Diff coeff parameters
10 : T = 1000+273 -- temperature
11 : D0sav = D00*math.exp(-E0/(0.026*T/300))
12 : D1sav = D10*math.exp(-E1/(0.026*T/300))
13 : ni = 7.14e18; L = 1.e-4 ; ul = 5e20; ur = 0; u = {}
14 :
15 : eq = function(fu,x,t,u,up,upp,ut) -- Equations
16 :     D = D0 + D1*u[1]/ni
17 :     fu[1] = ut[1] - D*upp[1] - D1*up[1]^2/ni
18 : end
19 : efl = function(fu,u,up) -- Left boundary
20 :     fu[1] = u[1] - ul
21 : end
22 : efr = function(fu,u,up) -- Right boundary
23 :     fu[1] = u[1] - ur
24 : end
25 :
26 : Nx = 200; Nt = 20 -- spatial steps, time steps
27 : x = xlg(0,L,L*1.e-4,Nx) -- Set up x and u arrays
28 : for i=1,Nx+1 do u[i] = 0.0 end; u[1] = ul
29 :
30 : D1=0; D0 = D0sav+D1sav*(ul/ni) -- Get initial trial solution
31 : pdeivbv({eq,efl,efr},{0,.001,1},x,u) -- Constant diffusion coeff
32 : D1=D1sav; D0 = D0sav -- Now real diffusion coeff
33 : sol = pdeivbvqs({eq,efl,efr},{0,{.01,1.e3},1,Nt},x,u)
34 : plot(sol[1],sol[#sol])
35 : write_data('list12_8.dat',sol)
```

Listing 12.8. Code segment for solving nonlinear diffusion PDE with parameters appropriate for Boron diffusion into Silicon.

Listing 12.8 shows code for solving this diffusion problem with a specific set of parameters on lines 9 through 13 that are applicable to Boron diffusion into Silicon. For this example the D_2 coefficient is taken to be zero and only a linear dependence of diffusion coefficient on concentration is included. The temperature for evaluating the parameters is taken as 1000 °C as seen on line 10 of the code. The sample thickness is taken as 1.e-4 cm and the impurity concentration at the origin is assumed to be held constant at 5×10^{20} which is a factor of about 70 above the intrinsic carrier density (n_i) at the given temperature. The functions defining the differential equation and boundary values on lines 15 through 24 should be very familiar to the reader by now. The main call to the PDE solver is on line 33 and uses the quick scan function pdeivbvqs() to cover a wide range in diffusion times from 0.01 to 1000 sec. Also, a logarithmically spaced spatial grid is again used as defined on line 27. The only surprising feature of the code is perhaps the initial call to the pdeivbv() solver on line 31 with a single time interval of 0 to 0.001 and line 30 which defines a constant diffusion coefficient before the call. This provides an initial smooth solution obtained with a linear equation in order to provide an initial guess at the solution of the nonlinear equation. Without this initial approximation, the code has convergence problems with the solution of the nonlinear equations. This is typical of many nonlinear problems where obtaining an initial starting solution is critical to obtaining a properly converged solution. In execution of the code, convergence problems will be manifest by observing that the printed number of iterations equals 50, the maximum default value.

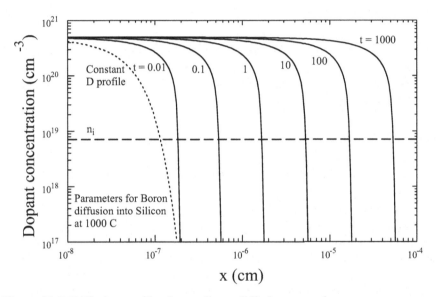

Figure 12.7. Diffusion profiles for nonlinear diffusion example.

Solutions obtained from Listing 12.8 are shown in Figure 12.7 covering the range of diffusion times from 0.01 to 1000 sec. There are several features of the solution worthy of note. First, the impurity profiles have essentially the same shape on the log-log scale and are characteristic of a diffusion front with a constant shape that moves through the spatial dimension advancing an order of magnitude in distance for every two orders of magnitude in time. This is similar to the constant diffusion coefficient case previously shown in Figure 12.2. The parallel curves in the figure suggest that the concentration profile is a function of $x/\sqrt{t}$ as is the known case for a constant diffusion coefficient. In semiconductor terms, diffusion into silicon with a constant background doping density would produce a p-n junction with a depth proportional to the square root of diffusion time. A second feature is the much more abrupt concentration profile as evidenced by comparing the solid curves with the dotted curve which is the profile for diffusion with a constant diffusion coefficient. Of course the exact details of the profile and depth will depend on the concentration relative to the intrinsic density (the n_i value) and as the concentration becomes much less than the intrinsic density the diffusion process would approach that of the constant diffusion coefficient case. The reader can vary the parameters and re-execute the code to observe changes in the diffusion profile.

One of the objectives of numerical solutions of partial differential equations is to explore the nature of solutions to physical problems. The observation above that the shape of the time dependent solution is independent of time leads to the conclusion that the solution to Eq. (12.31) should be of the form:

$$U(x,t) = U(0)f(x/2\sqrt{D_0 t}) = U(0)f(\eta)$$

$$\text{with } \eta = x/2\sqrt{D_0 t}$$

(12.32)

In this η is a dimensionless variable of distance and time. The inclusion of D_0 in the definition makes the variable dimensionless and the factor of 2 is included so that the solution for $D_1 = D_2 = 0$ will be the erfc(η) function. If a solution of this form exists then the partial differential equation can be converted into an ordinary differential equation in the variable η as:

$$(1+d_1 f + d_2 f^2)\frac{d^2 f}{d\eta^2} + \left[(d_1 + 2d_2 f)\frac{df}{d\eta} + 2\eta\right]\frac{df}{d\eta} = 0$$

(12.33)

where $d_1 = (D_1/D_0)(U(0)/n_i)$ and $d_2 = (D_2/D_0)(U(0)/n_i)^2$

By assuming the functional form of Eq. (12.32), the individual time and distance variables are eliminated from the equation and a single nonlinear differential equation is obtained in terms of the dimensionless variable η. Solutions of this equation with the boundary conditions $f(0) = 1$ and $f(\infty) = 0$ should provide a universal solution curve for a particular set of diffusion parameters expressed as d_1 and d_2. This one-dimensional boundary value problem can be solved by the techniques discussed in Chapter 10. A code segment for such a solution using the ode1fd() function is shown in Listing 12.9. The specific parameters used in the listing are the same as those used in Listing 12.8 which are for Boron diffusion into Silicon at 1000C. Lines 6 through 18 define parameters for the various solutions with optional parameters given for normalized distance and surface doping density in lines 11 through 18. Looking at the solutions in Figure 12.7, it can be seen that there is a very steep drop in diffusant concentration near the end of the diffusion profile when the surface concentration exceeds the intrinsic concentration.

This steep drop in concentration presents some difficulties in the numerical solutions with the code of Listing 12.9. For example attempting to obtain a solution much beyond the end of the abrupt drop in concentration can cause numerical instabilities unless small step sizes are taken near the end or range. However, the exact end of range is not known before solving the nonlinear equation. The approach used here has been to limit the solution range as expressed by L to some value just beyond the steep drop in solution value. The L values listed in the code have been obtained by some trial and error in running the code and discovering a value just beyond the steep drop in solution value. For the surface concentration of 5e20 on line 12 of the code, it can be estimated where the steep drop occurs using the data in Figure 12.7. It can be seen that for the 1000 sec curve the steep drop occurs at a depth of about 5.5×10^{-5} cm. This corresponds to the dimensionless value η of about 30 which is shown on line 12 of Listing 12.9. The universal differential equation is defined on lines 22 through 25 and the solution to the equation is obtained by a call to ode1fd() on line 30. Boundary functions are not used as the boundary conditions are fixed and the ode1fd() has a built-in set of fixed constant boundary conditions when boundary functions are not defined.

```
 1 : -- /* File list12_9.lua */
 2 : -- Programs to integrate nonlinear diffusion equation
 3 :
 4 : require'odebvfd'
 5 : -- Model equations to be solved
 6 : D00,D10,E0,E1 = 0.05, 0.95, 3.5, 3.5 -- Diff coeff parameters
 7 : T = 1000+273 -- temperature
 8 : D0 = D00*math.exp(-E0/(0.026*T/300))
 9 : D1 = D10*math.exp(-E1/(0.026*T/300))
10 : ni = 7.14e18; ur = 0.0; x,u = {},{}
11 : L = 43; ul = 1e21; ex = 'a' -- Use any of these
12 : --L = 30; ul = 5e20; ex = 'b'
13 : --L = 14; ul = 1e20; ex = 'c'
14 : --L = 10; ul = 5e19; ex = 'd'
15 : --L = 6; ul = 1e19; ex = 'e'
16 : --L = 4; ul = 5e18; ex = 'f'
17 : --L = 4; ul = 1e18; ex = 'g'
18 : --L = 4; ul = 1e17; ex = 'h'
19 : Dr = D1*ul/(D0*ni)
20 : print(1/(2*math.sqrt(D0*1e3)))
21 :
22 : eq = function(x,u,up,upp) -- Equations
23 :    D = 1 + Dr*u
24 :    return D*upp + (Dr*up + 2*x)*up
25 : end
26 :
27 : Nx = 10000 -- spatial steps, time steps
28 : -- Set up x and u arrays
29 : for i=1,Nx+1 do x[i] = (i-1)*L/Nx end
30 : u,nn,err1,err2 = ode1fd(eq,x,{1.0,0})
31 :
32 : plot(u); write_data('list12_9'..ex..'.dat',u)
```

Listing 12.9. Code for solving for universal diffusion curves for transient diffusion into a semiconductor. Parameters are for Boron diffusion into Silicon.

The set of universal solutions obtained from execution the code in Listing 12.9 are shown in Figure 12.8 for surface doping densities ranging from $1 \times 10^{17}/cm^3$ to $1 \times 10^{21}/cm^3$. At the lowest surface doping density the solution is close to the erfc() function which is the known solution for the linear diffusion equation. The solutions rapidly become much more abrupt as the surface density exceeds the intrinsic density and an approximately constant profile can be seen for surface densities above about $5X10^{18}/cm^3$. Although the shape remains approximately the same, the normalized depth of the diffusion profile continues to increase with increasing surface densities or increasing nonlinearity of the diffusion equation. For comparison, data points are shown corresponding to the transient solutions shown in Figure 12.7 illustrating that the universal diffusion curves do in fact correspond to the solution of the partial differential equation. A fairly large number of spatial points (1000) are used in the solutions as seen on line 27. This is an attempt to obtain sufficient spatial resolution in the rapidly decreasing solution near the end of the diffusion range. Even with this large number of steps, the numerical solution becomes difficult to obtain for the largest surface concentrations near the end of the range. One probably needs to use a non-uniform spatial step size with small step sizes near the end of the diffusion range. However, this is not easily implemented since one does not know the exact distance where the small step sizes are needed until a solution has been obtained.

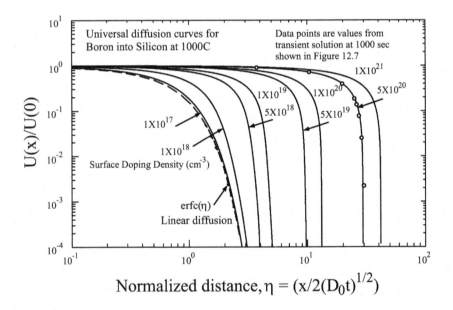

Figure 12.8. Universal diffusion curves for nonlinear diffusion. Parameters are for Boron diffusion into Silicon at 1000C. Data points compare the universal solutions with the transient solution at 1000 sec shown in Figure 12.7.

One could use an iterative approach to obtaining a more suitable step distribution, but such an approach will be left to the interested reader since the solutions obtained appear to agree very well with the full transient solution – at least for one surface doping density.

This look at the universal diffusion curves has been a little diversion from the topic of solving partial differential equations in two variables. However, as frequently occurs in real engineering problems, the pursuit of one problem frequently leads to consideration of another problem which in many cases can be a simpler problem or a more general solution to the original problem. In this case the ability to generate a set of universal diffusion curves for the nonlinear semiconductor diffusion problem provides a much more useful set of solutions than the originally pursued solutions which had to be expressed in terms of two variables instead of the one dimensionless variable combining both time and spatial dimension. Also numerical solutions of the universal curves in one dimension can be obtained much faster than the two dimensional solutions. Although the solutions for Listing 12.9 have taken a $D_2 = 0$ value, the example can obviously be expanded to include a nonzero value of this parameter as well. Such a second order term for example is needed for Phosphorus diffusion into Silicon. More examples of nonlinear diffusion are left to the interesed reader.

12.6.2 An Adiabatic Tubular Chemical Reactor Example

An example from the field of Chemical Engineering will now be given. This example is from the book, *Quasilinearization and Invariant Imbedding*, by E. S Lee (Academic Press, 1968). A set of equations is used to model the concentration of a particular reactant, in an adiabatic tubular chemical reactor with axial mixing:

$$\frac{1}{M}\frac{\partial^2 c}{\partial x^2} - \frac{\partial c}{\partial x} - \beta c^2 \exp(-E/RT) - \frac{\partial c}{\partial t} = 0$$

$$\frac{1}{M}\frac{\partial^2 T}{\partial x^2} - \frac{\partial T}{\partial x} + Q\beta c^2 \exp(-E/RT) - \frac{\partial T}{\partial t} = 0$$

$$0 < x < 48 = x_f$$

With Initial Conditions:

$$c(x) = 0.0; \quad T(x) = 1270 \tag{12.34}$$

and Boundary Conditions:

$$c(0) - \frac{1}{M}\frac{dc(0)}{dx} = c_e; \quad \mathrm{T}(0) - \frac{1}{M}\frac{dT(0)}{dx} = T_e$$

$$\frac{dc(x_f)}{dx} = 0; \quad \frac{dT(x_f)}{dx} = 0$$

These equations given in Eq. (12.34) can be compared to Eq. (11.65) which was previously given for the time independent case. The difference here is the presence of the time derivatives and the initial conditions on concentration and temperature. The reader is referred to Section 11.6 for a further discussion of this problem in the steady state. Computer code for solving this IV-BV problem is shown in Listing 12.10. The equation set and boundary conditions are defined on lines 10 through 21 of the listing in a manner very familiar by now to the reader. After setting up initial conditions and defining time parameters on lines 23 through 25, the call to the function pdeivbv() on line 26 returns an array of the solution values which are concentration and temperature along the spatial distance of the chemical reactor.

```
 1 : -- /* list12.10.lua */
 2 : -- Solution of nonlinear Chemical Engineering IV-BV problem
 3 : require"pdeivbv"
 4 : getfenv(pdeivbv).nprint = 1
 5 :
 6 : M,EdR,b,Q,ce,Te = 2, 22000, .5e8, 1000, 0.07, 1250
 7 : Ti = 1270 -- Initial temperature
 8 : L = 48; Nx = 200; Tm = 60; Nt = 6 -- spatial, time parameters
 9 :
10 : f = function(eqs,x,t,u,up,upp,ut) -- Differntial equation
11 :     eqs[1] = upp[1]/M-up[1]-b*u[1]^2*math.exp(-EdR/u[2])-ut[1]
12 :     eqs[2] = upp[2]/M-up[2]+Q*b*u[1]^2*math.exp(-EdR/u[2])-ut[2]
13 : end
14 : fl = function(eqs,u,up) -- Left boundary conditions
15 :     eqs[1] = u[1] - up[1]/M - ce
16 :     eqs[2] = u[2] -up[2]/M - Te
17 : end
18 : fr = function(eqs,u,up)
19 :     eqs[1] = up[1] -- Zero slopes
20 :     eqs[2] = up[2]
21 : end
22 :
23 : tvals = {0,Tm,Nt}
24 : x,u = {},{{},{}} -- Set initial values
25 : for i=1,Nx+1 do x[i],u[1][i],u[2][i] = L*(i-1)/Nx,0,1270 end
26 : s = pdeivbv({f,fl,fr},tvals,x,u) -- Solve equations
27 : plot(s[1],s[#s-1]); plot(s[1],s[#s])
28 : write_data('list12.10.dat',s)
Selected output:
Time = 1.000000e+001 Number of iterations in pdeivbv = 12
Time = 2.000000e+001 Number of iterations in pdeivbv = 10
Time = 3.000000e+001 Number of iterations in pdeivbv = 9
```

Listing 12.10. Code segment for solving for the transient response of a nonlinear tubular chemical reactor.

In order to monitor the solution, the nprint variable of the pdeivbv() routine is set to nonzero on line 4. This results in printed output such as the selected output which prints the maximum number of Newton iterations required to solve the set of equations for each time interval. For example for the 0 to 10 time interval, the maximum number of Newton iterations is printed as 12. The default number of time steps between the saved values is 10, so transient solutions were obtained for time values of 0, 1, 2, .. 10 and the largest number of Newton iterations taken for any of these solutions was 12. The output does not identify for which time interval the 12 iterations occurred. However, one would expect that it was probably the initial time interval since this represents the largest difference in the solution from the initial values. The output indicates that the number of Newton iterations decreases as the solution progresses in time and this is what one would expect if a good solution is being obtained as the solutions should approach time independent functions at large times. It is important to always print out the number of Newton iterations when starting a new problem as this is the only way a properly converged solution can be insured.

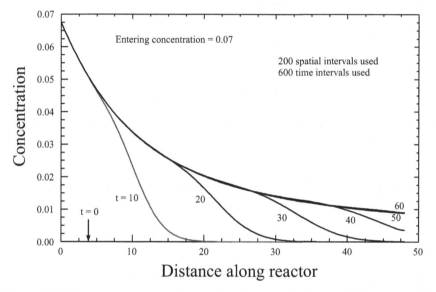

Figure 12.9. Transient concentration profiles for the tubular chemical reactor with parameters of Listing 12.10.

Graphs of the solution sets are shown in Figures 12.9 and 12.10. For the calculations, 200 spatial intervals and 600 time intervals have been used. The reader is encouraged to change the number of spatial or time intervals and re-execute the code and compare the results with that obtained for the present parameters. There should be little difference in the solutions for reasonable numbers of spatial and time points. The reader can also estimate the solution error by halving both the number of spatial and time intervals and using the h-2h algorithm as illustrated in Listing 12.7. The result of such a calculation is shown in Figure 12.11 for a time of t = 20 which is about midway through the transient solution time (code for this calculation is not shown). The solid curves show the calculated temperature and concentration while the dotted curves show the estimated error using the spatial and time intervals shown in the figure. The estimated accuracy of the temperature is seen to be about 4 decimal digits over the entire spatial range. The estimated

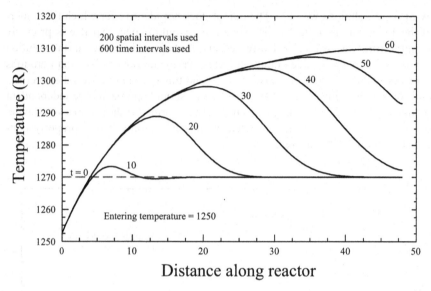

Figure 12.10. Transient temperature profiles for the tubular chemical reactor with parameters of Listing 12.10.

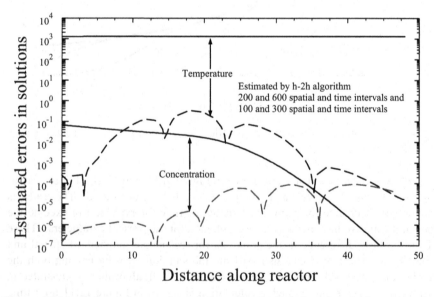

Figure 12.11. Estimated error in solutions for the tubular chemical reactor obtained by the h-2h algorithm at t = 20. Solid lines are solutions. Dotted curves are estimated errors.

accuracy for the concentration for x < 20 is also about 4 decimal digits. However, as the concentration value decreases, the accuracy degrades and at a distance of about 35 where the concentration is about 10^{-4} the estimated accuracy is about equal to the solution value. On the scale of the plots in Figures 12.9 and 12.10, the estimated errors would be too small to visually see. With these calculations it is expected that a very good indication of the transient solution of the tubular chemical reactor has been obtained for the parameters used in the numerical calculations. The reader can also compare the solutions with the steady state solution obtained in Section 11.6. The solutions for the t = 60 case are close to the steady state solutions as seen in Figure 11.29.

12.6.3 A Semiconductor PN Junction

Another example to be considered is the transient response of a p-n junction to a step change in applied voltage. Steady state solutions have previously been discussed in Section 11.6 and the reader is encouraged to review the discussion there and particular in connection with Eqs. (11.66) through (11.70). The one-dimensional geometry for a p-n junction is shown in Figure 12.12. Also shown at the terminal to the p-region is a step in applied voltage occurring at time t = 0. For a forward applied voltage that causes large current flow, the p-region is made positive with respect to the n-region.

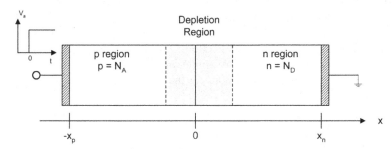

Figure 12.12. Geometry for transient response of p-n junction to step in applied voltage.

The set of partial differential equations are:

$$\frac{\partial^2 V}{\partial x^2} = \frac{q}{\varepsilon}(p - n + N_{Net}), \text{ with } N_{Net} = \begin{cases} N_D \text{ for } x > 0 \\ -N_A \text{ for } x < 0 \end{cases}$$

$$J_n = q\mu_n nE + qD_n \frac{\partial n}{\partial x}; \quad \frac{1}{q}\frac{\partial J_n}{\partial x} - U = \frac{\partial n}{\partial t} \tag{12.35}$$

$$J_p = q\mu_p pE - qD_p \frac{\partial p}{\partial x}; \quad -\frac{1}{q}\frac{\partial J_p}{\partial x} - U = \frac{\partial p}{\partial t}$$

$$\text{with } U = \frac{(pn - n_i^2)}{\tau_{no}(p + p_1) + \tau_{po}(n + n_1)}$$

These can be reduced to a set of three coupled differential equations which are first order in the time derivative and second order in the spatial derivative. As in Chapter 11 it is convenient to introduce the Quasi-Fermi potentials through the relationships:

$$n = (n_i^2 / N_A)\exp((V - \phi_n)/V_T)$$
$$p = N_A \exp((\phi_p - V)/V_T) \tag{12.36}$$

The three equations then become in terms of the electric potential and the two Quasi-Fermi potentials:

$$\frac{\partial^2 V}{\partial x^2} - \frac{q}{\varepsilon}(p - n + N_{Net}) = 0, \text{ with } N_{Net} = \begin{cases} N_D & \text{for } x > 0 \\ -N_A & \text{for } x < 0 \end{cases}$$

$$\frac{\partial V}{\partial t} - \frac{\partial \phi_p}{\partial t} + \mu_p \left[V_T \frac{\partial^2 \phi_p}{\partial x^2} + \frac{\partial \phi_p}{\partial x}\left(\frac{\partial V}{\partial x} - \frac{\partial \phi_p}{\partial x}\right) \right] - V_T \frac{U}{p} = 0 \tag{12.37}$$

$$\frac{\partial V}{\partial t} - \frac{\partial \phi_n}{\partial t} + \mu_n \left[V_T \frac{\partial^2 \phi_n}{\partial x^2} + \frac{\partial \phi_n}{\partial x}\left(\frac{\partial V}{\partial x} - \frac{\partial \phi_n}{\partial x}\right) \right] - V_T \frac{U}{n} = 0$$

To complete the mathematical formulation a set of initial conditions and boundary conditions is needed. For a p-n junction that is initially at equilibrium with no applied voltage, the two Quasi-Fermi levels are zero and the electric potential is a solution of Poisson's equation. Such a DC solution has previously been considered and obtained in Section 11.6. The reader is encouraged to review the material of that section before continuing with the present discussion. An appropriate set of boundary conditions for the left and right boundaries as developed in that section are:

Left boundary at $x = -x_p$

$$V(-x_p) = \phi_p(-x_p) = V_a = \text{Applied voltage}$$

$$\left(1 + \frac{\mu_n}{v_{sat}}\frac{\partial n(-x_p)}{\partial x}\right) - \exp((\phi_n(-x_p) - V(-x_p))/V_T) = 0$$

Right boundary at $x = x_n$ $\qquad\qquad\qquad\qquad$ (12.38)

$$V(x_n) = V_J = \text{Built-in junction voltage}$$

$$\left(1 + \frac{\mu_p}{v_{sat}}\frac{\partial p(x_n)}{\partial x}\right) - \exp((V(x_n) - \phi_p(x_n))/V_T) = 0$$

$$\phi_n(x_n) = 0$$

These boundary conditions are the same as those discussed in Chapter 11 for the steady state solution of a p-n junction problem. For reference v_{sat} is the maximum saturation velocity of the respective carriers (either electrons or holes).

```
 1 : -- /* File list12_11.lua */
 2 : -- Program for transient solution of p-n junction
 3 :
 4 : require"odefd"; require"pdeivbv"
 5 : getfenv(pdeivbv).nprint = 1
 6 : getfenv(ode2bvfd).umin = {5.e-4,5.e-4,5.e-4}
 7 : --getfenv(ode2bvfd).nprint = 1; getfenv(pdebivbv).nprint = 1
 8 :
 9 : -- Model equations to be solved
10 : L = 8.e-4; q = 1.6e-19; eps = 11.9*8.854e-14
11 : Na = 1e16; Nd = 1.e19 -- P and N type doping
12 : vth = .026; ni = 1.45e10
13 : tno = 1.e-8; tpo = 2.e-8
14 : unp = 1020 -- Minority carrier mobility in p region
15 : unn = 100 -- Majority carrier mobility in n region
16 : upn = 50 -- Minirity carrier mobility in n region
17 : upp = 400 -- Majority carrier mobility in p region
18 : vsat = 1.e7 -- Saturated velocity
19 : qdep = q/eps
20 : vj = vth*math.log(Na*Nd/ni^2)
21 : va = 0
22 :
23 : eq = function(fv,x,t,v,vp,vpp,vt) -- Equations with derivatives
24 :    if x<=0 then un,up,Nnet = unp,upp,-Na
25 :    else un,up,Nnet = unn,upn,Nd end
26 :    p = Na*math.exp((v[2]-v[1])/vth)
27 :    n = (ni^2/Na)*math.exp((v[1]-v[3])/vth)
28 :    U = (p*n-ni^2)/(tpo*(n+ni)+tno*(p+ni))
29 :    fv[1] = vpp[1] + qdep*(p - n + Nnet) -- V term
30 :    fv[2] = vt[1] - vt[2] + up*(vth*vpp[2] + vp[2]*(vp[2]-vp[1]))
31 :             - vth*U/p-- Up term for holes
31 :    fv[3] = vt[1] - vt[3] + un*(vth*vpp[3] + vp[3]*(vp[1]-vp[3]))
31 :             + vth*U/n-- Un term for electrons
32 : end
33 :
34 : efl = function(fv,v,vp) -- Using mixed boundary condition
35 :    fv[1] = v[1] - va
36 :    fv[2] = v[2] - va
37 :    fv[3] = (1+unp/vsat*(vp[3]))-math.exp((v[3]-v[1])/vth)
38 : end
39 : efr = function(fv,v,vp) -- Using mixed boundary condition
40 :    fv[1] = v[1] - vj
41 :    fv[2] = (1+upn/vsat*(vp[2]))-math.exp((v[1]-vj-v[2])/vth)
42 :    fv[3] = v[3]
43 : end
44 :
45 : ftzero = {0,0,0} -- Force zero time derivatives for Steady State
46 : eqss = function(fv,x,v,vx,vxx) -- zero time derivatives
47 :    eq(fv,x,0,v,vx,vxx,ftzero)
48 : end
49 : x1 = {} -- Set up nonuniform spatial array
50 : fact,step = 1.02,1.e-3; n1 =300; np1 = n1+1
51 : x1[1] = 0
52 : for i=2,np1 do -- Small to large steps
53 :    x1[i] = x1[i-1] + step; step = step*fact
54 : end
55 : j = np1
56 : for i=np1+1,2*n1+1 do -- Reverse array
```

```
57 :     x1[i] = x1[i-1] + x1[j] - x1[j-1]; j = j-1
58 : end
59 : x = {}; j = 1
60 : imax = #x1; xmax = x1[imax]
61 : for i=1,imax do -- Scale to -2L/3 to L/3
62 :     x[j] = (2*L/3)*(x1[j]/xmax - 1); j = j+1
63 : end
64 : for i=2,imax do
65 :     x[j] = (L/3)*x1[i]/xmax; j = j+1
66 : end
67 : nx = #x; nxm1 = nx-1 -- End of array definition
68 :
69 : v = {{},{},{}} -- Set initial voltage values
70 : for i=1,nx do
71 :     for j=1,3 do
72 :         if j==1 then
73 :             if x[i]<0 then v[1][i] = 0
74 :             else v[1][i] = vj end
75 :         else v[j][i] = 0 end
76 :     end
77 : end
78 : va = 0
79 : s= odefd({eqss,efl,efr},x,v) -- DC solution -- initial values
80 :
81 : va = 1 -- Step input voltage of 1 volt
82 : for i=1,nx do -- Add to existing voltages
83 :     duu = va-(x[i]-x[1])/(x[nx]-x[1])*va
84 :     for j=1,3 do v[j][i] = v[j][i] + duu end
85 : end
86 : --tvals = {0,{1.e-15,1.e-6},2,10} -- 20 steps/decade, save 2
87 : tvals = {0,{1.e-15,1.e-6},2,5} -- 10 steps/decade, save 2
88 : s = pdeivbvqs({eq,efl,efr},tvals,x,v)
89 : write_data('list12_11.dat',s)
```

Listing 12.11. Code for the transient response of a p-n junction with a step in applied voltage.

Listing 12.11 shows a code segment for solving for the transient response of such a p-n junction to a step change in applied voltage. In this case the applied voltage is assumed to be 1.0 Volts which is a rather large value of applied voltage as it exceeds the built-in junction potential and results in some interesting high carrier injection physics. The model equations and device parameters are specified on lines 9 through 32 of the code. The reader should have little difficulty in matching the defining equation set of lines 24 through 31 with the set in Eq. (12.37). Similarly the left and right boundary condition set on lines 34 through 43 are in the same form as Eq. (12.38). The code on lines 49 through 67 defines a non-uniform array of spatial points. The details will not be discussed here, but it will only be noted that the resulting array of spatial points has small spatial intervals near the left and right boundaries and small intervals near the doping step located at $x = 0$. If interested, the reader can plot out the resulting spatial distribution to observe the placement of small step sizes. The non-uniform spatial steps are defined such that there is a constant multiplicative factor between any adjacent pair of spatial points. As discussed in the previous chapter in connection with the p-n junction, it is expected that the changes in the solution variables will occur most rapidly near the junction interface and near the device boundaries and these are the spatial regions where small step sizes are placed.

To obtain a good set of initial spatial solutions, one needs to solve Poisson's equation or alternatively to solve the set of device equations with zero time derivatives. This initial solution set is obtained in the listing with lines 69 through 79 of the code with lines 70 through 76 setting an initial guess at the solution variables – a simple abrupt change is used for the electric potential at the junction interface. The call to function odefd() on line 79 returns the corresponding steady state solution for the solution variables. Note that this call uses the function eqss() defined on lines 46 through 48 of the listing. This function in turn simply calls the defining equation set after setting all the time derivatives to zero. In this manner the general equation set can be used for the initial solution so that a separate set of equations without the time derivative terms do not have to be defined.

The solution set obtained from the odefd() function on line 79 will be the solution just before the step in voltage is applied to the junction – this is often called the $t = 0^-$ value in engineering terminology. Immediately after the application of the applied voltage (of 1Volt in the code), the voltage solution will change across the device. In engineering terminology this is often called the $t = 0^+$ value. For the present problem, it is known that the carrier densities can not change abruptly (with infinite derivative) so that the internal charge distribution can not change abruptly as the voltage step is applied. The applied voltage must then give rise abruptly to a constant electric field increase across the device and a resulting linear increment in voltage and Quasi-Fermi potentials across the device. This linear potential increase is added to the steady-state solutions on lines 82 through 85 of the code. Finally the transient solution is obtained with the single line call to the quick scan function pdeivbvqs() on line 88 and the resulting solutions are saved to a file on line 89 of the code.

Before presenting the results, some discussion of the time scale used in the solution is in order. The tvals parameter defined on line 87 specifies that the solution will use a logarithmically spaced time scale beginning at 1×10^{-15} sec and continuing to 1×10^{-6} sec. This is a large range of over 9 orders of magnitude in time. More importantly how does one know what time interval is appropriate for a problem such as this? One approach is to simply explore various time intervals and see how the solution changes. If too small a time scale is selected, little change will occur in the solution variables and the time values can be increased. On the other hand, if too large a time scale is selected, the solution will change too rapidly in the initial time intervals and the time scale can be decreased. For the present problem the various time constants associated with the physical processes associated with the p-n junction have widely differing values and a large range of times is required to observe all the changes. To understand this it is necessary to consider possible time constants associated with a p-n junction problem. While this requires some knowledge of the physics of the problem, there are at least three time constants that might come into play. These are (a) the dielectric relaxation time, (b) the charging time of the junction capacitance and (c) the diffusion time of minority carriers across the semiconductor regions. Each of these will be considered in turn.

The dielectric relaxation time is the time required for charge redistribution in a conductive region and is given as:

$$\tau_{dr} = \varepsilon / \sigma = \varepsilon / (q\mu_n n) \text{ or } \varepsilon / (q\mu_p p) \tag{12.39}$$

Putting in appropriate parameter values, calculated values of 6.6×10^{-15} sec for the n-region and 1.6×10^{-12} sec for the p-region are obtained.

Next the junction capacitance charging time is the time required to change the voltage across the junction capacitance due to current flow through the bulk semiconductor regions (an RC time constant) and is given by:

$$\tau_{RC} = RC \approx ((x_p - W_{dep}) / q\mu_p p)(C / A) = (\varepsilon / q\mu_p p)((x_p - W_{dep}) / W_{deq})$$

$$W_{dep} \approx \sqrt{2\varepsilon V_J / qN_A} = \text{Depletion layer width} \tag{12.40}$$

The resistance and depletion layer width is dominated by the lightly doped p-side in this example. Interestingly, the RC time constant is the lightly doped region dielectric relaxation time increased by the ratio of the width of the conductive region to the depletion layer width. Putting values into this equation gives a value of 2.42×10^{-11} sec.

Finally the diffusion time can be estimated by:

$$\tau_{diff} = (x_p - W_{dep})^2 / D_{np} \text{ or } x_n^2 / D_{pn} \tag{12.41}$$

In these expressions, the width of the depletion layer on the heavily doped n^+ side is neglected and D is the appropriate minority carrier diffusion coefficient (related to the mobility by V_T). Numerical evaluation of these equations gives diffusion times of about 9.17×10^{-9} sec for the p-region and about 5.47×10^{-8} sec for the n-region. In summary then three major time constants can be evaluated for this problem ranging from about 6.6×10^{-15} sec to 2.4×10^{-11} sec and to 5.5×10^{-8} sec. Note that these are each separated by about three orders of magnitude in time and cover a total range of about 7 orders of magnitude in time.

Such a problem with widely varying time constants is most appropriate for solving with a logarithmically spaced time scale as is used in Listing 12.11 on line 86 where the statement tvals = {0,{1.e-15,1.e-6},2,5} occurs. This specifies an integration range beginning at zero time and employing a logarithmically spaced time range from 1.e-15 to 1.e-6 sec. Additionally, the 2,5 specification specifies that 2 solutions will be kept per decade in time and 5 additional solution will be obtained (but not kept) per saved solution. Thus a total of 10 solutions are specified per decade in time. The reader is encouraged to execute the code in Listing 12.11 and obtain the time solutions for this problem. The execution will take some time, so after starting the execution one may wish to take a coffee break and view the results after returning. More will be said later on the execution time for such problems. The commented statement on line 87 can be used to obtain 20 solutions per decade in time over the same time range.

Figure 12.13 shows a graph of one of the solution variables as obtained by Listing 12.11 and saved in the file list12_11.dat. This is for the hole Quasi-Fermi potential. The other two variables are not shown at this time. In general the plot is as one would expect, except for perhaps the solution curves between times of about $1X10^{-10}$ and $1X10^{-9}$ sec where there appear to be some oscillations in the spatial solutions. Whenever one sees such a solution, the immediate question should be are these wiggles in the solution real or artifacts of the numerical solution method? While such oscillations are possible in physical problems, it is know from much past experience that nature tends to prefer smoothly varying solutions and not solutions with many oscillations in either the spatial or time domains. Thus the accuracy of the solutions in this time interval should certainly be of question.

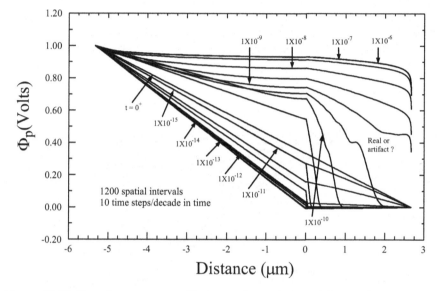

Figure 12.13. Transient response of hole quasi-Fermi potential around p-n junction for applied step voltage of 1 Volt with 10 time-steps per decade in time.

Since no explicit solutions are known with which to compare the numerical solutions, how is one to determine if the wiggles are real or an artifact of the numerical methods? One clue can come from the discussion in Section 10.3 where it was observed that the trapezoidal time integration algorithm could result in oscillations if a very rapid change occurred in the solution. From Figure 12.13, it is seen that the wiggles occur in the solution in a spatial and time region where the solution is changing very rapidly with time. Thus one is led to suspect that the wiggles are not real but simply due to time steps which are too large. To test this the code in Listing 12.11 can be re-executed increasing the number of time steps per decade in the solution. This can be done by commenting out line 87 of the listing and un-commenting line 86 which will then result in 20 time steps per decade in the solution.

The results of such an increase in the number of time steps is shown again for the hole Quasi-Fermi potential in Figure 12.14. The solution curves are now very similar to Figure 12.13 except for the time interval of about 1×10^{-10} and 1×10^{-9} where the wiggles in the solution are now missing and the spatial solution curves make very smooth transitions throughout the entire spatial regions. This is certainly more like what would be expected of the solutions for physical problems. The number of time steps can be further increased and it will be seen that little further change occurs in the solutions as the number of time steps is increased beyond about 20 time intervals per decade in time. Figures 12.15 and 12.16 now show the resulting solutions for the two other variables of electron Quasi-Fermi potential and electric potential. For most of the curves, only the solutions at each order of magnitude in time are shown. Two solutions are actually saved per order of magnitude in time and a few of these intermediate solutions are shown in Figures 12.14 and 12.15 where the variables are rapidly changing with time.

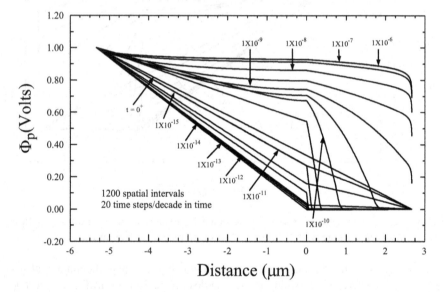

Figure 12.14. Transient response of hole quasi-Fermi potential around p-n junction for applied step voltage of 1 Volt with 20 time steps per decade in time.

To completely understand these solutions and their significance, would require a digression into semiconductor device physics which is beyond the scope of this work. However, the major time constants of the physics can be identified from a careful look at the solutions. Considering first the electric potential in Figure 12.16, it is seen that there is a rapid change in the potential in the n-region (x >0) over the time interval of 1×10^{-15} to 1×10^{-14} sec. This is the dielectric relaxation response in that region which rapidly reduces the electric field in that region to a very small value. Next one sees a significant change in the potential in the

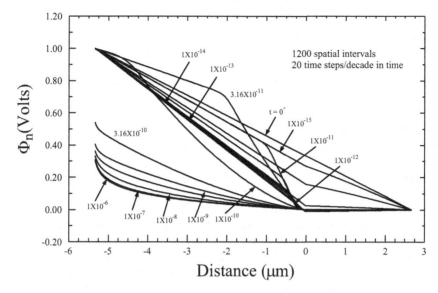

Figure 12.15. Transient response of electron quasi-Fermi potential around p-n junction for applied step voltage of 1 Volt with 20 time steps per decade in time.

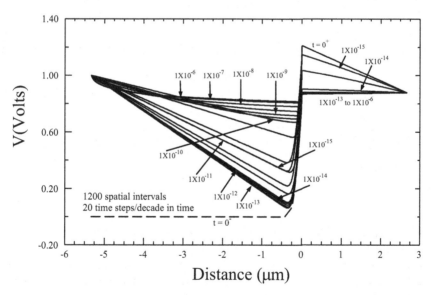

Figure 12.16. Transient response of electric potential around p-n junction for applied step voltage of 1 Volt with 20 time steps per decade in time.

p-region $(x < 0)$ over the time interval $1X10^{-12}$ to $1X10^{-10}$ sec which occurs due to the capacitance charging time of the p-n junction. Finally in Figures 12.14 and 12.15 one can see the minority carrier diffusion effects where the solutions do not reach a steady state value until about $1X10^{-8}$ sec for the electron Quasi-Fermi potential and about $1X10^{-7}$ sec for the hole Quasi-Fermi potential. These are consistent with the time constant estimates evaluated previously.

In addition to the spatial solutions at various times, in many cases of such partial differential equations, the time dependences of the solutions at various spatial points are also of interest. Such solutions can more readily show the various time constants of the problem than can the spatial solutions. No additional calculations are required to obtain such time based solutions, only additional arrays of solution variables need to be collected as the solution proceeds. For example in this problem one might be interested in the time evolution of the variables on each side of the p-n junction. For the call to pdeivbvqs() in Listing 12.11 using the tvals definition of line 86, twenty time solutions are actually calculated per decade in time. If these values are retained by the code at some specified spatial points, then the time evolution of the solutions can be readily visualized. A separate function pdeivbvqst() has been written to accomplish this task. Code for this will not be presented here since the core of the code is very similar to Listing 12.5 with the addition of code to collect time values at selected spatial points. The reader can explore the code in the pdeivbv.lua file. The only change in using the function is to specify an additional array of spatial points where one wishes to collect the time dependent data.

```
1 : -- /* File list12.12.lua */
2 : -- Program for transient solution of p-n junction
3 :
. . . . . Lines 4 through 85 are same as Listing 12.11.lua
. . . .
86 : tvals = {0,{1.e-15,1.e-6},2,10,{-2e-4,-1e-4,0,1e-4,2e-4}}
87 : s,st = pdeivbvqst({eq,efl,efr},tvals,x,v)
88 : write_data('list12.12.dat',s)
89 : write_data('list12.12t.dat',st)
```

Listing 12.12. Code example for saving time dependent data at selected spatial points. Only changes from Listing 12.11 are shown.

Code changes for the previous example using this function are shown in Listing 12.12. Only the changed lines are shown. The key to the additional specification is the last table entry in the tvals[] table which in this example requests that time dependent data be collected for the spatial points of -2e-4, -1e-4, 0, 1e-4, and 2e-4 which can be seen as entries in a final table in the tvals[] table. The call to the PDE solver, pdeivbvqst() then returns two arrays (s,st in the listing) with the first containing the requested spatial solutions and the second array containing the requested time dependent data. In the listing these are both saved to files for additional plotting and analysis on lines 88 and 89. The reader is encouraged to execute the code with these changes and observe the results. The execution time will not increase significantly since little time is needed to accumulate the time dependent arrays.

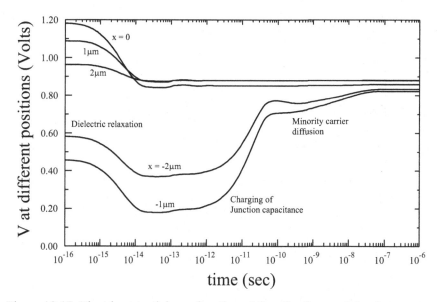

Figure 12.17. Electric potential as a function of time for five spatial points.

Some time dependent results are shown in Figures 12.17, 12.18 and 12.19 for the three solution variables at the five specified spatial points. First consider the electric potential in Figure 12.17. The transients due to the dielectric relaxation in the p-region are readily seen in the figure for all five spatial points in the region below about $1X10^{-14}$ sec. The potential changes on the n-side due to the charging of the junction capacitance are readily seen in the region from about $1X10^{-12}$ to about $1X10^{-10}$ sec. The final small changes in the potential out to about $1X10^{-7}$ sec are due to diffusion of electrons and holes away from the p-n junction. The time dependent changes in the Quasi-Fermi potentials almost mirror the electric potential changes until significant electron diffusion occurs beginning at about $1X10^{-11}$ sec for the $-1\mu m$ point in Figure 12.19. Finally one can see the changes in the hole Quasi-Fermi level in Figure 12.18 extending out to about $1X10^{-7}$ sec for the $2\mu m$ spatial point. This is due to minority carrier diffusion away from the junction interface.

In addition to the potentials for this problem, the time dependent carrier densities are of particular interest to semiconductor device researchers. These are related to the solution variables by the relationships in Eq. (12.36). These can be calculated from the saved results of the previous listings. The results of such calculations are shown in Figures 12.20 and 12.21 for holes and electrons. It can be seen that the carrier densities cover many orders of magnitude as the p-n junction is traversed.

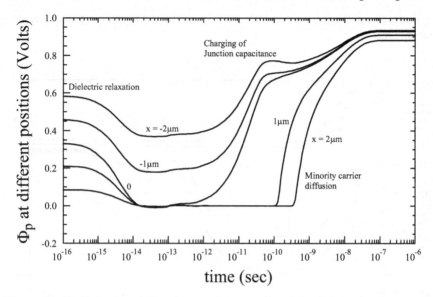

Figure 12.18. Hole Quasi-Fermi potential as a function of time for five spatial points.

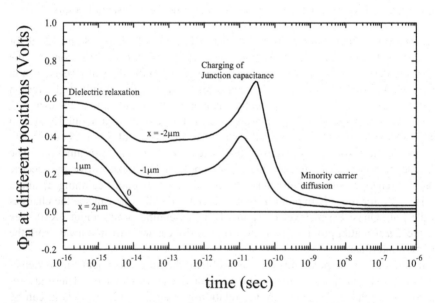

Figure 12.19. Electron Quasi-Fermi potential as a function of time for five spatial points.

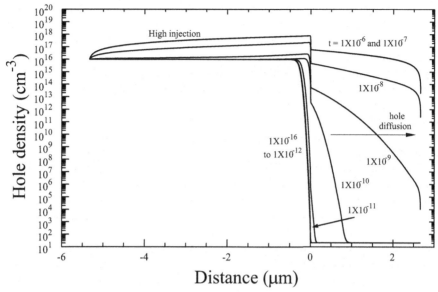

Figure 12.20. Spatial variations of hole densities at various times for p-n junction problem.

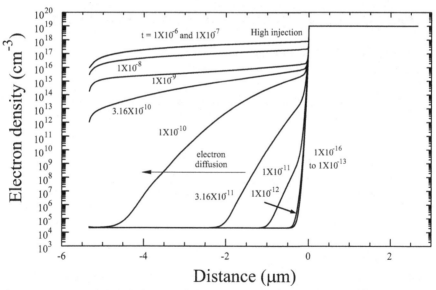

Figure 12.21. Spatial variations of electron densities at various times for p-n junction problem.

These large variations in carrier densities are one of the factors that make the solution of semiconductor problems especially challenging. On the linear spatial scale, the densities show very large changes over very small distances around the junction interface at $x = 0$. However, the use of a non-uniform spatial grid provides many spatial points used in the calculations over these small distances. Also the large drops in carrier densities near the boundaries of the device again correspond to regions with small spatial grid points. From the carrier densities and the potential graphs, it can be seen that steady state for the PDE occurs after about 1×10^{-7} sec which agrees with the previous estimate of the diffusion time for holes through the n-region.

The phenomena of minority carrier diffusion away from a forward biased p-n junction are readily seen in these figures. This is seen to begin to occur around 1×10^{-11} sec and is essentially completed by about 1×10^{-7} sec. This example is one which exhibits time constants spanning a very wide range in values. While one can obtain solutions for selected time intervals by using linear time intervals in the calculation, the use of a logarithmically spaced array of time values allows one to obtain accurate solutions over the wide range of time constants in one set of calculations. The problem has also demonstrated the use of a non-uniform spatial grid to place spatial solution points in regions where the solution is changing rapidly. The reader is encouraged to plot the distribution of spatial points and observe how the points are distributed around the junction and the end regions.

12.6.4 The Wave Equation for a Transmission Line

An example from Electrical Engineering is that of the wave equation describing a transmission line such as shown in Figure 12.22.

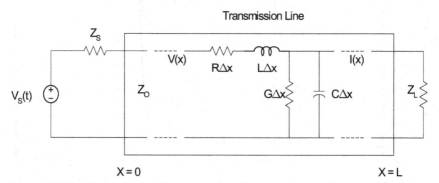

Figure 12.22. Transmission line example. Source voltage with source impedance and load impedance are shown. Incremental internal model is shown.

An incremental model for a section of the transmission line is shown as the R-L-G-C section in the box representing the transmission line. These have dimensions of resistance, inductance, conductance and capacitance per unit length of transmission line. Also shown is a possible time dependent voltage source with source impedance (Z_S) to excite the line and a load impedance (Z_L). The transmission line extends from x = 0 to some distance x = L. The physical variables of major importance along the transmission line are the voltage and current. These are described by the partial differential equations:

$$\frac{\partial V}{\partial x} + RI + L\frac{\partial I}{\partial t} = 0$$

$$\frac{\partial I}{\partial x} + GV + C\frac{\partial V}{\partial t} = 0$$

(12.42)

An ideal lossless transmission line is characterized by $R = 0$ and $G = 0$. The fundamental equations are first order in both the spatial variable and the time variable. The equations can be converted into uncoupled second order differential equations as:

$$\frac{\partial^2 V}{\partial x^2} - RGV - (LG + RC)\frac{\partial V}{\partial t} - LC\frac{\partial^2 V}{\partial t^2} = 0$$

$$\frac{\partial^2 I}{\partial x^2} - RGI - (LG + RC)\frac{\partial I}{\partial t} - LC\frac{\partial^2 I}{\partial t^2} = 0$$

(12.43)

Each variable satisfies the same second order partial differential equation. For a lossless transmission line these become the ideal wave equation (for V only):

$$\frac{\partial^2 V}{\partial x^2} - LC\frac{\partial^2 V}{\partial t^2} = 0 \text{ or}$$

$$\frac{\partial^2 V}{\partial x^2} - \frac{\partial^2 V}{\partial \tau^2} = 0 \text{ where } \tau = \frac{t}{\sqrt{LC}}$$

(12.44)

The solution of this linear wave equation is known to be of the form:

$$V = f(x - \tau) + g(x + \tau)$$

(12.45)

This general solution represents some function traveling to the right along the line (the $f()$ function) plus some function traveling to the left along the line (the $g()$ function). The velocity of the wave is +/- one x unit in one τ unit. The solutions travel without distortion along the line and the exact forms of the solutions are determined by the voltage (and current) waves set up at the boundaries of the transmission line. Since this is a classical problem of great practical importance, much is known about the solutions of the wave equation particularly for step excitations and sinusoidal excitations as one can find in Electrical Engineering textbooks.

The objective here is to illustrate the numerical solution of the wave equation for some simple cases. For the numerical solutions, it is convenient to use the normalized time τ in the basic set of Eq. (12.43). In addition it is convenient to introduce the concept of the basic transmission line impedance and two normalized line parameters as:

$$Z_0 = \sqrt{L/C} \text{ (in Ohm)}$$

$$R_R = R/Z_0; R_G = Z_0 G$$

(12.46)

In terms of these parameters, the basic line equations (from Eq. (12.42)) become:

$$\frac{\partial V}{\partial x} + Z_0 (R_R I + \frac{\partial I}{\partial \tau}) = 0$$

$$Z_0 \frac{\partial I}{\partial x} + R_G V + \frac{\partial V}{\partial \tau} = 0$$

(12.47)

It can be noted that in this formulation that $Z_0 I$ has the same dimensions as V. In a similar manner the two second order partial differential equations (from Eq. (12.43)) become:

$$\frac{\partial^2 V}{\partial x^2} - R_R R_G V - (R_R + R_G) \frac{\partial V}{\partial \tau} - \frac{\partial^2 V}{\partial \tau^2} = 0$$

$$\frac{\partial^2 I}{\partial x^2} - R_R R_G I - (R_R + R_G) \frac{\partial I}{\partial \tau} - \frac{\partial^2 I}{\partial \tau^2} = 0$$

(12.48)

In these equations the time is the normalized time. However, in the following this will be referred to simply as time with the understanding that this means "normalized time" with the normalization constant given in Eq. (12.44).

Listing 12.12 shows code for the numerical solution of the two coupled wave equations for the transmission line example. The functions required in the solution are contained in the odefd and pdeivbv modules and are brought into the solution on line 4 of the code. Lines 5 and 7 sets the nprint values for the solution codes so that the time iterations are printed and one can keep up with the solution as it progresses. The basic transmission line parameters are set on lines 11 through 17 of the code. The defining equation set is given by the eq() function on lines 19 through 24. The first two lines (20 and 21) show the defining equations for the second order set of equations while lines 22 and 23 which are commented out in the listing show the corresponding definitions for the first order equations. Boundary conditions are defined by the efl() and efr() functions on lines 26 through 35. For the left boundary condition, it is assumed that the excitation voltage source is a sinusoidal voltage source of magnitude vs and angular frequency w as defined on line 27 which is Kirchoff's voltage law as applied to the source voltage loop. This is an obvious boundary equation. However, for a set of two coupled differential equations, two boundary equations are required to uniquely determine a solution. A second boundary equation is not so readily identified. However, it is known that the basic set of first order equations as given by Eq. (12.47) must also be satisfied not only at each point along the transmission line but also at the line boundaries. However, which equation should be taken as the second boundary condition? Actually one can take either equation. This is because the other equation will be satisfied by the second order differential equation. For the efl() function the first of Eq. (12.47) is taken as the second boundary condition. The reader can change to the second equation and verify that either equation can be taken as the second boundary condition. It is noted that the line characteristic impedance Z_0 occurs in the second order formulation of the equation set only in the boundary conditions and not in the second order equations.

```
 1 : -- /* File list12_13.lua */
 2 : -- Program for transient solution of transmission line
 3 :
 4 : require"odefd"; require"pdeivbv"
 5 : getfenv(pdeivbv).nprint = 1
 6 : getfenv(ode2bvfd).umin = {5.e-4,5.e-4,5.e-4}
 7 : getfenv(ode2bvfd).nprint = 1
 8 :
 9 : tt = os.time()
10 : -- Model equations to be solved
11 : Zo = 50
12 : Rr = 0 -- Series resistance factor
13 : Rg = 0 -- Parallel conductance factor
14 : Rs = 100 -- Source resistance
15 : Rl = 100 -- Load resistance
16 : vs = 1.0 -- Step value of source voltage
17 : w = 4*math.pi; Rrg, Rrpg = Rr*Rg, Rr + Rg
18 :
19 : eq = function(fv,x,t,v,vp,vpp,vt,vtt) - Wave Equations
20 :    fv[1] = vpp[1] - Rrg*v[1] - Rrpg*vt[1] - vtt[1] - 2nd order
21 :    fv[2] = vpp[2] - Rrg*v[2] - Rrpg*vt[2] - vtt[2]
22 :    --fv[1] = vp[1]  + Zo*(Rr*v[2] + vt[2]) -- First order
23 :    --fv[2] = Zo*vp[2] +Rg*v[1]+ vt[1]
24 : end
25 :
26 : efl = function(fv,v,vp,t,vt) -- Boundary condition
27 :    fv[1] = vs*math.sin(w*t) - v[1] - Rs*v[2]
28 :    fv[2] = vp[1] + Zo*(Rr*v[2] + vt[2])
29 : end
30 : efr = function(fv,v,vp,t,vt) -- Boundary condition
31 :    fv[1] = v[2] -- Open circuit at load, I = 0
32 :    --fv[1] = v[1] -- Short circuit at load, V = 0
33 :    --fv[1] = v[1] - Rl*v[2] --Load resistace Rl at load,V = Rl*I
34 :    fv[2] = Zo*vp[2] + Rg*v[1] + vt[1]
35 : end
36 :
37 : x = {}; nx = 500; L = 1 -- Define x values
38 : for i=1,nx+1 do x[i] = L*(i-1)/nx end
39 :
40 : v = {{},{}} -- Set initial values, V = 0, I = 0 for all x values
41 : for i=1,nx+1 do
42 :    v[1][i] = 0; v[2][i] = 0 -- v[2] is current variable
43 : end
44 : tvals = {0,10,100,40,{0,L/4,L/2,3*L/4,L}}
45 : s,st = pdeivbvt({eq,efl,efr},tvals,x,v,vp,vpp)
46 : write_data(2,'list12_13.dat',s)
47 : write_data('list12_13t.dat',st)
48 : tt = os.time() - tt; print('time taken for calculation = ',tt)
```

Listing 12.13. Code segment for the numerical solution of the wave equations describing a transmission line.

For the right side set of boundary conditions, the equations depend on the termination of the transmission line. Three possible load conditions are shown as possible boundary conditions on lines 31 through 33: that of an open circuit at the load, a short circuit at the load and a load resistor of value Rl. The listing shows the open circuit case as used in the code with the other two possibilities commented out. The reader can execute the code with the other boundary conditions as desired. Also for the right boundary, the second boundary equation is taken as the second of the Eq. (12.47). Again this is arbitrary and is used simply to illustrate that either of the first order equations can be used as the second boundary condition.

The spatial array of points is defined by lines 37 and 38 while the initial values of the voltage and currents are defined as zero on lines 40 through 43. Finally, the desired time points for the solution are defined on line 44 by the statement tvals = {0, 10, 100, 40, {0, L/4, L/2, 3*L/4, L}}. This specifies a time (or normalized time) interval of 0 to 10 with solutions saved at 100 intervals over the range. This means that the spatial solutions will be saved at times of 0, .1, .2, --- 9.8, 9.9, 10. In addition, the 40 parameter specifies that calculations will be performed for 40 time intervals between each saved point. The net result is that the solution will be calculated for 100x40 = 400 time intervals (401 time points) over the time period of 0 to 10. Since the velocity of the transmission line wave is 1.0 (in normalized time units) this means that the solution will be calculated for 10 transits of a wave front across the transmission line. This is sufficient to approach a steady state solution. Finally, the {0, L/4, L/2, 3*L/4, L} parameter specifies spatial points along the line at which time dependent data will be collected. The numerical solution is then obtained by the call to the function pdeivbvt() on line 45. The last "t" character in the name specifies that the routine used in the solution will return the time solutions for the specified spatial points as indicated above. The last entry in the tvals array would not be used by a call to the pdeivbv() function, although the returned spatial solutions would by the same.

One final comment about the selected parameters concerns the selected frequency of the sinusoidal voltage source which is w = 4π on line 17. Since the wave transit time across the line is 1.0, this means that the line length corresponds to 2 wavelengths, or L = 2λ. Different line lengths in terms of wavelengths can be selected by varying the specified sinusoidal frequency or changing the line length. A value of w = 2π or an L = .5 would correspond to a line length of one wavelength. Since the calculation in Listing 12.13 involves many time solutions, the calculation will take some time to complete. This is another case where one might take a coffee break while the code executes.

Figure 12.23 shows spatial voltage variations across the transmission line for the sinusoidal excitation of Listing 12.13 for initial times between 0 and 1.0. The solution represents a sinusoidal voltage wave traveling along the line at a velocity of L/(unit time).

The amplitude of the voltage wave is $V_s Z_0/(Z_s + Z_0) = 0.33333$ Volts. The figure shows a sinusoidal wave traveling from the left side of the figure toward the right side. As seen in the figure, at time t = 1.0 the wavefront has just reached the load end of the line which in this example is an open circuit. Since the line is exactly 2 wavelengths long, 2 cycles of a sinusoidal voltage are present along the line. As the wavefront reaches the open circuited load, a reflected wave is set up traveling backwards along the line for t > 1.0.

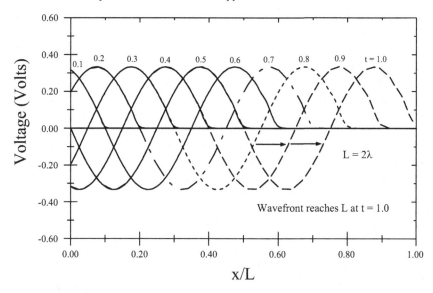

Figure 12.23. Initial voltage waveforms across transmission line for a sinusoidal voltage source.

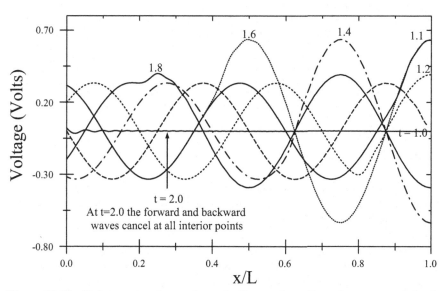

Figure 12.24. Voltage waveforms along the line as the reflected wave travels back along the line.

Solutions for larger time values from t = 1.0 to 2.0 are shown in Figure 12.24. This represents the time for a reflected wave to travel back to the source end of the line. The amplitude of the wave traveling backward along the line is equal to the amplitude of the initial forward traveling wave, resulting in a peak amplitude of the line voltage of twice the initial forward traveling wave. The solution for t = 2.0 is interesting in that the voltage is essentially zero throughout the entire transmission line. For this case the reflected wave at the open circuit exactly cancels the forward wave in amplitude and phase throughout the transmission line. The solution for the current variable, however, is not zero along the line so the line is storing energy in the inductance which for larger values of time results in a non-zero voltage waveform.

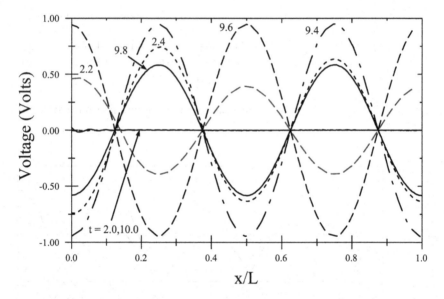

Figure 12.25. Voltage waveforms along the line as steady state is approached.

Finally, Figure 12.25 shows the line voltage waveforms as steady state is approached. In this case one sees a standing wave pattern established along the line with a peak amplitude of 1.0 Volts which is equal to the source voltage. The standing wave pattern goes through time values (multiples of 1.0 in normalized time) where the voltage along the line is zero as seen by the curve corresponding to t = 10.0 in the figure. One can see that for times as short as 2.4, the standing wave pattern is pretty well established across the line, although the amplitude has not reached a steady state value.

The line current should also approach a standing wave pattern with the current being zero at the ends of the line, but with non-zero values internal to the line as energy transfers back and forth between the line inductance and capacitance. This is seen in Figure 12.26 which shows $Z_0 I$ for the last few time points calculated. This is seen to have the same peak value as the voltage waveform along the line, illustrating that the current scales inversely with the line characteristic impedance. All of this is as expected from the known theory of linear transmission lines but it's nice to see that this is all consistent with the present numerical solutions.

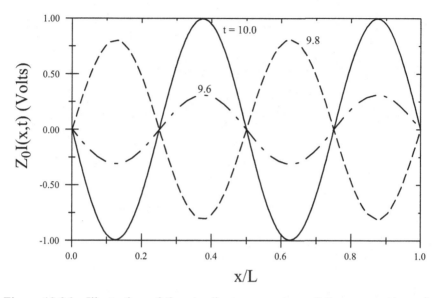

Figure 12.26. Illustration of the standing wave nature of the current along the transmission line as steady state is approached.

The time dependent voltages at the two ends of the transmission line are also of interest and these values are shown in Figures 12.27 and 12.28. The initial voltage at the source end of the line is determined by the line characteristic impedance and the source impedance and for this example is 0.333333 Volts as seen in Figure 12.27. This continues until t >= 2.0 which is sufficient time for the initial wave to travel the length of the line and a reflected wave to travel back to the source end of the line. The reflected wave causes the amplitude of the source end voltage to increase to 0.777777 Volts as again seen in Figure 12.27. This continues until t >= 4.0 at which time a second reflected wave at the load end has reached the source end of the line, resulting in another step in the amplitude of the source voltage. For t >= 8.0 sufficient time has elapsed for 4 forward and backward transits of the line and 4 steps in the voltage toward the steady state value of 1.0. After 4 round trips of waves along the line the amplitude is at 99.2% of its steady state value. The dashed curve in the Figure 12.28 represents the expected theoretical value of the amplitude of the voltage steps and very good agreement is seen with the numerical solution. However, an accurate solution approaching steady state does require many spatial solution points to sufficiently resolve the spatial waveforms and many time points to sufficiently resolve the time waveforms. The voltage at the load (or open circuited) end of the line also closely matches expected theoretical results as seen in Figure 12.28. The voltage is zero until the line wavefront has

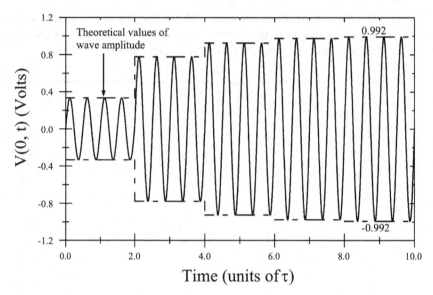

Figure 12.27. Time dependent voltage at source end of transmission line with sinusoidal excitation.

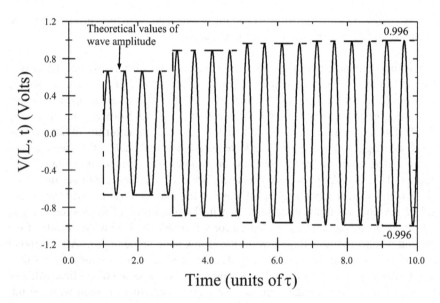

Figure 12.28. Time dependent voltage at load end of transmission line with sinusoidal excitation and with an open circuit load.

had time to reach the load end or for t <= 1.0. For 1.0 < t < 3.0 a reflected wave develops traveling back along the line and the load voltage becomes a sinusoid in time with a peak amplitude of 0.66666 Volts or twice the initial incident wave. The load voltage shows steps as the internal wavefront completes two complete transits along the line. For this example, the line voltage has reached 99.6% of its steady state value for t >= 9.0. Again the numerical solution agrees very well with the theoretical values of the sinusoidal amplitude as seen by the dotted curve in Figure 12.28.

From this example it can be seen that the solutions are within about 99% of the steady state values after about 8 line transit times or about 4 round trip transit times for a wave from the source to be reflected and return to the source. Since the exact solution for this problem is readily known, this provides an opportunity to explore the accuracy of the numerical solution for this particular example of spatial and time discretization. From Figure 12.23 some error can be seen in the solution near the load end of the line for t = 1.0 as the wave approaches the load. Ideally the solution should be a sinusoid with an abrupt change in the slope of the function at the leading edge of the wavefront. The numerical solution shows some rounding of the curve at the leading edge of the sinusoidal wave, resulting from the finite size of the spatial and time grid points. Before looking at the magnitude of the error consider another approach to the solution. It has previously been stated that the solution can be obtained by use of the two second order partial differential equations as given by lines 20 and 21 in Listing 12.13 as well as by the two first order partial differential equations as given by lines 22 and 23 of the listing.

The reader is encouraged to execute the code using these two functions and compare the results. One would expect that there might be some difference in the error in the numerical solution depending on whether one uses the first order or second order equations. Figure 12.29 shows the error in the numerical solution for the t = 1.0 case and for three different solution conditions. The upper two curves illustrate the error for the case of solving either first and second order equations and it is seen that the maximum errors are somewhat comparable, but the second order equations results is a slightly smaller maximum error. Finally the third and lower curve shows the error in the solution when using the second order equations but with an increase in the number of time and space increments by a factor of 2 (recall the h-2h algorithm for the previous chapter). If the solution error is second order in the mesh size, then the error would be expected to be a factor of 4 lower for this case. This is approximately the difference as can be seen from the two error curves, however the peak errors occur at different spatial locations for the two spatial grid sizes.

Figure 12.30 compares the solution errors for the time of t = 10.0 which is approaching the steady state. In this case the errors are more uniformly distributed over the length of the line as one would expect for a time independent boundary value problem. Again one sees a slight advantage of using the second order formulation as opposed to the first order equation formulation. Also the comparison of the h-2h cases more clearly illustrates the expected factor of 4 difference in maximum error. For the case of 500 spatial intervals and 400 time intervals, the maximum steady state error is about 4e-3 while the peak value is 0.333, representing an accuracy of about 1.2% for the worst case point.

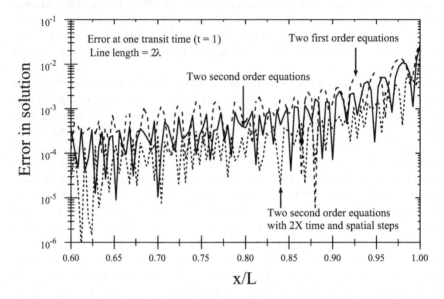

Figure 12.29. Error in the numerical solution for the example in Listing 12.13 for short times and long times.

While one result does not prove a general theorem, it is to be expected from a theoretical basis that the solution errors for other examples will also vary as the square of the number of time and spatial grid points. Doubling both of these, however, increases the execution time by about a factor of 4X, so a solution is always a tradeoff between accuracy and execution speed. One can always check the accuracy by doubling the time and spatial grid points and if one does not get essentially the same solution, one should be skeptical of a solution.

Results will now be shown for a few other combinations of parameters illustrated in Listing 12.13 but which are commented out in the listing. Replacing line 31 by line 32 in the function for the right boundary condition, results in specifying a short at the output of the transmission line instead of an open circuit. The reader is encouraged to perform this change and re-execute the example. The results for the time dependent voltage at the source end of the transmission line are shown in Figure 12.31. The voltage is seen to approach zero in steps every 4 cycles of the waveform as a wave has time to complete a round trip transit of the transmission line and the voltage is near zero after about 4 transits of the line as seen for t > 8. Again the voltage steps are as expected from known linear transmission line theory. Again standing waves of voltage and current exist along the line as steady state is approached with the voltage being zero at the ends of the line and the

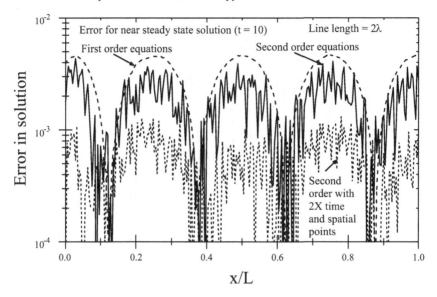

Figure 12.30. Errors in the numerical solution for the example in Listing 12.13 for times approaching steady state (t = 10.0).

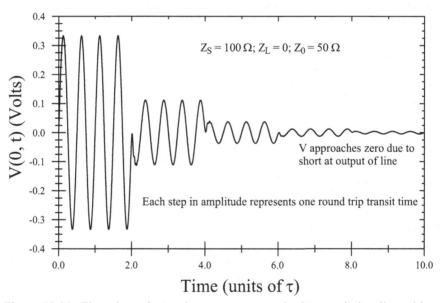

Figure 12.31. Time dependent voltage at source end of transmission line with a short at the output. Line is 2 wavelengths in length.

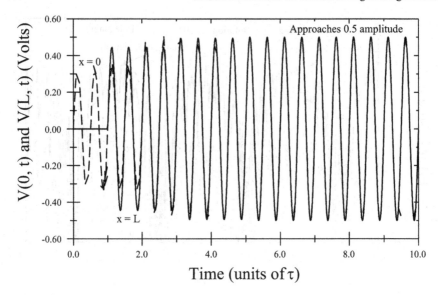

Figure 12.32. Time dependent voltage at source and load end of transmission line with a load resistance equal to the source resistance.

current having a maximum value at the ends of the line. The role of voltage and current being somewhat reversed in this case from the open circuit load voltage case.

The case of a load resistor of 100 Ω is shown in Figure 12.32 for both the source and load ends of the transmission line. In this case the steady state voltage is approached with both the source and load ends of the line having a peak sinusoidal amplitude of one-half of the source voltage (or 0.5 Volts). In this case the steady state amplitude is approached considerably faster than in the previous two examples.

The previous examples have all been for an ideal lossless transmission line with only internal inductance and capacitance. The basic equations include provisions for both a series resistance and a parallel conductance. These are easily included in the code of Listing 12.13 by changing the R_R and R_G parameters which are defined in Eq. (12.46). The reader is encouraged to change these parameters in the code and re-execute the examples above. One example is shown here in Figure 12.33 of the input voltage of a line with $R_R = 1$, $R_G = 0$ and for a line with a shorted output. This should be compared with Figure 12.31 for the corresponding response of a line with zero internal line resistance. In this case the steady state source voltage does not approach zero but the peak steady state voltage is about 0.188 Volts as can be seen from the figure. The reader can explore other examples of transmission lines with series resistance and parallel conductance.

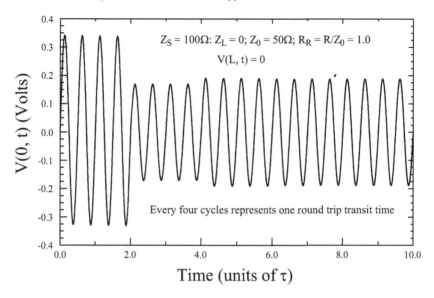

Figure 12.33. Voltage at source end of resistive transmission line with a short at the output. Line is 2 wavelengths in length.

The basic formulation of the transmission line equations can be used not only for linear problems as in these examples but also for nonlinear problems where the line equations and/or the load equations are nonlinear. Listing 12.14 shows one such example where a diode is used as the load resistance of a transmission line. The major change in the code is line 30 defining the diode equation for the right boundary condition. Other changes are the magnitude of the sinusoidal source voltage (5 Volts) and the source frequency (line 18) which makes the line length equal to one wavelength λ. The solution results for the source and load voltages are shown in Figure 12.34 and 12.35. As expected, one sees in Figure 12.35 that the diode limits the load voltage on the positive side to about 0.6 volts and acts like an open circuit for the negative voltages. Since the line length is a multiple of a wavelength, the voltage at the source side of the transmission line follows closely the load voltage when steady state is reached. This can bee seen be the equal values of the two solid curves in Figures 12.34 and 12.35 for t > 6.0 which corresponds to about three round trip transit times for the transmission line. Also shown in Figure 12.35 is $Z_0 I(L,t)$ which represents the diode current at the load scaled by the characteristic line impedance. As can be seen current flows only on the positive half cycles of the voltage pulse. As steady state is approached, the line voltage waveforms show interesting nonlinear waveforms as can be seen in Figure 12.36. The waveforms repeat as can be seen by the similar waveforms for times of 9 and 10 in the figure.

```
 1 : -- /* File list12_14.lua */
 2 : -- Program for transient solution of transmission line
 3 :
 4 : require"odefd"; require"pdeivbv"
 5 : getfenv(pdeivbv).nprint = 1
 6 : getfenv(ode2bvfd).umin = {5.e-4,5.e-4,5.e-4}
 7 : --getfenv(ode2bvfd).nprint = 1; getfenv(pdebivbv).nprint = 1
 8 :
 9 : tt = os.time()
10 : -- Model equations to be solved
11 : Zo = 50
12 : Vt, Is = .026, 1.e-12 -- Thermal voltage, Sat current
13 : Rr = 0 -- Series resistance factor
14 : Rg = 0 -- Parallel conductance factor
15 : Rs = 100 -- Source resistance
16 : Rl = 100 -- Load resistance
17 : vs = 5.0 -- Step value of source voltage
18 : w = 2*math.pi; Rrg, Rrpg = Rr*Rg, Rr + Rg
19 :
20 : eq = function(fv,x,t,v,vp,vpp,vt,vtt) -- Equations
21 :     fv[1] = vpp[1] - Rrg*v[1] - Rrpg*vt[1] - vtt[1] - 2nd order
22 :     fv[2] = vpp[2] - Rrg*v[2] - Rrpg*vt[2] - vtt[2]
23 :     end
24 :
25 : efl = function(fv,v,vp,t,vt) -- Using mixed boundary condition
26 :     fv[1] = vs*math.sin(w*t) - v[1] - Rs*v[2]
27 :     fv[2] = vp[1] + Zo*(Rr*v[2] + vt[2])
28 : end
29 : efr = function(fv,v,vp,t,vt) -- Using mixed boundary condition
30 :     fv[1] = v[2] - Is*(math.exp(v[1]/Vt) - 1) -- Diode at load
31 :     fv[2] = Zo*vp[2] + Rg*v[1] + vt[1]
32 : end
33 :
34 : x = {}; nx = 500; L = 1 -- Define x values
35 : for i=1,nx+1 do x[i] = L*(i-1)/nx end
36 :
37 : v = {{},{}} -- Set initial voltage values, V = 0, I = 0 for all
             x values
38 : for i=1,nx+1 do
39 :     v[1][i] = 0; v[2][i] = 0 -- v[2] is current value
40 : end
41 : tvals = {0,10,100,40,{0,.25,.5,.75,1}}
42 : -s, st = pdeivbvt({eq,efl,efr},tvals,x,v,vp,vpp)
43 : write_data(2,'list12_14.dat',s) -- Save same number of points as
             before
44 : write_data('list12_14t.dat',st)
45 : print('time taken for calculation = ',os.time()-tt);
             tt = os.time()
```

Listing 12.14. Code example for transmission line with a diode load and a sinusoidal source voltage.

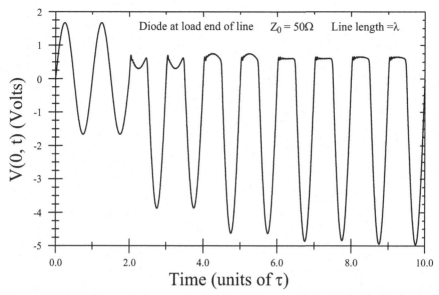

Figure 12.34. Voltage at source end of transmission line for a diode load.

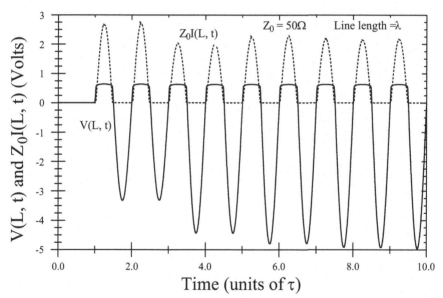

Figure 12.35 Voltage and current at load end of transmission line with a diode load.

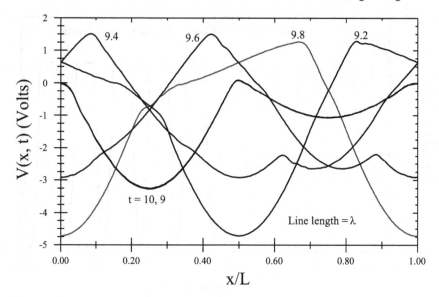

Figure 12.36. Steady state line voltage waveforms for transmission line with diode load.

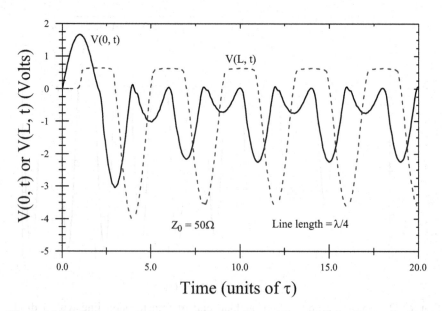

Figure 12.37. Example of short transmission line with a diode load. Line length is a quarter of a wavelength.

The type of response is dependent on the line length relative to the source signal wavelength. For example Figure 12.37 shows the voltage waveforms at the ends of the transmission line for the same example as above except that the line length is now $\lambda/4$ or a quarter of a wavelength. The code for this case is not shown but simply consists of changing line 18 of Listing 12.14 to w = 0.5*math.pi. In this case steady state is rapidly reached with the diode voltage somewhat similar to the longer line of Figure 12.34. However, the voltage at the source end of the line is very different from that shown in Figure 12.33. In this case the voltage waveform show two negative peaks per half cycle of the source voltage and the voltage values are all essentially negative with no positive going values. The reader is encouraged to explore other cases by re-executing the code in Listing 12.14 with changes in the source frequency and/or the other line parameters.

Some simple and important transmission line problems are virtually impossible to simulate numerically with any reasonable degree of engineering accuracy (such as 1% accuracy). It is thus important to understand the limitations of solving the wave equation with numerical techniques in such cases. One such simple and important case is that of a step (or pulse) voltage applied to a lossless transmission line. Listing 12.15 shows the only changes needed in Listing 12.13 for such an example where the code on line 27 substitutes a constant source voltage (of value vs) for the previous sinusoidal source.

```
 1 : -- /* File list12_15.lua */
 2 : -- Program for transient solution of transmission line
 3 :
 . . .
26 : efl = function(fv,v,vp,t,vt) -- Boundary condition
27 :    fv[1] = vs - v[1] - Rs*v[2]
28 :    fv[2] = vp[1] + Zo*(Rr*v[2] + vt[2])
29 : end
 . . .
44 : tvals = {0,10,100,400,{0,.25,.5,.75,1}}
45 : s,st = pdeivbvt({eq,efl,efr},tvals,x,v,vp,vpp)
 . . .
```

Listing 12.15. Code changes of Listing 12.13 for step voltage source at input of transmission line.

The time dependent numerical solution for this abrupt source voltage is shown in Figure 12.38 as the voltage wavefront propagates across the transmission line. Also shown with dotted lines are the ideal abrupt voltage wavefronts at various times. The numerical solutions are seen to have the general features of the exact solution and exhibit a sharp drop in voltage at progressive points along the transmission line as the wave propagates. However, the numerical solutions have an oscillatory artifact around the leading edge of the voltage wavefront and this oscillation is not damped out as time progresses. The key to identifying these oscillations as artifacts is the observation that the maximum and minimum points in the oscillation are separated in space by only one spatial grid point. This oscillation artifact arises from the limitations of the numerical techniques and is impossible to completely eliminate. For this example, spatial steps of 0.002 and time steps of 0.00025 were used in the simulations. Note that this is 10 times the number of time steps used in the previous calculation of Listing 12.14 (see 400 factor on line 44). With a time step of 0.0025 as in Listing 12.14, the oscillations will be somewhat larger than shown in the figure. The oscillations can be reduced (but not eliminated) by taking smaller time and spatial steps. The problem is of course the representation of an abrupt step in a variable (with infinite derivative) using a finite mesh of time and spatial grid points. With more spatial steps, the oscillations will occur at a higher frequency but will still be present. An accurate approximation requires an extremely fine array of time and spatial grid points. Even then there will always be some error in the approximation and some oscillations.

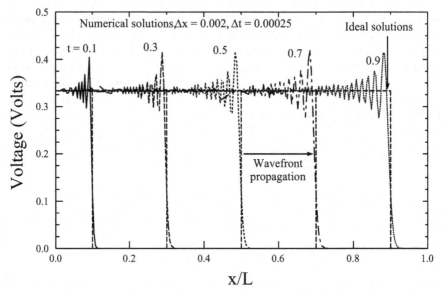

Figure 12.38. Numerical solution for ideal transmission line with a step in input voltage.

Thinking in terms of the frequency domain, an abrupt step in voltage contains frequency components out to infinite frequency. Any discrete array of time and spatial steps will be able to represent frequency components only out to some maximum frequency. To improve on the accuracy of the solution seen in Figure 12.38 would require a much finer array of spatial grid points around the wavefront. However, the wavefront moves with time across the entire range of spatial values so one needs an extremely fine mesh across the entire space or some means of moving a non-uniform spatial grid across the spatial dimension as the solution progresses. While it is possible to implement such a scheme it is not simple. In any case one also needs an extremely fine time grid as well to propagate the abrupt solution in time. So as a practical matter it can be ascertained that it is extremely difficult if not virtually impossible in reasonable computational times to achieve a high degree of accuracy for such problems of an abruptly propagating wavefront. The culprit of course is the abrupt change in source voltage. One could take the approach that abrupt changes never occur in real physical problems and always require a minimum of one time step for the voltage to change. Some numerical analysis programs (such as SPICE) take this approach. This minimizes the problem but even sources with an abrupt change in derivative can lead to solution artifacts around the leading edge of numerical solutions. This is seen to some extent in Figure 12.23 for the case of a sinusoidal voltage source turned on at t = 0. Such an example has an abrupt change in function derivative and it can be seen by a careful examination of Figure 12.23 that some numerically induced artifacts occur around the leading edge of the numerical solution. This is also the major source of the increasing error with distance seen in Figure 12.29 as the sinusoidal wavefront reaches the end of the transmission line. One of the lessons from this is that one should be very skeptical of a numerical solution that exhibits a rapid change in derivative and especially one that shows rapid oscillations around a rapidly changing region of the solution.

It is interesting to compare the wave equation with previous equations such as the diffusion equation. In Section 12.5.1 the diffusion equation was solved numerically for an abrupt step in source value at one of the boundaries. For that example it was found that non-uniform spatial and time grids were needed to achieve a high degree of accuracy, but small spatial steps were only needed at the initial boundary and small time steps were only needed to begin the solution. The difference with the diffusion equation is that the abrupt boundary condition rapidly (in time and space) becomes attenuated into a gradual change in the solution. This is in sharp contrast to the ideal wave equation where a solution propagates unattenuated or un-changed in shape through the spatial extent of the solution space making an accurate numerical solution exceedingly more difficult. While there is no readily available solution to this problem, the reader is advised to be very skeptical of numerical solutions of the wave equation for sources with abrupt changes in magnitude and/or derivative.

12.6.5 Nonlinear Wave Equations – Solitons

Nonlinear wave equations form an interesting branch of physical study and phenomena. Many such equations and solutions fall under the broad topic of solitons which are self-reinforcing waves representing a delicate balance between nonlinear and dispersive effects in a medium in which the wave travels. Although a precise definition of solitons is difficult they generally have the following properties: (1) they are waves of permanent shape, (2) they are localized and travel with some velocity and (3) they can interact strongly with other solitons, but emerge from the collision unchanged apart from a phase shift. One of the most studied nonlinear wave equations is the Korteweg-de Vries equation which in its simplified form is:

$$\frac{\partial^3 u}{\partial x^3} + 6u\frac{\partial u}{\partial x} + \frac{\partial u}{\partial t} = 0 \qquad (12.49)$$

This is third order in the spatial derivative and first order in the time derivative. One soliton solution to this nonlinear equation is known to be:

$$u = \frac{1}{2}c\frac{1}{\cosh^2((\sqrt{c}/2)(x-ct-a))} \qquad (12.50)$$

where a is a constant and c is the velocity of the soliton wave. The soliton wave amplitude peak is also seen to be $c/2$ so the soliton velocity depends directly on the amplitude of the soliton solution.

In order to explore numerical solutions of this nonlinear equation using the code segments in this chpater, the single third order equation needs to be converted into a set of equations with no more than second order derivatives. One such formulation is then the two coupled equations:

$$\frac{\partial u_1}{\partial x} - u_2 = 0$$

$$\frac{\partial^2 u_2}{\partial x^2} + 6u_1 u_2 + \frac{\partial u_1}{\partial t} = 0 \qquad (12.51)$$

In this formulation there is one first order equation and one second order equation in the spatial derivative. Another formulation would be in terms of three first order equations in the spatial derivative. However, the previously generated solution routines can handle equations with second order derivatives.

A code segment for solving this set of equations is shown in Listing 12.16. The code is similar to the past listings. The two defining partial differential equations are defined on lines 14 and 15. Boundary conditions of zero are taken at both boundaries for the two solution variables on lines 18 through 23. To begin the time solution, an initial spatial solution is assumed to exist as a triangular pulse of width 2 spatial units, centered at $x = 2$ and of peak value 5.0 is defined by the code on lines 29 through 35. The numerical time solution is obtained by the pdeivbvt() function on line 38. The tvals[] table on line 37 specifies a solution form $t = 0$ to t = 2.5 with 25 saved time solutions and with 20 additional solutions per saved interval. Also specified are spatial points at which time dependent data is collected.

A graph of the solution obtained by this code is shown in Figure 12.39. As can be seen the initial triangular pulse is rapidly transformed into a smoother pulse and this major pulse propagates along the distance axis with essentially a constant velocity and with essentially the same shape after some initial transformation of the triangle pulse into a smooth pulse. Some lower lever of signal is seen along the baseline but with no apparent consistent shape or propagation velocity. The major pulse peak is similar to the function described by Eq. (12.50) which is known to be a solution of the nonlinear wave equation. The wave peak is seen to travel from $x = 2$ to about $x = 18$ in $t = 2.0$ which is an average velocity of about 8.0. This is consistent with a peak pulse amplitude of around 4.0 and with Eq. (12.50) where the soliton velocity has a velocity of twice the pulse amplitude.

```
 1 : -- /* File list12_16.lua */
 2 : -- Program for transient solution of nonlinear transmission line
 3 :
 4 : require"odefd"; require"pdeivbv"; require"intp"
 5 : getfenv(pdeivbv).nprint = 1
 6 : getfenv(ode2bvfd).umin = {5.e-4,5.e-4,5.e-4}
 7 : --getfenv(ode2bvfd).nprint = 1; getfenv(pdebivbv).nprint = 1
 8 :
 9 : tt = os.time()
10 : -- Model equations to be solved
11 : us = 5.0 -- peak value of initial triangular wave
12 :
13 : eq = function(fu,x,t,u,up,upp,ut) -- Equations
14 :    fu[1] = up[1] - u[2]
15 :    fu[2] = upp[2] + 6*u[1]*u[2] + ut[1]
16 : end
17 :
18 : efl = function(fu,u,up,t,ut) -- Left boundary condition
19 :    fu[1], fu[2] = u[1], u[2]
20 : end
21 : efr = function(fu,u,up,t,ut) -- Right boundary condition
22 :    fu[1], fu[2] = u[1], u[2]
23 : end
24 :
25 : x = {}; nx = 1000; L = 20 -- Define x values
26 : for i=1,nx+1 do x[i] = L*(i-1)/nx end
27 :
28 : u = {{},{}} -- Set initial wave values,
29 : for i=1,nx+1 do -- Set initial triangular wave at x=2
30 :    if x[i]>3 then u[1][i] = 0
31 :    elseif x[i]>2 then u[1][i] = us*(3-x[i])
32 :    elseif x[i]>1 then u[1][i] = us*(x[i]-1)
33 :    else u[1][i] = 0 end
34 : end
35 : for i=1,nx+1 do u[2][i] = intp(x,u[1],x[i],1) end
36 :
37 : tvals = {0,2.5,25,20,{0,.25*L,.5*L,.75*L,L}}
38 : s,st = pdeivbvt({eq,efl,efr},tvals,x,u)
39 : write_data(2,'list12_16.dat',s) -- Save same number of points
40 : write_data('list12_16t.dat',st)
41 : print('time taken for calculation = ',os.time()-tt);
         tt = os.time()
```

Listing 12.16. Code segment for solution of soliton wave equation. Initial triangular pulse given at x = 2.

To explore the soliton solution further, Listing 12.17 shows code for setting up two ideal soliton solutions according to Eq. (12.50) but with different spatial locations and different amplitudes. The solution of the wave equation then simulates the time evolution of the solutions according to the nonlinear wave equation. The equations to be solved and the boundary conditions are the same as in the previous listing. In this case two functions are defined on lines 12 and 13, one for the cosh() function and one for the ideal soliton solution of Eq. (12.50). These are then used on lines 30 through 32 to set up an initial solution with an amplitude of 4 for the soliton at x = 4 and an amplitude of 1 for the soliton at x = 20. The pdeivbvt() function then called on line 35 simulates the development of the solutions from t = 0 to t = 5 at which point it is to be expected that the larger of the solitons will have traveled a distance of about 40 x units. To accommodate this traversal, the solution is simulated over an L = 40 range.

```
 1 : -- /* File list12_17.lua */
 2 : -- Program for transient solution of nonlinear transmission line
 3 :
 4 : require"odefd"; require"pdeivbv"
 5 : require"intp"
 6 : getfenv(pdeivbv).nprint = 1
 7 : getfenv(ode2bvfd).umin = {5.e-4,5.e-4,5.e-4}
 8 : --getfenv(ode2bvfd).nprint = 1; getfenv(pdebivbv).nprint = 1
 9 :
10 : tt = os.time()
11 : -- Model equations to be solved
12 : cosh = function(x) return (math.exp(x) + math.exp(-x))*.5 end
13 : ui = function(x,a,xo) return a/(cosh(0.5*math.sqrt(2*a)*
         (x-xo))^2) end
14 :
15 : eq = function(fu,x,t,u,up,upp,ut) -- Equations
16 :    fu[1] = up[1] - u[2]
17 :    fu[2] = upp[2] + 6*u[1]*u[2] + ut[1]
18 : end
19 : efl = function(fu,u,up,t,ut) -- Left boundary condition
20 :    fu[1], fu[2] = u[1], u[2]
21 : end
22 : efr = function(fu,u,up,t,ut) -- Right boundary condition
23 :    fu[1], fu[2] = u[1], u[2]
24 : end
25 :
26 : x = {}; nx = 1000; L = 40 -- Define x values
27 : for i=1,nx+1 do x[i] = L*(i-1)/nx end
28 :
29 : u = {{},{}} -- Set initial two solitons at x=4, x=20
30 : for i=1,nx+1 do u[1][i] = ui(x[i],4,4) end
31 : for i=1,nx+1 do u[1][i] = u[1][i] + ui(x[i],1,20) end
32 : for i=1,nx+1 do u[2][i] = intp(x,u[1],x[i],1) end
33 :
34 : tvals = {0,5,50,10,{0,.25*L,.5*L,.75*L,L}}
35 : s,st = pdeivbvt({eq,efl,efr},tvals,x,u)
36 : write_data(2,'list12_17.dat',s) -- Save same number of points
37 : write_data('list12_17t.dat',st)
38 : print('time taken for calculation = ',os.time()-tt)
```

Listing 12.17. Code segment for studying the propagation of two solitons with different amplitudes.

Results for selected times are shown in Figure 12.40. The solutions show several features as discussed above. First, the pulses appear to propagate with unchanged spatial shapes. Second, the larger amplitude pulse travels with the higher velocity and each velocity appears to be approximately twice the peak amplitude. A very interesting effect occurs when the two pulses coincide and this is seen to occur at a time of around 2.5. As seen in the figure the two pulses merge into a single solution of reduced height and broadened width. However, for longer times

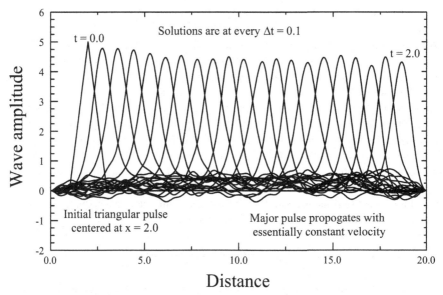

Figure 12.39. Solution waveform for the soliton wave equation of Listing 12.16.

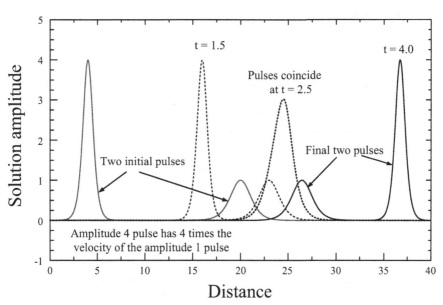

Figure 12.40. Propagation of two soliton pulses of different amplitudes and different velocities.

two pulses again emerge from the single broadened pulse with apparently unchanged height and width. It's as if the two soliton pulses have no interference with each other and simply pass each other and then continue on their way. However, the combined solution as the two pulses pass is not simply the sum of the two solutions as the peak amplitude would be increased and not decreased if this were the case. Also the sum of two individual solutions is not a solution of the nonlinear differential equation as can be verified from Eq. (12.49). The area under the solution curves remains constant as the two pulses pass and as the peak amplitude decreases and the width increases. A considerable literature exists on soliton solutions to nonlinear wave equations and especially on the Korteweg-de Vries nonlinear equation. The reader is encouraged to consult the literature and especially the web for more information and for further examples. This example is included here to illustrate the ease with which nonlinear wave equations can be simulated with the code segments developed in this chapter.

12.6.6 The Schrodinger Wave Equation

Schrodinger's wave equation is an especially important form of partial differential equation. It is the fundamental basis for much of modern physics as applied to engineering problems especially in the discipline of semiconductor engineering. The basic equation for the wavefunction associated with a particle in one spatial dimension and in time can be written as:

$$-i\hbar\frac{\partial\Psi}{\partial t} - \frac{\hbar^2}{2m}\frac{\partial^2\Psi}{\partial x^2} + V_p(x)\Psi = 0 \qquad (12.52)$$

where m is the mass of the particle being considered and $V_p(x)$ is the potential energy of the particle (as opposed to electrostatic potential). The equation is first order in time and second order in the spatial variable. This equation can be put into a normalized form by dividing by some energy reference (V_{po}) and making the following replacements:

$$\frac{V_{po}}{\hbar}t \to t; \quad \frac{\sqrt{2mV_{po}}}{\hbar}x \to x; \quad \frac{V_p}{V_{po}} \to V_p \qquad (12.53)$$

In terms of this new normalized set of variables, Schrodinger's wave equation becomes:

$$-j\frac{\partial\psi}{\partial t} - \frac{\partial^2\psi}{\partial x^2} + V_p(x)\psi = 0 \qquad (12.54)$$

Note that in this equation, the complex variable notation has been changed from the "i" used in most physics oriented works to the "j" commonly used in engineering work. This is in anticipation of using previously developed computer code which used the engineering notation. Some authors also note that the normalized equation is equivalent to working in a system of units in which $\hbar$ is equal to 1 and m is equal to 1/2.

It is also useful before considering solutions to the equation to further explore the normalization factors expressed by Eqs. (12.53). If one takes as a typical normalization energy of 1 eV, which might be a good number for an electron problem, then each normalized time unit in Eq. (12.53) will correspond to a real time of 6.58×10^{-16} sec $= 0.658$ fsec and each normalized distance unit will correspond to a real distance of $x = 1.95 \times 10^{-10}$ m $= 0.195$ nm, i.e. this is close to working in units of femto seconds and nanometers with energy expressed in eV.

The Schrodinger wave equation is functionally similar to the diffusion equation considered in Section 12.3. However, a significant difference here is the presence of the complex number j and the fact that the wavefunction is a complex number quantity as opposed to a real number quantity. An important connection to physically measured quantities is the magnitude squared of the wavefunction which gives the probability density function of finding the particle under consideration at point x. This requires that the integral of the wavefunction squared be equal to unity, or to the fact that the particle must be somewhere along the x axis. These results are summarized as:

$$\psi^*\psi = |\psi|^2 = \text{probability density function}$$

$$\int_{-\infty}^{\infty} \psi^*\psi \, dx = 1 \quad \text{normalization property} \tag{12.55}$$

A question that arises here is whether the previously developed numerical computer routines for initial value, boundary value problems can be used to solve Schrodinger's wave equation? One obvious change is the need to operate with complex algebra in the solution. For complex arithmetic operations this can be accomplished by calling into the code the complex math operation routines with the require"Complex" statement as previously used in several chapters (see Chapter 3 for some previous uses of complex numbers). If this change in the code is made and one attempts to execute the integration programs with complex numbers, errors are found in executing the math.abs() function. This function is used in various subroutines in order to establish convergence conditions looking at the absolute value of corrections to the solutions. This function does not operate properly when the calling argument is a complex number. However, a proper function is available for use with complex numbers as Complex.abs() which takes the magnitude of the complex variable with which it is called. Replacing the math.abs() function by this Complex.abs() function ensures that the corresponding function will only approach zero when both the real and imaginary parts of the corresponding complex number approach zero. This replacement is readily accomplished in computer code by the simple statement math.abs = Complex.abs. With these two modifications, the previously developed computer routines for solving PDE's of the initial value, boundary value type can be used for the solution of the Schrodinger wave equation.

For an initial demonstration example, the simplest Schrodinger problem is probably that of a particle in a box, where the potential energy becomes infinitely large at the boundaries of the box and the potential energy inside the box of some size L can be taken as zero. This can be expressed mathematically as:

$$V_p(x) = \begin{cases} 0 \text{ for } -L < x < L \\ \infty \text{ otherwise} \end{cases} \tag{12.56}$$

In this case the size of the box is taken as $2L$. In addition to the particle potential, some assumption must be made regarding the initial state of the particle wavefunction. For this example, it will be assumed that the particle is localized to some spatial region within the box and that it is traveling to the right initially with some velocity. For such an initial state, an appropriate initial wavefunction is:

$$\psi(x,0) = A \exp(-\alpha(x+x_o)^2) \exp(j\beta x) \tag{12.57}$$

This is an initial state centered at $x = -x_o$, with a Gaussian type probability density in the spatial domain and with a kinetic energy of $\beta^2 + \alpha$ and with a velocity of β in the positive x spatial direction. In order to properly normalize the wavefunction, the constant A must have the value $(2\alpha/\pi)^{1/4}$. A graph of the probability density for this function will subsequently be shown along with the time evolution of the solution.

Computer code for solving Schrodinger's equation for this particle in a box is shown in Listing 12.18. As discussed above, line 4 contains all the changes needed to use the previously developed code routines with a partial differential equation involving a complex variable. The PDE and boundary conditions are defined by the functions on lines 27 through 35. Recall that the complex quantity "j" is defined by the require "Complex" line and can be directly used in the differential equation on line 28. The spatial grid is defined on lines 37 through 40 and the initial wavefunction is defined on lines 44 through 48 with 1000 spatial intervals. In this initial defining equation (line 46) the constant A in Eq. (12.57) is taken as unity. Since this does not necessarily provide the proper normalization of the wavefunction, lines 50 and 51 calculate the area under the wavefunction squared and then properly normalize the wavefunction so that the integrated probability density is unity. The integrated probability value before and after normalization is printed to verify proper normalization. This makes uses of the sqarea() function on lines 18 through 24 to calculate the total square area under a wavefunction. An additional function tomagsq() is included on lines 11 through 17 to take as an argument a complex wavefunction array and return the corresponding magnitude squared of the wavefunction or the probability density function according to Eq. (12.55). Note that the maximum number of Newton iterations for the differential equation is set to 1 on line 9 which is possible since the Schrodinger equation is a linear equation.

For this example, the spatial interval is selected as -1 to +1 on a normalized scale. The initial wavefunction has a positive velocity of 100 units per unit of normalized time. Thus the time required to traverse the box is expected to be 0.02 normalized time units. So a time range of 0 to 0.06 was selected for the solution interval as this allows three traversals of the box. The major calling time intervals are set up on line 58 with the pdeivbvt() function called on line 59 to generate the time dependent solution. However, before this solution, another short time solution is generated on lines 54 through 56 by a call to the pdeivbvqs() function using

```
 1 : -- /* File list12_18.lua */
 2 : -- Program for transient solution of Schrodinger Wave Equation
 4 : require"Complex"; math.abs = Complex.abs -- Complex algebra
 5 : require"odefd"; require"pdeivbv"
 6 : require"intp"; require"intg"
 7 : getfenv(pdeivbv).nprint = 1 -- monitor progress
 8 : getfenv(pdebbviv).umin = {5.e-4,5.e-4,5.e-4}
 9 : getfenv(odefd).NMAX = 1
11 : tomagsq = function(s) -- Convert to magnitude squared
12 :     ns,nv = #s, #s[2]
13 :     for i=2,ns do
14 :             for k=1,nv do s[i][k] = math.abs(s[i][k])^2 end
15 :     end
16 :     return s
17 : end
18 : sqarea = function(x,y) -- Calculate area under wavefunction
19 :     local nx, sum = #x-1, 0
20 :     for i=1,nx     do -- works for nonumiform spatial grid
21 :         sum = sum + (x[i+1]-x[i])*(math.abs(y[i+1])^2+
    math.abs(y[i])^2)
22 :     end
23 :     return sum/2
24 : end
26 : -- Normalized Schrodinger equation to be solved
27 : eq = function(fv,x,t,v,vp,vpp,vt)
28 :     fv[1] = j*vt[1] + vpp[1] -- V =0 term
29 : end
30 : efl = function(fv,v,vp) -- Zero boundary value
31 :     fv[1] = v[1]
32 : end
33 : efr = function(fv,v,vp) -- Zero boundary value
34 :     fv[1] = v[1]
35 : end
37 : x = {}; nx = 1000 -- Set initial parameters
38 : L, alf, bet = 1, 75, 100
39 : xo = .5*L
40 : for i=1,nx+1 do x[i] = L*(-1 + (i-1)*2/nx) end
42 : v = {{}} -- Set initial voltage values
43 : vp,vpp = {{}}, {{}}
44 : for i=1,nx+1 do
45 :     xv = x[i]
46 :     v[1][i] = math.exp(-alf*(xv+xo)^2)*Complex.exp(bet*j*xv)
47 : end
48 : v[1][1] = Complex.new(0,0); v[1][nx+1] = v[1][1]
50 : sum =sqarea(x,v[1])^0.5; print('P = ',sum)
51 : for i=1,nx+1 do v[1][i] = v[1][i]/sum end
52 : print('Now P = ',sqarea(x,v[1]))
54 : tvals = {0,{1.e-6,1.e-5},1,5} -- Starter solution, to conserve P
55 : s = pdeivbvqs({eq,efl,efr},tvals,x,v,vp,vpp)
56 : sum = sqarea(s[1],s[3]); print('After initial short interval, P
= ',sum)
58 : tvals = {0,.08,160,10,{-.75,-.5,-.25,0,.25,.5,.75}}
59 : s,st = pdeivbvt({eq,efl,efr},tvals,x,v,vp,vpp)
60 : sum = sqarea(s[1],s[#s]);
   print('After final solution, P = ',sum)
62 : write_data('list12_18.dat',tomagsq(s)) -- Save as prob density
63 : write_data('list12_18t.dat',tomagsq(st))
Selected output:
P =    0.3804211494011
```

```
Now P =           1
After initial short interval, P =      0.9999989622412
After final solution, P =     0.99999896280236
```

Listing 12.18. Code segment for the time dependent solution of Schrodinger's wave equation in a box.

a very short time interval of 0 to 1.e-5 time units. The need for this is explained in the following paragraph.

An important feature of the time evolution of the Schrodinger wave equation is the conservation of probability. This means that the integral of the magnitude squared of the wavefunction over all space must be conserved as the wavefunction evolves in time. Any valid numerical solution of the wave equation must have this feature. It is known that the proper integration technique which preserves this feature is the trapezoidal method or as it is called when applied to PDE's, the Crank-Nicholson method. This is good in that the code developed in this chapter as Listing 12.1 and Listing 12.2 use the CN technique as the basic integration method. However, the CN (or trapezoidal) method can not be used to start the solution of a second order equation and the ode2bvfd() routine uses the backwards difference technique for an initial interval to begin a solution. This initial interval using the BD technique will not ensure the exact conservation of probability for the Schrodinger equation. However, if a short enough time interval is used, for all practical purposes, the BD technique will conserve the probability. Hence the use of an initial time interval using a log time scale from 0 to a time of 1.e-5 is set up on line 54 and used on line 55 with the pdeivbvqs() function. This initial calculation begins the solution with a time interval of 5e-8 and returns after a time interval of 1.e-5. The selected output shows that the initial integrated probability is exactly 1 as expected and that after this short time interval the integrated probability is 0.9999989622412 which is sufficiently close to unity for all practical purposes. The solution after this short time interval is then used as the initial solution for the longer time solution on lines 58 and 59. By including the vp and vpp calling arguments in the pdeivbvqs() function, this function returns arrays for the first and second time derivatives at the end of the short time solution. These arrays are then passed to the longer time solution in the pdeivbvt() function by again including these variables in the calling argument list. When these values are passed to the integration routine on line 59, the initial BD interval is not required and the solution technique begins directly with the CN technique. This eliminates errors in the magnitude squared solution for the second call to the solution routines.

What happens if the first short time interval solution is not included? In this case the call to the solution on line 59 would use an initial BD interval of 5e-5 before beginning the CN cycles. This much longer time interval with the BD technique will result in a much larger error in the initial solution and a much larger deviation from the ideal conservation of total probability. By using a very short initial time interval and then passing the appropriate time derivatives to the longer time interval solution, a complete solution can be obtained that properly conserves the total probability and which covers a much longer time interval. Because of this use of an initial BD technique to start the time solution for the pdeidbv() functions, one should always use an initial very short time interval solution to properly generate a set of time derivatives to then be used with a longer time interval solution when solving Schrodinger's equation with the routines of this chapter. The last printout in Listing 12.18 shows that the total probability for the last time solution is 0.99999896280236, indicating that the time evolution using the CN technique preserves the total probability over the total time of the solution. The reader can remove the initial short time solution on line 55 and observe the effect on the solution.

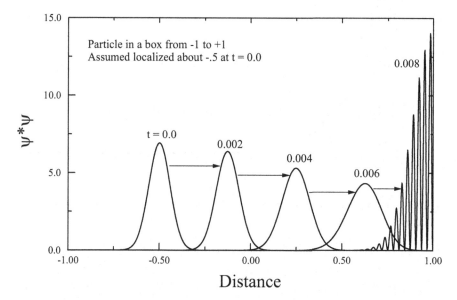

Figure 12.41. Probability density for particle in box. Localized particle moves for x = -.5 to x = 1 at time progresses for 0 to 0.008.

Some solutions for the time evolution of the probability density are shown in Figures 12.41 through 12.43. Figure 12.41 shows at t = 0.0 an initial Gaussian type wavefunction localized around x = -.5. As time progresses to t = 0.004, the wave function has progressed to about x = 0.25 and shows a slight spread in the width of the wavefunction. As the wavefunction encounters the right infinite potential boundary (see curve at t=0.008), a very rapid oscillation in the wavefunction is seen with the value at the boundary forced to zero. One should keep in mind the fact that the rapid oscillations are always present in the real and imaginary parts of the wavefunction. It is just that the magnitude squared does not show these rapid oscillations except when the particle encounters the box walls. How can one be sure that the oscillations in Figure 12.41 are a real physical effect as opposed to an artifact of the numerical solution such as observed in Figure 12.38? One of the keys is to examine the number of calculated spatial points within the oscillations. It will be found that there are a significant number of calculated spatial points within each cycle of the oscillation. For improved accuracy one might want to increase the number of spatial points used in the simulation.

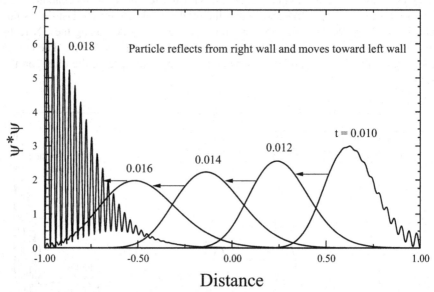

Figure 12.42. Probability density for particle in box. Particle moves from right boundary to left boundary as time progresses to t = 0.018.

Figure 12.42 illustrates the wavefunction as the particle bounces off the box wall and moves in the –x direction. For t = 0.016, the center of the wavefunction is back around x = -0.5 and shows essentially a Gaussian shape without the high frequency oscillations but with a larger mean spread in the density. The fact that the pulse regains the Gaussian shape after reflecting from the wall is another indication that the oscillations at the wall are not artifacts of the solution techniques. For t = 0.018 in Figure 12.42, the particle is strongly interacting with the left wall of the box and again rapid oscillations are sen in the magnitude squared wavefunction. Figure 12.43 shows the solution as the particle moves from near the left wall again back to the right wall. As time progresses the wavefunction is spreading more and more throughout the entire box and as time progresses further, it becomes more difficult to follow the development of the solution as the wavefunction fills the entire box with high frequency oscillations in the probability function. The final time point calculated by the code in Listing 12.18 is 0.08 and the probability function for this time is shown in Figure 12.44. In this case the high frequency nature of the wavefunction completely fills the box and there is little reminiscence of the localized nature of the initial wavefunction. The probability is approaching a more uniform probability of finding the particle anywhere within the box. Such is the nature of Quantum mechanics and the Schrodinger wave equation.

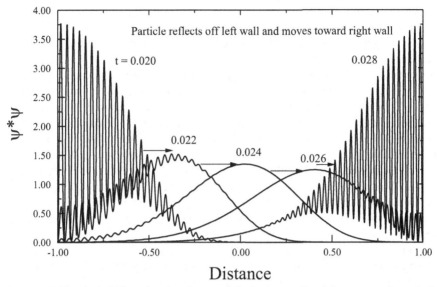

Figure 12.43. Probability density for particle in box. Particle moves from left boundary to right boundary as time progresses to t = 0.028.

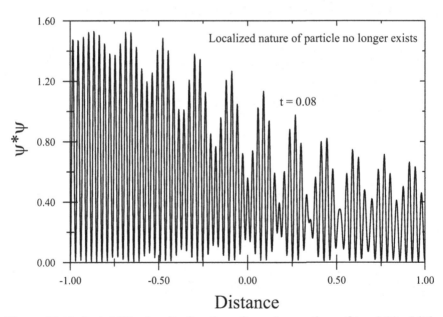

Figure 12.44. Probability density function after a longer time of t = 0.08. Little reminiscences are seen of the initial localized nature of the particle wavefunction.

Another classical example of the wave equation is that of a particle tunneling through an energy barrier where the particle does not have sufficient energy to classically get over the energy barrier. The previous example of a particle in a box can be extended to model this example problem by placing an energy barrier somewhere within the box and to the right of the initial particle location – say at x = 0. Since the normalized kinetic energy of the assumed initial wavefunction is about $1X10^4$ the height of the barrier must be somewhat larger and for an example has been selected as $1.2X10^4$ in normalized energy units. After experimenting some with a barrier width, it was decided that a width of about 0.02 in normalized distance units gives good results. The resulting code changes from Listing 12.18 are shown in Listing 12.19. Only a few changes are needed from the previous example. First the barrier properties are defined on line 26 and included in the PDE definition on lines 29 through 30. The barrier gives the PDE a dependency on the spatial variable. In order to increase the spatial resolution through the thin barrier, the number of spatial intervals is increased to 2000 on line 39. Finally the time for the calculation is set up on line 60 and the function pdeivbvt() is called on line 61 to return the solution. Values are saved for later graphing as desired. The time interval of 0 to 0.02 is much longer than needed to observe the tunneling of the initial pulse. Note that the height of the barrier is negative. This is a result of the signs taken on the normalized PDE.

```
 1 : -- /* File list12_19.lua */
 2 : -- Program for Schrodinger Wave Equation with barrier
 3 :
---
25 :
26 : xb, barrier = 0.02, -1.2e4
27 : -- Normalized Schrodinger equation to be solved
28 : eq = function(fv,x,t,v,vp,vpp,vt)
29 :    if x>=0 and x<xb then b = barrier else b = 0 end
30 :    fv[1] = j*vt[1] + vpp[1] + b*v[1]
31 : end
---
---
38 :
39 : x = {}; nx = 2000 -- Set initial parameters
---
59 :
60 : tvals = {0,.02,40,40,{-.75,-.5,-.25,0,.25,.5,.75}}
61 : s,st = pdeivbvt({eq,efl,efr},tvals,x,v,vp,vpp)
62 : sum=sqarea(s[1],s[#s]);print('After final solution,P = ',sum)
63 :
64 : write_data('list12_19.dat',tomagsq(s)) -- Save as prob density
65 : write_data('list12_19t.dat',tomagsq(st))
```

Listing 12.19. Code segment changes from Listing 12.18 for tunneling through a potential barrier.

Figure 12.45 shows some of the most interesting and important results of the transient calculation. The initial probability density as seen for the t = 0.0 curve is a Gaussian type distribution centered at x = -.5. The wavepacket moves with time to the right as in the previous example and begins to encounter the barrier between t = 0.001 and 0.002. By a time of 0.003 the pulse has interacted extensively with the barrier and one sees the beginning of a transmitted pulse and a reflected pulse. Again in the interaction region and in the reflected wave, the high frequency components of the wavefunction are strongly evident in the probability distribution at t = 0.003. By t = 0.004, there is a clearly developed transmitted pulse and a clearly defined reflected pulse both with essentially a Gaussian like probability density distribution. The transmitted pulse continues to the right and the reflected pulse continues to the left. As time continues these will encounter the walls of the enclosing box at x = +/- 1.0, being reflected again and again within the box and the barrier. However, the tunneling process is of interest here and this is adequately covered by the time scale shown in the figure.

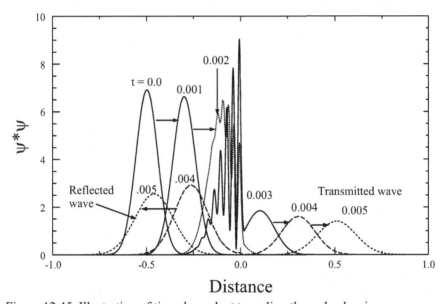

Figure 12.45. Illustration of time dependent tunneling through a barrier.

Figure 12.46 shows some more details of the numerical solution around the barrier location at a time of 0.003 where the wavefunction is strongly interacting with the barrier. The spatial scale is greatly expanded so the exponential like decay of the wavefunction within the barrier region can be seen. For this figure the barrier height has been scaled to fit on the vertical scale of the figure. Also spatial data points from the calculation are shown along the solution and it can be seen that about 10 calculated spatial data points occur within the energy barrier region. This is about the minimum that one would expect to give acceptable results and more spatial points would be desirable. However, if a uniform spatial grid is continued to be used, this would increase the computational time even further. A better approach would be to employ a non-uniform spatial grid with small step sizes around x = 0 where the barrier occurs. This will be left to the reader as an exercise to modify the code in Listing 12.19 in order to employ a non-uniform spatial grid. A finer spatial grid would certainly be required for cases where the transmission probability were considerable less than that in this example. Many practical engineering problems occur when the probability of tunneling for an individual particle is very small but observable effects are present because a large density of particles is incident on the barrier.

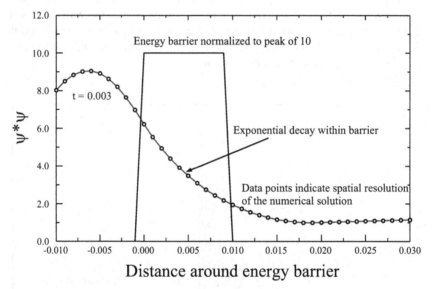

Figure 12.46. Detailed solution around the energy barrier for one particular time where strong interaction is occurring with the barrier.

From the areas under the two pulses – the transmitted and the reflected pulses – the transmission and reflection probabilities can be estimated. By such an integration process, values of 0.6457 and 0.3543 are obtained as the fraction of reflected and transmitted particles with an initial waveform as assumed in the calculation. Of course as dictated by quantum mechanics in any given measurement of such a system one would always observe a single particle as having been either reflected or transmitted with these probabilities. Such is the strange world of quantum mechanics.

The above solutions have been in terms of the normalized Schrodinger wave equation of Eq. (12.54). While the normalized equation is good for the numerical solutions, it is also appropriate to examine what actual physical parameters a particular solution might correspond to. This is going perhaps backwards from the more common case where one has a set of physical parameters which are then converted to a normalized set of parameters before the numerical solution. The normalized barrier height for the tunneling calculation has been taken as 1.2×10^4. If this were to correspond to an actual physical barrier height of perhaps 1.2eV =

$1.2 \times 1.6022 \times 10^{-19}$ joules then the normalization energy used in the normalized equations must be 0.0001eV. From this and Eq. (12.53) the corresponding normalizing factors for the spatial dimension and for time can be calculated as:

$$t_n = 6.582 \times 10^{-12} \text{ sec}$$
$$x_n = 1.952 \times 10^{-8} \text{ m} = 19.52 \text{ nm} \tag{12.58}$$

Thus the normalized tunneling results of Figure 12.45 would correspond to a physical problem of tunneling through a 1.2eV barrier by a particle with a kinetic energy of approximately 1.0eV and with the distance scale from -19.52 nm to 19.52 nm. The corresponding times on the figure would then be from 0 to $0.005 t_n = 3.29 \times 10^{-14}$ sec . Of course the vertical scale of the probability density would be much larger since the spatial scale is much smaller. The width of the tunneling barrier in actual dimensions would be 0.39 nm. One can verify that the above physical parameters would correspond to the solution given in Figure 12.45 by using the un-normalized Schrodinger equation in Listing 12.19 and using the above parameters for the spatial and time scales of the calculation. This is left as an exercise for the reader. In terms of the normalized equation, the same solution can correspond to different sets of physical parameters provided they correspond to the same set of normalized parameters. This is one advantage of working in dimensionless variables.

12.7 Boundary Value Problems (BVP) in Two Dimensions

Section 12.5 has explored partial differential equations in two variables of the initial value, boundary value type. These typically arise in physical problems involving one spatial variable and one time variable. Several examples have been given of such practical problems. The present section is devoted to PDEs in two dimensions where boundary values are specified in both dimensions. Typically these two dimensions are spatial dimensions. Perhaps the prototype BVP is Poisson's equation which in two dimensions is:

$$\frac{\partial^2 V}{\partial x^2} + \frac{\partial^2 V}{\partial y^2} = -\frac{\rho(x,y)}{\varepsilon} \tag{12.59}$$

where V is the electric potential and $\rho(x,y)$ is a spatial charge density. In a more general case, the charge density may also be a function of the potential. When the charge density is zero this is the well known Laplace equation.

A more general formulation of a two dimensional boundary value problem can be stated in terms of some general function which must be zero as in this notation:

$$F(x, y, U_{xx}, U_x, U, U_y, U_{yy}, U_{xy}) = 0 \tag{12.60}$$

where the various subscripts indicate partial derivatives with respect to the spatial dimensions. If the equation is nonlinear in the solution variable or in the partial derivatives, then a linearized version appropriate for Newton iterations can be obtained as previously discussed in Section (12.1). This general case will be considered here for code development. However, to introduce the topic, the simpler case of Poisson's equation as in Eq. (12.59) will first be considered. It will also be assumed in this introductory discussion that the spatial domain of interest is a rectangular area in the x and y coordinates.

If one has a uniform spatial grid in both dimensions of size Δx and Δy and finite differences are used to approximate the partial derivatives, then Poisson's equation becomes for a general spatial point:

$$\frac{V(j,i+1)-2V(j,i)+V(j,i-1)}{(\Delta x)^2}+\frac{V(j+1,i)-2V(j,i)+V(j-1,i)}{(\Delta y)^2}$$

$$+\frac{\rho(j,i)}{\varepsilon}=0 \qquad (12.61)$$

In this the notation j,i is used to indicate a general spatial point with j labeling the y points and i labeling the x points. When this discretization is applied to all spatial points, a set of coupled linear equations is obtained for the solution variable. The formulation of boundary values must also be considered before a complete set of equations can be described. However, for the internal points excluding the boundaries, the resulting equations couple the solution at some general point to the four neighboring points as illustrated in Figure 12.47 for the points around the (j,i) labeled general grid point. It is noted that a more general function involving first derivatives will also result in equations coupling the solution variable at the same neighboring points. It is to be noted that for a general rectangular two dimensional space as in Figure 12.47, there are several ways one can set up spatial dimensions and a spatial grid depending on which of the 4 corners of the space is

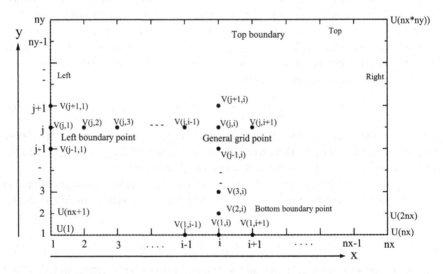

Figure 12.47. Illustration of coupled solution values for Poisson's equation.

taken as the origin of the coordinate system. For the purposes of discussion here, it will be assumed that the coordinate axes are taken as shown in the figure with the origin of both the x and y axes taken at the bottom left of the two dimensional space. This is more typical of standard directions than any of the other three possibilities. However, the nature of the set of equations does not depend on which corner is taken as the coordinate origins. Given that the origin of such a coordinate system is taken at the lower left corner, there is still the question of how to order the finite difference equation set. The equations can be ordered by counting first along the x axis and then along the y axis and this is the ordering shown in Figure 12.47. However, an equally valid ordering scheme would count along the y axis first and then along the x axis. These two ordering schemes are compared in Figures 12.48 and 12.49 for an array of 4 x axis intervals and 5 y axis intervals. Either of these schemes provides a valid and equivalent set of matrix equations.

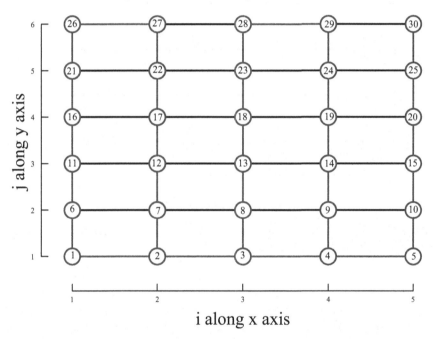

i along x axis

Figure 12.48. Illustration of numbering of nodes with incrementing along the x axis followed by incrementing along the y axis.

It is appropriate to discuss the choice of the integer i for labeling along the x axis and j for labeling along the y axis. Also note the notation in Figure 12.46 where the voltage at the nodes is written as $V(j,i)$ with the y axis integer specified first as opposed to the x axis integer. This is done to more closely identify with the labeling of rows and columns in the resulting equation matrix when the numbering scheme of Figure 12.48 is used. In this scheme, the matrix equation set corresponds to incrementing i first followed by incrementing j. Thus if the voltages are expressed as a matrix, the integer j corresponds to a row in the matrix while the integer i corresponds to a column in the matrix. This also corresponds to the storage notation in our computer language where V[j] corresponds to the jth row of the elements storing the values and V[j][i] corresponds to the ith element of the jth row.

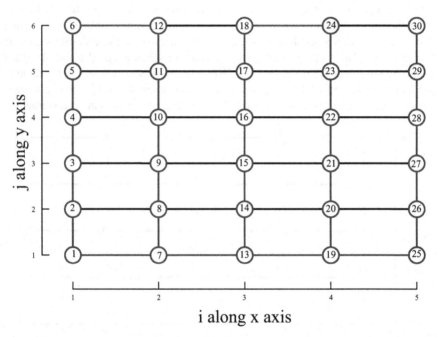

Figure 12.49. Illustration of numbering of nodes with incrementing along the y axis followed by incrementing along the x axis

In addition to the mode labeling schemes shown in Figures 12.48 and 12.49, other labeling schemes are certainly possible. Another frequently used scheme is to start at one corner and label the nodes sequentially along diagonal rows of nodes, ending at the opposite diagonal corner of the two dimensional space. The diagonal numbering scheme has certain advantages in solving the set of matrix equations and will be subsequently discussed.

The resulting sets of matrix equations resulting from the finite difference relationships of Eq. (12.61) have a special arrangement of coefficients as illustrated in Figure 12.50. This is for a particular simple form of boundary conditions which will be discussed subsequently. For the moment concentrate on the middle rows of the matrix. The dots indicate non-zero matrix elements and the other entries are zero. This form of the matrix has three non-zero rows along the diagonal corresponding to the i-1, i, and i+1 terms in the equation and two displaced diagonal rows corresponding to the j-1 and j+1 terms in the equation. This would correspond to numbering the spatial points sequentially along rows in the x direction as indicated in Figure 12.47. Numbering the spatial points sequentially along columns in the y direction will result in the three central diagonal rows corresponding to the j-1, j, and j+1 terms and the displaced diagonal rows corresponding to the i-1 and i+1 terms in the equation. The central three rows form a tri-diagonal collection of values. The separation between the main tri-diagonal elements and the off diagonal elements depends on the number of spatial grid points in the x (or y) dimension.

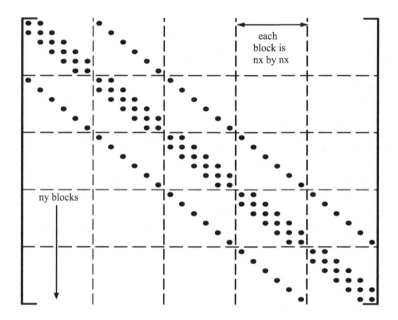

Figure 12.50. Matrix structure for coupled equation set based upon the finite difference method for a two dimensional boundary value problem. Matrix is of the "tri-diagonal with fringes" form.

The resulting equation set and form of the matrix is independent of whether one labels rows first or columns first. In either case this resulting particular form of a matrix is frequently called "tri-diagonal with fringes" with the tri-diagonal referring to the three central diagonal terms and the fringes referring to the displaced diagonal terms. As indicated in Figure 12.50 the matrix equation can also be considered as ny blocks of nx by nx matrix elements and the matrix has a tri-diagonal structure in terms of the nx by nx blocks of elements. The use of non-uniform spatial grids changes the coefficients associated with the various matrix elements, but does not modify the general structure of the matrix elements or give any additional elements. This particular matrix of 30 rows would correspond to only 6 spatial elements for the x dimension and 5 spatial elements for the y dimensions.

From this simple case, however, the structure of the equation matrix for larger numbers of spatial points can be readily visualized.

As indicated in the figure, there are two labeling sequences for the solution variable, one corresponding to the j, i spatial location and one corresponding to the location of the solution variable in the resulting set of matrix equations. The correspondence between these variables and locations is as follows:

$$U(m) = V(j,i) \text{ where } m = i + nx(j-1)$$
$$\text{with } 1 < i < nx, \ 1 < j < ny, \ 1 < m < nx * ny \tag{12.62}$$

The set of equations forms a sparse set of matrix equations. For example if the spatial grids give an array of 100 by 100 points with 10,000 resulting spatial points, the resulting matrix has less than $5 \times 10000 = 50,000$ non-zero elements as opposed to 100,000,000 elements in a completely full matrix. Less than .05% of the elements in a full matrix are thus non-zero. The elements of such a large matrix would never be stored in full form as shown in Figure 12.47 but stored by some sparse matrix storage technique. Even so the direct solution of such a sparse matrix is by no means a trivial task. For even a modest number of spatial grid points (100 by 100), the resulting matrix size and number of equations becomes very large. For this reason approximate solution methods are very frequently pursued as opposed to the direct solution of the resulting matrix equation set. Some of these approximate techniques will be discussed subsequently in this chapter. A complete discussion of such equations and solution methods can easily fill several books and the reader is referred to the literature for a more complete discussion of numerical techniques associated with the solution of sparse matrixes and with boundary value equations.

The equations appropriate for elements along the boundaries of the spatial region require special attention. Assuming again a rectangular spatial domain, there are four boundaries which can be designated as the top, bottom, left and right boundaries as illustrated in Figure 12.46. For discussion purposes, it will be assumed here that the bottom boundary corresponds to j = 1 for the y-axis labeling and that the left boundary corresponds to i = 1 for the x-axis labeling. The corner points can be associated with two possible boundaries. For example the i=1,j=1 corner point may be associated with either the left or bottom boundary. For this work, the corner points will be assumed to be associated with the x-axis boundaries so that the two lower corners will be assumed to be associated with the bottom boundary and the two upper corners will be assumed to be associated with the top boundary. If this is not the case these points can be treated as special cases in the equations defining the boundary points.

For the example of Posisson's equation, the form of the finite difference equations for the interior points with uniform spatial grids is as given by Eq. (12.61) and as shown in Figure 12.50. But how about the boundary points, what is the form of the equations for these points? The previous Figure 12.47 illustrates two boundary points – one along the bottom and one along the left boundary. For the simplest case of a known boundary value, the solution variable will be known along a boundary and only a diagonal element will be present in the resulting matrix equation. This is known as the Dirichlet boundary condition. The Neumann condition specifies the values of the normal gradients on the boundary and in general more complex boundary conditions, known as mixed boundary conditions, may exist involving both the value of the solution variable and the normal derivative. For the development here, it will be assumed that the various boundary conditions may be general functions of the solution variable and the first derivatives on the boundary. These can be specified in functional notation as:

$$\text{Bottom:} fb(x,V,\frac{\partial V}{\partial x},\frac{\partial V}{\partial y},i) = 0$$

$$\text{Top:} \quad ft(x,V,\frac{\partial V}{\partial x},\frac{\partial V}{\partial y},i) = 0$$

$$\text{Left:} \quad fl(y,V,\frac{\partial V}{\partial x},\frac{\partial V}{\partial y},j) = 0 \tag{12.63}$$

$$\text{Right:} \quad fr(y,V,\frac{\partial V}{\partial x},\frac{\partial V}{\partial y},j) = 0$$

The presence of the derivative terms couples the neighboring solution values into the boundary equations. For example along the lower boundary at point i the possible presence of the x-derivative will involve the $V(1,i-1)$, $V(1,i)$ and $V(1,i+1)$ terms as indicated in Figure 12.47. The possible presence of the y-directed or normal derivative will involve the $V(1,i)$, the $V(2,i)$ and possibly the $V(3,i)$ solution values. The simplest normal derivative would involve only the two vertical terms and force these values to be equal. However the use of a two point approximation, is not a very accurate approximation for the derivative and reasonably accuracy in the normal derivative dictates that three spatial points be used in the numerical derivative. However, this does complicate the resulting set of matrix equations somewhat as it introduces additional matrix terms not shown in the previous representation of Figure 12.48 for the form of the matrix equations. Including these possible terms along the four boundaries, gives the resulting form of the matrix equations as shown in Figure 12.51. The extra terms from boundary conditions are shown as open circles in the matrix. The resulting equation is still dominated by a central diagonal array of elements but in this case some additional elements are found along two extra diagonal rows within each sub block of the matrix making non-zero elements in five diagonal rows of the matrix. In addition, there are two additional fringe rows in the first and last block elements corresponding to the first and last nx sets of equations. Along the first nx equations these are readily seen to arise from the third row elements in Figure 12.47 used in approximating the normal derivative along the bottom boundary condition. Not all of these additional elements will be present in a given problem and they will only appear when the boundary conditions involve the normal derivative. However, in the most general case solution methods should be able to handle these additional elements. Many text books that discuss such boundary value problems do not include such terms and must therefore be assumed to be using a much simpler approximation for any normal

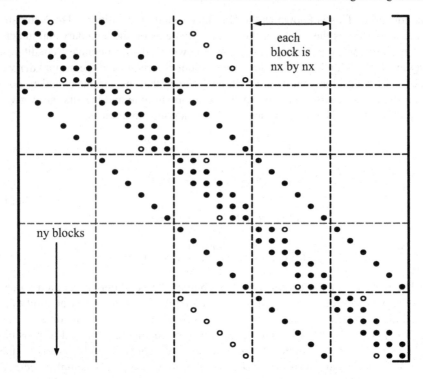

Figure 12.51. Structure of matrix equations with boundary equations involving three point derivative terms.

derivative boundary value conditions. With this as background we are not ready to formulate some computer code for such a BVP.

12.7.1 Formulation of Matrix Equations for BVP in Two Dimensions

The numerical solution of a two dimensional BVP can now be thought of as composed of two major parts: setting up the matrix equations and solving the resulting set of equations. This section will discuss generating the matrix equation set and the next section will discuss possible solution methods. Only the numbering scheme that numbers along the x axis first will be considered for formulating the matrix equations. If numbering along the y axis first is desired, a conversion function will be developed to convert from one form to the other. Developing a subroutine to generate the matrix equations is to a large extent a bookkeeping task to implement the math operators. Assuming that the differential equation is possibly nonlinear and of the form of Eq. (12.60) a linearization step can be performed as:

$$F(x, y, U_{xx}, U_x, U, U_y, U_{yy}) = 0$$

$$F + \frac{\partial F}{\partial U_{xx}} \frac{\partial^2 u}{\partial x^2} + \frac{\partial F}{\partial U_x} \frac{\partial u}{\partial x} + \frac{\partial F}{\partial U} u + \frac{\partial F}{\partial U_y} \frac{\partial u}{\partial y} + \frac{\partial F}{\partial U_{yy}} \frac{\partial^2 u}{\partial y^2} = 0 \qquad (12.64)$$

On the second line the function F and the partial derivatives of F are to be evaluated at some initial guess (U) of the solution and the function u is the correction to the initial guess. For a linear PDE, the initial guess is not important and can be taken as zero. For a nonlinear equation the solution must be iterated until a converged solution is obtained. The use of this form even for linear PDE's allows a common formulation for any type of boundary value problem and the computer can then set up the required matrix equations. Another numerical complexity can be the use of non-uniform spatial grids. This has previously been addressed in Section 11.5 dealing with single variable boundary value problems. The reader is referred back to that discussion for equations for the numerical first and second derivatives with a non-uniform spatial grid.

A code segment for generating the matrix equations for a two dimensional BVP is shown in Listing 12.20. The calling argument for this function is of the form setup2bveqs({fpde,fb,ft,fl,fr},x,y,u) where the first table entry is a listing of the functions involved in the solution: fpde() = the differential equation, fb() = bottom boundary equation, ft() = top boundary equation, fl() = left boundary equation and fr() = right boundary equation. The other arguments are the spatial x and y arrays and finally the initial approximation to the two dimensional solution. Some of the major features of the code will be briefly discussed. Lines 5 through 19 set up appropriate constants and arrays for the implementation. The code on lines 13 through 17 permit the input u array to be either a two dimensional array of solution values with the j, i labeling or a one dimensional array with the m labeling of Eq. (12.62). The a and b matrices are defined on line 10 with the two dimensional a matrix being defined by a = Spmat.new(nx,ny). This function simply returns a two dimensional array with ny rows and no entries in the columns. It also stores information regarding the nx and ny sizes so that the array can be further manipulated (such as reversing the x and y labeling of equation numbers). This information is stored as a table at index 0 as a[0] = {nx,ny,dir} where dir is a parameter indicating labeling as x first or y first. Lines 20 through 61 define local functions to interface to the equation and boundary functions. These are subsequently discussed. The code on Lines 62 through 69 attempts to find an appropriate probe factor to use in implementing the numerical partial derivative evaluations as indicated in Eq. (12.64). The probe factors are simple scaled to the maximum solution value on line 68.

The heart of the code is the loops over the y (or j) dimension from line 70 to line 124 and the x (or i) dimension from line 71 to line 122. The code on lines 72 through 99 set up factors (such as fx1, fx2, fx3, fxx1, fxx2, fxx3, etc.) to be used in evaluating the numerical derivatives. These factors are the terms needed to

```
 1 :  -- File pde2bv.lua --
 2 :  -- Code for PDE BV problems in 2 dimensions -- rectangular grid
 3 :  require'sdgauss'
 4 :  setup2bveqs = function(eqs,x,y,u,ndg) -- Set up 2D equation set
 5 :    -- u has the form u[j][i], j -> y grid and i -> x grid
 6 :    local u = u
 7 :    local uxx,ux,uij,uy,uyy,fxx,mxx,vx,xy
 8 :    local nx,ny,ndx,ndy = #x, #y, 0, 0
 9 :    local fctuxx,fctux,fctu,fctuy,fctuyy = FACT,FACT,FACT,
          FACT,FACT
10 :    local a, b = Spmat.new(nx,ny), Spmat.new(nx,-ny)
11 :    local sx,sy,ffxx,ffyy = {},{},{0,0,0}, {0,0,0}
12 :    local alfxi,alfyi,fx1,fx2,fx3,fy1,fy2,fy3
13 :    if #u~=nx*ny then -- for u in x,y format
14 :      local ux = {}
15 :      for j=1,ny do for i=1,nx do m = i+(j-1)*nx;
              ux[m] = u[j][i] end end
16 :      u = ux
17 :    end
18 :    for i=1,nx-1 do sx[i] = x[i+1] - x[i] end
19 :    for j=1,ny-1 do sy[j] = y[j+1] - y[j] end
20 :    local function setx(fxc,ffxc) -- Store x oriented elements
21 :      if fxc~=0.0 then for k=1,3 do
22 :        mxx = mx+k-2
23 :        if mxx==m and ndg~=nil then
              am[mxx][1] = am[mxx][1] + fxc*ffxc[k]
24 :        else am[mxx] = (am[mxx] or 0) + fxc*ffxc[k] end
25 :      end end
26 :    end
27 :    local function sety(fyc,ffyc) -- Store y oriented elements
28 :      if fyc~=0.0 then for k=1,3 do
29 :        mxx = my +(k-2)*nx
30 :        if mxx==m and ndg~=nil then
              am[mxx][2] = am[mxx][2] + fyc*ffyc[k]
31 :        else am[mxx] = (am[mxx] or 0) + fyc*ffyc[k] end
32 :      end end
33 :    end
34 :    local function bound(eq,x,uji,ux,uy,ij,nfeq)
35 :      if nfeq<4 then
36 :        fv = eq(x,uji,uy,ux,ij)
37 :        fu1 = (eq(x,uji,uy,ux+fctux,ij)-fv)/fctux
38 :        fu2 = (eq(x,uji,uy+fctuy,ux,ij)-fv)/fctuy
39 :        fu = (eq(x,uji+fctu,uy,ux,ij)-fv)/fctu
40 :      else
41 :        fv = eq(x,uji,ux,uy,ij)
42 :        fu1 = (eq(x,uji,ux+fctux,uy,ij)-fv)/fctux
43 :        fu2 = (eq(x,uji,ux,uy+fctuy,ij)-fv)/fctuy
44 :        fu = (eq(x,uji+fctu,ux,uy,ij)-fv)/fctu
45 :      end
46 :      return fv, 0.0, fu1, fu, fu2, 0.0
47 :    end
48 :    local function getfac(i,j,nfeq)
49 :      eq,xi,yj = eqs[nfeq], x[i], y[j]; uji = u[i+(j-1)*nx]
50 :      if nfeq==1 then
51 :        fv = eq(xi,yj,uxx,ux,uji,uy,uyy,i,j)
52 :        return fv, (eq(xi,yj,uxx+fctuxx,ux,uji,uy,uyy,i,j)-
              fv)/fctuxx,
53 :        (eq(xi,yj,uxx,ux+fctux,uji,uy,uyy,i,j)-fv)/fctux,
54 :        (eq(xi,yj,uxx,ux,uji+fctu,uy,uyy,i,j)-fv)/fctu,
```

```
55 :            (eq(xi,yj,uxx,ux,uji,uy+fctuy,uyy,i,j)-fv)/fctuy,
56 :            (eq(xi,yj,uxx,ux,uji,uy,uyy+fctuyy,i,j)-fv)/fctuyy
57 :       else
58 :            if nfeq<4 then return bound(eq,xi,uji,ux,uy,i,nfeq)
59 :            else return bound(eq,yj,uji,ux,uy,j,nfeq) end
60 :       end
61 :  end
62 :  mxx,vx,vy = 0.0, nx/abs(x[nx]-x[1]), ny/abs(y[ny]-y[1])
63 :  for j=1,ny do for i=1,nx do -- Find maximum solution value
64 :       mxx = max(mxx,abs(u[i+(j-1)*nx]))
65 :  end end
66 :  if mxx~=0.0 then -- Scale probe factors to max solution value
67 :       fctu = FACT*mxx
68 :       fctuxx,fctux,fctuy,fctuyy = fctu*vx^2,fctu*vx,fctu*vy,
                  fctu*vy^2
69 :  end
70 :  for j=1,ny do -- Loop over y-dimensions
71 :       for i=1,nx do -- Loop over x-dimensions
72 :            if j==1 then -- Bottom boundary row
73 :                 alfyi = sy[2]/sy[1]; fxx = 1/(alfyi*(sy[2]+sy[1]))
74 :                 fy1,fy2,fy3,ndy,nfeq = -alfyi*(2+alfyi)*fxx,
                         (1+alfyi)^2*fxx,-fxx,1,2
75 :            elseif j==ny then -- Top boundary
76 :                 alfyi = sy[ny-1]/sy[ny-2];
                         fxx = 1/(alfyi*(sy[ny-1]+sy[ny-2]))
77 :                 fy1,fy2,fy3,ndy,nfeq = alfyi^2*fxx,-
                         (1+alfyi)^2*fxx, (1+2*alfyi)*fxx,-1,3
78 :            else -- General interior point
79 :                 alfyi = sy[j]/sy[j-1];
                         fxx = 1/(alfyi*(sy[j]+sy[j-1]))
80 :                 fy1,fy2,fy3,ndy=-alfyi^2*fxx, (alfyi^2-1)*fxx,fxx,0
81 :                 fxx = 2/(sy[j]*(sy[j]+sy[j-1]))
82 :                 ffyy = {alfyi*fxx,-(alfyi+1)*fxx,fxx}
83 :                 if j>1 and j<ny then nfeq = 1 end
84 :            end
85 :            if i==1 then -- Left boundary
86 :                 alfxi = sx[2]/sx[1]; fxx = 1/(alfxi*(sx[2]+sx[1]))
87 :                 fx1,fx2,fx3,ndx = -alfxi*(2+alfxi)*fxx,
                         (1+alfxi)^2*fxx,-fxx,1
88 :                 if j>1 and j<ny then nfeq = 4 end
89 :            elseif i==nx then -- Right boundary
90 :                 alfxi = sx[nx-1]/sx[nx-2];
                         fxx = 1/(alfxi*(sx[nx-1]+sx[nx-2]))
91 :                 fx1,fx2,fx3,ndx = alfxi^2*fxx,-(1+alfxi)^2*fxx,
                         (1+2*alfxi)*fxx,-1
92 :                 if j>1 and j<ny then nfeq = 5 end
93 :            else -- General interior point
94 :                 alfxi = sx[i]/sx[i-1];
                         fxx = 1/(alfxi*(sx[i]+sx[i-1]))
95 :                 fx1,fx2,fx3,ndx = -alfxi^2*fxx,
                         (alfxi^2-1)*fxx,fxx,0
96 :                 fxx = 2/(sx[i]*(sx[i]+sx[i-1]))
97 :                 ffxx = {alfxi*fxx,-(alfxi+1)*fxx,fxx}
98 :                 if j>1 and j<ny then nfeq = 1 end
99 :            end
100:            -- Now evaluate derivatives
101:            if j==1 or j==ny then jj,ii = j+ndy, i
102:            elseif i==1 or i==nx then jj,ii = j, i+ndx
103:            else jj,ii = j, i end
```

```
104 :              m = ii + (jj-1)*nx
105 :              ujim,uji,ujip,ujmi,ujpi = u[m-1],u[m],u[m+1],
                       u[m-nx], u[m+nx]
106 :              ux,uxx = fx1*ujim + fx2*uji + fx3*ujip,
                       ffxx[1]*ujim + ffxx[2]*uji + ffxx[3]*ujip
107 :              uy,uyy = fy1*ujmi + fy2*uji + fy3*ujpi,
                       ffyy[1]*ujmi + ffyy[2]*uji + ffyy[3]*ujpi
108 :              -- Now probe equations for dependences
109 :              fv, fxx,fx,fu,fy,fyy = getfac(i,j,nfeq)
110 :              m = i + (j-1)*nx; b[m] = -fv -- diagonal at m,m
111 :              am = a[m] -- work on mth row
112 :              if ndg==nil then am[m] = 0 else am[m] = {0,0,0} end
113 :              if fu~=0.0 then -- Now store diagonal elements
114 :                  if ndg==nil then am[m] = fu
115 :                  else am[m][3] = 0.5*fu end
116 :              end
117 :              mx, my  = m+ndx, m+ndy*nx
118 :              ffx, ffy = {fx1,fx2,fx3}, {fy1,fy2,fy3}
119 :              setx(fxx,ffxx); setx(fx,ffx) -- Store x elements
120 :              sety(fy,ffy); sety(fyy,ffyy) -- Store y elements
121 :          end
122 :      end
123 :      return a,b -- Return matrix elements
124 : end
125 : setfenv(setup2bveqs,{table=table,Spmat=Spmat,abs=math.abs,
             ax=math.max,FACT=1.e-4,
126 : print=print})
```

Listing 12.20 Code segment for setting up matrix equations for two dimensional BVP on a rectangular grid.

evaluate the numerical partial derivatives for a non-uniform spatial grid. Special code is required if the node under consideration is on one of the four boundaries as opposed to a general inside node point. The comments indicate which sections apply to the boundaries and which apply to the general interior points. After setting up derivative factors, the numerical first and second derivatives are evaluated on lines 106 and 107. The function getfac() is then called on line 109. This function basically returns the function value and all the five partial derivatives identified in Eq. (12.64). The returned partial derivatives and the previously evaluated derivative factors are then used on lines 110 through 122 to set up the appropriate non-zero matrix elements in the a matrix. These are set by calling the local setx() and sety() functions as defined on lines 20 through 33 for setting the x oriented and y oriented derivative factors on lines 119 and 120.

Before each returned partial derivative is used in setx() or sety(), it is tested for a zero value and if found to be zero, no entry is made in the equation matrix. These tests are performed on lines 21 and 28. One perhaps surprising feature of the matrix storage (for a) is that under some conditions (when the ndg parameter is set) the diagonal elements are stored not as a single value but as a table of three values – see lines 23, 30 and 112 of the code. The contributions to the diagonal element from the x oriented derivatives are stored in the first element and the contributions from the y oriented derivatives are stored in the second element. Contributions from an extra dependency of the equation directly on the solution variable are stored in the third element of the diagonal term -- see line 115. This is done in anticipation of future approximate methods of solving the set of matrix equations where the individual contributions are needed in approximate techniques that interchange the role of the rows and columns of the matrix. For the discussion here this feature can be ignored but will be discussed again later.

Finally a brief discussion is in order for the getfac() function on lines 48 through 61. Basically this evaluates and returns all the various partial derivatives needed in the program. The function is passed a parameter nfeq which indicates if the evaluation is for an internal point or for a boundary point. A value of 1 for this parameter indicates an internal point and the derivatives are evaluated on lines 50 through 56. If the parameter value is greater than 1, indicating a boundary point, the extra function bound() is called on line 58 or 59 to set up the appropriate calls to the supplied boundary value functions. The bound() function on lines 34 through 47 then uses the appropriate user specified boundary function to evaluate the boundary elements terms for the matrix.

While the code in Listing12.20 is fairly long it only involves concepts previously used in this work such as the numerical partial derivatives and the appropriate factors for derivatives with a non-uniform spatial grid. There are the same factors used in Chapter 11 (Section 11.5) and the reader is referred to that discussion for the definitions of these terms. The code on lines 72 through 99 set up the various alfx and alfy factors for use in the numerical partial derivatives when using a non-uniform spatial grid. The setup2bveqs() function finally returns the a and b matrices on line 123 as needed for the matrix formulation au = b. All that remains after this is to solve the set of sparse matrix equations.

Before discussing methods to solve the equations it is appropriate to discuss the method used here to store the sparse matrix values. It is really quite simple and straightforward. As opposed to many computer languages such as C which require full storage of arrays, the language used here is very efficient is storing tables of any length. For example one can define t = {} as a table and then define t[100] = 100 and t[200] = 200. Memory will only be used for the two defined elements (plus some overhead for the table itself) so the sparse matrix storage requires that only the nonzero elements for a table be defined. Of course row tables must be established for each row in the two dimensional matrix. No new data structures or manipulations are required for storing the sparse matrix elements. As previously noted, the code in Listing 12.20 only places entries in the coefficient matrix if a partial derivative term is non-zero. As previously stated the diagonal matrix elements are stored as a table of two parts with the sum of the two values being the total diagonal element value.

As an example of the use of this setup2bveqs() function, Listing 12.21 shows a code segment for obtaining the matrix equations for perhaps the simplest of problems involving Poisson's equation. Lines 7 through 14 define the PDE and boundary conditions. The boundary conditions are u = 0 along the lower boundary, u = Vm along the upper boundary and zero normal derivatives along the left

```
 1 : -- File list12_21.lua --
 2 : -- Simple Example of BV problem in 2 D -- rectangular grid
 3 :
 4 : require"pde2bv"; require"spgauss"; require"sdgauss"
 5 : getfenv(spgauss).nprint=1; getfenv(spgauss).usage=2
 6 :
 7 : Vm = 1.0 -- Define equations to be solved
 8 : feq = function(x,y,uxx,ux,u,uy,uyy,i,j)
 9 :    return uxx + uyy -- Poisson's equation
10 : end
11 : fb = function(x,u,uy,ux,i) return u end -- 0 at bottom
12 : ft = function(x,u,uy,ux,i) return u - Vm end -- Vm at top
13 : fr = function(y,u,ux,uy,j) return ux end -- zero derivatives
14 : fl = function(y,u,ux,uy,j) return ux end
15 :
16 : x,y,u = {},{},{}; xmax,ymax = 1.0,1.0
17 : Nx,Ny = 40,40 -- 40 by 40 uniform grid
18 : nx,ny = Nx+1,Ny+1
19 : for i=1,nx do x[i] = xmax*(i-1)/Nx end
20 : for i = 1,ny do y[i] = ymax*(i-1)/Ny end
21 : for j = 1,ny do   -- Set zero initial values
22 :    u[j] = {}
23 :    for i = 1,nx do u[j][i] = 0 end
24 : end
25 : -- Set up matrix equations
26 : a,b = setup2bveqs({feq,fb,ft,fl,fr},x,y,u)
27 :
28 : _,_,nel = spgauss(a,b) -- usage statistics -- Try sdgauss()
29 : print('max #matrix elements = ',nel[2][#nel[2]])
30 : bb = to2darray(b) -- Convert to x,y array
31 : write_data("list12_21.dat",bb)
32 :
33 : udiff,mdiff,im,jm = {},0.0,0,0
34 : udiff[0] = b[0]; Vmx = b[#b]
35 : for j=1,ny do -- Calculate error in solution
36 :    for i=1,nx do
37 :        m = i + (j-1)*nx
38 :        udiff[m] = math.abs((Vmx*(j-1)/(ny-1)) - b[m])
39 :        if udiff[m]>mdiff then mdiff,im,jm = udiff[m],i,j end
40 :    end
41 : end
42 : print('max error = ',mdiff, '\nat j,i = ',jm,im)
43 : sdsav = to2darray(udiff) -- 2D Array of errors
44 : write_data('tmp12_21a.dat',sdsav) -- Save errors
45 : write_data('tmp12_21b.dat',nel) -- Save fill statistics
Selected output:
Completed row  100      in spgauss
Number of matrix elements =    10484
Completed row  200      in spgauss
Number of matrix elements =    13818
------
Completed row  1600     in spgauss
Number of matrix elements =    60261
max #matrix elements =         60216
max error =    5.7731597280508e-015
at j,i =        35       19
```

Listing12.21. Example code segment for solution of simple two dimensional Poisson equation problem.

and right boundaries. This is simply a parallel plate capacitor where the potential should vary linearly with distance between the upper and lower boundaries. The x and y grids are setup on lines 16 through 24 along with the initial solution array (with zero values). The set of finite difference equations in matrix form is obtained by calling the setup2bveqs() function on line 26. This set of equations is then solved by the spgauss() function on line 28. The remainder of the code is about storing the solution values and comparing with the known theoretical solution values. However before discussing the results, the next section considers various approaches to solving the set of matrix equations.

12.7.2 Solving Matrix Equations for BVP in Two Dimensions

The simplest and most straightforward method for solving the matrix equations is simply to use a direct solution method such as Gauss elimination. This is basically the method used in Listing 12.21 where the solution is obtained on line 28 by calling the spgauss() function which as discussed in Section 4.1 is a Gauss elimination method adapted for sparse matrices. As an alternative to direct solution there are many approximate methods that have been developed over the years for the iterative solution of large sets of linear equations. The basic question to be addressed is why would one use any method other than direct solution by Gauss elimination? The answer lies in the excessively large storage requirements and time requirements that can result from direct Gauss elimination when the spatial grid sizes become large. For example if one wants 100 by 100 x and y spatial intervals then one has an equation matrix of size 101X101 = 10,201 rows in the matrix. While the Gauss elimination process starts with only 5 (or fewer) non-zero elements per row, the number of non-zero elements grows rapidly as Gauss elimination proceeds. This is typically referred to as matrix "fill". By looking at the simple form of the matrix equation set in Figure 12.50 one can see that in the worst case, Gauss elimination will cause fill to occur completely between the major diagonal elements and the lower fringe row of elements. Since this is of width nx, in the worst case the total number of elements can be expressed as:

$$N_{el} <= ny(nx)^2 \tag{12.65}$$

For the above example of 100 by 100 spatial intervals this is 1,030,000 non-zero elements or just over 1 million elements. This can begin to tax the storage capacity of computers and can require long times for the direct solution. However it is stillmuch less than the full matrix size which is $nx^2 ny^2 = 104,060,401$.

However, the simplicity of the direct solution by Gauss elimination would dictate that this simple method be used whenever possible. The limits to a practical Gauss elimination process are not fixed but grow with each new generation of computer which gives increased storage capacity and execution speed. Certainly one should consider at present the use of Gauss elimination for any problem with a grid size of 100 by 100 or smaller. In a few years when the speed of computers is

1000X that of today and storage capacities are equivalently increased, this limit might be increased to 1000 by 1000 elements.

It is noted from Eq. (12.65) that one should take the x axis along the dimension with the smallest number of data points as this will give the smallest total value of matrix elements as expressed by the above equation. This would correspond to incrementing the equation count along the y axis first as illustrated in Figure 12.48 as opposed to incrementing first along the x axis as illustrated in Figure 12.47. However, there is a fairly simple transformation between the matrix elements in order to transfer from one labeling scheme to the other as illustrated in Figure 12.52. Elements along a particular row remain elements along a row with the axis transformation. However the matrix row associated with a particular x,y grid point changes as indicated in the figure from some m element to some mn element. Also the elements adjacent to the diagonal element and the fringe elements exchange places as indicated in the figure. This transformation is relatively easy to program and a code segment called rev2deqs() is included in the pde2bv package of functions to perform this axis reversal. Thus the approach taken here will be to always generate the set of matrix equations based upon the x axis first labeling and then to simply call the rev2deqs() function with the a,b arguments if the reversed axis labeling of the node points is desired. For minimum matrix fill this reversal should always be done if ny<nx.

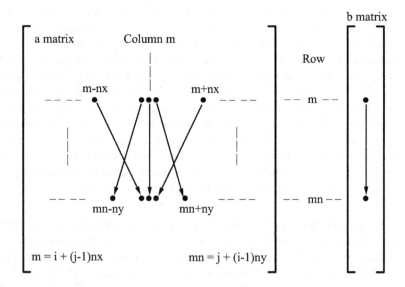

Figure 12.52. Illustration of matrix transformations for converting from x axis first labeling to y axis first labeling.

The topic of ordering the grid points in the matrix of equations is a topic that has received much attention over the years. In the most general case any given grid point can be labeled with any desired integer value between 1 and nx*ny as long as each integer is used only once in the labeling scheme. This produces an almost infinite set of possible labeling schemes. The x axis first or y axis first schemes discussed here are two of the most simple and straightforward labeling schemes. However, they are known to not be the most optimal labeling approaches if the goal is to keep the matrix fill to a minimum for many types of boundary value problems. Other possible ordering schemes go by such terms as diagonal ordering, alternating diagonal ordering, checkerboard ordering, one-way dissection, nested dissection, minimum degree ordering and Cuthill-Mckee ordering. In many cases the name gives some hints as to the node ordering approach used. A good review of the various methods has been given by Selberherr (in Analysis and Simulation of Semiconductor Devices, Springer-Verlag, 1984). Although in some cases there can be considerable savings in memory storage by minimizing matrix fill using these numbering schemes, most of these approaches will not be pursued here. Only the diagonal ordering is subsequently discussed as this is a simple extension that is easily implemented. The interested reader is encouraged to consult other references for other alternative node labeling approaches.

The spgauss() function used in Listing 12.21 and supplied by the require"spgauss" statement on line 26 was previously introduced in Chapter 4 and the reader is encouraged to examine the computer code for spgauss() in that chapter if desired. The program spgauss() provides additional feature if the "nprint" or "usage" parameters are set to non zero values as illustrated on line 5 of Listing 12.21. The nprint parameter causes a "completed row xxxx in spgauss" statement to be printed every 100 rows of the matrix as illustrated in the selected output shown in Listing 12.21. With this parameter one can follow the progress of the Gauss elimination rather than just see the computer go blank for some extended period. The usage parameter collects information on the matrix fill as the elimination process progresses. A 2 value for the usage parameter collects fill information as each row is eliminated while a 1 value collects information on the total number of matrix elements only at the end of the elimination phase, where presumably the largest storage requirement occurs.

The results obtained in Listing 12.21 for a 40 by 40 array of spatial points and included in the printed output, shows that spgauss() requires storage of 60,217 non zero elements at the end of the Gauss elimination process. This can be compared with the upper limit of 68,921 elements as given by Eq. (12.65). The simple upper limit is about 14% too high in this example. Figure 12.53 shows how the matrix fill increases as the Gauss elimination process proceeds with this example. This data was obtained from the nel array table returned by spgauss() on line 28 and stored in a saved file on line 45 in Listing 12.21. The first few rows and the last few rows cause essentially no increase in fill because only diagonal matrix elements exist for the fixed lower and upper boundary values. For the Gauss elimination steps for the other rows, the number of non-zero matrix elements increases essentially linearly with the row number that has been completed in the elimination process, reaching the final value of 60,217 elements. For this example, each step in the row elimination process increases the fill by approximately 33 elements which is the slope of the line in the figure. This is slightly less than the nx value of 41. This linear increase in fill is to be expected for the row or column ordering although there may be some variations, depending on the particular differential equation and the boundary conditions. Reversing the x and y axis will also give a slightly different total number of filled elements. The dotted curve labeled sdgauss showing a smaller matrix fill in the figure is subsequently discussed.

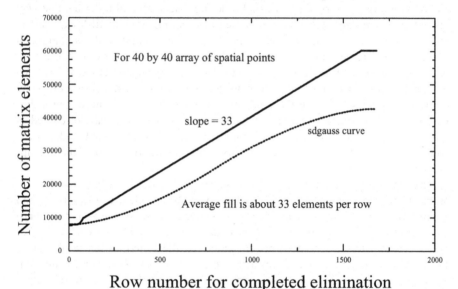

Row number for completed elimination
Figure 12.53. Increase in matrix fill with development of Gauss elimination.

The example of Listing 12.21 is perhaps the simplest example that can be simulated for Poisson's equation – no charge density and two opposite boundaries at fixed potentials, i.e. a parallel plate capacitor. However, some useful results regarding the solution of BVP's can be obtained from this example. The set of matrix equations is set up by the call to setup2dveqs() on line 26 and the spgauss() function is called on line 28 to solve the equation set. The returned solution set is a linear array of potential values with numbering from 1 to nx*ny. For most graphing programs, this needs to be converted back to a two dimensional array of potential values corresponding to the j,i values of the two dimensional spatial arrays. This is done by the call to the to2darray() function on line 30 of the listing. Details of this conversion function are not shown, but the reader can examine the code in the pde2bv.lua file if desired. Figure 12.54 shows a graph of the numerically calculated potential as a function of the two dimensions of the capacitor. This is for a spatial grid of 40 by 40 spatial intervals and the solution is just as expected with a linear variation between the two capacitor plates taken as being at

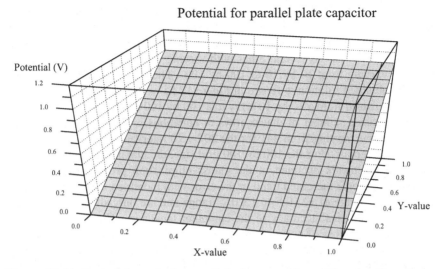

Figure 12.54. Potential graph for parallel plate capacitor. V varies linearly between capacitor plates.

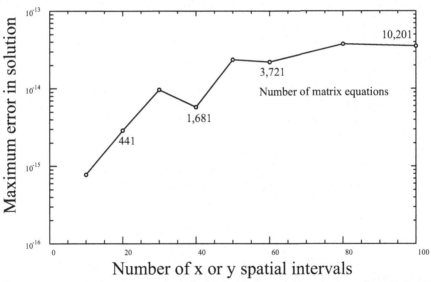

Figure 12.55. Maximum error in solution with increasing number of spatial grid points.

y = 0 and y = 1.0. While this is a simple example, the exact solution is known so the numerical results can be compared with the exact results and any errors due to the numerical techniques can be evaluated. To perform this comparison, the code in Listing 12.21 was run several times with varying numbers of spatial grid point intervals varying from 10 by 10 to 100 by 100. The maximum error in the numerically obtained solution is shown in Figure 12.55 as the number of spatial grid points is increased from a 10 by 10 array to a 100 by 100 array.

The increasing number of matrix rows or equations is shown by the numbers along the curve, varying from 441 for the 20 by 20 intervals to 10,201 for 100 by 100 spatial intervals. For the 100 by 100 grid size, the maximum number of non-zero matrix elements obtained in the Gauss elimination process was 975,546. While there is some increased trend in the error with increased matrix size, the errors are in the range of $1X10^{-15}$ to $1X10^{-14}$ range and are only slightly larger than the intrinsic numerical accuracy of the computer language. The results are basically a test of the loss of accuracy in the Gauss solution technique since for this problem the finite difference equations result in a set of exact equations for any grid size. For the linear solutions, there is no error in approximating the differential equation by finite difference equations. The results provide some confidence in the fact that the sparse Gauss elimination process introduces little error in the solution even for large numbers of equations. Hopefully this result will be carried over into more complex PDE's where an exact solution is not known. A final result of the accuracy studies is shown in Figure 12.56 which shows the details of the solution errors for the case of 100 by 100 spatial intervals. The error is plotted as a function of position in the x-y plane and it can be seen that the error is fairly uniformly distributed over the plane with no one region having an excessively large error.

The direct matrix solution by Gauss elimination as discussed above and as used in Listing 12.21 is the most straightforward approach to solving the set of coupled finite difference equations. However, for some problems, the number of equations and resulting storage requirements may be too large for this direct approach. Figure 12.57 shows how the storage requirement grows as the number of spatial intervals increases over the range of 10 to 100 intervals with equal numbers of x and y intervals. The requirement is seen to be close to the theoretical upper limit of the cube of the number of spatial intervals. So for example if a spatial grid is desired with 1000 by 1000 intervals then this would have slightly over $1X10^{6}$ equations and with fill around $9X10^{8}$ non zero matrix elements at the end of the Gauss reduction steps. At present this is too large to handle on PCs so another approach might be useful even at the expense of solution accuracy. Some approaches to approximate solution techniques will thus be discussed to complete this section. Such approaches typically require iteration over several approximations with the accuracy improving at each iterative step.

The matrix fill associated with solving a 2-dimensional PDE on a rectangular grid is dependent on the numbering technique for the spatial points. For different numbers of x and y oriented spatial points, this is obvious from the form of the matrix equations shown in Figure 12.51. For example with x first numbering, the

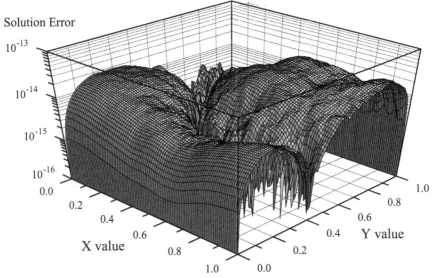

Figure 12.56. Solution error at various points in the x-y plane for a 100 by 100 interval spatial grid.

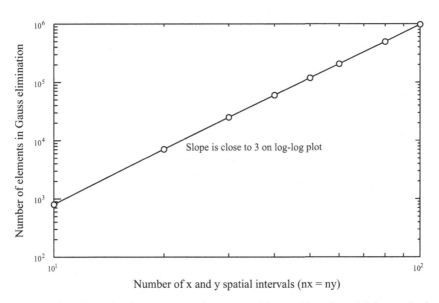

Figure 12.57. Growth of storage requirements with number of spatial intervals for sparse matrix Gauss elimination.

number of matrix elements in the Gauss reduction process is about 123,000 for a 40 by 80 grid size and is about 244,000 for an 80 by 40 grid. So for the minimum matrix fill the smallest number of grid elements should be taken along the x axis for x first numbering or along the y axis for y first numbering. Another possible numbering scheme is the diagonal numbering shown in Figure 12.58. This numbering starts at one corner (such as minimum x and y) and labels the points along a diagonal as shown by the arrow in the figure. Of course there are again two possible numbering schemes depending on whether the x or y axis is selected for the lowest numbers. This scheme in fact reduces the matrix fill when performing Gauss elimination, because it minimizes the distance of the fringe terms from the diagonal row of the matrix.

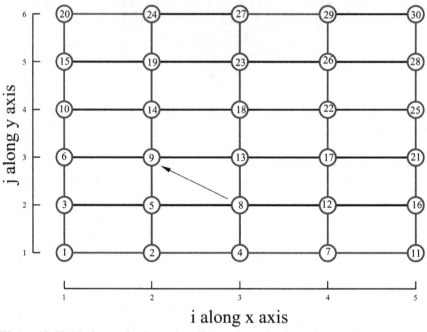

i along x axis

Figure 12.58. Node numbering using diagonal numbering scheme.

The node numbering transformation from an x or y first oriented scheme as shown in Figures 12.48 and 12.49 to the diagonal numbering shown in Figure 12.58 is relatively straightforward, requiring only two or three lines of code. Code for this transformation is available within the function sdgauss(a,b). This function is a direct replacement for spgauss(a,b) and solves a set of sparse matrix equations for a PDE problem returning the solution value and possible a table of matrix fill as the matrix is solved. The code in sdgauss() has basically four parts. First, a table is generated for the node transformation needed to implement the diagonal numbering shown in Figure 12.58. Second, this transformation is used on the a and b matrices to transform them to the diagonal numbering scheme. Third, the standard spgauss() is then called to solve the sparse matrix equation set. Finally, the solution set is transformed back to the x-y node numbering scheme before the function returns with the solved solution set. The code is not shown here as it is relatively straightforward and the reader can view it in the sdgauss.lua file if desired.

The PDE solution obtained in Listing 12.21 can be re-solved using the sparse diagonal Gauss solver by replacing the call to spgauss() on line 21 with a call to sdgauss(). The listing will also generate fill statistics for the new diagonal Gauss solution method. The reader is encouraged to make the change and re-execute the code observing the results. The maximum indicated error from such an execution will be 2.997e-015 which is again near the precision of the language. More importantly is the statistics on the matrix fill as the solution progresses. Some obtained results are shown in Figure 12.59 for the problem of Listing 12.21 with two spatial grid intervals – a 40 by 40 grid and an 80 by 80 grid. In order to compare the results over the large ranges involved, the results are shown on a log-log plot. This tends to compress the differences between the results, but in each case the maximum matrix fill for the diagonal numbering is about 70% of that for the x-y numbering. This ratio tends to be relatively constant for larger spatial grids with equal x and y intervals. The results for the 40 by 40 grid are the same as shown on a linear scale in Figure 12.53 with the dotted curve being the results for the sdgauss() function. For the 80 by 80 grid, the average fill with spgauss() is about 75 matrix elements per row of the Gauss reduction process while for sdgauss() the corresponding number is about 52 matrix elements per row.

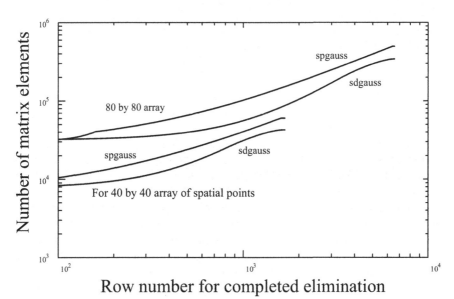

Figure 12.59. Comparison of matrix fill for spgauss() and sdgauss() functions with spatial grid sizes of 40 by 40 and 80 by 80.

For non equal spatial grid numbers, the difference between spgauss() and sdgauss() can be larger or smaller depending on the diagonal numbering relative to the smaller of the grid sides. The matrix fill for the diagonal numbering tends to be relatively independent of which side has the smaller number of spatial intervals while the fill with x-y numbering is reduced if the x side (with x first numbering) has the smallest number of spatial intervals. From this it can be concluded that the diagonal numbering has distinct advantages for the direct Gauss solution of a system of PDE generated matrix equations for a rectangular spatial grid region. Since there is no disadvantage of using the sdgauss() function in place of spgauss() for such problems, it will be used consistently in this chapter when a direct matrix solution of the matrix equations is to be used. It should be noted that the sdgauss() routine is specific to solving a system of sparse matrix equations with the structure generated by a two dimensional PDE problem. It can not be substituted for the spgauss() routine for the general case of a sparse matrix solution.

In addition to the direct solution of a set of matrix equations, several approximate solution methods involve what can be called "operator splitting" techniques. To begin this discussion, consider again the simple Poisson equation in the form:

$$\frac{\partial^2 V}{\partial x^2} + \frac{\partial^2 V}{\partial y^2} = -\frac{\rho}{\varepsilon} \text{ or in discrete form}$$

$$\frac{V(j,i+1) - 2V(j,i) + V(j,i-1)}{(\Delta x)^2} + \frac{V(j+1,i) - 2V(j,i) + V(j-1,i)}{(\Delta y)^2} \quad (12.66)$$

$$+ \frac{\rho(j,i)}{\varepsilon} = 0$$

One means of splitting the differential operator is along the dimensional axes which is achieved by writing the equations as:

$$\frac{\partial^2 V_{k+1}}{\partial x^2} = -(\frac{\partial^2 V_k}{\partial y^2} + \frac{\rho}{\varepsilon}) = F_x(x, y, V_k) \text{ or}$$

$$\frac{\partial^2 V_{k+1}}{\partial y^2} = -(\frac{\partial^2 V_k}{\partial x^2} + \frac{\rho}{\varepsilon}) = F_y(x, y, V_k) \quad (12.67)$$

In both of these forms it is assumed that the right hand side is some known function of position and if this is known then the problem is a much simpler single dimensional boundary value problem in x for the first equation and in y for the second form. The subscript k refers to some kth iteration values and k+1 refers to some new iteration values. Thus the iterative approach is to assume some initial approximation to the solution, to use this in the right hand side for the second derivative terms and then to calculate a new approximation by solving the resulting one dimensional BVP. The advantage of this is that the single dimensional finite difference equation set is a purely tri-diagonal set of equations which can be rapidly solved with no additional fill in the matrix equation set. A commonly used approach is to think in terms of a two step process whereby in the first step the equation set is formulated along the x direction as in the first of the equations and this is followed by a second step whereby the equation set is formulated along the y direction. This then constitutes one iterative step and is typically referred to as the alternating direction implicit or ADI approach.

An even simpler approach is to take all matrix terms in the set of coupled equations except the diagonal term to the right hand side and solve for each solution value using only the diagonal term. The process is then repeated until hopefully the solution converges to some correct solution. For equally spaced data and for Poisson's equation as in Eq. (12.66) this results in the iterative equation:

$$V_{k+1}(j,i) = \left[\frac{(\Delta x)^2 (\Delta y)^2}{2((\Delta x)^2 + (\Delta y)^2)} \right] \left[\frac{V_k(j,i+1) + V_k(j,i-1)}{(\Delta x)^2} + \right.$$
$$\left. \frac{V_k(j+1,i) + V_k(j-1,i)}{(\Delta y)^2} - \frac{\rho(j,i)}{\varepsilon} \right] \tag{12.68}$$

where the subscript k refers to the iteration index. For the simpler case of equal x and y spatial increments, this has the simpler form:

$$V_{k+1}(j,i) = \frac{1}{4} \left[V_k(j,i+1) + V_k(j,i-1) + V_k(j+1,i) + \right.$$
$$\left. V_k(j-1,i) - \rho(j,i)\Delta x \Delta y / \varepsilon \right] \tag{12.69}$$

In a space charge free region, the potential at point j,i is then simply the average of the four neighboring potentials. This iterative approach is known as Jacobi's method and dates back to the last century.

For the general case of non-uniform spatial grids, the resulting equations are somewhat more complicated but still straightforward. For such approaches as these operator splitting methods, use can still be made of the general function setup2bveqs() to generate the matrix coefficients. The possible use of such approaches is the reason the setup2bveqs() function has been coded with the possibility of separately storing the diagonal contributions from each of the spatial x and y variable derivatives. For the method of Eq. (12.68) only the total diagonal term is required and the central diagonal term can be used to incrementally update the solution and the process iterated any desired number of times. A slight modification of this method is known as the Gauss-Seidel method. The basis for this modification is to recognize that when the values are being updating for the solution value in row k for instance, the solution value for rows 1 through k-1 have already been updated and one can just go ahead and make use of these updated values rather than the original values for these node points. With this modification, the basic equation for updating a solution value is:

$$u^{new}[m] = \frac{1}{a[m][m]} \left[b[m] - \sum_{i=1}^{m-1} a[m][i]u^{new}[i] - \sum_{i=m+1}^{n} a[m][i]u^{old}[i] \right]$$
$$= u^{old}[m] + \frac{1}{a[m][m]} \left[b[m] - \sum_{i=1}^{m-1} a[m][i]u^{new}[i] - \sum_{i=m}^{n} a[m][i]u^{old}[i] \right] \tag{12.70}$$

The last form of the equation appears to require two arrays for the solution values, one for the new values and one for the old values. However, the replacement of the old solution values by the new solution values can be done in-place so only one array of solution values is required.

```
  1 : -- File pde2bv.lua --
  2 : -- Code for PDE BV problems in 2 dimensions -- rectangular grid
------
233 : sdsolve = function(a,b,u,lan) -- Single step of SOR
234 :   local n,epsm,bb,row,eps,jm,im = #b,0.0
235 :   local um = 0.0
236 :   lan = lan or 1
237 :   for j=1,n do -- Update each row in turn
238 :       row,eps = a[j],-b[j]
239 :       for i,v in pairs(row) do eps = eps + v*u[i] end
240 :       eps = eps/a[j][j]; u[j] = u[j] - lan*eps
241 :       if abs(eps)>abs(epsm) then epsm,jm,im = eps,j,i end
242 :       um = max(um,abs(u[j]))
243 :   end
244 :   return u, epsm, um, jm, im
245 : end
246 : setfenv(sdsolve,{table=table,type=type,abs=math.abs,
           max=math.max,pairs=pairs})
```

Listing 12.22 Code segment for implementing a single step in the Gauss-Seidel iterative solution of sparse matrix equation set.

A code segment for a single iterative step in such a process is shown in Listing 12.22 as a function sdsolve(), taking as input an a and b matrix of coefficients and an initial solution value u. Only a couple of features of the code are perhaps not straightforward. The parameter eps on line 239 accumulates the corrective term to be applied to the previous solution value at each node equation. At line 239 the eps term is the quantity in the brackets of the last form of Eq. (12.70). The maximum value of this correction term is found on line 241 along with the point at which it occurs. In addition the maximum value of the solution variable is found on line 242. These are returned by the function and can be used to monitor the progress of the solution and used, if desired, to terminate the iteration at some iterative step when the correction term is sufficiently small. A feature so far not discussed is the use of the lan term which is an input to the function and is used on line 240 to multiply the eqs correction term. This provides a means to either enhance or suppress the corrective term depending on whether this value is larger than or less than unity. For the present it will be considered as unity so the function simply returns the results of implementing the simple Jacobi method. This term is subsequently used to implement other types of solution algorithms.

Example code for implementing this iterative method into solving a BV problem is shown in Listing 12.23. The example problem is the same parallel plate capacitor problem previously considered and lines 1 through 24 simply define the functions and call setup2dbveqs() on line 24 to generate the matrix of coefficients. Two types of solution arrays are required – one a two dimensional array specified by j,i points (u in the listing) and a composite linear array corresponding to the equation numbering (ua in the listing). The setup2dbveqs() function returns the two dimensional array while the new function here, sdloop(), requires the linear array. The conversion between the 2-D array and the linear array of solution values is performed on lines 27 through 29 of the code. Lines 31 through 41 define a function which simply loops through the diagonal iteration function sdsolve() a specified number of times and returns the final solution array. The only other feature is the printing on lines 35 through 38 which prints an output line every 100 iterative loops so the user can monitor the progress of the iteration. This is most useful when one has large arrays (>50 by 50 arrays) and large numbers of iterations. The maximum correction value is also printed at each printing interval.

```lua
 1 : -- File list12_23.lua --
 2 : -- Example of BV problem in 2 D -- rectangular grid of points
 3 :
 4 : require"pde2bv"
 5 : Vm = 1.0 -- Define equations to be solved
 6 : feq = function(x,y,uxx,ux,u,uy,uyy,i,j)
 7 :    return uxx + uyy
 8 : end

 9 : fb = function(x,u,uy,ux,i) return u end
10 : ft = function(x,u,uy,ux,i) return u - Vm  end
11 : fr = function(y,u,ux,uy,j) return ux end
12 : fl = function(y,u,ux,uy,j) return ux end
13 :
14 : x,y,u = {},{},{}; xmax,ymax = 1.0,1.0
15 : Nx,Ny = 40,40
16 : nx,ny = Nx+1,Ny+1

17 : for i=1,nx do x[i] = xmax*(i-1)/Nx end
18 : for i = 1,ny do y[i] = ymax*(i-1)/Ny end
19 : for j = 1,ny do   -- Set zero initial values
20 :    u[j] = {}
21 :    for i = 1,nx do u[j][i] = 0 end
22 : end
23 :
24 : a,b = setup2bveqs({feq,fb,ft,fl,fr},x,y,u) -- get matrix eqs
25 :
26 : ua = {}
27 : for j=1,ny do -- convert u to single array
28 :    for i=1,nx do ua[i+(j-1)*nx] = u[j][i] end
29 : end
30 : function sdloop(a,b,ua,itt) -- Iterative loop
31 :    local jpold = 1
32 :    for k=1,itt do
33 :       ua,merr = sdsolve(a,b,ua) -- Single step
34 :       jprint = math.floor(k/100)
35 :       if jprint==jpold then
36 :           jpold=jprint+1
37 :           print("Completed iteration",k,
                     "with correction",merr);io.flush()
38 :       end
39 :    end
40 :    return ua
41 : end
42 :
43 : ua = sdloop(a,b,ua,2000)
44 : bb = to2darray(ua,nx,ny)
45 : write_data('list12_23.dat',bb)
```

Listing 12.23. Example code for solving BV problem using diagonal iteration and the sdloop() function of Listing 12.22.

Given enough iterations and time this diagonal iteration will converge to the true solution for this problem. Such a technique will converge when the problem is diagonally dominant. It is not the intent here to diverge into a discussion of when such a technique will converge and when it will not. It will not converge for all possible types of matrices. However, it does converge, but ever so slowly, for many important physical two dimensional problems. The intent here is to discuss some of the convergence rate issues using this simple example with known solution as the number of spatial intervals is increased. For small numbers of spatial intervals, the convergence can be reasonably fast. However, the solution effort must be compared with that of the direct matrix solution technique already discussed.

By executing the code in listing 12.23 for different spatial grid sizes and different numbers of iterations, the rate of convergence of the technique to the exact solution can be observed. Some interesting results are shown in Figure 12.60. When the maximum solution error is plotted vs. the number of iterations normalized by the number of equations being solved (nx*ny) the results as shown in the figure all fall on the same line. Thus if a given accuracy level is desired, the number of iterations must increase linearly with the number of equations or as the square of the number of spatial intervals (at least for equal numbers of x and y intervals). The curve shows that for an accuracy of about 1% one must have Nit/Neq = 0.84 and this ratio must be about 1.3 for an accuracy of 0.1%. This has important consequences with regard to the time required for the Gauss-Seidel

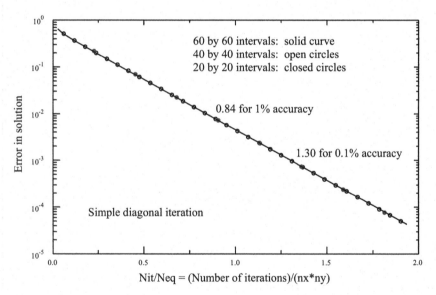

Figure 12.60. Error in solution of parallel plate capacitor for varying spatial grids and varying numbers of Gauss-Seidel iterations.

technique. The computational time required per iteration is also essentially directly proportional to the number of equations in the matrix. Thus the time required for a given accuracy level is:

$$T_{calc} = K(nx * ny)^2$$
$$= K(nx)^4 \text{ for } nx = ny$$

(12.71)

The constant of course depends on the speed of one's computer. However, a doubling of the number of spatial points requires an 8 fold increase in computer time for the same level of accuracy. While these results have been obtained with one particular simple BV problem, the general results can be expected to carry over to other problems. The convergence rate of this method is linear in the number of iterations and the convergence is very slow for large numbers of spatial intervals. This technique is primarily of historic interest and is not of practical use because of the slow convergence rate. However, it does form a basis for more practical modifications discussed below.

A much more practical algorithm is obtained if one makes an overcorrection to the Gauss-Seidel algorithm at each iterative step. As a preliminary discussion consider some iterative process that is proceeding with the calculated results of u_{k-1} and u_k at two consecutive steps. The quantity $u_k - u_{k-1}$ can be considered as an approximation to the expected change in the next step and these two values can be used to estimate the result at the next step by the equation:

$$u_{k+1} \simeq u_k + (u_k - u_{k-1}) = 2u_k - u_{k-1}$$
$$= \lambda u_k + (1-\lambda)u_{k-1} \text{ with } \lambda = 2$$
$$= u_{k-1} + \lambda(u_k - u_{k-1}) = u_{k-1} + \lambda \varepsilon_k$$

(12.72)

where the next term in the sequence is predicted by using the present term and the first order estimate of the derivative at the present step. This will be exact if the calculated result is changing exactly linearly with each iterative step. However, in the more likely case, the correction with $\lambda = 2$ slightly overestimates the change and a value somewhat less than 2 is probably a better value. This method of correcting a calculated iterative value is called relaxation and applied to a set of solution values would be implemented by the equations:

$$u_{k+1}(j,i) = \lambda u_{k+1}(j,i) + (1-\lambda)u_k(j,i)$$
$$u_{k+1}(j,i) = u_k(j,i) + \lambda(u_{k+1}(j,i) - u_k(j,i)) = u_k(j,i) + \lambda \varepsilon(j,i)$$

(12.73)

where the right hand side k value is the result of some present iteration and the k-1 value is the result of the previous iteration. When expressed in the second line form it is seen that the λ term is a weighting term applied to the correction to the kth solution value. Typical values of this parameter are between 0 and 2 which correspond to no correction and to a doubling of the calculated correction term. For values of the λ parameter below 1 this is know as under-relaxation and for values over 1 it is known as over-relaxation. From the simple discussion above an optimum value is expected to be close to 2. The convergence rate of an iterative process can in many cases be greatly accelerated when the optimum value of such as over-relaxation parameter is used. In practice for solving a set of matrix equations, the over-relaxation approach is applied to the Gauss-Seidel approach and is known as "simultaneous over-relaxation" (SOR). This results in the modified iterative equation:

$$u^{new}[m] = u^{old}[m] + \frac{\lambda}{a[m][m]} \left[b[m] - \sum_{i=1}^{m-1} a[m][i]u^{new}[i] - \sum_{i=m}^{n} a[m][i]u^{old}[i] \right]$$

(12.74)

where the over-relaxation parameter is applied to each step in the Gauss-Seidel iteration.

The code in Listing 12.22 provides for such an over-relaxation parameter as an input value. This can greatly accelerates the convergence rate and begins to give a practical iterative algorithm. However the selection of the most appropriate value of the acceleration parameter is not a straightforward task and some discussion of a selection method follows.

The optimum over-relaxation parameter is known only for a small class of linear problems and for select boundary conditions. The iteration matrix has eigenvalues each one of which reflects the factor by which the amplitude of an eigenmode of undesired residual is suppressed for each iterative step. Obviously the modulus of all these modes must be less than 1. The modulus of the factor with the largest amplitude is called the spectral radius and determines the overall long term convergence of the procedure for many iterative steps. If ρ_J is the spectral radius of the Jacobi iteration then the optimum value of λ is known to be:

$$\lambda = \frac{2}{1+\sqrt{1-\rho_J^2}}$$

(12.75)

Rather than using this optimum parameter from the start of the iterations, a standard procedure is to begin the iteration with $\lambda = 1$ (no over-relaxation) and to slowly approach the optimum value as the iteration proceeds according to the equations:

$$\lambda_1 = 1$$
$$\lambda_{k+1} = 1/(1 - \rho_J^2 \lambda_k / 4) \text{ with } k \geq 1$$
$$\lambda_k \to \frac{2}{1+\sqrt{1-\rho_J^2}} \text{ as } k \to \infty$$

(12.76)

(See *Numerical Recipes in C*, Cambridge Press, 1988 for a discussion).

A slight variation of this procedure is known as Chebyshev acceleration of SOR and consists of implementing the SOR procedure in two passes, one over the odd matrix elements and one over the even matrix elements with a modification to Eq. (12.76) for the first pass: The resulting iterative expressions being:

$$\lambda_0 = 1$$

$$\lambda_{1/2} = 1/(1 - \rho_J^2 / 2) \text{ for first half step}$$

$$\lambda_{k+1} = 1/(1 - \rho_J^2 \lambda_k / 4) \text{ for all other steps} \tag{12.77}$$

$$\lambda_k \rightarrow \frac{2}{1 + \sqrt{1 - \rho_J^2}} \text{ as } k \rightarrow \infty$$

Computer code for implementing one step of this odd/even Chebyshev algorithm is available as the function sdoesolve(a,b,ua,lan1,lan2) where lan1 and lan2 are the two required values for the odd and even half step iterations. This can be further embedded in a higher level function implementing a complete Chebyshev iterative solution. Code for this is not shown as it is similar to that in Listing 12.22 for sdsolve(). The reader can view the code for the sdoesolv() function and verify the implementation.

The Jacobi spectral radius is known for a few special structures and selected boundary conditions. On a rectangular grid with zero boundary conditions, it is known to be:

$$\rho_J = \frac{\cos(\pi / nx) + (\Delta x / \Delta y)^2 \cos(\pi / ny)}{1 + (\Delta x / \Delta y)^2} \tag{12.78}$$

where nx and ny are the number of x and y spatial points (as opposed to number of spatial intervals which are one less than these values). If the boundary conditions are periodic, then this replacement should be made in the equation: $\pi \rightarrow 2\pi$. For a square grid with equal numbers of spatial grid points this approaches $1 - (1/2)(\pi / nx)^2$ and $\lambda \rightarrow 2/(1 + \pi / nx)$.

To discuss these more practical iterative matrix solutions, it is convenient to have a problem for which both the exact solution is known and the Jacobi spectral radius is known. One such problem is the following form of Poisson's equation:

$$\frac{\partial^2 u}{\partial x^2} + \frac{\partial^2 u}{\partial y^2} + 2\pi^2 \sin(\pi x)\sin(\pi y) = 0 \tag{12.79}$$

$$\text{with solution } u = \sin(\pi x)\sin(\pi y), \ 0 \le x \le 1, 0 \le y \le 1$$

Listing 12.24 shows a code segment for solving this problem using the SOR approach and the sdsolve() routine with over-relaxation. The code on lines 1 through 31 should be very familiar by now. Line 24 calls the setup2bveqs() function to set up the matrix equations. Lines 32 through 34 define the Jacobi spectral radius and calculate the theoretical optimum over-relaxation factor. The sdloop() subroutine on lines 37 through 50 forms the heart of the calculation looping through the iterative solution of the matrix equations. Line 42 calls the sdsolv(a,b,ua,lan) function as defined in Listing 12.22 passing an over relaxation value, lan to the function. This in turn is updated between iterative calls by line 47 according to Eq, (12.76). Lines 43 through 46 simply print incremental values periodically so the progress of the solution can be monitored. Finally line 53 calls the sdloop() function requesting 250 iterations and the results are printed for later display. The selected output shows the maximum correction to the solution at the 25th iteration and 250th iteration as -0.01326 and 2.332e-013 respectively. These

```
 1 : -- File list12_24.lua --
 2 : -- Example of BV problem in 2 D -- rectangular grid of points
 3 :
 4 : require"pde2bv"
 5 : Vm = 1.0 -- Define equations to be solved
 6 : pi = math.pi;pi2 = 2*pi^2; sin=math.sin
 7 : feq = function(x,y,uxx,ux,u,uy,uyy,i,j)
 8 :    return uxx + uyy + Vm*pi2*sin(pi*x)*sin(pi*y)
 9 : end -- Now Zero boundary values
10 : fb = function(x,u,uy,ux,i) return u end
11 : ft = function(x,u,uy,ux,i) return u end
12 : fr = function(y,u,ux,uy,j) return u end
13 : fl = function(y,u,ux,uy,j) return u end
14 :
15 : x,y,u = {},{},{}; xmax,ymax = 1.0,1.0
16 : Nx,Ny = 50,50
17 : nx,ny = Nx+1,Ny+1; n = nx*ny
18 : for i=1,nx do x[i] = xmax*(i-1)/Nx end
19 : for i = 1,ny do y[i] = ymax*(i-1)/Ny end
20 : for j = 1,ny do   -- Set zero initial values
21 :    u[j] = {};      for i = 1,nx do u[j][i] = 0 end
22 : end
23 :
24 : a,b = setup2bveqs({feq,fb,ft,fl,fr},x,y,u)
25 :
26 : t = os.time()
27 : ua = {}
28 : for j=1,ny do
29 :    for i=1,nx do ua[i+(j-1)*nx] = u[j][i] end
30 : end
31 :
32 : rsp = ((x[2]-x[1])/(y[2]-y[1]))^2 -- Needed for lan
33 : rsp = ((math.cos(math.pi/nx)+rsp*math.cos(math.pi/ny))/(1+rsp))
34 : p4 = rsp^2/4; jprint=0; nprint = math.min(Nx,Ny)/4
35 : print('theoritical lan = ',2/(1+math.sqrt(1-rsp^2)))
36 :
37 : function sdloop(a,b,ua,itt)
38 :    local uold,jpold,lan = {},0,1
39 :    lan = 1.0
40 :    for k=1,itt do
41 :       ua,merr = sdsolve(a,b,ua,lan)
42 :       jprint = math.floor(k/nprint)
43 :       if jprint==jpold then
44 :          jpold=jpold+1
45 :          print("Completed iteration",k,"with correction",
                   merr,lan); io.flush()
46 :       end
47 :       lan = 1/(1- lan*p4) -- Update lan
48 :    end
49 :    return ua
50 : end
51 :
52 : ua = sdloop(a,b,ua,250)
53 : print('time = ',os.time()-t)
54 : bb = to2darray(ua,Nx,Ny)
55 : splot(bb); write_data('list12_24.dat',bb)
------
Selected output:
------
```

```
Completed iteration    25  with correction    -0.013264879594426
        1.8735713628921
------
Completed iteration    250 with correction    2.33239916270e-013
        1.8840181363533
```

Listing 12.24. Example code segment for SOR iterative solution of BVP.

are the corrective values and do not correspond exactly to the accuracy in the solution. The actual accuracy of the solution will be subsequently discussed. This particular calculation is for a 50 by 50 interval spatial grid.

Finally the solution generated by executing the code in Listing 12.24 is shown in Figure 12.61. A pop-up version of this graph is generated by line 55 of the code using the splot() function which takes as the argument a 2 dimensional array of solution values such as returned by the to2darray() function on line 54 of the code. The reader is encouraged to execute the code and observe the generated output and the graph.

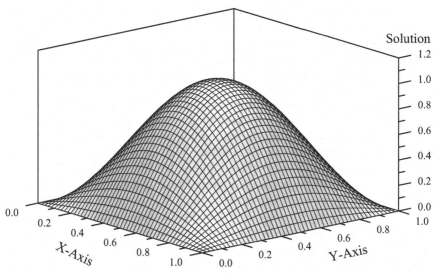

Figure 12.61. Solution of Poisson's equation in Listing 12.24 using the SOR iterative technique.

The code in Listing 12.24 can easily be packaged into a more general routine for implementing the iterative solution of the matrix equations. Also the SOR with Chebyshev enhancement needs to be explored. For these explorations, an iterative loop such as the one in Listing 12.24 has been implemented as pde2bvsor(a,b,u,rsp) which takes as input the matrices a and b and an initial guess at the solution u plus the spectral radius rsp and attempts to return an accurate solution to the matrix equations. An additional function pde2bvsoe(a,b,u,rsp) implements a similar solution using odd/even iteration with Chebyshev enhancements. In both cases the code is very much patterned after the sdloop() in Listing

12.24 with several enhancements. First if rsp is not input to the routines, a value is estimated from the number of x and y spatial points stored in the a matrix table (stored at a[0]). Second the routines attempt to achieve an error of 1.e-4 in the final solution value and then terminate. The code segment will execute some maximum number of iterative loops attempting to achieve this accuracy (default of 4000 iterations). A table of maximum correction terms can be saved, if desired, by the routines (by using getfenv(pde2bvsor).sverr=1). Finally, interim values can be periodically printed if desired (by setting an nprint parameter with getfenv(pde2bvsor).nprint = 1) so the user can monitor the progress of the solution. The functions return the final solution table plus an array of maximum correction values at each iterative step if this is requested.

An example of using the Chebyshev routine is shown in Listing 12.25. The defining code on lines 1 through 30 is identical to the code in Listing 12.24. After defining the functions and setting up the matrix a simple call to the function pde2bvcoe(a,b,ua) on line 32 returns the solved equation set plus an array showing the progress of the solution. It is noted that the solution converges to a maximum error correction of 4.70e-007 after 168 iterations. This is not the accuracy of the solution as will be subsequently discussed but considerably below the solution accuracy of 1.e-4 which the subroutine attempts to achieve. This particular calculation is for a 100 by 100 spatial interval grid. By comparing this back to Gauss-Seidel iteration it can be seen that the number of iterations is of order 100 times less with optimum over-relaxation.

Several iterative techniques have now been discussed for approximating the solution of the set of matrix equations associated with two dimensional BVPs. The most important of these are Gauss-Seidel, simple over-relaxation, successive over relaxation (SOR) and Chebyshev (with SOR). With the BVP in Listings 12.24 or 12.25 for which a closed form solution is known, both the convergence rate of the solution techniques and the accuracy of the solutions as a function of the number of iterative steps can be easily explored. Figure 12.62 shows the observed change in the maximum corrective value as a function of the number of iterations for Gauss-Seidel, Gauss-Seidel followed by over-relaxation, successive over relaxation (SOR) and Chebyshev enhanced over relaxation (COE). The figure illustrates the significant improvement in the maximum solution values obtained by the use of over relaxation as compared with simple Gauss-Seidel iteration.

Code for the Gauss-Seidel followed by over relaxation is not shown but the values are computed in the following manner. At each iterative step improved solution values are computed by the Gauss-Seidel technique of Eq. (12.70). After all new solution values are computed, simple over-relaxation is used to correct each solution value following Eq. (12.73) with the value of λ at each step increasing according to Eq. (12.76). This is in contrast to the successive over relaxation approach which applies over-relaxation to each diagonal element in turn as it is evaluated according to Eq. (12.74). The results show that for small numbers of iterations, simple over-relaxation provides the smallest corrective values while for larger numbers of iterations, the SOR approach gives smaller corrective values. Finally the Chebyshev enhancement approach of using odd-even reduction steps

```
 1 : -- File list12_25.lua --
 2 : -- Example of BV problem in 2 dimensions -- rectangular grid of
points
 3 :
 4 : require"pde2bv"; getfenv(pde2bvcoe).nprint=1; get-
fenv(pde2bvcoe).sverr=1
 5 : Vm = 1.0 -- Define equations to be solved
 6 : pi = math.pi;pi2 = 2*pi^2; sin=math.sin

 7 : feq = function(x,y,uxx,ux,u,uy,uyy,i,j)
 8 :    return uxx + uyy + Vm*pi2*sin(pi*x)*sin(pi*y)
 9 : end

10 : fb = function(x,u,uy,ux,i) return u end
11 : ft = function(x,u,uy,ux,i) return u end
12 : fr = function(y,u,ux,uy,j) return u end
13 : fl = function(y,u,ux,uy,j) return u end
14 :
15 : x,y,u = {},{},{}; xmax,ymax = 1.0,1.0
16 : Nx,Ny = 100,100
17 : nx,ny = Nx+1,Ny+1; n = nx*ny
18 : for i=1,nx do x[i] = xmax*(i-1)/Nx end
19 : for i = 1,ny do y[i] = ymax*(i-1)/Ny end
20 : for j = 1,ny do   -- Set zero initial values
21 :    u[j] = {}; for i = 1,nx do u[j][i] = 0 end
22 : end
23 :
24 : a,b = setup2bveqs({feq,fb,ft,fl,fr},x,y,u)
25 :
26 : t = os.time()
27 : ua = {}
28 : for j=1,ny do
29 :    for i=1,nx do ua[i+(j-1)*nx] = u[j][i] end
30 : end
31 :
32 : ua,errar = pde2bvcoe(a,b,ua)--,p)
33 : print('time = ',os.time()-t)
34 : bb = to2darray(ua,Nx,Ny); splot(bb)
35 : write_data('list12_25.dat',bb)
36 : write_data('list12_25_err.dat',errar)
Selected output:
Completed COE iteration        1        with correction    -
0.00098623005216721
Completed COE iteration        25       with correction    -
0.012497478127669
------
Exiting COE at iteration      168       with correction    -
4.7039880046323e-007
```

Listing 12.25. Illustration of iterative solution of BVP using Chebyshev enhanced SOR with function pde2bvcoe().

with simultaneous over correction provides the most rapid reduction in the corrective values. This is as expected from literature discussions of these techniques. At least for this example, the Chebyshev approach provides the most rapid reduction in the corrective terms.

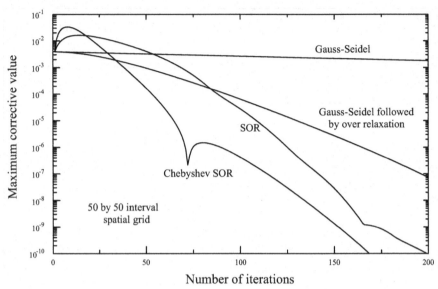

Figure 12.62. Examples of dependency of corrective error on number of iterations for various over relaxation methods.

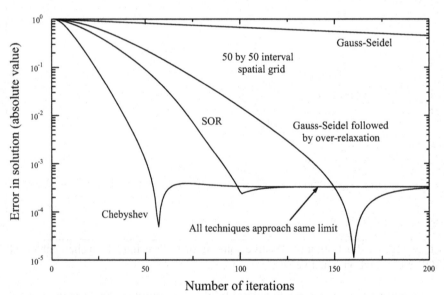

Figure 12.63. Maximum error in the solution for example in Listing 12.25 as a function of number of iterative steps in over-relaxation algorithms.

Not only is the value of the corrective term with increased iteration important, but even more important is the improvement in accuracy of the solution with increasing numbers of iterations and the improvement in solution accuracy is different from the decrease in the iterative correction term. In most cases the accuracy of the solution will not be known and one will have only the decrease in the value of the corrective terms on which to base an approximation to the overall solution accuracy. In this example however, the exact solution is known and thus the solution accuracy can be determined at each iteration step. Results of such an evaluation are shown in Figure 12.63 for the various iterative approaches. As the results show again the Chebyshev algorithm converges most rapidly toward the exact solution as evidenced by the smallest errors in the solution. Another interesting feature of the results is the fact that all three algorithms approach the same limiting solution accuracy for large numbers of iterations. This is in contrast to the values of the corrective terms seen in Figure 12.62 which continue to decrease with increasing numbers of iterations. In fact if one continues to increase the number of iterative steps, the correction terms in Figure 12.62 will eventually approach the 1.e-16 value which is the accuracy limit of the computer calculations. However, this level of accuracy is not seen in the accuracy of the generated solution. The reason for this is relatively straightforward. In addition to the accuracy in solving the set of matrix equations, there is a limit on the solution accuracy of the partial differential equation due to the use of the finite difference approximations for the continuous partial derivatives. In this case with a 50 by 50 interval spatial grid, this limit is an accuracy of about 3.3e-4 when the maximum value of the solution is 1.0, or an accuracy of about 0.033% of the peak value of the solution. It is very important to understand this limit because there is little to be gained in pushing the number of iterations beyond the value needed to achieve limits due to the finite number of spatial grid points. For this example this limit occurs at approximately 1Nx, 2Nx and 3Nx for the Chebyshev, SOR and simple-relaxation algorithms where Nx is the number of spatial intervals.

In order to explore these limits further, a series of calculations were made for the above example using varying numbers of spatial grid sizes. Some of the results are shown in Figure 12.64 for varying spatial interval sizes and for the Chebyshev algorithm. As can be seen for a given level of accuracy the number of required iterations increases as the number of spatial intervals increases. Also it can be sees that the limiting accuracy due to the finite differencing always occurs at large numbers of iterations and the limit varies very closely with the inverse square of the number of spatial intervals as first order theory would predict. For the 160 by 160 case, the limit of 3.22e-5 at 400 iterative steps is almost exactly 16 times smaller than the limit of 5.14e-4 for the 40 by 40 solution case.

Another important result that can be gleamed from the figure is the number of iterations required to reduce the solution error to a certain accuracy or to the limiting value as observed at large iteration values and as determined by the spatial grid size. Some of these relationships are shown in Figure 12.65, based upon this example and the data in Figure 12.64. First consider the lower of the curves in the figure which shows the number of iterations required to achieve an accuracy of

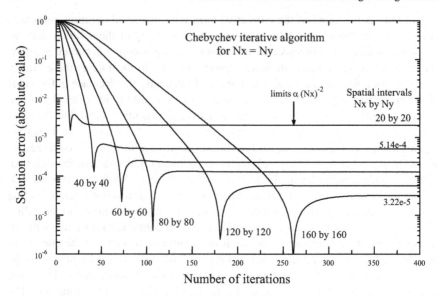

Figure 12.64. Solution error as spatial grid size and number of iterations are varied.

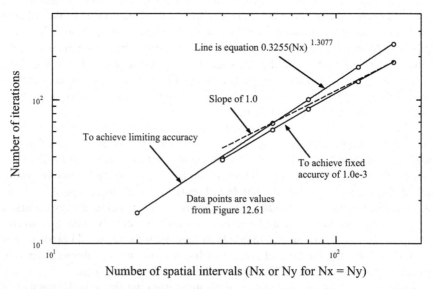

Figure 12.65. Observed relationship between number of spatial intervals and number of iterations required to reach limiting accuracy.

1.e-3. At large numbers of iterations the data appears to approach a linear relationship to the number of spatial intervals which is what one might expect. However, if one desires to achieve an accuracy limited by the intrinsic number of spatial intervals then more iterations are required as shown by the upper solid line in the figure. The data fits very consistently a power law relationship with an exponent of 1.3077. A more rapid increase to the limiting accuracy is to be expected if the number of iterations required to achieve a fixed accuracy varies linearly with the spatial grid size. The lower of the curves is probably of more significance since a fixed percentage error is usually desired in a calculation. The number of calculations associated with each iteration is expected to vary directly with the number of equations or with nx*ny. Then for a fixed accuracy one would expect the computational time to vary as:

$$T_{calc} = K(Nx * Ny)Nx$$
$$= K(Nx)^3 \text{ for } Nx = Ny$$
(12.80)

Doubling the number of spatial intervals then is expected to require a time increase of 8X in order to achieve the same relative solution accuracy. This is still a significant increase in calculation time but the time for the present example is not excessive for grid sizes up to several hundred by several hundred.

These results have been achieved with a simple Poisson equation on a square grid of spatial points. The advantage of this example is that an exact solution is available so considerable insight can be gained into the promise and limitations of these approximate matrix solution methods. However there is no absolute guarantee that the results will exactly carry over to other more complex problems with different second order equations, different numbers of x and y spatial intervals or with non-uniform spatial grids. However, the general trends are expected to be consistent with more complex problems. A major difficulty with all over relaxation techniques is selecting the relaxation factor or the Jacobi spectral radius. This is only known for a few simple grids and simple boundary conditions. In more real problems, the optimum value has to be determined in some empirical manner – for example by attempting several values to select the most optimal value that allows fast convergence but does not give a diverging correction. In the present example the iteration always converged. However, this is not true if the spectral radius factor is too large – the iterative process may simply diverge. Even with these difficulties, the iterative approach and especially the Chebyshev SOR enhancement becomes a viable iterative technique for solving two dimensional BVPs. For large array sizes, the limitation is the large number of iterations required and the long execution time. Further examples of applying these approximate matrix solution techniques to example problems will be considered in the next section.

One more example of a classical iterative method of solving a set of matrix equations will now be considered. For this approach take a look back at the operator splitting technique as presented in Eq. (12.67) for Poisson's equation. This technique splits the solution into two essentially one dimensional problems in terms of iterating on the matrix equations. When finite differences are applied to the left hand side of either of the terms in Eq. (12.67) essentially a tri-diagonal set of coupled equations is obtained. Such sets of equations are simple to solve and the reader is referred back to Section 11.5 for a discussion of such tri-diagonal sets of equations. The first question is what is needed to generate the sets of tri-diagonal equations? Can the previous equation setup function setup2bveqs() be of use? The answer is yes, this function has been programmed with just such an application in mind. The central diagonal term in the coefficient matrix has contributions from both the x oriented partial derivatives, the y oriented partial derivatives and possible from an explicit dependence of the differential equation on the solution variable. For direct matrix solution or for the previous iterative techniques, only the total value of the diagonal coefficient is required. However, for the operator splitting approach, the terms due to each individual derivative operator are required. Also one must decide how to distribute a term directly involving the solution variable between the two split operator versions of the equation. This is the reason the setup2bveqs() function in fact was originally written to store the diagonal term as a table of three values with just the parameters needed for the x and y oriented equations as given by Eq. (12.67) and for any direct contribution to the diagonal term This is another use of the concept of reusable computer code which can serve several purposes with a little foresight (or hindsight as the case might be). In any case, when writing a book one always claims that one had this in mind from the very beginning.

The basic idea of the alternating direction implicit (ADI) technique is to split the partial derivative operator into x and y components as in Eq. (12.67) and apply them in a two step process according to the following equations:

$$L_x V_{k+1/2} = \frac{\partial^2 V_{k+1/2}}{\partial x^2} = -(\frac{\partial^2 V_k}{\partial y^2} + \frac{\rho}{\varepsilon}) = F_x(x,y,V_k)$$

$$L_y V_{k+1} = \frac{\partial^2 V_{k+1}}{\partial y^2} = -(\frac{\partial^2 V_{k+1/2}}{\partial x^2} + \frac{\rho}{\varepsilon}) = F_y(x,y,V_{k+1/2})$$

$$(12.81)$$

In the more general case, the ρ term could be a function of the potential V which can be considered as the V_k or $V_{k+1/2}$ terms or as some combination of the potentials such as $(V_k + V_{k+1/2})/2$. In any case, in the first half step (first equation above) the x oriented derivatives are applied to the solution variable and the y oriented derivatives are collected with the known terms on the right hand side. The solution of the resulting equation set provides a new estimate of the solution at step k+1/2. This is then used in the second half step (second equation above) to approximate the x oriented derivatives and the y oriented derivatives applied to the solution variable to obtain a new approximate value of the derivative at the end of the two half steps or at step k+1. This two step process is then repeated until hopefully a solution is obtained to some desired degree of accuracy. The potential advantage of this ADI approach is that each step involves only a simple tri-diagonal set of matrix equations which can be very easily solved. Thus there is no growth of matrix elements as the method is applied. It does however require two formulations of the equation sets – one with primary node numbering along the x axis and one with primary node numbering along the y axis for use with the two direction oriented operators above. This numbering has been previously discussed and a routine reorder() developed to perform this reversal given one or the other formulation. With the reordering included, the operator splitting method is formulated as:

$$L_x U_{k+1/2} = \frac{\partial^2 U_{k+1/2}}{\partial x^2} = -(\frac{\partial^2 U_k}{\partial y^2} + \frac{\rho}{\varepsilon}) = F_x(x, y, U_k)$$

$$V_{k+1/2} = T_y(U_{k+1/2})$$

$$L_y V_{k+1} = \frac{\partial^2 V_{k+1}}{\partial y^2} = -(\frac{\partial^2 V_{k+1/2}}{\partial x^2} + \frac{\rho}{\varepsilon}) = F_y(x, y, V_{k+1/2})$$ (12.82)

$$U_{k+1} = T_x(V_{k+1})$$

In this $T_x()$ and $T_y()$ represent the transformations to x oriented row numbering and y oriented column numbering.

While this provides the basis of the operator splitting concept, the algorithm is typically not applied in this form because of numerical stability problems. The problem is typically reformulated into consideration of the time dependent two dimensional equation:

$$\frac{1}{D}\frac{\partial U}{\partial t} = \frac{\partial^2 U}{\partial x^2} + \frac{\partial^2 U}{\partial y^2} + \frac{\rho}{\varepsilon} = (L_x + L_y)U + \frac{\rho}{\varepsilon}$$ (12.83)

where D is some coefficient with dimensions of (distance)^2/time in order to make the equation dimensionally correct. This will in fact be recognized as the classical diffusion equation in time and two spatial dimensions. For splitting the operator into two dimensions, the time derivative is now approximated by a first order forward difference and for the two alternating directions the equations become:

$$\frac{U_{k+1/2} - U_k}{D\Delta t/2} = L_x U_{k+1/2} + L_y U_k + \frac{\rho}{\varepsilon}$$

$$\frac{U_{k+1} - U_{k+1/2}}{D\Delta t/2} = L_y U_{k+1} + L_x U_{k+1/2} + \frac{\rho}{\varepsilon}$$ (12.84)

where the transformations from x to y oriented numbering has been omitted.

When terms are collected these become:

$$(L_x - wI)U_{k+1/2} = -(L_y + wI)U_k - \frac{\rho}{\varepsilon}$$

$$(L_y - wI)U_{k+1} = -(L_x + wI)U_{k+1/2} - \frac{\rho}{\varepsilon}$$ (12.85)

where $w = \frac{2}{D\Delta t}$ and I is a unit matrix

Each diagonal term in the equation set now has an additional term which depends on the time step used in the iteration. It should be noted that some authors prefer to include a term (typically called r) with the L_x and L_y operators, in which case the definition of such a term is the reciprocal of the w factor used here. This modified operator splitting approach is what most authors refer to as the ADI technique. The original derivation of the equations can now be somewhat forgotten and consideration given to the optimal value of the w parameter for rapid convergence. Because of the stabilizing nature of the forward difference, this set of equations is known to be stable for all w under some restricted conditions such as the conditions that L_x and L_y commute, are symmetrical and have the same eigenvectors. While these conditions will not be true for all general problems and for nonlinear problems, it is clear that for large values of the w parameter, the diagonal terms on each side of the two equations in Eq. (12.85) will dominate and when these terms dominate there will be little change in the solution from iteration to iteration. From a purely operational point of view, the w term limits the change from iteration to iteration. The use of forward differencing is known to produce some error in a transient solution. However, the interest here is essentially in the limiting solution so the intermediate iterative values are not of major important – what is important is that the iterative procedure converge to the correct final solution values.

For a general formulation of a PDE in two dimensions as expressed in the form of Eq. (12.64) the Newton linearized form of the ADI equation set becomes:

$$\frac{\partial F}{\partial U_{xx}}\frac{\partial^2 u_{k+1/2}}{\partial x^2} + \frac{\partial F}{\partial U_x}\frac{\partial u_{k+1/2}}{\partial x} + (\frac{\partial F}{\partial U} - w)u_{k+1/2} =$$

$$-(\frac{\partial F}{\partial U_y}\frac{\partial u_k}{\partial y} + \frac{\partial F}{\partial U_{yy}}\frac{\partial^2 u_k}{\partial y^2} + wu_k + F)$$

$$\frac{\partial F}{\partial U_{yy}}\frac{\partial^2 u_{k+1}}{\partial y^2} + \frac{\partial F}{\partial U_y}\frac{\partial u_{k+1}}{\partial y} + (\frac{\partial F}{\partial U} - w)u_{k+1} =$$

$$-(\frac{\partial F}{\partial U_x}\frac{\partial u_{k+1/2}}{\partial x} + \frac{\partial F}{\partial U_{xx}}\frac{\partial^2 u_{k+1/2}}{\partial x^2} + wu_{k+1/2} + F)$$

(12.86)

In the first of these equations, the x-oriented derivatives are used to update the solution values while in the second of the equations the y-oriented derivatives are used to update the solution values. A complete ADI cycle applies both equations is sequence. The choice of associating the $\partial F / \partial u$ term entirely on the left hand side with the equation to be updated is open to debate. Another choice would be to associate part (perhaps half) of the term with both the old solution value and the equation to be updated. However the choice shown in Eq. (12.86) appears to be a good choice for the examples in this work.

Let's now consider appropriate values for the w parameter. The diagonal terms of the spatial L_x and L_y operators will be for a uniform grid of magnitude $2/(\Delta x)^2$ and $2/(\Delta y)^2$ or if the spatial dimension is normalized to 1 they will be of magnitude $2N_x^2$ or $2N_y^2$. A word is in order with respect to the sign of the w term in Eq. (12.85). For stability, the term must increase in magnitude the value of the diagonal matrix terms. For a positive second derivative term in an equation, the sign of the resulting diagonal matrix element is negative and thus the negative term on the left of Eq. (12.85) will be of the same sign as the diagonal operator term. This implies that one must select the sign of the w term to match the sign of the second derivative terms in a defining differential equation to be solved. The importance of the w terms in the equation will then be measured relative to the diagonal matrix terms contributed by a differential equation. If w is much smaller than these numbers, it will have a relatively small effect and one would expect the equations to converge more rapidly to a steady state or final solution. Of course a larger w value tends to be a more stabilizing factor in the solution. Again for a simple square spatial region with constant boundary conditions and for Poisson's equation, an optimal value of this factor is known. A detailed discussion of this theory would go too far beyond the intent here, but again the factor (or more exactly the reciprocal of w) is related to the eigenvalues of the L_x and L_y operators. Appropriate values of w for a model unit square with N intervals, span the range from $1/\lambda_{\max}$ to $1/\lambda_{\min}$, where:

$$\lambda_{\max} = 4\cos^2(\pi/2N) \text{ and } \lambda_{\min} = 4\sin^2(\pi/2N) \qquad (12.87)$$

For large N this range becomes $(1/4) < w < N^2/\pi^2 \sim 0.1N^2$. This is a rather wide range of possible values with vastly different convergence rates. The use of the upper limit for w would make the diagonal terms about 5% of the spatial operator terms while going to the other limit makes the contribution due to these terms much less important. This discussion provides some useful general rules about selecting an appropriate value of w. However for most problems, a value must be selected from trial and error based upon what value results in rapid convergence of the iterative process. There is considerable discussion in the literature about selecting the value of w and many authors suggest varying the value as the iterative solution process proceeds. Some examples will be discussed in the next section and in general for the work here a constant value of this parameter will be used. The reader should consult the literature for a more general discussion of procedures for varying this parameter.

Computer code for implementing this ADI algorithm is relatively straightforward but several functions are needed for the implementation. First code is needed for generating the set of matrix equations with separate accounting taken of the x and y oriented spatial operators. This is already available with the setup2bveqs() function in Listing 12.20 by including a non-nil final argument in the list of calling variables. Next some function is needed for transforming the matrix equations and any solution from the x oriented numbering to the y oriented numbering. This is just a reordering of the matrix terms or of the solution terms – primarily a book keeping process. A code segment named reorder() has been written for this reordering process and can be used in the form reorder(a,b) or reorder(a,b,w) where the call will reorder a square coefficient matrix a and a linear matrix b with a single call. At the same time the second form of the example with a w value will apply a w factor to the diagonal elements of the resulting coefficient a matrix as specified by Eq. (12.84). This is convenient as the w factor can be applied as the reordering process proceeds. More appropriately, a separate function updatew(a,w) or updatew(a,w1,w2) has been written to update the w value in the equation set. In the second form, the call can be used to change the w value in a set of matrix equations from the value w1 to w2. The second form of call is equivalent to the first form with updatew(a,w2-w1). Both of these are relatively straightforward algorithms and the reader can look at the code in the supplied pde2bv.lua set of code functions.

The final piece of code needed is a routine to solve a tri-diagonal set of matrix equations. The reader is referred back to Section 11.5 where this is discussed for a one dimensional boundary value problem. If one considers the x oriented operator at a fixed y value, then the present problem is essentially the same as a one dimensional case. Looking back at the structure of the matrix equations it can readily be seen that what is under consideration are the matrix elements clustered around the diagonal which are left when the off diagonal fringe elements are transferred to the known (or assumed known) side of the equation set. Many authors simply refer to solving the complete set of equations as a tri-diagonal set of equations. However, there are several subtle factors to be considered for the type of matrix found in this problem. First the boundary node equations are not a simple set of tri-diagonal equations. For example consider the equation set with x oriented numbering. For the boundary nodes (at y = 0 or y = ymax), the equations may have no x oriented terms. For example if the boundary condition is specified as some condition on the normal (or y oriented) derivative, then the collection of matrix terms for the L_x operator will all be zero. The normal derivative (or Dirichlet) boundary condition is certainly a common occurrence. Thus in solving the matrix equation along the x oriented operator, the first and last row in a solution set must be handled separately from the other rows as they may have no x oriented operator. The solution to this problem is to traverse these rows last after all other values have been updated and then all the values needed for any normal derivative terms have been calculated.

Another feature is that the solutions should be accomplished on a row by row (or column by column) basis and not by considering the complete matrix as a tridiagonal matrix. The need for this can be understood by considering the sub-matrix for a row of spatial points as illustrated in Figure 12.66 for the case of 10 spatial intervals along a row. If a three point derivative approximation is used for the end point normal derivatives, then the coefficient matrix has the possibility of extra matrix elements for the first and last rows associated with the sub-matrix as indicated by the open circles in the figure. This complicates slightly the solution for the tri-diagonal matrix, as the beginning and end points must be considered as special cases. This is the reason the solution must be developed on a row (or column) basis. Such end points were previously considered in Section 11.5 and the discussion here closely follows the development there.

Neglecting for the moment the end point equations, the equation for a general element in a row of the matrix can be expressed as:

$$a_k u_{k-1} + b_k u_k + c_k u_{k+1} + d_k = 0 \tag{12.88}$$

where subscript k labels the solution node and a, b and c are the three diagonal non-zero matrix elements. A straightforward method for solving this set of

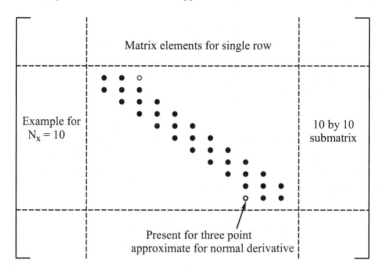

Figure 12.66. Example of matrix elements for a single row with the L_x operator.

tri-diagonal equations is to assume the solution can be written as:

$$u_{k-1} = e_k u_k + f_k \qquad (12.89)$$

When this is inserted into Eq. (12.88) and terms are rearranged one obtains the iterative relationship:

$$e_{k+1} = -\frac{c_k}{(a_k e_k + b_k)}; \quad f_{k+1} = -\frac{a_k f_k + d_k}{(a_k e_k + b_k)} \qquad (12.90)$$

Thus if starting values for the parameters are known, the iterative relationships can be evaluated across the matrix row and then the solution obtained by a transversal back across the structure applying Eq. (12.89). The only complicating part is obtaining a starting value for the parameters. The equations for the first two rows are:

$$b_1 u_1 + c_1 u_2 + \beta_1 u_3 + d_1 = 0$$
$$a_2 u_1 + b_2 u_2 + c_2 u_3 + d_2 = 0 \qquad (12.91)$$

where the β_1 term represents the additional open circle term in the first line of Figure 12.66 and is only present when the boundary condition involves the normal derivative. From these two equations, the first terms involving u_1 can be eliminated giving a relationship between u_2 and u_3 which then gives starting values for e_3 and f_3 as:

$$e_3 = \frac{(a_2 \beta_1 - b_1 c_2)}{(b_1 b_2 - a_2 c_1)}; \quad f_3 = \frac{(a_2 d_1 - b_1 d_2)}{(b_1 b_2 - a_2 c_1)} \qquad (12.92)$$

In applying the equations in reverse along the row, the first element value can be determined from Eq. (12.91) as:

$$u_1 = -\frac{(c_1 + b_2)u_2 + (\beta_1 + c_2)u_3 + (d_1 + d_2)}{(b_1 + a_2)} \tag{12.93}$$

The only remaining complexity is the evaluation of the end point solutions from the last two equations which can be written as:

$$a_{nx-1}u_{nx-2} + b_{nx-1}u_{nx-1} + c_{nx-1}u_{nx} + d_{nx-1} = 0$$
$$\beta_{nx}u_{nx-2} + a_{nx}u_{nx-1} + b_{nx}u_{nx} + d_{nx} = 0 \tag{12.94}$$

where in this case the β_{nx} term is the possible extra term along the last equation row in Figure 12.66 shown as an open circle. In these equations, the first terms can be eliminated using the relationship of Eq. (12.89) and then the resulting end point solution values can be solved for as:

$$u_{nx-1} = \frac{(c_{nx-1}d_{nx} - d_{nx-1}b_{nx}) + (c_{nx-1}\beta_{nx} - a_{nx-1}b_{nx})f_{nx-1}}{(a_{nx-1}b_{nx} - c_{nx-1}\beta_{nx})e_{nx-1} + (b_{nx-1}b_{nx} - c_{nx-1}a_{nx})}$$

$$u_{nx} = \frac{((a_{nx-1} + \beta_{nx})e_{nx-1} + b_{nx-1} + a_{nx})u_{nx-1} - (d_{nx-1} + d_{nx} + (a_{nx-1} + \beta_{nx})f_{nx-1})}{(c_{nx-1} + b_{nx})} \tag{12.95}$$

The two end point values can then be solved for in the order of u_{nx-1} followed by u_{nx}.

The solution procedure is now established: one begins with the initial definitions for the e and f coefficients at one side of the structure and evaluates these coefficients across a row of the structure using the iterative relationship of Eq. (12.90). At the end of the row, the last two solution values are evaluated according to Eq. (12.95) and then one steps back along the row evaluating all the solution values using the relationship of Eq. (12.89). The solution here is similar to that of Section 11.5 except for the treatment of the end point solution values.

Code for solving a set of such modified tri-diagonal equations is given as the subroutine trisolve(a,b,u) in the pde2bv.lua set of functions and can be accessed with a require 'pde2bv' program statement. The details of the implementation will not be presented here as it is a straightforward implementation of the equations outlined above. The reader is encouraged to examine the code in the pde2bv listing if desired. In the solution code the first and last rows are solved following the solution for the interior points as previously discussed and the internal node points are solved on a row by row basis as discussed above.

Listing 12.26 shows a code segment for using the ADI iterative approach for solving the matrix equations associated with a two dimensional PDE. The differential equation in this example is similar to that in Listing 12.25 but is:

$$\frac{\partial^2 u}{\partial x^2} + \frac{\partial^2 u}{\partial y^2} + V_m \pi^2 (\sin(\pi x)\sin(\pi y))^{20} = 0 \tag{12.96}$$

where the forcing function is much more sharply peaked near the center of the spatial grid because of the large power function (with exponent of 20). In this case an exact solution is not known. In Listing 12.26 the code for setting up the equation definitions and defining the spatial grid on lines 1 through 24 should be

```
 1 : -- File list12_26.lua --
 2 : -- Example of BV problem in 2 D -- rectangular grid of points
 3 :
 4 : require"pde2bv"
 5 : Vm = 5.0 -- Define equations to be solved
 6 : pi = math.pi;pi2 = 2*pi^2; sin=math.sin
 7 : feqs = { -- Table of functions
 8 :    function(x,y,uxx,ux,u,uy,uyy,i,j) -- General point
 9 :            return uxx + uyy +
                 Vm*pi2*(sin(pi*x/xmax)*sin(pi*y/ymax))^20
10 :    end,
11 :    function(x,u,uy,ux,i) return u end, -- Bottom boundary
12 :    function(x,u,uy,ux,i) return u end, -- Top boundary
13 :    function(y,u,ux,uy,j) return u end, -- Right boundary
14 :    function(y,u,ux,uy,j) return u end  -- Left boundary
15 : }  -- End general point and boundary values
16 :
17 : x,y,u = {},{},{}; xmax,ymax = 1.0,1.0 -- Try other values
18 : Nx,Ny = 80,80
19 : nx,ny = Nx+1,Ny+1; n = nx*ny
20 : for i = 1,nx do x[i] = xmax*(i-1)/Nx end
21 : for i = 1,ny do y[i] = ymax*(i-1)/Ny end
22 : for j = 1,ny do   -- Set x,y grid of initial values
23 :    u[j] = {}; for i = 1,nx do u[j][i] = 0 end
24 : end
25 : a,b = setup2bveqs(feqs,x,y,u,1) -- Set up equations, x first
26 :
27 : ar,br = reorder(a,b) -- Reorder with y first numbering
28 :
29 : wx,wy = 100,100 -- Set ADI factors -- Try other values
30 : updatew(ar,wy); updatew(a,wx)
31 :
32 : t = os.time()
33 : ua = Spmat.new(nx,-ny) -- Array for solution
34 : for j=1,ny do -- Linear array of solution values
35 :    for i=1,nx do ua[i+(j-1)*nx] = u[j][i] end
36 : end
37 : era,ier = {},1; nadi = 50
38 : for k=1,nadi do -- Loop over ADI solutions
39 :    ua,errm = trisolve(a,b,ua) -- X first labeling
40 :    print('k, first max corr =',k,errm);io.flush()
41 :    era[ier],ier = errm,ier+1
42 :    ua = reorder(ua) -- X first to Y first
43 :    ua,errm = trisolve(ar,br,ua) --Y first labeling
44 :    print('k, second max corr =',k,errm);io.flush()
45 :    era[ier],ier = errm,ier+1
46 :    ua = reorder(ua) -- Y first to X first
47 : end
48 : print('time = ',os.time()-t)
49 : print(' At end maximum correction =',errm)
50 : bb = to2darray(ua) -- Convert to 2D array for plots
51 : splot(bb); cplot(bb) -- Surface and coutour plots
52 : write_data('list12_26.dat',bb)
53 : write_data('list12_26a.dat',era)
54 : splot('list12_26.emf',bb)
```

Selected output:
```
k, first max corr =     49   1.4575868290123e-009
```

```
k, second max corr =   49        1.1627040441553e-009
k, first max corr  =       50    9.2912910698573e-010
k, second max corr =       50    8.6599016846378e-010
```

Listing 12.26. Example of ADI iterative solution of matrix equations for 2D PDE.

very familiar from previous examples. The call to setup2bveqs() on line 25 sets up the matrix equations with an x first ordering of the node equations and the call to reorder() on line 27 returns a set of matrix equations with a y first ordering of node equations. Note the use of a 1 or non-nil last parameter in the setup2bveqs() function argument list. This causes the returned coefficient matrix (the a matrix) to store the diagonal matrix values in three parts as previously discussed. Line 29 defines wx and wy factors corresponding to the 'w' factor in the ADI formulation of Eq. (12.85). In this case the possibility of two different 'w' factors for the two different ADI directions is included to accommodate possible differences if desired (more about this later). The calls to updatew() on line 30 insert the ADI 'w' factor into each set of matrix equations. Note that two solution arrays are defined – one (named u[]) with x,y ordering on lines 22 through 24 and one (named ua[]) with a single integer (x first) ordering on lines 34 through 36. Also two sets of coefficient matrixes are defined – one (named a,b on line 25) with x first ordering and one (named ar,br on line 27) with y first ordering.

The heart of the ADI iterative solution is the calculation loop between lines 38 and 47. The call to trisolve(a,b,ua) on line 39 solves the tri-diagonal equation set for the x first ordering and the call to trisolve(ar,br,ua) on line 43 solves the equation set for the y first ordering. Between calls to these solution sets the calls to reorder(ua) on lines 42 and 46 reorder the solution set (ua) between the two x and y oriented numbering schemes. Information is kept in the ua[0] table elements (ua[0][3] is either 1 or 2) regarding which node numbering scheme ua represents at any given time so that the function reorder() can properly reorder the solution values at each call to this function. In the present example, the ADI loop is executed a fixed number of times determined by the nadi parameter which is set at 50 on line 37 for this example of an 80 by 80 interval spatial grid. In order to monitor the progress of the solution, the maximum correction value to the solution is saved at each iterative step and printed to a saved file on line 53. The final solution is converted to a two dimensional array on line 50 by the to2darray() call and saved to a file on line 52. Note that the argument list to the to2darray() function has only a single value – the solution array. The function obtains spatial interval information from the ua[0] table entries. The selected output shows that the last correction values are in the range of 1e-9. However, this does not mean that the final calculated solution values are this accurate. The actual accuracy is discussed below.

The splot() and cplot() calls on line 51 of Listing 12.26 provide pop-up 2D surface graphs of the solution and contour plots for the solution variable. Finally the call to splot() on line 54 with a file name as the first argument saves the generated surface plot to the named file in an enhanced windows metafile (emf) format. These are useful for saving a copy of a pop-up graph. Figure 12.67 shows the

the previous solution of Figure 12.61 with a first order power on the sin() functions. The charge density for this solution is shown in Figure 12.68 and is very concentrated toward the center of the spatial area.

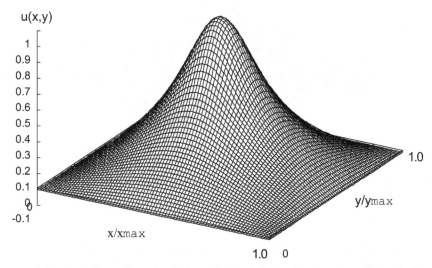

Figure 12.67. Surface plot of solution obtained from the ADI iterative solution technique for the example in Listing 12.26. The potential value is zero on all four boundaries.

In keeping with previous discussions, the accuracy of the implemented ADI iterative solution technique will be explored. While this is difficult to do for general problems, it can be explored for this straightforward example. Figure 12.69 shows the magnitude of the maximum correction to the solution at every ADI half step update. The lower curve labeled Correction is a plot of the data generated by the era[] array in Listing 12.26. Also shown is the absolute accuracy which is the difference between the ADI iterative solution and the exact solution to the set of matrix equations. And how do we know the exact solution since a closed form solution is not known for this example? For the exact solution, the same example was programmed using the spgauss() routine to solve the resulting set of matrix equations. While there is some error in the spgauss() solution relative to the differential equation due to the finite difference approach, the spgauss()solution should have very little error in regard to the solution of the set of matrix equations. The 'Correction' curve is for the worst case point in the two dimensional array so every correction value is at least as small as the values given by the curve. From the figure an initial fast drop in the corrective term is seen and this is followed by

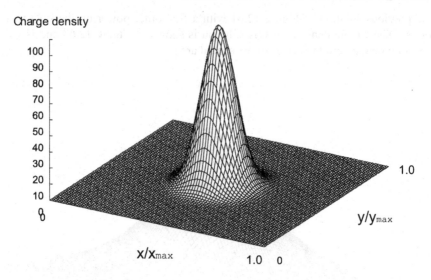

Figure 12.68. Charge density for the solution shown in Figure 12.67.

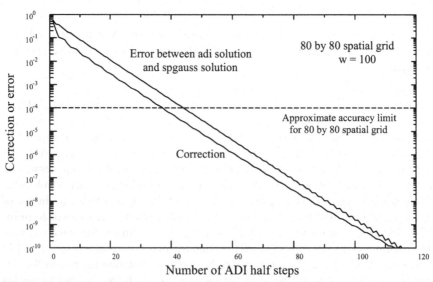

Figure 12.69. Error between the ADI iterative solution and the spgauss solution for each ADI half step along with the ADI correction at each half step.

an exponential decrease in the correction magnitude as a function of the number of ADI half steps. The solution error also follows an exponential decrease and is about one order of magnitude larger than the correction term at each iterative step. For this example at 80 half step iterations (equal to the spatial interval number) the worst case accuracy is seen to be about 0.01% of the peak solution value. This would probably be adequate for most applications. It can also be seen that the error can be made quite small if sufficient iterative steps are taken. Both the magnitude of the correction term and the error in solving the set of matrix equations is seen to approach very small values so the ADI approach is converging to an exact solution of the set of matrix equations.

Also shown in Figure 12.69 is a dotted line representing the approximate accuracy limit to the solution of the differential equation due to the finite spatial grid size as obtained in Figure 12.64. This is seen to occur at about 50 ADI half steps. What this means in terms of the accuracy of solving a 2D PDE is that any improved accuracy after about 25 full ADI steps in this example would be meaningless as this would be considerably below the accuracy limit imposed by the finite spatial grid size. Thus for spatial grid sizes that can be computed in reasonable computer times (several hundred by several hundred – perhaps 500 by 500) it is reasonable to terminate the iterative ADI solution after the correction error is on the order of 1.e-5 to 1.e-6. This conclusion also applies to any of the iterative approximate solution methods (identified as COE and SOR).

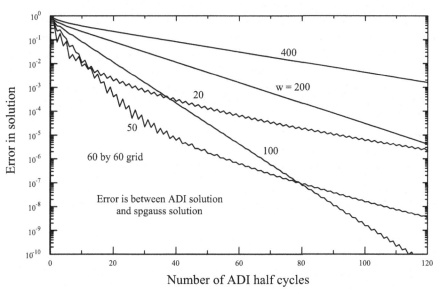

Figure 12.70. Convergence data for ADI technique with a 60 by 60 grid and for the example in Listing 12.26.

Also of interest is the convergence of the ADI technique for different 'w' factors. Some data on this is shown in Figure 12.70 for this example with a range of w values from 20 to 400. It can be seen that there is an optimal value for w which for this example is somewhere between 50 and 100 if 1.e-6 is the smallest correction factor of interest. At w values below 100 it is seen that accuracy for small numbers of iterative cycles is improved but the long term improvement in accuracy is considerably slower. This would suggest that one could probably use a variable w technique where a small initial value was transformed into a larger value as the iteration proceeds. However, this approach has not been explored and will be left to the reader to explore if desired.

Selecting an optimum value or sequence of values for w for an arbitrary linear or nonlinear problem is an unsolved problem. Some numerical experimentation is usually needed to obtain an appropriate value. By executing the code in Listing 12.26 over a range of spatial grid values (from 20 by 20 to 200 by 200) it was found for this example that the optimum w factor is essentially independent of spatial grid numbers (and has a value of about 87). Also it is found that the optimum value is about the same for non-equal spatial intervals on the x and y axis. Finally by changing the magnitude of the spatial range, it can be observed that the optimum w value scales with $(xmax - xmin)^{-2}$ or $(ymax-ymin)^{-2}$. The reader is encouraged to re-execute the code in Listing 12.26 and observe the solution and solution times for different grid ranges and spatial intervals.

The optimum w factor is also a function of the boundary conditions as can be verified by changing the boundary conditions in the example of Listing 12.26. If two opposite boundary conditions are changed to zero normal derivative instead of zero solution value on lines 11 through 14 of Listing 12.26 and calculations such as shown in Figure 12.70 are repeated, it is found that the optimum value of the w parameter is about 4 times the value (or about 360) obtained for the zero boundary case. This makes it more difficult to provide a default value for the w parameter that can be used with a wide variety of problems. For this example a value of w = 100 will provide reasonably fast convergence for both zero conditions on all four boundaries and for zero normal derivative on two boundary conditions. However this value is not optimum for either set of boundary conditions. For a general PDE the user should experiment with w values to determine an optimum value.

12.7.3 Putting it Together for BVPs in Two Dimensions

The previous section has discussed several techniques for solving the matrix equations associated with BVPs in two dimensions. This section will develop more general callable code routines which can be used to solve two dimensional BVPs. The approach is to develop a single routine that can use any of the discussed matrix solution techniques to solve both linear and nonlinear two dimensional BVPs. The basic ingredients for solving such problems are an array of spatial grid points, functions defining the differential equation and boundary conditions, and some criteria for determining the desired accuracy of the solution. If one has a nonlinear problem, the solution must be incorporated into a series of Newton iterative cycles using the "linearize and iterate" principle. So for non-linear problems this requires additional iterative cycles beyond any required in the iterative solution of the matrix equations. Thus the procedure will be that of an iterative cycle for solving the matrix equations within the non-linear Newton iterative cycles for solving the differential equation. There should be some way for the user to specify the desired method of solution – either the direct sparse matrix solution method or one of the iterative techniques. Finally an accuracy criterion must be specified for terminating the iterative cycles – for both the iterative matrix solution method and the Newton iterative cycles.

Listing 12.27 shows code for a function integrating the various matrix solution methods into a single more general function for solving a 2D PDE. The calling argument list includes a parameter 'tpsola' that is used to specify the desired solution method with these being (a) direct sparse matrix solution (SPM=1), (b) Chebyshev with odd/ever COE (COE=2), (c) SOR (SOR=3) and (d) ADI (ADI=4). The tests on lines 443 through 444 select the various solution methods based upon the input parameter. The core of the function is a Newton's method iterative cycle from line 461 to 481 that iteratively calls the selected solution method after updating the set of matrix equations on line 462. This iterative loop is to linearize and solve the basic PDE. At each Newton cycle, the solution is updated on line 476 along with a calculation of the maximum corrective value applied to the solution. When the maximum corrective error is a specified fraction of the peak solution value (with specified relative error of ERR = 1.e-5) the Newton loop exits at line 480 and the final solution is returned on line 487 along with the final maximum corrective term.

A few other bookkeeping sections in the code convert the input solution array from an x-y labeled two index matrix into a single column array if needed on lines 454 through 459 and a final loop converts the single column solution array into a j,i two index matrix of solution values on lines 486 through 488. The type of solution desired may be specified as a table of values in order to pass additional parameters to the PDE solver. If a table is specified for the typsola parameter, two additional parameters, rx and ry, can be extracted from the input list as seen on line 444. For a direct sparse matrix solution with the SPM parameter these are not used. For the COE and SOR methods, this feature can be used to specify a spectral radius value with only the rx parameter used. For the ADI method these two values can be used to specify wx and wy parameters. In all cases, default values of these parameters are available, if values are not input to the various functions.

The code calls other functions on lines 463 through 474 for the various solution methods. The most straightforward solution method is simply to solve the matrix equations by the diagonal numbering sparse matrix routine, sdgauss(), as on line 464. For the ADI method, the function pde2bvadi() is called on line 466. This function is essentially a callable function version of the code shown in Listing 12.26 from line 38 through 47 with additional testing of the maximum value of the correction term so that the function can return when the solution has reached a specified accuracy. The coded function has a default w value of 100 but the user

```
  1 : -- File pde2bv.lua -
442 : pde2bv = function(feqs,x,y,u,tpsola) -- 2 D PDE Solver
443 :   if type(tpsola)~='table' then tpsola = {tpsola} end
444 :   local tpsol,rx,ry=tpsola[1] or SPM,tpsola[2],
           tpsola[3] or tpsola[2]
445 :   local umx,errm,a,b,n,uxy,ndg = 0.0
446 :   local nx,ny = #x, #y
447 :   local uold,ua,n,itmx = {},{},nx*ny,0
448 :   local ur = Spmat.new(nx,-ny)
449 :   if linear==1 then nnmx=1 else nnmx=NMX end -- One cycle
450 :   if tpsol==ADI then ndg = 1 end
451 :   uold[0] = {u[0][1],u[0][2],u[0][3]}
452 :   ua[0] = {u[0][1],u[0][2],u[0][3]}
453 :   if #u==n then uxy=false else uxy=true end
454 :   for j=1,ny do
455 :     for i=1,nx do
456 :       k = i + (j-1)*nx
457 :       if uxy then ua[k] = u[j][i] else ua[k] = u[k] end
458 :     end
459 :   end
460 :   for k=1,n do uold[k],umx,ua[k] = ua[k],max(umx,ua[k]),0.0 end
461 :   for int=1,nnmx do -- Newton iterative loop
462 :     a,b = setup2bveqs(feqs,x,y,uold,ndg)
463 :     if tpsol==SPM then -- Solve with sparse matrix solver
464 :       sdgauss(a,b); ua = b -- b is new solution
465 :     elseif tpsol==COE then -- Solve with Chebychev SOR
466 :       ua = pde2bvcoe(a,b,ua,rx,umx)
467 :     elseif tpsol==SOR then -- Solve with SOR
468 :       ua = pde2bvsor(a,b,ua,rx,umx)
469 :     elseif tpsol==ADI then -- Solve with ADI
470 :       if rx==nil then rx = -abs(x[nx]-x[1]) end
471 :       if ry==nil then ry = -abs(y[ny]-y[1]) end
472 :       ua = pde2bvadi(a,b,ua,rx,ry,umx)
473 :     else print('Unknown type solution request:',tpsol,
           ' in pde2bv')
474 :     end
475 :     errm,umx,itmx = 0.0,0.0,itmx+1
476 :     for k=1,n do errm,uold[k] = max(errm,abs(ua[k])),
             uold[k]+ua[k] end
477 :     for k=1,n do umx,ua[k] = max(umx,abs(uold[k])), 0.0 end
478 :     if nprint~=0 then print('Completed Newton iteration',int,
             'with correction',errm); io.flush()
479 :     if seeplot~=0 then if seeplot==1 then
                   splot(to2darray(uold))
             else cplot(to2darray(uold)) end end end
480 :     if errm<ERR*umx then itmx = int; break end
481 :   end
482 :   if itmx==NMX then print('Maximum number of iterations
           exceeded in pde2bv!!')
483 :     io.flush() end
484 :   if uxy==false then return uold,errm,itmx
485 :   else
486 :     for j=1,ny do ur[j] = {}; for i=1,nx do k = i+(j-1)*nx;
               ur[j][i] = uold[k] end end
487 :     return ur,errm,itmx
488 :   end
489 : end
490 : setfenv(pde2bv,{sdgauss=sdgauss,max=math.max,abs=math.abs,
         table=table,
```

```
491 :  setup2bveqs=setup2bveqs,pde2bvcoe=pde2bvcoe,
          pde2bvsor=pde2bvsor,pde2bvadi=pde2bvadi,io=io,
          getfenv=getfenv,linear=0,SPM=1,COE=2,SOR=3,ADI=4,
          NMX=50,ERR=1.e-5,print=print,type=type,nprint=0,io=io,
          Spmat=Spmat,seeplot=0,cplot=cplot,splot=splot,
          to2darray=to2darray})
```

Listing 12.27. Code segment for solution of 2D PDE boundary value problem using one of 4 selected methods.

can change the value by specifying a value in calling the function. The COE and SOR functions pde2bvcoe() and pde2bvsor() have been previously discussed. The interested reader can view the code for these functions if desired. In all cases the approximate matrix solution techniques are iterated until the maximum relative correction error is less than 1.e-4. A final feature of the code is the test on line 449 for the value of a "linear" parameter. If this parameter is set to 1 by a statement getfenv(pde2bv).linear = 1, then the number of Newton iterations is set to 1 and a linear PDE is assumed. For a linear equation this can save computational time as otherwise, the code must execute 2 Newton cycles to determine that the equation set is a linear equation. For the direct sparse matrix solution method, this will reduce the execution time by a factor of 2. However, for the approximate solution methods, the time reduction will be relatively small as the extra call to the solution method will only require typically one additional approximate solution cycle.

An example of using this function for solving the previous Poisson's equation is shown in Listing 12.28. The function definitions are the same as in the previous Listing 12.26. In this case the spatial grid size is set at 100 by 100 intervals. One difference is the use of the setxy({xmin,xmax},{ymin,ymax},Nx,Ny) function on line 20 to set the x and y array values. This function is defined to simplify the definition of linear arrays from some min to max values for x and y with a specified number of spatial intervals. The new function pde2bv() is called on line 27 with the ADI solution method specified. The reader is encouraged to execute the code in the example changing the solution method and varying the number of spatial intervals. For this simple example, all the approximate solution methods rapidly converge to a specified correction accuracy of about 1.e-5 times the peak solution value.

The computer time taken by the solution is printed on line 30 so the user can compare the different solution methods. The relative times taken in solving this PDE problem with the code in Listing 12.28 was found to be in the order of COE:SOR:ADI:SPM = 1.0:1.40:1.47:24.0 where the COE solution method takes the shortest time and the other times are given relative to this method. However if the ADI parameter on line 29 is replaced by {ADI,87} which executes the ADI technique with the optimum w parameter value (as opposed to the default value of 100) then the ADI technique executes in approximately 50% less time and becomes the fastest solution method. Note that the direct sparse matrix solution technique takes much longer for this example than the approximate solution methods. Since this is a linear PDE, the SPM time can be cut in approximately half by

```
 1 : -- File list12_28.lua --
 2 : -- Example of BV problem in 2 D  using pde2bv() function
 3 :
 4 : require"pde2bv"
 5 : Vm = 5.0 -- Define equations to be solved
 6 : pi = math.pi;pi2 = 2*pi^2; sin=math.sin
 7 : xmax,ymax = 1,1
 8 : feqs = { -- Table of functions
 9 :     function(x,y,uxx,ux,u,uy,uyy,i,j) -- General point
10 :         return uxx + uyy +Vm*pi2*(sin(pi*x/xmax)*
                 sin(pi*y/ymax))^20
11 :     end,
12 :     function(x,u,uy,ux,i) return u end, -- Bottom boundary
13 :     function(x,u,uy,ux,i) return u end, -- Top boundary
14 :     function(y,u,ux,uy,j) return u end, -- Right boundary
15 :     function(y,u,ux,uy,j) return u end  -- Left boundary
16 : }   -- End general point and boundary values
17 :
18 : Nx,Ny = 100,100
19 : nx,ny = Nx+1,Ny+1
20 : x,y = setxy({0,xmax},{0,ymax},Nx,Ny)
21 : u = Spmat.new(nx,-ny)
22 : for j = 1,ny do   -- Set zero initial values
23 :     u[j] = {}; for i = 1,nx do u[j][i] = 0 end
24 : end
25 :
26 : SPM,COE,SOR,ADI = 1, 2, 3, 4 -- 4 solution types
27 : getfenv(pde2bv).nprint=1;getfenv(pde2bvadi).nprint=1
28 : t1 = os.clock()
29 : u,errm = pde2bv(feqs,x,y,u,ADI) -- Replace ADI as desired
30 : print('time =',os.clock()-t1)
31 : print(' At end maximum correction =',errm)
32 : splot(u); cplot(u)
33 : splot('list12_28.emf',u)
34 : write_data('list12_28.dat',u)
```

Listing 12.28. Example code for solving a 2D BVP with the pde2bv() function.

setting the "linear" parameter. However this would still execute about 12 times slower than the COE method.

If two of the zero boundary conditions on lines 12 through 15 are replaced by zero normal derivative boundary conditions, the approximate matrix solution techniques require more iterations to achieve the same level of accuracy and the execution times are increased by about a factor of 2. However, the order of the solution times remains the same with the COE providing the fastest solution time. The execution time for the SPM technique is obviously independent of the set of linear boundary conditions.

One of the reasons all of the approximate techniques are much faster than the direct matrix solution method is that the iterations can be terminated much before the ultimate accuracy of the approximate solution methods is achieved. For example, the SPM method solves the set of matrix equations with an accuracy approaching the internal machine accuracy. The approximate solution methods are on the other hand terminated when the relative accuracy of the correction terms is on the order of 1.e-5 which is many orders of magnitude less accurate than the

SPM matrix solution. Because of the accuracy limits of the finite difference spatial approximations to the PDE, additional accuracy is simply wasted in achieving the machine accuracy limit in solving the matrix equations. The ability of the approximate solution methods to terminate after achieving some relatively modest accuracy in the solution (relative to the machine accuracy) is one of the major reasons all the approximate solution methods are so much faster in this example than the direct matrix solution method. The approximate methods are also faster for a wide range of typical engineering problems. However, the approximate methods do not always converge so readily and may in fact not reach a converged solution for some problems. Also the SPM method can always be used to verify the accuracy of the various approximate methods. Several examples of 2D PDEs will be discussed in a following Section 12.8.

12.8 Two Dimensional BVPs with One Time Dimension

Before considering some additional examples of the finite difference method for two dimensional boundary value problems, it is useful to consider the possibility of adding another dimension to the problem and this is usually a time dimension. The prototypical example of such an equation is the time dependent diffusion equation:

$$\frac{\partial u}{\partial t} = \frac{\partial}{\partial x}(D\frac{\partial u}{\partial x}) + \frac{\partial}{\partial y}(D\frac{\partial u}{\partial y})$$

$$\frac{\partial u}{\partial t} = D(\frac{\partial^2 u}{\partial x^2} + \frac{\partial^2 u}{\partial y^2}) \text{ for constant } D$$

(12.97)

where D is a diffusion coefficient (dimensions of distance^2/time). For such time dependent physical problems, there are typically no mixed derivatives between the spatial variable and the time variable and this greatly simplifies the problem as opposed to a more general 3 dimensional problem where mixed time and derivatives might occur.

If the time derivative is approximated by the backwards differencing method for some time interval $h = \Delta t$ then the resulting equation becomes:

$$\frac{u_{t+h} - u_t}{h} = D(\frac{\partial^2 u_{t+h}}{\partial x^2} + \frac{\partial^2 u_{t+h}}{\partial y^2}) \text{ or}$$

$$\frac{\partial^2 u_{t+h}}{\partial x^2} + \frac{\partial^2 u_{t+h}}{\partial y^2} - \left(\frac{1}{Dh}\right)u_{t+h} + \left[\frac{u_t}{Dh}\right] = 0$$

(12.98)

The term is square brackets is the solution value at some time t and is assumed to be some known function of the spatial variables. For a given time and time step h, the form of this equation is identical to the general form of the two dimensional boundary value problem previously considered. As discussed in the previous section this is also very similar to the use of the w parameter in the ADI technique to stabilize the alternating direction iterative method. Thus at any given time and time step, the code routines already developed for the two dimensional BVP can be used essentially as already given to solve Eq. (12.98). Before delving into this however, it is known that the trapezoidal integration rule is a much more stable time integration algorithm than the backwards differencing method (see Section 10.1). With a little additional effort this can be implemented as:

$$\frac{u_{t+h} - u_t}{h} = \frac{u'_{t+h} + u'_t}{2} \text{ leading to}$$

$$\frac{\partial^2 u_{t+h}}{\partial x^2} + \frac{\partial^2 u_{t+h}}{\partial y^2} - \left(\frac{2}{Dh}\right)u_{t+h} + \left[\left(\frac{2u_t}{Dh}\right) - \left(\frac{\partial^2 u_t}{\partial x^2} + \frac{\partial^2 u_t}{\partial y^2}\right)\right] = 0 \qquad (12.99)$$

In this equation the term in brackets is some function of the spatial coordinates at time t and is known so the form of this equation is functionally the same as that of Eq. (12.98).

This is all basically identical to the case of one spatial dimension and one time dimension discussed in detail in Sections 10.1 and 10.7. The formulation can be extended to second derivative terms such as occur with the wave equation in two spatial dimensions. The net result is that at each time step the time derivatives can be replaced by the following equations:

$$u'_{t+h} = (u_{t+h} - un)/(h/2)$$
$$u''_{t+h} = (u_{t+h} - unn)/(h/2)^2$$
where   (12.100)
$$un = u_t + (h/2)u'_t = u_t + h_x u'_t$$
$$unn = u_t + (h)u'_t + (h/2)^2 u''_t = u_t + h_y u'_t + h_z u''_t$$

In these replacements, the quantities *un* and *unn* are known functions of the spatial dimensions assuming that the solution at time t has been obtained and the solution is being incremented from time t to a later time t + h.

With these replacements the formulism now exists for extending the two dimensional solvers in the previous section to solve for the time development of a two dimensional plus time PDE. There are, however, some practical considerations before discussing some code for performing this algorithm. One is the obvious fact that such solutions are going to be very time consuming if the solution is desired at a large number of time and spatial points. Second, the cumulative solution set can generate vast amounts of data. For example, a single solution for a 100 by 100 grid of spatial points generates a file of size around 172k bytes. If in addition one wishes to save the solution for 100 time points then a total of around 17.2M bytes of solution data will be generated. While one might calculate for 100 time points in order to achieve accuracy in a solution, it is doubtful that viewing a solution at 100 time points would provide more insight into a solution than a much smaller number of carefully selected time points. Other considerations are the additional storage requirements needed to generate the appropriate replacements for the time derivatives. Eq. (12.100) indicates that in addition to the solutions at a particular time it will be necessary to save arrays for the first and second time derivatives in order to effect the required replacements of the time derivatives at a new solution time. In addition for an initial value problem in time, the appropriate derivatives are not know at the initial starting point so an initial time interval will be required to start the trapezoidal algorithm that uses the simpler backwards differencing approach. This is identical to the approach used for time with one spatial dimension in Chapter 10.

With these considerations, computer code is shown in Listing 12.29 for implementing a series of time steps for solving a 2D PDE with time derivative. The code is very similar to Listing 12.1 that implements code for one time dimension and one spatial dimension. The major loop on time is from Line 548 to line 575. Within this loop the *un* and *unn* functions are evaluated for each data point on lines 553 and 554. The time is incremented on line 557 and the 2D solver pde2bv() is called on line 559 to obtain the solution at a given time value. The code on line 560 then checks for the maximum number of iterations and the first and second time derivative values are updated on line 562. After completing the specified time increments, within the time loop at line 569, the final solution value is saved in a 2 dimensional array on line 572 and returned by the function on line 573 or line 570 if no x-y array is needed. As used in Chapter 10 and in Listing 12.1 for solving initial value problems in time, if the second derivative is not input to the function, an initial interval of 4 time points is used in the calculation using the backwards differencing algorithm to begin the solution. Code to handle this task is contained in lines 542 through 547 and lines 565 through 566. The time derivatives are included in the argument list to pde1stp2bv1t() so that one can repeatedly call the function with increasing time steps and bypass the initial BD algorithm on all except the first call by passing the first and second derivative values back to the function on subsequent calls. This function named pde1stp2bv1t() is designed as a basic single or multiple time step solver and is basically designed to be called by an additional function which increments the time steps and saves any desired collection of solution values.

A final discussion is in order about the interface between this function and the 2D solver ped2bv() function which has no knowledge of time but solves only a set of equations in two spatial variables. For each time increment, Eq. (12.100) provide the required relationships between the time derivatives and the spatial variables. A local set of proxy functions implementing these relationships are defined in the code listing on lines 507 through 532. The pde2bv() function call on line 559 is passed this table of function (named feq) and it in turn calls these proxy functions. These proxy functions when called then evaluate the time derivatives as on lines 510, 514, 519, 524 and 529 and then in turn call the time dependent functions describing the differential equations including both spatial values and time values. References to these functions are passed into this function through the feqs table of functions in the calling arguments to the pde1stp2bv() function as the first parameter value. This is again an example of the reuse of a previously defined code segment to implement a higher level function.

```
  1 : -- File pde2bv.lua --
  2 : -- Code for PDE BV problems in 2 dimensions -- rectangular grid
----
----
495 : pde1stp2bv1t = function(feqs,tvals,x,y,ua,tpsola,ut,utt) --PDE
496 :   local j, neq, t, h, h2,h2sq,hs,hx,hy,hz -- Local variables
497 :   local unn,un,jfirst = {},{},0
498 :   local nit,nitt,errm = 0,0 -- Number of iterations
499 :   local nx,ny = #x, #y
500 :   local neq,uxy = nx*ny
501 :   local tmin,tmax,ntval = tvals[1],tvals[2],tvals[3]
502 :   local u,ur = Spmat.new(nx,-ny), Spmat.new(nx,-ny) -- arrays
503 :   ut = ut or {}; utt = utt or {}
504 :   if #ua==neq then uxy=false else uxy=true end
505 :   -- Functions to add next time values and time derivatives
506 :   local fpde,fbb,fbt,fbr,fbl = feqs[1],feqs[2],feqs[3],feqs[4],
            feqs[5]
507 :   feq = { -- Local functions to add time and time derivatives
508 :     function(x,y,uxx,ux,u,uy,uyy,i,j) -- General spatial point
509 :       local k = i + (j-1)*nx
510 :       local ut,utt = (u - un[k])/h2, (u - unn[k])/h2sq
511 :       return fpde(x,y,t,uxx,ux,u,uy,uyy,ut,utt,i,j)
512 :     end,
513 :     function(x,u,uy,ux,i) -- Bottom boundary
514 :       local ut,utt = (u - un[i])/h2, (u - unn[i])/h2sq
515 :       return fbb(x,t,u,uy,ux,ut,utt,i)
516 :     end,
517 :     function(x,u,uy,ux,i) -- Top boundary
518 :       local k = i+neq-nx
519 :       local ut,utt = (u - un[k])/h2, (u - unn[k])/h2sq
520 :       return fbt(x,t,u,uy,ux,ut,utt,i)
521 :     end,
522 :     function(y,u,ux,uy,j) -- Left boundary
523 :       local k = 1+(j-1)*nx
524 :       local ut,utt = (u - un[k])/h2, (u - unn[k])/h2sq
525 :       return fbr(y,t,u,ux,uy,ut,utt,j)
526 :     end,
527 :     function(y,u,ux,uy,j) -- Right boundary
528 :       local k = j*nx
529 :       local ut,utt = (u - un[k])/h2, (u - unn[k])/h2sq
530 :       return fbl(y,t,u,ux,uy,ut,utt,j)
531 :     end
532 :   }
533 :   for j=1,ny do -- Local array for solution in linear array
534 :     for i=1,nx do
535 :       k = i + (j-1)*nx; if uxy then u[k]=ua[j][i] else
                u[k]=ua[k] end
536 :     end
537 :   end
538 :   t = tmin -- Initial t value
539 :   hs = (tmax - t)/ntval -- Equal increments in t
540 :   -- If initial derivative not given, use BD for first 4 points
541 :   if #ut~=neq then for m=1,neq do ut[m] = ut[m] or 0 end end
542 :   if #utt~=neq then for m=1,neq do utt[m] = 0 end
543 :     jfirst,h = 0,0.25*hs; h2,h2sq,hy,hx,hz = h,h*h,h,0,0 -- BD
544 :   else
545 :     if bd~=false then jfirst,h = 4,hs;
              h2,h2sq,hy,hx,hz = h,h*h,h,0,0
546 :     else jfirst,h = 4,hs; h2,h2sq,hy,hx,hz = hs/2,h*h/4,h,h/2,
```

```
                    h*h/4 end
547 :    end
548 :    for k=1,ntval do -- Main loop for incrementing variable (t)
549 :        repeat -- Use BD for first 4 sub intervals of size h/4
550 :            jfirst = jfirst+1
551 :            -- Set up yn, and ynn arrays to solve equations
552 :            for m=1,neq do
553 :                un[m] = u[m] + hx*ut[m] -- hx = 0 or h/2
554 :                unn[m] = u[m] + hy*ut[m] + hz*utt[m] --
555 :                u[m] = u[m] + h*ut[m] -- Predicted value of u array
556 :            end
557 :            t = t + h -- Now increment t to next t value
558 :            -- Calculate new u values at next time step
559 :            u,errm,nitt = pde2bv(feq,x,y,u,tpsola) -- PDE at t
560 :            if nitt>nit then nit = nitt end -- Monitor max #
561 :            -- New derivative values, same function as in fnext
562 :            for m=1,neq do ut[m],utt[m] = (u[m] - un[m])/h2,
                        (u[m] - unn[m])/h2sq end
563 :        until jfirst>=4 -- End of first interval repeat using BD
564 :        if k==1 then
565 :            if bd~=false then jfirst,h = 4,hs; h2,h2sq,hy,hx,hz =
                    h,h*h,h,0,0
566 :            else jfirst,h=4,hs;h2,h2sq,hy,hx,hz=
                    hs/2,h*h/4,h,h/2,h*h/4 end
567 :        end
568 :        if nprint~=0 then print('Completed time =',t,
                    ' with correction',errm); io.flush() end
569 :    end -- End of main loop on t, now return solution array
570 :    if uxy==false then return u,errm,nit
571 :    else
572 :        for j=1,ny do ur[j] = {}; for i=1,nx do k = i+(j-1)*nx;
                    ur[j][i] = u[k] end end
573 :        return ur,errm,nit
574 :    end
575 : end -- End of pde1stp2bv1t
576 : setfenv(pde1stp2bv1t,{table=table,pde2bv=pde2bv,print=print,
            Spmat=Spmat,io=io,bd=false,nprint=0})
```

Listing 12.29. Code for incrementing a 2D PDE with time variable through one series of time steps.

To effectively use this time stepping function additional functions are needed which take desired time values for obtaining the solution and incrementally call this function to perform the detailed work. Two functions have been coded and provided as examples of this usage. The first is a function named pde2bv1t(feqs,tvals,x,y,uar,tpsola) and the second is the function pde2bv1tqs(feqs,tvals,x,y,uar,tpsola). Both of these have the same list of calling arguments and they differ in the fact that the first function uses time values linearly spaced in value and the second uses time values with a logarithmic spacing. This is similar to the "quick scan" time solutions discussed in Chapter 10 and the reader is referred back to that chapter for a more detailed discussion. The feqs argument is a table listing the functions in the order of differential equation, bottom boundary condition, top boundary condition, left boundary condition and right boundary condition. The uar variable is either an array of the initial values of the solution variable or a table of the initial value, the first derivative and the second derivative values. The possibility of including the two time derivatives is made so that one can repeatedly call the function and use the returned values of the time derivatives to pick up at some time value and continue a solution. The tpsola variable specifies the solution method and in it's simplest form is an integer specifying whether to use the SPM, the SOE, the ADI or the SOR solution methods. For all except the SPM method, additional parameters may be specified in a table to indicate spectral radius values or wx and wy values for the ADI technique.

The tvals argument is the most complicated of the input arguments and is a table listing the desired time solution values and has a similar form to that used for initial value problems in Chapter 10. The most general form of tvals for the pde2bv1t() function is: tvals = {tmin, {t1, t2, --- tn}, {np1, np2, --- npn}, {nt1, nt2, --- ntn}} where tmin is the initial time and the other parameters specify time intervals and numbers of solution points per time interval. For a time interval between tmin and t1, there will be np1 equally spaced solutions saved and there will be nt1 time calculations between each saved interval. For example if np1 = 4 and nt1 = 4 then there will be 16 = 4X4 solutions at equally spaced time points between tmin and t1 with the solution saved at 4 of the 16 time points. The same applies to the other intervals and the defining times. This allows the user to specify in very general terms how the solution is calculated and saved at various time values. For the number of saved time points and the extra time calculations, single numbers can be used which will apply to all time intervals and then the specification can be: tvals = {tmin, {t1, t2, --- tn}, np, nt}. Or a single time interval may be specified as in tvals = {tmin, tmax, np, nt}. Finally np and nt values may be omitted and default values of 10 and 10 will be used. Thus the simplest specification is tvals = {tmin, tmax} which will give the solution at 100 time points between tmin and tmax and save 10 of the solution values, equally spaced in time between tmin and tmax − well actually 11 values will be saved as the function also saves the initial value of the solution.

For the logarithmic or quick scan function, pde2bv1tsq(), the time specification is of the form tvals = {tmin, {t1, t2}, np, nt} where np and nt again specify number of saved time intervals and number of extra calculated time intervals between saved time points. In this case the time points between tmin and t1 are uniformly spaced and those between t1 and t2 are logarithmically spaced and the np and nt values specify numbers of time points per decade in time. As an example, tvals = {0, {1.e-4, 1.e0}, 2, 10} will lead to solutions saved at times 0, 5e-5, 1e-4 over the first interval and at 3.16e-4, 1.e-3, 3.16e-3, 1.e-2, 3.16e-2, 1.e-1, 3.16e-1 and 1.0 time values. For this example, the logarithmic range of time is over 3 decades in time. This is a very useful function and useful time specification as this function can cover a wide range of time values with minimal calculation expense. A somewhat similar wide time range can be achieved by specifying several ranges of time with different time intervals as in the previous paragraph. The pde2bv1tqs() function also provides one additional capability and this is the ability to specify specific spatial points and have the time dependent solution returned for these specified points. This is achieved by including a table of desired time points as the last argument in the calling argument list to the function. The points are specified in a table of the form {{x1,y1}, {x2,y2}, ---{xn,yn}}. The multi-step solvers pde2bv1t() and pde2bv1tqs() are essentially functions that do the bookkeeping needed to call the basic solver of Listing 12.29 multiple numbers of times and return a collected array of solution values. Code for these functions will not be shown but the reader is encouraged to inspect the code in the supplied files. Some examples will show the use of these functions.

Perhaps the simplest time dependent two dimensional PDE is the diffusion equation as given in Eq. (12.97). This equation has a range of engineering applications from temperature distribution in two dimensional plates to impurity diffusion in solids or semiconductors. Consider the diffusion problem as shown in Figure 12.71. The diffusion source is assumed to provide a constant boundary condition along the center half of the upper boundary. The normal derivative is taken to be zero along all the other boundary walls. This would approximate impurity diffusion in a semiconductor region that is enclosed by an oxide boundary on all sides except for a window in the oxide where the constant boundary condition occurs. The dimensions are assumed to be given in normalized dimensions. This can be thought of as replacing the diffusion coefficient in Eq. (12.97) by $D \rightarrow Dy_m^2$ where y_m is the maximum y dimension in the problem. The effective diffusion coefficient then has dimensions of time^{-1}. By defining a normalized time one can also eliminate D entirely and work in a dimensionless time variable. However this will not be done as one might want to consider a diffusion coefficient that depends on the concentration of diffusing species.

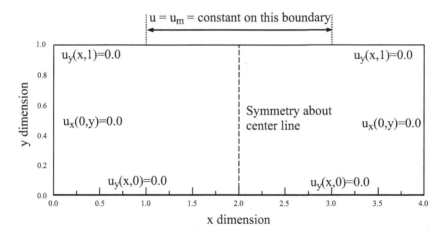

Figure 12.71. Spatial region for 2D transient diffusion problem.

It is readily seen that the problem as stated is symmetrical about the dotted center line in Figure 12.71. The diffusion problem can then be reduced to only the left (or right) half plane and this will increase the spatial resolution for a given number of spatial grid points. Since the problem is symmetrical across the dotted line, the normal derivative will also be zero along the dotted line. Many physical problems exhibit some type of symmetry and whenever possible this should be used to simplify the problem or as in this case to increase the solution accuracy for a given number of spatial grid points. In this case the solution will be obtained for only the left half of the structure running form $x = 0$ to $x = 2.0$. In terms of spatial dimensions, the x dimension is then twice the y dimension in this defined 2D space.

```
 1 : -- File list12_30.lua -- Example of time dependent BV problem
 2 :
 3 : require"pde2bv"
 4 : Um = 1.0 -- Define equations to be solved
 5 : D = 20 -- Constant diffusion coefficient
 6 : xmax,ymax = 2.0,1.0

 7 : feqs = { -- Table of functions
 8 :    function(x,y,t,uxx,ux,u,uy,uyy,ut,utt,i,j) -- General point
 9 :             return -ut + D*(uxx + uyy)
10 :    end,
11 :    function(x,t,u,uy,ux,i) return uy end, -- Bottom boundary
12 :    function(x,t,u,uy,ux,i) -- Top boundary
13 :             if x>=xmax/2 then return u-Um else return uy end end,
14 :    function(y,t,u,ux,uy,j) return ux end, -- Left boundary
15 :    function(y,t,u,ux,uy,j) return ux end  -- Right boundary
16 : }  -- End general point and boundary values
17 :
18 : Nx,Ny = 100,50 -- Change as desired
19 : nx,ny = Nx+1,Ny+1
20 : x,y = setxy({0,xmax},{0,ymax},Nx,Ny)
21 : u = {}; u[0] = {nx, ny, 1} -- Include size information
22 : for j = 1,ny do   -- Set zero initial values
23 :    u[j] = {}; for i = 1,nx do u[j][i] = 0 end
24 : end
25 : SPM,COE,SOR,ADI = 1, 2, 3, 4 -- possible solution methods
26 : t = os.time()
27 : getfenv(pde2bv1tqs).nprint=1
28 : getfenv(pde2bv1tqs).seeplot = 2 -- Uncomment for popup plots
29 : tvals = {0,{1.e-4,1.e-1},2,10,{{0,.5},{0,1},{2,0},{2,.75},
           {2,1},{1,.5},{1,.75},}}

30 : u,uxyt,errm = pde2bv1tqs(feqs,tvals,x,y,u,COE)
31 : --tvals = {0,{.001,.01},5} -- Use these for linear time
32 : --u,errm = pde2bv1t(feqs,tvals,x,y,u,COE)
33 :
34 : print('Time taken =',os.time()-t); io.flush()
35 : nsol = #u -- Number of time solutions saved
36 : for i=1,nsol do -- Save 2D data files
37 :    sfl = 'list12_30.'..i..'.dat'
38 :    write_data(sfl,reversexy(u[i])) -- reversexy() before save
39 : end
40 : splot('list12_30a.emf',reversexy(u[5]))
41 : cplot('list12_30b.emf',reversexy(u[5]))
42 : write_data('list12_30a.dat',uxyt) -- Save t dependent data
```

Listing 12.30. Code segment for solving for the transient diffusion into a 2D spatial region as defined in Figure 12.

Listing 12.30 shows computer code for simulating this 2D transient diffusion problem. The normalized diffusion constant is taken as 20 on line 5 and the maximum solution value is taken as a normalized value of 1.0 on line 4. Since the PDE is linear, the solution will scale linearly with any desired surface concentration. The defining PDE and boundary conditions are given on lines 7 through 16 with the constant concentration boundary specified along the top boundary evaluated on line 13. Along all other boundaries, the normal derivative is specified to be zero. The x and y grid point arrays are defined on lines 18 through 21 and the initial zero solution defined on lines 22 through 25. The logarithmic time step solver pde2bv1tqs() function is used to obtain the solution on line 30. The tvals array is defined appropriately on line 29 specifying an initial linear time step increment from 0 to 1.e-4 and then a logarithmic time step distribution from 1.e-4 to 1.e-1 or over 3 orders of magnitude in time. This allows one to observe the development of the solution from initial small times to close to the steady state with a minimal number of time points. For each of the time intervals, the 2,10 number specifications as the third and fourth table entries in tvals[] request that 2 solutions be saved for each time interval and that solutions be calculated for an additional 10 time intervals between the saved points. Thus calculations will be made for 20 time points per decade in time over the 1.e-4 to 0.1 time interval. It has been found that for a variety of problems using the logarithmic time spacing that 20 time intervals is sufficient to result in good solution accuracy. The final table in the tvals listing specifies a series of 6 spatial points ranging from {0, .5} to {1, .75} at which time dependent data will be collected and returned by the function.

The pde2bv1tqs() function returns two primary tables identified as u and uxyt on line 30. The first returned array (u here) is a table of tables giving the solutions u(y,x) at the specified time increments. For the tvals specification there will be 8 returned solution sets for times 0, 1.e-4, 3.16e-4, 1.e-3, 3.16e-3, 1.e-2, 3.16e-2, 1.e-1. The solution sets are saved into separate files on lines 36 through 39 of the listing. Note that the u(j,i) solution sets (j indicates y values and i indicates x values) are passed through the reversexy() function before saving to convert then to u(i,j) labeling. This is to accommodate most 2D graphing programs which expect this storage arrangement to produce a plot with the orientation of the structure in Figure 12.71. (This is true of the gnuplot routines used by the pop-up splot() and cplot() calls as well as the PSIPLOT commercial program.) This reversal is also performed on lines 40 and 41 before calling splot() and cplot(). With a file name specified in the calling arguments, the plots are saved to the named files for later use. Finally, The commented out lines of 31 and 32 can be used in place of lines 29 and 30 to experiment with the pde2bv1t() solver which uses a linear array of time points in obtaining the solutions. The reader is encouraged to experiment with different time formats for the listing.

Figures 12.72 and 12.73 show example saved plots from lines 40 and 41 of the code. These show surface and contour plots for a time of 3.16e-3. The surface plot is from the saved file on line 40. The contour plot is not exactly the saved file but is very similar to the saved file. The contour plot where each curve represents an increase in 0.05 of the peak concentration is perhaps the most easy to visualize of the two as the distance a given concentration has diffused into the structure can be readily seen. One can readily see the difference in the diffusion distance in the perpendicular direction from the diffusing surface as opposed to the shorter distance diffused along the top surface of the structure. The ratio of the lateral diffusion distance to vertical distance is approximately 20/27 = 0.74 (obtained from the 0.05 contour line).

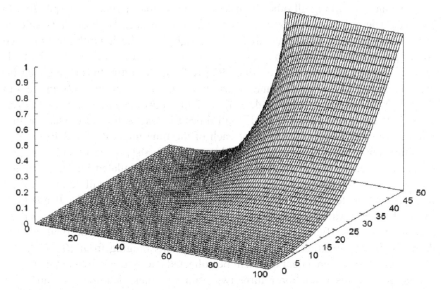

Figure 12.72. Surface profile plot for time of 3.16e-3 from the code in Listing 12.30. The x and y labels are in terms of grid point numbers (101 by 51 in value) corresponding to an x value of 2.0 and a y value of 1.0.

Another view of the solution can be obtained from the time dependent solutions at the set of selected spatial points defined by the tvals[] table and as saved to a file on line 42 of Listing 12.30. Selected data is shown in Figure 12.74 for 4 spatial points. The (2.0, .75) point is directly under the source surface and shows the most rapid diffusion solution. The (1, .75) point has a similar time solution with a somewhat slower diffusing front. The other points more remote from the source indicate longer times for the diffusing species to reach these spatial points. The increase in diffusion time for the remote points in this graph can be more readily seen in this graph than in the previous graphs. The plotted curves also show the logarithmic spacing of the solution times as the curves show similar shapes on the log time axis but are shifted in time for the various points. Such a graph can also be used to obtain a preliminary indication of possible numerical problems with such a solution. If any abrupt changes in the solution occur or any strange changes occur in the slope of the solution curves, one would be very suspicious of the solutions and such a situation would need to be followed up with a more careful study of the problem, possibly varying the number of spatial and/or time

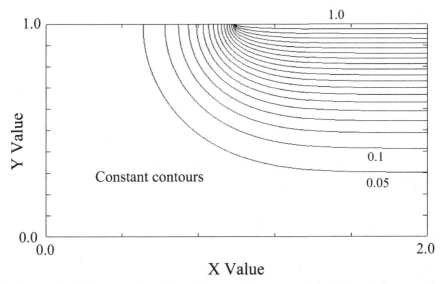

Figure 12.73. Contour plot of diffusion profile at time 1.36e-3 from the code in Listing 12.30.

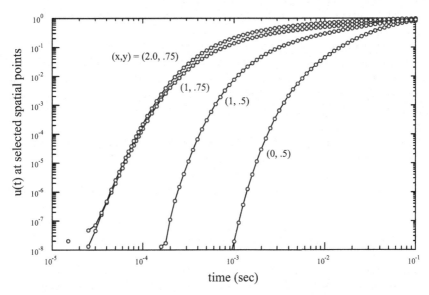

Figure 12.74. Time dependent values of the diffusion solution at selected spatial points from the data generated by Listing 12.30.

points. In the present case no unusual features are seen in the solutions as a function of time. A more detailed discussion of solution accuracy and means of exploring the accuracy of such numerical solutions are addressed in the next section.

Computer code has now been developed and discussed for several approaches to solving for the time independent and time dependent solutions of partial differential equations in two spatial dimensions and in possibly one additional time dimension. The next section explores the use of these approaches for several type of engineering problems.

12.9 Some Selected Examples of Two Dimensional BVPs

The preceding sections have developed a series of solution techniques for PDEs in two dimensions with the finite difference approach. This section uses the code segments developed to explore several examples. In addition the issue of accuracy of the solutions will be further explored and some techniques for estimating the accuracy developed.

12.9.1 Capacitor with offset bottom plate

A capacitor structure with an offset bottom plate is shown in Figure 12.75. Structures such as this occur in strip lines used to interconnect electrical components on circuit boards or integrated circuits. A typical structure would be symmetrical about a center line of the offset lower metal plate (shown as the cross hatched region in the figure). So the shown structure would more typically represent half of such a capacitor structure but this is all that is needed for analysis because of the symmetry of such a structure. Within the dielectric region the electric potential satisfies Poisson's equation while the top and bottom surfaces have constant potential values and the sides have zero normal derivative values.

Solutions for the potential distribution within the dielectric for such a structure are easily obtained from the subroutines developed in the previous section. A code segment for the numerical solution is shown in Listing 12.31. The code is straightforward with the function and boundary values defined on lines 7 through 19. The offset lower metal contact is implemented on line 10 of the code by simply setting the potential to a constant value of zero within this region. For this example 160 and 80 y and x spatial increments are used. The x dimension is taken as 5 times the y dimension in this example. The call to the 2D solver, pde2bv(), on line 34 uses the SOR iterative solution technique and the selected output shows that 280 iterations of the SOR algorithm are needed to achieve a correction error of -4.316e-7 which is probably more than sufficient for most engineering problems. Finally on line 35 pop-up surface and contour plots of the solution are called so a quick view of the solution can be obtained. The reader is encouraged to try obtaining the solution with some of the other solution methods. It will be found that the ADI technique will not properly execute with this particular problem because the solution node points within the metal contact have only diagonal matrix elements.

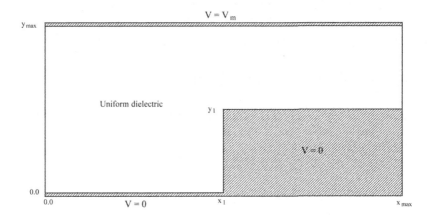

Figure 12.75. Capacitor structure with offset bottom plate.

One new feature of the code is the use of the surface electric field on lines 36 through 40 to calculate and print information regarding the capacitance of this physical structure. This requires some additional explanation. For a simple parallel plate capacitor the capacitance can be written as:

$$C = \frac{\varepsilon WL}{t} \tag{12.101}$$

where WL is the capacitor area and t is the dielectric thickness. For a more general structure such as considered here, the capacitance can be expressed as:

$$C = \frac{Q}{V} = \frac{W}{V} \int_0^{x_{max}} \varepsilon E_y dx = \varepsilon W \left[\int_0^{x_{max}} \frac{E_y}{V} dx \right] = \varepsilon W (L/t)_{eff}$$
$$\frac{C}{\varepsilon W} = \int_0^{x_{max}} \frac{E_y}{V} dx = (L/t)_{eff} \tag{12.102}$$

The capacitance per unit width into the page in Figure 12.75 is purely a function of the geometry of the structure and can be calculated as indicated by Eq. (12.102) or as implemented on lines 36 through 40 of Listing 12.31. The calculation on line 36 uses a grad() function that has been programmed to accept the x and y spatial grid along with a two-dimensional solution and return the results of applying the gradient operator to the solution. In this case:

$$u_x = \frac{\partial u}{\partial x}; \quad u_y = \frac{\partial u}{\partial y} \tag{12.103}$$

Code for the grad() function is not shown but is straightforward and the reader can view it if desired in the pde2bv.lua file. The printed result shows that the resulting normalized capacitance value is 7.7217. If the structure is approximated as two capacitors in parallel, then the resulting value would be (2.5/.5) + (2.5/1) = 7.50. The capacitance is increased somewhat above this by the fringing field that results from the vertical face of the stepped bottom contact.

```
 1 : -- File list12.31.lua --
 2 : -- Example of BV problem for capacitor with offset bottom plate
 3 :
 4 : require"pde2bv"
 5 : local xmax,ymax
 6 : Vm = 1
 7 : feqs = { -- Function and boundary conditions
 8 :     function(x,y,uxx,ux,u,uy,uyy,i,j)
 9 :         if x>x1 then
10 :             if y<=y1 then return u else return uxx+uyy end
11 :         else return uxx + uyy end
12 :     end,
13 :     function(x,u,uy,ux,i) return u end,
14 :     function(x,u,uy,ux,i) return u - Vm end,
15 :     function(y,u,ux,uy,j) return ux end,
16 :     function(y,u,ux,uy,j)
17 :         if y<=y1 then return u else return ux end
18 :     end
19 : }
20 :
21 : x,y = {},{}; Nx,Ny = 160,80
22 : nx,ny = Nx+1,Ny+1
23 : xmax,ymax = 5, 1
24 : x1,y1 = xmax/2, ymax/2
25 : for i=1,nx do x[i] = xmax*(i-1)/Nx end
26 : for j=1,ny do y[j] = ymax*(j-1)/Ny end
27 : u = Spmat.new(nx,-ny)
28 : for j = 1,ny do
29 :     u[j] = {}; for i = 1,nx do u[j][i] = 0 end
30 : end
31 :
32 : SPM,COE,SOR,ADI = 1, 2, 3, 4 -- 4 solution types
33 : getfenv(pde2bvsor).nprint=1
34 : u,errm = pde2bv(feqs,x,y,u,SOR) -- Replace SOR as desired
35 : ur = reversexy(u); splot(ur); cplot(ur)
36 : ux,uy = grad(x,y,u)
37 : sum = 0.5*(uy[ny][1] + uy[ny][nx])
38 : for i=2,nx-1 do sum = sum + uy[ny][i] end
39 : sum = sum*(x[2]-x[1])
40 : print('C/eps*W =',sum)
41 : write_data("list12.31.dat",ur); cplot("list12.31.emf",ur)
Selected output:
Exiting SOR at iteration  280 with correction  -4.3165156371483e-007
SOR spectral radius = 0.99924795250423
Exiting SOR at iteration   1  with correction  -4.1018179910054e-007
C/eps*W =       7.7216699737507
```

Listing 12.31. Code for capacitor BV problem with offset bottom plate.

The contour plot for this example is shown in Figure 12.76. It is readily seen that towards the ends of the structure the contour lines are essentially uniformly spaced as one would expect from a simple parallel plate capacitor. Near the top of the step in the bottom contact, the contour lines are closely spaced and this will result in high electric fields near the corner – a potential point of breakdown for the dielectric in such a structure. The algorithm used to generate the contour plots distorts slightly the lines near the corner of the structure.

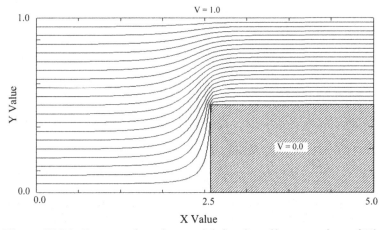

Figure 12.76. Contour plot of potential for the offset capacitor of Figure 12.75. Each contour is an increment of 0.05 Volts.

The reader is encouraged to explore this example by changing the solution method, by changing the geometrical ratios of the various regions and by changing the number of spatial grid points. It should also be noted that the discussion here is in terms of the electric potential of such a structure. However, the exact same formulation would be obtained if this were a problem in the mechanical engineering regime where one was interested in the temperature distribution between two plates with the cross hatched regions in Figure 12.75 held at constant temperatures. The temperature difference between the two plates would have exactly the same contour plots as in Figure 12.76. A thermal resistance for such a problem can also be ddefined somewhat similar to the capacitance in the present problem. This is left as an exercise for the reader.

12.9.2. Two or more different dielectrics in a structure involving Poisson's equation

Electrical problems involving dielectrics (or heat transfer problems with different materials) occur very frequently. For example in Figure 12.75 the lower part of the structure from 0 to y_1 might have a different dielectric constant from that in the upper part of the figure from y_1 to y_{max}. The question arises as how to handle such a case in the numerical approach. To consider this a simple parallel plate capacitor structure will be considered as shown in Figure 12.77 which has two different dielectrics in the upper and lower parts of the structure. Considering Poisson's equation in the y dimension one has:

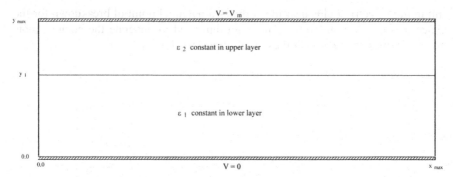

Figure 12.77. Capacitor with dielectrics having two different dielectric constants.

$$\frac{\partial D}{\partial y} = 0 = \frac{\partial}{\partial y}\left(\varepsilon\frac{\partial u}{\partial y}\right) = \frac{\partial u}{\partial y}\frac{\partial \varepsilon}{\partial y} + \varepsilon\frac{\partial^2 u}{\partial y^2} \qquad (12.104)$$

There is thus an extra contribution to the equation at the point in space where the two dielectrics with different dielectric constants meet. This is in fact a very sharply peaked derivative function and in the limit of an abrupt change in dielectric constant, becomes a delta function of spatial position. In many cases such a problem is approached by obtaining separate solutions in each of the regions and then joining the solutions across the boundary by the requirement that the D vector is constant across the boundary or in terms of electric fields that

$$\varepsilon_1 E_1 = \varepsilon_2 E_2 \ \text{ or } \ \varepsilon_1\frac{\partial u}{\partial y}\Big|_1 = \varepsilon_2\frac{\partial u}{\partial y}\Big|_2 \qquad (12.105)$$

Treating the regions as two separate problems and joining the solutions across the boundary is not so easily done with a numerical calculation so another approache will be considered here.

For this, consider the extra term near the interface in Eq. (12.104), which can be approximated as:

$$\frac{\partial u}{\partial y}\frac{\partial \varepsilon}{\partial y} \simeq \frac{\partial u}{\partial y}\frac{\varepsilon_2 - \varepsilon_1}{\Delta y} \simeq \varepsilon\frac{\partial u}{\partial y}\left[\frac{2(\varepsilon_2 - \varepsilon_1)}{(\varepsilon_2 + \varepsilon_1)\Delta y}\right] \qquad (12.106)$$

In this Δy is the spatial grid spacing at the interface and the dielectric constant at the interface has been taken as the average of the two values. Inserting this into Eq. (12.104) then gives a modified equation of:

$$\varepsilon\left(\frac{\partial^2 u}{\partial y^2} + \frac{\partial u}{\partial y}\left[\frac{2(\varepsilon_2 - \varepsilon_1)}{(\varepsilon_2 + \varepsilon_1)\Delta y}\right]\delta(y - y_1)\right) = 0 \qquad (12.107)$$

where the second term is to be applied only at the interface between the two dielectrics as indicated by the delta function in the equation.

This now provides a formulation for accounting for the changing dielectric constant. However, the question arises as to how well does it work in practice? Is it only an approximation or does it in fact result in the correct ratio of electric fields across the interface? To explore these questions, the code in Listing 12.32

```
 1 : -- File list12_32.lua --
 2 : -- Example of capacitor with two dielectrics
 3 :
 4 : require"pde2bv"
 5 :
 6 : feqs = {
 7 :    function(x,y,uxx,ux,u,uy,uyy,i,j)
 8 :       if j==nt then ext = fyext*uy else ext = 0.0 end
 9 :       return uxx + uyy + ext
10 :    end,
11 :    function(x,u,uy,ux,i) return u end,
12 :    function(x,u,uy,ux,i) return u-Vm end,
13 :    function(y,u,ux,uy,j) return ux end,
14 :    function(y,u,ux,uy,j) return ux end
15 : }
16 :
17 : ep1, ep2 = 3, 12
18 : Vm, xmax, ymax = 1.0, 1.0, 1.0
19 : Nx,Ny = 80,80
20 : nx,ny = Nx+1,Ny+1; nt = math.ceil(ny/2)
21 : x,y = setxy({0,xmax},{0,ymax},Nx,Ny)
22 : u = Spmat.new(nx,-ny)
23 : for j = 1,ny do   -- Set zero initial values
24 :    u[j] = {}; for i = 1,nx do u[j][i] = 0 end
25 : end
26 : fyext = 2*(ep2-ep1)/((ep2+ep1)*(y[nt]-y[nt-1]))
27 :
28 : SPM,COE,SOR,ADI = 1, 2, 3, 4 -- 4 solution types
29 : getfenv(pde2bvsor).nprint=1;getfenv(pde2bvcoe).nprint=1
30 : u,errm = pde2bv(feqs,x,y,u,SOR) -- Replace SOR as desired
31 : print('Dielectric change occurs at',nt)
32 : print(' At end maximum correction =',errm)
33 : ut = reversexy(u)
34 : splot(ut); cplot(ut); splot('list12_32a.emf',ut)
35 : write_data('list12_32a.dat',ut)
```

Listing 12.32. Code for capacitor structure with two different dielectrics.

shows this expression implemented for the parallel plate capacitor of Figure 12.77. For this example the interface between the two dielectrics is taken midway between the two capacitor plates with the integer value corresponding to that point being evaluated on line 20 of the code. The test for this integer value (j==nt) is on line 8 within the function describing the differential equation and the term ext is added to the equation only for this y value. The test is made on the integer value corresponding to the midpoint rather than the y value which would be 0.50. Because of roundoff errors in numerical calculations, the test of an exact equality of a floating point number may or may not succeed while the integer test will always put the extra term at exactly the correct boundary location. Problems such as this are the reason the code is written so that the call to the function defining the differential equation includes the integers j and i in addition to the y and x values. While this may seem redundant in most cases, it is vary valuable information in problems such as this. Line 26 calculates the value of the approximate derivative term to be applid exactly at the interface plane.

In this example the two dielectric constants are taken as 3 and 12 on line 17 with a ratio of 4 so that the exact answer for the internal electric fields can be easily evaluated and the fraction of the voltage dropped across each region of the dielectric is easily evaluated. For this ratio of dielectric constants one finds that $u(x,.5) = 0.8V_m$ or 80% of the voltage should be dropped across the region with the smallest dielectric constant. Figure 12.78 shows a surface plot of the potential solution resulting from executing the code in Listing 12.32 and which is saved in a file on line 34. While the results are in general what is to be expected, it is difficult to verify the exact accuracy of the results from this figure.

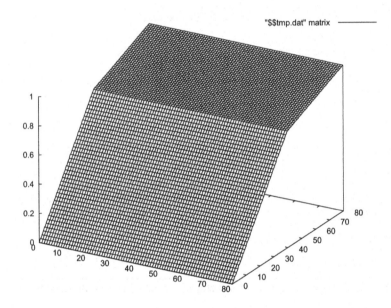

Figure 12.78. Surface plot of potential for parallel plate capacitor with two different dielectrics

A more detailed look at the solution through the dielectrics is shown in Figure 12.79 which shows the potential as a function of y along the centerline in the x dimension. The solid line shows the solution for an 80 by 80 interval spatial grid while the open circles show the solution points for a 20 by 20 interval spatial grid. As long as one uses an even number of spatial intervals, the j location of the interface in Listing 12.32 will occur at exactly the center of the dielectric. The results show that the abrupt transition in electric field (or slope of the lines in Figure 12.79) is accurately reproduced independently of the number of spatial intervals. In fact one can verify that using the extra term as identified in Eq. (12.106) at exactly the dielectric interface results in a set of difference equations across the interface that requires the electric fields to accurately reflect the desired physical ratio. This technique then provides a means for accurately modeling 2D structures

with different dielectric constants. For problems involving the temperature distribution with materials of different thermal properties, a similar approach can be
used to model such thermal problems. This is left as an exercise for the reader.

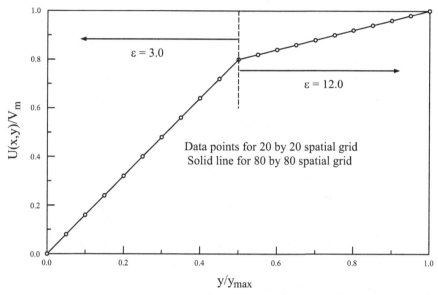

Figure 12.79. Detailed look at the solution through the two dielectrics for varying
spatial intervals.

It is also noted that if one takes the limits of very large and very small ratios of
the dielectric constants one gets a correction term of:

$$\frac{\partial u}{\partial y}\frac{\partial \varepsilon}{\partial y} \simeq \varepsilon \frac{\partial u}{\partial y}\left[\frac{2}{\Delta y}\right] \text{ for } \varepsilon_2 \gg \varepsilon_1 \text{ (also } E_2 \rightarrow 0)$$

$$\frac{\partial u}{\partial y}\frac{\partial \varepsilon}{\partial y} \simeq \varepsilon \frac{\partial u}{\partial y}\left[\frac{-2}{\Delta y}\right] \text{ for } \varepsilon_2 \ll \varepsilon_1 \text{ (also } E_1 \rightarrow 0)$$

(12.108)

For the first case the electric field in the second region will be forced to zero and
in the second case the electric field in the first region will be forced to zero. These
approximations can be used to model certain physical problems as will be shown
in the next section where one section of a physical structure is known to have zero
electric field at the interface.

12.9.3. Resistance of a Square Corner Resistor

In integrated circuits the layout of resistors frequently results in a resistor that in-
corporates a right angle square corner as illustrated in Figure 12.80. As seen in
this figure, this is a top view of a thin film resistor which has some uniform depth

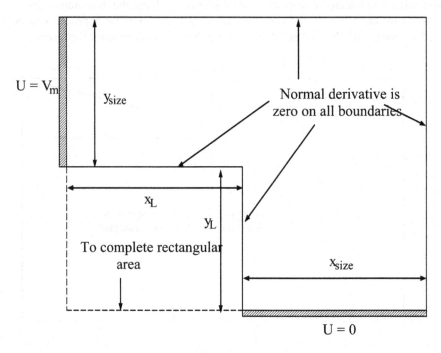

Figure 12.80. Resistor layout for square corner.

into the page in addition to the geometry shown in the figure. The cross hatched areas represent ohmic contacts to the region and current flows from one contact to the other around the corner when voltage is applied. The question to be addressed is the resistance of such a square corner resistor. This is usually specified in the number of "squares" represented by the structure. The resistance of a resistor can be expressed in simple form as

$$R = \rho \frac{L}{A} = \frac{\rho}{t} \frac{L}{W} = \rho_{\square} \frac{L}{W} \tag{12.109}$$

where ρ is the resistivity of the material and L and W are the length and width of the resistor and t is the thickness of the material (distance into page in Figure 12.80). Finally $\rho_{\square}$ is the resistivity in Ohms per square of material and L/W is the number of "squares" making up the resistor. A rectangular area of equal length and width constitutes "1 square".

For a more definite geometry consider the case where in Figure 12.80, $x_L = y_{size}$ and $y_L = x_{size}$, i.e. the structure is of equal widths in both directions and the lengths of the sides are equal. It is then expected that the equivalent size is 2 squares due to the side legs plus some contribution (of less than 1 square) from the square corner area. The potential within the resistor boundary will satisfy Laplace's equation (or Poisson's equation with zero charge density). However, this resistor area is not a rectangular region so it doesn't directly match the boundary requirements for the computer routines developed in this chapter. However, if the dotted region shown in the figure is added to the resistor then one does have a rectangular boundary region. But how can this be added without it effecting the potential distribution in the solid area and affecting the current flow and thus the equivalent resistance in some calculation? The answer lies in the use of Eq. (12.108) along the internal solid boundary of the region to force the normal derivative of the potential to be zero along the internal boundary as well as along the external boundaries. If the normal derivative (or electric field) is forced to zero along the internal boundary on the resistor side of the internal boundary then no current can flow across the solid line internal boundary and an accurate solution of the internal potential within the resistor can be obtained.

Code for solving for the potential in this structure and for evaluating the effective resistor squares is shown in Listing 12.33. Most of the code should be by now very familiar. The midpoint of the x and y ranges is evaluated on line 31 and used in the function defining the differential equation on lines 7 through 16 to insert a corrective term using fxext or fyext on lines 10 and 13 along the internal boundary forcing the normal derivative just inside the resistor structure to be zero. The boundary functions on lines 17 through 26 are used to force the correct voltages on the structure and to force the other normal derivatives to be zero. The pde2bv() function is called to solve the differential equation with boundary conditions on line 43. Finally after obtaining the solution, the effective number of resistive squares is evaluated on lines 47 through 51. The evaluation of the resistance is similar to the previous evaluation of capacitance except for an inverse calculation on line 51. A surface plot of the resulting solution is shown in Figure 12.81. For this figure only the solved potential inside the resistor boundaries are shown while the actual solution also gives potential values within the rectangular region from x,y = 0,0 to x,y = .5,.5. One can readily see the linear variations with spatial position near the contact boundaries indicating that the effects of the square corner have essentially been minimized at one unit distance from the corner. The numerically evaluated effective number of squares is 2.557 as printed on the output. Frequently this is considered as made up of one square from each of the legs on the structure and the remainder contributed by the square corner. This would place the contribution of the square corner at 0.557 squares. This can be compared with the value of 0.56 frequently used in the literature for such calculations.

A contour plot for the results is shown in Figure 12.80. Again only the values inside of the resistor boundary have meaning for this problem and only these values are shown. It is seen that the contour plots do in fact intersect the internal resistor boundary in a perpendicular manner, indicating that the condition of a zero normal derivative has in deed been achieved within the resistor boundary by the obtained solution. The pop-up plots and the saved plots generated by lines 46 and 52 of the code will contain plots for the complete rectangular region of the extended structure. The reader is encouraged to execute the code and observe the generated plots. Also the relative size of the resistor legs can be easily changed to

```
 1 :  -- File list12_33.lua --
 2 :  -- Example of square corner resistor
 3 :
 4 :  require"pde2bv"
 5 :
 6 :  feqs = {
 7 :     function(x,y,uxx,ux,u,uy,uyy,i,j)
 8 :        ext = 0
 9 :        if i==nxmid then
10 :             if j<=nymid then ext = fxext*ux end
11 :        end
12 :        if j==nymid then
13 :             if i<=nxmid then ext = ext + fyext*uy end
14 :        end
15 :        return uxx + uyy + ext
16 :     end,
17 :     function(x,u,uy,ux,i)
18 :        if i>nxmid-1 then return u
19 :        else return uy end
20 :     end,
21 :     function(x,u,uy,ux,i) return uy end,
22 :     function(y,u,ux,uy,j)
23 :        if j>nymid-1 then return u-Vm
24 :        else return ux end
25 :     end,
26 :     function(y,u,ux,uy,j) return ux end
27 :  }
28 :  Vm = 1.0
29 :  Nx,Ny = 80,80
30 :  nx,ny = Nx+1,Ny+1
31 :  nxmid,nymid = math.ceil(nx/2),math.ceil(ny/2)
32 :
33 :  x,y = setxy({0,1},{0,1},Nx,Ny)
34 :  u = Spmat.new(nx,-ny)
35 :  for j = 1,ny do   -- Set zero initial values
36 :     u[j] = {}; for i = 1,nx do u[j][i] = 0 end
37 :  end
38 :  fxext = 4/(x[nxmid+1]-x[nxmid-1])
39 :  fyext = 4/(y[nymid+1]-y[nymid-1])
40 :
41 :  SPM,COE,SOR,ADI = 1, 2, 3, 4 -- 4 solution types
42 :  getfenv(pde2bvsor).nprint=1;getfenv(pde2bvcoe).nprint=1
43 :  u,errm = pde2bv(feqs,x,y,u,COE) -- Replace COE as desired
44 :
45 :  print(' At end maximum correction =',errm)
46 :  ut = reversexy(u); splot(ut); cplot(ut)
47 :  ux,uy = grad(x,y,u)
48 :  sum = 0.5*(uy[1][nxmid] + uy[1][nx])
49 :  for i=nxmid+1,nx-1 do sum = sum + uy[1][i] end
50 :  sum = sum*(x[2]-x[1])
51 :  print('Eff squares =',1/sum)
52 :  splot('list12_33.emf',ut); cplot('list12_33a.emf',ut)
53 :  write_data('list12_33.dat',ut)
Selected output:
Eff squares =   2.5568718319112
```

Listing 12.33. Code for calculating resistance of square corner resistor.

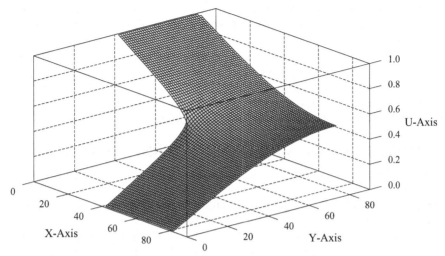

Figure 12.81 Surface plot of potential in square corner resistor.

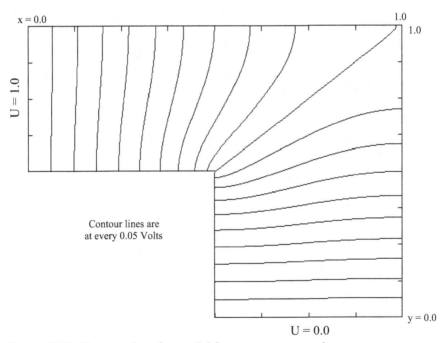

Figure 12.82. Contour plot of potential for square corner resistor.

observe effects on the number of effective squares. Finally the reader can solve the problem with the different approximation methods as well as the SPM exact matrix solution method. For this example all the 4 solution methods should converge to the same solution but with varying execution times. For spatial grids of 20 by 20 and 40 by 40 one should obtain effective squares of 2.545 and 2.554 respectively.

This example illustrates how with a little ingenuity the rectangular grid PDE code can be used to solve a variety of physical problems.

12.9.4 Semiconductor p-n junction – potential and carrier densities

The final example in this section is that of the potential about a p-n junction which is described by a nonlinear second order differential equation. A cross sectional view of a 2D p-n junction is shown in Figure 12.83. The p region would typically be formed by impurity diffusion through a surface oxide mask with the diffusion process similar to that discussed and modeled in Section 12.7. From the results shown in Figure 12.73 it can be seen that such a diffusion process gives a junction boundary with rounded corners and not the square corner as shown in Figure 12.83. However, this will be ignored here and a square corner junction considered just as shown in the figure. Such a junction can be formed in a semiconductor by the technique of etching a hole in the semiconductor and back filling with a doped epitaxial layer. In fact such processes are becoming used to form source-drain contacts for MOS devices, so the geometry with square junction corners is of some practical interest. It will be left to the reader to modify the problem if desired and consider junctions with rounded corners. The geometry is seen to be symmetrical about the center dotted line so again only half of the physical structure needs to be modeled and thereby the solution accuracy can be increased for a given number of spatial grid points.

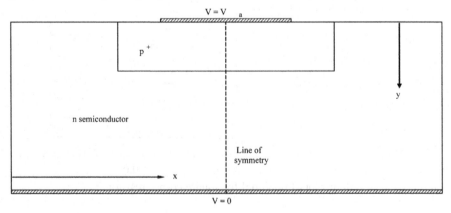

Figure 12.83. Cross section of semiconductor p-n junction.

The PDE corresponding to the equilibrium potential around the p-n junction is:

$$\frac{\partial^2 V}{\partial x^2} + \frac{\partial^2 V}{\partial y^2} = \frac{q}{\varepsilon}(p - n + N_{Net})$$

$$N_{Net} = N_D \text{ for } n \text{ region and } = -N_A \text{ for p region} \qquad (12.110)$$

$$p = N_A \exp(-V/V_T), \quad n = (n_i^2/N_D)\exp(V/V_T)$$

For this formulation, the potential is taken as zero deep within the p-type region where $p = N_A$. Since the thermal voltage is only 0.025 volts at room temperature, the exponential functions are rapidly varying functions of potential and this differential equation has a reasonably severe nonlinearity. This type of exponential variation is about as nonlinear a function as occurs in real physical engineering problems.

Code for implementing and solving these equations is shown in Listing 12.34. The doping densities are taken as 1.e19/cm^3 and 1.e17/cm^3 on the p and n sides of the junction. From previous knowledge, it is known that these densities give depletion regions with depths on the order of 1.e-4 cm or less so the dimensions of the structure are taken as 1.e-4 cm on each side, with the junction depth taken as 1/4 of the total structure thickness. Fundamental device and structure parameters are defined in the code on lines 4 through 9. The differential equation and boundary conditions are defined on lines 11 through 24. The top contact is taken to be at potential va = 0 and the bottom contact is taken to be at the junction built-in voltage as calculated on line 9. For an initial guess at the potential, the code on lines 30 through 37 set the initial potential everywhere in the p-region to zero and everywhere in the n-region to the built-in potential.

The call to pde2bv() on line 41 returns the array of solution values and with this code uses the COE iterative solution method. During the solution since the PDE is nonlinear, several Newton iterative loops are required with each step linearizing the PDE about the solution so far obtained. The selected output shows the progression of correction values at each Newton iterative step. The rapid convergence of the Newton iterations can be seen for iteration 4, 5 and 6 where the correction should become quadratic in the iterative step. The correction goes from a value in the third decimal place to a value in the 5th decimal place at steps 5 and 6. It will be recalled that within each Newton iterative step there are many COE iterative steps so this example has iterations within iterations to achieve the final solution. Fortunately the Newton iterative steps rapidly converge. The iterative steps of the COE solution can be observed if desired by un-commenting the statement getfenv(pde2bvcoe).nprint=1 on line 38.

Finally the code on lines 44 through 53 calculates the carrier densities, or rather the log10 of the carrier densities and stores these values in two arrays. The final potential and carrier densities are saved to data files on lines 55 and 56 for observing. Surface plots of the calculated potential and carrier densities for x and y spatial grids of 80 intervals are shown in Figures 12.84, 12.85 and 12.86. The results are as one would expect for a p-n junction. The reader is encouraged to exercise the code in Listing 12.34 by changing the dimensions, the number of spatial steps

```lua
 1 : -- File list12_34.lua --
 2 : -- Example of BV problem for p-n junction
 3 : require"pde2bv"
 4 : -- Material and device parameters
 5 : q = 1.6e-19; eps = 11.9*8.854e-14; L = 0.4e-4
 6 : vt = .026; ni = 1.45e10
 7 : Na = 1e19; Nd = 1.e17-- Doping densities
 8 : x1,y1,x2 = L/2, L/4, L*3/4; va = 0.0
 9 : qdep = q/eps; vj = vt*math.log(Na*Nd/ni^2); no = ni^2/Na
10 :
11 : feqs = {
12 :    function(x,y,uxx,ux,u,uy,uyy,i,j)
13 :       if x>=x1 and y<=y1 then Nnet = -Na
14 :       else Nnet = Nd end
15 :       p, n  = Na*math.exp(-u/vt), no*math.exp(u/vt)
16 :       return uxx + uyy + qdep*(Nnet + p - n)
17 :    end,
18 :    function(x,u,uy,ux,i)
19 :          if x>x2 then return u-va else return uy end
20 :    end,
21 :    function(x,u,uy,ux,i) return u-vj end,
22 :    function(y,u,ux,uy,j) return ux end,
23 :    function(y,u,ux,uy,j) return ux end
24 : }
25 :
26 : Nx,Ny = 80,80; nx,ny = Nx+1,Ny+1
27 : nyj = math.floor(Ny/4); nxj = math.floor(Nx/2)
28 : x,y = setxy({0,L},{0,L},Nx,Ny)
29 : u = Spmat.new(nx,-ny)
30 : for j = 1,ny do   -- Set initial values at 0 or Vj
31 :    yv = y[j]
32 :    u[j] = {}; for i = 1,nx do
33 :       xv = x[i]
34 :       if xv>=x1 and yv<=y1 then u[j][i] = 0.0
35 :       else u[j][i] = vj end
36 :    end
37 : end
38 : getfenv(pde2bv).nprint = 1; --getfenv(pde2bvcoe).nprint=1
39 : SPM,COE,SOR,ADI = 1, 2, 3, 4 -- 4 solution types
40 : t1 = os.clock()
41 : u,errm = pde2bv(feqs,x,y,u,COE) -- Replace COE as desired
42 : print('time =',os.clock()-t1)
43 :
44 : pa,na = {},{}; ut = reversexy(u)
45 : for i=1,nx do -- Calculate carrier densities
46 :    paa,naa = {}, {}
47 :    for j=1,ny do
48 :       ua = ut[i][j]
49 :       paa[j] = math.log10(Na*math.exp(-ua/vt))
50 :       naa[j] = math.log10(no*math.exp(ua/vt))
51 :    end
52 :    pa[i],na[i] = paa, naa
53 : end
54 : splot(ut); cplot(ut); splot('list12.34.emf',ut)
55 : write_data('list12.34u.dat',ut)
56 : write_data('list12.34p.dat',pa); write_data('list12.34n.dat',na)
```
Selected output:
```
Completed Newton iteration 4  with correction  0.029356356256872
Completed Newton iteration 5  with correction  0.0032467758599837
```

```
Completed Newton iteration 6  with correction  6.4289950325926e-005
```

Listing 12.34 Code segment for solving for potential and carrier densities at p-n junction.

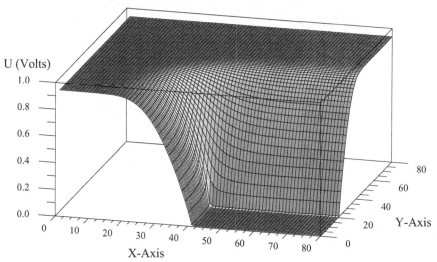

Figure 12.84. Surface plot of potential for p-n junction from example in Listing 12.34. Labeling along axes is in terms of spatial grid points (1 to 81).

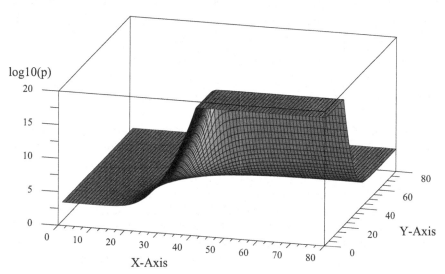

Figure 12.85. Surface plot of log10 of hole density about a p-n junction from Listing 12.34

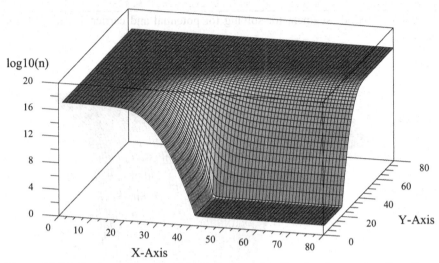

Figure 12.86. Surface plot of log10 of electron density about a p-n junction from Listing 12.34

and the carrier densities to observe changes in the solution. If the carrier densities are substantially reduced from the values shown, the dimensions of the region may have to be increased to completely observe the entire solution space. Other things of interest might be the use of nom-uniform grids to cover a wider range of distances but still maintain high resolution around the physical junction interface. The electric field within the device structure can also be calculated by use of the grad() function. The reader can also experiment with the various matrix solution techniques using the values defined on line 39 in place of the COE parametr on line 41.

12.9.5 Another Look at Accuracy of Solutions

The final topic for this chapter is another look at the accuracy of 2D PDE solutions obtained by the finite difference method. This was addressed somewhat in Section 12.6.1 for the case of an equation with a known solution. For the majority of PDEs, an exact solution is not known and that is the primary reason for resorting to numerical solutions. Especially with nonlinear equations such as that in the previous section, there are no known exact solutions with which to compare a numerical solution. One of the most useful techniques for exploring solution accuracy as has been discussed in previous chapters is to simply explore solutions for differing spatial grid sizes or differing time increments for time dependent equations. If the numerical solutions are "essentially" the same for varying grid sizes, then one can be somewhat confident that a valid solution has been obtained. Just what is meant by "essentially" the same is somewhat fuzzy. However, what this means can be defined in a somewhat more definite way. For example, it is known from the theory of finite differences, that the solution accuracy is expected to vary as the square of the number of spatial grid points. Thus the h-2h algorithm can be used to obtain an approximation of the solution accuracy for a given problem.

To explore the h-2h algorithm the PDE with boundary conditions must be solved for two spatial grid sizes varying the number of spatial step by a factor of 2, in both dimensions. This means that there will be a factor of 4 in the total number of spatial steps and the two solutions will vary in execution time by a factor of at least 4 (perhaps more depending on the approximate solution method used). To illustrate the h-2h algorithm with a PDE, the example of the p-n junction in the previous section was solved for spatial intervals of 160 by 160 and 80 by 80 and the resulting potential solutions saved in two files (list12_34u_80.dat and list12_34u_160.dat). Code for using these saved files and estimating the solution accuracy is shown in Listing12.35. First arrays are set up to hold spatial data and the two solution sets on lines 6 through 13. The previously calculated solution arrays are read into the program on lines 14 and 15. The heart of the calculation is line 20 which calls the previously developed odeerror() function (input with require"odeiv"). This function expects as input two arrays in the form of spatial values followed by solution values – one array for the h step sizes and one for the 2h step sizes. For this application, each row or solution values along a constant x (or y) axis can be used as the input arrays. So the loop from line 17 through 27 loops through the i values (or x values) of the solution arrays. Since there are twice as many x values in the 160 by 160 solution array as in the 80 by 80 solution array, the code on line 19 uses only every other i value in the 160 solution array. Lines 18 and 19 then set up the two arrays of x values and solution values for input to the odeerror() function which returns a table of two values – the x values followed by the error values. The odeerror()[2] statement selects only the error values to finally put in the saved error value array on line 26. The remaining code on lines 21 through 25 evaluates the maximum error in the potential solution and the spatial point where this occurs. The printed results show that for this example the maximum error in the solution is -0.00517 and occurs at spatial point 41, 39 which is basically at the corner of the p-n junction. This is certainly where the maximum error would be expected to occur.

A surface plot of the obtained error is shown in Figure 12.86 over the two dimensional array of solution values. The largest errors are concentrated around the p-n junction as expected with the largest value being the negative spike seen in the figure. Thus for this example the potential solution is expected to have an accuracy everywhere of less than about 0.5% of the peak potential value, which would probably be sufficient for most engineering work. To achieve a much higher accuracy would require a much larger number of spatial points or would require the use of a non-uniform spatial grid using smaller step sizes around the p-n junction material interface. Rounding the corner of the p-n junction would also probably lead to a more accurate solution as sharp corners are always difficult solution points for numerical techniques.

```
 1 : -- File list12_35.lua --
 2 : -- Example of h-2h algorithm for PDEs
 3 :
 4 : require"odeiv"
 5 :
 6 : u80, u160 = {},{} -- Define arrays for solutions
 7 : x80, x160 = {}, {} -- Linear spatial arrays
 8 : for i=1,81 do
 9 :    u80[i] = {}; x80[i] = (i-1)/81
10 : end
11 : for i=1,161 do
12 :    u160[i] = {}; x160[i] = (i-1)/160
13 : end
14 : read_data('list12_34u_80.dat',u80)
15 : read_data('list12_34u_160.dat',u160)
16 : err = {}; maxerr = 0.0
17 : for i=1,81 do
18 :    s80 = {x80, u80[i]}
19 :    s160 = {x160, u160[1+(i-1)*2]}
20 :    erri = odeerror(s80,s160)[2] -- column [2] for error values
21 :    for j=1,#erri do -- Maximum error
22 :       if math.abs(erri[j])>math.abs(maxerr) then
23 :          maxerr, im, jm = erri[j], i, j
24 :       end
25 :    end
26 :    err[i] = erri
27 : end
28 : splot(err); print('Maximum error =', maxerr, 'at i,j =', im,jm)
29 : write_data('list12_35.dat',err) -- Save error array
Selected output:
Maximum error = -0.0051744292251609   at i,j =        41      39
```

Listing 12.35. Code for estimating error of PDE solution by h-2h algorithm.

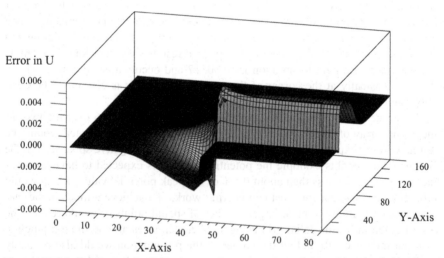

Figure 12.86. Surface plot of error around p-n junction obtained by the h-2h algorithm.

Although the potential solution is expected to be accurate to better than two decimal digits, the carrier densities are another matter. From Eq. (12.110) for n and p it is readily seen that

$$\delta p = -N_A \exp(-V/V_T)(\delta V/V_T) = -p(\delta V/V_T)$$

$$\delta n = (n_i^2/N_D)\exp(V/V_T)(\delta V/V_T) = n(\delta V/V_T)$$

(12.111)

In terms of percentage errors then these give:

$$\frac{\delta p}{p} = -\frac{\delta V}{V_T} = -(38.6/Volt)\delta V$$

$$\frac{\delta n}{n} = \frac{\delta V}{V_T} = (38.6/Volt)\delta V$$

(12.112)

Then in terms of the percentage uncertainty in carrier densities, the values are some 38.6 times the uncertainties in the potential. For this example an uncertainty in the potential of 0.00517 corresponds to about a 20% uncertainty in the carrier densities at the worst case point. This may still represent a very small percentage of the maximum carrier densities in the structure as the point where this occurs is deep within the depletion region where Poisson's equation is dominated not by the free carriers but by the fixed impurity charge density. This is the reason an accurate value of the potential can be obtained even in the presence of a rather high uncertainty in the free carrier densities – they are small contributions to Poisson's equation around the p-n junction interface. However, it is important to understand the accuracy with which any solution to a PDE can be obtained by any numerical technique.

The approach demonstrated here of using the h-2h algorithm can be applied to any PDE solution. A general function for doing this will not be given as the time involved in obtaining a solution probably means that one would prefer to obtain an estimate of the accuracy by performing calculations in several steps as done in this example. In general it is much more difficult to obtain high accuracy in the solution of 2D PDE's because of practical limits to the number of spatial intervals that can be used in a solution. Computational times can become excessively long with numbers of spatial intervals exceeding the 100 by 100 range. For previous one-dimensional equations, solutions with spatial numbers exceeding several thousand were easily obtained. An approach that should always be used with PDEs is to first perform calculations with small numbers of grid points to verify that the code is executing properly. Then the number of grid points can be increased and the code can even be executed overnight for higher accuracy since most of the time our computers remain inactive during the night.

12.10 Summary

This chapter has discussed the numerical solution of partial differential equations by the method of finite differences. This has made use of previously developed code for one dimensional equations. The FD method is most applicable to physical problems where the solution space matches a rectangular grid of independent variables. This is typically the case when one is dealing with problems involving one time dimension and one spatial dimension. These naturally match onto the finite difference method. The chapter contains several examples of such physical problems. The code routines developed for such problems handle nonlinear problems just as readily as linear problems. In keeping with the theme of this work, linear problems are treated as just special cases of the more general nonlinear problems.

For PDEs involving two spatial dimensions (plus perhaps time) the finite difference method does not provide as general a solution approach. This results from the fact that the boundaries of many real physical problems do not naturally match to a rectangular space of grid points. However, there are some problems where there is a natural match to such a rectangular space and several examples of such problems are included in this chapter. Also some techniques are shown that can be used to solve classes of physical problems where there is not an exact match to a rectangular set of boundary conditions. For the case of general non-rectangular boundary conditions, the method of finite elements discussed in the next chapter is in many cases a more appropriate solution method.

Considerably discussion has again been presented on the accuracy of the FD method as applied to PDEs. The accuracy is limited by the number of spatial grid elements that are used in the solution and this in turn is limited by the computational time for large numbers of grid points. A method using the h-2h algorithm has been demonstrated for estimating the accuracy of solving PDEs. To apply this algorithm to problems involving both spatial dimensions and time, the number of calculated points must be increased in both the spatial and time dimensions.

Several code segments have been developed in this chapter for aiding in the solution of PDEs and a brief summary of the most important of these is presented below;

1. pdeivbv() – Code for solving initial value, boundary value problem – usually in one time dimension and one spatial dimension.
2. pdeivbvt() – Code for solving initial value, boundary value problem with time dependent data collected for selected spatial points.
3. pdeivbvqs() – Code for solving initial value, boundary value problem, similar to pdeivbv() but with logarithmic spacing on time intervals.
4. pdeivbvsqt() – Code for solving initial value, boundary value problem with logarithmic time interval spacing and with time dependent data collected for selected spatial points.
5. setup2bveqs() – Code to set up matrix of finite difference equations in two dimensions for a PDE in two spatial dimensions.
6. sdgauss() – Code for the direct solution of system of sparse matrix equations of the form generated by PDE with diagonal node numbering. This uses spgauss() discussed in Chapter 4.
7. pde2bvsor() – Code for iterative solution of matrix equations by the SOR technique.

8. pde2bvcoe() – Code for iterative solution of matrix equations by the COE technique.
9. pde2bvadi() – Code for iterative solution of matrix equations by the ADI technique.
10. pde2bv() – Code for solving two dimensional PDE by finite difference technique using either SPM, SOR, COE or ADI technique.
11. pde2bv1t() – Code for solving PDE involving two spatial dimensions and one time dimension using either SPM, SOR, ÇOE or ADI techniques.
12. pde2bv1tqs() – Code for solving PDE involving two spatial dimensions and one time dimension with a logarithmic spacing on time intervals.
13. splot() – Pop-up surface plot of two dimensional solution.
14. cplot() – Pop-up contour plot of two dimensional solution.
15. grad() – Code for numerical calculation of two dimensional first derivatives or gradient of a spatial function.
16. intp2d() – Code for interpolation on a two dimensional surface.
17. to2darray() – Code for converting from a linear numbering scheme for grid points to a two dimensional numbering scheme.
18. reversexy() – Code for reversing the storage arrangements of a matrix from x first to y first or vice versa.
19. reorder() – Code for changing a linear grid numbering scheme from x axis first to y axis first or vice versa.

These code routines can be used for a variety of partial differential equations. These have been developed with a minimal of effort by making use of code segments developed in previous chapters. Using the examples in this chapter the reader should be able to explore an additional range of PDEs where the boundary conditions can be readily matched to the rectangular range.

The next chapter will discuss a second major technique for solving PDEs – that of the finite element method. Because this technique is much more amendable to physical problems with irregular boundaries, this final technique has found extensive application in a wide variety of engineering problems.

13 Partial Differential Equations: The Finite Element Method

The previous chapter has discussed the solution of partial differential equations using the classical finite difference approach. This method of solution is most appropriate for physical problems that match to a rectangular boundary area or that can be easily approximated by a rectangular boundary. One such class of PDEs is initial value problems in one spatial variable and one time variable. Other selected problems in two spatial dimensions are also amendable to this approach and selected examples are given in the previous chapter.

A large class of PDEs, however, involve spatial dimensions not confined to a rectangular geometry and for such problems the more recently developed finite element (FE) approach is in many cases much more appropriate. As in the previous chapter some of the important theory underlying the FE method will be discussed and selected computer code developed to implement a selected subset of FE approaches. The code is then used to illustrate the solution of selected PDEs by the FE approach.

13.1 An Introduction to the Finite Element Method

As with the finite difference (FD) approach, the FE approach seeks to obtain an approximate solution to a physical problem at a discrete set of spatial (or time) points. The approach was originally developed to study stresses in complex airframe structures, but has since been extended and applies to a broad range of physical phenomena. For a specific example consider the structure and problem shown in Figure 13.1. This shows an airfoil in the center (white area) of the figure within a rectangular chamber. This is typical of the problem of studying the air flow around an airfoil in the aircraft industry. The airfoil has a complex continually varying contour not readily approximated by a rectangular spatial grid structure. Shown surrounding the airfoil is a two dimensional spatial grid structure composed of an array of triangular spatial elements of varying size. Such an array of triangular elements is able to closely approximate the spatial boundary of the airfoil and the more triangular elements used with smaller dimensions around the airfoil the more closely the array can approximate any desired contour of the airfoil. In this example the "finite elements" are the triangular elements and the array of spatial grid points are the nodes of the triangles. Also noted in the figure is the continuous variation is size of the spatial elements from the airfoil boundary outward toward the rectangular boundaries. For this particular example a total of 2542 triangular elements are shown.

J.R. Hauser, *Numerical Methods for Nonlinear Engineering Models*, 883–987.
© Springer Science + Business Media B.V. 2009

Thus the FE approach envisions the spatial region as comprised of an array of spatial elements (triangles or other simple shapes) and the solution of a PDE is built up by interconnecting the solutions within these small subregions or elements to cover the entire spatial domain. The major advantages of this approach over the FD approach are: (1) the ability to approximate very complex boundary shapes and (2) the ability to gradually and continuously change the size of the spatial elements. These two advantages make the FE approach the preferred approach for a large range of real engineering problems.

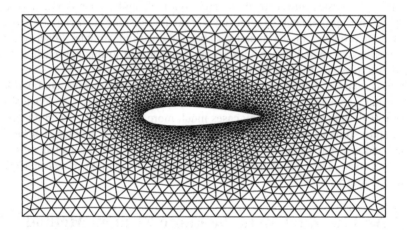

Figure 13.1. Spatial grid structure for example of an airfoil in a rectangular chamber.

Many researchers have contributed over the years to the theory and development of the FE approach and are two numerous to list here. For a history of the approach the reader is referred to the extensive literature on the method in books and on the web. An engineering library at a typical university will have over 100 books devoted entirely to the finite element method. The web has over 10,000,000 references to the FE method. Obviously a chapter devoted to the method can only cover some highlights of the method and introduce some of the more basic concepts and approaches of the method. As for a history of the approach it will simply be noted that the term "finite element method" became popular in the 1960s although many of the basic ideas of the method date back much further. As a practical engineering tool the method was developed primarily by engineers in the aircraft and automotive industries where complex spatial shapes are of primary importance. A good reference to the theory of the FE method is: *The Finite Element Method (5th Edition) Volume 1 - The Basis* by O. C. Zienkiewicz and R. L. Taylor, published by Elsevier. This is especially recommended because an online version is available at:

The FE method consists of at least the following steps: (1) discretizing the spatial regions of the problem into a series of finite elements, (2) selecting interpolation functions over the finite elements, (3) assembling of the element properties into a finite set of equations, (4) imposing boundary conditions, (5) solving the set of coupled equations and (6) assessing the solutions. Many of these steps are similar to the solution steps used in the FD method of the previous chapter. The major differences are in the details of the steps leading up to obtaining the set of equations to be solved – in this case for the properties of the finite elements. Each of these steps is now addressed in sections of this chapter.

13.2 Selecting and Determining Finite Elements

The first and perhaps the most basic step in the FE approach is the selection of finite elements to uniformly cover the spatial dimensions of a physical problem. While many possible shapes could be considered for the fundamental spatial elements, in practice, fairly simple shapes are typically used. For two spatial dimensions, the elements typically considered are triangular elements or general quadrilateral elements as illustrated in Figure 13.2. (The three dimensional case will be briefly considered in a subsequent section). Three triangular elements are shown in (a) – (c) with specified nodes of 3, 6 and 10 where nodes are identified by the solid points in the figure and are the points at which the solution variable is

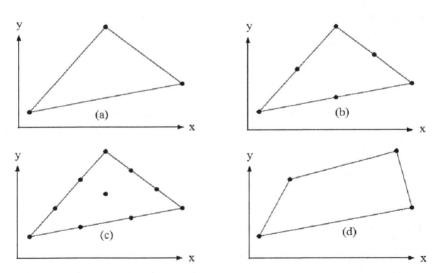

Figure 13.2 Examples of two-dimensional elements. (a) three node triangle (b) six node triangle (c) ten node triangle (d) general quadrilateral.

desired or assumed to be evaluated. Shown in (d) is a general quadrilateral element with four sides and 4 nodes. The simplest and most frequently used element is the three node triangle shown in Figure 13.2 (a). The six node triangle of (b) is also equivalent, in terms of nodes, to four, three node triangles as can be seen by embedding four triangles within the given triangle. In a great many cases of numerical methods, the simplest element repeated many times is not only the simplest approach, but also equivalent in accuracy for equivalent execution times to more complex algorithms. Thus, the three node triangle shown in Figure 13.2 will be the only spatial element considered in this work. As illustrated in Figure 13.1, it can be readily used to cover general spatial boundary surfaces.

With the selection of a basic spatial element, there is the major question of how to generate an appropriate grid of triangular elements such as shown in Figure 13.1 for a particular physical problem and geometry. For extremely simple problems the triangular grid might be generated manually but for general problems some automatic computer generation technique is required for practical applications. There are certain very desirable properties of a "good" spatial coverage of triangular elements. First the most desirable shape of a triangle is an equilateral triangle with all sides and angles equal. For such a case, each node of a triangle is surrounded by six equilateral triangles. A quick perusal of Figure 13.1 will show that the nodes of the triangles in this case are surrounded by five to seven triangles and the most common configuration is that of six triangles around a node. A second desirable feature is that all triangles have angles of 90° or less. Finally it is highly desirable that the size of the triangles vary slowly from one triangle to any adjacent triangle. Again a perusal of Figure 13.1 will verify that these features are present in that particular example.

A number of mesh generation programs are based on Delaunay triangulation in two dimensions. From the three nodes of a triangle, one can construct a circle passing through the three nodes. The Delaunay triangulation has the property that the circle so formed excludes all other nodes in the set of triangle nodes. The generation of a Delaunay triangulation for a given spatial domain is a somewhat specialized mathematical topic the details of which would take us far afield from the major emphasis of this work. However, this topic has fortunately been extensively pursued by others and several researchers have been very gracious in sharing theory and computer programs for generating such sets of Delaunay triangles. Two such programs available for free on the web with source code are named Triangle (http://www.cs.cmu.edu/~quake/triangle.html) and EasyMesh (http://www-dinma.univ.trieste.it/nirftc/research/easymesh/Default.htm). The reader is encouraged to investigate both programs but in the interest of time and effort only the EasyMesh program will be discussed and used here. This was the program used to generate the triangular mesh shown in Figure 13.1. Part of the reason for selecting this program is the ease with which it can be used.

EasyMesh is freely available and comes with a good description of its use and several examples. It requires one input file describing the spatial structure to be modeled and the program generates three output files that can be used in formulating a set of FE equations for some PDE. The program is executed on a PC command line as "EasyMesh name.d [options]" where name.d is the name of the input file. The most common option is +dfx which creates a drawing of the Delaunay (and Voronoi) mesh in .dxf (or Autodesk) format. The input file for Easymesh should have the following format:

- first line: `<number of nodes>`
- following lines: `<node number:> <x> <y> <triangle side> <boundary marker>`
- next line: `<number of segments>`
- following lines: `<segment number:> <start point> <end point> <boundary marker>`

The first major section of the input file contains a line by line listing of the node points defining the boundaries of the desired spatial region with a node number attached to each point followed by the `<x>` and `<y>` locations of the point. Other information that may be specified for a node is a `<triangle side>` value representing the desired length of the triangle sides which contain that node in the final triangulation. The smaller that value, the finer is the mesh around that point. Finally the `<boundary markers>` are tags used to identify which points and segments are associated with which boundary condition.

The major section of input is the "segments" section and describes how the nodes are connected to form a closed boundary region for the triangular elements. The boundaries are defined by a segment number with the starting and ending "node" numbers and a possible "boundary marker". Comments may also be included in the input file and these lines begin and end with a # character, and can be inserted anywhere inside the input file. The input file mane **must** have the extension `.d`. The input specification is perhaps best understood with an example to follow.

An example input file from the EasyMesh documentation is shown in Listing 13.1. The corresponding spatial geometry and mesh grid generated by this file is shown in Figure 13.3. This example has a hole in the structure along with a varying triangular size from the smallest triangles along the boundary of the hole to the largest along the outside boundary. In the input file a hole is distinguished from an external boundary by listing the nodes around the hole in a clockwise direction as opposed to a counterclockwise direction for the external boundary. An important feature of the input specification is the ability to specify an element size by the third entry used to define the critical boundary points. This can be seen by the 0.25 and 0.1 entries in the table following the node locations. Another feature of the EasyMesh program is the numbering of boundary points and boundary segments beginning with 0 as is common with C computer programs. This requires some translation when the results are interfaced to the Lua programming code as Lua typically defaults to tables beginning with 1. However, a beginning node number of 0 is required for the proper execution of the EasyMesh program.

```
#-----------#
# Example 1 #
#-----------#
#==========
| POINTS |
==========#
9 # number of points #

# Nodes which define the boundary #
0:    0    0    0.25    1
1:    5    0    0.25    1
2:    5    2    0.25    2
3:    4    3    0.25    3
4:    0    3    0.25    3

# Nodes which define the hole #
5:    1    1    0.1     4
6:    1    2    0.1     4
7:    2    2    0.1     4
8:    2    1    0.1     4

#=============
| SEGMENTS |
=============#
9 # Number of segments #

# Boundary segments #
0:    0    1    1
1:    1    2    2
2:    2    3    2
3:    3    4    3
4:    4    0    3

# Hole segments #
5:    5    6    4
6:    6    7    4
7:    7    8    4
8:    8    5    4
```

Listing 13.1. Example input file for EasyMesh program.

EasyMesh generates the following three output files with the following extensions and information about the triangles:

- .n   node file
- .e   element file
- .s   side file

These extensions are added to the input file name. If an input file is named NAME.d the following files will be created after executing the EasyMesh program:

Node file (NAME.n) has the following format:

- first line: <number of nodes>
- following lines: <node number:> <x> <y> <marker>
- the last two lines are comments inserted by the program to facilitate the reading of the node file

Element file (NAME.e) has the following format:

- first line   `<number of elements>`
- following lines: `<element number:> <i> <j> <k> <ei> <ej> <ek> <si> <sj> <sk> <xV> <yV> <marker>`
- the last two lines are comments inserted by the program to facilitate the reading of the element file

in the above, `i`, `j`, `k` are the three nodes belonging to the element, `ei`, `ej` and `ek` are the neighbouring elements, while `si`, `sj` and `sk` are the element sides. The entries `xV` and `yV` are the coordinates of the element circumcenter, which is also the Voronoi polygon node.

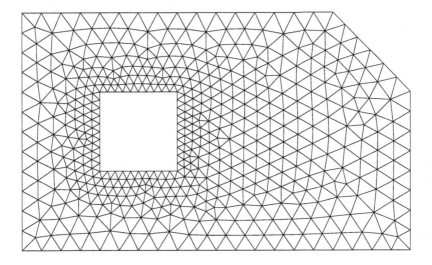

Figure 13.3. Finite element mesh generated by Easymesh for input file of Listing 13.1.

The `marker` is useful when the domain consists of more than one material. If `ei`, `ej` or `ek` is equal to -1, it means that the triangular element lies on the boundary.

Side file (NAME.s) has the following format:

- first line: `<number of sides>`
- following lines: `<side number:> <c> <d> <ea> <eb> <marker>`
- the last two lines are comments inserted by the program to facilitate the reading of the side file

where `c` and `d` are the starting and ending point of the side specified by node numbers, `ea` and `eb` are the elements on the left and on the right of the side. If `ea` or `eb` equals -1, it means that the left or right element does not exists, so the side is on the boundary.

These files provide essentially all the information ever needed to set up finite element equations for some physical problem associated with the triangular spatial grid and in many cases not all of the returned information is needed for setting up the equations for a PDE over the spatial domain. The output files generated by EasyMesh for all the examples in this chapter are provided in the supplied software so the user does not have to execute the EasyMesh program for the examples. However, the reader is strongly encouraged to download and execute the EasyMesh code in conjunction with the examples of this chapter. Also the reader will find valuable additional information in the examples and discussion files supplied with EasyMesh especially with regard to how to handle problems with several different materials embedded within the spatial domain.

13.3 Shape Functions and Natural Coordinates

Before discussing the formulation of physical equations over a set of finite elements some background material will be discusses regarding how solution variables can be expressed over a triangular spatial region. It is standard practice in FE analysis to represent physical variables as some function of the spatial coordinates over the area of a triangular region. If a triangle such as shown in Figure 13.2(a) is considered with three nodes and it is assumed that the physical variable is to be represented as some known values at the three node points than an appropriate variation of some quantity $u(x, y)$ over the triangular element is a linear variation such as:

$$u(x, y) = c_0 + c_1 x + c_2 y \qquad (13.1)$$

Since the solution is known at the three nodes and using the notation u_1, u_2 and u_3 for these three node values, the three constants in this equation can then be solved for in terms of the known node values. If six nodes are used over a triangular area as in Figure 13.2(b) then the variable can be expressed in terms of six unknown constants as for example in:

$$u(x, y) = c_0 + c_1 x + c_2 y + c_3 x^2 + c_4 y^2 + c_5 xy \qquad (13.2)$$

While this can more accurately represent a function over some spatial region, it leads to considerable more complexity in formulating a set of physical equations over a triangular grid. The reader can consult the literature for such a formulation. Only the simpler three point representation will be pursued here.

The constants in Eq. (13.1) can then be solved from the equation set:

$$\begin{bmatrix} 1 & x_1 & y_1 \\ 1 & x_2 & y_2 \\ 1 & x_3 & y_3 \end{bmatrix} \begin{bmatrix} c_0 \\ c_1 \\ c_2 \end{bmatrix} = \begin{bmatrix} u_1 \\ u_2 \\ u_3 \end{bmatrix} \qquad (13.3)$$

where the x, y pairs are the coordinates of the three node points of the triangle. The solution of this set of equations gives:

$$c_0 = \frac{u_1(x_2 y_3 - x_3 y_2) + u_2(x_3 y_1 - x_1 y_3) + u_3(x_1 y_2 - x_2 y_1)}{2\Delta}$$

$$c_1 = \frac{u_1(y_2 - y_3) + u_2(y_3 - y_1) + u_3(y_1 - y_2)}{2\Delta}$$

$$c_2 = \frac{u_1(x_3 - x_2) + u_2(x_1 - x_3) + u_3(x_2 - x_1)}{2\Delta} \quad (13.4)$$

$$\Delta = \text{area of triangle} = \frac{1}{2} \begin{vmatrix} 1 & x_1 & y_1 \\ 1 & x_2 & y_2 \\ 1 & x_3 & y_3 \end{vmatrix} =$$

$$(x_1 - x_3)(y_2 - y_3) - (x_2 - x_3)(y_1 - y_3)$$

To be more precise, Δ is the positive area of the triangle if the three nodes are labeled in a clockwise direction, otherwise it is the negative of the area if the nodes are numbered in a counter clockwise direction. The EasyMesh program labels the triangles in such a counter clockwise direction so this must be taken into account when computing the areas.

From this set of coefficients, Eq. (13.1) can then be expressed as:

$$u(x, y) = u_1 n_1(x, y) + u_2 n_2(x, y) + u_3 n_3(x, y) \quad (13.5)$$

where the n_1, n_2 and n_3 functions are known as shape functions and are given in terms of the triangle node points as:

$$n_1(x, y) = \frac{(x_2 y_3 - x_3 y_2) + (y_2 - y_3)x + (x_3 - x_2)y}{2\Delta}$$

$$n_2(x, y) = \frac{(x_3 y_1 - x_1 y_3) + (y_3 - y_1)x + (x_1 - x_3)y}{2\Delta} \quad (13.6)$$

$$n_3(x, y) = \frac{(x_1 y_2 - x_2 y_1) + (y_1 - y_2)x + (x_2 - x_1)y}{2\Delta}$$

The shape functions have the very useful property that each shape function in turn is equal to unity at the node corresponding to its subscript and is equal to zero at the other two nodes. The shape functions then describe faces of a triangular surface which is zero along one edge of the triangle and which linearly increases to a unity value at the node opposite to the zero edge. If the special case is considered where $u_1 = u_2 = u_3 = u(x, y) = $ constant then it is seen from Eq. (13.5) that this requires that $n_1 + n_2 + n_3 = 1$ which can also be verified by summing the expressions in Eq. (13.6). Figure 13.4 shows the surface of these three shape functions for a typical triangular element. The base triangle in the x-y plane is shown as a dotted surface. In each case the height of the shape function is unity at one of the nodes and goes to zero along the opposite boundary line. Each triangular element in the spatial domain of a given problem will have three associated shape functions. When needed to identify a particular triangular element, the notation n_{e1}, n_{e2} and n_{e3} will be used where the first subscript identifies the triangular element and the second subscript will identify the nodes with the node order specified by the *.e file returned by EasyMesh.

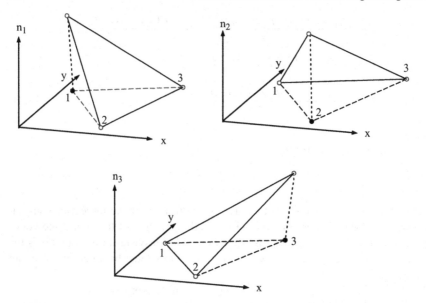

Figure 13.4. Illustration of three basic shape functions for a triangular element.

Another frequently used concept when dealing with finite elements is that of a set of natural coordinates, in this case corresponding to each of the sides of the triangle. Consider a set of new coordinates ℓ_1, ℓ_2 and ℓ_3 defined by the equations:

$$x = x_1\ell_1 + x_2\ell_2 + x_3\ell_3$$
$$y = y_1\ell_1 + y_2\ell_2 + y_3\ell_3 \qquad (13.7)$$

where each point in the x, y plane is represented as some weighted combination of the coordinates of the triangle nodes. Since there are only two spatial coordinates, there can not be three independent new coordinates, but this can be fixed by the additional requirement that $\ell_1 + \ell_2 + \ell_3 = 1$. One form of the transformation equations then becomes:

$$x = x_3 + (x_1 - x_3)\ell_1 + (x_2 - x_3)\ell_2$$
$$y = y_3 + (y_1 - y_3)\ell_1 + (y_2 - y_3)\ell_2 \qquad (13.8)$$

From these one can solve for the new coordinate sets in terms of the original x, y coordinates. This is straightforward and will be left as an exercise for the interested reader. The result is that it is found that the equations for ℓ_1, ℓ_2 and ℓ_3 are exactly the same as the equation set for the shape functions n_1, n_2 and n_3, i.e.

$$\ell_1(x, y) = n_1(x, y)$$
$$\ell_2(x, y) = n_2(x, y) \qquad (13.9)$$
$$\ell_3(x, y) = n_3(x.y)$$

As the shape functions vary from 0 to 1 across the triangular surface, the set of local coordinates vary in exactly the same manner from 0 to 1. In later discussions it will be convenient to think sometimes in terms of shape functions and some-times in terms of local coordinates in which case the h_i notation will be used for the shape functions and the ℓ_i notation will be used for the local coordinates but one should keep in mind that these are one and the same function of the x, y coor-dinates. Figure 13.5 illustrates the local coordinates within a triangular element. Lines of constant local coordinates are parallel to the triangle sides and at each node one of the local coordinates has value of 1 while the other local coordinates are of value 0. For any interior point, the local coordinates are also related to the three areas defined by the interior point and the three nodes of the triangle as illus-trated by the three areas shown in the figure. In the figure only lines of constant ℓ_1 are shown as dotted lines.

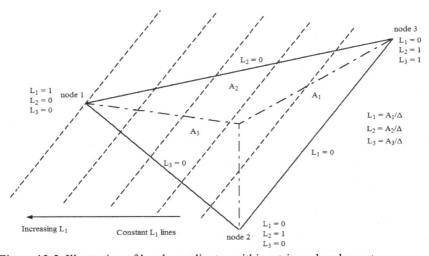

Figure 13.5. Illustration of local coordinates within a triangular element.

As will be subsequently developed, the shape functions and local coordinates play very important roles in formulating a set of matrix equations for a PDE over a spatial domain with the FE method while constant coordinate lines for ℓ_2 and ℓ_3 would be parallel to the other two sides of the triangle.

13.4 Formulation of Finite Element Equations

With the background material of the previous sections an approach for obtaining a set of difference equations approximating a partial differential equation over the spatial finite elements can now be discussed. There are two basic approaches used to obtain such a set of equations. These will first be illustrated with a specific PDE which will be taken as a form of the linear Poisson equation:

$$\frac{\partial^2 u}{\partial x^2} + \frac{\partial^2 u}{\partial y^2} + g(x, y) = 0 \qquad (13.10)$$

The first method derives from the calculus of variations and the known fact that for this equation the solution is a function that minimizes the functional:

$$I(u) = \iint_{\Omega} f(u, u_x, u_y, x, y) dx dy = \iint_{\Omega} \left[\frac{1}{2} \left(\frac{\partial u}{\partial x} \right)^2 + \frac{1}{2} \left(\frac{\partial u}{\partial y} \right)^2 - g(x, y) u \right] dx dy \quad (13.11)$$

The relationship between the differential equation and the functional f is the Euler-Lagrange equation:

$$\frac{\partial f}{\partial u} - \frac{\partial}{\partial x} \left(\frac{\partial f}{\partial u_x} \right) + \frac{\partial}{\partial y} \left(\frac{\partial f}{\partial u_y} \right) \quad (13.12)$$

It can be easily verified that the above formalism leads in reverse to the partial differential equation of Eq. (13.10). Thus the functional minimization approach is to seek an approximate solution that minimizes the value of the integral in Eq. (13.11).

To accomplish this one needs to parameterize the solution in terms of a finite set of solution values and then seek to minimize the functional with respect to the selected set of solution values. The set of solution values are taken as some set over each triangular element. The simplest sets of solution values to select are the three node point values as shown in Figure 13.2(a). However, a more complex set could be the six or ten values shown in Figure 13.2(b) and 13.2(c). After selecting the parameterizing set of values an interpolation method is then needed over the spatial domain. The conventional approach is to assume that the solution varies linearly between the triangular nodes and thus to use the shape function as derived in the previous section to approximate the spatial variation of the solution. When written in this form the solution becomes:

$$u(x, y) = u_1 N_1(x, y) + u_2 N_2(x, y) + \cdots u_K N_K(x, y)$$

$$= \sum_{k=1}^{K} u_k N_k(x, y) \text{ where } K = \text{ number of nodes}$$

$$N_k(x, y) = \sum_{i=1}^{N_e} n_{i,k}(x, y) \quad (13.13)$$

where N_e = number of elements around node k

This provides a linearly varying solution surface between all the triangle nodes. The interpolating functions, $N_k(x, y)$, require a little study and interpretation. They are functions that are zero in value except within the set of triangles surrounding the k node point as they are defined as the sum of the shape functions for all the triangles surrounding the node point. A visual view of these interpolation functions is shown in Figure 13.6 for a typical node with six surrounding triangles. This figure is drawn for simplicity for equilateral triangles, but one can envision the shape for non-equilateral triangles and for more or fewer triangles surrounding the node. There are typically 5 to 7 triangles arranged around each node point.

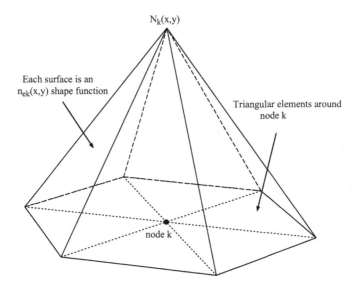

Figure 13.6. Spatial interpolation function around node k, composed of composite of local shape functions within adjoining triangular elements.

The set of u_i values which correspond to a minimum value of the integral in Eq. (13.11) are found by taking the differential of the integral with respect to each value and setting the result to zero. This gives the set of equations:

$$\frac{\partial I}{\partial u_i} = 0 = \iint_\Omega \left[\frac{\partial u}{\partial x} \frac{\partial}{\partial x} \left(\frac{\partial u}{\partial u_i} \right) + \frac{\partial u}{\partial y} \frac{\partial}{\partial y} \left(\frac{\partial u}{\partial u_i} \right) - g(x,y) \frac{\partial u}{\partial u_i} \right] dxdy$$

$$0 = \iint_\Omega \left[\frac{\partial u}{\partial x} \frac{\partial N_i}{\partial x} + \frac{\partial u}{\partial y} \frac{\partial N_i}{\partial y} - g(x,y) N_i \right] dxdy \qquad (13.14)$$

where $\dfrac{\partial u}{\partial u_i} = N_i$

Further making use of Eq. (13.13) for the solution and taking the spatial derivatives gives:

$$0 = \iint_{\Omega} \left[\sum_{k=1}^{K} u_k \frac{\partial N_k}{\partial x} \frac{\partial N_i}{\partial x} + \sum_{k=1}^{K} u_k \frac{\partial N_k}{\partial y} \frac{\partial N_i}{\partial y} - g(x,y)N_i \right] dxdy \qquad (13.15)$$

for $i = 1, 2, 3 \cdots K$

This provides one algebraic equation for each node in the spatial domain, so this formulation provides K equations in the K unknowns which are the solution values at the triangle nodes (where K is the number of triangle nodes). It should be noted that while the sums in the equations are denoted as being over all the nodes in the spatial domain, in applying the equation for any individual node, the sum is only non-zero over a small subset of nodes. For example for the ith equation, the interpolation function N_i as well as its derivative is only non-zero over the region of space occupied by the triangles immediately surrounding the ith node. This is a small subset of 5 to 7 triangles surrounding the ith node and contributions to the ith equation can only come from values at the nodes surrounding the ith node as illustrated in Figure 13.6. This means that in the resulting set of matrix equations, a given equation will have only a central diagonal term plus 5 to 7 off diagonal terms corresponding to the adjacent nodes. However, because the nodes in the triangular grid have no fixed numbering sequence, the resulting set of equations will not have any definite structure such as the banded structure obtained for the FD formulation.

Before continuing further the second major approach to obtaining an equation set will be discussed for this particular PDE. This approach is known as the "weighted residuals" approach. Since any numerical solution will only give an approximation to the exact solution, the differential equation of Eq. (13.10) is not expected to be identically zero at every point within the solution space. However, the ideal case of zero value can be approximated by multiplying any residual of this equation by some weighting function and integrating over the spatial domain and setting the result to zero as in the equation:

$$\iint_{\Omega} \left[\frac{\partial^2 u}{\partial x^2} + \frac{\partial^2 u}{\partial y^2} + g(x,y) \right] W(x,y) dxdy = 0 \qquad (13.16)$$

Standard practice is to consider a weighting function associated with each node and to make the weighting functions the same as the interpolating functions so that this becomes:

$$\iint_{\Omega} \left[\frac{\partial^2 u}{\partial x^2} + \frac{\partial^2 u}{\partial y^2} + g(x,y) \right] N_i dxdy = 0 \qquad (13.17)$$

In this the spatial notation has been dropped from the N_i function. Again since the N_i function is only non-zero over the triangular areas immediately surrounding the ith node as seen in Figure 13.6, the integral is only non-zero over a region of space immediately surrounding a particular node. This provides again the same number of equations as spatial nodes for obtaining a set of K equations in the unknowns which are again assumed to be expressible in terms of the values at the nodes and the interpolating functions. Standard practice is to also perform integration by parts on the second derivative terms in the form:

$$\iint_{\Omega} \frac{\partial^2 u}{\partial x^2} N_i dx dy = \int_{S_i} \frac{\partial u}{\partial x} N_i dy - \iint_{\Omega} \frac{\partial u}{\partial x} \frac{\partial N_i}{\partial x} dx dy$$

$$\iint_{\Omega} \frac{\partial^2 u}{\partial y^2} N_i dx dy = \int_{S_i} \frac{\partial u}{\partial y} N_i dx - \iint_{\Omega} \frac{\partial u}{\partial y} \frac{\partial N_i}{\partial y} dx dy$$

(13.18)

In each of these the first integrals is over the bounding surface of the triangles and the integrals are zero because the interpolating N_i functions are zero on the bounding surfaces. An exception to this would be any nodes on the bounding surface of a problem domain's spatial region. However, these are boundary nodes which in any case require special treatment and the equations for the boundary nodes will have to be handled separately – more on this later. So for the moment, neglecting the boundary nodes and putting these expressions back into Eq. (13.17) leads to exactly the same equation as before and as given in Eq. (13.14) (with only a reversed sign on all the terms).

In extending the theory to more general PDEs, either the functional minimization approach or the weighted residual approach can be used. For the set of classical PDEs, the minimization function for the Euler-Lagrange equation is known so one can apply the theory. However, such is not the case for general nonlinear PDEs. However the weighted residual approach can be used even for general nonlinear PDEs. Thus this is the path that will be pursued here. Consider a general two-dimensional PDE which is a function of the first and second derivatives as well as the spatial dimensions in the general form:

$$F(x, y, U_{xx}, U_x, U, U_y, U_{yy}) = 0$$

where $U_{xx} = \frac{\partial^2 U}{\partial x^2}, U_{yy} = \frac{\partial^2 U}{\partial y^2}, U_x = \frac{\partial U}{\partial x}, U_y = \frac{\partial U}{\partial y}$ (13.19)

If a first order Newton expansion about some initial solution such that $U \to U + u$ is now considered, a linear approximation to the equation is obtained in the usual manner as:

$$a_x \frac{\partial^2 u}{\partial x^2} + a_y \frac{\partial^2 u}{\partial y^2} + b_x \frac{\partial u}{\partial x} + b_y \frac{\partial u}{\partial y} + cu + F_0 = 0$$

where $a_x = \frac{\partial^2 F}{\partial U_{xx}^2}, a_y = \frac{\partial^2 F}{\partial U_{yy}^2}, b_x = \frac{\partial F}{\partial U_x}, b_y = \frac{\partial F}{\partial U_y}, c = \frac{\partial F}{\partial U}$ (13.20)

and F_0 is the function evaluated at the initial solution approximation. In this equation the coefficients on the various derivative terms may be general functions of the spatial variable. This is the standard technique for treating a nonlinear equation which has been extensively used in this work.

Now applying the weighted residual technique to this equation gives:

$$
\mathrm{Eq}(i) = \iint_{\Omega}\left[\, a_x \frac{\partial^2 u}{\partial x^2} + a_y \frac{\partial^2 u}{\partial y^2} + b_x \frac{\partial u}{\partial x} + b_y \frac{\partial u}{\partial y} + cu + F_0 \right] N_i \, dxdy = 0
$$

$$
= \int_L a_x \frac{\partial u}{\partial x} N_i \, dx + \int_L a_y \frac{\partial u}{\partial y} N_i \, dy + \iint_{\Omega}\left[\, -a_x \frac{\partial u}{\partial x}\frac{\partial N_i}{\partial x} - a_y \frac{\partial u}{\partial y}\frac{\partial N_i}{\partial y} \right] dxdy +
$$

$$
\iint_{\Omega}\left[\left(b_x - \frac{\partial a_x}{\partial x}\right)\frac{\partial u}{\partial x} + \left(b_y - \frac{\partial a_y}{\partial y}\right)\frac{\partial u}{\partial y} + cu + F_0 \right] N_i \, dxdy \qquad (13.21)
$$

$$
= \iint_{\Omega}\left[\, -a_x \frac{\partial u}{\partial x}\frac{\partial N_i}{\partial x} - a_y \frac{\partial u}{\partial y}\frac{\partial N_i}{\partial y} \right] dxdy +
$$

$$
\iint_{\Omega}\left[\left(b_x - \frac{\partial a_x}{\partial x}\right)\frac{\partial u}{\partial x} + \left(\dot{b}_y - \frac{\partial a_y}{\partial y}\right)\frac{\partial u}{\partial y} + cu + F_0 \right] N_i \, dxdy
$$

The second form of the equation is obtained by performing integration by parts on the second derivative terms. The third form is obtained by further neglecting the surface integrals which are zero for all internal nodes since the N_i function is zero on the boundaries of the integration domain. This equation is written as $\mathrm{Eq}(i)$ to emphasize that there is a separate equation for each internal node and this equation represents a set of matrix equations in the solution variables at the various triangle nodes. A final somewhat involved equation can be obtained by using the expansion in Eq. (13.13) for the solution variable in terms of the values at the nodes:

$$
\iint_{\Omega}\left[\, -a_x \sum_{k=1}^{K} u_k \frac{\partial N_k}{\partial x}\frac{\partial N_i}{\partial x} \right] dxdy + \iint_{\Omega}\left[\, -a_y \sum_{k=1}^{K} u_k \frac{\partial N_k}{\partial y}\frac{\partial N_i}{\partial y} \right] dxdy +
$$

$$
\iint_{\Omega}\left[\left(b_x - \frac{\partial a_x}{\partial x}\right)\sum_{k=1}^{K} u_k \frac{\partial N_k}{\partial x} N_i \right] dxdy +
$$

$$
\iint_{\Omega}\left[\left(b_y - \frac{\partial a_y}{\partial y}\right)\sum_{k=1}^{K} u_k \frac{\partial N_k}{\partial y} N_i \right] dxdy + \qquad (13.22)
$$

$$
\iint_{\Omega}\left[\, c \sum_{k=1}^{K} u_k N_k N_i \right] dxdy + \iint_{\Omega} F_0 N_i \, dxdy = 0
$$

This is a rather formidable looking equation; however, in application it is in fact simpler to apply than it may appear.

For each equation associated with each node there is only a small subset of all the node variables that appear in the equation and not the complete sum as indicated by the equation. This is because the N_i function is only non-zero within the spatial triangles immediately surrounding a given node. Thus as previously stated only node variables immediately surrounding a given node contribute to any given equation and this is between 5 to 7 node variables. Also the equations can be considered on a node by node basis or on a triangular element by element basis. On an element by element basis this amounts to reversing the order of the integration and summations in the equations and gives:

$$\sum_{k=1}^{K} u_k \sum_{\Delta_i} \left[\iint_{\Delta_i} \left(-a_x \frac{\partial N_k}{\partial x} \frac{\partial N_i}{\partial x} \right) dxdy + \iint_{\Delta_i} \left(-a_y \frac{\partial N_k}{\partial y} \frac{\partial N_i}{\partial y} \right) dxdy \right] +$$

$$\sum_{k=1}^{K} u_k \sum_{\Delta_i} \iint_{\Delta_i} \left[\left\{ \left(b_x - \frac{\partial a_x}{\partial x} \right) \frac{\partial N_k}{\partial x} N_i \right\} dxdy + \right.$$

$$\left. \iint_{\Delta_i} \left\{ \left(b_y - \frac{\partial a_y}{\partial y} \right) \frac{\partial N_k}{\partial y} N_i \right\} dxdy \right] \qquad (13.23)$$

$$\sum_{k=1}^{K} u_k \sum_{\Delta_i} \left[\iint_{\Delta_i} cN_k N_i dxdy \right] + \sum_{\Delta_i} \iint_{\Delta_i} F_0 N_i dxdy = 0$$

In this case the integrations are over the triangular elements with area Δ_i and the result is a set of matrix equations in the solution variable at the triangular nodes. When considered on an element by element basis, there are only three nodes to consider for each triangle and it is much simpler to evaluate the matrix equations on a triangular element by element bases. However, the evaluations over an individual triangle do not complete any one equation in the matrix and each triangle element only contributes a part of any complete node equation (three node equations in fact). On an element by element basis there are only four (or three finally) basic types of integrals that must be considered and these will be discussed in the next section.

13.5 Interpolation Functions and Integral Evaluation

In considering the evaluation of the integrals in Eq. (13.23) over one of the triangular elements, the N_k, N_i interpolation functions become simply the shape functions associated with the triangle under consideration, i.e. $N_k \rightarrow n_k$ and $N_i \rightarrow n_i$. Also for a given triangular element there are only three nodes and three shape functions to consider. The first type of integral comes from the first line in Eq. (13.23) and is of the form:

$$I1_{i,k} = -\iint_{\Delta} a_x(x,y) \frac{\partial n_k}{\partial x} \frac{\partial n_i}{\partial x} dxdy \text{ for } i,k = nd1, nd2 \text{ or } nd3 \qquad (13.24)$$

where $nd1, nd2$ and $nd3$ denote the three node numbers associated with a given triangular element. From Eq. (13.6) however, it can be seen that the shape functions are linear functions of x and y so over a given triangle, the partial derivatives are constant and can be taken out of the integral to give:

$$I1_{i,k} = -\left(\frac{\partial n_k}{\partial x} \right) \left(\frac{\partial n_i}{\partial x} \right) \iint_{\Delta} a_x(x,y) dxdy \qquad (13.25)$$

A similar expression results for the terms in the first line of Eq. (13.23) involving the y partial derivatives. This integral makes a contribution to row i and column k of the coefficient matrix. Before considering this further, the types of terms in the second and third lines of Eq. (13.23) will be considered.

For the second line terms, the form of the integral is:

$$I2_{i,k} = \iint_\Delta \left(b_x(x,y) - \frac{\partial a(x,y)}{\partial x} \right) \frac{\partial n_k}{\partial x} n_i \, dxdy$$

for $i,k = nd1, nd2$ or $nd3$ (13.26)

$$= \left(\frac{\partial n_k}{\partial x} \right) \iint_\Delta b_x'(x,y) n_i \, dxdy \text{ with } b_x'(x,y) = b_x(x,y) - \frac{\partial a(x,y)}{\partial x}$$

where the constant partial derivative has been taken out of the integral. A similar term results from the y derivative term on the second line. This term again makes a contribution to row i and column k of the coefficient matrix.

The remaining two integrals on the third line are of the form:

$$I3_{i,k} = \iint_\Delta c(x,y) n_k n_i \, dxdy$$

                                                                      (13.27)

$$I4_i = \iint_\Delta F_0(x,y) n_i \, dxdy$$

The third integral contributes to row i and column k of the coefficient matrix while the fourth term contributes to the right hand side of the matrix equations as it is independent of the node solution values. The fourth integral is in fact similar to the spatial integral for the second type of integral as indicated in Eq. (13.26).

The needed integrals then reduce to basically three types: one that simply integrates some spatial function over the triangular area (Eq. (13.25)), one that integrates a spatial function weighted by n_i (Eq. (13.26)) and finally one that integrates a spatial function weighted by $n_k n_i$ (first line of Eq. (13.27)). These integrals are most readily evaluated by transforming the integrations from the x,y coordinate system to the natural coordinate system as discussed in Section 13.3. In this coordinate system, the natural coordinates are identical to the shape functions, so the n_i values simply become the ℓ_i values, or the natural coordinate values. In transforming from the x,y coordinate system to the ℓ_1, ℓ_2 coordinate system one needs the transformation of variables equation: $dxdy = |J| d\ell_1 d\ell_2$ where J is the Jacobian matrix for the coordinate transformation, which is the matrix of coefficients in Eq. (13.8). When this is evaluated the result is: $dxdy = 2\Delta d\ell_1 d\ell_2$ where Δ is the area of the given triangle. Since there are 3 natural coordinates, one may choose to work with any 2 in the integration process. With this transformation, the three types of integrals become:

$$I1_{i,k} = -\left(\frac{\partial n_k}{\partial x} \right) \left(\frac{\partial n_i}{\partial x} \right) 2\Delta \iint_\Delta a_{\ell x}(\ell_1, \ell_2, \ell_3) dA_\ell$$

$$I2_{i,k} = \left(\frac{\partial n_k}{\partial x} \right) 2\Delta \iint_\Delta b_{\ell x}'(\ell_1, \ell_2, \ell_3) \ell_i dA_\ell \text{ for } i = 1,2 \text{ or } 3$$ (13.28)

$$I3_{i,k} = 2\Delta \iint_\Delta c_\ell(\ell_1, \ell_2, \ell_3) \ell_k \ell_i dA_\ell \text{ for } i,k = 1,2 \text{ or } 3$$

In each of these, the subscript ℓ has been added to the function to be integrated to indicate that it should be expressed in terms of the natural coordinate system using the relationships of Eq. (13.7). Also the integration is just indicated as dA_ℓ to indicate integration over the natural coordinate space which can be expressed in three equivalent forms as: $dA_\ell = d\ell_1 d\ell_2$ or $d\ell_2 d\ell_3$ or $d\ell_1 d\ell_3$. The total area in the natural coordinate space is 1/2 which cancels the multiplicative factor of 2. The general form of all the necessary integrals is:

$$I = 2\Delta \iint_\Delta f(\ell_1, \ell_2, \ell_3) dA_\ell \tag{13.29}$$

In the case of a PDE with constant coefficients, these can be very easily derived since it is known that

$$\iint_\Delta \ell_1^\alpha \ell_2^\beta \ell_3^\gamma dA_\ell = \frac{\alpha! \beta! \gamma!}{(\alpha + \beta + \gamma + 2)!} \tag{13.30}$$

and for constant coefficients, all the integrals are of this form.

For the more general case numerical integration can be used to approximate an integral such as in Eq. (13.29) where the function may be any arbitrary function of the normalized coordinates. Considerable past work has been done on numerical integration and especially on numerical integration as applied to triangular areas in previous FE work. Gaussian-Legendre numerical integration is a numerical method whereby a function is evaluated at selected points within an integration area or volume and the points summed with various weighting factors to give a numerical approximation to an integral. The technique is not limited to single variables, but can be applied to area (or volume) integrations as well. Figure 13.7 shows some typical points within a triangular area at which functions can be evaluated depending on the order of integration desired and with appropriate

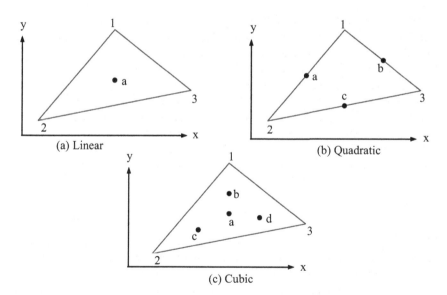

Figure 13.7. Function evaluation points for selected Gauss integration formulas over triangle areas. Points are exact for (a) linear (b) quadratic or (c) cubic function variations over the areas.

weighting. In (a) for example a single point evaluation at the center of the triangle with appropriate weighting gives an exact result for a function that varies linearly over the spatial domain. In (b) three points on the boundary lines are sufficient to give an exact result for a quadratic variation and for (c) the result is exact for a cubic spatial variation with four selected interior points.

Table 13.1. Numerical Integration Formulas for Triangles

Order	Points	Triangle Coordinates	Weights
(a) Linear	a	1/3, 1/3, 1/3	1
(b) Quadratic	a	1/2, 1/2, 0	1/3
	b	1/2, 0, 1/2	1/3
	c	0, 1/2, 1/2	1/3
(c) Cubic	a	1/3, 1/3, 1/3	-27/48
	b	0.6, 0.2, 0.2	25/48
	c	0.2, 0.6, 0.2	25/48
	d	0.2, 0.2, 0.6	25/48

The location of the points and the appropriate weighting factors are shown in Table 13.1. There are of course higher order integration formulas that use more evaluation points and achieve a higher degree of accuracy. For the present work if all the PDE equation coefficients are constant, then Eqs. (13.28) represent at most a quadratic variation over the triangle. However, the $I4$ integral in Eq. (13.27) could have a stronger variation with position. Thus the cubic formula associated with the points in Figure 13.7(c) appears to provide the best compromise for this work between simplicity and accuracy. It requires only one additional function evaluation above that required for the quadratic formula. With one of these integration formulas, the evaluation of Eq. (13.29) then becomes:

$$I = \Delta \sum_{j=1}^{n} W_j f(\ell_{1j}, \ell_{2j}, \ell_{3j}) \tag{13.31}$$

where n is the order of the integration formula and W_j are the weights given in Table 13.7 for the given triangle points. The 2 factor is missing from Eq. (13.31) because the area of a triangle in the natural coordinate space is 1/2.

Before putting it all together into programs for the FE method and showing some examples, one additional topic needs to be addressed and this in the subject of boundary conditions and how to handle these conditions in the FE method.

13.6 Boundary Conditions with the Finite Element Method

The previous section has considered the FE method as applied to internal nodes for a two dimensional problem. The triangle nodes located on the boundary of some physical region typically require special consideration. In formulating the equation set of Eq. (13.21) for the internal nodes, integral terms over a set of boundaries were neglected because the weighting functions are zero along all the internal boundaries. These integral terms must now be addressed and integrated into any additional specified boundary conditions. The neglected integrals are:

$$
\begin{aligned}
Eqb(i) &= \sum_{L_i} \left[\int_{L_i} a_x \frac{\partial u}{\partial x} N_i dx + \int_{L_i} a_y \frac{\partial u}{\partial y} N_i dy \right] \\
&= \sum_{L_i} \int_{L_i} \left[a_x \frac{\partial u}{\partial x} n_x + a_y \frac{\partial u}{\partial y} n_y \right] (1 - L / L_i) dL
\end{aligned}
\tag{13.32}
$$

where n_x and n_y are direction cosines of the outward normal to the surface and L_i is the length of a boundary side associated with the boundary node. In general there are two triangle sides associated with each boundary node for the triangular mesh case and the second form of the equation arises because the shape functions vary linearly with position along a boundary line.

Boundary conditions can be of several types. The most common boundary conditions are specifications of the value of the solution on the boundary nodes, the Dirichlet condition, or of the normal derivative on the boundary nodes, the Neuman condition. A more general set of boundary conditions known as mixed boundary conditions involve both the value and the normal derivative. A typical mixed boundary condition is of the form:

$$
a_x \frac{\partial u}{\partial x} n_x + a_y \frac{\partial u}{\partial y} n_y + g(x, y)u + h(x, y) = 0
\tag{13.33}
$$

This is a linear boundary condition, but in general such a mixed boundary condition could be nonlinear and of the form:

$$
\begin{aligned}
F_B(x, y, a_x \frac{\partial u}{\partial x} n_x &+ a_y \frac{\partial u}{\partial y} n_y, u) \\
&= F_B(x, y, a u_n, u) = 0
\end{aligned}
\tag{13.34}
$$

In this it is assumed that the normal surface derivatives always are accompanied by the a_x and a_y coefficients and this is indicated by the $a u_n$ notation in the second form of the expression. If this is the case, then the linearized version of the boundary condition will recover the form of Eq. (13.33) as:

$$\frac{\partial F_B}{\partial aU_n}\left[a_x\frac{\partial u}{\partial x}n_x+a_y\frac{\partial u}{\partial y}n_y\right]+\frac{\partial F_B}{\partial U}u+F_B=0$$

$$a_x\frac{\partial u}{\partial x}n_x+a_y\frac{\partial u}{\partial y}n_y+\left[\frac{F_{BU}}{F_{BUn}}\right]u+\left[\frac{F_B}{F_{BUn}}\right]=0 \tag{13.35}$$

$$\text{or } u=F_B/F_{BU} \text{ if } F_{BUn}=0$$

In the second form of the equation, the functional derivatives are denoted by subscripts and it is assumed that the coefficient of the normal derivative term is non-zero. If the normal derivative coefficient is in fact zero, then the boundary condition is that of a fixed value which can be easily solved for as in the last line of Eq. (13.35). In all cases, both of the functional derivatives can not be zero or a properly specified boundary condition does not exist.

With these conditions on the form of the boundary equation, the normal derivative terms in Eq. (13.32) can be replaced by that of Eq. (13.35) and the additional boundary equations that need to be evaluated are:

$$Eqb(i)=\sum_{L_i}\int_{L_i}\left[\left(\frac{F_{BU}}{F_{BUn}}\right)u+\left(\frac{F_B}{F_{BUn}}\right)\right](1-L/L_i)dL \tag{13.36}$$

This is readily evaluated by assuming that the potential varies linearly with distance along the boundary line between the boundary node for which the equation is being written and the two adjacent boundary nodes.

In order to implement this, one must know which adjacent triangle boundary nodes are associated with a given boundary node. This information can easily be gleamed from the sides file returned by the EasyMesh generation program. The format of this output file is: <side number:> <c> <d> <ea> <eb> <marker> where ,<c> and <d> are the nodes along a triangle side and <ea> and <eb> are the adjacent triangle elements. If either of these equals -1 then this identifies a non-existent triangle and thus identifies the side as a boundary line. From the other positive number, the triangle corresponding to the boundary side can be identified. Any triangle node on a boundary will have two sides associated with it as illustrated in Figure 13.8. The boundary is assumed to be to the left of the three triangle elements (T1,T2 and T3) shown in the figure. For some special cases, (such as the node at a right angle corner) the center triangle (T3) may not exist, but the other triangle elements will always exist for a boundary node. The normal derivatives are shown as directed outward from the spatial domain under consideration as is the normal convention.

Applying the relationships of Eq. (13.36) to the nodes in Figure 13.8 leads to:

$$u=u_1(1-L/L_1)+u_2(L/L_1) \text{ over } L_1 \ (0<L<L_1) \text{ and}$$
$$u=u_1(1-L/L_2)+u_3(L/L_2) \text{ over } L_2 \ (0<L<L_2) \tag{13.37}$$

The integral in Eq. (13.36) is then easily evaluated in terms of the solution values at the three boundary nodes under consideration. As seen in the figure, contributions to the equation for node 1 from the boundary conditions can possibly come from nodes 1, 2 and 3 in addition to the contributions from the internal nodes 4 and 5.

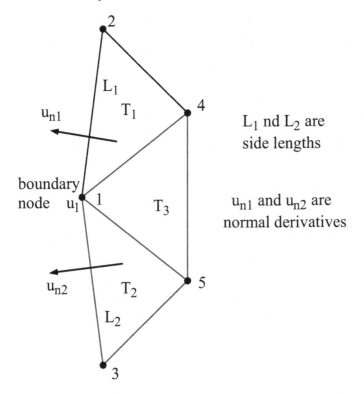

Figure 13.8. Illustration of boundary node and normal derivatives.

The procedure for formulating the boundary node equation set is then to sum over the interior triangles for all the boundary nodes in exactly the same manner as for the interior nodes. Then additional terms are added to the boundary node equations to implement the results of Eq. (13.36) if the boundary condition is that of a mixed type of if there is a non-zero normal derivative term. If the boundary condition is of the fixed value type, then the matrix equation diagonal element is simply set to the fixed value as given by the last line of Eq. (13.35).

There is however, one implementation detail that needs some further discussion. This is how to properly transition between two different normal boundary conditions along the set of boundary sides. The problem to be addressed can best be illustrated by some examples such as shown in Figure 13.9. Case (a) shows that of a transition from a region where the normal derivative (or mixed condition) is specified to a region where the solution value is specified. In this example the transition is shown at a sharp corner; however, it could just as easily occur along a relatively flat boundary region. A typical physical problem corresponding to this case would be a fixed solution value along the top and a zero normal derivative along the side. The specified boundary values could be functions of position along the boundary and not exactly fixed in value as indicated. The key point is a change in type of boundary condition at some transition point identified here as point k along the boundary. Case (b) corresponds to that of an abrupt transition between two different specified values of a normal or mixed boundary condition. This most commonly occurs for the case of a zero normal derivative on one boundary and that of a constant normal derivative (or constant material flux) along the other boundary. Again this could occur along a flat spatial boundary or at a sharp corner as indicated. The final case (c) is that of a change in mixed boundary condition but where a gradual transition in the condition is desired rather than an abrupt transition. While this case is not as likely to occur in physical problems, it is still of some interest and may occur in some problems.

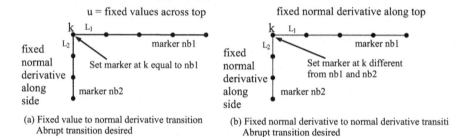

(a) Fixed value to normal derivative transition
Abrupt transition desired

(b) Fixed normal derivative to normal derivative transiti
Abrupt transition desired

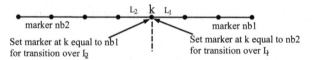

(c) Second case of fixed normal derivative to normal derivative transition
Gradual graded transition desired

Figure 13.9. Illustration of typical boundary value transition regions.

In all the cases it is assumed that the different boundary regions are identified by some integer boundary marker shown as nb1 and nb2 in the figure. This would be the boundary markers associated with an input file to EasyMesh as previously discussed and identified with boundary points and boundary segments in an input file to this program. The reader is referred back to Section 13.2 for a review of these markers. The question to be addressed here is how to designate the marker associated with the transition point so the desired transition in the specified boundary conditions can be uniquely identified and properly implemented. For case (a) the marking is relatively straightforward, any boundary node which does not involve a derivative term in specifying the boundary condition will simply be set in the matrix equation set to the appropriate value as in Eq. (13.35). Also the mixed boundary condition will be used in any boundary segment leading up to that node such as segment L_2 in Figure 13.9(a). An abrupt transition between the boundary specifications is easily implemented for this case.

The difficulty arises in distinguishing between cases (b) and (c) where one desires an abrupt change in the boundary equation or a gradual change in the boundary condition. This arises because any given boundary segment is associated with two boundary nodes, one to the right and one to the left of the segment. For case (b) it is desired to always use the boundary condition identified by nb1 for segment L_1 and boundary condition identified by nb2 for segment L_2. In case (c) however, it is desired to use the boundary condition identified by nb1 when dealing with the node to the right of a transition segment and boundary condition identified by nb2 when dealing with the node to the left of a transition segment. To properly accommodate these possibilities, the convention used here will be to use a unique boundary identifier for the transition node when an abrupt transition is desired as in case (b), i.e. the boundary marker associated with node k is taken to be different from either of the boundary markers on adjacent sides. This is like having an isolated boundary marker for one boundary node which is the transition node at which the mixed boundary condition abruptly changes. For case (c) no new boundary marker is used and the boundary region (either L_1 or L_2) will be a transition boundary region depending on whether the boundary marker for node k is set to nb1 or nb2.

The net result of this discussion is just basically one rule which is to use an isolated boundary marker for any boundary node where an abrupt transition is desired in specifying mixed boundary conditions. Otherwise the node numbering and marker numbering can follow any desired pattern. The extra marker is needed to uniquely identify the transition node. Otherwise with only two marker numbers, only a transition boundary interval can be identified. Examples of this will be given later and this will occur most frequently for abrupt corners where a flux of some physical quantity is specified on one boundary and a zero flux boundary condition is specified on the other adjoining boundary. Many physical PDEs have this type of boundary condition and this is probably more prevalent than that of the gradual transition of case (c). This convention of specifying boundary marker values will be taken into account when developing code for formulating the matrix equations. This convention is purely a convention adapted in this work and some other numbering convention could be selected.

13.7 The Complete Finite Element Method

The background has now been discussed for implementing a complete FE method for solving a general PDE. The implementation will of course be a subset of the general FE numerical method using selected features such as the use of triangular elements for a two dimensional spatial region. Also one of the simplest interpolation methods and numerical integration technique will be implemented. For a discussion of more general and complete FE approaches, the reader is referred to the extensive literature on this subject.

In keeping with the formulation of previous chapters, a series of code segments are desired so that the user can concentrate on simply composing a description of the differential equation to be solved and the boundary conditions. The user can then supply this information to a callable software function that sets up the appropriate matrix equations and solves the resulting set of matrix equations. This would be in addition to the use of a software package such as EasyMesh for generating a set of triangular elements for the spatial domain. Listing 13.2 shows a code segment for using such an approach for solving a PDE. This assumes that appropriate software functions have already been written and loaded with the "require'pde2fe'" statement on line 3 of the listing. Line 5 reads in the finite element information on the nodes, triangles and sides which is assumed to have been generated by EasyMesh and stored in list13_2.n, list13_2.e and list13_2.s files. Lines 11 through 17 define the differential equation to be solved and the boundary conditions as:

$$\frac{\partial^2 u}{\partial x^2} + \frac{\partial^2 u}{\partial y^2} + 2\pi^2 \sin(\pi x)\sin(\pi y) = 0$$

$$u(0, y) = u(1, y) = u(x, 0) = u(x, 1) = 0$$

(13.38)

This equation was also used in Chapter 12 as an example and has a known solution.

The PDE to be solved is defined on lines 11 through 14 by the feq() function and the zero boundary conditions are defined by the fb() function on lines 15 through 17. After setting up an array of zero initial solution values on line 19, the program calls the setup2feeqs() function on line 20 to formulate the matrix equations to be solved and the equations are solved by spgauss(), the sparse matrix solver introduced in Chapter 12, on line 23. The code on lines 25 through 28 convert the obtained solution variables to a set of solution values on a rectangular grid for subsequent plotting with splot() and cplot() on line 30. This approach to solving a PDE by the FE method is essentially identical to the approach developed in Chapter 12 for the FD method. The approach relies on two new functions readnts() and setup2feeqs() which need to be now discussed.

The function readnts() takes as input the output from the EasyMesh program and converts the file data into a form more appropriate for use by the setup2feeqs() program. One of the main functions of readnts() is to renumber the nodes, triangles and sides beginning with 1 instead of beginning with 0 as the EasyMesh program generates. The returned data is also stored as values in three tables (the nodes triangles and sides table on line 5). There is an additional transformation of the sides tabular data performed by the readnts() function. Data on the sides of the triangles is needed primarily for implementing appropriate boundary conditions on the PDE. Thus data is needed only for the triangle sides that constitute boundaries of the given spatial region. The readnts() function filters the sides data generated by EasyMesh and retains only the data involving triangle with sides on the boundary of the spatial region. In addition, the data is collected as a table indexed with the boundary node numbers as opposed to the sides labeling.

```
 1 : -- File list13_2.lua --
 2 :
 3 : require"spgauss"; require'pde2fe'
 4 :
 5 : nodes,triangles,sides = readnts('list13_2')
 6 :
 7 : nnds,nel = #nodes, #triangles
 8 : print('number nodes =',nnds)
 9 : print('number elements =',nel)
10 :
11 : pi = math.pi; pi2 = 2*pi^2; sin = math.sin
12 : function feq(x,y,uxx,ux,u,uy,uyy,ntr,mtn)
13 :    return uxx + uyy  + pi2*(sin(pi*x)*sin(pi*y))
14 : end
15 : function fb(nd,u,un,nbs)
16 :    return u
17 : end
18 :
19 : u = {}; for i=1,nnds do u[i] = 0.0 end
20 : a,b = setup2feeqs({feq,fb},{nodes,triangles,sides},u)
21 :
22 : getfenv(spgauss).nprint = 1
23 : spgauss(a,b)
24 :
25 : x,y = {},{}; NT = 21
26 : for j=1,NT do x[j] = (j-1)/(NT-1); y[j] = x[j] end
27 : sol,solxy = toxysol(b,x,y,nodes,triangles)
28 :
29 : write_data('list13_2a.dat',sol);write_data('list13_2b.dat',
         solxy)
30 : splot(solxy); cplot(solxy)
Selected Output:
number nodes = 781
number elements =      1460
```

Listing 13.2. Example of code segment for implementing FE solution method for
PDE.

The sides table returned by readnts() has two entries sides[i] = {side1, side2},
where side1 and side2 are tables describing the two boundary sides at boundary
node i in the form side1 = {c, d, ea, eb} where c and d are the starting and ending
nodes for the side and ea and eb are the numbers of the triangular elements on the
left and right sides of a boundary line. Since only the boundary sides are included,
either ea or eb will be -1 indicating a boundary side. The reader can view the code
for readnts() in the pde2fe.lua file for more details of the transformation.

The heart of applying the FE method in Listing 13.2 is the setup2feeqs() function on line 20 that sets up the matrix equations using the FE equations developed
in the preceding sections of this chapter. Computer code for this function is
shown in Listing 13.3 and will be briefly discussed. Lines 8 through 21 define local variables and constants to be used in the code with lines 16 through 18 defining the natural coordinates and weighting factors for the four point integration
technique defined in Table 13.1. Line 22 calls an external function calcsh() that
calculates the shape functions for all the triangle elements and returns the total

```
 1 : -- File pde2fe.lua -- Functions for 2D finite element analysis
 2 :
 3 : require"gauss"; require"spgauss"; require'pde2bv'
 4 : local NHF=13; local calcsh, deriv1tri, deriv2tri, ltoxy
 5 : local usave,utsave
 6 : setup2feeqs = function(eqs,nts,u) -- Set up 2D equations for FE
 7 :     -- u is of form u[k], nts = {nodes[], elements[], sides[]}
 8 :     local u,bxext,byext = u or {}, 0.0, 0.0
 9 :     local uxx,ux,uy,uyy,uxa,uya,ut
10 :     local nds,tri,sds = nts[1], nts[2], nts[3]
11 :     local nnds,nel,nsds = #nds, #tri, #sds
12 :     local fctuxx,fctux,fctu,fctuy,fctuyy = 1.0,FACT,FACT,FACT,1.0
13 :     local a, b = Spmat.new(nnds,-nnds), Spmat.new(nnds,-nnds)
14 :     local fv,fx,fu,fy,xi,yi,ui,xn,yn,mk = {},{},{},{},{},{},{},
            {},{},{}
15 :     local fxx,fyy,mxx,vx,vy,he,trik,n1,n2,n3,m1
16 :     local lipts = {{1/3,1/3,1/3},{.6,.2,.2},{.2,.6,.2},
            {.2,.2,.6}} --4 pts
17 :     local lkpts = {{1/3,.6,.2,.2},{1/3,.2,.6,.2},{1/3,.2,.2,.6}}
18 :     local fipts = {-27/48,25/48,25/48,25/48} --weighting Factors
19 :     if #u~=nnds then -- Make sure u is a proper table
20 :         local uxl = {}; for k=1,nnds do uxl[k] = 0.0 end; u = uxl
21 :     end
22 :     mxx = calcsh(tri,nds); vx = sqrt(1/mxx)
23 :     mxx,vy = 0.0, vx -- Find maximum solution value
24 :     for k=1,nnds do mxx = max(mxx, abs(u[k])) end
25 :     if mxx~=0.0 then -- Scale probe factors to maximum value
26 :         fctu = FACT*mxx
27 :         fctux,fctuy = fctu*vx, fctu*vy
28 :     end
29 :     uxa,uya = deriv1tri(tri,nds,u) -- Get first derivative arrays
30 :     eq,eqb = eqs[1], eqs[2] -- Differential, boundary equations
31 :     for k=1,nnds do a[k],b[k] = {}, 0 end
32 :     for k=1,nel do --Loop over triangles
33 :         trik = tri[k] -- one by one
34 :         n1,n2,n3,m1 = trik[1], trik[2], trik[3], trik[12]
35 :         ui[1],ui[2],ui[3] = u[n1], u[n2], u[n3] -- Three nodes
36 :         xi[1],xi[2],xi[3] = nds[n1][1], nds[n2][1], nds[n3][1]
37 :         yi[1],yi[2],yi[3] = nds[n1][2], nds[n2][2], nds[n3][2]
38 :         ns = {n1,n2,n3}
39 :         ux,uy,uxx,uyy = uxa[k],uya[k],0.0,0.0 -- Local derivatives
40 :         he = trik[NHF] -- shape factors for triangle
41 :         area,fxx,fyy = he[0], 0.0, 0.0 -- area factor
42 :         bxext,byext = 0.0,0.0
43 :         if truefd then -- extra first derivative terms?
44 :             for j=1,3 do -- loop over three triangle nodes
45 :                 xt,yt,ut = xi[j],yi[j],ui[j]
46 :                 lpts = {0,0,0}; lpts[j] = 1
47 :                 fvv = eq(xt,yt,uxx,ux,ut,uy,uyy,k,m1,ns,lpts)
48 :                 bxext=bxext-he[j][2]*(eq(xt,yt,uxx+fctuxx,ux,ut,
                        uy,uyy,k,m1,ns,lpts)-fvv)/fctuxx
49 :                 byext=byext-he[j][3]*(eq(xt,yt,uxx,ux,ut,uy,
                        uyy+fctuyy,k,m1,ns,lpts)-fvv)/fctuyy
50 :             end -- Now have da/dx and da/dy over triangle
51 :         end
52 :         for j=1,4 do -- loop over 4 integration points
53 :             lpts,fwt = lipts[j], fipts[j]; l1,l2,l3 =
                    lpts[1],lpts[2],lpts[3]
54 :             ut = ui[1]*l1 + ui[2]*l2 + ui[3]*l3 -- at intg points
```

```
 55 :          xt,yt = xi[1]*l1+xi[2]*l2+xi[3]*l3,
                   yi[1]*l1+yi[2]*l2+yi[3]*l3
 56 :          fvv = eq(xt,yt,uxx,ux,ut,uy,uyy,k,m1,ns,lpts) -- PDs
 57 :          fxx = fxx - fwt*(eq(xt,yt,uxx+fctuxx,ux,ut,
                   uy,uyy,k,m1,ns,lpts)-fvv)/fctuxx -- Sum ax
 58 :          fx[j] = fwt*((eq(xt,yt,uxx,ux+fctux,ut,uy,uyy,
                   k,m1,ns,lpts)-fvv)/fctux+bxext) -- bx terms
 59 :          fu[j] = fwt*(eq(xt,yt,uxx,ux,ut+fctu,uy,uyy,
                   k,m1,ns,lpts)-fvv)/fctu -- c terms
 60 :          fy[j] = fwt*((eq(xt,yt,uxx,ux,ut,uy+fctuy,uyy,
                   k,m1,ns,lpts)-fvv)/fctuy+byext) -- by terns
 61 :          fyy = fyy - fwt*(eq(xt,yt,uxx,ux,ut,uy,uyy+fctuyy,
                   k,m1,ns,lpts)-fvv)/fctuyy -- Sum ay terms
 62 :          fv[j] = fwt*(fvv + bxext*ux + byext*uy) -- Fo terms
 63 :      end
 64 :      fxx,fyy = fxx*area, fyy*area -- common area weighting
 65 :      for i=1,3 do -- loop over triangle nodes
 66 :          nnd,lk = trik[i], lkpts[i] -- primary node number
 67 :          fb,fbx,fby,fc = 0.0, 0.0, 0.0, 0.0 -- b,ux uy factors
 68 :          for j=1,4 do -- Step over itegration points
 69 :              fb = fb + lk[j]*fv[j] -- b matrix weighting
 70 :              fbx,fby = fbx + lk[j]*fx[j], fby + lk[j]*fy[j]
 71 :          end
 72 :          fbx,fby = fbx*area, fby*area -- common area weithting
 73 :          arow = a[nnd] -- Row of a matrix for inserting
 74 :          hx,hy = he[i][2], he[i][3] -- h factor derivatives
 75 :          b[nnd] = b[nnd] - fb*area - ux*hx*fxx - uy*hy*fyy
 76 :          for j=1,3 do -- step over 3 shape functions
 77 :              nc, hji,fc ,lj= trik[j], he[j], 0.0, lkpts[j]
 78 :              lx,ly = hji[2], hji[3] -- deriv of shape functions
 79 :              fa = hx*lx*fxx + hy*ly*fyy -- uxx and uyy factors
 80 :              if fa~=0.0 then arow[nc] = (arow[nc] or 0) + fa end
 81 :              if fbx~=0.0 then arow[nc] = (arow[nc] or 0) +
                       fbx*lx end
 82 :              if fby~=0.0 then arow[nc] = (arow[nc] or 0) +
                       fby*ly end
 83 :              for k=1,4 do fc = fc+fu[k]*lk[k]*lj[k] end --sum u
 84 :              if fc~=0.0 then arow[nc] = (arow[nc] or 0) +
                       fc*area end
 85 :          end
 86 :      end
 87 :  end
 88 :  for k,sda in pairs(sds) do -- Loop over boundary nodes
 89 :      arow,ut,ux = a[k], u[k], 0.0
 90 :      n1,n2,n3 = k,sda[1][2],sda[2][2]
 91 :      m1,mk[1],mk[2] = nds[k][3],nds[n2][3],nds[n3][3]--markers
 92 :      for j=1,2 do -- loop over two sides
 93 :          s1 = sda[j]; n1 = s1[2]
 94 :          if sda[1][3]<0 then nsgn = -1 else nsgn = 1 end
 95 :          xt,yt = nds[k][1] - nds[n1][1], nds[k][2] - nds[n1][2]
 96 :          fu[j] = sqrt(xt^2 + yt^2)
 97 :          xn[j],yn[j] = -yt*nsgn, xt*nsgn -- sides * length
 98 :      end
 99 :      lt = fu[1] + fu[2]
100 :      for j=1,2 do
101 :          s1 = sda[j]; if s1[3]<0 then ntri = s1[4] else
                   ntri = s1[3] end
102 :          ux = ux + (uxa[ntri]*xn[j]+uya[ntri]*yn[j])/lt --
103 :      end
```

```
104 :        fvv = eqb(k,ut,ux,m1,k)
105 :        fb  = (eqb(k,ut,ux+fctux,m1,k)-fvv)/fctux -- normal deriv
106 :        if fb==0.0 then -- No derivative term present in BV
107 :           arow = {}; a[k] = arow
108 :           arow[k]=(eqb(k,ut+fctu,ux,m1,k)-fvv)/fctu; b[k]=-fvv
109 :        else -- Mixed or normal boundary condition
110 :           if m1==mk[1] then mk[2] = m1 -- Test for equal markers
111 :           elseif m1==mk[2] then mk[1] = m1 end -- isolated node
112 :           for j=1,2 do -- Now get mixed boundary conditions
113 :              s1 = sda[j]; n1 = s1[2]
114 :              if s1[3]<0 then ntri = s1[4] else ntri = s1[3] end
115 :              ux = (uxa[ntri]*xn[j]+uya[ntri]*yn[j])/fu[j]
116 :              fvv =  eqb(k,ut,ux,mk[j],k)
117 :              fyy = (eqb(k,ut+fctu,ux,mk[j],k)-fvv)/fctu
118 :              fxx = (eqb(k,ut,ux+fctux,mk[j],k)-fvv)/fctux
119 :              if fyy~=0.0 then
120 :                 arow[k],arow[n1]=arow[k]-fyy*fu[j]/(fxx*3),
                         arow[n1]-fyy*fu[j]/(fxx*6)
121 :              end
122 :              b[k]  = b[k] + fvv*fu[j]/(fxx*2) - ux*fu[j]/2 +
                      (fyy*u[k]*fu[j]/(fxx*3)+yy*u[n1]*fu[j]/(fxx*6))
123 :           end
124 :        end
125 :     end
126 :     return a,b
127 : end
128 : setfenv(setup2feeqs,{FACT=1.e-4,NHF=NHF,table=table,
         Spmat=Spmat,sqrt=math.sqrt, max=math.max,abs=math.abs,
         calcsh=calcsh,deriv1tri=deriv1tri,deriv2tri=deriv2tri,
         he=he, pairs=pairs,truefd=false,print=print,table=table,
         whatis=whatis})
```

Listing 13.3. Code for setting up matrix equations for the FE method.

area of the spatial region occupied by the triangle set of elements. It is not clear from the function call where the shape function information is stored. However, data for the shape functions are stored as an entry in the table defining the triangles (the triangles array or the tri[] table in the listing). On entry to the calcsh() routine, the triangles array has 12 entries for each triangle as defined for the element file in Section 13.2 and as supplied as input to the program to the readnts() function. The shape functions are stored as an additional table in entry number 13 of the triangles file (NHF = 13 on line 4 defines this table entry number). How this data is stored and used will subsequently be discussed.

Lines 23 through 28 find the maximum value of the passed solution variable and then scale various probe factors, such as fctu, fctux and fctuy on lines 26 and 27, for the numerical evaluation of the functional derivatives of the defined PDE according to the maximum solution value. On line 29 an external function deriv1tre() is called to obtain values of the first derivatives of the input solution values. All of this is not necessary for a linear PDE but is included so that the code can also be used in a Newton loop for a nonlinear PDE.

The heart of the routine is the loop over the triangle regions from line 32 to line 88. The k'th triangle is selected on line 33 for processing and the x and y coordinates of the triangle nodes are evaluated on lines 36 and 37. The shape functions are selected on line 40 as the NHF'th entry in the k'th triangle table. The area of the triangle element is stored in the 0'th index for the shape function and recovered on line 41. The values at index 1, 2 and 3 for the shape function table are the constant, x dependent and y dependent coefficients of the shape function. For the moment skip over lines 43 through 51 of the code as this is not essential to understanding the function. A high level overview of the code is as follows. After selecting a particular triangle on line 33, the node numbers (line 34), solution values at the nodes (line 35) and coordinates of the nodes (lines 36 and 37) are evaluated. The for loop of lines 52 through 63 then step over the 4 internal points needed to apply the cubic Gauss integration approach of Table 13.1. Data is collected on the ax, ay, bx and by terms at the internal points – see the fxx, fx, fu, fy and fyy terms. Numerical derivatives are used within this loop to probe the PDE for the coefficients multiplying the second derivatives (lines 57 and 61), the first derivatives (lines 58 and 60) and the solution variable (line 59). For a nonlinear PDE this step provides the Newton linearization of the PDE. This loop collects data on the function values as needed for the subsequent evaluation of the integrals in Eq. (13.28). The code loop from line 65 through 87 then steps over the three nodes of the selected triangle setting up the contributions to the lines of the matrix equations associated with the three triangle nodes – equations number n1, n2 and n3 in the code notation. The appropriate row of the matrix under consideration within this loop is selected on line 73. For each node equation the selected triangle can make contributions to three columns of the coefficient matrix – the n1, n2 and n3 columns. The code loop from line 76 through 85 steps over these three columns with the appropriate shape factors and local coordinates associated with each column node. The contributions to the matrix elements are defined and set on lines 80 through 84 after appropriate weighting of the internal points according to Eq. (13.28). At the end of line 87and the loop over the triangles, the matrix elements for all the interior nodes have been evaluated. In addition part of the matrix elements for the boundary nodes have been evaluated.

As previously stated, the "sides" table is assumed to contain only data for elements associated with the boundary nodes and the loop from line 88 to 125 steps over these non-nil elements treating each boundary node in turn. The loop from 92 to 98 steps over the two boundary sides around a boundary node evaluating the length of the side and the normal vector to the boundary on line 97. A weighted value of the normal derivative of the solution along the boundary is evaluated on line 102. Lines 104 and 105 use numerical derivatives to probe the boundary condition function for the presence of a derivative term. If no boundary derivative term is found, then lines 107 through 108 complete the matrix definitions. Else the code from line 109 to 124 sets up the matrix equations implementing Eqs. (13.36) and (13.37) with a loop over the two adjacent sides from line 112 to line 123. The tests on lines 110 and 111 are used to determine if a boundary node has an isolated marker number indicating the use of an abrupt transition in mixed derivative values. The reader can observe the details of the equations and follow the implementation. Finally, the a and b matrix equations are returned by the setup2feeqs() code on line 126.

   Some discussion is needed about the calling arguments used to access user supplied functions describing the PDE to be solved, such as the call to eq() (the user supplied differential equation) on line 47 of Listing 13.3. The calling arguments are listed as: (xt,yt,uxx,ux,ut,uy, uyy,k,ml,xn,lpts). The arguments xt through uyy should be clear as these are the x and y locations to be evaluated and the function value (ut) plus the first and second derivative terms. The remaining terms are: k – triangle number, ml – material marker, xn – array of node numbers for the triangle and lpts – array of local coordinates for the spatial point being evaluated. For many PDE definitions, the equation can be clearly specified without any knowledge of these last four argument values and in fact these provide somewhat redundant information about the spatial point being probed. In principle, one should be able to define the PDE purely from the spatial coordinates and the function value and derivatives. However, the material marker information is especially useful for problems with abrupt changes in material properties at different spatial points. Finally the node numbers of the triangular region being probed and the local coordinates can be especially useful when one has a PDE where some function in the defining equation is known only at the spatial nodes of the triangles. Examples will be subsequently given where such information is needed to properly set up the PDE and boundary conditions. If this additional information is not needed for a particular PDE, the additional calling arguments can simply be omitted from the user defined equation for the PDE. Examples of this will also be given.

   This code segment as well as that in the previous chapter for setting up matrix equations for a PDE is one of the most involved in this work. However, it is relatively straightforward after one has mastered the required bookkeeping required to implement the equations on a triangle by triangle element basis. One detail of the code in Listing 13.3 is left unfinished and that is the "if" loop from line 43 through 51 described as an "extra first derivative term". To understand this loop, consider a common PDE of the form:

$$\frac{\partial}{\partial x}\left(a_x \frac{\partial U}{\partial x}\right) + \frac{\partial}{\partial y}\left(a_y \frac{\partial U}{\partial y}\right) + G(x,y,U_x,U_y,U) = 0 \tag{13.39}$$

Such an equation occurs very frequently in the physical world where the source of the equation involves the divergence of some vector flow problem. In the most general form the a_x and a_y terms may be functions of position and if this is expanded the equation becomes:

$$a_x \frac{\partial^2 U}{\partial x^2} + a_y \frac{\partial^2 U}{\partial y^2} + \left[\frac{\partial a_x}{\partial x}\frac{\partial U}{\partial x} + \frac{\partial a_y}{\partial y}\frac{\partial U}{\partial y}\right] + G(x,y,U_x,U_y,U) = 0 \tag{13.40}$$

where the term in square brackets represents extra first derivative terms in the equation. These will only be present if the a_x and a_y terms are functions of position. It should be noted that this assumes that the a_x and a_y terms are functions of position only and not functions of the solution variable. This will be a limitation of the code developed for solving this equation by the FE technique. If this form of the PDE is used and one goes back through the normal weighted residuals method and then perform integration by parts on the second derivative terms, it will be found that a set of negative first derivative terms is generated that exactly cancel the square bracket terms in this equations. This can also be verified by performing the weighted residual method on Eq. (13.39) and performing integration by parts on the second derivative terms.

Thus if one has an equation of the form of Eq. (13.39) converting to the exact form of Eq. (13.40) is not necessary provided one simply ignores any spatial dependency of the a_x and a_y coefficients in setting up the matrix equations. The setup2feeqs() code is programmed for this case and this is the purpose of the "if" loop from line 43 through 51. The variable truefd is set to a default value of false so the extra first derivative terms are not included in setting up the equations. Thus a standard procedure for coding a PDE of the form of Eq. (13.39) will be as follows:

$$\frac{\partial}{\partial x}\left(a_x(x,y)\frac{\partial U}{\partial x}\right) + \frac{\partial}{\partial y}\left(a_y(x,y)\frac{\partial U}{\partial y}\right) + G(x,y,U_x,U_y,U) = 0$$

Convert to: (13.41)

$$a_x(x,y)\frac{\partial^2 U}{\partial x^2} + a_y(x,y)\frac{\partial^2 U}{\partial y^2} + G(x,y,U_x,U_y,U) = 0$$

and neglect any spatial dependence of the coefficients. The code then generates correct equations, although the second form of the equation is not strictly mathematically correct! One may ask why go to this procedure? Why not just use the original formulation? However, in the original form of Eq. (13.39) the value of the a_x and a_y coefficients can not be evaluated by using numerical derivatives in any straightforward manner. The purpose of writing the equation in the form of the second line of Eq. (13.41) and coding in this manner is to allow the software to easily evaluate the a_x and a_y coefficients by use of numerical derivatives. An equally valid alternative approach is to include the additional first derivative terms in the defining PDE equations as in Eq. (13.40) and then to use a code statement getfenv(setup2feeqs).truefd = true before calling the setup2feeqs() function. The extra work of adding the first derivative terms will then be canceled by the added loop in the code. This is the purpose of the if loop from line 43 through 51.

It should be noted that the code developed for solving a PDE by the FE method as in Listing 13.3 has at least one very important limitation on the form of any possible nonlinearity in the PDE. The form of the a_x and a_y coefficients is limited to functions of only the spatial variables as indicated in Eq. (13.41). The code will not correctly formulate a PDE which has a nonlinear form of second derivatives, i.e. the second derivative coefficients can not depend on the solution variable or derivatives of the solution variable. The $G()$ function which represents functions of the solution variable and first derivative terms can have any nonlinear form in the defining PDE. This restriction comes about from the manner in which the second derivative term is converted to a first derivative term by use of the integration by parts algorithm. An example of how to properly reformulate a PDE with a nonlinear second derivative term will be subsequently illustrated.

Now that the matrix generation code has been discussed, the application in Listing 13.1 can be further discussed as an example of the FE method. The mesh for this problem was generated by the EasyMesh program with the input file shown in Listing 13.4. This is a simple square area of size 1 by 1 unit with a grid size of 0.04 specified on all four boundaries. The boundary sides are given labels of 1,2,3 and 4 by the last four lines in the file. These labels will come through in the output sides file as identifying labels for the sides that can be used to generate appropriate boundary conditions along the four sides and these numbers will be passed to the boundary function in the nbs parameter in the fb() function in Listing 13.2. For this particular example, the boundary condition is simply zero along all four boundaries and these side labeling values are not required. This is seen in Listing 13.2 where the nbs parameter of the fb() function is not used in the function.

```
# EasyMesh data for rectangular area #
# Generates 781 nodes and 1460 triangles #

4 # Number of nodes #

   0:    0      0      .04    1
   1:    1.0    0      .04    1
   2:    1.0    1.0    .04    3
   3:    0.0    1.0    .04    3

4 # Number of side segments #

   0:    0      1      1
   1:    1      2      2
   2:    2      3      3
   3:    3      0      4
```

Listing 13.4. Input data file to EasyMesh for example in Listing 13.2. This data is in file list13_2.d.

The discretization of the spatial domain generated by EasyMesh with Listing 13.4 as input is shown in Figure 13.10. The generated triangular mesh is almost a uniform grid over the spatial area with only a few areas deviating from the ideal of six triangles around each node point. The generated triangle set produces 781 nodes over the 1460 triangles with 2240 unique sides for the triangles. Note that the number of triangles is slightly less than twice the number of node points. This will always be the case with a triangular mesh as the 2X factor applies to an infinitely large array of points. The reader can peruse the output files from the EasyMesh program which have the names list13_2.n, list13_2.e and list13_2.s. These constitute the input files used by Listing 13.2 in the readnts() function call. Note that readnts() assumes that the standard file endings of .n, .e and .s are present.

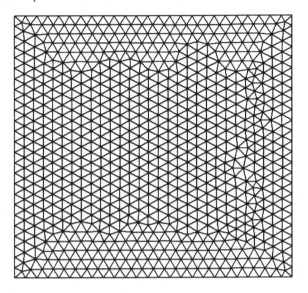

Figure 13.10 Mesh generated for square area by EasyMesh from Listing 13.4. The resulting area has 781 nodes and 1460 trianglular elements.

The execution of the code in Listing 13.2 with this spatial mesh generates a set of solution values at each of the triangle nodes. For graphing purposes, it is convenient to have a solution set corresponding to a uniform rectangular grid over the spatial domain. Thus the function toxysol(b,x,y,nodes,triangles) shown on line 27 of Listing 13.2 has been written. This takes as input a solution set b over a set of nodes and triangles and returns a solution set over an input x,y array of spatial points. This is an interesting problem of function interpolation over a spatial domain where the known solutions are at the nodes of the triangular spatial elements. Within each triangular element the solution is assumed to vary linearly with position (both x and y). However, for a given x and y position, it is not readily known within which triangle the point resides. Further, the numbering of the spatial triangles does not necessarily follow any definite pattern. Thus the problem of interpolating to a given x and y position comes down fundamentally to the problem of determining within which spatial triangle a point with coordinates x,y resides.

In terms of the natural coordinates for any given triangle, a general spatial point can be represented as in Eq. (13.8)

$$x = x_3 + (x_1 - x_3)\ell_1 + (x_2 - x_3)\ell_2$$
$$y = y_3 + (y_1 - y_3)\ell_1 + (y_2 - y_3)\ell_2$$

(13.42)

While this equation was originally developed to apply to spatial points within a given triangle, this can be applied to any general spatial point using the natural coordinates of any given triangle as a reference. These equations can be inverted to give the natural coordinates in terms of the triangle coordinates and some general spatial point as:

$$\ell_1 = \frac{(x-x_3)(y_2-y_3)+(y-y_3)(x_3-x_2)}{2\Delta}$$

$$\ell_2 = \frac{(x-x_3)(y_3-y_1)+(y-y_3)(x_1-x_3)}{2\Delta}$$

(13.43)

If the spatial point x,y resides within the given triangle then the following relationships hold:

$$0 \leq \ell_1 \leq 1$$

$$0 \leq \ell_2 \leq 1$$

$$\ell_1 + \ell_2 \leq 1$$

(13.44)

If these relationships are not satisfied, the spatial point lies outside the given triangle spatial domain. Thus a method for locating which triangle corresponds to a given spatial point is to step over all the triangles, applying Eq. (13.43) and checking the relationships of Eq. (13.44). As soon as these inequalities are satisfied the desired triangle has been found and the solution at the x,y spatial point can then be determined from the solution within the located triangle domain. This algorithm is implemented in the toxysol() function and the reader can inspect the code if desired. An enhancement to the algorithm is to save the number for the last found triangle and to begin the search for a new spatial point from that known triangle and step backwards and forward in the triangle numbering to find the triangle for a new set of spatial coordinates. For typical examples, this reduces considerably the time required to interpolate a set of spatial coordinates, since adjacent triangles tend to have locally close numbering. If all triangles are probed and the relationships of Eq. (13.44) are not satisfied within any of the triangles, then the x,y spatial point does not reside within the spatial domain of the solution. For such a case, the toxysol() function does not return any table entry for the spatial point. The toxysol() function actually returns two major arrays of solution values as indicated by the sol and solxy variables in Listing 13.2. The first array returns the solution set in the form of three columns of x,y,sol values with the x values first changing most rapidly. This is followed by a repeat set of values with the y values changing most rapidly. This is done so typical plotting programs can draw surface solution lines along both axes. The second array of solution values returned to solxy in Listing 13.2 is a two dimensional array of solution values with just the solution values and no data on the corresponding x and y coordinate values. This type of file is very useful in producing popup surface or contour plots as indicated by the last line in Listing 13.2. The reader should examine a typical output file to see the exact format of the generated data

The solution obtained from execution the code in Listing 13.2 is shown in Figure 13.11. This can be compared with Figure 12.61 which shows the solution for the same PDE obtained by the FD method. While the two solutions appear to be the same, one needs a more in-depth look at the accuracy of the solution to access the two methods for this particular problem. One of the useful aspects of this simple case is that an exact solution is known and an evaluation of the accuracy of the

FE method can be studied. This will be addressed in the next section. However, before that it is useful to embed the FE method into a higher level callable function that sets up the FE equations and solves the set of matrix equations.

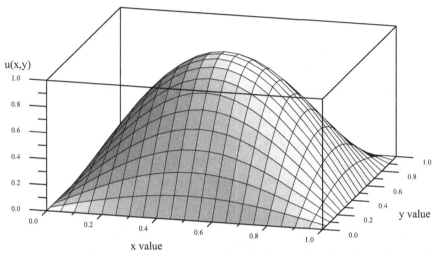

Figure 13.11. Solution for PDE in Listing 13.2 obtained by the FE method with the spatial grid in Figure 13.10.

Listing 13.5 shows a code segment for implementing such a higher level function named pde2fe(). This is similar to the pde2fd() function for the FD method in Chapter 12 and the code should be relatively straightforward to follow. Checks are made in the beginning (lines 133 through 140) to ensure appropriate input parameters. The main loop is from line 141 to 160 that implements a Newton's method for possible nonlinear equations. A single loop can be imposed for linear equations by the software statement getfenv(pde2fe).linear = 1 before using the pde2fe() function. The matrix equations are set up on line 142 by calling setup2feeqs() and solved by one of three possible methods. The most straightforward method is the direct matrix solution by spgauss() on line 144 which is the default method for the function. However, since the FE approach produces a set of matrix equations that are diagonally dominant, approximate solution methods can be used such as the COE and the SOR methods discussed in detail in Chapter 12. These can be used by setting the last calling argument (tpsola) in the argument list to the pde2fe() function to either 2 or 3. The remainder of the code (lines 154 through 159) checks for convergence of the Newton cycles and terminate the iterative loop when appropriate. In addition the code can print data on the progress of the solution if the nprint variable is set to a nonzero value.

```
  1 : -- File pde2fe.lua -- Functions for 2D finite element analysis
-----
132 : pde2fe = function(feqs,nts,u,tpsola) - 2D  PDE Solver
133 :   u = u or {}
134 :   if type(tpsola)~='table' then tpsola = {tpsola} end
135 :   local tpsol,rx,ry = tpsola[1] or SPM,tpsola[2],tpsola[3]
136 :   local umx,errm,krrm,a,b,uxy = 0.0
137 :   local uold,ua,n,itmx = {}, {}, #nts[1], 0
138 :   if #u~=n then for k=1,n do u[k] = u[k] or 0.0 end end141 :
139 :   if linear==1 then nnmx=1 else nnmx=NMX end -- Once for linear
140 :   for k=1,n do uold[k],umx,ua[k] = u[k], max(umx,u[k]), 0.0 end
141 :   for int=1,nnmx do
142 :       a,b = setup2feeqs(feqs,nts,uold)
143 :       if tpsol==SPM then -- Solve with sparse matrix solver
144 :           spgauss(a,b); ua = b -- b is new solution
145 :       elseif tpsol==COE then -- Solve with Chebychev SOR
146 :           if rx==nil then rx = cos(pi/sqrt(n)) end
147 :           ua = pde2bvcoe(a,b,ua,rx,umx)
148 :       elseif tpsol==SOR then -- Solve with SOR
149 :           if rx==nil then rx = cos(pi/sqrt(n)) end
150 :           ua = pde2bvsor(a,b,ua,rx,umx)
151 :       else
152 :           print('Unknown solution method specified in pde2fe.');
                    break
153 :       end
154 :       errm,krrm,umx,itmx = 0.0,0,0.0,itmx+1
155 :       for k=1,n do errm,uold[k] = max(errm,abs(ua[k])),
                    uold[k]+ua[k] end
156 :       for k=1,n do umx,ua[k] = max(umx,abs(uold[k])), 0.0 end
157 :       if nprint~=0 then print('Completed Newton iteration',int,
                    'with correction',errm)
158 :           io.flush() end
159 :       if abs(errm)<ERR*umx then itmx = int; break end
160 :   end
161 :   if itmx==NMX then print('Maximum number of iterations
            exceeded in pde2fe!!')
162 :   io.flush() end
163 :   return uold,errm,itmx
164 : end
165 : setfenv(pde2fe,{spgauss=spgauss,max=math.max,cos=math.cos,
          pi=math.pi,sqrt=math.sqrt, abs=math.abs,table=table,
          setup2feeqs=setup2feeqs,pde2bvcoe=pde2bvcoe,pde2bvsor=
          pde2bvsor,linear=0,SPM=1,COE=2,SOR=3,NMX=20,ERR=1.e-4,
          print=print,type=type,nprint=0,io=io,Spmat=Spmat,})
```

Listing 13.5 Code segment for implementing FE method with different possible matrix solution methods.

This function provides a simpler interface to the FE method and is functionally equivalent to the pde2fd() function implementing the FD method. The calling argument list is a table of functions describing the PDE and boundary conditions (feqs on line 132), a table of node/triangle/sides values (nts on line 132), an initial guess at the solution (u on line 132) and a parameter determining the type of solution method to be used (tpsola on line 132). The major difference in the calling arguments is the use of a table of nodes/triangle/sides values in the pde2fe() function in place of the x,y coordinate arrays passed to the pde2fd() function. The user is encouraged to solve similar problems by both approaches and compare the results. Of course this can only be done for problems with a rectangular geometry as this is all that can be handled by the FD method. Such problems will not be emphasized in this chapter as they do not make use of the major advantage of the FE method which is the ability to easily handle complex boundaries.

```lua
1 : -- File list13_6.lua --
 2 :
 3 : require'pde2fe'
 4 :
 5 : nts = {readnts('list13_2')}
 6 :
 7 : pi = math.pi; pi2 = 2*pi^2; sin = math.sin
 8 : feqs = {
 9 :    function(x,y,uxx,ux,u,uy,uyy,ntr,mtn)
10 :        return uxx + uyy  + pi2*(sin(pi*x)*sin(pi*y))
11 :    end,
12 :    function(nd,u,un,nbs)
13 :        return u
14 :    end
15 : }
16 :
17 : getfenv(spgauss).nprint = 1; getfenv(pde2fe).nprint=1
18 : getfenv(pde2fe).linear = 1
19 :
20 : u = pde2fe(feqs,nts)
21 : --u = pde2fe(feqs,nts,_,3) -- Try approximate methods
22 :
23 : x,y = {},{}; NT = 41
24 : for j=1,NT do x[j] = (j-1)/(NT-1); y[j] = x[j] end
25 : sol,solxy = toxysol(u,x,y,nts)
26 :
27 : write_data('list13_6a.dat',sol);
             write_data('list13_6b.dat',solxy)
28 : splot(solxy); cplot(solxy)
```

Listing 13.6. Example code segment for solving PDE with the pde2fe() function.

An example of the use of the pde2fe() function is shown in Listing 13.6. This is a repeat of the same problem as Listing 13.2 but with this higher level function. The call to the FE solving function is on line 20 and is simply pde2fe(feqs,nts) where feqs and nts are appropriate tables for the defining PDE functions and the spatial grid tables. Note that no initial solution value is passed to the function as a default table of zero values will be setup by the pde2fe() function. As can be seen this provides a simple interface to the FE method. The reader can execute the code in this listing and observe the same output as previously obtained. The reader is also encouraged to explore the use of the two approximate matrix solution methods with this example as illustrated by the commented statement on line 21. For a small number of nodes such as in this example (781 nodes) there is little advantage of the approximate methods over the direct matrix solution. However, for large numbers of nodes, the approximate methods are frequently considerably faster in terms of computer solution times.

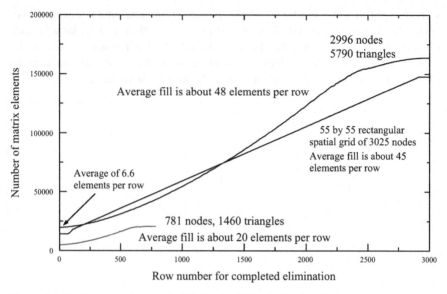

Figure 13.12. Data on the matrix fill for the FE solution in Listing 13.6.

It is interesting to compare the direct sparse matrix solution technique for the FE method with that for an equivalent FD method. By setting the "usage" parameter is spgauss() information on the matrix fill can be collected as the matrix is solved and passed back as a table from calling spgauss(). The reader can refer back to Listing 12.21 for an example and the discussion in connection with Figure 12.53. Figure 13.12 shows data on the matrix fill for the FE problem in Listing 13.2. Actually two sets of data are shown, one for the 781 nodes of Listing 13.2 and one for an increased number of nodes of 2996 and 5790 triangles. To generate the 2996 spatial grid, the EasyMesy program was executed with the input file shown in Listing13.4 but with the 0.04 entries changed to 0.02 in order to generate approximately 4 times as many grid points. At completion of the matrix reduction technique the average matrix fill is about 51 elements per row for the 2996 FE nodes and about 20 elements per row for the 781 FE nodes. For comparison the corresponding matrix fill curve for the FD method using a 54 by 54 grid of spatial x and y intervals, corresponds to 3025 nodes was generated by re-executing the code in Listing 12.21 with this rectangular spatial grid. Several features of the data and the comparison between the FE and FD matrix methods can be seen from Figure 13.12. First, the FE method has more matrix elements initially and ends with more fill than the FD method for equivalent numbers of nodes. This is to be expected since the FE method for this example has an average of about 6.6 non-zero elements per row in the matrix. This is consistent with the observation that each node has approximately 6 adjacent triangle elements and thus 6 nearest neighbor nodes to be entered into a matrix row. Second, the fill does not occur uniformly as the matrix solution progresses for the FE method but is approximately linear for the FD method. At the end of the matrix reduction process, the FE method fill exceeds that for the FD method, but the difference is not as great as might be expected. The FE method at the end used about 7% more non-zero matrix elements as compared with the FD method. The reader should be advised that these results should be taken as trends and not absolutes numbers. Many sparse matrix reduction packages perform row and column manipulations before the reduction process to minimize the fill during reduction. Such techniques have not been applied here and such an approach could reduce substantially the fill for certain problems. Also the fill is somewhat sensitive to the numbering scheme used for the nodes in both the FD and the FE approach. The numbering scheme returned by EasyMesh is employed here with no attempt to use a scheme that might reduce the matrix fill. However, this example would suggest that the numbering scheme used by EasyMesh is reasonably good for matrix fill – at least for this example.

13.8 Exploring the Accuracy of the Finite Element Method

Now that simple routines have been developed for the FE method of solving a PDE, it is appropriate to explore the accuracy of this approach for some simple problems for which an exact solution is known. One of the simplest such example is the PDE discussed in the previous section and used in Listing 13.2 and Listing 13.6 as well as previously used in Chapter 12 with the FD method. The error in the solution by the FE and FD methods are compared in Figure 13.13 for the case of 2996 nodes with the FE method and 3025 nodes with the FD method. The errors are shown as a function of the distance from the center of the square 1.0 by 1.0 area of the solution. The choice of this plotting technique is to obtain some trends in the error for the FE data. Plotting against node number produces a rather random plot. It can be seen that both the FE and FD methods show trends with respect to this plotting dimension. For the FD method, the largest errors occur near the center of the spatial domain. Also a rather smooth trend is seen in the error with spatial distance. For the FE method the largest errors occur at the largest distances from the center, i.e. in the corners of the spatial domain. Within the corners, the solution varies as the product of x and y spatial distance from the corners and it is within these regions were the assumption of a linear variation of the solution with position is least accurate and it will be recalled that a linear variation was assumed in deriving the fundamental FE equations. So the general trends for the FE method are understandable.

Since both sets of data have close to the same number of data points, it is readily seen that the FE method has a considerably smaller average error than does the FD method as the majority of the open circle points are below the solid points in the figure. The theoretical reason for this is not obvious since the error for both methods is expected to be proportional to the average spatial dimension of the fundamental spatial elements used in the solution. Finally it can be seen that the

FE errors are very scattered about some mean value and in fact the errors show a close to symmetrical distribution around 0 with both positive and negative errors. This is not shown in the figure since the magnitude of the errors is plotted. On the other hand the FD errors are of one sign only.

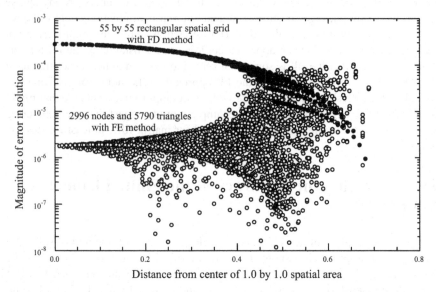

Figure 13.13. Comparison of solution errors with FE and FD methods for the example in Listing 13.2.

A comparison of the error in the FE method for two different values of total nodes is shown in Figure 13.14. The closed circles are for 781 nodes and the open circles are for 2996 nodes. The number of nodes for the two cases varies by approximately a factor of 4 which means that the average spatial dimension of the triangular elements should vary by approximately a factor of 2. One can easily see that the maximum errors for the two cases differ by at least a factor of 4 and the average error varies by an even larger value. In each case it appears that the overall accuracy could be improved by using smaller triangle elements within the corners of the spatial domain for the same number of total elements. This will be left to the reader to verify as it is not difficult to generate smaller triangular elements near various spatial points using the EasyMesh program.

While one should be careful to draw too many general conclusions from this one example, the results indicate that the FE method does provide good solution accuracy when compared with the FD method for the same number of node points. Because of the non-regular geometry of the fundamental elements in the FE method, it is difficult to develop any general model of the error in a PDE solution for the FE method.

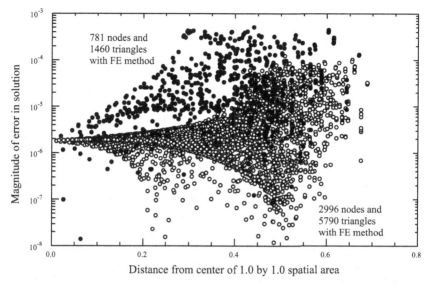

Figure 13.14. Comparison of error in solution by the FE method for different numbers of nodes and triangles in spatial domain. Example of Listing 13.2.

13.9 Two Dimensional BVPs with One Time Dimension

Many physical two dimensional boundary value problems also involve a time dimension. Just as with the FD technique, solving such problems with the FE technique represent an important application of the FE approach. The approach here is very similar to that used in Section 12.7 for the FD technique and the reader is referred to that section for a more detailed explanation of the approach. Again the diffusion equation provides a prototypical example of such a two dimensional boundary value problem with an additional time dimension:

$$
\frac{\partial u}{\partial t} = \frac{\partial}{\partial x}(D\frac{\partial u}{\partial x}) + \frac{\partial}{\partial y}(D\frac{\partial u}{\partial y})
$$

$$
\frac{\partial u}{\partial t} = D(\frac{\partial^2 u}{\partial x^2} + \frac{\partial^2 u}{\partial y^2}) \text{ for constant } D
$$

(13.45)

As in previous developments, the Crank-Nicholson (CN, also known as the trapezoidal or TP approximation to the time derivative) approach allows for the replacement of the time derivatives in terms of the solution values as:

$$u'_{t+h} = (u_{t+h} - un)/(h/2)$$

$$u''_{t+h} = (u_{t+h} - unn)/(h/2)^2$$

where   (13.46)

$$un = u_t + (h/2)u'_t = u_t + h_x u'_t$$

$$unn = u_t + (h)u'_t + (h/2)^2 u''_t = u_t + h_y u'_t + h_z u''_t$$

In these equations the primes indicate time derivatives at some time step t+h and the un and unn functions are to be evaluated at time step t where good approximations to the time derivatives are assumed to be known. In this manner the partial derivatives with respect to time can be eliminated from the equation and the solution can be bootstrapped along from one time increment to an additional time increment. This of course assumes that the problem is that of an initial value problem in the time dimension where a solution is known for some initial time at all points in the x-y space. If the problem involves a second time derivative, then the initial first time derivative must be known at all points in the x-y space. Such PDEs involving a boundary value problem in the spatial dimension and an initial value problem in time cover a broad range of practical engineering problems.

With this CN approximation to the time derivatives, the computer code developed for the two spatial dimensional boundary value problem in the previous section can be readily adapted to the initial value problem with the additional time dimension. Such a procedure has previously been developed in Section 12.7 for the FD method. One simply has to formulate computer code to interface between the existing code for the FE two dimensional method and user supplied code for defining the time dependent partial differential equation. Such a code segment is shown in Listing 13.7. This callable routine pde1stp2felt() implements one series of time steps with equal increments in the time variable. The input parameters for the function are feqs, tvals, nts, ua, tpsola, ut and utt where feqs is an array of functions defining the equation to be solved and the boundary values, tvals is an array of initial, final and increment values for the time variable, nts is the table of nodes, triangles and sides, tpsol is the type of solution method to be used, and finally ut and utt are arrays of first and second time derivative to possibly be input to the function. For an equation involving only the first time derivative, only the first 4 parameters are required. For an equation involving the second time derivative, all parameters expect utt are required as input. The function, code and calling arguments are very similar to the pde1stp2bv1t() function shown in Listing 12.29 for the FD approach. The only difference in calling arguments is the use of the nts array for node location information in place of x,y inputs for the FD routine.

The heart of the pde1stp2felt() function is the interface to the calling PDE and boundary function using the local proxy functions defined on lines 180 through 191. The proxy function defined on line 181 can be called by the previously defined two dimensional FE code and is devoid of any time information. The code from line 182 to 185 sets up the appropriate TP time derivatives as defined by Eq. (13.46) and then calls the user supplied function fpde() on line 186 which then

```lua
  1 : -- File pde2fe.lua -- Functions for 2D finite element analysis
170 : pde1stp2felt =function(feqs,tvals,nts,ua,tpsola,ut,utt)--2D,1T
171 :   local j, neq, t, h, h2,h2sq,hs,hx,hy,hz -- Local variables
172 :   local un,unn,unew = {},{},{}
173 :   local jfirst,nit,nitt,errm = 0,0,0 -- Number of iterations
174 :   local neq = #nts[1]
175 :   local tmin,tmax,ntval = tvals[1],tvals[2],tvals[3]
176 :   local u = ua or {} -- Single arrays
177 :   ut = ut or {}; utt = utt or {}
179 :   local fpde,fbv = feqs[1], feqs[2]
180 :   feq = { -- Local functions to add time and time derivatives
181 :     function(x,y,uxx,ux,up,uy,uyy,nt,m,nds,lds)--General point
182 :       local ut,utt,n1,n2,n3,l1,l2,l3 -- Local cord & nodes
183 :       n1,n2,n3,l1,l2,l3 = nds[1],nds[2],nds[3],lds[1],
                   lds[2],lds[3]
184 :       ut = (up-(un[n1]*l1+un[n2]*l2+un[n3]*l3))/h2 --un unn
185 :       utt = (up - (unn[n1]*l1+unn[n2]*l2+unn[n3]*l3))/h2sq
186 :       return fpde(x,y,t,uxx,ux,up,uy,uyy,ut,utt,
                   nt,m,nds,lds)--t added
187 :     end,
188 :     function(nd,u,und,nt,m) -- Boundary equations
189 :       local ut,utt = (u - un[nd])/h2, (u - unn[nd])/h2sq
190 :       return fbv(nd,t,u,und,ut,utt,nt,m) -- With added time
191 :     end }
193 :   t = tmin -- Initial t value
194 :   hs = (tmax - t)/ntval -- Equal steps in t used,
195 :   -- If initial deriv not given, use BD for first 4 points
196 :   if #ut~=neq then for m=1,neq do ut[m]=ut[m] or 0 end end --0?
197 :   if #utt~=neq then for m=1,neq do utt[m] = 0 end
198 :   jfirst,h = 0,0.25*hs; h2,h2sq,hy,hx,hz = h,h*h,h,0,0 --to BD
199 :   else -- Use TP or BD parameters
200 :     if bd~=false then jfirst,h = 4,hs;
                   h2,h2sq,hy,hx,hz = h,h*h,h,0,0
201 :     else jfirst,h = 4,hs; h2,h2sq,hy,hx,hz =
                   hs/2,h*h/4,h,h/2,h*h/4 end
202 :   end
203 :   for k=1,ntval do -- Main loop for incrementing variable (t)
204 :     repeat -- Use BD for first Dt with 4 sub intervals of h/4
205 :       jfirst = jfirst+1
207 :       for m=1,neq do
208 :         un[m] = u[m] + hx*ut[m] -- hx = 0 or h/2
209 :         unn[m] = u[m] + hy*ut[m] + hz*utt[m] -- hy=h, hz=0?
210 :         u[m] = u[m] + h*ut[m] -- Predicted value of u array
211 :       end
212 :       t = t + h -- Now increment t to next t value
214 :       u,errm,nitt = pde2fe(feq,nts,u,tpsola) -- Solve at t
215 :       if nitt>nit then nit = nitt end -- Monitor max #iter
216 :       -- New derivative values, same function as in feq()
217 :       for m=1,neq do ut[m],utt[m] = (u[m] - un[m])/h2,
                   (u[m] - unn[m])/h2sq end
218 :     until jfirst>=4 -- End of first interval repeat using BD
219 :     if k==1 then
220 :       if bd~=false then jfirst,h = 4,hs;
                   h2,h2sq,hy,hx,hz = h,h*h,h,0,0
221 :       else jfirst,h=4,hs;h2,h2sq,hy,hx,hz=hs/2,h*h/4,h,
                   h/2,h*h/4 end
222 :     end
223 :     if nprint~=0 then print('Completed time =',t,
                   ' with correction',errm); io.flush() end
```

```
224 :    end -- End of main loop on t, now return solution array
225 :    return u,errm,nit
226 : end -- End of pde1stp2felt
227 : setfenv(pde1stp2felt,{table=table,pde2fe=pde2fe,
              print=print,io=io,bd=false,nprint=0})
```

Listing 13.7. Code segment for implementing a 2D PDE with time variable through one time interval using the FE approach. (Some comment lines are missing from listing).

includes the first and second time derivative variables. Similarly the statements on lines 188 through 191 provide the required time derivative information for the user supplied boundary value equations.

The code from line 203 to 224 implements a loop over a series of time intervals at which the 2D PDE is solved by a call to pde2fe() on line 214 using the code in Listing 13.5. The un and unn functions are defined on lines 208 and 209 according to the formulas listed above. Note the use of information on the node numbers and local coordinates (nds and lds) on lines 183 through 185 to properly evaluate the un and unn functions at the spatial point being evaluated. This information is needed in this case because these functions are known only at the spatial nodes of the triangles and not at the triangle interior points where the functions are being evaluated. This is an example of the use of this passed information and without this information, the evaluation would take much longer as the x and y coordinate information would have to be used to essentially generate this information.

An additional feature of the code is the possible use of the BD time step approach for the first time increment with the repeat loop on line 204. This is needed to start an initial value problem for which some time derivative information is unknown. Finally a feature of the code is the ability to force the use of the BD algorithm for all time increments by setting the bd parameter to some value other than 'false', by the statement getfenv(pde1stp2felt).bd=true for example. This is primarily for comparison purposes and except for such comparisons it is recommended that the TP algorithm be used for all times except the initial time interval.

To provide a more user friendly interface to the code, two additional code segments have been defined and supplied. The first of these is the function pde2felt(feqs,tvals,nts,uar,tpsol) and the second is pde2feltqs(feqs,tvals,nts, uar,tpsol) which allow a more general specification of the time intervals for solving the 2D equation. Both of these have similar calling arguments. The feqs, nts and tpsol parameters are the same as previously defined. The uar parameter is either a simple table of initial values for the solution variable, or a table of three sub tables of initial values, first derivative values and second derivative values. The possibility of including the two time derivatives is made so that one can repeatedly call the function and use the returned values of the time derivatives to pick up at some later time and continue a solution. The tpsol variable specifies the solution method and in its simplest form is an integer specifying whether to use the SPM, COE or SOR matrix solution methods. The default method is the SPM method if this parameter is omitted. One may question the possible convergence of the COE and SOR approximation methods for the form of matrix obtained with the FE technique. Without attempting to discuss this issue, it will simply be found from the examples in this chapter that these approximate techniques appear to converge quite nicely – at least for the examples used here.

The tvals argument is the most complicated of the input arguments and is a table listing the desired time solution values and has a similar form to that used for FD initial value problems in Chapter 12. The most general form of tvals for the pde2bv1t() function is: tvals = {tmin, {t1, t2, --- tn}, {np1, np2, --- npn}, {nt1, nt2, --- ntn}} where tmin is the initial time and the other parameters specify time intervals and numbers of solution points per time interval. For a time interval between tmin and t1, there will be np1 equally spaced solutions saved and there will be nt1 time calculations between each saved interval. For example if np1 = 4 and nt1 = 4 then there will be 16 = 4X4 solutions at equally spaced time points with the solution saved at 4 of the 16 time points. The same applies to the other intervals and the defining times. This allows for a very general specification in how the solution is solved in time values. For the number of saved time points and the extra time calculations, single numbers can be used which will apply to all time intervals and then the specification may be of the form: tvals = {tmin, {t1, t2, --- tn}, np, nt}. Or a single time interval may be specified as in tvals = {tmin, tmax, np, nt}. Finally np and nt values may be omitted and default values of 10 and 10 will be used. Thus the simplest specification is tvals = {tmin, tmax} which will give the solution at 100 time points between tmin and tmax and save 10 of the solution values, equally spaced in time between tmin and tmax – well actually 11 values will be saved as the function also saves the initial value of the solution.

For the logarithmic or quick scan function, pde2fe1tsq(), the time specification is of the form tvals = {tmin, {t1, t2}, np, nt} where np and nt again specify saved time intervals and extra calculated time intervals between saved time points. In this case the time points between tmin and t1 are uniformly spaced and those between t1 and t2 are logarithmically spaced and the np and nt values specify numbers of time points per decade in time. As an example, tvals = {0, {1.e-4, 1.e0}, 2, 10} will lead to solutions saved at times 0, 5e-5, and 1e-4 over the first interval and at 3.16e-4, 1.e-3, 3.16e-3, 1.e-2, 3.16e-2, 1.e-1, 3.16e-1 and 1.0 time values. For this example the logarithmic range of time is over 3 decades in time. This is a very useful function and time specification for a solution can cover a wide range of time values with minimal calculation expense. A somewhat similar wide time range can be achieved by specifying several ranges of time with different time intervals as in the previous paragraph. The pde2fe1tqs() function also provides one additional capability and this is the ability to specify specific spatial points and have the time dependent solution returned for these specified points. This is achieved by including a table of desired time points as the last argument in the calling argument list to the function. The points are specified in a table of the form {{x1,y1}, {x2,y2}, ---{xn,yn}}. The multi-step solvers pde2fe1t() and pde2fe1tqs() are essentially functions that do the bookkeeping needed to call the basic solver of Listing 13.7 multiple times and return a collected array of solution values. Code for these functions will not be shown but the reader is encouraged to inspect the code in the supplied files. These functions and the format for the time intervals are essentially the same as the corresponding functions developed for the FD method in the previous chapter. Assuming the reader is familiar with that material, there should be little difficulty in applying the FE functions in this chapter and the use of these functions will be demonstrated with several examples.

An example will now be used to demonstrate the FE method with a 2D boundary value problem coupled with time as an initial value problem. In order to compare with the FD method of the previous chapter, the transient diffusion problem of Figure 12.71 will be repeated. This can provide not only a comparison with the FD approach but verify the accuracy of the FE method, assuming of course that the FD approach gives an accurate solution. The reader should be familiar with this example including the code in Listing 12.30 before proceeding here. The problem as defined in Figure 12.69 is that of diffusion from a constant surface concentration into a two-dimensional region with zero normal derivatives on the surface everywhere except for the window in which the diffusion occurs. The spatial region is taken to be of size 2.0 by 1.0 and the diffusion coefficient is taken to have the value 20.

```
# Example of transient diffusion #
5 # Boundary point section #

0: 0      0     .05   1
1: 2.0    0     .05   1
2: 2.0    1.0   .005  3
3: 1.0    1.0   .005  3
4: 0      1.0   .05   1

5 # Boundary sides section #

0: 0   1   2
1: 1   2   2
2: 2   3   3
3: 3   4   2
4: 4   0   2
```

Listing 13.8. EasyMesh definition file for transient diffusion problem.

For the FE method a triangular grid of spatial points must first be defined through use of the EasyMesh program. An input file for this example is shown in Listing 13.8. Five points are defined along the material boundaries with selected boundary condition markers of 1 for the end points with zero normal derivative and 3 for the constant concentration surface. Similarly 5 boundary sides are defined with markers of 2 and 3 for the zero normal derivative sides and the diffusing surface respectively. Note that the size of the desired triangles is specified as 0.005 at the edges of the diffusion surface and as 0.05 at the other material edges. When input to the EasyMesh program this generates the grid of spatial points shown in Figure 13.15. The part of the upper surface labeled as constant concentration is the window into which diffusion is assumed to occur. The much smaller triangular element sizes generated along this surface and in the region where diffusion will occur most rapidly are readily seen in the figure. It is also seen that the size of the triangles gradually change from the small triangles along the diffusing surface to the larger triangles along the bottom and side boundaries. The discretization generated by EasyMesh gives 2996 nodes and 5656 triangles.

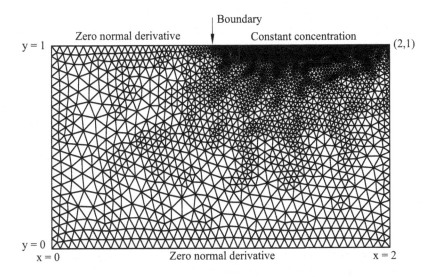

Figure 13.15. Triangular grid produced by EasyMesh for the input file of Listing 13.8.

This mesh can now be used to solve a two-dimensional diffusion problem with the same specifications as previously solved by the FD technique in Chapter 12 (Listing 12.30). A code segment for such a FE solution is shown in Listing 13.9. This can be compared on almost a line-by-line basis to the FD code in Listing 12.30. The PDE and boundary function definitions on lines 9 through 17 should be very familiar by now. The boundary condition is that of zero normal derivatives except for the boundary region with label 3 as implemented on line 14. This provides a good example of how the markers associated with the boundary sides can be so easily used to specify appropriate boundary conditions. The boundary marker values (with value of 2 or 3) as defined in the EasyMesh file of Listing 13.8 are passed to the boundary function on line 13 as the value of the nbs parameter. Arrays of rectangular x and y points are defined on lines 20 and 21 primarily for use in generating solution values for easy plotting of the results.

After setting up the desired time parameters on lines 29 and 30 (with tvals), the call to the quick scan pde2fe1tqs() function on line 31 returns the solution values in an array of solution values (by node number) at the requested solution times. Again the time specification table is somewhat complex and an explanation of the

```
 1 : -- File list13_9.lua -- time dependent BV problem in 2D
 2 :
 3 : require'pde2fe' -- Input FE code
 4 : nts ={readnts('list13_9')} -- Get node, triangle, sides data
 5 : print('Number of nodes, triangles =',#nts[1],#nts[2])
 6 : Um = 1.0 -- Define equations to be solved
 7 : D = 20 -- Constant diffusion coefficient
 8 : xmax,ymax = 2.0,1.0
 9 : feqs = { -- Table of functions
10 :     function(x,y,t,uxx,ux,u,uy,uyy,ut,utt) -- General point
11 :         return -ut + D*(uxx + uyy)
12 :     end,
13 :     function(nd,t,u,un,ut,utt,nbs)   -- Boundary values
14 :         if nbs==3 then return u - Um
15 :         else return un end
16 :     end
17 : }  -- End general point and boundary values
18 :
19 : u = {}; for k=1,#nts[1] do u[k] = 0.0 end
20 : x,y = {},{}; NT = 41 -- Change as desired
21 : for j=1,NT do y[j] = ymax*(j-1)/(NT-1); x[j] = 2*y[j] end
22 :
23 : SPM,COE,SOR = 1, 2, 3 -- 3 solution methods
24 : t = os.time() -- Time results
25 : getfenv(pde2feltqs).nprint=1 -- Observe progress in time?
26 : --getfenv(pde2bvcoe).nprint=1 -- Observe COE convergence?
27 : getfenv(pde2fe).nprint=1 -- Observe Newton results?
28 : --getfenv(pde2feltqs).seeplot = {x,y} -- See popup plots?
29 : tvals = {0,{1.e-4,1.e-1},2,10, -- Spatial points for saving
30 :     {{0,.5},{0,1},{2,0},{2,.75},{2,1},{1,.5},{1,.75}}}
31 : u,uxyt,errm = pde2feltqs(feqs,tvals,nts,u,COE)
32 :
33 : print('Time taken =',os.time()-t); io.flush()
34 : nsol = #u
35 : for i=1,nsol do -- Save 2D data files at saved times
36 :     sfl = 'list13_9.'..i..'.dat'
37 :     sol,solxy = toxysol(u[i],x,y,nts[1],nts[2]) -- to x-y arrays
38 :     write_data(sfl,sol) -- Save x-y arrays
39 :     if i==5 then -- Save popup plots at one time
40 :         splot('list13_9a.emf',solxy); cplot('list113_9b.emf',solxy)
41 :     end
42 : end
43 : write_data('list13_9a.dat',uxyt) -- Save time data
```

Listing 13.9. Code for solving a two-dimensional diffusion problem by the FE technique. Compare with FD technique in Listing 12.30.

values will be given. The first part of tvals is {0, {1.e-4, 1.e-1} 2, 10, --. This specifies a starting time of 0 and a first time interval of 0 to 1.e-4 over which 2 sets of solution values will be saved and the 10 time steps per saved interval will be uniformly spaced over the first time interval. Thus solutions will be generated at times of t = 0, .5e-5, 1.e-5, 1.5e-5, etc. and solutions will be saved at times of .5e-4 and 1.e-4. For the next time interval which is 1.e-4 to 1.e-1, the time steps will be logarithmically spaced with a factor of $10^{(1/20)} = 1.122$ per time step. Again 2 solutions will be saved per decade in time, one at values of 0.316 and one at 1.0 times the decades in time. The final table in tvals or the

{{0,.5},{0,1},{2,0},{2,.75},{2,1},{1,.5},{1,.75}} table indicates a series of spatial points in the form {xval, yval} where time solutions are to be saved for each time point calculated as the solution develops. For this example, the solution is generated for 80 time points (plus t = 0) so there will be 81 saved solution values for each of the specified spatial points.

The spatial solution values returned by pde2fe1tsq() are put into individual files for each of the saved times within the loop on lines 35 through 42. This makes use of the toxysol() function which takes the triangular node solution values and returns solution values for a range of specified x,y values, as generated by lines 20 and 21 for a rectangular array of 41 by 41 uniformly spaced x and y values. This is done to simplify the plotting for most programs that expect uniformly spaced data. Graphs from splot() and cplot() are saved to files on line 40 for the solution at time 3.16e-3 sec. Finally the time dependent data at the selected spatial points is saved into a file on line 43.

The reader should compare the results to now be presented with the FD solution obtained in Section 12.7. Figure 13.16 shows a surface plot of the diffusion profile at a time of 3.16e-3 sec as saved in the list13.9a.emf file. A comparison of this with the FD solution of Figure 12.70 shows essentially the same solution – at least to the extent that can be seen on this gross scale. The contour plot saved in file list13.9b.emf is shown in Figure 13.17 and this should be compared with Figure 12.71.

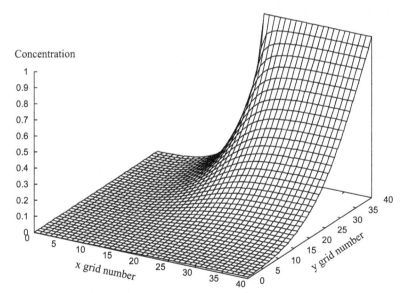

Figure 13.16. Surface profile plot for a time of 3.16e-3 from the code in Listing 13.9. The x and y labels are in terms of grid point numbers (41 by 41 in value) corresponding to x and y values of 2.0 and 1.0 respectively. (Compare with Figure 12.70).

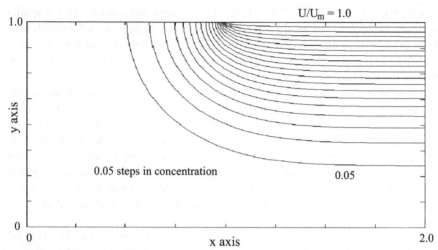

Figure 13.17. Contour plot of diffusion profile at a time of 3.16e-3 sec from the code in Listing 13.9. (Compare with Figure 12.71).

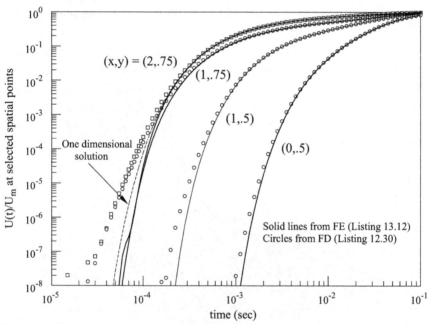

Figure 13.18. An expanded plot of the time dependent solution of the 2D diffusion problem at selected spatial points. The circles are points from Figure 12.72 obtained by the FD method.

A more detailed comparison of the FE and FD method is appropriate and this is illustrated in Figure 13.18 which shows the time dependence of the diffusing concentration at selected spatial points within the spatial domain. The solid lines are the results obtained here using the FE method with the triangular grid of Figure 13.15. The open circle and square points are from the FD method in Chapter 12 and shown in Figure 12.72. Very good agreement between the two methods is seen for the longer diffusion times, but some significant differences are seen at the shorter times and for low diffused concentrations. One would expect that the FE method gives the more accurate solution as the calculations were made using smaller fundamental spatial elements near the diffusion source. This expectation is verified to some extent by the dotted curve shown in comparison to the (2, .75) spatial point. This dotted curve is the known exact solution for a one-dimensional diffusion problem with a constant surface concentration. This should be a good approximation to the present 2D case where the diffusion profile of Figure 13.17 is approximately horizontal. The dotted curve is in fact seen to be closer to the FE generated solution. The difference in the solid and dotted curve is in the range of 1.e-7 to 1.e-6. The method used to solve the matrix elements in this example was the COE iterative technique. This was used as opposed to direct matrix inversion, because it is significantly faster in computer time. However, when executing the code in Listing 13.13 one will consistently see that the COE iterative technique terminates when the correction term is on the order of 1.e-7 to 1.e-6. Thus one would not expect greater accuracy in the solution that this and this is in fact observed. Whether one would achieve a greater accuracy using the exact matrix solution method (SPM) is left as an exercise for the reader.

This diffusion problem has now been solved twice by the FD and FE method and it could be argued that no new information has been generated by the FE method. This is certainly true in one sense. However, by comparing the two solution methods, a greater confidence is gained in the validity of the two techniques and especially in the FE method since the coding is considerably more involved than for the FD method. It is especially important to compare any new 2D method to other known 2D solutions since in most cases even for simple equations and simple geometries closed form exact solutions with which to compare the results do not exist. With this as a background, the FE approach both with and without the time dependency can be applied to more challenging example problems.

13.10 Selected Examples of PDEs with the Finite Element Method

13.10.1 Capacitor with Two Dielectric Materials

The first simple example to be considered is that of a capacitor with two different dielectric constants as used in the previous chapter and shown in Figure 12.75 and repeated here in Figure 13.19. While this is a simple example, it does illustrate the technique of different spatial material domains and how these are incorporated into the FE method. The spatial domain to be considered is a rectangular region from 0 to 2 in the x direction and from 0 to 1 in the y direction. Midway along the y direction (at y = 0.5) the dielectric constant changes value.

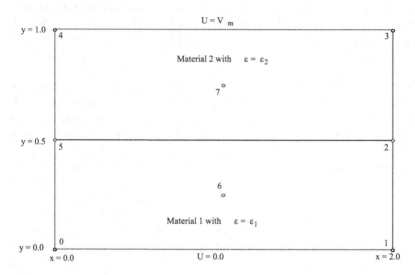

Figure 13.19. Spatial domain for example capacitor problem with identified points for input to EasyMesh

A mathematical formulation of the problem is:

$$\frac{\partial}{\partial x}(\varepsilon\frac{\partial U}{\partial x})+\frac{\partial}{\partial y}(\varepsilon\frac{\partial U}{\partial y}) = 0$$

$\varepsilon = \varepsilon_1$ for $0 \le y < 0.5$ and $\varepsilon = \varepsilon_2$ for $0.5 \le y < 1.0$ (13.47)

$U(x,0) = 0$ and $U(x,1.0) = V_m$

The first step is to set up a triangular spatial grid for the problem with a material boundary separating the regions of differing dielectric constant as in Figure 13.19. An appropriate input file for the EasyMesh grid generation program is shown in Listing 13.10. The spatial points corresponding to the input file are shown in Figure 13.19. Two extra points (2 and 5) are defined along the y boundaries to define the horizontal boundary between the two materials. In defining the spatial boundary, the outline of the entire region is first defined in the boundary lines section with the listing in a counter clockwise direction. Then the line separating the two materials is defined as the last line in Listing 13.10 (line 6:). Note the use of a marker number > 100 (400 used here) for the internal boundary line. This is not a requirement of the EasyMesh program but is imposed

```
# Example of capacitor with two dielectric materials #
8 # Boundary points section #

 0:    0    0     .04    1 # Boundary markers 1 to 4 #
 1:    2.0  0     .04    1
 2:    2.0  0.5   .04    2
 3:    2.0  1.0   .04    3
 4:    0.0  1.0   .04    3
 5:    0.0  0.5   .04    4

# Define internal points for two materials #
 6:    1.0  0.25 0     1 # Material 1 marker point #
 7:    1.0  0.75 0     2 # Material 2 marker point #

7 # Boundary lines section #

 0:    0    1    1 # Boundary markers 1 to 4, consistent with points #
 1:    1    2    2
 2:    2    3    2
 3:    3    4    3
 4:    4    5    4
 5:    5    0    4

# Define internal boundary line between two materials #
 6:    2    5    400 # Note, use marker number > 100 for internal
boundaries!!! #
```

Listing 13.10. Input file for EasyMesh defining example capacitor problem with two different materials.

here so that the software developed herein can distinguish an internal boundary point from an external boundary point. If used in this manner, the readnds() function will properly convert the EasyMesh output to proper format for use with the setup2feeqs() function. To uniquely define the two material regions, two isolated interior points (identified as 6: and 7:) are defined with a point in each material region. The last entry on these lines defines a material marker (1 or 2) which will be used to identify the two materials. These two markers will be output by EasyMesh in the *.e or element file to identify in which material the various triangles are located. These markers in turn will be passed by the setup2feeqs() function back through the calling arguments to the PDE defining function so the appropriate PDE can be used for each material. This will be subsequently seen in the code listing for solving this problem. Also note that the spatial size entry for the internal points is set at 0 which helps to identify these points as isolated internal nodes. Finally, note that no boundary lines are defined connecting to the isolated internal nodes.

The boundary conditions for this example are constant boundary values along the lower and upper boundary and zero normal derivative values along the vertical sides. Since transitions occur only between fixed boundary conditions along the top and bottom and fixed normal derivatives on the sides, no isolated boundary marker values are needed for this problem. On the two sides, the proper normal derivative boundary conditions will be used in the spatial regions directly adjacent to the corner points as desired.

The spatial grid generated by EasyMesh for this example is shown in Figure 13.20. By careful inspection it is seen that there is a definite horizontal boundary line traversing the center of the vertical structure as identified by the arrow in the figure and that no triangle boundaries cross this line separating the two materials. Otherwise the grid is rather unremarkable with 1572 nodes and 2990 triangular elements.

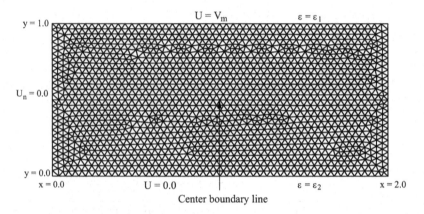

Figure 13.20. Spatial grid for example capacitor problem with two dielectric materials.

The code for solving for the potential in this capacitor structure is shown in Listing 13.11 using the FE grid structure shown above and contained in a set of list13_11.n, list13_11.e and list13_11.s files. The code for the differential equation on lines 10 through 14 illustrate how the material marker information from the EasyMesh file is used. The setup2feeqs() function in setting up the matrix equations passes the material marker for each triangle as the last argument into the user supplied function defining the differential equation as on line 10 (the mtn parameter). This marker can be readily used as on lines 11 and 12 to set the value of the dielectric constant for each of the material regions. The triangle number (the ntr parameter) is also passed to the function on line 10 but this is not used for the present problem. In a similar manner for the boundary function, the boundary marker (nbs) is passed to the defining boundary function on line 15 of the code and this is used to set the bottom and top fixed boundary conditions as well as the zero normal derivatives on the sides. It will note that the material markers and the boundary markers used in the function definitions of lines 10 through 19 match the corresponding values specified in the EasyMesh input file of Listing13.10.

After defining the PDE and boundary equations, the solution is obtained by a single line call to pde2fe() on line 26. Finally lines 28 through 29 use the returned solution set to obtain an x-y array of uniformly spaced solution values for easy plotting of the results. Figure 13.21 shows a surface plot of the resulting solution by the FE method. It will be recalled that this same problem was solved in Chapter 12 using the FD method and with a slight trick to handle the delta function derivative at the material boundary interface. The reader can compare the results here with those shown in Figure 12.76. Good agreement is seen from this graph with the expected result. However, a more detailed analysis is needed to access the exact accuracy of the solution. A comparison of the saved data file from Listing 13.11 shows that the maximum error in the FE numerical solution is about 1.0e-13, or at about the numerical limit of the machine accuracy. While this is not an absolute verification of the accuracy of the computer code for setting up the matrix equations and solving the equations, this does provide important additional confidence in the developed code and the FE method.

```
 1 : -- File list13_11.lua -- 2 dielectric capacitor
 2 :
 3 : require'pde2fe'
 4 :
 5 : nts = {readnts('list13_11')}
 6 :
 7 : Vm = 1.0
 8 : ep1,ep2 = 3, 12
 9 : feqs = {
10 :    function(x,y,uxx,ux,u,uy,uyy,ntr,mtn)
11 :       if mtn==1 then ep = ep1
12 :       else ep = ep2 end
13 :       return ep*(uxx + uyy)
14 :    end,
15 :    function(nd,u,un,nbs,kb)
16 :       if nbs==1 then return u -- Bottom boundary
17 :       elseif nbs==3 then return u - Vm -- Top boundary
18 :       else  return un end -- Sides
19 :    end
20 : }
21 :
22 : getfenv(spgauss).nprint = 1
23 : getfenv(pde2fe).nprint = 1
24 : getfenv(pde2fe).linear = 1
25 :
26 : u = pde2fe(feqs,nts)
27 :
28 : x,y = {},{}; NT = 21
29 : for j=1,NT do y[j] = (j-1)/(NT-1); x[j] = 2*y[j] end
30 : sol = toxysol(u,x,y,nts[1],nts[2])
31 :
32 : write_data('list13_11.dat',sol)
33 : splot(sol)
```

Listing 13.11. Code segment for solution of potential in capacitor structure with two different dielectric constant materials.

This example has been included partly to illustrate the fact that different material properties can be easily handled in the FE method and to illustrate how one specifies the different materials and boundaries in the input file to EasyMesh for proper interfacing with the FE numerical techniques. The approach used here is easily extendable to a wide variety of physical problems with complex geometries and different materials. It is noted that no special treatment is needed with the FE method to properly handle the abrupt change in dielectric constant at the material interface. This is related to the fact that an integration by parts is used in formulating the basic FE equations so that the second derivative form of Eq. (13.47) is actually converted to a first derivative term.

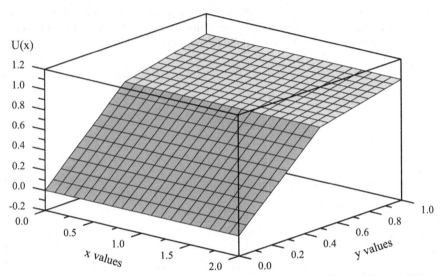

Figure 13.21 Surface plot of the potential solution for a capacitor with two dielectric constants with the FE method.

13.10.2 Square Corner Resistor

Another PED problem illustrated in Chapter 12 is the L-shaped (or square corner) resistor shown in Figure 12.78 and repeated here in Figure 13.22. This problem models electric current flow around a square corner resistor layout as in integrated circuits. One possible FE discretization is shown in Figure 13.22 for the case of 2996 nodes and 5760 triangular elements. This mesh was generated again by EasyMesh with the input file shown in Listing 13.12. The boundary markers used in this example are shown in both the input listing and in Figure 13.22. A unique boundary marker ranging from 1 to 6 has been used to identify each of the flat sides of the structure. These entries are in the "Boundary sides section" of the file and shown along the sides of the figure as the bm values. In addition each right angle corner boundary point is labeled with a separate boundary marker ranging in value from 7 to 10 and shown on the input file in the "Boundary points section". Each corner point then has a unique marker so that any desired mixed boundary condition can change abruptly at the sharp corner points. An additional feature is the use of a finer spatial grid around the sharp inner corner of the resistor (near x =

y = 1.0). This is achieved by specifying a spatial grid size of .01 at this corner point and a size of .04 for all other points.

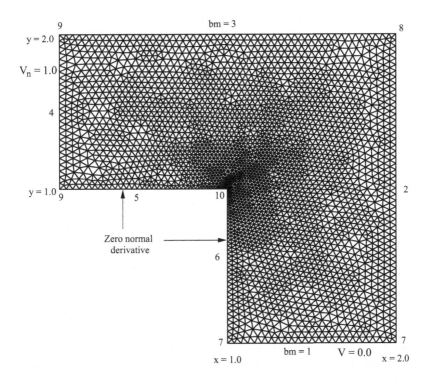

Figure 13.22. Geometry and spatial grid for L shaped square corner resistor example.

When this problem was solved in Chapter 12 by the FD method, the boundary conditions taken were that of a fixed voltage (of 0) along the bottom boundary and a fixed voltage (of 1.0) along the left boundary with zero normal derivatives along all other boundaries. This would again be an appropriate set of boundary conditions for use with the FE approach. However, to provide a little more variety and to illustrate additional approaches, for the present example a slightly different set of boundary conditions are taken at the left boundary for this repeat example. Instead of a fixed voltage at this boundary, a fixed normal derivative of the voltage is specified at this boundary. This is indicated in Figure 13.22 by the $V_n = 1.0$ value on the upper left side. Other boundary conditions are the same as before. Assuming a constant derivative is equivalent to assuming a constant current density across the left contact length. Such a constant current will result in a voltage at the contact and in fact the average value of the voltage (for a unit current density) will correspond exactly to the sheet resistance of the structure (in equivalent squares units). This is thus an easy way to evaluate the desired equivalent resistance of the structure. This also illustrates the method of handling a normal derivative boundary condition that changes abruptly at some spatial boundary point. In this case, the normal derivative should change abruptly from zero on the top and bottom sides to a finite value along the left boundary.

```
# Example of L shaped resistor #
6 # Boundary points section #

    0:    1     0     .04     7 # Boundary markers #
    1:    2     0     .04     7
    2:    2     2     .04     8
    3:    0     2     .04     9
    4:    0     1     .04     9
    5:    1     1     .01     10

6 # Boundary sides section #

    0:    0     1     1 # Lower contact marker #
    1:    1     2     2 # Normal derivative marker #
    2:    2     3     3
    3:    3     4     4 # Left contact marker #
    4:    4     5     5
    5:    5     0     6
```

Listing 13.12. Input file to EasyMesh for generation of spatial grid shown in Figure 13.22.

   For this particular problem with boundary conditions, not all of the various boundary markers are required. It is not essential that the corners labeled 7, 8 and 10 in Figure 13.22 have unique boundary markers since the lower boundary is that of a fixed value and the normal derivative boundary condition is the same on both sides of the corners labeled 8 and 10. Only the left corners labeled with a 9 need to have a unique boundary marker. If a fixed voltage is used for the left boundary there would also be no need for a unique boundary marker at the left corners. However, the use of extra corner markers at the lower boundary provides extra flexibility with respect to possible uses of other boundary conditions, such as specifying a normal derivative on the lower boundary. One must specify a specific voltage value at some point along the boundary surface or the PDE will have no unique solution. In this example the lower boundary is taken at zero potential.

   A code segment for solving the PDE associated with the square corner resistor is shown in Listing 13.13. The PDE definition on line 10 is simply Poisson's equation for the electric potential in the two dimensional surface. The boundary condition function is given on lines 12 through 17 implementing a zero solution value on the bottom boundary and a fixed normal derivative value of 1.0 on the left boundary and a zero normal derivative condition on all other boundary sides. The solution is obtained with the single line call to the pde2fe() function on line 22. The lines from 26 to 32 define a rectangular grid of points and interpolate the solution using the toxysol() function and saves the data for further plotting. Finally the effective number of squares of the corner resistor is evaluated on line 25 by obtaining the voltage at the center of the left contact and subtracting 2.0, the number of squares of the two legs leading to the corner resistor. The evaluated value is 0.5579 squares. This can be compared with the value of 0.5575 squares obtained in Chapter 12 by the FD technique.

```
 1 : -- File list13_13.lua -- FE analysis of Square Corner Resistor
 2 :
 3 : require'pde2fe'
 4 :
 5 : nts = {readnts('list13_13')}
 6 : Un = 1 -- Normal derivative on left boundary
 7 :
 8 : feqs = {
 9 :    function(x,y,uxx,ux,u,uy,uyy)
10 :       return uxx + uyy
11 :    end,
12 :    function(nd,u,un,nbs)
13 :       if nbs==1 then return u -- Bottom boundary
14 :       elseif nbs==7 then return u -- Bottom end points
15 :       elseif nbs==4 then return un - Un -- Left boundary
16 :       else  return un end -- Sides
17 :    end
18 : }
19 :
20 : getfenv(spgauss).nprint = 1 -- Just to see progress
21 : getfenv(pde2fe).linear = 1 -- Saves an extra Newton iteration
22 : u = pde2fe(feqs,nts)
23 :
24 : pt = {0,1.5} -- Point at center of left boundary
25 : print('Effective squares of corner =
                 ',intptri(u,nts[1],nts[2],pt)-2)
26 : x,y = {},{}; NT1 = 21; NT2 = 41 -- Arrays for plotting
27 : for j=1,NT1 do x[j] = 1+(j-1)/(NT1-1) end
28 : for j=1,NT2 do y[j] = 2*(j-1)/(NT2-1) end
29 : sol1 = toxysol(u,x,y,nts[1],nts[2]); splot(sol1)
30 : write_data('list13_13a.dat',sol1) -- Save for 1<x<2; 0<y<2
31 : for j=1,NT1 do x[j] = (j-1)/(NT1-1); y[j] = 1+x[j] end
32 : sol2 = toxysol(u,x,y,nts[1],nts[2]); splot(sol2)
33 : write_data('list13_13b.dat',sol2) -- Save for 0<x<1; 1<y<2
Selected output:
Effective squares of corner =        0 0.55794589640267
```

Listing 13.13. Code segment for solution of square corner resistor by FE method.

The obtained solution is shown in Figure 13.23. This can be compared with Figure 12.79 which was obtained by the FD method. The shape of the solution is the same, but the maximum solution value seen here is very close to 2.5579 Volts on the left boundary as opposed to the fixed value of 1.00 used in Chapter 12 for the FD method. This of course results from specifying the normal derivative at 1.0 in the present example rather than using a fixed boundary value for the potential at the contact. With a fixed normal derivative, there is some slight variation in the solution value along the left contact side. This is too small to be discerned in Figure 13.23 but can be seen by examining the output file of saved solution values.

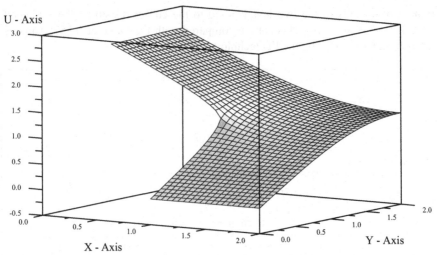

Figure 13.23. Surface plot of the voltage solution for the L shaped corner resistor of Figure 13.22.

This example also introduces a new function on line 25, the intptri() function that returns the solution value at a desired spatial point within the solution domain. The calling argument values (u,nts[1],nts[2],pt in the example) are the array of solution values, the array of triangle nodes, the array of triangles and the point (in format {x,y}) where the solution is to be evaluated. The function then returns the interpolated solution at the specified point, which in this example is the mid point of the left contact region.

The ability to vary the spatial size of the finite elements is important in this as well as many physical problems. From Figure 13.23 it is seen that the surface is rather smooth except near the inside corner at x = y = 1.0 where the potential varies much more rapidly. This can be seen more clearly in Figure 13.24 which plots the potential solution along the boundary of the resistor surface along the x = 1.0 and y = 1.0 boundaries. It is seen that the derivative becomes much larger near the inside corner where current crowding occurs. In anticipation of this type of solution, the spatial size of the triangular elements was reduced by a factor of 4 at the corner as can be seen in Figure 13.22. This should improve the accuracy of the solution in this critical region. However, more extensive calculations at varying grid sizes would be needed to verify if the grid size is sufficiently small in this region for achieving a highly accurate solution. The reader is encouraged to re-execute this example changing the boundary condition from that of a fixed normal derivative on the left side to that of a fixed potential and compare the solution with that obtained in Chapter 12. For this case, it will be necessary to numerically calculate the value of the derivative of the solution at the left boundary in order to obtain the number of equivalent squares for the resistor. An example of evaluating the first derivative within each triangular element is shown in the next example.

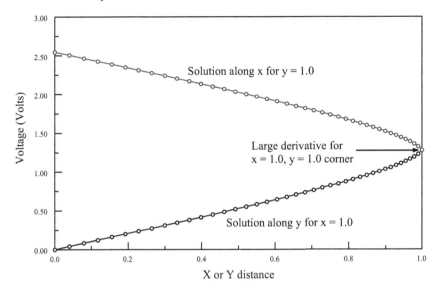

Figure 13.24. Plot of potential solution along the inter boundary and near the sharp corner.

For the code in Listing 13.13 the solution method for the set of obtained matrix equations will be the direct sparse matrix solution method. The SOR and COE iterative solution methods can also be used in this example to obtain a solution. This can be done by replacing line 22 of the code by the statement u = pde2fe(feqs,nts,u,2) for example for the COE method (or using 3 for the SOR method). Since the matrix equations are diagonally dominant, the approximation methods will converge to the proper solution values. The reader is encouraged to explore the solution with these approximate methods. However, it should be noted that the default method for selecting a spectral radius for use with the approximate methods is based upon the value for a rectangular array of uniformly spaced x and y grid elements. Such an approximation is probably not optimum for the geometry of this problem and for such triangular spatial elements. Never the less, as the reader can verify, the approximate methods will converge even for the default spectral radius values. However, with the default spectral radius value, the solution will require more computer time with the approximate methods than with the direct sparse matrix solution method. By experimenting with different spectral radius values, the reader can likely find an appropriate value that will result in a faster solution time for the approximate solution methods than the direct matrix solution method. Finding such an approach is left to the interested reader. For this example, the direct matrix solution method is sufficiently fast to give good results in reasonable computer times.

13.10.3 Temperature for Square Corner Resistor

In addition to the electric potential around a square corner resistor, any temperature increase along the resistor due to the electrical energy dissipation from the resistivity of the material might be of interest. The equation describing temperature (T) for such a problem is in the most general case:

$$\frac{\partial}{\partial x}(k_x \frac{\partial T}{\partial x}) + \frac{\partial}{\partial y}(k_y \frac{\partial T}{\partial y}) - \frac{1}{\rho c}\frac{\partial T}{\partial t} + q_{gen} = 0$$

where k_x, k_y are thermal conductances   (13.48)

ρ = material density, c = specific heat and

q_{gen} = power generation rate per unit volume

Now consider an isotropic material at steady state with a constant thermal conductivity and with the power generation due to an electric field. The equation then becomes:

$$\frac{\partial^2 T}{\partial x^2} + \frac{\partial^2 T}{\partial y^2} + \frac{\sigma}{k}E^2 = 0$$

where σ = Electrical conductivity and   (13.49)

E = Electric field

Thus an electric potential as calculated in Listing 13.13 can be used to evaluate the electric field within each of the triangular spatial elements and this can then subsequently be used in a second PDE calculation to evaluate a temperature change in a resistive element such as the square corner resistor.

The additional information needed for the numerical calculation concerns boundary values for the temperature. For this example, it will be assumed that the two ends of the resistor are fixed at a constant temperature (room temperature for example) and that no heat flux can flow out through the sides of the square corner resistor. The additional parameter is then the ratio of electrical to thermal conductivity. For this example this will be taken as 6.9 ($°C/W\Omega$) which is an appropriate value for 0.1 (Ωcm) silicon as the material of the resistor.

A code segment for this coupled electrical-thermal PDE problem is shown in Listing 13.14. A few features will be noted in the code. First the input data on nodes, triangles and sides using readnts() on line 6 has a second parameter L which is set to 1.e-2. This is a scale factor that scales the previously generated discretization file of list13_13 used in Listing 13.13 from a boundary size of 2 by 2 to a size of 1.e-2 by 1.e-2 in this case taken as a value in cm. This is the first time this optional parameter has been used but it may be used with any EasyMesh generated data set to scale the dimensions by a constant factor. Second, the electrical potential imposed on the resistor is set on line 5 to be 10.0 volts (by parameter Um). The remainder of the code should be readily understood. After defining the electrical equations on lines 10 through 21, the electrical potential is solved by the call to pde2fe() on line 22. This is the same as the calculation in Listing 13.13

```
 1 : -- File list13_14.lua -- Thermal heating of square corner R
 2 :
 3 : require'pde2fe'
 4 :
 5 : Um,K = 10.0,6.9; L = 1.e-2 -- L in cm
 6 : nts = {readnts('list13_13',L)} -- Read spatial grid; Scale to L
 7 : getfenv(spgauss).nprint = 1 -- just to see progress
 8 : getfenv(pde2fe).linear = 1 -- saves an extra Newton iteration
 9 :
10 : fels = { -- Equations for electric field
11 :    function(x,y,uxx,ux,u,uy,uyy,ntr,mtn)
12 :       return uxx + uyy
13 :    end,
14 :    function(nd,u,un,nbs,kb)
15 :       if nbs==1 then return u -- Bottom boundary
16 :       elseif nbs==7 then return u -- Bottom end points
17 :       elseif nbs==4 then return u - Um -- Left boundary
18 :       elseif nbs==9 then return u - Um
19 :       else  return un end -- All other sides
20 :    end
21 : }
22 : uel = pde2fe(fels,nts) -- Solve for electric potential
23 :
24 : efdx,efdy = derivtri(nts[2],nts[1],uel) -- Get electric field
25 : esq = {} -- Square of electric field
26 : for i=1,#efdx do esq[i] = efdx[i]^2+efdy[i]^2 end
27 :
28 : feqs = { -- table of PDE and boundary functions for temperature
29 :    function(x,y,uxx,ux,u,uy,uyy,ntr,mtn)
30 :       return uxx + uyy + K*esq[ntr]
31 :    end,
32 :    function(nd,u,un,nbs,kb)
33 :       if nbs==1 then return u -- Bottom boundary
34 :       elseif nbs==7 then return u -- Bottom end points
35 :       elseif nbs==4 then return u -- Left boundary
36 :       elseif nbs==9 then return u -- Left end points
37 :       else  return un end -- All other sides
38 :    end
39 : }
40 : T = pde2fe(feqs,nts) -- Solve equations for temperature
41 :
42 : x,y = {},{}; NT1 = 21; NT2 = 41 -- Rect array for plotting
43 : for j=1,NT1 do x[j] = L*(1+(j-1)/(NT1-1)) end
44 : for j=1,NT2 do y[j] = 2*L*(j-1)/(NT2-1) end

45 : sol1 = toxysol(T,x,y,nts[1],nts[2]); splot(sol1)
46 : for j=1,NT1 do x[j] = L*(j-1)/(NT1-1); y[j] = L+x[j] end
47 : sol2 = toxysol(T,x,y,nts[1],nts[2]); splot(sol2)

48 : Tm = 0; for i=1,#T do if T[i]>Tm then Tm,im = T[i],i end end
49 : print('Maximim T =',Tm,' at x,y =',nts[1][im][1],nts[1][im][2])
50 : write_data('list13_14a.dat',sol1)
51 : write_data('list13_14b.dat',sol2)
Selected Output:
Maximum T =    86.259927826694         at x,y =       0.02    0.02
```

Listing 13.14. Code segment for electrical-thermal square corner resistor PDE problem.

except for the fixed voltage boundary condition. The electric potential is then used with the derivtri() function on line 24 to evaluate the electric field within each triangular element with the results stored by triangle number. It will be recalled that one of the fundamental approximations of the FE method developed here is the assumption that the potential varies linearly with position within each of the spatial triangular elements and hence the electric field will be evaluated as constant within each triangular element. The electric field squared is then evaluated on line 26 and stored by triangle number in the esq[] table.

For the temperature calculation, the heat equation and boundary conditions are defined by the functions on lines 28 through 39 and the equations are solved for the temperature (T) by the pde2fe() call on line 40. In this case the temperature is actually the delta increase in temperature above the fixed boundary values as the value at the boundaries are taken as zero. The remainder of the code simply sets up x-y arrays and evaluates the temperature on such a grid for ease in plotting the results. It will be noted on line 30 that the triangular element number is used to access the electric field squared in the defining PDE. This is an example where a function is known within each triangle from a table of values as opposed to the code in setting up the time dependent equations in Listing 13.7 where functions were known at each triangle node from a table of values. The calling arguments to the user defined PDE provide sufficient input data to readily access the needed data for each of these cases. For this problem since only one material is present, information regarding the material number (the mtn entry) is not used. Listing 13.11 has provided an example of the use of this parameter.

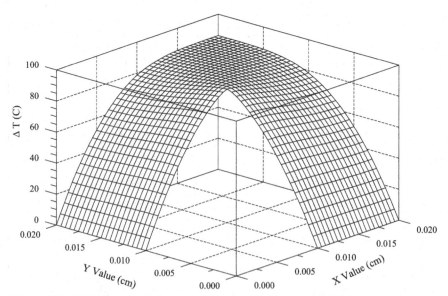

Figure 13.25. Temperature profile of square corner resistor with heat sinks at two ends of the resistor from calculation in Listing 13.14.

The computed temperature profile for the square corner resistor is shown in Figure 13.25 obtained from the saved files in Listing 13.14. The profile is symmetrical as expected about the corner of the resistor and for this example predicts a peak temperature of about 86.26 C above the temperature of the end heat sinks. The printed output shows that this peak value occurs at the maximum values of x and y as would be expected. This example illustrates how the results of one PDE calculation can be used as input to subsequent PDE calculations.

13.10.4 Air Velocity in Rectangular Duct

The variation of velocity within the cross section of a fluid such as air flowing in a long straight uniform duct is described by the equation:

$$\frac{\partial^2 u}{\partial x^2} + \frac{\partial^2 u}{\partial y^2} + 1 = 0 \tag{13.50}$$

where u is a normalized velocity related to other physical parameters by:

$$u = \frac{v}{2v_o f R_e} \tag{13.51}$$

Here v is the axial velocity of the fluid, v_o is the mean velocity, f is the Fanning friction factor and R_e is the Reynold's number. The boundary condition is that the velocity and hence the normalized velocity is zero on the boundaries of the duct. If the x and y dimensions are L_x and L_y and normalized dimensions are used this equation can be converted to the equivalent equation:

$$\frac{1}{L_x^2} \frac{\partial^2 u}{\partial x^2} + \frac{1}{L_y^2} \frac{\partial^2 u}{\partial y^2} + 1 = 0 \tag{13.52}$$

where now the spatial dimensions range from 0 to 1 over a square area. A triangular grid for such a square area has already been considered in Figure 13.10 and used in Listing 13.6.

A code segment for this PDE is shown in Listing 13.15. The x dimension is taken as twice the value in the y direction. The code is simple and straightforward, requiring little explanation. In this example the statement getfenv(pde2fe).linear = 1 is not included although the equation is obviously a linear PDE. This means that the software will assume that the PDE may possibly be a nonlinear equation requiring Newton iterations. The program output shows that the pde2fe() program does in fact perform 2 Newton iterations. However, the second Newton step is completed in one COE iterative step since the equation is in fact a linear PDE and the initial COE iterative solution provides a good approximation to the actual solution. Thus the omission of the linear statement results in 106 instead of 105 COE iterations. In addition, the program must formulate the equation set twice instead of once. With the COE iterative solution method, the additional time taken by these steps is a small fraction of the total solution time.

However, if the direct sparse matrix solution method is used in the solution, then the calculation will take approximately twice the time without indicating that the equation is linear as the SPM method will require a complete matrix solution at the second Newton iteration. The approximate COE iterative solution method is used here to illustrate another possible solution approach with the pde2fe() function.

```
 1 : -- File list13_15.lua -- Velocity in air duct
 2 :
 3 : require'pde2fe'
 4 : nts ={readnts('list13_15')} -- square area, 2996 nodes
 5 : Lx,Ly = 2,1 -- Ratio of lengths squared
 6 :
 7 : feqs = {
 8 :    function(x,y,uxx,ux,u,uy,uyy,ntr,mtn)
 9 :              return uxx/Lx^2 + uyy/Ly^2  + 1.0
10 :    end,
11 :    function(nd,u,un,nbs)
12 :              return u
13 :    end
14 : }
15 :
16 : getfenv(pde2bvcoe).nprint=1 -- Observe progress
17 : getfenv(pde2fe).nprint = 1
18 : SPM,COE,SOR = 1,2,3 -- Try different methods
19 :
20 : u = pde2fe(feqs,nts,u,COE) -- Solve equations
21 :
22 : x,y = {},{}; NT = 41-- x,y arrays for plotting
23 : for j=1,NT do x[j] = (j-1)/(NT-1); y[j] = x[j] end
24 : sol,soxy = toxysol(u,x,y,nts[1],nts[2])
25 : write_data('list13_15.dat',sol)
26 : splot(soxy);cplot(soxy) -- popup plots
Selected Output:
Exiting COE at iteration 105  with correction  8.2031290614486e-008
Completed Newton iteration 1  with correction  0.11386456145407
Exiting COE at iteration   1  with correction  6.4225615531569e-008
Completed Newton iteration 2  with correction  7.4339787610246e-008
```

Listing 13.15. Code for velocity profile of fluid flow in duct.

Two popup plots of the solution are obtained by executing the program so the reader can readily see the solution profiles. A contour plot of the solution is shown in Figure 13.26. The peak velocity at the center in normalized units is 0.1139 and the last contour line occurs at a value of 0.11. A surface plot is not shown but the reader can easily generate such a plot by executing the code in Listing 13.15. In this example, the triangular grid used on line 4 of the listing is similar to the grid shown in Figure 13.10, but has an increase in the number of nodes to 2996. This grid is actually the same as that used in Figure 13.12 and the reader is referred back to a discussion of that figure for the details of the grid structure. This example converges readily with any of the possible matrix solution methods and the reader is encouraged to explore other solution methods (SPM and SOR) by changing the parameter used on line 20 of Listing 13.15.

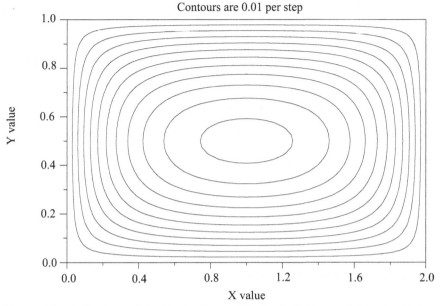

Figure 13.26. Contour plot of velocity profile in duct from solution in Listing 13.15.

13.10.5 Physical Deflection of Square Plate with Uniform Loading

The physical deflection of a square plate subject to an areal load q is described by the PDE:

$$\frac{\partial^4 z}{\partial x^4} + 2\frac{\partial^4 z}{\partial x^2 \partial y^2} + \frac{\partial^4 z}{\partial y^4} - \frac{q}{D} = 0$$

$$\text{where } D = \text{ flexural rigidity } = \frac{E\Delta z^3}{12(1-\sigma^2)}$$

(13.53)

In this E is the modulus of elasticity, Δz is the plate thickness and σ is Poisson's ratio. If a new variable is introduced, this can be converted into two second order PDEs as:

$$\frac{\partial^2 u}{\partial x^2} + \frac{\partial^2 u}{\partial y^2} - \frac{q}{D} = 0 \text{ and}$$

$$\frac{\partial^2 z}{\partial x^2} + \frac{\partial^2 z}{\partial y^2} - u = 0$$

(13.54)

Boundary conditions on both u and z are zero on all edges of the square plate.

Computer code for solving these two equations with boundary conditions is shown in Listing 13.16. Again since a square plate is being considered, the same

```
 1 :  -- File list13_16.lua -- Deflection of plate fixed at edges.
 2 :
 3 :  require'pde2fe'
 4 :  L = 2.0 -- Scale length to 2 by 2
 5 :  nts ={ readnts('list13_15',L)} -- Square area with 2996 nodes
 6 :  nds,tri = nts[1], nts[2] -- Node data
 7 :
 8 :  E,dz,sig,q = 2e11, 0.01, 0.3, 3.36e4
 9 :  D = E*dz^3/(12*(1-sig^2)); print('D = ',D)
10 :
11 :  feqs = {
12 :     function(x,y,uxx,ux,u,uy,uyy)
13 :        return uxx + uyy  - q/D
14 :     end,
15 :     function(nd,u,un,nbs)
16 :        return u
17 :     end
18 :  }
19 :
20 :  getfenv(spgauss).nprint = 1; getfenv(pde2fe).nprint = 1
21 :  getfenv(pde2fe).linear = 1 -- Saves Newton iterations
22 :
23 :  uv = pde2fe(feqs,nts) -- Solve equations for uz
24 :
25 :  feqz = {
26 :     function(x,y,uxx,ux,u,uy,uyy,ntr,mtn,nds,ltr)
27 :        local n1,n2,n3 = nds[1],nds[2],nds[3]
28 :        local l1,l2,l3 = ltr[1],ltr[2],ltr[3]
29 :        return uxx + uyy  - (uv[n1]*l1+uv[n2]*l2+uv[n3]*l3)
30 :     end,
31 :     function(nd,u,un,nbs)
32 :        return u
33 :     end
34 :  }
35 :  z = pde2fe(feqz,nts) -- Solve for deflection
36 :
37 :  x,y = {},{}; NT = 41 -- x - y grid for plotting
38 :  for j=1,NT do x[j] = L*(j-1)/(NT-1); y[j] = x[j] end

39 :  sol = toxysol(uv,x,y,nds,tri); solx = toxysol(z,x,y,nds,tri)
40 :  write_data('list13_16.dat',sol,solx)
41 :  splot(sol); splot(solx)
```

Listing 13.16. Code segment for calculating the deflection of a square plate fixed at the edges subject to a uniform loading.

triangular grid as used in Listing 13.15 is used in the solution (in list13.15 files) on line 5 with a scale factor of 2 so an area of 2 by 2 is modeled. The material and force parameters are defined on line 8 and these result in a flexural rigidity factor D of 18,315. Lines 11 through 23 define and solve the PDE for u while lines 25 through 35 define and solve for the z displacement. Finally lines 37 through 42 set up x and y grids for plotting and save the solution values. Popup plots allow the user to rapidly view the solution. The calls to the pde2fe() function in this example use the simplest possible forms of specifying only the equation set and the node-triangle-sides information. No initial arrays are provided as input and no solution method is requested. The default parameters will be used for these which for review are zero initial values for the solution variables and the full sparse matrix solution technique. The reader is encouraged to re-execute the code and explore the use of the COE and SOR approximate solution methods. While these execute somewhat faster, for this simple example the direct matrix solution method is also very fast.

This example illustrates several additional features of the software and the specification of parameters for a PDE. For the z deflection from Eq. (13.54) values of the u parameter must be evaluated. This variable in turn is obtained from the solution of a PDE as given on line 23. Thus values of the u parameter are known only at the node points of the triangular grid array. For the solution of the second PDE the differential equation needs to be evaluated at four spatial points that are internal to each triangular region. The reader is referred back to Figure 13.7(c) where these points are identified. The solution values must then be interpolated over the triangle using the values known at the triangle nodes. This evaluation is performed on line 29 of Listing 13.16 and uses the values at the nodes (n1, n2 and n3) along with the local coordinate values (l1, l2 and l3) for the calculation. As indicated on line 26 information on the triangle nodes and local coordinates are passed directly to the function used to define the PDE by the software that generates the matrix equations. While this information could be in principle be obtained from the x,y coordinate values, including this information in the calling argument list greatly simplifies the evaluation of the parameters for this PDE.

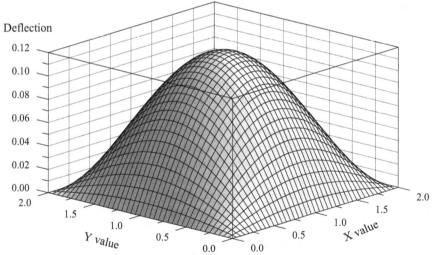

Figure 13.27. Deflection of uniformly loaded square plate obtained from PDE in Listing 13.16.

A surface plot of the calculated deflection is shown in Figure 13.27. The general shape of this is as one would expect with a maximum deflection at the center of the plate of 0.119 spatial units.

This is perhaps a good point to summarize the information passed to the PDE defining equation as on line 26. The passed information includes the x and y coordinates and in addition the number associated with the triangular region being evaluated (the ntr parameter), the number associated with the material marker (the mtn parameter), the three node numbers associated with the triangle (the nds parameter) and a table of the three local coordinates associated with the spatial x,y point. Examples have now been presented that use all of this information. Any given PDE may not require all of this information, but it is available if needed.

13.10.6 Physical Deflection of Circular Plate with Uniform Loading

This problem is similar to the previous problem except for being the deflection of a circular plate of radius 1.0 with a uniform loading. The equation set to be solved is the same as before and that of Eq. (13.54). The circular plate to be considered is illustrated in Figure 13.28 which shows the plate boundary as well as a triangular discretization obtained from executing the EasyMesh program. For input to this program a data file named list13_17.d was used. This file defines the circular boundary as a series of 80 straight line segments spaced at an angular spacing of $360/80 = 4.5°$. As can be seen in the figure the 80 straight line segments form a good approximation to a circular boundary region. In addition to the defining boundary region, the input data file defines four internal square regions which can be seen by close inspection of Figure 13.17. One of these squares located at coordinates +/-0.8, +/-0.8 is identified by the arrow in Figure 13.28. The other squares are located at +/-0.6, +/-0.4 and +/-0.2. These are used to force EasyMesh to generate small triangular elements in the interior of the circular region by specifying a mesh size of 0.04 along these interior square regions. Without defining these interior regions, EasyMesh will generate rather large triangular regions in the interior regions of the circle even though small element sizes are specified on the circular boundary. The generated triangular discretization is seen to be reasonably good but could be improved by forcing smaller element sizes at the center of the circle and by perhaps using six or eight sided boundaries for the four interior regions. The resulting input file (list13_17.d) for EasyMesh is rather large and is not shown here but the reader can examine the file and observe how the various regions are defined to control the size of the generated elements. The reader can also experiment with improving the grid with various changes to the input file. The lines connecting the interior squares are defined in a counter clockwise direction so holes are not generated within the circular region. The grid shown in Figure 13.28 consists of 1031 nodes and 1972 triangles.

A code segment for evaluating the deflection of the circular plate is shown in Listing 13.17. The code is essentially identical to that in Listing 13.16 for the deflection of a uniformly loaded square plate. The input file for nodes, triangles and sides is of course different on line 5. The other difference is the generation of the uniformly spaced x and y data arrays for plotting on line 38 which accounts for the axis values running from -1.0 to +1.0 for the circular plate as opposed to 0.0 to 2.0 for the square plate. The same material and loading parameters are used in both examples.

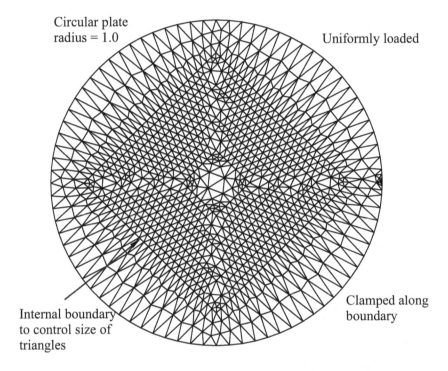

Figure 13.28. Generated triangular grid for circular plate clamped at edge of the circular region.

The calculated circular plate deflection is shown in Figure 13.29. This was plotted from the saved data file, but is very similar to the pop-up plot generated by execution the code. This can be compared with the square plate results in Figure 13.27. The solution shows the radial symmetry as expected for the circular plate. For the circular plate (diameter 2.0 units) the peak deflection is 0.0857 spatial units while the corresponding value for the square plate (2.0 by 2.0 units) was 0.119 spatial units. This agrees with expectations that the circular plate will deflect less than the square plate for the same uniform loading.

```
 1 : -- File list13_17.lua -- Deflection of circular plate
 2 :
 3 : require'pde2fe'
 4 : L = 1.0 -- Radius of circular plate
 5 : nts ={ readnts('list13_17') }
 6 : nds,tri = nts[1], nts[2] -- Node data
 7 :
 8 : E,dz,sig,q = 2e11, 0.01, 0.3, 3.36e4 -- Parameters
 9 : D = E*dz^3/(12*(1-sig^2)); print('D = ',D)
10 :
11 : feqs = {
12 :     function(x,y,uxx,ux,u,uy,uyy,ntr,mtn)
13 :         return uxx + uyy  - q/D
14 :     end,
15 :     function(nd,u,un,nbs)
16 :         return u
17 :     end
18 : }
19 :
20 : getfenv(spgauss).nprint = 1; getfenv(pde2fe).nprint = 1
21 : getfenv(pde2fe).linear = 1 -- Observe progress
22 : uv = pde2fe(feqs,nts) -- Solve equations, using SPM
23 :
24 : feqz = {
25 :     function(x,y,uxx,ux,u,uy,uyy,ntr,mtn,nds,ltr)
26 :         local n1,n2,n3 = nds[1],nds[2],nds[3]
27 :         local l1,l2,l3 = ltr[1],ltr[2],ltr[3]
28 :         return uxx + uyy  - (uv[n1]*l1+uv[n2]*l2+uv[n3]*l3)
29 :     end,
30 :     function(nd,u,un,nbs)
31 :         return u
32 :     end
33 : }
34 : z = pde2fe(feqz,nts) -- Try other solution methods?
35 :
36 : x,y = {},{}; NT = 41 -- x,y grid for plotting results
37 : for j=1,NT do x[j] = 2*L*(j-1)/(NT-1)-L; y[j] = x[j] end
38 : sol = toxysol(uv,x,y,nds,tri); solx = toxysol(z,x,y,nds,tri)
39 : write_data('list13_17.dat',sol,solx)
40 : print('Maximum deflection =',intptri(z,nds,tri,{0,0}))
41 : splot(sol); splot(solx)
Selected Output:
Maximum deflection =    0.085665044298009
```

Listing 13.17. Code segment for calculating the deflection of a uniformly loaded circular plate with fixed edges.

Some features of the values calculated for the rectangular grid on line 38 need some consideration. The x and y grid points evaluated on line 37 cover uniformly the entire two dimensional space from -1,-1 to 1,1 in x,y values. These array values are in turn passed to the toxysol() function on line 38 as points at which the solution is to be evaluated. However, the values returned by the toxysol() function contain only points within the circular spatial region as Figure 13.29 indicates and as can be seen by examining the stored data. The toxysol() function is able to determine that certain points in the input x,y arrays are outside the solution space and these points are omitted in the returned calculation set. This is a convenient feature of the toxysol() function that aids in plotting results from the FE method.

The triangular grid interpolation function intptri() is also used on line 40 to evaluate the maximum deflection at the center of the circular plate. It should also be noted that this function will return 'nil' if the requested point is outside the spatial region defined by the triangular grid structure.

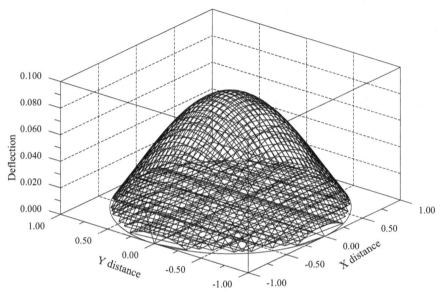

Figure 13.29. Deflection of uniformly loaded circular plate obtained from PDE in Listing 13.17.

13.10.7 Capacitor with Nonlinear Dielectric Material.

The previous examples have all involved PDE's where the coefficients of the second derivative terms have been constant (or piecewise constant). This example will begin to explore FE solutions with nonlinear terms involving the second derivative. Perhaps the simplest such example is that of a parallel plate capacitor such as illustrated in Figure 13.19 and with the spatial grid of Figure 13.20. Poisson's equation for such an example can be expressed as:

$$\frac{\partial}{\partial x}(\varepsilon\frac{\partial U}{\partial x}) + \frac{\partial}{\partial y}(\varepsilon\frac{\partial U}{\partial y}) = 0 \text{ with} \tag{13.55}$$

$$\varepsilon \text{ a possible function of } x, y, U$$

As shown the dielectric constant is assumed to be a possible function of the spatial coordinates as well as the potential. If the dielectric constant is only a function of the spatial coordinates, the equation is still a linear equation.

For a tractable example consider the dielectric constant as:

$$\varepsilon = \varepsilon_1 (1 + \gamma y)^2 (1 + \gamma U'')$$

$$\varepsilon(y = 0) = \varepsilon_1; \quad \varepsilon(y = 1) = \varepsilon_1 (1 + \gamma)^3$$

(13.56)

This is a somewhat artificial example as it would not be likely to find a real physical problem where the dielectric constant varied as in this equation. However, this is a good trial example to explore more complicated nonlinear PDEs since an exact solution can be easily obtained for this nonlinear dielectric constant with the boundary conditions of 0 and 1.0 for the potential on the y boundaries and a zero normal derivative along the x boundaries. The solution of Eq. (13.55) is then a function of the y dimension only and can be obtained from the equation:

$$\frac{U(1 + \gamma U'' / (n+1))}{(1 + \gamma / (n+1))} = \frac{(1+\gamma)}{\gamma} \frac{\gamma y}{(1 + \gamma y)}$$

(13.57)

For n = 1 this can be solved explicitly for the solution in terms of position y. For other values of n the equation provides y as a function of the solution variable U.

```
 1 : -- File list13_18.lua -- Capacitor with nonlinear dielectric
 2 :
 3 : require'pde2fe'
 4 :
 5 : nts = {readnts('list13_11')}
 6 :
 7 : Vm = 1.0
 8 : ep1,gr = 3,3
 9 : feqs = {
10 :     function(x,y,uxx,ux,u,uy,uyy,ntr,mtn)
11 :         ep = ep1*(1 + gr*y)^2*(1+gr*u^2)
12 :         return ep*(uxx + uyy) -- OK
13 :     end,
14 :     function(nd,u,un,nbs,kb)
15 :         if nbs==1 then return u -- Bottom boundary
16 :         elseif nbs==3 then return u - Vm -- Top boundary
17 :         else  return un end -- Sides
18 :     end
19 : }
20 :
21 : getfenv(pde2fe).nprint = 1 -- Observe progress
22 :
23 : u = pde2fe(feqs,nts)
24 :
25 : x,y = {},{}; NT = 21
26 : for j=1,NT do y[j] = (j-1)/(NT-1); x[j] = 2*y[j] end
27 : sol = toxysol(u,x,y,nts[1],nts[2])
28 :
29 : write_data('list13_18.dat',sol)
30 : splot(sol)
Selected output:
Completed Newton iteration 1  with correction  1.0000000000001
Completed Newton iteration 2  with correction  0.18241611326788
--------
Completed Newton iteration 7  with correction  0.00015327097061247
Completed Newton iteration 8  with correction  2.9376299328587e-005
```

Listing 13.18. Code segment for finite element solution of capacitor with nonlinear dielectric using pde2fe() function.

Computer code for solving this PDE for n = 2 is shown in Listing 13.18. The same spatial grid (line 5) is used as for the capacitor problem in Listing 13.11. After defining the PDE and boundary conditions on lines 9 through 19, the solution is obtained with a call to pde2fe() on line 23. This is about as simple a program as one can implement. Several features of the code will be seen. First no defining solution array or initial values are evaluated before the pde2fe() function is called and no input array is specified in the argument list. For such an argument list, the pde2fe() function will define an initial array and set all initial solution values to zero. Also no solution method is specified, so the pde2fe() function will use the default method which is to directly solve the set of matrix equations using the sparse matrix solution method. After obtaining the solution at the triangle grid points, an x-y array of solution values is obtained by lines 25 through 27 for ease in plotting the results. Some selected output is given which shows that the solution required 8 Newton iterations to solve the nonlinear PDE.

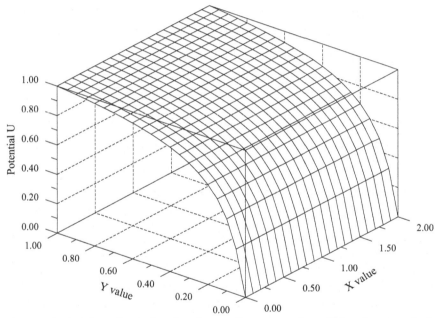

Figure 13.30. Plot of solution for the nonlinear capacitor dielectric example in Listing 13.18.

A plot of the resulting potential solution is shown in Figure 13.30. For this example the dielectric constant changes from a value of 3.0 at y = 0 to a value of 81.0 at y = 1. The potential changes rapidly in the region of low dielectric constant and changes very slowly in the region of high dielectric constant. This figure can be compared with Figure 13.17 which shows the potential solution for a dielectric with an abrupt change in the dielectric constant at the center of the structure. It can be seen from the spatial grid shown in Figure 13.20 that the grid structure has a boundary line in the center of the structure in the y dimension. This boundary line is not used in the present example, but the solution does not show any adverse effects of this defined boundary line for this example.

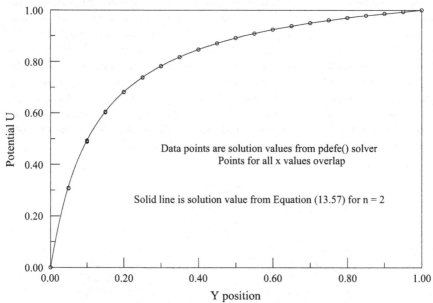

Figure 13.31. Comparison of numerical solution from Figure 13.30 with exact solution values as given by Eq. (13.57).

A comparison of the FE numerical solution with the exact solution values from Eq. (13.57) is shown in Figure 13.31. The data points in the figure are a composite collection of all the x valued solutions points with the results plotted as only a function of position y. A close examination of the figure reveals some slight difference in the potential values for any given y value, but the differences are small. The solid line is seen to be a close fit to the numerical solution values verifying the accuracy of the FE numerical solution. While this is a simple example, it is very important in developing software to have known solutions with which to compare the numerical results in order to verify the accuracy of any numerical computer code. While the accuracy of one solution for a nonlinear problem does not absolutely guarantee the accuracy for other nonlinear problems, the more such problems one solves, the more confidence can be obtained in any algorithm and computer program.

Since this is the first nonlinear PDE solved by the FE method, it is probably important to review again the technique for specifying the PDE to be solved. For this problem the defining equation is on line 12 of Listing 13.18 and is specified as:

$$\text{return ep*(uxx + uyy)} \qquad (13.58)$$

At first glance this would seem to be the same as specifying the equation uxx + uyy = 0 which is the equation for a material with a constant dielectric constant. However the obtained solution is obviously not that for a constant dielectric constant as that would result in a straight line variation with position. So the question arises as to what is being implemented here? The answer is that the formulation of the second derivative really means the following:

$$eq*(uxx + uyy) \text{ really means } \rightarrow \frac{\partial}{\partial x}(eq\frac{\partial u}{\partial x}) + \frac{\partial}{\partial y}(eq\frac{\partial u}{\partial y}) \qquad (13.59)$$

So in this case "What you see in **NOT** what you get". If this is confusing, the reader is referred back to the discussion in connection with Eq. (13.39) through (13.41) for a further clarification. This notation was adapted to have a simple input method for PDEs with the form of Eq. (13.59) which is the most common form of second derivative terms. The notation used provides the simplest code representation of the PDE that allows the software to evaluate the value of the eq parameter by use of numerical derivatives. If the statement getfenv(setup2feeqs).truefd = true is added to the code in Listing 13.18, then the computer code will give "What you see is what you get" results. However, then the additional first order derivative terms will have to be added to the defining equation set for Poisson's equation.

Perhaps to make the usage a little more clear, there are three ways to specify the PDE equation to be solved for this example where the dielectric constant can be a function of the spatial coordinates and the solution variable. These are:

$$(a) \; \varepsilon(U_{xx} + U_{yy}) = 0$$

$$(b) \; \varepsilon(U_{xx} + U_{yy}) + \frac{\partial \varepsilon}{\partial U}(U_x^2 + U_y^2) + \frac{\partial \varepsilon}{\partial x}U_y + \frac{\partial \varepsilon}{\partial y}U_y = 0 \qquad (13.60)$$

$$(c) \; (U_{xx} + U_{yy}) + \frac{1}{\varepsilon}\frac{\partial \varepsilon}{\partial U}(U_x^2 + U_y^2) + \frac{1}{\varepsilon}\frac{\partial \varepsilon}{\partial x}U_y + \frac{1}{\varepsilon}\frac{\partial \varepsilon}{\partial y}U_y = 0$$

where the explicit form of the dielectric constant and the derivative terms has not been shown. Form (b) is a direct result of expanding the partial derivative terms in Eq. (13.55) and has mathematically correct or 'true' first derivative terms. Form (c) is obtained from (b) by dividing by the dielectric constant and has a constant (unity) multiplying the second derivative terms. Finally form (a) is the simplified form omitting the expansion terms for the dielectric constant. While forms (b) or (c) would appear to be the most accurate expressions, form (a) is sufficient for specifying the PDE to the pde2fe() coded function. The reason for this is that the function (setup2feeqs()) used to formulate the equation set assumes that any spatial dependency of the factor multiplying the second derivative terms in form (a) resulted from the basic form of the equation as represented by Eq. (13.55). Further if form (b) is used to define the PDE, then the statement getfenv(setup2feeqs).truefd = true must be executed before calling the code to inform the software that the PDE is defined with the "true" first derivative terms. Form (c) can be used either with or without the setting of the truefd term since the factor multiplying the second derivative terms is a constant. The reader is encouraged to experiment with setting the truefd variable and with the three different formulations of the equation set as in Eq. (13.60).

Another factor to consider is the fact that the formulation of the equation set for Newton iterations in setup2feeqs() assumes that the factor multiplying the second derivative terms is only a function of the spatial variables and not a function of the solution variable. This means that in an example such as shown here, the linearized Newton equation set will be somewhat in error. However, if the second degree coefficient (ε in this case) varies sufficiently slowly with the solution variable, the Newton iterations can converge but may simply require a few more iterations. Provided the Newton iterations converge, the proper solution will be obtained in any case. For this example, it is found that the procedure does in fact converge in 8 Newton iterations to the expected solution even though there is a nonlinear term multiplying the second derivative.

13.10.8 Nonlinear Diffusion

In Section 12.5 the case of nonlinear diffusion in time and one spatial dimension was considered. Expanding this to two spatial dimensions, the PDE formulation is:

$$\frac{\partial}{\partial x}\left(D(U)\frac{\partial U}{\partial x} \right) + \frac{\partial}{\partial y}\left(D(U)\frac{\partial U}{\partial y} \right) - \frac{\partial U}{\partial t} = 0$$

(13.61)

$$D(U) = D_0 + D_1(U/ni) + D_2(U/ni)^2$$

In this equation the $D(U)$ function assumes a similar role to the $\varepsilon(x, y, U)$ function in the previous example. The reader is referred back to Section 12.5 for a discussion of this equation and the one dimensional time dependent solution. In the present chapter in Section 13.9 an example of linear diffusion into a two dimensional surface was presented. For that example, a triangular mesh array was developed and shown in Figure 13.15. The present example combines the nonlinear diffusion model of Section 12.5 with the FE mesh of Figure 13.15 to demonstrate a second nonlinear PDE solution using the FE approach. The reader should review this previous material as this section builds upon that material.

The code segment for this example is shown in Listing 13.19. In this example only the D_1 parameter is given a non-zero value and the parameters are customized for Boron diffusion into Silicon at 1000 C. The diffusion surface is assumed to be held at a fixed concentration of $5 \times 10^{20}/cm^3$. In the calculation, the solution is normalized to this concentration for simplicity in the solution. While the relative grid of Figure 13.15 is used in the calculation, the size of the rectangular area is changed by a factor of L = 1.e-4 as seen on line 4 of the listing. This is used on line 5 as the second argument of the readnts() function to change the dimensions to an area of 2.e-4 by 1.e-4 in place of the originally defined 2 by 1 area. Lines 7 through 13 define the material and diffusion parameters. This is followed on lines 14 through 23 by defining the PDE and boundary conditions in the standard way.

```
 1 : -- File list13_19.lua -- Time dependent nonlinear diffusion
 2 :
 3 : require'pde2fe' -- Input FE code
 4 : L = 1.e-4
 5 : nts ={readnts('list13_9',L)} -- Get node, triangle, sides data
 6 : -- Model equations to be solved
 7 : D00,D10,E0,E1 = 0.05, 0.95, 3.5, 3.5 -- Diffusion parameters
 8 : T = 1000+273 -- temperature
 9 : D0 = D00*math.exp(-E0/(0.026*T/300))
10 : D1 = D10*math.exp(-E1/(0.026*T/300))
11 : ni = 7.14e18; Un = 5e20
12 : Um = 1.0 -- Use normalized value
13 : xmax,ymax = 2*L, L
14 : feqs = { -- Table of functions
15 :    function(x,y,t,uxx,ux,u,uy,uyy,ut,utt) -- General point
16 :        D = D0 + D1*Un*u/ni
17 :        return D*(uxx + uyy) - ut
18 :    end,
19 :    function(nd,t,u,un,ut,utt,nbs)  -- Boundary values
20 :        if nbs==3 then return u - Um
21 :        else return un end
22 :    end
23 : }  -- End general point and boundary values
24 : u = {}
25 : for k=1,#nts[1] do
26 :    if nts[1][k][3] == 3 then u[k] = Um
27 :    else u[k] = 0.0 end
28 : end
29 :
30 : SPM,COE,SOR = 1, 2, 3 -- 3 solution methods
31 : t = os.time()
32 : getfenv(pde1stp2fe1t).nprint=1
33 : getfenv(pde2fe1tqs).nprint=1
34 : getfenv(pde2fe).nprint=1
35 : tvals = {0,{1,1000},2,20}
36 : u,errm = pde2fe1tqs(feqs,tvals,nts,u,SOR)
37 : print('Time taken =',os.time()-t); io.flush()
38 :
39 : x,y = {},{}; NT = 81 -- Change as desired
40 : for j=1,NT do y[j] = ymax*(j-1)/(NT-1); x[j] = 2*y[j] end
41 : nsol = #u

42 : for i=1,nsol do -- Save 2D data files
43 :    sfl = 'list13_19.'..i..'.dat'
44 :    sol,solxy = toxysol(u[i],x,y,nts[1],nts[2])
45 :    write_data(sfl,sol)
46 :    if i==nsol then
47 :            splot('list13_19a.emf',solxy);
cplot('list13_19b.emf',solxy)
48 :    end
49 : end

Selected output:
Time taken =    351
```

Listing 13.19. Code segment for FE solution of nonlinear diffusion problem.

Again the PDE to be solved on line 17 is not exactly as it appears but is a "proxy" equation for Eq. (13.61).

After defining time parameters on line 35, the time dependent solution is generated on line 36 by use of the quick scan pde2fe1tqs() function. After an initial time interval (from 0 to 1) using linearly spaced time values, the function increments time on a logarithmic basis from 1 to 1000. The 2,20 input specification in tvales[] on line 35 generates 2 saved solution values per decade in time and solve the equation set at an additional 20 time values between each saved solution. Thus a total of 40 time increments are used per decade in time or a total of 160 time solutions for the total calculation. Such a logarithmic time increment scheme is very useful for problems such as the diffusion equation where solution changes are very rapid for short times but become much slower as time progresses. The other parameter of note in the pde2fe1tqs() solution is the SOR (value of 3) parameter on line 36 which tells the routine to use the SOR iterative approximate method to solve the set of matrix equations.

The discretization for this problem generates 2996 nodes and thus there are 2996 equations to be solved for each time solution (or a 2996 by 2996 sparse matrix). Since the solution changes are expected to be small from time step to time step, the total solution time is expected to be reduced using the SOR technique as opposed to direct solution of the sparse matrix approach (using the SPM parameter). The reader can verify this by re-executing the code using the SPM parameter. The selected output line indicates that for the computer used to develop this code, the solution required 349 sec. (or about 6 min).

The printed output, not shown here, also indicates that 3 to 4 Newton iterative cycles are required at each time point to solve the nonlinear set of PDE equations. The net result of all this is that the set of 2996 by 2996 matrix equations must be solved a total of over 500 times (or about 0.7 sec per solution). The PDE solutions of this and the previous chapter are some of the longest running programs in this text. The reader is encouraged to re- execute the code in Listing 13.19 using the COE and SPM parameters and determine which approximate solution method is faster for this particular problem.

Figure 13.32 shows the computed concentration profile for the last time point calculated at t = 1000 sec. The plot is a log plot for the concentration profile indicating the sharp drop-off in concentration resulting from the concentration dependent diffusion profile. The solution appears to be in general as expected.

It would be very useful if the accuracy of the solution could be further verified as another check on the FE code. This can be done by comparing with the results of Section 12.5 and in particular Figure 12.7 which shows the concentration profile for a one-dimensional diffusion with the same parameters. To connect with the present calculation it can be noted that the diffusion along the x = 2.e-4 boundary should closely approximate the one dimensional diffusion profile. Figure 13.33 shows a comparison of the two generated solutions for times of 10, 100 and 1000 sec. In the figure the solid lines are the same as the curves in Figure 12.7 obtained as a one-dimensional solution of the nonlinear diffusion equation.

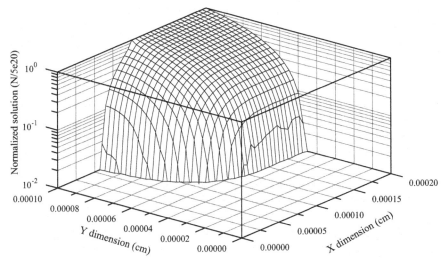

Figure 13.32. Calculated diffusion profile at a time of 1000 sec from Listing 13.19.

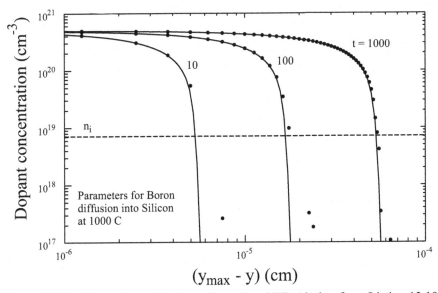

Figure 13.33. Comparison of the two dimensional FE solution from Listing 13.19 with the one dimensional solutions of Figure 12.7. Solid lines are from Figure 12.7. Data points are two-dimensional FE solution along the x = 2e-4 cm boundary.

The data points are from the two-dimensional solution of Listing 13.19 along the x = 2e-4 boundary. Relatively good agreement is seen for about two orders of magnitude in the solution variable. For solution values below the 1.e18 value, there is considerably difference especially for the shorter diffusion times. Also for the t = 10 sec case the spatial grid used in the x-y evaluation of the solution allows only five spatial points to be compared with the one-dimensional case. It can be recalled that the one-dimensional solution used a much finer spatial grid near the diffusion source to generate a much more accurate solution. It is obvious from Figure 13.33, that a finer spatial grid would be necessary to generate a more accurate solution for short times such as the 10 sec curve. The good agreement in the calculated values shown in Figure 13.33 however, are sufficient to conclude that the FE programs are generating accurate solutions but are of course fundamentally limited by the number of spatial intervals used in the calculation.

13.10.9 Semiconductor P-N Junction – Potential and Carrier Densities

This example is similar to that of Section 12.8.4 and the reader is referred back to that section for a more complete discussion. In particular the structural geometry for this example is shown in Figure 12.81. Again from the symmetry of the problem, only the left half of the structure will be considered. This is a rectangular spatial region of size 2.e-4 cm by 1.e-4 cm in the x and y dimensions. To compare with the numerical results of Section 12.8.4, the p-n junction will first be considered as having a rectangular shape as seen in Figure 12.81. The PDE corresponding to the equilibrium potential around the p-n junction is repeated here as:

$$\frac{\partial^2 V}{\partial x^2} + \frac{\partial^2 V}{\partial y^2} = \frac{q}{\varepsilon}(p - n + N_{Net})$$

$$N_{Net} = N_D \text{ for } n \text{ region and } = -N_A \text{ for p region} \qquad (13.62)$$

$$p = N_A \exp(-V/V_T), \quad n = (n_i^2/N_D)\exp(V/V_T)$$

The nonlinearity is in the exponential dependence of the carrier densities on the potential. An input file for generating a triangular spatial grid using EasyMesh is shown in Listing 13.20. The resulting spatial grid is shown in Figure 13.34. Note that in this figure the location of the p-n junction is near the bottom boundary as opposed to Figure 12.81 where the junction is near the top of the figure. Small spatial step sizes are generated along the p-n junction boundary and the detail is too small to be seen in this figure. The small triangular elements are specified in Listing 13.20 by the .01 parameter values associated with the points along the lines defining the junction boundary. The EasyMesh program using Listing 13.20 generates 2221 nodes and 4313 triangular elements.

```
# Example of p-n junction #
11 # Boundary point section #

  0: 0      0     .05  1
  1: .5     0     .01  1
  2: .75    0     .02  2
  3: 1.0    0     .02  2
  4: 1.0    .15   .02  2
  5: 1.0    .25   .01  1
  6: 1.0    1.0   .05  3
  7: 0      1.0   .05  3
  8: 0.5    0.25  .01  4

# define interior points for two materials #
  9: .5     0.5   0    1  # Material 1 marker #
  10: .75   0.1   0    2  # Material 2 marker #

10 # Boundary sides section #

  0: 0   1   1
  1: 1   2   1
  2: 2   3   2
  3: 3   4   2
  4: 4   5   1
  5: 5   6   1
  6: 6   7   3
  7: 7   0   1

  8: 5   8   400 # p-n junction boundary #
  9: 8   1   400
```

Listing 13.20. Input file for EasyMesh to generate spatial grid for p-n junction PDE.

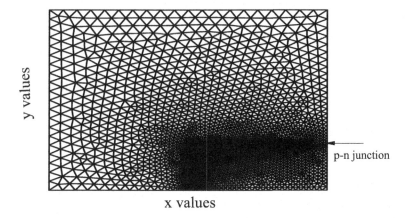

Figure 13.34. Spatial grid for p-n junction using input file from Listing 12.20.

```
 1 : -- File list13_21.lua --
 2 : -- Example of BV problem for p-n junction by FE method
 3 : require"pde2fe"; exp = math.exp
 4 :
 5 : -- Material and device parameters
 6 : q = 1.6e-19; eps = 11.9*8.854e-14; L = 0.4e-4
 7 : vt = .026; ni = 1.45e10
 8 : Na = 1e19; Nd = 1.e17-- Doping densities
 9 : x1,y1,x2 = L/2, L/4, L*3/4; va = 0.0
10 : qdep = q/eps; vj = vt*math.log(Na*Nd/ni^2); no = ni^2/Na
11 :
12 : nts = {readnts('list13_21',L)} -- Read spatial data
13 : nds = nts[1] -- nodes data
14 : npts,u = #nds, {} -- number of nodes
15 : for i=1,npts do -- set initial value & scale dimensions
16 :     xt,yt = nds[i][1],nds[i][2]
17 :     if xt>=0.5*L and yt<=0.25*L then u[i] = 0.0
18 :     else u[i] = vj end -- Set initial potential values
19 : end
20 :
21 : feqs = { -- Equation to be solved
22 :     function(x,y,uxx,ux,u,uy,uyy,ntr,mtn)
23 :             if mtn==2 then Nnet = -Na else Nnet = Nd end
24 :             p, n = Na*exp(-u/vt), no*exp(u/vt)
25 :             return uxx + uyy + qdep*(Nnet + p - n)
26 :     end,
27 :     function(nd,u,un,nbs,kb) -- Boundary values
28 :             if nbs==3 then return u-vj -- Top voltage contact
29 :             elseif nbs==2 then return u-va -- Bottom contact
30 :             else return un end -- Else zero normal derivative
31 :     end
32 : }
33 :
34 : getfenv(pde2fe).nprint = 1
35 : SPM,COE,SOR = 1, 2, 3 -- 3 solution methods
36 :
37 : u,errm = pde2fe(feqs,nts,u,COE) -- Solve by FE method
38 :
39 : x,y = {},{}; NT = 81
40 : for i=1,NT do -- Define uniform x-y grid for plotting
41 :     x[i] = (i-1)*L/(NT-1); y[i] = x[i]
42 : end
43 : ut = toxysol(u,x,y,nts[1],nts[2])
44 : pa,na = {},{} -- Calculate hole and electron densities
45 : for i=1,#ut[3] do
46 :     ua = ut[3][i]
47 :     pa[i] = math.log10(Na*exp(-ua/vt))
48 :     na[i] = math.log10(no*exp(ua/vt))
49 : end
50 : splot(ut); write_data('list13_21.dat',ut,pa,na)
```

Selected Output:

```
Completed Newton iteration 1  with correction  0.72172117308489
Completed Newton iteration 2  with correction  0.34879084122029
Completed Newton iteration 3  with correction  0.11186057092501
Completed Newton iteration 4  with correction  0.027747758446252
Completed Newton iteration 5  with correction  0.0029650148652629
Completed Newton iteration 6  with correction  5.3724198273396e-005
```

Listing 13.21. Code segment for solving p-n junction PDE using FE approach.

A code segment for solving the p-n junction PDE is shown in Listing 13.21. The code should be very familiar to the reader and only a brief explanation will be give. After defining material parameters and initial values through lines 19, the PDE and boundary conditions are defined by the two functions on lines 21 through 32. Fixed potentials are defined on the top and bottom contact areas and zero normal derivatives are defined on all other boundary surfaces. The PDE is solved by a call to the pde2fe() function on line 37. In this example, the COE approximate matrix solution method is used to solve the set of matrix equations. The iterative COE and SOR methods are somewhat faster, than the direct matrix solution method, although in this case, the solution is reasonable fast by either method since only 6 Newton iterations are required for the solution as observed from the printed output. The rapid convergence for the last two iterations is seen from the output. Finally, lines 39 through 50 define linear x-y arrays to save the solution for easier plotting of results.

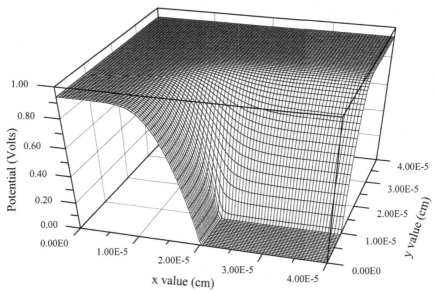

Figure 13.35. Plot of Potential for the p-n junction PDE in Listing 13.21 by the FE method.

A plot of the potential solution is shown in Figure 13.35. This should be compared with Figure 12.82 which is the solution for the same problem obtained by the FD method of the previous chapter. The figures appear to be very similar and in fact a more detailed look at the differences between the two solutions verifies the reasonably close agreement in the solutions. If the maximum difference in the potential solution are compared by the FE and FD method it is found that the largest difference is about 0.1 (volts) and occurs at the point where the x-y abrupt change in the p-n junction occurs, i.e. at x,y = 2.e-5,1.e-5. It would be expected that the FE method gives a more accurate solution since the spatial grid size is much smaller around the impurity interface than for the FD method. However, this can not be completely verified without other calculations at varying spatial step and increment sizes. This is left to the interested reader.

This example provides an opportunity to explore the rate of convergence of the solution as a function of the number of Newton iteration cycles and some results are shown in Figure 13.36. This data was obtained by temporarily specifying a very small error for terminating the series of Newton iterations. It will be recalled that for each Newton iterative cycle the nonlinear PDE is linearized about the approximate solution and then updated to a more accurate solution. Consider first the SPM curve. The figure shows that the maximum correction rapidly decreases in value and reaches a minimum correction value of around 1.e-16 which represents the limit of machine accuracy. Thus the FE solution approaches a static solution value after about 10 iterations with a relative error of about 1.e-16. This does not mean of course that the FE method results in a relative accuracy of 1.e-16 for the PDE as there is still an error due to the finite spatial grid size used in the solution. However, given the set of matrix equations, the solution accuracy is approaching that of the intrinsic software. To explore the true accuracy of the nonlinear PDE equation would require that the PDE be solved for a range of spatial grid sizes.

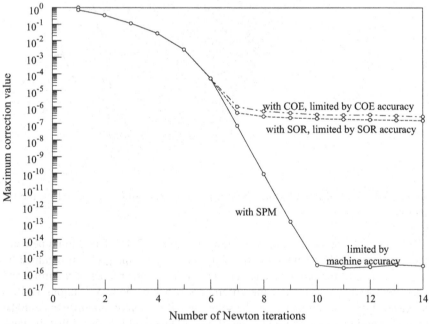

Figure 13.36. Maximum correction values with Newton iterations number for p-n junction PDE of Listing 13.21.

Now consider the COE and SOR labeled curves in Figure 13.36. The curves show that using these approximate solution methods for the matrix solution, the corrections at each Newton cycle is approximately the same until about the 7th Newton iteration. At this point the maximum correction is in the rage of 1.e-6 and the results become limited by the specified accuracy of the COE or SOR iterative method. For this example the maximum correction is close to the relative correction since the maximum value of the potential is close to unity. The default value in the pde2fe() code terminates the Newton iterations when the relative error is below 1.e-4. This is seen to be before the limits are reached for the COE or SOR iterative methods. This shows that any attempt to improve the accuracy of the Newton iterations should be accompanied by an improvement in the specified accuracy of the COE and SOR iterative methods. The existing default termination parameters should be sufficient for most engineering problems. The advantage of the approximate matrix solution methods is of course that the computer time required for the FE solution in Listing 13.21 can be greatly reduced by using either the COE or SOR iterative matrix solution methods as opposed to the direct matrix solution (or SPM).The user is encouraged to experiment with solving this problem with different methods and parameters.

The solution in Listing 13.21 also includes calculations of the electron and hole densities (or rather the log10 of these values) on lines 44 through 49. Although graphs of these are not shown here, the data is saved in an output file so the reader can plot and compare with similar results obtained in Chapter 12 by the FD technique.

This p-n junction example is useful in that it is directly comparable with the same example solved by the FD method. However, in the details of the example, it is not a very practical example as one would not typically encounter a p-n junction with a 90 degree angle between the x and y junction coordinates as in this example. From impurity diffusion, such as in Section 13.10.8, a typical p-n junction exhibits a circular arc at the edge of the junction such as shown in Figure 13.37. In this figure normalized dimensions are used so the relative structure can be used for various spatial region sizes and doping densities.

An appropriate input file for EasyMesh to generate a triangular grid of nodes and triangles for this example is shown in Listing 13.22. The input file is similar to that of Listing 13.20 except for the 5 defining line segments for the circular arc part of the p-n junction. Figure 13.37 is drawn with straight lines connecting these points so one can see that this defines reasonable well a circular arc. More line segments can be used if desired to more accurately define the circular arc section of the junction, but more segments will probably not produce any significantly different results. Structures of this general type are where the advantages of the FE method really shine as it is very easy to approximate general curved structures with straight line segments and use these in the EasyMesh program to generate an appropriate grid structure. In addition the input file defines two spatial points within the p and n type regions with material markers of 1 and 2 to aid in identifying the two types of semiconductor material.

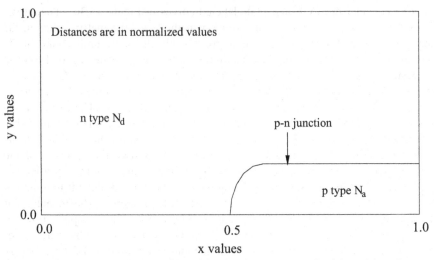

Figure 13.37. p-n junction profile with circular arc at end of junction. Due to symmetry only half of a junction is shown.

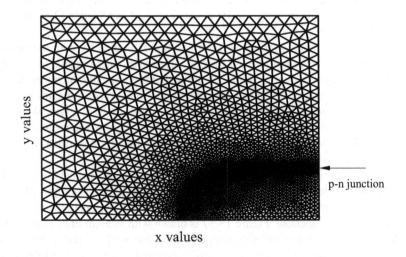

Figure 13.38. Triangular grid structure from EasyMesh for circular p-n junction boundary.

Figure 13.38 shows the triangular grid structure generated by EasyMesh for this input. The triangle size is very small around the physical p-n junction from the specification of the .01 parameter for the points of the circular arc and the triangular size increases as the triangles move away from the junction as desired to a specified size of .05 along the sides. This example generates 2090 nodes and 4051 triangles over the spatial region. The individual triangles are too small around the junction to be individually seen in Figure 13.38 but are approximately 1/5 of the length of those along the sides (or 1/25 the area).

```
# Example of p-n junction with circular boundary #
15 # Boundary point section #
 0: 0     0     .05  1
 1: .5    0     .01  1
 2: .75   0     .02  2
 3: 1.0   0     .02  2
 4: 1.0   .15   .02  2
 5: 1.0   .25   .01  1
 6: 1.0   1.0   .05  3
 7: 0     1.0   .05  3
 8: .5086 .0773 .01  4 # Circular arc #
 9: .5334 .1469 .01  4
10: .5721 .2023 .01  4
11: .6209 .2378 .01  4
12: .6750 .2500 .01  4
# define interior points for two materials #
13: .5    0.5   0    1 # Material 1 marker #
14: .75   0.1   0    2 # Material 2 marker #
14 # Boundary sides section #
 0: 0   1   1
 1: 1   2   1
 2: 2   3   2
 3: 3   4   2
 4: 4   5   1
 5: 5   6   1
 6: 6   7   3
 7: 7   0   1
 8: 1   8   400 # p-n junction boundary #
 9: 8   9   400
10: 9  10   400
11: 10 11   400
12: 11 12   400
13: 12 5    400
```

Listing 13.22. Input file for EasyMesh appropriate for the p-n junction structure shown in Figure 13.37.

Computer code for solving for the potential around the circular arc p-n junction using the FE method is shown in Listing 13.23. This code is almost identical to Listing 13.21. The major difference is in the input for the spatial grid structure which is obtained from the input file (in this case the "list13_23" files on line 12). The other significant difference is in setting up the initial values for the potential on lines 15 through 19 of the code. The triangle input date in field 12 of the array, provides a marker value (of 1 or 2) which identifies on which side of the p-n junction the triangle is located. This marker value is coded back in the EasyMesh input file by the interior point marker lines (lines 13: and 14:) of Listing 13.22. This

```
 1 : -- File list13_23.lua --
 2 : -- Example of BV problem for p-n junction by FE method
 3 : require"pde2fe"; exp = math.exp
 5 : -- Material and device parameters
 6 : q = 1.6e-19; eps = 11.9*8.854e-14; L = 0.4e-4
 7 : vt = .026; ni = 1.45e10
 8 : Na = 1e19; Nd = 1.e17-- Doping densities
 9 : x1,y1,x2 = L/2, L/4, L*3/4; va = 0.0
10 : qdep = q/eps; vj = vt*math.log(Na*Nd/ni^2); no = ni^2/Na
11 :
12 : nts = {readnts('list13_23',L)} -- Read spatial data
13 : ntr = nts[2] -- triangle data
14 : ntrs,u = #ntr, {} -- number of triangles
15 : for i=1,ntrs do -- set initial value & scale dimensions
16 :    tr = ntr[i] -- tr[12] is material number, 1 or 2
17 :    if tr[12]==2 then u[tr[1]],u[tr[2]],u[tr[3]] = 0,0,0
18 :    else u[tr[1]],u[tr[2]],u[tr[3]] = vj,vj,vj end
19 : end
20 :
21 : feqs = { -- Equation to be solved
22 :    function(x,y,uxx,ux,u,uy,uyy,ntr,mtn)
23 :       if mtn==2 then Nnet = -Na else Nnet = Nd end
24 :       p, n  = Na*exp(-u/vt), no*exp(u/vt)
25 :       return uxx + uyy + qdep*(Nnet + p - n)
26 :    end,
27 :    function(nd,u,un,nbs,kb) -- Boundary values
28 :       if nbs==3 then return u-vj -- Top voltage contact
29 :       elseif nbs==2 then return u-va -- Bottom contact
30 :       else return un end -- Else zero normal derivative
31 :    end
32 : }
33 :
34 : getfenv(pde2fe).nprint = 1
35 : SPM,COE,SOR = 1, 2, 3 -- 3 solution methods
36 :
37 : u,errm = pde2fe(feqs,nts,u,SOR) -- Solve by FE method
38 :
39 : x,y = {},{}; NT = 81
40 : for i=1,NT do -- Define uniform x-y grid for plotting
41 :    x[i] = (i-1)*L/(NT-1); y[i] = x[i]
42 : end
43 : ut = toxysol(u,x,y,nts[1],nts[2])
44 : pa,na = {},{} -- Calculate hole and electron densities
45 : for i=1,#ut[3] do
46 :    ua = ut[3][i]
47 :    pa[i] = math.log10(Na*exp(-ua/vt))
48 :    na[i] = math.log10(no*exp(ua/vt))
49 : end
50 : splot(ut); write_data('list13_23.dat',ut,pa,na)
Selected Output;
Completed Newton iteration 1  with correction  0.72985457097852
Completed Newton iteration 2  with correction  0.34899103517258
Completed Newton iteration 3  with correction  0.11163690124596
Completed Newton iteration 4  with correction  0.026996849735389
Completed Newton iteration 5  with correction  0.0027636904814619
Completed Newton iteration 6  with correction  4.5328056756516e-005
```

Listing 13.23. Code segment for solving for potential around p-n junction for Figure 13.37 with FE method.

value of the material marker is used on lines 17 and 18 of Listing 13.23 to set the values of the three triangle nodes to either 0.0 or vj. It will be noted that the nodes exactly on the junction boundary are associated with triangles located in both materials. Hence the nodes located exactly on the boundary, may be assigned to either of the initial voltage values. However this does not matter in the long run as this is all sorted out in the subsequent Newton iterations for the smoothly varying potential values around the p-n junction. For the initial approximations, it is only important that the values far from the junction be set to the appropriate boundary values. For this example the Newton iterations will not converge if one simply sets all the initial values to zero as the initial approximation is too far from the solution for a correct iterative convergence. This procedure could have been used in Listing 13.21 to set initial values and in this case the codes would be identical except for the input file defining the spatial nodes and triangles.

The listing shows that the SOR approximate matrix solution method is used on line 37 for the matrix solution method. This is used to simply illustrate another of the solution approaches. The reader should experiment with other possible methods (SPM and COE). In all cases the solution should converge in 6 Newton iterations with approximately the same correction factors as shown in the listing.

The potential solution and the calculated hole and electron densities are shown in Figures 13.39 through 13.41. These are from the saved files on line 50 of Listing 13.23. The computed results are evaluated on a rectangular grid of x-y points for easier plotting on lines 44 through 49 of the listing. Figure 13.39 can be compared directly with Figure 13.35 for the potential solution. For the carrier densities, the figures can be compared with Figures 12.83 and 12.84 for the square corner p-n junction case. In each case a much smoother transition in the solution

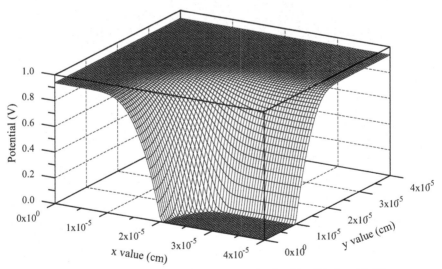

Figure 13.39. Potential around p-n junction as obtained by Listing 13.23.

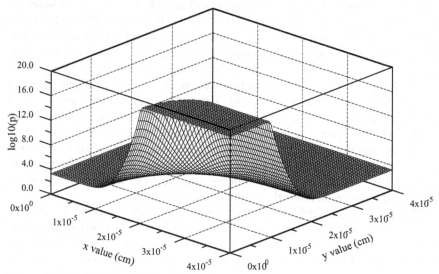

Figure 13.40 Calculated hole density (on a log10 scale) around p-n junction obtained from Listing 13.23

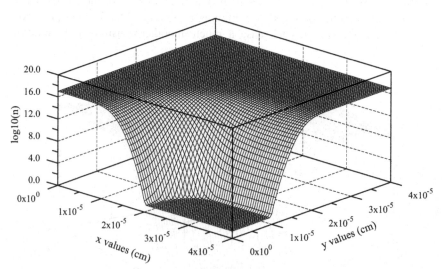

Figure 13.41 Calculated electron density (on a log10 scale) around p-n junction obtained from Listing 13.23.

variables are observed around the p-n junction edge. Since closed form solutions are not possible for such nonlinear PDEs, it is not possible to access exactly the accuracy of the obtained solutions. However, the previous comparisons with known solutions and the very smooth solution curves for the carrier densities that vary over many orders of magnitude, provide confidence that accurate solutions are being obtained. To probe further as to the accuracy would require additional calculations using different spatial grid sizes. The reader is encouraged to do this and compare results from using various spatial grids.

13.10.10 Vibrations of a clamped circular membrane

A circular plate clamped at the edges has previously been considered in Section 13.10.6. In that example the deflection of a plate under uniform loading was considered. The example here is somewhat similar but is that of the time dependent vibration of a clamped circular membrane such as a circular drum head. The describing PDE is of the form:

$$\frac{\partial^2 U}{\partial x^2} + \frac{\partial^2 U}{\partial y^2} - \frac{1}{c^2}\frac{\partial^2 U}{\partial t^2} - \frac{\gamma}{c}\frac{\partial U}{\partial t} = 0 \qquad (13.63)$$

In this c is the velocity of transverse waves and γ is an empirical "damping" coefficient. The velocity is related to more fundamental parameters by:

$$c = \sqrt{\frac{T_s}{\rho_s}}$$

$$T_s = \text{surface tension (N/m)} \qquad (13.64)$$

$$\rho_s = \text{surface mass density (kg/m}^2)$$

This example is thus that of a second order time derivative (and possible a first order one) which provides a final test of the FE time dependent code. (Previous examples have considered only first order derivatives).

This is a linear PDE problem that has been much studied in the classical literature. It is usually approached in the polar domain (or r and θ variables) and for the undamped case by converting the spatial derivatives to:

$$\frac{\partial^2 U}{\partial r^2} + \frac{1}{r}\frac{\partial U}{\partial r} + \frac{1}{r^2}\frac{\partial^2 U}{\partial \theta^2} - \frac{1}{c^2}\frac{\partial^2 U}{\partial t^2} \qquad (13.65)$$

with $U(a,\theta,t) = 0$ for fixed boundary at $r = a$

Solutions of this are know to exist in the form of product functions as:

$$U(r,\theta,t) = R(r)\Theta(\theta)T(t) \qquad (13.66)$$

Solutions for time ($T(t)$) are known to be sinusoidal in time and solutions for the angular dependence ($\Theta(\theta)$) are known to be sinusoidal in angle ($n\theta$ for example). Finally solutions for the radial function ($R(r)$) are known to be Bessel functions of integer order (n for example). Summarizing this, the solution for fundamental vibration modes is of the form:

$$U(r,\theta,t) = J_n(k_{m,n}r/a) \begin{Bmatrix} \cos(n\theta) \\ \sin(n\theta) \end{Bmatrix} \begin{Bmatrix} \cos(\omega t) \\ \sin(\omega t) \end{Bmatrix}$$

$$n = 1,2,3\ldots;\ k_{m,n} = m^{th}\ \text{root of}\ n^{th}\ \text{Bessel function}\ J_n(k_{m,n}) = 0 \qquad (13.67)$$

$$\text{and}\ \omega = k_{m,n}c/a$$

The meaning of the curly brackets in the above equation is that terms in the brackets may be multiplied by constants and added together for a complete solution. Finally a general solution may contain any number of fundamental modes multiplied by various constants. The solution space thus has a rich array of spatial and time responses. In order to narrow down the solution and have something to compare with the FE calculation, it will be assumed that the surface is initially excited with only the zero order Bessel function ($n = 0$). The first and second roots of J_0 are also known to occur at 2.40483 and 5.52008. If we then pick an initial excitation (at $t = 0$) corresponding to the second root, the response should be given by:

$$U(r,\theta,t) = J_0(5.52008r/a)\cos(5.52008t) \qquad (13.68)$$

With this initial condition, the FE numerical calculation can then be compared with a known expected time dependency.

For the FE analysis a discrete node and triangle spatial structure is needed and one such descretization has previously been defined for a circle of unit radius in Section 13.10.6 and as show in Figure 13.28. This triangular grid will also be used here for the FE analysis.

A code segment for the FE analysis is shown in Listing 13.24. The nodes and triangles are input on line 5 from the previous 'list13_17' files. The PDE and boundary condition are defined on lines 14 through 21 in the familiar form. Provisions are made for any value of C and gm although in this example these are taken as 1.0 and 0.0. The initial values for the solution variable are set using the loop on lines 24 through 26 over the nodes. First the r value is calculated on line 25 and then used on line 26 to set the initial solution value to the desired Bessel function which is obtained from the set of functions in the elemfunc file previously discussed in Section 7.8. The solution is obtained on line 32 by a call to the FE solver, pde2felt(), which includes one time dimension to the 2D FE equation set. Perhaps the only input needing additional explanation is the tvals parameter. The first four table values (0,8,200,1) specify a time interval of 0 to 8 sec (based upon the expected oscillation frequency). The 200,1 values specify 200 saved time interval values and only 1 time interval per saved time interval. In other words the input values request that all calculated time interval arrays be saved and returned by the pde2felt() function. Finally the fifth entry in tvals is a table specifying 7 spatial positions around the membrane at which time data will be collected at each calculated time point. Thus data should be collected at 200 (actually 201) time points during the FE calculation. Note that if the number of time points had been specified at 100,2 instead of the 200,1, calculations would be made at exactly the same number of time points, but spatial time dependent data would only be collected at the 100 saved time points. This is the reason for the rather large number of requested saved data points. In the loop of line 35 through 39, only every tenth saved file is actually written to a permanent file. Even then this calculation results in 21 saved data files. It's a pleasure for such a problem to have software that transparently manages data without having to allocate or deallocate storage space.

```lua
 1 : -- File list13_24.lua -- Vibration of circular membrane.
 2 :
 3 : require'pde2fe' ; require'elemfunc'
 4 : L = 1.0 -- Radius of circular plate
 5 : nts ={ readnts('list13_17') }
 6 : nds,tri = nts[1], nts[2] -- Node data
 7 :
 8 : C = 1; Csq = C^2-- Velocity -- change as desired
 9 : gm = 0.0 -- damping coefficient
10 : --gm = .2 -- change as desired
11 : x,y = {},{}; NT = 41
12 : for j=1,NT do x[j] = 2*L*(j-1)/(NT-1)-L; y[j] = x[j] end
13 :
14 : feqs = {
15 :     function(x,y,t,uxx,ux,u,uy,uyy,ut,utt)
16 :         return uxx + uyy  - utt/Csq - gm*ut/C -- 2D equation
17 :     end,
18 :     function(nd,t,u,un,ut,utt) -- Boundary value
19 :         return u
20 :     end
21 : }
22 :
23 : u = {}
24 : for k=1,#nds do
25 :     r = math.sqrt(nds[k][1]^2+nds[k][2]^2)
26 :     u[k] = elemfunc.J0(5.52008*r) -- Use Bessel function
27 : end
28 : SPM,COE,SOR = 1, 2, 3 -- 3 solution methods
29 : getfenv(pde2felt).nprint=1
30 : getfenv(pde2fe).linear = 1
31 : tvals = {0,8,200,1,{{0,0},{.5,0},{-.5,0},{.5,.5},
            {-.5,-.5},{.75,0},{0,.75}}}
32 : u,uxyt = pde2felt(feqs,tvals,nts,u,COE)
33 :
34 : nsol = #u
35 : for i=1,nsol,10 do -- Save 2D data files
36 :     sfl = 'list13_24.'..(1+(i-1)/10)..'.dat'
37 :     sol,solxy = toxysol(u[i],x,y,nts[1],nts[2])
38 :     write_data(sfl,sol); splot(sol)
39 : end
40 : write_data('list13_24a.dat',uxyt)
```

Listing 13.24. Code segment for FE analysis of vibration of circular membrane

Now for some calculated results. Figure 13.42 shows the initial (t = 0) membrane displacement which has the familiar shape of the J_0 Bessel function – kind of like a Mexican sombrero. According to Eq. (13.67), each spatial point should oscillate with the same frequency. This is verified in Figure 13.43 for three points by the saved time dependent data – saved on line 40 of the code. The three points

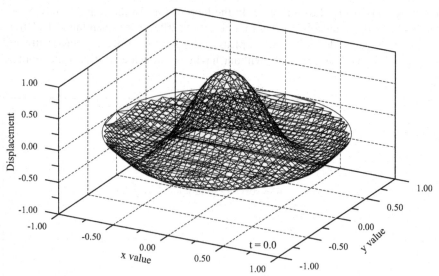

Figure 13.42. Initial displacement (t = 0.0) of circular membrane problem from Listing 13.24.

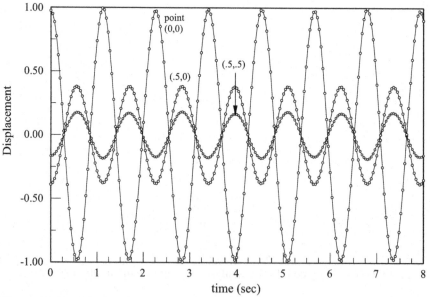

Figure 13.43. Calculated time dependent displacement for 3 points on the surface of the circular membrane problem from Listing 13.24.

are the center point and two points along the surface that have negative initial displacements. It can be seen that the three points do in fact have a sinusoidal oscillation and oscillate in phase with each passing through zero at the same time. The circular data points in the figure show the actually calculated and saved values. About 30 calculated time points are seen per cycle of the oscillation and this is about the minimum number of time points needed for a fairly accurate calculation. In this example more cycles of oscillation are calculated than are needed to observe the type of response. The reader is thus encouraged to repeat the calculation using a smaller time interval (perhaps two cycles) and keeping the same total number of time points. The results obtained can then be compared with a finer time resolution in the FE analysis.

A remaining question is with respect to the calculated oscillation frequency. A comparison of the center point oscillation with the theoretical value is shown in Figure 13.44. In this figure, the solid curve is the theoretically expected time dependency with the frequency given in Eq. (13.67) and the data points are the calculated values by the FE numerical analysis. A reasonably good agreement can be seen in the magnitude and frequency. However, a close examination of the results will indicate that the FE numerical analysis gives values that appear to have a slightly different frequency from the theoretical curve. This is a feature of any numerical calculation using the trapezoidal rule for time step approximation. This has been discussed in detail in Section 10.3 and the reader is encouraged to review that material for an explanation of this phenomena. A calculation with more time points per cycle of oscillation will cause the calculated points to agree more closely with the theoretical curve.

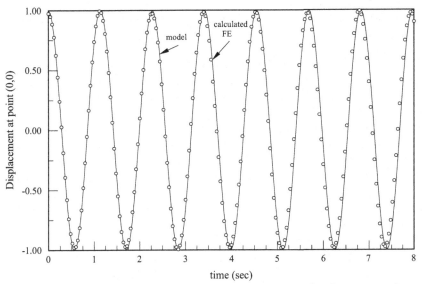

Figure 13.44. Comparison of the theoretical displacement for the center point with the calculated numerical values from the code in Listing 13.24.

To finish the discussion, three additional views of the membrane displacement are shown in Figures 13.45 through 13.47 at times where the peak in displacement at (0,0) is first negative (t = .4 sec) and then back at almost it's peak positive value (t = 1.2) and then finally at almost it's peak negative value (t = 4.0 sec). These four surface plots along with the time dependent data of Figure 13.43 should give the reader a good appreciation for the results of the FE analysis of the vibrating circular membrane with no damping.

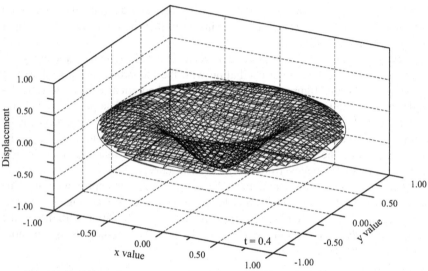

Fig 13.45. Membrane displacement at a time of 0.04 sec when the major peak has a negative value.

The presence of damping will cause the oscillations to damp out – using a finite gm in Listing 13.24. The results of such a calculation are compared in Figure 13.48 with the undamped case. For this example the code in Listing 13.24 was re-executed with a damping factor of 0.2. The results are as expected as the oscillations occur with the same natural frequency but appear damped out with time. For this example, damping is simply included in an empirical manner with a constant factor multiplying a first derivative time term in the PDE. Much information and many examples can be found on the web for the reader interested in further pursuit of the vibrating membrane example. From several web sites simulations of various fundamental modes of oscillation for a circular plate can be observed as described by Eq. (13.65).

This completes the examples to be given in this chapter of the use of the FE method for solving PDEs in two spatial dimensions and one time dimension. All parts of the developed code have been exercised and hopefully a high level of confidence has been developed with regard to the use of the method for a broad range of nonlinear engineering problems.

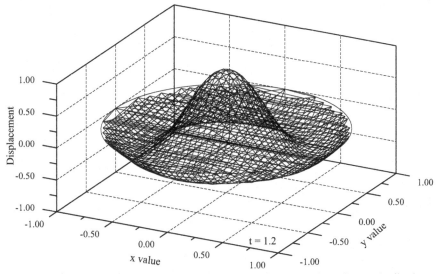

Fig 13.46. Membrane displacement at a time of 1.2 sec when the peak displacement is back at close to a peak of +1.0.

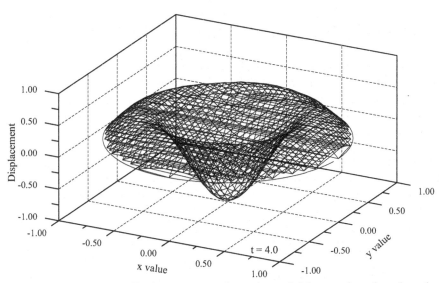

Fig 13.47. Membrane displacement at a later time of 4.0 sec when the when the major peak is at near a minimum value of -1.0.

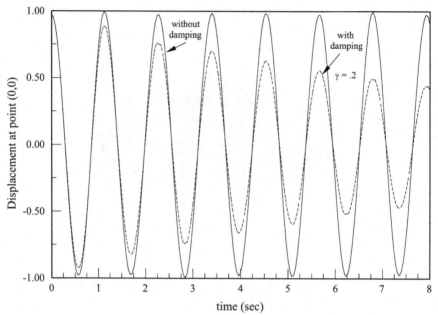

Figure 13.48 Comparison of circular membrane vibration with and without damping.

13.11 Some Final Considerations for the FE Method

The finite element method has developed into the most important approach for solving many engineering problems. Its premier advantage is the ability to readily model complex spatial geometries associated with material structures. This is in both two dimensions as covered in this work and three dimensions which are not considered in this work. A very extensive literature exists on this subject and many authors have written complete books on the subject. Obviously a single chapter can not hope to cover in depth the many important aspects of this subject. It is hoped that the material in this chapter provides the reader with a deeper appreciation for the approach and with some computer code that can be used for some of the simpler PDEs. A search of finite element resources in a library or on the web will result in a multitude of resources.

There are several ways in which the FE routines developed in this chapter could be extended and improved. Some of these will be briefly discussed for the interested reader. First the programs do not handle cases of mixed second partial derivatives in a PDE of the form:

$$a_x \frac{\partial^2 U}{\partial x^2} + a_{xy} \frac{\partial^2 U}{\partial x \partial y} + a_y \frac{\partial^2 U}{\partial y^2} + F(x, y, U_x, U_y, U) = 0 \qquad (13.69)$$

The code can be modified relatively easily to handle such an equation. However, mixed derivatives do not occur with most standard physical problems.

A second more extensive extension would be the development of programs to solve a number of coupled second order PDEs – for example 3 PDEs in three physical variables. These could also be time dependent. Such coupled PDEs occur frequently in physical problems. While extending the previous code for this case is relatively straightforward in principle, this is not a trivial task. Only some considerations for coupled equations will be considered here for N coupled equations. If one allows in the most general case, all ranges of derivatives to be expressed in the coupled equations, then one has N variables at each node and each defining PDE has N variables. Thus the number of node equations is increased by the factor N and the number of possible non-zero elements per row is increased by the factor N, giving a value of $NXN = N^2$ as the increased factor for the possible number of non-zero matrix elements. For the case of 3 coupled variables this is a factor of 9. Thus the computational time for coupled systems of equations can increase very fast.

For some physical problems with coupled variables, the approach of solving the equations in sequence and embedding the results in an iterative loop over the individual equations can be used to generate solutions. This works best when the coupling between the equations is weak. For the case of no coupling between two PDEs this sequence approach was used in Section 13.10.3 to solve in sequence for the current profile and then the temperature of a square corner resistor.

For many engineering PDE problems of significant complexity, one should consider using commercially available FE programs for obtaining solutions. Many such commercial programs have been developed and many have an emphasis on sub-disciplines of engineering. A brief listing of some of the more prevalent such commercial products is given here along with the engineering discipline that most frequently uses the software:

(a) ANSYS – Civil, Mechanical and Materials engineering
(b) ABAQUS – Civil, Mechanical and Materials engineering
(c) NASTRAN – Civil Mechanical and Materials engineering
(d) FEAP – Civil, Mechanical and Materials engineering
(e) FEM (with MATLAB) – General coupled physical equations
(f) SUPREM-IV – 2D semiconductor process models
(g) PISCES – 2D semiconductor device equations
(h) FIELDAY – 2D, 3D semiconductor device equations

If interested, the reader can find considerable information on each of these programs by searching the web. The last three entries are examples of FE programs and packages specialized to a particular engineering discipline, in this case the semiconductor device design discipline. The FE packages listed above find wide use in industry and should certainly be considered for very complex PDE problems. In general, however, they are not free as are the programs presented herein. In addition they require considerable effort to use as typically some type of specialized language must be learned in order to input problem definition information and to execute the programs. In most cases they are considerably more difficult to use than are the software code segments developed and discussed in this work.

13.12 Summary

This chapter has discussed the numerical solution of partial differential equations by the method of finite elements. For some problems this is a complementary method to the finite difference method of the previous chapter. For PDEs involving one spatial dimension and one time dimension, either of these approaches can usually be used to obtain accurate solutions. The finite element approach really shines when one has a PDE and boundary problem involving a non-rectangular spatial region. The more general spatial element allowed by the FE approach makes it easy to describe general spatial boundaries and boundary conditions associated with the boundaries.

The FE solvers developed in this chapter have made use of some of the approximate matrix solution techniques developed in the previous chapter. However, most of the code is new because of the different basic formulation of the finite element approach. In this work the method of weighted residuals has been used to formulate sets of FE node equations. This is one of the two basic methods typically used for this task. In addition, the development has been based upon the use of basic triangular spatial elements used to cover a two dimensional space. Other more general spatial elements have been sometimes used in the FE method. Finally the development has been restricted to two spatial dimensions and with possible an additional time dimension. The code has been developed in modular form so it can be easily applied to a variety of physical problems. In keeping with the nonlinear theme of this work, the FE analysis can be applied to either linear or nonlinear PDEs.

The developed code has been illustrated with a variety of physically based PDEs covering a range of geometries and physical disciplines. Not as much discussion on solution accuracy has been included as in previous chapters as the approach is not as amenable to a direct evaluation of solution accuracy, especially for nonlinear PDEs.

One of the keys to the easy solution of PDEs by the FE method is the generation of an appropriate coverage of various spatial domains by a set of triangle elements. For this task a freely available external program named EasyMesh was selected to be used in this chapter. Output from this program was then input into the developed programs for the FE analysis.

Several code segments have been developed in this chapter for the solution of PDEs and a brief summary of the most important of these is presented below:

1. setup2feeqs() – Code for setting up the matrix equations for a FE boundary
   value problem.
2. pde2fe() – Code for solving two dimensional PDE by the finite element technique.
3. pde1stp2fe1t() – Code for solving a two dimensional PDE plus time derivatives for one series of uniformly spaced time steps.
4. pde2fe1t() – Code for solving a two dimensional PDE plus time derivatives over a series of specified time intervals.
5. pde2fe1tqs() – Code for solving a two dimensional PDE plus time derivatives over logarithmically spaced time intervals.
6. toxysol() – Code for converting solution values from a triangular node set to an x-y rectangular grid set for aid in plotting results.
7. readnts() – Function for reading and converting node and triangle data from output of the EasyMesh program for use in FE solvers.
8. hsfunc() – Code for evaluating shape functions for a set of triangles and nodes.
9. tript() – Code for finding a triangle that contains a gives set of x,y coordinates.
10. intptri() – Code for interpolating a set of solution values at the triangle nodes to a desired x,y spatial point.
11. derivtri() – Code for evaluating the derivative of a solution set within the set of triangle spatial elements.

As illustrated in this chapter, these code routines can be used with a variety of partial differential equations from various fields of engineering. The programming interface and input required to solve PDE equations has been kept as simple as possible and as consistent as possible with that of previous chapters. With this background, the reader should be able to explore an additional range of PDEs with various spatial geometries within the restriction of nonlinear (or linear) PDEs involving two spatial dimensions plus one additional time dimension.

Appendix A: A Brief Summary of the Lua Programming Language

The programming language Lua is a relatively recent addition to the repertoire of computer programming languages. The first version (1.1) was released in July 1994. The most recent version of the language is version 5.1 (as of early 2008). Since its introduction it has assumed a dominant role as a favorite scripting language for computer games, to a large extent because of its small code size and fast execution speed. The major reasons for selecting this language for implementing the computer algorithms of this work are summarized in Chapter 2. The discussion here provides a more in depth summary of the most important features of the Lua language. However, the following two very excellent, in depth discussions are readily available on the web:

1. Lua Reference Manual -- http://www.lua.org/manual
2. Programming in Lua -- http://www.lua.org/pil

The material in this Appendix is a somewhat shortened version of the Lua Reference Manual and the reader is referred to this material for a much more complete discussion of the Lua language

Lua has been described by its developers as "an extension programming language designed to support general procedural programming with data description facilities". Many of the applications involve using Lua as a "scripting language" embedded within another language such as the C programming language. In fact Lua is written and compiled in clean C which is a common subset of ANSI C. Lua is a freely distributed programming language and a copy of the software is supplied on the disk accompanying this book. The user is advised to consult the Lua web site for possible later additions to the software (www.lua.org).

A.1 Lua Language Fundamentals

Names in Lua can be any string of letters, digits and underscores, not beginning with a digit. The exception to this rule is the set of following *keywords* that can not be used as object names:

and	break	do	else	elseif
end	false	for	function	if
in	local	nil	not	or
repeat	return	then	true	until
while				

Lua is case sensitive so that although end is a reserved word, End and END are perfectly valid names. As a convention, names beginning with an underscore followed by all capital letters are reserved for internal Lua usage and should not be used as naming conflicts may occur. Most of the reserved words are similar to reserved words in other computer languages and the list should be familiar to anyone familiar with another programming language.

In addition to the keywords, the following tokens also have special meaning in Lua:

```
+      -       *       /       %       ^       #
==     ~=      <=      >=      <       >       =
(      )       {       }       [       ]
;      :       ,       .       ..      ...
```

Again most of these symbols have the same meaning as in other languages. One exception is perhaps the use of ~= for "not equal".

Literal strings are delimited by matching single or double quotes and can contain C-like escape sequences such as '\b' (backspace), '\n' (newline) and '\t' (tab). A character in a string may also be specified by its numerical value using the escape sequence \ddd where ddd is a sequence of three decimal digits representing the character. Literal strings may also be defined using a long format of double brackets or double brackets with an intervening == string as for example [[text string]] or [==[text string]==].

Numerical constants consist of a sequence of numerical digits with an optional decimal part and an optional decimal exponent. Integer hexadecimal constants are also accepted if the digits are prefixed with 0x. Some valid numerical constants are:

```
5      5.00    0.005 3.33e-4       3.33E-4        0xff   0x4a
```

Internally Lua represents integers as simply double precision numerical numbers. Limits to the representation of numerical constants are discussed and explored in Chapter 2.

Lua comments begin with a double hyphen (--) anywhere outside a string. If the text following the double hyphen is not an opening long bracket, the comment is known as a *short comment* and extends until the end of the code line. Otherwise it is a *long comment* (beginning with --[) that runs until the corresponding closing long bracket (--]). Such long comments are very useful for extended comments in code or to temporarily disable sections of code extending over many lines of text.

Lua is a *dynamically typed language* and as such variables do not have types, only values have types. Because of this there are *no type definitions* in Lua and values carry their own type. For many users accompanied to other languages this takes some getting used to. However, it is one of the great features of the language as all values are *first-class values*. All values in the language may be stored in variable names, passed as arguments to other functions or returned as results.

Lua has eight basic types of values: *nil, Boolean, number, string, function, userdata, thread and table*. The type *nil* is that of the reserved work *nil* and usually represents the absence of a useful value, for example the value of an unassigned variable. A *Boolean* type has only the two values *false* and *true*. The types number and string should be familiar to anyone that has used a programming language. A variable of type *function* can be a code segment written in Lua or a function written in the C language. Functions are called with an argument list and functions may return any number of variables. The types *userdata* and *thread* are important types for embedded applications but are not used in the programs written in this work and thus are not discussed further here. The type of a value can be obtained by the function type(val) which returns a string identifying the type.

The fundamental data structure in Lua is the *table* and in fact this is the *only data structure in Lua*. While this is somewhat disconcerting to some new user of Lua it is one of the most useful features of Lua. The type table implements *associative arrays* which means that tables can be indexed not only with numbers but with any other value (except nil). Tables in Lua may be used to implement ordinary arrays, symbol tables, sets, records, graphs, trees, vectors, matrices, etc. In Lua tables are simply objects and the name of a table does not store the table – it simply contains a reference or pointer to the table. Tables can thus be readily passed to functions with a minimum of expense in computing resources.

Table constructors are expressions that create tables and are specially formatted expressions enclosed within braces ({ }). An example to create a table named tbl is:

$$tbl=\{3, 5.66, ['one'] = 1.0, ['name'] = 'John', [30] = 33\}$$

This creates a table with three integer fields (or indices) (1, 2 and 30) and with two string fields 'one' and 'name'. Examples of accessing the field values are: tbl[2] (value 5.66), tbl['one'] (value 1.0) and tbl['name'] (value 'John'). For table fields with string identifiers, Lua provides another more convenient mechanism of obtaining the value as for example tbl.one (value 1.0) or tbl.name = 'Joe' (changes string 'John' to string 'Joe'). All global variables in a Lua program live as fields in Lua tables called environment tables. A table field may reference another table and tables may be embedded within tables to any desired depth.

Variables in Lua are names of places that store values. They are of three different flavors: global variables, local variables and table fields. By convention, variables are global in scope unless limited in scope by the *local* keyword. Note that there is no *global* keyword as this is the default case.

A.2 Lua Statements

Statements in Lua follow almost the same form as in C and as other conventional programming languages. The basic unit of execution in Lua is called a *chunk* which is simply a sequence of Lua statements. Each statement in Lua may be optionally followed by a semicolon, although the normal convention in Lua programming is to only use the semicolon if an additional statement is placed on the same programming line. No empty statements are allowed, so that ; ; is a programming error.

Closely related to a chunk is the concept of a *block* which syntactically is the same as a chunk. Again a block is a list of Lua statements. A block may be explicitly delimited to produce a single statement of the form **do** block **end** where the **do** ... **end** structure represents a single statement. Such explicit blocks are useful to control the scope of variables as variables may be declared as *local* to the block of code.

Perhaps the most fundamental Lua statement is the assignment statement which contains a list of variables on the left side of an equal sign and a list of expressions on the right side of the equal sign. The assignment statement first evaluates all the expressions and only then performs the assignments. Some examples are:

$$i = 4$$

$$x, y, z = y, x, i$$

$$i, a[i] = i + 1, 10$$

In the second line the values stored in x and y are interchanged and z is set to the value of i. In the third line i is set to 5 and a[4] is set to 10. In an assignment statement the number of variables on the left may differ in number from the number of expressions on the right. If the number of variables is less than the number of expressions, the extra expressions are not evaluated. If the number of variables exceeds the number of expressions, the extra variables are set to **nil** values. The meaning of an assignment statement to a table field may be changed by use of a metatable as subsequently discussed.

Lua has several flow control structures that have similar meanings to those in other languages. The most important are:

> **while** *exp* **do** *block* **end**
> **repeat** *block* **until** *exp*
> **if** *exp* **then** *block* [**elseif** *exp* **then** *block*][**else** *block*] **end**

For the **if** statement, the expressions in brackets may or may not be present and any number of **elseif** clauses may be present. One of the differences with some other languages is the use of the **end** keyword to terminate the control structures. For the test expressions (*exp* terms in the above) any value may be evaluated by the expression with both **false** and **nil** considered as false. Any other evaluated value is considered as **true**. In the **repeat** structure the terminating *exp* can refer to local variables declared within the repeating block.

In addition Lua has two types of **for** control statements. One is a numeric for statement of the form:

> **for** var_name = *exp1*, *exp2* [*,exp3*] **do** *block* **end**

The block of code is repeated for var_name equal to *exp1* to *exp2* with the variable stepped by value *exp3* between each repeat execution. The default value of *exp3*, if omitted is 1. The control expressions are evaluated only once before the block is executed and they must all result in numbers. The loop variable var_name is considered local to the **for** loop and the value is not retained when the loop exits.

The generic **for** statement executes over functions called *iterators* and has the form:

for *namelist* **in** *explist* **do** *block* **end**

This type of **for** statement is not available in many common programming languages and takes some understanding for those first encountering Lua. An expression such as:

for *var_1, ..., var_n* **in** *explist* **do** *block* **end**

is best understood as equivalent to the following code segment:

```
do
    local f, s, var = explist
    while true do
        local var_1, ..., var_n = f(s, var)
        var = var_1
        if var == nil then break end
        block
    end
end
```

As the expanded equivalent indicates the evaluation of *explist* must return an *iterator* function (the *f* term), a *state variable* (the *s* term) and an *initial value* (the *var* term) for the *iterator variable*. One of the common uses of the generic **for** loop is to iterate over the values of a table and perform some operation on the table values. The reader is referred to the Lua programming manual for a more complete discussion of this control structure.

The keyword **break** may be used to exit control loops but must be the last statement in a block. Because of this restriction the construct **do break end** must sometimes be used to break from within the middle of a block of code.

A.3 Lua Expressions

Expressions constitute the fundamental construction unit for code in Lua. Such expressions have been indicated in the control structures of the preceding section. In its simplest form an expression consists of Lua keywords such as **true, false** or **nil**. Expressions may also be numbers, strings, functions, table constructors or variable names representing such values. Expressions may also be more complex units of code such as two expressions between a binary operator, such as *exp* =*exp1* + *exp2*. In general Lua expressions are similar to the fundamental expressions of the other languages.

The Lua evaluation of an expression typically results in one or more Lua object types being returned by the evaluation. The resulting evaluated items are then used by Lua as control parameters as in the previous section or as values to be set to other variables by use of the = construct as in *exp = exp1 / exp2*.

In Lua both function calls (see next section) and expressions may result in multiple values being evaluated. Lua has a specific set of rules as to how such multiple values are subsequently used by Lua code. This is necessitated by the fact that

Lua also allows multiple assignment statements. Some Lua statements and the results are shown below:

        a, b, c = x, y, z   -- a, b and c set to value of x, y and z
        a, b, c = x   -- a set to value of x, b and c set to nil
        a, b = x, y, z   -- a and b set to value of x and y, value of z is
                                  discarded

The more complex rules involve multiple values returned by function calls (such as f()). If a function call is used as the last (or only) element of a list of expressions then no adjustment is made to the number of returned values. In all other cases Lua adjusts the number of returned elements by a function call to one element, discarding all returned values except the first returned value. Some examples and the resulting adjustments are:

    f()   -- adjusted to 0 results when used as an isolated statement
    a, b, c = f()   -- f() is adjusted to 3 values adding nil values if needed
    a, b, c = f(), x   -- f() is adjusted to 1 value, b set to x value and c gets nil
    a, b, c = x, f()   -- f() is adjusted to 2 values, a is set to x value
    g(x, f())   -- g() gets x plus all returned values of f()
    g(f(), x)   -- f() is adjusted to 1 value and g() gets two parameters
    return f()   -- returns all results of f()
    return x, y, f()   -- returns x, y, and all results of f()
    return f(), x, y   -- f() is adjusted to 1 value and three values are returned
    {a, b, f()}   -- creates a list with elements a, b and all returned values of f()
    {f(), a, b}   -- f() is adjusted to 1 value and a list of three elements is created
    {f(), nil}   -- creates a list with 1 element, the first value returned by f()

An expression enclosed in parentheses always returns one value so that (f(x, y, z)) results in only one value regardless of the number of values returned by f().

Lua supports the usual array of arithmetic operators: + for addition, - for subtraction, * for multiplication, / for division, % for modulo, ^ for exponentiation and unitary – before any number. The supported relational operators are:

        ==   ~=   <   >   <=   >=

with the resulting evaluation resulting in a Lua **false** or **true** value. For equality the type of the operands must be the same. Objects such as tables and functions are compared by reference and two objects are considered equal only if they are the identically same object. For example two different tables with the same identical table values will not be considered as equal. The way that Lua compares tables can be changed by the use of metatables. The operator ~= is the negation of equality (the == operator).

The logical operators in Lua are the keywords **and, or** and **not**. These consider both **false** and **nil** as false and everything else as true. The **and** operator returns its first argument if it is **false** or **nil** and otherwise returns its second argument. The **or** operator returns its first argument if its value is different from **false** and **nil** and otherwise returns its second argument. Statements such as "a = a or 1" are frequently used in Lua to ensure that a variable (such as a) has either a defined value or is set to a default value of 1 if not previously defined.

For string concatenation Lua uses the '..' operator. If one of the operands is a number that can be converted to a string, Lua will perform the string conversion before applying the operator.

Lua has a special length operator, #, for obtaining the length of an object. The length of a string is the number of bytes. The length of a table t is defined as any integer index n such that t[n] is not **nil** and t[n+1] is **nil**. For a table with no **nil** values from 1 to n this gives the expected length of the table n. The operator causes some confusion when applied to tables with missing or **nil** values and for tables indexed by strings such as dictionary tables. Lua provides other mechanisms such as pairs() and ipairs() for stepping over such tables and the length operator should only be applied to tables indexed by numerical values with no missing elements.

Lua observes the following operator precedence with the order from lower to higher priority:

> **or**
> **and**
> < > <= >= ~= ==
> ..
> + -
> * / %
> **not** # - (unary)
> ^

Parentheses can be used to change the precedence of an expression.

A.4 Function Definitions and Function Calls

Functions are fundamental code blocks in any programming language leading to the ability to produce modular code segments that can be used in a variety of application. Two acceptable formats for function definitions are:

> fct = **function** (parlist) block **end**
> **function** fct (parlist) block **end**

where fct is a user supplied name for a function, parlist is a comma separated list of calling arguments and block is the Lua code to be executed by the function. When Lua encounters the function statement, the code is simply compiled by Lua and a later call to the function is used to explicitly execute the function. Function names in Lua are first class objects and can be stored in tables and passed to other functions just as any other variable in Lua. The keyword local can also be used to precede the function definition to limit the scope of the function name.

When Lua encounters a function definition and compiles the function it establishes an environment for the function that consists of the state of any global variables used within the function body. This means that different instances of the same function may refer to different external variables and may have different environmental tables. This is an important feature of Lua not available to statically compiled languages.

In defining a function the argument list (parlist above) is a list of comma separated Lua variable names. These names are treated as local variables to the function body and are initially set to the argument values passed to the function by a call to the function. The argument definition may (if desired) contain as the last argument in the list an entry consisting of three dots, i.e. '...'. This indicates that the function is a *vararg function* and can be passed a varying number of calling arguments. It is up to the called program to determine the number of such variable arguments.

In calling for the execution of a previously defined function the argument list used in calling the function is a list of comma separated statements that must evaluate to one of the eight basic Lua types of objects. For a function with a fixed number of calling arguments, the list of calling arguments is adjusted to match the number of calling arguments used in the defining statement. This is achieved by dropping arguments if the number exceeds the number in the defining code or adding nil arguments if the number is less than that in the defining code for the function. For a vararg function, all extra arguments are collected into a *vararg expression* that is also written as three dots. In this manner a function can accept a varying number of arguments in a calling statement. Lua makes no check on the type of argument used in calling a function to verify that they match the type of arguments used in defining the function. This is all left up to the user.

Object oriented programming in Lua typically involves the use of a table to store several functions that are associated with an object as well as storage of the object parameters with the same or another table. Recall that tables are the only data object defined in Lua. Lua provides a convenient means for references such named functions within tables. For example consider a table defined as C = {f1, f2, f3} where f1, f2 and f3 are names of functions that have previously been defined. Calls to the functions can then be made using the statements C.f1(), C.f2() or C.f3() with appropriate argument lists. The notation C.f1 is simply shorthand notation for C['f1'] which returns a reference to the f1 function. In defining methods for objects, Lua provides the syntactic sugar for the statement v:name(args) as equivalent to v.name(v, args). This is typically used in combination with metamethods for tables as described in the next section.

Values are returned by a function call with the statement **return** explist where explist is a comma separated list of Lua objects. As with break, the return statement must be the last statement of a block of code. In some cases this necessitates the **do return** explist **end** Lua construct to return from within a block. Lua implements function recursion and when the recursive function call is of the form **return** function_call() Lua implements proper tail calls where there is no limit on the number of nested tail calls that a program can execute.

Calls to evaluate a function may be made as simple single line statements in which case all values returned by the function are discarded. When a list of arguments is set equal to a function call, the number of returned arguments is adjusted to match the number of arguments in the left hand list of values.

A.5 Metatables

Every value in Lua may have an associated metatable that defines the behavior of the value under certain operations. Such metatables are typically used to define how table objects behave when used with certain inherent Lua operations on the table. For example if a non-numeric value is the operand in an addition statement, Lua checks for a metatable of the operand value with a table field named "__add". If such a field exists then Lua uses this function (__add()) to perform the addition function. The use of metatables is one of the features that makes Lua such an extensible language as tables can be used to define any type of object and metatables may be defined to indicate how basic language operations behave for the defined objects.

The keys in a metatable are called *events* and the values (typically functions) are called *metamethods*. For the example above the event would be the "add" operation and the metamethod would be the __add() function defined to handle the event. Values in a metatable can be set or changed by use of the Lua set-metatable(val, mtable) function and values may be queried through the get-metatable(val) function that returns the metatable itself.

Lua supports several standard metamethods for metatables. These are identified to Lua by a string name consisting of two initial underscores followed by a string identifier, such as "__add". The string must subsequently be the name of a defined function that handles the indicated operation for the Lua object with the given metatable. Lua supports the following metamethods:

"__add" : the addition or + operation
"__sub" : the subtraction or − operation
"__mul" : the multiplication or * operation
"__div" : the division or / operation
"__mod : the modulo or % operation
"__pow" : the power or ^ operation
"__unm" : the unary or − operation
"__concat" : the concatenation or .. operation
"__len" : the length or # operation
"__eq" : the equal or == operation
"__lt" : the less than or < operation
"__le" : the less than or equal or <= operation
"__index" : the table access or indexing operation (table[key] operation)
"__newindex" : the table assignment operation (table[key] = value operation)
"__call" ; the table call operation (used as table())

By defining a set of such functions for table objects, Lua provides an extension language for the operations indicated above to be automatically performed between newly defined language objects. This combined with the storing of functions as names within tables provides a powerful mechanism for object oriented programming in Lua.

A.6 Environments

Lua objects of type thread, function and userdata have another table associated with them called their *environment*. For the purpose here only function environments will be discussed. The environment table for a function defines all the global variables used to access values within the function. The environment of a function can be set with the statement "setfenv(fct, table)" and the environment table can be accessed by the statement "getfenv(fct)" which returns the environment table. An explicitly defined environmental table for a function is an excellent means of ensuring that the function does not define and export any undesirable global variables to a Lua program that uses the function.

A.7 Other Lua Language Features

Lua performs automatic garbage collection so the user does not have to worry with allocating and de-allocating memory as in many languages (such as C). None of the programs in this book worry with garbage collection but there are Lua functions for forcing garbage collection if needed. The reader is referred to the Lua manual for more details.

Lua supports coroutines that provide a form of *collaborative multithreading* of Lua code. By use of functions coroutine.yield() and coroutine.resume() Lua can be forced to suspend execution from one function and later resume the execution from within another Lua function. For some applications such an ability is very useful but this feature is not used in this work.

Lua has an extensive set of C Application Program Interface (API) functions for interfacing Lua with C language programs. C functions may be called from Lua programs or Lua programs may be called from C programs. The usefulness of this API is one of the Lua features that has made it the dominant scripting language for computer games. However, this feature is not used here so the reader is again referred to the Lua manual and Programming in Lua book for a discussion.

A.8 Lua Standard Libraries

In addition to the basic Lua language features, Lua provides several sets of standard libraries. Every Lua library is defined by an associative Lua table with the names of the functions as an index string into the table and the function is the value stored at the given index. The standard Lua libraries are grouped as follows:
- basic library (_G[])
- package library (package[])
- string manipulation (string[])

- table manipulation (table[])
- mathematical functions (math[])
- input and output functions (io[])
- operating system facilities (os[])
- debug facilities (debug[])

In the above the [] notation is used to indicate that the name[] is a table of values. These functions are implemented with the official C API and are provided as separate C modules. The standalone Lua execution program integrates all of these libraries into the executable program. The usage of Lua in this book assumes that these libraries are part of the Lua executable program although the debug facilities are not employed in this work.

The basic library functions are defined in a default Lua environment with the name _G. The table entries in the default environment may be observed from Lua by printing the entries in this table for example by execution the statement "for name, value in pairs(_G) do print(name) end". The following table entries are found in the _G table corresponding to the variable names in the standard Lua environment:

basic Lua environment names:

_G[], _VERSION<$, arg[], assert(), collectgarbage(), coroutine[], debug[], dofile(), error(), gcinfo(), getfenv(), getmetatable(), io[], ipairs(), load(), loadfile(), loadstring(), math[], module(), newproxy(), next(), os[], package[], pairs(), pcall(), print(), rawequal(), rawget(), rawset(), require(), select(), setfenv(), setmetatable(), string[], table[], tonumber(), tostring(), type(), unload(), unpack(), xpcall()

The names shown are not exactly the string names stored in the _G[] table but have been augmented at the end with a 2 character code to indicate which of the 8 fundamental Lua types of variable is stored in the table. For example the last two characters of [] indicate that the entry name refers to a table. The character codes and associated data types are: string = '<$', number = '<#', function = '()', table = '[]', userdata = '<@', thread = '<>', boolean = '<&'. The entries in the _G[] table are thus seen to be of type string, table or function. The use of most of the functions can be inferred from the names of the functions.

The functions and tables in the _G[] environmental table can be referenced in Lua by simply stating the name of the table entries. For example the version number can be printed by the statement "print(_VERSION)" and the type of a variable such as next() can be printed by the statement "print(type(next))".

The standard Lua libraries listed above are seen to be the names of entries in the _G[] table such as math for the math library. The functions supplied by any of the standard libraries can be observed by printing the corresponding table entries such as the statement "for name, value in pairs(math) do print(name) end" to observe the math table entries. The following is observed for the standard Lua library tables and corresponding functions:

- package library (package[])
  config<$, cpath<$, loaded[], loaders[], loadlib(), path<$, preload[], seeall()

- string manipulation library (string[])
  byte(), char(), dump(), find(), format(), gfind(), gmatch(), gsub(), len(), limit(), lower(), match(), rep(), reverse(), split(), sub(), trim(), upper()
- table manipulation library (table[])
  concat(), copy(), foreach(), foreachi(), insert(), inverse(), maxn(), remove(), reverse(), sort()}
- mathematical functions library (math[])
  abs(), acos(), asin(), atan(), atan2(), ceil(), cos(), cosh(), deg(), exp(), floor(), fmod(), frexp(), huge<#, ldexp(), log(), log10(), max(), min(), mod(), modf(), pi<#, pow(), rad(), random(), randomseed(), sin(), sinh(), sqrt(), tan(), tanh()
- input and output functions library (io[])
  close(), flush(), input(), lines(), open(), output(), popen(), read(), stderr<@, stdin<@, stdout<@, tmpfile(), type(), write()
- operating system facilities library (os[])
  clock(), date(), difftime(), execute(), exit(), getenv(), remove(), rename(), setlocale(), time(), tmpname()
- debug facilities library (debug[])
  debug(), getfenv(), gethook(), getinfo(), getlocal(), getmetatable(), getregistry(), getupvalue(), setfenv(), sethook(), setlocal(), setmetatable(), setupvalue(), traceback()

The majority of the library functions used in this work are reasonably self evident from the names of the functions or are explained in the text when encountered. Thus explanations of the various standard library functions will not be given here. The reader is referred to the Lua reference manuals for more details.

The conventional notation for referring to an entry in the _G[] table, such as the type() function would be the code _G.type(). This will certainly invoke the type() function. However, for simplicity Lua accepts references to entries in the _G[] table without the _G prefix so that the code statement type() can be used to invoke the function. For the other tables in the _G[] table, the standard Lua access methods must be used. For example to invoke the sin() function within the math[] table, the proper Lua code statement is "math.sin()". Another example is "os.clock()" to get information about the time for use in timing the execution of a block of code. To simplify notation one can define additional names such as sin = math.sin so that the sin function can be accessed by the statement: sin().

Appendix B: Software Installation

This Appendix describes the software supplied with this book and describes the typical installation and use of the software programs. This file is also supplied on the CD as the readme.txt file.

B.1 Introduction to Software on CD

The files supplied on the accompanying CD are arranged in directories as follows:

```
readme.txt
setup.exe
Examples/
        Chapter1/
        Chapter2/
        Chapter3/
        Chapter4/
        Chapter5/
        Chapter6/
        Chapter7/
        Chapter8/
        Chapter9/
        Chapter10/
        Chapter11/
        Chapter12/
        Chapter13/
Nonlinear Models/
        EasyMesh/
        gnuplot/
        Lua-5.1.3/
        SciTE174/
```

The readme.txt file is a text version of this Appendix.

The computer programs needed to execute the computer code for the various chapter of this book can be downloaded to the reader's computer by executing the setup.exe file on the CD. During the installation the user has the option of selecting a "compact" installation that installs only the Nonlinear Models directory or a "full" installation that in addition installs the files in the Examples directory. Unless the user's computer is low on disk space it is highly recommended that the Examples files also be copied as these contain the example computer code discussed in each chapter of the book. After the installation, the user's computer should be configured for easy access to these files.

The computer programs associated with each chapter of the book are organized into chapter directories under the Examples directory. The user can either access these from the CD or download these to the user's computer if desired using the "full" installation.

The user may also manually copy the appropriate files to a chosen directory on his/her computer. In this case the user will need to manually configure some of the files to properly access the features of the chapter programs. The user might wish to do this manual installation if the Lua language is already installed on the computer. This manual procedure is described in a later section of this file.

The files under the Nonlinear Models directory are programs and files required to execute the Lua programs as supplied with this book and as implemented in the text. The files are organized into 4 major directories that are now briefly discussed.

The EasyMesh directory contains all the files associated with the EasyMesh software. This software is used in Chapter 13 to generate triangular spatial grids for use in solving partial differential equations in two spatial dimensions. Programs in this directory are a copy of software that is freely available on the web at: http://www-dinma.univ.trieste.it/nirftc/research/easymesh/Default.htm. The files are supplied here as a convenience to the user of this book.

The gnuplot directory contains all the files associated with the gnuplot software which is public domain software for plotting graphs from data files. This software is used throughout this book to provide pop-up graphs of the curves generated by many of the computer programs. This software is freely available on the web at: http://www.gnuplot.info. The files are supplied here as a convenience to the user of this book.

The SciTE174 directory contains all the files associated with the SciTE software which is public domain software for a programming oriented text editor. This is the highly recommended text editor for use with the Lua programs developed and used in this book. Although any text editor may be used for editing Lua programs, the SciTE editor provides language highlighting and an easy interface for program development and execution. The software is freely available on the web at: http://scintilla.sourceforge.net/SciTE.html. The files are provided here as a convenience to the user of this book.

The Lua-5.1.3 directory contains all the files associated with Lua which is public domain software for the Lua language used as the basis for all the software examples in this book. Reasons for selecting this language are discussed in Chapter 2 of the text. The software is freely available on the web at: http://www.lua.org. The files are provided here as a convenience to the user of this book.

The Lua-5.1.3 directory is arranged into subdirectories as follows:

Lua-5.1.3/
    doc/
    etc/
    src/
           lua/
           stdlib/
    test/

The doc directory contains files with more detailed descriptions of the Lua language and the user can read the documentation for a more complete description of the Lua language. Even more complete descriptions of the language are available on the web at:

Lua Refernce Manual -- http://www.lua.org/manual and
Programming in Lua -- http://www.lua.org/pil

The src directory contains the computer code for the Lua language which is written in the C language. From the files in this directory an executable Lua program can be compiled. However, the user of this software does not have to perform this step as a lua.exe file is supplied in this directory. This lua.exe file is the program used to execute all the Lua programs developed in this book.

The lua directory under the src directory is an important directory for using the Lua programs in the various chapters of this book. This directory contains various code segments supplied with this book and that are required for executing the Lua programs in the various chapters of this book. All of the files supplied in the lua directory are programs developed specifically for use with this book and are not files supplied with the Lua language. They are collected within the src/lua directory because this is the standard directory searched by the lua.exe program for such files.

B.2 Full Installation of Software

For this discussion it is assumed that most users of this book and of the supplied software will not have the Lua language previously loaded on their computer. Thus this first description of installing the software is for users who do not have the Lua language on their computer and need the full installation of the software. For those who already have the Lua language installed, please skip to the next section on a partial installation of the software.

Executing the setup.exe program on the CD should provide the needed installation of the software required for execution of all the Lua programs associated with this book. The user will be asked during the installation to select a directory for the software. A suggested default directory is C:\Program Files. If this default is used the software programs will be copied into the C:\Program Files\Nonlinear Models directory with the configuration discussed in the previous section. Entries will also be made in the Registry such that the user can double click on a *.lua file and have the file automatically be loaded into the SciTE editor. The user can of course select a different major directory if desired, such as D:\Work in which case the software would be copied into the D:\Work\Nonlinear Models directory. The "full" software installation requires about 120MB of disk space on the selected drive while the "compact" installation requires about 35MB of disk space.

After installing the software with setup.exe on the disk, the user may verify the proper installation by the following procedure. Open MS Explorer (or another program) in the directory selected to download the disk files. Figure B.1 shows such a directory listing where the root directory is the default download directory of C:\Program Files\Nonlinear Models. The Folders listing on the left of the figure show the directories that should be created by the downloading process. The listing on the right shows the example code associated with Chapter 3 of the book with an arrow pointing to the list3_7.lua file.

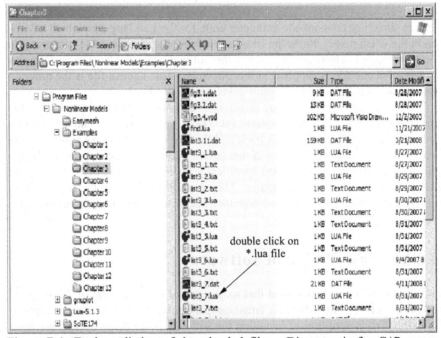

Figure B.1. Explorer listing of downloaded files. Directory is for C:\Program Files\Nonlinear Models\Examples\Chapter3.

If the software has downloaded properly and the computer is properly configured, a double mouse click (right button) on this *.lua file should bring up the SciTE editor with the Lua code from the file automatically loaded into in the editor. The expected result of such a double click on list3_7.lua is shown in Figure B.2. The left side of the figure shows the code from the list3_7.lua file. It is readily seen that the editor has syntax highlighting for Lua reserved words and other language structures. The right side shows the output from executing the Lua file. Program execution is readily performed from the SciTE editor from the Tools pull down menu. Right clicking on the Tools button along the top of the SciTE editor will give a drop down menu with the options of "compile" and "go". Selecting the "compile" option will execute the Lua compiler which provides a check of the program syntax without executing the code. The "go" option will compile and execute the Lua program and show any generated output from the program in the right half screen

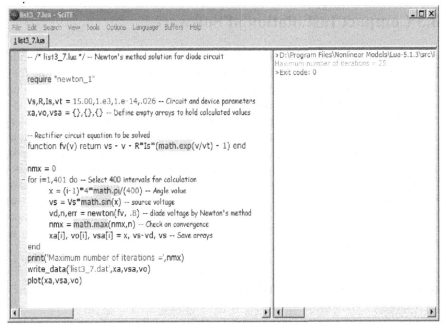

Figure B.2. Lua file list3_7.lua listed in the SciTE editor. Left side shows the Lua code file and the right side shows the result of executing the Lua file.

The Lua code on the left may be readily changed in the SciTE editor and the program re-executed without leaving the SciTE editor. This is the recommended way for execution the example code in this book.

If the software has been properly downloaded into the disk directories as indicated by Figure A.1 but a double click of the mouse on a *.lua file does not bring up the SciTE editor with the loaded code, it may be necessary to manual configure the software to associate the SciTE editor with a *.lua file. This can be performed as follows. Right click on any of the *.lua files to bring up a menu with an entry entitled "open with". Select this item to bring up the "Open With" menu and from this use the "Browse" button to select the SciTE.exe file in the Nonlinear Models\SciTE174\wscite directory. Also be sure to check the "Always use the selected program to open this kind of file" button. A double click on any *.lua file should then result in the *.lua file being loaded into the SciTE editor.

Of course any editor can be used to edit a *.lua file as it is just a pure text file. However, many simple text editors do not have syntax highlighting and automatic program execution. In any case one should use a programming language oriented editor with syntax highlighting and program execution mode such as the SciTE editor.

Assuming that the demonstration suggested above executes properly the download of the software has been successful and all the examples in the book should be easy to execute.

B.3 Compact Installation of Software

The "compact" software installation downloads all the software associated with the Nonlinear Models directory on the disk but does not download any of the Examples code for the book. This decreases the disk space required from about 120MB to about 35MB of disk space. In order to execute the Lua programs discussed in the book it is then necessary to separately download some of the chapter examples from the Examples directory. However, these can be downloaded on a chapter by chapter basis thereby saving disk space. If the user has sufficient disk space it is highly recommended that all the chapter examples simply be downloaded with the "full" installation. If the "compact" installation is used, and the download is successfully completed, the computer should be configured so that a double click on a *.lua file will properly load the file into the SciTE editor for changes or for execution. The reader is referred to the previous section for a discussion of the use of the software.

B.4 Manual Installation of Software

Under some circumstances the user may wish to manually install part or all of the supplied software. Cases where this might occur are when the user's computer already has some of the software already on the computer – for example already has Lua or SciTE on the computer. In this case the following procedure can be used to download selected parts of the supplied software. However, if sufficient space is available on the user's hard drive, it is highly recommended that the user simply download the extra copies of the software since the Lua examples are properly configured for use with software in the indicated directories. To download all of the supplied programs and examples requires about 120MB of disk space (or about 35MB if the examples are not downloaded). Downloading all the programs should not interfere with the user's execution of any previous software already on the user's computer, even if the user already has Lua, gnuplot or SciTE software on his/her computer. The following instructions are for the user who insists on downloading only selected parts of the software.

The user should select a disk drive and directory for storing the downloaded software. The default assumed directory is C:\Program Files\Nonlinear Models.

The user can select any other desired directory and for discussion purposes this directory will be referred to as the TopDir. The user can simply copy any of the programs on the software disk into this TopDir, such as the EasyMesh and gnuplot directories. For reference a brief description of the software in the supplied directories is:

EasyMesh – Programs used in Chapter 13 for generating a triangular mesh.
gnuplot – Programs for generating pop-up graphs of functions – used in all
                    chapters.
Lua-5.1.3 – Lua software for programming language used in book
SciTE174 – Programming text editor recommended for use with Lua

These directories and associated files can simply be copied to the TopDir of the user's computer.

For proper execution of the Lua programs in the Examples directory the Lua programs in the Nonlinear Models\Lua-5.1.3\src\lua directory on the disk must be accessible by the lua.exe program. If the user already has a version of Lua installed on his/her computer these files must be located in the src\lua directory of the user's Lua language directory. An experienced user of Lua should know how to achieve this. In any case even if the user has a copy of Lua already on his/her computer there is no harm in having another version of Lua in another directory if sufficient disk space is available (the Lua files on the disk require only 1.8MB of disk space). It is thus highly recommended that the user simply download the Lua files from the disk even if he/she already has Lua installed on his/her computer.

For a manual download of the files some configuration of the files is required for proper execution of the programs for each chapter. If the user selects TopDir = C:\Program Files\Nonlinear Models as the directory for the software, the downloaded programs already are configured for this directory and no modifications of the programs are required. The user can then skip the remainder of this section. Otherwise several programs must be modified so the software can properly locate needed files. First the software must know where (in what directory) the gnuplot software is located in order to produce the pop-up graphs used in the example programs. The program that must be modified is the lua.lua file in the "Lua-5.1.3\src\lua" directory. On the fourth line of this file should be the statement:

          local TopDir = "C:\\Program Files\\Nonlinear Models\\"

Using a text editor this statement should be modified to describe the directory into which the gnuplot software is located (yes the double \\ are required).

Second a file associated with the SciTE editor must be modified to properly tell the editor where the Lua executable code files are located. The file that must be modified is the lua.properties file in the "SciTE174\wscite" directory. Again using a text editor for this file, near the end of the file should be the statements:

```
command.compile.*.lua=C:\Program Files\Nonlinear
    Models\Lua-5.1.3\src\luac.exe -l lua.lua
    -o "$(FileName).luc" "$(FileNameExt)"
# Lua 5.0
command.go.*.lua=C:\Program Files\Nonlinear
    Models\Lua-5.1.3\src\lua.exe -l lua.lua
    "$(FileNameExt)"
```

The "C:\Program Files\Nonlinear Models" text must be changed to the directory in which the Lua files are located.

If the disk software is manually downloaded to the user's hard drive the above modifications must be manually made only if the user downloads the software to a directory different from the default of "C:\Program Files\Nonlinear Models". With the changes indicated above the software should be properly configured for executing the example programs. The final modification would be configuring the computer so that a double click on a *.lua file automatically loads the file into the SciTE editor. If the user is not proficient in this operation, the following procedure will perform the task. Right click on any of the *.lua files to bring up a menu with an entry entitled "open with". Select this item to bring up the "Open With" menu and from this use the "Browse" button to select the SciTE.exe file in the appropriate SciTE174\wscite directory. Also be sure to check the "Always use the selected program to open this kind of file" button. A double click on any *.lua file should then result in the *.lua file being loaded into the SciTE editor.

Subject Index